T0344965

Essential Mathematical Methods for the Physical Sciences

The mathematical methods that physical scientists need for solving substantial problems in their fields of study are set out clearly and simply in this tutorial-style textbook. Students will develop problem-solving skills through hundreds of worked examples, self-test questions and homework problems. Each chapter concludes with a summary of the main procedures and results and all assumed prior knowledge is summarized in one of the appendices. Over 300 worked examples show how to use the techniques and around 100 self-test questions in the footnotes act as checkpoints to build student confidence. Nearly 400 end-of-chapter problems combine ideas from the chapter to reinforce the concepts. Hints and outline answers to the odd-numbered problems are given at the end of each chapter, with fully worked solutions to these problems given in the accompanying *Student Solution Manual*. Fully worked solutions to all problems, password-protected for instructors, are available at www.cambridge.org/essential.

K. F. RILEY read mathematics at the University of Cambridge and proceeded to a Ph.D. there in theoretical and experimental nuclear physics. He became a Research Associate in elementary particle physics at Brookhaven, and then, having taken up a lectureship at the Cavendish Laboratory, Cambridge, continued this research at the Rutherford Laboratory and Stanford; in particular he was involved in the experimental discovery of a number of the early baryonic resonances. As well as having been Senior Tutor at Clare College, where he has taught physics and mathematics for over 40 years, he has served on many committees concerned with the teaching and examining of these subjects at all levels of tertiary and undergraduate education. He is also one of the authors of *200 Puzzling Physics Problems* (Cambridge University Press, 2001).

M. P. HOBSON read natural sciences at the University of Cambridge, specializing in theoretical physics, and remained at the Cavendish Laboratory to complete a Ph.D. in the physics of star formation. As a Research Fellow at Trinity Hall, Cambridge, and subsequently an Advanced Fellow of the Particle Physics and Astronomy Research Council, he developed an interest in cosmology, and in particular in the study of fluctuations in the cosmic microwave background. He was involved in the first detection of these fluctuations using a ground-based interferometer. Currently a University Reader at the Cavendish Laboratory, his research interests include both theoretical and observational aspects of cosmology, and he is the principal author of *General Relativity: An Introduction for Physicists* (Cambridge University Press, 2006). He is also a Director of Studies in Natural Sciences at Trinity Hall and enjoys an active role in the teaching of undergraduate physics and mathematics.

Essential Mathematical Methods for the Physical Sciences

K. F. RILEY
University of Cambridge

M. P. HOBSON
University of Cambridge

CAMBRIDGE
UNIVERSITY PRESS

University Printing House, Cambridge CB2 8BS, United Kingdom

One Liberty Plaza, 20th Floor, New York, NY 10006, USA

477 Williamstown Road, Port Melbourne, VIC 3207, Australia

314-321, 3rd Floor, Plot 3, Splendor Forum, Jasola District Centre, New Delhi - 110025, India

79 Anson Road, #06-04/06, Singapore 079906

Cambridge University Press is part of the University of Cambridge.

It furthers the University's mission by disseminating knowledge in the pursuit of
education, learning and research at the highest international levels of excellence.

www.cambridge.org
Information on this title: www.cambridge.org/9780521761147

First published 2011
Reprinted 2015
4th printing 2019

A catalogue record for this publication is available from the British Library

Library of Congress Cataloging in Publication data
Riley, K. F. (Kenneth Franklin), 1936–
Essential mathematical methods for the physical sciences : a tutorial guide / K.F. Riley, M.P. Hobson.
p. cm.
Includes index.
ISBN 978-0-521-76114-7
1. Mathematics – Textbooks. I. Hobson, M. P. (Michael Paul), 1967– II. Title.
QA37.3.R55 2011
510 – dc22 2010041509

ISBN 978-0-521-76114-7 Hardback

Additional resources for this publication at www.cambridge.org/essential

Contents

Preface

Since *Mathematical Methods for Physics and Engineering* (Cambridge: Cambridge University Press, 1998) by Riley, Hobson and Bence, hereafter denoted by *MMPE*, was first published, the range of material it covers has increased with each subsequent edition (2002 and 2006). Most of the additions have been in the form of introductory material covering polynomial equations, partial fractions, binomial expansions, coordinate geometry and a variety of basic methods of proof, though the third edition of *MMPE* also extended the range, but not the general level, of the areas to which the methods developed in the book could be applied. Recent feedback suggests that still further adjustments would be beneficial. In so far as content is concerned, the inclusion of some additional introductory material such as powers, logarithms, the sinusoidal and exponential functions, inequalities and the handling of physical dimensions, would make the starting level of the book better match that of some of its readers.

To incorporate these changes, and others to increase the user-friendliness of the text, into the current third edition of *MMPE* would inevitably produce a text that would be too ponderous for many students, to say nothing of the problems the physical production and transportation of such a large volume would entail. It is also the case that for students for whom a course on mathematical methods is *not* their first engagement with mathematics beyond high school level, all of the additional introductory material, as well as some of that presented in the early chapters of the original *MMPE*, would be ground already covered. For such students, typically those who have already taken two or three semesters of calculus, and perhaps an introductory course in ordinary differential equations, much of the first half of such an omnibus edition would be redundant.

For these reasons, we present under the current title, *Essential Mathematical Methods for the Physical Sciences*, an alternative edition of *MMPE*, one that focuses on the core of a putative extended third edition, omitting, except in summary form, all of the "mathematical tools" at one end, and some of the more specialized topics that appear in the third edition at the other. The emphasis is very much on developing the *methods* required by physical scientists before they can apply their knowledge of mathematical concepts to significant problems in their chosen fields.

For the record, we note that the more advanced topics in the third edition of *MMPE* that have fallen victim to this approach are quantum operators, tensors, group and representation theory, and numerical methods. The chapters on special functions, and the applications of complex variables have both been reduced to some extent, as have those on probability and statistics.

At the other end of the spectrum, the excised introductory material has not been altogether lost. Indeed, Appendix A of the present text consists entirely of summaries, in the style described in the penultimate paragraph of this Preface, of the material that is presumed to have been previously studied and mastered by the student. Clearly it can be used both as a reference/reminder and as an indicator of some missing background knowledge.

One aspect that has remained constant throughout the three editions of *MMPE*, is the general style of presentation of a topic – a qualitative introduction, physically based wherever possible, followed by a more formal presentation or proof, and finished with one or two full-worked examples. This format has been well received by reviewers, and there is no reason to depart from its basic structure.

In terms of style, many physical science students appear to be more comfortable with presentations that contain significant amounts of verbal explanation and comment, rather than with a series of mathematical equations the last line of which implies "job done". We have made changes that move the text in this direction. As is explained below, we also feel that if some of the advantages of small-group face-to-face teaching could be reflected in the written text, many students would find it beneficial.

One of the advantages of an oral approach to teaching, apparent to some extent in the lecture situation, and certainly in what are usually known as tutorials,[1] is the opportunity to follow the exposition of any particular point with an immediate short, but probing, question that helps to establish whether or not the student has grasped that point. This facility is not normally available when instruction is through a written medium, without having available at least the equipment necessary to access the contents of a storage disc.

In this book we have tried to go some way towards remedying this by making a non-standard use of footnotes. Some footnotes are used in traditional ways, to add a comment or a pertinent but not essential piece of additional information, to clarify a point by restating it in slightly different terms, or to make reference to another part of the text or an external source. However, about half of the nearly 300 footnotes in *this* book contain a question for the reader to answer or an instruction for them to follow; neither will call for a lengthy response, but in both cases an understanding of the associated material in the text will be required. This parallels the sort of follow-up a student might have to supply orally in a small-group tutorial, after a particular aspect of their written work has been discussed.

Naturally, students should attempt to respond to footnote questions using the skills and knowledge they have acquired, re-reading the relevant text if necessary, but if they are unsure of their answer, or wish to feel the satisfaction of having their correct response confirmed, they can consult the specimen answers given in Appendix H. Equally, footnotes in the form of observations will have served their purpose when students are consistently able to say to themselves "I didn't need that comment – I had already spotted and checked that particular point".

One further feature of the present volume is the inclusion at the end of each chapter, just before the problems begin, of a summary of the main results of that chapter. For some areas, this takes the form of a tabulation of the various case types that may arise in the context of the chapter; this should help the student to see the parallels between situations which in the main text are presented as a consecutive series of often quite lengthy pieces of mathematical development. It should be said that in such a summary it is not possible to state every detailed condition attached to each result, and the reader should consider

1 But in Cambridge are called "supervisions"!

the summaries as reminders and formulae providers, rather than as teaching text; that is the job of the main text and its footnotes.

Finally, we note, for the record, that the format and number of problems associated with the various remaining chapters have not been changed significantly, though problems based on excised topics have naturally been omitted. This means that hints or abbreviated solutions to all 200 odd-numbered problems appear in this text, and fully worked solutions of the same problems can be found in an accompanying volume, the *Student Solution Manual* for *Essential Mathematical Methods for the Physical Sciences*. Fully worked solutions to all problems, both odd- and even-numbered, are available to accredited instructors on the password-protected website www.cambridge.org/essential. Instructors wishing to have access to the website should contact solutions@cambridge.org for registration details.

Review of background topics

As explained in the Preface, this book is intended for those students who are pursuing a course in mathematical methods, but for whom it is not their first engagement with mathematics beyond high school level. Typically, such students will have already taken two or three semesters of calculus, and perhaps an introductory course in ordinary differential equations. The emphasis in the text is very much on developing the *methods* required by physical scientists before they can apply their knowledge of mathematical concepts to significant problems in their chosen fields; the basic mathematical "tools" that the student is presumed to have mastered are therefore not discussed in any detail.

However this introductory note and the associated appendix (Appendix A) are included both to act as a reference (or reminder) and to be an indicator of any presumed, but missing, topics in the student's background knowledge. The appendix consists of summary pages for ten major topic areas, ranging from powers and logarithms at one extreme to first-order ordinary differential equations at the other. The style they adopt is identical to that used for the chapter summary pages in the 17 main chapters of the book. It should be noted that in such summaries it is not possible to state every detailed condition attached to each result. In the areas covered in Appendix A, there are very few subtle situations to consider, but the reader should be aware that they may exist.

Naturally, being only summaries, the various sections of the appendix will not be sufficient for the student who needs to catch up in some area, to learn the particular topics from scratch. A more elementary text will clearly be needed; *Foundation Mathematics for the Physical Sciences* written by the current authors would be one such possibility.

1 Matrices and vector spaces

In so far as vector algebra is concerned (see the summary in Section A.9 of Appendix A), a *vector* can be considered as a geometrical object which has both a magnitude and a direction, and may be thought of as an arrow fixed in our familiar three-dimensional space. This space, in turn, may be defined by reference to, say, the fixed stars. This geometrical definition of a vector is both useful and important since it is *independent* of any coordinate system with which we choose to label points in space.

In most specific applications, however, it is necessary at some stage to choose a coordinate system and to break down a vector into its *component vectors* in the directions of increasing coordinate values. Thus for a particular Cartesian coordinate system (for example) the component vectors of a vector $\mathbf{a}$ will be $a_x\mathbf{i}$, $a_y\mathbf{j}$ and $a_z\mathbf{k}$ and the complete vector will be

$$\mathbf{a} = a_x\mathbf{i} + a_y\mathbf{j} + a_z\mathbf{k}. \tag{1.1}$$

Although for many purposes we need consider only real three-dimensional space, the notion of a vector may be extended to more abstract spaces, which in general can have an arbitrary number of dimensions N. We may still think of such a vector as an "arrow" in this abstract space, so that it is again *independent* of any (N-dimensional) coordinate system with which we choose to label the space. As an example of such a space, which, though abstract, has very practical applications, we may consider the description of a mechanical or electrical system. If the state of a system is uniquely specified by assigning values to a set of N variables, which could include angles or currents, for example, then that state can be represented by a vector in an N-dimensional space, the vector having those values as its components.[1]

In this chapter we first discuss general *vector spaces* and their properties. We then go on to consider the transformation of one vector into another by a linear operator. This leads naturally to the concept of a *matrix*, a two-dimensional array of numbers. The properties of matrices are then developed and we conclude with a discussion of how to use these properties to solve systems of linear equations and study some oscillatory systems.

1 This is an approach often used in control engineering.

1.1 Vector spaces

A set of objects (vectors) $\mathbf{a}$, $\mathbf{b}$, $\mathbf{c}$, ... is said to form a *linear vector space* V if:

(i) the set is closed under commutative and associative addition, so that

$$\mathbf{a} + \mathbf{b} = \mathbf{b} + \mathbf{a}, \tag{1.2}$$

$$(\mathbf{a} + \mathbf{b}) + \mathbf{c} = \mathbf{a} + (\mathbf{b} + \mathbf{c}); \tag{1.3}$$

(ii) the set is closed under multiplication by a scalar (any complex number) to form a new vector $\lambda\mathbf{a}$, the operation being both distributive and associative so that

$$\lambda(\mathbf{a} + \mathbf{b}) = \lambda\mathbf{a} + \lambda\mathbf{b}, \tag{1.4}$$

$$(\lambda + \mu)\mathbf{a} = \lambda\mathbf{a} + \mu\mathbf{a}, \tag{1.5}$$

$$\lambda(\mu\mathbf{a}) = (\lambda\mu)\mathbf{a}, \tag{1.6}$$

where λ and μ are arbitrary scalars;

(iii) there exists a *null vector* $\mathbf{0}$ such that $\mathbf{a} + \mathbf{0} = \mathbf{a}$ for all $\mathbf{a}$;

(iv) multiplication by unity leaves any vector unchanged, i.e. $1 \times \mathbf{a} = \mathbf{a}$;

(v) all vectors have a corresponding *negative vector* $-\mathbf{a}$ such that $\mathbf{a} + (-\mathbf{a}) = \mathbf{0}$. It follows from (1.5) with $\lambda = 1$ and $\mu = -1$ that $-\mathbf{a}$ is the same vector as $(-1) \times \mathbf{a}$.

We note that if we restrict all scalars to be real then we obtain a *real vector space* (an example of which is our familiar three-dimensional space); otherwise, in general, we obtain a *complex vector space*. We note that it is common to use the terms "vector space" and "space", instead of the more formal "linear vector space".

The *span* of a set of vectors $\mathbf{a}$, $\mathbf{b}$, ..., $\mathbf{s}$ is defined as the set of all vectors that may be written as a linear sum of the original set, i.e. all vectors

$$\mathbf{x} = \alpha\mathbf{a} + \beta\mathbf{b} + \cdots + \sigma\mathbf{s} \tag{1.7}$$

that result from the infinite number of possible values of the (in general complex) scalars $\alpha, \beta, \ldots, \sigma$. If $\mathbf{x}$ in (1.7) is equal to $\mathbf{0}$ for some choice of $\alpha, \beta, \ldots, \sigma$ (not *all* zero), i.e. if

$$\alpha\mathbf{a} + \beta\mathbf{b} + \cdots + \sigma\mathbf{s} = \mathbf{0}, \tag{1.8}$$

then the set of vectors $\mathbf{a}$, $\mathbf{b}$, ..., $\mathbf{s}$, is said to be *linearly dependent*. In such a set at least one vector is redundant, since it can be expressed as a linear sum of the others. If, however, (1.8) is not satisfied by *any* set of coefficients (other than the trivial case in which all the coefficients are zero) then the vectors are *linearly independent*, and no vector in the set can be expressed as a linear sum of the others.

If, in a given vector space, there exist sets of N linearly independent vectors, but no set of $N + 1$ linearly independent vectors, then the vector space is said to be N-dimensional. In this chapter we will limit our discussion to vector spaces of finite dimensionality.

1.1.1 Basis vectors

If V is an N-dimensional vector space then *any* set of N linearly independent vectors $\mathbf{e}_1, \mathbf{e}_2, \ldots, \mathbf{e}_N$ forms a *basis* for V. If $\mathbf{x}$ is an arbitrary vector lying in V then it can be written as a linear sum of these basis vectors:

$$\mathbf{x} = x_1\mathbf{e}_1 + x_2\mathbf{e}_2 + \cdots + x_N\mathbf{e}_N = \sum_{i=1}^{N} x_i\mathbf{e}_i, \tag{1.9}$$

for some set of coefficients x_i. Since any $\mathbf{x}$ lying in the span of V can be expressed in terms of the *basis* or *base vectors* $\mathbf{e}_i$, the latter are said to form a *complete* set.

The coefficients x_i are called the *components* of $\mathbf{x}$ with respect to the $\mathbf{e}_i$-basis. They are *unique*, since if both

$$\mathbf{x} = \sum_{i=1}^{N} x_i\mathbf{e}_i \quad \text{and} \quad \mathbf{x} = \sum_{i=1}^{N} y_i\mathbf{e}_i, \quad \text{then} \quad \sum_{i=1}^{N} (x_i - y_i)\mathbf{e}_i = \mathbf{0}. \tag{1.10}$$

Since the $\mathbf{e}_i$ are linearly independent, each coefficient in the final equation in (1.10) must be individually zero and so $x_i = y_i$ for all $i = 1, 2, \ldots, N$.

It follows from this that *any* set of N linearly independent vectors can form a basis for an N-dimensional space.[2] If we choose a different set $\mathbf{e}'_i$, $i = 1, \ldots, N$ then we can write $\mathbf{x}$ as

$$\mathbf{x} = x'_1\mathbf{e}'_1 + x'_2\mathbf{e}'_2 + \cdots + x'_N\mathbf{e}'_N = \sum_{i=1}^{N} x'_i\mathbf{e}'_i, \tag{1.11}$$

but this does not change the vector $\mathbf{x}$. The vector $\mathbf{x}$ (a geometrical entity) is independent of the basis – it is only the components of $\mathbf{x}$ that depend upon the basis.

1.1.2 The inner product

This subsection contains a working summary of the definition and properties of inner products; for a fuller mathematical treatment the reader is referred to Appendix B.

To describe how two vectors in a vector space "multiply" (as opposed to add or subtract) we define their *inner product*, denoted in general by $\langle \mathbf{a}|\mathbf{b}\rangle$. This is a scalar function of vectors $\mathbf{a}$ and $\mathbf{b}$, though it is not necessarily real. Alternative notations for $\langle \mathbf{a}|\mathbf{b}\rangle$ are $(\mathbf{a}, \mathbf{b})$, or simply $\mathbf{a} \cdot \mathbf{b}$.

The scalar or dot product, $\mathbf{a} \cdot \mathbf{b} \equiv |\mathbf{a}||\mathbf{b}| \cos\theta$, of two vectors in real three-dimensional space (where θ is the angle between the vectors) is an example of an inner product. In effect the notion of an inner product $\langle \mathbf{a}|\mathbf{b}\rangle$ is a generalization of the dot product to more abstract vector spaces. The inner product has the following properties (in which, as usual,

2 All bases contain *exactly* N base vectors. A (putative) alternative base with M ($< N$) vectors would imply that there is no set of more than M linearly independent vectors – but the original base is just such a set, giving a contradiction. Equally, $M > N$ would imply the existence of a linearly independent set with more than N members – contradicting the specification for the original base set. Hence $M = N$.

a * superscript denotes complex conjugation):[3]

$$\langle \mathbf{a}|\mathbf{b}\rangle = \langle \mathbf{b}|\mathbf{a}\rangle^*, \tag{1.12}$$

$$\langle \mathbf{a}|\lambda\mathbf{b} + \mu\mathbf{c}\rangle = \lambda\,\langle \mathbf{a}|\mathbf{b}\rangle + \mu\,\langle \mathbf{a}|\mathbf{c}\rangle, \tag{1.13}$$

$$\langle \lambda\mathbf{a} + \mu\mathbf{b}|\mathbf{c}\rangle = \lambda^*\,\langle \mathbf{a}|\mathbf{c}\rangle + \mu^*\,\langle \mathbf{b}|\mathbf{c}\rangle, \tag{1.14}$$

$$\langle \lambda\mathbf{a}|\mu\mathbf{b}\rangle = \lambda^*\mu\,\langle \mathbf{a}|\mathbf{b}\rangle. \tag{1.15}$$

Following the analogy with the dot product in three-dimensional real space, two vectors in a general vector space are defined to be *orthogonal* if $\langle \mathbf{a}|\mathbf{b}\rangle = 0$.

In the same way, the *norm* of a vector $\mathbf{a}$, defined by $||\mathbf{a}|| = \langle \mathbf{a}|\mathbf{a}\rangle^{1/2}$, is clearly a generalization of the length or modulus $|\mathbf{a}|$ of a vector $\mathbf{a}$ in three-dimensional space. In a general vector space $\langle \mathbf{a}|\mathbf{a}\rangle$ can be positive or negative; however, we will be concerned only with spaces in which $\langle \mathbf{a}|\mathbf{a}\rangle \geq 0$ and which are therefore said to have a *positive semi-definite norm*. In such a space $\langle \mathbf{a}|\mathbf{a}\rangle = 0$ implies $\mathbf{a} = \mathbf{0}$.

It is usual when working with an N-dimensional vector space to use a basis $\hat{\mathbf{e}}_1, \hat{\mathbf{e}}_2, \ldots, \hat{\mathbf{e}}_N$ that has the desirable property of being *orthonormal* (the basis vectors are mutually orthogonal and each has unit norm), i.e. a basis that has the property

$$\langle \hat{\mathbf{e}}_i\,|\,\hat{\mathbf{e}}_j\rangle = \delta_{ij}. \tag{1.16}$$

Here δ_{ij} is the *Kronecker delta* symbol, defined by the properties

$$\delta_{ij} = \begin{cases} 1 & \text{for } i = j, \\ 0 & \text{for } i \neq j. \end{cases}$$

Using the above basis, any two vectors $\mathbf{a}$ and $\mathbf{b}$ can be written as

$$\mathbf{a} = \sum_{i=1}^{N} a_i\,\hat{\mathbf{e}}_i \qquad \text{and} \qquad \mathbf{b} = \sum_{i=1}^{N} b_i\,\hat{\mathbf{e}}_i.$$

Furthermore, *in such an orthonormal basis* we have, for any $\mathbf{a}$,

$$\langle \hat{\mathbf{e}}_j|\mathbf{a}\rangle = \sum_{i=1}^{N} \langle \hat{\mathbf{e}}_j|a_i\,\hat{\mathbf{e}}_i\rangle = \sum_{i=1}^{N} a_i\,\langle \hat{\mathbf{e}}_j|\,\hat{\mathbf{e}}_i\rangle = a_j. \tag{1.17}$$

Thus the components of $\mathbf{a}$ are given by $a_i = \langle\,\hat{\mathbf{e}}_i\,|\mathbf{a}\rangle$. Note that this is *not* true unless the basis is orthonormal.

We can write the inner product of $\mathbf{a}$ and $\mathbf{b}$ in terms of their components in an orthonormal basis as

$$\begin{aligned}\langle \mathbf{a}|\mathbf{b}\rangle &= \langle a_1\,\hat{\mathbf{e}}_1 + a_2\,\hat{\mathbf{e}}_2 + \cdots + a_N\,\hat{\mathbf{e}}_N|b_1\,\hat{\mathbf{e}}_1 + b_2\,\hat{\mathbf{e}}_2 + \cdots + b_N\,\hat{\mathbf{e}}_N\rangle \\ &= \sum_{i=1}^{N} a_i^* b_i\,\langle \hat{\mathbf{e}}_i\,|\,\hat{\mathbf{e}}_i\rangle + \sum_{i=1}^{N}\sum_{j\neq i}^{N} a_i^* b_j\,\langle \hat{\mathbf{e}}_i\,|\,\hat{\mathbf{e}}_j\rangle \\ &= \sum_{i=1}^{N} a_i^* b_i,\end{aligned}$$

3 It is a useful exercise in close analysis to deduce properties (1.14) and (1.15), on a justified step-by-step basis, using only those given in (1.12) and (1.13) and the general properties of complex conjugation.

where the second equality follows from (1.15) and the third from (1.16). This is clearly a generalization of the expression for the dot product of vectors in three-dimensional space.

The extension of the above results to the case where the base vectors $\mathbf{e}_1, \mathbf{e}_2, \ldots, \mathbf{e}_N$ are *not* orthonormal is more mathematically complicated and given in Appendix B.

1.1.3 Some useful inequalities

For a set of objects (vectors) forming a linear vector space in which $\langle \mathbf{a}|\mathbf{a}\rangle \geq 0$ for all $\mathbf{a}$, there are a number of inequalities that often prove useful. Here we only list them; for the corresponding proofs the reader is referred to Appendix C.

(i) *Schwarz's inequality* states that

$$|\langle \mathbf{a}|\mathbf{b}\rangle| \leq ||\mathbf{a}||\,||\mathbf{b}||, \tag{1.18}$$

where the equality holds when $\mathbf{a}$ is a scalar multiple of $\mathbf{b}$, i.e. when $\mathbf{a} = \lambda\mathbf{b}$. It is important here to distinguish between the *absolute value* of a scalar, $|\lambda|$, and the *norm* of a vector, $||\mathbf{a}||$.

(ii) The *triangle inequality* states that

$$||\mathbf{a}+\mathbf{b}|| \leq ||\mathbf{a}|| + ||\mathbf{b}|| \tag{1.19}$$

and is the intuitive analogue of the observation that the length of any one side of a triangle cannot be greater than the sum of the lengths of the other two sides.

(iii) *Bessel's inequality* states that if $\hat{\mathbf{e}}_i$, $i = 1, 2, \ldots, N$ form an orthonormal basis in an N-dimensional vector space, then

$$||\mathbf{a}||^2 \geq \sum_i^M |\langle \hat{\mathbf{e}}_i|\mathbf{a}\rangle|^2, \tag{1.20}$$

where the equality holds if $M = N$. If $M < N$ then inequality results, unless the basis vectors omitted all have $a_i = 0$. This is the analogue of $|\mathbf{x}|^2$ for a three-dimensional vector $\mathbf{v}$ being equal to the sum of the squares of all its components, and if any are omitted the sum may fall short of $|\mathbf{x}|^2$.

To these inequalities can be added one equality that sometimes proves useful. The *parallelogram equality* reads

$$||\mathbf{a}+\mathbf{b}||^2 + ||\mathbf{a}-\mathbf{b}||^2 = 2\left(||\mathbf{a}||^2 + ||\mathbf{b}||^2\right), \tag{1.21}$$

and may be proved straightforwardly from the properties of the inner product.

1.2 Linear operators

We now discuss the action of *linear operators* on vectors in a vector space. A linear operator $\mathcal{A}$ associates with every vector $\mathbf{x}$ another vector

$$\mathbf{y} = \mathcal{A}\mathbf{x},$$

in such a way that, for two vectors $\mathbf{a}$ and $\mathbf{b}$,

$$\mathcal{A}(\lambda\mathbf{a} + \mu\mathbf{b}) = \lambda\mathcal{A}\mathbf{a} + \mu\mathcal{A}\mathbf{b},$$

where λ, μ are scalars. We say that $\mathcal{A}$ "operates" on $\mathbf{x}$ to give the vector $\mathbf{y}$. We note that the action of $\mathcal{A}$ is *independent* of any basis or coordinate system and may be thought of as "transforming" one geometrical entity (i.e. a vector) into another.

If we now introduce a basis $\mathbf{e}_i$, $i = 1, 2, \ldots, N$, into our vector space then the action of $\mathcal{A}$ on each of the basis vectors is to produce a linear combination of the latter; this may be written as

$$\mathcal{A}\mathbf{e}_j = \sum_{i=1}^{N} A_{ij}\mathbf{e}_i, \tag{1.22}$$

where A_{ij} is the ith component of the vector $\mathcal{A}\mathbf{e}_j$ in this basis; collectively the numbers A_{ij} are called the components of the linear operator in the $\mathbf{e}_i$-basis. *In this basis* we can express the relation $\mathbf{y} = \mathcal{A}\mathbf{x}$ in component form as

$$\mathbf{y} = \sum_{i=1}^{N} y_i\mathbf{e}_i = \mathcal{A}\left(\sum_{j=1}^{N} x_j\mathbf{e}_j\right) = \sum_{j=1}^{N} x_j \sum_{i=1}^{N} A_{ij}\mathbf{e}_i,$$

and hence, in purely component form, in this basis we have

$$y_i = \sum_{j=1}^{N} A_{ij}x_j. \tag{1.23}$$

If we had chosen a different basis $\mathbf{e}'_i$, in which the components of $\mathbf{x}$, $\mathbf{y}$ and $\mathcal{A}$ are x'_i, y'_i and A'_{ij} respectively then the geometrical relationship $\mathbf{y} = \mathcal{A}\mathbf{x}$ would be represented in this new basis by

$$y'_i = \sum_{j=1}^{N} A'_{ij}x'_j.$$

We have so far assumed that the vector $\mathbf{y}$ is in the same vector space as $\mathbf{x}$. If, however, $\mathbf{y}$ belongs to a different vector space, which may in general be M-dimensional ($M \neq N$) then the above analysis needs a slight modification. By introducing a basis set $\mathbf{f}_i$, $i = 1, 2, \ldots, M$, into the vector space to which $\mathbf{y}$ belongs we may generalize (1.22) as

$$\mathcal{A}\mathbf{e}_j = \sum_{i=1}^{M} A_{ij}\mathbf{f}_i,$$

where the components A_{ij} of the linear operator $\mathcal{A}$ relate to both of the bases $\mathbf{e}_j$ and $\mathbf{f}_i$.

The basic properties of linear operators, arising from their definition, are summarized as follows. If $\mathbf{x}$ is a vector and $\mathcal{A}$ and $\mathcal{B}$ are two linear operators then

$$(\mathcal{A} + \mathcal{B})\mathbf{x} = \mathcal{A}\mathbf{x} + \mathcal{B}\mathbf{x},$$
$$(\lambda\mathcal{A})\mathbf{x} = \lambda(\mathcal{A}\mathbf{x}),$$
$$(\mathcal{A}\mathcal{B})\mathbf{x} = \mathcal{A}(\mathcal{B}\mathbf{x}),$$

where in the last equality we see that the action of two linear operators in succession is associative. However, the product of two general linear operators is not commutative, i.e. $\mathcal{AB}\mathbf{x} \neq \mathcal{BA}\mathbf{x}$ in general.[4]

In an obvious way we define the null (or zero) and identity operators by

$$\mathcal{O}\mathbf{x} = \mathbf{0} \qquad \text{and} \qquad \mathcal{I}\mathbf{x} = \mathbf{x},$$

for any vector $\mathbf{x}$ in our vector space. Two operators $\mathcal{A}$ and $\mathcal{B}$ are equal if $\mathcal{A}\mathbf{x} = \mathcal{B}\mathbf{x}$ for all vectors $\mathbf{x}$. Finally, if there exists an operator $\mathcal{A}^{-1}$ such that

$$\mathcal{A}\mathcal{A}^{-1} = \mathcal{A}^{-1}\mathcal{A} = \mathcal{I}$$

then $\mathcal{A}^{-1}$ is the *inverse* of $\mathcal{A}$. Some linear operators do not possess an inverse and are called *singular*, whilst those operators that do have an inverse are termed *non-singular*.

1.3 Matrices

We have seen that in a particular basis $\mathbf{e}_i$ both vectors and linear operators can be described in terms of their components with respect to the basis. These components may be displayed as an array of numbers called a *matrix*. In general, if a linear operator $\mathcal{A}$ transforms vectors from an N-dimensional vector space, for which we choose a basis $\mathbf{e}_j$, $j = 1, 2, \ldots, N$, into vectors belonging to an M-dimensional vector space, with basis $\mathbf{f}_i$, $i = 1, 2, \ldots, M$, then we may represent the operator $\mathcal{A}$ by the matrix

$$\mathsf{A} = \begin{pmatrix} A_{11} & A_{12} & \ldots & A_{1N} \\ A_{21} & A_{22} & \ldots & A_{2N} \\ \vdots & \vdots & \ddots & \vdots \\ A_{M1} & A_{M2} & \ldots & A_{MN} \end{pmatrix}. \tag{1.24}$$

The *matrix elements* A_{ij} are the components of the linear operator with respect to the bases $\mathbf{e}_j$ and $\mathbf{f}_i$; the component A_{ij} of the linear operator appears in the ith row and jth column of the matrix. The array has M rows and N columns and is thus called an $M \times N$ matrix. If the dimensions of the two vector spaces are the same, i.e. $M = N$ (for example, if they are the same vector space) then we may represent $\mathcal{A}$ by an $N \times N$ or *square* matrix of *order* N. The component A_{ij}, which in general may be complex, is also commonly denoted by $(\mathsf{A})_{ij}$.

In a similar way we may denote a vector $\mathbf{x}$ in terms of its components x_i in a basis $\mathbf{e}_i$, $i = 1, 2, \ldots, N$, by the array

$$\mathsf{x} = \begin{pmatrix} x_1 \\ x_2 \\ \vdots \\ x_N \end{pmatrix},$$

4 Consider a two-dimensional linear vector space in which a typical vector is $\mathbf{x} = (x_1, x_2)$, with linear operators $\mathcal{A}$, $\mathcal{B}$ and $\mathcal{C}$ defined by $\mathcal{A}\mathbf{x} = (2x_1 + x_2, x_2)$, $\mathcal{B}\mathbf{x} = (x_1, x_1 + 2x_2)$ and $\mathcal{C}\mathbf{x} = (x_1 - x_2, 2x_2)$. Show that, although $\mathcal{A}$ and $\mathcal{C}$ commute, $\mathcal{A}$ and $\mathcal{B}$ do not.

which is a special case of (1.24) and is called a *column matrix* (or conventionally, and slightly confusingly, a *column vector* or even just a *vector* – strictly speaking the term "vector" refers to the geometrical entity $\mathbf{x}$). The column matrix x can also be written as

$$\mathsf{x} = (x_1 \quad x_2 \quad \cdots \quad x_N)^{\mathrm{T}},$$

which is the *transpose* of a *row matrix* (see Section 1.6).

We note that in a different basis $\mathbf{e}'_i$ the vector $\mathbf{x}$ would be represented by a *different* column matrix containing the components x'_i in the new basis, i.e.

$$\mathsf{x}' = \begin{pmatrix} x'_1 \\ x'_2 \\ \vdots \\ x'_N \end{pmatrix}.$$

Thus, we use x and x' to denote different column matrices which, in different bases $\mathbf{e}_i$ and $\mathbf{e}'_i$, represent the *same* vector $\mathbf{x}$. In many texts, however, this distinction is not made and $\mathbf{x}$ (rather than x) is equated to the corresponding column matrix; if we regard $\mathbf{x}$ as the geometrical entity, however, this can be misleading and so we explicitly make the distinction. A similar argument follows for linear operators; the same linear operator $\mathcal{A}$ is described in different bases by different matrices A and A', containing different matrix elements.

1.4 Basic matrix algebra

The basic algebra of matrices may be deduced from the properties of the linear operators that they represent. In a given basis the action of two linear operators $\mathcal{A}$ and $\mathcal{B}$ on an arbitrary vector $\mathbf{x}$ (see towards the end of Section 1.2), when written in terms of components using (1.23), is given by

$$\sum_j (\mathsf{A}+\mathsf{B})_{ij} x_j = \sum_j A_{ij} x_j + \sum_j B_{ij} x_j,$$
$$\sum_j (\lambda \mathsf{A})_{ij} x_j = \lambda \sum_j A_{ij} x_j,$$
$$\sum_j (\mathsf{AB})_{ij} x_j = \sum_k A_{ik} (\mathsf{Bx})_k = \sum_j \sum_k A_{ik} B_{kj} x_j.$$

Now, since $\mathbf{x}$ is arbitrary, we can immediately deduce the way in which matrices are added or multiplied, i.e.[5]

$$(\mathsf{A}+\mathsf{B})_{ij} = A_{ij} + B_{ij}, \tag{1.25}$$

5 Express the operators appearing in footnote 4 in matrix form and then use (1.27) to demonstrate their commutation or otherwise. Do operators $\mathcal{B}$ and $\mathcal{C}$ commute?

$$(\lambda \mathsf{A})_{ij} = \lambda A_{ij}, \tag{1.26}$$

$$(\mathsf{AB})_{ij} = \sum_k A_{ik} B_{kj}. \tag{1.27}$$

We note that a matrix element may, in general, be complex. We now discuss matrix addition and multiplication in more detail.

1.4.1 Matrix addition and multiplication by a scalar

From (1.25) we see that the sum of two matrices, $\mathsf{S} = \mathsf{A} + \mathsf{B}$, is the matrix whose elements are given by

$$S_{ij} = A_{ij} + B_{ij}$$

for every pair of subscripts i, j, with $i = 1, 2, \ldots, M$ and $j = 1, 2, \ldots, N$. For example, if A and B are 2×3 matrices then $\mathsf{S} = \mathsf{A} + \mathsf{B}$ is given by

$$\begin{pmatrix} S_{11} & S_{12} & S_{13} \\ S_{21} & S_{22} & S_{23} \end{pmatrix} = \begin{pmatrix} A_{11} & A_{12} & A_{13} \\ A_{21} & A_{22} & A_{23} \end{pmatrix} + \begin{pmatrix} B_{11} & B_{12} & B_{13} \\ B_{21} & B_{22} & B_{23} \end{pmatrix}$$
$$= \begin{pmatrix} A_{11}+B_{11} & A_{12}+B_{12} & A_{13}+B_{13} \\ A_{21}+B_{21} & A_{22}+B_{22} & A_{23}+B_{23} \end{pmatrix}. \tag{1.28}$$

Clearly, for the sum of two matrices to have any meaning, the matrices must have the same dimensions, i.e. both be $M \times N$ matrices.

From definition (1.28) it follows that $\mathsf{A} + \mathsf{B} = \mathsf{B} + \mathsf{A}$ and that the sum of a number of matrices can be written unambiguously without bracketing, i.e. matrix addition is *commutative* and *associative*.

The difference of two matrices is defined by direct analogy with addition. The matrix $\mathsf{D} = \mathsf{A} - \mathsf{B}$ has elements

$$D_{ij} = A_{ij} - B_{ij}, \qquad \text{for } i = 1, 2, \ldots, M,\ j = 1, 2, \ldots, N. \tag{1.29}$$

From (1.26) the product of a matrix A with a scalar λ is the matrix with elements λA_{ij}, for example

$$\lambda \begin{pmatrix} A_{11} & A_{12} & A_{13} \\ A_{21} & A_{22} & A_{23} \end{pmatrix} = \begin{pmatrix} \lambda A_{11} & \lambda A_{12} & \lambda A_{13} \\ \lambda A_{21} & \lambda A_{22} & \lambda A_{23} \end{pmatrix}. \tag{1.30}$$

Multiplication by a scalar is distributive and associative.

The following example illustrates these three elementary properties or definitions.

Example The matrices A, B and C are given by

$$\mathsf{A} = \begin{pmatrix} 2 & -1 \\ 3 & 1 \end{pmatrix}, \quad \mathsf{B} = \begin{pmatrix} 1 & 0 \\ 0 & -2 \end{pmatrix}, \quad \mathsf{C} = \begin{pmatrix} -2 & 1 \\ -1 & 1 \end{pmatrix}.$$

Find the matrix $\mathsf{D} = \mathsf{A} + 2\mathsf{B} - \mathsf{C}$.

Dealing separately with the elements in each particular position in the various matrices, we have

$$\mathsf{D} = \begin{pmatrix} 2 & -1 \\ 3 & 1 \end{pmatrix} + 2\begin{pmatrix} 1 & 0 \\ 0 & -2 \end{pmatrix} - \begin{pmatrix} -2 & 1 \\ -1 & 1 \end{pmatrix}$$
$$= \begin{pmatrix} 2+2\times 1-(-2) & -1+2\times 0-1 \\ 3+2\times 0-(-1) & 1+2\times(-2)-1 \end{pmatrix} = \begin{pmatrix} 6 & -2 \\ 4 & -4 \end{pmatrix}.$$

As a reminder, we note that for the question to have had any meaning, A, B and C all had to have the same dimensions, 2×2 in practice; the answer, D, is also 2×2. ◀

From the above considerations we see that the set of all, in general complex, $M \times N$ matrices (with fixed M and N) provide an example of a linear vector space – one whose elements have no obvious "arrow-like" qualities.

The space is of dimension MN. One basis for it is the set of $M \times N$ matrices $\mathsf{E}^{(p,q)}$ with the property that $E_{ij}^{(p,q)} = 1$ if $i = p$ and $j = q$ whilst $E_{ij}^{(p,q)} = 0$ for all other values of i and j, i.e. each matrix has only one non-zero entry, and that equals unity. Here the pair (p, q) is simply a label that picks out a particular one of the matrices $E^{(p,q)}$, the total number of which is MN.

1.4.2 Multiplication of matrices

Let us consider again the "transformation" of one vector into another, $\mathbf{y} = \mathcal{A}\mathbf{x}$, which, from (1.23), may be described in terms of components with respect to a particular basis as

$$y_i = \sum_{j=1}^{N} A_{ij} x_j \qquad \text{for } i = 1, 2, \ldots, M. \tag{1.31}$$

Writing this in matrix form as $\mathsf{y} = \mathsf{A}\mathsf{x}$ we have

$$\begin{pmatrix} y_1 \\ \boxed{y_2} \\ \vdots \\ y_M \end{pmatrix} = \begin{pmatrix} A_{11} & A_{12} & \ldots & A_{1N} \\ \hline A_{21} & A_{22} & \ldots & A_{2N} \\ \hline \vdots & \vdots & \ddots & \vdots \\ A_{M1} & A_{M2} & \ldots & A_{MN} \end{pmatrix} \begin{pmatrix} \boxed{\begin{matrix} x_1 \\ x_2 \\ \vdots \\ x_N \end{matrix}} \end{pmatrix} \tag{1.32}$$

where we have highlighted with boxes the components used to calculate the element y_2: using (1.31) for $i = 2$,

$$y_2 = A_{21}x_1 + A_{22}x_2 + \cdots + A_{2N}x_N.$$

All the other components y_i are calculated similarly.

If, instead, we operate with $\mathcal{A}$ on a basis vector $\mathbf{e}_j$ having all components zero except for the jth, which equals unity, then we find

$$\mathsf{A}\mathbf{e}_j = \begin{pmatrix} A_{11} & A_{12} & \ldots & A_{1N} \\ \boxed{A_{21} \quad A_{22} \quad \ldots \quad A_{2N}} \\ \vdots & \vdots & \ddots & \vdots \\ A_{M1} & A_{M2} & \ldots & A_{MN} \end{pmatrix} \boxed{\begin{pmatrix} 0 \\ 0 \\ \vdots \\ 1 \\ \vdots \\ 0 \end{pmatrix}} = \begin{pmatrix} A_{1j} \\ \boxed{A_{2j}} \\ \vdots \\ A_{Mj} \end{pmatrix},$$

and so confirm our identification of the matrix element A_{ij} as the ith component of $\mathsf{A}\mathbf{e}_j$ in this basis.

From (1.27) we can extend our discussion to the product of two matrices $\mathsf{P} = \mathsf{AB}$, where P is the matrix of the quantities formed by the operation of the rows of A on the columns of B, treating each column of B in turn as the vector $\mathbf{x}$ represented in component form in (1.31). It is clear that, for this to be a meaningful definition, the number of columns in A must equal the number of rows in B. Thus the product AB of an $M \times N$ matrix A with an $N \times R$ matrix B is itself an $M \times R$ matrix P, where

$$P_{ij} = \sum_{k=1}^{N} A_{ik} B_{kj} \qquad \text{for } i = 1, 2, \ldots, M, \quad j = 1, 2, \ldots, R.$$

For example, $\mathsf{P} = \mathsf{AB}$ may be written in matrix form

$$\begin{pmatrix} \boxed{P_{11}} & P_{12} \\ P_{21} & P_{22} \end{pmatrix} = \begin{pmatrix} \boxed{A_{11} \quad A_{12} \quad A_{13}} \\ A_{21} \quad A_{22} \quad A_{23} \end{pmatrix} \begin{pmatrix} \boxed{\begin{matrix} B_{11} \\ B_{21} \\ B_{31} \end{matrix}} & \begin{matrix} B_{12} \\ B_{22} \\ B_{32} \end{matrix} \end{pmatrix}$$

where

$$\begin{aligned} P_{11} &= A_{11}B_{11} + A_{12}B_{21} + A_{13}B_{31}, \\ P_{21} &= A_{21}B_{11} + A_{22}B_{21} + A_{23}B_{31}, \\ P_{12} &= A_{11}B_{12} + A_{12}B_{22} + A_{13}B_{32}, \\ P_{22} &= A_{21}B_{12} + A_{22}B_{22} + A_{23}B_{32}. \end{aligned}$$

Multiplication of more than two matrices follows naturally and is associative. So, for example,

$$\mathsf{A(BC)} \equiv \mathsf{(AB)C}, \qquad (1.33)$$

provided, of course, that all the products are defined.

As mentioned above, if A is an $M \times N$ matrix and B is an $N \times M$ matrix then two product matrices are possible, i.e.

$$\mathsf{P} = \mathsf{AB} \qquad \text{and} \qquad \mathsf{Q} = \mathsf{BA}.$$

These are clearly not the same, since P is an $M \times M$ matrix whilst Q is an $N \times N$ matrix. Thus, particular care must be taken to write matrix products in the intended order; $\mathsf{P} = \mathsf{AB}$

but $\mathsf{Q} = \mathsf{BA}$. We note in passing that A^2 means AA, A^3 means $\mathsf{A(AA)} = \mathsf{(AA)A}$ etc. Even if both A and B are square, in general

$$\mathsf{AB} \neq \mathsf{BA}, \tag{1.34}$$

i.e. the multiplication of matrices is not, in general, commutative. Consider the following.

Example Evaluate $\mathsf{P} = \mathsf{AB}$ and $\mathsf{Q} = \mathsf{BA}$ where

$$\mathsf{A} = \begin{pmatrix} 3 & 2 & -1 \\ 0 & 3 & 2 \\ 1 & -3 & 4 \end{pmatrix}, \qquad \mathsf{B} = \begin{pmatrix} 2 & -2 & 3 \\ 1 & 1 & 0 \\ 3 & 2 & 1 \end{pmatrix}.$$

As we saw for the 2×2 case above, the element P_{ij} of the matrix $\mathsf{P} = \mathsf{AB}$ is found by mentally taking the "scalar product" of the ith row of A with the jth column of B. For example, $P_{11} = 3 \times 2 + 2 \times 1 + (-1) \times 3 = 5$, $P_{12} = 3 \times (-2) + 2 \times 1 + (-1) \times 2 = -6$, etc. Thus

$$\mathsf{P} = \mathsf{AB} = \begin{pmatrix} 3 & 2 & -1 \\ 0 & 3 & 2 \\ 1 & -3 & 4 \end{pmatrix} \begin{pmatrix} 2 & -2 & 3 \\ 1 & 1 & 0 \\ 3 & 2 & 1 \end{pmatrix} = \begin{pmatrix} 5 & -6 & 8 \\ 9 & 7 & 2 \\ 11 & 3 & 7 \end{pmatrix},$$

and, similarly,

$$\mathsf{Q} = \mathsf{BA} = \begin{pmatrix} 2 & -2 & 3 \\ 1 & 1 & 0 \\ 3 & 2 & 1 \end{pmatrix} \begin{pmatrix} 3 & 2 & -1 \\ 0 & 3 & 2 \\ 1 & -3 & 4 \end{pmatrix} = \begin{pmatrix} 9 & -11 & 6 \\ 3 & 5 & 1 \\ 10 & 9 & 5 \end{pmatrix}.$$

These results illustrate that, in general, two matrices do not commute. ◀

The property that matrix multiplication is distributive over addition, i.e. that

$$(\mathsf{A} + \mathsf{B})\mathsf{C} = \mathsf{AC} + \mathsf{BC} \quad \text{and} \quad \mathsf{C}(\mathsf{A} + \mathsf{B}) = \mathsf{CA} + \mathsf{CB}, \tag{1.35}$$

follows directly from its definition.[6]

1.4.3 The null and identity matrices

Both the null matrix and the identity matrix are frequently encountered, and we take this opportunity to introduce them briefly, leaving their uses until later. The *null* or *zero* matrix $\mathsf{0}$ has all elements equal to zero, and so its properties are

$$\mathsf{A0} = \mathsf{0} = \mathsf{0A},$$
$$\mathsf{A} + \mathsf{0} = \mathsf{0} + \mathsf{A} = \mathsf{A}.$$

The *identity* matrix I has the property

$$\mathsf{AI} = \mathsf{IA} = \mathsf{A}.$$

6 But show that $(\mathsf{A} + \mathsf{B})(\mathsf{A} - \mathsf{B}) = \mathsf{A}^2 - \mathsf{B}^2$ if, and only if, A and B commute.

It is clear that, in order for the above products to be defined, the identity matrix must be square. The $N \times N$ identity matrix (often denoted by I_N) has the form

$$\mathsf{I}_N = \begin{pmatrix} 1 & 0 & \cdots & 0 \\ 0 & 1 & & \vdots \\ \vdots & & \ddots & 0 \\ 0 & \cdots & 0 & 1 \end{pmatrix}.$$

1.5 Functions of matrices

If a matrix A is *square* then, as mentioned above, one can define *powers* of A in a straightforward way. For example $\mathsf{A}^2 = \mathsf{AA}$, $\mathsf{A}^3 = \mathsf{AAA}$, or in the general case

$$\mathsf{A}^n = \mathsf{AA} \cdots \mathsf{A} \qquad (n \text{ times}),$$

where n is a positive integer. Having defined powers of a square matrix A, we may construct *functions* of A of the form

$$\mathsf{S} = \sum_n a_n \mathsf{A}^n,$$

where the a_k are simple scalars and the number of terms in the summation may be finite or infinite. In the case where the sum has an infinite number of terms, the sum has meaning only if it converges. A common example of such a function is the *exponential* of a matrix, which is defined by

$$\exp \mathsf{A} = \sum_{n=0}^{\infty} \frac{\mathsf{A}^n}{n!}. \qquad (1.36)$$

This definition can, in turn, be used to define other functions such as $\sin \mathsf{A}$ and $\cos \mathsf{A}$.[7]

1.6 The transpose of a matrix

In the next few sections we will consider some of the quantities that characterize any given matrix and also some other matrices that can be derived from the original. A tabulation of these derived quantities and matrices is given in the end-of-chapter Summary. We start with the transposed matrix.

We have seen that the components of a linear operator in a given coordinate system can be written in the form of a matrix A. We will also find it useful, however, to consider the different (but clearly related) matrix formed by interchanging the rows and columns of A. The matrix is called the *transpose* of A and is denoted by A^{T}. It is obvious that if A is an $M \times N$ matrix then its transpose A^{T} is an $N \times M$ matrix.

7 For the 3×3 matrix A that has $A_{11} = A_{33} = 1$, $A_{22} = -1$ and all other $A_{ij} = 0$, show that the trace of $\exp i\mathsf{A}$, i.e. the sum of its diagonal elements, is equal to $3 \cos 1 + i \sin 1$.

Example Find the transpose of the matrix

$$\mathsf{A} = \begin{pmatrix} 3 & 1 & 2 \\ 0 & 4 & 1 \end{pmatrix}.$$

By interchanging the rows and columns of A we immediately obtain

$$\mathsf{A}^{\mathrm{T}} = \begin{pmatrix} 3 & 0 \\ 1 & 4 \\ 2 & 1 \end{pmatrix}.$$

As it must be, given that A is a 2×3 matrix, A^{T} is a 3×2 matrix. ◀

As mentioned in Section 1.3, the transpose of a column matrix is a row matrix and vice versa. An important use of column and row matrices is in the representation of the inner product of two real vectors in terms of their components in a given basis. This notion is discussed fully in the next section, where it is extended to complex vectors.

The transpose of the product of two matrices, $(\mathsf{AB})^{\mathrm{T}}$, is given by the product of their transposes taken in the reverse order, i.e.

$$(\mathsf{AB})^{\mathrm{T}} = \mathsf{B}^{\mathrm{T}}\mathsf{A}^{\mathrm{T}}. \tag{1.37}$$

This is proved as follows:

$$\begin{aligned}(\mathsf{AB})^{\mathrm{T}}_{ij} = (\mathsf{AB})_{ji} &= \sum_k A_{jk}B_{ki} \\ &= \sum_k (\mathsf{A}^{\mathrm{T}})_{kj}(\mathsf{B}^{\mathrm{T}})_{ik} = \sum_k (\mathsf{B}^{\mathrm{T}})_{ik}(\mathsf{A}^{\mathrm{T}})_{kj} = (\mathsf{B}^{\mathrm{T}}\mathsf{A}^{\mathrm{T}})_{ij},\end{aligned}$$

and the proof can be extended to the product of several matrices to give[8]

$$(\mathsf{ABC}\cdots\mathsf{G})^{\mathrm{T}} = \mathsf{G}^{\mathrm{T}}\cdots\mathsf{C}^{\mathrm{T}}\mathsf{B}^{\mathrm{T}}\mathsf{A}^{\mathrm{T}}.$$

1.7 The complex and Hermitian conjugates of a matrix

Two further matrices that can be derived from a given general $M \times N$ matrix are the *complex conjugate*, denoted by A^*, and the *Hermitian conjugate*, denoted by $\mathsf{A}^\dagger$.

The complex conjugate of a matrix A is the matrix obtained by taking the complex conjugate of each of the elements of A, i.e.

$$(\mathsf{A}^*)_{ij} = (A_{ij})^*.$$

Obviously if a matrix is *real* (i.e. it contains only real elements) then $\mathsf{A}^* = \mathsf{A}$.

8 Convince yourself that, even if A, B, C, ..., G are not necessarily square matrices, but are compatible and the product $\mathsf{ABC}\cdots\mathsf{G}$ is meaningful, then their transposes are such that the product given on the RHS is also meaningful.

Example Find the complex conjugate of the matrix

$$\mathsf{A} = \begin{pmatrix} 1 & 2 & 3i \\ 1+i & 1 & 0 \end{pmatrix}.$$

By taking the complex conjugate of each element in turn,

$$\mathsf{A}^* = \begin{pmatrix} 1 & 2 & -3i \\ 1-i & 1 & 0 \end{pmatrix},$$

the complex conjugate of the whole matrix is obtained immediately. ◀

The Hermitian conjugate, or *adjoint*, of a matrix A is the transpose of its complex conjugate, or equivalently, the complex conjugate of its transpose, i.e.

$$\mathsf{A}^\dagger = (\mathsf{A}^*)^{\mathrm{T}} = (\mathsf{A}^{\mathrm{T}})^*.$$

We note that if A is real (and so $\mathsf{A}^* = \mathsf{A}$) then $\mathsf{A}^\dagger = \mathsf{A}^{\mathrm{T}}$, and taking the Hermitian conjugate is equivalent to taking the transpose. Following the previous line of argument for the transpose of the product of several matrices, the Hermitian conjugate of such a product can be shown to be given by

$$(\mathsf{AB}\cdots\mathsf{G})^\dagger = \mathsf{G}^\dagger\cdots\mathsf{B}^\dagger\mathsf{A}^\dagger. \tag{1.38}$$

Example Find the Hermitian conjugate of the matrix

$$\mathsf{A} = \begin{pmatrix} 1 & 2 & 3i \\ 1+i & 1 & 0 \end{pmatrix}.$$

Taking the complex conjugate of A from the previous example, and then forming its transpose, we find

$$\mathsf{A}^\dagger = \begin{pmatrix} 1 & 1-i \\ 2 & 1 \\ -3i & 0 \end{pmatrix}.$$

We could obtain the same result, of course, by first taking the transpose of A and then forming its complex conjugate. ◀

An important use of the Hermitian conjugate (or transpose in the real case) is in connection with the inner product of two vectors. Suppose that in a given orthonormal basis the vectors $\mathbf{a}$ and $\mathbf{b}$ may be represented by the column matrices

$$\mathsf{a} = \begin{pmatrix} a_1 \\ a_2 \\ \vdots \\ a_N \end{pmatrix} \quad \text{and} \quad \mathsf{b} = \begin{pmatrix} b_1 \\ b_2 \\ \vdots \\ b_N \end{pmatrix}. \tag{1.39}$$

Taking the Hermitian conjugate of a, to give a row matrix, and multiplying (on the right) by b we obtain

$$\mathsf{a}^\dagger\mathsf{b} = (a_1^* \; a_2^* \; \cdots \; a_N^*)\begin{pmatrix} b_1 \\ b_2 \\ \vdots \\ b_N \end{pmatrix} = \sum_{i=1}^{N} a_i^* b_i, \tag{1.40}$$

which is the expression for the inner product $\langle \mathbf{a}|\mathbf{b}\rangle$ in that basis.[9] We note that for real vectors (1.40) reduces to $\mathsf{a}^{\mathrm{T}}\mathsf{b} = \sum_{i=1}^{N} a_i b_i$.

1.8 The trace of a matrix

For a given matrix A, in the previous two sections we have considered various other matrices that can be derived from it. However, sometimes one wishes to derive a single number from a matrix. The simplest example is the *trace* (or *spur*) of a square matrix, which is denoted by $\mathrm{Tr}\,\mathsf{A}$. This quantity is defined as the sum of the diagonal elements of the matrix,

$$\mathrm{Tr}\,\mathsf{A} = A_{11} + A_{22} + \cdots + A_{NN} = \sum_{i=1}^{N} A_{ii}. \tag{1.41}$$

At this point, the definition may seem arbitrary, but as will be seen in this section, as well as later in the chapter, the trace of a matrix has properties that characterize the linear operator it represents, and are independent of the basis chosen for that representation. It is clear that taking the trace is a linear operation so that, for example,

$$\mathrm{Tr}(\mathsf{A} \pm \mathsf{B}) = \mathrm{Tr}\,\mathsf{A} \pm \mathrm{Tr}\,\mathsf{B}.$$

A very useful property of traces is that the trace of the product of two matrices is independent of the order of their multiplication; this result holds whether or not the matrices commute and is proved as follows:

$$\mathrm{Tr}\,\mathsf{AB} = \sum_{i=1}^{N}(\mathsf{AB})_{ii} = \sum_{i=1}^{N}\sum_{j=1}^{N} A_{ij}B_{ji} = \sum_{i=1}^{N}\sum_{j=1}^{N} B_{ji}A_{ij} = \sum_{j=1}^{N}(\mathsf{BA})_{jj} = \mathrm{Tr}\,\mathsf{BA}. \tag{1.42}$$

The result can be extended to the product of several matrices. For example, from (1.42), we immediately find

$$\mathrm{Tr}\,\mathsf{ABC} = \mathrm{Tr}\,\mathsf{BCA} = \mathrm{Tr}\,\mathsf{CAB},$$

which shows that the trace of a multiple product is invariant under cyclic permutations of the matrices in the product. Other easily derived properties of the trace are, for example, $\mathrm{Tr}\,\mathsf{A}^{\mathrm{T}} = \mathrm{Tr}\,\mathsf{A}$ and $\mathrm{Tr}\,\mathsf{A}^\dagger = (\mathrm{Tr}\,\mathsf{A})^*$.

9 It also follows that $\mathsf{a}^\dagger\mathbf{a} = \sum_{n=1}^{N} a_i^* a_i = \sum_{n=1}^{N} |a_i|^2$ is real for any vector $\mathbf{a}$, whether or not it has complex components.

1.9 The determinant of a matrix

For a given matrix A, the determinant det A (like the trace) is a single number (or algebraic expression) that depends upon the elements of A. Also like the trace, the determinant is defined only for *square* matrices. If, for example, A is a 3×3 matrix then its determinant, of *order* 3, is denoted by

$$\det \mathsf{A} = |\mathsf{A}| = \begin{vmatrix} A_{11} & A_{12} & A_{13} \\ A_{21} & A_{22} & A_{23} \\ A_{31} & A_{32} & A_{33} \end{vmatrix}, \tag{1.43}$$

i.e. the round or square brackets are replaced by vertical bars, similar to (large) modulus signs, but not to be confused with them.

In order to calculate the value of a general determinant of order n, we first define that of an order-1 determinant. We would not normally refer to an array with only one element as a matrix, but formally it is a 1×1 matrix, and it is useful to think of it as such for the present purposes. The determinant of such a matrix is *defined* to be the value of its single entry. Notice that, although when it is written in determinantal form it looks exactly like a modulus sign, $|a_{11}|$, it must not be treated as such, and, for example, a 1×1 matrix with a single entry -7 has determinant -7, not 7.

In order to define the determinant of an $n \times n$ matrix we will need to introduce the notions of the *minor* and the *cofactor* of an element of a matrix. We will then see that we can use the cofactors to write an order-3 determinant as the weighted sum of three order-2 determinants; these, in turn, will each be formally expanded in terms of two order-1 determinants.[10]

The minor M_{ij} of the element A_{ij} of an $N \times N$ matrix A is the determinant of the $(N-1) \times (N-1)$ matrix obtained by removing all the elements of the ith row and jth column of A; the associated cofactor, C_{ij}, is found by multiplying the minor by $(-1)^{i+j}$. The following example illustrates this.

Example Find the cofactor of the element A_{23} of the matrix

$$\mathsf{A} = \begin{pmatrix} A_{11} & A_{12} & A_{13} \\ A_{21} & A_{22} & A_{23} \\ A_{31} & A_{32} & A_{33} \end{pmatrix}.$$

Removing all the elements of the second row and third column of A and forming the determinant of the remaining terms gives the minor

$$M_{23} = \begin{vmatrix} A_{11} & A_{12} \\ A_{31} & A_{32} \end{vmatrix}.$$

10 Though in practice the values of order-2 determinants are nearly always computed directly.

Multiplying the minor by $(-1)^{2+3} = (-1)^5 = -1$ then gives

$$C_{23} = -\begin{vmatrix} A_{11} & A_{12} \\ A_{31} & A_{32} \end{vmatrix}$$

as the cofactor of A_{23}. ◀

We now define a determinant as *the sum of the products of the elements of any row or column and their corresponding cofactors*, e.g. $A_{21}C_{21} + A_{22}C_{22} + A_{23}C_{23}$ or $A_{13}C_{13} + A_{23}C_{23} + A_{33}C_{33}$. Such a sum is called a *Laplace expansion*. For example, in the first of these expansions, using the elements of the second row of the determinant defined by (1.43) and their corresponding cofactors, we write $|\mathsf{A}|$ as the Laplace expansion

$$\begin{aligned}|\mathsf{A}| &= A_{21}(-1)^{(2+1)}M_{21} + A_{22}(-1)^{(2+2)}M_{22} + A_{23}(-1)^{(2+3)}M_{23} \\ &= -A_{21}\begin{vmatrix} A_{12} & A_{13} \\ A_{32} & A_{33} \end{vmatrix} + A_{22}\begin{vmatrix} A_{11} & A_{13} \\ A_{31} & A_{33} \end{vmatrix} - A_{23}\begin{vmatrix} A_{11} & A_{12} \\ A_{31} & A_{32} \end{vmatrix}.\end{aligned}$$

We will see later that the value of the determinant is independent of the row or column chosen. Of course, we have not yet determined the value of $|\mathsf{A}|$ but, rather, written it as the weighted sum of three determinants of order 2. However, applying again the definition of a determinant, we can evaluate each of the order-2 determinants. As a typical example consider the first of these.

Example Evaluate the determinant

$$\begin{vmatrix} A_{12} & A_{13} \\ A_{32} & A_{33} \end{vmatrix}.$$

By considering the products of the elements of the first row in the determinant, and their corresponding cofactors (now order-1 determinants), we find

$$\begin{aligned}\begin{vmatrix} A_{12} & A_{13} \\ A_{32} & A_{33} \end{vmatrix} &= A_{12}(-1)^{(1+1)}|A_{33}| + A_{13}(-1)^{(1+2)}|A_{32}| \\ &= A_{12}A_{33} - A_{13}A_{32},\end{aligned}$$

where the values of the order-1 determinants $|A_{33}|$ and $|A_{32}|$ are, as defined earlier, A_{33} and A_{32} respectively. It must be remembered that the determinant is *not* necessarily the same as the modulus, e.g. $\det(-2) = |-2| = -2$, not 2. ◀

We can now combine all the above results to show that the value of the determinant (1.43) is given by

$$\begin{aligned}|\mathsf{A}| &= -A_{21}(A_{12}A_{33} - A_{13}A_{32}) + A_{22}(A_{11}A_{33} - A_{13}A_{31}) \\ &\quad - A_{23}(A_{11}A_{32} - A_{12}A_{31}) \qquad (1.44) \\ &= A_{11}(A_{22}A_{33} - A_{23}A_{32}) + A_{12}(A_{23}A_{31} - A_{21}A_{33}) \\ &\quad + A_{13}(A_{21}A_{32} - A_{22}A_{31}), \qquad (1.45)\end{aligned}$$

where the final expression gives the form in which the determinant is usually remembered and is the form that is obtained immediately by considering the Laplace expansion using the first row of the determinant. The last equality, which essentially rearranges a Laplace expansion using the second row into one using the first row, supports our assertion that the value of the determinant is unaffected by which row or column is chosen for the expansion. An alternative, but equivalent, view is contained in the next example.

Example Suppose the rows of a real 3×3 matrix A are interpreted as the components, in a given basis, of three (three-component) vectors **a**, **b** and **c**. Show that the determinant of A can be written as

$$|\mathsf{A}| = \mathbf{a} \cdot (\mathbf{b} \times \mathbf{c}).$$

If the rows of A are written as the components in a given basis of three vectors **a**, **b** and **c**, we have from (1.45) that

$$|\mathsf{A}| = \begin{vmatrix} a_1 & a_2 & a_3 \\ b_1 & b_2 & b_3 \\ c_1 & c_2 & c_3 \end{vmatrix} = a_1(b_2c_3 - b_3c_2) + a_2(b_3c_1 - b_1c_3) + a_3(b_1c_2 - b_2c_1).$$

From the general expression for a scalar triple product, it follows that we may write the determinant as

$$|\mathsf{A}| = \mathbf{a} \cdot (\mathbf{b} \times \mathbf{c}). \tag{1.46}$$

In other words, $|\mathsf{A}|$ is the volume of the parallelepiped defined by the vectors **a**, **b** and **c**. One could equally well interpret the *columns* of the matrix A as the components of three vectors, and result (1.46) would still hold.

This result provides a more memorable (and more meaningful) expression than (1.45) for the value of a 3×3 determinant. Indeed, using this geometrical interpretation, we see immediately that, if the vectors $\mathbf{a}_1, \mathbf{a}_2, \mathbf{a}_3$ are not linearly independent then the value of the determinant vanishes: $|\mathsf{A}| = 0$.[11] ◀

The evaluation of determinants of order greater than 3 follows the same general method as that presented above, in that it relies on successively reducing the order of the determinant by writing it as a Laplace expansion. Thus, a determinant of order 4 is first written as a sum of four determinants of order 3, which are then evaluated using the above method. For higher-order determinants, one cannot write down directly a simple geometrical expression for $|\mathsf{A}|$ analogous to that given in (1.46). Nevertheless, it is still true that if the rows or columns of the $N \times N$ matrix A are interpreted as the components in a given basis of N (N-component) vectors $\mathbf{a}_1, \mathbf{a}_2, \ldots, \mathbf{a}_N$, then the determinant $|\mathsf{A}|$ vanishes if these vectors are not all linearly independent.

1.9.1 Properties of determinants

A number of properties of determinants follow straightforwardly from the definition of det A; their use will often reduce the labor of evaluating a determinant. We present them here without specific proofs, though they all follow readily from the alternative form for a

11 Each can be expressed in terms of the other two; consequently, (i) they all lie in a plane, and (ii) the parallelepiped they define has zero volume.

determinant, given in equation (1.138) on p. 79, and expressed in terms of the Levi–Civita symbol ϵ_{ijk} (see Problem 1.37).

(i) *Determinant of the transpose.* The transpose matrix A^{T} (which, we recall, is obtained by interchanging the rows and columns of A) has the same determinant as A itself, i.e.

$$|\mathsf{A}^{\mathrm{T}}| = |\mathsf{A}|. \tag{1.47}$$

It follows that *any* theorem established for the rows of A will apply to the columns as well, and vice versa.

(ii) *Determinant of the complex and Hermitian conjugate.* It is clear that the matrix A^* obtained by taking the complex conjugate of each element of A has the determinant $|\mathsf{A}^*| = |\mathsf{A}|^*$. Combining this result with (1.47), we find that

$$|\mathsf{A}^{\dagger}| = |(\mathsf{A}^*)^{\mathrm{T}}| = |\mathsf{A}^*| = |\mathsf{A}|^*. \tag{1.48}$$

(iii) *Interchanging two rows or two columns.* If two rows (columns) of A are interchanged, its determinant changes sign but is unaltered in magnitude.

(iv) *Removing factors.* If all the elements of a single row (column) of A have a common factor, λ, then this factor may be removed; the value of the determinant is given by the product of the remaining determinant and λ. Clearly this implies that if all the elements of any row (column) are zero then $|\mathsf{A}| = 0$. It also follows that if every element of the $N \times N$ matrix A is multiplied by a constant factor λ then

$$|\lambda\mathsf{A}| = \lambda^N|\mathsf{A}|. \tag{1.49}$$

(v) *Identical rows or columns.* If any two rows (columns) of A are identical or are multiples of one another, then it can be shown that $|\mathsf{A}| = 0$.

(vi) *Adding a constant multiple of one row (column) to another.* The determinant of a matrix is unchanged in value by adding to the elements of one row (column) any fixed multiple of the elements of another row (column).

(vii) *Determinant of a product.* If A and B are square matrices of the same order then

$$|\mathsf{AB}| = |\mathsf{A}||\mathsf{B}| = |\mathsf{BA}|. \tag{1.50}$$

A simple extension of this property gives, for example,

$$|\mathsf{AB}\cdots\mathsf{G}| = |\mathsf{A}||\mathsf{B}|\cdots|\mathsf{G}| = |\mathsf{A}||\mathsf{G}|\cdots|\mathsf{B}| = |\mathsf{A}\cdots\mathsf{GB}|,$$

which shows that the determinant is invariant under permutation of the matrices in a multiple product.

1.9.2 Evaluation of determinants

There is no explicit procedure for using the above results in the evaluation of any given determinant, and judging the quickest route to an answer is a matter of experience. A general guide is to try to reduce all terms but one in a row or column to zero and hence in effect to obtain a determinant of smaller size. The steps taken in evaluating the determinant in the example below are certainly not the fastest, but they have been chosen in order to illustrate the use of most of the properties listed above.

Example Evaluate the determinant

$$|A| = \begin{vmatrix} 1 & 0 & 2 & 3 \\ 0 & 1 & -2 & 1 \\ 3 & -3 & 4 & -2 \\ -2 & 1 & -2 & -1 \end{vmatrix}.$$

Taking a factor 2 out of the third column and then adding the second column to the third gives

$$|A| = 2\begin{vmatrix} 1 & 0 & 1 & 3 \\ 0 & 1 & -1 & 1 \\ 3 & -3 & 2 & -2 \\ -2 & 1 & -1 & -1 \end{vmatrix} = 2\begin{vmatrix} 1 & 0 & 1 & 3 \\ 0 & 1 & 0 & 1 \\ 3 & -3 & -1 & -2 \\ -2 & 1 & 0 & -1 \end{vmatrix}.$$

Subtracting the second column from the fourth gives

$$|A| = 2\begin{vmatrix} 1 & 0 & 1 & 3 \\ 0 & 1 & 0 & 0 \\ 3 & -3 & -1 & 1 \\ -2 & 1 & 0 & -2 \end{vmatrix}.$$

We now note that the second row has only one non-zero element and so the determinant may conveniently be written as a Laplace expansion, i.e.

$$|A| = 2\times 1\times(-1)^{2+2}\begin{vmatrix} 1 & 1 & 3 \\ 3 & -1 & 1 \\ -2 & 0 & -2 \end{vmatrix} = 2\begin{vmatrix} 4 & 0 & 4 \\ 3 & -1 & 1 \\ -2 & 0 & -2 \end{vmatrix},$$

where the last equality follows by adding the second row to the first. It can now be seen that the first row is minus twice the third, and so the value of the determinant is zero, by property (v) above. ◀

1.10 The inverse of a matrix

Our first use of determinants will be in defining the *inverse* of a matrix. If we were dealing with ordinary numbers we would consider the relation $P = AB$ as equivalent to $B = P/A$, provided that $A \neq 0$. However, if A, B and P are matrices then this notation does not have an obvious meaning. What we really want to know is whether an explicit formula for B can be obtained in terms of A and P.

It will be shown that this is possible for those cases in which $|A| \neq 0$. A square matrix whose determinant is zero is called a *singular* matrix; otherwise it is *non-singular*. We will show that if A is non-singular we can define a matrix, denoted by A^{-1} and called the *inverse* of A, which has the property that if $AB = P$ then $B = A^{-1}P$. In words, B can be obtained by multiplying P from the left by A^{-1}. Analogously, if B is non-singular then, by multiplication from the right, $A = PB^{-1}$.

It is clear that

$$AI = A \quad\Rightarrow\quad I = A^{-1}A, \tag{1.51}$$

where I is the unit matrix, and so $\mathsf{A}^{-1}\mathsf{A} = \mathsf{I} = \mathsf{A}\mathsf{A}^{-1}$.[12] These statements are equivalent to saying that if we first multiply a matrix, B say, by A and then multiply by the inverse A^{-1}, we end up with the matrix we started with, i.e.

$$\mathsf{A}^{-1}\mathsf{A}\mathsf{B} = \mathsf{B}. \tag{1.52}$$

This justifies our use of the term "inverse". It is also clear that the inverse is only defined for square matrices.

So far we have only defined what we mean by the inverse of a matrix. Actually finding the inverse of a matrix A may be carried out in a number of ways. We will show that one method is to construct first the matrix C containing the cofactors of the elements of A, as discussed in Section 1.9. Then the required inverse A^{-1} can be found by forming the transpose of C and dividing by the determinant of A. Thus the elements of the inverse A^{-1} are given by

$$(\mathsf{A}^{-1})_{ik} = \frac{(\mathsf{C})^{\mathrm{T}}_{ik}}{|\mathsf{A}|} = \frac{C_{ki}}{|\mathsf{A}|}. \tag{1.53}$$

That this procedure does indeed result in the inverse may be seen by considering the components of $\mathsf{A}^{-1}\mathsf{A}$ with A^{-1} defined in this way, i.e.

$$(\mathsf{A}^{-1}\mathsf{A})_{ij} = \sum_k (\mathsf{A}^{-1})_{ik}(\mathsf{A})_{kj} = \sum_k \frac{C_{ki}}{|\mathsf{A}|}A_{kj} = \frac{|\mathsf{A}|}{|\mathsf{A}|}\delta_{ij}. \tag{1.54}$$

The last equality in (1.54) relies on the property

$$\sum_k C_{ki}A_{kj} = |\mathsf{A}|\delta_{ij}. \tag{1.55}$$

This can be proved by considering the matrix A' obtained from the original matrix A when the ith column of A is replaced by one of the other columns, say the jth; as an equation, $\mathsf{A}'_{ki} = \mathsf{A}_{kj}$. With this construction, A' is a matrix with two identical columns and so has zero determinant. However, replacing the ith column by another does not change the cofactors C_{ki} of the elements in the ith column, which are therefore the same in A and A', i.e. $C_{ki} = C'_{ki}$ for all k. Recalling the Laplace expansion of a general determinant, i.e.

$$|\mathsf{A}| = \sum_k A_{ki}C_{ki},$$

we obtain for the case $i \neq j$ that

$$\sum_k A_{kj}C_{ki} = \sum_k A'_{ki}C'_{ki} = |\mathsf{A}'| = 0.$$

The Laplace expansion itself deals with the case $i = j$, and the two together establish result (1.55).

12 It is not immediately obvious that $\mathsf{A}\mathsf{A}^{-1} = \mathsf{I}$, since A^{-1} has only been defined as a left inverse. Prove that the left inverse is also a right inverse by defining $\mathsf{A}_{\mathrm{R}}^{-1}$ by $\mathsf{A}\mathsf{A}_{\mathrm{R}}^{-1} = \mathsf{I}$ and then, by considering $\mathsf{A}^{-1}\mathsf{A}\mathsf{A}_{\mathrm{R}}^{-1}$, show that $\mathsf{A}_{\mathrm{R}}^{-1} = \mathsf{A}^{-1}$.

It is immediately obvious from (1.53) that the inverse of a matrix is not defined if the matrix is singular (i.e. if $|\mathsf{A}| = 0$).

Example Find the inverse of the matrix

$$\mathsf{A} = \begin{pmatrix} 2 & 4 & 3 \\ 1 & -2 & -2 \\ -3 & 3 & 2 \end{pmatrix}.$$

We first determine $|\mathsf{A}|$:

$$\begin{aligned} |\mathsf{A}| &= 2[-2(2) - (-2)3] + 4[(-2)(-3) - (1)(2)] + 3[(1)(3) - (-2)(-3)] \\ &= 11. \end{aligned} \tag{1.56}$$

This is non-zero and so an inverse matrix can be constructed. To do this we need the matrix of the cofactors, C, and hence C^{T}. We find[13]

$$\mathsf{C} = \begin{pmatrix} 2 & 4 & -3 \\ 1 & 13 & -18 \\ -2 & 7 & -8 \end{pmatrix} \quad \text{and} \quad \mathsf{C}^{\mathrm{T}} = \begin{pmatrix} 2 & 1 & -2 \\ 4 & 13 & 7 \\ -3 & -18 & -8 \end{pmatrix},$$

and hence

$$\mathsf{A}^{-1} = \frac{\mathsf{C}^{\mathrm{T}}}{|\mathsf{A}|} = \frac{1}{11}\begin{pmatrix} 2 & 1 & -2 \\ 4 & 13 & 7 \\ -3 & -18 & -8 \end{pmatrix}. \tag{1.57}$$

This result can be checked (somewhat tediously) by computing $\mathsf{A}^{-1}\mathsf{A}$. ◀

For a 2×2 matrix, the inverse has a particularly simple form. If the matrix is

$$\mathsf{A} = \begin{pmatrix} A_{11} & A_{12} \\ A_{21} & A_{22} \end{pmatrix}$$

then its determinant $|\mathsf{A}|$ is given by $|\mathsf{A}| = A_{11}A_{22} - A_{12}A_{21}$, and the matrix of cofactors is

$$\mathsf{C} = \begin{pmatrix} A_{22} & -A_{21} \\ -A_{12} & A_{11} \end{pmatrix}.$$

Thus the inverse of A is given by

$$\mathsf{A}^{-1} = \frac{\mathsf{C}^{\mathrm{T}}}{|\mathsf{A}|} = \frac{1}{A_{11}A_{22} - A_{12}A_{21}}\begin{pmatrix} A_{22} & -A_{12} \\ -A_{21} & A_{11} \end{pmatrix}. \tag{1.58}$$

It can be seen that the transposed matrix of cofactors for a 2×2 matrix is the same as the matrix formed by swapping the elements on the leading diagonal (A_{11} and A_{22}) and changing the signs of the other two elements (A_{12} and A_{21}). This is completely general for a 2×2 matrix and is easy to remember.

13 The reader should calculate at least some of the cofactors for themselves, paying particular attention to the sign of each.

The following are some further useful properties related to the inverse matrix and may be straightforwardly derived, as below.

(i) $(\mathsf{A}^{-1})^{-1} = \mathsf{A}$.
(ii) $(\mathsf{A}^{\mathrm{T}})^{-1} = (\mathsf{A}^{-1})^{\mathrm{T}}$.
(iii) $(\mathsf{A}^{\dagger})^{-1} = (\mathsf{A}^{-1})^{\dagger}$.
(iv) $(\mathsf{AB})^{-1} = \mathsf{B}^{-1}\mathsf{A}^{-1}$.
(v) $(\mathsf{AB}\cdots\mathsf{G})^{-1} = \mathsf{G}^{-1}\cdots\mathsf{B}^{-1}\mathsf{A}^{-1}$.

Example Prove the properties (i)–(v) stated above.

We begin by writing down the fundamental expression defining the inverse of a non-singular square matrix A:

$$\mathsf{AA}^{-1} = \mathsf{I} = \mathsf{A}^{-1}\mathsf{A}. \tag{1.59}$$

Property (i). This follows immediately from the expression (1.59).
Property (ii). Taking the transpose of each expression in (1.59) gives

$$(\mathsf{AA}^{-1})^{\mathrm{T}} = \mathsf{I}^{\mathrm{T}} = (\mathsf{A}^{-1}\mathsf{A})^{\mathrm{T}}.$$

Using the result (1.37) for the transpose of a product of matrices and noting that $\mathsf{I}^{\mathrm{T}} = \mathsf{I}$, we find

$$(\mathsf{A}^{-1})^{\mathrm{T}}\mathsf{A}^{\mathrm{T}} = \mathsf{I} = \mathsf{A}^{\mathrm{T}}(\mathsf{A}^{-1})^{\mathrm{T}}.$$

However, from (1.59), this implies $(\mathsf{A}^{-1})^{\mathrm{T}} = (\mathsf{A}^{\mathrm{T}})^{-1}$ and hence proves result (ii) above.

Property (iii). This may be proved in an analogous way to property (ii), by replacing the transposes in (ii) by Hermitian conjugates and using the result (1.38) for the Hermitian conjugate of a product of matrices.

Property (iv). Using (1.59), we may write

$$(\mathsf{AB})(\mathsf{AB})^{-1} = \mathsf{I} = (\mathsf{AB})^{-1}(\mathsf{AB}).$$

From the left-hand equality it follows, by multiplying on the left by A^{-1}, that

$$\mathsf{A}^{-1}\mathsf{AB}(\mathsf{AB})^{-1} = \mathsf{A}^{-1}\mathsf{I} \quad \text{and hence} \quad \mathsf{B}(\mathsf{AB})^{-1} = \mathsf{A}^{-1}.$$

Now multiplying on the left by B^{-1} gives

$$\mathsf{B}^{-1}\mathsf{B}(\mathsf{AB})^{-1} = \mathsf{B}^{-1}\mathsf{A}^{-1},$$

and hence the stated result.

Property (v). Finally, result (iv) may be extended to case (v) in a straightforward manner. For example, using result (iv) twice we find

$$(\mathsf{ABC})^{-1} = (\mathsf{BC})^{-1}\mathsf{A}^{-1} = \mathsf{C}^{-1}\mathsf{B}^{-1}\mathsf{A}^{-1}.$$

Clearly, this can then be further extended to cover the product of any finite number of matrices. ◀

We conclude this section by noting that the determinant $|\mathsf{A}^{-1}|$ of the inverse matrix can be expressed very simply in terms of the determinant $|\mathsf{A}|$ of the matrix itself. Again we start with the fundamental expression (1.59). Then, using the property (1.50) for the determinant of a product, we find

$$|AA^{-1}| = |A||A^{-1}| = |I|.$$

It is straightforward to show by Laplace expansion that $|I| = 1$, and so we arrive at the useful result

$$|A^{-1}| = \frac{1}{|A|}. \tag{1.60}$$

1.11 The rank of a matrix

The *rank* of a general $M \times N$ matrix is an important concept, particularly in the solution of sets of simultaneous linear equations, as discussed in the next section, and we now consider it in some detail. Like the trace and determinant, the rank of matrix A is a single number (or algebraic expression) that depends on the elements of A. Unlike the trace and determinant, however, the rank of a matrix can be defined even when A is not square. As we shall see, there are two *equivalent* definitions of the rank of a general matrix.

Firstly, the rank of a matrix may be defined in terms of the *linear independence* of vectors. Suppose that the columns of an $M \times N$ matrix are interpreted as the components in a given basis of N (M-component) vectors $\mathbf{v}_1, \mathbf{v}_2, \ldots, \mathbf{v}_N$, as follows:

$$A = \begin{pmatrix} \uparrow & \uparrow & & \uparrow \\ v_1 & v_2 & \ldots & v_N \\ \downarrow & \downarrow & & \downarrow \end{pmatrix}.$$

Then the *rank* of A, denoted by rank A or by $R(A)$, is defined as the number of *linearly independent* vectors in the set $\mathbf{v}_1, \mathbf{v}_2, \ldots, \mathbf{v}_N$, and equals the dimension of the vector space spanned by those vectors. Alternatively, we may consider the rows of A to contain the components in a given basis of the M (N-component) vectors $\mathbf{w}_1, \mathbf{w}_2, \ldots, \mathbf{w}_M$ as follows:

$$A = \begin{pmatrix} \leftarrow & w_1 & \rightarrow \\ \leftarrow & w_2 & \rightarrow \\ & \vdots & \\ \leftarrow & w_M & \rightarrow \end{pmatrix}.$$

It may then be shown[14] that the rank of A is also equal to the number of linearly independent vectors in the set $\mathbf{w}_1, \mathbf{w}_2, \ldots, \mathbf{w}_M$. From this definition it should be clear that the rank of A is unaffected by the exchange of two rows (or two columns) or by the multiplication of a row (or column) by a constant. Furthermore, suppose that a constant multiple of one row (column) is added to another row (column): for example, we might replace the row w_i by $w_i + cw_j$. This also has no effect on the number of linearly independent rows and so leaves the rank of A unchanged. We may use these properties to evaluate the rank of a given matrix.

14 For a fuller discussion, see, for example, C. D. Cantrell, *Modern Mathematical Methods for Physicists and Engineers* (Cambridge: Cambridge University Press, 2000), chapter 6.

A second (equivalent) definition of the rank of a matrix may be given and uses the concept of *submatrices*. A submatrix of A is any matrix that can be formed from the elements of A by ignoring one, or more than one, row or column. It may be shown that the rank of a general $M \times N$ matrix is equal to the size of the largest square submatrix of A whose determinant is non-zero. Therefore, if a matrix A has an $r \times r$ submatrix S with $|S| \neq 0$, but no $(r+1) \times (r+1)$ submatrix with non-zero determinant then the rank of the matrix is r. From either definition it is clear that the rank of A is less than or equal to the smaller of M and N.[15]

Example Determine the rank of the matrix

$$A = \begin{pmatrix} 1 & 1 & 0 & -2 \\ 2 & 0 & 2 & 2 \\ 4 & 1 & 3 & 1 \end{pmatrix}.$$

The largest possible square submatrices of A must be of dimension 3×3. Clearly, A possesses four such submatrices, the determinants of which are given by

$$\begin{vmatrix} 1 & 1 & 0 \\ 2 & 0 & 2 \\ 4 & 1 & 3 \end{vmatrix} = 0, \qquad \begin{vmatrix} 1 & 1 & -2 \\ 2 & 0 & 2 \\ 4 & 1 & 1 \end{vmatrix} = 0,$$

$$\begin{vmatrix} 1 & 0 & -2 \\ 2 & 2 & 2 \\ 4 & 3 & 1 \end{vmatrix} = 0, \qquad \begin{vmatrix} 1 & 0 & -2 \\ 0 & 2 & 2 \\ 1 & 3 & 1 \end{vmatrix} = 0.$$

In each case the determinant may be evaluated in the way described in Subsection 1.9.1. The fact that the determinants of all four 3×3 submatrices are zero implies that the rank of A is less than three.

The next largest square submatrices of A are of dimension 2×2. Consider, for example, the 2×2 submatrix formed by ignoring the third row and the third and fourth columns of A; this has determinant

$$\begin{vmatrix} 1 & 1 \\ 2 & 0 \end{vmatrix} = (1 \times 0) - (2 \times 1) = -2.$$

Since its determinant is non-zero, A is of rank 2 and we need not consider any other 2×2 submatrix. ◀

In the special case in which the matrix A is a *square* $N \times N$ matrix, by comparing either of the above definitions of rank with our discussion of determinants in Section 1.9, we see that $|A| = 0$ unless the rank of A is N. In other words, A is *singular* unless $R(A) = N$.

15 State the rank of an $N \times N$ matrix all of whose entries are equal to the non-zero value λ. Justify your answer by separate references to (a) the independence of its columns, (b) the determinant of any arbitrary 2×2 submatrix.

1.12 Simultaneous linear equations

In physical applications we often encounter sets of simultaneous linear equations. In general we may have M equations in N unknowns $x_1, x_2, \ldots, x_N$ of the form

$$\begin{aligned} A_{11}x_1 + A_{12}x_2 + \cdots + A_{1N}x_N &= b_1, \\ A_{21}x_1 + A_{22}x_2 + \cdots + A_{2N}x_N &= b_2, \\ &\vdots \\ A_{M1}x_1 + A_{M2}x_2 + \cdots + A_{MN}x_N &= b_M, \end{aligned} \tag{1.61}$$

where the A_{ij} and b_i have known values. If all the b_i are zero then the system of equations is called *homogeneous*, otherwise it is *inhomogeneous*. Depending on the given values, this set of equations for the N unknowns $x_1, x_2, \ldots, x_N$ may have either a unique solution, no solution or infinitely many solutions. Matrix analysis may be used to distinguish between the possibilities.

The set of equations may be expressed as a single matrix equation $\mathsf{A}\mathsf{x} = \mathsf{b}$, or, written out in full, as

$$\begin{pmatrix} A_{11} & A_{12} & \ldots & A_{1N} \\ A_{21} & A_{22} & \ldots & A_{2N} \\ \vdots & \vdots & \ddots & \vdots \\ A_{M1} & A_{M2} & \ldots & A_{MN} \end{pmatrix} \begin{pmatrix} x_1 \\ x_2 \\ \vdots \\ x_N \end{pmatrix} = \begin{pmatrix} b_1 \\ b_2 \\ \vdots \\ b_M \end{pmatrix}. \tag{1.62}$$

A fourth way of writing the same equations is to interpret the columns of A as the components, in some basis, of N (M-component) vectors $\mathbf{v}_1, \mathbf{v}_2 \ldots, \mathbf{v}_N$:

$$x_1\mathbf{v}_1 + x_2\mathbf{v}_2 + \cdots + x_N\mathbf{v}_N = \mathbf{b}. \tag{1.63}$$

In passing, we recall that the number of linearly independent vectors is equal to r, the rank of A.

1.12.1 The number of solutions

The rank of A has far-reaching consequences for the existence of solutions to sets of simultaneous linear equations such as (1.61). As just mentioned, these equations may have *no solution*, a *unique solution* or *infinitely many solutions*. We now discuss these three cases in turn.

No solution

The system of equations possesses no solution unless, as expressed in equation (1.63), $\mathbf{b}$ can be written as a linear combination of the columns of A; when it can, the $x_1, x_2, \ldots, x_N$ appearing in the combination give the solution. This in turn requires the set of vectors $\mathbf{b}, \mathbf{v}_1, \mathbf{v}_2, \ldots, \mathbf{v}_N$ to contain the same number of linearly independent vectors as the set $\mathbf{v}_1, \mathbf{v}_2, \ldots, \mathbf{v}_N$. In terms of matrices, this is equivalent to the requirement that the matrix A and the *augmented matrix*

$$\mathsf{M} = \begin{pmatrix} A_{11} & A_{12} & \ldots & A_{1N} & b_1 \\ A_{21} & A_{22} & \ldots & A_{2N} & b_1 \\ \vdots & & \ddots & & \vdots \\ A_{M1} & A_{M2} & \ldots & A_{MN} & b_M \end{pmatrix}$$

have the *same* rank r. If this condition is satisfied then the set of equations (1.61) will have either a unique solution or infinitely many solutions. If, however, A and M have different ranks, then there will be no solution.

A unique solution

If $\mathbf{b}$ can be expressed as in (1.63) and in addition $r = N$,[16] implying that the vectors $\mathbf{v}_1, \mathbf{v}_2, \ldots, \mathbf{v}_N$ are linearly independent, then the equations have a *unique solution* $x_1, x_2, \ldots, x_N$. The uniqueness follows from the uniqueness of the expansion of any vector in the vector space for which the $\mathbf{v}_i$ form a basis [see equation (1.10)].

Infinitely many solutions

If $\mathbf{b}$ can be expressed as in (1.63) but $r < N$ then only r of the vectors $\mathbf{v}_1, \mathbf{v}_2, \ldots, \mathbf{v}_N$ are linearly independent. We may therefore choose the coefficients of $n - r$ vectors in an arbitrary way, while still satisfying (1.63) for some set of coefficients $x_1, x_2, \ldots, x_N$; there are therefore *infinitely many solutions*.

We may use this result to investigate the special case of the solution of a *homogeneous* set of linear equations, for which $\mathbf{b} = \mathbf{0}$. Clearly the set *always* has the trivial solution $x_1 = x_2 = \cdots = x_N = 0$, and if $r = N$ this will be the only solution.

If $r < N$, however, there are infinitely many solutions; each will contain $N - r$ arbitrary components. In particular, we note that if $M < N$ (i.e. there are fewer equations than unknowns) then $r < N$ automatically. Hence a set of *homogeneous* linear equations with fewer equations than unknowns *always* has infinitely many solutions.

1.12.2 *N* simultaneous linear equations in *N* unknowns

A special case of (1.61) occurs when $M = N$. In this case the matrix A is *square* and we have the same number of equations as unknowns. Since A is square, the condition $r = N$ corresponds to $|\mathsf{A}| \neq 0$ and the matrix A is *non-singular*. The case $r < N$ corresponds to $|\mathsf{A}| = 0$, in which case A is *singular*.

As mentioned above, the equations will have a solution provided b can be written as in (1.63). If this is true then the equations will possess a unique solution when $|\mathsf{A}| \neq 0$ or infinitely many solutions when $|\mathsf{A}| = 0$. There exist several methods for obtaining the solution(s). Perhaps the most elementary method is *Gaussian elimination*; we will discuss this method first, and also address numerical subtleties such as equation interchange (pivoting). Following this, we will outline three further methods for solving a square set of simultaneous linear equations.

16 Note that M can be greater than N, but, if it is, then $M - N$ of the simultaneous equations must be expressible as linear combinations of the other N equations.

Gaussian elimination

This is probably one of the earliest techniques acquired by a student of algebra, namely the solving of simultaneous equations (initially only two in number) by the successive elimination of all the variables but one. This (known as *Gaussian elimination*) is achieved by using, at each stage, one of the equations to obtain an explicit expression for one of the remaining x_i in terms of the others and then substituting for that x_i in all other remaining equations. Eventually a single linear equation in just one of the unknowns is obtained. This is then solved and the result is resubstituted in previously derived equations (in reverse order) to establish values for all the x_i.

The method is probably very familiar to the reader, and so a specific example to illustrate this alone seems unnecessary. Instead, we will show how a calculation along such lines might be arranged so that the errors due to the inherent lack of precision in any calculating equipment do not become excessive. This can happen if the value of N is large and particularly (and we will merely state this) if the elements $A_{11}, A_{22}, \ldots, A_{NN}$ on the leading diagonal of the matrix of coefficients are small compared with the off-diagonal elements.

The process to be described is known as *Gaussian elimination with interchange.* The only, but essential, difference from straightforward elimination is that before each variable x_i is eliminated, the equations are reordered to put the largest (in modulus) remaining coefficient of x_i on the leading diagonal.

We will take as an illustration a straightforward three-variable example, which can in fact be solved perfectly well without any interchange since, with simple numbers and only two eliminations to perform, rounding errors do not have a chance to build up. However, the important thing is that the reader should appreciate how this would apply in (say) a computer program for a 1000-variable case, perhaps with unforeseeable zeros or very small numbers appearing on the leading diagonal.

Example Solve the simultaneous equations

$$\begin{array}{lrrrl} \text{(a)} & x_1 & +6x_2 & -4x_3 & = 8, \\ \text{(b)} & 3x_1 & -20x_2 & +x_3 & = 12, \\ \text{(c)} & -x_1 & +3x_2 & +5x_3 & = 3. \end{array} \tag{1.64}$$

Firstly, we interchange rows (a) and (b) to bring the term $3x_1$ onto the leading diagonal. In the following, we label the important equations (I), (II), (III), and the others alphabetically. A general (i.e. variable) label will be denoted by j.

$$\begin{array}{lrrrl} \text{(I)} & 3x_1 & -20x_2 & +x_3 & = 12, \\ \text{(d)} & x_1 & +6x_2 & -4x_3 & = 8, \\ \text{(e)} & -x_1 & +3x_2 & +5x_3 & = 3. \end{array}$$

For $(j) =$ (d) and (e), replace row (j) by

$$\text{row }(j) - \frac{a_{j1}}{3} \times \text{row (I)},$$

where a_{j1} is the coefficient of x_1 in row (j), to give the two equations

$$\begin{aligned}
&\text{(II)} \quad \left(6+\tfrac{20}{3}\right)x_2 \quad +\left(-4-\tfrac{1}{3}\right)x_3 \;=\; 8-\tfrac{12}{3},\\
&\text{(f)} \quad \left(3-\tfrac{20}{3}\right)x_2 \quad +\left(5+\tfrac{1}{3}\right)x_3 \;=\; 3+\tfrac{12}{3}.
\end{aligned}$$

Now $|6+\frac{20}{3}| > |3-\frac{20}{3}|$ and so no interchange is required before the next elimination. To eliminate x_2, replace row (f) by

$$\text{row (f)} - \frac{\left(-\frac{11}{3}\right)}{\frac{38}{3}} \times \text{row (II)}.$$

This gives

$$\text{(III)} \qquad \left[\tfrac{16}{3}+\tfrac{11}{38}\times\tfrac{(-13)}{3}\right]x_3 = 7+\tfrac{11}{38}\times 4.$$

Collecting together and tidying up the final equations, we have

$$\begin{aligned}
&\text{(I)} \quad 3x_1 \quad -20x_2 \quad +x_3 \;=\; 12,\\
&\text{(II)} \qquad\qquad 38x_2 \quad -13x_3 \;=\; 12,\\
&\text{(III)} \qquad\qquad\qquad\qquad x_3 \;=\; 2.
\end{aligned}$$

Starting with (III) and working backwards, it is now a simple matter to obtain

$$x_1 = 10, \qquad x_2 = 1, \qquad x_3 = 2$$

as the complete solution of the simultaneous equations. ◀

Direct inversion

Since A is square it will possess an inverse, provided $|\mathsf{A}| \neq 0$. Thus, if A is non-singular, we immediately obtain

$$\mathsf{x} = \mathsf{A}^{-1}\mathsf{b} \tag{1.65}$$

as the unique solution to the set of equations. However, if $\mathsf{b} = \mathbf{0}$ then we see immediately that the set of equations possesses only the trivial solution $\mathsf{x} = \mathbf{0}$. The direct inversion method has the advantage that, once A^{-1} has been calculated, one may obtain the solutions x corresponding to different vectors $\mathsf{b}_1, \mathsf{b}_2, \ldots$ on the RHS, with little further work.

Example Show that the set of simultaneous equations

$$\begin{aligned}
2x_1 + 4x_2 + 3x_3 &= 4,\\
x_1 - 2x_2 - 2x_3 &= 0,\\
-3x_1 + 3x_2 + 2x_3 &= -7,
\end{aligned} \tag{1.66}$$

has a unique solution, and find that solution.

The simultaneous equations can be represented by the matrix equation $\mathsf{Ax} = \mathsf{b}$, i.e.

$$\begin{pmatrix} 2 & 4 & 3 \\ 1 & -2 & -2 \\ -3 & 3 & 2 \end{pmatrix}\begin{pmatrix} x_1 \\ x_2 \\ x_3 \end{pmatrix} = \begin{pmatrix} 4 \\ 0 \\ -7 \end{pmatrix}.$$

As we have already shown that A^{-1} exists and have calculated it, see (1.57), it follows that $\mathsf{x} = \mathsf{A}^{-1}\mathsf{b}$ or, more explicitly, that

$$\begin{pmatrix} x_1 \\ x_2 \\ x_3 \end{pmatrix} = \frac{1}{11}\begin{pmatrix} 2 & 1 & -2 \\ 4 & 13 & 7 \\ -3 & -18 & -8 \end{pmatrix}\begin{pmatrix} 4 \\ 0 \\ -7 \end{pmatrix} = \begin{pmatrix} 2 \\ -3 \\ 4 \end{pmatrix}. \qquad (1.67)$$

Thus the unique solution is $x_1 = 2$, $x_2 = -3$, $x_3 = 4$. ◀

LU decomposition

Although conceptually simple, finding the solution by calculating A^{-1} can be computationally demanding, especially when N is large. In fact, as we shall now show, it is not necessary to perform the full inversion of A in order to solve the simultaneous equations $\mathsf{Ax} = \mathsf{b}$. Rather, we can perform a *decomposition* of the matrix into the product of a square *lower triangular* matrix L and a square *upper triangular* matrix U, which are such that[17]

$$\mathsf{A} = \mathsf{LU}, \qquad (1.68)$$

and then use the fact that triangular systems of equations can be solved very simply.

We must begin, therefore, by finding the matrices L and U such that (1.68) is satisfied. This may be achieved straightforwardly by writing out (1.68) in component form. For illustration, let us consider the 3×3 case. It is, in fact, always possible, and convenient, to take the diagonal elements of L as unity, so we have

$$\begin{aligned} \mathsf{A} &= \begin{pmatrix} 1 & 0 & 0 \\ L_{21} & 1 & 0 \\ L_{31} & L_{32} & 1 \end{pmatrix}\begin{pmatrix} U_{11} & U_{12} & U_{13} \\ 0 & U_{22} & U_{23} \\ 0 & 0 & U_{33} \end{pmatrix} \\ &= \begin{pmatrix} U_{11} & U_{12} & U_{13} \\ L_{21}U_{11} & L_{21}U_{12} + U_{22} & L_{21}U_{13} + U_{23} \\ L_{31}U_{11} & L_{31}U_{12} + L_{32}U_{22} & L_{31}U_{13} + L_{32}U_{23} + U_{33} \end{pmatrix}. \end{aligned} \qquad (1.69)$$

The nine unknown elements of L and U can now be determined by equating the nine elements of (1.69) to those of the 3×3 matrix A. This is done in the particular order illustrated in the example below.

Once the matrices L and U have been determined, one can use the decomposition to solve the set of equations $\mathsf{Ax} = \mathsf{b}$ in the following way. From (1.68), we have $\mathsf{LUx} = \mathsf{b}$, but this can be written as *two* triangular sets of equations

$$\mathsf{Ly} = \mathsf{b} \quad \text{and} \quad \mathsf{Ux} = \mathsf{y},$$

where y is another column matrix to be determined. One may easily solve the first triangular set of equations for y, which is then substituted into the second set. The required solution x is then obtained readily from the second triangular set of equations. We note that, as with direct inversion, once the LU decomposition has been determined, one can solve for various RHS column matrices $\mathsf{b}_1, \mathsf{b}_2, \ldots$, with little extra work.

17 Lower and upper triangular matrices are not formally defined and discussed until Subsection 1.13.2, but relevant aspects of their general structure will be apparent from the way they are used here.

Example Use LU decomposition to solve the set of simultaneous equations (1.66).

We begin the determination of the matrices L and U by equating the elements of the matrix in (1.69) with those of the matrix

$$\mathsf{A} = \begin{pmatrix} 2 & 4 & 3 \\ 1 & -2 & -2 \\ -3 & 3 & 2 \end{pmatrix}.$$

This is performed in the following order:

$$\begin{array}{llll}
\text{1st row:} & U_{11} = 2, & U_{12} = 4, & U_{13} = 3 \\
\text{1st column:} & L_{21}U_{11} = 1, & L_{31}U_{11} = -3 & \Rightarrow L_{21} = \frac{1}{2},\ L_{31} = -\frac{3}{2} \\
\text{2nd row:} & L_{21}U_{12} + U_{22} = -2, & L_{21}U_{13} + U_{23} = -2 & \Rightarrow U_{22} = -4,\ U_{23} = -\frac{7}{2} \\
\text{2nd column:} & L_{31}U_{12} + L_{32}U_{22} = 3 & & \Rightarrow L_{32} = -\frac{9}{4} \\
\text{3rd row:} & L_{31}U_{13} + L_{32}U_{23} + U_{33} = 2 & & \Rightarrow U_{33} = -\frac{11}{8}.
\end{array}$$

Thus we may write the matrix A as

$$\mathsf{A} = \mathsf{LU} = \begin{pmatrix} 1 & 0 & 0 \\ \frac{1}{2} & 1 & 0 \\ -\frac{3}{2} & -\frac{9}{4} & 1 \end{pmatrix} \begin{pmatrix} 2 & 4 & 3 \\ 0 & -4 & -\frac{7}{2} \\ 0 & 0 & -\frac{11}{8} \end{pmatrix}.$$

We must now solve the set of equations $\mathsf{Ly} = \mathsf{b}$, which read

$$\begin{pmatrix} 1 & 0 & 0 \\ \frac{1}{2} & 1 & 0 \\ -\frac{3}{2} & -\frac{9}{4} & 1 \end{pmatrix} \begin{pmatrix} y_1 \\ y_2 \\ y_3 \end{pmatrix} = \begin{pmatrix} 4 \\ 0 \\ -7 \end{pmatrix}.$$

Since this set of equations is triangular, we quickly find

$$y_1 = 4, \quad y_2 = 0 - (\tfrac{1}{2})(4) = -2, \quad y_3 = -7 - (-\tfrac{3}{2})(4) - (-\tfrac{9}{4})(-2) = -\tfrac{11}{2}.$$

These values must then be substituted into the equations $\mathsf{Ux} = \mathsf{y}$, which read

$$\begin{pmatrix} 2 & 4 & 3 \\ 0 & -4 & -\frac{7}{2} \\ 0 & 0 & -\frac{11}{8} \end{pmatrix} \begin{pmatrix} x_1 \\ x_2 \\ x_3 \end{pmatrix} = \begin{pmatrix} 4 \\ -2 \\ -\frac{11}{2} \end{pmatrix}.$$

This set of equations is also triangular, and, starting with the final row, we find the solution (in the given order)

$$x_3 = 4, \quad x_2 = -3, \quad x_1 = 2,$$

which agrees with the result found above by direct inversion. ◀

We note, in passing, that one can calculate both the inverse and the determinant of A from its LU decomposition. To find the inverse A^{-1}, one solves the system of equations $\mathsf{Ax} = \mathsf{b}$ repeatedly for the N different RHS column matrices $\mathsf{b} = \mathsf{e}_i$, $i = 1, 2, \ldots, N$, where e_i is the column matrix with its ith element equal to unity and the others equal to zero. The solution x in each case gives the corresponding column of A^{-1}. Evaluation of the determinant $|\mathsf{A}|$ is much simpler. From (1.68), we have

$$|\mathsf{A}| = |\mathsf{LU}| = |\mathsf{L}||\mathsf{U}|. \quad (1.70)$$

Since L and U are triangular, however, we see from (1.75) that their determinants are equal to the products of their diagonal elements. Since $L_{ii} = 1$ for all i, we thus find

$$|\mathsf{A}| = U_{11}U_{22}\cdots U_{NN} = \prod_{i=1}^{N} U_{ii}.$$

As an illustration, in the above example we find $|\mathsf{A}| = (2)(-4)(-11/8) = 11$, which, as it must, agrees with our earlier calculation (1.56).

Finally, a related but slightly different decomposition is possible if matrix A is what is known as *positive semi-definite*. This latter concept is discussed more fully in Section 1.18 in connection with quadratic and Hermitian forms, but for our present purposes we take it as meaning that the scalar quantity $\mathsf{x}^\dagger\mathsf{A}\mathsf{x}$ is real and greater than or equal to zero for *all* column matrices x. An alternative prescription is that all of the eigenvectors (see Section 1.14) of A are non-negative.

Given this definition, if the matrix A is symmetric and positive semi-definite then we can decompose it as

$$\mathsf{A} = \mathsf{L}\mathsf{L}^\dagger, \quad (1.71)$$

where L is a lower triangular matrix; this representation is known as a *Cholesky decomposition.*[18] We cannot set the diagonal elements of L equal to unity in this case, because we require the same number of independent elements in L as in A. The reason that the decomposition can only be applied to positive semi-definite matrices can be seen by considering the Hermitian form (or quadratic form in the real case)

$$\mathsf{x}^\dagger\mathsf{A}\mathsf{x} = \mathsf{x}^\dagger\mathsf{L}\mathsf{L}^\dagger\mathsf{x} = (\mathsf{L}^\dagger\mathsf{x})^\dagger(\mathsf{L}^\dagger\mathsf{x}).$$

Denoting the column matrix $\mathsf{L}^\dagger\mathsf{x}$ by y, we see that the last term on the RHS is $\mathsf{y}^\dagger\mathsf{y}$, which must be greater than or equal to zero. Thus, we require $\mathsf{x}^\dagger\mathsf{A}\mathsf{x} \geq 0$ for any arbitrary column matrix x.

As mentioned above, the requirement that a matrix be positive semi-definite is equivalent to demanding that all the eigenvalues of A are positive or zero. If one of the eigenvalues of A is zero, then, as will be shown in equation (1.104), $|\mathsf{A}| = 0$ and A is *singular*. Thus, if A is a non-singular matrix, it must be *positive definite* (rather than just positive semi-definite) for a Cholesky decomposition (1.71) to be possible. In fact, in this case, the inability to find a matrix L that satisfies (1.71) implies that A cannot be positive definite.

The Cholesky decomposition can be used in a way analogous to that in which the *LU* decomposition was employed earlier, but we will not explore this aspect further. Some practice decompositions are included in the problems at the end of this chapter.

18 In the special case where A is real, the decomposition becomes $\mathsf{A} = \mathsf{L}\mathsf{L}^{\mathrm{T}}$.

Cramer's rule

A further alternative method of solution is to use *Cramer's rule*, which also provides some insight into the nature of the solutions in the various cases. To illustrate this method let us consider a set of three equations in three unknowns,

$$\begin{aligned} A_{11}x_1 + A_{12}x_2 + A_{13}x_3 &= b_1, \\ A_{21}x_1 + A_{22}x_2 + A_{23}x_3 &= b_2, \\ A_{31}x_1 + A_{32}x_2 + A_{33}x_3 &= b_3, \end{aligned} \tag{1.72}$$

which may be represented by the matrix equation $\mathsf{Ax} = \mathsf{b}$. We wish either to find the solution(s) x to these equations or to establish that there are no solutions. From result (vi) of Subsection 1.9.1, the determinant $|\mathsf{A}|$ is unchanged by adding to its first column the combination

$$\frac{x_2}{x_1} \times (\text{second column of } \mathsf{A}) + \frac{x_3}{x_1} \times (\text{third column of } \mathsf{A}).$$

We thus obtain

$$|\mathsf{A}| = \begin{vmatrix} A_{11} & A_{12} & A_{13} \\ A_{21} & A_{22} & A_{23} \\ A_{31} & A_{32} & A_{33} \end{vmatrix} = \begin{vmatrix} A_{11} + (x_2/x_1)A_{12} + (x_3/x_1)A_{13} & A_{12} & A_{13} \\ A_{21} + (x_2/x_1)A_{22} + (x_3/x_1)A_{23} & A_{22} & A_{23} \\ A_{31} + (x_2/x_1)A_{32} + (x_3/x_1)A_{33} & A_{32} & A_{33} \end{vmatrix}.$$

Now the ith entry in the first column is simply b_i/x_1, with b_i as given by the original equations (1.72). Therefore substitution for the ith entry in the first column, yields

$$|\mathsf{A}| = \frac{1}{x_1} \begin{vmatrix} b_1 & A_{12} & A_{13} \\ b_2 & A_{22} & A_{23} \\ b_3 & A_{32} & A_{33} \end{vmatrix} = \frac{1}{x_1}\Delta_1.$$

The determinant Δ_1 is known as a *Cramer determinant*. Similar manipulations of the second and third columns of $|\mathsf{A}|$ yield x_2 and x_3, and so the full set of results reads

$$x_1 = \frac{\Delta_1}{|\mathsf{A}|}, \qquad x_2 = \frac{\Delta_2}{|\mathsf{A}|}, \qquad x_3 = \frac{\Delta_3}{|\mathsf{A}|}, \tag{1.73}$$

where

$$\Delta_1 = \begin{vmatrix} b_1 & A_{12} & A_{13} \\ b_2 & A_{22} & A_{23} \\ b_3 & A_{32} & A_{33} \end{vmatrix}, \quad \Delta_2 = \begin{vmatrix} A_{11} & b_1 & A_{13} \\ A_{21} & b_2 & A_{23} \\ A_{31} & b_3 & A_{33} \end{vmatrix}, \quad \Delta_3 = \begin{vmatrix} A_{11} & A_{12} & b_1 \\ A_{21} & A_{22} & b_2 \\ A_{31} & A_{32} & b_3 \end{vmatrix}.$$

It can be seen that each Cramer determinant Δ_i is simply $|\mathsf{A}|$ but with column i replaced by the RHS of the original set of equations. If $|\mathsf{A}| \neq 0$ then (1.73) gives the unique solution. The proof given here appears to fail if any of the solutions x_i is zero, but it can be shown that result (1.73) is valid even in such a case. The following example uses Cramer's method to solve the same set of equations as used in the previous two worked examples.

Example Use Cramer's rule to solve the set of simultaneous equations (1.66).

Let us again represent these simultaneous equations by the matrix equation $\mathsf{A}\mathbf{x} = \mathbf{b}$, i.e.

$$\begin{pmatrix} 2 & 4 & 3 \\ 1 & -2 & -2 \\ -3 & 3 & 2 \end{pmatrix} \begin{pmatrix} x_1 \\ x_2 \\ x_3 \end{pmatrix} = \begin{pmatrix} 4 \\ 0 \\ -7 \end{pmatrix}.$$

From (1.56), the determinant of A is given by $|\mathsf{A}| = 11$. Following the discussion given above, the three Cramer determinants are

$$\Delta_1 = \begin{vmatrix} 4 & 4 & 3 \\ 0 & -2 & -2 \\ -7 & 3 & 2 \end{vmatrix}, \quad \Delta_2 = \begin{vmatrix} 2 & 4 & 3 \\ 1 & 0 & -2 \\ -3 & -7 & 2 \end{vmatrix}, \quad \Delta_3 = \begin{vmatrix} 2 & 4 & 4 \\ 1 & -2 & 0 \\ -3 & 3 & -7 \end{vmatrix}.$$

These may be evaluated using the properties of determinants listed in Subsection 1.9.1 and we find $\Delta_1 = 22$, $\Delta_2 = -33$ and $\Delta_3 = 44$. From (1.73) the solution to the equations (1.66) is given by

$$x_1 = \frac{22}{11} = 2, \quad x_2 = \frac{-33}{11} = -3, \quad x_3 = \frac{44}{11} = 4,$$

which agrees with the solution found in the previous example. ◀

1.12.3 A geometrical interpretation

A helpful view of what is happening when simultaneous equations are solved, is to consider each of the equations as representing a surface in an N-dimensional space. This is most easily visualized in three (or two) dimensions. So, for example, we think of each of the three equations (1.72) as representing a plane in three-dimensional Cartesian coordinates. The sets of components of the vectors normal to the planes are (A_{11}, A_{12}, A_{13}), (A_{21}, A_{22}, A_{23}) and (A_{31}, A_{32}, A_{33}), and the perpendicular distances of the planes from the origin are given by

$$d_i = \frac{b_i}{\left(A_{i1}^2 + A_{i2}^2 + A_{i3}^2\right)^{1/2}} \quad \text{for } i = 1, 2, 3.$$

Finding the solution(s) to the simultaneous equations above corresponds to finding the point(s) of intersection of the planes.

If there is a unique solution the planes intersect at only a single point. This happens if their normals are linearly independent vectors. Since the rows of A represent the directions of these normals, this requirement is equivalent to $|\mathsf{A}| \neq 0$. If $\mathbf{b} = (0 \;\; 0 \;\; 0)^{\mathrm{T}} = \mathbf{0}$ then all the planes pass through the origin and, since there is only a single solution to the equations, the origin is that (trivial) solution.

Let us now turn to the cases where $|\mathsf{A}| = 0$. The simplest such case is that in which all three planes are parallel; this implies that the normals are all parallel and so A is of rank 1. Two possibilities exist:

(i) the planes are coincident, i.e. $d_1 = d_2 = d_3$, in which case there is an infinity of solutions;

(ii) the planes are not all coincident, i.e. $d_1 \neq d_2$ and/or $d_1 \neq d_3$ and/or $d_2 \neq d_3$, in which case there are no solutions.

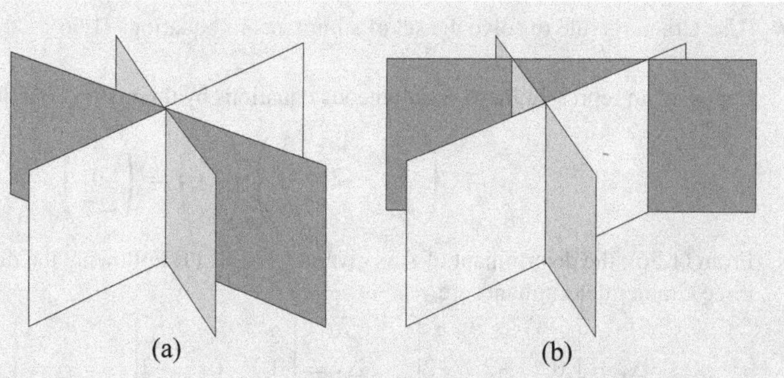

Figure 1.1 The two possible cases when A is of rank 2. In both cases all the normals lie in a horizontal plane but in (a) the planes all intersect on a single line (corresponding to an infinite number of solutions) whilst in (b) there are no common intersection points (no solutions).

It is apparent from (1.73) that case (i) occurs when all the Cramer determinants are zero and case (ii) occurs when at least one Cramer determinant is non-zero.

The most complicated cases with $|\mathsf{A}| = 0$ are those in which the normals to the planes themselves lie in a plane but are not parallel. In this case A has rank 2. Again two possibilities exist and these are shown in Figure 1.1. Just as in the rank-1 case, if all the Cramer determinants are zero then we get an infinity of solutions (this time on a line). Of course, in the special case in which $\mathsf{b} = \mathsf{0}$ (and the system of equations is homogeneous), the planes all pass through the origin and so they must intersect on a line through it. If at least one of the Cramer determinants is non-zero, we get no solution.

These rules may be summarized as follows.

(i) $|\mathsf{A}| \neq 0$, $\mathsf{b} \neq \mathsf{0}$: The three planes intersect at a single point that is not the origin, and so there is only one solution, given by both (1.65) and (1.73).

(ii) $|\mathsf{A}| \neq 0$, $\mathsf{b} = \mathsf{0}$: The three planes intersect at the origin only and there is only the trivial solution $\mathsf{x} = 0$.

(iii) $|\mathsf{A}| = 0$, $\mathsf{b} \neq \mathsf{0}$, Cramer determinants all zero: There is an infinity of solutions either on a line if A is rank 2, i.e. the cofactors are not all zero, or on a plane if A is rank 1, i.e. the cofactors are all zero.

(iv) $|\mathsf{A}| = 0$, $\mathsf{b} \neq \mathsf{0}$, Cramer determinants not all zero: No solutions.

(v) $|\mathsf{A}| = 0$, $\mathsf{b} = \mathsf{0}$: The three planes intersect on a line through the origin giving an infinity of solutions.

1.13 Special types of square matrix

Having examined some of the properties and uses of matrices, and of other matrices derived from them, we now consider some sets of square matrices that are characterized by a common structure or property possessed by their members; a summarizing table is given on p. 69. Matrices that are square, i.e. $N \times N$, appear in many physical applications and some special forms of square matrix are of particular importance.

1.13.1 Diagonal matrices

The unit matrix, which we have already encountered, is an example of a *diagonal* matrix. Such matrices are characterized by having non-zero elements only on the *leading diagonal*, i.e. only elements A_{ij} with $i = j$ may be non-zero. For example,

$$\mathsf{A} = \begin{pmatrix} 1 & 0 & 0 \\ 0 & 2 & 0 \\ 0 & 0 & -3 \end{pmatrix}$$

is a 3×3 diagonal matrix. Such a matrix is often denoted by $\mathsf{A} = \text{diag}\,(1, 2, -3)$. By performing a Laplace expansion, it is easily shown that the determinant of an $N \times N$ diagonal matrix is equal to the product of the diagonal elements.[19] Thus, if the matrix has the form $\mathsf{A} = \text{diag}(A_{11}, A_{22}, \ldots, A_{NN})$ then

$$|\mathsf{A}| = A_{11}A_{22}\cdots A_{NN}. \tag{1.74}$$

Moreover, it is also straightforward to show that the inverse of A is also a diagonal matrix given by

$$\mathsf{A}^{-1} = \text{diag}\left(\frac{1}{A_{11}}, \frac{1}{A_{22}}, \ldots, \frac{1}{A_{NN}}\right).$$

Finally, we note that, if two matrices A and B are *both* diagonal then they have the useful property that their product is commutative:

$$\mathsf{AB} = \mathsf{BA}.$$

Thus the set of all $N \times N$ diagonal matrices form a commuting set under matrix multiplication. This property is *not* shared by square matrices in general.

1.13.2 Lower and upper triangular matrices

We have already encountered triangular matrices in connection with *LU* and Cholesky decompositions, but we include them here for the sake of completeness.

A square matrix A is called *lower triangular* if all the elements *above* the principal diagonal are zero. For example, the general form for a 3×3 lower triangular matrix is

$$\mathsf{A} = \begin{pmatrix} A_{11} & 0 & 0 \\ A_{21} & A_{22} & 0 \\ A_{31} & A_{32} & A_{33} \end{pmatrix},$$

where the elements A_{ij} may be zero or non-zero. Similarly an *upper triangular* square matrix is one for which all the elements *below* the principal diagonal are zero. The general

19 Using this notation write down the form of the *most general, non-zero, singular, traceless, diagonal* 3×3 matrix.

3×3 form is thus

$$\mathsf{A} = \begin{pmatrix} A_{11} & A_{12} & A_{13} \\ 0 & A_{22} & A_{23} \\ 0 & 0 & A_{33} \end{pmatrix}.$$

By performing a Laplace expansion, it is straightforward to show that, in the general $N \times N$ case, the determinant of an upper or lower triangular matrix is equal to the product of its diagonal elements,

$$|\mathsf{A}| = A_{11}A_{22} \cdots A_{NN}. \tag{1.75}$$

Clearly property (1.74) of diagonal matrices is a special case of this more general result. Moreover, it may be shown that the inverse of a non-singular lower (upper) triangular matrix is also lower (upper) triangular.[20]

1.13.3 Symmetric and antisymmetric matrices

A square matrix A of order N with the property $\mathsf{A} = \mathsf{A}^{\mathrm{T}}$ is said to be *symmetric*. Similarly a matrix for which $\mathsf{A} = -\mathsf{A}^{\mathrm{T}}$ is said to be *anti-* or *skew*-symmetric and its diagonal elements $a_{11}, a_{22}, \ldots, a_{NN}$ are necessarily zero. Moreover, if A is (anti-)symmetric then so too is its inverse A^{-1}. This is easily proved by noting that if $\mathsf{A} = \pm\mathsf{A}^{\mathrm{T}}$ then

$$(\mathsf{A}^{-1})^{\mathrm{T}} = (\mathsf{A}^{\mathrm{T}})^{-1} = \pm\mathsf{A}^{-1}.$$

Any $N \times N$ matrix A can be written as the sum of a symmetric and an antisymmetric matrix, since we may write

$$\mathsf{A} = \tfrac{1}{2}(\mathsf{A} + \mathsf{A}^{\mathrm{T}}) + \tfrac{1}{2}(\mathsf{A} - \mathsf{A}^{\mathrm{T}}) = \mathsf{B} + \mathsf{C},$$

where clearly $\mathsf{B} = \mathsf{B}^{\mathrm{T}}$ and $\mathsf{C} = -\mathsf{C}^{\mathrm{T}}$. The matrix B is therefore called the symmetric part of A, and C is the antisymmetric part.

Example If A is an $N \times N$ antisymmetric matrix, show that $|\mathsf{A}| = 0$ if N is odd.

If A is antisymmetric then $\mathsf{A}^{\mathrm{T}} = -\mathsf{A}$. Using the properties of determinants (1.47) and (1.49), we have

$$|\mathsf{A}| = |\mathsf{A}^{\mathrm{T}}| = |-\mathsf{A}| = (-1)^N|\mathsf{A}|.$$

Thus, if N is odd then $|\mathsf{A}| = -|\mathsf{A}|$, which implies that $|\mathsf{A}| = 0$. ◀

1.13.4 Orthogonal matrices

A non-singular matrix with the property that its transpose is also its inverse,

$$\mathsf{A}^{\mathrm{T}} = \mathsf{A}^{-1}, \tag{1.76}$$

20 Determine where the following, clearly false, line of reasoning breaks down. Consider an upper triangular 3×3 matrix A which has unity for all its principal diagonal elements, $A_{12} = 0$, $A_{13} = a$ and $A_{23} = b$. It can be shown that $\mathsf{A} + \mathsf{A}^{-1} = 2\mathsf{I}$, and consequently (after multiplying through by A) we have $\mathsf{A}^2 - 2\mathsf{A} + \mathsf{I} = \mathsf{0}$. This can be written $(\mathsf{A} - \mathsf{I})(\mathsf{A} - \mathsf{I}) = (\mathsf{A} - \mathsf{I})^2 = \mathsf{0}$. Therefore $\mathsf{A} = \mathsf{I}$.

is called an *orthogonal matrix*. It follows immediately that the inverse of an orthogonal matrix is also orthogonal, since

$$(\mathsf{A}^{-1})^{\mathrm{T}} = (\mathsf{A}^{\mathrm{T}})^{-1} = (\mathsf{A}^{-1})^{-1}.$$

Moreover, since for an orthogonal matrix $\mathsf{A}^{\mathrm{T}}\mathsf{A} = \mathsf{I}$, we have

$$|\mathsf{A}^{\mathrm{T}}\mathsf{A}| = |\mathsf{A}^{\mathrm{T}}||\mathsf{A}| = |\mathsf{A}|^2 = |\mathsf{I}| = 1.$$

Thus the determinant of an orthogonal matrix must be $|\mathsf{A}| = \pm 1$.

An orthogonal matrix[21] represents, in a particular basis, a linear operator that leaves the norms (lengths) of real vectors unchanged, as we will now show. Suppose that $\mathbf{y} = \mathcal{A}\mathbf{x}$ is represented in some coordinate system by the matrix equation $\mathsf{y} = \mathsf{Ax}$; then $\langle \mathbf{y}|\mathbf{y}\rangle$ is given in this coordinate system by

$$\mathsf{y}^{\mathrm{T}}\mathsf{y} = \mathsf{x}^{\mathrm{T}}\mathsf{A}^{\mathrm{T}}\mathsf{A}\mathsf{x} = \mathsf{x}^{\mathrm{T}}\mathsf{x}.$$

Hence $\langle \mathbf{y}|\mathbf{y}\rangle = \langle \mathbf{x}|\mathbf{x}\rangle$, showing that the action of a linear operator represented by an orthogonal matrix does not change the norm of a real vector.

1.13.5 Hermitian and anti-Hermitian matrices

An *Hermitian* matrix is one that satisfies $\mathsf{A} = \mathsf{A}^\dagger$, where $\mathsf{A}^\dagger$ is the Hermitian conjugate discussed in Section 1.7. Similarly, if $\mathsf{A}^\dagger = -\mathsf{A}$, then A is called *anti-Hermitian*. A real (anti-)symmetric matrix is a special case of an (anti-)Hermitian matrix, in which all the elements of the matrix are real. Also, if A is an (anti-)Hermitian matrix then so too is its inverse A^{-1}, since

$$(\mathsf{A}^{-1})^\dagger = (\mathsf{A}^\dagger)^{-1} = \pm\mathsf{A}^{-1}.$$

Any $N \times N$ matrix A can be written as the sum of an Hermitian matrix and an anti-Hermitian matrix, since

$$\mathsf{A} = \tfrac{1}{2}(\mathsf{A} + \mathsf{A}^\dagger) + \tfrac{1}{2}(\mathsf{A} - \mathsf{A}^\dagger) = \mathsf{B} + \mathsf{C},$$

where clearly $\mathsf{B} = \mathsf{B}^\dagger$ and $\mathsf{C} = -\mathsf{C}^\dagger$. The matrix B is called the Hermitian part of A, and C is called the anti-Hermitian part.

1.13.6 Unitary matrices

A *unitary* matrix A is defined as one for which

$$\mathsf{A}^\dagger = \mathsf{A}^{-1}. \tag{1.77}$$

Clearly, if A is real then $\mathsf{A}^\dagger = \mathsf{A}^{\mathrm{T}}$, showing that a real orthogonal matrix is a special case of a unitary matrix, one in which all the elements are real.[22]

We note that the inverse A^{-1} of a unitary matrix is also unitary, since

$$(\mathsf{A}^{-1})^\dagger = (\mathsf{A}^\dagger)^{-1} = (\mathsf{A}^{-1})^{-1}.$$

21 A 2×2 matrix with both diagonal elements equal to $\cos\theta$ and off-diagonal elements $+\sin\theta$ and $-\sin\theta$ provides a practical example.

22 Three 2×2 matrices, S_x, S_y and S_z, are defined in Problem 1.10. Characterize each with respect to (a) reality, (b) symmetry, (c) Hermiticity, (d) orthogonality and (e) unitarity.

Moreover, since for a unitary matrix $\mathsf{A}^\dagger\mathsf{A} = \mathsf{I}$, we have

$$|\mathsf{A}^\dagger\mathsf{A}| = |\mathsf{A}^\dagger||\mathsf{A}| = |\mathsf{A}|^*|\mathsf{A}| = |\mathsf{I}| = 1.$$

Thus the determinant of a unitary matrix has unit modulus.

A unitary matrix represents, in a particular basis, a linear operator that leaves the norms (lengths) of complex vectors unchanged. If $\mathbf{y} = \mathcal{A}\mathbf{x}$ is represented in some coordinate system by the matrix equation $\mathsf{y} = \mathsf{A}\mathsf{x}$ then $\langle \mathbf{y}|\mathbf{y}\rangle$ is given in this coordinate system by

$$\mathsf{y}^\dagger\mathsf{y} = \mathsf{x}^\dagger\mathsf{A}^\dagger\mathsf{A}\mathsf{x} = \mathsf{x}^\dagger\mathsf{x}.$$

Hence $\langle \mathbf{y}|\mathbf{y}\rangle = \langle \mathbf{x}|\mathbf{x}\rangle$, showing that the action of the linear operator represented by a unitary matrix does not change the norm of a complex vector. The action of a unitary matrix on a complex column matrix thus parallels that of an orthogonal matrix acting on a real column matrix.

1.13.7 Normal matrices

A final important set of special matrices consists of the *normal* matrices, for which

$$\mathsf{A}\mathsf{A}^\dagger = \mathsf{A}^\dagger\mathsf{A},$$

i.e. a normal matrix is one that commutes with its Hermitian conjugate.

We can easily show that Hermitian matrices and unitary matrices (or symmetric matrices and orthogonal matrices in the real case) are examples of normal matrices. For an Hermitian matrix, $\mathsf{A} = \mathsf{A}^\dagger$ and so

$$\mathsf{A}\mathsf{A}^\dagger = \mathsf{A}\mathsf{A} = \mathsf{A}^\dagger\mathsf{A}.$$

Similarly, for a unitary matrix, $\mathsf{A}^{-1} = \mathsf{A}^\dagger$ and so

$$\mathsf{A}\mathsf{A}^\dagger = \mathsf{A}\mathsf{A}^{-1} = \mathsf{I} = \mathsf{A}^{-1}\mathsf{A} = \mathsf{A}^\dagger\mathsf{A}.$$

Finally, we note that, if A is normal then so too is its inverse A^{-1}, since

$$\mathsf{A}^{-1}(\mathsf{A}^{-1})^\dagger = \mathsf{A}^{-1}(\mathsf{A}^\dagger)^{-1} = (\mathsf{A}^\dagger\mathsf{A})^{-1} = (\mathsf{A}\mathsf{A}^\dagger)^{-1} = (\mathsf{A}^\dagger)^{-1}\mathsf{A}^{-1} = (\mathsf{A}^{-1})^\dagger\mathsf{A}^{-1}.$$

This broad class of matrices is formally important in the discussion of eigenvectors and eigenvalues (see the next section), as several general properties can be deduced purely on the basis that a matrix and its Hermitian conjugate commute. However, the corresponding general proofs tend to be more complicated than those treating only smaller classes of matrices and so, in the next sections, we have not pursued this broad approach.

1.14 Eigenvectors and eigenvalues

Suppose that a linear operator $\mathcal{A}$ transforms vectors $\mathbf{x}$ in an N-dimensional vector space into other vectors $\mathcal{A}\mathbf{x}$ in the same space. The possibility then arises that there exist vectors $\mathbf{x}$ each of which is transformed by $\mathcal{A}$ into a multiple of itself.[23] Such vectors would have

23 That is, after the transformation the vector still "points" in the same (or the directly opposite) direction in the vector space, even though it may have been changed in length.

to satisfy

$$\mathcal{A}\mathbf{x} = \lambda\mathbf{x}. \tag{1.78}$$

Any non-zero vector $\mathbf{x}$ that satisfies (1.78) for some value of λ is called an *eigenvector* of the linear operator $\mathcal{A}$, and λ is called the corresponding *eigenvalue*. As will be discussed below, in general the operator $\mathcal{A}$ has N independent eigenvectors $\mathbf{x}^i$, with eigenvalues λ_i. The λ_i are not necessarily all distinct.

If we choose a particular basis in the vector space, we can write (1.78) in terms of the components of $\mathcal{A}$ and $\mathbf{x}$ with respect to this basis as the matrix equation

$$\mathsf{A}\mathsf{x} = \lambda\mathsf{x}, \tag{1.79}$$

where A is an $N \times N$ matrix. The column matrices x that satisfy (1.79) obviously represent the eigenvectors $\mathbf{x}$ of $\mathcal{A}$ in our chosen coordinate system. Conventionally, these column matrices are also referred to as the *eigenvectors of the matrix* A.[24] Throughout this chapter we denote the ith eigenvector of a square matrix A by x^i and the corresponding eigenvalue by λ_i. This superscript notation for eigenvectors is used to avoid any confusion with components.

Clearly, if x is an eigenvector of A (with some eigenvalue λ) then any scalar multiple $\mu\mathsf{x}$ is also an eigenvector with the same eigenvalue; in other words, the factor by which the length of the vector is changed is independent of the original length. We therefore often use *normalized* eigenvectors, for which

$$\mathsf{x}^\dagger\mathsf{x} = 1$$

(note that $\mathsf{x}^\dagger\mathsf{x}$ corresponds to the inner product $\langle\mathbf{x}|\mathbf{x}\rangle$ in our basis). Any eigenvector x can be normalized by dividing all of its components by the scalar $(\mathsf{x}^\dagger\mathsf{x})^{1/2}$.

The problem of finding the eigenvalues and corresponding eigenvectors of a square matrix A plays an important role in many physical investigations. It is the standard basis for determining the normal modes of an oscillatory mechanical or electrical system, with applications ranging from the stability of bridges to the internal vibrations of molecules. It also provides the methodology for the particular formulation of quantum mechanics that is known as matrix mechanics.

We begin with an example that produces a simple deduction from the defining eigenvalue equation (1.79).

Example A non-singular matrix A has eigenvalues λ_i and eigenvectors x^i. Find the eigenvalues and eigenvectors of the inverse matrix A^{-1}.

The eigenvalues and eigenvectors of A satisfy

$$\mathsf{A}\mathsf{x}^i = \lambda_i\mathsf{x}^i.$$

24 In this context, when referring to linear combinations of eigenvectors x we will normally use the term "vector".

Left-multiplying both sides of this equation by A^{-1}, we find

$$\mathsf{A}^{-1}\mathsf{A}\mathsf{x}^i = \lambda_i \mathsf{A}^{-1}\mathsf{x}^i.$$

Since $\mathsf{A}^{-1}\mathsf{A} = \mathsf{I}$, dividing through by λ_i and interchanging the two sides of the equation, gives an eigenvalue equation for A^{-1}:

$$\mathsf{A}^{-1}\mathsf{x}^i = \frac{1}{\lambda_i}\mathsf{x}^i.$$

From this we see that each eigenvector $\mathbf{x}^i$ of A is also an eigenvector of A^{-1}, but that the corresponding eigenvalue is $1/\lambda_i$. As A and A^{-1} have the same dimensions, and hence the same number of independent eigenvectors, the two sets of eigenvectors are identical.[25,26] ◀

In the remainder of this section we will discuss some useful results concerning the eigenvectors and eigenvalues of certain special (though commonly occurring) square matrices. The results will be established for matrices whose elements may be complex; the corresponding properties for real matrices can be obtained as special cases.

1.14.1 Eigenvectors and eigenvalues of Hermitian and unitary matrices

We start by proving two powerful results about the eigenvalues and eigenfunctions of Hermitian matrices, namely:

(i) The eigenvalues of an Hermitian matrix are real.
(ii) The eigenvectors of an Hermitian matrix corresponding to different eigenvalues are orthogonal.

For the present we will assume that the eigenvalues of our Hermitian matrix A are distinct, and later show what modifications are needed when they are not.

Consider two eigenvalues λ_i and λ_j and their corresponding eigenvectors satisfying

$$\mathsf{A}\mathsf{x}^i = \lambda_i \mathsf{x}^i, \tag{1.80}$$

$$\mathsf{A}\mathsf{x}^j = \lambda_j \mathsf{x}^j. \tag{1.81}$$

Taking the Hermitian conjugate of (1.80) we find $(\mathsf{x}^i)^\dagger \mathsf{A}^\dagger = \lambda_i^*(\mathsf{x}^i)^\dagger$. Multiplying this on the right by x^j gives

$$(\mathsf{x}^i)^\dagger \mathsf{A}^\dagger \mathsf{x}^j = \lambda_i^*(\mathsf{x}^i)^\dagger \mathsf{x}^j,$$

and similarly multiplying (1.81) through on the left by $(\mathsf{x}^i)^\dagger$ yields

$$(\mathsf{x}^i)^\dagger \mathsf{A}\mathsf{x}^j = \lambda_j(\mathsf{x}^i)^\dagger \mathsf{x}^j.$$

Then, since $\mathsf{A}^\dagger = \mathsf{A}$, the two left-hand sides are equal and on subtraction we obtain

$$0 = (\lambda_i^* - \lambda_j)(\mathsf{x}^i)^\dagger \mathsf{x}^j. \tag{1.82}$$

To prove result (i) we need only set $j = i$. Then (1.82) reads

$$0 = (\lambda_i^* - \lambda_i)(\mathsf{x}^i)^\dagger \mathsf{x}^i.$$

25 If any of the λ_i are repeated, then linear combinations of the corresponding $\mathbf{x}^i$ may have to be formed.
26 Explain why, if one of the eigenvalues of A is 0, this does not imply that the inverse of A has an eigenvalue of ∞.

Now, since x is a non-zero vector, $(\mathsf{x}^i)^\dagger\mathsf{x}^i \neq 0$, implying that $\lambda_i^* = \lambda_i$, i.e. λ_i is real.

Result (ii) follows almost immediately because when $j \neq i$ in (1.82), and consequently $\lambda_j \neq \lambda_i = \lambda_i^*$, we must have $(\mathsf{x}^i)^\dagger\mathsf{x}^j = 0$, i.e. the relevant eigenvectors, x^i and x^j, are orthogonal.

We should also note at this point that, if A is anti-Hermitian (rather than Hermitian) and $\mathsf{A}^\dagger = -\mathsf{A}$ then the bracket in (1.82) reads $(\lambda_i^* + \lambda_j)$ and when j is set equal to i we conclude that $\lambda_i^* = -\lambda_i$, i.e. λ_i is purely imaginary. The previous conclusion about the orthogonality of the eigenvectors is unaltered.

As a reminder, we also recall that real symmetric matrices are special cases of Hermitian matrices, and so they too have real eigenvalues and mutually orthogonal eigenvectors.

The importance of result (i) for Hermitian matrices will be apparent to any student of quantum mechanics. In quantum mechanics the eigenvalues of operators correspond to measured values of observable quantities, e.g. energy, angular momentum, parity and so on, and these clearly must be real. If we use Hermitian operators to formulate the theories of quantum mechanics, the above property guarantees physically meaningful results.

We now turn our attention to unitary matrices and prove, by very similar means to those just employed, that the eigenvalues of a unitary matrix necessarily have unit modulus. A unitary matrix satisfies $\mathsf{A}^\dagger = \mathsf{A}^{-1}$ or, equivalently, $\mathsf{A}^\dagger\mathsf{A} = \mathsf{I}$.

Taking the Hermitian conjugate of (1.80) we have, as previously, that

$$(\mathsf{x}^i)^\dagger\mathsf{A}^\dagger = \lambda_i^*(\mathsf{x}^i)^\dagger, \tag{1.83}$$

whilst from (1.81)

$$\mathsf{A}\mathsf{x}^j = \lambda_j\mathsf{x}^j. \tag{1.84}$$

Now, right-multiplying the LHS of (1.83) by the LHS of (1.84), and correspondingly for the two RHSs, gives

$$\begin{aligned}(\mathsf{x}^i)^\dagger\mathsf{A}^\dagger\mathsf{A}\mathsf{x}^j &= \lambda_i^*(\mathsf{x}^i)^\dagger\lambda_j\mathsf{x}^j,\\ (\mathsf{x}^i)^\dagger\mathsf{x}^j &= \lambda_i^*\lambda_j(\mathsf{x}^i)^\dagger\mathsf{x}^j,\\ \left[1-\lambda_i^*\lambda_j\right](\mathsf{x}^i)^\dagger\mathsf{x}^j &= 0.\end{aligned}$$

Finally, setting $j = i$ and again noting that x^i is a non-zero vector, shows that

$$1 - |\lambda_i|^2 = 0.$$

Thus, the eigenvalues of a unitary matrix have unit modulus. The proof of the orthogonality property of its eigenvectors is as for Hermitian matrices. For completeness, we also note that a real orthogonal matrix is a special case of a unitary matrix; it too has eigenvectors of unit modulus.[27]

If some of the eigenvalues of a matrix are equal and one eigenvalue corresponds to two or more different eigenvectors (i.e. no two are simple multiples of each other), that eigenvalue is said to be *degenerate*. In this case further justification of the orthogonality of the eigenvectors is needed. The Gram–Schmidt orthogonalization procedure discussed in

27 In fact, for a real orthogonal matrix, the only possible eigenvalues are $\lambda = \pm 1$. Show this by proving that they must satisfy $\lambda^2 = 1$.

Appendix F provides a proof of, and a means of achieving, orthogonality, in that it shows how to construct a mutually orthogonal set of linear combinations of those eigenvectors that correspond to the degenerate eigenvalue. In practice, however, the method is laborious and the example in Subsection 1.15.1 gives a less rigorous but considerably quicker way of achieving the same end.

1.14.2 Eigenvectors and eigenvalues of a general square matrix

When an $N \times N$ matrix does not qualify for the broad, but nevertheless restricted, class of normal matrices (see Subsection 1.13.7), there are no general properties that can be ascribed to its eigenvalues and eigenvectors. In fact, in general it is not possible to find any orthogonal set of N eigenvectors or even to find *pairs* of orthogonal eigenvectors (except by chance in some cases). While its N non-orthogonal eigenvectors are usually linearly independent and hence form a basis for the N-dimensional vector space, even this is not necessarily so.

It may be shown (although we will not prove it) that any $N \times N$ matrix with *distinct* eigenvalues does have N linearly independent eigenvectors, which therefore do form a basis for the N-dimensional vector space. If a general square matrix has degenerate eigenvalues, however, then it may or may not have N linearly independent eigenvectors. A matrix whose eigenvectors are not linearly independent is said to be *defective*.

1.14.3 Simultaneous eigenvectors

We may now ask under what conditions two different Hermitian matrices can have a common set of eigenvectors. The result – that they do so if, and only if, they commute – has profound significance for the foundations of quantum mechanics.

To prove this important result let A and B be two $N \times N$ Hermitian matrices and x^i be the ith eigenvector of A corresponding to eigenvalue λ_i, i.e.

$$\mathsf{A}\mathsf{x}^i = \lambda_i \mathsf{x}^i \quad \text{for} \quad i = 1, 2, \ldots, N.$$

For the present we assume that the eigenvalues are all different.

(i) First suppose that A and B commute. Now consider

$$\mathsf{A}\mathsf{B}\mathsf{x}^i = \mathsf{B}\mathsf{A}\mathsf{x}^i = \mathsf{B}\lambda_i \mathsf{x}^i = \lambda_i \mathsf{B}\mathsf{x}^i,$$

where we have used the commutativity for the first equality and the eigenvector property for the second. It follows that $\mathsf{A}(\mathsf{B}\mathsf{x}^i) = \lambda_i(\mathsf{B}\mathsf{x}^i)$ and thus that $\mathsf{B}\mathsf{x}^i$ is an eigenvector of A corresponding to eigenvalue λ_i. But the eigenvector solutions of $(\mathsf{A} - \lambda_i \mathsf{I})\mathsf{x}^i = \mathbf{0}$ are unique to within a scale factor, and we therefore conclude that

$$\mathsf{B}\mathsf{x}^i = \mu_i \mathsf{x}^i$$

for some scale factor μ_i. However, this is just an eigenvector equation for B and shows that x^i is an eigenvector of B, in addition to being an eigenvector of A. By reversing the roles of A and B, it also follows that every eigenvector of B is an eigenvector of A. Thus the two sets of eigenvectors are identical.

(ii) Now suppose that A and B have all their eigenvectors in common, a typical one x^i satisfying both

$$\mathsf{A}\mathsf{x}^i = \lambda_i \mathsf{x}^i \quad \text{and} \quad \mathsf{B}\mathsf{x}^i = \mu_i \mathsf{x}^i.$$

As the eigenvectors span the N-dimensional vector space, any arbitrary vector x in the space can be written as a linear combination of the eigenvectors,

$$\mathsf{x} = \sum_{i=1}^{N} c_i \mathsf{x}^i.$$

Now consider both

$$\mathsf{A}\mathsf{B}\mathsf{x} = \mathsf{A}\mathsf{B}\sum_{i=1}^{N} c_i \mathsf{x}^i = \mathsf{A}\sum_{i=1}^{N} c_i \mu_i \mathsf{x}^i = \sum_{i=1}^{N} c_i \lambda_i \mu_i \mathsf{x}^i,$$

and

$$\mathsf{B}\mathsf{A}\mathsf{x} = \mathsf{B}\mathsf{A}\sum_{i=1}^{N} c_i \mathsf{x}^i = \mathsf{B}\sum_{i=1}^{N} c_i \lambda_i \mathsf{x}^i = \sum_{i=1}^{N} c_i \mu_i \lambda_i \mathsf{x}^i.$$

It follows that $\mathsf{A}\mathsf{B}\mathsf{x}$ and $\mathsf{B}\mathsf{A}\mathsf{x}$ are the same for any arbitrary x and hence that

$$(\mathsf{A}\mathsf{B} - \mathsf{B}\mathsf{A})\mathsf{x} = \mathsf{0}$$

for all x. That is, A and B *commute*.

This completes the proof that a necessary and sufficient condition for two Hermitian matrices to have a set of eigenvectors in common is that they commute. It should be noted that if an eigenvalue of A, say, is degenerate then not all of its possible sets of eigenvectors will also constitute a set of eigenvectors of B. However, provided that by taking linear combinations one set of joint eigenvectors can be found, the proof is still valid and the result still holds.

When extended to the case of Hermitian operators and continuous eigenfunctions (Sections 8.2 and 8.3) the connection between commuting matrices and a set of common eigenvectors plays a fundamental role in the postulatory basis of quantum mechanics. It draws the distinction between commuting and non-commuting observables and sets limits on how much information about a system can be known, even in principle, at any one time.

1.15 Determination of eigenvalues and eigenvectors

The next step is to show how the eigenvalues and eigenvectors of a given $N \times N$ matrix A are found. To do this we refer to (1.79) and, by replacing x by $\mathsf{I}\mathsf{x}$ where I is the unit matrix of order N, rewrite it as

$$\mathsf{A}\mathsf{x} - \lambda\mathsf{I}\mathsf{x} = (\mathsf{A} - \lambda\mathsf{I})\mathsf{x} = \mathsf{0}. \tag{1.85}$$

The point of doing this is immediate since (1.85) now has the form of a homogeneous set of simultaneous equations, the theory of which was developed in Section 1.12. What was proved there is that the equation $\mathsf{Bx} = \mathbf{0}$ only has a non-trivial solution x if $|\mathsf{B}| = 0$. Correspondingly, therefore, we must have in the present case that

$$|\mathsf{A} - \lambda \mathsf{I}| = 0, \tag{1.86}$$

if there are to be non-zero solutions x to (1.85).

Equation (1.86) is known as the *characteristic equation* for A and its LHS as the *characteristic* or *secular determinant* of A. The equation is a polynomial of degree N in the quantity λ. The N roots of this equation λ_i, $i = 1, 2, \ldots, N$, give the eigenvalues of A. Corresponding to each λ_i there will be a column vector x^i, which is the ith eigenvector of A and can be found by solving (1.85) for x.

It will be observed that when (1.86) is written out as a polynomial equation in λ, the coefficient of $-\lambda^{N-1}$ in the equation will be simply $A_{11} + A_{22} + \cdots + A_{NN}$, whilst that of λ^N will be unity. As discussed in Section 1.8, the quantity $\sum_{i=1}^{N} A_{ii}$ is the *trace* of A and, from the ordinary theory of polynomial equations will be equal to the sum of the roots of (1.86):

$$\sum_{i=1}^{N} \lambda_i = \operatorname{Tr} \mathsf{A}. \tag{1.87}$$

This can be used as one check that a computation of the eigenvalues λ_i has been done correctly. Unless equation (1.87) is satisfied by a computed set of eigenvalues, they have not been calculated correctly. However, that equation (1.87) is satisfied is a necessary, but not sufficient, condition for a correct computation. An alternative proof of (1.87) is given in Section 1.17. A straightforward example now follows.

Example Find the eigenvalues and normalized eigenvectors of the real symmetric matrix

$$\mathsf{A} = \begin{pmatrix} 1 & 1 & 3 \\ 1 & 1 & -3 \\ 3 & -3 & -3 \end{pmatrix}.$$

Using (1.86),

$$\begin{vmatrix} 1-\lambda & 1 & 3 \\ 1 & 1-\lambda & -3 \\ 3 & -3 & -3-\lambda \end{vmatrix} = 0.$$

Expanding out this determinant gives[28]

$$(1-\lambda)\left[(1-\lambda)(-3-\lambda) - (-3)(-3)\right] + 1\left[(-3)(3) - 1(-3-\lambda)\right] + 3\left[1(-3) - (1-\lambda)(3)\right] = 0, \tag{1.88}$$

28 This "head-on" method gives a cubic equation, the roots of which have to be obtained by inspiration. Obtain the same result by (i) adding the 2nd column to the 1st, and (ii) taking out a common factor $(2-\lambda)$. Then subtract the 1st row from the 2nd and obtain a quadratic expression in λ that can be factorized by inspection.

which simplifies to give

$$(1-\lambda)(\lambda^2+2\lambda-12)+(\lambda-6)+3(3\lambda-6)=0,$$
$$\Rightarrow \quad (\lambda-2)(\lambda-3)(\lambda+6)=0.$$

Hence the roots of the characteristic equation, which are the eigenvalues of A, are $\lambda_1=2$, $\lambda_2=3$, $\lambda_3=-6$. We note that, as expected,

$$\lambda_1+\lambda_2+\lambda_3=-1=1+1-3=A_{11}+A_{22}+A_{33}=\mathrm{Tr}\,\mathsf{A}.$$

For the first root, $\lambda_1=2$, a suitable eigenvector x^1, with elements x_1, x_2, x_3, must satisfy $\mathsf{A}\mathsf{x}^1=2\mathsf{x}^1$ or, equivalently,

$$\begin{aligned} x_1+x_2+3x_3&=2x_1,\\ x_1+x_2-3x_3&=2x_2,\\ 3x_1-3x_2-3x_3&=2x_3. \end{aligned} \tag{1.89}$$

These three equations are consistent (to ensure this, was the purpose behind finding the particular values of λ) and yield $x_3=0$, $x_1=x_2=k$, where k is any non-zero number. A suitable eigenvector would thus be

$$\mathsf{x}^1=(k\ \ k\ \ 0)^{\mathrm{T}}.$$

If we apply the normalization condition, we require $k^2+k^2+0^2=1$ or $k=1/\sqrt{2}$. Hence

$$\mathsf{x}^1=\left(\frac{1}{\sqrt{2}}\ \ \frac{1}{\sqrt{2}}\ \ 0\right)^{\mathrm{T}}=\frac{1}{\sqrt{2}}(1\ \ 1\ \ 0)^{\mathrm{T}}.$$

Repeating the last paragraph, but with the factor 2 on the RHS of (1.89) replaced successively by $\lambda_2=3$ and $\lambda_3=-6$, gives

$$\mathsf{x}^2=\frac{1}{\sqrt{3}}(1\ \ -1\ \ 1)^{\mathrm{T}} \text{ and } \mathsf{x}^3=\frac{1}{\sqrt{6}}(1\ \ -1\ \ -2)^{\mathrm{T}}$$

as two further normalized eigenvectors. ◀

In the above example, the three values of λ are all different and A is a real symmetric matrix. Thus we expect, and it is easily checked, that the three eigenvectors are mutually orthogonal, i.e.

$$\left(\mathsf{x}^1\right)^{\mathrm{T}}\mathsf{x}^2=\left(\mathsf{x}^1\right)^{\mathrm{T}}\mathsf{x}^3=\left(\mathsf{x}^2\right)^{\mathrm{T}}\mathsf{x}^3=0.$$

It will be apparent also that, as expected, the normalization of the eigenvectors has no effect on their orthogonality.

1.15.1 Degenerate eigenvalues

We now return to the case of degenerate eigenvalues, i.e. those that have two or more associated eigenvectors. We have shown already that it is always possible to construct an orthogonal set of eigenvectors for a normal matrix, see Appendix F; the following example, which exploits this natural or imposed mutual orthogonality, illustrates a heuristic method of finding such a set that is simpler than following the formal steps given in the appendix.

Example Construct an orthonormal set of eigenvectors for the matrix

$$\mathsf{A} = \begin{pmatrix} 1 & 0 & 3 \\ 0 & -2 & 0 \\ 3 & 0 & 1 \end{pmatrix}.$$

We first determine the eigenvalues using $|\mathsf{A} - \lambda \mathsf{I}| = 0$:

$$0 = \begin{vmatrix} 1-\lambda & 0 & 3 \\ 0 & -2-\lambda & 0 \\ 3 & 0 & 1-\lambda \end{vmatrix} = -(1-\lambda)^2(2+\lambda) + 3(3)(2+\lambda)$$
$$= (4-\lambda)(\lambda+2)^2.$$

Thus $\lambda_1 = 4$, $\lambda_2 = -2 = \lambda_3$. The normalized eigenvector $\mathsf{x}^1 = (x_1 \quad x_2 \quad x_3)^{\mathrm{T}}$ corresponding to the unrepeated eigenvalue is found from

$$\begin{pmatrix} 1 & 0 & 3 \\ 0 & -2 & 0 \\ 3 & 0 & 1 \end{pmatrix} \begin{pmatrix} x_1 \\ x_2 \\ x_3 \end{pmatrix} = 4 \begin{pmatrix} x_1 \\ x_2 \\ x_3 \end{pmatrix} \quad \Rightarrow \quad \mathsf{x}^1 = \frac{1}{\sqrt{2}} \begin{pmatrix} 1 \\ 0 \\ 1 \end{pmatrix}.$$

A general column vector that is orthogonal to x^1 is

$$\mathsf{x} = (a \quad b \quad -a)^{\mathrm{T}}, \qquad (1.90)$$

and it is easily shown that

$$\mathsf{A}\mathsf{x} = \begin{pmatrix} 1 & 0 & 3 \\ 0 & -2 & 0 \\ 3 & 0 & 1 \end{pmatrix} \begin{pmatrix} a \\ b \\ -a \end{pmatrix} = -2 \begin{pmatrix} a \\ b \\ -a \end{pmatrix} = -2\mathsf{x}.$$

Thus x is an eigenvector of A with associated eigenvalue -2. It is clear, however, that there is an infinite set of eigenvectors x all possessing the required property; the geometrical analogue is that there are an infinite number of corresponding vectors $\mathbf{x}$ lying in the plane that has $\mathbf{x}^1$ as its normal.

We do require that the two remaining eigenvectors are orthogonal to one another, but this still leaves an infinite number of possibilities. For x^2, therefore, let us choose a simple form of (1.90), suitably normalized, say,

$$\mathsf{x}^2 = (0 \quad 1 \quad 0)^{\mathrm{T}}.$$

The third eigenvector is then specified (to within an arbitrary multiplicative constant) by the requirement that it must be orthogonal to x^1 and x^2; thus x^3 may be found by evaluating the vector product of x^1 and x^2 and normalizing the result. This gives

$$\mathsf{x}^3 = \frac{1}{\sqrt{2}}(-1 \quad 0 \quad 1)^{\mathrm{T}},$$

corresponding to $a = -1$ and $b = 0$, and completes the construction of an orthonormal set of eigenvectors.[29] ◀

29 How would you find an orthonormal set if all three eigenvalues were equal?

1.16 Change of basis and similarity transformations

Throughout this chapter we have considered the vector $\mathbf{x}$ as a geometrical quantity that is independent of any basis (or coordinate system). If we introduce a basis $\mathbf{e}_i$, $i = 1, 2, \ldots, N$, into our N-dimensional vector space then we may write

$$\mathbf{x} = x_1\mathbf{e}_1 + x_2\mathbf{e}_2 + \cdots + x_N\mathbf{e}_N,$$

and represent $\mathbf{x}$ in this basis by the column matrix

$$\mathsf{x} = (x_1 \quad x_2 \quad \cdots \quad x_n)^{\mathrm{T}},$$

having components x_i. We now consider how these components change as a result of a prescribed change of basis. Let us introduce a new basis $\mathbf{e}'_i$, $i = 1, 2, \ldots, N$, which is related to the old basis by

$$\mathbf{e}'_j = \sum_{i=1}^{N} S_{ij}\mathbf{e}_i, \tag{1.91}$$

the coefficient S_{ij} being the ith component of $\mathbf{e}'_j$ with respect to the old (unprimed) basis. For an arbitrary vector $\mathbf{x}$ it follows that

$$\mathbf{x} = \sum_{i=1}^{N} x_i\mathbf{e}_i = \sum_{j=1}^{N} x'_j\mathbf{e}'_j = \sum_{j=1}^{N} x'_j \sum_{i=1}^{N} S_{ij}\mathbf{e}_i.$$

From this we derive the relationship between the components of $\mathbf{x}$ in the two coordinate systems as

$$x_i = \sum_{j=1}^{N} S_{ij}x'_j,$$

which we can write in matrix form as

$$\mathsf{x} = \mathsf{S}\mathsf{x}' \tag{1.92}$$

where S is the *transformation matrix* associated with the change of basis.

Furthermore, since the vectors $\mathbf{e}'_j$ are linearly independent, the matrix S is non-singular and so possesses an inverse S^{-1}. Multiplying (1.92) on the left by S^{-1} we find

$$\mathsf{x}' = \mathsf{S}^{-1}\mathsf{x}, \tag{1.93}$$

which relates the components of $\mathbf{x}$ in the new basis to those in the old basis. Comparing (1.93) and (1.91) we note that the components of $\mathbf{x}$ transform inversely to the way in which the basis vectors $\mathbf{e}_i$ themselves transform. This has to be so, as the vector $\mathbf{x}$ itself must remain unchanged.

We may also find the transformation law for the components of a linear operator under the same change of basis. The operator equation $\mathbf{y} = \mathcal{A}\mathbf{x}$ (which is basis independent) can be written as a matrix equation in each of the two bases as

$$\mathsf{y} = \mathsf{A}\mathsf{x}, \qquad \mathsf{y}' = \mathsf{A}'\mathsf{x}'. \tag{1.94}$$

But, using (1.92) to change from the unprimed to the primed basis, we may rewrite the first equation as

$$\mathsf{S}\mathsf{y}' = \mathsf{A}\mathsf{S}\mathsf{x}' \quad \Rightarrow \quad \mathsf{y}' = \mathsf{S}^{-1}\mathsf{A}\mathsf{S}\mathsf{x}'.$$

Comparing this with the second equation in (1.94) we find that the components of the linear operator $\mathcal{A}$ transform as

$$\mathsf{A}' = \mathsf{S}^{-1}\mathsf{A}\mathsf{S}. \tag{1.95}$$

Equation (1.95) is an example of a *similarity transformation* – a transformation that can be particularly useful in the conversion of matrices into convenient forms for computation.

Given a square matrix A, we may interpret it as representing a linear operator $\mathcal{A}$ in a given basis $\mathbf{e}_i$. From (1.95), however, we may also consider the matrix $\mathsf{A}' = \mathsf{S}^{-1}\mathsf{A}\mathsf{S}$, for any non-singular matrix S, as representing the same linear operator $\mathcal{A}$ but in a new basis $\mathbf{e}'_j$, related to the old basis by

$$\mathbf{e}'_j = \sum_i S_{ij}\mathbf{e}_i.$$

Therefore we would expect that any property of the matrix A that represents some (basis-independent) property of the linear operator $\mathcal{A}$ will also be a property of the matrix A′. We list a number of such properties below.

(i) If $\mathsf{A} = \mathsf{I}$ then $\mathsf{A}' = \mathsf{I}$, since, from (1.95),

$$\mathsf{A}' = \mathsf{S}^{-1}\mathsf{I}\mathsf{S} = \mathsf{S}^{-1}\mathsf{S} = \mathsf{I}. \tag{1.96}$$

(ii) The value of the determinant is unchanged:

$$|\mathsf{A}'| = |\mathsf{S}^{-1}\mathsf{A}\mathsf{S}| = |\mathsf{S}^{-1}||\mathsf{A}||\mathsf{S}| = |\mathsf{A}||\mathsf{S}^{-1}||\mathsf{S}| = |\mathsf{A}||\mathsf{S}^{-1}\mathsf{S}| = |\mathsf{A}|. \tag{1.97}$$

(iii) The characteristic determinant and hence the eigenvalues of A′ are the same as those of A: from (1.86),

$$\begin{aligned}|\mathsf{A}' - \lambda\mathsf{I}| &= |\mathsf{S}^{-1}\mathsf{A}\mathsf{S} - \lambda\mathsf{I}| = |\mathsf{S}^{-1}(\mathsf{A} - \lambda\mathsf{I})\mathsf{S}| \\ &= |\mathsf{S}^{-1}||\mathsf{S}||\mathsf{A} - \lambda\mathsf{I}| = |\mathsf{A} - \lambda\mathsf{I}|.\end{aligned} \tag{1.98}$$

(iv) The value of the trace is unchanged: this follows either from combining (1.87) and property (iii) above, or directly as follows,

$$\begin{aligned}\mathrm{Tr}\,\mathsf{A}' &= \sum_i A'_{ii} = \sum_i\sum_j\sum_k (\mathsf{S}^{-1})_{ij}A_{jk}S_{ki} \\ &= \sum_i\sum_j\sum_k S_{ki}(\mathsf{S}^{-1})_{ij}A_{jk} = \sum_j\sum_k \delta_{kj}A_{jk} = \sum_j A_{jj} \\ &= \mathrm{Tr}\,\mathsf{A}.\end{aligned} \tag{1.99}$$

An important class of similarity transformations is that for which S is a unitary matrix; in this case $\mathsf{A}' = \mathsf{S}^{-1}\mathsf{A}\mathsf{S} = \mathsf{S}^\dagger\mathsf{A}\mathsf{S}$. Unitary transformation matrices are particularly important, for the following reason. If the original basis $\mathbf{e}_i$ is orthonormal and the transformation matrix S is unitary then

$$\begin{aligned}\langle \mathbf{e}'_i|\mathbf{e}'_j\rangle &= \left\langle \sum_k S_{ki}\mathbf{e}_k \,\Big|\, \sum_r S_{rj}\mathbf{e}_r \right\rangle \\ &= \sum_k S^*_{ki} \sum_r S_{rj}\,\langle \mathbf{e}_k|\mathbf{e}_r\rangle \\ &= \sum_k S^*_{ki} \sum_r S_{rj}\delta_{kr} = \sum_k S^*_{ki}S_{kj} = \sum_k S^\dagger_{ik}S_{kj} = (\mathsf{S}^\dagger\mathsf{S})_{ij} = \delta_{ij},\end{aligned}$$

showing that the new basis is also orthonormal.

Furthermore, in addition to the properties of general similarity transformations, for unitary transformations the following hold.

(i) If A is Hermitian (anti-Hermitian) then A′ is Hermitian (anti-Hermitian), i.e. if $\mathsf{A}^\dagger = \pm\mathsf{A}$ then

$$(\mathsf{A}')^\dagger = (\mathsf{S}^\dagger\mathsf{A}\mathsf{S})^\dagger = \mathsf{S}^\dagger\mathsf{A}^\dagger\mathsf{S} = \pm\mathsf{S}^\dagger\mathsf{A}\mathsf{S} = \pm\mathsf{A}'. \qquad (1.100)$$

(ii) If A is unitary (so that $\mathsf{A}^\dagger = \mathsf{A}^{-1}$) then A′ is unitary, since

$$\begin{aligned}(\mathsf{A}')^\dagger\mathsf{A}' &= (\mathsf{S}^\dagger\mathsf{A}\mathsf{S})^\dagger(\mathsf{S}^\dagger\mathsf{A}\mathsf{S}) = \mathsf{S}^\dagger\mathsf{A}^\dagger\mathsf{S}\mathsf{S}^\dagger\mathsf{A}\mathsf{S} = \mathsf{S}^\dagger\mathsf{A}^\dagger\mathsf{A}\mathsf{S} \\ &= \mathsf{S}^\dagger\mathsf{I}\mathsf{S} = \mathsf{I}.\end{aligned} \qquad (1.101)$$

1.17 Diagonalization of matrices

Suppose that a linear operator $\mathcal{A}$ is represented in some basis $\mathbf{e}_i$, $i = 1, 2, \ldots, N$, by the matrix A. Consider a new basis $\mathbf{x}^j$ given by

$$\mathbf{x}^j = \sum_{i=1}^{N} S_{ij}\mathbf{e}_i,$$

where the $\mathbf{x}^j$ are chosen to be the eigenvectors of the linear operator $\mathcal{A}$, i.e.

$$\mathcal{A}\mathbf{x}^j = \lambda_j\mathbf{x}^j. \qquad (1.102)$$

In the new basis, $\mathcal{A}$ is represented by the matrix $\mathsf{A}' = \mathsf{S}^{-1}\mathsf{A}\mathsf{S}$, which has a particularly simple form, as we shall see shortly. The element S_{ij} of S is the ith component, in the old (unprimed) basis, of the jth eigenvector x^j of A, i.e. the columns of S are the eigenvectors of the matrix A:

$$\mathsf{S} = \begin{pmatrix} \uparrow & \uparrow & & \uparrow \\ \mathsf{x}^1 & \mathsf{x}^2 & \cdots & \mathsf{x}^N \\ \downarrow & \downarrow & & \downarrow \end{pmatrix},$$

that is, $S_{ij} = (\mathsf{x}^j)_i$. Therefore A' is given by

$$\begin{aligned}(\mathsf{S}^{-1}\mathsf{AS})_{ij} &= \sum_k \sum_l (\mathsf{S}^{-1})_{ik} A_{kl} S_{lj} \\ &= \sum_k \sum_l (\mathsf{S}^{-1})_{ik} A_{kl} (\mathsf{x}^j)_l \\ &= \sum_k (\mathsf{S}^{-1})_{ik} \lambda_j (\mathsf{x}^j)_k \\ &= \sum_k \lambda_j (\mathsf{S}^{-1})_{ik} S_{kj} = \lambda_j \delta_{ij}.\end{aligned}$$

So the matrix A' is diagonal with the eigenvalues of $\mathcal{A}$ as its diagonal elements, i.e.

$$\mathsf{A}' = \begin{pmatrix} \lambda_1 & 0 & \cdots & 0 \\ 0 & \lambda_2 & & \vdots \\ \vdots & & \ddots & 0 \\ 0 & \cdots & 0 & \lambda_N \end{pmatrix}.$$

Therefore, given a matrix A, if we construct the matrix S that has the eigenvectors of A as its columns then the matrix $\mathsf{A}' = \mathsf{S}^{-1}\mathsf{AS}$ is diagonal and has the eigenvalues of A as its diagonal elements. Since we require S to be non-singular ($|\mathsf{S}| \neq 0$), the N eigenvectors of A must be linearly independent and form a basis for the N-dimensional vector space. It may be shown that *any matrix with distinct eigenvalues* can be diagonalized by this procedure. If, however, a general square matrix has degenerate eigenvalues then it may, or may not, have N linearly independent eigenvectors. If it does not then it *cannot* be diagonalized.

For normal matrices (which include Hermitian, anti-Hermitian and unitary matrices)[30] the N eigenvectors are indeed linearly independent. Moreover, when normalized, these eigenvectors form an *orthonormal* set (or can be made to do so). Therefore the matrix S with these normalized eigenvectors as columns, i.e. whose elements are $S_{ij} = (\mathsf{x}^j)_i$, has the property

$$(\mathsf{S}^\dagger\mathsf{S})_{ij} = \sum_k (\mathsf{S}^\dagger)_{ik} (\mathsf{S})_{kj} = \sum_k S^*_{ki} S_{kj} = \sum_k (\mathsf{x}^i)^*_k (\mathsf{x}^j)_k = (\mathsf{x}^i)^\dagger \mathsf{x}^j = \delta_{ij}.$$

Hence S is unitary ($\mathsf{S}^{-1} = \mathsf{S}^\dagger$) and the original matrix A can be diagonalized by

$$\mathsf{A}' = \mathsf{S}^{-1}\mathsf{AS} = \mathsf{S}^\dagger\mathsf{AS}.$$

Therefore, any normal matrix A can be diagonalized by a similarity transformation using a *unitary* transformation matrix S.

30 See Subsection 1.13.7.

Example Diagonalize the matrix

$$\mathsf{A} = \begin{pmatrix} 1 & 0 & 3 \\ 0 & -2 & 0 \\ 3 & 0 & 1 \end{pmatrix}.$$

The matrix A is symmetric and so may be diagonalized by a transformation of the form $\mathsf{A}' = \mathsf{S}^\dagger \mathsf{A} \mathsf{S}$, where S has the normalized eigenvectors of A as its columns. We have already found these eigenvectors in Subsection 1.15.1, and so we can write straightaway

$$\mathsf{S} = \frac{1}{\sqrt{2}} \begin{pmatrix} 1 & 0 & -1 \\ 0 & \sqrt{2} & 0 \\ 1 & 0 & 1 \end{pmatrix}.$$

We note that although the eigenvalues of A are degenerate, its three eigenvectors are linearly independent and so A can still be diagonalized. Thus, calculating $\mathsf{S}^\dagger \mathsf{A} \mathsf{S}$ we obtain

$$\mathsf{S}^\dagger \mathsf{A} \mathsf{S} = \frac{1}{2} \begin{pmatrix} 1 & 0 & 1 \\ 0 & \sqrt{2} & 0 \\ -1 & 0 & 1 \end{pmatrix} \begin{pmatrix} 1 & 0 & 3 \\ 0 & -2 & 0 \\ 3 & 0 & 1 \end{pmatrix} \begin{pmatrix} 1 & 0 & -1 \\ 0 & \sqrt{2} & 0 \\ 1 & 0 & 1 \end{pmatrix}$$

$$= \begin{pmatrix} 4 & 0 & 0 \\ 0 & -2 & 0 \\ 0 & 0 & -2 \end{pmatrix},$$

which is the required diagonal matrix, and has, as expected, the eigenvalues of A as its diagonal elements. ◀

If a matrix A is diagonalized by the similarity transformation $\mathsf{A}' = \mathsf{S}^{-1}\mathsf{A}\mathsf{S}$, so that $\mathsf{A}' = \mathrm{diag}(\lambda_1, \lambda_2, \ldots, \lambda_N)$, then we have immediately

$$\mathrm{Tr}\,\mathsf{A}' = \mathrm{Tr}\,\mathsf{A} = \sum_{i=1}^{N} \lambda_i, \tag{1.103}$$

$$|\mathsf{A}'| = |\mathsf{A}| = \prod_{i=1}^{N} \lambda_i, \tag{1.104}$$

since the eigenvalues of the matrix are unchanged by the transformation.

1.18 Quadratic and Hermitian forms

Let us now introduce the concept of quadratic forms (and their complex analogues, Hermitian forms). A quadratic form Q is a scalar function of a real vector $\mathbf{x}$ given by

$$Q(\mathbf{x}) = \langle \mathbf{x} | \mathcal{A}\mathbf{x} \rangle, \tag{1.105}$$

for some real linear operator $\mathcal{A}$. In any given basis (coordinate system) we can write (1.105) in matrix form as

$$Q(\mathsf{x}) = \mathsf{x}^{\mathrm{T}} \mathsf{A} \mathsf{x}, \tag{1.106}$$

where A is a real matrix. In fact, as will be explained below, we need only consider the case where A is symmetric, i.e. $\mathsf{A} = \mathsf{A}^{\mathrm{T}}$. As an example in a three-dimensional space,

$$Q = \mathsf{x}^{\mathrm{T}}\mathsf{A}\mathsf{x} = \begin{pmatrix} x_1 & x_2 & x_3 \end{pmatrix} \begin{pmatrix} 1 & 1 & 3 \\ 1 & 1 & -3 \\ 3 & -3 & -3 \end{pmatrix} \begin{pmatrix} x_1 \\ x_2 \\ x_3 \end{pmatrix}$$
$$= x_1^2 + x_2^2 - 3x_3^2 + 2x_1x_2 + 6x_1x_3 - 6x_2x_3. \tag{1.107}$$

It is reasonable to ask whether a quadratic form $Q = \mathsf{x}^{\mathrm{T}}\mathsf{M}\mathsf{x}$, where M is any (possibly non-symmetric) real square matrix, is a more general definition. That this is not the case may be seen by expressing M in terms of a symmetric matrix $\mathsf{A} = \frac{1}{2}(\mathsf{M} + \mathsf{M}^{\mathrm{T}})$ and an antisymmetric matrix $\mathsf{B} = \frac{1}{2}(\mathsf{M} - \mathsf{M}^{\mathrm{T}})$ such that $\mathsf{M} = \mathsf{A} + \mathsf{B}$. We then have

$$Q = \mathsf{x}^{\mathrm{T}}\mathsf{M}\mathsf{x} = \mathsf{x}^{\mathrm{T}}\mathsf{A}\mathsf{x} + \mathsf{x}^{\mathrm{T}}\mathsf{B}\mathsf{x}. \tag{1.108}$$

However, Q is a scalar quantity and so

$$Q = Q^{\mathrm{T}} = (\mathsf{x}^{\mathrm{T}}\mathsf{A}\mathsf{x})^{\mathrm{T}} + (\mathsf{x}^{\mathrm{T}}\mathsf{B}\mathsf{x})^{\mathrm{T}} = \mathsf{x}^{\mathrm{T}}\mathsf{A}^{\mathrm{T}}\mathsf{x} + \mathsf{x}^{\mathrm{T}}\mathsf{B}^{\mathrm{T}}\mathsf{x} = \mathsf{x}^{\mathrm{T}}\mathsf{A}\mathsf{x} - \mathsf{x}^{\mathrm{T}}\mathsf{B}\mathsf{x}. \tag{1.109}$$

Comparing (1.108) and (1.109) shows that $\mathsf{x}^{\mathrm{T}}\mathsf{B}\mathsf{x} = 0$, and hence $\mathsf{x}^{\mathrm{T}}\mathsf{M}\mathsf{x} = \mathsf{x}^{\mathrm{T}}\mathsf{A}\mathsf{x}$, i.e. Q is unchanged by considering only the symmetric part of M. Hence, with no loss of generality, we may assume $\mathsf{A} = \mathsf{A}^{\mathrm{T}}$ in (1.106).

From its definition (1.105), Q is clearly a basis- (i.e. coordinate-) independent quantity. Let us therefore consider a new basis related to the old one by an orthogonal transformation matrix S, the components in the two bases of any vector $\mathbf{x}$ being related [as in (1.92)] by $\mathsf{x} = \mathsf{S}\mathsf{x}'$ or, equivalently, by $\mathsf{x}' = \mathsf{S}^{-1}\mathsf{x} = \mathsf{S}^{\mathrm{T}}\mathsf{x}$. We then have

$$Q = \mathsf{x}^{\mathrm{T}}\mathsf{A}\mathsf{x} = (\mathsf{x}')^{\mathrm{T}}\mathsf{S}^{\mathrm{T}}\mathsf{A}\mathsf{S}\mathsf{x}' = (\mathsf{x}')^{\mathrm{T}}\mathsf{A}'\mathsf{x}',$$

where (as expected) the matrix describing the linear operator $\mathcal{A}$ in the new basis is given by $\mathsf{A}' = \mathsf{S}^{\mathrm{T}}\mathsf{A}\mathsf{S}$ (since $\mathsf{S}^{\mathrm{T}} = \mathsf{S}^{-1}$).[31] But, from the previous section, if we choose as S the matrix whose columns are the *normalized* eigenvectors of A then $\mathsf{A}' = \mathsf{S}^{\mathrm{T}}\mathsf{A}\mathsf{S}$ is diagonal with the eigenvalues of A as the diagonal elements. In the new basis

$$Q = \mathsf{x}^{\mathrm{T}}\mathsf{A}\mathsf{x} = (\mathsf{x}')^{\mathrm{T}}\Lambda\mathsf{x}' = \lambda_1 x_1'^2 + \lambda_2 x_2'^2 + \cdots + \lambda_N x_N'^2, \tag{1.110}$$

where $\Lambda = \mathrm{diag}(\lambda_1, \lambda_2, \ldots, \lambda_N)$ and the λ_i are the eigenvalues of A. It should be noted that Q contains no cross-terms of the form $x_1'x_2'$.

Example Find an orthogonal transformation that takes the quadratic form (1.107) into the form

$$\lambda_1 x_1'^2 + \lambda_2 x_2'^2 + \lambda_3 x_3'^2.$$

31 Since A is symmetric, its normalized eigenvectors are orthogonal, or can be made so, and hence S is orthogonal with $\mathsf{S}^{-1} = \mathsf{S}^{\mathrm{T}}$.

The required transformation matrix S has the *normalized* eigenvectors of A as its columns. We have already found these in Section 1.15, and so we can write immediately

$$\mathsf{S} = \frac{1}{\sqrt{6}}\begin{pmatrix} \sqrt{3} & \sqrt{2} & 1 \\ \sqrt{3} & -\sqrt{2} & -1 \\ 0 & \sqrt{2} & -2 \end{pmatrix},$$

which is easily verified as being orthogonal. Since the eigenvalues of A are $\lambda = 2$, 3 and -6, the general result already proved shows that the transformation $\mathsf{x} = \mathsf{S}\mathsf{x}'$ will carry (1.107) into the form $2x_1'^2 + 3x_2'^2 - 6x_3'^2$. This may be verified most easily by writing out the inverse transformation $\mathsf{x}' = \mathsf{S}^{-1}\mathsf{x} = \mathsf{S}^{\mathrm{T}}\mathsf{x}$ and substituting. The inverse equations are

$$\begin{aligned} x_1' &= (x_1 + x_2)/\sqrt{2}, \\ x_2' &= (x_1 - x_2 + x_3)/\sqrt{3}, \\ x_3' &= (x_1 - x_2 - 2x_3)/\sqrt{6}. \end{aligned} \qquad (1.111)$$

If these are substituted into the form $Q = 2x_1'^2 + 3x_2'^2 - 6x_3'^2$ then the original expression (1.107) is recovered. ◀

In the definition of Q it was assumed that the components x_1, x_2, x_3 and the matrix A were real. It is clear that in this case the quadratic form $Q \equiv \mathsf{x}^{\mathrm{T}}\mathsf{A}\mathsf{x}$ is real also. Another, rather more general, expression that is also real is the *Hermitian form*

$$H(\mathsf{x}) \equiv \mathsf{x}^{\dagger}\mathsf{A}\mathsf{x}, \qquad (1.112)$$

where A is Hermitian (i.e. $\mathsf{A}^{\dagger} = \mathsf{A}$) and the components of x may now be complex. It is straightforward to show that H is real, since

$$H^* = (H^{\mathrm{T}})^* = \mathsf{x}^{\dagger}\mathsf{A}^{\dagger}\mathsf{x} = \mathsf{x}^{\dagger}\mathsf{A}\mathsf{x} = H.$$

With suitable generalization, the properties of quadratic forms apply also to Hermitian forms, but to keep the presentation simple we will restrict our discussion to quadratic forms.

A special case of a quadratic (Hermitian) form is one for which $Q = \mathsf{x}^{\mathrm{T}}\mathsf{A}\mathsf{x}$ is greater than zero for all column matrices x. By choosing as the basis the eigenvectors of A we have Q in the form

$$Q = \lambda_1 x_1^2 + \lambda_2 x_2^2 + \lambda_3 x_3^2.$$

The requirement that $Q > 0$ for all x means that all the eigenvalues λ_i of A must be positive. A symmetric (Hermitian) matrix A with this property is called *positive definite*. If, instead, $Q \geq 0$ for all x then it is possible that some of the eigenvalues are zero, and A is called *positive semi-definite*.

1.18.1 The stationary properties of the eigenvectors

Consider a quadratic form, such as $Q(\mathbf{x}) = \langle \mathbf{x}|\mathcal{A}\mathbf{x}\rangle$, equation (1.105), in a fixed basis. As the vector $\mathbf{x}$ is varied, through changes in its three components x_1, x_2 and x_3, the value of the quantity Q also varies. Because of the homogeneous form of Q we may restrict any investigation of these variations to vectors of unit length (since multiplying any vector $\mathbf{x}$ by any scalar k simply multiplies the value of Q by a factor k^2).

Of particular interest are any vectors $\mathbf{x}$ that make the value of the quadratic form a maximum or minimum. A necessary, but not sufficient, condition for this is that Q is stationary with respect to small variations $\Delta\mathbf{x}$ in $\mathbf{x}$, whilst $\langle\mathbf{x}|\mathbf{x}\rangle$ is maintained at a constant value (unity).

In the chosen basis the quadratic form is given by $Q = \mathsf{x}^{\mathrm{T}}\mathsf{A}\mathsf{x}$ and, using Lagrange undetermined multipliers to incorporate the variational constraints, we are led to seek solutions of

$$\Delta[\mathsf{x}^{\mathrm{T}}\mathsf{A}\mathsf{x} - \lambda(\mathsf{x}^{\mathrm{T}}\mathsf{x} - 1)] = 0. \tag{1.113}$$

This may be used directly, together with the fact that $(\Delta\mathsf{x}^{\mathrm{T}})\mathsf{A}\mathsf{x} = \mathsf{x}^{\mathrm{T}}\mathsf{A}\,\Delta\mathsf{x}$, since A is symmetric, to obtain

$$\mathsf{A}\mathsf{x} = \lambda\mathsf{x} \tag{1.114}$$

as the necessary condition that x must satisfy. If (1.114) is satisfied for some eigenvector x then the value of $Q(\mathsf{x})$ is given by

$$Q = \mathsf{x}^{\mathrm{T}}\mathsf{A}\mathsf{x} = \mathsf{x}^{\mathrm{T}}\lambda\mathsf{x} = \lambda. \tag{1.115}$$

However, if x and y are eigenvectors corresponding to different eigenvalues then they are (or can be chosen to be) orthogonal. Consequently the expression $\mathsf{y}^{\mathrm{T}}\mathsf{A}\mathsf{x}$ is necessarily zero, since

$$\mathsf{y}^{\mathrm{T}}\mathsf{A}\mathsf{x} = \mathsf{y}^{\mathrm{T}}\lambda\mathsf{x} = \lambda\mathsf{y}^{\mathrm{T}}\mathsf{x} = 0. \tag{1.116}$$

Summarizing, those column matrices x of unit magnitude that make the quadratic form Q stationary are eigenvectors of the matrix A, and the stationary value of Q is then equal to the corresponding eigenvalue. It is straightforward to see from the proof of (1.114) that, conversely, any eigenvector of A makes Q stationary.

Instead of maximizing or minimizing $Q = \mathsf{x}^{\mathrm{T}}\mathsf{A}\mathsf{x}$ subject to the constraint $\mathsf{x}^{\mathrm{T}}\mathsf{x} = 1$, an equivalent procedure is to extremize the function

$$\lambda(\mathsf{x}) = \frac{\mathsf{x}^{\mathrm{T}}\mathsf{A}\mathsf{x}}{\mathsf{x}^{\mathrm{T}}\mathsf{x}},$$

as we now show.

Example Show that if $\lambda(\mathsf{x})$ is stationary then x is an eigenvector of A and $\lambda(\mathsf{x})$ is equal to the corresponding eigenvalue.

We require $\Delta\lambda(\mathsf{x}) = 0$ with respect to small variations in x. Now

$$\begin{aligned}\Delta\lambda &= \frac{1}{(\mathsf{x}^{\mathrm{T}}\mathsf{x})^2}\left[(\mathsf{x}^{\mathrm{T}}\mathsf{x})\left(\Delta\mathsf{x}^{\mathrm{T}}\mathsf{A}\mathsf{x} + \mathsf{x}^{\mathrm{T}}\mathsf{A}\,\Delta\mathsf{x}\right) - \mathsf{x}^{\mathrm{T}}\mathsf{A}\mathsf{x}\left(\Delta\mathsf{x}^{\mathrm{T}}\mathsf{x} + \mathsf{x}^{\mathrm{T}}\Delta\mathsf{x}\right)\right]\\ &= \frac{2\Delta\mathsf{x}^{\mathrm{T}}\mathsf{A}\mathsf{x}}{\mathsf{x}^{\mathrm{T}}\mathsf{x}} - 2\left(\frac{\mathsf{x}^{\mathrm{T}}\mathsf{A}\mathsf{x}}{\mathsf{x}^{\mathrm{T}}\mathsf{x}}\right)\frac{\Delta\mathsf{x}^{\mathrm{T}}\mathsf{x}}{\mathsf{x}^{\mathrm{T}}\mathsf{x}},\end{aligned}$$

since $\mathsf{x}^{\mathrm{T}}\mathsf{A}\,\Delta\mathsf{x} = (\Delta\mathsf{x}^{\mathrm{T}})\mathsf{A}\mathsf{x}$ and $\mathsf{x}^{\mathrm{T}}\Delta\mathsf{x} = (\Delta\mathsf{x}^{\mathrm{T}})\mathsf{x}$. Thus

$$\Delta\lambda = \frac{2}{\mathsf{x}^{\mathrm{T}}\mathsf{x}}\Delta\mathsf{x}^{\mathrm{T}}[\mathsf{A}\mathsf{x} - \lambda(\mathsf{x})\mathsf{x}].$$

If $\Delta\lambda$ is to be zero, we must have that $\mathsf{A}\mathsf{x} = \lambda(\mathsf{x})\mathsf{x}$, i.e. x is an eigenvector of A with eigenvalue $\lambda(\mathsf{x})$. ◀

Thus the eigenvalues of a symmetric matrix A are the values of the function

$$\lambda(\mathsf{x}) = \frac{\mathsf{x}^{\mathrm{T}}\mathsf{A}\mathsf{x}}{\mathsf{x}^{\mathrm{T}}\mathsf{x}}$$

at its stationary points. The eigenvectors of A lie along those directions in space for which the quadratic form $Q = \mathsf{x}^{\mathrm{T}}\mathsf{A}\mathsf{x}$ has stationary values, given a fixed magnitude for the vector x. Similar results hold for Hermitian matrices.

1.18.2 Quadratic surfaces

The results of the previous subsection may be turned around to state that the surface given by

$$\mathsf{x}^{\mathrm{T}}\mathsf{A}\mathsf{x} = \text{constant} = 1 \text{ (say)} \qquad (1.117)$$

and called a *quadratic surface*, has stationary values of its radius (i.e. origin–surface distance) in those directions that are along the eigenvectors of A. More specifically, in three dimensions the quadratic surface $\mathsf{x}^{\mathrm{T}}\mathsf{A}\mathsf{x} = 1$ has its principal axes along the three mutually perpendicular eigenvectors of A, and the squares of the corresponding principal radii are given by λ_i^{-1}, $i = 1, 2, 3$.

As well as having this stationary property of the radius, a *principal axis* is characterized by the fact that any section of the surface perpendicular to it has some degree of symmetry about it. If the eigenvalues corresponding to any two principal axes are degenerate then the quadratic surface has rotational symmetry about the third principal axis and the choice of a pair of axes perpendicular to that axis is not uniquely defined.

Example Find the shape of the quadratic surface

$$x_1^2 + x_2^2 - 3x_3^2 + 2x_1x_2 + 6x_1x_3 - 6x_2x_3 = 1.$$

If, instead of expressing the quadratic surface in terms of x_1, x_2, x_3, as in (1.107), we were to use the new variables x_1', x_2', x_3' defined in (1.111), for which the coordinate axes are along the three mutually perpendicular eigenvector directions $(1, 1, 0)$, $(1, -1, 1)$ and $(1, -1, -2)$, then the equation of the surface would take the form (see (1.110))

$$\frac{x_1'^2}{(1/\sqrt{2})^2} + \frac{x_2'^2}{(1/\sqrt{3})^2} - \frac{x_3'^2}{(1/\sqrt{6})^2} = 1.$$

Thus, for example, a section of the quadratic surface in the plane $x_3' = 0$, i.e. $x_1 - x_2 - 2x_3 = 0$, is an ellipse, with semi-axes $1/\sqrt{2}$ and $1/\sqrt{3}$. Similarly a section in the plane $x_1' = x_1 + x_2 = 0$ is a hyperbola. ◀

Clearly the most general form of a quadratic surface, referred to its principal axes as coordinate axes, is

$$\pm\frac{x_1'^2}{a^2} \pm \frac{x_2'^2}{b^2} \pm \frac{x_3'^2}{c^2} = 1, \qquad (1.118)$$

where $\pm a^{-2}$, $\pm b^{-2}$ and $\pm c^{-2}$ are the three eigenvalues of the corresponding matrix A. For a real quadric surface, at least one of the signs on the LHS of (1.118) must be positive.

The simplest three-dimensional situation to visualize is that in which all of the eigenvalues are positive, since then the quadratic surface is an ellipsoid. If one eigenvalue is negative, as in the worked example, then the surface is ellipsoidal in some sections and hyperbolic in others, with the values of a, b and c and the value of the coordinate at which the section is taken determining where an ellipse terminates or a hyperbola begins.

The special case of one of the eigenvalues, λ_k say, being zero is worth mentioning. Then formally the corresponding principal radius becomes infinite or, more strictly, (1.118) becomes independent of x_k'. The corresponding quadratic surface is then a cylinder with its axis parallel to the x_k'-axis (i.e. in the direction of the corresponding eigenvector in the original coordinate system) and a cross-section given by (1.118) with no x_k' term; this will be an ellipse or hyperbola, depending on the relative sign of the other two eigenvalues.[32]

1.19 Normal modes

Any student of the physical sciences will encounter the subject of oscillations on many occasions and in a wide variety of circumstances, for example the voltage and current oscillations in an electric circuit, the vibrations of a mechanical structure and the internal motions of molecules. The matrices studied in this chapter provide a particularly simple way to approach what may appear, at first glance, to be difficult physical problems.

We will consider only systems for which a position-dependent potential exists, i.e. the potential energy of the system in any particular configuration depends upon the coordinates of the configuration, which need not be lengths, however; the potential must *not* depend upon the time derivatives (generalized velocities) of these coordinates.[33] A further restriction that we place is that the potential has a local minimum at the equilibrium point; physically, this is a necessary and sufficient condition for stable equilibrium. By suitably defining the origin of the potential, we may take its value at the equilibrium point as zero.

The coordinates chosen to describe a configuration of the system will be denoted by q_i, $i = 1, 2, \ldots, N$. The q_i need not be distances; some could be angles, for example. For convenience we can define the q_i so that they are all zero at the equilibrium point.[34]

32 What form do you expect the quadratic surface to take if two of the eigenvalues are zero (with the third one positive)?

33 So, for example, the potential $-q\mathbf{v}\cdot\mathbf{A}$ used in the Lagrangian description of a charged particle in an electromagnetic field is excluded.

34 Note that this does not mean that all coordinates have a common origin. For example, in molecular vibrations, the q_i are the displacements of the individual atoms from their equilibrium positions within the molecule; the individual origins therefore occupy different locations in space.

The instantaneous velocities of various parts of the system will depend upon the time derivatives of the q_i, denoted by $\dot{q}_i$. For small oscillations the velocities will be linear in the $\dot{q}_i$ and consequently the total kinetic energy T will be quadratic in them – and will include cross terms of the form $\dot{q}_i\dot{q}_j$ with $i \neq j$. The general expression for T can be written as the quadratic form

$$T = \sum_i \sum_j a_{ij}\dot{q}_i\dot{q}_j = \dot{\mathsf{q}}^{\mathrm{T}}\mathsf{A}\dot{\mathsf{q}}, \tag{1.119}$$

where $\dot{\mathsf{q}}$ is the column vector $(\dot{q}_1 \quad \dot{q}_2 \quad \cdots \quad \dot{q}_N)^{\mathrm{T}}$ and the $N \times N$ matrix A is real and may be chosen to be symmetric.[35] Furthermore, A, like any matrix corresponding to a kinetic energy, is positive definite; that is, whatever non-zero real values the $\dot{q}_i$ take, the quadratic form (1.119) has a value > 0.

Turning now to the potential energy, we may write its value for a configuration q by means of a Taylor expansion about the origin $\mathsf{q} = \mathsf{0}$,

$$V(\mathsf{q}) = V(\mathsf{0}) + \sum_i \frac{\partial V(\mathsf{0})}{\partial q_i} q_i + \frac{1}{2}\sum_i \sum_j \frac{\partial^2 V(\mathsf{0})}{\partial q_i \partial q_j} q_i q_j + \cdots .$$

However, we have chosen $V(\mathsf{0}) = 0$ and, since the origin is an equilibrium point, there is no force there and $\partial V(\mathsf{0})/\partial q_i = 0$. Consequently, to second order in the q_i we also have a quadratic form, but in the coordinates rather than in their time derivatives:

$$V = \sum_i \sum_j b_{ij} q_i q_j = \mathsf{q}^{\mathrm{T}}\mathsf{B}\mathsf{q}, \tag{1.120}$$

where B is, or can be made, symmetric.[36] In this case, and in general, the requirement that the potential is a minimum means that the potential matrix B, like the kinetic energy matrix A, is real and positive definite.

1.19.1 Typical oscillatory systems

We now introduce particular examples, although the results of this subsection are general, given the above restrictions, and the reader will find it easy to apply the results to many other instances.

Consider first a uniform rod of mass M and length l, attached by a light string also of length l to a fixed point P and executing small oscillations in a vertical plane. We choose as coordinates the angles θ_1 and θ_2 shown, with exaggerated magnitude, in Figure 1.2. In terms of these coordinates the center of gravity of the rod has, to *first order* in the θ_i, a velocity component in the x-direction equal to $l\dot{\theta}_1 + \frac{1}{2}l\dot{\theta}_2$ and in the y-direction equal to zero.[37] Adding in the rotational kinetic energy of the rod about its center of gravity we

35 For an electronic system consisting of capacitors and inductors, and for which the charges on the capacitors are used as the coordinates q_i, identify the quantity that corresponds to kinetic energy T in the mechanical case.

36 Write down the symmetric matrix B that is equivalent for the present purposes to the asymmetric matrix C whose entries are

$$(3,\ 4,\ -2,\ 0;\ -1,\ 5,\ 2,\ -3;\ 2,\ -4,\ -1,\ 6;\ 4,\ -3,\ 2,\ 0).$$

37 More strictly, that in the x-direction is the time derivative of $l\sin\theta_1 + \frac{1}{2}l\sin\theta_2$, the x-coordinate of the center of mass of the rod. The stated form is obtained either by using the chain rule, $dx/dt = dx/d\theta \cdot d\theta/dt$, followed by

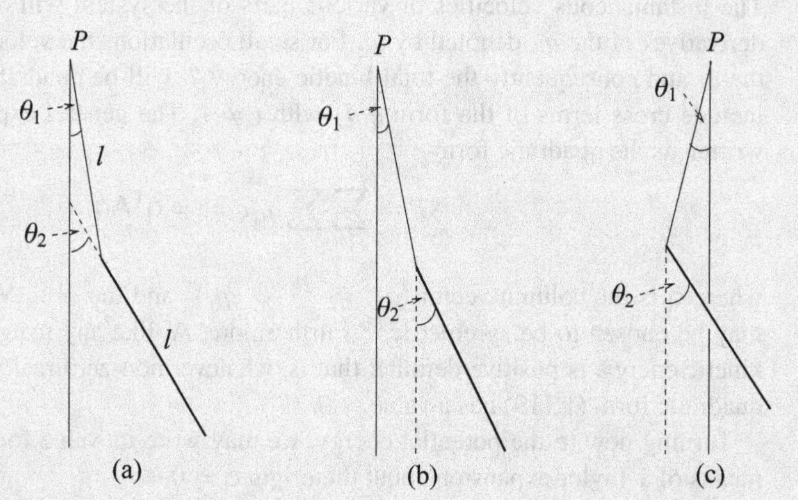

Figure 1.2 A uniform rod of length l attached to the fixed point P by a light string of the same length: (a) the general coordinate system; (b) approximation to the normal mode with lower frequency; (c) approximation to the mode with higher frequency.

obtain, to second order in the $\dot{\theta}_i$,

$$\begin{aligned} T &\approx \tfrac{1}{2}Ml^2(\dot{\theta}_1^2 + \tfrac{1}{4}\dot{\theta}_2^2 + \dot{\theta}_1\dot{\theta}_2) + \tfrac{1}{24}Ml^2\dot{\theta}_2^2 \\ &= \tfrac{1}{6}Ml^2\left(3\dot{\theta}_1^2 + 3\dot{\theta}_1\dot{\theta}_2 + \dot{\theta}_2^2\right) = \tfrac{1}{12}Ml^2\dot{\mathsf{q}}^{\mathrm{T}}\begin{pmatrix} 6 & 3 \\ 3 & 2 \end{pmatrix}\dot{\mathsf{q}}, \end{aligned} \qquad (1.121)$$

where $\dot{\mathsf{q}}^{\mathrm{T}} = (\dot{\theta}_1 \quad \dot{\theta}_2)$. The potential energy is given by

$$V = Mlg\left[(1-\cos\theta_1) + \tfrac{1}{2}(1-\cos\theta_2)\right] \qquad (1.122)$$

so that

$$V \approx \tfrac{1}{4}Mlg(2\theta_1^2 + \theta_2^2) = \tfrac{1}{12}Mlg\mathsf{q}^{\mathrm{T}}\begin{pmatrix} 6 & 0 \\ 0 & 3 \end{pmatrix}\mathsf{q}, \qquad (1.123)$$

where g is the acceleration due to gravity and $\mathsf{q} = (\theta_1 \quad \theta_2)^{\mathrm{T}}$; (1.123) is valid to second order in the θ_i.

With these expressions for T and V we now apply the conservation of energy,

$$\frac{d}{dt}(T+V) = 0, \qquad (1.124)$$

assuming that there are no external forces other than gravity. In matrix form (1.124) becomes

$$\frac{d}{dt}(\dot{\mathsf{q}}^{\mathrm{T}}\mathsf{A}\dot{\mathsf{q}} + \mathsf{q}^{\mathrm{T}}\mathsf{B}\mathsf{q}) = \ddot{\mathsf{q}}^{\mathrm{T}}\mathsf{A}\dot{\mathsf{q}} + \dot{\mathsf{q}}^{\mathrm{T}}\mathsf{A}\ddot{\mathsf{q}} + \dot{\mathsf{q}}^{\mathrm{T}}\mathsf{B}\mathsf{q} + \mathsf{q}^{\mathrm{T}}\mathsf{B}\dot{\mathsf{q}} = 0,$$

$\cos\theta \approx 1$ to first order in θ, or by using $\sin\theta \approx \theta$ to first order in θ, followed by differentiation with respect to time. The time derivative of $y = l(1-\cos\theta)$ contains the product of $\sin\theta$ and $d\theta/dt$ and is therefore of second order in θ.

which, using $\mathsf{A} = \mathsf{A}^{\mathrm{T}}$ and $\mathsf{B} = \mathsf{B}^{\mathrm{T}}$, gives

$$2\dot{\mathsf{q}}^{\mathrm{T}}(\mathsf{A}\ddot{\mathsf{q}} + \mathsf{B}\mathsf{q}) = 0.$$

We will assume, although it is not clear that this gives the only possible solution, that the above equation implies that the coefficient of each $\dot{q}_i$ is separately zero. Hence

$$\mathsf{A}\ddot{\mathsf{q}} + \mathsf{B}\mathsf{q} = 0. \tag{1.125}$$

For a rigorous derivation of this result, Lagrange's equations should be used, as in chapter 12.

Now we search for sets of coordinates q that *all* oscillate with the same period, i.e. the total motion repeats itself *exactly* after a *finite* interval. Solutions of this form will satisfy

$$\mathsf{q} = \mathsf{x}\cos\omega t; \tag{1.126}$$

the relative values of the elements of x in such a solution will indicate the extent to which each coordinate is involved in this special motion. In general there will be N values of ω if the matrices A and B are $N \times N$ and these values are known as *normal frequencies* or *eigenfrequencies*.

Putting (1.126) into (1.125) yields

$$-\omega^2\mathsf{A}\mathsf{x} + \mathsf{B}\mathsf{x} = (\mathsf{B} - \omega^2\mathsf{A})\mathsf{x} = 0. \tag{1.127}$$

Our work in Section 1.12 showed that this can have non-trivial solutions only if

$$|\mathsf{B} - \omega^2\mathsf{A}| = 0. \tag{1.128}$$

This is a form of characteristic equation for B, except that the unit matrix I has been replaced by A. It has the more familiar form if a choice of coordinates is made in which the kinetic energy T is a simple sum of squared terms, i.e. it has been diagonalized, and the scale of the new coordinates is then chosen to make each diagonal element unity.

However, even in the present case, (1.128) can be solved to yield ω_k^2 for $k = 1, 2, \ldots, N$, where N is the order of both A and B.[38] The values of ω_k can be used with (1.127) to find the corresponding column vector x^k and the initial (stationary) physical configuration that, on release, will execute motion with period $2\pi/\omega_k$.

In Subsection 1.14.1 we showed that the eigenvectors of a real symmetric matrix were, except in the case of degeneracy of the eigenvalues, mutually orthogonal. In the present situation an analogous, but not identical, result holds. It is shown in Subsection 1.19.2 that if x^1 and x^2 are two eigenvectors satisfying (1.127) for different values of ω^2 then they are orthogonal in the sense that

$$(\mathsf{x}^2)^{\mathrm{T}}\mathsf{A}\mathsf{x}^1 = 0 \quad \text{and} \quad (\mathsf{x}^2)^{\mathrm{T}}\mathsf{B}\mathsf{x}^1 = 0.$$

The direct "scalar product" $(\mathsf{x}^2)^{\mathrm{T}}\mathsf{x}^1$, formally equal to $(\mathsf{x}^2)^{\mathrm{T}}\mathsf{I}\,\mathsf{x}^1$, is not, in general, equal to zero.

38 Note that this implies that the number of normal frequencies of a system is equal to the number of coordinates needed to specify the system – though, of course, some may be repeated; clearly this will be the case if the "characteristic equation" (1.128), an Nth degree polynomial in ω^2, has repeated roots.

Returning to the suspended rod, we find from (1.128)

$$\left| \frac{Mlg}{12} \begin{pmatrix} 6 & 0 \\ 0 & 3 \end{pmatrix} - \frac{\omega^2 Ml^2}{12} \begin{pmatrix} 6 & 3 \\ 3 & 2 \end{pmatrix} \right| = 0.$$

Writing $\omega^2 l/g = \lambda$, this becomes

$$\begin{vmatrix} 6 - 6\lambda & -3\lambda \\ -3\lambda & 3 - 2\lambda \end{vmatrix} = 0 \quad \Rightarrow \quad \lambda^2 - 10\lambda + 6 = 0,$$

which has roots $\lambda = 5 \pm \sqrt{19}$. Thus we find that the two normal frequencies are given by $\omega_1 = (0.641g/l)^{1/2}$ and $\omega_2 = (9.359g/l)^{1/2}$. Putting the lower of the two values for ω^2, namely $(5 - \sqrt{19})g/l$, into (1.127) shows that for this mode

$$x_1 : x_2 = 3(5 - \sqrt{19}) : 6(\sqrt{19} - 4) = 1.923 : 2.153.$$

This corresponds to the case where the rod and string are almost straight out, i.e. they almost form a simple pendulum. Similarly it may be shown that the higher frequency corresponds to a solution where the string and rod are moving with opposite phases and $x_1 : x_2 = 9.359 : -16.718$. The two situations are shown in Figure 1.2.

In connection with quadratic forms it was shown in Section 1.18 how to make a change of coordinates such that the matrix for a particular form becomes diagonal. In Problem 1.42 a method is developed for diagonalizing simultaneously two quadratic forms (though the transformation matrix may not be orthogonal). If this process is carried out for A and B in a general system undergoing stable oscillations, the kinetic and potential energies in the new variables η_i take the forms

$$T = \sum_i \mu_i \dot{\eta}_i^2 = \dot{\eta}^{\mathrm{T}} \mathsf{M} \dot{\eta}, \quad \mathsf{M} = \mathrm{diag}\,(\mu_1, \mu_2, \ldots, \mu_N), \tag{1.129}$$

$$V = \sum_i \nu_i \eta_i^2 = \eta^{\mathrm{T}} \mathsf{N} \eta, \quad \mathsf{N} = \mathrm{diag}\,(\nu_1, \nu_2 \ldots, \nu_N), \tag{1.130}$$

and the equations of motion are the *uncoupled* equations

$$\mu_i \ddot{\eta}_i + \nu_i \eta_i = 0, \quad i = 1, 2, \ldots, N. \tag{1.131}$$

Clearly a simple renormalization of the η_i can be made that reduces all the μ_i in (1.129) to unity. When this is done the variables so formed are called *normal coordinates* and equations (1.131) the *normal equations*.

When a system is executing one of these simple harmonic motions it is said to be in a *normal mode*, and once started in such a mode it will repeat its motion exactly after each interval of $2\pi/\omega_i$. Any arbitrary motion of the system may be written as a superposition of the normal modes, and each component mode will execute harmonic motion with the corresponding eigenfrequency; however, unless by chance the eigenfrequencies are in integer relationship, the system will never return to its initial configuration after any finite time interval.[39]

As a second example we will consider a number of masses coupled together by springs. For this type of situation the potential and kinetic energies are automatically quadratic

39 For the rod on a string problem just considered, are there any non-trivial initial configurations, other than the two shown schematically in (b) and (c) of Figure 1.2, that repeat themselves after a finite time?

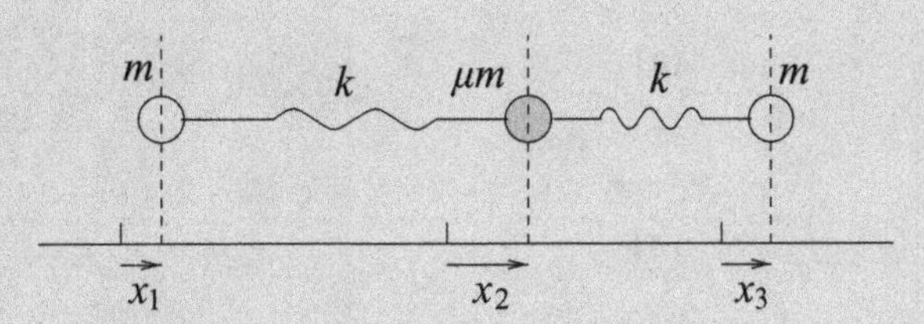

Figure 1.3 Three masses m, μm and m connected by two equal light springs of force constant k.

functions of the coordinates and their derivatives, provided the elastic limits of the springs are not exceeded, and the oscillations do not have to be vanishingly small for the analysis to be valid.

Example Find the normal frequencies and modes of oscillation of three particles of masses m, μm, m connected in that order in a straight line by two equal light springs of force constant k. (This arrangement could serve as a model for some linear molecules, e.g. CO_2.)

The situation is shown in Figure 1.3; the coordinates of the particles, x_1, x_2, x_3, are measured from their equilibrium positions, at which the springs are neither extended nor compressed.

The kinetic energy of the system is simply

$$T = \tfrac{1}{2}m\left(\dot{x}_1^2 + \mu\,\dot{x}_2^2 + \dot{x}_3^2\right),$$

whilst the potential energy stored in the springs is

$$V = \tfrac{1}{2}k\left[(x_2 - x_1)^2 + (x_3 - x_2)^2\right].$$

The kinetic- and potential-energy symmetric matrices are thus

$$\mathsf{A} = \frac{m}{2}\begin{pmatrix} 1 & 0 & 0 \\ 0 & \mu & 0 \\ 0 & 0 & 1 \end{pmatrix}, \qquad \mathsf{B} = \frac{k}{2}\begin{pmatrix} 1 & -1 & 0 \\ -1 & 2 & -1 \\ 0 & -1 & 1 \end{pmatrix}.$$

From (1.128), to find the normal frequencies we have to solve $|\mathsf{B} - \omega^2\mathsf{A}| = 0$. Thus, writing $m\omega^2/k = \lambda$, we have

$$\begin{vmatrix} 1-\lambda & -1 & 0 \\ -1 & 2-\mu\lambda & -1 \\ 0 & -1 & 1-\lambda \end{vmatrix} = 0,$$

which leads to $\lambda = 0$, 1 or $1 + 2/\mu$. The corresponding eigenvectors are respectively

$$\mathsf{x}^1 = \frac{1}{\sqrt{3}}\begin{pmatrix} 1 \\ 1 \\ 1 \end{pmatrix}, \quad \mathsf{x}^2 = \frac{1}{\sqrt{2}}\begin{pmatrix} 1 \\ 0 \\ -1 \end{pmatrix}, \quad \mathsf{x}^3 = \frac{1}{\sqrt{2 + (4/\mu^2)}}\begin{pmatrix} 1 \\ -2/\mu \\ 1 \end{pmatrix}.$$

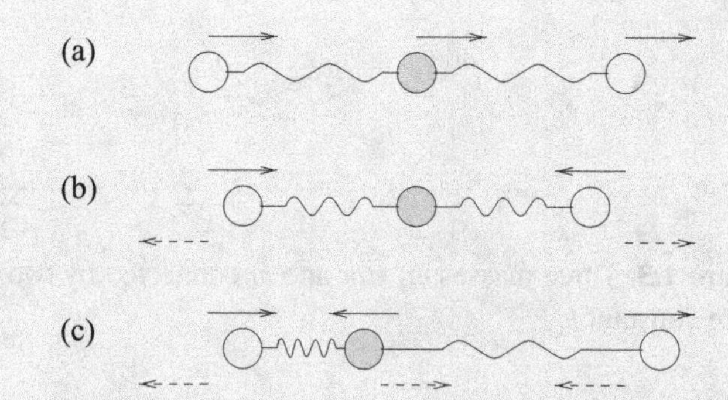

Figure 1.4 The normal modes of the masses and springs of a linear molecule such as CO_2. (a) $\omega^2 = 0$; (b) $\omega^2 = k/m$; (c) $\omega^2 = [(\mu + 2)/\mu](k/m)$.

The physical motions associated with these normal modes are illustrated in Figure 1.4. The first, with $\lambda = \omega = 0$ and all the x_i equal, merely describes bodily translation of the whole system, with no (i.e. zero-frequency) internal oscillations.[40]

In the second solution the central particle remains stationary, $x_2 = 0$, whilst the other two oscillate with equal amplitudes in antiphase with each other. This motion, which has frequency $\omega = (k/m)^{1/2}$, is illustrated in Figure 1.4(b).

The final and most complicated of the three normal modes has angular frequency $\omega = \{[(\mu + 2)/\mu](k/m)\}^{1/2}$, and involves a motion of the central particle which is in antiphase with that of the two outer ones and which has an amplitude $2/\mu$ times as great. In this motion [see Figure 1.4(c)] the two springs are compressed and extended in turn. We also note that in the second and third normal modes the center of mass of the molecule remains stationary. ◀

1.19.2 Rayleigh–Ritz method

We conclude this section with a discussion of the Rayleigh–Ritz method for estimating the eigenfrequencies of an oscillating system. We recall from the introduction to the section that for a system undergoing small oscillations the potential and kinetic energy are given by

$$V = \mathsf{q}^{\mathrm{T}}\mathsf{B}\mathsf{q} \quad \text{and} \quad T = \dot{\mathsf{q}}^{\mathrm{T}}\mathsf{A}\dot{\mathsf{q}},$$

where the components of q are the coordinates chosen to represent the configuration of the system and A and B are symmetric matrices (or may be chosen to be such). We also recall from (1.127) that the normal modes x^i and the eigenfrequencies ω_i are given by

$$(\mathsf{B} - \omega_i^2\mathsf{A})\mathsf{x}^i = 0. \tag{1.132}$$

It may be shown that the eigenvectors x^i corresponding to different normal modes are linearly independent and so form a complete set. Thus, any coordinate vector q can be

40 A zero eigenvalue, $\lambda = 0$, occurs whenever the determinant of the potential matrix B is zero, whatever the form of A, as can be seen from equation (1.128). This will happen for every system that has no outside forces acting upon it (reactions at anchor points are outside forces in this context) and will be typified by the sum of the coefficients in each column/row of B adding up to zero, thus automatically making $|\mathsf{B}| = 0$.

written $\mathsf{q} = \sum_j c_j \mathsf{x}^j$. We now consider the value of the generalized ratio of quadratic forms

$$\lambda(\mathsf{x}) = \frac{\mathsf{x}^{\mathrm{T}}\mathsf{B}\mathsf{x}}{\mathsf{x}^{\mathrm{T}}\mathsf{A}\mathsf{x}} = \frac{\sum_m (\mathsf{x}^m)^{\mathrm{T}} c_m^* \mathsf{B} \sum_i c_i \mathsf{x}^i}{\sum_j (\mathsf{x}^j)^{\mathrm{T}} c_j^* \mathsf{A} \sum_k c_k \mathsf{x}^k},$$

which, since both numerator and denominator are positive definite, is itself non-negative. Equation (1.132) can be used to replace $\mathsf{B}\mathsf{x}^i$, with the result that

$$\lambda(\mathsf{x}) = \frac{\sum_m (\mathsf{x}^m)^{\mathrm{T}} c_m^* \mathsf{A} \sum_i \omega_i^2 c_i \mathsf{x}^i}{\sum_j (\mathsf{x}^j)^{\mathrm{T}} c_j^* \mathsf{A} \sum_k c_k \mathsf{x}^k}$$

$$= \frac{\sum_m (\mathsf{x}^m)^{\mathrm{T}} c_m^* \sum_i \omega_i^2 c_i \mathsf{A}\mathsf{x}^i}{\sum_j (\mathsf{x}^j)^{\mathrm{T}} c_j^* \mathsf{A} \sum_k c_k \mathsf{x}^k}. \tag{1.133}$$

Now the eigenvectors x^i obtained by solving $(\mathsf{B} - \omega^2\mathsf{A})\mathsf{x} = \mathbf{0}$ are not mutually orthogonal unless either A or B is a multiple of the unit matrix. However, it may be shown that they do possess the desirable properties

$$(\mathsf{x}^j)^{\mathrm{T}}\mathsf{A}\mathsf{x}^i = 0 \quad \text{and} \quad (\mathsf{x}^j)^{\mathrm{T}}\mathsf{B}\mathsf{x}^i = 0 \quad \text{if } i \neq j. \tag{1.134}$$

This result is proved as follows. From (1.132) it is clear that, for general i and j,

$$(\mathsf{x}^j)^{\mathrm{T}}\left(\mathsf{B} - \omega_i^2\mathsf{A}\right)\mathsf{x}^i = 0. \tag{1.135}$$

But, by taking the transpose of (1.132) with i replaced by j and recalling that A and B are real and symmetric, we obtain

$$(\mathsf{x}^j)^{\mathrm{T}}\left(\mathsf{B} - \omega_j^2\mathsf{A}\right) = \mathbf{0}.$$

Forming the scalar product of this with x^i and subtracting the result from (1.135) gives

$$(\omega_j^2 - \omega_i^2)(\mathsf{x}^j)^{\mathrm{T}}\mathsf{A}\mathsf{x}^i = 0.$$

Thus, for $i \neq j$ and non-degenerate eigenvalues ω_i^2 and ω_j^2, we have that $(\mathsf{x}^j)^{\mathrm{T}}\mathsf{A}\mathsf{x}^i = 0$, and substituting this into (1.135) immediately establishes the corresponding result for $(\mathsf{x}^j)^{\mathrm{T}}\mathsf{B}\mathsf{x}^i$. Clearly, if either A or B is a multiple of the unit matrix then the eigenvectors are mutually orthogonal in the normal sense. The orthogonality relations (1.134) are derived again, and extended, in Problem 1.42.

Using the first of the relationships (1.134) to simplify (1.133), we find that

$$\lambda(\mathsf{x}) = \frac{\sum_i |c_i|^2 \omega_i^2 (\mathsf{x}^i)^{\mathrm{T}}\mathsf{A}\mathsf{x}^i}{\sum_k |c_k|^2 (\mathsf{x}^k)^{\mathrm{T}}\mathsf{A}\mathsf{x}^k}. \tag{1.136}$$

Now, if ω_0^2 is the lowest eigenfrequency then $\omega_i^2 \geq \omega_0^2$ for all i and, further, since $(\mathsf{x}^i)^{\mathrm{T}}\mathsf{A}\mathsf{x}^i \geq 0$ for all i the numerator of (1.136) is $\geq \omega_0^2 \sum_i |c_i|^2 (\mathsf{x}^i)^{\mathrm{T}}\mathsf{A}\mathsf{x}^i$. Hence

$$\lambda(\mathsf{x}) \equiv \frac{\mathsf{x}^{\mathrm{T}}\mathsf{B}\mathsf{x}}{\mathsf{x}^{\mathrm{T}}\mathsf{A}\mathsf{x}} \geq \omega_0^2, \tag{1.137}$$

for any x whatsoever (whether x is an eigenvector or not). Thus we are able to estimate the lowest eigenfrequency of the system by evaluating λ for a variety of vectors x, the components of which, it will be recalled, give the ratios of the coordinate amplitudes. This is sometimes a useful approach if many coordinates are involved and direct solution for the eigenvalues is not possible.

An additional result is that the maximum eigenfrequency $\omega_{\rm m}^2$ may also be estimated. It is obvious that if we replace the statement "$\omega_i^2 \geq \omega_0^2$ for all i" by "$\omega_i^2 \leq \omega_{\rm m}^2$ for all i", then $\lambda(\mathsf{x}) \leq \omega_{\rm m}^2$ for any x. Thus $\lambda(\mathsf{x})$ always lies between the lowest and highest eigenfrequencies of the system. Furthermore, $\lambda(\mathsf{x})$ has a *stationary* value, equal to ω_k^2, when x is the kth eigenvector (see Subsection 1.18.1).

Example Estimate the eigenfrequencies of the oscillating rod of Subsection 1.19.1.

Firstly we recall that

$$\mathsf{A} = \frac{Ml^2}{12}\begin{pmatrix} 6 & 3 \\ 3 & 2 \end{pmatrix} \quad \text{and} \quad \mathsf{B} = \frac{Mlg}{12}\begin{pmatrix} 6 & 0 \\ 0 & 3 \end{pmatrix}.$$

Physical intuition suggests that the slower mode will have a configuration approximating that of a simple pendulum (Figure 1.2), in which $\theta_1 = \theta_2$, and so we use this as a *trial vector*. Taking $\mathsf{x} = (\theta, \;\; \theta)^{\rm T}$,

$$\lambda(\mathsf{x}) = \frac{\mathsf{x}^{\rm T}\mathsf{B}\mathsf{x}}{\mathsf{x}^{\rm T}\mathsf{A}\mathsf{x}} = \frac{3Mlg\theta^2/4}{7Ml^2\theta^2/6} = \frac{9g}{14l} = 0.643\frac{g}{l},$$

and we conclude from (1.137) that the lower (angular) frequency is $\leq (0.643g/l)^{1/2}$. We have already seen on p. 62 that the true answer is $(0.641g/l)^{1/2}$ and so even a simple, but reasoned, guess has brought us very close to the precise answer.

Next we turn to the higher frequency. Here, a typical pattern of oscillation is not so obvious but, rather preempting the answer, we try $\theta_2 = -2\theta_1$; we then obtain $\lambda = 9g/l$ and so conclude that the higher eigenfrequency $\geq (9g/l)^{1/2}$. We have already seen that the exact answer is $(9.359g/l)^{1/2}$ and so again we have come close to it. ◀

A simplified version of the Rayleigh–Ritz method may be used to estimate the eigenvalues of a symmetric (or in general Hermitian) matrix B, the eigenvectors of which will be mutually orthogonal. By repeating the calculations leading to (1.136), A being replaced by the unit matrix I, it is easily verified that if

$$\lambda(\mathsf{x}) = \frac{\mathsf{x}^{\rm T}\mathsf{B}\mathsf{x}}{\mathsf{x}^{\rm T}\mathsf{x}}$$

is evaluated for *any* vector x then

$$\lambda_1 \leq \lambda(\mathsf{x}) \leq \lambda_{\rm m},$$

where $\lambda_1, \lambda_2, \ldots, \lambda_m$ are the eigenvalues of B in order of increasing size. A similar result holds for Hermitian matrices.[41]

1.20 The summation convention

In this chapter we have often needed to take a sum over a number of terms which are all of the same general form, and differ only in the value of an indexing subscript. Such a summation has been indicated by a summation sign, $\sum$, with the range of the subscript written above and below the sign. This very explicit notation has been deliberately adopted for the purposes of introducing the general procedures. However, the reader will, after a time, doubtless have felt that much of the notation is superfluous, particularly when there have been multiple sums appearing in a single expression, each with its own explicit summation sign; the derivation of equation (1.99) provides just such an example.

Such calculations can be significantly compacted, and in some cases simplified, if the Cartesian coordinates x, y and z are replaced symbolically by the indexed coordinates x_i, where i takes the values 1, 2 and 3, and the so-called *summation convention* is adopted. In this convention any *lower-case* alphabetic subscript that appears *exactly* twice in any term of an expression is understood to be summed over all the values that a subscript in that position can take (unless the contrary is specifically stated); there is no explicit summation sign.

The subscripted quantities may appear in the numerator and/or the denominator of a term in an expression. This naturally implies that any such pair of repeated subscripts must occur only in subscript positions that have the same range of values. Sometimes the ranges of values have to be specified, but usually they are apparent from the context.

As a basic example, in this notation

$$P_{ij} = \sum_{k=1}^{N} A_{ik} B_{kj}$$

becomes

$$P_{ij} = A_{ik} B_{kj} \qquad \text{i.e. without the explicit summation sign.}$$

In order to use the convention, partial differentiation with respect to Cartesian coordinates x, y and z is denoted by the generic symbol $\partial/\partial x_i$; this facilitates a compact and efficient notation for the development of vector calculus identities. These are studied in Chapter 2, though, for the same reasons that matrix algebra was first presented here without using the convention, vector calculus is initially developed there without recourse to it.

Further discussion of the summation convention, together with additional examples of it use, form the content of Appendix D. Considerable care is needed when using the convention, but mastering it is well worthwhile, as it considerably shortens many matrix algebra and vector calculus calculations.

41 A "continuous" version of this approach, using Hermitian operators (rather than Hermitian matrices) and wavefunctions (rather than vectors), is developed in Section 12.7 and in the problems at the end of Chapter 12.

SUMMARY

1. *Matrices and other quantities derived from an $M \times N$ matrix A*

Name	Symbol	How obtained	Notes
Trace	Tr A	Sum the elements on the leading diagonal	Needs $M = N$
2 × 2 determinant		$\begin{vmatrix} a_{11} & a_{12} \\ a_{21} & a_{22} \end{vmatrix} \equiv a_{11}a_{22} - a_{12}a_{21}$	Definition[a]
Determinant	$\lvert\mathsf{A}\rvert$	Make a Laplace expansion (Section 1.9) to reduce to a sum of 2 × 2 determinants	Needs $M = N$
Rank	$R(\mathsf{A})$	The largest value of r for which A has an $r \times r$ submatrix with a non-zero determinant	$R \le \min\{M, N\}$
Transpose	A^{T}	Interchange rows and columns: $(\mathsf{A}^{\mathrm{T}})_{ij} = A_{ji}$	A^{T} is $N \times M$
Complex conjugate	A^*	Take the complex conjugate of each element: $(\mathsf{A}^*)_{ij} = A^*_{ij}$	A^* is $M \times N$
Hermitian conjugate	$\mathsf{A}^\dagger$	Transpose the complex conjugate *or* complex conjugate the transpose: $(\mathsf{A}^\dagger)_{ij} = A^*_{ji}$	$\mathsf{A}^\dagger$ is $N \times M$
Minor	M_{ij}	Evaluate the determinant of the $(N-1) \times (N-1)$ matrix formed by deleting the ith row and the jth column	Needs $M = N$
Cofactor	C_{ij}	Multiply the minor M_{ij} by $(-1)^{i+j}$: $C_{ij} = (-1)^{i+j} M_{ij}$	Needs $M = N$
Inverse	A^{-1}	Divide each element of the transpose C^{T} of the matrix of cofactors C by the determinant of A; $(\mathsf{A}^{-1})_{ij} = C_{ji}/\lvert\mathsf{A}\rvert$	Needs $M = N$

[a] The formal definition of a 1 × 1 determinant is the value of its single entry (including its sign); it is not to be confused with $|a_{11}|$ meaning the (positive) modulus of a_{11}.

2. *Matrix algebra*
   - $(\mathsf{A} \pm \mathsf{B})_{ij} = A_{ij} \pm B_{ij}$.
   - $(\lambda\mathsf{A})_{ij} = \lambda A_{ij}$.
   - $(\mathsf{AB})_{ij} = \sum_k A_{ik} B_{kj}$.
   - $(\mathsf{A} + \mathsf{B})\mathsf{C} = \mathsf{AC} + \mathsf{BC}$ and $\mathsf{C}(\mathsf{A} + \mathsf{B}) = \mathsf{CA} + \mathsf{CB}$.
   - $\mathsf{AB} \neq \mathsf{BA}$, in general.

3. *Special types of square matrices*

Type	Symbolic property	Descriptive property
Real	$\mathsf{A}^* = \mathsf{A}$	Every element is real.
Imaginary	$\mathsf{A}^* = -\mathsf{A}$	Every element is imaginary or zero.
Diagonal	$A_{ij} = 0$ for $i \neq j$	Every off-diagonal element is zero.
Lower triangular	$A_{ij} = 0$ for $i < j$	Every element above the leading diagonal is zero.
Upper triangular	$A_{ij} = 0$ for $i > j$	Every element below the leading diagonal is zero.
Symmetric	$\mathsf{A}^{\mathrm{T}} = \mathsf{A}$	The matrix is equal to its transpose; $A_{ij} = A_{ji}$.
Antisymmetric *or* skew-symmetric	$\mathsf{A}^{\mathrm{T}} = -\mathsf{A}$	The matrix is equal to minus its transpose; $A_{ij} = -A_{ji}$. All of its diagonal elements must be zero.
Orthogonal	$\mathsf{A}^{\mathrm{T}} = \mathsf{A}^{-1}$	The transpose is equal to the inverse.
Hermitian	$\mathsf{A}^{\dagger} = \mathsf{A}$	The matrix is equal to its Hermitian conjugate; $A_{ij} = A_{ji}^*$.
Anti-Hermitian	$\mathsf{A}^{\dagger} = -\mathsf{A}$	The matrix is equal to minus its Hermitian conjugate; $A_{ij} = -A_{ji}^*$.
Unitary	$\mathsf{A}^{\dagger} = \mathsf{A}^{-1}$	The Hermitian conjugate is equal to the inverse.
Normal	$\mathsf{A}^{\dagger}\mathsf{A} = \mathsf{A}\mathsf{A}^{\dagger}$	The matrix commutes with its Hermitian conjugate.
Singular	$|\mathsf{A}| = 0$	The matrix has zero determinant (and no inverse).
Non-singular	$|\mathsf{A}| \neq 0$	The matrix has a non-zero determinant (and an inverse).
Defective		The $N \times N$ matrix has fewer than N linearly independent eigenvectors.

- Normal matrices include real symmetric, orthogonal, Hermitian and unitary matrices.
- $|\mathsf{A}^{\mathrm{T}}| = |\mathsf{A}|$; $|\mathsf{A}^{-1}| = |\mathsf{A}|^{-1}$.
- The determinant of an orthogonal matrix is equal to ± 1.

4. *Effects of matrix operations on matrix products*

Name	Effect on matrix product	Notes
Trace	$\mathrm{Tr}\,(\mathsf{AB}\ldots\mathsf{G}) = \mathrm{Tr}\,(\mathsf{B}\ldots\mathsf{GA})$	The product matrix $\mathsf{AB}\ldots\mathsf{G}$ must be square, though the individual matrices need not be. However, they must be compatible.
Determinant	$\lvert\mathsf{AB}\ldots\mathsf{G}\rvert = \lvert\mathsf{A}\rvert\lvert\mathsf{B}\rvert\ldots\lvert\mathsf{G}\rvert$	All matrices must be $N \times N$. Product is singular $\Leftrightarrow$ one or more of the individual matrices is singular.
Transpose	$(\mathsf{AB}\ldots\mathsf{G})^{\mathrm{T}} = \mathsf{G}^{\mathrm{T}}\ldots\mathsf{B}^{\mathrm{T}}\mathsf{A}^{\mathrm{T}}$	Matrices must be compatible but need not be square.
Complex conjugate	$(\mathsf{AB}\ldots\mathsf{G})^{*} = \mathsf{A}^{*}\mathsf{B}^{*}\ldots\mathsf{G}^{*}$	
Hermitian conjugate	$(\mathsf{AB}\ldots\mathsf{G})^{\dagger} = \mathsf{G}^{\dagger}\ldots\mathsf{B}^{\dagger}\mathsf{A}^{\dagger}$	Matrices must be compatible but need not be square.
Inverse	$(\mathsf{AB}\ldots\mathsf{G})^{-1} = \mathsf{G}^{-1}\ldots\mathsf{B}^{-1}\mathsf{A}^{-1}$	All matrices must be $N \times N$ and non-singular.

5. *Eigenvectors and eigenvalues*
   - The eigenvectors $\mathbf{x}^i$ and eigenvalues λ_i of a matrix A are defined by $\mathsf{A}\mathbf{x}^i = \lambda_i\mathbf{x}^i$.
   - The eigenvectors of a normal matrix corresponding to different eigenvalues are orthogonal.
   - The eigenvalues of an Hermitian (or real orthogonal) matrix are real.
   - The eigenvalues of a unitary matrix have unit modulus.
   - Two normal matrices commute $\Leftrightarrow$ they have a set of eigenvectors in common.
   - $\sum_i \lambda_i = \mathrm{Tr}\,\mathsf{A}$, $\prod_i \lambda_i = \lvert\mathsf{A}\rvert$.
   - A square matrix is singular $\Leftrightarrow$ at least one of its eigenvalues is zero.

6. *To find the eigenvalues and eigenvectors of a square matrix* A *and diagonalize it*
   (i) Solve the characteristic equation $\lvert\mathsf{A} - \lambda\mathsf{I}\rvert = 0$ for N values of λ, checking that $\sum_{i=1}^{N} \lambda_i = \mathrm{Tr}\,\mathsf{A}$.
   (ii) For each i, solve $\mathsf{A}\mathbf{x}^i = \lambda_i\mathbf{x}^i$ for $\mathbf{x}^i$.
   (iii) Construct the unitary matrix S whose columns are the normalized eigenvectors $\hat{\mathbf{x}}^i$ of A.
   (iv) Then $\mathsf{A}' = \mathsf{S}^{-1}\mathsf{AS} = \mathsf{S}^{\dagger}\mathsf{AS}$ is diagonal, with diagonal elements λ_i $(i = 1, \ldots, N)$.

7. *Quadratic forms and surfaces for $N = 3$*
   - The quadratic expression $Q(\mathbf{x}) = \mathbf{x}^{\mathrm{T}}\mathsf{A}\mathbf{x}$, with A symmetric, can be put in the form $\sum_{n=1}^{N} \lambda_i (x_i')^2$ using a real orthogonal change of basis $\mathbf{x} = \mathsf{S}\mathbf{x}'$, where S is as described in the previous section.
   - The equation $Q(\mathbf{x}) = 1$ represents a quadric surface whose principal axes lie in the directions of the eigenvectors $\mathbf{x}^i$ of A, and have lengths $(\sqrt{|\lambda_i|})^{-1}$.
   - If all the λ_i are positive the quadric surface is an ellipsoid; if one or two are negative, its cross-sections are a mixture of ellipses and hyperbolas. A zero eigenvalue gives rise to a "cylinder" whose axis lies along the corresponding eigenvector direction.

8. *Simultaneous linear equations,* $\mathsf{A}\mathbf{x} = \mathbf{b}$
   The Cramer determinant Δ_i is $|\mathsf{A}|$ but with the ith column of A replaced by the vector $\mathbf{b}$.

$\lvert\mathsf{A}\rvert$	$\mathbf{b}$	Δ_i	Number of solutions
$\neq 0$	$\neq \mathbf{0}$	–	one non-trivial, $\mathbf{x} = \mathsf{A}^{-1}\mathbf{b}$
	$= \mathbf{0}$	–	only trivial $\mathbf{x} = \mathbf{0}$
$= 0$	$\neq \mathbf{0}$	all $\Delta_i = 0$	infinite number
		at least one $\Delta_i \neq 0$	none
	$= \mathbf{0}$	–	infinite number

   Solution methods:
   - Direct inversion, $\mathbf{x} = \mathsf{A}^{-1}\mathbf{b}$.
   - *LU* decomposition: Find a lower diagonal matrix L and an upper diagonal matrix U such that $\mathsf{A} = \mathsf{LU}$. Then solve, successively, $\mathsf{L}\mathbf{y} = \mathbf{b}$ and $\mathsf{U}\mathbf{x} = \mathbf{y}$ to obtain $\mathbf{x}$.
   - Cholesky decomposition: If A is symmetric and positive definite ($\mathbf{x}^{\dagger}\mathsf{A}\mathbf{x} > 0$ for all non-zero $\mathbf{x}$) then find a lower diagonal matrix L such that $\mathsf{A} = \mathsf{L}\mathsf{L}^{\dagger}$ and proceed as in *LU* decomposition.
   - Cramer's rule: $x_i = \Delta_i / |\mathsf{A}|$.

9. *Normal modes definitions*
   For a system requiring N coordinates q_i (not necessarily distances) and their time derivatives $\dot{q}_i$ to specify its configuration, and with all $q_i = 0$ at equilibrium:
   - The kinetic energy quadratic form is $T = \dot{\mathsf{q}}^{\mathrm{T}}\mathsf{A}\mathsf{q} = \sum_{i,j} a_{ij}\dot{q}_i\dot{q}_j$.
   - The potential energy quadratic form is $V = \mathsf{q}^{\mathrm{T}}\mathsf{B}\mathsf{q} = \sum_{i,j} b_{ij}q_i q_j$.
   - Both A and B are symmetric positive definite $N \times N$ matrices.

10. *Normal frequencies and amplitudes*
    - The equation of motion is $\mathsf{A}\ddot{\mathsf{q}} + \mathsf{B}\mathsf{q} = 0$.
        (i) For periodic solutions $\mathsf{q} = \mathsf{x}\cos\omega t$, ω must satisfy $|\mathsf{B} - \omega^2\mathsf{A}| = 0$, thus giving (the squares of) the allowed normal frequencies.
        (ii) The corresponding vectors of amplitudes x^i are found, to within an overall scaling factor, by solving the linearly dependent set of equations $(\mathsf{B} - \omega^2\mathsf{A})\mathsf{x}^i = 0$.
        (iii) The amplitude vectors are orthogonal with respect to both A and B; $(\mathsf{x}^i)^{\mathrm{T}}\mathsf{A}\mathsf{x}^j = 0$ and $(\mathsf{x}^i)^{\mathrm{T}}\mathsf{B}\mathsf{x}^j = 0$ if $i \neq j$.
    - A value $\omega_i^2 = 0$ corresponds to bodily motion of the whole system, with no internal vibrations.

11. *Rayleigh–Ritz method*
    - The value of $\lambda(\mathsf{x}) = \dfrac{\mathsf{x}^{\mathrm{T}}\mathsf{B}\mathsf{x}}{\mathsf{x}^{\mathrm{T}}\mathsf{A}\mathsf{x}}$ always lies between (the square of) the lowest normal frequency ω_0 and that of the highest ω_{m}, however x is chosen.
    - *Any* evaluation, however contrived (or unjustified!) of $\lambda(\mathsf{x})$ gives an upper bound for (the square of) the lowest normal frequency.

PROBLEMS

1.1. Which of the following statements about linear vector spaces are true? Where a statement is false, give a counter-example to demonstrate this.

(a) Non-singular $N \times N$ matrices form a vector space of dimension N^2.
(b) Singular $N \times N$ matrices form a vector space of dimension N^2.
(c) Complex numbers form a vector space of dimension 2.
(d) Polynomial functions of x form an infinite-dimensional vector space.
(e) Series $\{a_0, a_1, a_2, \ldots, a_N\}$ for which $\sum_{n=0}^{N} |a_n|^2 = 1$ form an N-dimensional vector space.
(f) Absolutely convergent series form an infinite-dimensional vector space.
(g) Convergent series with terms of alternating sign form an infinite-dimensional vector space.

1.2. Consider the matrices

$$\text{(a)}\ \mathsf{B} = \begin{pmatrix} 0 & -i & i \\ i & 0 & -i \\ -i & i & 0 \end{pmatrix}, \qquad \text{(b)}\ \mathsf{C} = \frac{1}{\sqrt{8}} \begin{pmatrix} \sqrt{3} & -\sqrt{2} & -\sqrt{3} \\ 1 & \sqrt{6} & -1 \\ 2 & 0 & 2 \end{pmatrix}.$$

Are they (i) real, (ii) diagonal, (iii) symmetric, (iv) antisymmetric, (v) singular, (vi) orthogonal, (vii) Hermitian, (viii) anti-Hermitian, (ix) unitary, (x) normal?

1.3. By considering the matrices

$$\mathsf{A} = \begin{pmatrix} 1 & 0 \\ 0 & 0 \end{pmatrix}, \qquad \mathsf{B} = \begin{pmatrix} 0 & 0 \\ 3 & 4 \end{pmatrix},$$

show that $\mathsf{AB} = \mathsf{0}$ does *not* imply that either A or B is the zero matrix, but that it does imply that at least one of them is singular.

1.4. Evaluate the determinants

$$\text{(a)} \begin{vmatrix} a & h & g \\ h & b & f \\ g & f & c \end{vmatrix}, \qquad \text{(b)} \begin{vmatrix} 1 & 0 & 2 & 3 \\ 0 & 1 & -2 & 1 \\ 3 & -3 & 4 & -2 \\ -2 & 1 & -2 & 1 \end{vmatrix}$$

and

$$\text{(c)} \begin{vmatrix} gc & ge & a+ge & gb+ge \\ 0 & b & b & b \\ c & e & e & b+e \\ a & b & b+f & b+d \end{vmatrix}.$$

1.5. Using the properties of determinants, solve with a minimum of calculation the following equations for x:

$$\text{(a)} \begin{vmatrix} x & a & a & 1 \\ a & x & b & 1 \\ a & b & x & 1 \\ a & b & c & 1 \end{vmatrix} = 0, \qquad \text{(b)} \begin{vmatrix} x+2 & x+4 & x-3 \\ x+3 & x & x+5 \\ x-2 & x-1 & x+1 \end{vmatrix} = 0.$$

1.6. This problem considers a crystal whose unit cell has base vectors that are not necessarily mutually orthogonal.

(a) The basis vectors of the unit cell of a crystal, with the origin O at one corner, are denoted by $\mathbf{e}_1$, $\mathbf{e}_2$, $\mathbf{e}_3$. The matrix G has elements G_{ij}, where $G_{ij} = \mathbf{e}_i \cdot \mathbf{e}_j$ and H_{ij} are the elements of the matrix $\mathsf{H} \equiv \mathsf{G}^{-1}$. Show that the vectors $\mathbf{f}_i = \sum_j H_{ij}\mathbf{e}_j$ are the reciprocal vectors and that $H_{ij} = \mathbf{f}_i \cdot \mathbf{f}_j$.

(b) If the vectors $\mathbf{u}$ and $\mathbf{v}$ are given by

$$\mathbf{u} = \sum_i u_i \mathbf{e}_i, \qquad \mathbf{v} = \sum_i v_i \mathbf{f}_i,$$

obtain expressions for $|\mathbf{u}|$, $|\mathbf{v}|$, and $\mathbf{u} \cdot \mathbf{v}$.

(c) If the basis vectors are each of length a and the angle between each pair is $\pi/3$, write down G and hence obtain H.

(d) Calculate (i) the length of the normal from O onto the plane containing the points $p^{-1}\mathbf{e}_1$, $q^{-1}\mathbf{e}_2$, $r^{-1}\mathbf{e}_3$, and (ii) the angle between this normal and $\mathbf{e}_1$.

1.7. Prove the following results involving Hermitian matrices.

(a) If A is Hermitian and U is unitary then $\mathsf{U}^{-1}\mathsf{AU}$ is Hermitian.

(b) If A is anti-Hermitian then $i\mathsf{A}$ is Hermitian.

(c) The product of two Hermitian matrices A and B is Hermitian if and only if A and B commute.

(d) If S is a real antisymmetric matrix then $\mathsf{A} = (\mathsf{I} - \mathsf{S})(\mathsf{I} + \mathsf{S})^{-1}$ is orthogonal. If A is given by

$$\mathsf{A} = \begin{pmatrix} \cos\theta & \sin\theta \\ -\sin\theta & \cos\theta \end{pmatrix}$$

then find the matrix S that is needed to express A in the above form.

(e) If K is skew-Hermitian, i.e. $\mathsf{K}^\dagger = -\mathsf{K}$, then $\mathsf{V} = (\mathsf{I} + \mathsf{K})(\mathsf{I} - \mathsf{K})^{-1}$ is unitary.

1.8. A and B are real non-zero 3×3 matrices and satisfy the equation

$$(\mathsf{AB})^{\mathrm{T}} + \mathsf{B}^{-1}\mathsf{A} = \mathsf{0}.$$

(a) Prove that if B is orthogonal then A is antisymmetric.

(b) Without assuming that B is orthogonal, prove that A is singular.

1.9. The *commutator* $[\mathsf{X}, \mathsf{Y}]$ of two matrices is defined by the equation

$$[\mathsf{X}, \mathsf{Y}] = \mathsf{XY} - \mathsf{YX}.$$

Two anticommuting matrices A and B satisfy

$$\mathsf{A}^2 = \mathsf{I}, \qquad \mathsf{B}^2 = \mathsf{I}, \qquad [\mathsf{A}, \mathsf{B}] = 2i\mathsf{C}.$$

(a) Prove that $\mathsf{C}^2 = \mathsf{I}$ and that $[\mathsf{B}, \mathsf{C}] = 2i\mathsf{A}$.

(b) Evaluate $[[[\mathsf{A}, \mathsf{B}], [\mathsf{B}, \mathsf{C}]], [\mathsf{A}, \mathsf{B}]]$.

1.10. The four matrices S_x, S_y, S_z and I are defined by

$$\mathsf{S}_x = \begin{pmatrix} 0 & 1 \\ 1 & 0 \end{pmatrix}, \qquad \mathsf{S}_y = \begin{pmatrix} 0 & -i \\ i & 0 \end{pmatrix},$$
$$\mathsf{S}_z = \begin{pmatrix} 1 & 0 \\ 0 & -1 \end{pmatrix}, \qquad \mathsf{I} = \begin{pmatrix} 1 & 0 \\ 0 & 1 \end{pmatrix},$$

where $i^2 = -1$. Show that $\mathsf{S}_x^2 = \mathsf{I}$ and $\mathsf{S}_x\mathsf{S}_y = i\mathsf{S}_z$, and obtain similar results by permuting x, y and z. Given that $\mathbf{v}$ is a vector with Cartesian components (v_x, v_y, v_z), the matrix $\mathsf{S}(\mathbf{v})$ is defined as

$$\mathsf{S}(\mathbf{v}) = v_x\mathsf{S}_x + v_y\mathsf{S}_y + v_z\mathsf{S}_z.$$

Prove that, for general non-zero vectors $\mathbf{a}$ and $\mathbf{b}$,

$$\mathsf{S}(\mathbf{a})\mathsf{S}(\mathbf{b}) = \mathbf{a}\cdot\mathbf{b}\,\mathsf{I} + i\,\mathsf{S}(\mathbf{a}\times\mathbf{b}).$$

Without further calculation, deduce that $\mathsf{S}(\mathbf{a})$ and $\mathsf{S}(\mathbf{b})$ commute if and only if $\mathbf{a}$ and $\mathbf{b}$ are parallel vectors.

1.11. A general triangle has angles α, β and γ and corresponding opposite sides a, b and c. Express the length of each side in terms of the lengths of the other two sides and the relevant cosines, writing the relationships in matrix and vector form, using the vectors having components a, b, c and $\cos\alpha$, $\cos\beta$, $\cos\gamma$. Invert the matrix and hence deduce the cosine-law expressions involving α, β and γ.

1.12. Given a matrix

$$\mathsf{A} = \begin{pmatrix} 1 & \alpha & 0 \\ \beta & 1 & 0 \\ 0 & 0 & 1 \end{pmatrix},$$

where α and β are non-zero complex numbers, find its eigenvalues and eigenvectors. Find the respective conditions for (a) the eigenvalues to be real and (b) the eigenvectors to be orthogonal. Show that the conditions are jointly satisfied if and only if A is Hermitian.

1.13. Determine which of the matrices below are mutually commuting, and, for those that are, demonstrate that they have a complete set of eigenvectors in common:

$$\mathsf{A} = \begin{pmatrix} 6 & -2 \\ -2 & 9 \end{pmatrix}, \qquad \mathsf{B} = \begin{pmatrix} 1 & 8 \\ 8 & -11 \end{pmatrix},$$
$$\mathsf{C} = \begin{pmatrix} -9 & -10 \\ -10 & 5 \end{pmatrix}, \qquad \mathsf{D} = \begin{pmatrix} 14 & 2 \\ 2 & 11 \end{pmatrix}.$$

1.14. Do the following sets of equations have non-zero solutions? If so, find them.
(a) $3x + 2y + z = 0, \quad x - 3y + 2z = 0, \quad 2x + y + 3z = 0.$
(b) $2x = b(y + z), \quad x = 2a(y - z), \quad x = (6a - b)y - (6a + b)z.$

1.15. Solve the simultaneous equations

$$\begin{aligned} 2x + 3y + z &= 11, \\ x + y + z &= 6, \\ 5x - y + 10z &= 34. \end{aligned}$$

1.16. Solve the following simultaneous equations for x_1, x_2 and x_3, using matrix methods:

$$\begin{aligned} x_1 + 2x_2 + 3x_3 &= 1, \\ 3x_1 + 4x_2 + 5x_3 &= 2, \\ x_1 + 3x_2 + 4x_3 &= 3. \end{aligned}$$

1.17. Show that the following equations have solutions only if $\eta = 1$ or 2, and find them in these cases:

$$\begin{aligned} x + y + z &= 1, \\ x + 2y + 4z &= \eta, \\ x + 4y + 10z &= \eta^2. \end{aligned}$$

1.18. Find the condition(s) on α such that the simultaneous equations

$$\begin{aligned} x_1 + \alpha x_2 &= 1, \\ x_1 - x_2 + 3x_3 &= -1, \\ 2x_1 - 2x_2 + \alpha x_3 &= -2 \end{aligned}$$

have (a) exactly one solution, (b) no solutions, or (c) an infinite number of solutions; give all solutions where they exist.

1.19. Make an LU decomposition of the matrix

$$\mathsf{A} = \begin{pmatrix} 3 & 6 & 9 \\ 1 & 0 & 5 \\ 2 & -2 & 16 \end{pmatrix}$$

and hence solve $\mathsf{Ax} = \mathsf{b}$, where (i) $\mathsf{b} = (21 \quad 9 \quad 28)^{\mathrm{T}}$, (ii) $\mathsf{b} = (21 \quad 7 \quad 22)^{\mathrm{T}}$.

1.20. Make an LU decomposition of the matrix

$$\mathsf{A} = \begin{pmatrix} 2 & -3 & 1 & 3 \\ 1 & 4 & -3 & -3 \\ 5 & 3 & -1 & -1 \\ 3 & -6 & -3 & 1 \end{pmatrix}.$$

Hence solve $\mathsf{Ax} = \mathsf{b}$ for (i) $\mathsf{b} = (-4 \quad 1 \quad 8 \quad -5)^{\mathrm{T}}$, and (ii) $\mathsf{b} = (-10 \quad 0 \quad -3 \quad -24)^{\mathrm{T}}$.

Deduce that $\det \mathsf{A} = -160$ and confirm this by direct calculation.

1.21. Use the Cholesky decomposition method to determine whether the following matrices are positive definite. For each that is, determine the corresponding lower diagonal matrix L:

$$\mathsf{A} = \begin{pmatrix} 2 & 1 & 3 \\ 1 & 3 & -1 \\ 3 & -1 & 1 \end{pmatrix}, \qquad \mathsf{B} = \begin{pmatrix} 5 & 0 & \sqrt{3} \\ 0 & 3 & 0 \\ \sqrt{3} & 0 & 3 \end{pmatrix}.$$

1.22. Find the eigenvalues and a set of eigenvectors of the matrix

$$\begin{pmatrix} 1 & 3 & -1 \\ 3 & 4 & -2 \\ -1 & -2 & 2 \end{pmatrix}.$$

Verify that its eigenvectors are mutually orthogonal.

1.23. Find three real orthogonal column matrices, each of which is a simultaneous eigenvector of

$$\mathsf{A} = \begin{pmatrix} 0 & 0 & 1 \\ 0 & 1 & 0 \\ 1 & 0 & 0 \end{pmatrix} \quad \text{and} \quad \mathsf{B} = \begin{pmatrix} 0 & 1 & 1 \\ 1 & 0 & 1 \\ 1 & 1 & 0 \end{pmatrix}.$$

1.24. Use the results of the first worked example in Section 1.15 to evaluate, without repeated matrix multiplication, the expression $\mathsf{A}^6\mathsf{x}$, where $\mathsf{x} = (2 \quad 4 \quad -1)^{\mathrm{T}}$ and A is the matrix given in the example.

1.25. Given that A is a real symmetric matrix with normalized eigenvectors $\mathbf{e}^i$, obtain the coefficients α_i involved when column matrix x, which is the solution of

$$\mathsf{Ax} - \mu\mathsf{x} = \mathsf{v}, \qquad (*)$$

is expanded as $\mathsf{x} = \sum_i \alpha_i \mathbf{e}^i$. Here μ is a given constant and v is a given column matrix.

(a) Solve (*) when

$$\mathsf{A} = \begin{pmatrix} 2 & 1 & 0 \\ 1 & 2 & 0 \\ 0 & 0 & 3 \end{pmatrix},$$

$\mu = 2$ and $\mathsf{v} = (1 \quad 2 \quad 3)^{\mathrm{T}}$.

(b) Would (*) have a solution if $\mu = 1$ and (i) $\mathsf{v} = (1 \quad 2 \quad 3)^{\mathrm{T}}$, (ii) $\mathsf{v} = (2 \quad 2 \quad 3)^{\mathrm{T}}$? Where it does, find it.

1.26. Demonstrate that the matrix

$$\mathsf{A} = \begin{pmatrix} 2 & 0 & 0 \\ -6 & 4 & 4 \\ 3 & -1 & 0 \end{pmatrix}$$

is defective, i.e. does not have three linearly independent eigenvectors, by showing the following:

(a) its eigenvalues are degenerate and, in fact, all equal;

(b) any eigenvector has the form $(\mu \quad (3\mu - 2\nu) \quad \nu)^{\mathrm{T}}$;

(c) if two pairs of values, μ_1, ν_1 and μ_2, ν_2, define two independent eigenvectors v_1 and v_2, then *any* third similarly defined eigenvector v_3 can be written as a linear combination of v_1 and v_2, i.e.

$$\mathsf{v}_3 = a\mathsf{v}_1 + b\mathsf{v}_2,$$

where

$$a = \frac{\mu_3\nu_2 - \mu_2\nu_3}{\mu_1\nu_2 - \mu_2\nu_1} \quad \text{and} \quad b = \frac{\mu_1\nu_3 - \mu_3\nu_1}{\mu_1\nu_2 - \mu_2\nu_1}.$$

Illustrate (c) using the example $(\mu_1, \nu_1) = (1, 1)$, $(\mu_2, \nu_2) = (1, 2)$ and $(\mu_3, \nu_3) = (0, 1)$.

Show further that any matrix of the form

$$\begin{pmatrix} 2 & 0 & 0 \\ 6n - 6 & 4 - 2n & 4 - 4n \\ 3 - 3n & n - 1 & 2n \end{pmatrix}$$

is defective, with the same eigenvalues and eigenvectors as A.

1.27. By finding the eigenvectors of the Hermitian matrix

$$\mathsf{H} = \begin{pmatrix} 10 & 3i \\ -3i & 2 \end{pmatrix},$$

construct a unitary matrix U such that $\mathsf{U}^\dagger \mathsf{H} \mathsf{U} = \Lambda$, where Λ is a real diagonal matrix.

1.28. Use the stationary properties of quadratic forms to determine the maximum and minimum values taken by the expression

$$Q = 5x^2 + 4y^2 + 4z^2 + 2xz + 2xy$$

on the unit sphere, $x^2 + y^2 + z^2 = 1$. For what values of x, y, z do they occur?

1.29. Given that the matrix

$$\mathsf{A} = \begin{pmatrix} 2 & -1 & 0 \\ -1 & 2 & -1 \\ 0 & -1 & 2 \end{pmatrix}$$

has two eigenvectors of the form $(1 \quad y \quad 1)^{\mathrm{T}}$, use the stationary property of the expression $J(\mathsf{x}) = \mathsf{x}^{\mathrm{T}}\mathsf{A}\mathsf{x}/(\mathsf{x}^{\mathrm{T}}\mathsf{x})$ to obtain the corresponding eigenvalues. Deduce the third eigenvalue.

1.30. Find the lengths of the semi-axes of the ellipse

$$73x^2 + 72xy + 52y^2 = 100,$$

and determine its orientation.

1.31. The equation of a particular conic section is

$$Q \equiv 8x_1^2 + 8x_2^2 - 6x_1x_2 = 110.$$

Determine the type of conic section this represents, the orientation of its principal axes, and relevant lengths in the directions of these axes.

1.32. Show that the quadratic surface

$$5x^2 + 11y^2 + 5z^2 - 10yz + 2xz - 10xy = 4$$

is an ellipsoid with semi-axes of lengths 2, 1 and 0.5. Find the direction of its longest axis.

1.33. Find the direction of the axis of symmetry of the quadratic surface

$$7x^2 + 7y^2 + 7z^2 - 20yz - 20xz + 20xy = 3.$$

1.34. For the following matrices, find the eigenvalues and sufficient of the eigenvectors to be able to describe the quadratic surfaces associated with them:

$$\text{(a)} \begin{pmatrix} 5 & 1 & -1 \\ 1 & 5 & 1 \\ -1 & 1 & 5 \end{pmatrix}, \quad \text{(b)} \begin{pmatrix} 1 & 2 & 2 \\ 2 & 1 & 2 \\ 2 & 2 & 1 \end{pmatrix}, \quad \text{(c)} \begin{pmatrix} 1 & 2 & 1 \\ 2 & 4 & 2 \\ 1 & 2 & 1 \end{pmatrix}.$$

1.35. This problem demonstrates the reverse of the usual procedure of diagonalizing a matrix.

(a) Rearrange the result $\mathsf{A}' = \mathsf{S}^{-1}\mathsf{A}\mathsf{S}$ of Section 1.17 to express the original matrix A in terms of the unitary matrix S and the diagonal matrix A'. Hence show how to construct a matrix A that has given eigenvalues and given (orthogonal) column matrices as its eigenvectors.

(b) Find the matrix that has as eigenvectors $(1 \quad 2 \quad 1)^{\mathrm{T}}$, $(1 \quad -1 \quad 1)^{\mathrm{T}}$ and $(1 \quad 0 \quad -1)^{\mathrm{T}}$, with corresponding eigenvalues λ, μ and ν.

(c) Try a particular case, say $\lambda = 3$, $\mu = -2$ and $\nu = 1$, and verify by explicit solution that the matrix so found does have these eigenvalues.

1.36. Find an orthogonal transformation that takes the quadratic form

$$Q \equiv -x_1^2 - 2x_2^2 - x_3^2 + 8x_2x_3 + 6x_1x_3 + 8x_1x_2$$

into the form

$$\mu_1 y_1^2 + \mu_2 y_2^2 - 4y_3^2,$$

and determine μ_1 and μ_2 (see Section 1.18).

1.37. A more general form of expression for the determinant of a 3×3 matrix A than (1.45) is given by

$$|\mathsf{A}|\epsilon_{lmn} = A_{li}A_{mj}A_{nk}\epsilon_{ijk}. \tag{1.138}$$

The former could, as stated earlier in this chapter, have been written as

$$|\mathsf{A}| = \epsilon_{ijk}A_{i1}A_{j2}A_{k3}.$$

The more general form removes the explicit mention of 1, 2, 3 at the expense of an additional Levi–Civita symbol. As stated in the footnote on p. 790, the form of (1.138) can be readily extended to cover a general $N \times N$ matrix.

Use this more general form to prove properties (i), (iii), (v), (vi) and (vii) of determinants stated in Subsection 1.9.1. Property (iv) is obvious by inspection. For definiteness take $N = 3$, but convince yourself that your methods of proof would be valid for any positive integer N.

1.38. A double pendulum, smoothly pivoted at A, consists of two light rigid rods, AB and BC, each of length l, which are smoothly jointed at B and carry masses m and αm at B and C respectively. The pendulum makes small oscillations in one plane under gravity. At time t, AB and BC make angles $\theta(t)$ and $\phi(t)$,

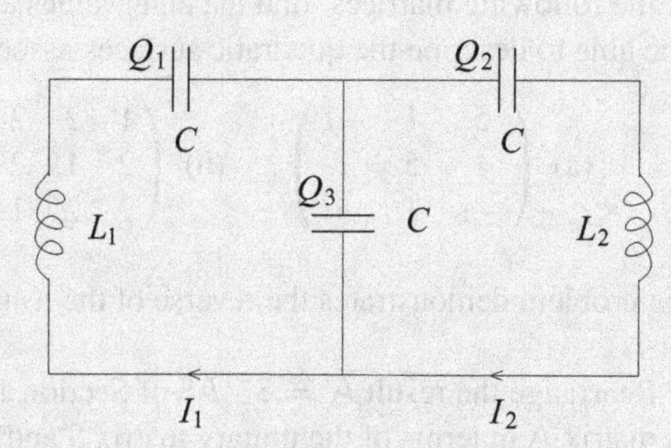

Figure 1.5 The circuit and notation for Problem 1.40.

respectively, with the downward vertical. Find quadratic expressions for the kinetic and potential energies of the system and hence show that the normal modes have angular frequencies given by

$$\omega^2 = \frac{g}{l}\left[1 + \alpha \pm \sqrt{\alpha(1+\alpha)}\right].$$

For $\alpha = 1/3$, show that in one of the normal modes the mid-point of BC does not move during the motion.

1.39. Three coupled pendulums swing perpendicularly to the horizontal line containing their points of suspension, and the following equations of motion are satisfied:

$$\begin{aligned}
-m\ddot{x}_1 &= cmx_1 + d(x_1 - x_2),\\
-M\ddot{x}_2 &= cMx_2 + d(x_2 - x_1) + d(x_2 - x_3),\\
-m\ddot{x}_3 &= cmx_3 + d(x_3 - x_2),
\end{aligned}$$

where x_1, x_2 and x_3 are measured from the equilibrium points; m, M and m are the masses of the pendulum bobs; and c and d are positive constants. Find the normal frequencies of the system and sketch the corresponding patterns of oscillation. What happens as $d \to 0$ or $d \to \infty$?

1.40. Consider the circuit consisting of three equal capacitors and two different inductors shown in Figure 1.5. For charges Q_i on the capacitors and currents I_i through the components, write down Kirchhoff's law for the total voltage change around each of two complete circuit loops. Note that, to within an unimportant constant, the conservation of current implies that $Q_3 = Q_1 - Q_2$. Express the loop equations in the form given in (1.125), namely

$$\mathsf{A}\ddot{\mathsf{Q}} + \mathsf{B}\mathsf{Q} = \mathsf{0}.$$

Use this to show that the normal frequencies of the circuit are given by

$$\omega^2 = \frac{1}{CL_1L_2}\left[L_1 + L_2 \pm (L_1^2 + L_2^2 - L_1L_2)^{1/2}\right].$$

Obtain the same matrices and result by finding the total energy stored in the various capacitors (typically $Q^2/(2C)$) and in the inductors (typically $LI^2/2$).

For the special case $L_1 = L_2 = L$ determine the relevant eigenvectors and so describe the patterns of current flow in the circuit.

1.41. Continue the worked example, modeling a linear molecule, discussed at the end of Subsection 1.19.1, for the case in which $\mu = 2$.

(a) Show that the eigenvectors derived there have the expected orthogonality properties with respect to both A and B.

(b) For the situation in which the atoms are released from rest with initial displacements $x_1 = 2\epsilon$, $x_2 = -\epsilon$ and $x_3 = 0$, determine their subsequent motions and maximum displacements.

1.42. *The simultaneous reduction to diagonal form of two real symmetric quadratic forms.*

Consider the two real symmetric quadratic forms $\mathsf{u}^{\mathrm{T}}\mathsf{A}\mathsf{u}$ and $\mathsf{u}^{\mathrm{T}}\mathsf{B}\mathsf{u}$, where u^{T} stands for the row matrix $(x \quad y \quad z)$, and denote by u^n those column matrices that satisfy

$$\mathsf{B}\mathsf{u}^n = \lambda_n \mathsf{A}\mathsf{u}^n, \tag{1.139}$$

in which n is a label and the λ_n are real, non-zero and all different.

(a) By multiplying (1.139) on the left by $(\mathsf{u}^m)^{\mathrm{T}}$, and the transpose of the corresponding equation for u^m on the right by u^n, show that $(\mathsf{u}^m)^{\mathrm{T}}\mathsf{A}\mathsf{u}^n = 0$ for $n \neq m$.

(b) By noting that $\mathsf{A}\mathsf{u}^n = (\lambda_n)^{-1}\mathsf{B}\mathsf{u}^n$, deduce that $(\mathsf{u}^m)^{\mathrm{T}}\mathsf{B}\mathsf{u}^n = 0$ for $m \neq n$.

(c) It can be shown that the u^n are linearly independent; the next step is to construct a matrix P whose columns are the vectors u^n.

(d) Make a change of variables $\mathsf{u} = \mathsf{P}\mathsf{v}$ such that $\mathsf{u}^{\mathrm{T}}\mathsf{A}\mathsf{u}$ becomes $\mathsf{v}^{\mathrm{T}}\mathsf{C}\mathsf{v}$, and $\mathsf{u}^{\mathrm{T}}\mathsf{B}\mathsf{u}$ becomes $\mathsf{v}^{\mathrm{T}}\mathsf{D}\mathsf{v}$. Show that C and D are diagonal by showing that $c_{ij} = 0$ if $i \neq j$, and similarly for d_{ij}.

Thus $\mathsf{u} = \mathsf{P}\mathsf{v}$ or $\mathsf{v} = \mathsf{P}^{-1}\mathsf{u}$ reduces both quadratics to diagonal form.

To summarize, the method is as follows:

(a) find the λ_n that allow (1.139) a non-zero solution, by solving $|\mathsf{B} - \lambda\mathsf{A}| = 0$;

(b) for each λ_n construct u^n;

(c) construct the non-singular matrix P whose columns are the vectors u^n;

(d) make the change of variable $\mathsf{u} = \mathsf{P}\mathsf{v}$.

1.43. It is shown in physics and engineering textbooks that circuits containing capacitors and inductors can be analyzed by replacing a capacitor of capacitance C by a "complex impedance" $1/(i\omega C)$ and an inductor of inductance L by an impedance $i\omega L$, where ω is the angular frequency of the currents flowing and $i^2 = -1$.

Use this approach and Kirchhoff's circuit laws to analyze the circuit shown in Figure 1.6 and obtain three linear equations governing the currents I_1, I_2 and I_3. Show that the only possible frequencies of self-sustaining currents satisfy either (a) $\omega^2 LC = 1$ or (b) $3\omega^2 LC = 1$. Find the corresponding current patterns and, in

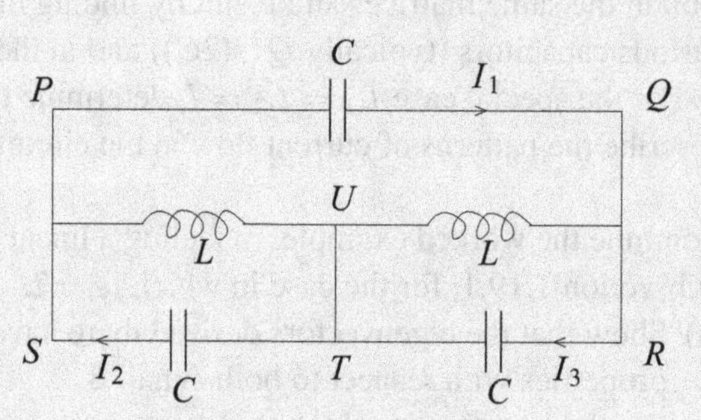

Figure 1.6 The circuit and notation for Problem 1.43.

each case, by identifying parts of the circuit in which no current flows, draw an equivalent circuit that contains only one capacitor and one inductor.

1.44. (*It is recommended that the reader does not attempt this question until Problem 1.42 has been studied.*)

Find a real linear transformation that simultaneously reduces the quadratic forms

$$3x^2 + 5y^2 + 5z^2 + 2yz + 6zx - 2xy,$$
$$5x^2 + 12y^2 + 8yz + 4zx$$

to diagonal form.

1.45. (*It is recommended that the reader does not attempt this question until Problem 1.42 has been studied.*)

If, in the pendulum system studied in Subsection 1.19.1, the string is replaced by a second rod identical to the first then the expressions for the kinetic energy T and the potential energy V become (to second order in the θ_i)

$$T \approx Ml^2\left(\tfrac{8}{3}\dot{\theta}_1^2 + 2\dot{\theta}_1\dot{\theta}_2 + \tfrac{2}{3}\dot{\theta}_2^2\right),$$
$$V \approx Mgl\left(\tfrac{3}{2}\theta_1^2 + \tfrac{1}{2}\theta_2^2\right).$$

Determine the normal frequencies of the system and find new variables ξ and η that will reduce these two expressions to diagonal form, i.e. to

$$a_1\dot{\xi}^2 + a_2\dot{\eta}^2 \qquad \text{and} \qquad b_1\xi^2 + b_2\eta^2.$$

1.46. Use the Rayleigh–Ritz method to estimate the lowest oscillation frequency of a heavy chain of N links, each of length a $(= L/N)$, which hangs freely from one end. Consider simple calculable configurations such as all links but one vertical, or all links collinear, etc.

1.47. Three particles of mass m are attached to a light horizontal string having fixed ends, the string being thus divided into four equal portions each of length a and under a tension T. Show that for small transverse vibrations the amplitudes x^i of the normal modes satisfy $\mathsf{B}\mathsf{x} = (ma\omega^2/T)\mathsf{x}$, where B is the matrix

$$\begin{pmatrix} 2 & -1 & 0 \\ -1 & 2 & -1 \\ 0 & -1 & 2 \end{pmatrix}.$$

Estimate the lowest and highest eigenfrequencies using trial vectors $(3 \quad 4 \quad 3)^{\mathrm{T}}$ and $(3 \quad -4 \quad 3)^{\mathrm{T}}$. Use also the exact vectors $\begin{pmatrix} 1 & \sqrt{2} & 1 \end{pmatrix}^{\mathrm{T}}$ and $\begin{pmatrix} 1 & -\sqrt{2} & 1 \end{pmatrix}^{\mathrm{T}}$ and compare the results.

HINTS AND ANSWERS

1.1. (a) False. $\mathbf{0}_N$, the $N \times N$ null matrix, is *not* non-singular.
(b) False. Consider the sum of $\begin{pmatrix} 1 & 0 \\ 0 & 0 \end{pmatrix}$ and $\begin{pmatrix} 0 & 0 \\ 0 & 1 \end{pmatrix}$.
(c) True.
(d) True.
(e) False. Consider $b_n = a_n + a_n$ for which $\sum_{n=0}^{N} |b_n|^2 = 4 \neq 1$, or note that there is no zero vector with unit norm.
(f) True.
(g) False. Consider the two series defined by

$$a_0 = \tfrac{1}{2}, \qquad a_n = 2(-\tfrac{1}{2})^n \quad \text{for} \quad n \geq 1; \qquad b_n = -(-\tfrac{1}{2})^n \quad \text{for} \quad n \geq 0.$$

The series that is the sum of $\{a_n\}$ and $\{b_n\}$ does not have alternating signs and so closure does not hold.

1.3. Use the property of the determinant of a matrix product.

1.5. (a) $x = a, b$ or c; (b) $x = -1$; the equation is linear in x.

1.7. (d) $\mathsf{S} = \begin{pmatrix} 0 & -\tan(\theta/2) \\ \tan(\theta/2) & 0 \end{pmatrix}$.
(e) Note that $(\mathsf{I} + \mathsf{K})(\mathsf{I} - \mathsf{K}) = \mathsf{I} - \mathsf{K}^2 = (\mathsf{I} - \mathsf{K})(\mathsf{I} + \mathsf{K})$.

1.9. (b) $32i\mathsf{A}$.

1.11. $a = b\cos\gamma + c\cos\beta$, and cyclic permutations; $a^2 = b^2 + c^2 - 2bc\cos\alpha$, and cyclic permutations.

1.13. C does not commute with the others; A, B and D have $(1 \quad -2)^{\mathrm{T}}$ and $(2 \quad 1)^{\mathrm{T}}$ as common eigenvectors.

1.15. $x = 3$, $y = 1$, $z = 2$.

1.17. First show that A is singular. $\eta = 1$, $x = 1 + 2z$, $y = -3z$; $\eta = 2$, $x = 2z$, $y = 1 - 3z$.

1.19. $L = (1, 0, 0; \tfrac{1}{3}, 1, 0; \tfrac{2}{3}, 3, 1)$, $\quad U = (3, 6, 9; 0, -2, 2; 0, 0, 4)$.

(i) $\mathsf{x} = (-1 \quad 1 \quad 2)^{\mathrm{T}}$. (ii) $\mathsf{x} = (-3 \quad 2 \quad 2)^{\mathrm{T}}$.

1.21. A is not positive definite as L_{33} is calculated to be $\sqrt{-6}$.
$\mathsf{B} = \mathsf{L}\mathsf{L}^{\mathrm{T}}$, where the non-zero elements of L are
$L_{11} = \sqrt{5},\ L_{31} = \sqrt{3/5},\ L_{22} = \sqrt{3},\ L_{33} = \sqrt{12/5}.$

1.23. For A: $(1 \quad 0 \quad -1)^{\mathrm{T}},\ (1 \quad \alpha_1 \quad 1)^{\mathrm{T}},\ (1 \quad \alpha_2 \quad 1)^{\mathrm{T}}$.
For B: $(1 \quad 1 \quad 1)^{\mathrm{T}},\ (\beta_1 \quad \gamma_1 \quad -\beta_1 - \gamma_1)^{\mathrm{T}},\ (\beta_2 \quad \gamma_2 \quad -\beta_2 - \gamma_2)^{\mathrm{T}}$.
The α_i, β_i and γ_i are arbitrary.
Simultaneous and orthogonal: $(1 \quad 0 \quad -1)^{\mathrm{T}},\ (1 \quad 1 \quad 1)^{\mathrm{T}},\ (1 \quad -2 \quad 1)^{\mathrm{T}}$.

1.25. $\alpha_j = (\mathsf{v} \cdot \mathsf{e}^{j*})/(\lambda_j - \mu)$, where λ_j is the eigenvalue corresponding to e^j.
(a) $\mathsf{x} = (2 \quad 1 \quad 3)^{\mathrm{T}}$.
(b) Since μ is equal to one of A's eigenvalues λ_j, the equation only has a solution if $\mathsf{v} \cdot \mathsf{e}^{j*} = 0$; (i) no solution; (ii) $\mathsf{x} = (1 \quad 1 \quad 3/2)^{\mathrm{T}}$.

1.27. $\mathsf{U} = (10)^{-1/2}(1, 3i; 3i, 1)$, $\Lambda = (1, 0; 0, 11)$.

1.29. $J = (2y^2 - 4y + 4)/(y^2 + 2)$ with stationary values at $y = \pm\sqrt{2}$ and corresponding eigenvalues $2 \mp \sqrt{2}$. From the trace property of A, the third eigenvalue equals 2.

1.31. Ellipse; $\theta = \pi/4$, $a = \sqrt{22}$; $\theta = 3\pi/4$, $b = \sqrt{10}$.

1.33. The direction of the eigenvector having the unrepeated eigenvalue is $(1, 1, -1)/\sqrt{3}$.

1.35. (a) $\mathsf{A} = \mathsf{S}\mathsf{A}'\mathsf{S}^{\dagger}$, where S is the matrix whose columns are the eigenvectors of the matrix A to be constructed, and $\mathsf{A}' = \mathrm{diag}\,(\lambda, \mu, \nu)$.
(b) $\mathsf{A} = (\lambda + 2\mu + 3\nu,\ 2\lambda - 2\mu,\ \lambda + 2\mu - 3\nu; 2\lambda - 2\mu,\ 4\lambda + 2\mu,\ 2\lambda - 2\mu;$
$\lambda + 2\mu - 3\nu,\ 2\lambda - 2\mu,\ \lambda + 2\mu + 3\nu)$.
(c) $\frac{1}{3}(1, 5, -2; 5, 4, 5; -2, 5, 1)$.

1.37. This solution is fuller than most. So, to save yet more additional text, the summation convention (Appendix D) is used and (1.138) denoted by $(*)$.
(i) We write the expression for $|\mathsf{A}^{\mathrm{T}}|$ using the given formalism, recalling that $(\mathsf{A}^{\mathrm{T}})_{ij} = (\mathsf{A})_{ji}$. We then multiply both sides by ϵ_{lmn} and sum over l, m and n:

$$\begin{aligned}
|\mathsf{A}^{\mathrm{T}}|\epsilon_{lmn} &= A_{il}A_{jm}A_{kn}\epsilon_{ijk}, \\
|\mathsf{A}^{\mathrm{T}}|\epsilon_{lmn}\epsilon_{lmn} &= A_{il}A_{jm}A_{kn}\epsilon_{lmn}\epsilon_{ijk} \\
&= |\mathsf{A}|\epsilon_{ijk}\epsilon_{ijk}, \\
|\mathsf{A}^{\mathrm{T}}| &= |\mathsf{A}|.
\end{aligned}$$

In the third line we have used the definition of $|\mathsf{A}|$ (with the roles of the sets of dummy variables $\{i, j, k\}$ and $\{l, m, n\}$ interchanged), and in the fourth line, we have canceled the scalar quantity $\epsilon_{lmn}\epsilon_{lmn} = \epsilon_{ijk}\epsilon_{ijk}$; the value of this scalar is $N(N-1)$, but that is irrelevant here.
(iii) Every non-zero term on the RHS of $(*)$ contains any particular row index once and only once. The same can be said for the Levi–Civita symbol on the LHS.

Thus interchanging two rows is equivalent to interchanging two of the subscripts of ϵ_{lmn} and thereby reversing its sign. Consequently, the whole RHS changes sign and the magnitude of $|\mathsf{A}|$ remains the same, though its sign is changed.

(v) If, say, $A_{pi} = \lambda A_{pj}$, for some particular pair of values i and j and all p, then in the (multiple) summation on the RHS of (∗) each A_{nk} appears multiplied by (with no summation over i and j)

$$\epsilon_{ijk} A_{li} A_{mj} + \epsilon_{jik} A_{lj} A_{mi} = \epsilon_{ijk} \lambda A_{lj} A_{mj} + \epsilon_{jik} A_{lj} \lambda A_{mj} = 0,$$

since $\epsilon_{ijk} = -\epsilon_{jik}$. Consequently, grouped in this way, all pairs of terms contribute nothing to the sum and $|\mathsf{A}| = 0$.

(vi) Consider the matrix B whose m, jth element is defined by $B_{mj} = A_{mj} + \lambda A_{pj}$, where $p \neq m$. The only case that needs detailed analysis is when l, m and n are all different. Since $p \neq m$ it must be the same as either l or n; suppose that $p = l$. The determinant of B is given by

$$\begin{aligned} |\mathsf{B}|\epsilon_{lmn} &= A_{li}(A_{mj} + \lambda A_{lj}) A_{nk} \epsilon_{ijk} \\ &= A_{li} A_{mj} A_{nk} \epsilon_{ijk} + \lambda A_{li} A_{lj} A_{nk} \epsilon_{ijk} \\ &= |\mathsf{A}|\epsilon_{lmn} + \lambda 0, \end{aligned}$$

where we have used the row equivalent of the intermediate result obtained for columns in (v). Thus we conclude that $|\mathsf{B}| = |\mathsf{A}|$.

(vii) If $\mathsf{X} = \mathsf{AB}$, then

$$|\mathsf{X}|\epsilon_{lmn} = A_{lx} B_{xi} A_{my} B_{yj} A_{nz} B_{zk} \epsilon_{ijk}.$$

Multiply both sides by ϵ_{lmn} and sum over l, m and n:

$$\begin{aligned} |\mathsf{X}|\epsilon_{lmn}\epsilon_{lmn} &= \epsilon_{lmn} A_{lx} A_{my} A_{nz} \epsilon_{ijk} B_{xi} B_{yj} B_{zk} \\ &= \epsilon_{xyz} |\mathsf{A}^{\mathrm{T}}| \epsilon_{xyz} |\mathsf{B}|, \\ \Rightarrow \quad |\mathsf{X}| &= |\mathsf{A}^{\mathrm{T}}|\,|\mathsf{B}| = |\mathsf{A}|\,|\mathsf{B}|, \quad \text{using result (i).} \end{aligned}$$

To obtain the last line we have canceled the non-zero scalar $\epsilon_{lmn}\epsilon_{lmn} = \epsilon_{xyz}\epsilon_{xyz}$ from both sides, as we did in the proof of result (i).

The extension to the product of any number of matrices is obvious. Replacing B by CD or by DC and applying the result just proved extends it to a product of three matrices. Extension to any higher number is done in the same way.

1.39. See Figure 1.7.

1.41. (b) $x_1 = \epsilon(\cos \omega t + \cos \sqrt{2}\omega t)$, $x_2 = -\epsilon \cos \sqrt{2}\omega t$, $x_3 = \epsilon(-\cos \omega t + \cos \sqrt{2}\omega t)$.

At various times the three displacements will reach 2ϵ, ϵ, 2ϵ respectively. For example, x_1 can be written as $2\epsilon \cos[(\sqrt{2} - 1)\omega t/2] \cos[(\sqrt{2} + 1)\omega t/2]$, i.e. an oscillation of angular frequency $(\sqrt{2} + 1)\omega/2$ and modulated amplitude $2\epsilon \cos[(\sqrt{2} - 1)\omega/2]$; the amplitude will reach 2ϵ after a time $\approx 4\pi/[\omega(\sqrt{2} - 1)]$.

1.43. As the circuit loops contain no voltage sources, the equations are homogeneous, and so for a non-trivial solution the determinant of coefficients must vanish.

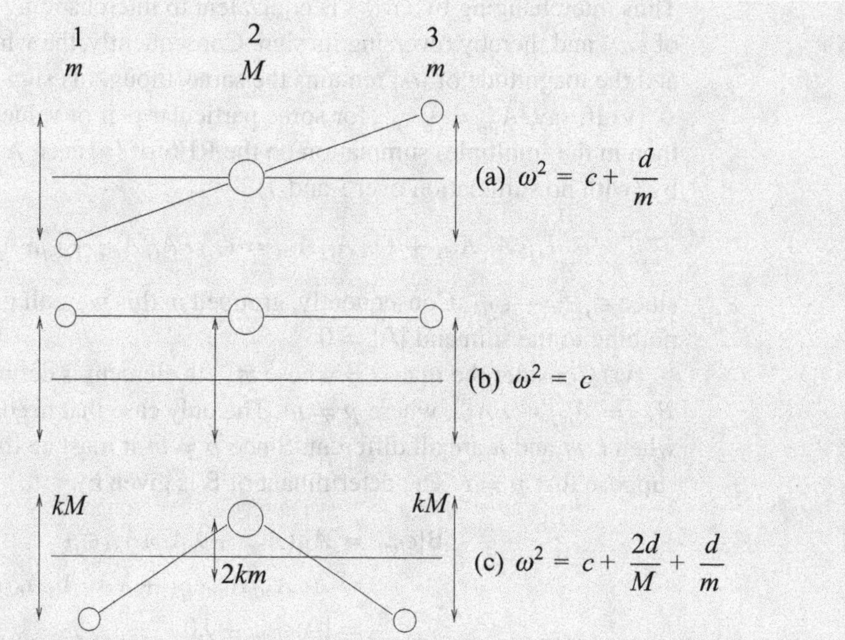

Figure 1.7 The normal modes, as viewed from above, of the coupled pendulums in Problem 1.39.

(a) $I_1 = 0$, $I_2 = -I_3$; no current in PQ; equivalent to two separate circuits of capacitance C and inductance L.

(b) $I_1 = -2I_2 = -2I_3$; no current in TU; capacitance $3C/2$ and inductance $2L$.

1.45. $\omega = (2.634g/l)^{1/2}$ or $(0.3661g/l)^{1/2}$; $\theta_1 = \xi + \eta$, $\theta_2 = 1.431\xi - 2.097\eta$.

1.47. Estimated, $10/17 < Ma\omega^2/T < 58/17$; exact, $2 - \sqrt{2} \le Ma\omega^2/T \le 2 + \sqrt{2}$.

2 Vector calculus

In Section A.9 of Appendix A we review the algebra of vectors, and in Chapter 1 we considered how to transform one vector into another using a linear operator. In this chapter and the next we discuss the calculus of vectors, i.e. the differentiation and integration both of vectors describing particular bodies, such as the velocity of a particle, and of vector fields, in which a vector is defined as a function of the coordinates throughout some volume (one-, two- or three-dimensional). Since the aim of this chapter is to develop methods for handling multi-dimensional physical situations, we will assume throughout that the functions with which we have to deal have sufficiently amenable mathematical properties, in particular that they are continuous and differentiable.

2.1 Differentiation of vectors

Let us consider a vector $\mathbf{a}$ that is a function of a scalar variable u. By this we mean that with each value of u we associate a vector $\mathbf{a}(u)$. For example, in Cartesian coordinates $\mathbf{a}(u) = a_x(u)\mathbf{i} + a_y(u)\mathbf{j} + a_z(u)\mathbf{k}$, where $a_x(u)$, $a_y(u)$ and $a_z(u)$ are scalar functions of u and are the components of the vector $\mathbf{a}(u)$ in the x-, y- and z-directions respectively. We note that if $\mathbf{a}(u)$ is continuous at some point $u = u_0$ then this implies that each of the Cartesian components $a_x(u)$, $a_y(u)$ and $a_z(u)$ is also continuous there.

Let us consider the derivative of the vector function $\mathbf{a}(u)$ with respect to u. The derivative of a vector function is defined in a similar manner to the ordinary derivative of a scalar function $f(x)$. The small change in the vector $\mathbf{a}(u)$ resulting from a small change Δu in the value of u is given by $\Delta\mathbf{a} = \mathbf{a}(u + \Delta u) - \mathbf{a}(u)$ (see Figure 2.1). The derivative of $\mathbf{a}(u)$ with respect to u is defined to be

$$\frac{d\mathbf{a}}{du} = \lim_{\Delta u \to 0} \frac{\mathbf{a}(u + \Delta u) - \mathbf{a}(u)}{\Delta u}, \tag{2.1}$$

assuming that the limit exists, in which case $\mathbf{a}(u)$ is said to be differentiable at that point. Note that $d\mathbf{a}/du$ is also a vector, which is not, in general, parallel to $\mathbf{a}(u)$. In Cartesian coordinates, the derivative of the vector $\mathbf{a}(u) = a_x\mathbf{i} + a_y\mathbf{j} + a_z\mathbf{k}$ is given by

$$\frac{d\mathbf{a}}{du} = \frac{da_x}{du}\mathbf{i} + \frac{da_y}{du}\mathbf{j} + \frac{da_z}{du}\mathbf{k}.$$

Perhaps the simplest application of the above is to finding the velocity and acceleration of a particle in classical mechanics. If the time-dependent position vector of the particle with respect to the origin in Cartesian coordinates is given by $\mathbf{r}(t) = x(t)\mathbf{i} + y(t)\mathbf{j} + z(t)\mathbf{k}$

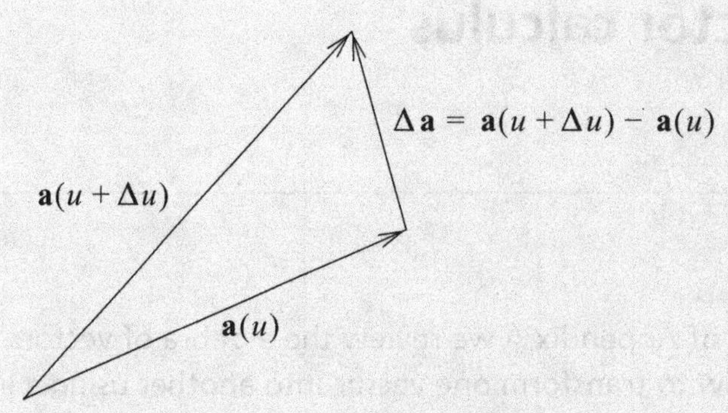

Figure 2.1 A small change in a vector $\mathbf{a}(u)$ resulting from a small change in u.

then the velocity of the particle is given by the vector

$$\mathbf{v}(t) = \frac{d\mathbf{r}}{dt} = \frac{dx}{dt}\mathbf{i} + \frac{dy}{dt}\mathbf{j} + \frac{dz}{dt}\mathbf{k}.$$

The direction of the velocity vector is along the tangent to the path $\mathbf{r}(t)$ at the instantaneous position of the particle, and its magnitude $|\mathbf{v}(t)|$ is equal to the speed of the particle. The acceleration of the particle is given in a similar manner by

$$\mathbf{a}(t) = \frac{d\mathbf{v}}{dt} = \frac{d^2x}{dt^2}\mathbf{i} + \frac{d^2y}{dt^2}\mathbf{j} + \frac{d^2z}{dt^2}\mathbf{k}.$$

These notions are illustrated in the following worked example.

Example The position vector of a particle at time t in Cartesian coordinates is given by $\mathbf{r}(t) = 2t^2\mathbf{i} + (3t - 2)\mathbf{j} + (3t^2 - 1)\mathbf{k}$. Find the speed of the particle at $t = 1$ and the component of its acceleration in the direction $\mathbf{s} = \mathbf{i} + 2\mathbf{j} + \mathbf{k}$.

The velocity and acceleration of the particle are given by

$$\mathbf{v}(t) = \frac{d\mathbf{r}}{dt} = 4t\mathbf{i} + 3\mathbf{j} + 6t\mathbf{k},$$

$$\mathbf{a}(t) = \frac{d\mathbf{v}}{dt} = 4\mathbf{i} + 6\mathbf{k}.$$

The speed of the particle at $t = 1$ is simply

$$|\mathbf{v}(1)| = \sqrt{4^2 + 3^2 + 6^2} = \sqrt{61}.$$

The acceleration of the particle is constant (i.e. independent of t), and its component in the direction $\mathbf{s}$ is given by

$$\mathbf{a} \cdot \hat{\mathbf{s}} = \frac{(4\mathbf{i} + 6\mathbf{k}) \cdot (\mathbf{i} + 2\mathbf{j} + \mathbf{k})}{\sqrt{1^2 + 2^2 + 1^2}} = \frac{5\sqrt{6}}{3}.$$

Note that the vector $\mathbf{s}$ had to be converted into the unit vector $\hat{\mathbf{s}}$, by dividing by its modulus, before it could be used to determine the component of $\mathbf{a}$ in its direction. ◀

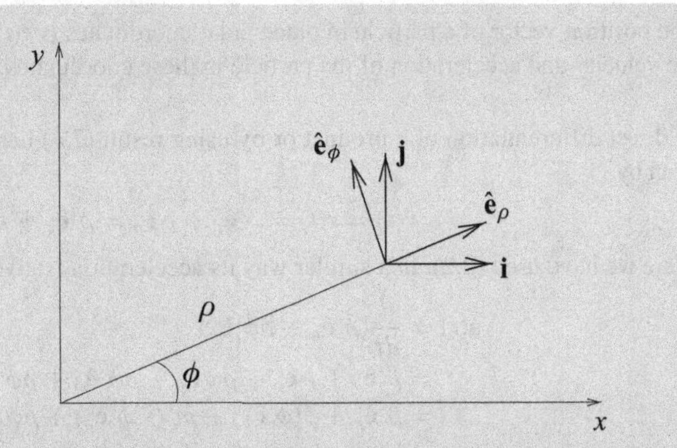

Figure 2.2 Unit basis vectors for two-dimensional Cartesian and plane polar coordinates.

In the case discussed above **i**, **j** and **k** are fixed, time-independent basis vectors. This may not be true of basis vectors in general; when we are not using Cartesian coordinates the basis vectors themselves must also be differentiated. We discuss basis vectors for non-Cartesian coordinate systems in detail in Section 2.9. Nevertheless, as a simple example, let us now consider two-dimensional plane polar coordinates ρ, ϕ.

Referring to Figure 2.2, imagine holding ϕ fixed and moving radially outwards, i.e. in the direction of increasing ρ. Let us denote the unit vector in this direction by $\hat{\mathbf{e}}_\rho$. Similarly, imagine keeping ρ fixed and moving around a circle of fixed radius in the direction of increasing ϕ. Let us denote the unit vector tangent to the circle by $\hat{\mathbf{e}}_\phi$. The two vectors $\hat{\mathbf{e}}_\rho$ and $\hat{\mathbf{e}}_\phi$ are the basis vectors for this two-dimensional coordinate system, just as **i** and **j** are basis vectors for two-dimensional Cartesian coordinates. All these basis vectors are shown in Figure 2.2.

An important difference between the two sets of basis vectors is that, while **i** and **j** are constant in magnitude *and direction*, the vectors $\hat{\mathbf{e}}_\rho$ and $\hat{\mathbf{e}}_\phi$ have constant magnitudes but their directions change as ρ and ϕ vary. Therefore, when calculating the derivative of a vector written in polar coordinates we must also differentiate the basis vectors. One way of doing this is to express $\hat{\mathbf{e}}_\rho$ and $\hat{\mathbf{e}}_\phi$ in terms of **i** and **j**. From Figure 2.2, we see that

$$\hat{\mathbf{e}}_\rho = \cos\phi\,\mathbf{i} + \sin\phi\,\mathbf{j},$$
$$\hat{\mathbf{e}}_\phi = -\sin\phi\,\mathbf{i} + \cos\phi\,\mathbf{j}.$$

Since **i** and **j** are constant vectors, we find that the derivatives of the basis vectors $\hat{\mathbf{e}}_\rho$ and $\hat{\mathbf{e}}_\phi$ with respect to t are given by

$$\frac{d\,\hat{\mathbf{e}}_\rho}{dt} = -\sin\phi\frac{d\phi}{dt}\mathbf{i} + \cos\phi\frac{d\phi}{dt}\mathbf{j} = \dot{\phi}\,\hat{\mathbf{e}}_\phi, \tag{2.2}$$

$$\frac{d\,\hat{\mathbf{e}}_\phi}{dt} = -\cos\phi\frac{d\phi}{dt}\mathbf{i} - \sin\phi\frac{d\phi}{dt}\mathbf{j} = -\dot{\phi}\,\hat{\mathbf{e}}_\rho, \tag{2.3}$$

where the overdot is the conventional notation for differentiation with respect to time.

Example The position vector of a particle in plane polar coordinates is $\mathbf{r}(t) = \rho(t)\,\hat{\mathbf{e}}_\rho$. Find expressions for the velocity and acceleration of the particle in these coordinates.

By direct differentiation of a product or by using result (2.4) below, the velocity of the particle is given by

$$\mathbf{v}(t) = \dot{\mathbf{r}}(t) = \dot{\rho}\,\hat{\mathbf{e}}_\rho + \rho\,\dot{\hat{\mathbf{e}}}_\rho = \dot{\rho}\,\hat{\mathbf{e}}_\rho + \rho\dot{\phi}\,\hat{\mathbf{e}}_\phi,$$

where we have used (2.2). In a similar way its acceleration is given by

$$\begin{aligned}\mathbf{a}(t) &= \frac{d}{dt}(\dot{\rho}\,\hat{\mathbf{e}}_\rho + \rho\dot{\phi}\,\hat{\mathbf{e}}_\phi)\\ &= \ddot{\rho}\,\hat{\mathbf{e}}_\rho + \dot{\rho}\,\dot{\hat{\mathbf{e}}}_\rho + \rho\dot{\phi}\,\dot{\hat{\mathbf{e}}}_\phi + \rho\ddot{\phi}\,\hat{\mathbf{e}}_\phi + \dot{\rho}\dot{\phi}\,\hat{\mathbf{e}}_\phi\\ &= \ddot{\rho}\,\hat{\mathbf{e}}_\rho + \dot{\rho}(\dot{\phi}\,\hat{\mathbf{e}}_\phi) + \rho\dot{\phi}(-\dot{\phi}\,\hat{\mathbf{e}}_\rho) + \rho\ddot{\phi}\,\hat{\mathbf{e}}_\phi + \dot{\rho}\dot{\phi}\,\hat{\mathbf{e}}_\phi\\ &= (\ddot{\rho} - \rho\dot{\phi}^2)\,\hat{\mathbf{e}}_\rho + (\rho\ddot{\phi} + 2\dot{\rho}\dot{\phi})\,\hat{\mathbf{e}}_\phi.\end{aligned}$$

Here, (2.2) and (2.3) were used to go from the second line to the third.[1] ◀

2.1.1 Differentiation of composite vector expressions

In composite vector expressions each of the vectors or scalars involved may be a function of some scalar variable u, as we have seen. The derivatives of such expressions are easily found using the definition (2.1) and the rules of ordinary differential calculus. They may be summarized by the following, in which we assume that **a** and **b** are differentiable vector functions of a scalar u and that ϕ is a differentiable scalar function of u:

$$\frac{d}{du}(\phi\mathbf{a}) = \phi\frac{d\mathbf{a}}{du} + \frac{d\phi}{du}\mathbf{a}, \tag{2.4}$$

$$\frac{d}{du}(\mathbf{a}\cdot\mathbf{b}) = \mathbf{a}\cdot\frac{d\mathbf{b}}{du} + \frac{d\mathbf{a}}{du}\cdot\mathbf{b}, \tag{2.5}$$

$$\frac{d}{du}(\mathbf{a}\times\mathbf{b}) = \mathbf{a}\times\frac{d\mathbf{b}}{du} + \frac{d\mathbf{a}}{du}\times\mathbf{b}. \tag{2.6}$$

The order of the factors in the terms on the RHS of (2.6) is, of course, just as important as it is in the original vector product.

Example A particle of mass m with position vector **r** relative to some origin O experiences a force **F**, which produces a torque (moment) $\mathbf{T} = \mathbf{r}\times\mathbf{F}$ about O. The angular momentum of the particle about O is given by $\mathbf{L} = \mathbf{r}\times m\mathbf{v}$, where **v** is the particle's velocity. Show that the rate of change of angular momentum is equal to the applied torque.

The rate of change of angular momentum is given by

$$\frac{d\mathbf{L}}{dt} = \frac{d}{dt}(\mathbf{r}\times m\mathbf{v}).$$

1 Apply this analysis to the case of a body of mass m moving with constant angular velocity ω in a circle of radius R centered on the origin, showing that the force needed to sustain the motion has magnitude $mR\omega^2$ and is directed towards the origin.

Using (2.6) we obtain

$$\frac{d\mathbf{L}}{dt} = \frac{d\mathbf{r}}{dt} \times m\mathbf{v} + \mathbf{r} \times \frac{d}{dt}(m\mathbf{v})$$
$$= \mathbf{v} \times m\mathbf{v} + \mathbf{r} \times \frac{d}{dt}(m\mathbf{v})$$
$$= \mathbf{0} + \mathbf{r} \times \mathbf{F} = \mathbf{T},$$

where in the last line we use Newton's second law, namely $\mathbf{F} = d(m\mathbf{v})/dt$. ◀

If a vector $\mathbf{a}$ is a function of a scalar variable s that is itself a function of u, so that $s = s(u)$, then the chain rule gives

$$\frac{d\mathbf{a}(s)}{du} = \frac{ds}{du}\frac{d\mathbf{a}}{ds}. \qquad (2.7)$$

The derivatives of more complicated vector expressions may be found by repeated application of the above equations.[2]

One further useful result can be derived by considering the derivative

$$\frac{d}{du}(\mathbf{a} \cdot \mathbf{a}) = 2\mathbf{a} \cdot \frac{d\mathbf{a}}{du};$$

since $\mathbf{a} \cdot \mathbf{a} = a^2$, where $a = |\mathbf{a}|$, we see that

$$\mathbf{a} \cdot \frac{d\mathbf{a}}{du} = 0 \quad \text{if } a \text{ is constant.} \qquad (2.8)$$

In other words, if a vector $\mathbf{a}(u)$ has a constant magnitude as u varies then it is perpendicular to the vector $d\mathbf{a}/du$.

2.1.2 Differential of a vector

As a final note on the differentiation of vectors, we can also define the *differential* of a vector, in a similar way to that of a scalar in ordinary differential calculus. In the definition of the vector derivative (2.1), we used the notion of a small change $\Delta\mathbf{a}$ in a vector $\mathbf{a}(u)$ resulting from a small change Δu in its argument. In the limit $\Delta u \to 0$, the change in $\mathbf{a}$ becomes infinitesimally small, and we denote it by the differential $d\mathbf{a}$. From (2.1) we see that the differential is given by

$$d\mathbf{a} = \frac{d\mathbf{a}}{du}\,du. \qquad (2.9)$$

Note that the differential of a vector is also a vector. As an example, the infinitesimal change in the position vector of a particle in an infinitesimal time dt is

$$d\mathbf{r} = \frac{d\mathbf{r}}{dt}\,dt = \mathbf{v}\,dt,$$

where $\mathbf{v}$ is the particle's velocity.[3]

2 Obtain an explicit cyclically invariant expression for the derivative with respect to u of the scalar triple product of the three vectors $\mathbf{a}(u)$, $\mathbf{b}(u)$ and $\mathbf{c}(u)$.

3 In the same way, the infinitesimal change in velocity in an infinitesimal time dt is given by $d\mathbf{v} = \mathbf{a}\,dt$, where $\mathbf{a}$ is the particle's acceleration.

2.2 Integration of vectors

The integration of a vector (or of an expression involving vectors that may itself be either a vector or scalar) with respect to a scalar u can be regarded as the inverse of differentiation. We must remember, however, that

(i) the integral has the same nature (vector or scalar) as the integrand,
(ii) the constant of integration for indefinite integrals must be of the same nature as the integral.

For example, if $\mathbf{a}(u) = d[\mathbf{A}(u)]/du$ then the indefinite integral of $\mathbf{a}(u)$ is given by

$$\int \mathbf{a}(u)\, du = \mathbf{A}(u) + \mathbf{b},$$

where $\mathbf{b}$ is a constant vector, of the same nature as $\mathbf{A}$. The definite integral of $\mathbf{a}(u)$ from $u = u_1$ to $u = u_2$ is given by

$$\int_{u_1}^{u_2} \mathbf{a}(u)\, du = \mathbf{A}(u_2) - \mathbf{A}(u_1).$$

Example A small particle of mass m orbits a much larger mass M centered at the origin O. According to Newton's law of gravitation, the position vector $\mathbf{r}$ of the small mass obeys the differential equation

$$m\frac{d^2\mathbf{r}}{dt^2} = -\frac{GMm}{r^2}\hat{\mathbf{r}}.$$

Show that the vector $\mathbf{r} \times d\mathbf{r}/dt$ is a constant of the motion.

Forming the vector product of the differential equation with $\mathbf{r}$, we obtain

$$\mathbf{r} \times \frac{d^2\mathbf{r}}{dt^2} = -\frac{GM}{r^2}\mathbf{r} \times \hat{\mathbf{r}}.$$

Since $\mathbf{r}$ and $\hat{\mathbf{r}}$ are collinear, $\mathbf{r} \times \hat{\mathbf{r}} = \mathbf{0}$ and therefore we have

$$\mathbf{r} \times \frac{d^2\mathbf{r}}{dt^2} = \mathbf{0}. \tag{2.10}$$

However,

$$\frac{d}{dt}\left(\mathbf{r} \times \frac{d\mathbf{r}}{dt}\right) = \mathbf{r} \times \frac{d^2\mathbf{r}}{dt^2} + \frac{d\mathbf{r}}{dt} \times \frac{d\mathbf{r}}{dt} = \mathbf{0},$$

since the first term is zero by (2.10), and the second is zero because it is the vector product of two parallel (in this case identical) vectors. Integrating, we obtain the required result

$$\mathbf{r} \times \frac{d\mathbf{r}}{dt} = \mathbf{c}, \tag{2.11}$$

where $\mathbf{c}$ is a constant vector.

As a further point of interest we may note that in an infinitesimal time dt the change in the position vector of the small mass is $d\mathbf{r}$ and the element of area swept out by the position vector of the particle is simply $dA = \frac{1}{2}|\mathbf{r} \times d\mathbf{r}|$. Dividing both sides of this equation by dt, we conclude that

$$\frac{dA}{dt} = \frac{1}{2}\left|\mathbf{r} \times \frac{d\mathbf{r}}{dt}\right| = \frac{|\mathbf{c}|}{2},$$

and that the physical interpretation of the above result (2.11) is that the position vector $\mathbf{r}$ of the small mass sweeps out equal areas in equal times. This result is in fact valid for motion under any force that acts along the line joining the two particles. ◀

2.3 Vector functions of several arguments

The concept of the derivative of a vector is easily extended to cases where the vectors (or scalars) are functions of more than one independent scalar variable, $u_1, u_2, \ldots, u_n$. In this case, the results of Subsection 2.1.1 are still valid, except that the derivatives become partial derivatives $\partial\mathbf{a}/\partial u_i$ defined as in ordinary differential calculus. For example, in Cartesian coordinates,

$$\frac{\partial \mathbf{a}}{\partial u_r} = \frac{\partial a_x}{\partial u_r}\mathbf{i} + \frac{\partial a_y}{\partial u_r}\mathbf{j} + \frac{\partial a_z}{\partial u_r}\mathbf{k}.$$

In particular, (2.7) generalizes to the chain rule of partial differentiation. If $\mathbf{a} = \mathbf{a}(u_1, u_2, \ldots, u_n)$ and each of the u_i is also a function $u_i(v_1, v_2, \ldots, v_n)$ of the variables v_i then the generalization is

$$\frac{\partial \mathbf{a}}{\partial v_i} = \frac{\partial \mathbf{a}}{\partial u_1}\frac{\partial u_1}{\partial v_i} + \frac{\partial \mathbf{a}}{\partial u_2}\frac{\partial u_2}{\partial v_i} + \cdots + \frac{\partial \mathbf{a}}{\partial u_n}\frac{\partial u_n}{\partial v_i} = \sum_{j=1}^{n} \frac{\partial \mathbf{a}}{\partial u_j}\frac{\partial u_j}{\partial v_i}. \tag{2.12}$$

A special case of this rule arises when $\mathbf{a}$ is an explicit function of some variable v, as well as of scalars $u_1, u_2, \ldots, u_n$ that are themselves functions of v; then we have

$$\frac{d\mathbf{a}}{dv} = \frac{\partial \mathbf{a}}{\partial v} + \sum_{j=1}^{n} \frac{\partial \mathbf{a}}{\partial u_j}\frac{\partial u_j}{\partial v}. \tag{2.13}$$

We may also extend the concept of the differential of a vector given in (2.9) to vectors dependent on several variables $u_1, u_2, \ldots, u_n$:

$$d\mathbf{a} = \frac{\partial \mathbf{a}}{\partial u_1} du_1 + \frac{\partial \mathbf{a}}{\partial u_2} du_2 + \cdots + \frac{\partial \mathbf{a}}{\partial u_n} du_n = \sum_{j=1}^{n} \frac{\partial \mathbf{a}}{\partial u_j} du_j. \tag{2.14}$$

As an example, the infinitesimal change in an electric field $\mathbf{E}$ in moving from a position $\mathbf{r}$ to a neighboring one $\mathbf{r} + d\mathbf{r}$ is given by

$$d\mathbf{E} = \frac{\partial \mathbf{E}}{\partial x} dx + \frac{\partial \mathbf{E}}{\partial y} dy + \frac{\partial \mathbf{E}}{\partial z} dz. \tag{2.15}$$

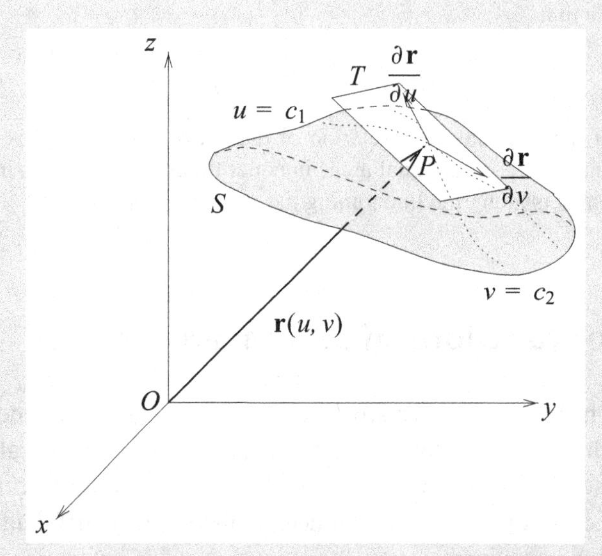

Figure 2.3 The tangent plane T to a surface S at a particular point P; $u = c_1$ and $v = c_2$ are the coordinate curves, shown by dotted lines, that pass through P. The broken line shows some particular parametric curve $\mathbf{r} = \mathbf{r}(\lambda)$ lying in the surface.

2.4 Surfaces

A surface S in space can be described by the vector $\mathbf{r}(u, v)$ joining the origin O of a coordinate system to a point on the surface (see Figure 2.3). As the parameters u and v vary, the end-point of the vector moves over the surface.

In Cartesian coordinates the surface is given by

$$\mathbf{r}(u, v) = x(u, v)\mathbf{i} + y(u, v)\mathbf{j} + z(u, v)\mathbf{k},$$

where $x = x(u, v)$, $y = y(u, v)$ and $z = z(u, v)$ are the parametric equations of the surface. We can also represent a surface by $z = f(x, y)$ or $g(x, y, z) = 0$. Either of these representations can be converted into the parametric form. For example, if $z = f(x, y)$ then by setting $u = x$ and $v = y$ the surface can be represented in parametric form by

$$\mathbf{r}(u, v) = u\mathbf{i} + v\mathbf{j} + f(u, v)\mathbf{k}.$$

Any curve $\mathbf{r}(\lambda)$, where λ is a parameter, on the surface S can be represented by a pair of equations relating the parameters u and v, for example $u = f(\lambda)$ and $v = g(\lambda)$. A parametric representation of the curve can easily be found by straightforward substitution, i.e. $\mathbf{r}(\lambda) = \mathbf{r}(u(\lambda), v(\lambda))$. Using (2.12) for the case where the vector is a function of a single variable λ so that the LHS becomes a total derivative, the tangent to the curve $\mathbf{r}(\lambda)$ at any point is given by

$$\frac{d\mathbf{r}}{d\lambda} = \frac{\partial \mathbf{r}}{\partial u}\frac{du}{d\lambda} + \frac{\partial \mathbf{r}}{\partial v}\frac{dv}{d\lambda}. \tag{2.16}$$

The two curves $u =$ constant and $v =$ constant passing through any point P on S are called *coordinate curves*. For the curve $u =$ constant, for example, we have $du/d\lambda = 0$, and so from (2.16) its tangent vector is in the direction $\partial \mathbf{r}/\partial v$. Similarly, the tangent vector to the curve $v =$ constant is in the direction $\partial \mathbf{r}/\partial u$.

If the surface is smooth then at any point P on S the vectors $\partial \mathbf{r}/\partial u$ and $\partial \mathbf{r}/\partial v$ are linearly independent and define the *tangent plane* T at the point P (see Figure 2.3). A vector normal to the surface at P is given by

$$\mathbf{n} = \frac{\partial \mathbf{r}}{\partial u} \times \frac{\partial \mathbf{r}}{\partial v}. \tag{2.17}$$

In the neighborhood of P, an infinitesimal vector displacement $d\mathbf{r}$ is written

$$d\mathbf{r} = \frac{\partial \mathbf{r}}{\partial u}\, du + \frac{\partial \mathbf{r}}{\partial v}\, dv.$$

The *element of area* at P, an infinitesimal parallelogram whose sides are the coordinate curves, has magnitude

$$dS = \left| \frac{\partial \mathbf{r}}{\partial u}\, du \times \frac{\partial \mathbf{r}}{\partial v}\, dv \right| = \left| \frac{\partial \mathbf{r}}{\partial u} \times \frac{\partial \mathbf{r}}{\partial v} \right| du\, dv = |\mathbf{n}|\, du\, dv. \tag{2.18}$$

Thus the total area of the surface is

$$A = \iint_R \left| \frac{\partial \mathbf{r}}{\partial u} \times \frac{\partial \mathbf{r}}{\partial v} \right| du\, dv = \iint_R |\mathbf{n}|\, du\, dv, \tag{2.19}$$

where R is the region in the uv-plane corresponding to the range of parameter values that define the surface.

Example Find the element of area on the surface of a sphere of radius a, and hence calculate the total surface area of the sphere.

We can represent a point $\mathbf{r}$ on the surface of the sphere in terms of the two parameters θ and ϕ:

$$\mathbf{r}(\theta, \phi) = a \sin\theta \cos\phi\, \mathbf{i} + a \sin\theta \sin\phi\, \mathbf{j} + a \cos\theta\, \mathbf{k},$$

where θ and ϕ are the polar and azimuthal angles respectively. At any point P, vectors tangent to the coordinate curves $\theta =$ constant and $\phi =$ constant are

$$\frac{\partial \mathbf{r}}{\partial \theta} = a \cos\theta \cos\phi\, \mathbf{i} + a \cos\theta \sin\phi\, \mathbf{j} - a \sin\theta\, \mathbf{k},$$

$$\frac{\partial \mathbf{r}}{\partial \phi} = -a \sin\theta \sin\phi\, \mathbf{i} + a \sin\theta \cos\phi\, \mathbf{j}.$$

A normal $\mathbf{n}$ to the surface at this point is then given by

$$\mathbf{n} = \frac{\partial \mathbf{r}}{\partial \theta} \times \frac{\partial \mathbf{r}}{\partial \phi} = \begin{vmatrix} \mathbf{i} & \mathbf{j} & \mathbf{k} \\ a\cos\theta\cos\phi & a\cos\theta\sin\phi & -a\sin\theta \\ -a\sin\theta\sin\phi & a\sin\theta\cos\phi & 0 \end{vmatrix}$$

$$= a^2\sin\theta(\sin\theta\cos\phi\,\mathbf{i} + \sin\theta\sin\phi\,\mathbf{j} + \cos\theta\,\mathbf{k}),$$

which has a magnitude of $a^2\sin\theta$. Therefore, the element of area at P is, from (2.18),

$$dS = a^2\sin\theta\, d\theta\, d\phi,$$

and the total surface area of the sphere is given by

$$A = \int_0^{\pi} d\theta \int_0^{2\pi} d\phi\, a^2 \sin\theta = 4\pi a^2.$$

This familiar result can, of course, be proved by much simpler methods![4] ◄

2.5 Scalar and vector fields

We now turn to the case where a particular scalar or vector quantity is defined not just at a point in space but continuously as a *field* throughout some region of space R (which is often the whole space). Although the concept of a field is valid for spaces with an arbitrary number of dimensions, in the remainder of this chapter we will restrict our attention to the familiar three-dimensional case. A *scalar field* $\phi(x, y, z)$ associates a scalar with each point in R, while a *vector field* $\mathbf{a}(x, y, z)$ associates a vector with each point. In what follows, we will assume that the variation in the scalar or vector field from point to point is both continuous and differentiable in R.

Simple examples of scalar fields include the pressure at each point in a fluid and the electrostatic potential at each point in space in the presence of an electric charge. Vector fields relating to the same physical systems are the velocity vector in a fluid (giving the local speed and direction of the flow) and the electric field.

With the study of continuously varying scalar and vector fields there arises the need to consider their derivatives and also the integration of field quantities along lines, over surfaces and throughout volumes in the field. We defer the discussion of line, surface and volume integrals until the next chapter, and in the remainder of this chapter we concentrate on the definitions of vector differential operators and their properties.

2.6 Vector operators

Certain differential operations may be performed on scalar and vector fields and have wide-ranging applications in the physical sciences. The most important operations are

4 Use a similar method to show that the surface element on the paraboloid of revolution given in cylindrical polar coordinates by $\rho^2 = 4az$ is $dS = (2a)^{-1}(\rho^2 + 4a^2)^{1/2}\, d\rho\, d\phi$.

those of finding the *gradient* of a scalar field and the *divergence* and *curl* of a vector field. It is usual to define these operators from a strictly mathematical point of view, as we do below. In the following chapter, however, we will discuss their geometrical definitions, which rely on the concept of integrating vector quantities along lines and over surfaces.

Central to all these differential operations is the vector operator ∇, which is called *del* (or sometimes *nabla*) and in Cartesian coordinates is defined by

$$\nabla \equiv \mathbf{i}\frac{\partial}{\partial x} + \mathbf{j}\frac{\partial}{\partial y} + \mathbf{k}\frac{\partial}{\partial z}. \tag{2.20}$$

The form of this operator in non-Cartesian coordinate systems is discussed in Sections 2.8 and 2.9.

2.6.1 Gradient of a scalar field

The *gradient* of a scalar field $\phi(x, y, z)$ is defined by

$$\text{grad}\,\phi = \nabla\phi = \mathbf{i}\frac{\partial \phi}{\partial x} + \mathbf{j}\frac{\partial \phi}{\partial y} + \mathbf{k}\frac{\partial \phi}{\partial z}. \tag{2.21}$$

Clearly, $\nabla\phi$ is a vector field whose x-, y- and z-components are the first partial derivatives of $\phi(x, y, z)$ with respect to x, y and z respectively. Also note that the vector field $\nabla\phi$ should not be confused with $\phi\nabla$, which has components $(\phi\,\partial/\partial x, \phi\,\partial/\partial y, \phi\,\partial/\partial z)$, and is a vector operator.[5]

Example Find the gradient of the scalar field $\phi = xy^2z^3$.

From (2.21) the gradient of ϕ, obtained by differentiating with respect to x, y and z in turn, is given by

$$\nabla\phi = y^2z^3\mathbf{i} + 2xyz^3\mathbf{j} + 3xy^2z^2\mathbf{k}.$$

Note that each component can be a function of some or all of the coordinates. ◀

The gradient of a scalar field ϕ has some interesting geometrical properties. Let us first consider the problem of *calculating the rate of change of ϕ in some particular direction.* For an infinitesimal vector displacement $d\mathbf{r}$, forming its scalar product with $\nabla\phi$ we obtain

$$\begin{aligned}\nabla\phi \cdot d\mathbf{r} &= \left(\mathbf{i}\frac{\partial \phi}{\partial x} + \mathbf{j}\frac{\partial \phi}{\partial y} + \mathbf{k}\frac{\partial \phi}{\partial z}\right) \cdot (\mathbf{i}\,dx + \mathbf{j}\,dy + \mathbf{k}\,dx), \\ &= \frac{\partial \phi}{\partial x}dx + \frac{\partial \phi}{\partial y}dy + \frac{\partial \phi}{\partial z}dz, \\ &= d\phi, \end{aligned} \tag{2.22}$$

which is the infinitesimal change in ϕ in going from position $\mathbf{r}$ to $\mathbf{r} + d\mathbf{r}$. In particular, if $\mathbf{r}$ depends on some parameter u such that $\mathbf{r}(u)$ defines a curve in space, then the total

5 Distinguish between (i) $(\phi\nabla)\psi$, (ii) $(\nabla\phi)\psi$ and (iii) $\nabla(\phi\psi)$ and determine the x-component of each if $\phi(x, y, z) = x^2y^2z^2$ and $\psi(x, y, z) = x^2 + y^2 + z^2$.

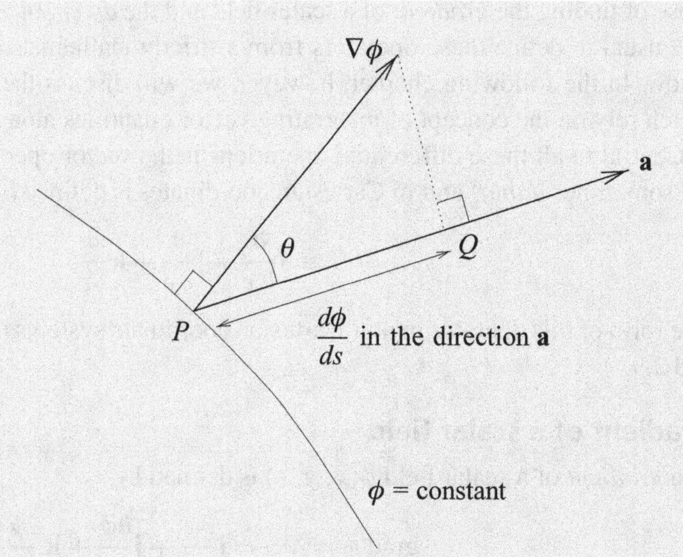

Figure 2.4 Geometrical properties of $\nabla\phi$. PQ gives the value of $d\phi/ds$ in the direction $\mathbf{a}$.

derivative of ϕ with respect to u along the curve is simply

$$\frac{d\phi}{du} = \nabla\phi \cdot \frac{d\mathbf{r}}{du}. \tag{2.23}$$

In the particular case where the parameter u is the arc length s along the curve, the total derivative of ϕ with respect to s along the curve is given by

$$\frac{d\phi}{ds} = \nabla\phi \cdot \hat{\mathbf{t}}, \tag{2.24}$$

where $\hat{\mathbf{t}}$ is the unit tangent to the curve at the given point.

In general, the rate of change of ϕ with respect to the distance s in a particular direction $\mathbf{a}$ is given by

$$\frac{d\phi}{ds} = \nabla\phi \cdot \hat{\mathbf{a}} \tag{2.25}$$

and is called the directional derivative. Since $\hat{\mathbf{a}}$ is a unit vector we have

$$\frac{d\phi}{ds} = |\nabla\phi|\cos\theta,$$

where θ is the angle between $\hat{\mathbf{a}}$ and $\nabla\phi$ as shown in Figure 2.4. Clearly $\nabla\phi$ lies in the direction of the fastest increase in ϕ, and $|\nabla\phi|$ is the largest possible value of $d\phi/ds$. Similarly, the largest rate of decrease of ϕ is $d\phi/ds = -|\nabla\phi|$ in the direction of $-\nabla\phi$.

Example For the function $\phi = x^2y + yz$ at the point $(1, 2, -1)$, find its rate of change with distance in the direction $\mathbf{a} = \mathbf{i} + 2\mathbf{j} + 3\mathbf{k}$. At this same point, what is the greatest possible rate of change with distance and in which direction does it occur?

The gradient of ϕ is given by (2.21):

$$\begin{aligned}\nabla\phi &= 2xy\mathbf{i} + (x^2 + z)\mathbf{j} + y\mathbf{k},\\ &= 4\mathbf{i} + 2\mathbf{k} \qquad \text{at the point } (1, 2, -1).\end{aligned}$$

The unit vector in the direction of $\mathbf{a}$ is $\hat{\mathbf{a}} = \frac{1}{\sqrt{14}}(\mathbf{i} + 2\mathbf{j} + 3\mathbf{k})$, so the rate of change of ϕ with distance s in this direction is, using (2.25),

$$\frac{d\phi}{ds} = \nabla\phi \cdot \hat{\mathbf{a}} = \frac{1}{\sqrt{14}}(4 + 6) = \frac{10}{\sqrt{14}}.$$

From the above discussion, at the point $(1, 2, -1)$ $d\phi/ds$ will be greatest in the direction of $\nabla\phi = 4\mathbf{i} + 2\mathbf{k}$ and has the value $|\nabla\phi| = \sqrt{20}$ in this direction. ◀

We can extend the above analysis to find the rate of change of a vector field (rather than a scalar field as above) in a particular direction. The scalar differential operator $\hat{\mathbf{a}} \cdot \nabla$ can be shown to give the rate of change with distance in the direction $\hat{\mathbf{a}}$ of the quantity (vector or scalar) on which it acts. In Cartesian coordinates it may be written as

$$\hat{\mathbf{a}} \cdot \nabla = a_x\frac{\partial}{\partial x} + a_y\frac{\partial}{\partial y} + a_z\frac{\partial}{\partial z}. \tag{2.26}$$

Thus we can write the infinitesimal change in an electric field in moving from $\mathbf{r}$ to $\mathbf{r} + d\mathbf{r}$ given in (2.15) as $d\mathbf{E} = (d\mathbf{r} \cdot \nabla)\mathbf{E}$.

A second interesting geometrical property of $\nabla\phi$ may be found by considering the surface defined by $\phi(x, y, z) = c$, where c is some constant. If $\hat{\mathbf{t}}$ is a unit tangent to this surface at some point then clearly $d\phi/ds = 0$ in this direction and from (2.24) we have $\nabla\phi \cdot \hat{\mathbf{t}} = 0$. In other words, *$\nabla\phi$ is a vector normal to the surface $\phi(x, y, z) = c$ at every point*, as shown in Figure 2.4.[6]

If $\hat{\mathbf{n}}$ is a unit normal to the surface in the direction of increasing $\phi(x, y, z)$, then the gradient is sometimes written

$$\nabla\phi \equiv \frac{\partial\phi}{\partial n}\hat{\mathbf{n}}, \tag{2.27}$$

where $\partial\phi/\partial n \equiv |\nabla\phi|$ is the rate of change of ϕ in the direction $\hat{\mathbf{n}}$ and is called the *normal derivative*.

6 If $\phi(x, y, z) = Ar^{-n}$, with $A > 0$ and $r^2 = x^2 + y^2 + z^2$, identify the surfaces of constant ϕ and hence the direction of $\nabla\phi$. Confirm your conclusion by explicit calculation, working in Cartesian coordinates and using the chain rule to evaluate the required derivatives.

Example Find expressions for the equations of the tangent plane and the line normal to the surface $\phi(x, y, z) = c$ at the point P with coordinates x_0, y_0, z_0. Use the results to find the equations of the tangent plane and the line normal to the surface of the sphere $\phi = x^2 + y^2 + z^2 = a^2$ at the point $(0, 0, a)$.

A vector normal to the surface $\phi(x, y, z) = c$ at the point P is simply $\nabla\phi$ evaluated at that point; we denote it by $\mathbf{n}_0$. If $\mathbf{r}_0$ is the position vector of the point P relative to the origin, and $\mathbf{r}$ is the position vector of any point on the tangent plane, then the vector equation of the tangent plane is

$$(\mathbf{r} - \mathbf{r}_0) \cdot \mathbf{n}_0 = 0.$$

Similarly, if $\mathbf{r}$ is the position vector of any point on the straight line passing through P (with position vector $\mathbf{r}_0$) in the direction of the normal $\mathbf{n}_0$ then the vector equation of this line is

$$(\mathbf{r} - \mathbf{r}_0) \times \mathbf{n}_0 = \mathbf{0}.$$

For the surface of the sphere $\phi = x^2 + y^2 + z^2 = a^2$,

$$\begin{aligned} \nabla\phi &= 2x\mathbf{i} + 2y\mathbf{j} + 2z\mathbf{k} \\ &= 2a\mathbf{k} \qquad \text{at the point } (0, 0, a). \end{aligned}$$

Therefore the equation of the tangent plane to the sphere at this point is

$$(\mathbf{r} - \mathbf{r}_0) \cdot 2a\mathbf{k} = 0.$$

This gives $2a(z - a) = 0$ or $z = a$, as expected. The equation of the line normal to the sphere at the point $(0, 0, a)$ is

$$(\mathbf{r} - \mathbf{r}_0) \times 2a\mathbf{k} = \mathbf{0},$$

which gives $2ay\mathbf{i} - 2ax\mathbf{j} = \mathbf{0}$ or $x = y = 0$, i.e. the z-axis, as expected. Figure 2.5 shows the tangent plane and normal to the surface of the sphere at this point. ◀

Further properties of the gradient operation, which are analogous to those of the ordinary derivative, are listed in Subsection 2.7.1 and may be easily proved. In addition to these, we note that the gradient operation also obeys the chain rule as in ordinary differential calculus, i.e. if ϕ and ψ are scalar fields in some region R then[7]

$$\nabla\left[\phi(\psi)\right] = \frac{\partial\phi}{\partial\psi}\nabla\psi.$$

2.6.2 Divergence of a vector field

The *divergence* of a vector field $\mathbf{a}(x, y, z)$ is defined by

$$\operatorname{div}\mathbf{a} = \nabla \cdot \mathbf{a} = \frac{\partial a_x}{\partial x} + \frac{\partial a_y}{\partial y} + \frac{\partial a_z}{\partial z}, \qquad (2.28)$$

where a_x, a_y and a_z are the x-, y- and z-components of $\mathbf{a}$. Clearly, $\nabla \cdot \mathbf{a}$ is a scalar field. Any vector field $\mathbf{a}$ for which $\nabla \cdot \mathbf{a} = 0$ is said to be *solenoidal*; this property may apply to the whole field, or only to particular regions or points of it.

7 Evaluate both sides of this equation in the particular case that $\psi(x, y, z) = z(x^2 + y^2)$ and $\phi(\psi) = \psi^2$ and verify that they are the same function of x, y and z.

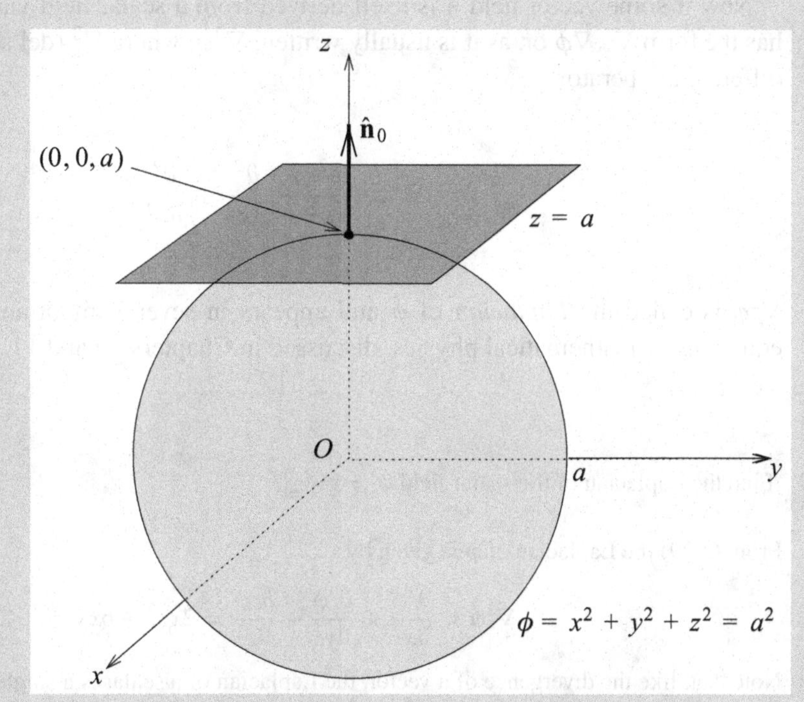

Figure 2.5 The tangent plane and the normal to the surface of the sphere $\phi = x^2 + y^2 + z^2 = a^2$ at the point $\mathbf{r}_0$ with coordinates (0, 0, a).

Example Find the divergence of the vector field $\mathbf{a} = x^2y^2\mathbf{i} + y^2z^2\mathbf{j} + x^2z^2\mathbf{k}$.

From (2.28) the divergence of $\mathbf{a}$ is given by

$$\nabla \cdot \mathbf{a} = 2xy^2 + 2yz^2 + 2x^2z = 2(xy^2 + yz^2 + x^2z).$$

Although this expression contains three terms, they are simply added together and the expression is a scalar, not a vector. ◀

The geometric definition of divergence and its physical meaning will be discussed in the next chapter. For the moment, we merely note that the divergence can be considered as a quantitative measure of how much a vector field diverges (spreads out) or converges at any given point. For example, if we consider the vector field $\mathbf{v}(x, y, z)$ describing the local velocity at any point in a fluid then $\nabla \cdot \mathbf{v}$ is equal to the net rate of outflow of fluid per unit volume, evaluated at a point (by letting a small volume at that point tend to zero).

Now if some vector field $\mathbf{a}$ is itself derived from a scalar field via $\mathbf{a} = \nabla\phi$ then $\nabla \cdot \mathbf{a}$ has the form $\nabla \cdot \nabla\phi$ or, as it is usually written, $\nabla^2\phi$, where ∇^2 (del squared) is the scalar differential operator

$$\nabla^2 \equiv \frac{\partial^2}{\partial x^2} + \frac{\partial^2}{\partial y^2} + \frac{\partial^2}{\partial z^2}. \tag{2.29}$$

$\nabla^2\phi$ is called the *Laplacian* of ϕ and appears in several important partial differential equations of mathematical physics, discussed in Chapters 10 and 11.

Example Find the Laplacian of the scalar field $\phi = xy^2z^3$.

From (2.29) the Laplacian of ϕ is given by

$$\nabla^2\phi = \frac{\partial^2\phi}{\partial x^2} + \frac{\partial^2\phi}{\partial y^2} + \frac{\partial^2\phi}{\partial z^2} = 2xz^3 + 6xy^2z.$$

Note that, like the divergence of a vector, the Laplacian of a scalar is a single quantity (i.e. another scalar), even though the general expression for it contains more than one term. ◀

2.6.3 Curl of a vector field

The *curl* of a vector field $\mathbf{a}(x, y, z)$ is defined by

$$\text{curl}\,\mathbf{a} = \nabla \times \mathbf{a} = \left(\frac{\partial a_z}{\partial y} - \frac{\partial a_y}{\partial z}\right)\mathbf{i} + \left(\frac{\partial a_x}{\partial z} - \frac{\partial a_z}{\partial x}\right)\mathbf{j} + \left(\frac{\partial a_y}{\partial x} - \frac{\partial a_x}{\partial y}\right)\mathbf{k},$$

where a_x, a_y and a_z are the x-, y- and z-components of $\mathbf{a}$. The RHS can be written in a more memorable form as a determinant:

$$\nabla \times \mathbf{a} = \begin{vmatrix} \mathbf{i} & \mathbf{j} & \mathbf{k} \\ \dfrac{\partial}{\partial x} & \dfrac{\partial}{\partial y} & \dfrac{\partial}{\partial z} \\ a_x & a_y & a_z \end{vmatrix}, \tag{2.30}$$

where it is understood that, on expanding the determinant, the partial derivatives in the second row act on the components of $\mathbf{a}$ in the third row. Clearly, $\nabla \times \mathbf{a}$ is itself a vector field. Any vector field $\mathbf{a}$ for which $\nabla \times \mathbf{a} = \mathbf{0}$ is said to be *irrotational*; this property, like that of being solenoidal, may apply to the whole field, or only to particular regions or points of it.

Example Find the curl of the vector field $\mathbf{a} = x^2y^2z^2\mathbf{i} + y^2z^2\mathbf{j} + x^2z^2\mathbf{k}$.

The curl of $\mathbf{a}$ is given by

$$\begin{aligned}
\nabla \times \mathbf{a} &= \begin{vmatrix} \mathbf{i} & \mathbf{j} & \mathbf{k} \\ \dfrac{\partial}{\partial x} & \dfrac{\partial}{\partial y} & \dfrac{\partial}{\partial z} \\ x^2y^2z^2 & y^2z^2 & x^2z^2 \end{vmatrix} \\
&= \left[\frac{\partial}{\partial y}(x^2z^2z^2) - \frac{\partial}{\partial z}(y^2z^2)\right]\mathbf{i} + \left[\frac{\partial}{\partial z}(x^2y^2z^2) - \frac{\partial}{\partial x}(x^2z^2)\right]\mathbf{j} \\
&\quad + \left[\frac{\partial}{\partial x}(y^2z^2) - \frac{\partial}{\partial y}(x^2y^2z^2)\right]\mathbf{k}, \\
&= -2\left[y^2z\mathbf{i} + (xz^2 - x^2y^2z)\mathbf{j} + x^2yz^2\mathbf{k}\right].
\end{aligned}$$

As with any general vector, each of the components of the curl of a vector can depend on some or all of the coordinates.[8] ◀

For a vector field $\mathbf{v}(x, y, z)$ describing the local velocity at any point in a fluid, $\nabla \times \mathbf{v}$ is a measure of the angular velocity of the fluid in the neighborhood of that point. If a small paddle wheel were placed at various points in the fluid then it would tend to rotate in regions where $\nabla \times \mathbf{v} \neq \mathbf{0}$, while it would not rotate in regions where $\nabla \times \mathbf{v} = \mathbf{0}$.

Another insight into the physical interpretation of the curl operator is gained by considering the vector field $\mathbf{v}$ describing the velocity at any point in a rigid body rotating about some axis with angular velocity ω. If $\mathbf{r}$ is the position vector of the point with respect to some origin on the axis of rotation then the velocity of the point is given by $\mathbf{v} = \boldsymbol{\omega} \times \mathbf{r}$. Without any loss of generality, we may take $\boldsymbol{\omega}$ to lie along the z-axis of our coordinate system, so that $\boldsymbol{\omega} = \omega\mathbf{k}$. The velocity field is then $\mathbf{v} = -\omega y\,\mathbf{i} + \omega x\,\mathbf{j}$. The curl of this vector field is easily found to be

$$\nabla \times \mathbf{v} = \begin{vmatrix} \mathbf{i} & \mathbf{j} & \mathbf{k} \\ \dfrac{\partial}{\partial x} & \dfrac{\partial}{\partial y} & \dfrac{\partial}{\partial z} \\ -\omega y & \omega x & 0 \end{vmatrix} = 2\omega\mathbf{k} = 2\boldsymbol{\omega}. \tag{2.31}$$

Therefore the curl of the velocity field is a vector equal to twice the angular velocity vector of the rigid body about its axis of rotation. We give a full geometrical discussion of the curl of a vector in the next chapter.

2.7 Vector operator formulae

In the same way as for ordinary vectors, certain identities involving vector operators exist. In addition to these, there are various relations involving the action of vector operators on sums and products of scalar and vector fields. Some of these relations have been mentioned

8 For the field considered here, where is the field irrotational?

Table 2.1 *Vector operators acting on sums and products. The operator ∇ is defined in (2.20); ϕ and ψ are scalar fields, **a** and **b** are vector fields*

$$\begin{aligned}
\nabla(\phi+\psi) &= \nabla\phi + \nabla\psi \\
\nabla\cdot(\mathbf{a}+\mathbf{b}) &= \nabla\cdot\mathbf{a} + \nabla\cdot\mathbf{b} \\
\nabla\times(\mathbf{a}+\mathbf{b}) &= \nabla\times\mathbf{a} + \nabla\times\mathbf{b} \\
\nabla(\phi\psi) &= \phi\nabla\psi + \psi\nabla\phi \\
\nabla(\mathbf{a}\cdot\mathbf{b}) &= \mathbf{a}\times(\nabla\times\mathbf{b}) + \mathbf{b}\times(\nabla\times\mathbf{a}) + (\mathbf{a}\cdot\nabla)\mathbf{b} + (\mathbf{b}\cdot\nabla)\mathbf{a} \\
\nabla\cdot(\phi\mathbf{a}) &= \phi\nabla\cdot\mathbf{a} + \mathbf{a}\cdot\nabla\phi \\
\nabla\cdot(\mathbf{a}\times\mathbf{b}) &= \mathbf{b}\cdot(\nabla\times\mathbf{a}) - \mathbf{a}\cdot(\nabla\times\mathbf{b}) \\
\nabla\times(\phi\mathbf{a}) &= \nabla\phi\times\mathbf{a} + \phi\nabla\times\mathbf{a} \\
\nabla\times(\mathbf{a}\times\mathbf{b}) &= \mathbf{a}(\nabla\cdot\mathbf{b}) - \mathbf{b}(\nabla\cdot\mathbf{a}) + (\mathbf{b}\cdot\nabla)\mathbf{a} - (\mathbf{a}\cdot\nabla)\mathbf{b}
\end{aligned}$$

earlier, but we list all the most important ones here for convenience. The validity of these relations may be easily verified by direct calculation; in most cases, the quickest and most compact way of doing this is to use the notation and results discussed in Appendix E.

Although some of the following vector relations are expressed in Cartesian coordinates, it may be proved that they are all independent of the choice of coordinate system. This is to be expected since grad, div and curl all have clear geometrical definitions, which are discussed more fully in the next chapter and which do not rely on any particular choice of coordinate system.

2.7.1 Vector operators acting on sums and products

Let ϕ and ψ be scalar fields and **a** and **b** be vector fields. Assuming these fields are differentiable, the action of grad, div and curl on various sums and products of them is presented in Table 2.1.

These relations can be proved by direct calculation. For example, the penultimate entry is proved as follows.

Example Show that

$$\nabla\times(\phi\mathbf{a}) = \nabla\phi\times\mathbf{a} + \phi\nabla\times\mathbf{a}.$$

The x-component of the LHS is

$$\begin{aligned}
\frac{\partial}{\partial y}(\phi a_z) - \frac{\partial}{\partial z}(\phi a_y) &= \phi\frac{\partial a_z}{\partial y} + \frac{\partial\phi}{\partial y}a_z - \phi\frac{\partial a_y}{\partial z} - \frac{\partial\phi}{\partial z}a_y, \\
&= \phi\left(\frac{\partial a_z}{\partial y} - \frac{\partial a_y}{\partial z}\right) + \left(\frac{\partial\phi}{\partial y}a_z - \frac{\partial\phi}{\partial z}a_y\right), \\
&= \phi(\nabla\times\mathbf{a})_x + (\nabla\phi\times\mathbf{a})_x,
\end{aligned}$$

where, for example, $(\nabla\phi\times\mathbf{a})_x$ denotes the x-component of the vector $\nabla\phi\times\mathbf{a}$. Incorporating the y- and z-components, which can be similarly found, we obtain the stated result. ◀

An alternative proof using the methods of Appendix E and the summation convention is

$$[\nabla \times (\phi \mathbf{a})]_i = \epsilon_{ijk}\frac{\partial(\phi a_k)}{\partial x_j} = \epsilon_{ijk}\frac{\partial \phi}{\partial x_j} a_k + \epsilon_{ijk}\phi\frac{\partial a_k}{\partial x_j} = [\nabla\phi \times \mathbf{a}]_i + [\phi(\nabla \times \mathbf{a})]_i.$$

Some useful special cases of the relations in Table 2.1 are worth noting. If $\mathbf{r}$ is the position vector relative to some origin and $r = |\mathbf{r}|$, then

$$\nabla\phi(r) = \frac{d\phi}{dr}\hat{\mathbf{r}},$$

$$\nabla \cdot [\phi(r)\mathbf{r}] = 3\phi(r) + r\frac{d\phi(r)}{dr},$$

$$\nabla^2\phi(r) = \frac{d^2\phi(r)}{dr^2} + \frac{2}{r}\frac{d\phi(r)}{dr},$$

$$\nabla \times [\phi(r)\mathbf{r}] = \mathbf{0}.$$

These results may be proved straightforwardly using Cartesian coordinates[9] but far more simply using spherical polar coordinates, which are discussed in Subsection 2.8.2. Particular cases of these results are

$$\nabla r = \hat{\mathbf{r}}, \qquad \nabla \cdot \mathbf{r} = 3, \qquad \nabla \times \mathbf{r} = \mathbf{0},$$

together with

$$\nabla\left(\frac{1}{r}\right) = -\frac{\hat{\mathbf{r}}}{r^2},$$

$$\nabla \cdot \left(\frac{\hat{\mathbf{r}}}{r^2}\right) = -\nabla^2\left(\frac{1}{r}\right) = 4\pi\delta(\mathbf{r}),$$

where $\delta(\mathbf{r})$ is the Dirac delta function, discussed in Chapter 5. The last equation is important in the solution of certain partial differential equations and is discussed further in Chapter 10.

2.7.2 Combinations of grad, div and curl

We now consider the action of two vector operators in succession on a scalar or vector field. We can immediately discard four of the nine obvious combinations of grad, div and curl, since they clearly do not make sense. If ϕ is a scalar field and $\mathbf{a}$ is a vector field, these four combinations are grad(grad ϕ), div(div $\mathbf{a}$), curl(div $\mathbf{a}$) and grad(curl $\mathbf{a}$). In each case the second (outer) vector operator is acting on the wrong type of field, i.e. scalar instead of vector or vice versa. In grad(grad ϕ), for example, grad acts on grad ϕ, which is a vector field, but we know that grad only acts on scalar fields (although in fact it is possible to form the *outer product* of the del operator with a vector to give what is known as a tensor, but that need not concern us here).

9 Prove the second result using Cartesian coordinates. Use the chain rule and recall that $\partial r/\partial x = x/r$, etc.

Of the five valid combinations of grad, div and curl, two are identically zero, namely[10]

$$\text{curl grad}\,\phi = \nabla \times \nabla\phi = \mathbf{0}, \tag{2.32}$$

$$\text{div curl}\,\mathbf{a} = \nabla \cdot (\nabla \times \mathbf{a}) = 0. \tag{2.33}$$

From (2.32), we see that if $\mathbf{a}$ is derived from the gradient of some scalar function such that $\mathbf{a} = \nabla\phi$ then it is necessarily irrotational ($\nabla \times \mathbf{a} = 0$). We also note that if $\mathbf{a}$ is an irrotational vector field then another irrotational vector field is $\mathbf{a} + \nabla\phi + \mathbf{c}$, where ϕ is any scalar field and $\mathbf{c}$ is any constant vector. This follows since

$$\nabla \times (\mathbf{a} + \nabla\phi + \mathbf{c}) = \nabla \times \mathbf{a} + \nabla \times \nabla\phi = \mathbf{0}.$$

Similarly, from (2.33) we may infer that if $\mathbf{b}$ is the curl of some vector field $\mathbf{a}$ such that $\mathbf{b} = \nabla \times \mathbf{a}$ then $\mathbf{b}$ is solenoidal ($\nabla \cdot \mathbf{b} = 0$). Obviously, if $\mathbf{b}$ is solenoidal and $\mathbf{c}$ is any constant vector then $\mathbf{b} + \mathbf{c}$ is also solenoidal.

The three remaining combinations of grad, div and curl are

$$\text{div grad}\,\phi = \nabla \cdot \nabla\phi = \nabla^2\phi = \frac{\partial^2\phi}{\partial x^2} + \frac{\partial^2\phi}{\partial y^2} + \frac{\partial^2\phi}{\partial z^2}, \tag{2.34}$$

$$\begin{aligned} \text{grad div}\,\mathbf{a} &= \nabla(\nabla \cdot \mathbf{a}), \\ &= \left(\frac{\partial^2 a_x}{\partial x^2} + \frac{\partial^2 a_y}{\partial x \partial y} + \frac{\partial^2 a_z}{\partial x \partial z}\right)\mathbf{i} + \left(\frac{\partial^2 a_x}{\partial y \partial x} + \frac{\partial^2 a_y}{\partial y^2} + \frac{\partial^2 a_z}{\partial y \partial z}\right)\mathbf{j} \\ &\quad + \left(\frac{\partial^2 a_x}{\partial z \partial x} + \frac{\partial^2 a_y}{\partial z \partial y} + \frac{\partial^2 a_z}{\partial z^2}\right)\mathbf{k}, \end{aligned} \tag{2.35}$$

$$\text{curl curl}\,\mathbf{a} = \nabla \times (\nabla \times \mathbf{a}) = \nabla(\nabla \cdot \mathbf{a}) - \nabla^2\mathbf{a}, \tag{2.36}$$

where (2.34) and (2.35) are expressed in Cartesian coordinates. In (2.36), the term $\nabla^2\mathbf{a}$ has the linear differential operator ∇^2 acting on a vector [as opposed to a scalar as in (2.34)], which of course consists of a sum of unit vectors multiplied by components. Two cases arise.

(i) If the unit vectors are constants (i.e. they are independent of the values of the coordinates) then the differential operator gives a non-zero contribution only when acting upon the components, the unit vectors being merely multipliers.
(ii) If the unit vectors vary as the values of the coordinates change (i.e. are not constant in direction throughout the whole space) then the derivatives of these vectors appear as contributions to $\nabla^2\mathbf{a}$.

Cartesian coordinates are an example of the first case in which each component satisfies $(\nabla^2\mathbf{a})_i = \nabla^2 a_i$. In this case (2.36) can be applied to each component separately:

$$[\nabla \times (\nabla \times \mathbf{a})]_i = [\nabla(\nabla \cdot \mathbf{a})]_i - \nabla^2 a_i. \tag{2.37}$$

10 Prove these results by using the summation convention, showing that the two LHSs take the forms

$$\epsilon_{ijk}\frac{\partial^2\phi}{\partial x_j \partial x_k} \quad \text{and} \quad \epsilon_{ijk}\frac{\partial^2 a_k}{\partial x_i \partial x_j},$$

and then considering the effects of interchanging j and k in the first case, and i and j in the second.

However, cylindrical and spherical polar coordinates come in the second class. For them (2.36) is still true, but the further step to (2.37) cannot be made.

More complicated vector operator relations may be proved using combinations of the relations given above. The following example shows that the vector product of two vectors each of which has been derived as the gradient of a scalar is *always* solenoidal.

Example Show that

$$\nabla \cdot (\nabla\phi \times \nabla\psi) = 0,$$

where ϕ and ψ are scalar fields.

From the previous section we have

$$\nabla \cdot (\mathbf{a} \times \mathbf{b}) = \mathbf{b} \cdot (\nabla \times \mathbf{a}) - \mathbf{a} \cdot (\nabla \times \mathbf{b}).$$

If we let $\mathbf{a} = \nabla\phi$ and $\mathbf{b} = \nabla\psi$ then we obtain

$$\nabla \cdot (\nabla\phi \times \nabla\psi) = \nabla\psi \cdot (\nabla \times \nabla\phi) - \nabla\phi \cdot (\nabla \times \nabla\psi) = 0, \tag{2.38}$$

since $\nabla \times \nabla\phi = 0 = \nabla \times \nabla\psi$, from (2.32). ◀

2.8 Cylindrical and spherical polar coordinates

The operators we have discussed in this chapter, i.e. grad, div, curl and ∇^2, have all been defined in terms of Cartesian coordinates, but for many physical situations other coordinate systems are more natural. For example, many systems, such as an isolated charge in space, have spherical symmetry and spherical polar coordinates would be the obvious choice. For axisymmetric systems, such as fluid flow in a pipe, cylindrical polar coordinates are the natural choice. The physical laws governing the behavior of the systems are often expressed in terms of the vector operators we have been discussing, and so it is necessary to be able to express these operators in these other, non-Cartesian, coordinates. We first consider the two most common non-Cartesian coordinate systems, i.e. cylindrical and spherical polars, and then go on to discuss general curvilinear coordinates in the next section.

2.8.1 Cylindrical polar coordinates

As shown in Figure 2.6, the position of a point in space P having Cartesian coordinates x, y, z may be expressed in terms of cylindrical polar coordinates ρ, ϕ, z, where

$$x = \rho\cos\phi, \quad y = \rho\sin\phi, \quad z = z, \tag{2.39}$$

and $\rho \geq 0$, $0 \leq \phi < 2\pi$ and $-\infty < z < \infty$. The position vector of P may therefore be written

$$\mathbf{r} = \rho\cos\phi\,\mathbf{i} + \rho\sin\phi\,\mathbf{j} + z\,\mathbf{k}. \tag{2.40}$$

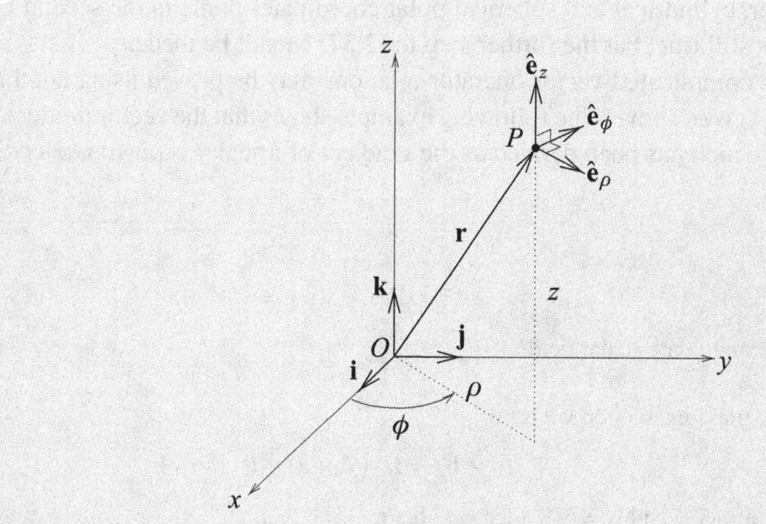

Figure 2.6 Cylindrical polar coordinates ρ, ϕ, z.

If we take the partial derivatives of $\mathbf{r}$ with respect to ρ, ϕ and z respectively then we obtain the three vectors

$$\mathbf{e}_\rho = \frac{\partial \mathbf{r}}{\partial \rho} = \cos\phi\,\mathbf{i} + \sin\phi\,\mathbf{j}, \tag{2.41}$$

$$\mathbf{e}_\phi = \frac{\partial \mathbf{r}}{\partial \phi} = -\rho\sin\phi\,\mathbf{i} + \rho\cos\phi\,\mathbf{j}, \tag{2.42}$$

$$\mathbf{e}_z = \frac{\partial \mathbf{r}}{\partial z} = \mathbf{k}. \tag{2.43}$$

These vectors lie in the directions of increasing ρ, ϕ and z respectively but are not all of unit length.[11] Although $\mathbf{e}_\rho$, $\mathbf{e}_\phi$ and $\mathbf{e}_z$ form a useful set of basis vectors in their own right (we will see in Section 2.9 that such a basis is sometimes the *most* useful), it is usual to work with the corresponding *unit* vectors, which are obtained by dividing each vector by its modulus to give

$$\hat{\mathbf{e}}_\rho = \mathbf{e}_\rho = \cos\phi\,\mathbf{i} + \sin\phi\,\mathbf{j}, \tag{2.44}$$

$$\hat{\mathbf{e}}_\phi = \frac{1}{\rho}\mathbf{e}_\phi = -\sin\phi\,\mathbf{i} + \cos\phi\,\mathbf{j}, \tag{2.45}$$

$$\hat{\mathbf{e}}_z = \mathbf{e}_z = \mathbf{k}. \tag{2.46}$$

These three unit vectors, like the Cartesian unit vectors **i**, **j** and **k**, form an orthonormal triad[12] at each point in space, i.e. the basis vectors are mutually orthogonal and of unit length (see Figure 2.6). Unlike the fixed vectors **i**, **j** and **k**, however, $\hat{\mathbf{e}}_\rho$ and $\hat{\mathbf{e}}_\phi$ change direction as P moves.

11 $\mathbf{e}_\rho$ and $\mathbf{e}_z$ *are* of unit length, but $\mathbf{e}_\phi$ has length ρ, which, moreover, varies with the position of P.
12 Taken in the order given, ρ, ϕ, z, they form a right-handed set, as the reader should verify.

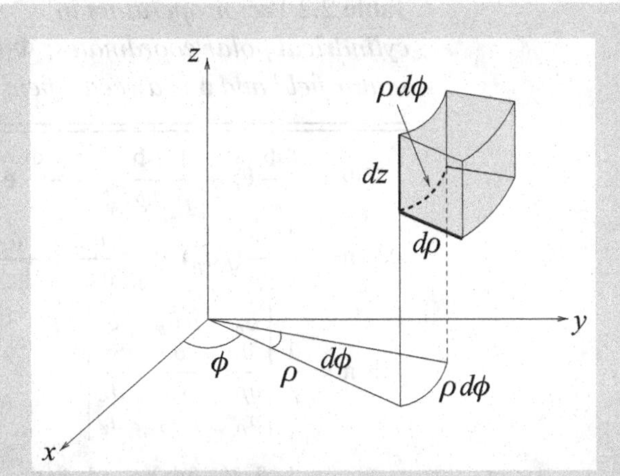

Figure 2.7 The element of volume in cylindrical polar coordinates is given by $\rho\, d\rho\, d\phi\, dz$.

The expression for a general infinitesimal vector displacement $d\mathbf{r}$ in the position of P is given, from (2.14), by

$$\begin{aligned} d\mathbf{r} &= \frac{\partial \mathbf{r}}{\partial \rho}\, d\rho + \frac{\partial \mathbf{r}}{\partial \phi}\, d\phi + \frac{\partial \mathbf{r}}{\partial z}\, dz \\ &= d\rho\, \mathbf{e}_\rho + d\phi\, \mathbf{e}_\phi + dz\, \mathbf{e}_z \\ &= d\rho\ \hat{\mathbf{e}}_\rho + \rho\, d\phi\ \hat{\mathbf{e}}_\phi + dz\ \hat{\mathbf{e}}_z. \end{aligned} \tag{2.47}$$

This expression illustrates an important difference between Cartesian and cylindrical polar coordinates (or non-Cartesian coordinates in general). In Cartesian coordinates, the distance moved in going from x to $x + dx$, with y and z held constant, is simply $ds = dx$. However, in cylindrical polars, if ϕ changes by $d\phi$, with ρ and z held constant, then the distance moved is *not* $d\phi$, but $ds = \rho\, d\phi$. Factors, such as the ρ in $\rho\, d\phi$, that multiply the coordinate differentials to give distances are known as *scale factors*. From (2.47), the scale factors for the ρ-, ϕ- and z-coordinates are therefore 1, ρ and 1 respectively.

The magnitude ds of the displacement $d\mathbf{r}$ is given in cylindrical polar coordinates by

$$(ds)^2 = d\mathbf{r} \cdot d\mathbf{r} = (d\rho)^2 + \rho^2 (d\phi)^2 + (dz)^2,$$

where in the second equality we have used the fact that the basis vectors are orthonormal. We can also find the volume element in a cylindrical polar system (see Figure 2.7) by calculating the volume of the infinitesimal parallelepiped defined by the vectors $d\rho\ \hat{\mathbf{e}}_\rho$, $\rho\, d\phi\ \hat{\mathbf{e}}_\phi$ and $dz\ \hat{\mathbf{e}}_z$. As stated in point 4 of Section A.9, this is given by the scalar triple product of the three vectors:

$$dV = \left| d\rho\ \hat{\mathbf{e}}_\rho \cdot (\rho\, d\phi\ \hat{\mathbf{e}}_\phi \times dz\ \hat{\mathbf{e}}_z) \right| = \rho\, d\rho\, d\phi\, dz,$$

which again uses the fact that the basis vectors are orthonormal. For a simple coordinate system such as cylindrical polars the expressions for $(ds)^2$ and dV are obvious from the geometry.

Table 2.2 *Vector operators in cylindrical polar coordinates; Φ is a scalar field and* **a** *is a vector field*

$$\nabla\Phi = \frac{\partial\Phi}{\partial\rho}\,\hat{\mathbf{e}}_\rho + \frac{1}{\rho}\frac{\partial\Phi}{\partial\phi}\,\hat{\mathbf{e}}_\phi + \frac{\partial\Phi}{\partial z}\,\hat{\mathbf{e}}_z$$

$$\nabla\cdot\mathbf{a} = \frac{1}{\rho}\frac{\partial}{\partial\rho}(\rho a_\rho) + \frac{1}{\rho}\frac{\partial a_\phi}{\partial\phi} + \frac{\partial a_z}{\partial z}$$

$$\nabla\times\mathbf{a} = \frac{1}{\rho}\begin{vmatrix} \hat{\mathbf{e}}_\rho & \rho\,\hat{\mathbf{e}}_\phi & \hat{\mathbf{e}}_z \\ \dfrac{\partial}{\partial\rho} & \dfrac{\partial}{\partial\phi} & \dfrac{\partial}{\partial z} \\ a_\rho & \rho a_\phi & a_z \end{vmatrix}$$

$$\nabla^2\Phi = \frac{1}{\rho}\frac{\partial}{\partial\rho}\left(\rho\frac{\partial\Phi}{\partial\rho}\right) + \frac{1}{\rho^2}\frac{\partial^2\Phi}{\partial\phi^2} + \frac{\partial^2\Phi}{\partial z^2}$$

We will now express the vector operators discussed in this chapter in terms of cylindrical polar coordinates. Let us consider a vector field $\mathbf{a}(\rho, \phi, z)$ and a scalar field $\Phi(\rho, \phi, z)$, where we use Φ for the scalar field to avoid confusion with the azimuthal angle ϕ. We must first write the vector field in terms of the basis vectors of the cylindrical polar coordinate system, i.e.

$$\mathbf{a} = a_\rho\,\hat{\mathbf{e}}_\rho + a_\phi\,\hat{\mathbf{e}}_\phi + a_z\,\hat{\mathbf{e}}_z,$$

where a_ρ, a_ϕ and a_z are the components of **a** in the ρ-, ϕ- and z-directions respectively. The expressions for grad, div, curl and ∇^2 can then be calculated and are given in Table 2.2. Since the derivations of these expressions are rather complicated we leave them until our discussion of general curvilinear coordinates in the next section; the reader could well postpone examination of these formal proofs until some experience of using the expressions has been gained.

Example Express the vector field $\mathbf{a} = yz\,\mathbf{i} - y\,\mathbf{j} + xz^2\,\mathbf{k}$ in cylindrical polar coordinates, and hence calculate its divergence. Show that the same result is obtained by evaluating the divergence in Cartesian coordinates.

The basis vectors of the cylindrical polar coordinate system are given in (2.44)–(2.46). Solving these equations simultaneously for **i**, **j** and **k** we obtain

$$\begin{aligned}\mathbf{i} &= \cos\phi\,\hat{\mathbf{e}}_\rho - \sin\phi\,\hat{\mathbf{e}}_\phi,\\ \mathbf{j} &= \sin\phi\,\hat{\mathbf{e}}_\rho + \cos\phi\,\hat{\mathbf{e}}_\phi,\\ \mathbf{k} &= \hat{\mathbf{e}}_z.\end{aligned}$$

Substituting these relations and (2.39) into the expression for **a** we find

$$\begin{aligned}\mathbf{a} &= z\rho\sin\phi\,(\cos\phi\,\hat{\mathbf{e}}_\rho - \sin\phi\,\hat{\mathbf{e}}_\phi) - \rho\sin\phi\,(\sin\phi\,\hat{\mathbf{e}}_\rho + \cos\phi\,\hat{\mathbf{e}}_\phi) + z^2\rho\cos\phi\,\hat{\mathbf{e}}_z\\ &= (z\rho\sin\phi\cos\phi - \rho\sin^2\phi)\,\hat{\mathbf{e}}_\rho - (z\rho\sin^2\phi + \rho\sin\phi\cos\phi)\,\hat{\mathbf{e}}_\phi + z^2\rho\cos\phi\,\hat{\mathbf{e}}_z.\end{aligned}$$

From this expression for **a**, the individual components a_ρ, a_ϕ and a_z can be read off and substituted into the formula for $\nabla \cdot \mathbf{a}$ given in Table 2.2. When the partial differentiations indicated there are carried out,[13] the result is

$$\begin{aligned}\nabla \cdot \mathbf{a} &= 2z \sin\phi \cos\phi - 2\sin^2\phi - 2z\sin\phi\cos\phi - \cos^2\phi + \sin^2\phi + 2z\rho\cos\phi \\ &= 2z\rho\cos\phi - 1.\end{aligned}$$

Alternatively, and much more quickly in this case, we can calculate the divergence directly in Cartesian coordinates. We obtain

$$\nabla \cdot \mathbf{a} = \frac{\partial a_x}{\partial x} + \frac{\partial a_y}{\partial y} + \frac{\partial a_z}{\partial z} = 0 + (-1) + 2xz = 2zx - 1,$$

which on substituting $x = \rho\cos\phi$ yields the same result as the calculation in cylindrical polars. ◀

Finally, we note that similar results can be obtained for (two-dimensional) polar coordinates in a plane by omitting the z-dependence. For example, $(ds)^2 = (d\rho)^2 + \rho^2(d\phi)^2$, while the element of volume is replaced by the element of area $dA = \rho\, d\rho\, d\phi$.

2.8.2 Spherical polar coordinates

As shown in Figure 2.8, the position of a point in space P, with Cartesian coordinates x, y, z, may be expressed in terms of spherical polar coordinates r, θ, ϕ, where

$$x = r\sin\theta\cos\phi, \qquad y = r\sin\theta\sin\phi, \qquad z = r\cos\theta, \tag{2.48}$$

and $r \geq 0$, $0 \leq \theta \leq \pi$ and $0 \leq \phi < 2\pi$. The position vector of P may therefore be written as

$$\mathbf{r} = r\sin\theta\cos\phi\,\mathbf{i} + r\sin\theta\sin\phi\,\mathbf{j} + r\cos\theta\,\mathbf{k}.$$

If, in a similar manner to that used in the previous section for cylindrical polars, we find the partial derivatives of **r** with respect to r, θ and ϕ respectively and divide each of the resulting vectors by its modulus then we obtain the unit basis vectors

$$\begin{aligned}\hat{\mathbf{e}}_r &= \sin\theta\cos\phi\,\mathbf{i} + \sin\theta\sin\phi\,\mathbf{j} + \cos\theta\,\mathbf{k}, \\ \hat{\mathbf{e}}_\theta &= \cos\theta\cos\phi\,\mathbf{i} + \cos\theta\sin\phi\,\mathbf{j} - \sin\theta\,\mathbf{k}, \\ \hat{\mathbf{e}}_\phi &= -\sin\phi\,\mathbf{i} + \cos\phi\,\mathbf{j}.\end{aligned}$$

These unit vectors are in the directions of increasing r, θ and ϕ respectively and are the orthonormal basis set for spherical polar coordinates, as shown in Figure 2.8.

A general infinitesimal vector displacement in spherical polars is, from (2.14),

$$d\mathbf{r} = dr\,\hat{\mathbf{e}}_r + r\,d\theta\,\hat{\mathbf{e}}_\theta + r\sin\theta\,d\phi\,\hat{\mathbf{e}}_\phi; \tag{2.49}$$

thus the scale factors for the r-, θ- and ϕ-coordinates are 1, r and $r\sin\theta$ respectively. The magnitude ds of the displacement $d\mathbf{r}$ is given by

$$(ds)^2 = d\mathbf{r}\cdot d\mathbf{r} = (dr)^2 + r^2(d\theta)^2 + r^2\sin^2\theta(d\phi)^2,$$

13 Doing this for yourself gives useful practice.

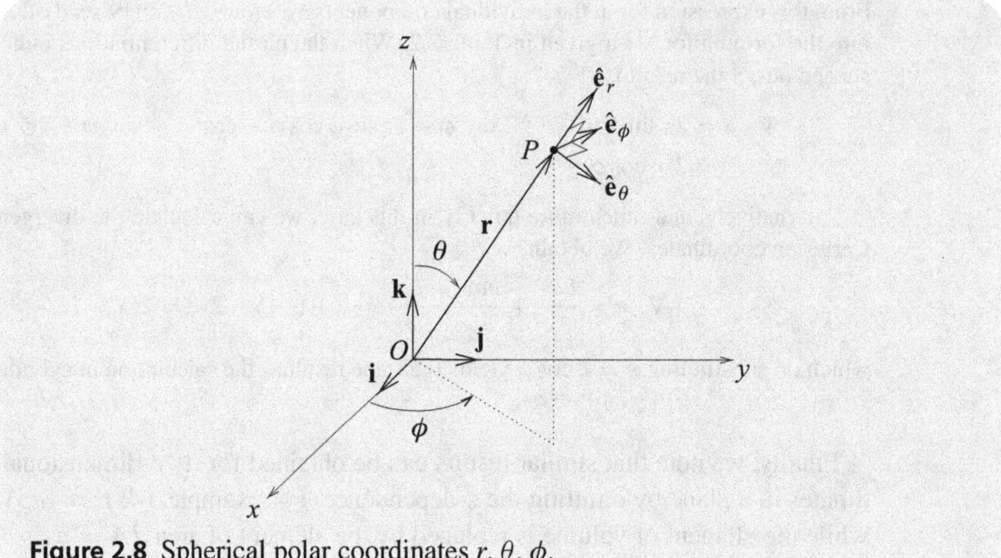

Figure 2.8 Spherical polar coordinates r, θ, ϕ.

since the basis vectors form an orthonormal set. The element of volume in spherical polar coordinates (see Figure 2.9) is the volume of the infinitesimal parallelepiped defined by the vectors $dr\ \hat{\mathbf{e}}_r$, $r\,d\theta\ \hat{\mathbf{e}}_\theta$ and $r\sin\theta\,d\phi\ \hat{\mathbf{e}}_\phi$ and is given by

$$dV = \left|dr\ \hat{\mathbf{e}}_r \cdot (r\,d\theta\ \hat{\mathbf{e}}_\theta\ \times\ r\sin\theta\,d\phi\ \hat{\mathbf{e}}_\phi)\right| = r^2\sin\theta\,dr\,d\theta\,d\phi,$$

where again we use the fact that the basis vectors are orthonormal. The same expressions for $(ds)^2$ and dV could be obtained by visual examination of the geometry of the spherical polar coordinate system.

We will now express the standard vector operators in spherical polar coordinates, using the same techniques as for cylindrical polar coordinates. We consider a scalar field $\Phi(r, \theta, \phi)$ and a vector field $\mathbf{a}(r, \theta, \phi)$. The latter may be written in terms of the basis vectors of the spherical polar coordinate system as

$$\mathbf{a} = a_r\ \hat{\mathbf{e}}_r + a_\theta\ \hat{\mathbf{e}}_\theta + a_\phi\ \hat{\mathbf{e}}_\phi,$$

where a_r, a_θ and a_ϕ are the components of $\mathbf{a}$ in the r-, θ- and ϕ- directions respectively. The expressions for grad, div, curl and ∇^2 are given in Table 2.3. The derivations of these results are given in the next section.

As a final note we mention that in the expression for $\nabla^2\Phi$ given in Table 2.3 we can rewrite the first term on the RHS as follows:[14]

$$\frac{1}{r^2}\frac{\partial}{\partial r}\left(r^2\frac{\partial\Phi}{\partial r}\right) = \frac{1}{r}\frac{\partial^2}{\partial r^2}(r\Phi).$$

This alternative expression can sometimes be useful in shortening calculations.

14 Show that both expressions are equal to $\partial^2\Phi/\partial r^2 + (2/r)\,\partial\Phi/\partial r$.

Table 2.3 *Vector operators in spherical polar coordinates; Φ is a scalar field and* **a** *is a vector field*

$$\nabla\Phi = \frac{\partial\Phi}{\partial r}\hat{\mathbf{e}}_r + \frac{1}{r}\frac{\partial\Phi}{\partial\theta}\hat{\mathbf{e}}_\theta + \frac{1}{r\sin\theta}\frac{\partial\Phi}{\partial\phi}\hat{\mathbf{e}}_\phi$$

$$\nabla\cdot\mathbf{a} = \frac{1}{r^2}\frac{\partial}{\partial r}(r^2 a_r) + \frac{1}{r\sin\theta}\frac{\partial}{\partial\theta}(\sin\theta\, a_\theta) + \frac{1}{r\sin\theta}\frac{\partial a_\phi}{\partial\phi}$$

$$\nabla\times\mathbf{a} = \frac{1}{r^2\sin\theta}\begin{vmatrix} \hat{\mathbf{e}}_r & r\,\hat{\mathbf{e}}_\theta & r\sin\theta\,\hat{\mathbf{e}}_\phi \\ \dfrac{\partial}{\partial r} & \dfrac{\partial}{\partial\theta} & \dfrac{\partial}{\partial\phi} \\ a_r & ra_\theta & r\sin\theta\, a_\phi \end{vmatrix}$$

$$\nabla^2\Phi = \frac{1}{r^2}\frac{\partial}{\partial r}\left(r^2\frac{\partial\Phi}{\partial r}\right) + \frac{1}{r^2\sin\theta}\frac{\partial}{\partial\theta}\left(\sin\theta\frac{\partial\Phi}{\partial\theta}\right) + \frac{1}{r^2\sin^2\theta}\frac{\partial^2\Phi}{\partial\phi^2}$$

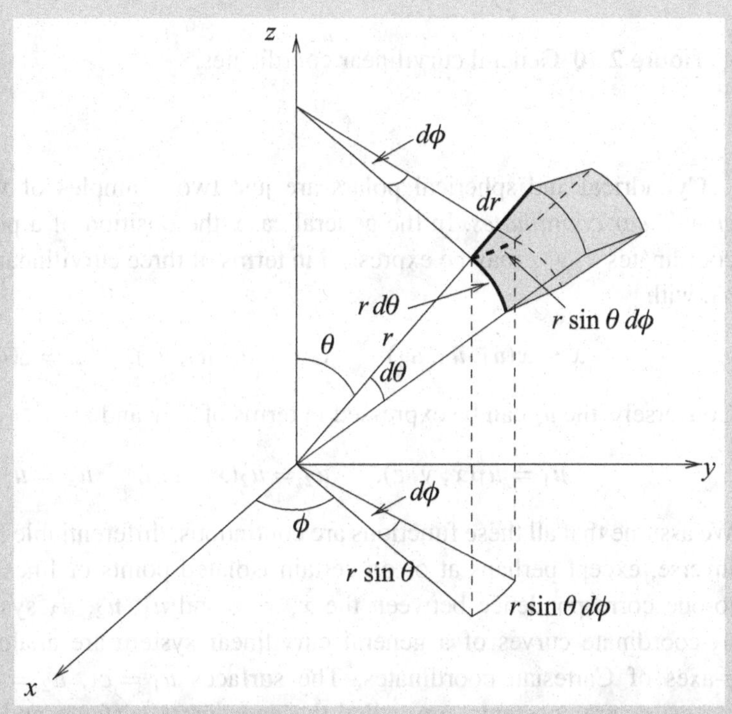

Figure 2.9 The element of volume in spherical polar coordinates is given by $r^2 \sin\theta\, dr\, d\theta\, d\phi$.

2.9 General curvilinear coordinates

As indicated earlier, the contents of this section are more formal and technically complicated than hitherto. The section could be omitted until the reader has had some experience of using its results.

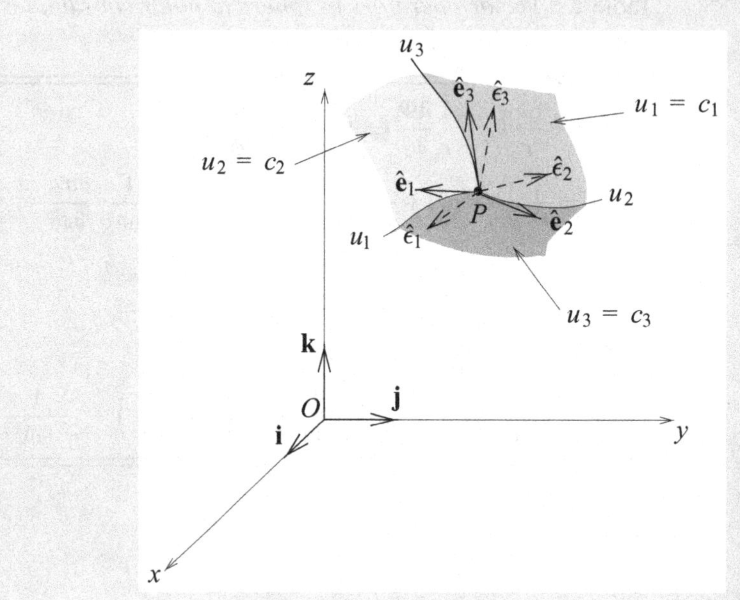

Figure 2.10 General curvilinear coordinates.

Cylindrical and spherical polars are just two examples of what are called *general curvilinear coordinates*. In the general case, the position of a point P having Cartesian coordinates x, y, z may be expressed in terms of three curvilinear coordinates, u_1, u_2 and u_3, with

$$x = x(u_1, u_2, u_3), \qquad y = y(u_1, u_2, u_3), \qquad z = z(u_1, u_2, u_3).$$

Conversely, the u_i can be expressed in terms of x, y and z:

$$u_1 = u_1(x, y, z), \qquad u_2 = u_2(x, y, z), \qquad u_3 = u_3(x, y, z).$$

We assume that all these functions are continuous, differentiable and have a single-valued inverse, except perhaps at or on certain isolated points or lines, so that there is a one-to-one correspondence between the x, y, z and u_1, u_2, u_3 systems. The u_1-, u_2- and u_3-coordinate curves of a general curvilinear system are analogous to the x-, y- and z-axes of Cartesian coordinates. The surfaces $u_1 = c_1$, $u_2 = c_2$ and $u_3 = c_3$, where c_1, c_2, c_3 are constants, are called the *coordinate surfaces* and each pair of these surfaces has its intersection in a curve called a *coordinate curve* or *line* (see Figure 2.10).

As an example that is already familiar, in the spherical polar coordinate system the coordinate surfaces are spheres, cones and a "sheaf" of half-planes containing the z-axis. The coordinate curves defined by the intersections of spheres and cones are circles, on which $u_3 = \phi$ identifies any particular point P; the curves determined by spheres and half-planes are semi-circular arcs (on which $u_2 = \theta$ defines P); the cones and half-planes meet in radial lines, on which the value of $u_1 = r$ picks out any particular point.

If at each point in space the three coordinate surfaces passing through the point meet at right angles then the curvilinear coordinate system is called *orthogonal*. In our example of spherical polars, the three coordinate surfaces passing through the point (R, Θ, Φ) are the sphere $r = R$, the circular cone $\theta = \Theta$ and the plane $\phi = \Phi$, which intersect at right angles at that point. Therefore spherical polars form an orthogonal coordinate system (as do cylindrical polars[15]).

If $\mathbf{r}(u_1, u_2, u_3)$ is the position vector of the point P then $\mathbf{e}_1 = \partial\mathbf{r}/\partial u_1$ is a vector tangent to the u_1-curve at P (for which u_2 and u_3 are constants) in the direction of increasing u_1. Similarly, $\mathbf{e}_2 = \partial\mathbf{r}/\partial u_2$ and $\mathbf{e}_3 = \partial\mathbf{r}/\partial u_3$ are vectors tangent to the u_2- and u_3-curves at P in the direction of increasing u_2 and u_3 respectively. Denoting the lengths of these vectors by h_1, h_2 and h_3, the *unit* vectors in each of these directions are given by

$$\hat{\mathbf{e}}_1 = \frac{1}{h_1}\frac{\partial\mathbf{r}}{\partial u_1}, \qquad \hat{\mathbf{e}}_2 = \frac{1}{h_2}\frac{\partial\mathbf{r}}{\partial u_2}, \qquad \hat{\mathbf{e}}_3 = \frac{1}{h_3}\frac{\partial\mathbf{r}}{\partial u_3},$$

where $h_1 = |\partial\mathbf{r}/\partial u_1|$, $h_2 = |\partial\mathbf{r}/\partial u_2|$ and $h_3 = |\partial\mathbf{r}/\partial u_3|$.

The quantities h_1, h_2, h_3 are the scale factors of the curvilinear coordinate system. The element of distance associated with an infinitesimal change du_i in one of the coordinates is $h_i\, du_i$. In the previous section we found that the scale factors for cylindrical and spherical polar coordinates were

$$\text{for cylindrical polars} \quad h_\rho = 1, \quad h_\phi = \rho, \quad h_z = 1,$$
$$\text{for spherical polars} \quad h_r = 1, \quad h_\theta = r, \quad h_\phi = r\sin\theta.$$

Although the vectors $\mathbf{e}_1, \mathbf{e}_2, \mathbf{e}_3$ form a perfectly good basis for the curvilinear coordinate system, it is usual to work with the corresponding unit vectors $\hat{\mathbf{e}}_1$, $\hat{\mathbf{e}}_2$, $\hat{\mathbf{e}}_3$. For an orthogonal curvilinear coordinate system these unit vectors form an orthonormal basis.[16]

An infinitesimal vector displacement in general curvilinear coordinates is given by, from (2.14),

$$d\mathbf{r} = \frac{\partial\mathbf{r}}{\partial u_1}du_1 + \frac{\partial\mathbf{r}}{\partial u_2}du_2 + \frac{\partial\mathbf{r}}{\partial u_3}du_3 \tag{2.50}$$
$$= du_1\,\mathbf{e}_1 + du_2\,\mathbf{e}_2 + du_3\,\mathbf{e}_3 \tag{2.51}$$
$$= h_1\,du_1\,\hat{\mathbf{e}}_1 + h_2\,du_2\,\hat{\mathbf{e}}_2 + h_3\,du_3\,\hat{\mathbf{e}}_3. \tag{2.52}$$

In the case of *orthogonal* curvilinear coordinates, where the $\hat{\mathbf{e}}_i$ are mutually perpendicular, the element of arc length is given by[17]

$$(ds)^2 = d\mathbf{r}\cdot d\mathbf{r} = h_1^2(du_1)^2 + h_2^2(du_2)^2 + h_3^2(du_3)^2. \tag{2.53}$$

The volume element for the coordinate system is the volume of the infinitesimal parallelepiped defined by the vectors $(\partial\mathbf{r}/\partial u_i)\,du_i = du_i\,\mathbf{e}_i = h_i\,du_i\,\hat{\mathbf{e}}_i$, for $i = 1, 2, 3$. For

15 Identify the shapes of the coordinate surfaces for cylindrical polar coordinates.

16 Though, in general, their directions in space depend upon where they are located – unlike the situation in a Cartesian system, where the directions are always the same.

17 Remember that $\hat{\mathbf{e}}_i\cdot\hat{\mathbf{e}}_i = 1$ and that $\hat{\mathbf{e}}_i\cdot\hat{\mathbf{e}}_j = 0$ if $i \neq j$. Consequently, there are no cross-terms of the form du_1du_2, say, in the expression for $(ds)^2$.

orthogonal coordinates this is given by

$$\begin{aligned} dV &= \left|du_1\,\mathbf{e}_1 \cdot (du_2\,\mathbf{e}_2 \times du_3\,\mathbf{e}_3)\right| \\ &= \left|h_1\,\hat{\mathbf{e}}_1 \cdot (h_2\,\hat{\mathbf{e}}_2 \times h_3\,\hat{\mathbf{e}}_3)\right| du_1\,du_2\,du_3 \\ &= h_1 h_2 h_3\,du_1\,du_2\,du_3. \end{aligned}$$

Now, in addition to the set $\{\hat{\mathbf{e}}_i\}$, $i = 1, 2, 3$, there exists another set of three unit basis vectors at P. Since ∇u_1 is a vector normal to the surface $u_1 = c_1$, a unit vector in this direction is $\hat{\boldsymbol{\epsilon}}_1 = \nabla u_1/|\nabla u_1|$. Similarly, $\hat{\boldsymbol{\epsilon}}_2 = \nabla u_2/|\nabla u_2|$ and $\hat{\boldsymbol{\epsilon}}_3 = \nabla u_3/|\nabla u_3|$ are unit vectors normal to the surfaces $u_2 = c_2$ and $u_3 = c_3$ respectively.

Therefore at each point P in a curvilinear coordinate system, there exist, in general, two sets of unit vectors: $\{\hat{\mathbf{e}}_i\}$, tangent to the coordinate curves, and $\{\hat{\boldsymbol{\epsilon}}_i\}$, normal to the coordinate surfaces. A vector $\mathbf{a}$ can be written in terms of either set of unit vectors:

$$\mathbf{a} = a_1\,\hat{\mathbf{e}}_1 + a_2\,\hat{\mathbf{e}}_2 + a_3\,\hat{\mathbf{e}}_3 = A_1\hat{\boldsymbol{\epsilon}}_1 + A_2\hat{\boldsymbol{\epsilon}}_2 + A_3\hat{\boldsymbol{\epsilon}}_3,$$

where a_1, a_2, a_3 and A_1, A_2, A_3 are the components of $\mathbf{a}$ in the two systems.

However, it is intuitively the case, and it may be shown mathematically, that if the coordinate system is an orthogonal one, then the two bases are the same. In other words, in a system in which the coordinate surfaces passing through any one point meet at right angles, the direction in which the position vector moves when only u_1, say, is changed, is the same direction as that of the normal to the particular constant-u_1 surface that passes through the point.

Non-orthogonal coordinates are difficult to work with and beyond the scope of this book, and so from now on we will consider only orthogonal systems and not need to distinguish between the two sets of base vectors; for the rest of our discussion we will use the set $\{\hat{\mathbf{e}}_i\}$.

We next derive expressions for the standard vector operators in *orthogonal* curvilinear coordinates. The expressions for the vector operators in cylindrical and spherical polar coordinates given in Tables 2.2 and 2.3 respectively can be found from those derived below by inserting the appropriate scale factors.

2.9.1 Gradient

As a total differential, the change $d\Phi$ in a scalar field Φ resulting from changes du_1, du_2, du_3 in the coordinates u_1, u_2, u_3 is given by

$$d\Phi = \frac{\partial\Phi}{\partial u_1}\,du_1 + \frac{\partial\Phi}{\partial u_2}\,du_2 + \frac{\partial\Phi}{\partial u_3}\,du_3.$$

For orthogonal curvilinear coordinates u_1, u_2, u_3 we find from (2.52), and comparison with (2.22), that we can write this as

$$d\Phi = \nabla\Phi \cdot d\mathbf{r}, \tag{2.54}$$

where $\nabla\Phi$ is given by

$$\nabla\Phi = \frac{1}{h_1}\frac{\partial\Phi}{\partial u_1}\,\hat{\mathbf{e}}_1 + \frac{1}{h_2}\frac{\partial\Phi}{\partial u_2}\,\hat{\mathbf{e}}_2 + \frac{1}{h_3}\frac{\partial\Phi}{\partial u_3}\,\hat{\mathbf{e}}_3. \tag{2.55}$$

This implies that the del operator can be written

$$\nabla = \frac{\hat{\mathbf{e}}_1}{h_1}\frac{\partial}{\partial u_1} + \frac{\hat{\mathbf{e}}_2}{h_2}\frac{\partial}{\partial u_2} + \frac{\hat{\mathbf{e}}_3}{h_3}\frac{\partial}{\partial u_3}.$$

A particular result that we will need below is obtained by setting $\Phi = u_i$ in (2.55); this leads immediately to

$$\nabla u_i = \frac{\hat{\mathbf{e}}_i}{h_i} \qquad \text{for } i = 1, 2, 3. \tag{2.56}$$

2.9.2 Divergence

In order to derive the expression for the divergence of a vector field in orthogonal curvilinear coordinates, we must first write the vector field in terms of the basis vectors of the coordinate system:

$$\mathbf{a} = a_1\,\hat{\mathbf{e}}_1 + a_2\,\hat{\mathbf{e}}_2 + a_3\,\hat{\mathbf{e}}_3.$$

The divergence is then given by

$$\nabla\cdot\mathbf{a} = \frac{1}{h_1h_2h_3}\left[\frac{\partial}{\partial u_1}(h_2h_3a_1) + \frac{\partial}{\partial u_2}(h_3h_1a_2) + \frac{\partial}{\partial u_3}(h_1h_2a_3)\right], \tag{2.57}$$

as we now prove.

Example Prove the expression for $\nabla\cdot\mathbf{a}$ in orthogonal curvilinear coordinates.

Let us first consider the sub-expression $\nabla\cdot(a_1\,\hat{\mathbf{e}}_1)$. Using (2.56) twice, we can write $\hat{\mathbf{e}}_1 = \hat{\mathbf{e}}_2\times\hat{\mathbf{e}}_3 = h_2\nabla u_2 \times h_3\nabla u_3$, and so

$$\begin{aligned}\nabla\cdot(a_1\,\hat{\mathbf{e}}_1) &= \nabla\cdot(a_1h_2h_3\nabla u_2\times\nabla u_3),\\ &= \nabla(a_1h_2h_3)\cdot(\nabla u_2\times\nabla u_3) + a_1h_2h_3\nabla\cdot(\nabla u_2\times\nabla u_3).\end{aligned}$$

Here we have used the sixth vector identity in Table 2.1, with the product $a_1h_2h_3$ replacing ϕ and the vector product $\nabla u_2\times\nabla u_3$ replacing the vector. However, both u_2 and u_3 are scalar fields (as well as being coordinates) and therefore, from (2.38), $\nabla\cdot(\nabla u_2\times\nabla u_3) = 0$. So, using (2.56) again, we obtain

$$\nabla\cdot(a_1\,\hat{\mathbf{e}}_1) = \nabla(a_1h_2h_3)\cdot\left(\frac{\hat{\mathbf{e}}_2}{h_2}\times\frac{\hat{\mathbf{e}}_3}{h_3}\right) = \nabla(a_1h_2h_3)\cdot\frac{\hat{\mathbf{e}}_1}{h_2h_3}.$$

Using (2.55) to find the gradient of $a_1h_2h_3$, but retaining only the $\hat{\mathbf{e}}_1$ component (because of the scalar product with $\hat{\mathbf{e}}_1$ in the above equation), we find on substitution that

$$\nabla\cdot(a_1\,\hat{\mathbf{e}}_1) = \frac{1}{h_1h_2h_3}\frac{\partial}{\partial u_1}(a_1h_2h_3).$$

Repeating the analysis for $\nabla\cdot(a_2\,\hat{\mathbf{e}}_2)$ and $\nabla\cdot(a_3\,\hat{\mathbf{e}}_3)$, and adding the results, we obtain the general expression for the divergence of $\mathbf{a}$ as stated in (2.57). ◀

2.9.3 Laplacian

In expression (2.57) for the divergence, we now replace **a** by $\nabla\Phi$ as given by (2.55), i.e. we set $a_i = h_i^{-1}\partial\Phi/\partial u_i$. When this is done we obtain

$$\nabla^2\Phi = \frac{1}{h_1h_2h_3}\left[\frac{\partial}{\partial u_1}\left(\frac{h_2h_3}{h_1}\frac{\partial\Phi}{\partial u_1}\right) + \frac{\partial}{\partial u_2}\left(\frac{h_3h_1}{h_2}\frac{\partial\Phi}{\partial u_2}\right) + \frac{\partial}{\partial u_3}\left(\frac{h_1h_2}{h_3}\frac{\partial\Phi}{\partial u_3}\right)\right],$$

which is the general expression for the Laplacian of Φ in orthogonal curvilinear coordinates.

2.9.4 Curl

The curl of a vector field $\mathbf{a} = a_1\,\hat{\mathbf{e}}_1 + a_2\,\hat{\mathbf{e}}_2 + a_3\,\hat{\mathbf{e}}_3$ in orthogonal curvilinear coordinates is given by

$$\nabla \times \mathbf{a} = \frac{1}{h_1h_2h_3}\begin{vmatrix} h_1\,\hat{\mathbf{e}}_1 & h_2\,\hat{\mathbf{e}}_2 & h_3\,\hat{\mathbf{e}}_3 \\ \dfrac{\partial}{\partial u_1} & \dfrac{\partial}{\partial u_2} & \dfrac{\partial}{\partial u_3} \\ h_1a_1 & h_2a_2 & h_3a_3 \end{vmatrix}. \tag{2.58}$$

The proof of this is similar to that for the divergence operator, and again we give it as a worked example.

Example Prove the expression for $\nabla \times \mathbf{a}$ in orthogonal curvilinear coordinates.

Let us first consider the sub-expression $\nabla \times (a_1\,\hat{\mathbf{e}}_1)$. Since $\hat{\mathbf{e}}_1 = h_1\nabla u_1$, we have, using the penultimate entry in Table 2.1, that

$$\begin{aligned}\nabla \times (a_1\,\hat{\mathbf{e}}_1) &= \nabla \times (a_1h_1\nabla u_1),\\ &= \nabla(a_1h_1) \times \nabla u_1 \;+\; a_1h_1(\nabla \times \nabla u_1).\end{aligned}$$

But $\nabla \times \nabla u_1 = 0$, so we obtain

$$\nabla \times (a_1\,\hat{\mathbf{e}}_1) = \nabla(a_1h_1) \times \frac{\hat{\mathbf{e}}_1}{h_1}.$$

Letting $\Phi = a_1h_1$ in (2.55) and substituting into the above equation, we find

$$\nabla \times (a_1\,\hat{\mathbf{e}}_1) = \frac{\hat{\mathbf{e}}_2}{h_3h_1}\frac{\partial}{\partial u_3}(a_1h_1) - \frac{\hat{\mathbf{e}}_3}{h_1h_2}\frac{\partial}{\partial u_2}(a_1h_1).$$

Notice that it is the $\hat{\mathbf{e}}_3$ component of $\nabla(a_1h_1)$ that produces the $\hat{\mathbf{e}}_2$ component of the above expression, and vice versa. The corresponding analysis of $\nabla \times (a_2\,\hat{\mathbf{e}}_2)$ produces terms in $\hat{\mathbf{e}}_3$ and $\hat{\mathbf{e}}_1$, whilst that of $\nabla \times (a_3\,\hat{\mathbf{e}}_3)$ produces terms in $\hat{\mathbf{e}}_1$ and $\hat{\mathbf{e}}_2$. When the three results are added together, the coefficients multiplying $\hat{\mathbf{e}}_1$, $\hat{\mathbf{e}}_2$ and $\hat{\mathbf{e}}_3$ are the same as those obtained by writing out (2.58) explicitly, thus proving the stated result. ◀

The general expressions for the vector operators in orthogonal curvilinear coordinates are shown for reference in Table 2.4. The explicit results for cylindrical and spherical

Table 2.4 *Vector operators in orthogonal curvilinear coordinates* u_1, u_2, u_3. *Φ is a scalar field and* **a** *is a vector field*

$$\nabla\Phi = \frac{1}{h_1}\frac{\partial\Phi}{\partial u_1}\hat{\mathbf{e}}_1 + \frac{1}{h_2}\frac{\partial\Phi}{\partial u_2}\hat{\mathbf{e}}_2 + \frac{1}{h_3}\frac{\partial\Phi}{\partial u_3}\hat{\mathbf{e}}_3$$

$$\nabla\cdot\mathbf{a} = \frac{1}{h_1h_2h_3}\left[\frac{\partial}{\partial u_1}(h_2h_3a_1) + \frac{\partial}{\partial u_2}(h_3h_1a_2) + \frac{\partial}{\partial u_3}(h_1h_2a_3)\right]$$

$$\nabla\times\mathbf{a} = \frac{1}{h_1h_2h_3}\begin{vmatrix} h_1\hat{\mathbf{e}}_1 & h_2\hat{\mathbf{e}}_2 & h_3\hat{\mathbf{e}}_3 \\ \dfrac{\partial}{\partial u_1} & \dfrac{\partial}{\partial u_2} & \dfrac{\partial}{\partial u_3} \\ h_1a_1 & h_2a_2 & h_3a_3 \end{vmatrix}$$

$$\nabla^2\Phi = \frac{1}{h_1h_2h_3}\left[\frac{\partial}{\partial u_1}\left(\frac{h_2h_3}{h_1}\frac{\partial\Phi}{\partial u_1}\right) + \frac{\partial}{\partial u_2}\left(\frac{h_3h_1}{h_2}\frac{\partial\Phi}{\partial u_2}\right) + \frac{\partial}{\partial u_3}\left(\frac{h_1h_2}{h_3}\frac{\partial\Phi}{\partial u_3}\right)\right]$$

polar coordinates, given in Tables 2.2 and 2.3 respectively, are obtained by substituting the appropriate set of scale factors in each case.

SUMMARY

1. *Derivatives of products*
   - $\dfrac{d}{du}(\phi\mathbf{a}) = \phi\dfrac{d\mathbf{a}}{du} + \dfrac{d\phi}{du}\mathbf{a},$
   - $\dfrac{d}{du}(\mathbf{a}\cdot\mathbf{b}) = \mathbf{a}\cdot\dfrac{d\mathbf{b}}{du} + \dfrac{d\mathbf{a}}{du}\cdot\mathbf{b},$
   - $\dfrac{d}{du}(\mathbf{a}\times\mathbf{b}) = \mathbf{a}\times\dfrac{d\mathbf{b}}{du} + \dfrac{d\mathbf{a}}{du}\times\mathbf{b}.$

2. *Integrals of vectors with respect to scalars*
   (i) The integral has the same nature (vector or scalar) as the integrand.
   (ii) The constant of integration for indefinite integrals must be of the same nature as the integral.

3. *For a surface given by* $\mathbf{r} = \mathbf{r}(u, v)$
   - If $u = u(\lambda)$ and $v = v(\lambda)$ is a curve $\mathbf{r} = \mathbf{r}(\lambda)$ lying in the surface, then the tangent vector to the curve is

$$\mathbf{t} = \frac{d\mathbf{r}}{d\lambda} = \frac{\partial\mathbf{r}}{\partial u}\frac{du}{d\lambda} + \frac{\partial\mathbf{r}}{\partial v}\frac{dv}{d\lambda}.$$

   - A normal to the tangent plane and the size of an element of area are

$$\mathbf{n} = \frac{\partial\mathbf{r}}{\partial u}\times\frac{\partial\mathbf{r}}{\partial v} \quad\text{and}\quad dS = |\mathbf{n}|\,du\,dv.$$

4. *Vector operators acting on fields, and their Cartesian forms*
   - Gradient of a scalar:

$$\text{grad}\,\phi = \nabla\phi = \mathbf{i}\frac{\partial\phi}{\partial x} + \mathbf{j}\frac{\partial\phi}{\partial y} + \mathbf{k}\frac{\partial\phi}{\partial z} \quad \text{with} \quad \nabla\phi\cdot d\mathbf{r} = d\phi.$$

   - The rate of change in the direction of **a** is given by the operator

$$\hat{\mathbf{a}}\cdot\nabla = \hat{a}_x\frac{\partial}{\partial x} + \hat{a}_y\frac{\partial}{\partial y} + \hat{a}_z\frac{\partial}{\partial z}.$$

   - Divergence of a vector:

$$\text{div}\,\mathbf{a} = \nabla\cdot\mathbf{a} = \frac{\partial a_x}{\partial x} + \frac{\partial a_y}{\partial y} + \frac{\partial a_z}{\partial z}.$$

   - Curl of a vector:

$$\text{curl}\,\mathbf{a} = \nabla\times\mathbf{a} = \begin{vmatrix} \mathbf{i} & \mathbf{j} & \mathbf{k} \\ \frac{\partial}{\partial x} & \frac{\partial}{\partial y} & \frac{\partial}{\partial z} \\ a_x & a_y & a_z \end{vmatrix}.$$

   - For the actions of the operators on sums and products, see Table 2.1 on p. 104.

5. *Combinations of vector operators, and their Cartesian forms*

Name	Symbolic form	Value or Cartesian form
del$^2\,\phi$	$\nabla^2\phi = \nabla\cdot\nabla\phi$	$\frac{\partial^2\phi}{\partial x^2} + \frac{\partial^2\phi}{\partial y^2} + \frac{\partial^2\phi}{\partial z^2}$
curl grad ϕ	$\nabla\times\nabla\phi$	$\mathbf{0}$
div curl **a**	$\nabla\cdot(\nabla\times\mathbf{a})$	0
grad div **a**	$\nabla(\nabla\cdot\mathbf{a})$	$\left(\frac{\partial^2 a_x}{\partial x^2} + \frac{\partial^2 a_y}{\partial x\partial y} + \frac{\partial^2 a_z}{\partial x\partial z}\right)\mathbf{i} + (\cdots)\mathbf{j} + (\cdots)\mathbf{k}$
curl curl **a**	$\nabla\times(\nabla\times\mathbf{a}) = \nabla(\nabla\cdot\mathbf{a}) - \nabla^2\mathbf{a}$	$\left(\frac{\partial^2 a_y}{\partial x\partial y} + \frac{\partial^2 a_z}{\partial x\partial z} - \frac{\partial^2 a_x}{\partial y^2} - \frac{\partial^2 a_x}{\partial z^2}\right)\mathbf{i} + (\cdots)\mathbf{j} + (\cdots)\mathbf{k}$

   - In Cartesian coordinates, $(\nabla^2\mathbf{a})_i = \nabla^2 a_i$.
   - $\nabla\cdot(\nabla\phi\times\nabla\psi) = 0$ if ϕ and ψ are scalar fields.

6. *Vector operators in polar coordinates*
   - For cylindrical polars, see Table 2.2 on p. 110.
   - For spherical polars, see Table 2.3 on p. 113.

7. *General orthogonal curvilinear coordinates with* $\mathbf{r} = \mathbf{r}(u_1, u_2, u_3)$
   - Scale factors and unit vectors
$$h_i = \left|\frac{\partial \mathbf{r}}{\partial u_i}\right|, \qquad \hat{\mathbf{e}}_i = \frac{1}{h_i}\frac{\partial \mathbf{r}}{\partial u_i}.$$
   - For cylindrical polars $h_1 = 1, h_2 = \rho, h_3 = 1$; for spherical polars $h_1 = 1, h_2 = r, h_3 = r\sin\theta$.
   - $(ds)^2 = d\mathbf{r}\cdot d\mathbf{r} = h_1^2(du_1)^2 + h_2^2(du_2)^2 + h_3^2(du_3)^2$ with no cross-terms of the form $du_1\,du_2$.
   - $dV = h_1h_2h_3\,du_1\,du_2\,du_3$.
   - $\nabla = \dfrac{\hat{\mathbf{e}}_1}{h_1}\dfrac{\partial}{\partial u_1} + \dfrac{\hat{\mathbf{e}}_2}{h_2}\dfrac{\partial}{\partial u_2} + \dfrac{\hat{\mathbf{e}}_3}{h_3}\dfrac{\partial}{\partial u_3}$.
   - For vector operators, see Table 2.4 on p. 119.

PROBLEMS

2.1. Evaluate the integral
$$\int \left[\mathbf{a}(\dot{\mathbf{b}}\cdot\mathbf{a} + \mathbf{b}\cdot\dot{\mathbf{a}}) + \dot{\mathbf{a}}(\mathbf{b}\cdot\mathbf{a}) - 2(\dot{\mathbf{a}}\cdot\mathbf{a})\mathbf{b} - \dot{\mathbf{b}}|\mathbf{a}|^2\right] dt$$
in which $\dot{\mathbf{a}}$, $\dot{\mathbf{b}}$ are the derivatives of the real vectors $\mathbf{a}$, $\mathbf{b}$ with respect to t.

2.2. At time $t = 0$, the vectors $\mathbf{E}$ and $\mathbf{B}$ are given by $\mathbf{E} = \mathbf{E}_0$ and $\mathbf{B} = \mathbf{B}_0$, where the unit vectors, $\mathbf{E}_0$ and $\mathbf{B}_0$ are fixed and orthogonal. The equations of motion are
$$\frac{d\mathbf{E}}{dt} = \mathbf{E}_0 + \mathbf{B}\times\mathbf{E}_0,$$
$$\frac{d\mathbf{B}}{dt} = \mathbf{B}_0 + \mathbf{E}\times\mathbf{B}_0.$$
Find $\mathbf{E}$ and $\mathbf{B}$ at a general time t, showing that after a long time the directions of $\mathbf{E}$ and $\mathbf{B}$ have almost interchanged.

2.3. The general equation of motion of a (non-relativistic) particle of mass m and charge q when it is placed in a region where there is a magnetic field $\mathbf{B}$ and an electric field $\mathbf{E}$ is
$$m\ddot{\mathbf{r}} = q(\mathbf{E} + \dot{\mathbf{r}}\times\mathbf{B});$$
here $\mathbf{r}$ is the position of the particle at time t and $\dot{\mathbf{r}} = d\mathbf{r}/dt$, etc. Write this as three separate equations in terms of the Cartesian components of the vectors involved.

For the simple case of crossed uniform fields $\mathbf{E} = E\mathbf{i}$, $\mathbf{B} = B\mathbf{j}$, in which the particle starts from the origin at $t = 0$ with $\dot{\mathbf{r}} = v_0\mathbf{k}$, find the equations of motion and show the following:

(a) if $v_0 = E/B$ then the particle continues its initial motion;

(b) if $v_0 = 0$ then the particle follows the space curve given in terms of the parameter ξ by

$$x = \frac{mE}{B^2 q}(1 - \cos\xi), \qquad y = 0, \qquad z = \frac{mE}{B^2 q}(\xi - \sin\xi).$$

Interpret this curve geometrically and relate ξ to t. Show that the total distance traveled by the particle after time t is given by

$$\frac{2E}{B}\int_0^t \left|\sin\frac{Bqt'}{2m}\right| dt'.$$

2.4. Use vector methods to find the maximum angle to the horizontal at which a stone may be thrown so as to ensure that it is always moving away from the thrower.

2.5. If two systems of coordinates with a common origin O are rotating with respect to each other, the measured accelerations differ in the two systems. Denoting by $\mathbf{r}$ and $\mathbf{r}'$ position vectors in frames $OXYZ$ and $OX'Y'Z'$, respectively, the connection between the two is

$$\ddot{\mathbf{r}}' = \ddot{\mathbf{r}} + \dot{\boldsymbol{\omega}} \times \mathbf{r} + 2\boldsymbol{\omega} \times \dot{\mathbf{r}} + \boldsymbol{\omega} \times (\boldsymbol{\omega} \times \mathbf{r}),$$

where $\boldsymbol{\omega}$ is the angular velocity vector of the rotation of $OXYZ$ with respect to $OX'Y'Z'$ (taken as fixed). The third term on the RHS is known as the Coriolis acceleration, whilst the final term gives rise to a centrifugal force.

Consider the application of this result to the firing of a shell of mass m from a stationary ship on the steadily rotating earth, working to the first order in ω ($= 7.3 \times 10^{-5}$ rad s^{-1}). If the shell is fired with velocity $\mathbf{v}$ at time $t = 0$ and only reaches a height that is small compared with the radius of the earth, show that its acceleration, as recorded on the ship, is given approximately by

$$\ddot{\mathbf{r}} = \mathbf{g} - 2\boldsymbol{\omega} \times (\mathbf{v} + \mathbf{g}t),$$

where $m\mathbf{g}$ is the weight of the shell measured on the ship's deck.

The shell is fired at another stationary ship (a distance $\mathbf{s}$ away) and $\mathbf{v}$ is such that the shell would have hit its target had there been no Coriolis effect.

(a) Show that without the Coriolis effect the time of flight of the shell would have been $\tau = -2\mathbf{g} \cdot \mathbf{v}/g^2$.

(b) Show further that when the shell actually hits the sea it is off-target by approximately

$$\frac{2\tau}{g^2}[(\mathbf{g} \times \boldsymbol{\omega}) \cdot \mathbf{v}](\mathbf{g}\tau + \mathbf{v}) - (\boldsymbol{\omega} \times \mathbf{v})\tau^2 - \frac{1}{3}(\boldsymbol{\omega} \times \mathbf{g})\tau^3.$$

(c) Estimate the order of magnitude $\boldsymbol{\Delta}$ of this miss for a shell for which the initial speed v is 300 m s^{-1}, firing close to its maximum range ($\mathbf{v}$ makes an angle of $\pi/4$ with the vertical) in a northerly direction, whilst the ship is stationed at latitude 45° North.

2.6. Find the areas of the given surfaces using parametric coordinates.

(a) Using the parameterization $x = u\cos\phi$, $y = u\sin\phi$, $z = u\cot\Omega$, find the sloping surface area of a right circular cone of semi-angle Ω whose base has radius a. Verify that it is equal to $\frac{1}{2}\times$ perimeter of the base $\times$ slope height.

(b) Using the same parameterization as in (a) for x and y, and an appropriate choice for z, find the surface area between the planes $z = 0$ and $z = Z$ of the paraboloid of revolution $z = \alpha(x^2 + y^2)$.

2.7. Parameterizing the hyperboloid

$$\frac{x^2}{a^2} + \frac{y^2}{b^2} - \frac{z^2}{c^2} = 1$$

by $x = a\cos\theta\sec\phi$, $y = b\sin\theta\sec\phi$, $z = c\tan\phi$, show that an area element on its surface is

$$dS = \sec^2\phi\left[c^2\sec^2\phi\left(b^2\cos^2\theta + a^2\sin^2\theta\right) + a^2b^2\tan^2\phi\right]^{1/2}\,d\theta\,d\phi.$$

Use this formula to show that the area of the curved surface $x^2 + y^2 - z^2 = a^2$ between the planes $z = 0$ and $z = 2a$ is

$$\pi a^2\left(6 + \frac{1}{\sqrt{2}}\sinh^{-1}2\sqrt{2}\right).$$

2.8. For the function

$$z(x, y) = (x^2 - y^2)e^{-x^2-y^2},$$

find the location(s) at which the steepest gradient occurs. What are the magnitude and direction of that gradient? The algebra involved is easier if plane polar coordinates are used.

2.9. Verify by direct calculation that

$$\nabla\cdot(\mathbf{a}\times\mathbf{b}) = \mathbf{b}\cdot(\nabla\times\mathbf{a}) - \mathbf{a}\cdot(\nabla\times\mathbf{b}).$$

2.10. In the following problems, **a**, **b** and **c** are vector fields.

(a) Simplify

$$\nabla\times\mathbf{a}(\nabla\cdot\mathbf{a}) \;+\; \mathbf{a}\times[\nabla\times(\nabla\times\mathbf{a})] \;+\; \mathbf{a}\times\nabla^2\mathbf{a}.$$

(b) By explicitly writing out the terms in Cartesian coordinates, prove that

$$[\mathbf{c}\cdot(\mathbf{b}\cdot\nabla) - \mathbf{b}\cdot(\mathbf{c}\cdot\nabla)]\,\mathbf{a} = (\nabla\times\mathbf{a})\cdot(\mathbf{b}\times\mathbf{c}).$$

(c) Prove that $\mathbf{a}\times(\nabla\times\mathbf{a}) = \nabla(\frac{1}{2}a^2) - (\mathbf{a}\cdot\nabla)\mathbf{a}$.

2.11. Evaluate the Laplacian of the function

$$\psi(x, y, z) = \frac{zx^2}{x^2 + y^2 + z^2}$$

(a) directly in Cartesian coordinates, and (b) after changing to a spherical polar coordinate system. Verify that, as they must, the two methods give the same result.

2.12. Verify that (2.37) is valid for each component separately when $\mathbf{a}$ is the Cartesian vector $x^2y\,\mathbf{i} + xyz\,\mathbf{j} + z^2y\,\mathbf{k}$, by showing that each side of the equation is equal to $z\,\mathbf{i} + (2x + 2z)\,\mathbf{j} + x\,\mathbf{k}$.

2.13. The (Maxwell) relationship between a time-independent magnetic field $\mathbf{B}$ and the current density $\mathbf{J}$ (measured in SI units in $\mathrm{A\,m^{-2}}$) producing it,

$$\nabla \times \mathbf{B} = \mu_0 \mathbf{J},$$

can be applied to a long cylinder of conducting ionized gas which, in cylindrical polar coordinates, occupies the region $\rho < a$.

(a) Show that a uniform current density $(0, C, 0)$ and a magnetic field $(0, 0, B)$, with B constant $(=B_0)$ for $\rho > a$ and $B = B(\rho)$ for $\rho < a$, are consistent with this equation. Given that $B(0) = 0$ and that $\mathbf{B}$ is continuous at $\rho = a$, obtain expressions for C and $B(\rho)$ in terms of B_0 and a.

(b) The magnetic field can be expressed as $\mathbf{B} = \nabla \times \mathbf{A}$, where $\mathbf{A}$ is known as the vector potential. Show that a suitable $\mathbf{A}$ that has only one non-vanishing component, $A_\phi(\rho)$, can be found, and obtain explicit expressions for $A_\phi(\rho)$ for both $\rho < a$ and $\rho > a$. Like $\mathbf{B}$, the vector potential is continuous at $\rho = a$.

(c) The gas pressure $p(\rho)$ satisfies the hydrostatic equation $\nabla p = \mathbf{J} \times \mathbf{B}$ and vanishes at the outer wall of the cylinder. Find a general expression for p.

2.14. Evaluate the Laplacian of a vector field using two different coordinate systems as follows.

(a) For cylindrical polar coordinates ρ, ϕ, z, evaluate the derivatives of the three unit vectors with respect to each of the coordinates, showing that only $\partial \hat{\mathbf{e}}_\rho/\partial\phi$ and $\partial \hat{\mathbf{e}}_\phi/\partial\phi$ are non-zero.

(i) Hence evaluate $\nabla^2\mathbf{a}$ when $\mathbf{a}$ is the vector $\hat{\mathbf{e}}_\rho$, i.e. a vector of unit magnitude everywhere directed radially outwards and expressed by $a_\rho = 1, a_\phi = a_z = 0$.

(ii) Note that it is trivially obvious that $\nabla \times \mathbf{a} = \mathbf{0}$ and hence that equation (2.36) requires that $\nabla(\nabla \cdot \mathbf{a}) = \nabla^2\mathbf{a}$.

(iii) Evaluate $\nabla(\nabla \cdot \mathbf{a})$ and show that the latter equation holds, but that

$$[\nabla(\nabla \cdot \mathbf{a})]_\rho \neq \nabla^2 a_\rho.$$

(b) Rework the same problem in Cartesian coordinates (where, as it happens, the algebra is more complicated).

2.15. Maxwell's equations for electromagnetism in free space (i.e. in the absence of charges, currents and dielectric or magnetic media) can be written

$$\text{(i)}\ \nabla \cdot \mathbf{B} = 0, \qquad \text{(ii)}\ \nabla \cdot \mathbf{E} = 0,$$

$$\text{(iii)}\ \nabla \times \mathbf{E} + \frac{\partial \mathbf{B}}{\partial t} = \mathbf{0}, \quad \text{(iv)}\ \nabla \times \mathbf{B} - \frac{1}{c^2}\frac{\partial \mathbf{E}}{\partial t} = \mathbf{0}.$$

A vector $\mathbf{A}$ is defined by $\mathbf{B} = \nabla \times \mathbf{A}$, and a scalar ϕ by $\mathbf{E} = -\nabla\phi - \partial\mathbf{A}/\partial t$. Show that if the condition

$$\text{(v)}\ \nabla \cdot \mathbf{A} + \frac{1}{c^2}\frac{\partial \phi}{\partial t} = 0$$

is imposed (this is known as choosing the Lorentz gauge), then $\mathbf{A}$ and ϕ satisfy wave equations as follows:

$$\text{(vi)}\ \nabla^2\phi - \frac{1}{c^2}\frac{\partial^2 \phi}{\partial t^2} = 0,$$

$$\text{(vii)}\ \nabla^2\mathbf{A} - \frac{1}{c^2}\frac{\partial^2 \mathbf{A}}{\partial t^2} = \mathbf{0}.$$

The reader is invited to proceed as follows.

(a) Verify that the expressions for $\mathbf{B}$ and $\mathbf{E}$ in terms of $\mathbf{A}$ and ϕ are consistent with (i) and (iii).

(b) Substitute for $\mathbf{E}$ in (ii) and use the derivative with respect to time of (v) to eliminate $\mathbf{A}$ from the resulting expression. Hence obtain (vi).

(c) Substitute for $\mathbf{B}$ and $\mathbf{E}$ in (iv) in terms of $\mathbf{A}$ and ϕ. Then use the gradient of (v) to simplify the resulting equation and so obtain (vii).

2.16. For a description using spherical polar coordinates with axial symmetry, of the flow of a very viscous fluid, the components of the velocity field $\mathbf{u}$ are given in terms of the *stream function* ψ by

$$u_r = \frac{1}{r^2\sin\theta}\frac{\partial \psi}{\partial \theta}, \qquad u_\theta = \frac{-1}{r\sin\theta}\frac{\partial \psi}{\partial r}.$$

Find an explicit expression for the differential operator E defined by

$$E\psi = -(r\sin\theta)(\nabla \times \mathbf{u})_\phi.$$

The stream function satisfies the equation of motion $E^2\psi = 0$ and, for the flow of a fluid past a sphere, takes the form $\psi(r,\theta) = f(r)\sin^2\theta$. Show that $f(r)$ satisfies the (ordinary) differential equation

$$r^4 f^{(4)} - 4r^2 f'' + 8rf' - 8f = 0.$$

2.17. Paraboloidal coordinates u, v, ϕ are defined in terms of Cartesian coordinates by

$$x = uv\cos\phi, \qquad y = uv\sin\phi, \qquad z = \tfrac{1}{2}(u^2 - v^2).$$

Identify the coordinate surfaces in the u, v, ϕ system. Verify that each coordinate surface (u = constant, say) intersects every coordinate surface on which one of the other two coordinates (v, say) is constant. Show further that the system of coordinates is an orthogonal one and determine its scale factors. Prove that the u-component of $\nabla \times \mathbf{a}$ is given by

$$\frac{1}{(u^2+v^2)^{1/2}}\left(\frac{a_\phi}{v} + \frac{\partial a_\phi}{\partial v}\right) - \frac{1}{uv}\frac{\partial a_v}{\partial \phi}.$$

2.18. In a Cartesian system, A and B are the points $(0, 0, -1)$ and $(0, 0, 1)$ respectively. In a new coordinate system a general point P is given by (u_1, u_2, u_3) with $u_1 = \frac{1}{2}(r_1 + r_2)$, $u_2 = \frac{1}{2}(r_1 - r_2)$, $u_3 = \phi$; here r_1 and r_2 are the distances AP and BP and ϕ is the angle between the plane ABP and $y = 0$.

(a) Express z and the perpendicular distance ρ from P to the z-axis in terms of u_1, u_2, u_3.
(b) Evaluate $\partial x/\partial u_i$, $\partial y/\partial u_i$, $\partial z/\partial u_i$, for $i = 1, 2, 3$.
(c) Find the Cartesian components of $\hat{\mathbf{u}}_j$ and hence show that the new coordinates are mutually orthogonal. Evaluate the scale factors and the infinitesimal volume element in the new coordinate system.
(d) Determine and sketch the forms of the surfaces $u_i =$ constant.
(e) Find the most general function f of u_1 only that satisfies $\nabla^2 f = 0$.

2.19. Hyperbolic coordinates u, v, ϕ are defined in terms of Cartesian coordinates by

$$x = \cosh u \cos v \cos\phi, \qquad y = \cosh u \cos v \sin\phi, \qquad z = \sinh u \sin v.$$

Sketch the coordinate curves in the $\phi = 0$ plane, showing that far from the origin they become concentric circles and radial lines. In particular, identify the curves $u = 0$, $v = 0$, $v = \pi/2$ and $v = \pi$. Calculate the tangent vectors at a general point, show that they are mutually orthogonal and deduce that the appropriate scale factors are

$$h_u = h_v = (\cosh^2 u - \cos^2 v)^{1/2}, \qquad h_\phi = \cosh u \cos v.$$

Find the most general function $\psi(u)$ of u only that satisfies Laplace's equation $\nabla^2\psi = 0$.

HINTS AND ANSWERS

2.1. Group the terms so that they form the total derivatives of compound vector expressions. The integral has the value $\mathbf{a} \times (\mathbf{a} \times \mathbf{b}) + \mathbf{h}$.

2.3. For crossed uniform fields $\ddot{x} + (Bq/m)^2 x = q(E - Bv_0)/m$, $\ddot{y} = 0$, $m\dot{z} = qBx + mv_0$;
(b) $\xi = Bqt/m$; the path is a cycloid in the plane $y = 0$; $ds = [(dx/dt)^2 + (dz/dt)^2]^{1/2}\,dt$.

2.5. $\mathbf{g} = \ddot{\mathbf{r}}' - \boldsymbol{\omega} \times (\boldsymbol{\omega} \times \mathbf{r})$, where $\ddot{\mathbf{r}}'$ is the shell's acceleration measured by an observer fixed in space. To first order in ω, the direction of $\mathbf{g}$ is radial, i.e. parallel to $\ddot{\mathbf{r}}'$.
(a) Note that $\mathbf{s}$ is orthogonal to $\mathbf{g}$.
(b) If the actual time of flight is T, use $(\mathbf{s} + \boldsymbol{\Delta}) \cdot \mathbf{g} = 0$ to show that

$$T \approx \tau(1 + 2g^{-2}(\mathbf{g} \times \boldsymbol{\omega}) \cdot \mathbf{v} + \cdots).$$

In the Coriolis terms, it is sufficient to put $T \approx \tau$.

(c) For this situation $(\mathbf{g} \times \boldsymbol{\omega}) \cdot \mathbf{v} = 0$ and $\boldsymbol{\omega} \times \mathbf{v} = \mathbf{0}$; $\tau \approx 43$ s and $\Delta = 10$–15 m to the East.

2.7. To integrate $\sec^2\phi(\sec^2\phi + \tan^2\phi)^{1/2}\,d\phi$ put $\tan\phi = 2^{-1/2}\sinh\psi$.

2.9. Work in Cartesian coordinates, regrouping the terms obtained by evaluating the divergence on the LHS

2.11. (a) $2z(x^2 + y^2 + z^2)^{-3}[(y^2 + z^2)(y^2 + z^2 - 3x^2) - 4x^4]$;
(b) $2r^{-1}\cos\theta\,(1 - 5\sin^2\theta\cos^2\phi)$; both are equal to $2zr^{-4}(r^2 - 5x^2)$.

2.13. Use the formulae given in Table 2.2.
(a) $C = -B_0/(\mu_0 a)$; $B(\rho) = B_0\rho/a$.
(b) $B_0\rho^2/(3a)$ for $\rho < a$, and $B_0[\rho/2 - a^2/(6\rho)]$ for $\rho > a$.
(c) $[B_0^2/(2\mu_0)][1 - (\rho/a)^2]$.

2.15. Recall that $\nabla \times \nabla\phi = \mathbf{0}$ for any scalar ϕ and that $\partial/\partial t$ and ∇ act on different variables.

2.17. Two sets of paraboloids of revolution about the z-axis and the sheaf of planes containing the z-axis. For constant u, $-\infty < z < u^2/2$; for constant v, $-v^2/2 < z < \infty$. The scale factors are $h_u = h_v = (u^2 + v^2)^{1/2}$, $h_\phi = uv$.

2.19. The tangent vectors are as follows: for $u = 0$, the line joining $(1, 0, 0)$ and $(-1, 0, 0)$; for $v = 0$, the line joining $(1, 0, 0)$ and $(\infty, 0, 0)$; for $v = \pi/2$, the line $(0, 0, z)$; for $v = \pi$, the line joining $(-1, 0, 0)$ and $(-\infty, 0, 0)$. $\psi(u) = 2\tan^{-1} e^u + c$, derived from $\partial[\cosh u(\partial\psi/\partial u)]/\partial u = 0$.

3 Line, surface and volume integrals

In the previous chapter we encountered continuously varying scalar and vector fields and discussed the action of various differential operators on them. There is often a need to consider, not only these differential operations, but also the integration of field quantities along lines, over surfaces and throughout volumes. In general the integrand may be scalar or vector in nature, but the evaluation of such integrals involves their reduction to one or more scalar integrals, which are then evaluated. In the case of surface and volume integrals this requires the evaluation of double and triple integrals.

3.1 Line integrals

In this section we discuss *line* or *path integrals*, in which some quantity related to the field is integrated between two given points in space, A and B, along a prescribed curve C that joins them. In general, we may encounter line integrals of the forms

$$\int_C \phi\, d\mathbf{r}, \qquad \int_C \mathbf{a}\cdot d\mathbf{r}, \qquad \int_C \mathbf{a}\times d\mathbf{r}, \tag{3.1}$$

where ϕ is a scalar field and $\mathbf{a}$ is a vector field. The three integrals themselves are respectively vector, scalar and vector in nature. As we will see below, in physical applications line integrals of the second type are by far the most common.

The formal definition of a line integral closely follows that of ordinary integrals and can be considered as the limit of a sum. We may divide the path C joining the points A and B into N small line elements $\Delta\mathbf{r}_p$, $p = 1, \ldots, N$. If (x_p, y_p, z_p) is any point on the line element $\Delta\mathbf{r}_p$ then the second type of line integral in (3.1), for example, is defined as

$$\int_C \mathbf{a}\cdot d\mathbf{r} = \lim_{N\to\infty} \sum_{p=1}^{N} \mathbf{a}(x_p, y_p, z_p)\cdot \Delta\mathbf{r}_p,$$

where it is assumed that all $|\Delta\mathbf{r}_p| \to 0$ as $N \to \infty$.

Each of the line integrals in (3.1) is evaluated over some curve C that may be either open (A and B being distinct points) or closed (the curve C forms a loop, so that A and B are coincident). In the case where C is closed, the line integral is written $\oint_C$ to indicate this. The curve may be given either parametrically by $\mathbf{r}(u) = x(u)\mathbf{i} + y(u)\mathbf{j} + z(u)\mathbf{k}$ or by means of simultaneous equations relating x, y, z for the given path (in Cartesian coordinates).

In general, the value of the line integral depends not only on the end-points A and B but also on the path C joining them. For a closed curve we must also specify the direction around the loop in which the integral is taken. It is usually taken to be such that a person walking around the loop C in this direction always has the region R on his/her left; this is equivalent to traversing C in the anticlockwise direction (as viewed from above).

3.1.1 Evaluating line integrals

The method of evaluating a line integral is to reduce it to a set of scalar integrals. It is usual to work in Cartesian coordinates, in which case $d\mathbf{r} = dx\,\mathbf{i} + dy\,\mathbf{j} + dz\,\mathbf{k}$. The first type of line integral in (3.1) then becomes simply

$$\int_C \phi\, d\mathbf{r} = \mathbf{i}\int_C \phi(x, y, z)\, dx + \mathbf{j}\int_C \phi(x, y, z)\, dy + \mathbf{k}\int_C \phi(x, y, z)\, dz.$$

The three integrals on the RHS are ordinary scalar integrals that can be evaluated in the usual way once the path of integration C has been specified. Note that in the above we have used relations of the form

$$\int \phi\, \mathbf{i}\, dx = \mathbf{i}\int \phi\, dx,$$

which is allowable since the Cartesian unit vectors are of constant magnitude and direction and hence may be taken out of the integral. If we had been using a different coordinate system, such as spherical polars, then, as we saw in the previous chapter, the unit basis vectors would not be constant. In that case the basis vectors could not be factorized out of the integral.

The second and third line integrals in (3.1) can also be reduced to a set of scalar integrals by writing the vector field $\mathbf{a}$ in terms of its Cartesian components as $\mathbf{a} = a_x\mathbf{i} + a_y\mathbf{j} + a_z\mathbf{k}$, where a_x, a_y and a_z are each (in general) functions of x, y and z. The second line integral in (3.1), for example, can then be written as

$$\begin{aligned}\int_C \mathbf{a}\cdot d\mathbf{r} &= \int_C (a_x\mathbf{i} + a_y\mathbf{j} + a_z\mathbf{k})\cdot(dx\,\mathbf{i} + dy\,\mathbf{j} + dz\,\mathbf{k}) \\ &= \int_C (a_x\, dx + a_y\, dy + a_z\, dz) \\ &= \int_C a_x\, dx + \int_C a_y\, dy + \int_C a_z\, dz. \end{aligned} \quad (3.2)$$

A similar procedure may be followed for the third type of line integral in (3.1), which involves a cross product.[1]

Line integrals have properties that are analogous to those of ordinary integrals. In particular, the following are useful properties (which we illustrate using the second form of line integral in (3.1) but which are valid for all three types).

1 Write out this integral explicitly in Cartesian coordinates.

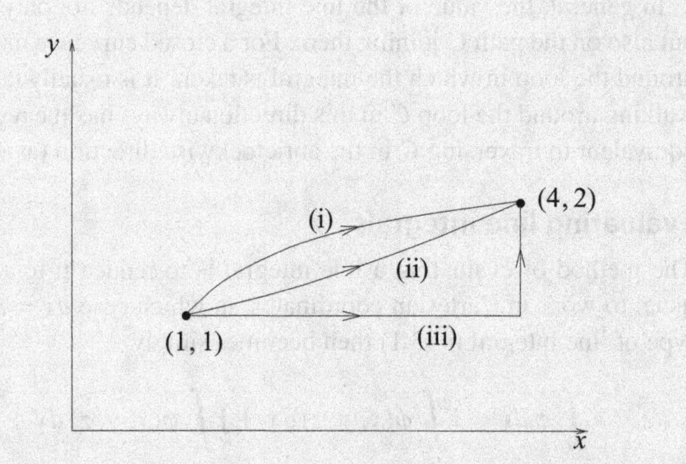

Figure 3.1 Different possible paths between the points (1, 1) and (4, 2).

(i) Reversing the path of integration changes the sign of the integral. If the path C along which the line integrals are evaluated has A and B as its end-points then

$$\int_A^B \mathbf{a} \cdot d\mathbf{r} = -\int_B^A \mathbf{a} \cdot d\mathbf{r}.$$

This implies that if the path C is a loop then integrating around the loop in the opposite direction changes the sign of the integral.

(ii) If the path of integration is subdivided into smaller segments then the sum of the separate line integrals along each segment is equal to the line integral along the whole path. So, if P is any point on the path of integration that lies between the path's end-points A and B then

$$\int_A^B \mathbf{a} \cdot d\mathbf{r} = \int_A^P \mathbf{a} \cdot d\mathbf{r} + \int_P^B \mathbf{a} \cdot d\mathbf{r}.$$

Example Evaluate the line integral $I = \int_C \mathbf{a} \cdot d\mathbf{r}$, where $\mathbf{a} = (x + y)\mathbf{i} + (y - x)\mathbf{j}$, along each of the paths in the xy-plane shown in Figure 3.1, namely

(i) the parabola $y^2 = x$ from (1, 1) to (4, 2),
(ii) the curve $x = 2u^2 + u + 1$, $y = 1 + u^2$ from (1, 1) to (4, 2),
(iii) the line $y = 1$ from (1, 1) to (4, 1), followed by the line $x = 4$ from (4, 1).

Since each of the paths lies entirely in the xy-plane, we have $d\mathbf{r} = dx\,\mathbf{i} + dy\,\mathbf{j}$. We can therefore write the line integral as

$$I = \int_C \mathbf{a} \cdot d\mathbf{r} = \int_C [(x + y)\,dx + (y - x)\,dy]. \tag{3.3}$$

We must now evaluate this line integral along each of the prescribed paths.

Case (i). Along the parabola $y^2 = x$ we have $2y\,dy = dx$. Substituting for x in (3.3) and using just the limits on y, we obtain

$$I = \int_{(1,1)}^{(4,2)} [(x+y)\,dx + (y-x)\,dy] = \int_1^2 [(y^2+y)2y + (y-y^2)]\,dy = 11\tfrac{1}{3}.$$

Note that we could just as easily have substituted for y and obtained an integral in x, which would have given the same result.[2]

Case (ii). The second path is given in terms of a parameter u. We could eliminate u between the two equations to obtain a relationship between x and y directly and proceed as above, but it is usually quicker to write the line integral in terms of the parameter u. Along the curve $x = 2u^2 + u + 1$, $y = 1 + u^2$ we have $dx = (4u+1)\,du$ and $dy = 2u\,du$. Substituting for x and y in (3.3) and writing the correct limits on u, we obtain

$$\begin{aligned} I &= \int_{(1,1)}^{(4,2)} [(x+y)\,dx + (y-x)\,dy] \\ &= \int_0^1 [(3u^2+u+2)(4u+1) - (u^2+u)2u]\,du = 10\tfrac{2}{3}. \end{aligned}$$

Case (iii). For the third path the line integral must be evaluated along the two line segments separately and the results added together. First, along the line $y = 1$ we have $dy = 0$. Substituting this into (3.3) and using just the limits on x for this segment, we obtain

$$\int_{(1,1)}^{(4,1)} [(x+y)\,dx + (y-x)\,dy] = \int_1^4 (x+1)\,dx = 10\tfrac{1}{2}.$$

Next, along the line $x = 4$ we have $dx = 0$. Substituting this into (3.3) and using just the limits on y for this segment, we obtain

$$\int_{(4,1)}^{(4,2)} [(x+y)\,dx + (y-x)\,dy] = \int_1^2 (y-4)\,dy = -2\tfrac{1}{2}.$$

The value of the line integral along the whole path is just the sum of the values of the line integrals along each segment, and is given by $I = 10\tfrac{1}{2} - 2\tfrac{1}{2} = 8$. ◀

When calculating a line integral along some curve C, which is given in terms of x, y and z, we are sometimes faced with the problem that the curve C is such that x, y and z are not single-valued functions of one another over the entire length of the curve. This is a particular problem for closed loops in the xy-plane (and also for some open curves). In such cases the path may be subdivided into shorter line segments along which one coordinate is a single-valued function of the other two. The sum of the line integrals along these segments is then equal to the line integral along the entire curve C. A better solution, however, is to represent the curve in a parametric form $\mathbf{r}(u)$ that is valid for its entire length.

2 Show that this is so.

Example Evaluate the line integral $I = \oint_C x\,dy$, where C is the circle in the xy-plane defined by $x^2 + y^2 = a^2$, $z = 0$.

Adopting the usual convention mentioned above, the circle C is to be traversed in the anticlockwise direction. Taking the circle as a whole means x is not a single-valued function of y. We must therefore divide the path into two parts with $x = +\sqrt{a^2 - y^2}$ for the semi-circle lying to the right of $x = 0$, and $x = -\sqrt{a^2 - y^2}$ for the semi-circle lying to the left of $x = 0$. The required line integral is then the sum of the integrals along the two semi-circles. Substituting for x, and then setting $y = a\sin\theta$, it is given by

$$\begin{aligned} I = \oint_C x\,dy &= \int_{-a}^{a} \sqrt{a^2 - y^2}\,dy + \int_{a}^{-a} \left(-\sqrt{a^2 - y^2}\right) dy \\ &= 4\int_0^a \sqrt{a^2 - y^2}\,dy \\ &= 4a^2 \int_0^{\pi/2} \sqrt{1 - \sin^2\theta}\cos\theta\,d\theta = 4a^2 \int_0^{\pi/2} \cos^2\theta\,d\theta = \pi a^2. \end{aligned}$$

Alternatively, we can represent the entire circle parametrically, in terms of the azimuthal angle ϕ, so that $x = a\cos\phi$ and $y = a\sin\phi$ with ϕ running from 0 to 2π. The integral can now be evaluated over the whole circle at once. Noting that $dy = a\cos\phi\,d\phi$, we can rewrite the line integral completely in terms of the parameter ϕ and obtain

$$I = \oint_C x\,dy = a^2 \int_0^{2\pi} \cos^2\phi\,d\phi = \pi a^2.$$

The final evaluation used the fact that the integral with respect to ϕ of $\cos^2\lambda\phi$ (or $\sin^2\lambda\phi$) over any range that is $m\pi$ in length, is equal to $\frac{1}{2}$ times the length of the range if λm is an integer.[3]

◀

3.1.2 Physical examples of line integrals

There are many physical examples of line integrals, but perhaps the most common is the expression for the total work done by a force $\mathbf{F}$ when it moves its point of application from a point A to a point B along a given curve C. We allow the magnitude and direction of $\mathbf{F}$ to vary along the curve. Let the force act at a point $\mathbf{r}$ and consider a small displacement $d\mathbf{r}$ along the curve; then the small amount of work done is given by the scalar product $dW = \mathbf{F}\cdot d\mathbf{r}$ (note that dW can be either positive or negative). Therefore, the total work done in traversing the path C is

$$W_C = \int_C \mathbf{F}\cdot d\mathbf{r}.$$

Naturally, other physical quantities can be expressed in such a way. For example, the electrostatic potential energy gained by moving a charge q along a path C in an electric field $\mathbf{E}$ is $-q\int_C \mathbf{E}\cdot d\mathbf{r}$. We may also note that Ampère's law concerning the magnetic

3 A result worth remembering, since the squares of sinusoids occur throughout the mathematics of physics and engineering.

field $\mathbf{B}$ associated with a current-carrying wire can be written as

$$\oint_C \mathbf{B} \cdot d\mathbf{r} = \mu_0 I,$$

where I is the current enclosed by a closed path C traversed in a right-handed sense with respect to the current direction.

Magnetostatics also provides a physical example of the third type of line integral in (3.1). If a loop of wire C carrying a current I is placed in a magnetic field $\mathbf{B}$ then the force $d\mathbf{F}$ on a small length $d\mathbf{r}$ of the wire is given by $d\mathbf{F} = I\, d\mathbf{r} \times \mathbf{B}$, and so the total (vector) force on the loop is

$$\mathbf{F} = I \oint_C d\mathbf{r} \times \mathbf{B}.$$

3.1.3 Line integrals with respect to a scalar

In addition to those listed in (3.1), we can form other types of line integral, which depend on a particular curve C but for which we integrate with respect to a scalar du, rather than the vector differential $d\mathbf{r}$. This distinction is somewhat arbitrary, however, since we can always rewrite line integrals containing the vector differential $d\mathbf{r}$ as a line integral with respect to some scalar parameter. If the path C along which the integral is taken is described parametrically by $\mathbf{r}(u)$ then

$$d\mathbf{r} = \frac{d\mathbf{r}}{du}\, du,$$

and the second type of line integral in (3.1), for example, can be written as

$$\int_C \mathbf{a} \cdot d\mathbf{r} = \int_C \mathbf{a} \cdot \frac{d\mathbf{r}}{du}\, du.$$

A similar procedure can be followed for the other types of line integral in (3.1).

Commonly occurring special cases of line integrals with respect to a scalar are

$$\int_C \phi\, ds, \qquad \int_C \mathbf{a}\, ds,$$

where s is the arc length along the curve C. We can always represent C parametrically by $\mathbf{r}(u)$, and since

$$(ds)^2 = (dx)^2 + (dy)^2 + (dz)^2 = d\mathbf{r} \cdot d\mathbf{r},$$

we have

$$\left(\frac{ds}{du}\right)^2 = \frac{d\mathbf{r}}{du} \cdot \frac{d\mathbf{r}}{du}.$$

Consequently we may write

$$ds = \sqrt{\frac{d\mathbf{r}}{du} \cdot \frac{d\mathbf{r}}{du}}\, du.$$

The line integrals can therefore be expressed entirely in terms of the parameter u and thence evaluated.

Example Evaluate the line integral $I = \int_C (x-y)^2\,ds$, where C is the semi-circle of radius a running from $A = (a, 0)$ to $B = (-a, 0)$ and for which $y \geq 0$.

The semi-circular path from A to B can be described in terms of the azimuthal angle ϕ (measured from the x-axis) by

$$\mathbf{r}(\phi) = a\cos\phi\,\mathbf{i} + a\sin\phi\,\mathbf{j},$$

where ϕ runs from 0 to π. Therefore the element of arc length is given by

$$ds = \sqrt{\frac{d\mathbf{r}}{d\phi}\cdot\frac{d\mathbf{r}}{d\phi}}\,d\phi = a[(-\sin\phi)^2 + (\cos\phi)^2]\,d\phi = a\,d\phi.$$

Since $(x-y)^2 = a^2(\cos^2\phi - 2\sin\phi\cos\phi + \sin^2\phi) = a^2(1 - \sin 2\phi)$, the line integral becomes

$$I = \int_C (x-y)^2\,ds = \int_0^\pi a^3(1 - \sin 2\phi)\,d\phi = \pi a^3.$$

As a further illustration of the importance of the integral path chosen, we note that the integral directly along the x-axis, between the same end-points and with the same integrand, has the negative value of $-\frac{2}{3}a^3$ (as the reader may wish to verify). ◀

Finally in this section, we note the form for an element of arc length in three dimensions. As discussed in the previous chapter, the expression (2.53) for its square in general three-dimensional orthogonal curvilinear coordinates u_1, u_2, u_3 is

$$(ds)^2 = h_1^2\,(du_1)^2 + h_2^2\,(du_2)^2 + h_3^2\,(du_3)^2,$$

where h_1, h_2, h_3 are the scale factors of the coordinate system. If a curve C in three dimensions is given parametrically by the equations $u_i = u_i(\lambda)$ for $i = 1, 2, 3$ then the element of arc length along the curve is[4]

$$ds = \sqrt{h_1^2\left(\frac{du_1}{d\lambda}\right)^2 + h_2^2\left(\frac{du_2}{d\lambda}\right)^2 + h_3^2\left(\frac{du_3}{d\lambda}\right)^2}\,d\lambda.$$

3.2 Connectivity of regions

In physical systems it is usual to define a scalar or vector field in some region R. In the next and some later sections we will need the concept of the *connectivity* of such a region in both two and three dimensions.

We begin by discussing planar regions. A plane region R is said to be *simply connected* if every simple closed curve within R can be continuously shrunk to a point without leaving the region [see Figure 3.2(a)]. If, however, the region R contains a hole then there exist simple closed curves that cannot be shrunk to a point without leaving R [see

4 Express the relationship between ds and $d\phi$ for a spiral given in cylindrical polar coordinates by $\rho = k\phi^2$ and $z = \mu\phi^3$.

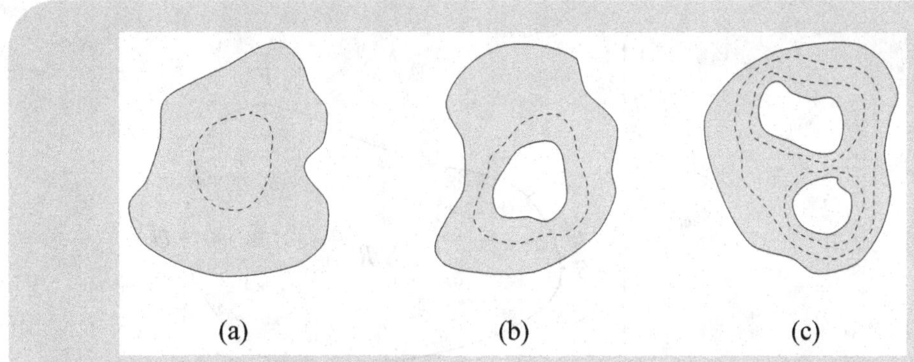

Figure 3.2 (a) A simply connected region; (b) a doubly connected region; (c) a triply connected region.

Figure 3.2(b)]. Such a region is said to be doubly connected, since its boundary has two distinct parts. Similarly, a region with $n - 1$ holes is said to be *n-fold connected*, or *multiply connected* (the region in Figure 3.2(c) is triply connected).

These ideas can be extended to regions that are not planar, such as general three-dimensional surfaces and volumes. The same criteria concerning the shrinking of closed curves to a point also apply when deciding the connectivity of such regions. In these cases, however, the curves must lie in the surface or volume in question. For example, the interior of a torus is not simply connected, since there exist closed curves in the interior that cannot be shrunk to a point without leaving the torus. The region between two concentric spheres of different radii is simply connected.[5]

3.3 Green's theorem in a plane

In Subsection 3.1.1 we considered (amongst other things) the evaluation of line integrals for which the path C is closed and lies entirely in the xy-plane. Since the path is closed it will enclose a region R of the plane. We now show how to express the line integral around the loop as a double integral over the enclosed region R.

Suppose the functions $P(x, y)$, $Q(x, y)$ and their partial derivatives are single-valued, finite and continuous inside and on the boundary C of some simply connected region R in the xy-plane. *Green's theorem in a plane* (sometimes called the divergence theorem in two dimensions) then states

$$\oint_C (P\,dx + Q\,dy) = \iint_R \left(\frac{\partial Q}{\partial x} - \frac{\partial P}{\partial y} \right) dx\,dy, \tag{3.4}$$

and so relates the line integral around C to a double integral over the enclosed region R. This theorem may be proved straightforwardly in the following way. Consider the simply

5 Are the following simply or multiply connected: (a) the glass of a wine glass, (b) the clay of a coffee cup, and (c) the clay of a Pythagorean cup with the connecting hole blocked with clay (consult Wikipedia if you are not familiar with one)?

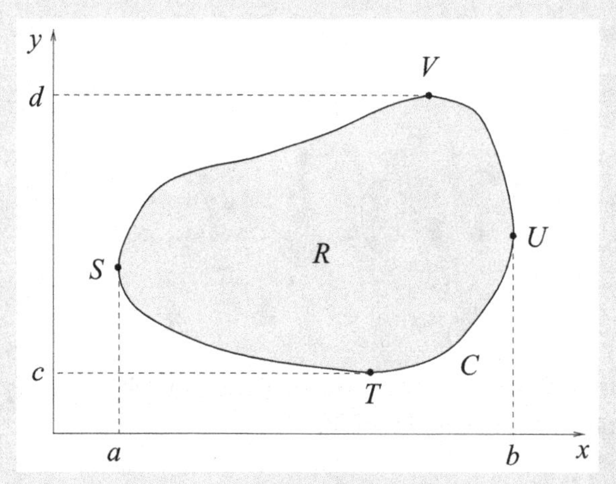

Figure 3.3 A simply connected region R bounded by the curve C.

connected region R in Figure 3.3, and let $y = y_1(x)$ and $y = y_2(x)$ be the equations of the curves STU and SVU respectively. We then write

$$\begin{aligned}\iint_R \frac{\partial P}{\partial y}\,dx\,dy &= \int_a^b dx \int_{y_1(x)}^{y_2(x)} dy\,\frac{\partial P}{\partial y} = \int_a^b dx\Big[P(x,y)\Big]_{y=y_1(x)}^{y=y_2(x)}\\ &= \int_a^b \Big[P(x,y_2(x)) - P(x,y_1(x))\Big]\,dx\\ &= -\int_a^b P(x,y_1(x))\,dx - \int_b^a P(x,y_2(x))\,dx = -\oint_C P\,dx.\end{aligned}$$

If we now let $x = x_1(y)$ and $x = x_2(y)$ be the equations of the curves TSV and TUV respectively, we can similarly show that

$$\begin{aligned}\iint_R \frac{\partial Q}{\partial x}\,dx\,dy &= \int_c^d dy \int_{x_1(y)}^{x_2(y)} dx\,\frac{\partial Q}{\partial x} = \int_c^d dy\Big[Q(x,y)\Big]_{x=x_1(y)}^{x=x_2(y)}\\ &= \int_c^d \Big[Q(x_2(y),y) - Q(x_1(y),y)\Big]\,dy\\ &= \int_d^c Q(x_1,y)\,dy + \int_c^d Q(x_2,y)\,dy = \oint_C Q\,dy.\end{aligned}$$

Subtracting these two results gives Green's theorem in a plane.[6]

6 Notice that there is no necessary connection between $P(x, y)$ and $Q(x, y)$ and that each result could stand alone. The difference in sign is simply the combined result of the conventional choices for (i) the x- and y-axes and (ii) the positive direction of traversing a closed contour.

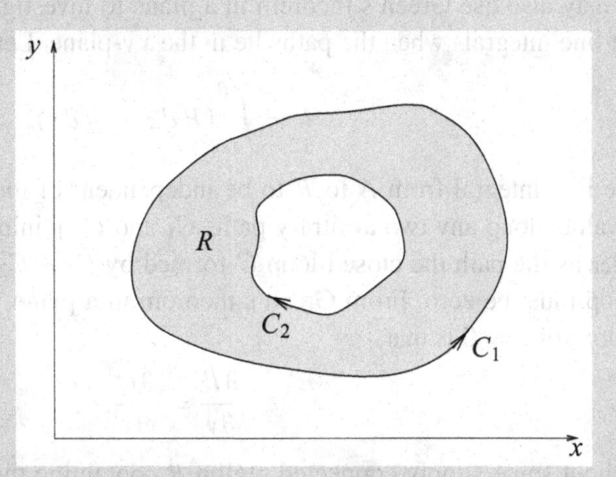

Figure 3.4 A doubly connected region R bounded by the curves C_1 and C_2.

Example Show that the area of a region R enclosed by a simple closed curve C is given by $A = \frac{1}{2}\oint_C (x\,dy - y\,dx) = \oint_C x\,dy = -\oint_C y\,dx$. Hence calculate the area of the ellipse $x = a\cos\phi$, $y = b\sin\phi$.

In Green's theorem (3.4) put $P = -y$ and $Q = x$; then

$$\oint_C (x\,dy - y\,dx) = \iint_R (1+1)\,dx\,dy = 2\iint_R dx\,dy = 2A.$$

Therefore the area of the region is $A = \frac{1}{2}\oint_C (x\,dy - y\,dx)$. Alternatively, we could put $P = 0$ and $Q = x$ and obtain $A = \oint_C x\,dy$, or put $P = -y$ and $Q = 0$, which gives $A = -\oint_C y\,dx$.

The area of the ellipse $x = a\cos\phi$, $y = b\sin\phi$ is given by

$$A = \frac{1}{2}\oint_C (x\,dy - y\,dx) = \frac{1}{2}\int_0^{2\pi} ab(\cos^2\phi + \sin^2\phi)\,d\phi$$
$$= \frac{ab}{2}\int_0^{2\pi} d\phi = \pi ab.$$

The parameterization used here is that for the standard form of an ellipse, $x^2/a^2 + y^2/b^2 = 1$. ◄

It may further be shown that Green's theorem in a plane is also valid for multiply connected regions. In this case, the line integral must be taken over all the distinct boundaries of the region. Furthermore, each boundary must be traversed in the positive direction, so that a person traveling along it in this direction always has the region R on their left. In order to apply Green's theorem to the region R shown in Figure 3.4, the line integrals must be taken over both boundaries, C_1 and C_2, in the directions indicated, and the results added together.[7]

7 A coin has the form of a uniform circular disc of radius a with a central circular hole of radius b removed from it. By setting $Q = xy^2$ and $P = -x^2y$, show that its moment of inertia about a central axis perpendicular to its plane is $\frac{1}{2}m(a^2 + b^2)$, where m is the mass of the coin.

We may also use Green's theorem in a plane to investigate the path independence (or not) of line integrals when the paths lie in the xy-plane. Let us consider the line integral

$$I = \int_A^B (P\,dx + Q\,dy).$$

For the line integral from A to B to be independent of the path taken, it must have the same value along any two arbitrary paths C_1 and C_2 joining the points. Moreover, if we consider as the path the closed loop C formed by $C_1 - C_2$ then the line integral around this loop must be zero. From Green's theorem in a plane, (3.4), we see that a *sufficient* condition for $I = 0$ is that

$$\frac{\partial P}{\partial y} = \frac{\partial Q}{\partial x}, \tag{3.5}$$

throughout some simply connected region R containing the loop, where we assume that these partial derivatives are continuous in R.

It may be shown that (3.5) is also a *necessary* condition for $I = 0$ and is equivalent to requiring $P\,dx + Q\,dy$ to be an exact differential of some function $\psi(x, y)$ such that $P\,dx + Q\,dy = d\psi$. It follows that $\int_A^B (P\,dx + Q\,dy) = \psi(B) - \psi(A)$ and that $\oint_C (P\,dx + Q\,dy)$ around any closed loop C in the region R is identically zero.[8] These results are special cases of the general results for paths in three dimensions, which are discussed in the next section.

Example Evaluate the line integral

$$I = \oint_C \left[(e^x y + \cos x \sin y)\,dx + (e^x + \sin x \cos y)\,dy\right],$$

around the ellipse $x^2/a^2 + y^2/b^2 = 1$.

Clearly, it is not straightforward to calculate this line integral directly. However, if we let

$$P = e^x y + \cos x \sin y \qquad \text{and} \qquad Q = e^x + \sin x \cos y,$$

then $\partial P/\partial y = e^x + \cos x \cos y = \partial Q/\partial x$, and so $P\,dx + Q\,dy$ is an exact differential (it is actually the differential of the function $f(x, y) = e^x y + \sin x \sin y$). From the above discussion, we can conclude immediately that $I = 0$. ◀

3.4 Conservative fields and potentials

So far we have made the point that, in general, the value of a line integral between two points A and B depends on the path C taken from A to B. In the previous section, however, we saw that, for paths in the xy-plane, line integrals whose integrands have

8 Show that this is the case if $Q = xy^2$ and $P = x^2 y$ (compare with footnote 7) and C is any circular contour of radius r centered on the origin. Identify ψ in this case.

certain properties are independent of the path taken. We now extend that discussion to the full three-dimensional case.

For line integrals of the form $\int_C \mathbf{a}\cdot d\mathbf{r}$, there exists a class of vector fields for which the line integral between two points is *independent* of the path taken. Such vector fields are called *conservative*. A vector field $\mathbf{a}$ that has continuous partial derivatives in a simply connected region R is conservative if, and only if, any of the following is true.[9]

(i) The integral $\int_A^B \mathbf{a}\cdot d\mathbf{r}$, where A and B lie in the region R, is independent of the path from A to B. Hence the integral $\oint_C \mathbf{a}\cdot d\mathbf{r}$ around any closed loop in R is zero.
(ii) There exists a single-valued function ϕ of position such that $\mathbf{a} = \nabla\phi$.
(iii) $\nabla \times \mathbf{a} = \mathbf{0}$.
(iv) $\mathbf{a}\cdot d\mathbf{r}$ is an exact differential.

The validity or otherwise of any of these statements implies the same for the other three, as we will now show.

First, let us assume that (i) above is true. If the line integral from A to B is independent of the path taken between the points then its value must be a function only of the positions of A and B. We may therefore write

$$\int_A^B \mathbf{a}\cdot d\mathbf{r} = \phi(B) - \phi(A), \tag{3.6}$$

which defines a single-valued scalar function of position ϕ. If the points A and B are separated by an infinitesimal displacement $d\mathbf{r}$ then (3.6) becomes

$$\mathbf{a}\cdot d\mathbf{r} = d\phi,$$

which shows that we require $\mathbf{a}\cdot d\mathbf{r}$ to be an exact differential: condition (iv). From (2.22) we can write $d\phi = \nabla\phi\cdot d\mathbf{r}$, and so we have

$$(\mathbf{a} - \nabla\phi)\cdot d\mathbf{r} = 0.$$

Since $d\mathbf{r}$ is arbitrary, we find that $\mathbf{a} = \nabla\phi$; this immediately implies $\nabla\times\mathbf{a} = \mathbf{0}$, condition (iii) [see (2.32)].

Alternatively, if we suppose that there exists a single-valued function of position ϕ such that $\mathbf{a} = \nabla\phi$ then $\nabla\times\mathbf{a} = \mathbf{0}$ follows as before. The line integral around a closed loop then becomes

$$\oint_C \mathbf{a}\cdot d\mathbf{r} = \oint_C \nabla\phi\cdot d\mathbf{r} = \oint d\phi.$$

Since we defined ϕ to be single-valued, this integral is zero as required.

Now suppose $\nabla\times\mathbf{a} = \mathbf{0}$. From Stoke's theorem, which is discussed in Section 3.9, we immediately obtain $\oint_C \mathbf{a}\cdot d\mathbf{r} = 0$; then $\mathbf{a} = \nabla\phi$ and $\mathbf{a}\cdot d\mathbf{r} = d\phi$ follow as above.

9 It may be helpful to keep in mind the physical example of a charge q in an electric field $\mathbf{E}$. Then the electrostatic potential is $-\phi$ with $\mathbf{E} = \nabla\phi$, the integral $q\int_A^B \mathbf{E}\cdot d\mathbf{r}$ is the work done on the charge as it moves from A to B (by any route), and $\phi(A) - \phi(B)$ is the increase (or decrease if negative) in the potential energy of the charge.

Finally, let us suppose $\mathbf{a}\cdot d\mathbf{r} = d\phi$. Then immediately we have $\mathbf{a} = \nabla\phi$, and the other results follow as above.

Example Evaluate the line integral $I = \int_A^B \mathbf{a}\cdot d\mathbf{r}$, where $\mathbf{a} = (xy^2 + z)\mathbf{i} + (x^2y + 2)\mathbf{j} + x\mathbf{k}$, A is the point (c, c, h) and B is the point $(2c, c/2, h)$, along the different paths

(i) C_1, given by $x = cu$, $y = c/u$, $z = h$,

(ii) C_2, given by $2y = 3c - x$, $z = h$.

Show that the vector field $\mathbf{a}$ is in fact conservative, and find ϕ such that $\mathbf{a} = \nabla\phi$.

Expanding out the integrand, we have

$$I = \int_{(c,\,c,\,h)}^{(2c,\,c/2,\,h)} \left[(xy^2 + z)\,dx + (x^2y + 2)\,dy + x\,dz\right], \tag{3.7}$$

which we must evaluate along each of the paths C_1 and C_2.

(i) Along C_1 we have $dx = c\,du$, $dy = -(c/u^2)\,du$, $dz = 0$, and on substituting in (3.7) and finding the limits on u, we obtain

$$I = \int_1^2 c\left(h - \frac{2}{u^2}\right) du = c(h-1).$$

(ii) Along C_2 we have $2\,dy = -dx$, $dz = 0$ and, on substituting in (3.7) and using the limits on x, we obtain

$$I = \int_c^{2c} \left(\tfrac{1}{2}x^3 - \tfrac{9}{4}cx^2 + \tfrac{9}{4}c^2x + h - 1\right) dx = c(h-1).$$

Hence the line integral has the same value along paths C_1 and C_2. Taking the curl of $\mathbf{a}$, we have

$$\nabla\times\mathbf{a} = (0-0)\mathbf{i} + (1-1)\mathbf{j} + (2xy - 2xy)\mathbf{k} = \mathbf{0},$$

so $\mathbf{a}$ is a conservative vector field, and the line integral between two points must be independent of the path taken. Since $\mathbf{a}$ is conservative, we can write $\mathbf{a} = \nabla\phi$. Therefore, ϕ must satisfy

$$\frac{\partial\phi}{\partial x} = xy^2 + z,$$

which implies that $\phi = \frac{1}{2}x^2y^2 + zx + f(y, z)$ for some function f. Secondly, we require

$$\frac{\partial\phi}{\partial y} = x^2y + \frac{\partial f}{\partial y} = x^2y + 2,$$

which implies $f = 2y + g(z)$. Finally, since

$$\frac{\partial\phi}{\partial z} = x + \frac{\partial g}{\partial z} = x,$$

we have $g = \text{constant} = k$. It can be seen that we have explicitly constructed the function $\phi = \frac{1}{2}x^2y^2 + zx + 2y + k$. ◀

The quantity ϕ that figures so prominently in this section is called the *scalar potential function* of the conservative vector field $\mathbf{a}$ (which satisfies $\nabla\times\mathbf{a} = \mathbf{0}$), and is unique up to an arbitrary additive constant. Scalar potentials that are multivalued functions of position (but in simple ways) are also of value in describing some physical situations, the most obvious example being the scalar magnetic potential associated with a current-carrying wire. When the integral of a field quantity around a closed loop is considered, provided the loop does not enclose a net current, the potential is single-valued and all the above results still hold. If the loop does enclose a net current, however, our analysis is no longer valid and extra care must be taken.

If, instead of being conservative, a vector field $\mathbf{b}$ satisfies $\nabla \cdot \mathbf{b} = 0$ (i.e. $\mathbf{b}$ is solenoidal) then it is both possible and useful, for example in the theory of electromagnetism, to define a *vector field* $\mathbf{a}$ such that $\mathbf{b} = \nabla \times \mathbf{a}$. It may be shown that such a vector field $\mathbf{a}$ always exists. Further, if $\mathbf{a}$ is one such vector field then $\mathbf{a}' = \mathbf{a} + \nabla\psi + \mathbf{c}$, where ψ is any scalar function and $\mathbf{c}$ is any constant vector, also satisfies the above relationship, i.e. $\mathbf{b} = \nabla \times \mathbf{a}'$. This was discussed more fully in Subsection 2.7.2.

3.5 Surface integrals

As with line integrals, integrals over surfaces can involve vector and scalar fields and, equally, can result in either a vector or a scalar. The simplest case involves entirely scalars and is of the form

$$\int_S \phi \, dS. \tag{3.8}$$

As analogues of the line integrals listed in (3.1), we may also encounter surface integrals involving vectors, namely

$$\int_S \phi \, d\mathbf{S}, \qquad \int_S \mathbf{a} \cdot d\mathbf{S}, \qquad \int_S \mathbf{a} \times d\mathbf{S}. \tag{3.9}$$

All the above integrals are taken over some surface S, which may be either open or closed, and are therefore, in general, double integrals. Following the notation for line integrals, for surface integrals over a closed surface $\int_S$ is replaced by $\oint_S$.

The vector differential $d\mathbf{S}$ in (3.9) represents a vector area element of the surface S. It may also be written $d\mathbf{S} = \hat{\mathbf{n}}\, dS$, where $\hat{\mathbf{n}}$ is a unit normal to the surface at the position of the element and dS is the scalar area of the element used in (3.8). The convention for the direction of the normal $\hat{\mathbf{n}}$ to a surface depends on whether the surface is open or closed. A closed surface, see Figure 3.5(a), does not have to be simply connected (for example, the surface of a torus is not), but it does have to enclose a volume V, which may be of infinite extent. The direction of $\hat{\mathbf{n}}$ is taken to point outwards from the enclosed volume as shown. An open surface, see Figure 3.5(b), spans some perimeter curve C. The direction of $\hat{\mathbf{n}}$ is then given by the right-hand sense with respect to the direction in which the perimeter is traversed, i.e. follows the right-hand screw rule. An open surface does not have to be simply connected but for our purposes it must be two-sided (a Möbius strip is an example of a one-sided surface).

The formal definition of a surface integral is very similar to that of a line integral. We divide the surface S into N elements of area ΔS_p, $p = 1, 2, \ldots, N$, each with a unit normal $\hat{\mathbf{n}}_p$. If (x_p, y_p, z_p) is any point in ΔS_p then the second type of surface integral in

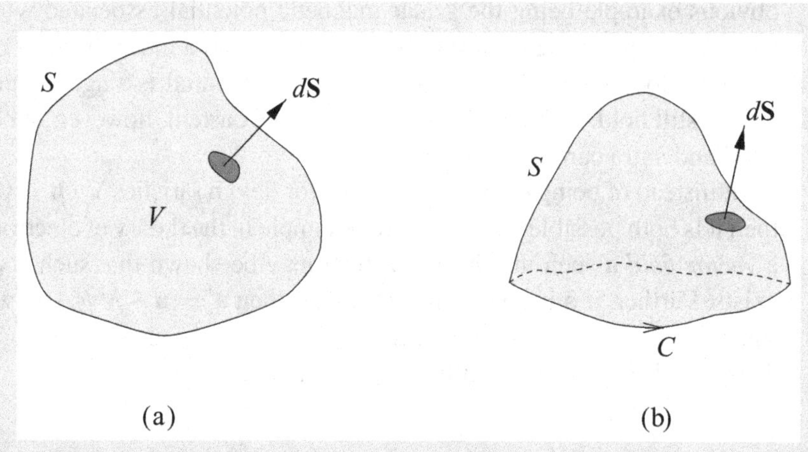

Figure 3.5 (a) A closed surface and (b) an open surface. In each case a normal to the surface is shown: $d\mathbf{S} = \hat{\mathbf{n}}\, dS$.

(3.9), for example, is defined as

$$\int_S \mathbf{a} \cdot d\mathbf{S} = \lim_{N\to\infty} \sum_{p=1}^{N} \mathbf{a}(x_p, y_p, z_p) \cdot \hat{\mathbf{n}}_p \Delta S_p,$$

where it is required that all $\Delta S_p \to 0$ as $N \to \infty$.

3.5.1 Evaluating surface integrals

We now consider how to evaluate surface integrals over some general surface. This involves writing the scalar area element dS in terms of the coordinate differentials of our chosen coordinate system. In some particularly simple cases this is very straightforward. For example, if S is the surface of a sphere of radius a (or some part thereof) then using spherical polar coordinates θ, ϕ on the sphere we have $dS = a^2 \sin\theta\, d\theta\, d\phi$. For a general surface, however, it is not usually possible to represent the surface in a simple way in any particular coordinate system. In such cases, it is usual to work in Cartesian coordinates and consider the projections of the surface onto the coordinate planes.

Consider a surface (or part of a surface) S as in Figure 3.6. The surface S is projected onto a region R of the xy-plane, so that an element of surface area dS projects onto the area element dA. From the figure, we see that $dA = |\cos\alpha|\, dS$, where α is the angle between the unit vector $\mathbf{k}$ in the z-direction and the unit normal $\hat{\mathbf{n}}$ to the surface at P. So, at any given point of S, we have simply

$$dS = \frac{dA}{|\cos\alpha|} = \frac{dA}{|\hat{\mathbf{n}} \cdot \mathbf{k}|}.$$

Now, if the surface S is given by the equation $f(x, y, z) = 0$ then, as shown in Subsection 2.6.1, the unit normal at any point of the surface is given by $\hat{\mathbf{n}} = \nabla f / |\nabla f|$ evaluated at that point, cf. (2.27). The scalar element of surface area then becomes

$$dS = \frac{dA}{|\hat{\mathbf{n}} \cdot \mathbf{k}|} = \frac{|\nabla f|\, dA}{\nabla f \cdot \mathbf{k}} = \frac{|\nabla f|\, dA}{\partial f/\partial z}, \tag{3.10}$$

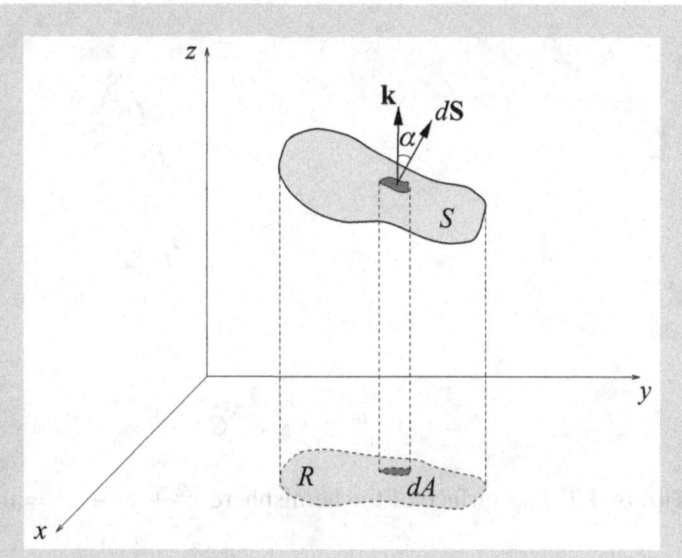

Figure 3.6 A surface S (or part thereof) projected onto a region R in the xy-plane; $d\mathbf{S}$ is a surface element.

where $|\nabla f|$ and $\partial f/\partial z$ are evaluated on the surface S. We can therefore express any surface integral over S as a double integral over the region R in the xy-plane.

In the following example both the specific and more general approaches are illustrated; not surprisingly, because of its more universal applicability, the latter is the longer.

Example Evaluate the surface integral $I = \int_S \mathbf{a} \cdot d\mathbf{S}$, where $\mathbf{a} = x\mathbf{i}$ and S is the surface of the hemisphere $x^2 + y^2 + z^2 = a^2$ with $z \geq 0$.

The surface of the hemisphere is shown in Figure 3.7. In this case dS may be easily expressed in spherical polar coordinates as $dS = a^2 \sin\theta \, d\theta \, d\phi$, and the unit normal to the surface at any point is simply $\hat{\mathbf{r}}$. On the surface of the hemisphere we have $x = a \sin\theta \cos\phi$ and so

$$\mathbf{a} \cdot d\mathbf{S} = x\,(\mathbf{i} \cdot \hat{\mathbf{r}})\, dS = (a \sin\theta \cos\phi)(\sin\theta \cos\phi)(a^2 \sin\theta \, d\theta \, d\phi).$$

Therefore, inserting the correct limits on θ and ϕ, we have

$$I = \int_S \mathbf{a} \cdot d\mathbf{S} = a^3 \int_0^{\pi/2} d\theta \, \sin^3\theta \int_0^{2\pi} d\phi \, \cos^2\phi = \frac{2\pi a^3}{3}.$$

We could, however, follow the general prescription above and project the hemisphere S onto the region R in the xy-plane that is a circle of radius a centered at the origin. Writing the equation of the surface of the hemisphere as $f(x, y) = x^2 + y^2 + z^2 - a^2 = 0$ and using (3.10), we have

$$I = \int_S \mathbf{a} \cdot d\mathbf{S} = \int_S x\,(\mathbf{i} \cdot \hat{\mathbf{r}})\, dS = \int_R x\,(\mathbf{i} \cdot \hat{\mathbf{r}}) \frac{|\nabla f|\, dA}{\partial f/\partial z}.$$

Now $\nabla f = 2x\mathbf{i} + 2y\mathbf{j} + 2z\mathbf{k} = 2\mathbf{r}$, so on the surface S we have $|\nabla f| = 2|\mathbf{r}| = 2a$. On S we also have $\partial f/\partial z = 2z = 2\sqrt{a^2 - x^2 - y^2}$ and $\mathbf{i} \cdot \hat{\mathbf{r}} = x/a$. Therefore, the integral

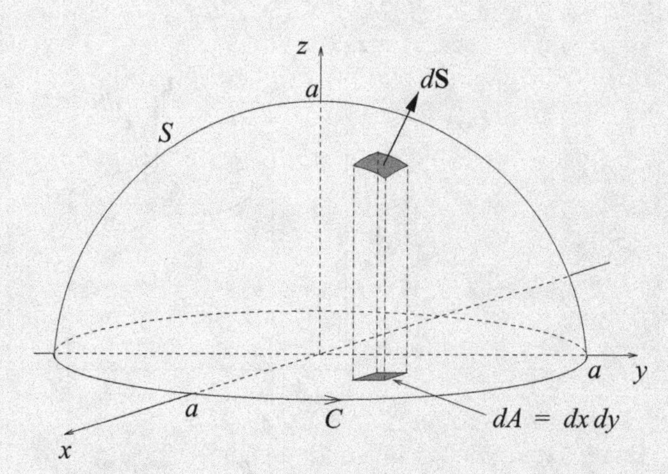

Figure 3.7 The surface of the hemisphere $x^2 + y^2 + z^2 = a^2$, $z \geq 0$.

becomes

$$I = \iint_R \frac{x^2}{\sqrt{a^2 - x^2 - y^2}}\, dx\, dy.$$

Although this integral may be evaluated directly, it is quicker to transform to plane polar coordinates:

$$\begin{aligned} I &= \iint_{R'} \frac{\rho^2 \cos^2\phi}{\sqrt{a^2 - \rho^2}}\, \rho\, d\rho\, d\phi \\ &= \int_0^{2\pi} \cos^2\phi\, d\phi \int_0^a \frac{\rho^3\, d\rho}{\sqrt{a^2 - \rho^2}}. \end{aligned}$$

Making the substitution $\rho = a \sin u$, we finally obtain

$$I = \int_0^{2\pi} \cos^2\phi\, d\phi \int_0^{\pi/2} a^3 \sin^3 u\, du = \frac{2\pi a^3}{3}.$$

The first integral contributes a factor π to the final answer and the second contributes $2a^3/3$. ◀

In the above discussion we assumed that any line parallel to the z-axis intersects S only once. If this is not the case, we must split up the surface into smaller surfaces S_1, S_2, etc. that are of this type. The surface integral over S is then the sum of the surface integrals over S_1, S_2 and so on. This is always necessary for closed surfaces.

Sometimes we may need to project a surface S (or some part of it) onto the zx- or yz-plane, rather than the xy-plane; for such cases, the above analysis is easily modified.

3.5.2 Vector areas of surfaces

The vector area of a surface S is defined as

$$\mathbf{S} = \int_S d\mathbf{S},$$

where the surface integral may be evaluated as above.

Example Find the vector area of the surface of the hemisphere $x^2 + y^2 + z^2 = a^2$ with $z \geq 0$.

As in the previous example, $d\mathbf{S} = a^2 \sin\theta \, d\theta \, d\phi \, \hat{\mathbf{r}}$ in spherical polar coordinates. Therefore the vector area is given by

$$\mathbf{S} = \iint_S a^2 \sin\theta \, \hat{\mathbf{r}} \, d\theta \, d\phi.$$

Now, since $\hat{\mathbf{r}}$ varies over the surface S, it also must be integrated. This is most easily achieved by writing $\hat{\mathbf{r}}$ in terms of the constant Cartesian basis vectors. On S we have

$$\hat{\mathbf{r}} = \sin\theta \cos\phi \, \mathbf{i} + \sin\theta \sin\phi \, \mathbf{j} + \cos\theta \, \mathbf{k},$$

so the expression for the vector area becomes

$$\begin{aligned}
\mathbf{S} &= \mathbf{i}\left(a^2 \int_0^{2\pi} \cos\phi \, d\phi \int_0^{\pi/2} \sin^2\theta \, d\theta\right) + \mathbf{j}\left(a^2 \int_0^{2\pi} \sin\phi \, d\phi \int_0^{\pi/2} \sin^2\theta \, d\theta\right) \\
&\quad + \mathbf{k}\left(a^2 \int_0^{2\pi} d\phi \int_0^{\pi/2} \sin\theta \cos\theta \, d\theta\right) \\
&= \mathbf{0} + \mathbf{0} + \pi a^2 \mathbf{k} = \pi a^2 \mathbf{k}.
\end{aligned}$$

Note that the magnitude of $\mathbf{S}$ is the projected area of the hemisphere onto the xy-plane, and not the surface area of the hemisphere. ◀

The hemispherical shell discussed above is an example of an open surface. For a closed surface, however, the vector area is always zero.[10] This may be seen by projecting the surface down onto each Cartesian coordinate plane in turn. For each projection, every positive element of area on the upper surface is canceled by the corresponding negative element on the lower surface. Therefore, each component of $\mathbf{S} = \oint_S d\mathbf{S}$ vanishes.

An important corollary of this result is that the vector area of an open surface depends only on its perimeter, or boundary curve, C. This may be proved as follows. If surfaces S_1 and S_2 have the same perimeter then $S_1 - S_2$ is a closed surface, for which

$$\oint d\mathbf{S} = \int_{S_1} d\mathbf{S} - \int_{S_2} d\mathbf{S} = \mathbf{0}.$$

Hence $\mathbf{S}_1 = \mathbf{S}_2$. Moreover, we may derive an expression for the vector area of an open surface S solely in terms of a line integral around its perimeter C. Since we may choose any surface with perimeter C, we will consider a cone with its vertex at the origin (see Figure 3.8). The vector area of the elementary triangular region shown in the figure is

10 Use this result to reduce the solution to the previous worked example to a single sentence.

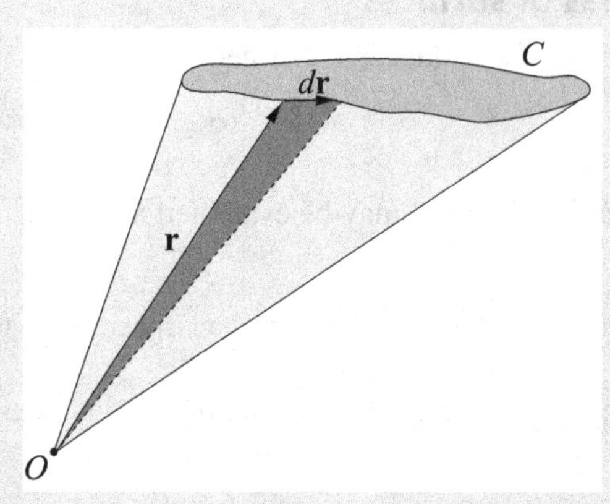

Figure 3.8 The conical surface spanning the perimeter C and having its vertex at the origin.

$d\mathbf{S} = \frac{1}{2}\mathbf{r} \times d\mathbf{r}$.[11] Therefore, the vector area of the cone, and hence of *any* open surface with perimeter C, is given by the line integral[12]

$$\mathbf{S} = \frac{1}{2}\oint_C \mathbf{r} \times d\mathbf{r}.$$

For a surface confined to the xy-plane, $\mathbf{r} = x\mathbf{i} + y\mathbf{j}$ and $d\mathbf{r} = dx\,\mathbf{i} + dy\,\mathbf{j}$, and so, applying the above prescription, we obtain for this special case that the area is $A = \frac{1}{2}\oint_C (x\,dy - y\,dx)$; this is as we found in Section 3.3.

Example Find the vector area of the surface of the hemisphere $x^2 + y^2 + z^2 = a^2$, $z \geq 0$, by evaluating the line integral $\mathbf{S} = \frac{1}{2}\oint_C \mathbf{r} \times d\mathbf{r}$ around its perimeter.

The perimeter C of the hemisphere is the circle $x^2 + y^2 = a^2$, on which we have

$$\mathbf{r} = a\cos\phi\,\mathbf{i} + a\sin\phi\,\mathbf{j}, \qquad d\mathbf{r} = -a\sin\phi\,d\phi\,\mathbf{i} + a\cos\phi\,d\phi\,\mathbf{j}.$$

Therefore the cross product $\mathbf{r} \times d\mathbf{r}$ is given by

$$\mathbf{r} \times d\mathbf{r} = \begin{vmatrix} \mathbf{i} & \mathbf{j} & \mathbf{k} \\ a\cos\phi & a\sin\phi & 0 \\ -a\sin\phi\,d\phi & a\cos\phi\,d\phi & 0 \end{vmatrix} = a^2(\cos^2\phi + \sin^2\phi)\,d\phi\,\mathbf{k} = a^2\,d\phi\,\mathbf{k},$$

and the vector area becomes

$$\mathbf{S} = \tfrac{1}{2}a^2\mathbf{k}\int_0^{2\pi} d\phi = \pi a^2\,\mathbf{k},$$

in agreement with the result of the previous worked example and footnote 10. ◀

11 Note that in this case $\hat{\mathbf{n}}$ points into what, in Figure 3.8, would normally be described as the "interior" of the hollow cone; however, its direction is in agreement with the convention described on p. 141.

12 Note that the value obtained for $\mathbf{S}$ does *not* depend upon the position of the surface's perimeter relative to the origin.

3.5.3 Physical examples of surface integrals

There are many examples of surface integrals in the physical sciences. Surface integrals of the form (3.8) occur in computing the total electric charge on a surface or the mass of a shell, $\int_S \rho(\mathbf{r})\,dS$, given the charge or mass density $\rho(\mathbf{r})$. For surface integrals involving vectors, the second form in (3.9) is the most common. For a vector field $\mathbf{a}$, the surface integral $\int_S \mathbf{a}\cdot d\mathbf{S}$ is called the *flux* of $\mathbf{a}$ through S. Examples of physically important flux integrals are numerous.[13] For example, let us consider a surface S in a fluid with density $\rho(\mathbf{r})$ that has a velocity field $\mathbf{v}(\mathbf{r})$. The mass of fluid crossing an element of surface area $d\mathbf{S}$ in time dt is $dM = \rho\mathbf{v}\cdot d\mathbf{S}\,dt$. Therefore the *net* total mass flux of fluid crossing S is $M = \int_S \rho(\mathbf{r})\mathbf{v}(\mathbf{r})\cdot d\mathbf{S}$. As another example, the electromagnetic flux of energy out of a given volume V bounded by a surface S is $\oint_S(\mathbf{E}\times\mathbf{H})\cdot d\mathbf{S}$.

The solid angle, to be defined below, subtended at a point O by a surface (closed or otherwise) can also be represented by an integral of this form, although it is not strictly a flux integral (unless we imagine isotropic rays radiating from O). The integral

$$\Omega = \int_S \frac{\mathbf{r}\cdot d\mathbf{S}}{r^3} = \int_S \frac{\hat{\mathbf{r}}\cdot d\mathbf{S}}{r^2} \tag{3.11}$$

gives the *solid angle* Ω *subtended at* O *by a surface* S if $\mathbf{r}$ is the position vector measured from O of an element of the surface.[14] A little thought will show that (3.11) takes account of all three relevant factors: the size of the element of surface, its inclination to the line joining the element to O, and the distance from O. Such a general expression is often useful for computing solid angles when the three-dimensional geometry is complicated. Note that (3.11) remains valid when the surface S is not convex and when a single ray from O in certain directions would cut S in more than one place (but we exclude multiply connected regions). In particular, when the surface is closed $\Omega = 0$ if O is outside S and $\Omega = 4\pi$ if O is an interior point.

Surface integrals resulting in vectors occur less frequently. An example is afforded, however, by the total resultant force experienced by a body immersed in a stationary fluid in which the hydrostatic pressure is given by $p(\mathbf{r})$. The pressure is everywhere inwardly directed and the resultant force is $\mathbf{F} = -\oint_S p\,d\mathbf{S}$, taken over the whole surface.

3.6 Volume integrals

Volume integrals are defined in an obvious way and are generally simpler than line or surface integrals since the element of volume dV is a scalar quantity. We may encounter volume integrals of the forms

$$\int_V \phi\,dV, \qquad \int_V \mathbf{a}\,dV. \tag{3.12}$$

Clearly, the first form results in a scalar, whereas the second form yields a vector. Two closely related physical examples, one of each kind, are provided by the total mass of

13 Probably the most familiar is Gauss's theorem, which can be written as $\int_S \mathbf{E}\cdot d\mathbf{S} = \epsilon_0^{-1}\sum_i q_i$ for a system of charges q_i in a vacuum that are contained within a surface S.

14 Use this result to find an expression for the solid angle enclosed by a cone of half-angle α.

a fluid contained in a volume V, given by $\int_V \rho(\mathbf{r})\,dV$, and the total linear momentum of that same fluid, given by $\int_V \rho(\mathbf{r})\mathbf{v}(\mathbf{r})\,dV$ where $\mathbf{v}(\mathbf{r})$ is the velocity field in the fluid. As a slightly more complicated example of a volume integral we may consider the following.

Example Find an expression for the angular momentum of a solid body rotating with angular velocity $\boldsymbol{\omega}$ about an axis through the origin.

Consider a small volume element dV situated at position $\mathbf{r}$; its linear momentum is $\rho\,dV\dot{\mathbf{r}}$, where $\rho = \rho(\mathbf{r})$ is the density distribution, and its angular momentum about O is $\mathbf{r} \times \rho\dot{\mathbf{r}}\,dV$. Thus for the whole body the angular momentum $\mathbf{L}$ is

$$\mathbf{L} = \int_V (\mathbf{r} \times \dot{\mathbf{r}})\rho\,dV.$$

Putting $\dot{\mathbf{r}} = \boldsymbol{\omega} \times \mathbf{r}$ yields

$$\mathbf{L} = \int_V [\mathbf{r} \times (\boldsymbol{\omega} \times \mathbf{r})]\,\rho\,dV = \int_V \boldsymbol{\omega} r^2 \rho\,dV - \int_V (\mathbf{r} \cdot \boldsymbol{\omega})\mathbf{r}\rho\,dV.$$

It should be noted that both integrals produce vectors; the first is necessarily positive and in the direction of $\boldsymbol{\omega}$, but the second could be in any direction. ◀

The first type of volume integral in (3.12) is a standard multiple integral with a non-constant integrand, and evaluation of the second type follows directly from it since we can write

$$\int_V \mathbf{a}\,dV = \mathbf{i}\int_V a_x\,dV + \mathbf{j}\int_V a_y\,dV + \mathbf{k}\int_V a_z\,dV, \tag{3.13}$$

where a_x, a_y, a_z are the Cartesian components of $\mathbf{a}$. Of course, we could have written $\mathbf{a}$ in terms of the basis vectors of some other coordinate system (e.g. spherical polars) but, since such basis vectors are not, in general, constant, they cannot be taken out of the integral sign as in (3.13) and must be included as part of the integrand.

The volume of a three-dimensional region V can obviously be expressed as $V = \int_V dV$, and this integral may be evaluated directly once the limits of integration have been found. However, the volume of the region equally obviously depends only on the surface S that bounds it. We should therefore be able to express the volume V in terms of a surface integral over S. This is indeed possible, and the appropriate expression may be derived as follows. Referring to Figure 3.9, let us suppose that the origin O is contained within V. The volume of the small shaded cone is $dV = \frac{1}{3}\mathbf{r} \cdot d\mathbf{S}$; the total volume of the region is thus given by

$$V = \frac{1}{3}\oint_S \mathbf{r} \cdot d\mathbf{S}.$$

It may be shown that this expression is valid even when O is not contained in V. Although this surface integral form is available, in practice it is usually simpler to evaluate the volume integral directly.

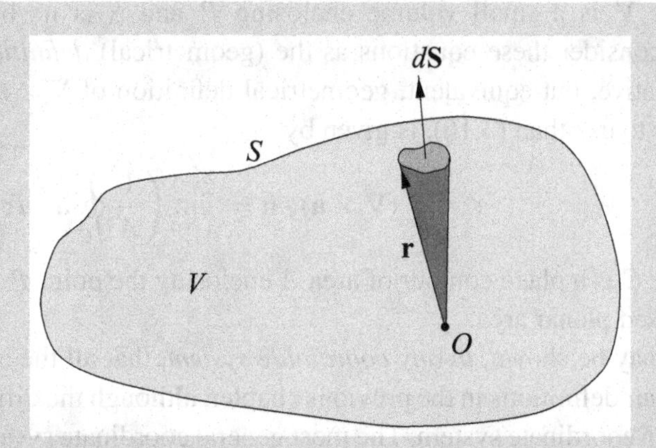

Figure 3.9 A general volume V containing the origin and bounded by the closed surface S.

Example Find the volume enclosed between a sphere of radius a centered on the origin and a circular cone of half-angle α with its vertex at the origin.

The element of vector area $d\mathbf{S}$ on the surface of the sphere is given in spherical polar coordinates by $a^2 \sin\theta \, d\theta \, d\phi \, \hat{\mathbf{r}}$. Now taking the axis of the cone to lie along the z-axis (from which θ is measured) the required volume is given by

$$V = \frac{1}{3}\oint_S \mathbf{r}\cdot d\mathbf{S} = \frac{1}{3}\int_0^{2\pi} d\phi \int_0^{\alpha} a^2 \sin\theta \, \mathbf{r}\cdot\hat{\mathbf{r}} \, d\theta$$

$$= \frac{1}{3}\int_0^{2\pi} d\phi \int_0^{\alpha} a^3 \sin\theta \, d\theta = \frac{2\pi a^3}{3}(1-\cos\alpha).$$

If the cone is formally "turned inside out", i.e. α is set equal to π, then the formula for the volume of a complete sphere is recovered. ◀

3.7 Integral forms for grad, div and curl

In the previous chapter we defined the vector operators grad, div and curl in purely mathematical terms, which depended on the coordinate system in which they were expressed. An interesting application of line, surface and volume integrals is the expression of grad, div and curl in coordinate-free, geometrical terms.

If ϕ is a scalar field and $\mathbf{a}$ is a vector field then it may be shown that at any point P

$$\nabla\phi = \lim_{V\to 0}\left(\frac{1}{V}\oint_S \phi \, d\mathbf{S}\right), \tag{3.14}$$

$$\nabla\cdot\mathbf{a} = \lim_{V\to 0}\left(\frac{1}{V}\oint_S \mathbf{a}\cdot d\mathbf{S}\right), \tag{3.15}$$

$$\nabla\times\mathbf{a} = \lim_{V\to 0}\left(\frac{1}{V}\oint_S d\mathbf{S}\times\mathbf{a}\right), \tag{3.16}$$

where V is a small volume enclosing P and S is its bounding surface. Indeed, we may consider these equations as the (geometrical) *definitions* of grad, div and curl. An alternative, but equivalent, geometrical definition of $\nabla \times \mathbf{a}$ at a point P, which is often easier to use than (3.16), is given by

$$(\nabla \times \mathbf{a}) \cdot \hat{\mathbf{n}} = \lim_{A \to 0} \left(\frac{1}{A} \oint_C \mathbf{a} \cdot d\mathbf{r} \right), \tag{3.17}$$

where C is a plane contour of area A enclosing the point P and $\hat{\mathbf{n}}$ is the unit normal to the enclosed planar area.

It may be shown, *in any coordinate system*, that all the above equations are consistent with our definitions in the previous chapter, although the difficulty of proof depends on the chosen coordinate system. The most general coordinate system encountered in that chapter was one with orthogonal curvilinear coordinates u_1, u_2, u_3, of which Cartesians, cylindrical polars and spherical polars are all special cases. Although it may be shown that (3.14) leads to the usual expression for grad in curvilinear coordinates, the proof requires complicated manipulations of the derivatives of the basis vectors with respect to the coordinates and is not presented here. In Cartesian coordinates, however, the proof is quite simple.

Example Show that the geometrical definition of *grad* leads to the usual expression for $\nabla\phi$ in Cartesian coordinates.

Consider the surface S of a small rectangular volume element $\Delta V = \Delta x\, \Delta y\, \Delta z$ that has its faces parallel to the x-, y-, and z-coordinate surfaces; the point P (see above) is at one corner. We must calculate the surface integral (3.14) over each of its six faces. Remembering that the normal to the surface points outwards from the volume on each face, the two faces with $x =$ constant have areas $\Delta \mathbf{S} = -\mathbf{i}\, \Delta y\, \Delta z$ and $\Delta \mathbf{S} = \mathbf{i}\, \Delta y\, \Delta z$ respectively. Furthermore, over each small surface element, we may take ϕ to be constant,[15] so that the net contribution to the surface integral from these two faces is, to first order in Δx,

$$\begin{aligned}[(\phi + \Delta\phi) - \phi]\, \Delta y\, \Delta z\, \mathbf{i} &= \left(\phi + \frac{\partial \phi}{\partial x} \Delta x - \phi \right) \Delta y\, \Delta z\, \mathbf{i} \\ &= \frac{\partial \phi}{\partial x} \Delta x\, \Delta y\, \Delta z\, \mathbf{i}.\end{aligned}$$

The surface integral over the pairs of faces with $y =$ constant and $z =$ constant respectively may be found in a similar way, and we obtain

$$\oint_S \phi\, d\mathbf{S} = \left(\frac{\partial \phi}{\partial x}\mathbf{i} + \frac{\partial \phi}{\partial y}\mathbf{j} + \frac{\partial \phi}{\partial z}\mathbf{k} \right) \Delta x\, \Delta y\, \Delta z.$$

Therefore $\nabla\phi$ at the point P is given by

$$\begin{aligned}\nabla\phi &= \lim_{\Delta x, \Delta y, \Delta z \to 0} \left[\frac{1}{\Delta x\, \Delta y\, \Delta z} \left(\frac{\partial \phi}{\partial x}\mathbf{i} + \frac{\partial \phi}{\partial y}\mathbf{j} + \frac{\partial \phi}{\partial z}\mathbf{k} \right) \Delta x\, \Delta y\, \Delta z \right] \\ &= \frac{\partial \phi}{\partial x}\mathbf{i} + \frac{\partial \phi}{\partial y}\mathbf{j} + \frac{\partial \phi}{\partial z}\mathbf{k},\end{aligned}$$

which is the same expression as the purely mathematical one for $\nabla\phi$. ◀

15 But, in general, different on the two faces.

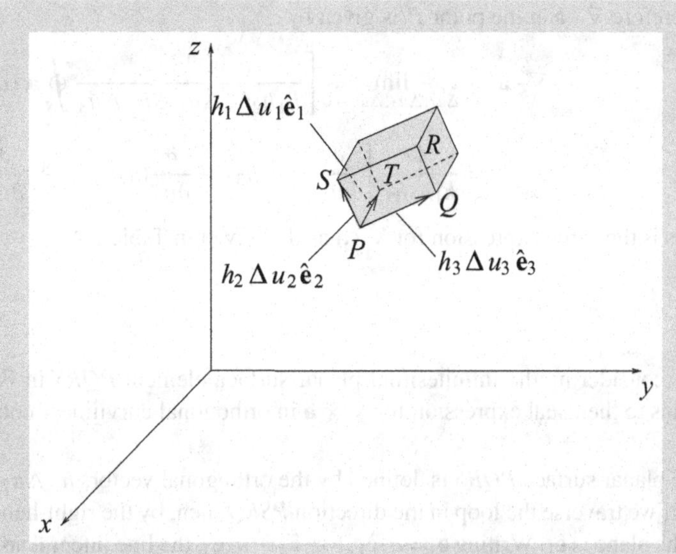

Figure 3.10 A general volume ΔV in orthogonal curvilinear coordinates u_1, u_2, u_3. PT gives the vector $h_1\,\Delta u_1\,\hat{\mathbf{e}}_1$, PS gives $h_2\,\Delta u_2\,\hat{\mathbf{e}}_2$ and PQ gives $h_3\,\Delta u_3\,\hat{\mathbf{e}}_3$.

We now turn to (3.15) and (3.17). These geometrical definitions may be shown straightforwardly to lead to the usual expressions for div and curl in orthogonal curvilinear coordinates.

Example By considering the infinitesimal volume element $dV = h_1h_2h_3\,\Delta u_1\,\Delta u_2\,\Delta u_3$ shown in Figure 3.10, show that (3.15) leads to the usual expression for $\nabla\cdot\mathbf{a}$ in orthogonal curvilinear coordinates.

Let us write the vector field in terms of its components with respect to the basis vectors of the curvilinear coordinate system as $\mathbf{a} = a_1\hat{\mathbf{e}}_1 + a_2\hat{\mathbf{e}}_2 + a_3\hat{\mathbf{e}}_3$. We consider first the contribution to the RHS of (3.15) from the two faces with $u_1 =$ constant, i.e. $PQRS$ and the face opposite it (see Figure 3.10). Now, the volume element is formed from the orthogonal vectors $h_1\,\Delta u_1\,\hat{\mathbf{e}}_1$, $h_2\,\Delta u_2\,\hat{\mathbf{e}}_2$ and $h_3\,\Delta u_3\,\hat{\mathbf{e}}_3$ at the point P and so for $PQRS$ we have[16]

$$\Delta\mathbf{S} = h_2h_3\,\Delta u_2\,\Delta u_3\,\hat{\mathbf{e}}_3\times\hat{\mathbf{e}}_2 = -h_2h_3\,\Delta u_2\,\Delta u_3\,\hat{\mathbf{e}}_1.$$

Reasoning along the same lines as in the previous example, we conclude that the contribution to the surface integral of $\mathbf{a}\cdot d\mathbf{S}$ over $PQRS$ and its opposite face taken together is given by[17]

$$\frac{\partial}{\partial u_1}(\mathbf{a}\cdot\Delta\mathbf{S})\,\Delta u_1 = \frac{\partial}{\partial u_1}(a_1h_2h_3)\,\Delta u_1\,\Delta u_2\,\Delta u_3.$$

The surface integrals over the pairs of faces with $u_2 =$ constant and $u_3 =$ constant respectively may be found in a similar way, and we obtain

$$\oint_S \mathbf{a}\cdot d\mathbf{S} = \left[\frac{\partial}{\partial u_1}(a_1h_2h_3) + \frac{\partial}{\partial u_2}(a_2h_3h_1) + \frac{\partial}{\partial u_3}(a_3h_1h_2)\right]\Delta u_1\,\Delta u_2\,\Delta u_3.$$

16 Recall that $\Delta\mathbf{S}$ is in the direction of the outward normal to the volume; hence $\hat{\mathbf{e}}_3\times\hat{\mathbf{e}}_2$, and not $\hat{\mathbf{e}}_2\times\hat{\mathbf{e}}_3$.

17 Note that, since $\Delta\mathbf{S}$ is (anti-)parallel to $\hat{\mathbf{e}}_1$, only the a_1 component of $\mathbf{a}$ contributes to $\mathbf{a}\cdot\Delta\mathbf{S}$.

Therefore $\nabla \cdot \mathbf{a}$ at the point P is given by

$$\nabla \cdot \mathbf{a} = \lim_{\Delta u_1, \Delta u_2, \Delta u_3 \to 0} \left[\frac{1}{h_1 h_2 h_3 \, \Delta u_1 \, \Delta u_2 \, \Delta u_3} \oint_S \mathbf{a} \cdot d\mathbf{S} \right]$$

$$= \frac{1}{h_1 h_2 h_3} \left[\frac{\partial}{\partial u_1}(a_1 h_2 h_3) + \frac{\partial}{\partial u_2}(a_2 h_3 h_1) + \frac{\partial}{\partial u_3}(a_3 h_1 h_2) \right].$$

This is the same expression for $\nabla \cdot \mathbf{a}$ as that given in Table 2.4. ◀

Example By considering the infinitesimal planar surface element *PQRS* in Figure 3.10, show that (3.17) leads to the usual expression for $\nabla \times \mathbf{a}$ in orthogonal curvilinear coordinates.

The planar surface *PQRS* is defined by the orthogonal vectors $h_2 \, \Delta u_2 \, \hat{\mathbf{e}}_2$ and $h_3 \, \Delta u_3 \, \hat{\mathbf{e}}_3$ at the point P. If we traverse the loop in the direction *PSRQ* then, by the right-hand convention, the unit normal to the plane is $\hat{\mathbf{e}}_1$. Writing $\mathbf{a} = a_1 \hat{\mathbf{e}}_1 + a_2 \hat{\mathbf{e}}_2 + a_3 \hat{\mathbf{e}}_3$, the line integral around the loop in this direction is given by the sum of four scalar products, each of which has a non-zero contribution from only one of the components of $\mathbf{a}$. The contributions are, in order, $h_2 a_2$, $h_3 a_3$ evaluated at $u_2 + \Delta u_2$, $-h_2 a_2$ evaluated at $u_3 + \Delta u_3$, and $-h_3 a_3$; the negative signs in the final two contributions arise because along *RQ* and *QP* the direction of traversal is in the negative $\hat{\mathbf{e}}_2$ and $\hat{\mathbf{e}}_3$ directions, respectively. The line integral is thus

$$\oint_{PSRQ} \mathbf{a} \cdot d\mathbf{r} = a_2 h_2 \, \Delta u_2 + \left[a_3 h_3 + \frac{\partial}{\partial u_2}(a_3 h_3) \, \Delta u_2 \right] \Delta u_3$$

$$- \left[a_2 h_2 + \frac{\partial}{\partial u_3}(a_2 h_2) \, \Delta u_3 \right] \Delta u_2 - a_3 h_3 \, \Delta u_3$$

$$= \left[\frac{\partial}{\partial u_2}(a_3 h_3) - \frac{\partial}{\partial u_3}(a_2 h_2) \right] \Delta u_2 \, \Delta u_3.$$

Therefore from (3.17) the component of $\nabla \times \mathbf{a}$ in the direction $\hat{\mathbf{e}}_1$ at P is given by

$$(\nabla \times \mathbf{a})_1 = \lim_{\Delta u_2, \Delta u_3 \to 0} \left[\frac{1}{h_2 h_3 \, \Delta u_2 \, \Delta u_3} \oint_{PSRQ} \mathbf{a} \cdot d\mathbf{r} \right]$$

$$= \frac{1}{h_2 h_3} \left[\frac{\partial}{\partial u_2}(h_3 a_3) - \frac{\partial}{\partial u_3}(h_2 a_2) \right].$$

The other two components are found by cyclically permuting the subscripts 1, 2, 3. Each of the components so found is in accord with the determinantal expression for $\nabla \times \mathbf{a}$ given in Table 2.4. ◀

Finally, we note that we can also write the ∇^2 operator as a surface integral by setting $\mathbf{a} = \nabla \phi$ in (3.15), to obtain

$$\nabla^2 \phi = \nabla \cdot \nabla \phi = \lim_{V \to 0} \left(\frac{1}{V} \oint_S \nabla \phi \cdot d\mathbf{S} \right).$$

3.8 Divergence theorem and related theorems

The divergence theorem relates the total flux of a vector field out of a closed surface S to the integral of the divergence of the vector field over the enclosed volume V; it follows almost immediately from our geometrical definition of divergence (3.15).

Imagine a volume V, in which a vector field $\mathbf{a}$ is continuous and differentiable, to be divided up into a large number of small volumes V_i. Using (3.15), we have for each small volume

$$(\nabla \cdot \mathbf{a})V_i \approx \oint_{S_i} \mathbf{a} \cdot d\mathbf{S},$$

where S_i is the surface of the small volume V_i. Summing over i we find that contributions from surface elements interior to S cancel since each surface element appears in two terms with opposite signs, the outward normals in the two terms being equal and opposite. Only contributions from surface elements that are also parts of S survive. If each V_i is allowed to tend to zero then we obtain the *divergence theorem*,

$$\int_V \nabla \cdot \mathbf{a}\, dV = \oint_S \mathbf{a} \cdot d\mathbf{S}. \tag{3.18}$$

We note that the divergence theorem holds for both simply and multiply connected surfaces, provided that they are closed and enclose some non-zero volume V.

The theorem finds most use as a tool in formal manipulations, but sometimes it is of value in transforming surface integrals of the form $\int_S \mathbf{a} \cdot d\mathbf{S}$ into volume integrals or vice versa. For example, setting $\mathbf{a} = \mathbf{r}$ we immediately obtain

$$\int_V \nabla \cdot \mathbf{r}\, dV = \int_V 3\, dV = 3V = \oint_S \mathbf{r} \cdot d\mathbf{S},$$

which gives the expression for the volume of a region found in Section 3.6. The use of the divergence theorem is further illustrated in the following example.

Example Evaluate the surface integral $I = \int_S \mathbf{a} \cdot d\mathbf{S}$, where $\mathbf{a} = (y - x)\mathbf{i} + x^2 z\mathbf{j} + (z + x^2)\mathbf{k}$ and S is the open surface of the hemisphere $x^2 + y^2 + z^2 = a^2$, $z \geq 0$.

We could evaluate this surface integral directly, but the algebra is somewhat lengthy. We will therefore evaluate it by use of the divergence theorem. Since the latter only holds for closed surfaces enclosing a non-zero volume V, let us first consider the closed surface $S' = S + S_1$, where S_1 is the circular area in the xy-plane given by $x^2 + y^2 \leq a^2$, $z = 0$; S' then encloses a hemispherical volume V. By the divergence theorem we have

$$\int_V \nabla \cdot \mathbf{a}\, dV = \oint_{S'} \mathbf{a} \cdot d\mathbf{S} = \int_S \mathbf{a} \cdot d\mathbf{S} + \int_{S_1} \mathbf{a} \cdot d\mathbf{S}.$$

Now $\nabla \cdot \mathbf{a} = -1 + 0 + 1 = 0$, so we can write

$$\int_S \mathbf{a} \cdot d\mathbf{S} = -\int_{S_1} \mathbf{a} \cdot d\mathbf{S}.$$

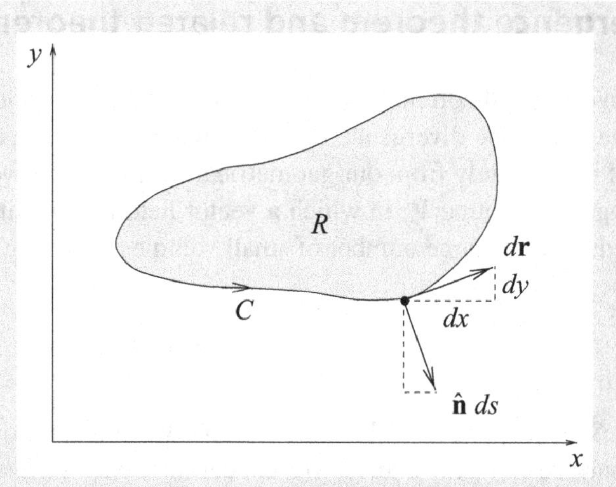

Figure 3.11 A closed curve C in the xy-plane bounding a region R. Vectors tangent and normal to the curve at a given point are also shown.

The surface integral over S_1 is easily evaluated. Remembering that the normal to the surface points outward from the volume, a surface element on S_1 is simply $d\mathbf{S} = -\mathbf{k}\,dx\,dy$. On S_1 we also have $\mathbf{a} = (y - x)\,\mathbf{i} + x^2\,\mathbf{k}$, so that

$$I = -\int_{S_1} \mathbf{a} \cdot d\mathbf{S} = \iint_R x^2\,dx\,dy,$$

where R is the circular region in the xy-plane given by $x^2 + y^2 \leq a^2$. Transforming to plane polar coordinates we have

$$I = \iint_{R'} \rho^2 \cos^2\phi\ \rho\,d\rho\,d\phi = \int_0^{2\pi} \cos^2\phi\,d\phi \int_0^a \rho^3\,d\rho = \frac{\pi a^4}{4}.$$

Thus the integral $\int \mathbf{a} \cdot d\mathbf{S}$ over a curved surface, for an intricate vector field $\mathbf{a}$, has been evaluated by computing the integral of a much simpler field over an easily specified plane surface. ◀

It is also interesting to consider the two-dimensional version of the divergence theorem. As an example, let us consider a two-dimensional planar region R in the xy-plane bounded by some closed curve C (see Figure 3.11). At any point on the curve the vector $d\mathbf{r} = dx\,\mathbf{i} + dy\,\mathbf{j}$ is a tangent to the curve and the vector $\hat{\mathbf{n}}\,ds = dy\,\mathbf{i} - dx\,\mathbf{j}$ is a normal pointing out of the region R. If the vector field $\mathbf{a}$ is continuous and differentiable in R then the two-dimensional divergence theorem in Cartesian coordinates gives

$$\iint_R \left(\frac{\partial a_x}{\partial x} + \frac{\partial a_y}{\partial y}\right) dx\,dy = \oint \mathbf{a} \cdot \hat{\mathbf{n}}\,ds = \oint_C (a_x\,dy - a_y\,dx).$$

Letting $P = -a_y$ and $Q = a_x$, we recover Green's theorem in a plane, which was discussed in Section 3.3.

3.8.1 Green's theorems

Consider two scalar functions ϕ and ψ that are continuous and differentiable in some volume V bounded by a surface S. Applying the divergence theorem to the vector field

$\phi\nabla\psi$ we obtain

$$\oint_S \phi\nabla\psi \cdot d\mathbf{S} = \int_V \nabla\cdot(\phi\nabla\psi)\,dV$$
$$= \int_V \left[\phi\nabla^2\psi + (\nabla\phi)\cdot(\nabla\psi)\right]\,dV. \tag{3.19}$$

Reversing the roles of ϕ and ψ in (3.19) and subtracting the two equations gives

$$\oint_S (\phi\nabla\psi - \psi\nabla\phi)\cdot d\mathbf{S} = \int_V (\phi\nabla^2\psi - \psi\nabla^2\phi)\,dV. \tag{3.20}$$

Equation (3.19) is usually known as Green's first theorem and (3.20) as his second. Green's second theorem is useful in the development of the Green's functions used in the solution of partial differential equations (see Chapter 11).

3.8.2 Other related integral theorems

There exist two other integral theorems which are closely related to the divergence theorem and which are of some use in physical applications. If ϕ is a scalar field and $\mathbf{b}$ is a vector field and both ϕ and $\mathbf{b}$ satisfy our usual differentiability conditions in some volume V bounded by a closed surface S then

$$\int_V \nabla\phi\,dV = \oint_S \phi\,d\mathbf{S}, \tag{3.21}$$

$$\int_V \nabla\times\mathbf{b}\,dV = \oint_S d\mathbf{S}\times\mathbf{b}. \tag{3.22}$$

The first of these is proved in the following example.

Example Use the divergence theorem to prove equation (3.21).

In the divergence theorem (3.18) let $\mathbf{a} = \phi\mathbf{c}$, where $\mathbf{c}$ is an arbitrary constant vector. We then have

$$\int_V \nabla\cdot(\phi\mathbf{c})\,dV = \oint_S \phi\mathbf{c}\cdot d\mathbf{S}.$$

Expanding out the integrand on the LHS we have

$$\nabla\cdot(\phi\mathbf{c}) = \phi\nabla\cdot\mathbf{c} + \mathbf{c}\cdot\nabla\phi = \mathbf{c}\cdot\nabla\phi,$$

since $\mathbf{c}$ is constant. Also, $\phi\mathbf{c}\cdot d\mathbf{S} = \mathbf{c}\cdot\phi d\mathbf{S}$, so we obtain

$$\int_V \mathbf{c}\cdot(\nabla\phi)\,dV = \oint_S \mathbf{c}\cdot\phi\,d\mathbf{S}.$$

Since $\mathbf{c}$ is constant we may take it out of both integrals to give

$$\mathbf{c}\cdot\int_V \nabla\phi\,dV = \mathbf{c}\cdot\oint_S \phi\,d\mathbf{S},$$

and since $\mathbf{c}$ is arbitrary we obtain the stated result (3.21).[18] ◀

18 Provide a formal proof of "$\mathbf{c}\cdot\mathbf{P} = \mathbf{c}\cdot\mathbf{Q}$ implies that $\mathbf{P} = \mathbf{Q}$ if $\mathbf{c}$ is arbitrary".

Equation (3.22) may be proved in a similar way by letting $\mathbf{a} = \mathbf{b} \times \mathbf{c}$ in the divergence theorem, where $\mathbf{c}$ is again a constant vector.

3.8.3 Physical applications of the divergence theorem

The divergence theorem is useful in deriving many of the most important partial differential equations in physics (see Chapter 10). The basic idea is to use the divergence theorem to convert an integral form, often derived from observation, into an equivalent differential form (used in theoretical statements).

Example For a compressible fluid with time-varying position-dependent density $\rho(\mathbf{r}, t)$ and velocity field $v(\mathbf{r}, t)$, in which fluid is neither being created nor destroyed, show that

$$\frac{\partial \rho}{\partial t} + \nabla \cdot (\rho \mathbf{v}) = 0.$$

For an arbitrary volume V in the fluid, the conservation of mass tells us that the rate of increase or decrease of the mass M of fluid in the volume must equal the net rate at which fluid is entering or leaving the volume, i.e.

$$\frac{dM}{dt} = -\oint_S \rho \mathbf{v} \cdot d\mathbf{S},$$

where S is the surface bounding V. But the mass of fluid in V is simply $M = \int_V \rho \, dV$, so we have

$$\frac{d}{dt} \int_V \rho \, dV + \oint_S \rho \mathbf{v} \cdot d\mathbf{S} = 0.$$

Taking the derivative inside the first integral[19] on the LHS and using the divergence theorem to rewrite the second integral, we obtain

$$\int_V \frac{\partial \rho}{\partial t} dV + \int_V \nabla \cdot (\rho \mathbf{v}) \, dV = \int_V \left[\frac{\partial \rho}{\partial t} + \nabla \cdot (\rho \mathbf{v}) \right] dV = 0.$$

Since the volume V is arbitrary, the integrand (which is assumed continuous) must be identically zero, so we obtain

$$\frac{\partial \rho}{\partial t} + \nabla \cdot (\rho \mathbf{v}) = 0.$$

This is known as the *continuity equation*. It can also be applied to other systems, for example those in which ρ is the density of electric charge or the heat content, etc. For the flow of an incompressible fluid, $\rho =$ constant and the continuity equation becomes simply $\nabla \cdot \mathbf{v} = 0$. ◀

In the previous example, we assumed that there were no sources or sinks in the volume V, i.e. that there was no part of V in which fluid was being created or destroyed. We now consider the case where a finite number of *point* sources and/or sinks are present in an incompressible fluid. Let us first consider the simple case where a single source is located at the origin, out of which a quantity of fluid flows radially at a rate Q ($\text{m}^3\ \text{s}^{-1}$). The

19 The derivative is with respect to time while the integral is with respect to space, and so this interchange is permissible.

velocity field is given by

$$\mathbf{v} = \frac{Q\mathbf{r}}{4\pi r^3} = \frac{Q\hat{\mathbf{r}}}{4\pi r^2}.$$

Now, for a sphere S_1 of radius r centered on the source, the flux across S_1 is

$$\oint_{S_1} \mathbf{v} \cdot d\mathbf{S} = |\mathbf{v}| 4\pi r^2 = Q.$$

Since $\mathbf{v}$ has a singularity at the origin it is not differentiable there, i.e. $\nabla \cdot \mathbf{v}$ is not defined there, but at all other points $\nabla \cdot \mathbf{v} = 0$, as required for an incompressible fluid. Therefore, from the divergence theorem, for any closed surface S_2 that does not enclose the origin we have

$$\oint_{S_2} \mathbf{v} \cdot d\mathbf{S} = \int_V \nabla \cdot \mathbf{v}\, dV = 0.$$

Thus we see that the surface integral $\oint_S \mathbf{v} \cdot d\mathbf{S}$ has value Q or zero depending on whether or not S encloses the source. In order that the divergence theorem is valid for *all* surfaces S, irrespective of whether they enclose the source, we write

$$\nabla \cdot \mathbf{v} = Q\delta(\mathbf{r}),$$

where $\delta(\mathbf{r})$ is the three-dimensional Dirac delta function. The properties of this function are discussed fully in Chapter 5, but for the moment we note that it is defined in such a way that

$$\delta(\mathbf{r} - \mathbf{a}) = 0 \qquad \text{for } \mathbf{r} \neq \mathbf{a},$$

$$\int_V f(\mathbf{r})\delta(\mathbf{r} - \mathbf{a})\, dV = \begin{cases} f(\mathbf{a}) & \text{if } \mathbf{a} \text{ lies in } V \\ 0 & \text{otherwise} \end{cases}$$

for any well-behaved function $f(\mathbf{r})$. Therefore, for any volume V containing the source at the origin, we have

$$\int_V \nabla \cdot \mathbf{v}\, dV = Q \int_V \delta(\mathbf{r})\, dV = Q,$$

which is consistent with $\oint_S \mathbf{v} \cdot d\mathbf{S} = Q$ for a closed surface enclosing the source. Hence, by introducing the Dirac delta function the divergence theorem can be made valid even for non-differentiable point sources.

The generalization to several sources and sinks is straightforward. For example, if a source is located at $\mathbf{r} = \mathbf{a}$ and a sink of equal strength at $\mathbf{r} = \mathbf{b}$, then the velocity field is

$$\mathbf{v} = \frac{(\mathbf{r} - \mathbf{a})Q}{4\pi |\mathbf{r} - \mathbf{a}|^3} - \frac{(\mathbf{r} - \mathbf{b})Q}{4\pi |\mathbf{r} - \mathbf{b}|^3}$$

and its divergence is given by

$$\nabla \cdot \mathbf{v} = Q\delta(\mathbf{r} - \mathbf{a}) - Q\delta(\mathbf{r} - \mathbf{b}).$$

Therefore, the integral $\oint_S \mathbf{v} \cdot d\mathbf{S}$ has the value Q if S encloses the source, $-Q$ if S encloses the sink, and 0 if S encloses neither the source nor sink or encloses them both. This analysis also applies to other physical systems – for example, in electrostatics we can regard the sources and sinks as positive and negative point charges respectively and replace $\mathbf{v}$ by the electric field $\mathbf{E}$.

3.9 Stokes' theorem and related theorems

Stokes' theorem is the "curl analogue" of the divergence theorem and relates the integral of the curl of a vector field over an open surface S to the line integral of the vector field around the perimeter C bounding the surface.

Following the same lines as for the derivation of the divergence theorem, we can divide the surface S into many small areas S_i with boundaries C_i and unit normals $\hat{\mathbf{n}}_i$. Using (3.17), we have for each small area

$$(\nabla \times \mathbf{a}) \cdot \hat{\mathbf{n}}_i \; S_i \approx \oint_{C_i} \mathbf{a} \cdot d\mathbf{r}.$$

Summing over i we find that on the RHS all parts of all interior boundaries that are not part of C are included twice, being traversed in opposite directions on each occasion and thus contributing canceling contributions. Only contributions from line elements that are also parts of C survive. If each S_i is allowed to tend to zero then we obtain Stokes' theorem,

$$\int_S (\nabla \times \mathbf{a}) \cdot d\mathbf{S} = \oint_C \mathbf{a} \cdot d\mathbf{r}. \tag{3.23}$$

We note that Stokes' theorem holds for both simply and multiply connected open surfaces, provided that they are two-sided.

Just as the divergence theorem (3.18) can be used to relate volume and surface integrals for certain types of integrand, Stokes' theorem can be used in evaluating surface integrals of the form $\oint_S (\nabla \times \mathbf{a}) \cdot d\mathbf{S}$ as line integrals or vice versa.

Example Given the vector field $\mathbf{a} = y\,\mathbf{i} - x\,\mathbf{j} + z\,\mathbf{k}$, verify Stokes' theorem for the hemispherical surface $x^2 + y^2 + z^2 = a^2$, $z \geq 0$.

Let us first evaluate the surface integral

$$\int_S (\nabla \times \mathbf{a}) \cdot d\mathbf{S}$$

over the hemisphere. It is easily shown that $\nabla \times \mathbf{a} = -2\,\mathbf{k}$, and the surface element is $d\mathbf{S} = a^2 \sin\theta \, d\theta \, d\phi \, \hat{\mathbf{r}}$ in spherical polar coordinates. Therefore

$$\begin{aligned}
\int_S (\nabla \times \mathbf{a}) \cdot d\mathbf{S} &= \int_0^{2\pi} d\phi \int_0^{\pi/2} d\theta \left(-2a^2 \sin\theta\right) \hat{\mathbf{r}} \cdot \mathbf{k} \\
&= -2a^2 \int_0^{2\pi} d\phi \int_0^{\pi/2} \sin\theta \left(\frac{z}{a}\right) d\theta \\
&= -2a^2 \int_0^{2\pi} d\phi \int_0^{\pi/2} \sin\theta \cos\theta \, d\theta = -2\pi a^2.
\end{aligned}$$

We now evaluate the line integral around the perimeter curve C of the surface, which is the circle $x^2 + y^2 = a^2$ in the xy-plane. This is given by

$$\oint_C \mathbf{a} \cdot d\mathbf{r} = \oint_C (y\,\mathbf{i} - x\,\mathbf{j} + z\,\mathbf{k}) \cdot (dx\,\mathbf{i} + dy\,\mathbf{j} + dz\,\mathbf{k})$$

$$= \oint_C (y\,dx - x\,dy).$$

Using plane polar coordinates, on C we have $x = a\cos\phi$, $y = a\sin\phi$ so that $dx = -a\sin\phi\,d\phi$, $dy = a\cos\phi\,d\phi$, and the line integral becomes

$$\oint_C (y\,dx - x\,dy) = -a^2 \int_0^{2\pi} (\sin^2\phi + \cos^2\phi)\,d\phi = -a^2 \int_0^{2\pi} d\phi = -2\pi a^2.$$

Since the surface and line integrals have the same value,[20] we have verified Stokes' theorem in this case. ◄

The two-dimensional version of Stokes' theorem also yields Green's theorem in a plane. Consider the region R in the xy-plane shown in Figure 3.11, in which a vector field $\mathbf{a}$ is defined. Since $\mathbf{a} = a_x\,\mathbf{i} + a_y\,\mathbf{j}$, we have $\nabla \times \mathbf{a} = (\partial a_y/\partial x - \partial a_x/\partial y)\,\mathbf{k}$, and Stokes' theorem becomes

$$\iint_R \left(\frac{\partial a_y}{\partial x} - \frac{\partial a_x}{\partial y} \right) dx\,dy = \oint_C (a_x\,dx + a_y\,dy).$$

Letting $P = a_x$ and $Q = a_y$ we recover Green's theorem in a plane, (3.4).

3.9.1 Related integral theorems

As for the divergence theorem, there exist two other integral theorems that are closely related to Stokes' theorem. If ϕ is a scalar field and $\mathbf{b}$ is a vector field, and both ϕ and $\mathbf{b}$ satisfy our usual differentiability conditions on some two-sided open surface S bounded by a closed perimeter curve C, then

$$\int_S d\mathbf{S} \times \nabla\phi = \oint_C \phi\,d\mathbf{r}, \tag{3.24}$$

$$\int_S (d\mathbf{S} \times \nabla) \times \mathbf{b} = \int_S [\nabla(\mathbf{b} \cdot d\mathbf{S}) - (\nabla \cdot \mathbf{b})d\mathbf{S}] = \oint_C d\mathbf{r} \times \mathbf{b}. \tag{3.25}$$

Example Use Stokes' theorem to prove equation (3.24).

In Stokes' theorem, (3.23), let $\mathbf{a} = \phi\mathbf{c}$, where $\mathbf{c}$ is a constant vector. We then have

$$\int_S [\nabla \times (\phi\mathbf{c})] \cdot d\mathbf{S} = \oint_C \phi\mathbf{c} \cdot d\mathbf{r}. \tag{3.26}$$

Expanding out the integrand on the LHS we have

$$\nabla \times (\phi\mathbf{c}) = \nabla\phi \times \mathbf{c} + \phi\nabla \times \mathbf{c} = \nabla\phi \times \mathbf{c},$$

20 Note that, since *any* open surface with boundary C will do, the value of the surface integral can be written down immediately if the plane surface $x^2 + y^2 = a^2$, $z = 0$ is used.

since $\mathbf{c}$ is constant, and the scalar triple product on the LHS of (3.26) can therefore be written

$$[\nabla \times (\phi \mathbf{c})] \cdot d\mathbf{S} = (\nabla\phi \times \mathbf{c}) \cdot d\mathbf{S} = \mathbf{c} \cdot (d\mathbf{S} \times \nabla\phi).$$

Substituting this into (3.26) and taking $\mathbf{c}$ out of both integrals because it is constant, we find

$$\mathbf{c} \cdot \int_S d\mathbf{S} \times \nabla\phi = \mathbf{c} \cdot \oint_C \phi \, d\mathbf{r}.$$

Since $\mathbf{c}$ is an *arbitrary* constant vector, result (3.24) follows. ◀

Equation (3.25) may be proved in a similar way, by letting $\mathbf{a} = \mathbf{b} \times \mathbf{c}$ in Stokes' theorem, where $\mathbf{c}$ is again a constant vector. The equality between the two integrands for the surface integral is most easily shown using the summation convention notation (Appendix E).

We also note that by setting $\mathbf{b} = \mathbf{r}$ in (3.25) we find

$$\int_S (d\mathbf{S} \times \nabla) \times \mathbf{r} = \oint_C d\mathbf{r} \times \mathbf{r}.$$

Expanding out the integrand on the LHS gives

$$(d\mathbf{S} \times \nabla) \times \mathbf{r} = \nabla(d\mathbf{S} \cdot \mathbf{r}) - d\mathbf{S}(\nabla \cdot \mathbf{r}) = d\mathbf{S} - 3\,d\mathbf{S} = -2\,d\mathbf{S}.$$

Therefore, as we found in Subsection 3.5.2, the vector area of an open surface S is given by

$$\mathbf{S} = \int_S d\mathbf{S} = \frac{1}{2}\oint_C \mathbf{r} \times d\mathbf{r}.$$

3.9.2 Physical applications of Stokes' theorem

Like the divergence theorem, Stokes' theorem is useful for converting integral equations into differential ones.

Example From Ampère's law derive Maxwell's equation in the case where the currents are steady, i.e. $\nabla \times \mathbf{B} - \mu_0 \mathbf{J} = \mathbf{0}$.

Ampère's rule for a distributed current with current density $\mathbf{J}$ is

$$\oint_C \mathbf{B} \cdot d\mathbf{r} = \mu_0 \int_S \mathbf{J} \cdot d\mathbf{S},$$

for any circuit C bounding a surface S. Using Stokes' theorem, the LHS can be transformed into $\int_S (\nabla \times \mathbf{B}) \cdot d\mathbf{S}$; hence

$$\int_S (\nabla \times \mathbf{B} - \mu_0 \mathbf{J}) \cdot d\mathbf{S} = 0$$

for *any* surface S. This can only be so if $\nabla \times \mathbf{B} - \mu_0 \mathbf{J} = \mathbf{0}$, which is the required relation. Similarly, from Faraday's law of electromagnetic induction we can derive Maxwell's equation $\nabla \times \mathbf{E} = -\partial \mathbf{B}/\partial t$. ◀

In Subsection 3.8.3 we discussed the flow of an incompressible fluid in the presence of several sources and sinks. Let us now consider *vortex* flow in an incompressible fluid with a velocity field

$$\mathbf{v} = \frac{1}{\rho}\hat{\mathbf{e}}_\phi,$$

in cylindrical polar coordinates ρ, ϕ, z. For this velocity field $\nabla \times \mathbf{v}$ equals zero everywhere except on the axis $\rho = 0$, where $\mathbf{v}$ has a singularity. Therefore $\oint_C \mathbf{v} \cdot d\mathbf{r}$ equals zero for any path C that does not enclose the vortex line on the axis and 2π if C does enclose the axis. In order for Stokes' theorem to be valid for all paths C, we therefore set

$$\nabla \times \mathbf{v} = 2\pi\delta(\rho),$$

where $\delta(\rho)$ is the Dirac delta function, to be discussed in Subsection 5.2. Now, since $\nabla \times \mathbf{v} = \mathbf{0}$, except on the axis $\rho = 0$, there exists a scalar potential ψ such that $\mathbf{v} = \nabla\psi$. It may easily be shown that $\psi = \phi$, the azimuthal angle. Therefore, if C does not enclose the axis then

$$\oint_C \mathbf{v} \cdot d\mathbf{r} = \oint d\phi = 0,$$

and if C does enclose the axis

$$\oint_C \mathbf{v} \cdot d\mathbf{r} = \Delta\phi = 2\pi n,$$

where n is the number of times we traverse C. Thus ϕ is a multivalued potential.

Similar analyses are valid for other physical systems – for example, in magnetostatics we may replace the vortex lines by current-carrying wires and the velocity field $\mathbf{v}$ by the magnetic field $\mathbf{B}$.

SUMMARY

1. *Line integral types*

   Scalar type: $\int_C \mathbf{a} \cdot d\mathbf{r}$; vector type: $\int_C \phi\, d\mathbf{r}$ or $\int_C \mathbf{a} \times d\mathbf{r}$.

2. *Green's theorem in a plane*

$$\oint_C (P\,dx + Q\,dy) = \iint_R \left(\frac{\partial Q}{\partial x} - \frac{\partial P}{\partial y}\right) dx\,dy.$$

   The theorem is valid for a multiply connected region provided C includes all boundaries and they are traversed in the positive direction.

3. *Conservative fields*

   A vector field $\mathbf{a}$ that has continuous partial derivatives in a simply connected region R is conservative if, and only if, any of the following is true (each implies the other three).

(i) The integral $\int_A^B \mathbf{a} \cdot d\mathbf{r}$, where A and B lie in the region R, is independent of the path from A to B. Hence the integral $\oint_C \mathbf{a} \cdot d\mathbf{r}$ around any closed loop in R is zero.

(ii) There exists a single-valued function ϕ of position such that $\mathbf{a} = \nabla\phi$.

(iii) $\nabla \times \mathbf{a} = \mathbf{0}$.

(iv) $\mathbf{a} \cdot d\mathbf{r}$ is an exact differential.

4. *Solenoidal fields*
   If $\nabla \cdot \mathbf{b} = 0$, then it is always possible to find infinitely many vector fields $\mathbf{a}$ such that $\mathbf{b} = \nabla \times \mathbf{a}$. If $\mathbf{a}$ is one such field, then $\mathbf{a}' = \mathbf{a} + \nabla\psi + \mathbf{c}$ is another, for any scalar ψ and any constant vector $\mathbf{c}$.

5. *Surface integrals*
   - Scalar type: $\int_S \mathbf{a} \cdot d\mathbf{S}$; vector type: $\int_S \phi\, d\mathbf{S}$ or $\int_S \mathbf{a} \times d\mathbf{S}$.
   - The scalar element of area dS on the surface $f(x, y, z) = 0$ is related to its projection dA on the xy-plane by $dS = \dfrac{|\nabla f|}{\partial f/\partial z}\, dA$.
   - The vector area of a surface, $\mathbf{S} = \int d\mathbf{S}$, is always zero for a closed surface.
   - The vector area of an open surface depends only on its boundary curve C and is given by $\mathbf{S} = \dfrac{1}{2}\oint_C \mathbf{r} \times d\mathbf{r}$.
   - The solid angle Ω subtended at the origin by a surface S is given by $\Omega = \displaystyle\int_S \frac{\hat{\mathbf{r}} \cdot d\mathbf{S}}{r^2}$.

6. *Theorems for surface integrals*
   - Stokes' theorem: $\int_S (\nabla \times \mathbf{a}) \cdot d\mathbf{S} = \oint_C \mathbf{a} \cdot d\mathbf{r}$.
   - Other theorems:

$$\int_S d\mathbf{S} \times \nabla\phi = \oint_C \phi\, d\mathbf{r} \quad \text{and} \quad \int_S [\nabla(\mathbf{b} \cdot d\mathbf{S}) - (\nabla \cdot \mathbf{b})d\mathbf{S}] = \oint_C d\mathbf{r} \times \mathbf{b}.$$

7. *Volume integrals*
   - Scalar type: $\int_V \phi\, dV$; vector type: $\int_V \mathbf{a}\, dV$.
   - The volume of a closed region depends only on its bounding surface S and is given by $V = \dfrac{1}{3}\oint_S \mathbf{r} \cdot d\mathbf{S}$.
   - Grad, div and curl can be represented/defined by integrals over the surface of a (vanishingly) small volume:

$$\nabla\phi = \lim_{V \to 0}\left(\frac{1}{V}\oint_S \phi\, d\mathbf{S}\right), \qquad \nabla \cdot \mathbf{a} = \lim_{V \to 0}\left(\frac{1}{V}\oint_S \mathbf{a} \cdot d\mathbf{S}\right),$$

$$\nabla \times \mathbf{a} = \lim_{V \to 0}\left(\frac{1}{V}\oint_S d\mathbf{S} \times \mathbf{a}\right).$$

8. *Theorems for volume integrals*
   - Divergence theorem: $\int_V \nabla \cdot \mathbf{a}\, dV = \oint_S \mathbf{a} \cdot d\mathbf{S}$.
   - Green's 1st theorem: $\oint_S \phi \nabla \psi \cdot d\mathbf{S} = \int_V [\phi \nabla^2 \psi + (\nabla \phi) \cdot (\nabla \psi)]\, dV$.
   - Green's 2nd theorem: $\oint_S (\phi \nabla \psi - \psi \nabla \phi) \cdot d\mathbf{S} = \int_V (\phi \nabla^2 \psi - \psi \nabla^2 \phi)\, dV$.
   - Other theorems: $\int_V \nabla \phi\, dV = \oint_S \phi\, d\mathbf{S}$ and $\int_V \nabla \times \mathbf{b}\, dV = \oint_S d\mathbf{S} \times \mathbf{b}$.

PROBLEMS

3.1. The vector field $\mathbf{F}$ is defined by

$$\mathbf{F} = 2xz\mathbf{i} + 2yz^2\mathbf{j} + (x^2 + 2y^2z - 1)\mathbf{k}.$$

Calculate $\nabla \times \mathbf{F}$ and deduce that $\mathbf{F}$ can be written $\mathbf{F} = \nabla \phi$. Determine the form of ϕ.

3.2. A vector field $\mathbf{Q}$ is defined as

$$\mathbf{Q} = \left[3x^2(y+z) + y^3 + z^3\right]\mathbf{i} + \left[3y^2(z+x) + z^3 + x^3\right]\mathbf{j} + \left[3z^2(x+y) + x^3 + y^3\right]\mathbf{k}.$$

Show that $\mathbf{Q}$ is a conservative field, construct its potential function and hence evaluate the integral $J = \int \mathbf{Q} \cdot d\mathbf{r}$ along any line connecting the point A at $(1, -1, 1)$ to B at $(2, 1, 2)$.

3.3. A vector field $\mathbf{F}$ is given by $xy^2\mathbf{i} + 2\mathbf{j} + x\mathbf{k}$, and L is a path parameterized by $x = ct$, $y = c/t$, $z = d$ for the range $1 \le t \le 2$. Evaluate the three integrals (a) $\int_L \mathbf{F}\, dt$, (b) $\int_L \mathbf{F}\, dy$ and (c) $\int_L \mathbf{F} \cdot d\mathbf{r}$.

3.4. By making an appropriate choice for the functions $P(x, y)$ and $Q(x, y)$ that appear in Green's theorem in a plane, show that the integral of $x - y$ over the upper half of the unit circle centered on the origin has the value $-\frac{2}{3}$. Show the same result by direct integration in Cartesian coordinates.

3.5. Determine the point of intersection P, in the first quadrant, of the two ellipses

$$\frac{x^2}{a^2} + \frac{y^2}{b^2} = 1 \quad \text{and} \quad \frac{x^2}{b^2} + \frac{y^2}{a^2} = 1.$$

Taking $b < a$, consider the contour L that bounds the area in the first quadrant that is common to the two ellipses. Show that the parts of L that lie along the coordinate axes contribute nothing to the line integral around L of $x\, dy - y\, dx$.

Using a parameterization of each ellipse similar to that employed in the example in Section 3.3, evaluate the two remaining line integrals and hence find the total area common to the two ellipses.

3.6. By using parameterizations of the form $x = a\cos^n\theta$ and $y = a\sin^n\theta$ for suitable values of n, find the area bounded by the curves

$$x^{2/5} + y^{2/5} = a^{2/5} \quad \text{and} \quad x^{2/3} + y^{2/3} = a^{2/3}.$$

3.7. Evaluate the line integral

$$I = \oint_C \left[y(4x^2 + y^2)\,dx + x(2x^2 + 3y^2)\,dy \right]$$

around the ellipse $x^2/a^2 + y^2/b^2 = 1$.

3.8. Criticize the following "proof" that $\pi = 0$.

(a) Apply Green's theorem in a plane to the functions $P(x, y) = \tan^{-1}(y/x)$ and $Q(x, y) = \tan^{-1}(x/y)$, taking the region R to be the unit circle centered on the origin.

(b) The RHS of the equality so produced is

$$\iint_R \frac{y - x}{x^2 + y^2}\,dx\,dy,$$

which, either from symmetry considerations or by changing to plane polar coordinates, can be shown to have zero value.

(c) In the LHS of the equality, set $x = \cos\theta$ and $y = \sin\theta$, yielding $P(\theta) = \theta$ and $Q(\theta) = \pi/2 - \theta$. The line integral becomes

$$\int_0^{2\pi} \left[\left(\frac{\pi}{2} - \theta \right) \cos\theta - \theta \sin\theta \right] d\theta,$$

which has the value 2π.

(d) Thus $2\pi = 0$ and the stated result follows.

3.9. A single-turn coil C of arbitrary shape is placed in a magnetic field $\mathbf{B}$ and carries a current I. Show that the couple acting upon the coil can be written as

$$\mathbf{M} = I \int_C (\mathbf{B} \cdot \mathbf{r})\,d\mathbf{r} - I \int_C \mathbf{B}(\mathbf{r} \cdot d\mathbf{r}).$$

For a planar rectangular coil of sides $2a$ and $2b$ placed with its plane vertical and at an angle ϕ to a uniform horizontal field $\mathbf{B}$, show that $\mathbf{M}$ is, as expected, $4abBI\cos\phi\,\mathbf{k}$.

3.10. Find the vector area $\mathbf{S}$ of the part of the curved surface of the hyperboloid of revolution

$$\frac{x^2}{a^2} - \frac{y^2 + z^2}{b^2} = 1$$

that lies in the region $z \geq 0$ and $a \leq x \leq \lambda a$.

3.11. An axially symmetric solid body with its axis AB vertical is immersed in an incompressible fluid of density ρ_0. Use the following method to show that, whatever the shape of the body, for $\rho = \rho(z)$ in cylindrical polars the Archimedean upthrust is, as expected, $\rho_0 g V$, where V is the volume of the body.

Express the vertical component of the resultant force on the body, $-\int p \, d\mathbf{S}$, where p is the pressure, in terms of an integral; note that $p = -\rho_0 g z$ and that for an annular surface element of width dl, $\mathbf{n} \cdot \mathbf{n}_z \, dl = -d\rho$. Integrate by parts and use the fact that $\rho(z_A) = \rho(z_B) = 0$.

3.12. Show that the expression below is equal to the solid angle subtended by a rectangular aperture, of sides $2a$ and $2b$, at a point on the normal through its center, and at a distance c from the aperture:

$$\Omega = 4 \int_0^b \frac{ac}{(y^2 + c^2)(y^2 + c^2 + a^2)^{1/2}} \, dy.$$

By setting $y = (a^2 + c^2)^{1/2} \tan \phi$, change this integral into the form

$$\int_0^{\phi_1} \frac{4ac \cos \phi}{c^2 + a^2 \sin^2 \phi} \, d\phi,$$

where $\tan \phi_1 = b/(a^2 + c^2)^{1/2}$, and hence show that

$$\Omega = 4 \tan^{-1} \left[\frac{ab}{c(a^2 + b^2 + c^2)^{1/2}} \right].$$

3.13. A vector field $\mathbf{a}$ is given by $-zxr^{-3}\mathbf{i} - zyr^{-3}\mathbf{j} + (x^2 + y^2)r^{-3}\mathbf{k}$, where $r^2 = x^2 + y^2 + z^2$. Establish that the field is conservative (a) by showing that $\nabla \times \mathbf{a} = \mathbf{0}$, and (b) by constructing its potential function ϕ.

3.14. A vector field $\mathbf{a}$ is given by $(z^2 + 2xy)\,\mathbf{i} + (x^2 + 2yz)\,\mathbf{j} + (y^2 + 2zx)\,\mathbf{k}$. Show that $\mathbf{a}$ is conservative and that the line integral $\int \mathbf{a} \cdot d\mathbf{r}$ along any line joining (1, 1, 1) and (1, 2, 2) has the value 11.

3.15. A force $\mathbf{F}(\mathbf{r})$ acts on a particle at $\mathbf{r}$. In which of the following cases can $\mathbf{F}$ be represented in terms of a potential? Where it can, find the potential.

(a) $\mathbf{F} = F_0 \left[\mathbf{i} - \mathbf{j} - \frac{2(x-y)}{a^2} \mathbf{r} \right] \exp\left(-\frac{r^2}{a^2} \right);$

(b) $\mathbf{F} = \frac{F_0}{a} \left[z\mathbf{k} + \frac{(x^2 + y^2 - a^2)}{a^2} \mathbf{r} \right] \exp\left(-\frac{r^2}{a^2} \right);$

(c) $\mathbf{F} = F_0 \left[\mathbf{k} + \frac{a(\mathbf{r} \times \mathbf{k})}{r^2} \right].$

3.16. One of Maxwell's electromagnetic equations states that all magnetic fields $\mathbf{B}$ are solenoidal (i.e. $\nabla \cdot \mathbf{B} = 0$). Determine whether each of the following vectors could represent a real magnetic field; where it could, try to find a suitable vector potential $\mathbf{A}$, i.e. such that $\mathbf{B} = \nabla \times \mathbf{A}$. (*Hint*: seek a vector potential that is parallel to $\nabla \times \mathbf{B}$.)

(a) $\dfrac{B_0 b}{r^3}[(x-y)z\,\mathbf{i}+(x-y)z\,\mathbf{j}+(x^2-y^2)\,\mathbf{k}]$ in Cartesians with $r^2=x^2+y^2+z^2$.

(b) $\dfrac{B_0 b^3}{r^3}[\cos\theta\,\cos\phi\,\hat{\mathbf{e}}_r-\sin\theta\,\cos\phi\,\hat{\mathbf{e}}_\theta+\sin 2\theta\,\sin\phi\,\hat{\mathbf{e}}_\phi]$ in spherical polars.

(c) $B_0 b^2\left[\dfrac{z\rho}{(b^2+z^2)^2}\,\hat{\mathbf{e}}_\rho+\dfrac{1}{b^2+z^2}\,\hat{\mathbf{e}}_z\right]$ in cylindrical polars.

3.17. The vector field $\mathbf{f}$ has components $y\mathbf{i}-x\mathbf{j}+\mathbf{k}$ and γ is a curve given parametrically by

$$\mathbf{r}=(a-c+c\cos\theta)\mathbf{i}+(b+c\sin\theta)\mathbf{j}+c^2\theta\mathbf{k},\qquad 0\le\theta\le 2\pi.$$

Describe the shape of the path γ and show that the line integral $\int_\gamma \mathbf{f}\cdot d\mathbf{r}$ vanishes. Does this result imply that $\mathbf{f}$ is a conservative field?

3.18. A vector field $\mathbf{a}=f(r)\mathbf{r}$ is spherically symmetric and everywhere directed away from the origin. Show that $\mathbf{a}$ is irrotational, but that it is also solenoidal only if $f(r)$ is of the form Ar^{-3}.

3.19. Evaluate the surface integral $\int \mathbf{r}\cdot d\mathbf{S}$, where $\mathbf{r}$ is the position vector, over that part of the surface $z=a^2-x^2-y^2$ for which $z\ge 0$, by each of the following methods.

(a) Parameterize the surface as $x=a\sin\theta\,\cos\phi$, $y=a\sin\theta\,\sin\phi$, $z=a^2\cos^2\theta$, and show that

$$\mathbf{r}\cdot d\mathbf{S}=a^4(2\sin^3\theta\cos\theta+\cos^3\theta\sin\theta)\,d\theta\,d\phi.$$

(b) Apply the divergence theorem to the volume bounded by the surface and the plane $z=0$.

3.20. Obtain an expression for the value ϕ_P at a point P of a scalar function ϕ that satisfies $\nabla^2\phi=0$, in terms of its value and normal derivative on a surface S that encloses it, by proceeding as follows.

(a) In Green's second theorem, take ψ at any particular point Q as $1/r$, where r is the distance of Q from P. Show that $\nabla^2\psi=0$, except at $r=0$.

(b) Apply the result to the doubly connected region bounded by S and a small sphere Σ of radius δ centered on P.

(c) Apply the divergence theorem to show that the surface integral over Σ involving $1/\delta$ vanishes, and prove that the term involving $1/\delta^2$ has the value $4\pi\phi_P$.

(d) Conclude that

$$\phi_P=-\frac{1}{4\pi}\int_S \phi\frac{\partial}{\partial n}\left(\frac{1}{r}\right)\,dS+\frac{1}{4\pi}\int_S\frac{1}{r}\frac{\partial\phi}{\partial n}\,dS.$$

This important result shows that the value at a point P of a function ϕ that satisfies $\nabla^2\phi=0$ everywhere within a closed surface S that encloses P may be expressed *entirely* in terms of its value and normal derivative on S. This matter is taken up more generally in connection with Green's functions in Chapter 11 and in connection with functions of a complex variable in Section 14.10.

3.21. Use result (3.21), together with an appropriately chosen scalar function ϕ, to prove that the position vector $\bar{\mathbf{r}}$ of the center of mass of an arbitrarily shaped body of volume V and uniform density can be written

$$\bar{\mathbf{r}} = \frac{1}{V}\oint_S \tfrac{1}{2}r^2\, d\mathbf{S}.$$

3.22. A rigid body of volume V and surface S rotates with angular velocity $\boldsymbol{\omega}$. Show that

$$\boldsymbol{\omega} = -\frac{1}{2V}\oint_S \mathbf{u}\times d\mathbf{S},$$

where $\mathbf{u}(\mathbf{x})$ is the velocity of the point $\mathbf{x}$ on the surface S.

3.23. Demonstrate the validity of the divergence theorem:
(a) by calculating the flux of the vector

$$\mathbf{F} = \frac{\alpha\mathbf{r}}{(r^2+a^2)^{3/2}}$$

through the spherical surface $|\mathbf{r}| = \sqrt{3}a$;
(b) by showing that

$$\nabla\cdot\mathbf{F} = \frac{3\alpha a^2}{(r^2+a^2)^{5/2}}$$

and evaluating the volume integral of $\nabla\cdot\mathbf{F}$ over the interior of the sphere $|\mathbf{r}| = \sqrt{3}a$. The substitution $r = a\tan\theta$ will prove useful in carrying out the integration.

3.24. Prove equation (3.22) and, by taking $\mathbf{b} = zx^2\mathbf{i} + zy^2\mathbf{j} + (x^2-y^2)\mathbf{k}$, show that the two integrals

$$I = \int x^2\, dV \quad \text{and} \quad J = \int \cos^2\theta\sin^3\theta\cos^2\phi\, d\theta\, d\phi,$$

both taken over the unit sphere, must have the same value. Evaluate both directly to show that the common value is $4\pi/15$.

3.25. In a uniform conducting medium with unit relative permittivity, charge density ρ, current density $\mathbf{J}$, electric field $\mathbf{E}$ and magnetic field $\mathbf{B}$, Maxwell's electromagnetic equations take the form (with $\mu_0\epsilon_0 = c^{-2}$)
(i) $\nabla\cdot\mathbf{B} = 0$, (ii) $\nabla\cdot\mathbf{E} = \rho/\epsilon_0$,
(iii) $\nabla\times\mathbf{E} + \dot{\mathbf{B}} = \mathbf{0}$, (iv) $\nabla\times\mathbf{B} - (\dot{\mathbf{E}}/c^2) = \mu_0\mathbf{J}$.

The density of stored energy in the medium is given by $\frac{1}{2}(\epsilon_0 E^2 + \mu_0^{-1} B^2)$. Show that the rate of change of the total stored energy in a volume V is equal to

$$-\int_V \mathbf{J} \cdot \mathbf{E}\, dV - \frac{1}{\mu_0} \oint_S (\mathbf{E} \times \mathbf{B}) \cdot d\mathbf{S},$$

where S is the surface bounding V.

[The first integral gives the ohmic heating loss, whilst the second gives the electromagnetic energy flux out of the bounding surface. The vector $\mu_0^{-1}(\mathbf{E} \times \mathbf{B})$ is known as the Poynting vector.]

3.26. A vector field $\mathbf{F}$ is defined in cylindrical polar coordinates ρ, θ, z by

$$\mathbf{F} = F_0 \left(\frac{x \cos \lambda z}{a} \mathbf{i} + \frac{y \cos \lambda z}{a} \mathbf{j} + (\sin \lambda z) \mathbf{k} \right)$$

$$\equiv \frac{F_0 \rho}{a} (\cos \lambda z) \mathbf{e}_\rho + F_0 (\sin \lambda z) \mathbf{k},$$

where $\mathbf{i}$, $\mathbf{j}$ and $\mathbf{k}$ are the unit vectors along the Cartesian axes and $\mathbf{e}_\rho$ is the unit vector $(x/\rho)\mathbf{i} + (y/\rho)\mathbf{j}$.

(a) Calculate, as a surface integral, the flux of $\mathbf{F}$ through the closed surface bounded by the cylinders $\rho = a$ and $\rho = 2a$ and the planes $z = \pm a\pi/2$.

(b) Evaluate the same integral using the divergence theorem.

3.27. The vector field $\mathbf{F}$ is given by

$$\mathbf{F} = (3x^2yz + y^3z + xe^{-x})\mathbf{i} + (3xy^2z + x^3z + ye^x)\mathbf{j} + (x^3y + y^3x + xy^2z^2)\mathbf{k}.$$

Calculate (a) directly, and (b) by using Stokes' theorem the value of the line integral $\int_L \mathbf{F} \cdot d\mathbf{r}$, where L is the (three-dimensional) closed contour $OABCDEO$ defined by the successive vertices $(0, 0, 0)$, $(1, 0, 0)$, $(1, 0, 1)$, $(1, 1, 1)$, $(1, 1, 0)$, $(0, 1, 0)$, $(0, 0, 0)$.

3.28. A vector force field $\mathbf{F}$ is defined in Cartesian coordinates by

$$\mathbf{F} = F_0 \left[\left(\frac{y^3}{3a^3} + \frac{y}{a} e^{xy/a^2} + 1 \right) \mathbf{i} + \left(\frac{xy^2}{a^3} + \frac{x + y}{a} e^{xy/a^2} \right) \mathbf{j} + \frac{z}{a} e^{xy/a^2} \mathbf{k} \right].$$

Use Stokes' theorem to calculate

$$\oint_L \mathbf{F} \cdot d\mathbf{r},$$

where L is the perimeter of the rectangle $ABCD$ given by $A = (0, a, 0)$, $B = (a, a, 0)$, $C = (a, 3a, 0)$ and $D = (0, 3a, 0)$.

HINTS AND ANSWERS

3.1. Show that $\nabla \times \mathbf{F} = \mathbf{0}$. The potential $\phi_F(\mathbf{r}) = x^2z + y^2z^2 - z$.

3.3. (a) $c^3 \ln 2\,\mathbf{i} + 2\,\mathbf{j} + (3c/2)\mathbf{k}$; (b) $(-3c^4/8)\mathbf{i} - c\,\mathbf{j} - (c^2 \ln 2)\mathbf{k}$; (c) $c^4 \ln 2 - c$.

3.5. For P, $x = y = ab/(a^2 + b^2)^{1/2}$. The relevant limits are $0 \leq \theta_1 \leq \tan^{-1}(b/a)$ and $\tan^{-1}(a/b) \leq \theta_2 \leq \pi/2$. The total common area is $4ab\tan^{-1}(b/a)$.

3.7. Show that, in the notation of Section 3.3, $\partial Q/\partial x - \partial P/\partial y = 2x^2$; $I = \pi a^3 b/2$.

3.9. $\mathbf{M} = I\int_C \mathbf{r} \times (d\mathbf{r} \times \mathbf{B})$. Show that the horizontal sides in the first term and the whole of the second term contribute nothing to the couple.

3.11. Note that, if $\hat{\mathbf{n}}$ is the outward normal to the surface, $\hat{\mathbf{n}}_z \cdot \hat{\mathbf{n}}\, dl$ is equal to $-d\rho$

3.13. (b) $\phi = c + z/r$.

3.15. (a) Yes, $F_0(x - y)\exp(-r^2/a^2)$; (b) yes, $-F_0[(x^2 + y^2)/(2a)]\exp(-r^2/a^2)$; (c) no, $\nabla \times \mathbf{F} \neq \mathbf{0}$.

3.17. A spiral of radius c with its axis parallel to the z-direction and passing through (a, b). The pitch of the spiral is $2\pi c^2$. No, because (i) γ is not a closed loop and (ii) the line integral must be zero for *every* closed loop, not just for a particular one. In fact $\nabla \times \mathbf{f} = -2\mathbf{k} \neq \mathbf{0}$ shows that $\mathbf{f}$ is not conservative.

3.19. (a) $d\mathbf{S} = (2a^3\cos\theta\sin^2\theta\cos\phi\,\mathbf{i} + 2a^3\cos\theta\sin^2\theta\sin\phi\,\mathbf{j} + a^2\cos\theta\sin\theta\,\mathbf{k})\, d\theta\, d\phi$.
(b) $\nabla \cdot \mathbf{r} = 3$; over the plane $z = 0$, $\mathbf{r} \cdot d\mathbf{S} = 0$.
The necessarily common value is $3\pi a^4/2$.

3.21. Write $\mathbf{r}$ as $\nabla(\frac{1}{2}r^2)$.

3.23. The answer is $3\sqrt{3}\pi\alpha/2$ in each case.

3.25. Identify the expression for $\nabla \cdot (\mathbf{E} \times \mathbf{B})$ and use the divergence theorem.

3.27. (a) The successive contributions to the integral are:
$1 - 2e^{-1}$, 0, $2 + \frac{1}{2}e$, $-\frac{7}{3}$, $-1 + 2e^{-1}$, $-\frac{1}{2}$.
(b) $\nabla \times \mathbf{F} = 2xyz^2\mathbf{i} - y^2z^2\mathbf{j} + ye^x\mathbf{k}$. Show that the contour is equivalent to the sum of two plane square contours in the planes $z = 0$ and $x = 1$, the latter being traversed in the negative sense. Integral $= \frac{1}{6}(3e - 5)$.

4 Fourier series

The reader will be familiar with how, through Taylor series (see Section A.6 of Appendix A), complicated functions may be expressed as power series. However, this is not the only way in which a function may be represented as a series, and the subject of this chapter is the expression of functions as a sum of sine and cosine terms. Such a representation is called a *Fourier series*. Unlike Taylor series, a Fourier series can describe functions that are not everywhere continuous and/or differentiable. There are also other advantages in using trigonometric terms. They are easy to differentiate and integrate, their moduli are easily taken and each term contains only one characteristic frequency. This last point is important because, as we shall see later, Fourier series are often used to represent the response of a system to a periodic input, and this response often depends directly on the frequency content of the input.[1] Fourier series are used in a wide variety of such physical situations, including the vibrations of a finite string, the scattering of light by a diffraction grating and the transmission of an input signal by an electronic circuit.

4.1 The Dirichlet conditions

We have already mentioned that Fourier series may be used to represent some functions for which a Taylor series expansion is not possible. The particular conditions that a function $f(x)$ must fulfill in order that it may be expanded as a Fourier series are known as the *Dirichlet conditions*, and may be summarized by the following four points:

(i) the function must be periodic;
(ii) it must be single-valued and continuous, except possibly at a finite number of finite discontinuities;
(iii) it must have only a finite number of maxima and minima within one period;
(iv) the integral over one period of $|f(x)|$ must converge.

If the above conditions are satisfied then the Fourier series converges to $f(x)$ at all points where $f(x)$ is continuous. The convergence of the Fourier series at points of discontinuity is discussed in Section 4.4. The last three Dirichlet conditions are almost always met in real

1 Recall, for example, that the angle through which a glass prism refracts a ray of light depends upon the frequency of that light.

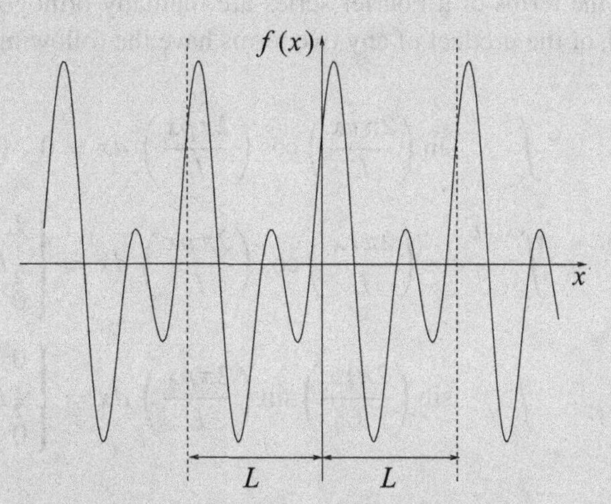

Figure 4.1 An example of a function that may be represented as a Fourier series without modification.

applications, but not all functions are periodic and hence do not fulfill the first condition. It may be possible, however, to represent a non-periodic function as a Fourier series by manipulation of the function into a periodic form. This is discussed in Section 4.5. An example of a function that may, without modification, be represented as a Fourier series is shown in Figure 4.1.

We have stated without proof that any function that satisfies the Dirichlet conditions may be represented as a Fourier series, i.e. it can be expressed as a linear sum of sine and cosine terms. Let us now show why this is plausible, though we will not give a proof that would satisfy a strict mathematician.

The first thing to note is that both sines and cosines are needed. We could not use only sine terms since sine functions, being odd functions of their arguments [i.e. a function for which $f(-x) = -f(x)$], could not represent an even function [i.e. functions for which $f(-x) = f(x)$]. This is obvious when we try to express a function $f(x)$ that takes a non-zero value at $x = 0$. Clearly, since $\sin nx = 0$ for all values of n, we could not represent $f(x)$ at $x = 0$ by a sine series. Similarly odd functions could not be represented by a cosine series since cosine is an even function. Nevertheless, it is possible to represent *all* odd functions by a sine series and *all* even functions by a cosine series. Now, since all functions may be written as the sum of an odd and an even part,

$$\begin{aligned} f(x) &= \tfrac{1}{2}[f(x) + f(-x)] + \tfrac{1}{2}[f(x) - f(-x)] \\ &= f_{\text{even}}(x) + f_{\text{odd}}(x), \end{aligned}$$

we can write any function as the sum of a sine series and a cosine series.[2]

2 Separate the function $\sin(2x + \alpha)$ into its odd and even parts.

All the terms of a Fourier series are mutually orthogonal, i.e. the integrals, over one period, of the product of any two terms have the following properties:

$$\int_{x_0}^{x_0+L} \sin\left(\frac{2\pi rx}{L}\right)\cos\left(\frac{2\pi px}{L}\right)dx = 0 \quad \text{for all } r \text{ and } p, \tag{4.1}$$

$$\int_{x_0}^{x_0+L} \cos\left(\frac{2\pi rx}{L}\right)\cos\left(\frac{2\pi px}{L}\right)dx = \begin{cases} L & \text{for } r = p = 0, \\ \frac{1}{2}L & \text{for } r = p > 0, \\ 0 & \text{for } r \neq p, \end{cases} \tag{4.2}$$

$$\int_{x_0}^{x_0+L} \sin\left(\frac{2\pi rx}{L}\right)\sin\left(\frac{2\pi px}{L}\right)dx = \begin{cases} 0 & \text{for } r = p = 0, \\ \frac{1}{2}L & \text{for } r = p > 0, \\ 0 & \text{for } r \neq p, \end{cases} \tag{4.3}$$

where r and p are integers greater than or equal to zero; these formulae are easily derived using the trigonometric addition results summarized in Section A.1. A full discussion of why it is possible to expand a function as a sum of mutually orthogonal functions is given in Chapter 8.

The Fourier series expansion of the function $f(x)$ is conventionally written

$$f(x) = \frac{a_0}{2} + \sum_{r=1}^{\infty}\left[a_r \cos\left(\frac{2\pi rx}{L}\right) + b_r \sin\left(\frac{2\pi rx}{L}\right)\right], \tag{4.4}$$

where a_0, a_r, b_r are constants called the *Fourier coefficients*. These coefficients are analogous to those in a power series expansion and the determination of their numerical values is the essential step in writing a function as a Fourier series.

This chapter continues with a discussion of how to find the Fourier coefficients for particular functions. We then discuss simplifications to the general Fourier series that may save considerable effort in calculations. This is followed by the alternative representation of a function as a complex Fourier series, and we conclude with a discussion of Parseval's theorem.

4.2 The Fourier coefficients

We have indicated that a series that satisfies the Dirichlet conditions may be written in the form (4.4), and now consider how to find the Fourier coefficients for any particular function. To this end, and throughout the mathematics of the physical sciences, the following set of special values of sinusoidal functions will prove extremely useful, and should be committed to memory by the reader. For integer n:

$$\sin n\pi = 0, \quad \sin\left(n + \tfrac{1}{2}\right)\pi = (-1)^n, \tag{4.5}$$

$$\cos n\pi = (-1)^n, \quad \cos\left(n + \tfrac{1}{2}\right)\pi = 0. \tag{4.6}$$

For a periodic function $f(x)$ of period L we will find that the Fourier coefficients are given by

$$a_r = \frac{2}{L}\int_{x_0}^{x_0+L} f(x)\cos\left(\frac{2\pi rx}{L}\right) dx, \tag{4.7}$$

$$b_r = \frac{2}{L}\int_{x_0}^{x_0+L} f(x)\sin\left(\frac{2\pi rx}{L}\right) dx, \tag{4.8}$$

where x_0 is arbitrary but is often taken as 0 or $-L/2$. The apparently arbitrary factor $\frac{1}{2}$ that appears in the a_0 term in (4.4) is included so that (4.7) may apply for $r = 0$ as well as for $r > 0$. The relations (4.7) and (4.8) may be derived as follows.

Suppose the Fourier series expansion of $f(x)$ can be written as in (4.4),

$$f(x) = \frac{a_0}{2} + \sum_{r=1}^{\infty}\left[a_r\cos\left(\frac{2\pi rx}{L}\right) + b_r\sin\left(\frac{2\pi rx}{L}\right)\right].$$

Then, multiplying by $\cos(2\pi px/L)$, integrating over one full period in x and changing the order of the summation and integration, we get

$$\begin{aligned}\int_{x_0}^{x_0+L} f(x)\cos\left(\frac{2\pi px}{L}\right) dx &= \frac{a_0}{2}\int_{x_0}^{x_0+L}\cos\left(\frac{2\pi px}{L}\right) dx \\ &\quad + \sum_{r=1}^{\infty} a_r\int_{x_0}^{x_0+L}\cos\left(\frac{2\pi rx}{L}\right)\cos\left(\frac{2\pi px}{L}\right) dx \\ &\quad + \sum_{r=1}^{\infty} b_r\int_{x_0}^{x_0+L}\sin\left(\frac{2\pi rx}{L}\right)\cos\left(\frac{2\pi px}{L}\right) dx.\end{aligned} \tag{4.9}$$

We can now find the Fourier coefficients by considering (4.9) as p takes different values. Using the orthogonality conditions (4.1)–(4.3) of the previous section, we find that when $p = 0$ (4.9) becomes

$$\int_{x_0}^{x_0+L} f(x)dx = \frac{a_0}{2}L.$$

When $p \neq 0$ the only non-vanishing term on the RHS of (4.9) occurs when $r = p$, and so

$$\int_{x_0}^{x_0+L} f(x)\cos\left(\frac{2\pi rx}{L}\right) dx = \frac{a_r}{2}L.$$

The other Fourier coefficients b_r may be found by repeating the above process but multiplying by $\sin(2\pi px/L)$ instead of $\cos(2\pi px/L)$ (see Problem 4.2).

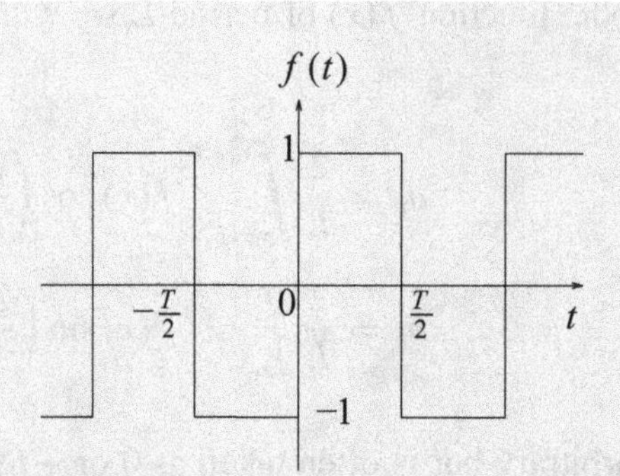

Figure 4.2 A square-wave function.

Example Express the square-wave function illustrated in Figure 4.2 as a Fourier series.

Physically this might represent the input to an electrical circuit that switches between a high and a low state with time period T. The square wave may be represented by

$$f(t) = \begin{cases} -1 & \text{for } -\frac{1}{2}T \le t < 0, \\ +1 & \text{for } 0 \le t < \frac{1}{2}T. \end{cases}$$

In deriving the Fourier coefficients, we note firstly that the function is an odd function and so the series will contain only sine terms (this simplification is discussed further in the following section). To evaluate the coefficients in the sine series we use (4.8). Hence

$$\begin{aligned} b_r &= \frac{2}{T}\int_{-T/2}^{T/2} f(t)\sin\left(\frac{2\pi rt}{T}\right) dt \\ &= \frac{4}{T}\int_{0}^{T/2} \sin\left(\frac{2\pi rt}{T}\right) dt \\ &= \frac{2}{\pi r}\left[1-(-1)^r\right]. \end{aligned}$$

Thus the sine coefficients are zero if r is even and equal to $4/(\pi r)$ if r is odd. Hence the Fourier series for the square-wave function may be written as

$$f(t) = \frac{4}{\pi}\left(\sin\omega t + \frac{\sin 3\omega t}{3} + \frac{\sin 5\omega t}{5} + \cdots\right), \tag{4.10}$$

where $\omega = 2\pi/T$ is called the *angular frequency*. ◀

4.3 Symmetry considerations

The example in the previous section employed the useful property that since the function to be represented was odd, all the cosine terms of the Fourier series were absent. It is often the case that the function we wish to express as a Fourier series has a particular symmetry, which we can exploit to reduce the calculational labor of evaluating Fourier coefficients. Functions that are symmetric or antisymmetric about the origin (i.e. even and odd functions respectively) admit particularly useful simplifications. Functions that are odd in x have no cosine terms (see Section 4.1) and all the a-coefficients are equal to zero. Similarly, functions that are even in x have no sine terms and all the b-coefficients are zero. Since the Fourier series of odd or even functions contain only half the coefficients required for a general periodic function, there is a considerable reduction in the algebra needed to find a Fourier series.

The consequences of symmetry or antisymmetry of the function about the quarter period (i.e. about $L/4$) are a little less obvious. Furthermore, the results are not used as often as those above and the remainder of this section can be omitted on a first reading without loss of continuity. The following argument gives the required results.

Suppose that $f(x)$ has even or odd symmetry about $L/4$, i.e. $f(L/4 - x) = \pm f(x - L/4)$. For convenience, we make the substitution $s = x - L/4$ and hence $f(-s) = \pm f(s)$. We can now see that

$$b_r = \frac{2}{L}\int_{x_0}^{x_0+L} f(s)\sin\left(\frac{2\pi rs}{L} + \frac{\pi r}{2}\right) ds,$$

where the limits of integration have been left unaltered since f is, of course, periodic in s as well as in x. If we use the expansion

$$\sin\left(\frac{2\pi rs}{L} + \frac{\pi r}{2}\right) = \sin\left(\frac{2\pi rs}{L}\right)\cos\left(\frac{\pi r}{2}\right) + \cos\left(\frac{2\pi rs}{L}\right)\sin\left(\frac{\pi r}{2}\right),$$

we can see immediately, using the results given in (4.5) and (4.6), that the trigonometric part of the integrand is an odd function of s if r is even and an even function of s if r is odd. Hence if $f(s)$ is even and r is even then the integral is zero, as is also the case if $f(s)$ is odd and r is odd. Similar results can be derived for the Fourier a-coefficients and we conclude that

(i) if $f(x)$ is even about $L/4$ then $a_{2r+1} = 0$ and $b_{2r} = 0$,
(ii) if $f(x)$ is odd about $L/4$ then $a_{2r} = 0$ and $b_{2r+1} = 0$.

All the above results follow automatically when the Fourier coefficients are evaluated in any particular case, but prior knowledge of them will often enable some coefficients to be set equal to zero on inspection and so substantially reduce the computational labor. As an example, the square-wave function shown in Figure 4.2 is (i) an odd function of t, so that all $a_r = 0$, and (ii) even about the point $t = T/4$, so that $b_{2r} = 0$. Thus we can say immediately that only sine terms of odd harmonics will be present and therefore will need to be calculated; this is confirmed in the expansion (4.10).

4.4 Discontinuous functions

The Fourier series expansion usually works well for functions that are discontinuous in the required range. However, the series itself does not produce a discontinuous function and we state without proof that the value of the expanded $f(x)$ at a discontinuity will be half-way between the upper and lower values. Expressing this more mathematically, at a point of finite discontinuity, x_d, the Fourier series converges to

$$\tfrac{1}{2}\lim_{\epsilon\to 0}[f(x_d+\epsilon)+f(x_d-\epsilon)].$$

Very close to a discontinuity, the Fourier series representation of the function will overshoot its value (at the discontinuity). Although as more terms are included the maximum overshoot moves in position arbitrarily close to the discontinuity, it never disappears even in the limit of an infinite number of terms. This behavior is known as *Gibbs' phenomenon*. A full discussion is not pursued here but suffice it to say that the size of the overshoot is proportional to the magnitude of the discontinuity.

Example Find the value to which the Fourier series of the square-wave function discussed in Section 4.2 converges at $t = 0$.

It can be seen that the function is discontinuous at $t = 0$ and, by the above rule, we expect the series to converge to a value half-way between the upper and lower values, in other words to converge to zero in this case. Considering the Fourier series of this function, (4.10), we see that all the terms are zero and hence the Fourier series converges to zero as expected. The Gibbs phenomenon for the square-wave function is shown in Figure 4.3. ◀

4.5 Non-periodic functions

We have already mentioned that a Fourier representation may sometimes be used for non-periodic functions. If we wish to find the Fourier series of a non-periodic function only within a fixed range then we may *continue* the function outside the range so as to make it periodic. The Fourier series of this periodic function would then correctly represent the non-periodic function in the desired range. Since we are often at liberty to extend the function in a number of ways, we can sometimes make it odd or even and so reduce the amount of calculation required. Figure 4.4(b) shows the simplest extension to the function shown in Figure 4.4(a). However, this extension has no particular symmetry. Figures 4.4(c), (d) show extensions as odd and even functions respectively with the benefit that only sine or cosine terms appear in the resulting Fourier series. We note that these last two extensions each give a function of period $2L$.

In view of the result of Section 4.4, it must be added that the continuation must not be discontinuous at the end-points of the interval of interest; if it is, the series will not converge to the required value there. The requirement that the series converges appropriately may thus reduce the choice of continuations. This aspect is discussed further at the end of the following example.

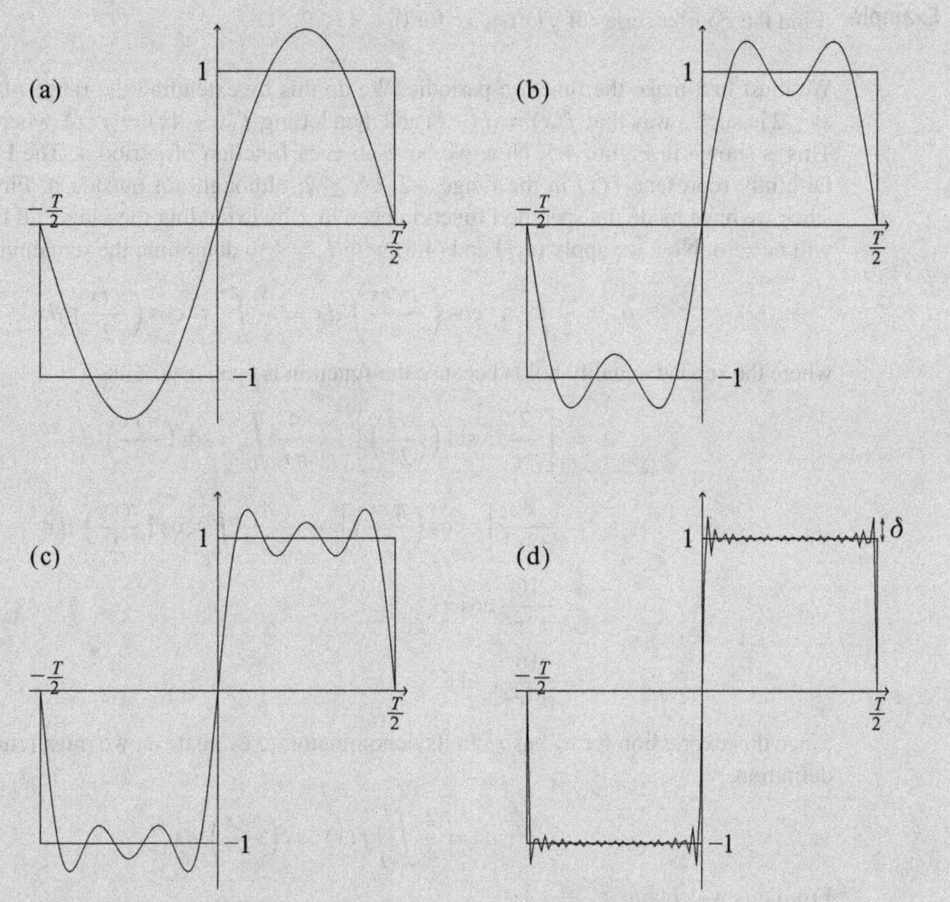

Figure 4.3 The convergence of a Fourier series expansion of a square-wave function, including (a) one term, (b) two terms, (c) three terms and (d) 20 terms. The overshoot δ is shown in (d).

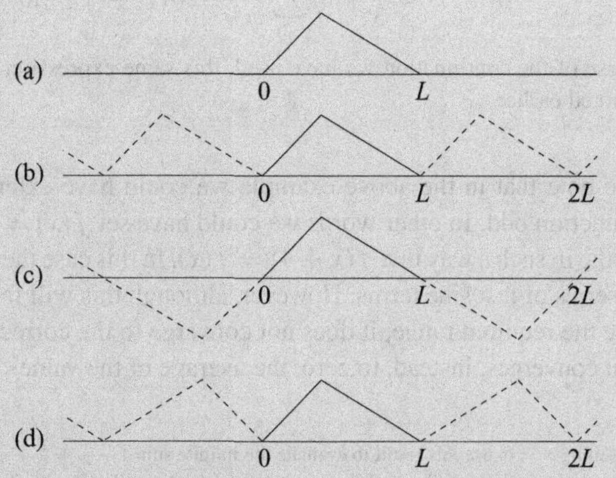

Figure 4.4 Possible periodic extensions of a function. See the main text.

Example Find the Fourier series of $f(x) = x^2$ for $0 < x \leq 2$.

We must first make the function periodic. We do this by extending the range of interest to $-2 < x \leq 2$ in such a way that $f(x) = f(-x)$ and then letting $f(x + 4k) = f(x)$, where k is any integer. This is shown in Figure 4.5. Now we have an even function of period 4. The Fourier series will faithfully represent $f(x)$ in the range $-2 < x \leq 2$, although not outside it. Firstly we note that since we have made the specified function even in x by extending the range, all the coefficients b_r will be zero. Now we apply (4.7) and (4.8) with $L = 4$ to determine the remaining coefficients:

$$a_r = \frac{2}{4}\int_{-2}^{2} x^2 \cos\left(\frac{2\pi rx}{4}\right) dx = \frac{4}{4}\int_0^2 x^2 \cos\left(\frac{\pi rx}{2}\right) dx,$$

where the second equality holds because the function is even in x. Thus

$$\begin{aligned} a_r &= \left[\frac{2}{\pi r}x^2 \sin\left(\frac{\pi rx}{2}\right)\right]_0^2 - \frac{4}{\pi r}\int_0^2 x \sin\left(\frac{\pi rx}{2}\right) dx \\ &= \frac{8}{\pi^2 r^2}\left[x \cos\left(\frac{\pi rx}{2}\right)\right]_0^2 - \frac{8}{\pi^2 r^2}\int_0^2 \cos\left(\frac{\pi rx}{2}\right) dx \\ &= \frac{16}{\pi^2 r^2}\cos \pi r \\ &= \frac{16}{\pi^2 r^2}(-1)^r. \end{aligned}$$

Since this expression for a_r has r^2 in its denominator, to evaluate a_0 we must return to the original definition,

$$a_r = \frac{2}{4}\int_{-2}^{2} f(x) \cos\left(\frac{\pi rx}{2}\right) dx.$$

From this we obtain

$$a_0 = \frac{2}{4}\int_{-2}^{2} x^2\, dx = \frac{4}{4}\int_0^2 x^2\, dx = \frac{8}{3}.$$

The final expression for $f(x)$ is then[3]

$$x^2 = \frac{4}{3} + 16\sum_{r=1}^{\infty}\frac{(-1)^r}{\pi^2 r^2}\cos\left(\frac{\pi rx}{2}\right) \quad \text{for } 0 < x \leq 2.$$

Because of the continuation we have used, this same expression is also valid for $-2 \leq x \leq 0$, as was noted earlier. ◀

We note that in the above example we could have extended the range so as to make the function odd. In other words we could have set $f(x) = -f(-x)$ and then made $f(x)$ periodic in such a way that $f(x + 4) = f(x)$. In this case the resulting Fourier series would be a series of just sine terms. However, although this will faithfully represent the function inside the required range, it does not converge to the correct values of $f(x) = \pm 4$ at $x = \pm 2$; it converges, instead, to zero, the average of the values at the two ends of the range.[4]

3 By setting $x = 0$, use this result to evaluate the infinite sum $1 - \frac{1}{4} + \frac{1}{9} - \frac{1}{16} + \frac{1}{25} + \cdots$.

4 Show that a further drawback of this particular extension is that the coefficients have only a r^{-1} convergence, as opposed to the r^{-2} convergence for the extension actually used.

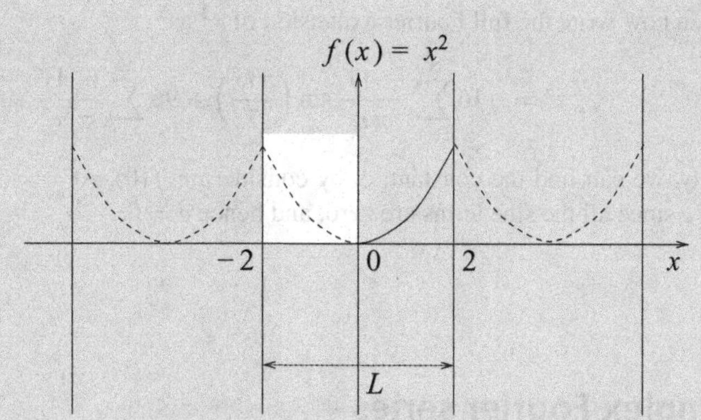

Figure 4.5 $f(x) = x^2, 0 < x \leq 2$, with the range extended to give periodicity.

4.6 Integration and differentiation

It is sometimes possible to find the Fourier series of a function by integration or differentiation of another Fourier series. If the Fourier series of $f(x)$ is integrated term by term then the resulting Fourier series converges to the integral of $f(x)$. Clearly, when integrating in such a way there is a constant of integration that must be found.

If $f(x)$ is a continuous function of x for all x and $f(x)$ is also periodic then the Fourier series that results from differentiating term by term converges to $f'(x)$, provided that $f'(x)$ itself satisfies the Dirichlet conditions.

These two properties of Fourier series can sometimes prove useful for calculating complicated Fourier series; simple Fourier series can be evaluated (or found from standard tables) and the more complicated series may then be built up by integration and/or differentiation, as in the following example.

Example Find the Fourier series of $f(x) = x^3$ for $0 < x \leq 2$.

In the example discussed in the previous section we found the Fourier series for $f(x) = x^2$ in the required range. So, if we *integrate* this term by term, we obtain

$$\frac{x^3}{3} = \frac{4}{3}x + 32\sum_{r=1}^{\infty}\frac{(-1)^r}{\pi^3 r^3}\sin\left(\frac{\pi r x}{2}\right) + c,$$

where c is, so far, an arbitrary constant. We have not yet found the Fourier series for x^3 because the term $\frac{4}{3}x$ appears in the expansion. However, by now *differentiating* the same initial expression for x^2 we obtain

$$2x = -8\sum_{r=1}^{\infty}\frac{(-1)^r}{\pi r}\sin\left(\frac{\pi r x}{2}\right).$$

We can now write the full Fourier expansion of x^3 as[5]

$$x^3 = -16\sum_{r=1}^{\infty}\frac{(-1)^r}{\pi r}\sin\left(\frac{\pi r x}{2}\right) + 96\sum_{r=1}^{\infty}\frac{(-1)^r}{\pi^3 r^3}\sin\left(\frac{\pi r x}{2}\right) + c.$$

Finally, we can find the constant, c, by considering $f(0)$. At $x = 0$, our Fourier expansion gives $x^3 = c$ since all the sine terms are zero, and hence $c = 0$. ◀

4.7 Complex Fourier series

As a Fourier series expansion in general contains both sine and cosine parts, it may be written more compactly using a complex exponential expansion. This simplification makes use of the property that $\exp(irx) = \cos rx + i \sin rx$. The complex Fourier series expansion is written

$$f(x) = \sum_{r=-\infty}^{\infty} c_r \exp\left(\frac{2\pi irx}{L}\right), \tag{4.11}$$

where the Fourier coefficients are given by[6]

$$c_r = \frac{1}{L}\int_{x_0}^{x_0+L} f(x)\exp\left(-\frac{2\pi irx}{L}\right)dx. \tag{4.12}$$

This relation can be derived, in a similar manner to that of Section 4.2, by multiplying (4.11) by $\exp(-2\pi ipx/L)$ before integrating and using the orthogonality relation

$$\int_{x_0}^{x_0+L}\exp\left(-\frac{2\pi ipx}{L}\right)\exp\left(\frac{2\pi irx}{L}\right)dx = \begin{cases} L & \text{for } r = p, \\ 0 & \text{for } r \neq p. \end{cases}$$

The complex Fourier coefficients in (4.11) have the following relations to the real Fourier coefficients:

$$\begin{aligned} c_r &= \tfrac{1}{2}(a_r - ib_r), \\ c_{-r} &= \tfrac{1}{2}(a_r + ib_r). \end{aligned} \tag{4.13}$$

Note that if $f(x)$ is real then $c_{-r} = c_r^*$, where the asterisk represents complex conjugation. As a particular case, c_0 is real.[7]

5 Do you expect the series obtained for (a) x^2 in the previous example, and (b) x^3 in this example, to show the Gibbs phenomenon?

6 Note the minus sign in the exponent and that the multiplying factor is $1/L$, and not $2/L$.

7 Identify what the value of c_0 represents in general.

Example Find a complex Fourier series for $f(x) = x$ in the range $-2 < x < 2$.

Using (4.12), for $r \neq 0$,

$$c_r = \frac{1}{4}\int_{-2}^{2} x \exp\left(-\frac{\pi i r x}{2}\right) dx$$

$$= \left[-\frac{x}{2\pi i r}\exp\left(-\frac{\pi i r x}{2}\right)\right]_{-2}^{2} + \int_{-2}^{2} \frac{1}{2\pi i r}\exp\left(-\frac{\pi i r x}{2}\right) dx$$

$$= -\frac{1}{\pi i r}\left[\exp(-\pi i r) + \exp(\pi i r)\right] + \left[\frac{1}{r^2\pi^2}\exp\left(-\frac{\pi i r x}{2}\right)\right]_{-2}^{2}$$

$$= \frac{2i}{\pi r}\cos \pi r - \frac{2i}{r^2\pi^2}\sin \pi r = \frac{2i}{\pi r}(-1)^r. \tag{4.14}$$

For $r = 0$, we find by simple direct integration that $c_0 = 0$ (as expected, see footnote 7) and hence

$$x = \sum_{\substack{r=-\infty \\ r\neq 0}}^{\infty} \frac{2i(-1)^r}{r\pi}\exp\left(\frac{\pi i r x}{2}\right).$$

We note that the Fourier series derived for x in Section 4.6 gives $a_r = 0$ for all r and

$$b_r = -\frac{4(-1)^r}{\pi r},$$

and so, using (4.13), we confirm that c_r and c_{-r} have the forms derived above. It is also apparent that the relationship $c_r^* = c_{-r}$ holds, as we expect since $f(x)$ is real. ◀

4.8 Parseval's theorem

Parseval's theorem gives a useful way of relating the Fourier coefficients to the function that they describe. Essentially a conservation law, it states that

$$\frac{1}{L}\int_{x_0}^{x_0+L} |f(x)|^2 dx = \sum_{r=-\infty}^{\infty} |c_r|^2$$

$$= \left(\tfrac{1}{2}a_0\right)^2 + \tfrac{1}{2}\sum_{r=1}^{\infty}\left(a_r^2 + b_r^2\right). \tag{4.15}$$

In a more memorable form, this says that the sum of the moduli squared of the complex Fourier coefficients is equal to the average value of $|f(x)|^2$ over one period. Parseval's theorem can be proved straightforwardly by writing $f(x)$ as a Fourier series and evaluating the required integral, but the algebra is messy. Therefore, we shall use an alternative method, for which the algebra is simple and which, in fact, leads to a more general form of the theorem.

Let us consider two functions $f(x)$ and $g(x)$, which are (or can be made) periodic with period L and which have Fourier series (expressed in complex form)

$$f(x) = \sum_{r=-\infty}^{\infty} c_r \exp\left(\frac{2\pi irx}{L}\right),$$

$$g(x) = \sum_{r=-\infty}^{\infty} \gamma_r \exp\left(\frac{2\pi irx}{L}\right),$$

where c_r and γ_r are the complex Fourier coefficients of $f(x)$ and $g(x)$ respectively. Thus

$$f(x)g^*(x) = \sum_{r=-\infty}^{\infty} c_r g^*(x) \exp\left(\frac{2\pi irx}{L}\right).$$

Integrating this equation with respect to x over the interval $(x_0, x_0 + L)$ and dividing by L, we find

$$\begin{aligned}\frac{1}{L}\int_{x_0}^{x_0+L} f(x)g^*(x)\,dx &= \sum_{r=-\infty}^{\infty} c_r \frac{1}{L}\int_{x_0}^{x_0+L} g^*(x)\exp\left(\frac{2\pi irx}{L}\right) dx \\ &= \sum_{r=-\infty}^{\infty} c_r \left[\frac{1}{L}\int_{x_0}^{x_0+L} g(x)\exp\left(\frac{-2\pi irx}{L}\right) dx\right]^* \\ &= \sum_{r=-\infty}^{\infty} c_r \gamma_r^*,\end{aligned}$$

where the last equality uses (4.12). Finally, if we let $g(x) = f(x)$ then we obtain Parseval's theorem (4.15).[8] This result can be proved in a similar manner using the sine and cosine form of the Fourier series, but the algebra is slightly more complicated.

Parseval's theorem is sometimes used to sum series. However, if one is presented with a series to sum, it is not usually possible to decide which Fourier series should be used to evaluate it. Rather, useful summations are nearly always found serendipitously. The following example shows the evaluation of a sum by a Fourier series method.

8 Use the coefficients obtained previously for $f(x) = x$ in $-2 < x < 2$ to show that $\sum_{r=1}^{\infty} r^{-2} = \pi^2/6$.

Example Using Parseval's theorem and the Fourier series for $f(x) = x^2$ found in Section 4.5, calculate the sum $\sum_{r=1}^{\infty} r^{-4}$.

Firstly we find the average value of $[f(x)]^2$ over the interval $-2 < x \leq 2$:

$$\frac{1}{4}\int_{-2}^{2} x^4\,dx = \frac{16}{5}.$$

Now we evaluate the right-hand side of (4.15), noting that there are no b_r terms:

$$\left(\tfrac{1}{2}a_0\right)^2 + \tfrac{1}{2}\sum_{1}^{\infty} a_r^2 + \tfrac{1}{2}\sum_{1}^{\infty} b_r^2 = \left(\tfrac{4}{3}\right)^2 + \tfrac{1}{2}\sum_{r=1}^{\infty} \frac{(16)^2}{\pi^4 r^4}.$$

Equating the two expressions we find

$$\sum_{r=1}^{\infty} \frac{1}{r^4} = \frac{2\pi^4}{(16)^2}\left(\frac{16}{5} - \frac{16}{9}\right) = \frac{\pi^4}{90},$$

a result that can easily be verified numerically to any reasonable required accuracy, because of the rapid convergence of the series. ◀

SUMMARY

1. *Dirichlet conditions*
   (i) The function $f(x)$ must be periodic.
   (ii) It must be single-valued and continuous, except possibly at a finite number of finite discontinuities.
   (iii) It must have only a finite number of maxima and minima within one period L.
   (iv) The integral over one period of $|f(x)|$ must converge.

2. *Fourier expansion and coefficients*

$$f(x) = \frac{a_0}{2} + \sum_{r=1}^{\infty}\left[a_r \cos\left(\frac{2\pi rx}{L}\right) + b_r \sin\left(\frac{2\pi rx}{L}\right)\right],$$

where

$$a_r = \frac{2}{L}\int_{x_0}^{x_0+L} f(x)\cos\left(\frac{2\pi rx}{L}\right)dx, \quad b_r = \frac{2}{L}\int_{x_0}^{x_0+L} f(x)\sin\left(\frac{2\pi rx}{L}\right)dx,$$

and x_0 is arbitrary, but is often taken as 0 or $-L/2$.

- Where a function is discontinuous, the Fourier series converges to the average of the two limiting values at the discontinuity.
- At a discontinuity the Gibbs phenomenon moves closer to the point of discontinuity as the number of terms is increased, but never disappears.
- For a function that is defined over a finite range and then extended to make a periodic function, the period of the latter is often not that of the original, but some multiple of it.

3. *Symmetry considerations*
   (i) If $f(x)$ is even about $x = 0$ then all $b_r = 0$.
   (ii) If $f(x)$ is odd about $x = 0$ then all $a_r = 0$.
   (iii) If $f(x)$ is even about $x = L/4$ then $a_{2r+1} = 0$ and $b_{2r} = 0$.
   (iv) If $f(x)$ is odd about $x = L/4$ then $a_{2r} = 0$ and $b_{2r+1} = 0$.

4. *Manipulation of series*
   - The term-by-term integral of the Fourier series for $f(x)$ converges to the Fourier series of $\int^x f(u)\,du$ to within a constant of integration.
   - The term-by-term derivative of the Fourier series for $f(x)$ converges to the Fourier series of df/dx, provided the latter satisfies the Dirichlet conditions.

5. *Complex Fourier series*
   - The complex series expansion is

$$f(x) = \sum_{r=-\infty}^{\infty} c_r \exp\left(\frac{2\pi irx}{L}\right), \text{ with } c_r = \frac{1}{L}\int_{x_0}^{x_0+L} f(x)e^{-2\pi irx/L}\,dx.$$

   - The connections between the coefficients c_r and those of the corresponding real series are $c_r = \frac{1}{2}(a_r - ib_r)$ and $c_{-r} = \frac{1}{2}(a_r + ib_r)$.

6. *Integral theorems*
   - Parseval's theorem

$$\frac{1}{L}\int_{x_0}^{x_0+L} |f(x)|^2 dx = \sum_{r=-\infty}^{\infty} |c_r|^2 = \left(\tfrac{1}{2}a_0\right)^2 + \tfrac{1}{2}\sum_{r=1}^{\infty}(a_r^2 + b_r^2).$$

   - If $f(x) = \sum_{r=-\infty}^{\infty} c_r e^{2\pi irx/L}$ and $g(x) = \sum_{r=-\infty}^{\infty} \gamma_r e^{2\pi irx/L}$, then

$$\frac{1}{L}\int_{x_0}^{x_0+L} f(x)g^*(x)\,dx = \sum_{r=-\infty}^{\infty} c_r\gamma_r^*.$$

   - Parseval's theorem is a special case in which $f(x) = g(x)$.

PROBLEMS

4.1. Prove the orthogonality relations stated in Section 4.1.

4.2. Derive the Fourier coefficients b_r in a similar manner to the derivation of the a_r in Section 4.2.

4.3. Which of the following functions of x could be represented by a Fourier series over the range indicated?

(a) $\tanh^{-1}(x)$, $\quad -\infty < x < \infty$;
(b) $\tan x$, $\quad -\infty < x < \infty$;
(c) $|\sin x|^{-1/2}$, $\quad -\infty < x < \infty$;
(d) $\cos^{-1}(\sin 2x)$, $\quad -\infty < x < \infty$;
(e) $x \sin(1/x)$, $\quad -\pi^{-1} < x \leq \pi^{-1}$, cyclically repeated.

4.4. By moving the origin of t to the center of an interval in which $f(t) = +1$, i.e. by changing to a new independent variable $t' = t - \frac{1}{4}T$, express the square-wave function in the example in Section 4.2 as a cosine series. Calculate the Fourier coefficients involved (a) directly and (b) by changing the variable in result (4.10).

4.5. Find the Fourier series of the function $f(x) = x$ in the range $-\pi < x \leq \pi$. Hence show that

$$1 - \frac{1}{3} + \frac{1}{5} - \frac{1}{7} + \cdots = \frac{\pi}{4}.$$

4.6. For the function

$$f(x) = 1 - x, \qquad 0 \leq x \leq 1,$$

find (a) the Fourier sine series and (b) the Fourier cosine series. Which would be better for numerical evaluation? Relate your answer to the relevant periodic continuations.

4.7. For the continued functions used in Problem 4.6 and the derived corresponding series, consider (i) their derivatives and (ii) their integrals. Do they give meaningful equations? You will probably find it helpful to sketch all the functions involved.

4.8. The function $y(x) = x \sin x$ for $0 \leq x \leq \pi$ is to be represented by a Fourier series of period 2π that is either even or odd. By sketching the function and considering its derivative, determine which series will have the more rapid convergence. Find the full expression for the better of these two series, showing that the convergence $\sim n^{-3}$ and that alternate terms are missing.

4.9. Find the Fourier coefficients in the expansion of $f(x) = \exp x$ over the range $-1 < x < 1$. What value will the expansion have when $x = 2$?

4.10. By integrating term by term the Fourier series found in the previous question and using the Fourier series for $f(x) = x$ found in Section 4.6, show that $\int \exp x \, dx = \exp x + c$. Why is it not possible to show that $d(\exp x)/dx = \exp x$ by differentiating the Fourier series of $f(x) = \exp x$ in a similar manner?

4.11. Consider the function $f(x) = \exp(-x^2)$ in the range $0 \le x \le 1$. Show how it should be continued to give as its Fourier series a series (the actual form is not wanted) (a) with only cosine terms, (b) with only sine terms, (c) with period 1 and (d) with period 2.

Would there be any difference between the values of the last two series at (i) $x = 0$, (ii) $x = 1$?

4.12. Find, without calculation, which terms will be present in the Fourier series for the periodic functions $f(t)$, of period T, that are given in the range $-T/2$ to $T/2$ by:

(a) $f(t) = 2$ for $0 \le |t| < T/4$, $f = 1$ for $T/4 \le |t| < T/2$;

(b) $f(t) = \exp[-(t - T/4)^2]$;

(c) $f(t) = -1$ for $-T/2 \le t < -3T/8$ and $3T/8 \le t < T/2$, $f(t) = 1$ for $-T/8 \le t < T/8$; the graph of f is completed by two straight lines in the remaining ranges so as to form a continuous function.

4.13. Consider the representation as a Fourier series of the displacement of a string lying in the interval $0 \le x \le L$ and fixed at its ends, when it is pulled aside by y_0 at the point $x = L/4$. Sketch the continuations for the region outside the interval that will

(a) produce a series of period L,

(b) produce a series that is antisymmetric about $x = 0$, and

(c) produce a series that will contain only cosine terms.

(d) What are (i) the periods of the series in (b) and (c) and (ii) the value of the "a_0-term" in (c)?

(e) Show that a typical term of the series obtained in (b) is

$$\frac{32y_0}{3n^2\pi^2} \sin \frac{n\pi}{4} \sin \frac{n\pi x}{L}.$$

4.14. Show that the Fourier series for the function $y(x) = |x|$ in the range $-\pi \le x < \pi$ is

$$y(x) = \frac{\pi}{2} - \frac{4}{\pi} \sum_{m=0}^{\infty} \frac{\cos(2m+1)x}{(2m+1)^2}.$$

By integrating this equation term by term from 0 to x, find the function $g(x)$ whose Fourier series is

$$\frac{4}{\pi} \sum_{m=0}^{\infty} \frac{\sin(2m+1)x}{(2m+1)^3}.$$

Deduce the value of the sum S of the series

$$1 - \frac{1}{3^3} + \frac{1}{5^3} - \frac{1}{7^3} + \cdots.$$

4.15. Using the result of Problem 4.14, determine, as far as possible by inspection, the forms of the functions of which the following are the Fourier series:

(a) $\cos\theta + \frac{1}{9}\cos 3\theta + \frac{1}{25}\cos 5\theta + \cdots;$

(b) $\sin\theta + \frac{1}{27}\sin 3\theta + \frac{1}{125}\sin 5\theta + \cdots;$

(c) $\frac{L^2}{3} - \frac{4L^2}{\pi^2}\left[\cos\frac{\pi x}{L} - \frac{1}{4}\cos\frac{2\pi x}{L} + \frac{1}{9}\cos\frac{3\pi x}{L} - \cdots\right].$

(You may find it helpful to first set $x = 0$ in the quoted result and so obtain values for $S_o = \sum(2m+1)^{-2}$ and other sums derivable from it.)

4.16. By finding a cosine Fourier series of period 2 for the function $f(t)$ that takes the form $f(t) = \cosh(t-1)$ in the range $0 \le t \le 1$, prove that

$$\sum_{n=1}^{\infty} \frac{1}{n^2\pi^2 + 1} = \frac{1}{e^2 - 1}.$$

Deduce values for the sums $\sum(n^2\pi^2 + 1)^{-1}$ over odd n and even n separately.

4.17. Find the (real) Fourier series of period 2 for $f(x) = \cosh x$ and $g(x) = x^2$ in the range $-1 \le x \le 1$. By integrating the series for $f(x)$ twice, prove that

$$\sum_{n=1}^{\infty} \frac{(-1)^{n+1}}{n^2\pi^2(n^2\pi^2 + 1)} = \frac{1}{2}\left(\frac{1}{\sinh 1} - \frac{5}{6}\right).$$

4.18. Express the function $f(x) = x^2$ as a Fourier sine series in the range $0 < x \le 2$ and show that it converges to zero at $x = \pm 2$.

4.19. Demonstrate explicitly for the square-wave function discussed in Section 4.2 that Parseval's theorem (4.15) is valid. You will need to use the relationship

$$\sum_{m=0}^{\infty} \frac{1}{(2m+1)^2} = \frac{\pi^2}{8}.$$

Show that a filter that transmits frequencies only up to $8\pi/T$ will still transmit more than 90% of the power in such a square-wave voltage signal.

4.20. Show that the Fourier series for $|\sin\theta|$ in the range $-\pi \le \theta \le \pi$ is given by

$$|\sin\theta| = \frac{2}{\pi} - \frac{4}{\pi}\sum_{m=1}^{\infty} \frac{\cos 2m\theta}{4m^2 - 1}.$$

By setting $\theta = 0$ and $\theta = \pi/2$, deduce values for

$$\sum_{m=1}^{\infty} \frac{1}{4m^2 - 1} \quad \text{and} \quad \sum_{m=1}^{\infty} \frac{1}{16m^2 - 1}.$$

4.21. Find the complex Fourier series for the periodic function of period 2π defined in the range $-\pi \leq x \leq \pi$ by $y(x) = \cosh x$. By setting $x = 0$ prove that

$$\sum_{n=1}^{\infty} \frac{(-1)^n}{n^2 + 1} = \frac{1}{2}\left(\frac{\pi}{\sinh \pi} - 1\right).$$

4.22. The repeating output from an electronic oscillator takes the form of a sine wave $f(t) = \sin t$ for $0 \leq t \leq \pi/2$; it then drops instantaneously to zero and starts again. The output is to be represented by a complex Fourier series of the form

$$\sum_{n=-\infty}^{\infty} c_n e^{4nti}.$$

Sketch the function and find an expression for c_n. Verify that $c_{-n} = c_n^*$. Demonstrate that setting $t = 0$ and $t = \pi/2$ produces differing values for the sum

$$\sum_{n=1}^{\infty} \frac{1}{16n^2 - 1}.$$

Determine the correct value and check it using the result of Problem 4.20.

4.23. Apply Parseval's theorem to the series found in the previous problem and so derive a value for the sum of the series

$$\frac{17}{(15)^2} + \frac{65}{(63)^2} + \frac{145}{(143)^2} + \cdots + \frac{16n^2 + 1}{(16n^2 - 1)^2} + \cdots.$$

4.24. A string, anchored at $x = \pm L/2$, has a fundamental vibration frequency of $2L/c$, where c is the speed of transverse waves on the string. It is pulled aside at its center point by a distance y_0 and released at time $t = 0$. Its subsequent motion can be described by the series

$$y(x,t) = \sum_{n=1}^{\infty} a_n \cos \frac{n\pi x}{L} \cos \frac{n\pi ct}{L}.$$

Find a general expression for a_n and show that only the odd harmonics of the fundamental frequency are present in the sound generated by the released string. By applying Parseval's theorem, find the sum S of the series $\sum_0^{\infty} (2m+1)^{-4}$.

4.25. Show that Parseval's theorem for two real functions whose Fourier expansions have cosine and sine coefficients a_n, b_n and α_n, β_n takes the form

$$\frac{1}{L}\int_0^L f(x)g(x)\,dx = \frac{1}{4}a_0\alpha_0 + \frac{1}{2}\sum_{n=1}^{\infty}(a_n\alpha_n + b_n\beta_n).$$

(a) Demonstrate that for $g(x) = \sin mx$ or $\cos mx$ this reduces to the definition of the Fourier coefficients.
(b) Explicitly verify the above result for the case in which $f(x) = x$ and $g(x)$ is the square-wave function, both in the interval $-1 \leq x \leq 1$.
[Note that $g = g^*$, and it is the integral of fg^* that will have to be formally evaluated using the complex Fourier series representations of the two functions.]

4.26. An odd function $f(x)$ of period 2π is to be approximated by a Fourier sine series having only m terms. The error in this approximation is measured by the square deviation

$$E_m = \int_{-\pi}^{\pi}\left[f(x) - \sum_{n=1}^{m} b_n \sin nx\right]^2 dx.$$

By differentiating E_m with respect to the coefficients b_n, find the values of b_n that minimize E_m.

Sketch the graph of the function $f(x)$, where

$$f(x) = \begin{cases} -x(\pi + x) & \text{for } -\pi \leq x < 0, \\ x(x - \pi) & \text{for } 0 \leq x < \pi. \end{cases}$$

If $f(x)$ is to be approximated by the first three terms of a Fourier sine series, what values should the coefficients have so as to minimize E_3? What is the resulting value of E_3?

HINTS AND ANSWERS

4.1. Note that the only integral of a sinusoid around a complete cycle of length L that is not zero is the integral of $\cos(2\pi nx/L)$ when $n = 0$.

4.3. Only (c). In terms of the Dirichlet conditions (Section 4.1), the others fail as follows: (a) (i); (b) (ii); (d) (ii); (e) (iii).

4.5. $f(x) = 2\sum_1^{\infty}(-1)^{n+1}n^{-1}\sin nx$; set $x = \pi/2$.

4.7. (i) Series (a) from Problem 4.6 does not converge and cannot represent the function $y(x) = -1$. Series (b) reproduces the square-wave function of equation (4.10).

(ii) Series (a) gives the series for $y(x) = -x - \frac{1}{2}x^2 - \frac{1}{2}$ in the range $-1 \leq x \leq 0$ and for $y(x) = x - \frac{1}{2}x^2 - \frac{1}{2}$ in the range $0 \leq x \leq 1$. Series (b) gives the series for $y(x) = x + \frac{1}{2}x^2 + \frac{1}{2}$ in the range $-1 \leq x \leq 0$ and for $y(x) = x - \frac{1}{2}x^2 + \frac{1}{2}$ in the range $0 \leq x \leq 1$.

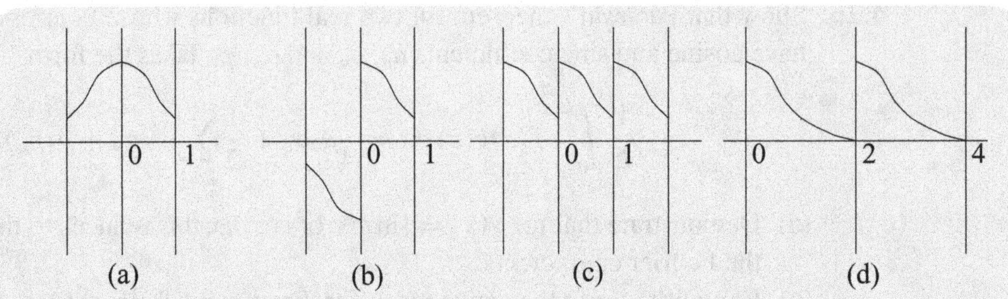

Figure 4.6 Continuations of $\exp(-x^2)$ in $0 \le x \le 1$ to give: (a) cosine terms only; (b) sine terms only; (c) period 1; (d) period 2.

4.9. $f(x) = (\sinh 1)\{1 + 2\sum_1^\infty (-1)^n (1 + n^2\pi^2)^{-1}[\cos(n\pi x) - n\pi \sin(n\pi x)]\}$; the series will converge to the same value as it does at $x = 0$, i.e. $f(0) = 1$.

4.11. See Figure 4.6. (c) (i) $(1 + e^{-1})/2$, (ii) $(1 + e^{-1})/2$; (d) (i) $(1 + e^{-4})/2$, (ii) e^{-1}.

4.13. (d) (i) The periods are both $2L$; (ii) $y_0/2$.

4.15. $S_o = \pi^2/8$. If $S_e = \sum(2m)^{-2}$ then $S_e = \frac{1}{4}(S_e + S_o)$, yielding $S_o - S_e = \pi^2/12$ and $S_e + S_o = \pi^2/6$.
(a) $(\pi/4)(\pi/2 - |\theta|)$; (b) $(\pi\theta/4)(\pi/2 - |\theta|/2)$ from integrating (a). (c) Even function; average value $L^2/3$; $y(0) = 0$; $y(L) = L^2$; probably $y(x) = x^2$. Compare with the worked example in Section 4.5.

4.17. $\cosh x = (\sinh 1)[1 + 2\sum_{n=1}^\infty (-1)^n (\cos n\pi x)/(n^2\pi^2 + 1)]$ and after integrating twice this form must be recovered. Use $x^2 = \frac{1}{3} + 4\sum(-1)^n(\cos n\pi x)/(n^2\pi^2)]$ to eliminate the quadratic term arising from the constants of integration; there is no linear term.

4.19. $C_{\pm(2m+1)} = \mp 2i/[(2m+1)\pi]$; $\sum |C_n|^2 = (4/\pi^2) \times 2 \times (\pi^2/8)$; the values $n = \pm 1, \pm 3$ contribute $> 90\%$ of the total.

4.21. $c_n = [(-1)^n \sinh \pi]/[\pi(1 + n^2)]$. Having set $x = 0$, separate out the $n = 0$ term and note that $(-1)^n = (-1)^{-n}$.

4.23. $(\pi^2 - 8)/16$.

4.25. (b) All a_n and α_n are zero; $b_n = 2(-1)^{n+1}/(n\pi)$ and $\beta_n = 4/(n\pi)$. You will need the result quoted in Problem 4.19.

5 Integral transforms

In the previous chapter we encountered the Fourier series representation of a periodic function in a fixed interval as a superposition of sinusoidal functions. It is often desirable, however, to obtain such a representation for functions that are defined over an infinite interval and have no particular periodicity. Such a representation is called a *Fourier transform* and is one of a class of representations called *integral transforms.*

We begin by considering Fourier transforms as a generalization of Fourier series. We then go on to discuss the properties of the Fourier transform and its applications. In the second part of the chapter we present an analogous discussion of the closely related *Laplace transform*.

5.1 Fourier transforms

The Fourier transform provides a representation of functions defined over an infinite interval and having no particular periodicity, in terms of a superposition of sinusoidal functions. It may thus be considered as a generalization of the Fourier series representation of periodic functions. Since Fourier transforms are often used to represent time-varying functions, we shall present much of our discussion in terms of $f(t)$, rather than $f(x)$, although in some spatial examples $f(x)$ will be the more natural notation and we shall use it as appropriate. Our only requirement on $f(t)$ will be that $\int_{-\infty}^{\infty} |f(t)|\, dt$ is finite.

In order to develop the transition from Fourier series to Fourier transforms, we first recall that a function of period T may be represented as a complex Fourier series, cf. (4.11),

$$f(t) = \sum_{r=-\infty}^{\infty} c_r\, e^{2\pi irt/T} = \sum_{r=-\infty}^{\infty} c_r\, e^{i\omega_r t}, \tag{5.1}$$

where $\omega_r = 2\pi r/T$. As the period T tends to infinity, the "frequency quantum" $\Delta\omega = 2\pi/T$ becomes vanishingly small and the spectrum of allowed frequencies ω_r becomes a continuum. Thus, the infinite sum of terms in the Fourier series becomes an integral, and the coefficients c_r become functions of the *continuous* variable ω, as follows.

We recall, cf. (4.12), that the coefficients c_r in (5.1) are given by

$$c_r = \frac{1}{T}\int_{-T/2}^{T/2} f(t)\, e^{-2\pi irt/T}\, dt = \frac{\Delta\omega}{2\pi}\int_{-T/2}^{T/2} f(t)\, e^{-i\omega_r t}\, dt, \tag{5.2}$$

where we have written the integral in two alternative forms and, for convenience, made one period run from $-T/2$ to $+T/2$ rather than from 0 to T. Substituting from (5.2) into

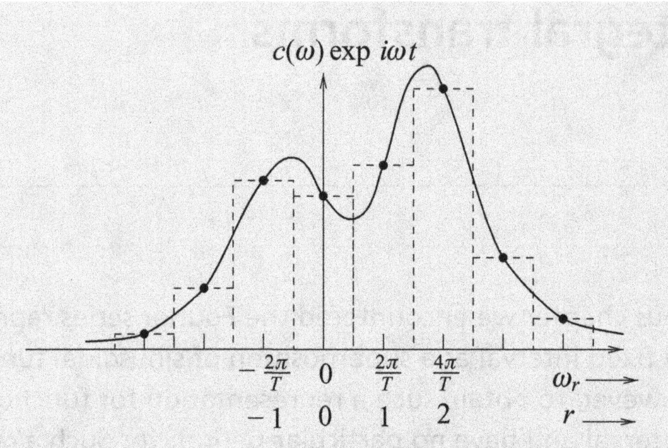

Figure 5.1 The relationship between the Fourier terms for a function of period T and the Fourier integral (the area below the solid line) of the function.

(5.1) gives

$$f(t) = \sum_{r=-\infty}^{\infty} \frac{\Delta\omega}{2\pi} \int_{-T/2}^{T/2} f(u)\, e^{-i\omega_r u}\, du\, e^{i\omega_r t}. \tag{5.3}$$

At this stage ω_r is still a discrete function of r equal to $2\pi r/T$.

The solid points in Figure 5.1 are a plot of (say, the real part of) $c_r\, e^{i\omega_r t}$ as a function of r (or equivalently of ω_r) and it is clear that $(2\pi/T)c_r\, e^{i\omega_r t}$ gives the area of the rth broken-line rectangle. If T tends to ∞ then $\Delta\omega$ $(= 2\pi/T)$ becomes infinitesimal, the width of the rectangles tends to zero and, from the mathematical definition of an integral,

$$\sum_{r=-\infty}^{\infty} \frac{\Delta\omega}{2\pi} g(\omega_r)\, e^{i\omega_r t} \quad \to \quad \frac{1}{2\pi} \int_{-\infty}^{\infty} g(\omega)\, e^{i\omega t}\, d\omega.$$

In this particular case

$$g(\omega_r) = \int_{-T/2}^{T/2} f(u)\, e^{-i\omega_r u}\, du,$$

and (5.3) becomes

$$f(t) = \frac{1}{2\pi} \int_{-\infty}^{\infty} d\omega\, e^{i\omega t} \int_{-\infty}^{\infty} du\, f(u)\, e^{-i\omega u}. \tag{5.4}$$

This result is known as *Fourier's inversion theorem*.

From it we may define the *Fourier transform* of $f(t)$ by

$$\widetilde{f}(\omega) = \frac{1}{\sqrt{2\pi}} \int_{-\infty}^{\infty} f(t)\, e^{-i\omega t}\, dt, \tag{5.5}$$

and its inverse by

$$f(t) = \frac{1}{\sqrt{2\pi}} \int_{-\infty}^{\infty} \widetilde{f}(\omega)\, e^{i\omega t}\, d\omega. \tag{5.6}$$

Including the constant $1/\sqrt{2\pi}$ in the definition of $\widetilde{f}(\omega)$ (whose mathematical existence as $T \to \infty$ is assumed here without proof) is clearly arbitrary, the only requirement being that the product of the constants in (5.5) and (5.6) should equal $1/(2\pi)$. Our definition is chosen to be as symmetric as possible.[1]

We first illustrate the general procedure with a straightforward example.

Example Find the Fourier transform of the exponential decay function $f(t) = 0$ for $t < 0$ and $f(t) = A\,e^{-\lambda t}$ for $t \geq 0$ $(\lambda > 0)$.

Using the definition (5.5) and separating the integral into two parts,

$$\begin{aligned}\widetilde{f}(\omega) &= \frac{1}{\sqrt{2\pi}}\int_{-\infty}^{0}(0)\,e^{-i\omega t}\,dt + \frac{A}{\sqrt{2\pi}}\int_{0}^{\infty}e^{-\lambda t}\,e^{-i\omega t}\,dt\\ &= 0 + \frac{A}{\sqrt{2\pi}}\left[-\frac{e^{-(\lambda+i\omega)t}}{\lambda+i\omega}\right]_0^\infty\\ &= \frac{A}{\sqrt{2\pi}(\lambda+i\omega)},\end{aligned}$$

which is the required transform. It is clear that the multiplicative constant A does not affect the form of the transform, merely its amplitude. This transform may be verified by resubstitution of the above result into (5.6) to recover $f(t)$, but evaluation of the integral requires the use of complex-variable contour integration (Chapter 14). ◀

5.1.1 The uncertainty principle

An important function that appears in many areas of physical science, either precisely or as an approximation to a physical situation, is the *Gaussian* or *normal* distribution. Its Fourier transform is of importance both in itself and also because, when interpreted statistically, it readily illustrates a form of *uncertainty principle*. Its general form is found as follows.

Example Find the Fourier transform of the normalized Gaussian distribution

$$f(t) = \frac{1}{\tau\sqrt{2\pi}}\exp\left(-\frac{t^2}{2\tau^2}\right), \qquad -\infty < t < \infty.$$

This Gaussian distribution is centered on $t = 0$ and has a root mean square deviation $\Delta t = \tau$. (Any reader who is unfamiliar with this interpretation of the distribution should refer to Chapter 16.)

1 The normal practice in engineering is to have a unit constant for the Fourier transform, and a factor of $1/(2\pi)$ in the inverse transform. Either choice is, of course, satisfactory so long as it is applied consistently, though some formulae, e.g. the convolution theorems discussed later, are affected by it.

Using the definition (5.5), the Fourier transform of $f(t)$ is given by

$$\begin{aligned}\widetilde{f}(\omega) &= \frac{1}{\sqrt{2\pi}}\int_{-\infty}^{\infty}\frac{1}{\tau\sqrt{2\pi}}\exp\left(-\frac{t^2}{2\tau^2}\right)\exp(-i\omega t)\,dt \\ &= \frac{1}{\sqrt{2\pi}}\int_{-\infty}^{\infty}\frac{1}{\tau\sqrt{2\pi}}\exp\left\{-\frac{1}{2\tau^2}\left[t^2+2\tau^2 i\omega t+(\tau^2 i\omega)^2-(\tau^2 i\omega)^2\right]\right\}dt,\end{aligned}$$

where the quantity $-(\tau^2 i\omega)^2/(2\tau^2)$ has been both added and subtracted in the exponent in order to allow the factors involving the variable of integration t to be expressed as a complete square. Hence the expression can be written

$$\widetilde{f}(\omega) = \frac{\exp(-\frac{1}{2}\tau^2\omega^2)}{\sqrt{2\pi}}\left\{\frac{1}{\tau\sqrt{2\pi}}\int_{-\infty}^{\infty}\exp\left[-\frac{(t+i\tau^2\omega)^2}{2\tau^2}\right]dt\right\}.$$

The quantity inside the braces is the normalization integral for the Gaussian and equals unity, although to show this strictly needs results from complex variable theory (Chapter 14). That it is equal to unity can be made plausible by changing the variable to $s = t + i\tau^2\omega$ and assuming that the imaginary parts introduced into the integration path and limits (where the integrand goes rapidly to zero anyway) make no difference.

We are left with the result that

$$\widetilde{f}(\omega) = \frac{1}{\sqrt{2\pi}}\exp\left(\frac{-\tau^2\omega^2}{2}\right), \tag{5.7}$$

which is another Gaussian distribution, centered on zero and with a root mean square deviation $\Delta\omega = 1/\tau$. It is interesting to note, and an important property, that the Fourier transform of a Gaussian is another Gaussian. ◀

In the above example the root mean square deviation in t was τ, and so it is seen that the deviations or "spreads" in t and in ω are inversely related:

$$\Delta\omega\,\Delta t = 1,$$

independently of the value of τ. In physical terms, the narrower in time is, say, an electrical impulse, the greater the spread of frequency components it must contain. Similar physical statements are valid for other pairs of Fourier-related variables, such as spatial position and wave number. In an obvious notation, $\Delta k\Delta x = 1$ for a Gaussian wave packet.

The uncertainty relations as usually expressed in quantum mechanics can be related to this if the de Broglie and Einstein relationships for momentum and energy are introduced; they are

$$p = \hbar k \qquad \text{and} \qquad E = \hbar\omega.$$

Here $\hbar$ is Planck's constant h divided by 2π; p and E are the momentum and energy of the system, respectively. In a quantum mechanics setting $f(t)$ is a wavefunction and the distribution of the wave intensity in time is given by $|f|^2$ (also a Gaussian). Similarly, the intensity distribution in frequency is given by $|\widetilde{f}|^2$. These two distributions have respective root mean square deviations of $\tau/\sqrt{2}$ and $1/(\sqrt{2}\tau)$, giving, after incorporation

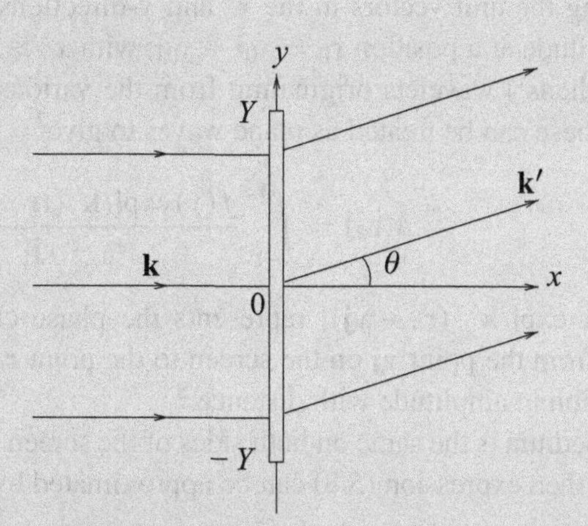

Figure 5.2 Diffraction grating of width $2Y$ with light of wavelength $2\pi/k$ being diffracted through an angle θ.

of the above relations,

$$\Delta E\,\Delta t = \hbar/2 \qquad \text{and} \qquad \Delta p\,\Delta x = \hbar/2.$$

The factors of $1/2$ that appear are specific to the Gaussian form, but any distribution $f(t)$ produces for the product $\Delta E \Delta t$ a quantity $\lambda\hbar$ in which λ is strictly positive (in fact, the Gaussian value of $1/2$ is the minimum possible).

5.1.2 Fraunhofer diffraction

We take our final example of the Fourier transform from the field of optics. The pattern of transmitted light produced by a partially opaque (or phase-changing) object upon which a coherent beam of radiation falls is called a *diffraction pattern* and, in particular, when the cross-section of the object is small compared with the distance at which the light is observed the pattern is known as a *Fraunhofer* diffraction pattern.

We will consider only the case in which the light is monochromatic with wavelength λ. The direction of the incident beam of light can then be described by the *wave vector* $\mathbf{k}$; the magnitude of this vector is given by the *wave number* $k = 2\pi/\lambda$ of the light. The essential quantity in a Fraunhofer diffraction pattern is the dependence of the observed amplitude (and hence intensity) on the angle θ between the viewing direction $\mathbf{k}'$ and the direction $\mathbf{k}$ of the incident beam. This is entirely determined by the spatial distribution of the amplitude and phase of the light at the object, the transmitted intensity in a particular direction $\mathbf{k}'$ being determined by the corresponding Fourier component of this spatial distribution.

As an example, we take as an object a simple two-dimensional screen of width $2Y$ on which light of wave number k is incident normally; see Figure 5.2. We suppose that at the position $(0, y)$ the amplitude of the transmitted light is $f(y)$ per unit length in the y-direction [$f(y)$ may be complex]. The function $f(y)$ is called an *aperture function*. Both the screen and beam are assumed infinite in the z-direction.

Denoting the unit vectors in the x- and y-directions by $\mathbf{i}$ and $\mathbf{j}$ respectively, the total light amplitude at a position $\mathbf{r}_0 = x_0\mathbf{i} + y_0\mathbf{j}$, with $x_0 > 0$, will be the superposition of all the (Huyghens') wavelets originating from the various parts of the screen. For large r_0 $(= |\mathbf{r}_0|)$, these can be treated as plane waves to give[2]

$$A(\mathbf{r}_0) = \int_{-Y}^{Y} \frac{f(y)\exp[i\mathbf{k}' \cdot (\mathbf{r}_0 - y\mathbf{j})]}{|\mathbf{r}_0 - y\mathbf{j}|}\,dy. \tag{5.8}$$

The factor $\exp[i\mathbf{k}' \cdot (\mathbf{r}_0 - y\mathbf{j})]$ represents the phase change undergone by the light in traveling from the point $y\mathbf{j}$ on the screen to the point $\mathbf{r}_0$, and the denominator represents the reduction in amplitude with distance.[3]

If the medium is the same on both sides of the screen then $\mathbf{k}' = k\cos\theta\,\mathbf{i} + k\sin\theta\,\mathbf{j}$, and if $r_0 \gg Y$ then expression (5.8) can be approximated by

$$A(\mathbf{r}_0) = \frac{\exp(i\mathbf{k}' \cdot \mathbf{r}_0)}{r_0} \int_{-\infty}^{\infty} f(y)\exp(-iky\sin\theta)\,dy. \tag{5.9}$$

We have used the fact that $f(y) = 0$ for $|y| > Y$ to extend the integral to infinite limits. The intensity in the direction θ is then given by

$$I(\theta) = |A|^2 = \frac{2\pi}{{r_0}^2}|\tilde{f}(q)|^2, \tag{5.10}$$

where $q = k\sin\theta$. We now consider a specific case.

Example Evaluate $I(\theta)$ for an aperture consisting of two long slits each of width $2b$ whose centers are separated by a distance $2a$, $a > b$; the slits are illuminated by light of wavelength λ.

The aperture function is plotted in Figure 5.3. We first need to find $\tilde{f}(q)$:

$$\begin{aligned}
\tilde{f}(q) &= \frac{1}{\sqrt{2\pi}}\int_{-a-b}^{-a+b} e^{-iqx}\,dx + \frac{1}{\sqrt{2\pi}}\int_{a-b}^{a+b} e^{-iqx}\,dx \\
&= \frac{1}{\sqrt{2\pi}}\left[-\frac{e^{-iqx}}{iq}\right]_{-a-b}^{-a+b} + \frac{1}{\sqrt{2\pi}}\left[-\frac{e^{-iqx}}{iq}\right]_{a-b}^{a+b} \\
&= \frac{-1}{iq\sqrt{2\pi}}\left[e^{-iq(-a+b)} - e^{-iq(-a-b)} + e^{-iq(a+b)} - e^{-iq(a-b)}\right].
\end{aligned}$$

After some manipulation we obtain

$$\tilde{f}(q) = \frac{4\cos qa\sin qb}{q\sqrt{2\pi}}.$$

2 This is the approach first used by Fresnel. For simplicity we have omitted from the integral a multiplicative inclination factor that depends on angle θ and decreases as θ increases.

3 Recall that the system is infinite in the z-direction and so the "spreading" is effectively in two dimensions only.

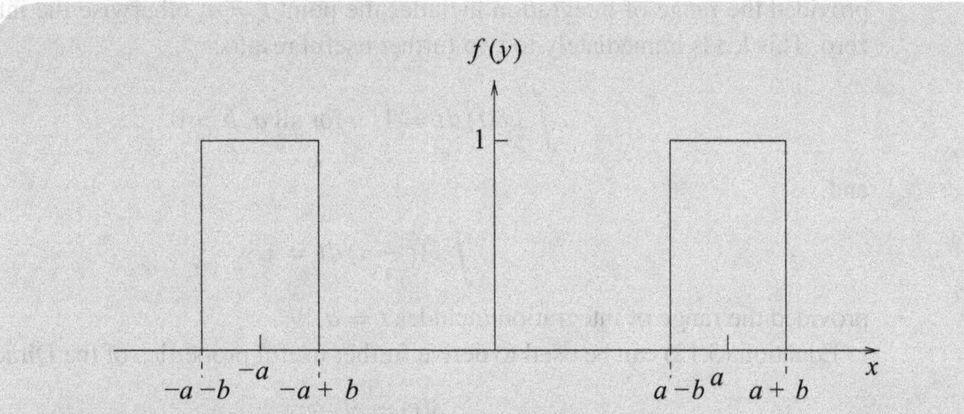

Figure 5.3 The aperture function $f(y)$ for two wide slits.

Now applying (5.10) we find

$$I(\theta) = \frac{16\cos^2 qa \sin^2 qb}{q^2 r_0{}^2},$$

where r_0 is the distance from the center of the aperture. Remembering that $q = (2\pi \sin\theta)/\lambda$, and hence varies as θ varies, we see that the illumination on a distant viewing screen will show a complicated series of maxima and minima.[4] ◀

5.2 The Dirac δ-function

Before going on to consider further properties of Fourier transforms we make a digression to discuss the Dirac δ-function and its relation to Fourier transforms. The δ-function is different from most functions encountered in the physical sciences but we will see that a rigorous mathematical definition exists; the utility of the δ-function will be demonstrated throughout the remainder of this chapter. It can be visualized as a very sharp narrow pulse (in space, time, density, etc.) which produces an integrated effect having a definite magnitude. The formal properties of the δ-function may be summarized as follows.

The Dirac δ-function has the property that

$$\delta(t) = 0 \quad \text{for } t \neq 0, \tag{5.11}$$

but its fundamental defining property is

$$\int f(t)\delta(t-a)\,dt = f(a), \tag{5.12}$$

4 Do you expect a maximum or minimum in the "straight ahead" direction, $\theta = 0$? If the former, how does its intensity vary with (i) a, and (ii) b?

provided the range of integration includes the point $t = a$; otherwise the integral equals zero. This leads immediately to two further useful results:

$$\int_{-a}^{b} \delta(t)\,dt = 1 \qquad \text{for all } a, b > 0 \tag{5.13}$$

and

$$\int \delta(t-a)\,dt = 1, \tag{5.14}$$

provided the range of integration includes $t = a$.

Equation (5.12) can be used to derive further useful properties of the Dirac δ-function:

$$\delta(t) = \delta(-t), \tag{5.15}$$

$$\delta(at) = \frac{1}{|a|}\delta(t), \tag{5.16}$$

$$t\delta(t) = 0. \tag{5.17}$$

We now prove the second of these.

Example Prove that $\delta(bt) = \delta(t)/|b|$.

Let us first consider the case where $b > 0$. It follows that

$$\int_{-\infty}^{\infty} f(t)\delta(bt)\,dt = \int_{-\infty}^{\infty} f\left(\frac{t'}{b}\right)\delta(t')\frac{dt'}{b} = \frac{1}{b}f(0) = \frac{1}{b}\int_{-\infty}^{\infty} f(t)\delta(t)\,dt,$$

where we have made the substitution $t' = bt$. But $f(t)$ is arbitrary and so we immediately see that $\delta(bt) = \delta(t)/b = \delta(t)/|b|$ for $b > 0$.

Now consider the case where $b = -c < 0$. It follows that

$$\int_{-\infty}^{\infty} f(t)\delta(bt)\,dt = \int_{\infty}^{-\infty} f\left(\frac{t'}{-c}\right)\delta(t')\left(\frac{dt'}{-c}\right) = \int_{-\infty}^{\infty} \frac{1}{c} f\left(\frac{t'}{-c}\right)\delta(t')\,dt'$$

$$= \frac{1}{c}f(0) = \frac{1}{|b|}f(0) = \frac{1}{|b|}\int_{-\infty}^{\infty} f(t)\delta(t)\,dt,$$

where we have made the substitution $t' = bt = -ct$. But $f(t)$ is arbitrary and so

$$\delta(bt) = \frac{1}{|b|}\delta(t),$$

for all b, which establishes the result. ◀

Furthermore, by considering an integral of the form

$$\int f(t)\delta(h(t))\,dt,$$

and making a change of variables to $z = h(t)$, we may show that

$$\delta(h(t)) = \sum_i \frac{\delta(t-t_i)}{|h'(t_i)|}, \tag{5.18}$$

where the t_i are those values of t for which $h(t) = 0$ and $h'(t)$ stands for dh/dt.[5]

The derivative of the delta function, $\delta'(t)$, is defined by

$$\int_{-\infty}^{\infty} f(t)\delta'(t)\,dt = \Big[f(t)\delta(t)\Big]_{-\infty}^{\infty} - \int_{-\infty}^{\infty} f'(t)\delta(t)\,dt$$
$$= -f'(0), \tag{5.19}$$

and similarly for higher derivatives.[6]

For many practical purposes, effects that are not strictly described by a δ-function may be analyzed as such, if they take place in an interval much shorter than the response interval of the system on which they act. For example, the idealized notion of an impulse of magnitude J applied at time t_0 can be represented by

$$j(t) = J\delta(t - t_0). \tag{5.20}$$

Many physical situations are described by a δ-function in space rather than in time. Moreover, we often require the δ-function to be defined in more than one dimension. For example, the charge density of a point charge q at a point $\mathbf{r}_0$ may be expressed as a three-dimensional δ-function

$$\rho(\mathbf{r}) = q\delta(\mathbf{r} - \mathbf{r}_0) = q\delta(x - x_0)\delta(y - y_0)\delta(z - z_0), \tag{5.21}$$

so that a discrete "quantum" is expressed as if it were a continuous distribution. From (5.21) we see that (as expected) the total charge enclosed in a volume V is given by

$$\int_V \rho(\mathbf{r})\,dV = \int_V q\delta(\mathbf{r} - \mathbf{r}_0)\,dV = \begin{cases} q & \text{if } \mathbf{r}_0 \text{ lies in } V, \\ 0 & \text{otherwise.} \end{cases}$$

Closely related to the Dirac δ-function is the *Heaviside* or *unit step function* $H(t)$, for which

$$H(t) = \begin{cases} 1 & \text{for } t > 0, \\ 0 & \text{for } t < 0. \end{cases} \tag{5.22}$$

This function is clearly discontinuous at $t = 0$ and it is usual to take $H(0) = 1/2$. A combination of Heaviside functions can be used to describe a function that has a constant value over a limited range. For example,

$$f(t) = 3[H(t - a) - H(t - b)]$$

would describe a function that has the value 3 for $a < t < b$ but is zero outside this range.[7]

5 Result (5.16) is a particular example, in which $h(t) = bt$.

6 Give an integral expression, involving the function $f(x)$ and the delta function and/or its derivatives, that is equal to the nth derivative of $f(x)$ evaluated at $x = a$.

7 Using a "Morse code" in which dots are represented by unit δ-functions and dashes by a value of $+1$ maintained for three time units, construct the function of t that corresponds to the international distress signal "SOS" ($\cdots - - - \cdots$) starting at $t = 1$. The space between "sounds" should be one time unit and that between letters should be three units.

The Heaviside function is related to the delta function by

$$H'(t) = \delta(t). \tag{5.23}$$

Example Prove relation (5.23).

Considering, for an *arbitrary* function $f(x)$, the integral

$$\begin{aligned}\int_{-\infty}^{\infty} f(t)H'(t)\,dt &= \Big[f(t)H(t)\Big]_{-\infty}^{\infty} - \int_{-\infty}^{\infty} f'(t)H(t)\,dt \\ &= f(\infty) - \int_{0}^{\infty} f'(t)\,dt \\ &= f(\infty) - \Big[f(t)\Big]_{0}^{\infty} = f(0),\end{aligned}$$

and comparing it with (5.12) when $a = 0$ immediately shows that $H'(t) = \delta(t)$. ◀

5.2.1 Relation of the δ-function to Fourier transforms

In the previous section we introduced the Dirac δ-function as a way of representing very sharp narrow pulses, but in no way related it to Fourier transforms. We now show that the δ-function can equally well be defined in a way that more naturally relates it to the Fourier transform.

Referring back to the Fourier inversion theorem (5.4), we have

$$\begin{aligned}f(t) &= \frac{1}{2\pi}\int_{-\infty}^{\infty} d\omega\, e^{i\omega t}\int_{-\infty}^{\infty} du\; f(u)\,e^{-i\omega u} \\ &= \int_{-\infty}^{\infty} du\; f(u)\left\{\frac{1}{2\pi}\int_{-\infty}^{\infty} e^{i\omega(t-u)}\,d\omega\right\}.\end{aligned}$$

Comparison of this with (5.12) shows that we may write the δ-function as[8]

$$\delta(t-u) = \frac{1}{2\pi}\int_{-\infty}^{\infty} e^{i\omega(t-u)}\,d\omega. \tag{5.24}$$

Considered as a Fourier transform, this representation shows that a very narrow time peak at $t = u$ results from the superposition of a complete spectrum of harmonic waves, all frequencies having the same amplitude and all waves being in phase at $t = u$. This suggests that the δ-function may also be represented as the limit of the transform of a uniform distribution of unit height as the width of this distribution becomes infinite.

8 Note that the multiplicative factor $1/(2\pi)$ is completely determined and does *not* depend on the arbitrary choice of scaling in the definition of a Fourier transform. See footnote 1.

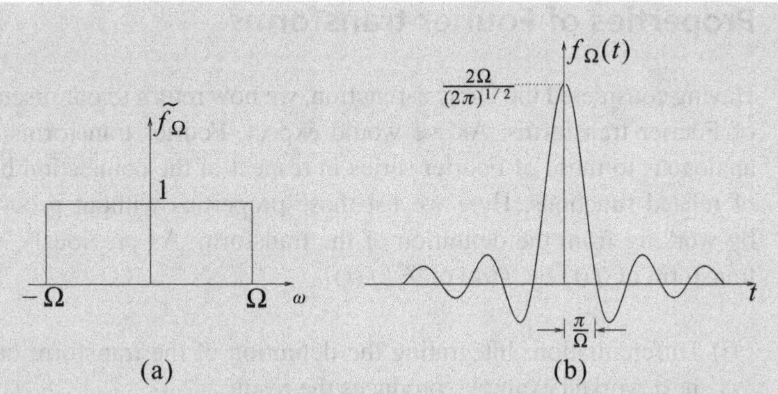

Figure 5.4 (a) A Fourier transform showing a rectangular distribution of frequencies between $\pm\Omega$; (b) the function of which it is the transform, which is proportional to $t^{-1}\sin\Omega t$.

Consider the rectangular distribution of frequencies shown in Figure 5.4(a). From (5.6), taking the inverse Fourier transform,

$$f_\Omega(t) = \frac{1}{\sqrt{2\pi}} \int_{-\Omega}^{\Omega} 1 \times e^{i\omega t}\, d\omega$$

$$= \frac{2\Omega}{\sqrt{2\pi}} \frac{\sin \Omega t}{\Omega t}. \tag{5.25}$$

This function is illustrated in Figure 5.4(b) and it is apparent that, for large Ω, it becomes very large at $t = 0$ and also very narrow about $t = 0$, as we qualitatively expect and require. We also note that, in the limit $\Omega \to \infty$, $f_\Omega(t)$, as defined by the inverse Fourier transform, tends to $(2\pi)^{1/2}\delta(t)$ by virtue of (5.24). Hence we may conclude that the δ-function can also be represented by

$$\delta(t) = \lim_{\Omega\to\infty} \left(\frac{\sin \Omega t}{\pi t} \right). \tag{5.26}$$

Several other function representations are equally valid, e.g. the limiting cases of rectangular, triangular or Gaussian distributions; the only essential requirements are a knowledge of the area under such a curve and that undefined operations such as dividing by zero are not inadvertently carried out on the δ-function whilst some non-explicit representation is being employed.

We also note that the Fourier transform definition of the delta function, (5.24), shows that the latter is real since

$$\delta^*(t) = \frac{1}{2\pi} \int_{-\infty}^{\infty} e^{-i\omega t}\, d\omega = \delta(-t) = \delta(t).$$

Finally, the Fourier transform of a δ-function is simply

$$\widetilde{\delta}(\omega) = \frac{1}{\sqrt{2\pi}} \int_{-\infty}^{\infty} \delta(t)\, e^{-i\omega t}\, dt = \frac{1}{\sqrt{2\pi}}. \tag{5.27}$$

5.3 Properties of Fourier transforms

Having considered the Dirac δ-function, we now return to our discussion of the properties of Fourier transforms. As we would expect, Fourier transforms have many properties analogous to those of Fourier series in respect of the connection between the transforms of related functions. Here we list these properties without proof; they can be verified by working from the definition of the transform. As previously, we denote the Fourier transform of $f(t)$ by $\widetilde{f}(\omega)$ or $\mathcal{F}[f(t)]$.

(i) Differentiation: Integrating the definition of the transform once by parts, as in the next worked example, produces the result

$$\mathcal{F}\left[f'(t)\right] = i\omega\widetilde{f}(\omega). \tag{5.28}$$

This may be extended to higher derivatives, using repeated integration by parts, and yields

$$\mathcal{F}\left[f''(t)\right] = i\omega\mathcal{F}\left[f'(t)\right] = -\omega^2\widetilde{f}(\omega),$$

and so on, each additional differentiation introducing a further factor of $i\omega$ on the RHS.

(ii) Integration: Again, integration by parts is used, but with the indefinite integral being the factor that is differentiated. The result is

$$\mathcal{F}\left[\int^t f(s)\,ds\right] = \frac{1}{i\omega}\widetilde{f}(\omega) + 2\pi c\delta(\omega), \tag{5.29}$$

where the term $2\pi c\delta(\omega)$ represents the Fourier transform of the constant of integration associated with the indefinite integral.

(iii) Scaling:

$$\mathcal{F}[f(at)] = \frac{1}{a}\widetilde{f}\left(\frac{\omega}{a}\right). \tag{5.30}$$

(iv) Translation:

$$\mathcal{F}[f(t+a)] = e^{ia\omega}\widetilde{f}(\omega). \tag{5.31}$$

(v) Exponential multiplication:

$$\mathcal{F}\left[e^{\alpha t}f(t)\right] = \widetilde{f}(\omega + i\alpha), \tag{5.32}$$

where α may be real, imaginary or complex.

Example Prove relation (5.28).

Calculating the Fourier transform of $f'(t)$ directly, we obtain

$$\begin{aligned}\mathcal{F}\left[f'(t)\right] &= \frac{1}{\sqrt{2\pi}}\int_{-\infty}^{\infty} f'(t)\,e^{-i\omega t}\,dt \\ &= \frac{1}{\sqrt{2\pi}}\left[e^{-i\omega t} f(t)\right]_{-\infty}^{\infty} + \frac{1}{\sqrt{2\pi}}\int_{-\infty}^{\infty} i\omega\, e^{-i\omega t} f(t)\,dt \\ &= i\omega \tilde{f}(\omega),\end{aligned}$$

provided $f(t) \to 0$ as $t \to \pm\infty$. This it must do, since $\int_{-\infty}^{\infty} |f(t)|\,dt$ is finite. ◀

To illustrate a use and also a proof of (5.32), let us consider an amplitude-modulated radio wave. Suppose a message to be broadcast is represented by $f(t)$. The message can be added electronically to a constant signal a of magnitude such that $a + f(t)$ is never negative, and then the sum can be used to modulate the amplitude of a carrier signal of frequency ω_c. Using a complex exponential notation, the transmitted amplitude is now

$$g(t) = A\,[a + f(t)]\,e^{i\omega_c t}. \tag{5.33}$$

Ignoring in the present context the effect of the term $Aa\exp(i\omega_c t)$, which gives a contribution to the transmitted spectrum only at $\omega = \omega_c$, we obtain for the new spectrum

$$\begin{aligned}\tilde{g}(\omega) &= \frac{1}{\sqrt{2\pi}} A \int_{-\infty}^{\infty} f(t)\,e^{i\omega_c t}\,e^{-i\omega t}\,dt \\ &= \frac{1}{\sqrt{2\pi}} A \int_{-\infty}^{\infty} f(t)\,e^{-i(\omega-\omega_c)t}\,dt \\ &= A\tilde{f}(\omega - \omega_c),\end{aligned} \tag{5.34}$$

which is simply a shift of the whole spectrum by the carrier frequency. The use of different carrier frequencies enables signals to be separated.

5.3.1 Odd and even functions

If $f(t)$ is odd or even, then we may derive alternative forms of Fourier's inversion theorem, which lead to the definition of different transform pairs. Let us first consider an odd function $f(t) = -f(-t)$, whose Fourier transform is given by

$$\begin{aligned}\tilde{f}(\omega) &= \frac{1}{\sqrt{2\pi}}\int_{-\infty}^{\infty} f(t)\,e^{-i\omega t}\,dt \\ &= \frac{1}{\sqrt{2\pi}}\int_{-\infty}^{\infty} f(t)(\cos\omega t - i\sin\omega t)\,dt \\ &= \frac{-2i}{\sqrt{2\pi}}\int_{0}^{\infty} f(t)\sin\omega t\,dt,\end{aligned}$$

where in the last line we use the fact that $f(t)$ and $\sin\omega t$ are odd, whereas $\cos\omega t$ is even.

We note that $\widetilde{f}(-\omega) = -\widetilde{f}(\omega)$, i.e. $\widetilde{f}(\omega)$ is an odd function of ω. Hence

$$f(t) = \frac{1}{\sqrt{2\pi}} \int_{-\infty}^{\infty} \widetilde{f}(\omega)\, e^{i\omega t}\, d\omega = \frac{2i}{\sqrt{2\pi}} \int_{0}^{\infty} \widetilde{f}(\omega) \sin \omega t\, d\omega$$
$$= \frac{2}{\pi} \int_{0}^{\infty} d\omega\, \sin \omega t \left\{ \int_{0}^{\infty} f(u) \sin \omega u\, du \right\}.$$

Thus we may define the *Fourier sine transform pair* for odd functions:

$$\widetilde{f}_s(\omega) = \sqrt{\frac{2}{\pi}} \int_{0}^{\infty} f(t)\, \sin \omega t\, dt, \tag{5.35}$$

$$f(t) = \sqrt{\frac{2}{\pi}} \int_{0}^{\infty} \widetilde{f}_s(\omega)\, \sin \omega t\, d\omega. \tag{5.36}$$

Note that although the Fourier sine transform pair was derived by considering an odd function $f(t)$ defined over all t, the definitions (5.35) and (5.36) only require $f(t)$ and $\widetilde{f}_s(\omega)$ to be defined for positive t and ω respectively. For an even function, i.e. one for which $f(t) = f(-t)$, we can define the *Fourier cosine transform pair* in a similar way, but with $\sin \omega t$ replaced by $\cos \omega t$.

5.3.2 Convolution and deconvolution

It is apparent that any attempt to measure the value of a physical quantity is limited, to some extent, by the finite resolution of the measuring apparatus used. On the one hand, the physical quantity we wish to measure will be in general a function of an independent variable, x say, i.e. the true function to be measured takes the form $f(x)$. On the other hand, the apparatus we are using does not give the true output value of the function; a resolution function $g(y)$ is involved. By this we mean that the probability that an output value $y = 0$ will be recorded instead as being between y and $y + dy$ is given by $g(y)\, dy$.

Some possible resolution functions of this sort are shown in Figure 5.5. To obtain good results we wish the resolution function to be as close to a δ-function as possible [case (a)]. A typical piece of apparatus has a resolution function of finite width, although if it is accurate the mean is centered on the true value [case (b)]. However, some apparatuses may show biases that tend to shift observations to higher or lower values than the true ones [cases (c) and (d)], thereby exhibiting systematic errors.

Given that the true distribution is $f(x)$ and the resolution function of our measuring apparatus is $g(y)$, we wish to calculate what the observed distribution $h(z)$ will be. The symbols x, y and z all refer to the same physical variable (e.g. length or angle), but are denoted differently because the variable appears in the analysis in three different roles.

The probability that a true reading lying between x and $x + dx$, and so having probability $f(x)\, dx$ of being selected by the experiment, will be moved by the instrumental resolution by an amount $z - x$ into a small interval of width dz is $g(z - x)\, dz$. Hence the combined probability that the interval dx will give rise to an observation appearing in the interval dz is $f(x)\, dx\, g(z - x)\, dz$. Adding together the contributions from all values of x that can lead to an observation in the range z to $z + dz$, we find that the observed

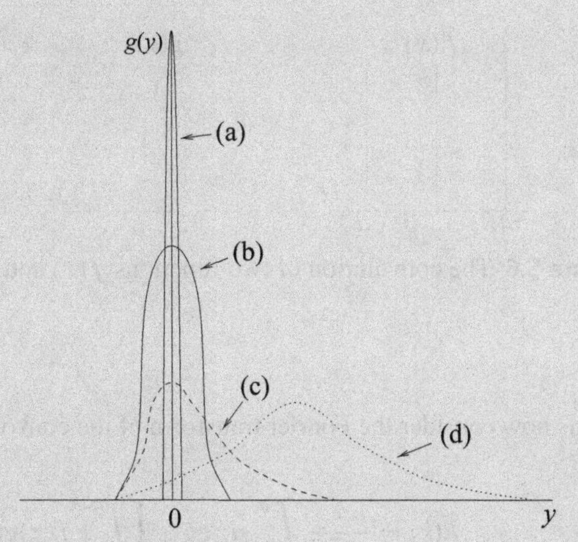

Figure 5.5 Resolution functions: (a) ideal δ-function; (b) typical unbiased resolution; (c) and (d) biases tending to shift observations to higher values than the true one.

distribution is given by

$$h(z) = \int_{-\infty}^{\infty} f(x)g(z-x)\,dx. \tag{5.37}$$

The integral in (5.37) is called the *convolution* of the functions f and g and is often written $f * g$. The convolution defined above is commutative ($f * g = g * f$), associative and distributive. The observed distribution is thus the convolution of the true distribution and the experimental resolution function. The result will be that the observed distribution is broader and smoother than the true one and, if $g(y)$ has a bias, the maxima will normally be displaced from their true positions. It is also obvious from (5.37) that if the resolution is the ideal δ-function, $g(y) = \delta(y)$ then $h(z) = f(z)$ and the observed distribution is the true one.

It is interesting to note, and a very important property, that the convolution of any function $g(y)$ with a number of delta functions leaves a copy of $g(y)$ at the position of each of the delta functions, as is illustrated in the next worked example.

Example Find the convolution of the function $f(x) = \delta(x+a) + \delta(x-a)$ with the function $g(y)$ plotted in Figure 5.6.

Using the convolution integral (5.37)

$$h(z) = \int_{-\infty}^{\infty} f(x)g(z-x)\,dx = \int_{-\infty}^{\infty} [\delta(x+a) + \delta(x-a)]g(z-x)\,dx$$
$$= g(z+a) + g(z-a).$$

This convolution $h(z)$ is plotted in Figure 5.6. ◀

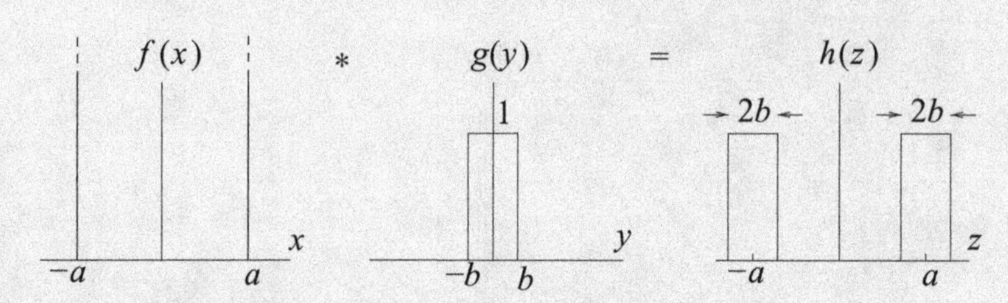

Figure 5.6 The convolution of two functions $f(x)$ and $g(y)$.

Let us now consider the Fourier transform of the convolution (5.37); this is given by

$$\begin{aligned}\widetilde{h}(k) &= \frac{1}{\sqrt{2\pi}}\int_{-\infty}^{\infty} dz\, e^{-ikz}\left\{\int_{-\infty}^{\infty} f(x)g(z-x)\,dx\right\}\\ &= \frac{1}{\sqrt{2\pi}}\int_{-\infty}^{\infty} dx\ f(x)\left\{\int_{-\infty}^{\infty} g(z-x)\,e^{-ikz}\,dz\right\}.\end{aligned}$$

If we let $u = z - x$ in the second integral we have

$$\begin{aligned}\widetilde{h}(k) &= \frac{1}{\sqrt{2\pi}}\int_{-\infty}^{\infty} dx\ f(x)\left\{\int_{-\infty}^{\infty} g(u)\,e^{-ik(u+x)}\,du\right\}\\ &= \frac{1}{\sqrt{2\pi}}\int_{-\infty}^{\infty} f(x)\,e^{-ikx}\,dx\int_{-\infty}^{\infty} g(u)\,e^{-iku}\,du\\ &= \frac{1}{\sqrt{2\pi}}\times\sqrt{2\pi}\,\widetilde{f}(k)\times\sqrt{2\pi}\,\widetilde{g}(k) = \sqrt{2\pi}\,\widetilde{f}(k)\widetilde{g}(k).\end{aligned}\qquad(5.38)$$

Hence the Fourier transform of a convolution $f * g$ is equal to the product of the separate Fourier transforms multiplied by $\sqrt{2\pi}$; this result is called the *convolution theorem.*

It may be proved similarly that the converse is also true, namely that the Fourier transform of the product $f(x)g(x)$ is given by[9]

$$\mathcal{F}\left[f(x)g(x)\right] = \frac{1}{\sqrt{2\pi}}\widetilde{f}(k) * \widetilde{g}(k).\qquad(5.39)$$

9 Using the "engineering notation" (see footnote 1), there is no $\sqrt{2\pi}$ in (5.38) and the factor in (5.39) is $1/(2\pi)$, rather than $1/\sqrt{2\pi}$.

Example Find the Fourier transform of the function in Figure 5.3 representing two wide slits by considering the Fourier transforms of (i) two δ-functions, at $x = \pm a$, (ii) a rectangular function of height 1 and width $2b$ centered on $x = 0$.

(i) The Fourier transform of the two δ-functions is given by

$$\tilde{f}(q) = \frac{1}{\sqrt{2\pi}} \int_{-\infty}^{\infty} \delta(x-a)\, e^{-iqx}\, dx + \frac{1}{\sqrt{2\pi}} \int_{-\infty}^{\infty} \delta(x+a)\, e^{-iqx}\, dx$$

$$= \frac{1}{\sqrt{2\pi}} \left(e^{-iqa} + e^{iqa}\right) = \frac{2\cos qa}{\sqrt{2\pi}}.$$

(ii) The Fourier transform of the broad slit is

$$\tilde{g}(q) = \frac{1}{\sqrt{2\pi}} \int_{-b}^{b} e^{-iqx}\, dx = \frac{1}{\sqrt{2\pi}} \left[\frac{e^{-iqx}}{-iq}\right]_{-b}^{b}$$

$$= \frac{-1}{iq\sqrt{2\pi}} (e^{-iqb} - e^{iqb}) = \frac{2\sin qb}{q\sqrt{2\pi}}.$$

We have already seen that the convolution of these functions is the required function representing two wide slits (see Figure 5.6). So, using the convolution theorem, the Fourier transform of the convolution is $\sqrt{2\pi}$ times the product of the individual transforms, i.e. $4\cos qa\ \sin qb/(q\sqrt{2\pi})$. This is, of course, the same result as that obtained in the example in Subsection 5.1.2.

◀

The inverse of convolution, called *deconvolution*, allows us to find a true distribution $f(x)$ given an observed distribution $h(z)$ and a resolution function $g(y)$.

Example An experimental quantity $f(x)$ is measured using apparatus with a known resolution function $g(y)$ to give an observed distribution $h(z)$. How may $f(x)$ be extracted from the measured distribution?

From the convolution theorem (5.38), the Fourier transform of the measured distribution is

$$\tilde{h}(k) = \sqrt{2\pi}\, \tilde{f}(k)\tilde{g}(k),$$

from which we obtain

$$\tilde{f}(k) = \frac{1}{\sqrt{2\pi}} \frac{\tilde{h}(k)}{\tilde{g}(k)}.$$

Then on inverse Fourier transforming we find

$$f(x) = \frac{1}{\sqrt{2\pi}} \mathcal{F}^{-1}\left[\frac{\tilde{h}(k)}{\tilde{g}(k)}\right].$$

In words, to extract the true distribution, we divide the Fourier transform of the observed distribution by that of the resolution function for each value of k and then take the inverse Fourier transform of the function so generated.

◀

This explicit method of extracting true distributions is straightforward for exact functions but, in practice, because of experimental and statistical uncertainties in the experimental data or because data over only a limited range are available, it is often not very precise, involving as it does three (numerical) transforms each requiring in principle an integral over an infinite range.

5.3.3 Parseval's theorem

Just as there is a connection between the integral of the squared modulus of an original function and the sum of the squared moduli of the Fourier coefficients that represent it in a Fourier series, so there is a connection between the same integral of the function and an *integral* over the squared amplitude of its Fourier transform representation. Like its series counterpart, this connection is known as *Parseval's theorem*. It takes the form

$$\int_{-\infty}^{\infty} |f(x)|^2\, dx = \int_{-\infty}^{\infty} |\widetilde{f}(k)|^2\, dk, \tag{5.40}$$

and can be proved by substituting the inverse Fourier transform representations of $f(x)$ and $f^*(x)$, its complex conjugate, into the integral on the LHS, carrying out the x integration first, and then using the δ-function interpretation (5.24) of an infinite integral with a purely exponential integrand.

When f is a physical amplitude these integrals relate to the total intensity involved in some physical process.

Example The displacement of a damped harmonic oscillator as a function of time is given by

$$f(t) = \begin{cases} 0 & \text{for } t < 0, \\ e^{-t/\tau} \sin \omega_0 t & \text{for } t \geq 0. \end{cases}$$

Find the Fourier transform of this function and so give a physical interpretation of Parseval's theorem.

Using the usual definition for the Fourier transform we find

$$\widetilde{f}(\omega) = \int_{-\infty}^{0} 0 \times e^{-i\omega t}\, dt + \int_{0}^{\infty} e^{-t/\tau} \sin \omega_0 t \, e^{-i\omega t}\, dt.$$

Writing $\sin \omega_0 t$ as $(e^{i\omega_0 t} - e^{-i\omega_0 t})/2i$ we obtain

$$\begin{aligned} \widetilde{f}(\omega) &= 0 + \frac{1}{2i} \int_0^{\infty} \left[e^{-it(\omega - \omega_0 - i/\tau)} - e^{-it(\omega + \omega_0 - i/\tau)} \right] dt \\ &= \frac{1}{2} \left[\frac{1}{\omega + \omega_0 - i/\tau} - \frac{1}{\omega - \omega_0 - i/\tau} \right], \end{aligned}$$

which is the required Fourier transform. The physical interpretation of $|\widetilde{f}(\omega)|^2$ is the energy content per unit frequency interval (i.e. the *energy spectrum*) whilst $|f(t)|^2$ is proportional to the sum of the kinetic and potential energies of the oscillator. Hence (to within a constant) Parseval's theorem shows the equivalence of these two alternative specifications for the total energy. ◀

5.3.4 Fourier transforms in higher dimensions

The concept of the Fourier transform can be extended naturally to more than one dimension. For example, in three dimensions we can define the (spatial) Fourier transform of $f(x, y, z)$ as

$$\widetilde{f}(k_x, k_y, k_z) = \frac{1}{(2\pi)^{3/2}} \iiint f(x, y, z)\, e^{-ik_x x} e^{-ik_y y} e^{-ik_z z}\, dx\, dy\, dz, \tag{5.41}$$

and its inverse as

$$f(x, y, z) = \frac{1}{(2\pi)^{3/2}} \iiint \widetilde{f}(k_x, k_y, k_z)\, e^{ik_x x} e^{ik_y y} e^{ik_z z}\, dk_x\, dk_y\, dk_z. \tag{5.42}$$

Denoting the vector with components $k_x,\ k_y,\ k_z$ by $\mathbf{k}$ and that with components $x,\ y,\ z$ by $\mathbf{r}$, we can write the Fourier transform pair (5.41), (5.42) as

$$\widetilde{f}(\mathbf{k}) = \frac{1}{(2\pi)^{3/2}} \int f(\mathbf{r})\, e^{-i\mathbf{k}\cdot\mathbf{r}}\, d^3\mathbf{r}, \tag{5.43}$$

$$f(\mathbf{r}) = \frac{1}{(2\pi)^{3/2}} \int \widetilde{f}(\mathbf{k})\, e^{i\mathbf{k}\cdot\mathbf{r}}\, d^3\mathbf{k}. \tag{5.44}$$

From these relations we may deduce that the three-dimensional Dirac δ-function can be written as

$$\delta(\mathbf{r}) = \frac{1}{(2\pi)^3} \int e^{i\mathbf{k}\cdot\mathbf{r}}\, d^3\mathbf{k}. \tag{5.45}$$

Similar relations to (5.43), (5.44) and (5.45) exist for spaces of other dimensionalities.

Example In three-dimensional space a function $f(\mathbf{r})$ possesses spherical symmetry, so that $f(\mathbf{r}) = f(r)$. Find the Fourier transform of $f(\mathbf{r})$ as a one-dimensional integral.

Let us choose spherical polar coordinates in which the vector $\mathbf{k}$ of the Fourier transform lies along the polar axis ($\theta = 0$). This we can do since $f(\mathbf{r})$ is spherically symmetric. We then have

$$d^3\mathbf{r} = r^2 \sin\theta\, dr\, d\theta\, d\phi \qquad \text{and} \qquad \mathbf{k}\cdot\mathbf{r} = kr\cos\theta,$$

where $k = |\mathbf{k}|$. The Fourier transform is then given by

$$\begin{aligned}
\widetilde{f}(\mathbf{k}) &= \frac{1}{(2\pi)^{3/2}} \int f(\mathbf{r})\, e^{-i\mathbf{k}\cdot\mathbf{r}}\, d^3\mathbf{r} \\
&= \frac{1}{(2\pi)^{3/2}} \int_0^\infty dr \int_0^\pi d\theta \int_0^{2\pi} d\phi\, f(r) r^2 \sin\theta\, e^{-ikr\cos\theta} \\
&= \frac{1}{(2\pi)^{3/2}} \int_0^\infty dr\, 2\pi f(r) r^2 \int_0^\pi d\theta\, \sin\theta\, e^{-ikr\cos\theta}.
\end{aligned}$$

The integral over θ may be straightforwardly evaluated by noting that

$$\frac{d}{d\theta}(e^{-ikr\cos\theta}) = ikr\sin\theta\, e^{-ikr\cos\theta}.$$

Therefore

$$\begin{aligned}\widetilde{f}(\mathbf{k}) &= \frac{1}{(2\pi)^{3/2}}\int_0^\infty dr\, 2\pi f(r)r^2\left[\frac{e^{-ikr\cos\theta}}{ikr}\right]_{\theta=0}^{\theta=\pi}\\ &= \frac{1}{(2\pi)^{3/2}}\int_0^\infty 4\pi r^2 f(r)\left(\frac{\sin kr}{kr}\right)dr.\end{aligned}$$

It will be noted that when $\mathbf{k} = \mathbf{0}$ and so $k \to 0$, the factor in large parentheses tends to unity and the Fourier transform becomes effectively the normalization integral of $f(\mathbf{r})$. ◀

A similar result may be obtained for two-dimensional Fourier transforms in which $f(\mathbf{r}) = f(\rho)$, i.e. $f(\mathbf{r})$ is independent of azimuthal angle ϕ. In this case, using the integral representation of the Bessel function $J_0(x)$ given at the very end of Subsection 9.5.3, we find

$$\widetilde{f}(\mathbf{k}) = \frac{1}{2\pi}\int_0^\infty 2\pi\rho f(\rho)J_0(k\rho)\,d\rho. \tag{5.46}$$

5.4 Laplace transforms

Often we are interested in functions $f(t)$ for which the Fourier transform does not exist because $f \not\to 0$ as $t \to \infty$, and so the integral defining $\widetilde{f}$ does not converge. This would be the case for the function $f(t) = t$, which does not possess a Fourier transform. Furthermore, we might be interested in a given function only for $t > 0$, for example when we are given its value at $t = 0$ in an initial-value problem. This leads us to consider the Laplace transform, $\bar{f}(s)$ or $\mathcal{L}[\,f(t)\,]$, of $f(t)$, which is defined by

$$\bar{f}(s) \equiv \int_0^\infty f(t)e^{-st}\,dt, \tag{5.47}$$

provided that the integral exists. We assume here that s is real, but complex values would have to be considered in a more detailed study.[10] In practice, for a given function $f(t)$ there will be some real number s_0 such that the integral in (5.47) exists for $s > s_0$ but diverges for $s \le s_0$.

Through (5.47) we define a *linear* transformation $\mathcal{L}$ that converts functions of the variable t to functions of a new variable s. Its linearity is expressed by

$$\mathcal{L}[af_1(t) + bf_2(t)] = a\mathcal{L}[f_1(t)] + b\mathcal{L}[f_2(t)] = a\bar{f}_1(s) + b\bar{f}_2(s). \tag{5.48}$$

10 It will be clear that, so far as the convergence of the integral is concerned, it is only the real part of any complex s that matters; any imaginary part could be considered as a part of a redefined $f(t)$, taking the form of a phase factor of unit modulus.

Example Find the Laplace transforms of the functions (i) $f(t) = 1$, (ii) $f(t) = e^{at}$, (iii) $f(t) = t^n$, for $n = 0, 1, 2, \ldots$.

(i) By direct application of the definition of a Laplace transform (5.47), we find

$$\mathcal{L}[1] = \int_0^\infty e^{-st}\,dt = \left[\frac{-1}{s}e^{-st}\right]_0^\infty = \frac{1}{s}, \qquad \text{if } s > 0,$$

where the restriction $s > 0$ is required for the integral to exist.

(ii) Again using (5.47) directly, we find

$$\begin{aligned}\bar{f}(s) &= \int_0^\infty e^{at}e^{-st}\,dt = \int_0^\infty e^{(a-s)t}\,dt \\ &= \left[\frac{e^{(a-s)t}}{a-s}\right]_0^\infty = \frac{1}{s-a} \qquad \text{if } s > a.\end{aligned}$$

(iii) Once again using the definition (5.47), we have

$$\bar{f}_n(s) = \int_0^\infty t^n e^{-st}\,dt.$$

Integrating by parts we find

$$\begin{aligned}\bar{f}_n(s) &= \left[\frac{-t^n e^{-st}}{s}\right]_0^\infty + \frac{n}{s}\int_0^\infty t^{n-1}e^{-st}\,dt \\ &= 0 + \frac{n}{s}\bar{f}_{n-1}(s), \qquad \text{if } s > 0.\end{aligned}$$

We now have a recursion relation between successive transforms and by calculating $\bar{f}_0$ we can infer $\bar{f}_1$, $\bar{f}_2$, etc. Since $t^0 = 1$, (i) above gives

$$\bar{f}_0 = \frac{1}{s}, \qquad \text{if } s > 0, \tag{5.49}$$

and

$$\bar{f}_1(s) = \frac{1}{s^2}, \quad \bar{f}_2(s) = \frac{2!}{s^3}, \quad \ldots, \quad \bar{f}_n(s) = \frac{n!}{s^{n+1}} \qquad \text{if } s > 0.$$

Thus, in each case (i)–(iii), direct application of the definition of the Laplace transform (5.47) yields the required result.[11] ◀

Unlike that for the Fourier transform, the inversion of the Laplace transform is not an easy operation to perform, since an explicit formula for $f(t)$, given $\bar{f}(s)$, is not straightforwardly obtained from (5.47). The general method for obtaining an inverse Laplace transform makes use of complex variable theory and is not discussed until

11 Verify the linearity of the Laplace transform operation as follows. Write an equation expressing e^{at}, with $a > 0$, as an infinite sum and take the transforms of both sides. Then show, using the binomial theorem, that the resulting equation in s is valid. Verify also that the condition on s for all transforms to be defined is the same as that for the validity of the binomial expansion.

Table 5.1 *Standard Laplace transforms. The transforms are valid for $s > s_0$*

$f(t)$	$\bar{f}(s)$	s_0
c	c/s	0
ct^n	$cn!/s^{n+1}$	0
$\sin bt$	$b/(s^2+b^2)$	0
$\cos bt$	$s/(s^2+b^2)$	0
e^{at}	$1/(s-a)$	a
$t^n e^{at}$	$n!/(s-a)^{n+1}$	a
$\sinh at$	$a/(s^2-a^2)$	$\|a\|$
$\cosh at$	$s/(s^2-a^2)$	$\|a\|$
$e^{at}\sin bt$	$b/[(s-a)^2+b^2]$	a
$e^{at}\cos bt$	$(s-a)/[(s-a)^2+b^2]$	a
$t^{1/2}$	$\frac{1}{2}(\pi/s^3)^{1/2}$	0
$t^{-1/2}$	$(\pi/s)^{1/2}$	0
$\delta(t-t_0)$	e^{-st_0}	0
$H(t-t_0)=\begin{cases}1 & \text{for } t\geq t_0\\ 0 & \text{for } t<t_0\end{cases}$	e^{-st_0}/s	0

Chapter 15. However, some progress can be made without having to find an *explicit* inverse, since we can prepare from (5.47) a "dictionary" of the Laplace transforms of common functions and, when faced with an inversion to carry out, hope to find the given transform (together with its parent function) in the listing. Such a list is given in Table 5.1.

When finding inverse Laplace transforms using Table 5.1, it is useful to note that for all practical purposes the inverse Laplace transform is unique[12] and linear and so

$$\mathcal{L}^{-1}\left[a\bar{f}_1(s)+b\bar{f}_2(s)\right]=af_1(t)+bf_2(t). \tag{5.50}$$

In many practical problems, the function of s, of which the inverse Laplace transform is to be found, is the ratio of two polynomials. In these cases, the method of partial fractions can be used to express the function in terms of entries that appear in the table, as is illustrated below.

12 This is not strictly true, since two functions can differ from one another at a finite number of isolated points but have the *same* Laplace transform.

Example Using Table 5.1 find $f(t)$ if

$$\bar{f}(s) = \frac{s+3}{s(s+1)}.$$

Using partial fractions $\bar{f}(s)$ may be written

$$\bar{f}(s) = \frac{3}{s} - \frac{2}{s+1}.$$

Comparing this with the standard Laplace transforms in Table 5.1, we find that the inverse transform of $3/s$ is 3 for $s > 0$ and the inverse transform of $2/(s+1)$ is $2e^{-t}$ for $s > -1$, and so

$$f(t) = 3 - 2e^{-t},$$

but only for $s > 0$, so that both conditions on s are satisfied. ◀

5.4.1 Laplace transforms of derivatives and integrals

One of the main uses of Laplace transforms is in solving differential equations. Differential equations are the subject of the next six chapters and we will return to the application of Laplace transforms to their solution in Chapter 6. In the meantime we will derive some of the required basic results, in particular the Laplace transforms of general derivatives and the indefinite integral.

The Laplace transform of the first derivative of $f(t)$ is given by

$$\begin{aligned}\mathcal{L}\left[\frac{df}{dt}\right] &= \int_0^\infty \frac{df}{dt} e^{-st}\, dt \\ &= \left[f(t)e^{-st}\right]_0^\infty + s\int_0^\infty f(t)e^{-st}\, dt \\ &= -f(0) + s\bar{f}(s), \qquad \text{for } s > 0. \end{aligned} \tag{5.51}$$

The evaluation relies only on integration by parts; as this can be repeated, higher-order derivatives may be found in a similar manner.

Example Find the Laplace transform of d^2f/dt^2.

Using the definition of the Laplace transform and integrating by parts we obtain

$$\begin{aligned}\mathcal{L}\left[\frac{d^2f}{dt^2}\right] &= \int_0^\infty \frac{d^2f}{dt^2} e^{-st}\, dt \\ &= \left[\frac{df}{dt}e^{-st}\right]_0^\infty + s\int_0^\infty \frac{df}{dt} e^{-st}\, dt \\ &= -\frac{df}{dt}(0) + s[s\bar{f}(s) - f(0)], \qquad \text{for } s > 0,\end{aligned}$$

where (5.51) has been substituted for the integral. This can be written more neatly as

$$\mathcal{L}\left[\frac{d^2 f}{dt^2}\right] = s^2 \bar{f}(s) - sf(0) - \frac{df}{dt}(0), \qquad \text{for } s > 0.$$

It should be noted that the Laplace transform of the second derivative of $f(t)$ automatically has the initial ($t = 0$) values of the function and its first derivative built into it. ◀

In general the Laplace transform of the nth derivative is given by

$$\mathcal{L}\left[\frac{d^n f}{dt^n}\right] = s^n \bar{f} - s^{n-1} f(0) - s^{n-2}\frac{df}{dt}(0) - \cdots - \frac{d^{n-1} f}{dt^{n-1}}(0), \qquad \text{for } s > 0. \quad (5.52)$$

Again, the initial values of lower derivatives are built into the transform.

We now turn to integration, which is much more straightforward. From the definition (5.47),

$$\mathcal{L}\left[\int_0^t f(u)\,du\right] = \int_0^\infty dt\, e^{-st} \int_0^t f(u)\,du$$

$$= \left[-\frac{1}{s}e^{-st}\int_0^t f(u)\,du\right]_0^\infty + \int_0^\infty \frac{1}{s}e^{-st} f(t)\,dt.$$

The first term on the RHS vanishes at both limits,[13] and so

$$\mathcal{L}\left[\int_0^t f(u)\,du\right] = \frac{1}{s}\mathcal{L}[\,f\,]. \quad (5.53)$$

5.4.2 Other properties of Laplace transforms

From Table 5.1 it will be apparent that multiplying a function $f(t)$ by e^{at} has the effect on its transform that s is replaced by $s - a$. This is easily proved generally:

$$\mathcal{L}\left[e^{at} f(t)\right] = \int_0^\infty f(t) e^{at} e^{-st}\,dt$$

$$= \int_0^\infty f(t) e^{-(s-a)t}\,dt$$

$$= \bar{f}(s-a). \quad (5.54)$$

As it were, multiplying $f(t)$ by e^{at} moves the origin of s by an amount a.

We may now consider the effect of multiplying the Laplace transform $\bar{f}(s)$ by e^{-bs} ($b > 0$). From the definition (5.47),

$$e^{-bs}\bar{f}(s) = \int_0^\infty e^{-s(t+b)} f(t)\,dt$$

$$= \int_b^\infty e^{-sz} f(z-b)\,dz,$$

13 Explain why.

on putting $t + b = z$. Thus $e^{-bs}\bar{f}(s)$ is the Laplace transform of a function $g(t)$ defined by

$$g(t) = \begin{cases} 0 & \text{for } 0 < t \le b, \\ f(t-b) & \text{for } t > b. \end{cases}$$

In other words, the function f has been translated to "later" t (larger values of t) by an amount b.

Further properties of Laplace transforms can be proved in similar ways and are listed below.

(i) $$\mathcal{L}[f(at)] = \frac{1}{a}\bar{f}\left(\frac{s}{a}\right), \qquad (5.55)$$

(ii) $$\mathcal{L}\left[t^n f(t)\right] = (-1)^n \frac{d^n \bar{f}(s)}{ds^n}, \qquad \text{for } n = 1, 2, 3, \ldots, \qquad (5.56)$$

(iii) $$\mathcal{L}\left[\frac{f(t)}{t}\right] = \int_s^\infty \bar{f}(u)\,du, \qquad (5.57)$$

provided $\lim_{t\to 0}[f(t)/t]$ exists.

Additional results can be obtained by combining two or more of the properties derived so far, as is now illustrated.

Example Find an expression for the Laplace transform of $t\,d^2f/dt^2$.

From the definition of the Laplace transform we have

$$\begin{aligned} \mathcal{L}\left[t\frac{d^2f}{dt^2}\right] &= \int_0^\infty e^{-st}t\frac{d^2f}{dt^2}\,dt \\ &= -\frac{d}{ds}\int_0^\infty e^{-st}\frac{d^2f}{dt^2}\,dt \\ &= -\frac{d}{ds}[s^2\bar{f}(s) - sf(0) - f'(0)] \\ &= -s^2\frac{d\bar{f}}{ds} - 2s\bar{f} + f(0). \end{aligned}$$

Clearly, any general result, such as this one, is of some value, but here the transform is not very convenient for future manipulation as it contains a derivative with respect to s. ◀

Finally we mention the convolution theorem for Laplace transforms (which is analogous to that for Fourier transforms discussed in Subsection 5.3.2). If the functions f and g have Laplace transforms $\bar{f}(s)$ and $\bar{g}(s)$ then

$$\mathcal{L}\left[\int_0^t f(u)g(t-u)\,du\right] = \bar{f}(s)\bar{g}(s), \qquad (5.58)$$

where the integral in the brackets on the LHS is the *convolution* of f and g, denoted by $f * g$. As in the case of Fourier transforms, the convolution defined above is commutative,

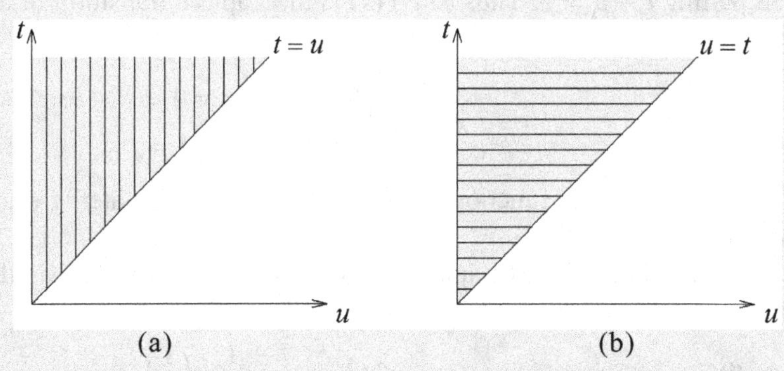

Figure 5.7 Two representations of the Laplace transform convolution (see text).

i.e. $f * g = g * f$, and is associative and distributive. From (5.58) we also see that

$$\mathcal{L}^{-1}\left[\bar{f}(s)\bar{g}(s)\right] = \int_0^t f(u)g(t-u)\,du = f * g.$$

The proof of (5.58) is given in the following worked example.

Example Prove the convolution theorem for Laplace transforms.

From the definition (5.58),

$$\begin{aligned}\bar{f}(s)\bar{g}(s) &= \int_0^\infty e^{-su} f(u)\,du \int_0^\infty e^{-sv} g(v)\,dv \\ &= \int_0^\infty du \int_0^\infty dv\, e^{-s(u+v)} f(u)g(v).\end{aligned}$$

Now letting $u + v = t$ changes the limits on the integrals, with the result that

$$\bar{f}(s)\bar{g}(s) = \int_0^\infty du\, f(u) \int_u^\infty dt\, g(t-u)\, e^{-st}.$$

As shown in Figure 5.7(a) the shaded area of integration may be considered as the sum of vertical strips. However, we may instead integrate over this area by summing over horizontal strips as shown in Figure 5.7(b). Then the integral can be written as

$$\begin{aligned}\bar{f}(s)\bar{g}(s) &= \int_0^t du\, f(u) \int_0^\infty dt\, g(t-u)\, e^{-st} \\ &= \int_0^\infty dt\, e^{-st}\left\{\int_0^t f(u)g(t-u)\,du\right\} \\ &= \mathcal{L}\left[\int_0^t f(u)g(t-u)\,du\right],\end{aligned}$$

as given in equation (5.58). ◀

The properties of the Laplace transform derived in this section can sometimes be useful in finding the Laplace transforms of particular functions.

Example Find the Laplace transform of $f(t) = t \sin bt$.

Although we could calculate the Laplace transform directly, we can use (5.56) to give

$$\bar{f}(s) = (-1)\frac{d}{ds}\mathcal{L}[\sin bt] = -\frac{d}{ds}\left(\frac{b}{s^2+b^2}\right) = \frac{2bs}{(s^2+b^2)^2}, \qquad \text{for } s > 0.$$

The direct method of integration by parts yields that same result, as the reader may care to verify. ◀

5.5 Concluding remarks

In this chapter we have discussed Fourier and Laplace transforms in some detail. Both are examples of *integral transforms*, which can be considered in a more general context.

A general integral transform of a function $f(t)$ takes the form

$$F(\alpha) = \int_a^b K(\alpha, t) f(t)\, dt, \tag{5.59}$$

where $F(\alpha)$ is the transform of $f(t)$ with respect to the *kernel* $K(\alpha, t)$, and α is the transform variable. For example, in the Laplace transform case $K(s, t) = e^{-st}$, $a = 0$ and $b = \infty$, whilst in the one-dimensional Fourier transform $K(k, x) = e^{-ikx}$, $a = -\infty$ and $b = \infty$.

Very often the inverse transform can also be written straightforwardly and we obtain a transform pair similar to that encountered in Fourier transforms. Examples of such pairs are

(i) the Hankel transform

$$F(k) = \int_0^\infty f(x) J_n(kx) x\, dx,$$

$$f(x) = \int_0^\infty F(k) J_n(kx) k\, dk,$$

where the J_n are Bessel functions of order n, and

(ii) the Mellin transform

$$F(z) = \int_0^\infty t^{z-1} f(t)\, dt,$$

$$f(t) = \frac{1}{2\pi i}\int_{-i\infty}^{i\infty} t^{-z} F(z)\, dz.$$

Although we do not have the space to discuss their general properties, the reader should at least be aware of this wider class of integral transforms.

SUMMARY

1. *Dirac δ-function*
   - Definition: $\int f(t)\delta(t-a)\,dt = f(a)$ if the integration range includes $t = a$; otherwise the integral is zero.
   - $\int \delta(t-a)\,dt = 1$ if the integration range includes $t = a$.
   - $\delta(-t) = \delta(t), \quad \delta(at) = \frac{1}{|a|}\delta(t), \quad t\delta(t) = 0.$
   - $\delta(h(t)) = \sum_i \frac{\delta(t - t_i)}{|h'(t_i)|}$, where the t_i are the zeros of $h(t)$.
   - The derivatives $\delta^{(n)}(t)$ of the δ-function are defined by

$$\int_{-\infty}^{\infty} f(t)\delta^{(n)}(t)\,dt = (-1)^n f^{(n)}(0).$$

   - The Heaviside function $H(t)$, which is defined as $H(t) = 1$ for $t > 0$ and $H(t) = 0$ for $t < 0$, has the property $H'(t) = \delta(t)$.
   - Integral representation:

$$\delta(t-u) = \frac{1}{2\pi}\int_{-\infty}^{\infty} e^{i\omega(t-u)}\,d\omega, \qquad \delta(\mathbf{r}) = \frac{1}{(2\pi)^3}\int e^{i\mathbf{k}\cdot\mathbf{r}}\,d^3\mathbf{k}.$$

2. *Fourier and Laplace transforms*
   A Fourier transform $\mathcal{F}[f(t)]$ is a linear transformation $f(t) \to \widetilde{f}(\omega)$ given by

$$\widetilde{f}(\omega) \equiv \frac{1}{\sqrt{2\pi}}\int_{-\infty}^{\infty} f(t)\,e^{-i\omega t}\,dt, \text{ with inverse } f(t) = \frac{1}{\sqrt{2\pi}}\int_{-\infty}^{\infty} \widetilde{f}(\omega)\,e^{i\omega t}\,d\omega.$$

   - Fourier-related variables satisfy "uncertainty principles", e.g.

$$\Delta k\,\Delta x \geq c > 0 \quad \text{and} \quad \Delta\omega\,\Delta t \geq c' > 0,$$

   where Δu is some measure of the spread of u about its mean.
   - Parseval's theorem: $\int_{-\infty}^{\infty} |f(t)|^2\,dt = \int_{-\infty}^{\infty} |\widetilde{f}(\omega)|^2\,d\omega.$

   A Laplace transform $\mathcal{L}[f(t)]$ is a linear transformation $f(t) \to \bar{f}(s)$ given by

$$\bar{f}(s) \equiv \int_0^{\infty} f(t)e^{-st}\,dt \text{ for } s > s_0 \text{ with } s_0 \text{ depending on the form of } f(t).$$

   - For standard Laplace transforms, see Table 5.1 on p. 212.
   - $\mathcal{L}\left[t^n f(t)\right] = (-1)^n \frac{d^n \bar{f}(s)}{ds^n}$, for $n = 1, 2, 3\ldots.$

3. *Fourier and Laplace transforms of related functions*
   For Fourier transforms $f(t)$ can be non-zero for all values of t, but for Laplace transforms $f(t)$ is understood to be, more explicitly, $H(t)f(t)$.

Related function	Fourier transform	Laplace transform
$f(at)$	$\frac{1}{a}\tilde{f}\left(\frac{\omega}{a}\right)$	$\frac{1}{a}\bar{f}\left(\frac{s}{a}\right)$
$f(t-b)$	$e^{-ib\omega}\tilde{f}(\omega)$	$e^{-bs}\bar{f}(s)$
$e^{\alpha t}f(t)$	$\tilde{f}(\omega+i\alpha)$	$\bar{f}(s-\alpha)$
$f'(t)$	$i\omega\tilde{f}(\omega)$	$s\bar{f}(s)-f(0)$
$f''(t)$	$-\omega^2\tilde{f}(\omega)$	$s^2\bar{f}(s)-sf(0)-\frac{df}{dt}(0)$
$f^{(n)}(t)$	$(i)^n\omega^n\tilde{f}(\omega)$	$s^n\bar{f}-s^{n-1}f(0)-s^{n-2}\frac{df}{dt}(0)-\cdots-\frac{d^{n-1}f}{dt^{n-1}}(0)$
$\int^t f(u)\,du$	$\frac{1}{i\omega}\tilde{f}(\omega)+2\pi c\delta(\omega)$	$\frac{1}{s}\bar{f}(s)$
$f(t)*g(t)$	$\sqrt{2\pi}\,\tilde{f}(\omega)\tilde{g}(\omega)$	$\bar{f}(s)\bar{g}(s)$
$f(t)g(t)$	$\frac{1}{\sqrt{2\pi}}\tilde{f}(\omega)*\tilde{g}(\omega)$	–

PROBLEMS

5.1. Find the Fourier transform of the function $f(t)=\exp(-|t|)$.

(a) By applying Fourier's inversion theorem prove that

$$\frac{\pi}{2}\exp(-|t|)=\int_0^\infty \frac{\cos\omega t}{1+\omega^2}\,d\omega.$$

(b) By making the substitution $\omega=\tan\theta$, demonstrate the validity of Parseval's theorem for this function.

5.2. Use the general definition and properties of Fourier transforms to show the following.

(a) If $f(x)$ is periodic with period a then $\tilde{f}(k)=0$, unless $ka=2\pi n$ for integer n.

(b) The Fourier transform of $tf(t)$ is $i\,d\tilde{f}(\omega)/d\omega$.

(c) The Fourier transform of $f(mt+c)$ is

$$\frac{e^{i\omega c/m}}{m}\tilde{f}\left(\frac{\omega}{m}\right).$$

5.3. Find the Fourier transform of $H(x-a)e^{-bx}$, where $H(x)$ is the Heaviside function.

5.4. Prove that the Fourier transform of the function $f(t)$ defined in the tf-plane by straight-line segments joining $(-T, 0)$ to $(0, 1)$ to $(T, 0)$, with $f(t) = 0$ outside $|t| < T$, is

$$\tilde{f}(\omega) = \frac{T}{\sqrt{2\pi}} \operatorname{sinc}^2\left(\frac{\omega T}{2}\right),$$

where sinc x is defined as $(\sin x)/x$.

Use the general properties of Fourier transforms to determine the transforms of the following functions, graphically defined by straight-line segments and equal to zero outside the ranges specified:

(a) $(0, 0)$ to $(0.5, 1)$ to $(1, 0)$ to $(2, 2)$ to $(3, 0)$ to $(4.5, 3)$ to $(6, 0)$;
(b) $(-2, 0)$ to $(-1, 2)$ to $(1, 2)$ to $(2, 0)$;
(c) $(0, 0)$ to $(0, 1)$ to $(1, 2)$ to $(1, 0)$ to $(2, -1)$ to $(2, 0)$.

5.5. By taking the Fourier transform of the equation

$$\frac{d^2\phi}{dx^2} - K^2\phi = f(x),$$

show that its solution, $\phi(x)$, can be written as

$$\phi(x) = \frac{-1}{\sqrt{2\pi}} \int_{-\infty}^{\infty} \frac{e^{ikx}\tilde{f}(k)}{k^2 + K^2}\, dk,$$

where $\tilde{f}(k)$ is the Fourier transform of $f(x)$.

5.6. By differentiating the definition of the Fourier sine transform $\tilde{f}_s(\omega)$ of the function $f(t) = t^{-1/2}$ with respect to ω, and then integrating the resulting expression by parts, find an elementary differential equation satisfied by $\tilde{f}_s(\omega)$. Hence show that this function is its own Fourier sine transform, i.e. $\tilde{f}_s(\omega) = Af(\omega)$, where A is a constant. Show that it is also its own Fourier cosine transform. Assume that the limit as $x \to \infty$ of $x^{1/2} \sin \alpha x$ can be taken as zero.

5.7. Find the Fourier transform of the unit rectangular distribution

$$f(t) = \begin{cases} 1 & |t| < 1, \\ 0 & \text{otherwise.} \end{cases}$$

Determine the convolution of f with itself and, without further integration, determine its transform. Deduce that

$$\int_{-\infty}^{\infty} \frac{\sin^2 \omega}{\omega^2}\, d\omega = \pi, \qquad \int_{-\infty}^{\infty} \frac{\sin^4 \omega}{\omega^4}\, d\omega = \frac{2\pi}{3}.$$

5.8. Calculate the Fraunhofer spectrum produced by a diffraction grating, uniformly illuminated by light of wavelength $2\pi/k$, as follows. Consider a grating with $4N$ equal strips each of width a and alternately opaque and transparent. The aperture function is then

$$f(y) = \begin{cases} A & \text{for } (2n+1)a \le y \le (2n+2)a, \quad -N \le n < N, \\ 0 & \text{otherwise.} \end{cases}$$

(a) Show, for diffraction at angle θ to the normal to the grating, that the required Fourier transform can be written

$$\tilde{f}(q) = (2\pi)^{-1/2} \sum_{r=-N}^{N-1} \exp(-2iarq) \int_a^{2a} A \exp(-iqu)\, du,$$

where $q = k \sin\theta$.

(b) Evaluate the integral and sum to show that

$$\tilde{f}(q) = (2\pi)^{-1/2} \exp(-iqa/2) \frac{A \sin(2qaN)}{q \cos(qa/2)},$$

and hence that the intensity distribution $I(\theta)$ in the spectrum is proportional to

$$\frac{\sin^2(2qaN)}{q^2 \cos^2(qa/2)}.$$

(c) For large values of N, the numerator in the above expression has very closely spaced maxima and minima as a function of θ and effectively takes its mean value, 1/2, giving a low-intensity background. Much more significant peaks in $I(\theta)$ occur when $\theta = 0$ or the cosine term in the denominator vanishes. Show that the corresponding values of $|\tilde{f}(q)|$ are

$$\frac{2aNA}{(2\pi)^{1/2}} \quad \text{and} \quad \frac{4aNA}{(2\pi)^{1/2}(2m+1)\pi}, \qquad \text{with } m \text{ integral.}$$

Note that the constructive interference makes the maxima in $I(\theta) \propto N^2$, not N. Of course, observable maxima only occur for $0 \le \theta \le \pi/2$.

5.9. By finding the complex Fourier *series* for its LHS show that either side of the equation

$$\sum_{n=-\infty}^{\infty} \delta(t + nT) = \frac{1}{T} \sum_{n=-\infty}^{\infty} e^{-2\pi nit/T}$$

can represent a periodic train of impulses. By expressing the function $f(t + nX)$, in which X is a constant, in terms of the Fourier *transform* $\tilde{f}(\omega)$ of $f(t)$, show that

$$\sum_{n=-\infty}^{\infty} f(t + nX) = \frac{\sqrt{2\pi}}{X} \sum_{n=-\infty}^{\infty} \tilde{f}\left(\frac{2n\pi}{X}\right) e^{2\pi nit/X}.$$

This result is known as the *Poisson summation formula.*

5.10. In many applications in which the frequency spectrum of an analogue signal is required, the best that can be done is to sample the signal $f(t)$ a finite number of times at fixed intervals, and then use a *discrete Fourier transform* F_k to estimate discrete points on the (true) frequency spectrum $\tilde{f}(\omega)$.

(a) By an argument that is essentially the converse of that given in Section 5.1, show that, if N samples f_n, beginning at $t = 0$ and spaced τ apart, are taken, then $\tilde{f}(2\pi k/(N\tau)) \approx F_k \tau$ where

$$F_k = \frac{1}{\sqrt{2\pi}} \sum_{n=0}^{N-1} f_n e^{-2\pi nki/N}.$$

(b) For the function $f(t)$ defined by

$$f(t) = \begin{cases} 1 & \text{for } 0 \le t < 1 \\ 0 & \text{otherwise,} \end{cases}$$

from which eight samples are drawn at intervals of $\tau = 0.25$, find a formula for $|F_k|$ and evaluate it for $k = 0, 1, \ldots, 7$.

(c) Find the exact frequency spectrum of $f(t)$ and compare the actual and estimated values of $\sqrt{2\pi}|\tilde{f}(\omega)|$ at $\omega = k\pi$ for $k = 0, 1, \ldots, 7$. Note the relatively good agreement for $k < 4$ and the lack of agreement for larger values of k.

5.11. For a function $f(t)$ that is non-zero only in the range $|t| < T/2$, the full frequency spectrum $\tilde{f}(\omega)$ can be constructed, in principle exactly, from values at discrete sample points $\omega = n(2\pi/T)$. Prove this as follows.

(a) Show that the coefficients of a complex Fourier *series* representation of $f(t)$ with period T can be written as

$$c_n = \frac{\sqrt{2\pi}}{T} \tilde{f}\left(\frac{2\pi n}{T}\right).$$

(b) Use this result to represent $f(t)$ as an infinite sum in the defining integral for $\tilde{f}(\omega)$, and hence show that

$$\tilde{f}(\omega) = \sum_{n=-\infty}^{\infty} \tilde{f}\left(\frac{2\pi n}{T}\right) \operatorname{sinc}\left(n\pi - \frac{\omega T}{2}\right),$$

where sinc x is defined as $(\sin x)/x$.

5.12. A signal obtained by sampling a function $x(t)$ at regular intervals T is passed through an electronic filter, whose response $g(t)$ to a unit δ-function input is represented in a tg-plot by straight lines joining $(0, 0)$ to $(T, 1/T)$ to $(2T, 0)$ and is zero for all other values of t. The output of the filter is the convolution of the input, $\sum_{-\infty}^{\infty} x(t)\delta(t - nT)$, with $g(t)$.

Using the convolution theorem, and the result given in Problem 5.4, show that the output of the filter can be written

$$y(t) = \frac{1}{2\pi} \sum_{n=-\infty}^{\infty} x(nT) \int_{-\infty}^{\infty} \operatorname{sinc}^2\left(\frac{\omega T}{2}\right) e^{-i\omega[(n+1)T-t]} d\omega.$$

5.13. Find the Fourier transform specified in part (a) and then use it to answer part (b).

(a) Find the Fourier transform of

$$f(\gamma, p, t) = \begin{cases} e^{-\gamma t} \sin pt & t > 0, \\ 0 & t < 0, \end{cases}$$

where γ (> 0) and p are constant parameters.

(b) The current $I(t)$ flowing through a certain system is related to the applied voltage $V(t)$ by the equation

$$I(t) = \int_{-\infty}^{\infty} K(t-u)V(u)\,du,$$

where

$$K(\tau) = a_1 f(\gamma_1, p_1, \tau) + a_2 f(\gamma_2, p_2, \tau).$$

The function $f(\gamma, p, t)$ is as given in (a) and all the a_i, γ_i (> 0) and p_i are fixed parameters. By considering the Fourier transform of $I(t)$, find the relationship that must hold between a_1 and a_2 if the total net charge Q passed through the system (over a very long time) is to be zero for an arbitrary applied voltage.

5.14. Prove the equality

$$\int_0^{\infty} e^{-2at} \sin^2 at\,dt = \frac{1}{\pi} \int_0^{\infty} \frac{a^2}{4a^4 + \omega^4}\,d\omega.$$

5.15. A linear amplifier produces an output that is the convolution of its input and its response function. The Fourier transform of the response function for a particular amplifier is

$$\tilde{K}(\omega) = \frac{i\omega}{\sqrt{2\pi}(\alpha + i\omega)^2}.$$

Determine the time variation of its output $g(t)$ when its input is the Heaviside step function. [Consider the Fourier transform of a decaying exponential function and the result of Problem 5.2(b).]

5.16. In quantum mechanics, two equal-mass particles having momenta $\mathbf{p}_j = \hbar\mathbf{k}_j$ and energies $E_j = \hbar\omega_j$ and represented by plane wavefunctions $\phi_j = \exp[i(\mathbf{k}_j \cdot \mathbf{r}_j - \omega_j t)]$, $j = 1, 2$, interact through a potential $V = V(|\mathbf{r}_1 - \mathbf{r}_2|)$. In first-order perturbation theory the probability of scattering to a state with momenta and energies $\mathbf{p}'_j$, E'_j is determined by the modulus squared of the quantity

$$M = \iiint \psi_{\rm f}^* V \psi_{\rm i}\,d\mathbf{r}_1\,d\mathbf{r}_2\,dt.$$

The initial state, $\psi_{\rm i}$, is $\phi_1\phi_2$ and the final state, $\psi_{\rm f}$, is $\phi'_1\phi'_2$.

(a) By writing $\mathbf{r}_1 + \mathbf{r}_2 = 2\mathbf{R}$ and $\mathbf{r}_1 - \mathbf{r}_2 = \mathbf{r}$ and assuming that $d\mathbf{r}_1\,d\mathbf{r}_2 = d\mathbf{R}\,d\mathbf{r}$, show that M can be written as the product of three one-dimensional integrals.

(b) From two of the integrals deduce energy and momentum conservation in the form of δ-functions.

(c) Show that M is proportional to the Fourier transform of V, i.e. to $\widetilde{V}(\mathbf{k})$ where $2\hbar\mathbf{k} = (\mathbf{p}_2 - \mathbf{p}_1) - (\mathbf{p}_2' - \mathbf{p}_1')$ or, alternatively, $\hbar\mathbf{k} = \mathbf{p}_1' - \mathbf{p}_1$.

5.17. For some ion–atom scattering processes, the potential V of the previous problem may be approximated by $V = |\mathbf{r}_1 - \mathbf{r}_2|^{-1}\exp(-\mu|\mathbf{r}_1 - \mathbf{r}_2|)$. Show, using the result of the worked example in Subsection 5.3.4, that the probability that the ion will scatter from, say, $\mathbf{p}_1$ to $\mathbf{p}_1'$ is proportional to $(\mu^2 + k^2)^{-2}$, where $k = |\mathbf{k}|$ and $\mathbf{k}$ is as given in part (c) of the previous problem.

5.18. The equivalent duration and bandwidth, T_e and B_e, of a signal $x(t)$ are defined in terms of the latter and its Fourier transform $\tilde{x}(\omega)$ by

$$T_\text{e} = \frac{1}{x(0)}\int_{-\infty}^{\infty} x(t)\,dt,$$

$$B_\text{e} = \frac{1}{\tilde{x}(0)}\int_{-\infty}^{\infty} \tilde{x}(\omega)\,d\omega,$$

where neither $x(0)$ nor $\tilde{x}(0)$ is zero. Show that the product $T_\text{e}B_\text{e} = 2\pi$ (this is a form of uncertainty principle), and find the equivalent bandwidth of the signal

$$x(t) = \exp(-|t|/T).$$

For this signal, determine the fraction of the total energy that lies in the frequency range $|\omega| < B_\text{e}/4$. You will need the indefinite integral with respect to x of $(a^2 + x^2)^{-2}$, which is

$$\frac{x}{2a^2(a^2 + x^2)} + \frac{1}{2a^3}\tan^{-1}\frac{x}{a}.$$

5.19. Prove the expressions given in Table 5.1 for the Laplace transforms of $t^{-1/2}$ and $t^{1/2}$, by setting $x^2 = ts$ in the result

$$\int_0^{\infty} \exp(-x^2)\,dx = \tfrac{1}{2}\sqrt{\pi}.$$

5.20. Find the functions $y(t)$ whose Laplace transforms are the following:

(a) $1/(s^2 - s - 2)$;

(b) $2s/[(s+1)(s^2+4)]$;

(c) $e^{-(\gamma+s)t_0}/[(s+\gamma)^2 + b^2]$.

5.21. Use the properties of Laplace transforms to prove the following without evaluating any Laplace integrals explicitly:

(a) $\mathcal{L}\left[t^{5/2}\right] = \frac{15}{8}\sqrt{\pi}s^{-7/2}$;

(b) $\mathcal{L}\left[(\sinh at)/t\right] = \frac{1}{2}\ln\left[(s+a)/(s-a)\right], \quad s > |a|$;

(c) $\mathcal{L}\left[\sinh at\cos bt\right] = a(s^2 - a^2 + b^2)[(s-a)^2 + b^2]^{-1}[(s+a)^2 + b^2]^{-1}$.

5.22. Find the solution (the so-called *impulse response* or *Green's function*) of the equation

$$T\frac{dx}{dt} + x = \delta(t)$$

by proceeding as follows.

(a) Show by substitution that

$$x(t) = A(1 - e^{-t/T})H(t)$$

is a solution, for which $x(0) = 0$, of

$$T\frac{dx}{dt} + x = AH(t), \qquad (*)$$

where $H(t)$ is the Heaviside step function.

(b) Construct the solution when the RHS of $(*)$ is replaced by $AH(t - \tau)$, with $dx/dt = x = 0$ for $t < \tau$, and hence find the solution when the RHS is a rectangular pulse of duration τ.

(c) By setting $A = 1/\tau$ and taking the limit as $\tau \to 0$, show that the impulse response is $x(t) = T^{-1}e^{-t/T}$.

(d) Obtain the same result much more directly by taking the Laplace transform of each term in the original equation, solving the resulting algebraic equation and then using the entries in Table 5.1.

5.23. This problem is concerned with the limiting behavior of Laplace transforms.

(a) If $f(t) = A + g(t)$, where A is a constant and the indefinite integral of $g(t)$ is bounded as its upper limit tends to ∞, show that

$$\lim_{s\to 0} s\bar{f}(s) = A.$$

(b) For $t > 0$, the function $y(t)$ obeys the differential equation

$$\frac{d^2y}{dt^2} + a\frac{dy}{dt} + by = c\cos^2\omega t,$$

where a, b and c are positive constants. Find $\bar{y}(s)$ and show that $s\bar{y}(s) \to c/2b$ as $s \to 0$. Interpret the result in the t-domain.

5.24. By writing $f(x)$ as an integral involving the δ-function $\delta(\xi - x)$ and taking the Laplace transforms of both sides, show that the transform of the solution of the equation

$$\frac{d^4y}{dx^4} - y = f(x)$$

for which y and its first three derivatives vanish at $x = 0$ can be written as

$$\bar{y}(s) = \int_0^\infty f(\xi)\frac{e^{-s\xi}}{s^4 - 1}\,d\xi.$$

Use the properties of Laplace transforms and the entries in Table 5.1 to show that

$$y(x) = \frac{1}{2}\int_0^x f(\xi)\,[\sinh(x-\xi) - \sin(x-\xi)]\,d\xi.$$

5.25. The function $f_a(x)$ is defined as unity for $0 < x < a$ and zero otherwise. Find its Laplace transform $\bar{f}_a(s)$ and deduce that the transform of $xf_a(x)$ is

$$\frac{1}{s^2}\left[1 - (1+as)e^{-sa}\right].$$

Write $f_a(x)$ in terms of Heaviside functions and hence obtain an explicit expression for

$$g_a(x) = \int_0^x f_a(y) f_a(x-y)\,dy.$$

Use the expression to write $\bar{g}_a(s)$ in terms of the functions $\bar{f}_a(s)$ and $\bar{f}_{2a}(s)$, and their derivatives, and hence show that $\bar{g}_a(s)$ is equal to the square of $\bar{f}_a(s)$, in accordance with the convolution theorem.

5.26. Show that the Laplace transform of $f(t-a)H(t-a)$, where $a \geq 0$, is $e^{-as}\bar{f}(s)$ and that, if $g(t)$ is a periodic function of period T, $\bar{g}(s)$ can be written as

$$\frac{1}{1-e^{-sT}}\int_0^T e^{-st} g(t)\,dt.$$

(a) Sketch the periodic function defined in $0 \leq t \leq T$ by

$$g(t) = \begin{cases} 2t/T & 0 \leq t < T/2 \\ 2(1 - t/T) & T/2 \leq t \leq T, \end{cases}$$

and, using the previous result, find its Laplace transform.

(b) Show, by sketching it, that

$$\frac{2}{T}[tH(t) + 2\sum_{n=1}^{\infty}(-1)^n(t - \tfrac{1}{2}nT)H(t - \tfrac{1}{2}nT)]$$

is another representation of $g(t)$ and hence derive the relationship

$$\tanh x = 1 + 2\sum_{n=1}^{\infty}(-1)^n e^{-2nx}.$$

HINTS AND ANSWERS

5.1. Note that the integrand has different analytic forms for $t < 0$ and $t \geq 0$. $(2/\pi)^{1/2}(1+\omega^2)^{-1}$.

5.3. $(1/\sqrt{2\pi})[(b-ik)/(b^2+k^2)]e^{-a(b+ik)}$.

5.5. Use or derive $\widetilde{\phi''}(k) = -k^2\tilde{\phi}(k)$ to obtain an algebraic equation for $\tilde{\phi}(k)$ and then use the Fourier inversion formula.

5.7. $(2/\sqrt{2\pi})(\sin\omega/\omega)$.
The convolution is $2 - |t|$ for $|t| < 2$, zero otherwise. Use the convolution theorem.
$(4/\sqrt{2\pi})(\sin^2\omega/\omega^2)$.
Apply Parseval's theorem to f and to $f * f$.

5.9. The Fourier coefficient is T^{-1}, independent of n. Make the changes of variables $t \to \omega$, $n \to -n$ and $T \to 2\pi/X$ and apply the translation theorem.

5.11. (b) Recall that the infinite integral involved in defining $\tilde{f}(\omega)$ has a non-zero integrand only in $|t| < T/2$.

5.13. (a) $(1/\sqrt{2\pi})\{p/[(\gamma + i\omega)^2 + p^2]\}$.
(b) Show that $Q = \sqrt{2\pi}\,\tilde{I}(0)$ and use the convolution theorem. The required relationship is $a_1p_1/(\gamma_1^2 + p_1^2) + a_2p_2/(\gamma_2^2 + p_2^2) = 0$.

5.15. $\tilde{g}(\omega) = 1/[\sqrt{2\pi}(\alpha + i\omega)^2]$, leading to $g(t) = te^{-\alpha t}$.

5.17. $\tilde{V}(\mathbf{k}) \propto [-2\pi/(ik)]\int\{\exp[-(\mu - ik)r] - \exp[-(\mu + ik)r]\}\,dr$.

5.19. Prove the result for $t^{1/2}$ by integrating that for $t^{-1/2}$ by parts.

5.21. (a) Use (5.56) with $n = 2$ on $\mathcal{L}\left[\sqrt{t}\,\right]$; (b) use (5.57);
(c) consider $\mathcal{L}\left[\exp(\pm at)\cos bt\right]$ and use the translation property, Subsection 5.4.2.

5.23. (a) Note that $|\lim\int g(t)e^{-st}\,dt| \le |\lim\int g(t)\,dt|$.
(b) $(s^2 + as + b)\bar{y}(s) = \{c(s^2 + 2\omega^2)/[s(s^2 + 4\omega^2)]\} + (a + s)y(0) + y'(0)$.
For this damped system, at large t (corresponding to $s \to 0$) rates of change are negligible and the equation reduces to $by = c\cos^2\omega t$. The average value of $\cos^2\omega t$ is $\frac{1}{2}$.

5.25. $s^{-1}[1 - \exp(-sa)]$; $g_a(x) = x$ for $0 < x < a$, $g_a(x) = 2a - x$ for $a \le x \le 2a$, $g_a(x) = 0$ otherwise.

6 Higher-order ordinary differential equations

Differential equations are the group of equations that contain derivatives. Chapters 6–11 discuss a variety of differential equations, starting in this chapter with those ordinary differential equations (ODEs) that have closed-form solutions. As its name suggests, an ODE contains only ordinary derivatives (no partial derivatives) and describes the relationship between these derivatives of the *dependent variable*, usually called y, with respect to the *independent variable*, usually called x. The solution to such an ODE is therefore a function of x and is written $y(x)$. For an ODE to have a closed-form solution, it must be possible to express $y(x)$ in terms of the standard elementary functions such as x^2, $\sqrt{x}$, $\exp x$, $\ln x$, $\sin x$, etc. The solutions of some differential equations cannot, however, be written in closed form, but only as an infinite series that carry no special names; these are discussed in Chapter 7.

Ordinary differential equations may be separated conveniently into different categories according to their general characteristics. The primary grouping adopted here is by the *order* of the equation. The order of an ODE is simply the order of the highest derivative it contains. Thus equations containing dy/dx, but no higher derivatives, are called first order, those containing d^2y/dx^2 are called second order and so on.

Ordinary differential equations may be classified further according to *degree*. The degree of an ODE is the power to which the highest-order derivative is raised, after the equation has been rationalized to contain only integer powers of derivatives. Hence the ODE

$$\frac{d^3y}{dx^3} + x\left(\frac{dy}{dx}\right)^{3/2} + x^2y = 0$$

is of third order and second degree, since after rationalization it contains the term $(d^3y/dx^3)^2$.

As explained in the Preface, it is assumed that the reader has had some previous experience with differential equations and is familiar with the standard procedures for dealing with first-order ODEs. These procedures are summarized for reference purposes in Section A.10 of Appendix A. We, therefore, begin our current study of differential equations with second-order ODEs, starting with a brief review of general considerations that affect all ODEs, and then moving on to equations with more specific structures.

6.1 General considerations

The *general solution* to an ODE is the most general function $y(x)$ that satisfies the equation; it will contain *constants of integration* which may be determined by the application of some suitable *boundary conditions*. For example, we may be told that for a certain first-order differential equation, the solution $y(x)$ is equal to zero when the independent variable x is equal to unity; this allows us to determine the value of the constant of integration.

The *general solutions* to nth-order ODEs will contain n (essential) arbitrary constants of integration and therefore we will need n (independent and self-consistent) boundary conditions if these constants are to be determined (see Subsection 6.1.1). When the boundary conditions have been applied, and the constants found, we are left with a *particular solution* to the ODE, which obeys the given boundary conditions. Some ODEs of degree greater than unity also possess *singular solutions*, which are solutions that contain no arbitrary constants and cannot be found from the general solution; some first-order equation types that might give rise to singular solutions are indicted in the summary table in Section A.10. When any solution to an ODE has been found, it is always possible to check its validity by substitution into the original equation and verification that any given boundary conditions are met.

In this chapter, we discuss various types of first-degree ODEs and then go on to examine those higher-degree equations that can be solved in closed form. At the outset, however, we discuss the general form of the solutions of ODEs; this discussion is relevant to both first- and higher-order ODEs.

6.1.1 General form of solution

It is helpful when considering the general form of the solution to an ODE to consider the inverse process, namely that of obtaining an ODE from a given group of functions, each one of which is a solution of the ODE. Suppose the members of the group can be written as

$$y = f(x, a_1, a_2, \ldots, a_n), \tag{6.1}$$

each member being specified by a different set of values[1] of the parameters a_i. For example, consider the group of functions

$$y = a_1 \sin x + a_2 \cos x; \tag{6.2}$$

here $n = 2$.

Since an ODE is required for which *any* of the group is a solution, it clearly must not contain any of the a_i. As there are n of the a_i in expression (6.1), we must obtain $n + 1$ equations involving them in order that, by elimination, we can obtain one final equation without them.

Initially we have only (6.1), but if this is differentiated n times, a total of $n + 1$ equations is obtained from which (in principle) all the a_i can be eliminated, to give one ODE satisfied

1 This does not preclude some values being the same in two different sets, but does require that at least one of the a_i is different for any pair of members.

by all the group. As a result of the n differentiations, $d^n y/dx^n$ will be present in one of the $n+1$ equations and hence in the final equation, which will therefore be of nth order.

In the case of (6.2), we have

$$\frac{dy}{dx} = a_1 \cos x - a_2 \sin x,$$

$$\frac{d^2y}{dx^2} = -a_1 \sin x - a_2 \cos x.$$

Here the elimination of a_1 and a_2 is trivial (because of the similarity of the forms of y and d^2y/dx^2), resulting in

$$\frac{d^2y}{dx^2} + y = 0,$$

a second-order equation.[2]

Thus, to summarize, a group of functions (6.1) with n parameters satisfies an nth-order ODE in general (although in some degenerate cases an ODE of less than nth order is obtained). The intuitive converse of this is that the general solution of an nth-order ODE contains n arbitrary parameters (constants); for our purposes, this will be assumed to be valid although a totally general proof is difficult.

As mentioned earlier, external factors affect a system described by an ODE, by fixing the values of the dependent variables for particular values of the independent ones. These externally imposed (or *boundary*) conditions on the solution are thus the means of determining the parameters and so of specifying precisely which function is the required solution. It is apparent that the number of boundary conditions should match the number of parameters and hence the order of the equation, if a unique solution is to be obtained. Fewer independent boundary conditions than this will lead to a number of undetermined parameters in the solution, whilst an excess will usually mean that no acceptable solution is possible.

For an nth-order equation the required n boundary conditions can take many forms, for example the value of y at n different values of x, or the value of any $n-1$ of the n derivatives $dy/dx, d^2y/dx^2, \ldots, d^ny/dx^n$ together with that of y, all for the same value of x, or many intermediate combinations.

6.1.2 Linear equations

The development in the rest of this chapter is largely concerned with second-order ODEs and is divided into three main sections. In the first of these we discuss linear equations with constant coefficients. This is followed in the second section by an investigation of linear equations with variable coefficients. Finally, in the third section, we discuss a few methods that may be of use in solving general ODEs, both linear and non-linear. However, we start by considering some general points relating to *all* linear ODEs, whatever the nature of the coefficients appearing in them.

2 Find the differential equation satisfied by all functions of the form $y(x) = ax^3 + be^{-x}$, where a and b are arbitrary constants. Verify your answer by re-substituting $y = x^3$ and $y = e^{-x}$ separately, formally corresponding to the cases $a = 1, b = 0$ and $a = 0, b = 1$, respectively.

Linear equations are of paramount importance in the description of physical processes. Moreover, it is an empirical fact that, when put into mathematical form, many natural processes appear as higher-order linear ODEs, most often as second-order equations. Although we could restrict our attention to these second-order equations, the generalization to nth-order equations requires little extra work, and so we will consider this more general case.

A linear ODE of general order n has the form

$$a_n(x)\frac{d^n y}{dx^n} + a_{n-1}(x)\frac{d^{n-1} y}{dx^{n-1}} + \cdots + a_1(x)\frac{dy}{dx} + a_0(x)y = f(x). \tag{6.3}$$

If $f(x) = 0$ then the equation is called *homogeneous*; otherwise it is *inhomogeneous*. As discussed above, the general solution to (6.3) will contain n arbitrary constants, which may be determined if n boundary conditions are also provided.[3]

In order to solve any equation of the form (6.3), we need first to find the general solution of the *complementary equation*, i.e. the equation formed by setting $f(x) = 0$:

$$a_n(x)\frac{d^n y}{dx^n} + a_{n-1}(x)\frac{d^{n-1} y}{dx^{n-1}} + \cdots + a_1(x)\frac{dy}{dx} + a_0(x)y = 0. \tag{6.4}$$

To determine the general solution of (6.4), we must find n linearly independent functions that satisfy it. Once we have found these solutions, the general solution is given by a linear superposition of these n functions. In other words, if the n solutions of (6.4) are $y_1(x),\ y_2(x), \ldots, y_n(x)$, then the general solution is given by the linear superposition

$$y_c(x) = c_1 y_1(x) + c_2 y_2(x) + \cdots + c_n y_n(x), \tag{6.5}$$

where the c_m are arbitrary constants that may be determined if n boundary conditions are provided.[4] The linear combination $y_c(x)$ is called the *complementary function* of (6.3).

The question naturally arises how we establish that any n individual solutions to (6.4) are indeed linearly independent.[5] For n functions to be linearly independent over an interval, there must not exist *any* set of constants $c_1,\ c_2, \ldots, c_n$ such that

$$c_1 y_1(x) + c_2 y_2(x) + \cdots + c_n y_n(x) = 0 \tag{6.6}$$

over the interval in question, except for the trivial case $c_1 = c_2 = \cdots = c_n = 0$.

A statement equivalent to (6.6), which is perhaps more useful for the practical determination of linear independence, can be found by repeatedly differentiating (6.6), $n-1$ times in all, to obtain n simultaneous equations for $c_1,\ c_2, \ldots, c_n$:

$$\begin{aligned} c_1 y_1(x) + c_2 y_2(x) + \cdots + c_n y_n(x) &= 0 \\ c_1 y_1'(x) + c_2 y_2'(x) + \cdots + c_n y_n'(x) &= 0 \\ &\ \vdots \\ c_1 y_1^{(n-1)}(x) + c_2 y_2^{(n-1)} + \cdots + c_n y_n^{(n-1)}(x) &= 0, \end{aligned} \tag{6.7}$$

3 They must be both independent and self-consistent.

4 See footnote 3.

5 For $n = 2$ this is trivial, as the requirement is that one is not a simple multiple of the other. However, for higher values of n, determination by inspection becomes increasingly more difficult and a mechanistic procedure is needed.

where the primes denote differentiation with respect to x. Referring to the discussion of simultaneous linear equations given in Chapter 1, if the determinant of the coefficients of $c_1, c_2, \ldots, c_n$ is non-zero then the only solution to equations (6.7) is the trivial solution $c_1 = c_2 = \cdots = c_n = 0$. In other words, the n functions $y_1(x), y_2(x), \ldots, y_n(x)$ are linearly independent over an interval if

$$W(y_1, y_2, \ldots, y_n) = \begin{vmatrix} y_1 & y_2 & \ldots & y_n \\ y_1{}' & y_2{}' & & \vdots \\ \vdots & & \ddots & \vdots \\ y_1^{(n-1)} & \ldots & \ldots & y_n^{(n-1)} \end{vmatrix} \neq 0 \tag{6.8}$$

over that interval; $W(y_1, y_2, \ldots, y_n)$ is called the *Wronskian* of the set of functions. It should be noted, however, that the vanishing of the Wronskian does not guarantee that the functions are linearly dependent.[6]

If the original equation (6.3) has $f(x) = 0$ (i.e. it is homogeneous) then of course the complementary function $y_c(x)$ in (6.5) is already the general solution. If, however, the equation has $f(x) \neq 0$ (i.e. it is inhomogeneous) then $y_c(x)$ is only one part of the solution. The general solution of (6.3) is then given by

$$y(x) = y_c(x) + y_p(x), \tag{6.9}$$

where $y_p(x)$ is the *particular integral*, which can be *any* function that satisfies (6.3) directly, provided it is linearly independent of $y_c(x)$. It should be emphasized that, for practical purposes, *any* such function, no matter how simple (or complicated), is equally valid in forming the general solution (6.9).

It is important to realize that the above method for finding the general solution to an ODE by superposing particular solutions assumes crucially that the ODE is linear. For non-linear equations, discussed in Section 6.6, this method cannot be used, and indeed it is often impossible to find closed-form solutions to such equations.

Before we leave the general properties of linear equations, there is an essential point to be made in connection with fitting boundary conditions for inhomogeneous equations. Making the general solution fit the given boundary conditions determines the unknown constants that appear as part of the complementary function. However, it is crucial that the conditions are incorporated *after* the particular integral has been included in the solution. As an illustration of this, consider the following example (in which the statements made about the forms of solutions may be checked by re-substitution).

The complementary function solution of the equation

$$\frac{d^2y}{dx^2} - \frac{dy}{dx} - 2y = x$$

is $y_c(x) = Ae^{2x} + Be^{-x}$ and a particular integral is $y_p(x) = \frac{1}{4} - \frac{1}{2}x$. Suppose that the given boundary conditions are $y(0) = 1$ and $y'(0) = 0$.

6 Consider the functions $f(x) = x^5$ and $g(x) = |x^5|$, defined as x^5 for $x \geq 0$ and $-x^5$ for $x < 0$. Show by considering the solutions of $af(x) + bg(x) = 0$ at $x = \pm 1$ that they are linearly independent, but, by evaluating it, that their Wronskian is everywhere zero.

If these conditions are (mistakenly) fitted to the complementary function alone, we obtain

$$A + B = 1 \text{ and } 2A - B = 0 \quad \Rightarrow \quad A = \tfrac{1}{3} \text{ and } B = \tfrac{2}{3}.$$

The subsequent addition of the particular integral then yields as the (incorrect) full solution

$$y(x) = \tfrac{1}{3}e^{2x} + \tfrac{2}{3}e^{-x} + \tfrac{1}{4} - \tfrac{1}{2}x.$$

Re-substitution will confirm that, although this $y(x)$ is a solution of the differential equation, it does *not* satisfy the boundary conditions.

The correct procedure is to take the general solution, including the particular integral,

$$y(x) = Ae^{2x} + Be^{-x} + \tfrac{1}{4} - \tfrac{1}{2}x,$$

and make *this* fit the boundary conditions. They then require

$$A + B + \tfrac{1}{4} = 1 \quad \text{and} \quad 2A - B - \tfrac{1}{2} = 0 \quad \Rightarrow \quad A = \tfrac{5}{12} \quad \text{and} \quad B = \tfrac{1}{3}.$$

The correct full solution is therefore

$$y(x) = \tfrac{5}{12}e^{2x} + \tfrac{1}{3}e^{-x} + \tfrac{1}{4} - \tfrac{1}{2}x,$$

as can be confirmed, if necessary, by calculating $y(0)$ and $y'(0)$.

6.2 Linear equations with constant coefficients

If the a_m in (6.3) are constants rather than functions of x then we have

$$a_n \frac{d^n y}{dx^n} + a_{n-1} \frac{d^{n-1} y}{dx^{n-1}} + \cdots + a_1 \frac{dy}{dx} + a_0 y = f(x). \tag{6.10}$$

Equations of this sort are very common throughout the physical sciences and engineering, and the method for their solution falls into two parts as discussed in the previous section, i.e. finding the complementary function $y_c(x)$ and finding the particular integral $y_p(x)$. If $f(x) = 0$ in (6.10) then we do not have to find a particular integral, and the complementary function is by itself the general solution.[7]

6.2.1 Finding the complementary function $y_c(x)$

The complementary function must satisfy

$$a_n \frac{d^n y}{dx^n} + a_{n-1} \frac{d^{n-1} y}{dx^{n-1}} + \cdots + a_1 \frac{dy}{dx} + a_0 y = 0 \tag{6.11}$$

and contain n arbitrary constants [see equation (6.5)]. The standard method for finding $y_c(x)$ is to try a solution of the form $y = Ae^{\lambda x}$, substituting this into (6.11). After dividing the resulting equation through by $Ae^{\lambda x}$, we are left with a polynomial equation in λ of order n; this is the *auxiliary equation* and reads

$$a_n \lambda^n + a_{n-1}\lambda^{n-1} + \cdots + a_1\lambda + a_0 = 0. \tag{6.12}$$

7 Formally, we can think of the solution $y(x) = 0$ for all x as a (particularly simple, but perfectly acceptable) particular integral $y_p(x)$. This is then added to $y_c(x)$ to give the general solution.

In general the auxiliary equation has n roots, say $\lambda_1, \lambda_2, \ldots, \lambda_n$. In certain cases, some of these roots may be repeated and some may be complex. The three main cases are as follows.

(i) *All roots real and distinct.* In this case the n solutions to (6.11) are $y(x) = \exp(\lambda_m x)$ for $m = 1$ to n. It is easily shown by calculating the Wronskian (6.8) of these functions that if all the λ_m are distinct then these solutions are linearly independent. We can therefore linearly superpose them, as in (6.5), to form the complementary function

$$y_c(x) = c_1 e^{\lambda_1 x} + c_2 e^{\lambda_2 x} + \cdots + c_n e^{\lambda_n x}. \tag{6.13}$$

(ii) *Some roots complex.* For the special (but usual) case that all the coefficients a_m in (6.11) are real, if one of the roots of the auxiliary equation (6.12) is complex, say $\alpha + i\beta$, then its complex conjugate $\alpha - i\beta$ is also a root. In this case we can write

$$\begin{aligned} c_1 e^{(\alpha+i\beta)x} + c_2 e^{(\alpha-i\beta)x} &= e^{\alpha x}(d_1 \cos \beta x + d_2 \sin \beta x) \\ &= A e^{\alpha x} \left\{ \begin{matrix} \sin \\ \cos \end{matrix} \right\} (\beta x + \phi), \end{aligned} \tag{6.14}$$

where A and ϕ are arbitrary constants.

(iii) *Some roots repeated.* If, for example, λ_1 occurs k times ($k > 1$) as a root of the auxiliary equation, then we have not found n linearly independent solutions of (6.11); formally the Wronskian (6.8) of these solutions, having two or more identical columns, is equal to zero. We must therefore find $k - 1$ further solutions that are linearly independent of those already found and also of each other. By direct substitution into (6.11) it is found that[8]

$$x e^{\lambda_1 x}, \quad x^2 e^{\lambda_1 x}, \quad \ldots, \quad x^{k-1} e^{\lambda_1 x}$$

are also solutions, and by calculating the Wronskian it can be shown that they, together with the solutions already found, form a linearly independent set of n functions. Therefore the complementary function is given by

$$y_c(x) = (c_1 + c_2 x + \cdots + c_k x^{k-1}) e^{\lambda_1 x} + c_{k+1} e^{\lambda_{k+1} x} + c_{k+2} e^{\lambda_{k+2} x} + \cdots + c_n e^{\lambda_n x}. \tag{6.15}$$

If more than one root is repeated the above argument is easily extended. For example, suppose as before that λ_1 is a k-fold root of the auxiliary equation and, further, that λ_2 is an l-fold root (of course, $k > 1$ and $l > 1$). Then, from the above argument, the complementary function reads

$$\begin{aligned} y_c(x) = {} & (c_1 + c_2 x + \cdots + c_k x^{k-1}) e^{\lambda_1 x} \\ & + (c_{k+1} + c_{k+2} x + \cdots + c_{k+l} x^{l-1}) e^{\lambda_2 x} \\ & + c_{k+l+1} e^{\lambda_{k+l+1} x} + c_{k+l+2} e^{\lambda_{k+l+2} x} + \cdots + c_n e^{\lambda_n x}. \end{aligned} \tag{6.16}$$

The following is a simple example.

8 A general algebraic proof of this is rather messy, as it involves Leibnitz' theorem and multiple summations.

Example Find the complementary function of the equation

$$\frac{d^2y}{dx^2} - 2\frac{dy}{dx} + y = e^x. \tag{6.17}$$

Setting the RHS to zero, substituting $y = Ae^{\lambda x}$ and dividing through by $Ae^{\lambda x}$ we obtain the auxiliary equation

$$\lambda^2 - 2\lambda + 1 = 0.$$

The root $\lambda = 1$ occurs twice and so, although e^x is a solution to (6.17), we must find a further solution to the equation that is linearly independent of e^x. From the above discussion, we deduce that xe^x is such a solution, and so the full complementary function is given by the linear superposition

$$y_c(x) = (c_1 + c_2 x)e^x.$$

This can be checked by re-substitution; only the c_2xe^x actually needs to be checked, as the c_1e^x term is a proved solution, rather than merely a stated one. ◄

Solution method. *Set the RHS of the ODE to zero (if it is not already so), and substitute* $y = Ae^{\lambda x}$. *After dividing through the resulting equation by* $Ae^{\lambda x}$, *obtain an nth-order polynomial equation in* λ *[the auxiliary equation, see (6.12)]. Solve the auxiliary equation to find the n roots,* $\lambda_1, \lambda_2, \ldots, \lambda_n$, *say. If all these roots are real and distinct then* $y_c(x)$ *is given by (6.13). If, however, some of the roots are complex or repeated then* $y_c(x)$ *is given by (6.14) or (6.15), or the extension (6.16) of the latter, respectively.*

6.2.2 Finding the particular integral $y_p(x)$

There is no generally applicable method for finding the particular integral $y_p(x)$ but, for linear ODEs with constant coefficients and a simple RHS, $y_p(x)$ can often be found by inspection or by assuming a parameterized form similar to $f(x)$. The latter method is sometimes called the *method of undetermined coefficients*. If $f(x)$ contains only polynomial, exponential, or sine and cosine terms then, by assuming a trial function for $y_p(x)$ of similar form but one which contains a number of undetermined parameters and substituting this trial function into (6.11), the parameters can be found and an acceptable $y_p(x)$ deduced.[9] Standard trial functions are as follows.

(i) If $f(x) = ae^{rx}$ then try

$$y_p(x) = be^{rx}.$$

(ii) If $f(x) = a_1 \sin rx + a_2 \cos rx$ (a_1 or a_2 may be zero) then try

$$y_p(x) = b_1 \sin rx + b_2 \cos rx.$$

9 It should always be borne in mind that *any* valid particular integral will do; the difference between that and any other particular integral can always be made up for by a different choice of constants in the complementary function. Show this more symbolically by writing the solution $y(x)$ as both $\sum c_i y_i(x) + y_{p1}(x)$ and $\sum d_i y_i(x) + y_{p2}(x)$, thus obtaining an expression for the difference $y_{p2}(x) - y_{p1}(x)$ in the particular solutions.

(iii) If $f(x) = a_0 + a_1 x + \cdots + a_N x^N$ (some a_m may be zero) then try

$$y_p(x) = b_0 + b_1 x + \cdots + b_N x^N.$$

(iv) If $f(x)$ is the sum or product of any of the above then try $y_p(x)$ as the sum or product of the corresponding individual trial functions.

It should be noted that this method fails if any term in the assumed trial function is also contained within the complementary function $y_c(x)$. In such a case the trial function should be multiplied by the smallest integer power of x such that it will then contain no term that already appears in the complementary function. The undetermined coefficients in the trial function can now be found by substitution into (6.10).[10]

The next worked example illustrates this point – in duplicate, it may be said!

Example Find a particular integral of the equation

$$\frac{d^2y}{dx^2} - 2\frac{dy}{dx} + y = e^x.$$

From the above discussion our first guess at a trial particular integral would be $y_p(x) = be^x$. However, since the complementary function of this equation is $y_c(x) = (c_1 + c_2 x)e^x$ (as in the previous subsection), we see that e^x is already contained in it, as indeed is xe^x. Multiplying our first guess by the lowest integer power of x such that the result does not appear in $y_c(x)$, we therefore try $y_p(x) = bx^2e^x$. Substituting this into the ODE, we find that $b = 1/2$, so the particular integral is given by $y_p(x) = x^2e^x/2$. ◀

Three further methods that are useful in finding the particular integral $y_p(x)$ are those based on Green's functions, the variation of parameters, and a change in the dependent variable using knowledge of the complementary function. However, since these methods are also applicable to equations with variable coefficients, a discussion of them is postponed until Section 6.5.

Solution method. *If the RHS of an ODE contains only functions mentioned at the start of this subsection then the appropriate trial function should be substituted into it, thereby fixing the undetermined parameters. If, however, the RHS of the equation is not of this form then one of the more general methods outlined in Subsections 6.5.3–6.5.5 should be used; perhaps the most straightforward of these is the variation-of-parameters method.*

6.2.3 Constructing the general solution $y_c(x) + y_p(x)$

As stated earlier, the full solution to the ODE (6.10) is found by adding together the complementary function and any particular integral. In order to illustrate further the material discussed in the last two subsections, let us find the general solution to a new example, starting from the beginning.

10 It is important to recognize that the coefficient in the particular integral is *not* arbitrary, unlike those in the complementary function. Thus the number of boundary conditions needed to determine the unknown coefficients in a general solution is not altered by the inclusion of a particular integral.

Example Solve

$$\frac{d^2y}{dx^2} + 4y = x^2 \sin 2x. \tag{6.18}$$

First we set the RHS to zero and assume the trial solution $y = Ae^{\lambda x}$. Substituting this into (6.18) leads to the auxiliary equation

$$\lambda^2 + 4 = 0 \quad \Rightarrow \quad \lambda = \pm 2i. \tag{6.19}$$

Therefore the complementary function is given by

$$y_c(x) = c_1 e^{2ix} + c_2 e^{-2ix} = d_1 \cos 2x + d_2 \sin 2x. \tag{6.20}$$

We must now turn our attention to the particular integral $y_p(x)$. Consulting the list of standard trial functions in the previous subsection, we find that a first guess at a suitable trial function for this case should be

$$(ax^2 + bx + c)\sin 2x + (dx^2 + ex + f)\cos 2x. \tag{6.21}$$

However, we see that this trial function contains terms in $\sin 2x$ and $\cos 2x$, both of which already appear in the complementary function (6.20). We must therefore multiply (6.21) by the smallest integer power of x which ensures that none of the resulting terms appears in $y_c(x)$. Since multiplying by x will suffice, we finally assume the trial function

$$(ax^3 + bx^2 + cx)\sin 2x + (dx^3 + ex^2 + fx)\cos 2x. \tag{6.22}$$

Substituting this into (6.18) to fix the constants appearing in (6.22), we find the particular integral to be[11]

$$y_p(x) = -\frac{x^3}{12}\cos 2x + \frac{x^2}{16}\sin 2x + \frac{x}{32}\cos 2x. \tag{6.23}$$

The general solution to (6.18) then reads

$$\begin{aligned} y(x) &= y_c(x) + y_p(x) \\ &= d_1 \cos 2x + d_2 \sin 2x - \frac{x^3}{12}\cos 2x + \frac{x^2}{16}\sin 2x + \frac{x}{32}\cos 2x, \end{aligned}$$

with d_1 and d_2 undetermined until two boundary conditions are imposed. ◀

6.3 Linear recurrence relations

Before continuing our discussion of higher-order ODEs, we take this opportunity to introduce the discrete analogues of differential equations, which are called *recurrence relations* (or sometimes *difference equations*). Whereas a differential equation gives a prescription, in terms of current values, for the new value of a dependent variable at a point only infinitesimally far away, a recurrence relation describes how the next in a sequence of values u_n, defined only at (non-negative) integer values of the "independent variable" n, is to be calculated.

11 Carry out this substitution, using Leibnitz' theorem to obtain the second derivatives, and show that the equations to be satisfied are $12a = 0$, $8b + 6d = 0$, $4c = 0$, $-12d = 1$, $6a - 8e = 0$ and $2b - 4f = 0$.

In its most general form a recurrence relation expresses the way in which u_{n+1} is to be calculated from all the preceding values $u_0, u_1, \ldots, u_n$. Just as the most general differential equations are intractable, so are the most general recurrence relations, and we will limit ourselves to analogues of the types of differential equations studied earlier in this chapter, namely those that are linear, have constant coefficients and possess simple functions on the RHS. Such equations occur over a broad range of engineering and statistical physics as well as in the realms of finance, business planning and gambling! They form the basis of many numerical methods, particularly those concerned with the numerical solution of ordinary and partial differential equations.

A general recurrence relation is exemplified by the formula

$$u_{n+1} = \sum_{r=0}^{N-1} a_r u_{n-r} + k, \tag{6.24}$$

where N and the a_r are fixed and k is a constant or a simple function of n. Such an equation, involving terms of the series whose indices differ by up to N (ranging from $n - N + 1$ to n), is called an Nth-order recurrence relation. It is clear that, given values for $u_0, u_1, \ldots, u_{N-1}$, this is a definitive scheme for generating the series and therefore has a unique solution.

Paralleling the nomenclature of differential equations, if the term not involving any u_n is absent, i.e. $k = 0$, then the recurrence relation is called *homogeneous*. The parallel continues with the form of the general solution of (6.24). If v_n is the general solution of the homogeneous relation, and w_n is *any* solution of the full relation, then

$$u_n = v_n + w_n$$

is the most general solution of the complete recurrence relation. This is straightforwardly verified as follows:

$$\begin{aligned} u_{n+1} &= v_{n+1} + w_{n+1} \\ &= \sum_{r=0}^{N-1} a_r v_{n-r} + \sum_{r=0}^{N-1} a_r w_{n-r} + k \\ &= \sum_{r=0}^{N-1} a_r (v_{n-r} + w_{n-r}) + k \\ &= \sum_{r=0}^{N-1} a_r u_{n-r} + k. \end{aligned}$$

Of course, if $k = 0$ then $w_n = 0$ for all n is a trivial particular solution and the complementary solution, v_n, is itself the most general solution.

6.3.1 First-order recurrence relations

First-order relations, for which $N = 1$, are exemplified by

$$u_{n+1} = a u_n + k, \tag{6.25}$$

with u_0 specified. The solution to the homogeneous relation is immediate,

$$u_n = Ca^n,$$

and, if k is a constant, the particular solution is equally straightforward: $w_n = K$ for all n, provided K is chosen to satisfy

$$K = aK + k,$$

i.e. $K = k(1-a)^{-1}$. This will be sufficient unless $a = 1$, in which case $u_n = u_0 + nk$ is obvious by inspection.

Thus the general solution of (6.25) is

$$u_n = \begin{cases} Ca^n + k/(1-a) & a \neq 1, \\ u_0 + nk & a = 1. \end{cases} \tag{6.26}$$

If u_0 is specified for the case of $a \neq 1$ then C must be chosen as $C = u_0 - k/(1-a)$, resulting in the equivalent form

$$u_n = u_0 a^n + k\frac{1-a^n}{1-a}. \tag{6.27}$$

We now illustrate this method with a worked example.

Example A house-buyer borrows capital B from a bank that charges a fixed annual rate of interest $R\%$. If the loan is to be repaid over Y years, at what value should the fixed annual payments P, made at the end of each year, be set? For a loan over 25 years at 6%, what percentage of the first year's payment goes towards paying off the capital?

Let u_n denote the outstanding debt at the end of year n, and write $R/100 = r$. Then the relevant recurrence relation is

$$u_{n+1} = u_n(1+r) - P$$

with $u_0 = B$. From (6.27) we have

$$u_n = B(1+r)^n - P\frac{1-(1+r)^n}{1-(1+r)}.$$

As the loan is to be repaid over Y years, $u_Y = 0$ and thus

$$P = \frac{Br(1+r)^Y}{(1+r)^Y - 1}.$$

The first year's interest is rB and so the fraction of the first year's payment going towards capital repayment is $(P - rB)/P$, which, using the above expression for P, is equal to $(1+r)^{-Y}$. With the given figures, this is (only) 23%. ◀

With only small modifications, the method just described can be adapted to handle recurrence relations in which the constant k in (6.25) is replaced by $k\alpha^n$, i.e. the relation is

$$u_{n+1} = au_n + k\alpha^n. \tag{6.28}$$

As for an inhomogeneous linear differential equation (see Subsection 6.2.2), we may try as a potential particular solution a form which resembles the term that makes the equation inhomogeneous. Here, the presence of the term $k\alpha^n$ indicates that a particular solution of the form $u_n = A\alpha^n$ should be tried. Substituting this into (6.28) gives

$$A\alpha^{n+1} = aA\alpha^n + k\alpha^n,$$

from which it follows that $A = k/(\alpha - a)$ and that there is a particular solution having the form $u_n = k\alpha^n/(\alpha - a)$, provided $\alpha \neq a$. For the special case $\alpha = a$, the reader can readily verify that a particular solution of the form $u_n = An\alpha^n$ is appropriate. This mirrors the corresponding situation for linear differential equations when the RHS of the differential equation is contained in the complementary function of its LHS.

In summary, the general solution to (6.28) is

$$u_n = \begin{cases} C_1 a^n + k\alpha^n/(\alpha - a) & \alpha \neq a, \\ C_2 a^n + kn\alpha^{n-1} & \alpha = a, \end{cases} \tag{6.29}$$

with $C_1 = u_0 - k/(\alpha - a)$ and $C_2 = u_0$.

6.3.2 Second-order recurrence relations

We consider next recurrence relations that involve u_{n-1} in the prescription for u_{n+1} and treat the general case in which the intervening term, u_n, is also present. A typical equation is thus

$$u_{n+1} = au_n + bu_{n-1} + k. \tag{6.30}$$

As previously, the general solution of this is $u_n = v_n + w_n$, where v_n satisfies

$$v_{n+1} = av_n + bv_{n-1} \tag{6.31}$$

and w_n is *any* particular solution of (6.30); the proof follows the same lines as that given earlier.

We have already seen for a first-order recurrence relation that the solution to the homogeneous equation is given by terms forming a geometric series, and we consider a corresponding series of powers in the present case. Setting $v_n = A\lambda^n$ in (6.31) for some λ, as yet undetermined, gives the requirement that λ should satisfy

$$A\lambda^{n+1} = aA\lambda^n + bA\lambda^{n-1}.$$

Dividing through by $A\lambda^{n-1}$ (assumed non-zero) shows that λ could be either of the roots, λ_1 and λ_2, of

$$\lambda^2 - a\lambda - b = 0, \tag{6.32}$$

which is known as the *characteristic equation* of the recurrence relation.

That there are two possible series of terms of the form $A\lambda^n$ is consistent with the fact that two initial values (boundary conditions) have to be provided before the series can be calculated by repeated use of (6.30). These two values are sufficient to determine the appropriate coefficient A for each of the series. Since (6.31) is both linear and homogeneous, and is satisfied by both $v_n = A\lambda_1^n$ and $v_n = B\lambda_2^n$, its general solution is

$$v_n = A\lambda_1^n + B\lambda_2^n,$$

for arbitrary values of A and B.[12]

If the coefficients a and b are such that (6.32) has two equal roots, i.e. $a^2 = -4b$, then, as in the analogous case of repeated roots for differential equations [see Subsection 6.2.1(iii)], the second term of the general solution is replaced by $Bn\lambda_1^n$ to give

$$v_n = (A + Bn)\lambda_1^n.$$

A further possibility is that the roots of the characteristic equation are complex, in which case the general solution of the homogeneous equation takes the form

$$v_n = A\mu^n e^{in\theta} + B\mu^n e^{-in\theta} = \mu^n(C\cos n\theta + D\sin n\theta).$$

Finding a particular solution is straightforward if k is a constant: a trivial but adequate solution is $w_n = k(1 - a - b)^{-1}$ for all n. As with first-order equations, particular solutions can be found for other simple forms of k by trying functions similar to k itself. Thus particular solutions for the cases $k = Cn$ and $k = D\alpha^n$ can be found by trying $w_n = E + Fn$ and $w_n = G\alpha^n$ respectively.

Example Find the value of u_{16} if the series u_n satisfies

$$u_{n+1} + 4u_n + 3u_{n-1} = n$$

for $n \geq 1$, with $u_0 = 1$ and $u_1 = -1$.

We first solve the characteristic equation,

$$\lambda^2 + 4\lambda + 3 = 0,$$

to obtain the roots $\lambda = -1$ and $\lambda = -3$. Thus the complementary function is

$$v_n = A(-1)^n + B(-3)^n.$$

In view of the form of the RHS of the original relation, we try

$$w_n = E + Fn$$

as a particular solution and obtain

$$E + F(n+1) + 4(E + Fn) + 3[E + F(n-1)] = n,$$

yielding $F = 1/8$ and $E = 1/32$.

Thus the complete general solution is

$$u_n = A(-1)^n + B(-3)^n + \frac{n}{8} + \frac{1}{32},$$

12 Of which second-order recurrence relation and initial values would $u_n = 3(-2)^n - 2(-3)^n$ be the unique solution? Evaluate u_4, (a) directly using your recurrence relation, and (b) by using the given solution.

and now using the given values for u_0 and u_1 determines A as $7/8$ and B as $3/32$. Thus

$$u_n = \frac{1}{32}\left[28(-1)^n + 3(-3)^n + 4n + 1\right].$$

Finally, substituting $n = 16$ gives $u_{16} = 4\,035\,633$, a value the reader may (or may not) wish to verify by repeated application of the initial recurrence relation. ◀

6.3.3 Higher-order recurrence relations

It will be apparent that linear recurrence relations of order $N > 2$ do not present any additional difficulty in principle, though two obvious practical difficulties are (i) that the characteristic equation is of order N and in general will not have roots that can be written in closed form and (ii) that a correspondingly large number of given values is required to determine the N otherwise arbitrary constants in the solution. The algebraic labor needed to solve the set of simultaneous linear equations that determines them increases rapidly with N. We do not give specific examples here, but some are included in the problems at the end of this chapter.

6.4 Laplace transform method

Having briefly discussed recurrence relations, we now return to the main topic of this chapter, i.e. methods for obtaining solutions to higher-order ODEs. One such method is that of Laplace transforms, which is very useful for solving linear ODEs with constant coefficients. Taking the Laplace transform of such an equation transforms it into a purely *algebraic* equation in terms of the Laplace transform of the required solution. Once the algebraic equation has been solved for this Laplace transform, the general solution to the original ODE can be obtained by performing an inverse Laplace transform. One advantage of this method is that, for given boundary conditions, it provides the solution in just one step, instead of having to find the complementary function and particular integral separately.

In order to apply the method we need only two results from Laplace transform theory (see Section 5.4). First, the Laplace transform of a function $f(x)$ is defined by

$$\bar{f}(s) \equiv \int_0^\infty e^{-sx} f(x)\,dx, \qquad (6.33)$$

from which we can derive the second useful relation. This concerns the Laplace transform of the nth derivative of $f(x)$:

$$\overline{f^{(n)}}(s) = s^n \bar{f}(s) - s^{n-1} f(0) - s^{n-2} f'(0) - \cdots - s f^{(n-2)}(0) - f^{(n-1)}(0), \qquad (6.34)$$

where the primes and superscripts in parentheses denote differentiation with respect to x. Using these relations, along with Table 5.1, on p. 212, which gives Laplace transforms of standard functions, we are in a position to solve a linear ODE with constant coefficients by this method.

Example Solve

$$\frac{d^2y}{dx^2} - 3\frac{dy}{dx} + 2y = 2e^{-x}, \tag{6.35}$$

subject to the boundary conditions $y(0) = 2$, $y'(0) = 1$.

Taking the Laplace transform of (6.35) and using the table of standard results we obtain

$$s^2\bar{y}(s) - sy(0) - y'(0) - 3\left[s\bar{y}(s) - y(0)\right] + 2\bar{y}(s) = \frac{2}{s+1},$$

which, after the boundary-condition values have been explicitly included, reduces to

$$(s^2 - 3s + 2)\bar{y}(s) - 2s + 5 = \frac{2}{s+1}. \tag{6.36}$$

Solving this algebraic equation for $\bar{y}(s)$, the Laplace transform of the required solution to (6.35), we obtain

$$\bar{y}(s) = \frac{2s^2 - 3s - 3}{(s+1)(s-1)(s-2)} = \frac{1}{3(s+1)} + \frac{2}{s-1} - \frac{1}{3(s-2)}, \tag{6.37}$$

where in the final step we have used partial fractions. Taking the inverse Laplace transform of (6.37), again using Table 5.1, we find the specific solution to (6.35) to be

$$y(x) = \tfrac{1}{3}e^{-x} + 2e^{x} - \tfrac{1}{3}e^{2x}.$$

Clearly, the first term in the solution corresponds to the particular integral in the general method, and the final two terms to the complementary function. As noted, the Laplace transform method finds them both at the same time.

It should be noted that if the boundary conditions in a problem are given as symbols, rather than just numbers, then the step involving partial fractions can often involve a considerable amount of algebra. For such cases, the method loses some of its attractiveness. ◀

The Laplace transform method is usually very convenient for solving sets of *simultaneous* linear ODEs with constant coefficients, as we now illustrate.

Example Two electrical circuits, both of negligible resistance, each consist of a coil having self-inductance L and a capacitor having capacitance C. The mutual inductance of the two circuits is M. There is no source of e.m.f. in either circuit. Initially the second capacitor is given a charge CV_0, the first capacitor being uncharged, and at time $t = 0$ a switch in the second circuit is closed to complete the circuit. Find the subsequent current in the first circuit.

Subject to the initial conditions $q_1(0) = \dot{q}_1(0) = \dot{q}_2(0) = 0$ and $q_2(0) = CV_0 = V_0/G$, say, we have to solve

$$L\ddot{q}_1 + M\ddot{q}_2 + Gq_1 = 0,$$

$$M\ddot{q}_1 + L\ddot{q}_2 + Gq_2 = 0.$$

On taking the Laplace transform of the above equations, we obtain

$$(Ls^2 + G)\bar{q}_1 + Ms^2\bar{q}_2 = sMV_0C,$$
$$Ms^2\bar{q}_1 + (Ls^2 + G)\bar{q}_2 = sLV_0C.$$

Eliminating $\bar{q}_2$ and rewriting as an equation for $\bar{q}_1$, we find

$$\bar{q}_1(s) = \frac{MV_0 s}{[(L+M)s^2 + G][(L-M)s^2 + G]}$$
$$= \frac{V_0}{2G}\left[\frac{(L+M)s}{(L+M)s^2 + G} - \frac{(L-M)s}{(L-M)s^2 + G}\right].$$

Using Table 5.1,

$$q_1(t) = \tfrac{1}{2}V_0C(\cos\omega_1 t - \cos\omega_2 t),$$

where $\omega_1^2(L+M) = G$ and $\omega_2^2(L-M) = G$. Thus the current is given by

$$i_1(t) = \dot{q}_1(t) = \tfrac{1}{2}V_0C(\omega_2\sin\omega_2 t - \omega_1\sin\omega_1 t).$$

As expected, and required, both the initial charge on the first capacitor and the initial current in the first circuit are zero. ◀

Solution method. *Perform a Laplace transform, as defined in (6.33), on the entire equation, using (6.34) to calculate the transform of the derivatives. Then solve the resulting algebraic equation for $\bar{y}(s)$, the Laplace transform of the required solution to the ODE. By using the method of partial fractions and consulting a table of Laplace transforms of standard functions, calculate the inverse Laplace transform. The resulting function $y(x)$ is the solution of the ODE that obeys the given boundary conditions.*

6.5 Linear equations with variable coefficients

There is no generally applicable method of solving equations with coefficients that are functions of x. Nevertheless, there are certain cases in which a solution is possible. Some of the methods discussed in this section are also useful for finding the general solution or particular integral for equations with constant coefficients that have proved impenetrable by the techniques discussed earlier.

6.5.1 The Legendre and Euler linear equations

Legendre's linear equation has the form

$$a_n(\alpha x + \beta)^n \frac{d^n y}{dx^n} + \cdots + a_1(\alpha x + \beta)\frac{dy}{dx} + a_0 y = f(x), \qquad (6.38)$$

where α, β and the a_n are constants, and may be solved by making the substitution $\alpha x + \beta = e^t$ and using t as the new independent variable.[13] We then have

$$\frac{dy}{dx} = \frac{dt}{dx}\frac{dy}{dt} = \frac{\alpha}{\alpha x + \beta}\frac{dy}{dt},$$

$$\frac{d^2y}{dx^2} = \frac{d}{dx}\frac{dy}{dx} = \frac{\alpha^2}{(\alpha x + \beta)^2}\left(\frac{d^2y}{dt^2} - \frac{dy}{dt}\right),$$

and so on for higher derivatives. Therefore we can write the terms of (6.38) as

$$\begin{aligned}
(\alpha x + \beta)\frac{dy}{dx} &= \alpha\frac{dy}{dt},\\
(\alpha x + \beta)^2\frac{d^2y}{dx^2} &= \alpha^2\frac{d}{dt}\left(\frac{d}{dt} - 1\right)y,\\
&\vdots\\
(\alpha x + \beta)^n\frac{d^ny}{dx^n} &= \alpha^n\frac{d}{dt}\left(\frac{d}{dt} - 1\right)\cdots\left(\frac{d}{dt} - n + 1\right)y.
\end{aligned} \tag{6.39}$$

Substituting equations (6.39) into the original equation (6.38), the latter becomes a linear ODE with constant coefficients, i.e.

$$a_n\alpha^n\frac{d}{dt}\left(\frac{d}{dt} - 1\right)\cdots\left(\frac{d}{dt} - n + 1\right)y + \cdots + a_1\alpha\frac{dy}{dt} + a_0 y = f\left(\frac{e^t - \beta}{\alpha}\right),$$

which can be solved by the methods of Section 6.2.

A special case of Legendre's linear equation, for which $\alpha = 1$ and $\beta = 0$, is *Euler's equation*,

$$a_n x^n\frac{d^ny}{dx^n} + \cdots + a_1 x\frac{dy}{dx} + a_0 y = f(x); \tag{6.40}$$

it may be solved in a similar manner to the above by substituting $x = e^t$.

If, in (6.40), $f(x) = 0$, or even if it is not but we are seeking the complementary function, then substituting $y = x^\lambda$ leads to a simple algebraic equation in λ, which can be solved to yield the solution to (6.40). This is more straightforward than the e^t change of variable, as there is no need to calculate new derivatives. In the event that the algebraic equation for λ has repeated roots, extra care is needed. If λ_1 is a k-fold root ($k > 1$) then the k linearly independent solutions corresponding to this root are x^{λ_1}, $x^{\lambda_1}\ln x, \ldots, x^{\lambda_1}(\ln x)^{k-1}$.

13 For t to be real requires that $\alpha x + \beta$ is never negative. This is effectively the same restriction as requiring equation (6.38) to have no singular points, i.e. the equation has coefficients that are everywhere finite when the coefficient of its highest derivative is made unity. See Chapter 7 for a fuller discussion of singular points.

Example Solve

$$x^2\frac{d^2y}{dx^2}+x\frac{dy}{dx}-4y=0 \qquad (6.41)$$

by both of the methods discussed above.

First we make the substitution $x = e^t$, which, after canceling e^t, gives an equation with constant coefficients, i.e.

$$\frac{d}{dt}\left(\frac{d}{dt}-1\right)y+\frac{dy}{dt}-4y=0 \quad\Rightarrow\quad \frac{d^2y}{dt^2}-4y=0. \qquad (6.42)$$

Using the methods of Section 6.2, the general solution of (6.42), and therefore of (6.41), is given by

$$y = c_1e^{2t}+c_2e^{-2t}=c_1x^2+c_2x^{-2}.$$

Since the RHS of (6.41) is zero, we can reach the same solution by substituting $y = x^\lambda$ into (6.41). This gives

$$\lambda(\lambda-1)x^\lambda+\lambda x^\lambda-4x^\lambda=0,$$

which reduces to

$$(\lambda^2-4)x^\lambda=0.$$

This has the solutions $\lambda = \pm 2$, and so we obtain

$$y = c_1x^2+c_2x^{-2}$$

as the general solution, in agreement with our previous result. ◀

Solution method. *If the ODE is of the Legendre form (6.38) then substitute $\alpha x+\beta = e^t$. This results in an equation of the same order but with constant coefficients, which can be solved by the methods of Section 6.2. If the ODE is of the Euler form (6.40) with a non-zero RHS then substitute $x = e^t$; this again leads to an equation of the same order but with constant coefficients. If, however, $f(x) = 0$ in the Euler equation (6.40) then the equation may also be solved by substituting $y = x^\lambda$. This leads to an algebraic equation whose solution gives the allowed values of λ; the general solution is then the linear superposition of these functions.*

6.5.2 Exact equations

Sometimes an ODE may be merely the derivative of another ODE of one order lower. If this is the case then the ODE is called *exact*. The nth-order linear ODE

$$a_n(x)\frac{d^ny}{dx^n}+\cdots+a_1(x)\frac{dy}{dx}+a_0(x)y=f(x) \qquad (6.43)$$

is exact if the LHS can be written as a simple derivative, i.e. if

$$a_n(x)\frac{d^ny}{dx^n}+\cdots+a_0(x)y=\frac{d}{dx}\left[b_{n-1}(x)\frac{d^{n-1}y}{dx^{n-1}}+\cdots+b_0(x)y\right]. \qquad (6.44)$$

It may be shown that, for (6.44) to hold, we require

$$a_0(x) - a_1'(x) + a_2''(x) - \cdots + (-1)^n a_n^{(n)}(x) = 0, \tag{6.45}$$

where the prime again denotes differentiation with respect to x. If (6.45) is satisfied then straightforward integration leads to a new equation of one order lower. If this simpler equation can be solved then a solution to the original equation is obtained. Of course, if the above process leads to an equation that is itself exact then the analysis can be repeated to reduce the order still further.

Example Solve

$$(1-x^2)\frac{d^2y}{dx^2} - 3x\frac{dy}{dx} - y = 1. \tag{6.46}$$

Comparing with (6.43), we have $a_2 = 1 - x^2$, $a_1 = -3x$ and $a_0 = -1$. It is easily shown that $a_0 - a_1' + a_2'' = 0$; so (6.46) is exact and can therefore be written in the form

$$\frac{d}{dx}\left[b_1(x)\frac{dy}{dx} + b_0(x)y\right] = 1. \tag{6.47}$$

Expanding the LHS of (6.47) we find

$$\frac{d}{dx}\left(b_1\frac{dy}{dx} + b_0 y\right) = b_1\frac{d^2y}{dx^2} + (b_1' + b_0)\frac{dy}{dx} + b_0' y. \tag{6.48}$$

Comparing (6.46) and (6.48) we find

$$b_1 = 1 - x^2, \qquad b_1' + b_0 = -3x, \qquad b_0' = -1.$$

These relations integrate consistently to give $b_1 = 1 - x^2$ and $b_0 = -x$, so (6.46) can be written as

$$\frac{d}{dx}\left[(1-x^2)\frac{dy}{dx} - xy\right] = 1. \tag{6.49}$$

Integrating (6.49) gives us directly the first-order linear ODE

$$\frac{dy}{dx} - \left(\frac{x}{1-x^2}\right)y = \frac{x + c_1}{1-x^2},$$

which can be solved by multiplying through by an integrating factor of $\sqrt{1-x^2}$ and has

$$y = \frac{c_1 \sin^{-1} x + c_2}{\sqrt{1-x^2}} - 1$$

as its solution. ◀

It is worth noting that, even if an original higher-order ODE is not exact in its given form, it may sometimes be made exact by multiplying through by some suitable integrating factor. Unfortunately, no straightforward standard method for finding such integrating factors exists and one often has to rely on inspection or experience.

Example Solve

$$x(1-x^2)\frac{d^2y}{dx^2} - 3x^2\frac{dy}{dx} - xy = x. \qquad (6.50)$$

It is easily shown that (6.50) is not exact, but we also see immediately that by multiplying it through by $1/x$ we recover (6.46), which is exact and has already been solved.[14] ◀

Another important point is that an ODE need not be linear to be exact, although no simple rule such as (6.45) exists if it is not linear. Nevertheless, it is often worth exploring the possibility that a non-linear equation is exact, since it could then be reduced in order by one and may lead to a soluble equation.

Solution method. *For a linear ODE of the form (6.43) check whether it is exact using equation (6.45). If it is not, then attempt to find an integrating factor which when multiplying the equation makes it exact. Once the equation is exact write the LHS as a derivative as in (6.44) and, by expanding this derivative and comparing with the LHS of the ODE, determine the functions $b_m(x)$ in (6.44). Integrate the resulting equation to yield another ODE, of one order lower. This may be solved or simplified further if the new ODE is itself exact or can be made so.*

6.5.3 Partially known complementary function

Suppose we wish to solve the nth-order linear ODE

$$a_n(x)\frac{d^ny}{dx^n} + \cdots + a_1(x)\frac{dy}{dx} + a_0(x)y = f(x), \qquad (6.51)$$

and we happen to know that $u(x)$ is a solution of (6.51) when the RHS is set to zero, i.e. $u(x)$ is one part of the complementary function. By making the substitution $y(x) = u(x)v(x)$, we can transform (6.51) into an equation of order $n-1$ in dv/dx. This simpler equation may prove soluble. In particular, if the original equation is of second order then we obtain a first-order equation in dv/dx, which may be soluble using the methods summarized in Section A.10.[15]

This particular approach gives both the remaining term in the complementary function and a particular integral. The method therefore provides a useful way of calculating particular integrals for second-order equations with variable (or constant) coefficients.

14 As a further example, show that the equation

$$2x^2\frac{d^2y}{dx^2} + x(2x-1)\frac{dy}{dx} + y = 0$$

is not exact as it stands, but can be made so by dividing all through by λx^2, where λ is any non-zero constant.

15 Given that $u(x) = x$ is one solution of $(1-x^2)y'' - 2xy' + 2y = 0$, show that $y(x) = xv(x)$, where $v'(x) = Ax^{-2}(1-x^2)^{-1}$ is another.

Example Solve

$$\frac{d^2y}{dx^2} + y = \csc x. \tag{6.52}$$

We see that the RHS does not fall into any of the categories listed in Subsection 6.2.2, and so we are at an initial loss as to how to find the particular integral. However, the complementary function of (6.52) is

$$y_c(x) = c_1 \sin x + c_2 \cos x,$$

and so let us choose the solution $u(x) = \cos x$ (we could equally well choose $\sin x$) and make the substitution $y(x) = v(x)u(x) = v(x) \cos x$ into (6.52). This gives

$$\cos x \frac{d^2v}{dx^2} - 2 \sin x \frac{dv}{dx} = \csc x, \tag{6.53}$$

which is a first-order linear ODE in dv/dx and may be solved by multiplying through by a suitable integrating factor. Writing (6.53) as

$$\frac{d^2v}{dx^2} - 2 \tan x \frac{dv}{dx} = \frac{\csc x}{\cos x}, \tag{6.54}$$

we see that the required integrating factor is given by

$$\exp\left\{-2 \int \tan x \, dx\right\} = \exp\left[2 \ln(\cos x)\right] = \cos^2 x.$$

Multiplying both sides of (6.54) by the integrating factor $\cos^2 x$ we obtain

$$\frac{d}{dx}\left(\cos^2 x \frac{dv}{dx}\right) = \cot x,$$

which integrates to give

$$\cos^2 x \frac{dv}{dx} = \ln(\sin x) + c_1.$$

After rearranging and integrating again, this becomes

$$\begin{aligned} v &= \int \sec^2 x \ln(\sin x) \, dx + c_1 \int \sec^2 x \, dx \\ &= \tan x \ln(\sin x) - x + c_1 \tan x + c_2. \end{aligned}$$

Therefore the general solution to (6.52) is given by $y = uv = v \cos x$, i.e.

$$y = c_1 \sin x + c_2 \cos x + \sin x \ln(\sin x) - x \cos x,$$

which contains the full complementary function and the particular integral. ◀

Solution method. *If $u(x)$ is a known solution of the nth-order equation (6.51) with $f(x) = 0$, then make the substitution $y(x) = u(x)v(x)$ in (6.51). This leads to an equation of order $n - 1$ in dv/dx, which might be soluble.*

6.5.4 Variation of parameters

The method of variation of parameters proves useful in finding particular integrals for linear ODEs with variable (and constant) coefficients. However, it requires knowledge of the entire complementary function, not just of one part of it as in the previous subsection.

Suppose we wish to find a particular integral of the equation

$$a_n(x)\frac{d^n y}{dx^n} + \cdots + a_1(x)\frac{dy}{dx} + a_0(x)y = f(x), \tag{6.55}$$

and the complementary function $y_c(x)$ (the general solution of (6.55) with $f(x) = 0$) is

$$y_c(x) = c_1 y_1(x) + c_2 y_2(x) + \cdots + c_n y_n(x),$$

where the functions $y_m(x)$ are known. We now assume that a particular integral of (6.55) can be expressed in a form similar to that of the complementary function, but with the constants c_m replaced by functions of x, i.e. we assume a particular integral of the form

$$y_p(x) = k_1(x)y_1(x) + k_2(x)y_2(x) + \cdots + k_n(x)y_n(x). \tag{6.56}$$

This will no longer satisfy the complementary equation (i.e. (6.55) with the RHS set to zero) but might, with suitable choices of the functions $k_i(x)$, be made equal to $f(x)$, thus producing not a complementary function but a particular integral.

Since we have n arbitrary functions $k_1(x), k_2(x), \ldots, k_n(x)$, but only one restriction on them (namely the ODE), we may impose a further $n - 1$ constraints. We can choose these constraints to be as convenient as possible, and the simplest choice is given by

$$\begin{aligned}
k_1'(x)y_1(x) + k_2'(x)y_2(x) + \cdots + k_n'(x)y_n(x) &= 0,\\
k_1'(x)y_1'(x) + k_2'(x)y_2'(x) + \cdots + k_n'(x)y_n'(x) &= 0,\\
&\ \ \vdots\\
k_1'(x)y_1^{(n-2)}(x) + k_2'(x)y_2^{(n-2)}(x) + \cdots + k_n'(x)y_n^{(n-2)}(x) &= 0,\\
k_1'(x)y_1^{(n-1)}(x) + k_2'(x)y_2^{(n-1)}(x) + \cdots + k_n'(x)y_n^{(n-1)}(x) &= \frac{f(x)}{a_n(x)},
\end{aligned} \tag{6.57}$$

where the primes denote differentiation with respect to x. The last of these equations is not a freely chosen constraint; given the previous $n - 1$ constraints and the original ODE, it is essential that it be satisfied.

This choice of constraints is easily justified (although the algebra is quite messy). Differentiating (6.56) with respect to x, we obtain

$$y_p' = k_1 y_1' + k_2 y_2' + \cdots + k_n y_n' + \left[k_1' y_1 + k_2' y_2 + \cdots + k_n' y_n\right],$$

where, for the moment, we drop the explicit x-dependence of these functions. Since we are free to choose our constraints as we wish, let us define the expression in square brackets to be zero, giving the first equation in (6.57). Differentiating again we find

$$y_p'' = k_1 y_1'' + k_2 y_2'' + \cdots + k_n y_n'' + \left[k_1' y_1' + k_2' y_2' + \cdots + k_n' y_n'\right].$$

Once more we can choose the expression in brackets to be zero, giving the second equation in (6.57). We can repeat this procedure, choosing the corresponding expression in each case to be zero. This yields the first $n-1$ equations in (6.57). The mth derivative of $y_{\rm p}$ for $m<n$ is then given by

$$y_{\rm p}^{(m)} = k_1 y_1^{(m)} + k_2 y_2^{(m)} + \cdots + k_n y_n^{(m)}.$$

Differentiating $y_{\rm p}$ once more we find that its nth derivative is given by

$$y_{\rm p}^{(n)} = k_1 y_1^{(n)} + k_2 y_2^{(n)} + \cdots + k_n y_n^{(n)} + \left[k_1' y_1^{(n-1)} + k_2' y_2^{(n-1)} + \cdots + k_n' y_n^{(n-1)}\right].$$

Substituting the expressions for $y_{\rm p}^{(m)}$, $m=0$ to n, into the original ODE (6.55), we obtain

$$\sum_{m=0}^{n} a_m\left[k_1 y_1^{(m)} + k_2 y_2^{(m)} + \cdots + k_n y_n^{(m)}\right] + a_n\left[k_1' y_1^{(n-1)} + k_2' y_2^{(n-1)} + \cdots + k_n' y_n^{(n-1)}\right] = f(x),$$

i.e.

$$\sum_{m=0}^{n} a_m \sum_{j=1}^{n} k_j y_j^{(m)} + a_n\left[k_1' y_1^{(n-1)} + k_2' y_2^{(n-1)} + \cdots + k_n' y_n^{(n-1)}\right] = f(x).$$

Rearranging the order of summation on the LHS, we find

$$\sum_{j=1}^{n} k_j\left[a_n y_j^{(n)} + \cdots + a_1 y_j' + a_0 y_j\right] + a_n\left[k_1' y_1^{(n-1)} + k_2' y_2^{(n-1)} + \cdots + k_n' y_n^{(n-1)}\right] = f(x). \qquad (6.58)$$

But since the functions y_j are solutions of the complementary equation of (6.55) we have (for all j)

$$a_n y_j^{(n)} + \cdots + a_1 y_j' + a_0 y_j = 0.$$

Therefore (6.58) becomes

$$a_n\left[k_1' y_1^{(n-1)} + k_2' y_2^{(n-1)} + \cdots + k_n' y_n^{(n-1)}\right] = f(x),$$

which is the final equation given in (6.57).

Considering (6.57) to be a set of simultaneous equations in the set of unknowns $k_1'(x), k_2', \ldots, k_n'(x)$, we see that the determinant of the coefficients of these functions is equal to the Wronskian $W(y_1, y_2, \ldots, y_n)$, which is non-zero since the solutions $y_m(x)$ are linearly independent; see equation (6.8). Therefore (6.57) can be solved for the functions $k_m'(x)$, which in turn can be integrated, setting all constants of integration equal to zero, to give $k_m(x)$. The general solution to (6.55) is then given by

$$y(x) = y_{\rm c}(x) + y_{\rm p}(x) = \sum_{m=1}^{n}[c_m + k_m(x)]y_m(x).$$

Note that if non-zero constants of integration are included in the $k_m(x)$ then, as well as finding the particular integral, we redefine the arbitrary constants c_m in the complementary function.

We now re-solve the worked example from the previous subsection, using this alternative method. We also include some defined boundary conditions.

Example Use the variation-of-parameters method to solve

$$\frac{d^2y}{dx^2} + y = \csc x, \tag{6.59}$$

subject to the boundary conditions $y(0) = y(\pi/2) = 0$.

The complementary function of (6.59) is again

$$y_c(x) = c_1 \sin x + c_2 \cos x.$$

We therefore assume a particular integral of the form

$$y_p(x) = k_1(x) \sin x + k_2(x) \cos x,$$

and impose the additional constraints of (6.57), i.e.

$$k_1'(x) \sin x + k_2'(x) \cos x = 0,$$

$$k_1'(x) \cos x - k_2'(x) \sin x = \csc x.$$

Solving these equations for $k_1'(x)$ and $k_2'(x)$ gives

$$k_1'(x) = \cos x \ \csc x = \cot x,$$

$$k_2'(x) = -\sin x \ \csc x = -1.$$

Hence, ignoring the constants of integration, $k_1(x)$ and $k_2(x)$ are given by

$$k_1(x) = \ln(\sin x),$$

$$k_2(x) = -x.$$

The general solution to the ODE (6.59) is therefore

$$y(x) = [c_1 + \ln(\sin x)] \sin x + (c_2 - x) \cos x,$$

which is identical to the solution found in Subsection 6.5.3. Applying the boundary conditions $y(0) = y(\pi/2) = 0$ we find $c_1 = c_2 = 0$ and so

$$y(x) = \ln(\sin x) \sin x - x \cos x.$$

It will be apparent that, although establishing the general variation-of-parameters result for arbitrary n is algebraically demanding, for any specific case the calculations are reasonably tractable, provided the integrations of the k_i' can be carried out. ◀

Solution method. *If the complementary function of (6.55) is known then assume a particular integral of the same form but with the constants replaced by functions of x. Impose the constraints in (6.57) and solve the resulting system of equations for the unknowns $k_1'(x), k_2'(x), \ldots, k_n'(x)$. Integrate these functions, setting constants of integration equal to zero, to obtain $k_1(x), k_2(x), \ldots, k_n(x)$ and hence the particular integral.*

6.5.5 Green's functions

The Green's function method of solving linear ODEs bears a striking resemblance to the method of variation of parameters discussed in the previous subsection; it too requires knowledge of the entire complementary function in order to find the particular integral and therefore the general solution. The Green's function approach is different, however, because once the Green's function for a particular LHS of (6.3) and particular boundary conditions has been found, then the solution for *any* RHS, i.e. for any $f(x)$, can be written down immediately, albeit in the form of an integral.

Although the Green's function method can be approached by considering the superposition of eigenfunctions of the equation (see Chapter 8) and is also applicable to the solution of partial differential equations (see Chapter 11), this section adopts a more utilitarian approach based on the properties of the Dirac delta function (see Subsection 5.2) and deals only with the use of Green's functions in solving ODEs.

Let us again consider the equation

$$a_n(x)\frac{d^n y}{dx^n} + \cdots + a_1(x)\frac{dy}{dx} + a_0(x)y = f(x), \tag{6.60}$$

but for the sake of brevity we now denote the LHS by $\mathcal{L}y(x)$, i.e. as a linear differential operator acting on $y(x)$. Thus (6.60) now reads

$$\mathcal{L}y(x) = f(x). \tag{6.61}$$

Let us suppose that a function $G(x, z)$ (the *Green's function*) exists such that the general solution to (6.61), which obeys some set of imposed boundary conditions in the range $a \leq x \leq b$, is given by

$$y(x) = \int_a^b G(x, z) f(z)\, dz, \tag{6.62}$$

where z is an integration variable. If we apply the linear differential operator $\mathcal{L}$ to both sides of (6.62) and use (6.61) then we obtain

$$\mathcal{L}y(x) = \int_a^b \left[\mathcal{L}G(x, z)\right] f(z)\, dz = f(x). \tag{6.63}$$

Comparison of (6.63) with a standard property of the Dirac delta function (see Subsection 5.2), namely

$$f(x) = \int_a^b \delta(x - z) f(z)\, dz,$$

for $a \leq x \leq b$, shows that for (6.63) to hold for any arbitrary function $f(x)$, we require (for $a \leq x \leq b$) that

$$\mathcal{L}G(x, z) = \delta(x - z), \tag{6.64}$$

i.e. the Green's function $G(x, z)$ *must satisfy the original ODE with the RHS set equal to a delta function.* $G(x, z)$ may be thought of physically as the response to a unit impulse at $x = z$, of a system subject to the imposed boundary conditions.

In addition to (6.64), we must impose two further sets of restrictions on $G(x, z)$. The first is the requirement that the general solution $y(x)$ in (6.62) obeys the boundary conditions.

For *homogeneous* boundary conditions, in which $y(x)$ and/or its derivatives are required to be *zero* at specified points, this is most simply arranged by demanding that $G(x, z)$ itself obeys the boundary conditions when it is considered as a function of x alone; if, for example, we require $y(a) = y(b) = 0$ then we should also demand $G(a, z) = G(b, z) = 0$. Situations involving inhomogeneous boundary conditions are discussed at the end of this subsection.

The second set of restrictions concerns the continuity or discontinuity of $G(x, z)$ and its derivatives at $x = z$ and can be found by integrating (6.64) with respect to x over the small interval $[z - \epsilon,\ z + \epsilon]$ and taking the limit as $\epsilon \to 0$. We then obtain

$$\lim_{\epsilon\to 0} \sum_{m=0}^{n} \int_{z-\epsilon}^{z+\epsilon} a_m(x) \frac{d^m G(x, z)}{dx^m}\, dx = \lim_{\epsilon\to 0} \int_{z-\epsilon}^{z+\epsilon} \delta(x - z)\, dx = 1. \tag{6.65}$$

Since $d^n G/dx^n$ exists at $x = z$ but with value infinity, the $(n - 1)$th-order derivative must have a finite discontinuity there, whereas all the lower-order derivatives, $d^m G/dx^m$ for $m < n - 1$, must be continuous at this point. Therefore the terms containing these derivatives cannot contribute to the value of the integral on the LHS of (6.65). Noting that, apart from an arbitrary additive constant, $\int (d^m G/dx^m)\, dx = d^{m-1} G/dx^{m-1}$, and integrating the terms on the LHS of (6.65) by parts we find

$$\lim_{\epsilon\to 0} \int_{z-\epsilon}^{z+\epsilon} a_m(x) \frac{d^m G(x, z)}{dx^m}\, dx = 0 \tag{6.66}$$

for $m = 0$ to $n - 1$. Thus, since only the term containing $d^n G/dx^n$ contributes to the integral in (6.65), we conclude, after performing an integration by parts, that

$$\lim_{\epsilon\to 0} \left[a_n(x) \frac{d^{n-1} G(x, z)}{dx^{n-1}} \right]_{z-\epsilon}^{z+\epsilon} = 1. \tag{6.67}$$

Thus we have the further n constraints that $G(x, z)$ and its derivatives up to order $n - 2$ are continuous at $x = z$ but that $d^{n-1} G/dx^{n-1}$ has a discontinuity of $1/a_n(z)$ at $x = z$.

Thus the properties of the Green's function $G(x, z)$ for an nth-order linear ODE may be summarized by the following.

(i) $G(x, z)$ obeys the original ODE but with $f(x)$ on the RHS set equal to a delta function $\delta(x - z)$.
(ii) When considered as a function of x alone $G(x, z)$ obeys the specified (homogeneous) boundary conditions on $y(x)$.
(iii) The derivatives of $G(x, z)$ with respect to x up to order $n - 2$ are continuous at $x = z$, but the $(n - 1)$th-order derivative has a discontinuity of $1/a_n(z)$ at this point.

To illustrate the Green's function method, we now solve a (by now) familiar equation for a third time.

Example Use Green's functions to solve

$$\frac{d^2y}{dx^2} + y = \csc x, \tag{6.68}$$

subject to the boundary conditions $y(0) = y(\pi/2) = 0$.

From (6.64) we see that the Green's function $G(x, z)$ must satisfy

$$\frac{d^2G(x,z)}{dx^2} + G(x, z) = \delta(x - z). \tag{6.69}$$

Now it is clear that for $x \neq z$ the RHS of (6.69) is zero, and we are left with the task of finding the general solution to the homogeneous equation, i.e. the complementary function. The complementary function of (6.69) consists of a linear superposition of $\sin x$ and $\cos x$ and *must* consist of different superpositions on either side of $x = z$, since its $(n-1)$th derivative (i.e. the first derivative in this case) is required to have a discontinuity there. Therefore we assume the form of the Green's function to be

$$G(x, z) = \begin{cases} A(z)\sin x + B(z)\cos x & \text{for } x < z, \\ C(z)\sin x + D(z)\cos x & \text{for } x > z. \end{cases}$$

Note that we have performed a similar (but not identical) operation to that used in the variation-of-parameters method, i.e. we have replaced the constants in the complementary function with functions (this time of z).

We must now impose the relevant restrictions on $G(x, z)$ in order to determine the functions $A(z), \ldots, D(z)$. The first of these is that $G(x, z)$ should itself obey the homogeneous boundary conditions $G(0, z) = G(\pi/2, z) = 0$. This leads to the conclusion that $B(z) = C(z) = 0$, so we now have

$$G(x, z) = \begin{cases} A(z)\sin x & \text{for } x < z, \\ D(z)\cos x & \text{for } x > z. \end{cases}$$

The second restriction is the continuity conditions given in equations (6.66), (6.67), namely that, for this second-order equation, $G(x, z)$ is continuous at $x = z$ and dG/dx has a discontinuity of $1/a_2(z) = 1$ at this point. Applying these two constraints we have

$$D(z)\cos z - A(z)\sin z = 0,$$

$$-D(z)\sin z - A(z)\cos z = 1.$$

Solving these equations for $A(z)$ and $D(z)$, we find

$$A(z) = -\cos z, \qquad D(z) = -\sin z.$$

Thus we have

$$G(x, z) = \begin{cases} -\cos z \sin x & \text{for } x < z, \\ -\sin z \cos x & \text{for } x > z. \end{cases}$$

Therefore, from (6.62), the general solution to (6.68) that obeys the boundary conditions $y(0) = y(\pi/2) = 0$ is given by[16]

$$\begin{aligned} y(x) &= \int_0^{\pi/2} G(x,z) \csc z \, dz \\ &= -\cos x \int_0^x \sin z \csc z \, dz - \sin x \int_x^{\pi/2} \cos z \csc z \, dz \\ &= -x \cos x + \sin x \ln(\sin x), \end{aligned}$$

which agrees with the result obtained in the previous subsections. ◀

As mentioned earlier, once a Green's function has been obtained for a given LHS and boundary conditions, it can be used to find a general solution for any RHS; thus, the solution of $d^2y/dx^2 + y = f(x)$, with $y(0) = y(\pi/2) = 0$, is given immediately by

$$\begin{aligned} y(x) &= \int_0^{\pi/2} G(x,z) f(z) \, dz \\ &= -\cos x \int_0^x \sin z \; f(z) \, dz - \sin x \int_x^{\pi/2} \cos z \; f(z) \, dz. \end{aligned} \tag{6.70}$$

As an example, the reader may wish to verify that if $f(x) = \sin 2x$ then (6.70) gives $y(x) = (-\sin 2x)/3$, a solution easily verified by direct substitution. In general, analytic integration of (6.70) for arbitrary $f(x)$ will prove intractable; then the integrals must be evaluated numerically.

A further useful aspect of the Green's function method is that, although above it was used to provide a general solution, it can also be employed to find a particular integral if the complementary function is known. This is easily seen since in (6.70) the constant integration limits 0 and $\pi/2$ lead merely to constant values by which the factors $\sin x$ and $\cos x$ are multiplied; thus the complementary function is reconstructed. The rest of the general solution, i.e. the particular integral, comes from the variable integration limit x appearing in both integrals. Therefore by changing $\int_x^{\pi/2}$ to $-\int^x$, and so dropping the constant integration limits, we can find just the particular integral. For example, a particular integral of $d^2y/dx^2 + y = f(x)$ that satisfies the above boundary conditions is given by

$$y_{\rm p}(x) = -\cos x \int^x \sin z \; f(z) \, dz + \sin x \int^x \cos z \; f(z) \, dz.$$

A very important point to understand about the Green's function method is that a particular $G(x,z)$ applies to a given LHS of an ODE *and* the imposed boundary conditions, i.e. *the same equation with different boundary conditions will have a different Green's function.* To illustrate this point, let us consider again the ODE solved in (6.70), but with different boundary conditions.

16 Note very carefully which part of the Green's function is used in which part of the integral; the integration is over z, not over x. For the integration from 0 to x, the integration variable z is less than x and the second form given for $G(x,z)$, namely $-\sin z \cos x$, is the appropriate one. Conversely, for the integral from x to $\pi/2$, $z > x$ and the first form is the one to use.

Example Use Green's functions to solve

$$\frac{d^2y}{dx^2} + y = f(x), \tag{6.71}$$

subject to the one-point boundary conditions $y(0) = y'(0) = 0$.

We first note that the relevant range is now $0 < x < \infty$. Again (6.69) is required to hold and so we again assume a Green's function of the form

$$G(x,z) = \begin{cases} A(z)\sin x + B(z)\cos x & \text{for } x < z, \\ C(z)\sin x + D(z)\cos x & \text{for } x > z. \end{cases}$$

However, we now require $G(x, z)$ to obey the boundary conditions $G(0, z) = G'(0, z) = 0$, which imply $A(z) = B(z) = 0$. Therefore we have

$$G(x,z) = \begin{cases} 0 & \text{for } x < z, \\ C(z)\sin x + D(z)\cos x & \text{for } x > z. \end{cases}$$

Applying the continuity conditions on $G(x, z)$ as before now gives

$$C(z)\sin z + D(z)\cos z = 0,$$

$$C(z)\cos z - D(z)\sin z = 1,$$

which are solved to give

$$C(z) = \cos z, \qquad D(z) = -\sin z.$$

Recognizing that $C(z)\sin x + D(z)\cos x = \cos z \sin x - \sin z \cos x$ can be written more compactly as $\sin(x - z)$, we can write the full Green's function as

$$G(x,z) = \begin{cases} 0 & \text{for } x < z, \\ \sin(x-z) & \text{for } x > z, \end{cases}$$

and the general solution to (6.71) that obeys the boundary conditions $y(0) = y'(0) = 0$ is

$$y(x) = \int_0^\infty G(x,z)f(z)\,dz$$

$$= \int_0^x \sin(x-z)f(z)\,dz,$$

where we have used the fact that $G(x, z)$ is zero for all $z > x$ to reduce the upper limit of the integral from ∞ to x. This form of solution is in line with the physical notion of "causality" in that, if x represented time, we would not expect, for a system that has started "from rest" $[y(0) = y'(0) = 0]$, that its response, $y(x)$, at time x would be affected by the value of f at a time z greater than x. The same considerations do not apply to systems with two-point boundary conditions since some property of y at the upper boundary is pre-ordained. ◀

Finally, we consider how to deal with inhomogeneous boundary conditions such as $y(a) = \alpha$, $y(b) = \beta$ or $y(0) = y'(0) = \gamma$, where α, β, γ are non-zero. The simplest method of solution in this case is to make a change of variable such that the boundary conditions in the new variable, u say, are homogeneous, i.e. $u(a) = u(b) = 0$ or $u(0) = u'(0) = 0$, etc. For nth-order equations we generally require n boundary conditions to fix the solution, but these n boundary conditions can be of various types: we could have the n-point boundary conditions $y(x_m) = y_m$ for $m = 1$ to n, or the one-point boundary conditions $y(x_0) = y'(x_0) = \cdots = y^{(n-1)}(x_0) = y_0$, or something in between. In all cases a suitable change of variable is

$$u = y - h(x),$$

where $h(x)$ is an $(n-1)$th-order polynomial that obeys the boundary conditions.

For example, if we are considering the second-order case with boundary conditions $y(a) = \alpha$, $y(b) = \beta$ then a suitable change of variable is

$$u = y - (mx + c),$$

where $y = mx + c$ is the straight line through the points (a, α) and (b, β), for which $m = (\alpha - \beta)/(a - b)$ and $c = (\beta a - \alpha b)/(a - b)$. Alternatively, if the boundary conditions for our second-order equation are $y(0) = y'(0) = \gamma$ then we would make the same change of variable, but this time $y = mx + c$ would be the straight line through $(0, \gamma)$ with slope γ, i.e. $m = c = \gamma$.

Solution method. *Require that the Green's function $G(x, z)$ obeys the original ODE, but with the RHS set to a delta function $\delta(x - z)$. This is equivalent to assuming that $G(x, z)$ is given by the complementary function of the original ODE, with the constants replaced by functions of z; these functions are different for $x < z$ and $x > z$. Now require also that $G(x, z)$ obeys the given homogeneous boundary conditions and impose the continuity conditions given in (6.66) and (6.67). The general solution to the original ODE is then given by (6.62). For inhomogeneous boundary conditions, make the change of dependent variable $u = y - h(x)$, where $h(x)$ is a polynomial obeying the given boundary conditions.*

6.6 General ordinary differential equations

In this section, we discuss miscellaneous methods for simplifying general ODEs. These methods are applicable to both linear and non-linear equations and in some cases may lead to a solution. More often than not, however, finding a closed-form solution to a general non-linear ODE proves impossible.

6.6.1 Dependent variable absent

If an ODE does not contain the dependent variable y explicitly, but only its derivatives, then the change of variable $p = dy/dx$ leads to an equation of one order lower. As a first example consider the following.

Example Solve

$$\frac{d^2y}{dx^2} + 2\frac{dy}{dx} = 4x. \tag{6.72}$$

This is transformed by the substitution $p = dy/dx$ to the first-order equation

$$\frac{dp}{dx} + 2p = 4x. \tag{6.73}$$

The solution to (6.73) is then found using a standard method for first-order ODEs and reads[17]

$$p = \frac{dy}{dx} = ae^{-2x} + 2x - 1,$$

where a is a constant. Thus by direct integration the solution to (6.72) is

$$y(x) = c_1 e^{-2x} + x^2 - x + c_2,$$

which, as expected for a second-order differential equation, contains two arbitrary constants. ◀

An extension to the above method is appropriate if an ODE contains only derivatives of y that are of order m and greater. Then the substitution $p = d^m y/dx^m$ reduces the order of the ODE by m.

Solution method. *If the ODE contains only derivatives of y that are of order m and greater, then the substitution $p = d^m y/dx^m$ reduces the order of the equation by m.*

6.6.2 Independent variable absent

If an ODE does not contain the independent variable x explicitly, except in d/dx, d^2/dx^2, etc., then as in the previous subsection we make the substitution $p = dy/dx$ but also write

$$\frac{d^2y}{dx^2} = \frac{dp}{dx} = \frac{dy}{dx}\frac{dp}{dy} = p\frac{dp}{dy},$$

$$\frac{d^3y}{dx^3} = \frac{d}{dx}\left(p\frac{dp}{dy}\right) = \frac{dy}{dx}\frac{d}{dy}\left(p\frac{dp}{dy}\right) = p^2\frac{d^2p}{dy^2} + p\left(\frac{dp}{dy}\right)^2, \tag{6.74}$$

and so on for higher-order derivatives. This leads to an equation of one order lower. This time, our worked example is a non-linear equation.

17 Try to derive this without having to look up the method.

Example Solve

$$1 + y\frac{d^2y}{dx^2} + \left(\frac{dy}{dx}\right)^2 = 0. \qquad (6.75)$$

Making the substitutions $dy/dx = p$ and $d^2y/dx^2 = p(dp/dy)$ we obtain the first-order ODE

$$1 + yp\frac{dp}{dy} + p^2 = 0.$$

This is separable in p and y, and may be solved in the normal way to obtain

$$(1 + p^2)y^2 = c_1.$$

This equation can, in its turn, be solved to yield p, which may now be rewritten in terms of y to give

$$p = \frac{dy}{dx} = \pm\sqrt{\frac{c_1^2 - y^2}{y^2}}.$$

Again a separable equation is obtained and it may be integrated to give[18]

$$(x + c_2)^2 + y^2 = c_1^2$$

as the general solution of (6.75). ◀

Solution method. *If the ODE does not contain x explicitly then substitute $p = dy/dx$, along with the relations for higher derivatives given in (6.74), to obtain an equation of one order lower, which may prove easier to solve.*

6.6.3 Equations homogeneous in *x* or *y* alone

One of the standard methods for the solution of first-order differential equations (see Section A.10) deals with equations that are homogeneous in x and y in the sense that dy/dx can be expressed purely in terms of the ratio y/x. Here we consider differential equations that are homogeneous in x or y alone; by this we mean that if x, say, were replaced by λx, then every term in the equation would be multiplied by the same power of λ.[19] Thus x^2, $x^3\,dy/dx$, $yx^4\,d^2y/dx^2$ and $x\,dx/dy$ could all form part of the same homogeneous equation in x alone, but none of $x\,dy/dx$, d^2x/dy^2 and $x\,d^2y/dx^2$ could be part of that same equation if it were to remain homogeneous. An example of an equation homogeneous in x alone might be

$$x\frac{d^2y}{dx^2} + (1 - y)\frac{dy}{dx} = 0,$$

18 (a) Identify geometrically the family of solutions generated as c_1 and c_2 are varied. (b) Using the general expressions for the radius of curvature of a curve and for the angle ψ that the tangent to a curve makes with the x-axis, show that equation (6.75) expresses a simple geometric property of a typical member of the family of solutions.

19 A more informal specification might be that each term in the equation contains the same "net power" of x, treating x almost as if it were a physical dimension and requiring dimensional consistency, as for acceptable physical equations. The "net power" of x is more technically known as the "weight" of x.

whilst one homogeneous in y alone could be

$$y\frac{d^2y}{dx^2} + (1-x^2)\left(\frac{dy}{dx}\right)^2 + ny^2 = 0.$$

We note that the Euler equation of Subsection 6.5.1 is a special, linear example of an equation homogeneous in x alone.

Equations homogeneous in x can be simplified by the substitution $x = e^t$, in that this leads to an equation in which the new independent variable t occurs only in the form d/dt. This happens because any factor x^n becomes a factor e^{nt} and each differential operation d/dx contributes a factor $e^{-t}d/dt$; in each term of the homogeneous equation, the net power of e^t introduced is the same and it can be canceled throughout. Similarly, if an equation is homogeneous in y alone, then substituting $y = e^v$ leads to an equation in which the new dependent variable, v, occurs only in the form d/dv.

Our worked example is homogeneous in x alone.

Example Solve

$$x^2\frac{d^2y}{dx^2} + x\frac{dy}{dx} + \frac{2}{y^3} = 0.$$

Since this equation is homogeneous in x alone, we substitute $x = e^t$ and obtain

$$e^{2t}e^{-t}\frac{d}{dt}\left(e^{-t}\frac{dy}{dt}\right) + e^te^{-t}\frac{dy}{dt} + \frac{2}{y^3} = \frac{d^2y}{dt^2} + \frac{2}{y^3} = 0,$$

which does not contain the new independent variable t except as d/dt. Such equations may often be solved by the method of Subsection 6.6.2, but in this case we can multiply through by dy/dt and then integrate directly to obtain

$$\frac{dy}{dt} = \sqrt{2(c_1 + 1/y^2)}.$$

This equation is separable, and we find

$$\int \frac{dy}{\sqrt{2(c_1 + 1/y^2)}} = t + c_2.$$

By multiplying the numerator and denominator of the integrand on the LHS by y, we find the solution

$$\frac{\sqrt{c_1y^2 + 1}}{\sqrt{2}c_1} = t + c_2.$$

Remembering that $t = \ln x$, we finally obtain

$$\frac{\sqrt{c_1y^2 + 1}}{\sqrt{2}c_1} = \ln x + c_2.$$

Note that we must *not* replace $\sqrt{2}c_1$ by a third arbitrary constant c_3; the two appearances of c_1 in the final solution must be maintained, and they must have the same value in each place. ◀

Solution method. *If the "weight" of x taken alone is the same in every term in the ODE then the substitution $x = e^t$ leads to an equation in which the new independent variable t is absent except in the form d/dt. If the "weight" of y taken alone is the same in every term then the substitution $y = e^v$ leads to an equation in which the new dependent variable v is absent except in the form d/dv.*

SUMMARY

1. *General considerations*
   - A set of n functions are linearly independent over an interval if their Wronskian

$$W(y_1, y_2, \ldots, y_n) = \begin{vmatrix} y_1 & y_2 & \cdots & y_n \\ y_1{}' & y_2{}' & & \vdots \\ \vdots & & \ddots & \vdots \\ y_1^{(n-1)} & \cdots & \cdots & y_n^{(n-1)} \end{vmatrix}$$

     is not identically zero over that interval; the vanishing of the Wronskian does *not* guarantee linear dependence.
   - An nth-order homogeneous linear ODE has n linearly independent solutions, $y_i(x)$ for $i = 1, 2, \ldots, n$ and the complementary function (CF) is $y_c(x) = \sum_i c_i y_i(x)$.
   - The complete solution to an inhomogeneous linear equation is $y(x) = y_c(x) + y_p(x)$, where $y_p(x)$ is *any* particular solution (however simple) of the ODE.
   - An nth-order equation requires n independent and self-consistent boundary conditions (BC) for a unique solution.
   - *Warning*: The BC must be applied to $y_c(x) + y_p(x)$ as a whole (and *not* to y_c alone, with y_p added later).

2. *Linear equations with constant coefficients*

$$a_n \frac{d^n y}{dx^n} + a_{n-1} \frac{d^{n-1} y}{dx^{n-1}} + \cdots + a_1 \frac{dy}{dx} + a_0 y = f(x). \qquad (*)$$

   - With $f(x)$ set equal to zero, a trial solution of the form $y = e^{\lambda x}$ gives an nth-degree polynomial in λ with
     (i) each real distinct root λ_i giving a solution $e^{\lambda_i x}$,
     (ii) pairs of complex roots $\alpha \pm i\beta$ giving solutions $e^{\alpha x}(d_1 \cos \beta x + d_2 \sin \beta x)$,
     (iii) a k-repeated root λ_i giving k (of the n) solutions as $e^{\lambda_i x}, xe^{\lambda_i x}, \ldots, x^{k-1}e^{\lambda_i x}$.
     The CF is a linear combination of the solutions so found.
   - A particular integral (PI) can be found by trying a multiple of $f(x)$. If $f(x)$ is proportional to a term in the CF, then the PI is $y_m(x) = Ax^m f(x)$, where m is the lowest positive integer such that y_m does not appear in the CF.

- Taking the Laplace transform of $(*)$ converts it to an algebraic equation for $\bar{y}(s)$, which can often be inverse transformed to $y(x)$, using partial fractions and a table of Laplace transforms.

3. *Linear recurrence relations with constant coefficients*
The general Nth-order recurrence relation is

$$u_{n+1} = \sum_{r=0}^{N-1} a_r u_{n-r} + k(n). \qquad (**)$$

- If v_n is the general solution of $(**)$ when $k = 0$, and w_n is *any* solution of $(**)$, then the general solution is $u_n = v_n + w_n$.
- Setting $v_n = A\lambda^n$ in $(**)$ with $k = 0$ gives the characteristic equation $\lambda^N = \Sigma_{r=0}^{N-1} a_r \lambda^{N-1-r}$, an Nth-degree polynomial equation.
- If the N roots of the characteristic equation are λ_i, then $v_n = \sum_{i=1}^{N} A_i \lambda_i^n$.
- If two of the roots are complex conjugates $\alpha \pm i\beta$, then two of the terms in v_n are replaced by $r^n(A\cos n\phi + B\sin n\phi)$ where $\tan\phi = \beta/\alpha$ and $r = \sqrt{\alpha^2 + \beta^2}$.
- If λ_i is a k-fold root, k of the terms in v_n are replaced by $(A_1 + A_2 n + \cdots + A_{k-1} n^{k-1})\lambda_i^n$.
- A particular solution w_n is sought by trying forms similar to $k(n)$.
- The coefficients in $u_n = v_n + w_n$ are determined by the given initial values u_0, $u_1, \ldots, u_N$.

4. *Linear equations with variable coefficients*

$$a_n(x)\frac{d^n y}{dx^n} + \cdots + a_1(x)\frac{dy}{dx} + a_0(x)y = f(x). \qquad (***)$$

- Legendre's linear equation, $\Sigma_{i=0}^{n} a_i(\alpha x + \beta)^i \dfrac{d^i y}{dx^i} = f(x)$, can be reduced to one with constant coefficients by setting $\alpha x + \beta = e^t$.
- Euler's linear equation is a special case of Legendre's, with $\alpha = 1$ and $\beta = 0$; it can usually be solved more easily by setting $y = x^\lambda$ and obtaining an nth-degree polynomial equation for the allowed values of λ.
- If $a_0(x) - a_1'(x) + a_2''(x) - \cdots + (-1)^n a_n^{(n)}(x) = 0$, then $(***)$ is exact and can be integrated once without modification.
- If one solution $u(x)$ of $(***)$ is known, then setting $y(x) = u(x)v(x)$ gives an equation of order $n - 1$ for dv/dx. A PI is also generated when re-substitution for v is made.
- The method of variation of parameters (see p. 250) can be used to find a PI.
- If, for $a \le x \le b$, a (Green's) function $G(x, z)$
  (i) obeys the original ODE but with $f(x)$ on the RHS set equal to a delta function $\delta(x - z)$,

(ii) when considered as a function of x alone it obeys the specified (homogeneous) BCs on $y(x)$ at $x = a$ and $x = b$,

(iii) its derivatives with respect to x up to order $n - 2$ are continuous at $x = z$, but the $(n - 1)$th-order derivative has a discontinuity of $1/a_n(z)$ at this point,

then

$$y(x) = \int_a^b G(x, z)f(z)\,dz$$

is the required solution of $(***)$ for any $f(x)$ and the same BC.

5. *Miscellaneous methods*
   - If the ODE contains only derivatives of y that are of order m and greater, then the substitution $p = d^m y/dx^m$ reduces the order of the equation by m.
   - If the ODE does not contain x explicitly, set $dy/dx = p$, $d^2y/dx^2 = p(dp/dy)$, ... (see p. 259) and solve for $p = p(y)$. Then integrate $dy/dx = p(y)$.
   - If an equation is homogeneous in x alone, substitute $x = e^t$. This leads to an equation in which t occurs only in the form d/dt. Similarly, for an equation homogeneous in y alone set $y = e^v$.

PROBLEMS

6.1. A simple harmonic oscillator, of mass m and natural frequency ω_0, experiences an oscillating driving force $f(t) = ma\cos\omega t$. Therefore, its equation of motion is

$$\frac{d^2x}{dt^2} + \omega_0^2 x = a\cos\omega t,$$

where x is its position. Given that at $t = 0$ we have $x = dx/dt = 0$, find the function $x(t)$. Describe the solution if ω is approximately, but not exactly, equal to ω_0.

6.2. Find the roots of the auxiliary equation for the following. Hence solve them for the boundary conditions stated.

(a) $\dfrac{d^2f}{dt^2} + 2\dfrac{df}{dt} + 5f = 0$, with $f(0) = 1$, $f'(0) = 0$.

(b) $\dfrac{d^2f}{dt^2} + 2\dfrac{df}{dt} + 5f = e^{-t}\cos 3t$, with $f(0) = 0$, $f'(0) = 0$.

6.3. The theory of bent beams shows that at any point in the beam the "bending moment" is given by K/ρ, where K is a constant (that depends upon the beam material and cross-sectional shape) and ρ is the radius of curvature at that point. Consider a light beam of length L whose ends, $x = 0$ and $x = L$, are supported at the same vertical height and which has a weight W suspended from its center.

Verify that at any point x ($0 \le x \le L/2$ for definiteness) the net magnitude of the bending moment (bending moment = force × perpendicular distance) due to the weight and support reactions, evaluated on either side of x, is $Wx/2$.

If the beam is only slightly bent, so that $(dy/dx)^2 \ll 1$, where $y = y(x)$ is the downward displacement of the beam at x, show that the beam profile satisfies the approximate equation

$$\frac{d^2y}{dx^2} = -\frac{Wx}{2K}.$$

By integrating this equation twice and using physically imposed conditions on your solution at $x = 0$ and $x = L/2$, show that the downward displacement at the center of the beam is $WL^3/(48K)$.

6.4. Solve the differential equation

$$\frac{d^2f}{dt^2} + 6\frac{df}{dt} + 9f = e^{-t},$$

subject to the conditions $f = 0$ and $df/dt = \lambda$ at $t = 0$.

Find the equation satisfied by the positions of the turning points of $f(t)$ and hence, by drawing suitable sketch graphs, determine the number of turning points the solution has in the range $t > 0$ if (a) $\lambda = 1/4$, and (b) $\lambda = -1/4$.

6.5. The function $f(t)$ satisfies the differential equation

$$\frac{d^2f}{dt^2} + 8\frac{df}{dt} + 12f = 12e^{-4t}.$$

For the following sets of boundary conditions determine whether it has solutions, and, if so, find them:
(a) $f(0) = 0, \quad f'(0) = 0, \quad f(\ln\sqrt{2}) = 0;$
(b) $f(0) = 0, \quad f'(0) = -2, \quad f(\ln\sqrt{2}) = 0.$

6.6. Determine the values of α and β for which the following four functions are linearly dependent:

$$\begin{aligned} y_1(x) &= x\cosh x + \sinh x, \\ y_2(x) &= x\sinh x + \cosh x, \\ y_3(x) &= (x+\alpha)e^x, \\ y_4(x) &= (x+\beta)e^{-x}. \end{aligned}$$

You will find it convenient to work with those linear combinations of the $y_i(x)$ that can be written the most compactly.

6.7. A solution of the differential equation

$$\frac{d^2y}{dx^2} + 2\frac{dy}{dx} + y = 4e^{-x}$$

takes the value 1 when $x = 0$ and the value e^{-1} when $x = 1$. What is its value when $x = 2$?

6.8. The two functions $x(t)$ and $y(t)$ satisfy the simultaneous equations

$$\frac{dx}{dt} - 2y = -\sin t,$$
$$\frac{dy}{dt} + 2x = 5\cos t.$$

Find explicit expressions for $x(t)$ and $y(t)$, given that $x(0) = 3$ and $y(0) = 2$. Sketch the solution trajectory in the xy-plane for $0 \le t < 2\pi$, showing that the trajectory crosses itself at $(0, 1/2)$ and passes through the points $(0, -3)$ and $(0, -1)$ in the negative x-direction.

6.9. Find the general solutions of

(a) $\dfrac{d^3y}{dx^3} - 12\dfrac{dy}{dx} + 16y = 32x - 8,$

(b) $\dfrac{d}{dx}\left(\dfrac{1}{y}\dfrac{dy}{dx}\right) + (2a\coth 2ax)\left(\dfrac{1}{y}\dfrac{dy}{dx}\right) = 2a^2,$

where a is a constant.

6.10. Use the method of Laplace transforms to solve

(a) $\dfrac{d^2f}{dt^2} + 5\dfrac{df}{dt} + 6f = 0, \qquad f(0) = 1,\ f'(0) = -4,$

(b) $\dfrac{d^2f}{dt^2} + 2\dfrac{df}{dt} + 5f = 0, \qquad f(0) = 1,\ f'(0) = 0.$

6.11. The quantities $x(t)$, $y(t)$ satisfy the simultaneous equations

$$\ddot{x} + 2n\dot{x} + n^2x = 0,$$
$$\ddot{y} + 2n\dot{y} + n^2y = \mu\dot{x},$$

where $x(0) = y(0) = \dot{y}(0) = 0$ and $\dot{x}(0) = \lambda$. Show that

$$y(t) = \tfrac{1}{2}\mu\lambda t^2\left(1 - \tfrac{1}{3}nt\right)\exp(-nt).$$

6.12. Use Laplace transforms to solve, for $t \ge 0$, the differential equations

$$\ddot{x} + 2x + y = \cos t,$$
$$\ddot{y} + 2x + 3y = 2\cos t,$$

which describe a coupled system that starts from rest at the equilibrium position. Show that the subsequent motion takes place along a straight line in the xy-plane. Verify that the frequency at which the system is driven is equal to one of the resonance frequencies of the system; explain why there is *no* resonant behavior in the solution you have obtained.

6.13. Two unstable isotopes A and B and a stable isotope C have the following decay rates per atom present: $A \to B$, $3\,\text{s}^{-1}$; $A \to C$, $1\,\text{s}^{-1}$; $B \to C$, $2\,\text{s}^{-1}$. Initially a quantity x_0 of A is present, but there are no atoms of the other two types. Using Laplace transforms, find the amount of C present at a later time t.

6.14. For a lightly damped ($\gamma < \omega_0$) harmonic oscillator driven at its undamped resonance frequency ω_0, the displacement $x(t)$ at time t satisfies the equation

$$\frac{d^2x}{dt^2} + 2\gamma\frac{dx}{dt} + \omega_0^2 x = F \sin \omega_0 t.$$

Use Laplace transforms to find the displacement at a general time if the oscillator starts from rest at its equilibrium position.

(a) Show that ultimately the oscillation has amplitude $F/(2\omega_0\gamma)$, with a phase lag of $\pi/2$ relative to the driving force per unit mass F.

(b) By differentiating the original equation, conclude that if $x(t)$ is expanded as a power series in t for small t, then the first non-vanishing term is $F\omega_0 t^3/6$. Confirm this conclusion by expanding your explicit solution.

6.15. The "golden mean", which is said to describe the most esthetically pleasing proportions for the sides of a rectangle (e.g. the ideal picture frame), is given by the limiting value of the ratio of successive terms of the Fibonacci series u_n, which is generated by

$$u_{n+2} = u_{n+1} + u_n,$$

with $u_0 = 0$ and $u_1 = 1$. Find an expression for the general term of the series and verify that the golden mean is equal to the larger root of the recurrence relation's characteristic equation.

6.16. In a particular scheme for numerically modeling one-dimensional fluid flow, the successive values, u_n, of the solution are connected for $n \geq 1$ by the difference equation

$$c(u_{n+1} - u_{n-1}) = d(u_{n+1} - 2u_n + u_{n-1}),$$

where c and d are positive constants. The boundary conditions are $u_0 = 0$ and $u_M = 1$. Find the solution to the equation, and show that successive values of u_n will have alternating signs if $c > d$.

6.17. The first few terms of a series u_n, starting with u_0, are $1, 2, 2, 1, 6, -3$. The series is generated by a recurrence relation of the form

$$u_n = Pu_{n-2} + Qu_{n-4},$$

where P and Q are constants. Find an expression for the general term of the series and show that, in fact, the series consists of two interleaved series given by

$$u_{2m} = \tfrac{2}{3} + \tfrac{1}{3}4^m,$$
$$u_{2m+1} = \tfrac{7}{3} - \tfrac{1}{3}4^m,$$

for $m = 0, 1, 2, \ldots$.

6.18. Find an explicit expression for the u_n satisfying

$$u_{n+1} + 5u_n + 6u_{n-1} = 2^n,$$

given that $u_0 = u_1 = 1$. Deduce that $2^n - 26(-3)^n$ is divisible by 5 for all non-negative integers n.

6.19. Find the general expression for the u_n satisfying

$$u_{n+1} = 2u_{n-2} - u_n$$

with $u_0 = u_1 = 0$ and $u_2 = 1$, and show that they can be written in the form

$$u_n = \frac{1}{5} - \frac{2^{n/2}}{\sqrt{5}} \cos\left(\frac{3\pi n}{4} - \phi\right),$$

where $\tan\phi = 2$.

6.20. Consider the seventh-order recurrence relation

$$u_{n+7} - u_{n+6} - u_{n+5} + u_{n+4} - u_{n+3} + u_{n+2} + u_{n+1} - u_n = 0.$$

Find the most general form of its solution, and show that:
(a) if only the four initial values $u_0 = 0$, $u_1 = 2$, $u_2 = 6$ and $u_3 = 12$ are specified, then the relation has one solution that cycles repeatedly through this set of four numbers;
(b) but if, in addition, it is required that $u_4 = 20$, $u_5 = 30$ and $u_6 = 42$ then the solution is unique, with $u_n = n(n+1)$.

6.21. Find the general solution of

$$x^2\frac{d^2y}{dx^2} - x\frac{dy}{dx} + y = x,$$

given that $y(1) = 1$ and $y(e) = 2e$.

6.22. Find the general solution of

$$(x+1)^2\frac{d^2y}{dx^2} + 3(x+1)\frac{dy}{dx} + y = x^2.$$

6.23. Prove that the general solution of

$$(x-2)\frac{d^2y}{dx^2}+3\frac{dy}{dx}+\frac{4y}{x^2}=0$$

is given by

$$y(x)=\frac{1}{(x-2)^2}\left[k\left(\frac{2}{3x}-\frac{1}{2}\right)+cx^2\right].$$

6.24. Use the method of variation of parameters to find the general solutions of (a) $\dfrac{d^2y}{dx^2}-y=x^n$, (b) $\dfrac{d^2y}{dx^2}-2\dfrac{dy}{dx}+y=2xe^x$.

6.25. Find the Green's function that satisfies

$$\frac{d^2G(x,\xi)}{dx^2}-G(x,\xi)=\delta(x-\xi)\qquad\text{with}\qquad G(0,\xi)=G(1,\xi)=0.$$

6.26. Consider the equation

$$F(x,y)=x(x+1)\frac{d^2y}{dx^2}+(2-x^2)\frac{dy}{dx}-(2+x)y=0.$$

(a) Given that $y_1(x)=1/x$ is one of its solutions, find a second linearly independent one by setting $y_2(x)=y_1(x)u(x)$.
(b) Hence, using the variation of parameters method, find the general solution of

$$F(x,y)=(x+1)^2.$$

6.27. Show generally that if $y_1(x)$ and $y_2(x)$ are linearly independent solutions of

$$\frac{d^2y}{dx^2}+p(x)\frac{dy}{dx}+q(x)y=0,$$

with $y_1(0)=0$ and $y_2(1)=0$, then the Green's function $G(x,\xi)$ for the interval $0\le x,\xi\le 1$ and with $G(0,\xi)=G(1,\xi)=0$ can be written in the form

$$G(x,\xi)=\begin{cases}y_1(x)y_2(\xi)/W(\xi) & 0<x<\xi,\\ y_2(x)y_1(\xi)/W(\xi) & \xi<x<1,\end{cases}$$

where $W(x)=W[y_1(x),y_2(x)]$ is the Wronskian of $y_1(x)$ and $y_2(x)$.

6.28. Use the result of the previous problem to find the Green's function $G(x,\xi)$ that satisfies

$$\frac{d^2G}{dx^2}+3\frac{dG}{dx}+2G=\delta(x-x),$$

in the interval $0 \le x, \xi \le 1$, with $G(0,\xi) = G(1,\xi) = 0$. Hence obtain integral expressions for the solution of

$$\frac{d^2y}{dx^2} + 3\frac{dy}{dx} + 2y = \begin{cases} 0 & 0 < x < x_0, \\ 1 & x_0 < x < 1, \end{cases}$$

distinguishing between the cases (a) $x < x_0$, and (b) $x > x_0$.

6.29. The equation of motion for a driven damped harmonic oscillator can be written

$$\ddot{x} + 2\dot{x} + (1 + \kappa^2)x = f(t),$$

with $\kappa \ne 0$. If it starts from rest with $x(0) = 0$ and $\dot{x}(0) = 0$, find the corresponding Green's function $G(t,\tau)$ and verify that it can be written as a function of $t - \tau$ only. Find the explicit solution when the driving force is the unit step function, i.e. $f(t) = H(t)$. Confirm your solution by taking the Laplace transforms of both it and the original equation.

6.30. Show that the Green's function for the equation

$$\frac{d^2y}{dx^2} + \frac{y}{4} = f(x),$$

subject to the boundary conditions $y(0) = y(\pi) = 0$, is given by

$$G(x,z) = \begin{cases} -2\cos\frac{1}{2}x\sin\frac{1}{2}z & 0 \le z \le x, \\ -2\sin\frac{1}{2}x\cos\frac{1}{2}z & x \le z \le \pi. \end{cases}$$

6.31. Find the Green's function $x = G(t, t_0)$ that solves

$$\frac{d^2x}{dt^2} + \alpha\frac{dx}{dt} = \delta(t - t_0)$$

under the initial conditions $x = dx/dt = 0$ at $t = 0$. Hence solve

$$\frac{d^2x}{dt^2} + \alpha\frac{dx}{dt} = f(t),$$

where $f(t) = 0$ for $t < 0$. Evaluate your answer explicitly for $f(t) = Ae^{-at}$ $(t > 0)$.

6.32. Consider the equation

$$\frac{d^2y}{dx^2} + f(y) = 0,$$

where $f(y)$ can be any function.

(a) By multiplying through by dy/dx, obtain the general solution relating x and y.

(b) A mass m, initially at rest at the point $x = 0$, is accelerated by a force

$$f(x) = A(x_0 - x)\left[1 + 2\ln\left(1 - \frac{x}{x_0}\right)\right].$$

Its equation of motion is $m\,d^2x/dt^2 = f(x)$. Find x as a function of time, and show that ultimately the particle has traveled a distance x_0.

6.33. Solve

$$2y\frac{d^3y}{dx^3} + 2\left(y + 3\frac{dy}{dx}\right)\frac{d^2y}{dx^2} + 2\left(\frac{dy}{dx}\right)^2 = \sin x.$$

6.34. Find the general solution of the equation

$$x\frac{d^3y}{dx^3} + 2\frac{d^2y}{dx^2} = Ax.$$

6.35. Confirm that the equation

$$2x^2y\frac{d^2y}{dx^2} + y^2 = x^2\left(\frac{dy}{dx}\right)^2 \qquad (*)$$

is homogeneous in both x and y separately. Make two successive transformations that exploit this fact, starting with a substitution for x, to obtain an equation of the form

$$2\frac{d^2v}{dt^2} + \left(\frac{dv}{dt}\right)^2 - 2\frac{dv}{dt} + 1 = 0.$$

By writing $dv/dt = p$, solve this equation for $v = v(t)$ and hence find the solution to (*).

HINTS AND ANSWERS

6.1. The function is $a(\omega_0^2 - \omega^2)^{-1}(\cos\omega t - \cos\omega_0 t)$; for moderate t, $x(t)$ is a sine wave of linearly increasing amplitude $(t\sin\omega_0 t)/(2\omega_0)$; for large t it shows beats of maximum amplitude $2(\omega_0^2 - \omega^2)^{-1}$.

6.3. Ignore the term y'^2, compared with 1, in the expression for ρ. $y = 0$ at $x = 0$. From symmetry, $dy/dx = 0$ at $x = L/2$.

6.5. General solution $f(t) = Ae^{-6t} + Be^{-2t} - 3e^{-4t}$. (a) No solution, inconsistent boundary conditions; (b) $f(t) = 2e^{-6t} + e^{-2t} - 3e^{-4t}$.

6.7. The auxiliary equation has repeated roots and the RHS is contained in the complementary function. The solution is $y(x) = (A + Bx)e^{-x} + 2x^2e^{-x}$. $y(2) = 5e^{-2}$.

6.9. (a) The auxiliary equation has roots $2, 2, -4$; $(A + Bx)\exp 2x + C\exp(-4x)$ $+2x + 1$; (b) multiply through by $\sinh 2ax$ and note that $\int \operatorname{cosech} 2ax\,dx = (2a)^{-1}\ln(|\tanh ax|)$; $y = B(\sinh 2ax)^{1/2}(|\tanh ax|)^A$.

6.11. Use Laplace transforms; write $s(s+n)^{-4}$ as $(s+n)^{-3} - n(s+n)^{-4}$.

6.13. $\mathcal{L}[C(t)] = x_0(s+8)/[s(s+2)(s+4)]$, yielding $C(t) = x_0[1 + \frac{1}{2}\exp(-4t) - \frac{3}{2}\exp(-2t)]$.

6.15. The characteristic equation is $\lambda^2 - \lambda - 1 = 0$.
$u_n = [(1+\sqrt{5})^n - (1-\sqrt{5})^n]/(2^n\sqrt{5})$.

6.17. From u_4 and u_5, $P = 5$, $Q = -4$. $u_n = 3/2 - 5(-1)^n/6 + (-2)^n/4 + 2^n/12$.

6.19. The general solution is $A + B2^{n/2}\exp(i3\pi n/4) + C2^{n/2}\exp(i5\pi n/4)$. The initial values imply that $A = 1/5$, $B = (\sqrt{5}/10)\exp[i(\pi - \phi)]$ and $C = (\sqrt{5}/10)\exp[i(\pi + \phi)]$.

6.21. This is Euler's equation; setting $x = \exp t$ produces $d^2z/dt^2 - 2\,dz/dt + z = \exp t$, with complementary function $(A + Bt)\exp t$ and particular integral $t^2(\exp t)/2$; $y(x) = x + [x \ln x(1 + \ln x)]/2$.

6.23. After multiplication through by x^2 the coefficients are such that this is an exact equation. The resulting first-order equation, in standard form, needs an integrating factor $(x-2)^2/x^2$.

6.25. Given the boundary conditions, it is better to work with $\sinh x$ and $\sinh(1-x)$ than with $e^{\pm x}$; $G(x,\xi) = -[\sinh(1-\xi)\sinh x]/\sinh 1$ for $x < \xi$ and $-[\sinh(1-x)\sinh\xi]/\sinh 1$ for $x > \xi$.

6.27. Follow the method of Subsection 6.5.5, but using general rather than specific functions.

6.29. $G(t,\tau) = 0$ for $t < \tau$ and $\kappa^{-1}e^{-(t-\tau)}\sin[\kappa(t-\tau)]$ for $t > \tau$. For a unit step input, $x(t) = (1+\kappa^2)^{-1}(1 - e^{-t}\cos\kappa t - \kappa^{-1}e^{-t}\sin\kappa t)$. Both transforms are equivalent to $s[(s+1)^2 + \kappa^2]\bar{x} = 1$.

6.31. Use continuity and the step condition on $\partial G/\partial t$ at $t = t_0$ to show that
$G(t, t_0) = \alpha^{-1}\{1 - \exp[\alpha(t_0 - t)]\}$ for $0 \le t_0 \le t$;
$x(t) = A(\alpha - a)^{-1}\{a^{-1}[1 - \exp(-at)] - \alpha^{-1}[1 - \exp(-\alpha t)]\}$.

6.33. The LHS of the equation is exact for two stages of integration and then needs an integrating factor $\exp x$; $2y\,d^2y/dx^2 + 2y\,dy/dx + 2(dy/dx)^2$;
$2y\,dy/dx + y^2 = d(y^2)/dx + y^2$; $y^2 = A\exp(-x) + Bx + C - (\sin x - \cos x)/2$.

6.35. Set $x = e^t$ to obtain $2yd^2y/dt^2 - 2ydy/dt - (dy/dt)^2 + y^2 = 0$ and then set $y = e^v$. After one integration $(p-1)^{-1} = \frac{1}{2}t + A$. After the second, $v = 2\ln(A + \frac{1}{2}t) + t + B$, leading to $y = x(C + D\ln x)^2$.

7 Series solutions of ordinary differential equations

In the previous chapter the solution of both homogeneous and non-homogeneous linear ODEs of order ≥ 2 was discussed. In particular we developed methods for solving some equations in which the coefficients were not constant but functions of the independent variable x. In each case we were able to write the solutions to such equations in terms of elementary functions, or as integrals. In general, however, the solutions of equations with variable coefficients cannot be written in this way, and we must consider alternative approaches.

In this chapter we discuss a method for obtaining solutions to linear ODEs in the form of convergent series. Such series can be evaluated numerically, and those occurring most commonly are named and tabulated. There is in fact no distinct borderline between this and the previous chapter, since solutions in terms of elementary functions may equally well be written as convergent series (i.e. the relevant Taylor series). Indeed, it is partly because some series occur so frequently that they are given special names such as $\sin x$, $\cos x$ or $\exp x$.

Since, in this chapter, we shall be concerned principally with second-order linear ODEs we begin with a discussion of this type of equation, and obtain some general results that will prove useful when we come to discuss series solutions.

7.1 Second-order linear ordinary differential equations

Any homogeneous second-order linear ODE can be written in the form

$$y'' + p(x)y' + q(x)y = 0, \tag{7.1}$$

where $y' = dy/dx$ and $p(x)$ and $q(x)$ are given functions of x. From the previous chapter, we recall that the most general form of the solution to (7.1) is

$$y(x) = c_1 y_1(x) + c_2 y_2(x), \tag{7.2}$$

where $y_1(x)$ and $y_2(x)$ are *linearly independent* solutions of (7.1), and c_1 and c_2 are constants that are fixed by the boundary conditions (if supplied).

A full discussion of the linear independence of sets of functions was given at the beginning of the previous chapter, but for just two functions y_1 and y_2 to be linearly independent we simply require that y_2 is not a multiple of y_1. Equivalently, y_1 and y_2 must be such that the equation

$$c_1 y_1(x) + c_2 y_2(x) = 0$$

is *only* satisfied for $c_1 = c_2 = 0$. Therefore the linear independence of $y_1(x)$ and $y_2(x)$ can usually be deduced by inspection, but in any case can always be verified by the evaluation of the Wronskian of the two solutions,

$$W(x) = \begin{vmatrix} y_1 & y_2 \\ y_1' & y_2' \end{vmatrix} = y_1 y_2' - y_2 y_1'. \tag{7.3}$$

If $W(x) \neq 0$ anywhere in a given interval then y_1 and y_2 are linearly independent in that interval.[1]

An alternative expression for $W(x)$, of which we will make use later, may be derived by differentiating (7.3) with respect to x to give

$$W' = y_1 y_2'' + y_1' y_2' - y_2 y_1'' - y_2' y_1' = y_1 y_2'' - y_1'' y_2.$$

Since both y_1 and y_2 satisfy (7.1), we may substitute for y_1'' and y_2'' to obtain

$$W' = -y_1(p y_2' + q y_2) + (p y_1' + q y_1) y_2 = -p(y_1 y_2' - y_1' y_2) = -pW.$$

Integrating, we find

$$W(x) = C \exp\left\{ -\int^x p(u)\, du \right\}, \tag{7.4}$$

where C is a constant.[2] We note further that in the special case $p(x) \equiv 0$ we obtain $W =$ constant.

Example The functions $y_1 = \sin x$ and $y_2 = \cos x$ are both solutions of the equation $y'' + y = 0$. Evaluate the Wronskian of these two solutions, and hence show that they are linearly independent.

The Wronskian of y_1 and y_2 is given by

$$W = y_1 y_2' - y_2 y_1' = -\sin^2 x - \cos^2 x = -1.$$

Since $W \neq 0$ the two solutions are linearly independent. We also note that $y'' + y = 0$ is a special case of (7.1) with $p(x) = 0$. We therefore expect, from (7.4), that W will be a constant, as is indeed the case. ◀

From the previous chapter we recall that, once we have obtained the general solution to the homogeneous second-order ODE (7.1) in the form (7.2), the general solution to the *inhomogeneous* equation

$$y'' + p(x)y' + q(x)y = f(x) \tag{7.5}$$

can be written as the sum of the solution to the homogeneous equation $y_c(x)$ (the complementary function) and *any* function $y_p(x)$ (the particular integral) that satisfies (7.5) and is linearly independent of $y_c(x)$. We have therefore

$$y(x) = c_1 y_1(x) + c_2 y_2(x) + y_p(x). \tag{7.6}$$

1 Use the Wronskian test to show that the set of functions $y_1 = \tan x$, $y_2 = \sec x$ and $y_3 = 1$ are not linearly dependent, but that the set $y_1 = \tan^2 x$, $y_2 = \sec^2 x$ and $y_3 = 1$ are.

2 For the two functions $y_1 = e^{3x}$ and $y_2 = e^{-2x}$, find the second-order equation of which they are the independent solutions and calculate their Wronskian. Verify that (7.4) is satisfied.

General methods for obtaining y_p, that are applicable to equations with variable coefficients, such as the variation of parameters or Green's functions, were discussed in the previous chapter. An alternative description of the Green's function method for solving inhomogeneous equations is given in the next chapter. For the present, however, we will restrict our attention to the solutions of homogeneous ODEs in the form of convergent series.

7.2 Ordinary and singular points of an ODE

So far we have implicitly assumed that $y(x)$ is a *real* function of a *real* variable x. However, this is not always the case, and in the remainder of this chapter we broaden our discussion by generalizing to a *complex* function $y(z)$ of a *complex* variable z.

Let us therefore consider the second-order linear homogeneous ODE

$$y'' + p(z)y' + q(z) = 0, \tag{7.7}$$

where now $y' = dy/dz$; this is a straightforward generalization of (7.1). A full discussion of complex functions and differentiation with respect to a complex variable z is given in Chapter 14, but for the purposes of the present chapter we need not concern ourselves with many of the subtleties that exist. In particular, we may treat differentiation with respect to z in a way analogous to ordinary differentiation with respect to a real variable x.

In (7.7), if, at some point $z = z_0$, the functions $p(z)$ and $q(z)$ are finite and can be expressed as complex power series, i.e.

$$p(z) = \sum_{n=0}^{\infty} p_n(z - z_0)^n, \qquad q(z) = \sum_{n=0}^{\infty} q_n(z - z_0)^n,$$

then $p(z)$ and $q(z)$ are said to be *analytic* at $z = z_0$, and this point is called an *ordinary point* of the ODE. If, however, $p(z)$ or $q(z)$, or both, diverge at $z = z_0$ then it is called a *singular point* of the ODE.

Even if an ODE is singular at a given point $z = z_0$, it may still possess a non-singular (finite) solution at that point. In fact, the necessary and sufficient condition[3] for such a solution to exist is that $(z - z_0)p(z)$ and $(z - z_0)^2q(z)$ are both analytic at $z = z_0$. Singular points that have this property are called *regular singular points*, whereas any singular point not satisfying both these criteria is termed an *irregular* or *essential* singularity.

Example Legendre's equation has the form

$$(1 - z^2)y'' - 2zy' + \ell(\ell + 1)y = 0, \tag{7.8}$$

where ℓ is a constant. Show that $z = 0$ is an ordinary point and $z = \pm 1$ are regular singular points of this equation.

3 See, for example, H. Jeffreys and B. S. Jeffreys, *Methods of Mathematical Physics*, 3rd edn (Cambridge: Cambridge University Press, 1966), p. 479.

Firstly, divide through by $1 - z^2$ to put the equation into our standard form (7.7):

$$y'' - \frac{2z}{1-z^2}y' + \frac{\ell(\ell+1)}{1-z^2}y = 0.$$

Comparing this with (7.7), we identify $p(z)$ and $q(z)$ as

$$p(z) = \frac{-2z}{1-z^2} = \frac{-2z}{(1+z)(1-z)}, \qquad q(z) = \frac{\ell(\ell+1)}{1-z^2} = \frac{\ell(\ell+1)}{(1+z)(1-z)}.$$

By inspection, both $p(z)$ and $q(z)$ are analytic at $z = 0$, which is therefore an ordinary point, but both diverge for $z = \pm 1$, which are thus singular points. However, at $z = 1$ we see that both $(z-1)p(z)$ and $(z-1)^2 q(z)$ are analytic and hence $z = 1$ is a regular singular point. Similarly, at $z = -1$ both $(z+1)p(z)$ and $(z+1)^2 q(z)$ are analytic, and it too is a regular singular point. ◀

So far we have assumed that z_0 is finite. However, we may sometimes wish to determine the nature of the point $|z| \to \infty$. This may be achieved straightforwardly by substituting $w = 1/z$ into the equation and investigating the behavior at $w = 0$.

Example Show that Legendre's equation has a regular singularity at $|z| \to \infty$.

Letting $w = 1/z$, the derivatives with respect to z become

$$\frac{dy}{dz} = \frac{dy}{dw}\frac{dw}{dz} = -\frac{1}{z^2}\frac{dy}{dw} = -w^2\frac{dy}{dw},$$

$$\frac{d^2y}{dz^2} = \frac{dw}{dz}\frac{d}{dw}\left(\frac{dy}{dz}\right) = -w^2\left(-2w\frac{dy}{dw} - w^2\frac{d^2y}{dw^2}\right) = w^3\left(2\frac{dy}{dw} + w\frac{d^2y}{dw^2}\right).$$

If we substitute these derivatives into Legendre's equation (7.8) we obtain

$$\left(1 - \frac{1}{w^2}\right)w^3\left(2\frac{dy}{dw} + w\frac{d^2y}{dw^2}\right) + 2\frac{1}{w}w^2\frac{dy}{dw} + \ell(\ell+1)y = 0,$$

which simplifies to give

$$w^2(w^2-1)\frac{d^2y}{dw^2} + 2w^3\frac{dy}{dw} + \ell(\ell+1)y = 0.$$

Dividing through by $w^2(w^2-1)$ to put the equation into standard form, and comparing with (7.7), we identify $p(w)$ and $q(w)$ as

$$p(w) = \frac{2w}{w^2-1}, \qquad q(w) = \frac{\ell(\ell+1)}{w^2(w^2-1)}.$$

At $w = 0$, $p(w)$ is analytic but $q(w)$ diverges, and so the point $|z| \to \infty$ is a singular point of Legendre's equation. However, since wp and w^2q are both analytic at $w = 0$, $|z| \to \infty$ is a regular singular point.[4] ◀

4 Considering the results of this and the previous worked example taken together, and without consulting Table 7.1, state how many singular points Legendre's equation has in total. How many of them are regular singular points?

Table 7.1 *Important second-order linear ODEs in the physical sciences and engineering*

Equation	Regular singularities	Essential singularities
Hypergeometric $z(1-z)y'' + [c-(a+b+1)z]y' - aby = 0$	$0, 1, \infty$	–
Legendre $(1-z^2)y'' - 2zy' + \ell(\ell+1)y = 0$	$-1, 1, \infty$	–
Associated Legendre $(1-z^2)y'' - 2zy' + \left[\ell(\ell+1) - \frac{m^2}{1-z^2}\right]y = 0$	$-1, 1, \infty$	–
Chebyshev $(1-z^2)y'' - zy' + \nu^2 y = 0$	$-1, 1, \infty$	–
Confluent hypergeometric $zy'' + (c-z)y' - ay = 0$	0	∞
Bessel $z^2y'' + zy' + (z^2-\nu^2)y = 0$	0	∞
Laguerre $zy'' + (1-z)y' + \nu y = 0$	0	∞
Associated Laguerre $zy'' + (m+1-z)y' + (\nu-m)y = 0$	0	∞
Hermite $y'' - 2zy' + 2\nu y = 0$	–	∞
Simple harmonic oscillator $y'' + \omega^2 y = 0$	–	∞

Table 7.1 lists the singular points of several second-order linear ODEs that play important roles in the analysis of many problems in physics and engineering. A full discussion of the solutions to each of the equations in Table 7.1 and their properties is left until Chapter 9; exceptions to this are the hypergeometric and confluent hypergeometric equations, for which discussion of their solutions is beyond the scope of this book. We now develop the general methods by which series solutions may be obtained.

7.3 Series solutions about an ordinary point

If $z = z_0$ is an ordinary point of (7.7) then it may be shown that *every* solution $y(z)$ of the equation is also analytic at $z = z_0$. From now on we will take z_0 as the origin, i.e. $z_0 = 0$. If this is not already the case, then a substitution $Z = z - z_0$ will make it so.[5] Since every solution is analytic, $y(z)$ can be represented by a power series of the form (see

5 Chebyshev's equation (see Table 7.1) has a singularity at $z = 1$. Rewrite the equation in terms of a new variable such that this singularity is situated at the new origin. Where are the other singularities of this new equation?

Section 14.11)

$$y(z) = \sum_{n=0}^{\infty} a_n z^n. \qquad (7.9)$$

Moreover, it may be shown that such a power series converges for $|z| < R$, where R is the radius of convergence and is equal to the distance from $z = 0$ to the nearest singular point of the ODE (see Chapter 14). At the radius of convergence, however, the series may or may not converge (though it will diverge at at least one point on it).

Since every solution of (7.7) is analytic at an ordinary point, it is always possible to obtain two *independent* solutions (from which the general solution (7.2) can be constructed) of the form (7.9). The derivatives of y with respect to z are given by

$$y' = \sum_{n=0}^{\infty} n a_n z^{n-1} = \sum_{n=0}^{\infty} (n+1) a_{n+1} z^n, \qquad (7.10)$$

$$y'' = \sum_{n=0}^{\infty} n(n-1) a_n z^{n-2} = \sum_{n=0}^{\infty} (n+2)(n+1) a_{n+2} z^n. \qquad (7.11)$$

Note that, in each case, in the first equality the sum can still start at $n = 0$ since the first term in (7.10) and the first two terms in (7.11) are automatically zero. The second equality in each case is obtained by shifting the summation index so that the sum can be written in terms of coefficients of z^n. By substituting (7.9)–(7.11) into the ODE (7.7), and requiring that the coefficients of each power of z sum to zero, we obtain a *recurrence relation* expressing each a_n in terms of the previous a_r $(0 \le r \le n-1)$.

Our first worked example tackles a very familiar equation, for which the solution is nearly always given in terms of named functions.

Example Find the series solutions, about $z = 0$, of

$$y'' + y = 0.$$

By inspection, $z = 0$ is an ordinary point of the equation, and so we may obtain two independent solutions by making the substitution $y = \sum_{n=0}^{\infty} a_n z^n$. Using (7.9) and (7.11) we find

$$\sum_{n=0}^{\infty} (n+2)(n+1) a_{n+2} z^n + \sum_{n=0}^{\infty} a_n z^n = 0,$$

which may be written as

$$\sum_{n=0}^{\infty} [(n+2)(n+1) a_{n+2} + a_n] z^n = 0.$$

For this equation to be satisfied we require that the coefficient of each power of z vanishes *separately*, and so we obtain the two-term recurrence relation

$$a_{n+2} = -\frac{a_n}{(n+2)(n+1)} \qquad \text{for } n \ge 0.$$

Using this relation, we can calculate, say, the even coefficients $a_2,\ a_4,\ a_6$ and so on, for a given a_0. Alternatively, starting with a_1, we obtain the odd coefficients $a_3,\ a_5$, etc. Two independent solutions of the ODE can be obtained by setting either $a_0 = 0$ or $a_1 = 0$. Firstly, if we set $a_1 = 0$ and choose $a_0 = 1$ then we obtain the solution

$$y_1(z) = 1 - \frac{z^2}{2!} + \frac{z^4}{4!} - \cdots = \sum_{n=0}^{\infty} \frac{(-1)^n}{(2n)!} z^{2n}.$$

Secondly, if we set $a_0 = 0$ and choose $a_1 = 1$ then we obtain a second, *independent*, solution

$$y_2(z) = z - \frac{z^3}{3!} + \frac{z^5}{5!} - \cdots = \sum_{n=0}^{\infty} \frac{(-1)^n}{(2n+1)!} z^{2n+1}.$$

Recognizing these two series as $\cos z$ and $\sin z$, we can write the general solution as

$$y(z) = c_1 \cos z + c_2 \sin z,$$

where c_1 and c_2 are arbitrary constants that are fixed by boundary conditions (if supplied). We note that both solutions converge for all z, as might be expected since the ODE possesses no singular points (except $|z| \to \infty$). ◀

Solving the above example was quite straightforward and the resulting series were easily recognized and written in *closed form* (i.e. in terms of elementary functions); *this is not usually the case.* Another simplifying feature of the previous example was that we obtained a two-term recurrence relation relating a_{n+2} and a_n, so that the odd- and even-numbered coefficients were independent of one another. In general, the recurrence relation expresses a_n in terms of any number of the previous a_r $(0 \le r \le n-1)$. The following example illustrates this point.

Example Find the series solutions, about $z = 0$, of

$$y'' - \frac{2}{(1-z)^2} y = 0.$$

By inspection, $z = 0$ is an ordinary point, and therefore we may find two independent solutions by substituting $y = \sum_{n=0}^{\infty} a_n z^n$. Using (7.10) and (7.11), and multiplying through by $(1-z)^2$, we find

$$(1 - 2z + z^2) \sum_{n=0}^{\infty} n(n-1) a_n z^{n-2} - 2 \sum_{n=0}^{\infty} a_n z^n = 0,$$

which leads to

$$\sum_{n=0}^{\infty} n(n-1) a_n z^{n-2} - 2 \sum_{n=0}^{\infty} n(n-1) a_n z^{n-1} + \sum_{n=0}^{\infty} n(n-1) a_n z^n - 2 \sum_{n=0}^{\infty} a_n z^n = 0.$$

In order to write all these series in terms of the coefficients of z^n, we must shift the summation index in the first two sums, obtaining

$$\sum_{n=0}^{\infty} (n+2)(n+1) a_{n+2} z^n - 2 \sum_{n=0}^{\infty} (n+1) n a_{n+1} z^n + \sum_{n=0}^{\infty} (n^2 - n - 2) a_n z^n = 0,$$

which can be written as

$$\sum_{n=0}^{\infty} (n+1)[(n+2) a_{n+2} - 2n a_{n+1} + (n-2) a_n] z^n = 0.$$

By demanding that the coefficients of each power of z vanish separately, we obtain the three-term recurrence relation

$$(n+2)a_{n+2} - 2na_{n+1} + (n-2)a_n = 0 \qquad \text{for } n \geq 0,$$

which determines a_n for $n \geq 2$ in terms of a_0 and a_1. From a three-term (or more) recurrence relation, it is, in general, difficult to find a_n in explicit form.[6] This particular recurrence relation, however, has two straightforward solutions. One solution is $a_n = a_0$ for all n, in which case (choosing $a_0 = 1$) we find

$$y_1(z) = 1 + z + z^2 + z^3 + \cdots = \frac{1}{1-z}.$$

The other solution to the recurrence relation is $a_1 = -2a_0$, $a_2 = a_0$ and $a_n = 0$ for $n > 2$, so that (again choosing $a_0 = 1$) we obtain a *polynomial* solution to the ODE:

$$y_2(z) = 1 - 2z + z^2 = (1-z)^2.$$

The linear independence of y_1 and y_2 is obvious but can be checked by computing the Wronskian

$$W = y_1 y_2' - y_1' y_2 = \frac{1}{1-z}[-2(1-z)] - \frac{1}{(1-z)^2}(1-z)^2 = -3.$$

Since $W \neq 0$, the two solutions y_1 and y_2 are indeed linearly independent. The general solution of the ODE is therefore

$$y(z) = \frac{c_1}{1-z} + c_2(1-z)^2.$$

We observe that y_1 (and hence the general solution) is singular at $z = 1$, which is the singular point of the ODE nearest to $z = 0$, but the polynomial solution, y_2, is valid for all finite z. ◀

The above example illustrates the possibility that, in some cases, we may find that the recurrence relation leads to $a_n = 0$ for $n > N$, for one or both of the two solutions; we then obtain a *polynomial* solution to the equation. Polynomial solutions are discussed more fully in Section 7.6, but one obvious property of such solutions is that they converge for all finite z. By contrast, as mentioned above, for solutions in the form of an infinite series the circle of convergence extends only as far as the singular point nearest to that about which the solution is being obtained.

7.4 Series solutions about a regular singular point

From Table 7.1 we see that several of the most important second-order linear ODEs in physics and engineering have regular singular points in the finite complex plane. We must extend our discussion, therefore, to obtaining series solutions to ODEs about such points. In what follows we assume that the regular singular point about which the solution is required is at $z = 0$, since, as we have seen, if this is not already the case then a substitution of the form $Z = z - z_0$ will make it so.

6 Though, of course, any particular coefficient can, in principle, be calculated by repeated application of the recurrence relation, given a sufficient number of early values.

If $z = 0$ is a regular singular point of the equation

$$y'' + p(z)y' + q(z)y = 0$$

then at least one of $p(z)$ and $q(z)$ is not analytic at $z = 0$, and in general we should not expect to find a power series solution of the form (7.9). We must therefore extend the method to include a more general form for the solution. In fact, it may be shown (Fuch's theorem) that there exists *at least one* solution to the above equation,[7] of the form

$$y = z^\sigma \sum_{n=0}^{\infty} a_n z^n, \tag{7.12}$$

where the exponent σ is a number that may be real or complex and where $a_0 \neq 0$ (since, if it were otherwise, σ could be redefined as $\sigma + 1$ or $\sigma + 2$ or $\cdots$ so as to make $a_0 \neq 0$). Such a series is called a generalized power series or *Frobenius series.* As in the case of a simple power series solution, the radius of convergence of the Frobenius series is, in general, equal to the distance from the expansion point to the next nearest singularity of the ODE.

Since $z = 0$ is a regular singularity of the ODE, it follows that $zp(z)$ and $z^2q(z)$ are analytic at $z = 0$, so that we may write

$$zp(z) \equiv s(z) = \sum_{n=0}^{\infty} s_n z^n,$$

$$z^2q(z) \equiv t(z) = \sum_{n=0}^{\infty} t_n z^n,$$

where we have defined the analytic functions $s(z)$ and $t(z)$ for later convenience. The original ODE therefore becomes

$$y'' + \frac{s(z)}{z}y' + \frac{t(z)}{z^2}y = 0.$$

Let us substitute the Frobenius series (7.12) into this equation. The derivatives of (7.12) with respect to z are given by

$$y' = \sum_{n=0}^{\infty} (n+\sigma)a_n z^{n+\sigma-1}, \tag{7.13}$$

$$y'' = \sum_{n=0}^{\infty} (n+\sigma)(n+\sigma-1)a_n z^{n+\sigma-2}, \tag{7.14}$$

and we obtain

$$\sum_{n=0}^{\infty} (n+\sigma)(n+\sigma-1)a_n z^{n+\sigma-2} + s(z)\sum_{n=0}^{\infty} (n+\sigma)a_n z^{n+\sigma-2} + t(z)\sum_{n=0}^{\infty} a_n z^{n+\sigma-2} = 0.$$

7 But, of course, not more than two.

Dividing this equation through by $z^{\sigma-2}$, we find

$$\sum_{n=0}^{\infty} [(n+\sigma)(n+\sigma-1) + s(z)(n+\sigma) + t(z)]\, a_n z^n = 0. \tag{7.15}$$

Setting $z = 0$, all terms in the sum with $n > 0$ vanish, implying that

$$[\sigma(\sigma-1) + s(0)\sigma + t(0)]a_0 = 0,$$

which, since we require $a_0 \neq 0$, yields the *indicial equation*

$$\sigma(\sigma-1) + s(0)\sigma + t(0) = 0. \tag{7.16}$$

This equation is a quadratic in σ and in general has two roots, the nature of which determines the forms of possible series solutions.

The two roots of the indicial equation, σ_1 and σ_2, are called the *indices* of the regular singular point. By substituting each of these roots into (7.15) in turn and requiring that the coefficients of each power of z vanish separately,[8] we obtain a recurrence relation (for each root) expressing each a_n as a function of the previous a_r $(0 \le r \le n-1)$.

We will see that the larger root of the indicial equation *always* yields a solution to the ODE in the form of a Frobenius series (7.12). The form of the second solution depends, however, on the relationship between the two indices σ_1 and σ_2. There are three possible general cases: (i) distinct roots not differing by an integer; (ii) repeated roots; (iii) distinct roots differing by a non-zero integer. Below, we discuss each of these in turn.

Before continuing, however, we note that, as was the case for solutions in the form of a simple power series, it is always worth investigating whether a Frobenius series found as a solution to a problem is summable in closed form or expressible in terms of known functions. We illustrate this point below, but the reader should avoid gaining the impression that this is always so or that, if one worked hard enough, a closed-form solution could always be found without using the series method. As mentioned earlier, this is *not* the case, and very often an infinite series solution is the best one can do.

7.4.1 Distinct roots not differing by an integer

If the roots of the indicial equation, σ_1 and σ_2, differ by an amount that is not an integer then the recurrence relations corresponding to each root lead to two linearly independent solutions of the ODE:

$$y_1(z) = z^{\sigma_1} \sum_{n=0}^{\infty} a_n z^n, \qquad y_2(z) = z^{\sigma_2} \sum_{n=0}^{\infty} b_n z^n,$$

with both solutions taking the form of a Frobenius series. The linear independence of these two solutions follows from the fact that y_2/y_1 is not a constant since $\sigma_1 - \sigma_2$ is not

8 Before doing so, it is most advisable to multiply the differential equation all through by whatever is needed to remove all inverse powers of any factor containing z (other than the ones responsible for the singularity) that appear in $p(z)$ or $q(z)$. If this is not done, even a simple factor such as $(1-z)^{-1}$ has to be expanded by the binomial theorem and results in the product of two (or more) infinite series, from which picking out particular powers of z is extremely difficult.

an integer. Because y_1 and y_2 are linearly independent, we may use them to construct the general solution $y = c_1 y_1 + c_2 y_2$.

We also note that this case includes complex conjugate roots where $\sigma_2 = \sigma_1^*$, since $\sigma_1 - \sigma_2 = \sigma_1 - \sigma_1^* = 2i \,\mathrm{Im}\, \sigma_1$ cannot be equal to a real integer.

Example Find the power series solutions about $z = 0$ of

$$4zy'' + 2y' + y = 0.$$

Dividing through by $4z$ to put the equation into standard form, we obtain

$$y'' + \frac{1}{2z}y' + \frac{1}{4z}y = 0, \qquad (7.17)$$

and on comparing with (7.7) we identify $p(z) = 1/(2z)$ and $q(z) = 1/(4z)$. Clearly $z = 0$ is a singular point of (7.17), but since $zp(z) = 1/2$ and $z^2 q(z) = z/4$ are finite there, it is a regular singular point. We therefore substitute the Frobenius series $y = z^\sigma \sum_{n=0}^{\infty} a_n z^n$ into (7.17). Using (7.13) and (7.14), we obtain

$$\sum_{n=0}^{\infty} (n+\sigma)(n+\sigma-1)a_n z^{n+\sigma-2} + \frac{1}{2z}\sum_{n=0}^{\infty}(n+\sigma)a_n z^{n+\sigma-1} + \frac{1}{4z}\sum_{n=0}^{\infty} a_n z^{n+\sigma} = 0,$$

which, on dividing through by $z^{\sigma-2}$, gives

$$\sum_{n=0}^{\infty} \left[(n+\sigma)(n+\sigma-1) + \tfrac{1}{2}(n+\sigma) + \tfrac{1}{4}z\right] a_n z^n = 0. \qquad (7.18)$$

If we set $z = 0$ then all terms in the sum with $n > 0$ vanish, and we obtain the indicial equation

$$\sigma(\sigma - 1) + \tfrac{1}{2}\sigma = 0,$$

which has roots $\sigma = 1/2$ and $\sigma = 0$. Since these roots do not differ by an integer, we expect to find two independent solutions to (7.17), in the form of Frobenius series.

Demanding that the coefficients of z^n vanish separately in (7.18), we obtain the recurrence relation

$$(n+\sigma)(n+\sigma-1)a_n + \tfrac{1}{2}(n+\sigma)a_n + \tfrac{1}{4}a_{n-1} = 0. \qquad (7.19)$$

If we choose the larger root, $\sigma = 1/2$, of the indicial equation, then (7.19) becomes

$$(4n^2 + 2n)a_n + a_{n-1} = 0 \quad \Rightarrow \quad a_n = \frac{-a_{n-1}}{2n(2n+1)}.$$

Setting $a_0 = 1$, we find $a_n = (-1)^n/(2n+1)!$, and so the solution to (7.17) is given by

$$\begin{aligned} y_1(z) &= \sqrt{z}\sum_{n=0}^{\infty} \frac{(-1)^n}{(2n+1)!} z^n \\ &= \sqrt{z} - \frac{(\sqrt{z})^3}{3!} + \frac{(\sqrt{z})^5}{5!} - \cdots = \sin\sqrt{z}. \end{aligned}$$

To obtain the second solution we set $\sigma = 0$ (the smaller root of the indicial equation) in (7.19), which gives

$$(4n^2 - 2n)a_n + a_{n-1} = 0 \quad \Rightarrow \quad a_n = -\frac{a_{n-1}}{2n(2n-1)}.$$

Setting $a_0 = 1$ now gives $a_n = (-1)^n/(2n)!$, and so the second (independent) solution to (7.17) is

$$y_2(z) = \sum_{n=0}^{\infty} \frac{(-1)^n}{(2n)!} z^n = 1 - \frac{(\sqrt{z})^2}{2!} + \frac{(\sqrt{4})^4}{4!} - \cdots = \cos\sqrt{z}.$$

We may check that $y_1(z)$ and $y_2(z)$ are indeed linearly independent by computing the Wronskian as follows:

$$\begin{aligned} W &= y_1 y_2' - y_2 y_1' \\ &= \sin\sqrt{z}\left(-\frac{1}{2\sqrt{z}}\sin\sqrt{z}\right) - \cos\sqrt{z}\left(\frac{1}{2\sqrt{z}}\cos\sqrt{z}\right) \\ &= -\frac{1}{2\sqrt{z}}\left(\sin^2\sqrt{z} + \cos^2\sqrt{z}\right) = -\frac{1}{2\sqrt{z}} \neq 0. \end{aligned}$$

Since $W \neq 0$, the solutions $y_1(z)$ and $y_2(z)$ are linearly independent. Hence, the general solution to (7.17) is given by

$$y(z) = c_1 \sin\sqrt{z} + c_2 \cos\sqrt{z},$$

i.e. by any linear combination of the two independent series solutions that have been found. ◀

7.4.2 Repeated root of the indicial equation

If the indicial equation has a repeated root, so that $\sigma_1 = \sigma_2 = \sigma$, then obviously only one solution in the form of a Frobenius series (7.12) may be found as described above, i.e.

$$y_1(z) = z^\sigma \sum_{n=0}^{\infty} a_n z^n.$$

Methods for obtaining a second, linearly independent, solution are discussed in Section 7.5.

7.4.3 Distinct roots differing by an integer

Whatever the roots of the indicial equation, the recurrence relation corresponding to the larger of the two always leads to a solution of the ODE. However, if the roots of the indicial equation differ by an integer then the recurrence relation corresponding to the smaller root may or may not lead to a second linearly independent solution, depending on the ODE under consideration. Note that for complex roots of the indicial equation, the "larger" root is taken to be the one with the larger real part.

Example Find the power series solutions about $z = 0$ of

$$z(z-1)y'' + 3zy' + y = 0. \quad (7.20)$$

Dividing through by $z(z-1)$ to put the equation into standard form, we obtain

$$y'' + \frac{3}{(z-1)}y' + \frac{1}{z(z-1)}y = 0, \quad (7.21)$$

and on comparing with (7.7) we identify $p(z) = 3/(z-1)$ and $q(z) = 1/[z(z-1)]$. We immediately see that $z = 0$ is a singular point of (7.21), but since $zp(z) = 3z/(z-1)$ and $z^2q(z) = z/(z-1)$

are finite there, it is a regular singular point and we expect to find at least one solution in the form of a Frobenius series. We therefore substitute $y = z^\sigma \sum_{n=0}^{\infty} a_n z^n$ into (7.21) and, using (7.13) and (7.14), we obtain

$$\sum_{n=0}^{\infty}(n+\sigma)(n+\sigma-1)a_n z^{n+\sigma-2} + \frac{3}{z-1}\sum_{n=0}^{\infty}(n+\sigma)a_n z^{n+\sigma-1} + \frac{1}{z(z-1)}\sum_{n=0}^{\infty} a_n z^{n+\sigma} = 0,$$

which, on dividing through by $z^{\sigma-2}$, gives

$$\sum_{n=0}^{\infty}\left[(n+\sigma)(n+\sigma-1) + \frac{3z}{z-1}(n+\sigma) + \frac{z}{z-1}\right]a_n z^n = 0.$$

Although we could use this expression to find the indicial equation and recurrence relations, the working is simpler if we now multiply through by $z-1$ to give[9]

$$\sum_{n=0}^{\infty}[(z-1)(n+\sigma)(n+\sigma-1) + 3z(n+\sigma) + z]\,a_n z^n = 0. \tag{7.22}$$

If we set $z = 0$ then all terms in the sum with the exponent of z greater than zero vanish, and we obtain the indicial equation

$$\sigma(\sigma-1) = 0,$$

which has the roots $\sigma = 1$ and $\sigma = 0$. Since the roots differ by an integer (unity), it may not be possible to find two linearly independent solutions of (7.21) in the form of Frobenius series. We are guaranteed, however, to find one such solution corresponding to the larger root, $\sigma = 1$.

Demanding that the coefficients of z^n vanish separately in (7.22), we obtain the recurrence relation[10]

$$(n-1+\sigma)(n-2+\sigma)a_{n-1} - (n+\sigma)(n+\sigma-1)a_n + 3(n-1+\sigma)a_{n-1} + a_{n-1} = 0,$$

which can be simplified to give

$$(n+\sigma-1)a_n = (n+\sigma)a_{n-1}. \tag{7.23}$$

On substituting $\sigma = 1$ into this expression, we obtain

$$a_n = \left(\frac{n+1}{n}\right)a_{n-1},$$

and on setting $a_0 = 1$ we find $a_n = n+1$; so one solution to (7.21) is given by

$$y_1(z) = z\sum_{n=0}^{\infty}(n+1)z^n = z(1+2z+3z^2+\cdots)$$

$$= \frac{z}{(1-z)^2}. \tag{7.24}$$

9 See footnote 8.

10 For practice, derive this recurrence relation for yourself.

If we attempt to find a second solution (corresponding to the smaller root of the indicial equation) by setting $\sigma = 0$ in (7.23), we find

$$a_n = \left(\frac{n}{n-1}\right) a_{n-1}.$$

But we require $a_0 \neq 0$, so a_1 is formally infinite and the method fails. We discuss how to find a second linearly independent solution in the next section. ◀

One particular case is worth mentioning. If the point about which the solution is required, i.e. $z = 0$, is in fact an ordinary point of the ODE, rather than a wrongly presumed regular singular point, then substitution of the Frobenius series (7.12) leads to an indicial equation with roots $\sigma = 0$ and $\sigma = 1$. Although these roots differ by an integer (unity), the recurrence relations corresponding to the two roots yield two linearly independent power series solutions (one for each root), as expected from Section 7.3.

7.5 Obtaining a second solution

Whilst attempting to construct solutions to an ODE in the form of Frobenius series about a regular singular point, we found in the previous section that when the indicial equation has a repeated root, or roots differing by an integer, we can (in general) find only one solution of this form. In order to construct the general solution to the ODE, however, we require two linearly independent solutions y_1 and y_2. We now consider some methods for obtaining a second solution in this case.

7.5.1 The Wronskian method

If y_1 and y_2 are two linearly independent solutions of the standard equation

$$y'' + p(z)y' + q(z)y = 0$$

then the Wronskian of these two solutions is given by $W(z) = y_1 y_2' - y_2 y_1'$. Dividing the Wronskian by y_1^2 we find that we obtain an expression that can be written in the form of a total derivative:

$$\frac{W}{y_1^2} = \frac{y_2'}{y_1} - \frac{y_1'}{y_1^2} y_2 = \frac{y_2'}{y_1} + \left[\frac{d}{dz}\left(\frac{1}{y_1}\right)\right] y_2 = \frac{d}{dz}\left(\frac{y_2}{y_1}\right).$$

Since this RHS is a total derivative, the equation can be integrated, and gives

$$y_2(z) = y_1(z) \int^z \frac{W(u)}{y_1^2(u)}\, du.$$

Now using the alternative expression for $W(z)$ given in (7.4) with $C = 1$ (since we are not concerned with this normalizing factor), we find

$$y_2(z) = y_1(z) \int^z \frac{1}{y_1^2(u)} \exp\left\{-\int^u p(v)\, dv\right\} du. \tag{7.25}$$

Hence, given y_1, we can in principle compute y_2. Note that the lower limits of integration have been omitted. If constant lower limits are included then they merely lead to a constant times the first solution.

Example Find a second solution of equation (7.21) using the Wronskian method.

For the ODE (7.21) we have $p(z) = 3/(z-1)$, and from (7.24) we see that one solution to (7.21) is $y_1 = z/(1-z)^2$. Substituting for p and y_1 in (7.25) we have

$$\begin{aligned}
y_2(z) &= \frac{z}{(1-z)^2}\int^z \frac{(1-u)^4}{u^2}\exp\left(-\int^u \frac{3}{v-1}\,dv\right)du \\
&= \frac{z}{(1-z)^2}\int^z \frac{(1-u)^4}{u^2}\exp\left[-3\ln(u-1)\right]du \\
&= \frac{z}{(1-z)^2}\int^z \frac{u-1}{u^2}\,du \\
&= \frac{z}{(1-z)^2}\left(\ln z + \frac{1}{z}\right).
\end{aligned}$$

By calculating the Wronskian of y_1 and y_2 it could easily be shown that, as expected, the two solutions are linearly independent. In fact, as the Wronskian has already been evaluated as $W(u) = \exp[-3\ln(u-1)]$, i.e. as $W(z) = (z-1)^{-3}$, no further calculation is needed. ◀

An alternative (but equivalent) method of finding a second solution is simply to assume that the second solution has the form $y_2(z) = u(z)y_1(z)$ for some function $u(z)$ to be determined (this method was discussed more fully in Subsection 6.5.3). From (7.25), we see that the second solution derived from the Wronskian is indeed of this form. Substituting $y_2(z) = u(z)y_1(z)$ into the ODE leads to a first-order ODE in which u' is the dependent variable; this may then be solved, at least formally if not analytically.

7.5.2 The derivative method

The derivative method of finding a second solution begins with the derivation of a recurrence relation for the coefficients a_n in a Frobenius series solution, as in the previous section. However, rather than putting $\sigma = \sigma_1$ in this recurrence relation to evaluate the first series solution, we now keep σ as a variable parameter. This means that the computed a_n are functions of σ and the computed solution is now a function of z and σ:

$$y(z, \sigma) = z^\sigma \sum_{n=0}^{\infty} a_n(\sigma) z^n. \tag{7.26}$$

Of course, if we put $\sigma = \sigma_1$ in this, we obtain immediately the first series solution, but for the moment we leave σ as a parameter.

For brevity let us denote the differential operator on the LHS of our standard ODE (7.7) by $\mathcal{L}$, so that

$$\mathcal{L} = \frac{d^2}{dz^2} + p(z)\frac{d}{dz} + q(z),$$

and examine the effect of $\mathcal{L}$ on the series $y(z,\sigma)$ in (7.26). It is clear that the series $\mathcal{L}y(z,\sigma)$ will consist of a single term in z^σ, since the recurrence relation defining the $a_n(\sigma)$ was deliberately constructed so as to make the coefficients of higher powers of z in $\mathcal{L}y(z,\sigma)$ vanish. But the coefficient of z^σ is simply the LHS of the indicial equation. Therefore, if the roots of the indicial equation are $\sigma=\sigma_1$ and $\sigma=\sigma_2$, it follows that we can write $\mathcal{L}y(z,\sigma)$ in the form

$$\mathcal{L}y(z,\sigma)=a_0(\sigma-\sigma_1)(\sigma-\sigma_2)z^\sigma. \tag{7.27}$$

And so, as in the previous section, we see that for $y(z,\sigma)$ to be a solution of the ODE $\mathcal{L}y=0$, σ must equal σ_1 or σ_2; this applies whether the two values are the same, differ by an integer, or are unrelated. For simplicity we shall set $a_0=1$ in the following discussion.

Let us first consider the case in which the two roots of the indicial equation are equal, i.e. $\sigma_2=\sigma_1$. From (7.27) we then have

$$\mathcal{L}y(z,\sigma)=(\sigma-\sigma_1)^2z^\sigma.$$

Differentiating this equation with respect to σ we obtain[11]

$$\frac{\partial}{\partial\sigma}\left[\mathcal{L}y(z,\sigma)\right]=(\sigma-\sigma_1)^2z^\sigma\ln z+2(\sigma-\sigma_1)z^\sigma.$$

This is equal to zero if $\sigma=\sigma_1$, as would be expected for the derivative of a quadratic expression that has a repeated zero.

Now, since $\partial/\partial\sigma$ and $\mathcal{L}$ are operators that differentiate with respect to different variables, we can reverse their order and conclude that

$$\mathcal{L}\left[\frac{\partial}{\partial\sigma}y(z,\sigma)\right]=0 \qquad \text{at } \sigma=\sigma_1.$$

Hence, the function in square brackets, evaluated at $\sigma=\sigma_1$ and denoted by

$$\left[\frac{\partial}{\partial\sigma}y(z,\sigma)\right]_{\sigma=\sigma_1}, \tag{7.28}$$

is also a solution of the original ODE $\mathcal{L}y=0$, and is in fact the second linearly independent solution that we were looking for.

The case in which the roots of the indicial equation differ by an integer is rather more complicated, but can be treated in a similar way.[12] We give here only the final result, which is that the second linearly independent solution is

$$\left\{\frac{\partial}{\partial\sigma}\left[(\sigma-\sigma_2)y(z,\sigma)\right]\right\}_{\sigma=\sigma_2}, \tag{7.29}$$

where σ_2 is the smaller of the two roots, i.e. the one for which the straightforward series solution method fails. We now go back to complete a previous worked example for which this was the case.

11 Make sure that you can show that $dz^\sigma/d\sigma=z^\sigma\ln z$.

12 For a discussion of the method see, for example, K. F. Riley, *Mathematical Methods for the Physical Sciences* (Cambridge: Cambridge University Press, 1974), pp. 158–9.

Example Find a second solution of equation (7.21) using the derivative method.

From (7.23) the recurrence relation (with σ as a parameter) is given by

$$(n+\sigma-1)a_n = (n+\sigma)a_{n-1}.$$

Setting $a_0 = 1$ we find that the coefficients have the particularly simple form $a_n(\sigma) = (\sigma+n)/\sigma$. We therefore consider the function

$$y(z,\sigma) = z^\sigma \sum_{n=0}^{\infty} a_n(\sigma) z^n = z^\sigma \sum_{n=0}^{\infty} \frac{\sigma+n}{\sigma} z^n.$$

The smaller root of the indicial equation for (7.21) is $\sigma_2 = 0$, and so from (7.29) a second, linearly independent, solution to the ODE is given by

$$\left\{\frac{\partial}{\partial\sigma}[\sigma y(z,\sigma)]\right\}_{\sigma=0} = \left\{\frac{\partial}{\partial\sigma}\left[z^\sigma \sum_{n=0}^{\infty}(\sigma+n)z^n\right]\right\}_{\sigma=0}.$$

The derivative with respect to σ is given by

$$\frac{\partial}{\partial\sigma}\left[z^\sigma \sum_{n=0}^{\infty}(\sigma+n)z^n\right] = z^\sigma \ln z \sum_{n=0}^{\infty}(\sigma+n)z^n + z^\sigma \sum_{n=0}^{\infty} z^n,$$

which on setting $\sigma = 0$ gives the second solution

$$\begin{aligned} y_2(z) &= \ln z \sum_{n=0}^{\infty} n z^n + \sum_{n=0}^{\infty} z^n \\ &= \frac{z}{(1-z)^2}\ln z + \frac{1}{1-z} \\ &= \frac{z}{(1-z)^2}\left(\ln z + \frac{1}{z} - 1\right). \end{aligned}$$

This second solution is the same as that obtained by the Wronskian method in the previous subsection except for the addition of some of the first solution. ◀

7.5.3 Series form of the second solution

Using any of the methods discussed above, we can find the general form of the second solution to the ODE. Usually, this form is most easily found using the derivative method. Let us first consider the case where the two solutions of the indicial equation are equal. In this case a second solution is given by (7.28), which may be written as

$$\begin{aligned} y_2(z) &= \left[\frac{\partial y(z,\sigma)}{\partial\sigma}\right]_{\sigma=\sigma_1} \\ &= (\ln z) z^{\sigma_1} \sum_{n=0}^{\infty} a_n(\sigma_1) z^n + z^{\sigma_1} \sum_{n=1}^{\infty}\left[\frac{da_n(\sigma)}{d\sigma}\right]_{\sigma=\sigma_1} z^n \\ &= y_1(z)\ln z + z^{\sigma_1}\sum_{n=1}^{\infty} b_n z^n, \end{aligned} \qquad (7.30)$$

where $b_n = [da_n(\sigma)/d\sigma]_{\sigma=\sigma_1}$. One could equally obtain the coefficients b_n by direct substitution of the form (7.30) into the original ODE.

In the case where the roots of the indicial equation differ by an integer (not equal to zero), then from (7.29) a second solution is given by

$$y_2(z) = \left\{ \frac{\partial}{\partial \sigma} [(\sigma - \sigma_2) y(z, \sigma)] \right\}_{\sigma=\sigma_2}$$

$$= \ln z \left[(\sigma - \sigma_2) z^{\sigma} \sum_{n=0}^{\infty} a_n(\sigma) z^n \right]_{\sigma=\sigma_2} + z^{\sigma_2} \sum_{n=0}^{\infty} \left[\frac{d}{d\sigma} (\sigma - \sigma_2) a_n(\sigma) \right]_{\sigma=\sigma_2} z^n.$$

But, it can be shown[13] that $[(\sigma - \sigma_2) y(z, \sigma)]$ at $\sigma = \sigma_2$ is just a multiple of the first solution $y(z, \sigma_1)$. Therefore the second solution is of the form

$$y_2(z) = c y_1(z) \ln z + z^{\sigma_2} \sum_{n=0}^{\infty} b_n z^n, \tag{7.31}$$

where c is a constant. In some cases, however, c might be zero, and so the second solution would not contain the term in $\ln z$ and could be written simply as a Frobenius series. Clearly this corresponds to the case in which the substitution of a Frobenius series into the original ODE yields two solutions automatically. In either case, the coefficients b_n may also be found by direct substitution of the form (7.31) into the original ODE.

7.6 Polynomial solutions

We have seen that the evaluation of successive terms of a series solution to a differential equation is carried out by means of a recurrence relation. The form of the relation for a_n depends upon n, the previous values of a_r $(r < n)$ and the parameters of the equation. It may happen, as a result of this, that for some value of $n = N + 1$ the computed value a_{N+1} is zero and that all higher a_r also vanish. If this is so, and the corresponding solution of the indicial equation σ is a positive integer or zero, then we are left with a finite polynomial of degree $N' = N + \sigma$ as a solution of the ODE:

$$y(z) = \sum_{n=0}^{N} a_n z^{n+\sigma}. \tag{7.32}$$

In many applications in theoretical physics (particularly in quantum mechanics) the termination of a potentially infinite series after a finite number of terms is of crucial importance in establishing physically acceptable descriptions and properties of systems. The condition under which such a termination occurs is therefore of considerable importance. Consider the following example.

13 See footnote 12.

Example Find power series solutions about $z = 0$ of

$$y'' - 2zy' + \lambda y = 0. \tag{7.33}$$

For what values of λ does the equation possess a polynomial solution? Find such a solution for $\lambda = 4$.

Clearly $z = 0$ is an ordinary point of (7.33) and so we look for solutions of the form $y = \sum_{n=0}^{\infty} a_n z^n$. Substituting this into the ODE and multiplying through by z^2 we find

$$\sum_{n=0}^{\infty} [n(n-1) - 2z^2 n + \lambda z^2] a_n z^n = 0.$$

By demanding that the coefficients of each power of z vanish separately we derive the recurrence relation[14]

$$n(n-1)a_n - 2(n-2)a_{n-2} + \lambda a_{n-2} = 0,$$

which may be rearranged to give

$$a_n = \frac{2(n-2) - \lambda}{n(n-1)} a_{n-2} \qquad \text{for } n \geq 2. \tag{7.34}$$

This recurrence relation connects only alternate a_n, and so the odd and even coefficients are independent of one another, and two solutions to (7.33) may be derived. We either set $a_1 = 0$ and $a_0 = 1$ to obtain

$$y_1(z) = 1 - \lambda \frac{z^2}{2!} - \lambda(4 - \lambda)\frac{z^4}{4!} - \lambda(4 - \lambda)(8 - \lambda)\frac{z^6}{6!} - \cdots \tag{7.35}$$

or set $a_0 = 0$ and $a_1 = 1$ to obtain

$$y_2(z) = z + (2 - \lambda)\frac{z^3}{3!} + (2 - \lambda)(6 - \lambda)\frac{z^5}{5!} + (2 - \lambda)(6 - \lambda)(10 - \lambda)\frac{z^7}{7!} + \cdots.$$

Now, from the recurrence relation (7.34) (or in this case from the expressions for y_1 and y_2 themselves) we see that for the ODE to possess a polynomial solution we require $\lambda = 2(n - 2)$ for $n \geq 2$ or, more simply, $\lambda = 2n$ for $n \geq 0$, i.e. λ must be an even positive integer. If $\lambda = 4$ then from (7.35) the ODE has the polynomial solution

$$y_1(z) = 1 - \frac{4z^2}{2!} = 1 - 2z^2.$$

This can be confirmed trivially by re-substitution. ◀

A simpler method of obtaining finite polynomial solutions is to *assume* a solution of the form (7.32), where $a_N \neq 0$. Instead of starting with the lowest power of z, as we have done up to now, this time we start by considering the coefficient of the highest power z^N; such a power now exists because of our assumed form of solution.[15]

14 Do this for yourself, noting how, for any fixed power of z, the nth power in this instance, the powers of z that appear explicitly in the expression in square brackets affect the value r of the subscript of the appropriate coefficient, a_r.

15 Of course, if, in fact, the equation has no polynomial solutions, then a non-integer, or zero, or negative value is found for N.

Example By assuming a polynomial solution find the values of λ in (7.33) for which such a solution exists.

We assume a polynomial solution to (7.33) of the form $y = \sum_{n=0}^{N} a_n z^n$. Substituting this form into (7.33) we find

$$\sum_{n=0}^{N} \left[n(n-1)a_n z^{n-2} - 2zna_n z^{n-1} + \lambda a_n z^n \right] = 0.$$

Now, instead of starting with the lowest power of z, we start with the highest. Thus, demanding that the coefficient of z^N vanishes, we require $-2N + \lambda = 0$, i.e. $\lambda = 2N$, as we found in the previous example. By demanding that the coefficient of a general power of z is zero, the same recurrence relation as above may be derived and the solutions found. ◀

SUMMARY

With $\mathcal{L}y \equiv y'' + p(z)y' + q(z)y = 0$ as the standard form.

1. *Ordinary and singular points of the equation*
   The function $f(z)$ is analytic at the point $z = z_0$ if it can be expressed as a power series $f(z) = \sum_{n=0}^{\infty} a_n (z - z_0)^n$.

Point type at z_0	Analytic	Not analytic
Ordinary	$p(z)$ and $q(z)$	–
Singular	–	$p(z)$ or $q(z)$
Regular singularity	$(z - z_0)p(z)$ and $(z - z_0)^2 q(z)$	$p(z)$ or $q(z)$
Essential singularity	–	$(z - z_0)p(z)$ or $(z - z_0)^2 q(z)$

   - The equation has a singular point at infinity if the equation obtained by making the substitution $w = 1/z$ has a singular point at $w = 0$.
   - For the singularities of some important equations, see Table 7.1 on p. 277.

2. *Series solutions about the origin* ($z_0 = 0$)
   If the expansion point is not the origin, make it so, using the substitution $Z = z - z_0$.
   - If the origin is an ordinary point, there are two linearly independent solutions of $\mathcal{L}y = 0$ of the form $y(z) = \sum_{n=0}^{\infty} a_n z^n$, corresponding to two different pairs of values for a_0 and a_1.

- If the origin is a regular singular point, there is at least one (but not more than two) solutions of the form $y(z,\sigma) = z^\sigma \sum_{n=0}^{\infty} a_n z^n$ with $a_0 \neq 0$. Substituting this into $\mathcal{L}y = 0$ and separately equating to zero the coefficient of each power of z yields:
  (i) The indicial equation, a quadratic equation in σ (from the coefficient of the lowest power of z present).
  (ii) The recurrence relation involving two or more successive expansion coefficients a_n (from the coefficient of a general power of z).
  (iii) The radius of convergence of the series solution (from the ratio of successive terms actually present in the series). This is always equal to the distance from the origin of the (next) nearest singular point.
  (iv) An indication of whether or not the equation has polynomial solutions (i.e. the recurrence relation shows that $a_n = 0$ for all n greater than some N). Whether this happens usually depends on the value of a parameter contained in $q(z)$.

3. *Forms of the second solution in particular cases*

The general form is $y(z,\sigma) \equiv z^\sigma \sum_{n=0}^{\infty} a_n z^n$. The first solution is $y_1(z) \equiv y(z,\sigma_1)$. Here $\sigma_1 \geq \sigma_2$ and m is a positive integer.

Case	Calculation of $y_2(z)$	Form of $y_2(z)$
$\sigma_1 \neq \sigma_2 + m$	$y(z,\sigma_2)$	$z^{\sigma_2} \sum_{n=0}^{\infty} a_n z^n$
$\sigma_1 = \sigma_2$	$\left[\dfrac{\partial y(z,\sigma)}{\partial \sigma}\right]_{\sigma=\sigma_1}$	$y_1(z)\ln z + z^{\sigma_1} \sum_{n=1}^{\infty} b_n z^n$
$\sigma_1 = \sigma_2 + m$	$\left\{\dfrac{\partial[(\sigma-\sigma_2)y(z,\sigma)]}{\partial \sigma}\right\}_{\sigma=\sigma_2}$	$c y_1(z)\ln z + z^{\sigma_2} \sum_{n=0}^{\infty} b_n z^n$
$\sigma_1 = \sigma_2$ or $\sigma_1 = \sigma_2 + m$	Wronskian method	$y_1(z) \int^z \dfrac{g(u)}{y_1^2(u)}\, du$, where $g(u) = \exp\left\{-\int^u p(v)\, dv\right\}$

PROBLEMS

7.1. Find two power series solutions about $z = 0$ of the differential equation

$$(1 - z^2)y'' - 3zy' + \lambda y = 0.$$

Deduce that the value of λ for which the corresponding power series becomes an Nth-degree polynomial $U_N(z)$ is $N(N+2)$. Construct $U_2(z)$ and $U_3(z)$.

7.2. Find solutions, as power series in z, of the equation

$$4zy'' + 2(1 - z)y' - y = 0.$$

Identify one of the solutions and verify it by direct substitution.

7.3. Find power series solutions in z of the differential equation

$$zy'' - 2y' + 9z^5 y = 0.$$

Identify closed forms for the two series, calculate their Wronskian, and verify that they are linearly independent. Compare the Wronskian with that calculated from the differential equation.

7.4. Change the independent variable in the equation

$$\frac{d^2 f}{dz^2} + 2(z - a)\frac{df}{dz} + 4f = 0 \qquad (*)$$

from z to $x = z - \alpha$, and find two independent series solutions, expanded about $x = 0$, of the resulting equation. Deduce that the general solution of $(*)$ is

$$f(z, \alpha) = A(z - \alpha)e^{-(z-\alpha)^2} + B\sum_{m=0}^{\infty} \frac{(-4)^m m!}{(2m)!}(z - \alpha)^{2m},$$

with A and B arbitrary constants.

7.5. Investigate solutions of Legendre's equation at one of its singular points as follows.

(a) Verify that $z = 1$ is a regular singular point of Legendre's equation and that the indicial equation for a series solution in powers of $(z - 1)$ has a double root at $\sigma = 0$.

(b) Obtain the corresponding recurrence relation and show that a polynomial solution is obtained if ℓ is a positive integer.

(c) Determine the radius of convergence R of the $\sigma = 0$ series and relate it to the positions of the singularities of Legendre's equation.

7.6. Verify that $z = 0$ is a regular singular point of the equation

$$z^2 y'' - \tfrac{3}{2}zy' + (1 + z)y = 0,$$

and that the indicial equation has roots 2 and 1/2. Show that the general solution is given by

$$y(z) = 6a_0 z^2 \sum_{n=0}^{\infty} \frac{(-1)^n (n+1)2^{2n} z^n}{(2n+3)!}$$

$$+ b_0\left(z^{1/2} + 2z^{3/2} - \frac{z^{1/2}}{4}\sum_{n=2}^{\infty} \frac{(-1)^n 2^{2n} z^n}{n(n-1)(2n-3)!}\right).$$

7.7. Use the derivative method to obtain, as a second (independent) solution of Bessel's equation for the case when $\nu = 0$, the following expression:

$$J_0(z)\ln z - \sum_{n=1}^{\infty} \frac{(-1)^n}{(n!)^2} \left(\sum_{r=1}^{n} \frac{1}{r} \right) \left(\frac{z}{2} \right)^{2n},$$

given that the first solution is $J_0(z)$, as specified by (9.76).

7.8. Consider a series solution of the equation

$$zy'' - 2y' + yz = 0 \qquad (*)$$

about its regular singular point.

(a) Show that its indicial equation has roots that differ by an integer but that the two roots nevertheless generate linearly independent solutions

$$y_1(z) = 3a_0 \sum_{n=1}^{\infty} \frac{(-1)^{n+1}\, 2n z^{2n+1}}{(2n+1)!},$$

$$y_2(z) = a_0 \sum_{n=0}^{\infty} \frac{(-1)^{n+1}(2n-1)z^{2n}}{(2n)!}.$$

(b) Show that $y_1(z)$ is equal to $3a_0(\sin z - z\cos z)$ by expanding the sinusoidal functions. Then, using the Wronskian method, find an expression for $y_2(z)$ in terms of sinusoids. You will need to write z^2 as $(z/\sin z)(z\sin z)$ and integrate by parts to evaluate the integral involved.

(c) Confirm that the two solutions are linearly independent by showing that their Wronskian is equal to $-z^2$, as would be expected from the form of $(*)$.

7.9. Find series solutions of the equation $y'' - 2zy' - 2y = 0$. Identify one of the series as $y_1(z) = \exp z^2$ and verify this by direct substitution. By setting $y_2(z) = u(z)y_1(z)$ and solving the resulting equation for $u(z)$, find an explicit form for $y_2(z)$ and deduce that

$$\int_0^x e^{-v^2}\,dv = e^{-x^2} \sum_{n=0}^{\infty} \frac{n!}{2(2n+1)!}(2x)^{2n+1}.$$

7.10. Find the radius of convergence of a series solution about the origin for the equation $(z^2 + az + b)y'' + 2y = 0$ in the following cases:

(a) $a = 5,\ b = 6$; (b) $a = 5,\ b = 7$.

Show that if a and b are real and $4b > a^2$, then the radius of convergence is always given by $b^{1/2}$.

7.11. For the equation $y'' + z^{-3}y = 0$, show that the origin becomes a regular singular point if the independent variable is changed from z to $x = 1/z$. Hence find a series solution of the form $y_1(z) = \sum_0^\infty a_n z^{-n}$. By setting $y_2(z) = u(z)y_1(z)$ and expanding the resulting expression for du/dz in powers of z^{-1}, show that $y_2(z)$ has the asymptotic form

$$y_2(z) = c\left[z + \ln z - \tfrac{1}{2} + O\left(\frac{\ln z}{z}\right)\right],$$

where c is an arbitrary constant.

7.12. Prove that the Laguerre equation,

$$z\frac{d^2y}{dz^2} + (1-z)\frac{dy}{dz} + \lambda y = 0,$$

has polynomial solutions $L_N(z)$ if λ is a non-negative integer N, and determine the recurrence relationship for the polynomial coefficients. Hence show that an expression for $L_N(z)$, normalized in such a way that $L_N(0) = N!$, is

$$L_N(z) = \sum_{n=0}^{N} \frac{(-1)^n (N!)^2}{(N-n)!(n!)^2} z^n.$$

Evaluate $L_3(z)$ explicitly.

7.13. The origin is an ordinary point of the Chebyshev equation,

$$(1-z^2)y'' - zy' + m^2 y = 0,$$

which therefore has series solutions of the form $z^\sigma \sum_0^\infty a_n z^n$ for $\sigma = 0$ and $\sigma = 1$.

(a) Find the recurrence relationships for the a_n in the two cases and show that there exist polynomial solutions $T_m(z)$:

(i) for $\sigma = 0$, when m is an even integer, the polynomial having $\frac{1}{2}(m+2)$ terms;

(ii) for $\sigma = 1$, when m is an odd integer, the polynomial having $\frac{1}{2}(m+1)$ terms.

(b) $T_m(z)$ is normalized so as to have $T_m(1) = 1$. Find explicit forms for $T_m(z)$ for $m = 0, 1, 2, 3$.

(c) Show that the corresponding non-terminating series solutions $S_m(z)$ have as their first few terms

$$S_0(z) = a_0\left(z + \frac{1}{3!}z^3 + \frac{9}{5!}z^5 + \cdots\right),$$

$$S_1(z) = a_0\left(1 - \frac{1}{2!}z^2 - \frac{3}{4!}z^4 - \cdots\right),$$

$$S_2(z) = a_0\left(z - \frac{3}{3!}z^3 - \frac{15}{5!}z^5 - \cdots\right),$$

$$S_3(z) = a_0\left(1 - \frac{9}{2!}z^2 + \frac{45}{4!}z^4 + \cdots\right).$$

7.14. Obtain the recurrence relations for the solution of Legendre's equation (9.1) in *inverse* powers of z, i.e. set $y(z) = \sum a_n z^{\sigma-n}$, with $a_0 \neq 0$. Deduce that, if ℓ is an integer, then the series with $\sigma = \ell$ will terminate and hence converge for all z, whilst the series with $\sigma = -(\ell+1)$ does not terminate and hence converges only for $|z| > 1$.

HINTS AND ANSWERS

7.1. Note that $z = 0$ is an ordinary point of the equation.
For $\sigma = 0$, $a_{n+2}/a_n = [n(n+2) - \lambda]/[(n+1)(n+2)]$ and, correspondingly, for $\sigma = 1$, $U_2(z) = a_0(1 - 4z^2)$ and $U_3(z) = a_0(z - 2z^3)$.

7.3. $\sigma = 0$ and 3; $a_{6m}/a_0 = (-1)^m/(2m)!$ and $a_{6m}/a_0 = (-1)^m/(2m+1)!$, respectively. $y_1(z) = a_0 \cos z^3$ and $y_2(z) = a_0 \sin z^3$. The Wronskian is $\pm 3a_0^2 z^2 \neq 0$.

7.5. (b) $a_{n+1}/a_n = [\,\ell(\ell+1) - n(n+1)\,]/[\,2(n+1)^2\,]$.
(c) $R = 2$, equal to the distance between $z = 1$ and the closest singularity at $z = -1$.

7.7. A typical term in the series for $y(\sigma, z)$ is $\dfrac{(-1)^n z^{2n}}{[\,(\sigma+2)(\sigma+4)\cdots(\sigma+2n)\,]^2}$.

7.9. The origin is an ordinary point. Determine the constant of integration by examining the behavior of the related functions for small x.
$y_2(z) = (\exp z^2)\int_0^z \exp(-x^2)\,dx$.

7.11. The transformed equation is $xy'' + 2y' + y = 0$; $a_n = (-1)^n(n+1)^{-1}(n!)^{-2}a_0$; $du/dz = A[\,y_1(z)\,]^{-2}$.

7.13. (a) (i) $a_{n+2} = [a_n(n^2 - m^2)]/[(n+2)(n+1)]$,
(ii) $a_{n+2} = \{a_n[(n+1)^2 - m^2]\}/[(n+3)(n+2)]$; (b) $1, z, 2z^2 - 1, 4z^3 - 3z$.

8 Eigenfunction methods for differential equations

In the two previous chapters we dealt with the solution of differential equations of order n by two different methods. In one method, we found n independent solutions of the equation and then combined them, weighted with coefficients determined by the boundary conditions; in the other we found solutions in terms of series whose coefficients were related by (in general) an n-term recurrence relation and thence fixed by the boundary conditions. For both approaches the linearity of the equation was an important or essential factor in the utility of the method, and in this chapter our aim will be to exploit the superposition properties of linear differential equations even further.

We will be concerned with the solution of equations of the inhomogeneous form

$$\mathcal{L}y(x) = f(x), \tag{8.1}$$

where $f(x)$ is a prescribed or general function *and* the boundary conditions to be satisfied by the solution $y = y(x)$, for example at the limits $x = a$ and $x = b$, are given. The expression $\mathcal{L}y(x)$ stands for a linear differential operator $\mathcal{L}$ acting upon the function $y(x)$.[1]

In general, unless $f(x)$ is both known and simple, it will not be possible to find particular integrals of (8.1), even if complementary functions can be found that satisfy $\mathcal{L}y = 0$. The idea is therefore to exploit the linearity of $\mathcal{L}$ by building up the required solution $y(x)$ as a *superposition*, generally containing an infinite number of terms, of some set of functions $\{y_i(x)\}$ that each individually satisfy the boundary conditions. Clearly this brings in a quite considerable complication but since, within reason, we may select the set of functions to suit ourselves, we can obtain sizeable compensation for this complication. Indeed, if the set chosen is one containing functions that, when acted upon by $\mathcal{L}$, produce particularly simple results then we can "show a profit" on the operation. In particular, if the set consists of those functions y_i which satisfy the boundary conditions, and for which

$$\mathcal{L}y_i(x) = \lambda_i y_i(x), \tag{8.2}$$

where λ_i is a constant, then a distinct advantage may be obtained from the maneuver because all the differentiation will have disappeared from (8.1).

1 For example, in the Legendre equation $(1 - x^2)y'' - 2xy' + \lambda y = f(x)$, the operator $\mathcal{L}$ is the differential *operator* $(1 - x^2)d^2/dx^2 - 2x\,d/dx + \lambda$, though, as will be discussed shortly, only the terms generating derivatives are essential, as the constant term λ can be treated as part of an eigenvalue.

Equation (8.2) is clearly reminiscent of the equation satisfied by the *eigenvectors* $\mathbf{x}^i$ of a linear operator $\mathcal{A}$, namely

$$\mathcal{A}\mathbf{x}^i = \lambda_i \mathbf{x}^i, \tag{8.3}$$

where λ_i is a constant and is called the *eigenvalue* associated with $\mathbf{x}^i$. By analogy, in the context of differential equations a function $y_i(x)$ satisfying (8.2) is called an *eigenfunction* of the operator $\mathcal{L}$ (under the imposed boundary conditions) and λ_i is then called the eigenvalue associated with the eigenfunction $y_i(x)$. Clearly, the eigenfunctions $y_i(x)$ of $\mathcal{L}$ are only determined up to an arbitrary scale factor by (8.2).

Probably the most familiar equation of the form (8.2) is that which describes a simple harmonic oscillator, i.e.

$$\mathcal{L}y \equiv -\frac{d^2y}{dt^2} = \omega^2 y, \quad \text{where } \mathcal{L} \equiv -d^2/dt^2. \tag{8.4}$$

Imposing the boundary condition that the solution is periodic with period T, the eigenfunctions in this case are given by $y_n(t) = A_n e^{i\omega_n t}$, where $\omega_n = 2\pi n/T$, $n = 0, \pm 1, \pm 2, \ldots$ and the A_n are constants. The eigenvalues[2] are $\omega_n^2 = n^2\omega_1^2 = n^2(2\pi/T)^2$.

We may discuss a somewhat wider class of differential equations by considering a slightly more general form of (8.2), namely

$$\mathcal{L}y_i(x) = \lambda_i \rho(x) y_i(x), \tag{8.5}$$

where $\rho(x)$ is a *weight function*. In many applications $\rho(x)$ is unity for all x, in which case (8.2) is recovered; in general, though, it is a function determined by the choice of coordinate system used in describing a particular physical situation. The only requirement on $\rho(x)$ is that it is real and does not change sign in the range $a \le x \le b$, so that it can, without loss of generality, be taken to be non-negative throughout; of course, $\rho(x)$ must be the same function for all values of λ_i. A function $y_i(x)$ that satisfies (8.5) is called an eigenfunction of the operator $\mathcal{L}$ with respect to the weight function $\rho(x)$.

This chapter will not cover methods used to determine the eigenfunctions of (8.2) or (8.5), since we have discussed those in previous chapters, but, rather, will use the properties of the eigenfunctions to solve inhomogeneous equations of the form (8.1). We shall see later that the sets of eigenfunctions $y_i(x)$ of a particular class of operators called *Hermitian operators* (the operator in the simple harmonic oscillator equation is an example) have particularly useful properties and these will be studied in detail. It turns out that many of the interesting differential operators met within the physical sciences are Hermitian.

Before continuing our investigation of the eigenfunctions of Hermitian operators, however, we discuss in the next section some properties of general sets of functions. The material discussed is somewhat more formal than that contained in most of this book and could be omitted on a first reading.

2 Sometimes ω_n is referred to as the eigenvalue of this equation, but we will avoid this confusing terminology.

8.1 Sets of functions

In Chapter 1 we discussed the definition of a vector space but concentrated on spaces of finite dimensionality. We consider now the *infinite*-dimensional space of all reasonably well-behaved functions $f(x)$, $g(x)$, $h(x)$, ... on the interval $a \le x \le b$. That these functions form a linear vector space is shown by noting the following properties. The set is closed under

(i) addition, which is commutative and associative, i.e.

$$f(x) + g(x) = g(x) + f(x),$$
$$[f(x) + g(x)] + h(x) = f(x) + [g(x) + h(x)],$$

(ii) multiplication by a scalar, which is distributive and associative, i.e.

$$\lambda [f(x) + g(x)] = \lambda f(x) + \lambda g(x),$$
$$\lambda [\mu f(x)] = (\lambda\mu) f(x),$$
$$(\lambda + \mu) f(x) = \lambda f(x) + \mu f(x).$$

Furthermore, in such a space

(iii) there exists a "null vector" 0 such that $f(x) + 0 = f(x)$,
(iv) multiplication by unity leaves any function unchanged, i.e. $1 \times f(x) = f(x)$,
(v) each function has an associated negative function $-f(x)$ such that $f(x) + [-f(x)] = 0$.

By analogy with finite-dimensional vector spaces we now introduce a set of linearly independent basis functions $y_n(x)$, $n = 0, 1, \ldots, \infty$, such that *any* "reasonable" function in the interval $a \le x \le b$ (i.e. it obeys the Dirichlet conditions discussed in Chapter 4) can be expressed as the linear sum of these functions:

$$f(x) = \sum_{n=0}^{\infty} c_n y_n(x).$$

Clearly if a different set of linearly independent basis functions $u_n(x)$ is chosen then the function can be expressed in terms of the new basis,

$$f(x) = \sum_{n=0}^{\infty} d_n u_n(x),$$

where the d_n are a different set of coefficients. In each case, provided the basis functions are linearly independent, the coefficients are unique.

We may also define an *inner product* on our function space by

$$\langle f|g\rangle = \int_a^b f^*(x) g(x) \rho(x)\, dx, \qquad (8.6)$$

where $\rho(x)$ is the weight function, which we require to be real and non-negative in the interval $a \le x \le b$. As mentioned above, $\rho(x)$ is often unity for all x. Two functions are said to be *orthogonal* [with respect to the weight function $\rho(x)$] on the interval $[a, b]$ if

$$\langle f|g\rangle = \int_a^b f^*(x)g(x)\rho(x)\,dx = 0, \tag{8.7}$$

and the *norm* of a function is defined as

$$||f|| = \langle f|f\rangle^{1/2} = \left[\int_a^b f^*(x)f(x)\rho(x)\,dx\right]^{1/2} = \left[\int_a^b |f(x)|^2\rho(x)\,dx\right]^{1/2}. \tag{8.8}$$

It is also common practice to define a *normalized* function by $\hat{f} = f/\,||f||$, which has unit norm.

An infinite-dimensional vector space of functions, for which an inner product is defined, is called a *Hilbert space*. Using the concept of the inner product, we can choose a basis of linearly independent functions $\hat{\phi}_n(x)$, $n = 0, 1, 2, \ldots$ that are orthonormal, i.e. such that

$$\left\langle\hat{\phi}_i|\hat{\phi}_j\right\rangle = \int_a^b \hat{\phi}_i^*(x)\hat{\phi}_j(x)\rho(x)\,dx = \delta_{ij}. \tag{8.9}$$

If $y_n(x)$, $n = 0, 1, 2, \ldots$, are a linearly independent, but not orthonormal, basis for the Hilbert space then an orthonormal set of basis functions $\hat{\phi}_n$ may be produced using the Gram–Schmidt procedure (i.e. in a similar manner to that used in the construction of a set of orthogonal eigenvectors of an Hermitian matrix; see Chapter 1 and Appendix F) as follows:

$$\begin{aligned}
\phi_0 &= y_0,\\
\phi_1 &= y_1 - \hat{\phi}_0\left\langle\hat{\phi}_0|y_1\right\rangle,\\
\phi_2 &= y_2 - \hat{\phi}_1\left\langle\hat{\phi}_1|y_2\right\rangle - \hat{\phi}_0\left\langle\hat{\phi}_0|y_2\right\rangle,\\
&\vdots\\
\phi_n &= y_n - \hat{\phi}_{n-1}\left\langle\hat{\phi}_{n-1}|y_n\right\rangle - \cdots - \hat{\phi}_0\left\langle\hat{\phi}_0|y_n\right\rangle,\\
&\vdots
\end{aligned}$$

It is straightforward to check that each ϕ_n is orthogonal to all its predecessors ϕ_i, $i = 0, 1, 2, \ldots, n-1$ and so the functions ϕ_n form an orthogonal set. In general they do not have unit norms, but can clearly be made to do so using individual normalization factors, thus generating an orthonormal set.

Example Starting from the linearly independent functions $y_n(x) = x^n$, $n = 0, 1, \ldots$, construct three orthonormal functions over the range $-1 < x < 1$, assuming a weight function of unity.

The first unnormalized function ϕ_0 is simply equal to the first of the original functions, i.e.

$$\phi_0 = 1.$$

The normalization is carried out by dividing by

$$\langle\phi_0|\phi_0\rangle^{1/2} = \left(\int_{-1}^{1} 1 \times 1\, du\right)^{1/2} = \sqrt{2},$$

with the result that the first normalized function $\hat{\phi}_0$ is given by

$$\hat{\phi}_0 = \frac{\phi_0}{\sqrt{2}} = \sqrt{\frac{1}{2}}.$$

The second unnormalized function is found by applying the above Gram–Schmidt orthogonalization procedure, i.e.

$$\phi_1 = y_1 - \hat{\phi}_0 \langle\hat{\phi}_0|y_1\rangle.$$

It can easily be shown that $\langle\hat{\phi}_0|y_1\rangle = 0$, and so $\phi_1 = x$. Normalizing then gives

$$\hat{\phi}_1 = \phi_1 \left(\int_{-1}^{1} u \times u\, du\right)^{-1/2} = \sqrt{\tfrac{3}{2}}x.$$

The third unnormalized function is similarly given by

$$\begin{aligned}\phi_2 &= y_2 - \hat{\phi}_1 \langle\hat{\phi}_1|y_2\rangle - \hat{\phi}_0 \langle\hat{\phi}_0|y_2\rangle \\ &= x^2 - 0 - \tfrac{1}{3},\end{aligned}$$

which, on normalizing, gives

$$\hat{\phi}_2 = \phi_2 \left(\int_{-1}^{1} \left(u^2 - \tfrac{1}{3}\right)^2 du\right)^{-1/2} = \tfrac{1}{2}\sqrt{\tfrac{5}{2}}(3x^2 - 1).$$

Comparison of the functions $\hat{\phi}_0$, $\hat{\phi}_1$ and $\hat{\phi}_2$ with the list in Subsection 9.1.1, shows that this procedure generates (multiples of)[3] the first three Legendre polynomials. ◀

If a function is expressed in terms of an *orthonormal* basis $\hat{\phi}_n(x)$ as

$$f(x) = \sum_{n=0}^{\infty} c_n \hat{\phi}_n(x) \tag{8.10}$$

then the coefficients c_n are given by

$$c_n = \langle\hat{\phi}_n|f\rangle = \int_a^b \hat{\phi}_n^*(x) f(x) \rho(x)\, dx. \tag{8.11}$$

Note that this is true only if the basis is orthonormal.

Since for a Hilbert space $\langle f|f\rangle \geq 0$, the inequalities discussed in Subsection 1.1.3 hold. The proofs are not repeated here, but the relationships are listed for completeness.

(i) The Schwarz inequality states that

$$|\langle f|g\rangle| \leq \langle f|f\rangle^{1/2} \langle g|g\rangle^{1/2}, \tag{8.12}$$

3 For largely historical reasons, the normalization of the Legendre polynomials is not the "natural" one found in this example, but one determined by the requirement that $P_\ell(1) = 1$. The "naturally normalized" functions are a factor of $\sqrt{(2\ell + 1)/2}$ greater than the conventionally defined Legendre polynomials.

where the equality holds when $f(x)$ is a scalar multiple of $g(x)$, i.e. when they are linearly dependent.

(ii) The triangle inequality states that

$$||f+g|| \leq ||f|| + ||g||, \tag{8.13}$$

where again equality holds when $f(x)$ is a scalar multiple of $g(x)$.

(iii) Bessel's inequality requires the introduction of an *orthonormal* basis $\hat{\phi}_n(x)$ so that any function $f(x)$ can be written as

$$f(x) = \sum_{n=0}^{\infty} c_n \hat{\phi}_n(x),$$

where $c_n = \langle \hat{\phi}_n | f \rangle$. Bessel's inequality then states that

$$\langle f | f \rangle \geq \sum_n |c_n|^2. \tag{8.14}$$

The equality holds if the summation is over all the basis functions. If some values of n are omitted from the sum then the inequality results (unless, of course, the c_n happen to be zero for all values of n omitted, in which case the equality remains).

8.2 Adjoint, self-adjoint and Hermitian operators

Having discussed general sets of functions, we now return to the discussion of eigenfunctions of linear operators. We begin by introducing the *adjoint* of an operator $\mathcal{L}$, denoted by $\mathcal{L}^\dagger$, which is defined by

$$\int_a^b f^*(x) \left[\mathcal{L}g(x)\right] dx = \int_a^b [\mathcal{L}^\dagger f(x)]^* g(x)\, dx + \text{boundary terms}, \tag{8.15}$$

where the boundary terms are evaluated at the end-points of the interval $[a, b]$. Thus, for any given linear differential operator $\mathcal{L}$, the adjoint operator $\mathcal{L}^\dagger$ can be found by repeated integration by parts; this is so because each additional integration by parts transfers one further differentiation from $g(x)$ to $f(x)$. If the highest order differentiation appearing in $\mathcal{L}$ is the nth, then after n integrations all of the differentiations will be acting upon f. The definite integrals in the successive integrations generate the boundary value terms in (8.15).

An operator is said to be *self-adjoint* if $\mathcal{L}^\dagger = \mathcal{L}$. If, in addition, certain boundary conditions are met by the functions f and g on which a self-adjoint operator acts, or by the operator itself, such that the boundary terms in (8.15) vanish, then the operator is said to be *Hermitian* over the interval $a \leq x \leq b$. Thus, in this case,

$$\int_a^b f^*(x) \left[\mathcal{L}g(x)\right] dx = \int_a^b [\mathcal{L}f(x)]^* g(x)\, dx. \tag{8.16}$$

A little careful study will reveal the similarity between the definition of an Hermitian operator and the definition of an Hermitian matrix given in Chapter 1.

Example Show that the linear operator $\mathcal{L} = d^2/dt^2$ is self-adjoint, and determine the required boundary conditions for the operator to be Hermitian over the interval t_0 to $t_0 + T$.

Substituting into the LHS of the definition of the adjoint operator (8.15) and integrating by parts gives

$$\int_{t_0}^{t_0+T} f^* \frac{d^2 g}{dt^2}\, dt = \left[f^* \frac{dg}{dt} \right]_{t_0}^{t_0+T} - \int_{t_0}^{t_0+T} \frac{df^*}{dt} \frac{dg}{dt}\, dt.$$

Integrating the second term on the RHS by parts once more yields

$$\int_{t_0}^{t_0+T} f^* \frac{d^2 g}{dt^2}\, dt = \left[f^* \frac{dg}{dt} \right]_{t_0}^{t_0+T} + \left[-\frac{df^*}{dt} g \right]_{t_0}^{t_0+T} + \int_{t_0}^{t_0+T} g \frac{d^2 f^*}{dt^2}\, dt,$$

which, by comparison with (8.15), proves that $\mathcal{L}$ is a self-adjoint operator. Moreover, from (8.16), we see that $\mathcal{L}$ is an Hermitian operator over the required interval provided

$$\left[f^* \frac{dg}{dt} \right]_{t_0}^{t_0+T} = \left[\frac{df^*}{dt} g \right]_{t_0}^{t_0+T}.$$

This would be the case if, for example, the set of functions to which f and g belonged were all those of the form $Ae^{-\alpha t^2}$ with $\alpha > 0$ over the range defined by $t_0 = 0$ and $T = \infty$.[4] ◀

We showed in Chapter 1 that the eigenvalues of Hermitian matrices are real and that their eigenvectors can be chosen to be orthogonal. Similarly, the eigenvalues of Hermitian operators are real and their eigenfunctions can be chosen to be orthogonal; we will prove these properties in the following section. Hermitian operators (or matrices) are often used in the formulation of quantum mechanics. The eigenvalues then give the possible measured values of an observable quantity such as energy or angular momentum, and the physical requirement that such quantities must be real is ensured by the reality of these eigenvalues.

Furthermore, the infinite set of eigenfunctions of an Hermitian operator form a complete basis set over the relevant interval, so that it is possible to expand any function $y(x)$ obeying the appropriate conditions in an eigenfunction series over this interval:

$$y(x) = \sum_{n=0}^{\infty} c_n y_n(x), \tag{8.17}$$

where the choice of suitable values for the c_n will make the sum arbitrarily close to $y(x)$.[5] These useful properties provide the motivation for a detailed study of Hermitian operators.

4 Suggest a set of functions that would make $\mathcal{L}$ Hermitian if $t_0 = 0$ and $T = 2\pi$.

5 The proof of the completeness of the eigenfunctions of an Hermitian operator is beyond the scope of this book. The reader should refer, for example, to R. Courant and D. Hilbert, *Methods of Mathematical Physics* (New York: Interscience, 1953).

8.3 Properties of Hermitian operators

We now provide proofs of some of the useful properties of Hermitian operators. Again much of the analysis is similar to that for Hermitian matrices in Chapter 1, although the present section stands alone. (Here, and throughout the remainder of this chapter, we will write out inner products in full. We note, however, that the inner product notation often provides a neat form in which to express results.)

8.3.1 Reality of the eigenvalues

Consider an Hermitian operator for which (8.5) is satisfied by at least two eigenfunctions $y_i(x)$ and $y_j(x)$, which have corresponding eigenvalues λ_i and λ_j, so that

$$\mathcal{L}y_i = \lambda_i \rho(x) y_i, \tag{8.18}$$

$$\mathcal{L}y_j = \lambda_j \rho(x) y_j, \tag{8.19}$$

where we have allowed for the presence of a weight function $\rho(x)$. Multiplying (8.18) by y_j^* and (8.19) by y_i^* and then integrating gives

$$\int_a^b y_j^* \mathcal{L} y_i \, dx = \lambda_i \int_a^b y_j^* y_i \rho \, dx, \tag{8.20}$$

$$\int_a^b y_i^* \mathcal{L} y_j \, dx = \lambda_j \int_a^b y_i^* y_j \rho \, dx. \tag{8.21}$$

Remembering that we have required $\rho(x)$ to be real, the complex conjugate of (8.20) becomes

$$\int_a^b y_j (\mathcal{L} y_i)^* \, dx = \lambda_i^* \int_a^b y_i^* y_j \rho \, dx, \tag{8.22}$$

and using the definition of an Hermitian operator (8.16) it follows that the LHS of (8.22) is equal to the LHS of (8.21). Thus

$$(\lambda_i^* - \lambda_j) \int_a^b y_i^* y_j \rho \, dx = 0. \tag{8.23}$$

If $i = j$ then $\lambda_i = \lambda_i^*$ (since $\int_a^b y_i^* y_i \rho \, dx \neq 0$), which is a statement that the eigenvalue λ_i is real.

8.3.2 Orthogonality and normalization of the eigenfunctions

From (8.23), it is immediately apparent that two eigenfunctions y_i and y_j that correspond to different eigenvalues, i.e. such that $\lambda_i \neq \lambda_j$, satisfy

$$\int_a^b y_i^* y_j \rho \, dx = 0, \tag{8.24}$$

which is a statement of the orthogonality of y_i and y_j.

If one (or more) of the eigenvalues is degenerate, however, we have different eigenfunctions corresponding to the same eigenvalue, and the proof of orthogonality is not so straightforward. Nevertheless, an orthogonal set of eigenfunctions may be constructed using the *Gram–Schmidt orthogonalization* method mentioned earlier in this chapter, described in Appendix F, and used in Chapter 1 to construct a set of orthogonal eigenvectors of an Hermitian matrix. We repeat the analysis here for completeness.

Suppose, for the sake of our proof, that λ_0 is k-fold degenerate, i.e.

$$\mathcal{L}y_i = \lambda_0 \rho y_i \qquad \text{for } i = 0, 1, \ldots, k-1, \tag{8.25}$$

but that λ_0 is different from any of λ_k, λ_{k+1}, etc. Then any linear combination of these y_i is also an eigenfunction with eigenvalue λ_0 since

$$\mathcal{L}z \equiv \mathcal{L}\sum_{i=0}^{k-1} c_i y_i = \sum_{i=0}^{k-1} c_i \mathcal{L}y_i = \sum_{i=0}^{k-1} c_i \lambda_0 \rho y_i = \lambda_0 \rho z. \tag{8.26}$$

If the y_i defined in (8.25) are not already mutually orthogonal then consider the new eigenfunctions z_i constructed by the following procedure, in which each of the new functions z_i is to be normalized, to give $\hat{z}_i$, before proceeding to the construction of the next one:[6]

$$\begin{aligned}
z_0 &= y_0, \\
z_1 &= y_1 - \left(\hat{z}_0 \int_a^b \hat{z}_0^* y_1 \rho \, dx\right), \\
z_2 &= y_2 - \left(\hat{z}_1 \int_a^b \hat{z}_1^* y_2 \rho \, dx\right) - \left(\hat{z}_0 \int_a^b \hat{z}_0^* y_2 \rho \, dx\right), \\
&\ \ \vdots \\
z_{k-1} &= y_{k-1} - \left(\hat{z}_{k-2} \int_a^b \hat{z}_{k-2}^* y_{k-1} \rho \, dx\right) - \cdots - \left(\hat{z}_0 \int_a^b \hat{z}_0^* y_{k-1} \rho \, dx\right).
\end{aligned}$$

Each of the integrals is just a number and thus each new function z_i is, as can be shown from (8.26), an eigenvector of $\mathcal{L}$ with eigenvalue λ_0. It is straightforward to check that each z_i is orthogonal to all its predecessors. Thus, by this explicit construction we have shown that an orthogonal set of eigenfunctions of an Hermitian operator $\mathcal{L}$ can be obtained. Clearly the orthogonal set obtained, z_i, is not unique, since a different set could be obtained simply by relabeling the original $y_i(x)$ and carrying out the same procedure.

In general, since $\mathcal{L}$ is linear, the normalization of its eigenfunctions $y_i(x)$ is arbitrary. It is often convenient, however, to work in terms of the normalized eigenfunctions $\hat{y}_i(x)$ defined by $\int_a^b \hat{y}_i^* \hat{y}_i \rho \, dx = 1$. These then form an orthonormal set and we can write

$$\int_a^b \hat{y}_i^* \hat{y}_j \rho \, dx = \delta_{ij}, \tag{8.27}$$

a relationship valid for all pairs of values i, j.

6 The normalization can be carried out by dividing the eigenfunction z_i by $(\int_a^b z_i^* z_i \rho \, dx)^{1/2}$.

8.3.3 Completeness of the eigenfunctions

As noted earlier, the eigenfunctions of an Hermitian operator may be shown to form a complete basis set over the relevant interval. One may thus expand any (reasonable) function $y(x)$ obeying appropriate boundary conditions in an eigenfunction series over the interval, as in (8.17). Working in terms of the normalized eigenfunctions $\hat{y}_n(x)$, we may thus write

$$f(x) = \sum_n \hat{y}_n(x) \int_a^b \hat{y}_n^*(z) f(z) \rho(z)\, dz$$
$$= \int_a^b f(z)\rho(z) \sum_n \hat{y}_n(x)\hat{y}_n^*(z)\, dz.$$

Since this is true for any $f(x)$, we must have that

$$\rho(z) \sum_n \hat{y}_n(x)\hat{y}_n^*(z) = \delta(x - z). \tag{8.28}$$

This is called the *completeness* or *closure* property of the eigenfunctions. It defines a complete set. If the spectrum of eigenvalues of $\mathcal{L}$ is anywhere continuous then the eigenfunction $y_n(x)$ must be treated as $y(n, x)$ and an integration carried out over n.

We also note that the RHS of (8.28) is a δ-function and so is only non-zero when $z = x$; thus $\rho(z)$ on the LHS can be replaced by $\rho(x)$ if required, i.e.

$$\rho(z) \sum_n \hat{y}_n(x)\hat{y}_n^*(z) = \rho(x) \sum_n \hat{y}_n(x)\hat{y}_n^*(z). \tag{8.29}$$

8.3.4 Construction of real eigenfunctions

Recall that the eigenfunction y_i satisfies

$$\mathcal{L}y_i = \lambda_i \rho y_i \tag{8.30}$$

and that, for a real Hermitian operator, the complex conjugate of this gives

$$\mathcal{L}y_i^* = \lambda_i^* \rho y_i^* = \lambda_i \rho y_i^*, \tag{8.31}$$

where the last equality follows because the eigenvalues are real, i.e. $\lambda_i = \lambda_i^*$. Thus, y_i and y_i^* are eigenfunctions corresponding to the same eigenvalue and hence, because of the linearity of $\mathcal{L}$, at least one of $y_i^* + y_i$ and $i(y_i^* - y_i)$, which are both real, is a non-zero eigenfunction corresponding to that eigenvalue.[7] It follows that the eigenfunctions of a real Hermitian differential operator can always be made real by taking suitable linear combinations, though taking such linear combinations will only be necessary in cases where a particular λ is degenerate, i.e. corresponds to more than one linearly independent eigenfunction.

7 Taking the particular example of $y_j(x) = e^{ik_jx}$, identify the appropriate Hermitian operator $\mathcal{L}$ and the real eigenfunction(s) that are generated by this procedure.

8.4 Sturm–Liouville equations

One of the most important applications of our discussion of Hermitian operators is to the study of *Sturm–Liouville equations*, which take the general form

$$p(x)\frac{d^2y}{dx^2} + r(x)\frac{dy}{dx} + q(x)y + \lambda\rho(x)y = 0, \qquad \text{where } r(x) = \frac{dp(x)}{dx} \tag{8.32}$$

and p, q and r are real functions of x.[8] A variational approach to the Sturm–Liouville equation, which is useful in estimating the eigenvalues λ for a given set of boundary conditions on y, is discussed in Chapter 12. For now, however, we concentrate on demonstrating that solutions of the Sturm–Liouville equation that satisfy appropriate boundary conditions are the eigenfunctions of an Hermitian operator.

It is clear that (8.32) can be written

$$\mathcal{L}y = \lambda\rho(x)y, \qquad \text{where } \mathcal{L} \equiv -\left[p(x)\frac{d^2}{dx^2} + r(x)\frac{d}{dx} + q(x)\right]. \tag{8.33}$$

Using the condition that $r(x) = p'(x)$, it will be seen that the general Sturm–Liouville equation (8.32) can also be rewritten as

$$(py')' + qy + \lambda\rho y = 0, \tag{8.34}$$

where primes denote differentiation with respect to x. Using (8.33) this may also be written $\mathcal{L}y \equiv -(py')' - qy = \lambda\rho y$, which defines a more useful form for the Sturm–Liouville linear operator, namely

$$\mathcal{L} \equiv -\left[\frac{d}{dx}\left(p(x)\frac{d}{dx}\right) + q(x)\right]. \tag{8.35}$$

8.4.1 Hermitian nature of the Sturm–Liouville operator

As we show in the next worked example, the linear operator of the Sturm–Liouville equation (8.35) is self-adjoint. Moreover, the operator is Hermitian over the range $[a, b]$ provided certain boundary conditions are met, namely that any two eigenfunctions y_i and y_j of (8.33) must satisfy

$$\left[y_i^* p y_j'\right]_{x=a} = \left[y_i^* p y_j'\right]_{x=b} \qquad \text{for all } i, j. \tag{8.36}$$

Rearranging (8.36), we can write

$$\left[y_i^* p y_j'\right]_{x=a}^{x=b} = 0 \tag{8.37}$$

as an equivalent statement of the required boundary conditions. These boundary conditions are in fact not too restrictive and are met, for instance, by the sets $y(a) = y(b) = 0$; $y(a) = y'(b) = 0$; $p(a) = p(b) = 0$ and by many other sets. It is important to note that in order to satisfy (8.36) and (8.37) one boundary condition must be specified at each end of the range.

8 We note that sign conventions vary in this expression for the general Sturm–Liouville equation; some authors use $-\lambda\rho(x)y$ on the LHS of (8.32).

Example Prove that the Sturm–Liouville operator is Hermitian over the range $[a, b]$ and under the boundary conditions (8.37).

Putting the Sturm–Liouville form $\mathcal{L}y = -(py')' - qy$ into the definition (8.16) of an Hermitian operator, the LHS may be written as a sum of two terms, i.e.

$$-\int_a^b [y_i^*(py_j')' + y_i^* q y_j]\,dx = -\int_a^b y_i^*(py_j')'\,dx - \int_a^b y_i^* q y_j\,dx.$$

The first term may be integrated by parts to give

$$-\Big[y_i^* p y_j'\Big]_a^b + \int_a^b (y_i^*)' p y_j'\,dx.$$

The boundary-value term in this is zero because of the boundary conditions, and so integrating by parts again yields

$$\Big[(y_i^*)' p y_j\Big]_a^b - \int_a^b ((y_i^*)' p)' y_j\,dx.$$

Again, the boundary-value term is zero, leaving us with

$$-\int_a^b [y_i^*(py_j')' + y_i^* q y_j]\,dx = -\int_a^b [y_j(p(y_i^*)')' + y_j q y_i^*]\,dx,$$

which proves that the Sturm–Liouville operator is Hermitian over the prescribed interval. The proof shows that, even if the boundary-value terms were not zero, or did not cancel each other, the S–L operator would still be self-adjoint; this property is clearly directly related to the structure of the operator (see Problem 8.7). ◀

It is also worth noting that, since $p(a) = p(b) = 0$ is a valid set of boundary conditions, many Sturm–Liouville equations possess a "natural" interval $[a, b]$ over which the corresponding differential operator $\mathcal{L}$ is Hermitian *irrespective* of the boundary conditions satisfied by its eigenfunctions at $x = a$ and $x = b$ (the only requirement being that they are regular, i.e. not infinite, at these end-points).

8.4.2 Transforming an equation into Sturm–Liouville form

Many of the second-order differential equations encountered in physical problems are examples of the Sturm–Liouville equation (8.34). Moreover, *any* second-order differential equation of the form

$$p(x)y'' + r(x)y' + q(x)y + \lambda\rho(x)y = 0 \tag{8.38}$$

can be converted into Sturm–Liouville form by multiplying through by a suitable integrating factor, which is given by

$$F(x) = \exp\left\{\int^x \frac{r(u) - p'(u)}{p(u)}\,du\right\}. \tag{8.39}$$

It is easily verified that (8.38) then takes the Sturm–Liouville form,

$$[F(x)p(x)y']' + F(x)q(x)y + \lambda F(x)\rho(x)y = 0, \tag{8.40}$$

Table 8.1 *The Sturm–Liouville form (8.34) for important ODEs in the physical sciences and engineering. The asterisk denotes that, for Bessel's equation, a change of variable $x \to x/\alpha$ is required to give the conventional normalization used here, but is not needed for the transformation into Sturm–Liouville form*

Equation	$p(x)$	$q(x)$	λ	$\rho(x)$
Hypergeometric	$x^c(1-x)^{a+b-c+1}$	0	$-ab$	$x^{c-1}(1-x)^{a+b-c}$
Legendre	$1-x^2$	0	$\ell(\ell+1)$	1
Associated Legendre	$1-x^2$	$-m^2/(1-x^2)$	$\ell(\ell+1)$	1
Chebyshev	$(1-x^2)^{1/2}$	0	ν^2	$(1-x^2)^{-1/2}$
Confluent hypergeometric	$x^c e^{-x}$	0	$-a$	$x^{c-1}e^{-x}$
Bessel*	x	$-\nu^2/x$	α^2	x
Laguerre	xe^{-x}	0	ν	e^{-x}
Associated Laguerre	$x^{m+1}e^{-x}$	0	ν	$x^m e^{-x}$
Hermite	e^{-x^2}	0	2ν	e^{-x^2}
Simple harmonic	1	0	ω^2	1

with a different, but still non-negative, weight function $F(x)\rho(x)$. Table 8.1 summarizes the Sturm–Liouville form (8.34) for several of the equations listed in Table 7.1. These forms can be determined using (8.39), as illustrated in the following example.

Example Put the following equations into Sturm–Liouville (SL) form:

(i) $(1-x^2)y'' - xy' + \nu^2 y = 0$ (Chebyshev equation);
(ii) $xy'' + (1-x)y' + \nu y = 0$ (Laguerre equation);
(iii) $y'' - 2xy' + 2\nu y = 0$ (Hermite equation).

(i) From (8.39), the required integrating factor is

$$F(x) = \exp\left(\int^x \frac{u}{1-u^2}\,du\right) = \exp\left[-\tfrac{1}{2}\ln(1-x^2)\right] = (1-x^2)^{-1/2}.$$

Thus, the Chebyshev equation becomes

$$(1-x^2)^{1/2}y'' - x(1-x^2)^{-1/2}y' + \nu^2(1-x^2)^{-1/2}y = \left[(1-x^2)^{1/2}y'\right]' + \nu^2(1-x^2)^{-1/2}y = 0,$$

which is in SL form with $p(x) = (1-x^2)^{1/2}$, $q(x) = 0$, $\rho(x) = (1-x^2)^{-1/2}$ and $\lambda = \nu^2$.

(ii) From (8.39), the required integrating factor is

$$F(x) = \exp\left(\int^x -1\,du\right) = \exp(-x).$$

Thus, the Laguerre equation becomes

$$xe^{-x}y'' + (1-x)e^{-x}y' + \nu e^{-x}y = (xe^{-x}y')' + \nu e^{-x}y = 0,$$

which is in SL form with $p(x) = xe^{-x}$, $q(x) = 0$, $\rho(x) = e^{-x}$ and $\lambda = \nu$.

(iii) From (8.39), the required integrating factor is

$$F(x) = \exp\left(\int^x -2u\, du\right) = \exp(-x^2).$$

Thus, the Hermite equation becomes

$$e^{-x^2}y'' - 2xe^{-x^2}y' + 2\nu e^{-x^2}y = (e^{-x^2}y')' + 2\nu e^{-x^2}y = 0,$$

which is in SL form with $p(x) = e^{-x^2}$, $q(x) = 0$, $\rho(x) = e^{-x^2}$ and $\lambda = 2\nu$. ◀

From the $p(x)$ entries in Table 8.1, we may read off the natural interval over which the corresponding Sturm–Liouville operator (8.35) is Hermitian; in each case this is given by $[a, b]$, where $p(a) = p(b) = 0$. Thus, the natural interval for the Legendre equation, the associated Legendre equation and the Chebyshev equation is $[-1, 1]$; for the Laguerre and associated Laguerre equations the interval is $[0, \infty]$; and for the Hermite equation it is $[-\infty, \infty]$. In addition, from (8.37), one sees that for the simple harmonic equation one requires only that $[a, b] = [x_0, x_0 + 2\pi]$. We also note that, as required, the weight function in each case is finite and non-negative over the natural interval. Occasionally, a little more care is required in determining the conditions for a Sturm–Liouville operator of the form (8.35) to be Hermitian over some natural interval, as is illustrated in the following example.

Example Express the hypergeometric equation,

$$x(1-x)y'' + [\,c - (a+b+1)x\,]y' - aby = 0,$$

in Sturm–Liouville form. Hence determine the natural interval over which the resulting Sturm–Liouville operator is Hermitian and the corresponding conditions that one must impose on the parameters a, b and c.

As usual for an equation not already in SL form, we first determine the appropriate integrating factor. This is given, as in equation (8.39), by

$$\begin{aligned} F(x) &= \exp\left[\int^x \frac{c - (a+b+1)u - 1 + 2u}{u(1-u)}\, du\right] \\ &= \exp\left[\int^x \frac{c - 1 - (a+b-1)u}{u(1-u)}\, du\right] \\ &= \exp\left[\int^x \left(\frac{c-1}{1-u} + \frac{c-1}{u} - \frac{a+b-1}{1-u}\, du\right)\right] \\ &= \exp\left[\,(a+b-c)\ln(1-x) + (c-1)\ln x\,\right] \\ &= x^{c-1}(1-x)^{a+b-c}. \end{aligned}$$

When the equation is multiplied through by $F(x)$ it takes the form

$$\left[x^c(1-x)^{a+b-c+1}y'\right]' - abx^{c-1}(1-x)^{a+b-c}y = 0.$$

Now, for the corresponding Sturm–Liouville operator to be Hermitian, the conditions to be imposed are as follows.

(i) The boundary condition (8.37); if $c > 0$ and $a + b - c + 1 > 0$, this is satisfied automatically for $0 \le x \le 1$, which is thus the natural interval in this case.
(ii) The weight function $x^{c-1}(1-x)^{a+b-c}$ must be finite and not change sign in the interval $0 \le x \le 1$. This means that both exponents in it must be positive, i.e. $c - 1 > 0$ and $a + b - c > 0$.

Putting together the conditions on the parameters gives the double inequality $a + b > c > 1$. ◀

Finally, we consider Bessel's equation,

$$x^2 y'' + xy' + (x^2 - \nu^2)y = 0,$$

which may be converted into Sturm–Liouville form, but only in a somewhat unorthodox fashion. It is conventional first to divide the Bessel equation by x and then to change variables to $\bar{x} = x/\alpha$. In this case, it becomes

$$\bar{x}y''(\alpha\bar{x}) + y'(\alpha\bar{x}) - \frac{\nu^2}{\bar{x}}y(\alpha\bar{x}) + \alpha^2\bar{x}y(\alpha\bar{x}) = 0, \tag{8.41}$$

where a prime now indicates differentiation with respect to $\bar{x}$. Dropping the bars on the independent variable, we thus have

$$[xy'(\alpha x)]' - \frac{\nu^2}{x}y(\alpha x) + \alpha^2 xy(\alpha x) = 0, \tag{8.42}$$

which is in SL form with $p(x) = x$, $q(x) = -\nu^2/x$, $\rho(x) = x$ and $\lambda = \alpha^2$. It should be noted, however, that in this case the eigenvalue (actually its square root) appears in the argument of the dependent variable.

8.5 Superposition of eigenfunctions: Green's functions

We have already seen that if

$$\mathcal{L}y_n(x) = \lambda_n \rho(x) y_n(x), \tag{8.43}$$

where $\mathcal{L}$ is an Hermitian operator, then the eigenvalues λ_n are real and the eigenfunctions $y_n(x)$ are orthogonal, or can be made so. Let us assume that we know the eigenfunctions $y_n(x)$ of $\mathcal{L}$ that individually satisfy (8.43) as well as some imposed boundary conditions that make $\mathcal{L}$ Hermitian.

Now consider the problem of solving the inhomogeneous differential equation

$$\mathcal{L}y(x) = f(x), \tag{8.44}$$

subject to the same boundary conditions. Since the eigenfunctions of $\mathcal{L}$ form a complete set, the full solution, $y(x)$, to (8.44) may be written as a superposition of eigenfunctions, i.e.

$$y(x) = \sum_{n=0}^{\infty} c_n y_n(x), \tag{8.45}$$

for some choice of the constants c_n. Making full use of the linearity of $\mathcal{L}$, we have

$$f(x) = \mathcal{L}y(x) = \mathcal{L}\left(\sum_{n=0}^{\infty} c_n y_n(x)\right) = \sum_{n=0}^{\infty} c_n \mathcal{L} y_n(x) = \sum_{n=0}^{\infty} c_n \lambda_n \rho(x) y_n(x). \tag{8.46}$$

Multiplying the first and last terms of (8.46) by y_j^* and integrating, we obtain

$$\int_a^b y_j^*(z) f(z)\, dz = \sum_{n=0}^{\infty} \int_a^b c_n \lambda_n y_j^*(z) y_n(z) \rho(z)\, dz, \tag{8.47}$$

where we have used z as the integration variable for later convenience. Finally, using the orthogonality condition (8.27), we see that the integrals on the RHS are zero unless $n = j$, and so obtain

$$c_n = \frac{1}{\lambda_n} \frac{\int_a^b y_n^*(z) f(z)\, dz}{\int_a^b y_n^*(z) y_n(z) \rho(z)\, dz}. \tag{8.48}$$

Thus, if we can find all the eigenfunctions of a differential operator then (8.48) can be used to find the weighting coefficients for the superposition, to give as the full solution

$$y(x) = \sum_{n=0}^{\infty} \frac{1}{\lambda_n} \frac{\int_a^b y_n^*(z) f(z)\, dz}{\int_a^b y_n^*(z) y_n(z) \rho(z)\, dz} y_n(x). \tag{8.49}$$

If we work with normalized eigenfunctions $\hat{y}_n(x)$, so that

$$\int_a^b \hat{y}_n^*(z) \hat{y}_n(z) \rho(z)\, dz = 1 \qquad \text{for all } n,$$

and we assume that we may interchange the order of summation and integration, then (8.49) can be written as

$$y(x) = \int_a^b \left\{ \sum_{n=0}^{\infty} \left[\frac{1}{\lambda_n} \hat{y}_n(x) \hat{y}_n^*(z) \right] \right\} f(z)\, dz.$$

The quantity in braces, which is a function of x and z only, is usually written $G(x, z)$, and is the *Green's function* for the problem. With this notation,

$$y(x) = \int_a^b G(x, z) f(z)\, dz, \tag{8.50}$$

where

$$G(x, z) = \sum_{n=0}^{\infty} \frac{1}{\lambda_n} \hat{y}_n(x) \hat{y}_n^*(z). \tag{8.51}$$

We note that $G(x, z)$ is determined entirely by the boundary conditions and the eigenfunctions $\hat{y}_n$, and hence by $\mathcal{L}$ itself, and that $f(z)$ depends purely on the RHS of the inhomogeneous equation (8.44). Thus, for a given $\mathcal{L}$ and boundary conditions we can establish, once and for all, a function $G(x, z)$ that will enable us to solve the inhomogeneous equation for *any* RHS. From (8.51) we also note that

$$G(x, z) = G^*(z, x). \tag{8.52}$$

We have already met the Green's function in the solution of second-order differential equations in Chapter 6, as the function that both satisfies the equation $\mathcal{L}[G(x, z)] = \delta(x - z)$ and meets the boundary conditions. The formulation given above is an alternative, though equivalent, one.

Example Find an appropriate Green's function for the equation

$$y'' + \tfrac{1}{4}y = f(x),$$

with boundary conditions $y(0) = y(\pi) = 0$. Hence, solve for (i) $f(x) = \sin 2x$ and (ii) $f(x) = x/2$.

One approach to solving this problem is to use the methods of Chapter 6 and find a complementary function and particular integral. However, in order to illustrate the techniques developed in the present chapter we will use the superposition of eigenfunctions, which, as may easily be checked, produces the same solution.

The operator on the LHS of this equation is already Hermitian under the given boundary conditions, and so we seek its eigenfunctions. These satisfy the equation

$$y'' + \tfrac{1}{4}y = \lambda y.$$

This equation has the familiar solution

$$y(x) = A \sin\left(\sqrt{\tfrac{1}{4} - \lambda}\right)x + B \cos\left(\sqrt{\tfrac{1}{4} - \lambda}\right)x.$$

Now, the boundary conditions require that $B = 0$ and $\sin\left(\sqrt{\tfrac{1}{4} - \lambda}\right)\pi = 0$, and so

$$\sqrt{\tfrac{1}{4} - \lambda} = n, \qquad \text{where } n = 0, \pm 1, \pm 2, \ldots.$$

Therefore, the independent eigenfunctions that satisfy the boundary conditions are

$$y_n(x) = A_n \sin nx,$$

where n is any non-negative integer, and the corresponding eigenvalues are $\lambda_n = \tfrac{1}{4} - n^2$. The normalization condition further requires

$$\int_0^\pi A_n^2 \sin^2 nx\, dx = 1 \quad \Rightarrow \quad A_n = \left(\frac{2}{\pi}\right)^{1/2}.$$

Comparison with (8.51) shows that the appropriate Green's function is therefore given by

$$G(x, z) = \frac{2}{\pi} \sum_{n=0}^{\infty} \frac{\sin nx \sin nz}{\tfrac{1}{4} - n^2}.$$

Case (i). Using (8.50), the solution with $f(x) = \sin 2x$ is given by

$$y(x) = \frac{2}{\pi}\int_0^{\pi}\left(\sum_{n=0}^{\infty}\frac{\sin nx \sin nz}{\frac{1}{4}-n^2}\right)\sin 2z\,dz = \frac{2}{\pi}\sum_{n=0}^{\infty}\frac{\sin nx}{\frac{1}{4}-n^2}\int_0^{\pi}\sin nz \sin 2z\,dz.$$

Now the integral is zero unless $n = 2$, in which case it is

$$\int_0^{\pi}\sin^2 2z\,dz = \frac{\pi}{2}.$$

Thus the single term

$$y(x) = -\frac{2}{\pi}\frac{\sin 2x}{15/4}\frac{\pi}{2} = -\frac{4}{15}\sin 2x$$

is the full solution for $f(x) = \sin 2x$. This is, of course, exactly the solution found by using the methods of Chapter 6.

Case (ii). The solution with $f(x) = x/2$ is given by

$$y(x) = \int_0^{\pi}\left(\frac{2}{\pi}\sum_{n=0}^{\infty}\frac{\sin nx \sin nz}{\frac{1}{4}-n^2}\right)\frac{z}{2}\,dz = \frac{1}{\pi}\sum_{n=0}^{\infty}\frac{\sin nx}{\frac{1}{4}-n^2}\int_0^{\pi}z\sin nz\,dz.$$

The integral may be evaluated by integrating by parts. For $n \neq 0$,

$$\begin{aligned}\int_0^{\pi}z\sin nz\,dz &= \left[-\frac{z\cos nz}{n}\right]_0^{\pi} + \int_0^{\pi}\frac{\cos nz}{n}\,dz \\ &= \frac{-\pi\cos n\pi}{n} + \left[\frac{\sin nz}{n^2}\right]_0^{\pi} \\ &= -\frac{\pi(-1)^n}{n}.\end{aligned}$$

For $n = 0$ the integral is zero, and thus the infinite series

$$y(x) = \sum_{n=1}^{\infty}(-1)^{n+1}\frac{\sin nx}{n\left(\frac{1}{4}-n^2\right)}$$

is the full solution for $f(x) = x/2$. Using the methods of Subsection 6.2.2, the solution is found to be $y(x) = 2x - 2\pi\sin(x/2)$, which may be shown to be equal to the above solution by expanding $2x - 2\pi\sin(x/2)$ as a Fourier sine series. ◀

SUMMARY

1. *Linear differential operators*
   - In the equation $\mathcal{L}y_i(x) = \lambda_i\rho(x)y_i(x)$, the function $y_i(x)$ is an *eigenfunction* of the *linear differential operator* $\mathcal{L}$ with respect to the *weight function* ρ with *eigenvalue* λ_i.
   - The adjoint of $\mathcal{L}$, denoted by $\mathcal{L}^\dagger$, is defined by

$$\int_a^b f^*(x)\left[\mathcal{L}g(x)\right]dx = \int_a^b [\mathcal{L}^\dagger f(x)]^* g(x)\,dx + \text{boundary terms.}$$

- The operator $\mathcal{L}$ is self-adjoint if $\mathcal{L}^\dagger = \mathcal{L}$, and Hermitian if, in addition, the boundary terms vanish.

2. *Properties of Hermitian operators*
   - Their eigenvalues are real.
   - Their eigenfunctions corresponding to different eigenvalues are orthogonal with respect to ρ.
   - Even if some eigenvalues are degenerate, an orthonormal set of eigenfunctions can be constructed. They satisfy $\int_a^b \hat{y}_i^* \hat{y}_j \rho \, dx = \delta_{ij}$.

3. *Sturm–Liouville equations*
   SL equations have the general form $(py')' + qy + \lambda \rho y = 0$.
   - The corresponding operator is $\mathcal{L} \equiv -\left[\dfrac{d}{dx}\left(p(x)\dfrac{d}{dx}\right) + q(x)\right]$.
   - The operator is Hermitian if $[y_i^* p y_j']_a^b = 0$.
   - If $p(a) = p(b) = 0$, then $[a, b]$ is a natural interval for the operator.
   - For some important SL equations, see Table 8.1 on p. 310.
   - The general second-order ODE $p(x)y'' + r(x)y' + q(x)y + \lambda \rho(x) y = 0$ can be converted to SL form by multiplying through by the integrating factor

$$F(x) = \exp\left\{\int^x \frac{r(u) - p'(u)}{p(u)} \, du\right\}.$$

4. *Green's function to solve $\mathcal{L}(y) = f(x)$*
   The solution is given by

$$y(x) = \int_a^b G(x, z) f(z) \, dz, \quad \text{where} \quad G(x, z) = \sum_{n=0}^{\infty} \frac{1}{\lambda_n} \hat{y}_n(x) \hat{y}_n^*(z),$$

   the summation being over those normalized eigenfunctions that satisfy the boundary conditions.

PROBLEMS

8.1. By considering $\langle h|h \rangle$, where $h = f + \lambda g$ with λ real, prove that, for two functions f and g,

$$\langle f|f \rangle \langle g|g \rangle \geq \tfrac{1}{4}[\langle f|g \rangle + \langle g|f \rangle]^2.$$

The function $y(x)$ is real and positive for all x. Its Fourier cosine transform $\tilde{y}_c(k)$ is defined by

$$\tilde{y}_c(k) = \int_{-\infty}^{\infty} y(x) \cos(kx) \, dx,$$

and it is given that $\tilde{y}_c(0) = 1$. Prove that

$$\tilde{y}_c(2k) \geq 2[\tilde{y}_c(k)]^2 - 1.$$

8.2. Write the homogeneous Sturm–Liouville eigenvalue equation for which $y(a) = y(b) = 0$ as

$$\mathcal{L}(y; \lambda) \equiv (py')' + qy + \lambda \rho y = 0,$$

where $p(x)$, $q(x)$ and $\rho(x)$ are continuously differentiable functions. Show that if $z(x)$ and $F(x)$ satisfy $\mathcal{L}(z; \lambda) = F(x)$, with $z(a) = z(b) = 0$, then

$$\int_a^b y(x)F(x)\,dx = 0.$$

Demonstrate the validity of this general result by direct calculation for the specific case in which $p(x) = \rho(x) = 1$, $q(x) = 0$, $a = -1$, $b = 1$ and $z(x) = 1 - x^2$.

8.3. Consider the real eigenfunctions $y_n(x)$ of a Sturm–Liouville equation,

$$(py')' + qy + \lambda \rho y = 0, \qquad a \leq x \leq b,$$

in which $p(x)$, $q(x)$ and $\rho(x)$ are continuously differentiable real functions and $p(x)$ does not change sign in $a \leq x \leq b$. Take $p(x)$ as positive throughout the interval, if necessary by changing the signs of all eigenvalues. For $a \leq x_1 \leq x_2 \leq b$, establish the identity

$$(\lambda_n - \lambda_m) \int_{x_1}^{x_2} \rho y_n y_m \,dx = \left[y_n\, p\, y_m' - y_m\, p\, y_n' \right]_{x_1}^{x_2}.$$

Deduce that if $\lambda_n > \lambda_m$ then $y_n(x)$ must change sign between two successive zeros of $y_m(x)$.

[The reader may find it helpful to illustrate this result by sketching the first few eigenfunctions of the system $y'' + \lambda y = 0$, with $y(0) = y(\pi) = 0$, and the Legendre polynomials $P_n(z)$ for $n = 2, 3, 4, 5$.]

8.4. Show that the equation

$$y'' + a\delta(x)y + \lambda y = 0,$$

with $y(\pm\pi) = 0$ and a real, has a set of eigenvalues λ satisfying

$$\tan(\pi\sqrt{\lambda}) = \frac{2\sqrt{\lambda}}{a}.$$

Investigate the conditions under which negative eigenvalues, $\lambda = -\mu^2$, with μ real, are possible.

8.5. Use the properties of Legendre polynomials to solve the following problems.
(a) Find the solution of $(1 - x^2)y'' - 2xy' + by = f(x)$, valid in the range $-1 \leq x \leq 1$ and finite at $x = 0$, in terms of Legendre polynomials.

(b) If $b = 14$ and $f(x) = 5x^3$, find the explicit solution and verify it by direct substitution.

[The first six Legendre polynomials are listed in Subsection 9.1.1.]

8.6. Starting from the linearly independent functions $1, x, x^2, x^3, \ldots,$ in the range $0 \le x < \infty$, find the first three orthogonal functions ϕ_0, ϕ_1 and ϕ_2, with respect to the weight function $\rho(x) = e^{-x}$. By comparing your answers with the Laguerre polynomials generated by the recurrence relation (9.112), deduce the form of $\phi_3(x)$.

8.7. Consider the set of functions, $\{f(x)\}$, of the real variable x, defined in the interval $-\infty < x < \infty$, that $\to 0$ at least as quickly as x^{-1} as $x \to \pm\infty$. For unit weight function, determine whether each of the following linear operators is Hermitian when acting upon $\{f(x)\}$:

$$\text{(a)}\ \frac{d}{dx} + x; \quad \text{(b)}\ -i\frac{d}{dx} + x^2; \quad \text{(c)}\ ix\frac{d}{dx}; \quad \text{(d)}\ i\frac{d^3}{dx^3}.$$

8.8. A particle moves in a parabolic potential in which its natural angular frequency of oscillation is $\frac{1}{2}$. At time $t = 0$ it passes through the origin with velocity v. It is then suddenly subjected to an additional acceleration, of $+1$ for $0 \le t \le \pi/2$, followed by -1 for $\pi/2 < t \le \pi$. At the end of this period it is again at the origin. Apply the results of the worked example in Section 8.5 to show that

$$v = -\frac{8}{\pi}\sum_{m=0}^{\infty}\frac{1}{(4m+2)^2 - \frac{1}{4}} \approx -0.81.$$

8.9. Find an eigenfunction expansion for the solution, with boundary conditions $y(0) = y(\pi) = 0$, of the inhomogeneous equation

$$\frac{d^2y}{dx^2} + \kappa y = f(x),$$

where κ is a constant and

$$f(x) = \begin{cases} x & 0 \le x \le \pi/2, \\ \pi - x & \pi/2 < x \le \pi. \end{cases}$$

8.10. Consider the following two approaches to constructing a Green's function.

(a) Find those eigenfunctions $y_n(x)$ of the self-adjoint linear differential operator d^2/dx^2 that satisfy the boundary conditions $y_n(0) = y_n(\pi) = 0$, and hence construct its Green's function $G(x, z)$.

(b) Construct the same Green's function using a method based on the complementary function of the appropriate differential equation and the boundary conditions to be satisfied at the position of the δ-function, showing that it is

$$G(x,z) = \begin{cases} x(z-\pi)/\pi & 0 \le x \le z, \\ z(x-\pi)/\pi & z \le x \le \pi. \end{cases}$$

(c) By expanding the function given in (b) in terms of the eigenfunctions $y_n(x)$, verify that it is the same function as that derived in (a).

8.11. The differential operator $\mathcal{L}$ is defined by

$$\mathcal{L}y = -\frac{d}{dx}\left(e^x \frac{dy}{dx}\right) - \tfrac{1}{4}e^x y.$$

Determine the eigenvalues λ_n of the problem

$$\mathcal{L}y_n = \lambda_n e^x y_n \qquad 0 < x < 1,$$

with boundary conditions

$$y(0) = 0, \qquad \frac{dy}{dx} + \tfrac{1}{2}y = 0 \quad \text{at} \quad x = 1.$$

(a) Find the corresponding unnormalized y_n, and also a weight function $\rho(x)$ with respect to which the y_n are orthogonal. Hence, select a suitable normalization for the y_n.

(b) By making an eigenfunction expansion, solve the equation

$$\mathcal{L}y = -e^{x/2}, \quad 0 < x < 1,$$

subject to the same boundary conditions as previously.

8.12. Show that the linear operator

$$\mathcal{L} \equiv \tfrac{1}{4}(1+x^2)^2\tfrac{d^2}{dx^2} + \tfrac{1}{2}x(1+x^2)\tfrac{d}{dx} + a,$$

acting upon functions defined in $-1 \le x \le 1$ and vanishing at the end-points of the interval, is Hermitian with respect to the weight function $(1+x^2)^{-1}$.

By making the change of variable $x = \tan(\theta/2)$, find two even eigenfunctions, $f_1(x)$ and $f_2(x)$, of the differential equation

$$\mathcal{L}u = \lambda u.$$

8.13. By substituting $x = \exp t$, find the normalized eigenfunctions $y_n(x)$ and the eigenvalues λ_n of the operator $\mathcal{L}$ defined by

$$\mathcal{L}y = x^2y'' + 2xy' + \tfrac{1}{4}y, \qquad 1 \le x \le e,$$

with $y(1) = y(e) = 0$. Find, as a series $\sum a_n y_n(x)$, the solution of $\mathcal{L}y = x^{-1/2}$.

8.14. Express the solution of Poisson's equation in electrostatics,

$$\nabla^2\phi(\mathbf{r}) = -\rho(\mathbf{r})/\epsilon_0,$$

where ρ is the non-zero charge density over a finite part of space, in the form of an integral and hence identify the Green's function for the ∇^2 operator.

8.15. In the quantum-mechanical study of the scattering of a particle by a potential, a Born-approximation solution can be obtained in terms of a function $y(\mathbf{r})$ that satisfies an equation of the form

$$(-\nabla^2 - K^2)y(\mathbf{r}) = F(\mathbf{r}).$$

Assuming that $y_{\mathbf{k}}(\mathbf{r}) = (2\pi)^{-3/2}\exp(i\mathbf{k}\cdot\mathbf{r})$ is a suitably normalized eigenfunction of $-\nabla^2$ corresponding to eigenvalue k^2, find a suitable Green's function $G_K(\mathbf{r},\mathbf{r}')$. By taking the direction of the vector $\mathbf{r}-\mathbf{r}'$ as the polar axis for a $\mathbf{k}$-space integration, show that $G_K(\mathbf{r},\mathbf{r}')$ can be reduced to

$$\frac{1}{4\pi^2|\mathbf{r}-\mathbf{r}'|}\int_{-\infty}^{\infty}\frac{w\sin w}{w^2 - w_0^2}\,dw,$$

where $w_0 = K|\mathbf{r}-\mathbf{r}'|$.

[This integral can be evaluated using a contour integration (Chapter 14) to give $(4\pi|\mathbf{r}-\mathbf{r}'|)^{-1}\exp(iK|\mathbf{r}-\mathbf{r}'|)$.]

HINTS AND ANSWERS

8.1. Express the condition $\langle h|h\rangle \geq 0$ as a quadratic equation in λ and then apply the condition for no real roots, noting that $\langle f|g\rangle + \langle g|f\rangle$ is real. To put a limit on $\int y\cos^2 kx\,dx$, set $f = y^{1/2}\cos kx$ and $g = y^{1/2}$ in the inequality.

8.3. Follow an argument similar to that used for proving the reality of the eigenvalues, but integrate from x_1 to x_2, rather than from a to b. Take x_1 and x_2 as two successive zeros of $y_m(x)$ and note that, if the sign of y_m is α then the sign of $y_m'(x_1)$ is α whilst that of $y_m'(x_2)$ is $-\alpha$. Now assume that $y_n(x)$ does not change sign in the interval and has a constant sign β; show that this leads to a contradiction between the signs of the two sides of the identity.

8.5. (a) $y = \sum a_n P_n(x)$ with

$$a_n = \frac{n+1/2}{b-n(n+1)}\int_{-1}^{1} f(z)P_n(z)\,dz;$$

(b) $5x^3 = 2P_3(x) + 3P_1(x)$, giving $a_1 = 1/4$ and $a_3 = 1$, leading to $y = 5(2x^3 - x)/4$.

8.7. (a) No, $\int g f^{*\prime} dx \neq 0$; (b) yes; (c) no, $i\int f^* g\,dx \neq 0$; (d) yes.

8.9. The normalized eigenfunctions are $(2/\pi)^{1/2}\sin nx$, with n an integer.
$y(x) = (4/\pi)\sum_{n\text{ odd}}[(-1)^{(n-1)/2}\sin nx]/[n^2(\kappa - n^2)]$.

8.11. $\lambda_n = (n+1/2)^2\pi^2$, $n = 0, 1, 2, \ldots$.

(a) Since $y_n(1)y'_m(1) \neq 0$, the Sturm–Liouville boundary conditions are not satisfied and the appropriate weight function has to be justified by inspection. The normalized eigenfunctions are $\sqrt{2}e^{-x/2}\sin[(n+1/2)\pi x]$, with $\rho(x) = e^x$.

(b) $y(x) = (-2/\pi^3)\sum_{n=0}^{\infty} e^{-x/2}\sin[(n+1/2)\pi x]/(n+1/2)^3$.

8.13. $y_n(x) = \sqrt{2}x^{-1/2}\sin(n\pi \ln x)$ with $\lambda_n = -n^2\pi^2$;

$$a_n = \begin{cases} -(n\pi)^{-2}\int_1^e \sqrt{2}x^{-1}\sin(n\pi \ln x)\,dx = -\sqrt{8}(n\pi)^{-3} & \text{for } n \text{ odd,} \\ 0 & \text{for } n \text{ even.} \end{cases}$$

8.15. Use the form of Green's function that is the integral over all eigenvalues of the "outer product" of two eigenfunctions corresponding to the same eigenvalue, but with arguments $\mathbf{r}$ and $\mathbf{r}'$.

9 Special functions

In the previous two chapters, we introduced the most important second-order linear ODEs in physics and engineering, listing their regular and irregular singular points in Table 7.1 and their Sturm–Liouville forms in Table 8.1. These equations occur with such frequency that solutions to them, which obey particular commonly occurring boundary conditions, have been extensively studied and given special names.

In this chapter, we discuss these so-called "special functions" and their properties. Inevitably, for each set of functions in turn, the discussion has to cover the differential equation they satisfy, their polynomial or power series form with some particular examples, their orthogonality and normalization properties, and their recurrence relations. In addition, as first introduced in this chapter, most sets possess a Rodrigues' formula and a generating function.

Although each of these aspects needs to be treated in sufficient detail for the enquiring reader to be satisfied about the validity of the results stated, their serial presentation, for one set of functions after another, tends to become rather overwhelming. Consequently it is suggested that once the reader has become familiar with the general nature of each of the aspects, by studying, say, Sections 9.1 to 9.3 on Legendre functions, associated Legendre functions and spherical harmonics, he or she may treat other sets of functions more lightly, turning in the first instance to the summary beginning on p. 377, and only referring to the detailed derivations, proofs and worked examples in Sections 9.4 to 9.9 when specific needs arise.

To end the chapter, we also study some special functions that are not derived from solutions of important second-order ODEs, namely the gamma function and related functions. These convenient functions appear in a number of contexts, and so in Section 9.10 we gather together some of their properties, with a minimum of formal proofs.

9.1 Legendre functions

Legendre's differential equation has the form

$$(1 - x^2)y'' - 2xy' + \ell(\ell + 1)y = 0, \qquad (9.1)$$

and has three regular singular points, at $x = -1, 1, \infty$. It occurs in numerous physical applications and particularly in problems with axial symmetry that involve the ∇^2 operator, when they are expressed in spherical polar coordinates. In normal usage the variable x

in Legendre's equation is the cosine of the polar angle in spherical polars, and thus $-1 \leq x \leq 1$. The parameter ℓ is a given real number, and any solution of (9.1) is called a *Legendre function.*

In Subsection 7.2, we showed that $x = 0$ is an ordinary point of (9.1), and so we expect to find two linearly independent solutions of the form $y = \sum_{n=0}^{\infty} a_n x^n$. Substituting, we find

$$\sum_{n=0}^{\infty} \left[n(n-1)a_n x^{n-2} - n(n-1)a_n x^n - 2na_n x^n + \ell(\ell+1)a_n x^n\right] = 0,$$

which on collecting terms gives

$$\sum_{n=0}^{\infty} \{(n+2)(n+1)a_{n+2} - [n(n+1) - \ell(\ell+1)]a_n\} x^n = 0.$$

The recurrence relation is therefore

$$a_{n+2} = \frac{[n(n+1) - \ell(\ell+1)]}{(n+1)(n+2)} a_n, \tag{9.2}$$

for $n = 0, 1, 2, \ldots$. If we choose $a_0 = 1$ and $a_1 = 0$ then we obtain the solution

$$y_1(x) = 1 - \ell(\ell+1)\frac{x^2}{2!} + (\ell-2)\ell(\ell+1)(\ell+3)\frac{x^4}{4!} - \cdots, \tag{9.3}$$

whereas on choosing $a_0 = 0$ and $a_1 = 1$ we find a second solution

$$y_2(x) = x - (\ell-1)(\ell+2)\frac{x^3}{3!} + (\ell-3)(\ell-1)(\ell+2)(\ell+4)\frac{x^5}{5!} - \cdots. \tag{9.4}$$

By applying the ratio test to these series, we find that both series converge for $|x| < 1$, and so their radius of convergence is unity. This is as expected, as it is the distance from the expansion point, $x = 0$, of the nearest singular point of the equation (here, both $x = 1$ and $x = -1$). Since (9.3) contains only even powers of x and (9.4) contains only odd powers, these two solutions cannot be proportional to one another, and are therefore linearly independent. Hence, the general solution to (9.1) for $|x| < 1$ is

$$y(x) = c_1 y_1(x) + c_2 y_2(x).$$

9.1.1 Legendre functions for integer ℓ

In many physical applications the parameter ℓ in Legendre's equation (9.1) is an integer, i.e. $\ell = 0, 1, 2, \ldots$. In this case, the recurrence relation (9.2) gives

$$a_{\ell+2} = \frac{[\ell(\ell+1) - \ell(\ell+1)]}{(\ell+1)(\ell+2)} a_\ell = 0,$$

i.e. the series terminates and we obtain a polynomial solution of order ℓ. In particular, if ℓ is even, then $y_1(x)$ in (9.3) reduces to a polynomial, whereas if ℓ is odd the same is true of $y_2(x)$ in (9.4). These solutions, when suitably normalized, are called the *Legendre polynomials* of order ℓ; they are written $P_\ell(x)$ and are valid for all finite x. It is conventional to normalize $P_\ell(x)$ in such a way that $P_\ell(1) = 1$, and, since for ℓ even/odd the polynomial

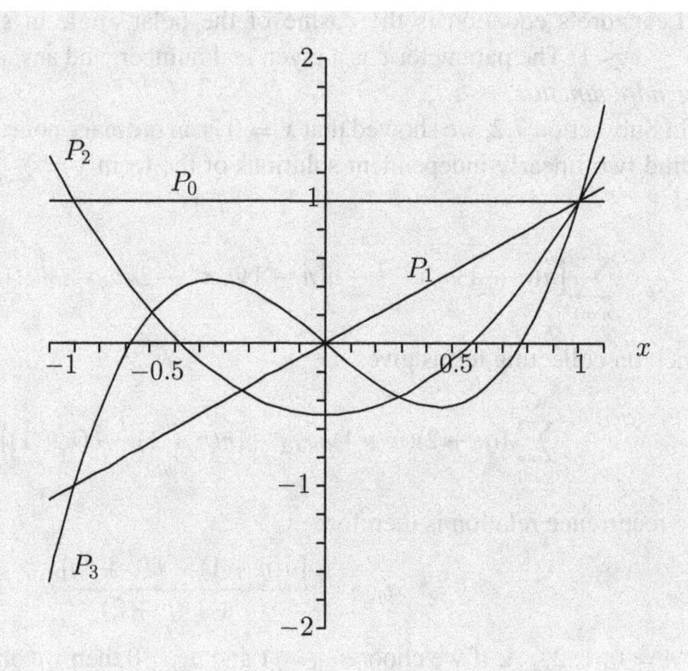

Figure 9.1 The first four Legendre polynomials.

consists only of even/odd powers of x, it follows that $P_\ell(-1) = (-1)^\ell$. The first few Legendre polynomials are easily constructed and are given by

$$P_0(x) = 1, \qquad P_1(x) = x,$$
$$P_2(x) = \tfrac{1}{2}(3x^2 - 1), \qquad P_3(x) = \tfrac{1}{2}(5x^3 - 3x),$$
$$P_4(x) = \tfrac{1}{8}(35x^4 - 30x^2 + 3), \quad P_5(x) = \tfrac{1}{8}(63x^5 - 70x^3 + 15x).$$

The first four Legendre polynomials are plotted in Figure 9.1.

Although, according to whether ℓ is an even or odd integer, respectively, either $y_1(x)$ in (9.3) or $y_2(x)$ in (9.4) terminates to give a multiple of the corresponding Legendre polynomial $P_\ell(x)$, the other series in each case does not terminate and therefore converges only for $|x| < 1$. According to whether ℓ is even or odd, we define *Legendre functions of the second kind* as $Q_\ell(x) = \alpha_\ell y_2(x)$ or $Q_\ell(x) = \beta_\ell y_1(x)$, respectively, where the constants α_ℓ and β_ℓ are conventionally taken to have the values

$$\alpha_\ell = \frac{(-1)^{\ell/2} 2^\ell [(\ell/2)!]^2}{\ell!} \qquad \text{for } \ell \text{ even,} \tag{9.5}$$

$$\beta_\ell = \frac{(-1)^{(\ell+1)/2} 2^{\ell-1} \{[(\ell-1)/2]!\}^2}{\ell!} \qquad \text{for } \ell \text{ odd.} \tag{9.6}$$

These complicated normalization factors are chosen so that the $Q_\ell(x)$ obey the same recurrence relations as the $P_\ell(x)$ (see Subsection 9.1.2).

The general solution of Legendre's equation for *integer* ℓ is therefore

$$y(x) = c_1 P_\ell(x) + c_2 Q_\ell(x), \tag{9.7}$$

where $P_\ell(x)$ is a polynomial of order ℓ, and so converges for all x, and $Q_\ell(x)$ is an infinite series that converges only for $|x| < 1$.[1]

By using the Wronskian method, Section 7.5, we may obtain closed forms for the $Q_\ell(x)$. That for $\ell = 0$ is found in the next worked example.

Example Use the Wronskian method to find a closed-form expression for $Q_0(x)$.

From (7.25) a second solution to Legendre's equation (9.1), with $\ell = 0$, is[2]

$$\begin{aligned} y_2(x) &= P_0(x) \int^x \frac{1}{[P_0(u)]^2} \exp\left(\int^u \frac{2v}{1-v^2}\, dv \right) du \\ &= \int^x \exp\left[-\ln(1-u^2) \right] du \\ &= \int^x \frac{du}{(1-u^2)} = \frac{1}{2} \ln\left(\frac{1+x}{1-x} \right), \end{aligned} \tag{9.8}$$

where, in the second line, we have used the fact that $P_0(x) = 1$, and in the final line have expressed the u integrand in partial fractions before integrating.

All that remains is to adjust the normalization of this solution so that it agrees with (9.5). Expanding the logarithm in (9.8) as the difference between two Maclaurin series we obtain

$$y_2(x) = x + \frac{x^3}{3} + \frac{x^5}{5} + \cdots .$$

Comparing this with the expression for $Q_0(x)$, using (9.4) with $\ell = 0$ and the normalization (9.5), we find that $y_2(x)$ is already correctly normalized, and so

$$Q_0(x) = \frac{1}{2} \ln\left(\frac{1+x}{1-x} \right).$$

Of course, we might have recognized the series (9.4) for $\ell = 0$, but to do so for larger ℓ would prove progressively more difficult. ◀

Using the above method for $\ell = 1$, we find[3]

$$Q_1(x) = \frac{x}{2} \ln\left(\frac{1+x}{1-x} \right) - 1.$$

Closed forms for higher-order $Q_\ell(x)$ may now be found using the recurrence relation (9.27) derived in the next subsection. The first few Legendre functions of the second kind are plotted in Figure 9.2.

1 It is possible, in fact, to find a second solution in terms of an infinite series of *negative* powers of x that is finite for $|x| > 1$ (see Problem 7.14).

2 Note that the integral over v in the round brackets is actually independent of the value of ℓ.

3 Carry through this calculation, noting that $P_1(x) = x$. Assume, as is the case here, that the required partial fraction expansion takes the form $Au^{-2} + B(1-u)^{-1} + C(1+u)^{-1}$, i.e. there is no u^{-1} term. Can you explain why this must be so?

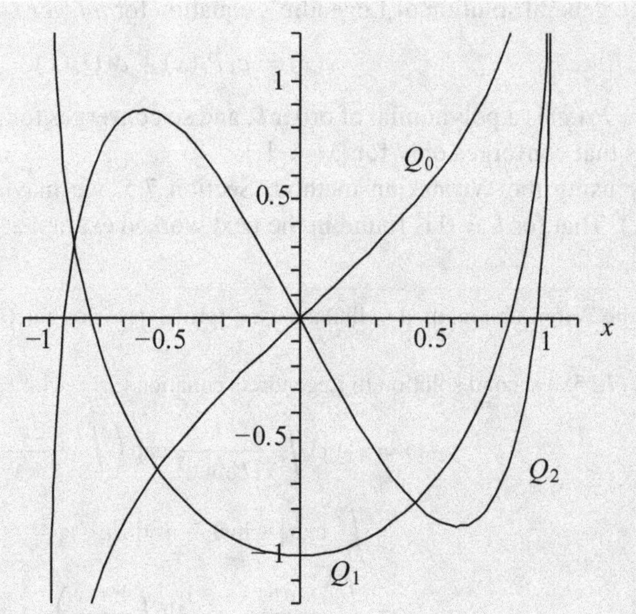

Figure 9.2 The first three Legendre functions of the second kind.

9.1.2 Properties of Legendre polynomials

As stated earlier, when encountered in physical problems the variable x in Legendre's equation is usually the cosine of the polar angle θ in spherical polar coordinates, and we then require the solution $y(x)$ to be regular at $x = \pm 1$, which corresponds to $\theta = 0$ or $\theta = \pi$. For this to occur we require the equation to have a polynomial solution, and so ℓ must be an integer. Furthermore, we also require the coefficient c_2 of the function $Q_\ell(x)$ in (9.7) to be zero, since $Q_\ell(x)$ is singular at $x = \pm 1$; the overall consequence of these requirements is that the general solution is simply a multiple of the relevant Legendre polynomial $P_\ell(x)$. For this reason, in this section we will study the properties of the Legendre polynomials $P_\ell(x)$ in some detail.

Rodrigues' formula

As an aid to establishing further properties of the Legendre polynomials we now develop Rodrigues' representation of these functions.[4] Rodrigues' formula for the $P_\ell(x)$ is

$$P_\ell(x) = \frac{1}{2^\ell \ell!} \frac{d^\ell}{dx^\ell}(x^2 - 1)^\ell. \tag{9.9}$$

4 Rodrigues' (not Rodrigue's) formula is a general device for generating a set of orthonormal polynomials $y_n(x)$ that satisfy a given differential equation of the form $p(x)y'' + r(x)y' + \lambda y = 0$ for a set of values of λ that are a particular (equation-dependent) function of n. Here $p(x)$ is a polynomial of degree 2 or less, and $r(x)$ is linear in x or a constant. If a non-negative weight function is defined by $w(x) = [p(x)]^{-1} \exp\{\int^x [r(u)/p(u)]\,du\}$, then the nth-order polynomial defined by $[w(x)]^{-1}$ times the nth derivative of $w(x)[p(x)]^n$ is proportional to $y_n(x)$. The polynomials so generated are independent and mutually orthogonal with respect to $w(x)$. They can be normalized by constants that, for any given equation, have a defined dependence on n.

To prove that this is a representation we let $u = (x^2 - 1)^\ell$, so that $u' = 2\ell x(x^2 - 1)^{\ell-1}$ and

$$(x^2 - 1)u' - 2\ell x u = 0.$$

If we differentiate this expression $\ell + 1$ times using Leibnitz' theorem, we obtain

$$\left[(x^2 - 1)u^{(\ell+2)} + 2x(\ell + 1)u^{(\ell+1)} + \ell(\ell + 1)u^{(\ell)}\right] - 2\ell\left[xu^{(\ell+1)} + (\ell + 1)u^{(\ell)}\right] = 0,$$

which reduces to

$$(x^2 - 1)u^{(\ell+2)} + 2xu^{(\ell+1)} - \ell(\ell + 1)u^{(\ell)} = 0.$$

Changing the sign all through, we recover Legendre's equation (9.1) with $u^{(\ell)}$ as the dependent variable. Since, from (9.9), ℓ is an integer and $u^{(\ell)}$ is regular at $x = \pm 1$, we may make the identification

$$u^{(\ell)}(x) = c_\ell P_\ell(x), \tag{9.10}$$

for some constant c_ℓ that depends on ℓ. To establish the value of c_ℓ we note that the only term in the expression for the ℓth derivative of $(x^2 - 1)^\ell$ that does not contain a factor $x^2 - 1$, and therefore does not vanish at $x = 1$, is $(2x)^\ell \ell!(x^2 - 1)^0$. Putting $x = 1$ in (9.10) and recalling that $P_\ell(1) = 1$, therefore shows that $c_\ell = 2^\ell \ell!$, thus completing the proof of Rodrigues' formula (9.9).

Example Use Rodrigues' formula to show that

$$I_\ell = \int_{-1}^{1} P_\ell(x)P_\ell(x)\,dx = \frac{2}{2\ell + 1}. \tag{9.11}$$

The result is trivially obvious for $\ell = 0$ and so we assume $\ell \geq 1$. Then, by Rodrigues' formula,

$$I_\ell = \frac{1}{2^{2\ell}(\ell!)^2} \int_{-1}^{1} \left[\frac{d^\ell(x^2 - 1)^\ell}{dx^\ell}\right]\left[\frac{d^\ell(x^2 - 1)^\ell}{dx^\ell}\right] dx.$$

Repeated integration by parts, with all boundary terms vanishing, reduces this to

$$I_\ell = \frac{(-1)^\ell}{2^{2\ell}(\ell!)^2} \int_{-1}^{1} (x^2 - 1)^\ell \frac{d^{2\ell}}{dx^{2\ell}}(x^2 - 1)^\ell\,dx$$

$$= \frac{(2\ell)!}{2^{2\ell}(\ell!)^2} \int_{-1}^{1} (1 - x^2)^\ell\,dx.$$

If we write

$$K_\ell = \int_{-1}^{1} (1 - x^2)^\ell\,dx,$$

then integration by parts (taking a factor 1 as the second part) gives

$$K_\ell = \int_{-1}^{1} 2\ell x^2(1 - x^2)^{\ell-1}\,dx.$$

Writing $2\ell x^2$ as $2\ell - 2\ell(1-x^2)$ we obtain

$$K_\ell = 2\ell \int_{-1}^{1} (1-x^2)^{\ell-1}\,dx - 2\ell \int_{-1}^{1} (1-x^2)^{\ell}\,dx$$
$$= 2\ell K_{\ell-1} - 2\ell K_\ell$$

and hence the recurrence relation $(2\ell+1)K_\ell = 2\ell K_{\ell-1}$. We therefore find

$$K_\ell = \frac{2\ell}{2\ell+1}\frac{2\ell-2}{2\ell-1}\cdots\frac{2}{3}K_0 = 2^\ell \ell!\frac{2^\ell \ell!}{(2\ell+1)!}2 = \frac{2^{2\ell+1}(\ell!)^2}{(2\ell+1)!},$$

which, when substituted into the expression for I_ℓ, establishes the required result. ◀

Mutual orthogonality

In Section 8.4, we noted that Legendre's equation was of Sturm–Liouville form with $p = 1 - x^2$, $q = 0$, $\lambda = \ell(\ell+1)$ and $\rho = 1$, and that its natural interval was $[-1, 1]$. Since the Legendre polynomials $P_\ell(x)$ are regular at the end-points $x = \pm 1$, they must be mutually orthogonal over this interval, i.e.

$$\int_{-1}^{1} P_\ell(x)P_k(x)\,dx = 0 \qquad \text{if } \ell \neq k. \tag{9.12}$$

Although this result follows from the general considerations of the previous chapter,[5] it may also be proved directly, as shown in the following example.

Example Prove directly that the Legendre polynomials $P_\ell(x)$ are mutually orthogonal over the interval $-1 < x < 1$.

Since the $P_\ell(x)$ satisfy Legendre's equation we may write

$$\left[(1-x^2)P_\ell'\right]' + \ell(\ell+1)P_\ell = 0,$$

where $P_\ell' = dP_\ell/dx$. Multiplying through by P_k and integrating from $x = -1$ to $x = 1$, we obtain

$$\int_{-1}^{1} P_k\left[(1-x^2)P_\ell'\right]'\,dx + \int_{-1}^{1} P_k\ell(\ell+1)P_\ell\,dx = 0.$$

Integrating the first term by parts and noting that the boundary contribution vanishes at both limits because of the factor $1 - x^2$, we find

$$-\int_{-1}^{1} P_k'(1-x^2)P_\ell'\,dx + \int_{-1}^{1} P_k\ell(\ell+1)P_\ell\,dx = 0.$$

Now, if we reverse the roles of ℓ and k and subtract one expression from the other, we conclude that

$$[k(k+1) - \ell(\ell+1)]\int_{-1}^{1} P_k P_\ell\,dx = 0,$$

5 Or from the establishment of the Legendre polynomials by Rodrigues' formula, taken together with the stated (but unproven) properties given in the previous footnote.

and therefore, since $k \neq \ell$, we must have the result (9.12). As a particular case, we note that if we put $k = 0$ we obtain

$$\int_{-1}^{1} P_\ell(x)\,dx = 0 \qquad \text{for } \ell \neq 0,$$

i.e. every Legendre polynomial, except $P_0(x)$, has zero average value over the range $-1 \leq x \leq 1$. The exception, having value 1 for all x, clearly has a unit average. ◀

As we discussed in the previous chapter, the mutual orthogonality (and completeness) of the $P_\ell(x)$ means that any reasonable function $f(x)$ (i.e. one obeying the Dirichlet conditions discussed at the start of Chapter 4) can be expressed in the interval $|x| < 1$ as an infinite sum of Legendre polynomials,

$$f(x) = \sum_{\ell=0}^{\infty} a_\ell P_\ell(x), \tag{9.13}$$

where the coefficients a_ℓ are given by

$$a_\ell = \frac{2\ell + 1}{2} \int_{-1}^{1} f(x) P_\ell(x)\,dx, \tag{9.14}$$

as is proved below. For polynomial functions $f(x)$, the sum has only a finite number of terms, the highest value of ℓ needed being equal to the degree of the polynomial.[6]

Example Prove the expression (9.14) for the coefficients in the Legendre polynomial expansion of a function $f(x)$.

If we multiply (9.13) by $P_k(x)$ and integrate from $x = -1$ to $x = 1$ then we obtain

$$\int_{-1}^{1} P_k(x) f(x)\,dx = \sum_{\ell=0}^{\infty} a_\ell \int_{-1}^{1} P_k(x) P_\ell(x)\,dx$$

$$= a_k \int_{-1}^{1} P_k(x) P_k(x)\,dx = \frac{2a_k}{2k+1},$$

where we have used the orthogonality property (9.12) and the normalization property (9.11). ◀

Generating function

A useful device for manipulating and studying sequences of functions or quantities labeled by an integer variable (here, the Legendre polynomials $P_\ell(x)$ labeled by ℓ) is a *generating function*. The generating function has perhaps its greatest utility in the area of probability theory (see Chapter 16). However, it is also a great convenience in our present study.

6 Express the function $f(x) = 1 + x + x^2 + x^3$ as a sum of Legendre polynomials. Check your expansion against the function at $x = +1$ and $x = -1$. In retrospect, can you see how to easily determine the final two a_ℓ needed for the expansion of a general polynomial, once the rest have been calculated? Confirm that your proposal works for the given example.

The generating function for, say, a series of functions $f_n(x)$ for $n = 0, 1, 2, \ldots$ is a function $G(x, h)$ containing, as well as x, a dummy variable h such that

$$G(x, h) = \sum_{n=0}^{\infty} f_n(x)h^n,$$

i.e. $f_n(x)$ is the coefficient of h^n in the expansion of G in powers of h. The utility of the device lies in the fact that sometimes it is possible to find a closed form for $G(x, h)$.

For our study of Legendre polynomials let us consider the functions $P_n(x)$ defined by the equation

$$G(x, h) = (1 - 2xh + h^2)^{-1/2} = \sum_{n=0}^{\infty} P_n(x)h^n. \tag{9.15}$$

As we show below, the functions so defined are identical to the Legendre polynomials and the function $(1 - 2xh + h^2)^{-1/2}$ is in fact the generating function for them. In the process we will also deduce several useful relationships between the various polynomials and their derivatives.

Example Show that the functions $P_n(x)$ defined by equation (9.15) satisfy Legendre's equation.

In the following $dP_n(x)/dx$ will be denoted by P_n'. Firstly, we differentiate the defining equation (9.15) with respect to x and get

$$h(1 - 2xh + h^2)^{-3/2} = \sum P_n' h^n. \tag{9.16}$$

Also, we differentiate (9.15) with respect to h to yield

$$(x - h)(1 - 2xh + h^2)^{-3/2} = \sum n P_n h^{n-1}. \tag{9.17}$$

Equation (9.16) can then be written, using (9.15), as

$$h \sum P_n h^n = (1 - 2xh + h^2) \sum P_n' h^n,$$

and equating the coefficients of h^{n+1} we obtain the recurrence relation

$$P_n = P_{n+1}' - 2xP_n' + P_{n-1}'. \tag{9.18}$$

Equations (9.16) and (9.17) can be combined as

$$(x - h) \sum P_n' h^n = h \sum n P_n h^{n-1},$$

from which the coefficient of h^n yields a second recurrence relation,

$$xP_n' - P_{n-1}' = nP_n; \tag{9.19}$$

eliminating P_{n-1}' between (9.18) and (9.19) then gives the further result

$$(n + 1)P_n = P_{n+1}' - xP_n'. \tag{9.20}$$

If we now take the result (9.20) with n replaced by $n-1$ and add x times (9.19) to it we obtain

$$(1-x^2)P_n' = n(P_{n-1} - xP_n). \tag{9.21}$$

Finally, differentiating both sides with respect to x and using (9.19) again, we find

$$\begin{aligned}(1-x^2)P_n'' - 2xP_n' &= n[(P_{n-1}' - xP_n') - P_n] \\ &= n(-nP_n - P_n) = -n(n+1)P_n,\end{aligned}$$

and so the P_n defined by (9.15) do indeed satisfy Legendre's equation. ◀

The above example shows that the functions $P_n(x)$ defined by (9.15) satisfy Legendre's equation with $\ell = n$ (an integer) and, also from (9.15), these functions are regular at $x = \pm 1$. Thus P_n must be some multiple of the nth Legendre polynomial. It therefore remains only to verify the normalization. This is easily done at $x = 1$, when G becomes

$$G(1, h) = [(1-h)^2]^{-1/2} = 1 + h + h^2 + \cdots,$$

and we can see that all the P_n so defined have $P_n(1) = 1$ as required, and are thus identical to the Legendre polynomials.[7]

A particular use of the generating function (9.15) is in representing the inverse distance between two points in three-dimensional space in terms of Legendre polynomials. If two points $\mathbf{r}$ and $\mathbf{r}'$ are at distances r and r', respectively, from the origin, with $r' < r$, then

$$\begin{aligned}\frac{1}{|\mathbf{r}-\mathbf{r}'|} &= \frac{1}{(r^2 + r'^2 - 2rr'\cos\theta)^{1/2}} \\ &= \frac{1}{r[1 - 2(r'/r)\cos\theta + (r'/r)^2]^{1/2}} \\ &= \frac{1}{r}\sum_{\ell=0}^{\infty}\left(\frac{r'}{r}\right)^{\ell} P_\ell(\cos\theta),\end{aligned} \tag{9.22}$$

where θ is the angle between the two position vectors $\mathbf{r}$ and $\mathbf{r}'$. If $r' > r$, however, r and r' must be exchanged in (9.22) or the series would not converge. This result may be used, for example, to write down the electrostatic potential at a point $\mathbf{r}$ due to a charge q at the point $\mathbf{r}'$. Thus, in the case $r' < r$, this is given by

$$V(\mathbf{r}) = \frac{q}{4\pi\epsilon_0 r}\sum_{\ell=0}^{\infty}\left(\frac{r'}{r}\right)^{\ell} P_\ell(\cos\theta).$$

We note that in the special case where the charge is at the origin, and $r' = 0$, only the $\ell = 0$ term in the series is non-zero and the expression reduces correctly to the familiar form $V(\mathbf{r}) = q/(4\pi\epsilon_0 r)$.

7 By using the generating function and considering the case of $x = 0$, show that the constant term in the polynomial expression for $P_\ell(x)$ is zero if ℓ is odd, and equal to $[(-1)^r(2r)!]/[4^r(r!)^2]$ if $\ell = 2r$.

Recurrence relations

In our discussion of the generating function above, we derived several useful recurrence relations satisfied by the Legendre polynomials $P_n(x)$. In particular, from (9.18), we have the four-term recurrence relation

$$P'_{n+1} + P'_{n-1} = P_n + 2xP'_n.$$

Also, from (9.19)–(9.21), we have the three-term recurrence relations

$$P'_{n+1} = (n+1)P_n + xP'_n, \tag{9.23}$$

$$P'_{n-1} = -nP_n + xP'_n, \tag{9.24}$$

$$(1-x^2)P'_n = n(P_{n-1} - xP_n), \tag{9.25}$$

$$(2n+1)P_n = P'_{n+1} - P'_{n-1}, \tag{9.26}$$

where the final relation is obtained immediately by subtracting the second from the first. Many other useful recurrence relations can be derived from those given above and from the generating function. We now derive one that contains no derivatives.

Example Prove the recurrence relation

$$(n+1)P_{n+1} = (2n+1)xP_n - nP_{n-1}. \tag{9.27}$$

Substituting from (9.15) into (9.17), we find

$$(x-h)\sum P_n h^n = (1 - 2xh + h^2)\sum nP_n h^{n-1}.$$

Equating coefficients of h^n we obtain

$$xP_n - P_{n-1} = (n+1)P_{n+1} - 2xnP_n + (n-1)P_{n-1},$$

which on rearrangement gives the stated result. ◀

The recurrence relation derived in the above example is particularly convenient for determining $P_n(x)$, either numerically or algebraically. One starts with $P_0(x) = 1$ and $P_1(x) = x$ and iterates the recurrence relation until $P_n(x)$ is obtained.[8]

In summary, the situation concerning Legendre polynomials is as follows. There are three possible starting points, which have been shown to be equivalent: the defining equation (9.1) together with the condition $P_n(1) = 1$; Rodrigues' formula (9.9); and the generating function (9.15). In addition there are a variety of relationships and recurrence relations (not particularly memorable, but collectively useful) and, as will be apparent from the work of Chapter 10, they together form a powerful tool for use in axially symmetric situations in which the ∇^2 operator is involved and spherical polar coordinates are employed.

8 Calculate $P_4(x)$ in this way, showing that it is given by $\frac{1}{8}(35x^4 - 30x^2 + 3)$.

9.2 Associated Legendre functions

The associated Legendre equation has the form

$$(1-x^2)y'' - 2xy' + \left[\ell(\ell+1) - \frac{m^2}{1-x^2}\right]y = 0; \tag{9.28}$$

it has three regular singular points, at $x = -1, 1$, and ∞, and reduces to Legendre's equation (9.1) when $m = 0$. It occurs in physical applications involving the operator ∇^2, when the latter is expressed in spherical polars. In such cases, $-\ell \leq m \leq \ell$ and m is restricted to integer values, a situation which we will assume from here on. As was the case for Legendre's equation, in normal usage the variable x is the cosine of the polar angle in spherical polars, and thus $-1 \leq x \leq 1$. Any solution of (9.28) is called an *associated Legendre function*.

The point $x = 0$ is an ordinary point of (9.28), and one could obtain series solutions of the form $y = \sum_{n=0} a_n x^n$ in the same manner as that used for Legendre's equation. In this case, however, it is more instructive to note that if $u(x)$ is a solution of Legendre's equation (9.1), then

$$y(x) = (1-x^2)^{|m|/2}\frac{d^{|m|}u}{dx^{|m|}} \tag{9.29}$$

is a solution of the associated equation (9.28), as we now prove.

Example Prove that if $u(x)$ is a solution of Legendre's equation, then $y(x)$ given in (9.29) is a solution of the associated equation.

For simplicity, let us begin by assuming that m is non-negative. Legendre's equation for u reads

$$(1-x^2)u'' - 2xu' + \ell(\ell+1)u = 0,$$

and, on differentiating this equation m times using Leibnitz' theorem, we obtain

$$(1-x^2)v'' - 2x(m+1)v' + (\ell-m)(\ell+m+1)v = 0, \tag{9.30}$$

where $v(x) = d^m u/dx^m$. On setting

$$y(x) = (1-x^2)^{m/2}v(x),$$

the derivatives v' and v'' may be written as

$$v' = (1-x^2)^{-m/2}\left(y' + \frac{mx}{1-x^2}y\right),$$

$$v'' = (1-x^2)^{-m/2}\left[y'' + \frac{2mx}{1-x^2}y' + \frac{m}{1-x^2}y + \frac{m(m+2)x^2}{(1-x^2)^2}y\right].$$

Substituting these expressions into (9.30) and simplifying, we obtain

$$(1-x^2)y'' - 2xy' + \left[\ell(\ell+1) - \frac{m^2}{1-x^2}\right]y = 0,$$

which shows that y is a solution of the associated Legendre equation (9.28). Finally, we note that if m is negative, the value of m^2 is unchanged, and so a solution for positive m is also a solution for the corresponding negative value of m.[9] ◀

Thus, by applying (9.29) to the two linearly independent series solutions of Legendre's equation given in (9.3) and (9.4), which we now denote by $u_1(x)$ and $u_2(x)$, we obtain two linearly independent series solutions $y_1(x)$ and $y_2(x)$ of the associated equation. From the general convergence properties of power series and their derivatives, we see that both $y_1(x)$ and $y_2(x)$ will, like $u_1(x)$ and $u_2(x)$, converge for $|x| < 1$. Hence the general solution to (9.28) in this range is given by

$$y(x) = c_1 y_1(x) + c_2 y_2(x).$$

9.2.1 Associated Legendre functions for integer ℓ

If ℓ and m are both integers, as is the case in most physical applications, then the general solution to (9.28) is denoted by

$$y(x) = c_1 P_\ell^m(x) + c_2 Q_\ell^m(x), \tag{9.31}$$

where $P_\ell^m(x)$ and $Q_\ell^m(x)$ are associated Legendre functions of the first and second kind, respectively. For non-negative values of m, these functions are related to the ordinary Legendre functions for integer ℓ by

$$P_\ell^m(x) = (1 - x^2)^{m/2} \frac{d^m P_\ell}{dx^m}, \qquad Q_\ell^m(x) = (1 - x^2)^{m/2} \frac{d^m Q_\ell}{dx^m}. \tag{9.32}$$

We see immediately that, as required, the associated Legendre functions reduce to the ordinary Legendre functions when $m = 0$. Since it is m^2 that appears in the associated Legendre equation (9.28), the associated Legendre functions for negative m values must be proportional to the corresponding function for non-negative m. The constant of proportionality is a matter of convention. For the $P_\ell^m(x)$ it is usual to regard the definition (9.32) as being valid also for negative m values. Although differentiating a negative number of times is not defined, when $P_\ell(x)$ is expressed in terms of the Rodrigues' formula (9.9), this problem does not occur for $-\ell \le m \le \ell$.[10] In this case,

$$P_\ell^{-m}(x) = (-1)^m \frac{(\ell - m)!}{(\ell + m)!} P_\ell^m(x). \tag{9.33}$$

9 Note that prescription (9.29) is expressed in terms of $|m|$ and specifies exactly the same actions, whether m is positive or negative.

10 Some authors define $P_\ell^{-m}(x) = P_\ell^m(x)$, and similarly for the $Q_\ell^m(x)$, in which case m is replaced by $|m|$ in the definitions (9.32). It should be noted that, in this case, many of the results presented in this section also require m to be replaced by $|m|$.

Example Prove result (9.33) for associated Legendre functions with negative values of m.

From (9.32) and the Rodrigues' formula (9.9) for the Legendre polynomials, we have

$$P_\ell^m(x) = \frac{1}{2^\ell \ell!}(1-x^2)^{m/2}\frac{d^{\ell+m}}{dx^{\ell+m}}(x^2-1)^\ell,$$

and, without loss of generality, we may assume that m is non-negative. It is convenient to write $(x^2-1) = (x+1)(x-1)$ and use Leibnitz' theorem to evaluate the derivative, which yields

$$P_\ell^m(x) = \frac{1}{2^\ell \ell!}(1-x^2)^{m/2}\sum_{r=0}^{\ell+m}\frac{(\ell+m)!}{r!(\ell+m-r)!}\frac{d^r(x+1)^\ell}{dx^r}\frac{d^{\ell+m-r}(x-1)^\ell}{dx^{\ell+m-r}}.$$

Considering the two derivative factors in a term in the summation, we note that the first is non-zero only for $r \le \ell$ and the second is non-zero for $\ell+m-r \le \ell$. Combining these conditions yields $m \le r \le \ell$. Performing the derivatives, we thus obtain

$$P_\ell^m(x) = \frac{1}{2^\ell \ell!}(1-x^2)^{m/2}\sum_{r=m}^{\ell}\frac{(\ell+m)!}{r!(\ell+m-r)!}\frac{\ell!(x+1)^{\ell-r}}{(\ell-r)!}\frac{\ell!(x-1)^{r-m}}{(r-m)!}$$

$$= (-1)^{m/2}\frac{\ell!(\ell+m)!}{2^\ell}\sum_{r=m}^{\ell}\frac{(x+1)^{\ell-r+\frac{m}{2}}(x-1)^{r-\frac{m}{2}}}{r!(\ell+m-r)!(\ell-r)!(r-m)!}. \qquad (9.34)$$

Repeating the above calculation for $P_\ell^{-m}(x)$ and identifying once more those terms in the sum that are non-zero, we find

$$P_\ell^{-m}(x) = (-1)^{-m/2}\frac{\ell!(\ell-m)!}{2^\ell}\sum_{r=0}^{\ell-m}\frac{(x+1)^{\ell-r-\frac{m}{2}}(x-1)^{r+\frac{m}{2}}}{r!(\ell-m-r)!(\ell-r)!(r+m)!}$$

$$= (-1)^{-m/2}\frac{\ell!(\ell-m)!}{2^\ell}\sum_{\bar{r}=m}^{\ell}\frac{(x+1)^{\ell-\bar{r}+\frac{m}{2}}(x-1)^{\bar{r}-\frac{m}{2}}}{(\bar{r}-m)!(\ell-\bar{r})!(\ell+m-\bar{r})!\bar{r}!}, \qquad (9.35)$$

where, in the second equality, we have rewritten the summation in terms of the new index $\bar{r} = r+m$. Comparing (9.34) and (9.35), we immediately arrive at the required result (9.33). ◀

Since $P_\ell(x)$ is a polynomial of order ℓ, we have $P_\ell^m(x) = 0$ for $|m| > \ell$. From its definition, it is clear that $P_\ell^m(x)$ is also a polynomial of order ℓ if m is even, but contains the factor $(1-x^2)$ to a fractional power if m is odd. In either case, $P_\ell^m(x)$ is regular at $x = \pm 1$. The first few associated Legendre functions of the first kind are easily constructed and are given by (omitting the $m = 0$ cases)[11]

$$P_1^1(x) = (1-x^2)^{1/2}, \qquad P_2^1(x) = 3x(1-x^2)^{1/2},$$

$$P_2^2(x) = 3(1-x^2), \qquad P_3^1(x) = \tfrac{3}{2}(5x^2-1)(1-x^2)^{1/2},$$

$$P_3^2(x) = 15x(1-x^2), \qquad P_3^3(x) = 15(1-x^2)^{3/2}.$$

Finally, we note that the associated Legendre functions of the second kind $Q_\ell^m(x)$, like $Q_\ell(x)$, are singular at $x = \pm 1$.

11 Taking $x = \cos\theta$, as in most physical examples, express the functions in terms of $\cos\theta$ and $\sin\theta$ and note how the powers of $\sin\theta$ vary with m for any given ℓ.

9.2.2 Properties of associated Legendre functions $P_\ell^m(x)$

When encountered in physical problems the variable x in the associated Legendre equation (as in the ordinary Legendre equation) is usually the cosine of the polar angle θ in spherical polar coordinates, and we then require the solution $y(x)$ to be regular at $x = \pm 1$ (corresponding to $\theta = 0$ or $\theta = \pi$). For this to occur, we require ℓ to be an integer and the coefficient c_2 of the function $Q_\ell^m(x)$ in (9.31) to be zero, since $Q_\ell^m(x)$ is singular at $x = \pm 1$, with the result that the general solution is simply some multiple of one of the associated Legendre functions of the first kind, $P_\ell^m(x)$. We will study the further properties of these functions in the remainder of this subsection.

Mutual orthogonality

As noted in Section 8.4, the associated Legendre equation is of Sturm–Liouville form $(py)' + qy + \lambda\rho y = 0$, with $p = 1 - x^2$, $q = -m^2/(1-x^2)$, $\lambda = \ell(\ell+1)$ and $\rho = 1$, and its natural interval is thus $[-1, 1]$. Since the associated Legendre functions $P_\ell^m(x)$ are regular at the end-points $x = \pm 1$, they must be mutually orthogonal with respect to weight function ρ over this interval for a fixed value of m, i.e.

$$\int_{-1}^{1} P_\ell^m(x) P_k^m(x)\, dx = 0 \qquad \text{if } \ell \neq k. \tag{9.36}$$

This result may also be proved directly in a manner similar to that used for demonstrating the orthogonality of the Legendre polynomials $P_\ell(x)$ in Subsection 9.1.2. Note that the value of m must be the same for the two associated Legendre functions for (9.36) to hold. The normalization condition when $\ell = k$ may be obtained using the Rodrigues' formula, as shown in the following example.

Example Show that

$$I_{\ell m} \equiv \int_{-1}^{1} P_\ell^m(x) P_\ell^m(x)\, dx = \frac{2}{2\ell+1}\frac{(\ell+m)!}{(\ell-m)!}. \tag{9.37}$$

From the definition (9.32) and the Rodrigues' formula (9.9) for $P_\ell(x)$, we may write

$$I_{\ell m} = \frac{1}{2^{2\ell}(\ell!)^2}\int_{-1}^{1}\left[(1-x^2)^m \frac{d^{\ell+m}(x^2-1)^\ell}{dx^{\ell+m}}\right]\left[\frac{d^{\ell+m}(x^2-1)^\ell}{dx^{\ell+m}}\right] dx,$$

where the square brackets identify the factors to be used when integrating by parts. Performing the integration by parts $\ell + m$ times, and noting that all boundary terms vanish, we obtain

$$I_{\ell m} = \frac{(-1)^{\ell+m}}{2^{2\ell}(\ell!)^2}\int_{-1}^{1}(x^2-1)^\ell \frac{d^{\ell+m}}{dx^{\ell+m}}\left[(1-x^2)^m \frac{d^{\ell+m}(x^2-1)^\ell}{dx^{\ell+m}}\right] dx.$$

Using Leibnitz' theorem, the second factor in the integrand may be written as

$$\frac{d^{\ell+m}}{dx^{\ell+m}}\left[(1-x^2)^m \frac{d^{\ell+m}(x^2-1)^\ell}{dx^{\ell+m}}\right] = \sum_{r=0}^{\ell+m}\frac{(\ell+m)!}{r!(\ell+m-r)!}\frac{d^r(1-x^2)^m}{dx^r}\frac{d^{2\ell+2m-r}(x^2-1)^\ell}{dx^{2\ell+2m-r}}.$$

Considering the two derivative factors in a term in the summation on the RHS, we see that the first is non-zero only for $r \le 2m$, whereas the second is non-zero only for $2\ell + 2m - r \le 2\ell$. Combining these conditions, we find that the only non-zero term in the sum is that for which $r = 2m$. Thus, we may write

$$I_{\ell m} = \frac{(-1)^{\ell+m}}{2^{2\ell}(\ell!)^2}\frac{(\ell+m)!}{(2m)!(\ell-m)!}\int_{-1}^{1}(1-x^2)^{\ell}\frac{d^{2m}(1-x^2)^m}{dx^{2m}}\frac{d^{2\ell}(1-x^2)^{\ell}}{dx^{2\ell}}\,dx.$$

Since $d^{2\ell}(1-x^2)^{\ell}/dx^{2\ell} = (-1)^{\ell}(2\ell)!$, and noting that $(-1)^{2\ell+2m} = 1$, we have

$$I_{\ell m} = \frac{1}{2^{2\ell}(\ell!)^2}\frac{(2\ell)!(\ell+m)!}{(\ell-m)!}\int_{-1}^{1}(1-x^2)^{\ell}\,dx.$$

We have already shown in Subsection 9.1.2 that

$$K_{\ell} \equiv \int_{-1}^{1}(1-x^2)^{\ell}\,dx = \frac{2^{2\ell+1}(\ell!)^2}{(2\ell+1)!},$$

and so we obtain the final result

$$I_{\ell m} = \frac{2}{2\ell+1}\frac{(\ell+m)!}{(\ell-m)!}.$$

As expected, for $m = 0$ this reduces to the corresponding result for Legendre polynomials. ◀

The orthogonality and normalization conditions, (9.36) and (9.37) respectively, mean that the associated Legendre functions $P_{\ell}^m(x)$, with m fixed, may be used in a similar way to the Legendre polynomials to expand any reasonable function $f(x)$ on the interval $|x| < 1$ in a series of the form

$$f(x) = \sum_{k=0}^{\infty} a_{m+k}P_{m+k}^m(x), \tag{9.38}$$

where, in this case, the coefficients are given by

$$a_{\ell} = \frac{2\ell+1}{2}\frac{(\ell-m)!}{(\ell+m)!}\int_{-1}^{1} f(x)P_{\ell}^m(x)\,dx.$$

We note that the series takes the form (9.38) because $P_{\ell}^m(x) = 0$ for $m > \ell$.

Finally, it is worth noting that the associated Legendre functions $P_{\ell}^m(x)$ must also obey a second orthogonality relationship. This comes about because one may equally well write the associated Legendre equation (9.28) in Sturm–Liouville form $(py)' + qy + \lambda\rho y = 0$, with $p = 1 - x^2$, $q = \ell(\ell+1)$, $\lambda = -m^2$ and $\rho = (1-x^2)^{-1}$; once again the natural interval is $[-1, 1]$. Since the associated Legendre functions $P_{\ell}^m(x)$ are regular at the end-points $x = \pm 1$, they must therefore be mutually orthogonal with respect to the weight function $(1-x^2)^{-1}$ over this interval for a fixed value of ℓ, i.e.

$$\int_{-1}^{1} P_{\ell}^m(x)P_{\ell}^k(x)(1-x^2)^{-1}\,dx = 0 \quad \text{if } |m| \ne |k|. \tag{9.39}$$

One may also show straightforwardly that the corresponding normalization condition when $m = k$ is given by

$$\int_{-1}^{1} P_\ell^m(x)P_\ell^m(x)(1-x^2)^{-1}\,dx = \frac{(\ell+m)!}{m(\ell-m)!}.$$

In solving physical problems, however, the orthogonality condition (9.39) is not of any practical use.

Generating function

The generating function for associated Legendre functions can be easily derived by combining their definition (9.32) with the generating function for the Legendre polynomials given in (9.15). We find that

$$G(x,h) = \frac{(2m)!(1-x^2)^{m/2}}{2^m m!(1-2hx+h^2)^{m+1/2}} = \sum_{n=0}^{\infty} P_{n+m}^m(x)h^n. \tag{9.40}$$

Example Derive expression (9.40) for the associated Legendre generating function.

The generating function (9.15) for the Legendre polynomials reads

$$\sum_{n=0}^{\infty} P_n h^n = (1-2xh+h^2)^{-1/2}.$$

Differentiating both sides of this result m times (assuming m to be non-negative), multiplying through by $(1-x^2)^{m/2}$ and using the definition (9.32) of the associated Legendre functions, we obtain

$$\sum_{n=0}^{\infty} P_n^m h^n = (1-x^2)^{m/2}\frac{d^m}{dx^m}(1-2xh+h^2)^{-1/2}.$$

Performing the derivatives on the RHS gives

$$\sum_{n=0}^{\infty} P_n^m h^n = \frac{1\cdot 3\cdot 5\cdots(2m-1)(1-x^2)^{m/2}h^m}{(1-2xh+h^2)^{m+1/2}}.$$

Dividing through by h^m, re-indexing the summation on the LHS and noting that, quite generally,

$$1\cdot 3\cdot 5\cdots(2r-1) = \frac{1\cdot 2\cdot 3\cdots 2r}{2\cdot 4\cdot 6\cdots 2r} = \frac{(2r)!}{2^r r!},$$

we obtain the final result (9.40).[12] ◀

Recurrence relations

As one might expect, the associated Legendre functions satisfy certain recurrence relations. Indeed, the presence of the two indices n and m means that a much wider range of recurrence relations may be derived. Here we shall content ourselves with quoting just

12 Use the generating function to calculate $P_3^1(x)$ as given on p. 335.

four of the more useful ones:

$$P_n^{m+1} = \frac{2mx}{(1-x^2)^{1/2}} P_n^m + [m(m-1) - n(n+1)] P_n^{m-1}, \tag{9.41}$$

$$(2n+1)x P_n^m = (n+m) P_{n-1}^m + (n-m+1) P_{n+1}^m, \tag{9.42}$$

$$(2n+1) P_n^m = (P_{n+1}^{m+1} - P_{n-1}^{m+1})(1-x^2)^{-1/2}, \tag{9.43}$$

$$2(1-x^2)^{1/2}(P_n^m)' = P_n^{m+1} - (n+m)(n-m+1) P_n^{m-1}. \tag{9.44}$$

We note that, by virtue of our adopted definition (9.32), these recurrence relations are equally valid for negative and non-negative values of m. These relations may be derived in a number of ways, such as using the generating function (9.40) or, as shown below, by differentiation of the recurrence relations for the Legendre polynomials $P_\ell(x)$.

Example Use the recurrence relation $(2n+1)P_n = P'_{n+1} - P'_{n-1}$ for Legendre polynomials to derive the result (9.43).

Differentiating the recurrence relation for the Legendre polynomials m times, we have

$$(2n+1)\frac{d^m P_n}{dx^m} = \frac{d^{m+1} P_{n+1}}{dx^{m+1}} - \frac{d^{m+1} P_{n-1}}{dx^{m+1}}.$$

Multiplying through by $(1-x^2)^{(m+1)/2}$ and using the definition (9.32) immediately gives the result (9.43). ◀

9.3 Spherical harmonics

The associated Legendre functions discussed in the previous section occur most commonly when obtaining solutions in spherical polar coordinates of Laplace's equation $\nabla^2 u = 0$ (see Subsection 11.3.1). In particular, one finds that, for solutions that are finite on the polar axis, the angular part of the solution is given by

$$\Theta(\theta)\Phi(\phi) = P_\ell^m(\cos\theta)(C\cos m\phi + D\sin m\phi),$$

where ℓ and m are integers with $-\ell \le m \le \ell$. This general form is sufficiently common that particular functions of θ and ϕ called *spherical harmonics* are defined and tabulated. The spherical harmonics $Y_\ell^m(\theta,\phi)$ are defined by

$$Y_\ell^m(\theta,\phi) = (-1)^m \left[\frac{2\ell+1}{4\pi}\frac{(\ell-m)!}{(\ell+m)!}\right]^{1/2} P_\ell^m(\cos\theta)\exp(im\phi). \tag{9.45}$$

Using (9.33), we note that

$$Y_\ell^{-m}(\theta,\phi) = (-1)^m \left[Y_\ell^m(\theta,\phi)\right]^*,$$

where the asterisk denotes complex conjugation. The first few spherical harmonics

$Y_\ell^m(\theta, \phi) \equiv Y_\ell^m$ are as follows:

$$Y_0^0 = \sqrt{\frac{1}{4\pi}}, \qquad Y_1^0 = \sqrt{\frac{3}{4\pi}} \cos\theta,$$

$$Y_1^{\pm 1} = \mp\sqrt{\frac{3}{8\pi}} \sin\theta \exp(\pm i\phi), \qquad Y_2^0 = \sqrt{\frac{5}{16\pi}}(3\cos^2\theta - 1),$$

$$Y_2^{\pm 1} = \mp\sqrt{\frac{15}{8\pi}} \sin\theta \cos\theta \exp(\pm i\phi), \quad Y_2^{\pm 2} = \sqrt{\frac{15}{32\pi}} \sin^2\theta \exp(\pm 2i\phi).$$

Since they contain as their θ-dependent part the solution P_ℓ^m to the associated Legendre equation, the Y_ℓ^m are mutually orthogonal when integrated from -1 to $+1$ over $d(\cos\theta)$. Their mutual orthogonality with respect to ϕ $(0 \le \phi \le 2\pi)$ is even more obvious. The numerical factor in (9.45) is chosen to make the Y_ℓ^m an orthonormal set, i.e.

$$\int_{-1}^{1} \int_{0}^{2\pi} \left[Y_\ell^m(\theta, \phi)\right]^* Y_{\ell'}^{m'}(\theta, \phi)\, d\phi\, d(\cos\theta) = \delta_{\ell\ell'}\delta_{mm'}. \tag{9.46}$$

In addition, the spherical harmonics form a complete set in that any reasonable function (i.e. one that is likely to be met in a physical situation) of θ and ϕ can be expanded as a sum of such functions,

$$f(\theta, \phi) = \sum_{\ell=0}^{\infty} \sum_{m=-\ell}^{\ell} a_{\ell m} Y_\ell^m(\theta, \phi), \tag{9.47}$$

the constants $a_{\ell m}$ being given by

$$a_{\ell m} = \int_{-1}^{1} \int_{0}^{2\pi} \left[Y_\ell^m(\theta, \phi)\right]^* f(\theta, \phi)\, d\phi\, d(\cos\theta). \tag{9.48}$$

This is in exact analogy with a Fourier series and is a particular example of the general property of Sturm–Liouville solutions.

Aside from the orthonormality condition (9.46), the most important relationship obeyed by the Y_ℓ^m is the *spherical harmonic addition theorem*. This reads

$$P_\ell(\cos\gamma) = \frac{4\pi}{2\ell + 1} \sum_{m=-\ell}^{\ell} Y_\ell^m(\theta, \phi)[Y_\ell^m(\theta', \phi')]^*, \tag{9.49}$$

where (θ, ϕ) and (θ', ϕ') denote two different directions in our spherical polar coordinate system that are separated by an angle γ. Spherical trigonometry (or vector methods) shows that the connection between these angles is

$$\cos\gamma = \cos\theta\, \cos\theta' + \sin\theta\, \sin\theta'\, \cos(\phi - \phi'). \tag{9.50}$$

The proof of (9.49) is somewhat lengthy and of little help when it comes to applying the theorem, and so we do not give it here.[13]

13 But verify that (9.49) is satisfied in the particular case of $\ell = 1$ and two diametrically opposed directions (θ, ϕ) and $(\pi - \theta, \phi + \pi)$.

9.4 Chebyshev functions

Chebyshev's equation has the form

$$(1-x^2)y'' - xy' + \nu^2 y = 0, \tag{9.51}$$

and has three regular singular points, at $x = -1, 1, \infty$. By comparing it with (9.1), we see that the Chebyshev equation is very similar in form to Legendre's equation. Despite this similarity, equation (9.51) does not occur very often in physical problems, though its solutions are of considerable importance in numerical analysis. The parameter ν is a given real number, but in nearly all practical applications it takes an integer value. From here on we thus assume that $\nu = n$, where n is a non-negative integer. As was the case for Legendre's equation, in normal usage the variable x is the cosine of an angle, and so $1 \le x \le 1$. Any solution of (9.51) is called a *Chebyshev function*.

The point $x = 0$ is an ordinary point of (9.51), and so we expect to find two linearly independent solutions of the form $y = \sum_{m=0}^{\infty} a_m x^m$. One could find the recurrence relations for the coefficients a_m in a similar manner to that used for Legendre's equation in Section 9.1 (see Problem 7.13). For Chebyshev's equation, however, it is easier and more illuminating to take a different approach. In particular, we note that on making the substitution $x = \cos\theta$, and consequently $d/dx = (-1/\sin\theta)\,d/d\theta$, Chebyshev's equation becomes (with $\nu = n$)

$$\frac{d^2y}{d\theta^2} + n^2 y = 0,$$

which is the simple harmonic equation with solutions $\cos n\theta$ and $\sin n\theta$. The corresponding linearly independent solutions of Chebyshev's equation are thus given by

$$T_n(x) = \cos(n\cos^{-1}x) \quad \text{and} \quad V_n(x) = \sin(n\cos^{-1}x). \tag{9.52}$$

It is straightforward to show that the $T_n(x)$ are *polynomials* of order n, whereas the $V_n(x)$ are *not* polynomials. This we now do.

Example Find explicit forms for the series expansions of $T_n(x)$ and $V_n(x)$.

Writing $x = \cos\theta$, it is convenient first to form the complex superposition

$$\begin{aligned} T_n(x) + iV_n(x) &= \cos n\theta + i\sin n\theta \\ &= (\cos\theta + i\sin\theta)^n \\ &= \left(x + i\sqrt{1-x^2}\right)^n \qquad \text{for } |x| \le 1. \end{aligned}$$

Then, on expanding out the last expression using the binomial theorem, we obtain

$$T_n(x) = x^n - {}^nC_2 x^{n-2}(1-x^2) + {}^nC_4 x^{n-4}(1-x^2)^2 - \cdots, \tag{9.53}$$

$$V_n(x) = \sqrt{1-x^2}\left[{}^nC_1 x^{n-1} - {}^nC_3 x^{n-3}(1-x^2) + {}^nC_5 x^{n-5}(1-x^2)^2 - \cdots\right], \tag{9.54}$$

where ${}^nC_r = n!/[r!(n-r)!]$ is a binomial coefficient. We thus see that $T_n(x)$ is a polynomial of order n, but $V_n(x)$ is not a polynomial. ◀

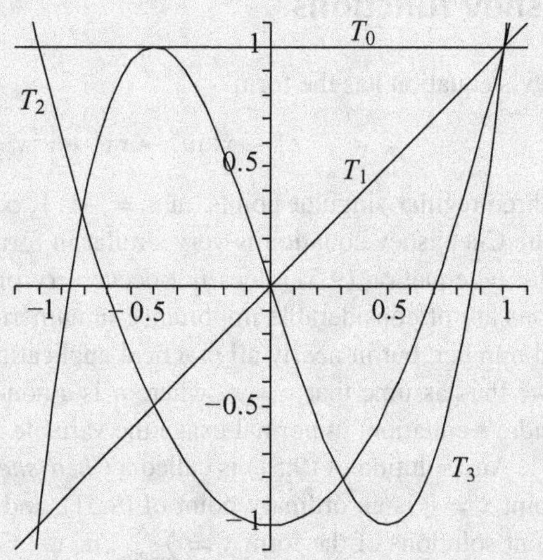

Figure 9.3 The first four Chebyshev polynomials of the first kind.

It is conventional to define the additional functions

$$W_n(x) = (1 - x^2)^{-1/2} T_{n+1}(x) \quad \text{and} \quad U_n(x) = (1 - x^2)^{-1/2} V_{n+1}(x). \tag{9.55}$$

From (9.53) and (9.54), we see immediately that $U_n(x)$ is a *polynomial* of order n, but that $W_n(x)$ is *not* a polynomial. In practice, it is usual to work entirely in terms of $T_n(x)$ and $U_n(x)$, which are known, respectively, as *Chebyshev polynomials of the first and second kind*. In particular, we note that the general solution to Chebyshev's equation can be written in terms of these polynomials as

$$y(x) = \begin{cases} c_1 T_n(x) + c_2 \sqrt{1 - x^2}\, U_{n-1}(x) & \text{for } n = 1, 2, 3, \ldots, \\ c_1 + c_2 \sin^{-1} x & \text{for } n = 0. \end{cases}$$

The $n = 0$ solution could also be written as $d_1 + c_2 \cos^{-1} x$ with $d_1 = c_1 + \frac{1}{2}\pi c_2$.

The first few Chebyshev polynomials of the first kind are easily constructed and are given by[14]

$$T_0(x) = 1, \qquad T_1(x) = x,$$
$$T_2(x) = 2x^2 - 1, \qquad T_3(x) = 4x^3 - 3x,$$
$$T_4(x) = 8x^4 - 8x^2 + 1, \qquad T_5(x) = 16x^5 - 20x^3 + 5x.$$

The functions $T_0(x)$, $T_1(x)$, $T_2(x)$ and $T_3(x)$ are plotted in Figure 9.3.
In general, the Chebyshev polynomials $T_n(x)$ satisfy $T_n(-x) = (-1)^n T_n(x)$, which is easily deduced from (9.53). Similarly, it is straightforward to deduce the following special

14 $T_n(x)$ is simply the expression for $\cos n\theta$ when written in terms of $\cos\theta = x$. Thus, $\cos 2\theta = 2\cos^2\theta - 1 \;\Rightarrow\; T_2(x) = 2x^2 - 1$, etc. Similarly, $U_n(x)$ is $[\sin(n+1)\theta]/[\sin\theta]$ written in terms of $\cos\theta$. See equations (9.62) and (9.63).

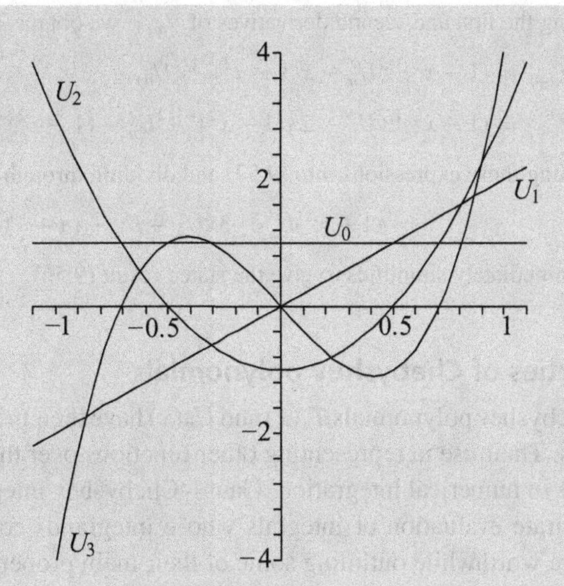

Figure 9.4 The first four Chebyshev polynomials of the second kind.

values:

$$T_n(1) = 1, \qquad T_n(-1) = (-1)^n, \qquad T_{2n}(0) = (-1)^n, \qquad T_{2n+1}(0) = 0.$$

The first few Chebyshev polynomials of the second kind are also easily found and read

$$\begin{aligned} U_0(x) &= 1, & U_1(x) &= 2x, \\ U_2(x) &= 4x^2 - 1, & U_3(x) &= 8x^3 - 4x, \\ U_4(x) &= 16x^4 - 12x^2 + 1, & U_5(x) &= 32x^5 - 32x^3 + 6x. \end{aligned}$$

The functions $U_0(x)$, $U_1(x)$, $U_2(x)$ and $U_3(x)$ are plotted in Figure 9.4. As may be deduced from (9.54) and (9.55), the Chebyshev polynomials $U_n(x)$ satisfy $U_n(-x) = (-1)^n U_n(x)$; they also have the special values:

$$U_n(1) = n + 1, \qquad U_n(-1) = (-1)^n(n+1), \qquad U_{2n}(0) = (-1)^n, \qquad U_{2n+1}(0) = 0.$$

The equation that the derived functions $U_n(x)$ satisfy is found in the next worked example.

Example Show that the Chebyshev polynomials $U_n(x)$ satisfy the differential equation

$$(1 - x^2)U_n''(x) - 3xU_n'(x) + n(n+2)U_n(x) = 0. \tag{9.56}$$

From (9.55), we have $V_{n+1} = (1 - x^2)^{1/2} U_n$ and these functions satisfy the Chebyshev equation (9.51) with $\nu = n + 1$, namely

$$(1 - x^2)V_{n+1}'' - xV_{n+1}' + (n+1)^2 V_{n+1} = 0. \tag{9.57}$$

Evaluating the first and second derivatives of V_{n+1}, we obtain

$$V'_{n+1} = (1-x^2)^{1/2}U'_n - x(1-x^2)^{-1/2}U_n,$$

$$V''_{n+1} = (1-x^2)^{1/2}U''_n - 2x(1-x^2)^{-1/2}U'_n - (1-x^2)^{-1/2}U_n - x^2(1-x^2)^{-3/2}U_n.$$

Substituting these expressions into (9.57) and dividing through by $(1-x^2)^{1/2}$, we find

$$(1-x^2)U''_n - 3xU'_n - U_n + (n+1)^2U_n = 0,$$

which immediately simplifies to give the stated result (9.56). ◀

9.4.1 Properties of Chebyshev polynomials

The Chebyshev polynomials $T_n(x)$ and $U_n(x)$ have their principal applications in numerical analysis. Their use in representing other functions over the range $|x| < 1$ plays an important role in numerical integration; Gauss–Chebyshev integration is of particular value for the accurate evaluation of integrals whose integrands contain factors $(1-x^2)^{\pm 1/2}$. It is therefore worthwhile outlining some of their main properties.

Rodrigues' formula

The Chebyshev polynomials $T_n(x)$ and $U_n(x)$ may be expressed in terms of a Rodrigues' formula, in a similar way to that used for the Legendre polynomials discussed in Subsection 9.1.2. For the Chebyshev polynomials, we have

$$T_n(x) = \frac{(-1)^n\sqrt{\pi}(1-x^2)^{1/2}}{2^n(n-\frac{1}{2})!}\frac{d^n}{dx^n}(1-x^2)^{n-\frac{1}{2}},$$

$$U_n(x) = \frac{(-1)^n\sqrt{\pi}(n+1)}{2^{n+1}(n+\frac{1}{2})!(1-x^2)^{1/2}}\frac{d^n}{dx^n}(1-x^2)^{n+\frac{1}{2}}.$$

These Rodrigues' formulae may be proved in an analogous manner to that used in Subsection 9.1.2 when establishing the corresponding expression for the Legendre polynomials.

Mutual orthogonality

In Section 8.4, we noted that Chebyshev's equation could be put into Sturm–Liouville form with $p = (1-x^2)^{1/2}$, $q = 0$, $\lambda = n^2$ and $\rho = (1-x^2)^{-1/2}$, and its natural interval is thus $[-1, 1]$. Since the Chebyshev polynomials of the first kind, $T_n(x)$, are solutions of the Chebyshev equation and are regular at the end-points $x = \pm 1$, they must be mutually orthogonal over this interval with respect to the weight function $\rho = (1-x^2)^{-1/2}$, i.e.

$$\int_{-1}^{1} T_n(x)T_m(x)(1-x^2)^{-1/2}\,dx = 0 \qquad \text{if } n \neq m. \tag{9.58}$$

The normalization, when $m = n$, is easily found by making the substitution $x = \cos\theta$ and using (9.52). We immediately obtain

$$\int_{-1}^{1} T_n(x)T_n(x)(1-x^2)^{-1/2}\,dx = \begin{cases} \pi & \text{for } n = 0, \\ \pi/2 & \text{for } n = 1, 2, 3, \ldots. \end{cases} \tag{9.59}$$

The orthogonality and normalization conditions mean that any (reasonable) function $f(x)$ can be expanded over the interval $|x| < 1$ in a series of the form

$$f(x) = \tfrac{1}{2}a_0 + \textstyle\sum_{n=1}^{\infty} a_n T_n(x),$$

where the coefficients in the expansion are given by[15]

$$a_n = \frac{2}{\pi}\int_{-1}^{1} f(x)T_n(x)(1-x^2)^{-1/2}\,dx.$$

For the Chebyshev polynomials of the second kind, $U_n(x)$, we see from (9.55) that $(1-x^2)^{1/2}U_n(x) = V_{n+1}(x)$ satisfies Chebyshev's equation (9.51) with $\nu = n+1$. Thus, the orthogonality relation for the $U_n(x)$, obtained by replacing $T_i(x)$ by $V_{i+1}(x)$ in equation (9.58), reads

$$\int_{-1}^{1} U_n(x)U_m(x)(1-x^2)^{1/2}\,dx = 0 \qquad \text{if } n \neq m.$$

The corresponding normalization condition, when $n = m$, can again be found by making the substitution $x = \cos\theta$, as illustrated in the following example.

Example Show that

$$I \equiv \int_{-1}^{1} U_n(x)U_n(x)(1-x^2)^{1/2}\,dx = \frac{\pi}{2}.$$

From (9.55), we see that

$$I = \int_{-1}^{1} V_{n+1}(x)V_{n+1}(x)(1-x^2)^{-1/2}\,dx,$$

which, on substituting $x = \cos\theta$, gives

$$I = \int_{\pi}^{0} \sin(n+1)\theta\,\sin(n+1)\theta\,\frac{1}{\sin\theta}(-\sin\theta)\,d\theta = \frac{\pi}{2}.$$

At the final step, we have used the standard result about the integral of the square of a sinusoid. ◀

The above orthogonality and normalization conditions allow one to expand any (reasonable) function in the interval $|x| < 1$ in a series of the form

$$f(x) = \sum_{n=0}^{\infty} a_n U_n(x),$$

in which the coefficients a_n are given by

$$a_n = \frac{2}{\pi}\int_{-1}^{1} f(x)U_n(x)(1-x^2)^{1/2}\,dx.$$

15 Express the function $f(x) = 1 + x + x^2 + x^3$ as a sum of Chebyshev polynomials. It is easier in this case to rewrite $f(x)$ directly in terms of the $T_n(x)$ by adding and subtracting terms, starting from the highest power of x present. Compare with footnote 6.

Generating functions

The generating functions for the Chebyshev polynomials of the first and second kinds are given, respectively, by

$$G_{\mathrm{I}}(x, h) = \frac{1 - xh}{1 - 2xh + h^2} = \sum_{n=0}^{\infty} T_n(x)h^n, \tag{9.60}$$

$$G_{\mathrm{II}}(x, h) = \frac{1}{1 - 2xh + h^2} = \sum_{n=0}^{\infty} U_n(x)h^n. \tag{9.61}$$

These prescriptions may be proved in a manner similar to that used in Subsection 9.1.2 for the generating function of the Legendre polynomials. For the Chebyshev polynomials, however, the generating functions are of less practical use, since most of the useful results can be obtained more easily by taking advantage of the trigonometric forms (9.52), as illustrated below.

Recurrence relations

There exist many useful recurrence relationships for the Chebyshev polynomials $T_n(x)$ and $U_n(x)$. They are most easily derived by setting $x = \cos\theta$ and using (9.52) and (9.55) to write

$$T_n(x) = T_n(\cos\theta) = \cos n\theta, \tag{9.62}$$

$$U_n(x) = U_n(\cos\theta) = \frac{\sin(n+1)\theta}{\sin\theta}. \tag{9.63}$$

One may then use standard formulae for the trigonometric functions to derive a wide variety of recurrence relations. Of particular use are the trigonometric identities

$$\cos(n \pm 1)\theta = \cos n\theta\ \cos\theta \mp \sin n\theta\ \sin\theta, \tag{9.64}$$

$$\sin(n \pm 1)\theta = \sin n\theta\ \cos\theta \pm \cos n\theta\ \sin\theta. \tag{9.65}$$

Example Show that the Chebyshev polynomials satisfy the recurrence relations

$$T_{n+1}(x) - 2xT_n(x) + T_{n-1}(x) = 0, \tag{9.66}$$

$$U_{n+1}(x) - 2xU_n(x) + U_{n-1}(x) = 0. \tag{9.67}$$

Adding the result (9.64) with the plus sign to the corresponding result with a minus sign gives

$$\cos(n+1)\theta + \cos(n-1)\theta = 2\cos n\theta\ \cos\theta.$$

Using (9.62) and setting $x = \cos\theta$ immediately gives a rearrangement of the required result (9.66). Similarly, adding the plus and minus cases of result (9.65) gives

$$\sin(n+1)\theta + \sin(n-1)\theta = 2\sin n\theta\ \cos\theta.$$

Dividing through on both sides by $\sin\theta$ and using (9.63) yields (9.67). ◀

The recurrence relations (9.66) and (9.67) are extremely useful in the practical computation of Chebyshev polynomials. For example, given the values of $T_0(x)$ and $T_1(x)$ at some point x, the result (9.66) may be used iteratively to obtain the value of any $T_n(x)$ at that point; similarly, (9.67) may be used to calculate the value of any $U_n(x)$ at some point x, given the values of $U_0(x)$ and $U_1(x)$ at that point.

Further recurrence relations satisfied by the Chebyshev polynomials are

$$T_n(x) = U_n(x) - xU_{n-1}(x), \tag{9.68}$$

$$(1-x^2)U_n(x) = xT_{n+1}(x) - T_{n+2}(x), \tag{9.69}$$

which establish useful relationships between the two sets of polynomials $T_n(x)$ and $U_n(x)$. The relation (9.68) follows immediately from (9.65), whereas (9.69) follows from (9.64), with n replaced by $n+1$, on noting that $\sin^2\theta = 1 - x^2$. Additional useful results concerning the derivatives of Chebyshev polynomials may be obtained from (9.62) and (9.63), as illustrated in the following example.

Example Show that

$$T_n'(x) = nU_{n-1}(x),$$
$$(1-x^2)U_n'(x) = xU_n(x) - (n+1)T_{n+1}(x).$$

These results are most easily derived from the expressions (9.62) and (9.63) by noting that $d/dx = (-1/\sin\theta)\,d/d\theta$. Thus,

$$T_n'(x) = -\frac{1}{\sin\theta}\frac{d(\cos n\theta)}{d\theta} = \frac{n\sin n\theta}{\sin\theta} = nU_{n-1}(x).$$

Similarly, we find

$$U_n'(x) = -\frac{1}{\sin\theta}\frac{d}{d\theta}\left[\frac{\sin(n+1)\theta}{\sin\theta}\right] = \frac{\sin(n+1)\theta\,\cos\theta}{\sin^3\theta} - \frac{(n+1)\cos(n+1)\theta}{\sin^2\theta}$$
$$= \frac{x\,U_n(x)}{1-x^2} - \frac{(n+1)T_{n+1}(x)}{1-x^2},$$

which rearranges immediately to yield the stated result. ◀

9.5 Bessel functions

Bessel's equation has the form

$$x^2y'' + xy' + (x^2 - \nu^2)y = 0, \tag{9.70}$$

which has a regular singular point at $x = 0$ and an essential singularity at $x = \infty$. The parameter ν is a given number, which we may take as ≥ 0 with no loss of generality. The equation arises from physical situations similar to those involving Legendre's equation, but when cylindrical, rather than spherical, polar coordinates are employed. The variable

x in Bessel's equation is usually a multiple of a radial distance and therefore ranges from 0 to ∞.

We shall seek solutions to Bessel's equation in the form of infinite series. Writing (9.70) in the standard form used in Chapter 7, we have

$$y'' + \frac{1}{x}y' + \left(1 - \frac{\nu^2}{x^2}\right)y = 0. \tag{9.71}$$

By inspection, $x = 0$ is a regular singular point; hence we try a solution of the form $y = x^\sigma \sum_{n=0}^{\infty} a_n x^n$. Substituting this into (9.71) and multiplying the resulting equation by $x^{2-\sigma}$, we obtain

$$\sum_{n=0}^{\infty}\left[(\sigma+n)(\sigma+n-1)+(\sigma+n)-\nu^2\right]a_n x^n + \sum_{n=0}^{\infty} a_n x^{n+2} = 0,$$

which simplifies to

$$\sum_{n=0}^{\infty}\left[(\sigma+n)^2-\nu^2\right]a_n x^n + \sum_{n=0}^{\infty} a_n x^{n+2} = 0.$$

Considering the coefficient of x^0, we obtain the indicial equation

$$\sigma^2 - \nu^2 = 0,$$

and so $\sigma = \pm\nu$. For coefficients of higher powers of x we find

$$\left[(\sigma+1)^2-\nu^2\right]a_1 = 0, \tag{9.72}$$

$$\left[(\sigma+n)^2-\nu^2\right]a_n + a_{n-2} = 0 \quad \text{for } n \geq 2. \tag{9.73}$$

Substituting $\sigma = \pm\nu$ into (9.72) and (9.73), we obtain the recurrence relations

$$(1 \pm 2\nu)a_1 = 0, \tag{9.74}$$

$$n(n \pm 2\nu)a_n + a_{n-2} = 0 \quad \text{for } n \geq 2. \tag{9.75}$$

We consider now the form of the general solution to Bessel's equation (9.70) for two cases: the case for which ν is not an integer and that for which it is (including zero).

9.5.1 Bessel functions for non-integer ν

If ν is a non-integer then, in general, the two roots of the indicial equation, $\sigma_1 = \nu$ and $\sigma_2 = -\nu$, will not differ by an integer, and we may obtain two linearly independent solutions in the form of Frobenius series. Special considerations do arise, however, when $\nu = m/2$ for $m = 1, 3, 5, \ldots$, and $\sigma_1 - \sigma_2 = 2\nu = m$ is an (odd positive) integer. When this happens, we may always obtain a solution in the form of a Frobenius series corresponding to the larger root, $\sigma_1 = \nu = m/2$, as described above. However, for the smaller root, $\sigma_2 = -\nu = -m/2$, we must determine whether a second Frobenius series solution is possible by examining the recurrence relation (9.75), which reads

$$n(n-m)a_n + a_{n-2} = 0 \quad \text{for } n \geq 2.$$

Since m is an *odd* positive integer in this case, we can use this recurrence relation (starting with $a_0 \neq 0$) to calculate $a_2, a_4, a_6, \ldots$ in the knowledge that all these terms will remain finite. It is possible in this case, therefore, to find a second solution in the form of a Frobenius series, one that corresponds to the smaller root σ_2.

Thus, in general, for non-integer ν we have from (9.74) and (9.75)

$$\begin{aligned} a_n &= -\frac{1}{n(n \pm 2\nu)} a_{n-2} \quad \text{for } n = 2, 4, 6, \ldots, \\ &= 0 \quad \text{for } n = 1, 3, 5, \ldots. \end{aligned}$$

Setting $a_0 = 1$ in each case, we obtain the two solutions

$$y_{\pm\nu}(x) = x^{\pm\nu}\left[1 - \frac{x^2}{2(2 \pm 2\nu)} + \frac{x^4}{(2 \times 4)(2 \pm 2\nu)(4 \pm 2\nu)} - \cdots\right].$$

It is customary, however, to set

$$a_0 = \frac{1}{2^{\pm\nu}\Gamma(1 \pm \nu)},$$

where $\Gamma(x)$ is the *gamma function*, described in Subsection 9.10.1; it may be regarded as the generalization of the factorial function to non-integer and/or negative arguments.[16] The two solutions of (9.70) are then written as $J_\nu(x)$ and $J_{-\nu}(x)$, where

$$\begin{aligned} J_\nu(x) &= \frac{1}{\Gamma(\nu+1)}\left(\frac{x}{2}\right)^\nu \left[1 - \frac{1}{\nu+1}\left(\frac{x}{2}\right)^2 + \frac{1}{(\nu+1)(\nu+2)}\frac{1}{2!}\left(\frac{x}{2}\right)^4 - \cdots\right] \\ &= \sum_{n=0}^{\infty} \frac{(-1)^n}{n!\Gamma(\nu+n+1)}\left(\frac{x}{2}\right)^{\nu+2n}; \end{aligned} \tag{9.76}$$

replacing ν by $-\nu$ gives $J_{-\nu}(x)$. The functions $J_\nu(x)$ and $J_{-\nu}(x)$ are called *Bessel functions of the first kind, of order* ν.

Since ν is not an integer, $\Gamma(-\nu + n + 1)$ is finite, and so the first term of each series is a finite non-zero multiple of x^ν and $x^{-\nu}$, respectively. Consequently, we can deduce that $J_\nu(x)$ and $J_{-\nu}(x)$ are linearly independent; this may be confirmed by calculating the Wronskian of these two functions. Therefore, for non-integer ν the general solution of Bessel's equation (9.70) is given by

$$y(x) = c_1 J_\nu(x) + c_2 J_{-\nu}(x). \tag{9.77}$$

We note that Bessel functions of half-integer order are expressible in closed form in terms of trigonometric functions, as illustrated in the following example.

16 In particular, $\Gamma(n+1) = n!$ for $n = 0, 1, 2, \ldots$, and $\Gamma(n)$ is infinite if n is any integer ≤ 0.

Example Find the general solution of

$$x^2y'' + xy' + (x^2 - \tfrac{1}{4})y = 0.$$

This is Bessel's equation with $\nu = 1/2$, so from (9.77) the general solution is simply

$$y(x) = c_1 J_{1/2}(x) + c_2 J_{-1/2}(x).$$

However, Bessel functions of half-integral order can be expressed in terms of trigonometric functions. To show this, we note from (9.76) that

$$J_{\pm 1/2}(x) = x^{\pm 1/2} \sum_{n=0}^{\infty} \frac{(-1)^n x^{2n}}{2^{2n \pm 1/2} n! \Gamma(1 + n \pm \frac{1}{2})}.$$

Using the fact that $\Gamma(x+1) = x\Gamma(x)$ and $\Gamma(\frac{1}{2}) = \sqrt{\pi}$, we find that, for $\nu = 1/2$,

$$\begin{aligned}
J_{1/2}(x) &= \frac{(\frac{1}{2}x)^{1/2}}{\Gamma(\frac{3}{2})} - \frac{(\frac{1}{2}x)^{5/2}}{1!\Gamma(\frac{5}{2})} + \frac{(\frac{1}{2}x)^{9/2}}{2!\Gamma(\frac{7}{2})} - \cdots \\
&= \frac{(\frac{1}{2}x)^{1/2}}{(\frac{1}{2})\sqrt{\pi}} - \frac{(\frac{1}{2}x)^{5/2}}{1!(\frac{3}{2})(\frac{1}{2})\sqrt{\pi}} + \frac{(\frac{1}{2}x)^{9/2}}{2!(\frac{5}{2})(\frac{3}{2})(\frac{1}{2})\sqrt{\pi}} - \cdots \\
&= \frac{(\frac{1}{2}x)^{1/2}}{(\frac{1}{2})\sqrt{\pi}}\left(1 - \frac{x^2}{3!} + \frac{x^4}{5!} - \cdots\right) = \frac{(\frac{1}{2}x)^{1/2}}{(\frac{1}{2})\sqrt{\pi}} \frac{\sin x}{x} = \sqrt{\frac{2}{\pi x}} \sin x,
\end{aligned}$$

whereas for $\nu = -1/2$ we obtain

$$\begin{aligned}
J_{-1/2}(x) &= \frac{(\frac{1}{2}x)^{-1/2}}{\Gamma(\frac{1}{2})} - \frac{(\frac{1}{2}x)^{3/2}}{1!\Gamma(\frac{3}{2})} + \frac{(\frac{1}{2}x)^{7/2}}{2!\Gamma(\frac{5}{2})} - \cdots \\
&= \frac{(\frac{1}{2}x)^{-1/2}}{\sqrt{\pi}}\left(1 - \frac{x^2}{2!} + \frac{x^4}{4!} - \cdots\right) = \sqrt{\frac{2}{\pi x}} \cos x.
\end{aligned}$$

Therefore the general solution we require is

$$y(x) = c_1 J_{1/2}(x) + c_2 J_{-1/2}(x) = c_1\sqrt{\frac{2}{\pi x}} \sin x + c_2\sqrt{\frac{2}{\pi x}} \cos x.$$

It is worth noting that if a solution finite at $x = 0$ is required, then $c_2 = 0$. ◀

9.5.2 Bessel functions for integer ν

The definition of the Bessel function $J_\nu(x)$ given in (9.76) is, of course, valid for all values of ν, but, as we shall see, in the case of integer ν the general solution of Bessel's equation cannot be written in the form (9.77). Firstly, let us consider the case $\nu = 0$, so that the two solutions to the indicial equation are equal, and we clearly obtain only one solution in the form of a Frobenius series. From (9.76), this is given

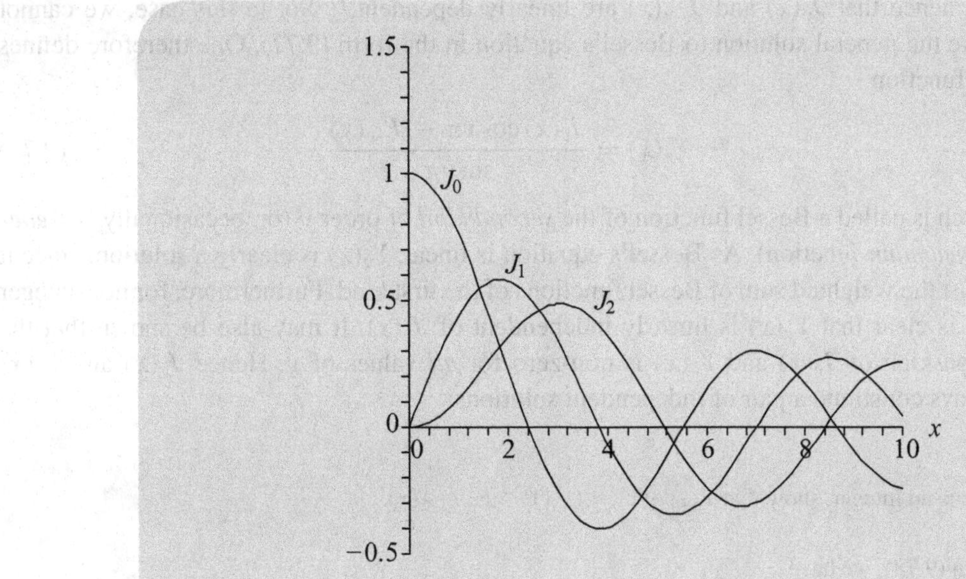

Figure 9.5 The first three integer-order Bessel functions of the first kind.

by

$$J_0(x) = \sum_{n=0}^{\infty} \frac{(-1)^n x^{2n}}{2^{2n} n! \Gamma(1+n)}$$
$$= 1 - \frac{x^2}{2^2} + \frac{x^4}{2^2 4^2} - \frac{x^6}{2^2 4^2 6^2} + \cdots.$$

In general, however, if ν is a positive integer then the solutions of the indicial equation differ by an integer. For the larger root, $\sigma_1 = \nu$, we may find a solution $J_\nu(x)$, for $\nu = 1, 2, 3, \ldots$, in the form of the Frobenius series given by (9.76). Graphs of $J_0(x)$, $J_1(x)$ and $J_2(x)$ are plotted in Figure 9.5 for real x.

For the smaller root, $\sigma_2 = -\nu$, however, the recurrence relation (9.75) becomes

$$n(n-m)a_n + a_{n-2} = 0 \quad \text{for } n \geq 2,$$

where $m = 2\nu$ is now an *even* positive integer, i.e. $m = 2, 4, 6, \ldots$. Starting with $a_0 \neq 0$ we may then calculate $a_2, a_4, a_6, \ldots$, but we see that when $n = m$ the coefficient a_n is formally infinite, and the method fails to produce a second solution in the form of a Frobenius series.

In fact, by replacing ν by $-\nu$ in the definition of $J_\nu(x)$ given in (9.76), it can be shown that, for integer ν,

$$J_{-\nu}(x) = (-1)^\nu J_\nu(x),$$

and hence that $J_\nu(x)$ and $J_{-\nu}(x)$ are linearly dependent.[17] So, in this case, we cannot write the general solution to Bessel's equation in the form (9.77). One therefore defines the function

$$Y_\nu(x) = \frac{J_\nu(x)\cos\nu\pi - J_{-\nu}(x)}{\sin\nu\pi}, \tag{9.78}$$

which is called a Bessel function of the *second kind* of order ν (or, occasionally, a *Weber* or *Neumann* function). As Bessel's equation is linear, $Y_\nu(x)$ is clearly a solution, since it is just the weighted sum of Bessel functions of the first kind. Furthermore, for non-integer ν it is clear that $Y_\nu(x)$ is linearly independent of $J_\nu(x)$. It may also be shown that the Wronskian of $J_\nu(x)$ and $Y_\nu(x)$ is non-zero for *all* values of ν. Hence $J_\nu(x)$ and $Y_\nu(x)$ always constitute a pair of independent solutions.

Example If n is an integer, show that $Y_{n+1/2}(x) = (-1)^{n+1}J_{-n-1/2}(x)$.

From (9.78), we have

$$Y_{n+1/2}(x) = \frac{J_{n+1/2}(x)\cos(n+\frac{1}{2})\pi - J_{-n-1/2}(x)}{\sin(n+\frac{1}{2})\pi}.$$

If n is an integer, $\cos(n+\frac{1}{2})\pi = 0$ and $\sin(n+\frac{1}{2})\pi = (-1)^n$, and so we immediately obtain $Y_{n+1/2}(x) = (-1)^{n+1}J_{-n-1/2}(x)$, as required. ◀

When ν is an integer, the expression (9.78) becomes an indeterminate form 0/0 because, for integer ν, we have $\sin\nu\pi = 0$, $\cos\nu\pi = (-1)^\nu$ and $J_{-\nu}(x) = (-1)^\nu J_\nu(x)$. However, this indeterminate form can be evaluated using l'Hôpital's rule. And so for integer ν, we define $Y_\nu(x)$ as

$$Y_\nu(x) = \lim_{\mu\to\nu}\left[\frac{J_\mu(x)\cos\mu\pi - J_{-\mu}(x)}{\sin\mu\pi}\right], \tag{9.79}$$

which gives a linearly independent second solution for this case. Thus, we may write the general solution of Bessel's equation, valid for *all* ν, as

$$y(x) = c_1J_\nu(x) + c_2Y_\nu(x). \tag{9.80}$$

The functions $Y_0(x)$, $Y_1(x)$ and $Y_2(x)$ are plotted in Figure 9.6.

Finally, we note that, in some applications, it is convenient to work with complex linear combinations of Bessel functions of the first and second kinds given by

$$H_\nu^{(1)}(x) = J_\nu(x) + iY_\nu(x), \qquad H_\nu^{(2)}(x) = J_\nu(x) - iY_\nu(x);$$

these are called, respectively, *Hankel functions* of the first and second kind of order ν.

17 Prove this. Note that for $-\nu+n+1 \le 0$, $\Gamma(-\nu+n+1) = \infty$. Change the summation index to (the integer) $s = n-\nu$ and note that $(\nu+s)!\,\Gamma(s+1)$ can be written as $\Gamma(\nu+s+1)\,s!$.

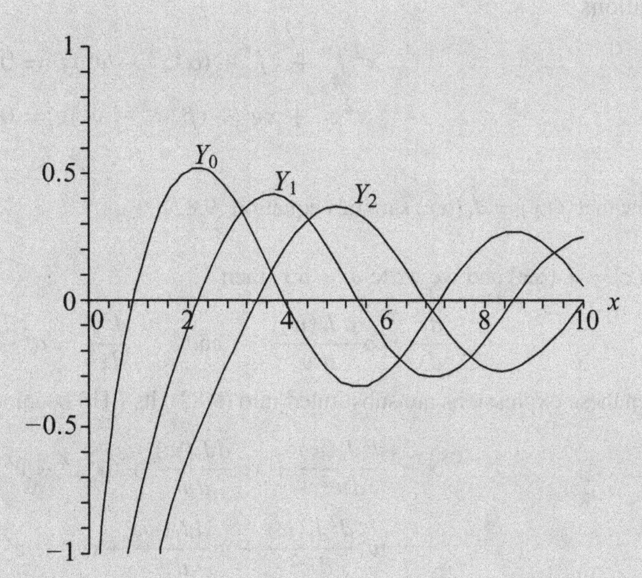

Figure 9.6 The first three integer-order Bessel functions of the second kind.

9.5.3 Properties of Bessel functions $J_\nu(x)$

In physical applications, we often require that the solution is regular at $x = 0$, but, from its definition (9.78) or (9.79), it is clear that $Y_\nu(x)$ is singular at the origin, and so in such physical situations the coefficient c_2 in (9.80) must be set to zero; the solution is then simply some multiple of $J_\nu(x)$. These Bessel functions of the first kind have various useful properties that are worthy of further discussion. Unless otherwise stated, the results presented in this section apply to Bessel functions $J_\nu(x)$ of integer and non-integer order.

Mutual orthogonality

In Section 8.4, we noted that Bessel's equation (9.70) could be put into conventional Sturm–Liouville form with $p = x$, $q = -\nu^2/x$, $\lambda = \alpha^2$ and $\rho = x$, provided αx is the argument of y. From the form of p, we see that there is no natural interval over which one would expect the solutions of Bessel's equation corresponding to different eigenvalues λ (but fixed ν) to be automatically orthogonal. Nevertheless, provided the Bessel functions satisfied appropriate boundary conditions, we would expect them to obey an orthogonality relationship over some interval $[a, b]$ of the form

$$\int_a^b x J_\nu(\alpha x) J_\nu(\beta x)\, dx = 0 \qquad \text{for } \alpha \neq \beta. \tag{9.81}$$

To determine the boundary conditions required for this result to hold, and hence find the acceptable combinations of values of α, β, a and b, let us consider the functions $f(x) = J_\nu(\alpha x)$ and $g(x) = J_\nu(\beta x)$, which, as is proved below, respectively satisfy the equations

$$x^2 f'' + xf' + (\alpha^2 x^2 - \nu^2) f = 0, \tag{9.82}$$

$$x^2 g'' + xg' + (\beta^2 x^2 - \nu^2) g = 0. \tag{9.83}$$

Example Show that $f(x) = J_\nu(\alpha x)$ satisfies equation (9.82).

If $f(x) = J_\nu(\alpha x)$ and we write $w = \alpha x$, then

$$\frac{df}{dx} = \alpha \frac{dJ_\nu(w)}{dw} \qquad \text{and} \qquad \frac{d^2 f}{dx^2} = \alpha^2 \frac{d^2 J_\nu(w)}{dw^2}.$$

When these expressions are substituted into (9.82), its LHS becomes

$$\begin{aligned} &x^2\alpha^2 \frac{d^2 J_\nu(w)}{dw^2} + x\alpha \frac{dJ_\nu(w)}{dw} + (\alpha^2 x^2 - \nu^2) J_\nu(w) \\ &= w^2 \frac{d^2 J_\nu(w)}{dw^2} + w \frac{dJ_\nu(w)}{dw} + (w^2 - \nu^2) J_\nu(w). \end{aligned}$$

But, from Bessel's equation itself, this final expression is equal to zero, thus verifying that $f(x)$ does satisfy (9.82). ◀

Now multiplying (9.83) by $f(x)$ and (9.82) by $g(x)$, subtracting them, and dividing through by x gives

$$\frac{d}{dx}[x(fg' - gf')] = (\alpha^2 - \beta^2) x f g, \tag{9.84}$$

where we have used the fact that

$$\frac{d}{dx}[x(fg' - gf')] = x(fg'' - gf'') + (fg' - gf').$$

By integrating (9.84) over any given range $x = a$ to $x = b$, we obtain

$$\int_a^b x f(x) g(x)\, dx = \frac{1}{\alpha^2 - \beta^2} \Big[x f(x) g'(x) - x g(x) f'(x) \Big]_a^b,$$

which, on setting $f(x) = J_\nu(\alpha x)$ and $g(x) = J_\nu(\beta x)$, becomes

$$\int_a^b x J_\nu(\alpha x) J_\nu(\beta x)\, dx = \frac{1}{\alpha^2 - \beta^2} \Big[\beta x J_\nu(\alpha x) J_\nu'(\beta x) - \alpha x J_\nu(\beta x) J_\nu'(\alpha x) \Big]_a^b. \tag{9.85}$$

If $\alpha \neq \beta$, and the interval $[a, b]$ is such that the expression on the RHS of (9.85) equals zero, then we obtain the orthogonality condition (9.81). This happens, for example, if $J_\nu(\alpha x)$ and $J_\nu(\beta x)$ vanish at $x = a$ and $x = b$, or if $J_\nu'(\alpha x)$ and $J_\nu'(\beta x)$ vanish at $x = a$ and $x = b$, or for many more general conditions. It should be noted that the boundary term is automatically zero at the point $x = 0$, as one might expect from the fact that the Sturm–Liouville form of Bessel's equation has $p(x) = x$.

If $\alpha = \beta$, the RHS of (9.85) takes the indeterminate form $0/0$. But it may still be evaluated using l'Hôpital's rule, or alternatively we may calculate the relevant integral directly, as follows.

Example Evaluate the integral

$$\int_a^b J_\nu^2(\alpha x) x \, dx.$$

Ignoring the integration limits for the moment,

$$\int J_\nu^2(\alpha x) x \, dx = \frac{1}{\alpha^2} \int J_\nu^2(u) u \, du,$$

where $u = \alpha x$. Integrating by parts yields

$$I = \int J_\nu^2(u) u \, du = \tfrac{1}{2} u^2 J_\nu^2(u) - \int J_\nu(u) J_\nu'(u) u^2 \, du.$$

Now Bessel's equation (9.70) can be rearranged as

$$u^2 J_\nu(u) = \nu^2 J_\nu(u) - u J_\nu'(u) - u^2 J_\nu''(u),$$

which, on substitution into the expression for I, gives

$$I = \tfrac{1}{2} u^2 J_\nu^2(u) - \int J_\nu'(u)[\nu^2 J_\nu(u) - u J_\nu'(u) - u^2 J_\nu''(u)] \, du$$
$$= \tfrac{1}{2} u^2 J_\nu^2(u) - \tfrac{1}{2} \nu^2 J_\nu^2(u) + \tfrac{1}{2} u^2 [J_\nu'(u)]^2 + c.$$

Since $u = \alpha x$, the required integral is given by

$$\int_a^b J_\nu^2(\alpha x) x \, dx = \frac{1}{2} \left[\left(x^2 - \frac{\nu^2}{\alpha^2} \right) J_\nu^2(\alpha x) + x^2 [J_\nu'(\alpha x)]^2 \right]_a^b, \qquad (9.86)$$

which gives the normalization condition for Bessel functions of the first kind. ◀

Since the Bessel functions $J_\nu(x)$ possess the orthogonality property (9.85), we may expand any reasonable function $f(x)$, i.e. one obeying the Dirichlet conditions discussed in Chapter 4, in the interval $0 \le x \le b$ in terms of them. The interval is taken to be $0 \le x \le b$, as then one need only ensure that the appropriate boundary condition is satisfied at $x = b$, since the boundary condition at $x = 0$ is met automatically. The expansion as a sum of Bessel functions of a given (non-negative) order ν, takes the form

$$f(x) = \sum_{n=0}^{\infty} c_n J_\nu(\alpha_n x), \qquad (9.87)$$

provided that the α_n are chosen such that $J_\nu(\alpha_n b) = 0$, so as to make the RHS of (9.85) equal to zero.[18] The coefficients c_n are then given by

$$c_n = \frac{2}{b^2 J_{\nu+1}^2(\alpha_n b)} \int_0^b f(x) J_\nu(\alpha_n x) x \, dx. \qquad (9.88)$$

18 Notice that it is the allowed values of α and β, as two of the α_n, that are being determined for a given b by this boundary value requirement; it is *not* the value of b being determined by given values of α and β.

The manipulation of the normalization constant from the form containing $J'_\nu(\alpha x)$, that appears in (9.86), into the form given above is shown as part of the next worked example.

Example Prove expression (9.88) for the expansion coefficients c_n.

If we multiply (9.87) by $x J_\nu(\alpha_m x)$ and integrate from $x = 0$ to $x = b$ then we obtain

$$\begin{aligned}\int_0^b x J_\nu(\alpha_m x) f(x)\, dx &= \sum_{n=0}^{\infty} c_n \int_0^b x J_\nu(\alpha_m x) J_\nu(\alpha_n x)\, dx \\ &= c_m \int_0^b J_\nu^2(\alpha_m x) x\, dx \\ &= \tfrac{1}{2} c_m b^2 J'^2_\nu(\alpha_m b) = \tfrac{1}{2} c_m b^2 J^2_{\nu+1}(\alpha_m b),\end{aligned}$$

where in the last two lines we have used (9.85) with $\alpha_m = \alpha \neq \beta = \alpha_n$, (9.86), the fact that $J_\nu(\alpha_m b) = 0$ and (9.92), which is proved below. ◀

Recurrence relations

The recurrence relations enjoyed by Bessel functions of the first kind, $J_\nu(x)$, can be derived directly from the power series definition (9.76). For example, to prove the recurrence relation

$$\frac{d}{dx}[x^\nu J_\nu(x)] = x^\nu J_{\nu-1}(x) \tag{9.89}$$

we start from the power series definition (9.76) of $J_\nu(x)$ and obtain

$$\begin{aligned}\frac{d}{dx}[x^\nu J_\nu(x)] &= \frac{d}{dx} \sum_{n=0}^{\infty} \frac{(-1)^n x^{2\nu+2n}}{2^{\nu+2n} n!\Gamma(\nu+n+1)} \\ &= \sum_{n=0}^{\infty} \frac{(-1)^n x^{2\nu+2n-1}}{2^{\nu+2n-1} n!\Gamma(\nu+n)} \\ &= x^\nu \sum_{n=0}^{\infty} \frac{(-1)^n x^{(\nu-1)+2n}}{2^{(\nu-1)+2n} n!\Gamma((\nu-1)+n+1)} = x^\nu J_{\nu-1}(x).\end{aligned}$$

It may similarly be shown that

$$\frac{d}{dx}[x^{-\nu} J_\nu(x)] = -x^{-\nu} J_{\nu+1}(x). \tag{9.90}$$

From (9.89) and (9.90) the remaining recurrence relations may be derived. Expanding out the derivative on the LHS of (9.89) and dividing through by $x^{\nu-1}$, we obtain the relation

$$x J'_\nu(x) + \nu J_\nu(x) = x J_{\nu-1}(x). \tag{9.91}$$

Similarly, by expanding out the derivative on the LHS of (9.90), and multiplying through by $x^{\nu+1}$, we find

$$x J'_\nu(x) - \nu J_\nu(x) = -x J_{\nu+1}(x). \tag{9.92}$$

Adding (9.91) and (9.92) and dividing through by x gives

$$J_{\nu-1}(x) - J_{\nu+1}(x) = 2J'_\nu(x). \tag{9.93}$$

Finally, subtracting (9.92) from (9.91) and dividing by x gives

$$J_{\nu-1}(x) + J_{\nu+1}(x) = \frac{2\nu}{x} J_\nu(x). \tag{9.94}$$

Example Given that $J_{1/2}(x) = (2/\pi x)^{1/2} \sin x$ and that $J_{-1/2}(x) = (2/\pi x)^{1/2} \cos x$, express $J_{3/2}(x)$ and $J_{-3/2}(x)$ in terms of trigonometric functions.

From (9.92) we have

$$\begin{aligned} J_{3/2}(x) &= \frac{1}{2x} J_{1/2}(x) - J'_{1/2}(x) \\ &= \frac{1}{2x}\left(\frac{2}{\pi x}\right)^{1/2} \sin x - \left(\frac{2}{\pi x}\right)^{1/2} \cos x + \frac{1}{2x}\left(\frac{2}{\pi x}\right)^{1/2} \sin x \\ &= \left(\frac{2}{\pi x}\right)^{1/2} \left(\frac{1}{x} \sin x - \cos x\right). \end{aligned}$$

Similarly, from (9.91) we have

$$\begin{aligned} J_{-3/2}(x) &= -\frac{1}{2x} J_{-1/2}(x) + J'_{-1/2}(x) \\ &= -\frac{1}{2x}\left(\frac{2}{\pi x}\right)^{1/2} \cos x - \left(\frac{2}{\pi x}\right)^{1/2} \sin x - \frac{1}{2x}\left(\frac{2}{\pi x}\right)^{1/2} \cos x \\ &= \left(\frac{2}{\pi x}\right)^{1/2} \left(-\frac{1}{x} \cos x - \sin x\right). \end{aligned}$$

We see that, by repeated use of these recurrence relations, all Bessel functions $J_\nu(x)$ of half-integer order may be expressed in terms of trigonometric functions. From their definition (9.78), Bessel functions of the second kind, $Y_\nu(x)$, of half-integer order can be similarly expressed. ◀

Finally, we note that the relations (9.89) and (9.90) may be rewritten in integral form as

$$\int x^\nu J_{\nu-1}(x)\, dx = x^\nu J_\nu(x),$$

$$\int x^{-\nu} J_{\nu+1}(x)\, dx = -x^{-\nu} J_\nu(x).$$

If ν is an integer, the recurrence relations of this section may be proved using the generating function for Bessel functions discussed below. It may be shown that Bessel functions of the second kind, $Y_\nu(x)$, also satisfy the recurrence relations derived above.

Generating function

The Bessel functions $J_\nu(x)$, where $\nu = n$ is an integer, can be described by a generating function in a way similar to that discussed for Legendre polynomials in Subsection 9.1.2.

The generating function for Bessel functions of integer order is given by

$$G(x,h) = \exp\left[\frac{x}{2}\left(h - \frac{1}{h}\right)\right] = \sum_{n=-\infty}^{\infty} J_n(x)h^n. \tag{9.95}$$

By expanding the exponential as a power series, it is straightforward to verify that the functions $J_n(x)$ defined by (9.95) are indeed Bessel functions of the first kind, as given by (9.76).

The generating function (9.95) is useful for finding, for Bessel functions of integer order, properties that can often be extended to the non-integer case. In particular, the Bessel function recurrence relations may be derived.

Example Use the generating function to prove, for integer ν, the recurrence relation (9.94), i.e.

$$J_{\nu-1}(x) + J_{\nu+1}(x) = \frac{2\nu}{x} J_\nu(x).$$

Differentiating $G(x,h)$ with respect to h we obtain

$$\frac{\partial G(x,h)}{\partial h} = \frac{x}{2}\left(1 + \frac{1}{h^2}\right) G(x,h) = \sum_{n=-\infty}^{\infty} nJ_n(x)h^{n-1},$$

which can be written using (9.95) again as

$$\frac{x}{2}\left(1 + \frac{1}{h^2}\right) \sum_{n=-\infty}^{\infty} J_n(x)h^n = \sum_{n=-\infty}^{\infty} nJ_n(x)h^{n-1}.$$

Equating coefficients of h^n we obtain

$$\frac{x}{2}[J_n(x) + J_{n+2}(x)] = (n+1)J_{n+1}(x),$$

which, on replacing n by $\nu - 1$, gives the required recurrence relation. ◀

Integral representations

The generating function (9.95) can also be used to derive *integral representations* of Bessel functions of integer order.

Example Show that for integer n the Bessel function $J_n(x)$ is given by

$$J_n(x) = \frac{1}{\pi}\int_0^\pi \cos(n\theta - x\sin\theta)\, d\theta. \tag{9.96}$$

By expanding out the cosine term in the integrand in (9.96) we obtain the integral

$$I = \frac{1}{\pi}\int_0^\pi [\cos(x\sin\theta)\cos n\theta + \sin(x\sin\theta)\sin n\theta]\, d\theta. \tag{9.97}$$

Now, we may express $\cos(x\sin\theta)$ and $\sin(x\sin\theta)$ in terms of Bessel functions by setting $h = \exp i\theta$ in (9.95) to give

$$\exp\left[\frac{x}{2}(\exp i\theta - \exp(-i\theta))\right] = \exp(ix\sin\theta) = \sum_{m=-\infty}^{\infty} J_m(x)\exp im\theta.$$

Using de Moivre's theorem, $\exp i\theta = \cos\theta + i\sin\theta$, we then obtain

$$\exp(ix\sin\theta) = \cos(x\sin\theta) + i\sin(x\sin\theta) = \sum_{m=-\infty}^{\infty} J_m(x)(\cos m\theta + i\sin m\theta).$$

Equating the real and imaginary parts of this expression gives

$$\cos(x\sin\theta) = \sum_{m=-\infty}^{\infty} J_m(x)\cos m\theta,$$

$$\sin(x\sin\theta) = \sum_{m=-\infty}^{\infty} J_m(x)\sin m\theta.$$

Substituting these expressions into (9.97) then yields

$$I = \frac{1}{\pi}\sum_{m=-\infty}^{\infty}\int_0^{\pi}[J_m(x)\cos m\theta\cos n\theta + J_m(x)\sin m\theta\sin n\theta]\,d\theta.$$

However, using the orthogonality of the trigonometric functions [see equations (4.1)–(4.3)], we obtain

$$I = \frac{1}{\pi}\frac{\pi}{2}[J_n(x) + J_n(x)] = J_n(x),$$

which proves the integral representation (9.96). ◀

Finally, we mention the special case of the integral representation (9.96) for $n = 0$. Recalling that $\cos(-x\sin\theta) = \cos(x\sin\theta)$, we have

$$J_0(x) = \frac{1}{\pi}\int_0^{\pi}\cos(x\sin\theta)\,d\theta = \frac{1}{2\pi}\int_0^{2\pi}\cos(x\sin\theta)\,d\theta,$$

since $\cos(x\sin\theta)$ repeats itself in the range $\theta = \pi$ to $\theta = 2\pi$. However, $\sin(x\sin\theta)$ changes sign in this range and so

$$\frac{1}{2\pi}\int_0^{2\pi}\sin(x\sin\theta)\,d\theta = 0.$$

Using de Moivre's theorem, we can therefore write

$$J_0(x) = \frac{1}{2\pi}\int_0^{2\pi}\exp(ix\sin\theta)\,d\theta = \frac{1}{2\pi}\int_0^{2\pi}\exp(ix\cos\theta)\,d\theta.$$

There are in fact many other integral representations of Bessel functions; they can be derived from those given.

9.6 Spherical Bessel functions

When obtaining solutions of Helmholtz' equation $(\nabla^2 + k^2)u = 0$ in spherical polar coordinates (see Subsection 11.3.2), one finds that, for solutions that are finite on the polar axis, the radial part $R(r)$ of the solution must satisfy the equation

$$r^2 R'' + 2rR' + [k^2r^2 - \ell(\ell+1)]R = 0, \tag{9.98}$$

where ℓ is an integer. This equation looks very much like Bessel's equation and can in fact be reduced to it by writing $R(r) = r^{-1/2}S(r)$, in which case $S(r)$ then satisfies

$$r^2 S'' + rS' + \left[k^2r^2 - \left(\ell + \tfrac{1}{2}\right)^2\right] S = 0.$$

On making the change of variable $x = kr$ and letting $y(x) = S(kr)$, we obtain

$$x^2 y'' + xy' + \left[x^2 - \left(\ell + \tfrac{1}{2}\right)^2\right] y = 0,$$

where the primes now denote d/dx. This is Bessel's equation of order $\ell + \frac{1}{2}$ and has as its solutions $y(x) = J_{\ell+1/2}(x)$ and $Y_{\ell+1/2}(x)$. The general solution of (9.98) can therefore be written

$$R(r) = r^{-1/2}[c_1 J_{\ell+1/2}(kr) + c_2 Y_{\ell+1/2}(kr)],$$

where c_1 and c_2 are constants that may be determined from the boundary conditions on the solution. In particular, for solutions that are finite at the origin we require $c_2 = 0$.

The functions $x^{-1/2}J_{\ell+1/2}(x)$ and $x^{-1/2}Y_{\ell+1/2}(x)$, when suitably normalized, are called *spherical Bessel functions* of the first and second kind, respectively, and are denoted as follows:

$$j_\ell(x) = \sqrt{\frac{\pi}{2x}} J_{\ell+1/2}(x), \tag{9.99}$$

$$n_\ell(x) = \sqrt{\frac{\pi}{2x}} Y_{\ell+1/2}(x). \tag{9.100}$$

For integer ℓ, we also note that $Y_{\ell+1/2}(x) = (-1)^{\ell+1} J_{-\ell-1/2}(x)$, as discussed in Subsection 9.5.2. Moreover, in Subsection 9.5.1, we noted that Bessel functions of the first kind, $J_\nu(x)$, of half-integer order are expressible in closed form in terms of trigonometric functions. Thus, all spherical Bessel functions of both the first and second kinds may be expressed in such a form. In particular, using the results of the worked example in Subsection 9.5.1, we find that

$$j_0(x) = \frac{\sin x}{x}, \tag{9.101}$$

$$n_0(x) = -\frac{\cos x}{x}. \tag{9.102}$$

Expressions for higher-order spherical Bessel functions are most easily obtained by repeated use of a recurrence relation based on (9.90). Although we do not prove it here, this reads

$$f_\ell(x) = (-1)^\ell x^\ell \left(\frac{1}{x}\frac{d}{dx}\right)^\ell f_0(x), \tag{9.103}$$

where $f_\ell(x)$ denotes either $j_\ell(x)$ or $n_\ell(x)$.

Using (9.103) and the expressions (9.101) and (9.102), one quickly finds, for example, that[19]

$$j_1(x) = \frac{\sin x}{x^2} - \frac{\cos x}{x}, \qquad j_2(x) = \left(\frac{3}{x^3} - \frac{1}{x}\right)\sin x - \frac{3\cos x}{x^2},$$

$$n_1(x) = -\frac{\cos x}{x^2} - \frac{\sin x}{x}, \qquad n_2(x) = -\left(\frac{3}{x^3} - \frac{1}{x}\right)\cos x - \frac{3\sin x}{x^2}.$$

Finally, we note that the orthogonality properties of the spherical Bessel functions follow directly from the orthogonality condition (9.85) for Bessel functions of the first kind.

9.7 Laguerre functions

Laguerre's equation has the form

$$xy'' + (1 - x)y' + \nu y = 0; \tag{9.104}$$

it has a regular singularity at $x = 0$ and an essential singularity at $x = \infty$. The parameter ν is a given real number, although it nearly always takes an integer value in physical applications. The Laguerre equation appears in the description of the wavefunction of the hydrogen atom. Any solution of (9.104) is called a *Laguerre function*.

Since the point $x = 0$ is a regular singularity, we may find at least one solution in the form of a Frobenius series (see Section 7.4):

$$y(x) = \sum_{m=0}^{\infty} a_m x^{m+\sigma}. \tag{9.105}$$

Substituting this series into (9.104) and dividing through by $x^{\sigma-1}$, we obtain

$$\sum_{m=0}^{\infty} [(m+\sigma)(m+\sigma-1) + (1-x)(m+\sigma) + \nu x]\, a_m x^m = 0. \tag{9.106}$$

Setting $x = 0$, so that only the $m = 0$ term remains, we obtain the indicial equation $\sigma^2 = 0$, which trivially has $\sigma = 0$ as its repeated root. Thus, Laguerre's equation has only one solution of the form (9.105), and it, in fact, reduces to a simple power series. Substituting $\sigma = 0$ into (9.106) and demanding that the coefficient of x^{m+1} vanishes, we obtain the recurrence relation

$$a_{m+1} = \frac{m - \nu}{(m+1)^2} a_m.$$

19 Derive the four results given, so as to ensure that the notation has been understood.

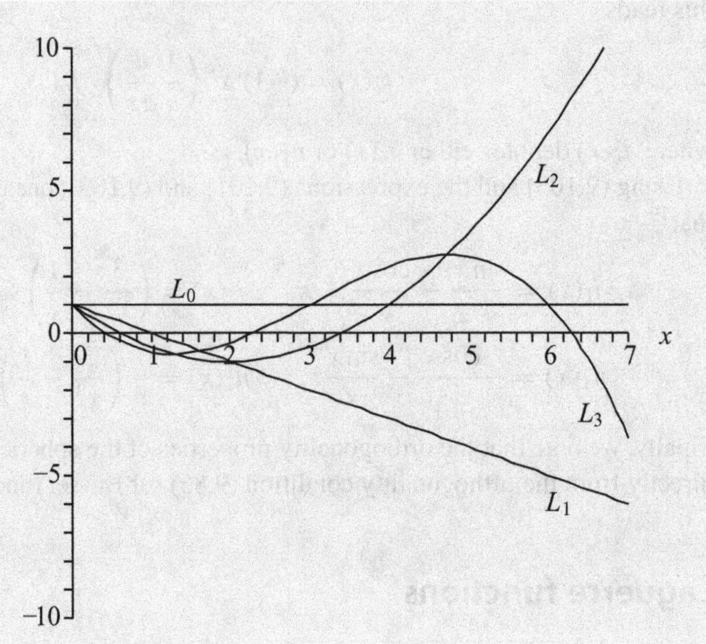

Figure 9.7 The first four Laguerre polynomials.

As mentioned above, in nearly all physical applications, the parameter ν takes integer values. Therefore, if $\nu = n$, where n is a non-negative integer, we see that $a_{n+1} = a_{n+2} = \cdots = 0$, and so our solution to Laguerre's equation is a polynomial of order n. It is conventional to choose $a_0 = 1$, so that the solution, written with the highest power of x as the first term, is given by

$$L_n(x) = \frac{(-1)^n}{n!}\left[x^n - \frac{n^2}{1!}x^{n-1} + \frac{n^2(n-1)^2}{2!}x^{n-2} - \cdots + (-1)^n n!\right] \tag{9.107}$$

$$= \sum_{m=0}^{n}(-1)^m \frac{n!}{(m!)^2(n-m)!}x^m, \tag{9.108}$$

where $L_n(x)$ is called the nth *Laguerre polynomial*. We note in particular that $L_n(0) = 1$. The first few Laguerre polynomials are given by

$$L_0(x) = 1, \qquad 3!L_3(x) = -x^3 + 9x^2 - 18x + 6,$$

$$L_1(x) = -x + 1, \qquad 4!L_4(x) = x^4 - 16x^3 + 72x^2 - 96x + 24,$$

$$2!L_2(x) = x^2 - 4x + 2, \qquad 5!L_5(x) = -x^5 + 25x^4 - 200x^3 + 600x^2 - 600x + 120.$$

The functions $L_0(x)$, $L_1(x)$, $L_2(x)$ and $L_3(x)$ are plotted in Figure 9.7.

9.7.1 Properties of Laguerre polynomials

The Laguerre polynomials and functions derived from them are important in the analysis of the quantum mechanical behavior of some physical systems. We therefore briefly outline their useful properties in this section.

Rodrigues' formula

The Laguerre polynomials can be expressed in terms of a Rodrigues' formula given by

$$L_n(x) = \frac{e^x}{n!}\frac{d^n}{dx^n}\left(x^n e^{-x}\right), \tag{9.109}$$

which may be proved straightforwardly by calculating the nth derivative explicitly using Leibnitz' theorem and comparing the result with (9.108). This is illustrated in the following example.

Example Prove that the expression (9.109) yields the nth Laguerre polynomial.

Evaluating the nth derivative in (9.109) using Leibnitz' theorem, we find

$$\begin{aligned} L_n(x) &= \frac{e^x}{n!}\sum_{r=0}^{n} {}^nC_r \frac{d^r x^n}{dx^r}\frac{d^{n-r}e^{-x}}{dx^{n-r}} \\ &= \frac{e^x}{n!}\sum_{r=0}^{n}\frac{n!}{r!(n-r)!}\frac{n!}{(n-r)!}x^{n-r}(-1)^{n-r}e^{-x} \\ &= \sum_{r=0}^{n}(-1)^{n-r}\frac{n!}{r!(n-r)!(n-r)!}x^{n-r}. \end{aligned}$$

Relabeling the summation using the index $m = n - r$, we obtain

$$L_n(x) = \sum_{m=0}^{n}(-1)^m\frac{n!}{(m!)^2(n-m)!}x^m,$$

which is precisely the expression (9.108) for the nth Laguerre polynomial. ◀

Mutual orthogonality

In Section 8.4, we noted that Laguerre's equation could be put into Sturm–Liouville form with $p = xe^{-x}$, $q = 0$, $\lambda = \nu$ and $\rho = e^{-x}$, and its natural interval is thus $[0, \infty]$. Since the Laguerre polynomials $L_n(x)$ are solutions of the equation and are regular at the end-points, they must be mutually orthogonal over this interval with respect to the weight function $\rho = e^{-x}$, i.e.[20]

$$\int_0^\infty L_n(x)L_k(x)e^{-x}\,dx = 0 \qquad \text{if } n \neq k.$$

20 This specific form of the weight function means that for the numerical integration from 0 to ∞ of integrands containing a factor $e^{-\alpha x}$, a simple scaling of the integration variable will cast it into a form for which Gauss–Laguerre integration is particularly suitable. This integration scheme, based on the Laguerre polynomials, effectively handles this exponential factor analytically.

This result may also be proved directly using the Rodrigues' formula (9.109).[21] Indeed, as we show below, the normalization of the Laguerre polynomials is most easily verified using this method with k set equal to n.

Example Show that

$$I \equiv \int_0^\infty L_n(x)L_n(x)e^{-x}\,dx = 1. \tag{9.110}$$

Using the Rodrigues' formula (9.109) to replace the second Laguerre factor, we may write

$$I = \frac{1}{n!}\int_0^\infty L_n(x)\frac{d^n}{dx^n}(x^n e^{-x})\,dx = \frac{(-1)^n}{n!}\int_0^\infty \frac{d^n L_n}{dx^n}x^n e^{-x}\,dx,$$

where, in the second equality, we have integrated by parts n times and used the fact that the boundary terms all vanish. When $d^n L_n/dx^n$ is evaluated using (9.108), only the derivative of the $m = n$ term survives and that has the value $[(-1)^n n!\,n!\,]/[(n!)^2\,0!] = (-1)^n$. Thus we have

$$I = \frac{1}{n!}\int_0^\infty x^n e^{-x}\,dx = 1,$$

where, in the second equality, we use the expression (9.133) defining the gamma function (see Section 9.10). ◀

The above orthogonality and normalization conditions allow us to expand any (reasonable) function in the interval $0 \le x < \infty$ in a series of the form

$$f(x) = \sum_{n=0}^{\infty} a_n L_n(x),$$

in which the coefficients a_n are given by

$$a_n = \int_0^\infty f(x)L_n(x)e^{-x}\,dx.$$

We note that it is sometimes convenient to define the *orthonormal Laguerre functions* $\phi_n(x) = e^{-x/2}L_n(x)$, which may also be used to produce a series expansion of a function in the interval $0 \le x < \infty$.

Generating function

The generating function for the Laguerre polynomials is given by

$$G(x,h) = \frac{e^{-xh/(1-h)}}{1-h} = \sum_{n=0}^{\infty} L_n(x)h^n. \tag{9.111}$$

We may prove this result by differentiating the generating function with respect to x and h, respectively, to obtain recurrence relations for the Laguerre polynomials, which may

21 Verify this result by direct calculation when $n = 1$ and $k = 2$.

then be combined to show that the functions $L_n(x)$ in (9.111) do indeed satisfy Laguerre's equation (as discussed in the next Subsection).[22]

Recurrence relations

The Laguerre polynomials obey a number of useful recurrence relations. The three most important relations are as follows:

$$(n+1)L_{n+1}(x) = (2n+1-x)L_n(x) - nL_{n-1}(x), \tag{9.112}$$

$$L_{n-1}(x) = L'_{n-1}(x) - L'_n(x), \tag{9.113}$$

$$xL'_n(x) = nL_n(x) - nL_{n-1}(x). \tag{9.114}$$

The first two relations can be derived from the generating function (9.111) – this is done below – and may be combined to yield the third result.[23]

Example Derive the recurrence relations (9.112) and (9.113).

Differentiating the generating function (9.111) with respect to h, we find

$$\frac{\partial G}{\partial h} = \frac{(1-x-h)e^{-xh/(1-h)}}{(1-h)^3} = \sum nL_n h^{n-1}.$$

Thus, we may write

$$(1-x-h)\sum L_n h^n = (1-h)^2 \sum nL_n h^{n-1},$$

and, on equating coefficients of h^n on each side, we obtain

$$(1-x)L_n - L_{n-1} = (n+1)L_{n+1} - 2nL_n + (n-1)L_{n-1},$$

which trivially rearranges to give the recurrence relation (9.112).

To obtain the recurrence relation (9.113), we begin by differentiating the generating function (9.111) with respect to x, which yields

$$\frac{\partial G}{\partial x} = -\frac{he^{-xh/(1-h)}}{(1-h)^2} = \sum L'_n h^n,$$

and thus we have

$$-h\sum L_n h^n = (1-h)\sum L'_n h^n.$$

Equating coefficients of h^n on each side then gives

$$-L_{n-1} = L'_n - L'_{n-1},$$

which immediately simplifies to give (9.113). ◀

22 Show that the generating function gives the correct value for $L_n(0)$.

23 It is easier algebraically to first change $n-1$ to n (and n to $n+1$) in (9.113) and multiply the equation through by x. Then substitution from (9.114) for each term on the RHS gives an equation that can be rearranged to give (9.112). This shows that the three equations are consistent and that the third could, by suitable manipulation, be derived from the first two.

9.8 Associated Laguerre functions

The associated Laguerre equation has the form

$$xy'' + (m + 1 - x)y' + ny = 0: \tag{9.115}$$

it has a regular singularity at $x = 0$ and an essential singularity at $x = \infty$. We restrict our attention to the situation in which the parameters n and m are both non-negative integers, as is the case in nearly all physical problems. The associated Laguerre equation occurs most frequently in quantum mechanical applications. Any solution of (9.115) is called an *associated Laguerre function*.

Solutions of (9.115) for non-negative integers n and m are given by the *associated Laguerre polynomials*

$$L_n^m(x) = (-1)^m \frac{d^m}{dx^m} L_{n+m}(x), \tag{9.116}$$

where $L_n(x)$ are the ordinary Laguerre polynomials.[24]

Example Show that the functions $L_n^m(x)$ defined in (9.116) are solutions of (9.115).

Since the Laguerre polynomials $L_n(x)$ are solutions of Laguerre's equation (9.104), we have

$$xL''_{n+m} + (1 - x)L'_{n+m} + (n + m)L_{n+m} = 0.$$

Differentiating this equation m times using Leibnitz' theorem and rearranging, we find

$$xL_{n+m}^{(m+2)} + (m + 1 - x)L_{n+m}^{(m+1)} + nL_{n+m}^{(m)} = 0.$$

On multiplying through by $(-1)^m$ and setting $L_n^m = (-1)^m L_{n+m}^{(m)}$, in accord with (9.116), we obtain

$$x(L_n^m)'' + (m + 1 - x)(L_n^m)' + nL_n^m = 0,$$

which shows that the functions L_n^m are indeed solutions of (9.115). ◀

In particular, we note that $L_n^0(x) = L_n(x)$. As discussed in the previous section, $L_n(x)$ is a polynomial of order n and so it follows that $L_n^m(x)$ is also. The first few associated Laguerre polynomials are easily found using (9.116):

$$L_0^m(x) = 1,$$
$$L_1^m(x) = -x + m + 1,$$
$$2!L_2^m(x) = x^2 - 2(m + 2)x + (m + 1)(m + 2),$$
$$3!L_3^m(x) = -x^3 + 3(m + 3)x^2 - 3(m + 2)(m + 3)x + (m + 1)(m + 2)(m + 3).$$

24 Note that some authors define the associated Laguerre polynomials as $\mathcal{L}_n^m(x) = (d^m/dx^m)L_n(x)$, which is thus related to our expression (9.116) by $L_n^m(x) = (-1)^m \mathcal{L}_{n+m}^m(x)$.

Indeed, in the general case, one may show straightforwardly from the definition (9.116) and the expression (9.108) for the ordinary Laguerre polynomials that

$$L_n^m(x) = \sum_{k=0}^{n} (-1)^k \frac{(n+m)!}{k!(n-k)!(k+m)!} x^k. \tag{9.117}$$

9.8.1 Properties of associated Laguerre polynomials

The properties of the associated Laguerre polynomials follow directly from those of the ordinary Laguerre polynomials through the definition (9.116). We shall therefore only briefly outline the most useful results here.

Rodrigues' formula

A Rodrigues' formula for the associated Laguerre polynomials is given by

$$L_n^m(x) = \frac{e^x x^{-m}}{n!} \frac{d^n}{dx^n}(x^{n+m} e^{-x}). \tag{9.118}$$

It can be proved by evaluating the nth derivative using Leibnitz' theorem (see Problem 9.7).

Mutual orthogonality

In Section 8.4, we noted that the associated Laguerre equation could be transformed into a Sturm–Liouville one with $p = x^{m+1}e^{-x}$, $q = 0$, $\lambda = n$ and $\rho = x^m e^{-x}$, and its natural interval is thus $[0, \infty]$. Since the associated Laguerre polynomials $L_n^m(x)$ are solutions of the equation and are regular at the end-points, those with the same m but differing values of the eigenvalue $\lambda = n$ must be mutually orthogonal over this interval with respect to the weight function $\rho = x^m e^{-x}$, i.e.

$$\int_0^\infty L_n^m(x) L_k^m(x) x^m e^{-x}\, dx = 0 \qquad \text{if } n \neq k.$$

This result may also be proved directly using the Rodrigues' formula (9.118), as may the normalization condition when $k = n$.

Example Show that

$$I \equiv \int_0^\infty L_n^m(x) L_n^m(x) x^m e^{-x}\, dx = \frac{(n+m)!}{n!}. \tag{9.119}$$

Using the Rodrigues' formula (9.118), we may write

$$I = \frac{1}{n!}\int_0^\infty L_n^m(x) \frac{d^n}{dx^n}(x^{n+m} e^{-x})\, dx = \frac{(-1)^n}{n!}\int_0^\infty \frac{d^n L_n^m}{dx^n} x^{n+m} e^{-x}\, dx,$$

where, in the second equality, we have integrated by parts n times and used the fact that the boundary terms all vanish. From (9.117) we see that $d^n L_n^m/dx^n = (-1)^n$. Thus we have

$$I = \frac{1}{n!}\int_0^\infty x^{n+m} e^{-x}\, dx = \frac{(n+m)!}{n!},$$

where, in the second equality, we use the expression (9.133) defining the gamma function (see Section 9.10). ◀

The above orthogonality and normalization conditions allow us to expand any (reasonable) function in the interval $0 \le x < \infty$ in a series of the form

$$f(x) = \sum_{n=0}^{\infty} a_n L_n^m(x),$$

in which the coefficients a_n are given by

$$a_n = \frac{n!}{(n+m)!} \int_0^{\infty} f(x) L_n^m(x) x^m e^{-x}\, dx.$$

We note that it is sometimes convenient to define the *orthogonal associated Laguerre functions* $\phi_n^m(x) = x^{m/2} e^{-x/2} L_n^m(x)$, which may also be used to produce a series expansion of a function in the interval $0 \le x < \infty$.

Generating function

The generating function for the associated Laguerre polynomials is given by

$$G(x,h) = \frac{e^{-xh/(1-h)}}{(1-h)^{m+1}} = \sum_{n=0}^{\infty} L_n^m(x) h^n. \tag{9.120}$$

This can be obtained by differentiating the generating function (9.111) for the ordinary Laguerre polynomials m times with respect to x, and using (9.116). As an example of its direct use, we can set $x = 0$ and obtain an expression for $L_n^m(0)$:

$$\begin{aligned}\sum_{n=0}^{\infty} L_n^m(0) h^n &= \frac{1}{(1-h)^{m+1}} \\ &= 1 + (m+1)h + \frac{(m+1)(m+2)}{2!}h^2 + \cdots \\ &\quad + \frac{(m+1)(m+2)\cdots(m+n)}{n!}h^n + \cdots,\end{aligned}$$

where, in the second equality, we have expanded the RHS using the binomial theorem. On equating coefficients of h^n, we immediately obtain

$$L_n^m(0) = \frac{(n+m)!}{n!\,m!}.$$

Recurrence relations

The various recurrence relations satisfied by the associated Laguerre polynomials may be derived by differentiating the generating function (9.120) with respect to either or both of x and h, or by differentiating with respect to x the recurrence relations obeyed by the ordinary Laguerre polynomials, discussed in Subsection 9.7.1. Of the many recurrence relations satisfied by the associated Laguerre polynomials, two of the most useful are as follows:

$$(n+1)L_{n+1}^m(x) = (2n+m+1-x)L_n^m(x) - (n+m)L_{n-1}^m(x), \tag{9.121}$$

$$x(L_n^m)'(x) = nL_n^m(x) - (n+m)L_{n-1}^m(x). \tag{9.122}$$

For proofs of these relations the reader is referred to Problem 9.7.

9.9 Hermite functions

Hermite's equation has the form

$$y'' - 2xy' + 2\nu y = 0, \tag{9.123}$$

and has an essential singularity at $x = \infty$. The parameter ν is a given real number, although it nearly always takes an integer value in physical applications. The Hermite equation appears in the description of the wavefunction of the harmonic oscillator. Any solution of (9.123) is called a *Hermite function.*

Since $x = 0$ is an ordinary point of the equation, we may find two linearly independent solutions in the form of a power series (see Section 7.3):

$$y(x) = \sum_{m=0}^{\infty} a_m x^m. \tag{9.124}$$

Substituting this series into (9.104) yields

$$\sum_{m=0}^{\infty} [(m+1)(m+2)a_{m+2} + 2(\nu - m)a_m] \, x^m = 0.$$

Demanding that the coefficient of each power of x vanishes, we obtain the recurrence relation

$$a_{m+2} = -\frac{2(\nu - m)}{(m+1)(m+2)} a_m.$$

As mentioned above, in nearly all physical applications, the parameter ν takes integer values. Therefore, if $\nu = n$, where n is a non-negative integer, we see that $a_{n+2} = a_{n+4} = \cdots = 0$, and so one solution of Hermite's equation is a polynomial of order n. For even n, it is conventional to choose $a_0 = (-1)^{n/2} n!/(n/2)!$, whereas for odd n one takes $a_1 = (-1)^{(n-1)/2} 2n!/[\frac{1}{2}(n-1)]!$. These choices allow a general solution to be written as

$$H_n(x) = (2x)^n - n(n-1)(2x)^{n-1} + \frac{n(n-1)(n-2)(n-3)}{2!}(2x)^{n-4} - \cdots \tag{9.125}$$

$$= \sum_{m=0}^{[n/2]} (-1)^m \frac{n!}{m!(n-2m)!}(2x)^{n-2m}, \tag{9.126}$$

where $H_n(x)$ is called the nth *Hermite polynomial* and the notation $[n/2]$ denotes the integer part of $n/2$. We note in particular that $H_n(-x) = (-1)^n H_n(x)$. The first few Hermite polynomials are given by

$$H_0(x) = 1, \qquad H_3(x) = 8x^3 - 12x,$$
$$H_1(x) = 2x, \qquad H_4(x) = 16x^4 - 48x^2 + 12,$$
$$H_2(x) = 4x^2 - 2, \qquad H_5(x) = 32x^5 - 160x^3 + 120x.$$

The functions $H_0(x)$, $H_1(x)$, $H_2(x)$ and $H_3(x)$ are plotted in Figure 9.8.

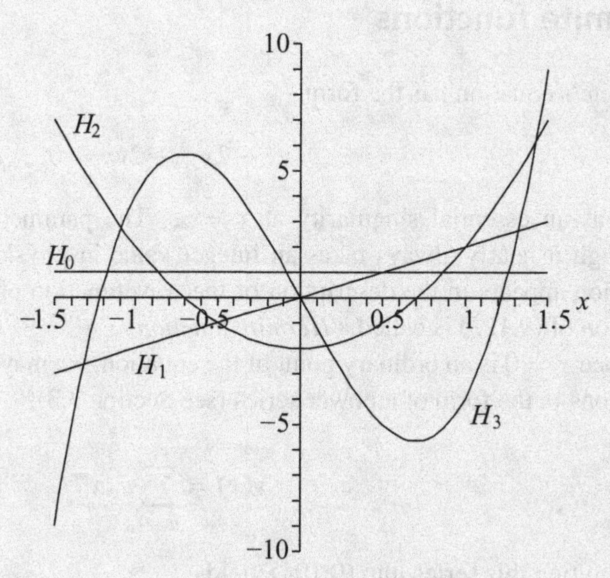

Figure 9.8 The first four Hermite polynomials.

9.9.1 Properties of Hermite polynomials

The Hermite polynomials and functions derived from them are important in the analysis of the quantum mechanical behavior of some physical systems. We therefore briefly outline their useful properties in this section.

Rodrigues' formula

The Rodrigues' formula for the Hermite polynomials is given by

$$H_n(x) = (-1)^n e^{x^2} \frac{d^n}{dx^n}(e^{-x^2}). \tag{9.127}$$

This can be proved using Leibnitz' theorem, as follows.

Example Prove the Rodrigues' formula (9.127) for the Hermite polynomials.

Letting $u = e^{-x^2}$ and differentiating with respect to x, we quickly find that

$$u' + 2xu = 0.$$

Differentiating this equation $n+1$ times using Leibnitz' theorem then gives

$$u^{(n+2)} + 2xu^{(n+1)} + 2(n+1)u^{(n)} = 0,$$

which, on introducing the new variable $v = (-1)^n u^{(n)}$, reduces to

$$v'' + 2xv' + 2(n+1)v = 0. \tag{9.128}$$

Now letting $y = e^{x^2} v$, we may write the derivatives of v as

$$v' = e^{-x^2}(y' - 2xy),$$
$$v'' = e^{-x^2}(y'' - 4xy' + 4x^2y - 2y).$$

Substituting these expressions into (9.128), and dividing through by e^{-x^2}, finally yields Hermite's equation,

$$y'' - 2xy + 2ny = 0,$$

thus demonstrating that $y = (-1)^n e^{x^2} d^n(e^{-x^2})/dx^n$ is indeed a solution. Moreover, since this solution is clearly a polynomial of order n, it must be some multiple of $H_n(x)$. The normalization is easily checked by noting that, from (9.127), the highest-order term is $(2x)^n$, which agrees with the expression (9.125). ◀

Mutual orthogonality

We saw in Section 8.4 that Hermite's equation could be cast in Sturm–Liouville form with $p = e^{-x^2}$, $q = 0$, $\lambda = 2n$ and $\rho = e^{-x^2}$, and its natural interval is thus $[-\infty, \infty]$. Since the Hermite polynomials $H_n(x)$ are solutions of the equation and are regular at the end-points, they must be mutually orthogonal over this interval with respect to the weight function $\rho = e^{-x^2}$, i.e.

$$\int_{-\infty}^{\infty} H_n(x)H_k(x)e^{-x^2}\,dx = 0 \qquad \text{if } n \neq k.$$

This result may also be proved directly using the Rodrigues' formula (9.127).[25] Indeed, the normalization, when $k = n$, is most easily found in this way.

Example Show that

$$I \equiv \int_{-\infty}^{\infty} H_n(x)H_n(x)e^{-x^2}\,dx = 2^n n!\sqrt{\pi}. \tag{9.129}$$

Using the Rodrigues' formula (9.127), we may write

$$I = (-1)^n \int_0^{\infty} H_n(x)\frac{d^n}{dx^n}(e^{-x^2})\,dx = \int_{-\infty}^{\infty} \frac{d^n H_n}{dx^n} e^{-x^2}\,dx,$$

where, in the second equality, we have integrated by parts n times and used the fact that the boundary terms all vanish. From (9.125) we see that $d^n H_n/dx^n = 2^n n!$. Thus we have

$$I = 2^n n! \int_{-\infty}^{\infty} e^{-x^2}\,dx = 2^n n!\sqrt{\pi},$$

where, in the second equality, we use the standard result for the area under a Gaussian curve. ◀

25 This result forms the basis of Gauss–Hermite numerical integration between $-\infty$ and ∞, of particular value for integrands containing factors of the form $e^{-\alpha x^2}$. See footnote 20.

The above orthogonality and normalization conditions allow any (reasonable) function in the interval $-\infty \leq x < \infty$ to be expanded in a series of the form

$$f(x) = \sum_{n=0}^{\infty} a_n H_n(x),$$

in which the coefficients a_n are given by

$$a_n = \frac{1}{2^n n! \sqrt{\pi}} \int_{-\infty}^{\infty} f(x) H_n(x) e^{-x^2}\, dx.$$

We note that it is sometimes convenient to define the *orthogonal Hermite functions* $\phi_n(x) = e^{-x^2/2} H_n(x)$; they also may be used to produce a series expansion of a function in the interval $-\infty \leq x < \infty$. Indeed, $\phi_n(x)$ is proportional to the wavefunction of a particle in the nth energy level of a quantum harmonic oscillator.

Generating function

The generating function equation for the Hermite polynomials reads

$$G(x, h) = e^{2hx - h^2} = \sum_{n=0}^{\infty} \frac{H_n(x)}{n!} h^n, \qquad (9.130)$$

a result that will now be proved using the Rodrigues' formula (9.127).

Example Show that the functions $H_n(x)$ in (9.130) are the Hermite polynomials.

It is often more convenient to write the generating function (9.130) as

$$G(x, h) = e^{x^2} e^{-(x-h)^2} = \sum_{n=0}^{\infty} \frac{H_n(x)}{n!} h^n.$$

Differentiating this form k times with respect to h gives

$$\sum_{n=k}^{\infty} \frac{H_n}{(n-k)!} h^{n-k} = \frac{\partial^k G}{\partial h^k} = e^{x^2} \frac{\partial^k}{\partial h^k} e^{-(x-h)^2} = (-1)^k e^{x^2} \frac{\partial^k}{\partial x^k} e^{-(x-h)^2}.$$

Relabeling the summation on the LHS using the new index $m = n - k$, we obtain

$$\sum_{m=0}^{\infty} \frac{H_{m+k}}{m!} h^m = (-1)^k e^{x^2} \frac{\partial^k}{\partial x^k} e^{-(x-h)^2}.$$

Setting $h = 0$ in this equation, we find

$$H_k(x) = (-1)^k e^{x^2} \frac{d^k}{dx^k}(e^{-x^2}),$$

which is the Rodrigues' formula (9.127) for the Hermite polynomials. ◀

The generating function (9.130) is also useful for determining special values of the Hermite polynomials. In particular, it is straightforward to show that $H_{2n}(0) = (-1)^n (2n)!/n!$ and $H_{2n+1}(0) = 0$.

Recurrence relations

The two most useful recurrence relations satisfied by the Hermite polynomials are given by

$$H_{n+1}(x) = 2xH_n(x) - 2nH_{n-1}(x), \tag{9.131}$$

$$H_n'(x) = 2nH_{n-1}(x). \tag{9.132}$$

The first relation provides a simple iterative way of evaluating the nth Hermite polynomials at some point $x = x_0$, given the values of $H_0(x)$ and $H_1(x)$ at that point. For proofs of these recurrence relations, see Problem 9.5.

9.10 The gamma function and related functions

Many times in this chapter, and often throughout the rest of the book, we have made mention of the gamma function and related functions such as the error functions. Although not derived as the solutions of important second-order ODEs, these convenient functions appear in a number of contexts, and so here we gather together some of their properties. This final section should be regarded merely as a reference containing some useful relations obeyed by these functions; a minimum of formal proofs is given.

9.10.1 The gamma function

The *gamma function* $\Gamma(n)$ is defined by

$$\Gamma(n) = \int_0^\infty x^{n-1}e^{-x}\,dx, \tag{9.133}$$

which converges for $n > 0$, where in general n is a real number. Replacing n by $n+1$ in (9.133) and integrating the RHS by parts, we find

$$\begin{aligned}\Gamma(n+1) &= \int_0^\infty x^n e^{-x}\,dx \\ &= \left[-x^n e^{-x}\right]_0^\infty + \int_0^\infty n x^{n-1} e^{-x}\,dx \\ &= n\int_0^\infty x^{n-1}e^{-x}\,dx,\end{aligned}$$

from which we obtain the important result

$$\Gamma(n+1) = n\Gamma(n). \tag{9.134}$$

From (9.133), we see that $\Gamma(1) = 1$, and so, if n is a positive integer,

$$\Gamma(n+1) = n!. \tag{9.135}$$

In fact, equation (9.135) serves as a definition of the factorial function even for non-integer n. For negative n the factorial function is defined by

$$n! = \frac{(n+m)!}{(n+m)(n+m-1)\cdots(n+1)}, \tag{9.136}$$

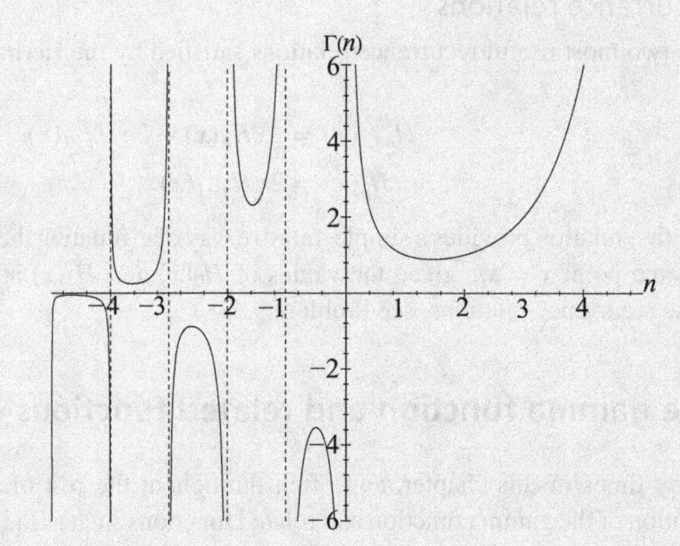

Figure 9.9 The gamma function $\Gamma(n)$.

where m is any positive integer that makes $n + m > 0$. Different choices of m ($> -n$) do not lead to different values for $n!$. A plot of the gamma function is given in Figure 9.9, where it can be seen that the function is infinite for negative integer values of n, in accordance with (9.136). For an extension of the factorial function to complex arguments, see Problem 9.11.

By letting $x = y^2$ in (9.133), we immediately obtain another useful representation of the gamma function given by

$$\Gamma(n) = 2\int_0^{\infty} y^{2n-1} e^{-y^2}\, dy. \tag{9.137}$$

Setting $n = \frac{1}{2}$ we find the result

$$\Gamma\left(\tfrac{1}{2}\right) = 2\int_0^{\infty} e^{-y^2}\, dy = \int_{-\infty}^{\infty} e^{-y^2}\, dy = \sqrt{\pi},$$

where we have used the standard value for the Gaussian integral. From this result, $\Gamma(n)$ for half-integral n can be found using (9.134). Some immediately derivable factorial values of half integers are

$$\left(-\tfrac{3}{2}\right)! = -2\sqrt{\pi}, \quad \left(-\tfrac{1}{2}\right)! = \sqrt{\pi}, \quad \left(\tfrac{1}{2}\right)! = \tfrac{1}{2}\sqrt{\pi}, \quad \left(\tfrac{3}{2}\right)! = \tfrac{3}{4}\sqrt{\pi}.$$

Moreover, it may be shown for non-integral n that the gamma function satisfies the important identity

$$\Gamma(n)\Gamma(1-n) = \frac{\pi}{\sin n\pi}. \tag{9.138}$$

It can also be shown that the gamma function is given by

$$\Gamma(n+1) = \sqrt{2\pi n}\, n^n e^{-n}\left(1 + \frac{1}{12n} + \frac{1}{288n^2} - \frac{139}{51\,840n^3} + \cdots\right) = n!, \tag{9.139}$$

which is known as *Stirling's asymptotic series*. For large n the first term dominates, and so

$$n! \approx \sqrt{2\pi n}\, n^n e^{-n}; \tag{9.140}$$

this is known as *Stirling's approximation*. This approximation is particularly useful in statistical thermodynamics, when arrangements of a large number of particles are to be considered.[26]

Example Prove Stirling's approximation $n! \approx \sqrt{2\pi n}\, n^n e^{-n}$ for large n.

From (9.133), the extended definition of the factorial function (which is valid for $n > -1$) is given by

$$n! = \int_0^\infty x^n e^{-x}\, dx = \int_0^\infty e^{n\ln x - x}\, dx. \tag{9.141}$$

If we let $x = n + y$, then

$$\ln x = \ln n + \ln\left(1 + \frac{y}{n}\right)$$
$$= \ln n + \frac{y}{n} - \frac{y^2}{2n^2} + \frac{y^3}{3n^3} - \cdots.$$

Substituting this result into (9.141), we obtain

$$n! = \int_{-n}^\infty \exp\left[n\left(\ln n + \frac{y}{n} - \frac{y^2}{2n^2} + \cdots\right) - n - y\right] dy.$$

Thus, when n is sufficiently large, we may approximate $n!$ by

$$n! \approx e^{n\ln n - n}\int_{-\infty}^\infty e^{-y^2/(2n)}\, dy = e^{n\ln n - n}\sqrt{2\pi n} = \sqrt{2\pi n}\, n^n e^{-n},$$

which is Stirling's approximation (9.140). ◀

9.10.2 The incomplete gamma function

In the definition (9.133) of the gamma function, we may divide the range of integration into two parts and write

$$\Gamma(n) = \int_0^x u^{n-1}e^{-u}\, du + \int_x^\infty u^{n-1}e^{-u}\, du \equiv \gamma(n, x) + \Gamma(n, x), \tag{9.142}$$

whereby we have defined the *incomplete gamma functions* $\gamma(n, x)$ and $\Gamma(n, x)$, respectively. The choice of which of these two functions to use is merely a matter of convenience.

26 Make a spreadsheet to generate Stirling's approximation, including the first correction term, for $1 \le n \le 30$, and show that it has an accuracy of the order of 0.1% even for such small numbers.

Example Show that if n is a positive integer

$$\Gamma(n, x) = (n-1)!e^{-x} \sum_{k=0}^{n-1} \frac{x^k}{k!}.$$

From (9.142), on integrating by parts we find

$$\Gamma(n, x) = \int_x^\infty u^{n-1}e^{-u}du = x^{n-1}e^{-x} + (n-1)\int_x^\infty u^{n-2}e^{-u}du$$
$$= x^{n-1}e^{-x} + (n-1)\Gamma(n-1, x),$$

which is valid for arbitrary n. If n is an integer, however, by repeated resubstitution we obtain

$$\Gamma(n, x) = x^{n-1}e^{-x} + (n-1)\{x^{n-2}e^{-x} + (n-2)[x^{n-3}e^{-x} + \cdots]\}$$
$$= e^{-x}[x^{n-1} + (n-1)x^{n-2} + (n-1)(n-2)x^{n-3} + \cdots + (n-1)!]$$
$$= (n-1)!\, e^{-x} \sum_{k=0}^{n-1} \frac{x^k}{k!},$$

which is the required result. ◀

We note that it is common to define, in addition, "normalized" functions:

$$P(a, x) \equiv \frac{\gamma(a, x)}{\Gamma(a)}, \qquad Q(a, x) \equiv \frac{\Gamma(a, x)}{\Gamma(a)},$$

They are also often called incomplete gamma functions and care is needed if names rather than symbols are used; it is clear that $Q(a, x) = 1 - P(a, x)$.

9.10.3 The error function

Finally, we mention the *error function*, which is encountered in probability theory and in the solutions of some partial differential equations. The error function is a particular case of the incomplete gamma function, $\text{erf}(x) = \gamma(\frac{1}{2}, x^2)/\sqrt{\pi}$, and is thus given by[27]

$$\text{erf}(x) = \frac{2}{\sqrt{\pi}} \int_0^x e^{-u^2}\, du = 1 - \frac{2}{\sqrt{\pi}} \int_x^\infty e^{-u^2}\, du. \tag{9.143}$$

From this definition we can easily see that

$$\text{erf}(0) = 0, \qquad \text{erf}(\infty) = 1, \qquad \text{erf}(-x) = -\text{erf}(x).$$

By making the substitution $y = \sqrt{2}u$ in (9.143), we find

$$\text{erf}(x) = \sqrt{\frac{2}{\pi}} \int_0^{\sqrt{2}x} e^{-y^2/2}\, dy.$$

27 Show this connection by using definition (9.142) and the substitution $u = v^2$.

The cumulative probability function $\Phi(x)$ for the standard Gaussian distribution may be written in terms of the error function as follows:

$$\Phi(x) = \frac{1}{\sqrt{2\pi}} \int_{-\infty}^{x} e^{-y^2/2}\, dy$$
$$= \frac{1}{2} + \frac{1}{\sqrt{2\pi}} \int_{0}^{x} e^{-y^2/2}\, dy$$
$$= \frac{1}{2} + \frac{1}{2}\mathrm{erf}\left(\frac{x}{\sqrt{2}}\right).$$

It is also sometimes useful to define the *complementary error function*

$$\mathrm{erfc}(x) = 1 - \mathrm{erf}(x) = \frac{2}{\sqrt{\pi}} \int_{x}^{\infty} e^{-u^2}\, du = \frac{\Gamma(\frac{1}{2}, x^2)}{\sqrt{\pi}}. \quad (9.144)$$

SUMMARY

1. *Equations in Sturm–Liouville form*

$$(py')' + qy + \lambda\rho y = 0$$

For the forms of $p(x)$, $q(x)$, λ and ρ for important ODEs in the physical sciences see Table 8.1 on p. 310. The natural interval is denoted by $[a, b]$.

Properties of commonly used polynomial solutions of SL equations

Name and symbol	Natural interval	$\rho(x)\,dx$	$\int_a^b y_k(x)y_{k'}(x)\rho(x)\,dx$	Main application
				Spherical polars, ∇^2
Legendre, $P_\ell(x)$	$[-1, 1]$	dx	$\frac{2}{2\ell+1}\delta_{\ell\ell'}$	Azimuthally symmetric
Associated Legendre, $P_\ell^m(x)$	$[-1, 1]$	dx	$\frac{2}{2\ell+1}\frac{(\ell+m)!}{(\ell-m)!}\delta_{\ell\ell'}\delta_{mm'}$	Azimuthally asymmetric
Spherical harmonics, $Y_\ell^m(\theta, \phi)$	$\theta\,[0, \pi]$ $\phi\,[0, 2\pi]$	$\sin\theta\, d\theta\, d\phi$	$\delta_{\ell\ell'}\delta_{mm'}$	Azimuthally asymmetric
Chebyshev, $T_n(x)$	$[-1, 1]$	$\frac{dx}{\sqrt{1-x^2}}$	$\begin{cases} \pi\,\delta_{nn'} & n = 0 \\ (\pi/2)\,\delta_{nn'} & n > 0 \end{cases}$	Numerical analysis
Chebyshev, $U_n(x)$	$[-1, 1]$	$\sqrt{1-x^2}\,dx$	$(\pi/2)\,\delta_{nn'}$	Numerical analysis

(*cont.*)

(*cont.*)

Name and symbol	Natural interval	$\rho(x)\,dx$	$\int_a^b y_k(x)y_{k'}(x)\rho(x)\,dx$	Main application
Laguerre, $L_n(x)$	$[0,\infty]$	$e^{-x}\,dx$	$\delta_{nn'}$	Quantum hydrogen atom
Associated Laguerre, $L_n^m(x)$	$[0,\infty]$	$x^m e^{-x}\,dx$	$\frac{(n+m)!}{n!}\delta_{nn'}$	Quantum hydrogen atom
Hermite, $H_n(x)$	$[-\infty,\infty]$	$e^{-x^2}\,dx$	$2^n\,n!\,\sqrt{\pi}\,\delta_{nn'}$	Quantum oscillators

Bessel and spherical Bessel functions, used principally in cylindrical polar solutions of equations involving ∇^2, are not finite polynomials. They do not have simple normalization and orthogonality properties, though they do possess a straightforward generating function and many integral representations. The generating function for Bessel functions is

$$\exp\left[\frac{x}{2}\left(h-\frac{1}{h}\right)\right]=\sum_{-\infty}^{\infty}J_n(x)h^n.$$

The spherical Bessel function $j_\ell(x)$ is defined as $\sqrt{\frac{\pi}{2x}}J_{\ell+1/2}(x)$. For further details, see pp. 347–361.

Calculation of commonly used polynomial solutions of SL equations

The generating functions given are for $\sum_{n=0}^{\infty}y_n(x)h^n$, except for that for the associated Legendre functions which generates $\sum_{n=0}^{\infty}P_{n+m}^m(x)h^n$. In the expressions for the generating functions of those polynomials that are marked with an asterisk (*), the function $f(x,h)\equiv(1-2xh+h^2)$.

Name and symbol	Rodrigues' formula or definition†	Generating function
*Legendre, $P_\ell(x)$	$\frac{1}{2^\ell\,\ell!}\frac{d^\ell}{dx^\ell}[(x^2-1)^\ell]$	$\frac{1}{f(x,h)^{1/2}}$
*Associated Legendre, $P_\ell^m(x)$	$\frac{(1-x^2)^{m/2}}{2^\ell\,\ell!}\frac{d^{\ell+m}}{dx^{\ell+m}}[(x^2-1)^\ell]$	$\frac{(2m)!\,(1-x^2)^{m/2}}{2^m\,m!\,f(x,h)^{m+1/2}}$

(*cont.*)

Name and symbol	Rodrigues' formula or definition†	Generating function
†Spherical harmonics, $Y_\ell^m(\theta,\phi)$	$(-1)^m \left[\frac{2\ell+1}{4\pi}\frac{(\ell-m)!}{(\ell+m)!}\right]^{1/2} P_\ell^m(\cos\theta)e^{im\phi}$	–
*Chebyshev, $T_n(x)$	$\frac{(-1)^n\sqrt{\pi}\,(1-x^2)^{1/2}}{2^n(n-\frac{1}{2})!}\frac{d^n}{dx^n}(1-x^2)^{n-1/2}$	$\frac{1-xh}{f(x,h)}$
*Chebyshev, $U_n(x)$	$\frac{(-1)^n\sqrt{\pi}\,(n+1)}{2^{n+1}(n+\frac{1}{2})!\,(1-x^2)^{1/2}}\frac{d^n}{dx^n}(1-x^2)^{n+1/2}$	$\frac{1}{f(x,h)}$
Laguerre, $L_n(x)$	$\frac{e^x}{n!}\frac{d^n}{dx^n}(x^n e^{-x})$	$\frac{e^{-xh/(1-h)}}{1-h}$
Associated Laguerre, $L_n^m(x)$	$\frac{e^x}{n!\,x^m}\frac{d^n}{dx^n}(x^{n+m}e^{-x})$	$\frac{e^{-xh/(1-h)}}{(1-h)^{m+1}}$
Hermite, $H_n(x)$	$(-1)^n e^{x^2}\frac{d^n}{dx^n}\left(e^{-x^2}\right)$	e^{2hx-h^2}

Recurrence relations for Bessel functions and commonly used polynomial solutions of SL equations

For all of the functions considered, there are many recurrence relations that involve derivatives of the corresponding functions. However, the list below has been restricted to those relations that can be used to find the next function in a series without having to calculate derivatives.

Name and symbol	Recurrence relation for $y_{n+1}(x)$
Legendre, $P_\ell(x)$	$P_{n+1} = \frac{(2n+1)xP_n - nP_{n-1}}{n+1}$
Associated Legendre, $P_\ell^m(x)$	$P_{n+1}^m = \frac{(2n+1)xP_n^m - (n+m)P_{n-1}^m}{n-m+1}$
	$P_n^{m+1} = \frac{2mx}{(1-x^2)^{1/2}}P_n^m + [m(m-1)-n(n+1)]P_n^{m-1}$
Chebyshev, $T_n(x)$	$T_{n+1}(x) = 2xT_n(x) - T_{n-1}(x)$
Chebyshev, $U_n(x)$	$U_{n+1}(x) = 2xU_n(x) - U_{n-1}(x)$
Bessel, $J_\nu(x)$	$J_{\nu+1} = \frac{2\nu}{x}J_\nu - J_{\nu-1}$
Spherical Bessel, $j_\ell(x)$	$j_{\ell+1} = \frac{2\ell+1}{x}j_\ell - j_{\ell-1}$

(*cont.*)

(*cont.*)

Name and symbol	Recurrence relation for $y_{n+1}(x)$
Laguerre, $L_n(x)$	$L_{n+1} = \dfrac{(2n+1-x)L_n - nL_{n-1}}{n+1}$
Associated Laguerre, $L_n^m(x)$	$L_{n+1}^m = \dfrac{(2n+m+1-x)L_n^m - (n+m)L_{n-1}^m}{n+1}$
Hermite, $H_n(x)$	$H_{n+1} = 2xH_n - 2nH_{n-1}$

2. *The gamma function*
   Defined by $\Gamma(n) = \int_0^\infty x^{n-1}e^{-x}\,dx$.
   - $\Gamma(n+1) = n\Gamma(n)$.
   - $n! \equiv \Gamma(n+1)$, for all positive n.
   - $n! \equiv \dfrac{(n+m)!}{(n+m)(n+m-1)\cdots(n+1)}$, where integer $m > -n$ for negative non-integer n.
   - $n! = \infty$ if n is a negative integer.
   - Stirling's approximation: $n! \approx \sqrt{2\pi n}\, n^n e^{-n}\left(1 + \dfrac{1}{12n} + \cdots\right)$.
   - Incomplete gamma functions:
   $$\gamma(n,x) \equiv \int_0^x u^{n-1}e^{-u}\,du \quad \text{and} \quad \Gamma(n,x) \equiv \int_x^\infty u^{n-1}e^{-u}\,du$$
   with $\gamma(n,x) + \Gamma(n,x) = \Gamma(n)$ for any x.
3. *The error function*
   Defined by $\text{erf}(x) = \dfrac{2}{\sqrt{\pi}}\int_0^x e^{-u^2}\,du$.
   - $\text{erf}(0) = 0$, $\text{erf}(\infty) = 1$, and $\text{erf}(-x) = -\text{erf}(x)$.
   - Gaussian cumulative probability
   $$\Phi(x) \equiv \frac{1}{\sqrt{2\pi}}\int_{-\infty}^x e^{-u^2/2}\,du = \frac{1}{2} + \frac{1}{2}\text{erf}\left(\frac{x}{\sqrt{2}}\right).$$
   - Complementary error function $\text{erfc}(x) = 1 - \text{erf}(x) = \dfrac{1}{\sqrt{\pi}}\Gamma(\tfrac{1}{2}, x^2)$.

PROBLEMS

9.1. Use the explicit expressions

$$Y_0^0 = \sqrt{\tfrac{1}{4\pi}}, \qquad Y_1^0 = \sqrt{\tfrac{3}{4\pi}}\cos\theta,$$

$$Y_1^{\pm1} = \mp\sqrt{\tfrac{3}{8\pi}}\sin\theta\exp(\pm i\phi), \qquad Y_2^0 = \sqrt{\tfrac{5}{16\pi}}(3\cos^2\theta - 1),$$

$$Y_2^{\pm1} = \mp\sqrt{\tfrac{15}{8\pi}}\sin\theta\cos\theta\exp(\pm i\phi), \quad Y_2^{\pm2} = \sqrt{\tfrac{15}{32\pi}}\sin^2\theta\exp(\pm 2i\phi),$$

to verify for $\ell = 0, 1, 2$ that

$$\sum_{m=-\ell}^{\ell} \left|Y_\ell^m(\theta, \phi)\right|^2 = \frac{2\ell + 1}{4\pi},$$

and so is independent of the values of θ and ϕ. This is true for any ℓ, but a general proof is more involved. This result helps to reconcile intuition with the apparently arbitrary choice of polar axis in a general quantum mechanical system.

9.2. Express the function

$$f(\theta, \phi) = \sin\theta[\sin^2(\theta/2)\cos\phi + i\cos^2(\theta/2)\sin\phi] + \sin^2(\theta/2)$$

as a sum of spherical harmonics.

9.3. Use the generating function for the Legendre polynomials $P_n(x)$ to show that

$$\int_0^1 P_{2n+1}(x)\,dx = (-1)^n \frac{(2n)!}{2^{2n+1}n!(n+1)!}$$

and that, except for the case $n = 0$,

$$\int_0^1 P_{2n}(x)\,dx = 0.$$

9.4. Carry through the following procedure as a proof of the result

$$I_n = \int_{-1}^{1} P_n(z)P_n(z)\,dz = \frac{2}{2n+1}.$$

(a) Square both sides of the generating-function definition of the Legendre polynomials,

$$(1 - 2zh + h^2)^{-1/2} = \sum_{n=0}^{\infty} P_n(z)h^n.$$

(b) Express the RHS as a sum of powers of h, obtaining expressions for the coefficients.
(c) Integrate the RHS from -1 to 1 and use the orthogonality property of the Legendre polynomials.
(d) Similarly integrate the LHS and expand the result in powers of h.
(e) Compare coefficients.

9.5. The Hermite polynomials $H_n(x)$ may be defined by

$$\Phi(x, h) = \exp(2xh - h^2) = \sum_{n=0}^{\infty} \frac{1}{n!} H_n(x)h^n.$$

Show that

$$\frac{\partial^2\Phi}{\partial x^2} - 2x\frac{\partial\Phi}{\partial x} + 2h\frac{\partial\Phi}{\partial h} = 0,$$

and hence that the $H_n(x)$ satisfy the Hermite equation

$$y'' - 2xy' + 2ny = 0,$$

where n is an integer ≥ 0.
Use Φ to prove that
(a) $H_n'(x) = 2nH_{n-1}(x)$,
(b) $H_{n+1}(x) - 2xH_n(x) + 2nH_{n-1}(x) = 0$.

9.6. A charge $+2q$ is situated at the origin and charges of $-q$ are situated at distances $\pm a$ from it along the polar axis. By relating it to the generating function for the Legendre polynomials, show that the electrostatic potential Φ at a point (r, θ, ϕ) with $r > a$ is given by

$$\Phi(r, \theta, \phi) = \frac{2q}{4\pi\epsilon_0 r}\sum_{s=1}^{\infty}\left(\frac{a}{r}\right)^{2s} P_{2s}(\cos\theta).$$

9.7. For the associated Laguerre polynomials, carry out the following:
(a) Prove the Rodrigues' formula

$$L_n^m(x) = \frac{e^x x^{-m}}{n!}\frac{d^n}{dx^n}(x^{n+m}e^{-x}),$$

taking the polynomials to be defined by

$$L_n^m(x) = \sum_{k=0}^{n}(-1)^k \frac{(n+m)!}{k!(n-k)!(k+m)!}x^k.$$

(b) Prove the recurrence relations

$$(n+1)L_{n+1}^m(x) = (2n+m+1-x)L_n^m(x) - (n+m)L_{n-1}^m(x),$$

$$x(L_n^m)'(x) = nL_n^m(x) - (n+m)L_{n-1}^m(x),$$

but this time taking the polynomial as defined by

$$L_n^m(x) = (-1)^m \frac{d^m}{dx^m}L_{n+m}(x)$$

or the generating function.

9.8. The quantum mechanical wavefunction for a one-dimensional simple harmonic oscillator in its nth energy level is of the form

$$\psi(x) = \exp(-x^2/2)H_n(x),$$

where $H_n(x)$ is the nth Hermite polynomial. The generating function for the polynomials is

$$G(x, h) = e^{2hx-h^2} = \sum_{n=0}^{\infty}\frac{H_n(x)}{n!}h^n.$$

(a) Find $H_i(x)$ for $i = 1, 2, 3, 4$.

(b) Evaluate by direct calculation

$$\int_{-\infty}^{\infty} e^{-x^2} H_p(x) H_q(x)\, dx,$$

(i) for $p = 2, q = 3$; (ii) for $p = 2, q = 4$; (iii) for $p = q = 3$. Check your answers against the expected values $2^p\, p!\sqrt{\pi}\,\delta_{pq}$.

[You will find it convenient to use

$$\int_{-\infty}^{\infty} x^{2n} e^{-x^2}\, dx = \frac{(2n)!\sqrt{\pi}}{2^{2n} n!}$$

for integer $n \geq 0$.]

9.9. By initially writing $y(x)$ as $x^{1/2} f(x)$ and then making subsequent changes of variable, reduce Stokes' equation,

$$\frac{d^2y}{dx^2} + \lambda xy = 0,$$

to Bessel's equation. Hence show that a solution that is finite at $x = 0$ is a multiple of $x^{1/2} J_{1/3}(\frac{2}{3}\sqrt{\lambda x^3})$.

9.10. By choosing a suitable form for h in their generating function,

$$G(z, h) = \exp\left[\frac{z}{2}\left(h - \frac{1}{h}\right)\right] = \sum_{n=-\infty}^{\infty} J_n(z) h^n,$$

show that integral representations of the Bessel functions of the first kind are given, for integral m, by

$$J_{2m}(z) = \frac{(-1)^m}{2\pi} \int_0^{2\pi} \cos(z\cos\theta)\cos 2m\theta\, d\theta \qquad m \geq 1,$$

$$J_{2m+1}(z) = \frac{(-1)^m}{2\pi} \int_0^{2\pi} \sin(z\cos\theta)\cos(2m+1)\theta\, d\theta \qquad m \geq 0.$$

9.11. The complex function $z!$ is defined by

$$z! = \int_0^{\infty} u^z e^{-u}\, du \qquad \text{for Re } z > -1.$$

For Re $z \leq -1$ it is defined by

$$z! = \frac{(z+n)!}{(z+n)(z+n-1)\cdots(z+1)},$$

where n is any (positive) integer $> -\text{Re } z$. Being the ratio of two polynomials, $z!$ is analytic everywhere in the finite complex plane except at the poles that occur when z is a negative integer.

(a) Show that the definition of $z!$ for Re $z \leq -1$ is independent of the value of n chosen.

(b) Prove that the residue[28] of $z!$ at the pole $z = -m$, where m is an integer > 0, is $(-1)^{m-1}/(m-1)!$.

9.12. Show, from its definition, that the Bessel function of the second kind, and of integral order ν, can be written as

$$Y_\nu(z) = \frac{1}{\pi}\left[\frac{\partial J_\mu(z)}{\partial \mu} - (-1)^\nu \frac{\partial J_{-\mu}(z)}{\partial \mu}\right]_{\mu=\nu}.$$

Using the explicit series expression for $J_\mu(z)$, show that $\partial J_\mu(z)/\partial \mu$ can be written as

$$J_\nu(z)\ln\left(\frac{z}{2}\right) + g(\nu, z),$$

and deduce that $Y_\nu(z)$ can be expressed as

$$Y_\nu(z) = \frac{2}{\pi} J_\nu(z)\ln\left(\frac{z}{2}\right) + h(\nu, z),$$

where $h(\nu, z)$, like $g(\nu, z)$, is a power series in z.

9.13. The integral

$$I = \int_{-\infty}^{\infty} \frac{e^{-k^2}}{k^2 + a^2}\, dk, \qquad (*)$$

in which $a > 0$, occurs in some statistical mechanics problems. By first considering the integral

$$J = \int_0^\infty e^{iu(k+ia)}\, du,$$

and a suitable variation of it, show that $I = (\pi/a)\exp(a^2)\,\mathrm{erfc}(a)$, where $\mathrm{erfc}(x)$ is the complementary error function.

9.14. Consider two series expansions of the error function as follows.

(a) Obtain a series expansion of the error function $\mathrm{erf}(x)$ in ascending powers of x. How many terms are needed to give a value correct to four significant figures for $\mathrm{erf}(1)$?

(b) Obtain an asymptotic expansion that can be used to estimate $\mathrm{erfc}(x)$ for large x (> 0) in the form of a series

$$\mathrm{erfc}(x) = R(x) = e^{-x^2} \sum_{n=0}^{\infty} \frac{a_n}{x^n}.$$

Consider what bounds can be put on the estimate and at what point the infinite series should be terminated in a practical estimate. In particular, estimate $\mathrm{erfc}(1)$ and test the answer for compatibility with that in part (a).

28 If you are not (yet) familiar with the notion of a residue in complex variable theory, treat this part of the problem as evaluating the limit as $z \to -m$ of $(z+m)z!$.

9.15. Prove two of the properties of the incomplete gamma function $P(a, x^2)$ as follows.

(a) By considering its form for a suitable value of a, show that the error function can be expressed as a particular case of the incomplete gamma function.

(b) The Fresnel integrals, of importance in the study of the diffraction of light, are given by

$$C(x) = \int_0^x \cos\left(\frac{\pi}{2}t^2\right) dt, \qquad S(x) = \int_0^x \sin\left(\frac{\pi}{2}t^2\right) dt.$$

Show that they can be expressed in terms of the error function by

$$C(x) + iS(x) = A\,\mathrm{erf}\left[\frac{\sqrt{\pi}}{2}(1-i)x\right],$$

where A is a (complex) constant, which you should determine. Hence express $C(x) + iS(x)$ in terms of the incomplete gamma function.

HINTS AND ANSWERS

9.1. Note that taking the square of the modulus eliminates all mention of ϕ.

9.3. Integrate both sides of the generating function definition from $x = 0$ to $x = 1$, and then expand the resulting term, $(1 + h^2)^{1/2}$, using a binomial expansion. Show that ${}^{1/2}C_m$ can be written as $[(-1)^{m-1}(2m-2)!]/[2^{2m-1}m!(m-1)!]$.

9.5. Prove the stated equation using the explicit closed form of the generating function. Then substitute the series and require the coefficient of each power of h to vanish. (b) Differentiate result (a) and then use (a) again to replace the derivatives.

9.7. (a) Write the result of using Leibnitz' theorem on the product of x^{n+m} and e^{-x} as a finite sum, evaluate the separated derivatives, and then re-index the summation. (b) For the first recurrence relation, differentiate the generating function with respect to h and then use the generating function again to replace the exponential. Equating coefficients of h^n then yields the result. For the second, differentiate the corresponding relationship for the ordinary Laguerre polynomials m times.

9.9. $x^2 f'' + xf' + (\lambda x^3 - \frac{1}{4})f = 0$. Then, in turn, set $x^{3/2} = u$, and $\frac{2}{3}\lambda^{1/2}u = v$; then v satisfies Bessel's equation with $\nu = \frac{1}{3}$.

9.11. (a) Show that the ratio of two definitions based on m and n, with $m > n > -\mathrm{Re}\, z$, is unity, independent of the actual values of m and n.
(b) Consider the limit as $z \to -m$ of $(z + m)z!$, with the definition of $z!$ based on n where $n > m$.

9.13. Express the integrand in partial fractions and use J, as given, and $J' = \int_0^\infty \exp[-iu(k - ia)]\,du$ to express I as the sum of two double integral expressions. Reduce them using the standard Gaussian integral, and then make a change of variable $2v = u + 2a$.

9.15. (a) If the dummy variable in the incomplete gamma function is t, make the change of variable $y = +\sqrt{t}$. Now choose a so that $2(a - 1) + 1 = 0$; $\text{erf}(x) = P(\frac{1}{2}, x^2)$.
(b) Change the integration variable u in the standard representation of the RHS to s, given by $u = \frac{1}{2}\sqrt{\pi}(1 - i)s$, and note that $(1 - i)^2 = -2i$. $A = (1 + i)/2$. From part (a), $C(x) + iS(x) = \frac{1}{2}(1 + i)P(\frac{1}{2}, -\frac{1}{2}\pi i\, x^2)$.

10 Partial differential equations

In this chapter and the next, the solution of differential equations of types typically encountered in the physical sciences and engineering is extended to situations involving more than one independent variable. A partial differential equation (PDE) is an equation relating an unknown function (the dependent variable) of two or more variables to its partial derivatives with respect to those variables. The most commonly occurring independent variables are those describing position and time, and so we will couch our discussion and examples in notation appropriate to them.

As in the rest of this book, we will focus our attention on the equations that arise most often in physical situations. We will restrict our discussion, therefore, to linear PDEs, i.e. those of first degree in the dependent variable. Furthermore, we will discuss primarily second-order equations. The solution of first-order PDEs will necessarily be involved in treating these, and some of the methods discussed can be extended without difficulty to third- and higher-order equations. We shall also see that many ideas developed for ODEs can be carried over directly into the study of PDEs.

Initially, in the current chapter, we will concentrate on general solutions of PDEs in terms of arbitrary functions of particular combinations of the independent variables, and on the solutions that may be derived from them in the presence of boundary conditions. We also discuss the existence and uniqueness of the solutions to PDEs under given boundary conditions.

In the following chapter the methods most commonly used in practice for obtaining solutions to PDEs subject to given boundary conditions will be considered. These methods include the separation of variables, integral transforms and Green's functions. It will become apparent that some of the results of the present chapter, based on combining the independent variables, are in fact the same solutions as those found using separated variables, but arrived at by a different approach.

10.1 Important partial differential equations

Most of the important PDEs of physics are second-order and linear. In order to gain familiarity with their general forms, some of them will now be briefly discussed. These equations apply to a wide variety of different physical systems.

Since, in general, the PDEs listed below describe three-dimensional situations, the independent variables are $\mathbf{r}$ and t, where $\mathbf{r}$ is the position vector and t is time. The actual variables used to specify the position vector $\mathbf{r}$ are dictated by the coordinate system

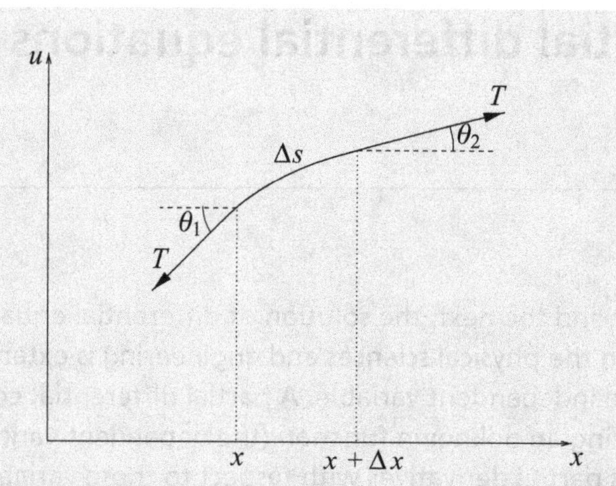

Figure 10.1 The forces acting on an element of a string under uniform tension T.

in use. For example, in Cartesian coordinates the independent variables of position are x, y and z, whereas in spherical polar coordinates they are r, θ and ϕ. The equations may be written in a coordinate-independent manner, however, by using the Laplacian operator ∇^2.

10.1.1 The wave equation

The wave equation

$$\nabla^2 u = \frac{1}{c^2}\frac{\partial^2 u}{\partial t^2} \tag{10.1}$$

describes as a function of position and time the displacement from equilibrium, $u(\mathbf{r}, t)$, of a vibrating string or membrane or a vibrating solid, gas or liquid. The equation also occurs in electromagnetism, where u may be a component of the electric or magnetic field in an electromagnetic wave or the current or voltage along a transmission line. The quantity c is the speed of propagation of the waves. Our first two worked examples are the constructions, rather than the solutions, of partial differential equations; we begin with the wave equation.

Example Find the equation satisfied by small transverse displacements $u(x, t)$ of a uniform string of mass per unit length ρ held under a uniform tension T, assuming that the string is initially located along the x-axis in a Cartesian coordinate system.

Figure 10.1 shows the forces acting on an elemental length Δs of the string. If the tension T in the string is uniform along its length then the net upward vertical force on the element is

$$\Delta F = T\sin\theta_2 - T\sin\theta_1.$$

Assuming that the angles θ_1 and θ_2 are both small, we may make the approximation $\sin\theta \approx \tan\theta$. Since at any point on the string the slope $\tan\theta = \partial u/\partial x$, the force can be written

$$\Delta F = T\left[\frac{\partial u(x+\Delta x, t)}{\partial x} - \frac{\partial u(x,t)}{\partial x}\right] \approx T\frac{\partial^2 u(x,t)}{\partial x^2}\Delta x,$$

where we have used the definition of the partial derivative to simplify the RHS.

This upward force may be equated, by Newton's second law, to the product of the mass of the element and its upward acceleration. The element has a mass $\rho\,\Delta s$, which is approximately equal to $\rho\,\Delta x$ if the vibrations of the string are small, and so we have

$$\rho\,\Delta x\frac{\partial^2 u(x,t)}{\partial t^2} = T\frac{\partial^2 u(x,t)}{\partial x^2}\Delta x.$$

Dividing both sides by Δx we obtain, for the vibrations of the string, the one-dimensional wave equation

$$\frac{\partial^2 u}{\partial x^2} = \frac{1}{c^2}\frac{\partial^2 u}{\partial t^2},$$

where $c^2 = T/\rho$. ◀

The longitudinal vibrations of an elastic rod obey a very similar equation to that derived in the above example, namely

$$\frac{\partial^2 u}{\partial x^2} = \frac{\rho}{E}\frac{\partial^2 u}{\partial t^2};$$

here ρ is the mass per unit volume and E is Young's modulus. Note that in this example the displacement u is along the rod, and not perpendicular to it.

The wave equation can be generalized slightly. For example, in the case of the vibrating string, there could also be an external upward vertical force $f(x,t)$ per unit length acting on the string at time t. The transverse vibrations would then satisfy the equation

$$T\frac{\partial^2 u}{\partial x^2} + f(x,t) = \rho\frac{\partial^2 u}{\partial t^2},$$

which is clearly of the form "upward force per unit length = mass per unit length × upward acceleration".

Similar examples, but involving two or three spatial dimensions rather than one, are provided by the equation governing the transverse vibrations of a stretched membrane subject to an external vertical force density $f(x,y,t)$,

$$T\left(\frac{\partial^2 u}{\partial x^2} + \frac{\partial^2 u}{\partial y^2}\right) + f(x,y,t) = \rho(x,y)\frac{\partial^2 u}{\partial t^2},$$

where ρ is the mass per unit area of the membrane and T is the tension per unit length within it.

10.1.2 The diffusion equation

The diffusion equation

$$\kappa\nabla^2 u = \frac{\partial u}{\partial t} \qquad (10.2)$$

describes the temperature u in a thermally conducting region containing no heat sources or sinks; it also applies to the diffusion of a chemical that has a concentration $u(\mathbf{r}, t)$. The constant κ is called the diffusivity. The equation is clearly second order in the three spatial variables, but first order in time.

Example Derive the equation satisfied by the temperature $u(\mathbf{r}, t)$ at time t for a material of uniform thermal conductivity k, specific heat capacity s and density ρ. Express the equation in Cartesian coordinates.

Let us consider an arbitrary volume V lying within the solid and bounded by a surface S (this may coincide with the surface of the solid if so desired). At any point in the solid the rate of heat flow per unit area in any given direction $\hat{\mathbf{r}}$ is proportional to minus the component of the temperature gradient in that direction and so is given by $(-k\nabla u) \cdot \hat{\mathbf{r}}$. The total flux of heat *out* of the volume V per unit time is given by

$$\begin{aligned} -\frac{dQ}{dt} &= \int_S (-k\nabla u) \cdot \hat{\mathbf{n}}\, dS \\ &= \int_V \nabla \cdot (-k\nabla u)\, dV, \end{aligned} \tag{10.3}$$

where Q is the total heat energy in V at time t and $\hat{\mathbf{n}}$ is the outward-pointing unit normal to S; note that we have used the divergence theorem to convert the surface integral into a volume integral.

We can also express Q as a volume integral over V,

$$Q = \int_V s\rho u\, dV,$$

and its rate of change is then given by

$$\frac{dQ}{dt} = \int_V s\rho \frac{\partial u}{\partial t}\, dV, \tag{10.4}$$

where we have taken the derivative with respect to time inside the integral (using Leibnitz' rule).

Comparing (10.3) and (10.4), and remembering that the volume V is arbitrary, we obtain the three-dimensional diffusion equation

$$\kappa \nabla^2 u = \frac{\partial u}{\partial t},$$

where the diffusion coefficient $\kappa = k/(s\rho)$.[1] If we write ∇^2 in terms of x, y and z we obtain

$$\kappa \left(\frac{\partial^2 u}{\partial x^2} + \frac{\partial^2 u}{\partial y^2} + \frac{\partial^2 u}{\partial z^2} \right) = \frac{\partial u}{\partial t},$$

and so express the equation explicitly in Cartesian coordinates. ◀

The diffusion equation just derived can be generalized to

$$k\nabla^2 u + f(\mathbf{r}, t) = s\rho \frac{\partial u}{\partial t}.$$

1 Note that if the thermal conductivity k or the combination $s\rho$ varied with position, we would not be able to characterize the material with a single diffusion coefficient – not even with one that was allowed to vary with position. This is because, as noted below, the equation contains $\nabla.(k\nabla u)$ and hence would generate additional terms of the form $(\partial k/\partial x)(\partial u/\partial x)$.

The second term, $f(\mathbf{r}, t)$, represents a varying density of heat sources throughout the material but is often not required in physical applications. In the most general case, k, s and ρ may depend on position $\mathbf{r}$, in which case the first term becomes $\nabla \cdot (k\nabla u)$. However, in the simplest application the heat flow is one-dimensional with no heat sources, and the equation becomes (in Cartesian coordinates)

$$\frac{\partial^2 u}{\partial x^2} = \frac{s\rho}{k}\frac{\partial u}{\partial t}.$$

10.1.3 Laplace's equation

Laplace's equation,

$$\nabla^2 u = 0, \tag{10.5}$$

may be obtained by setting $\partial u/\partial t = 0$ in the diffusion equation (10.2), and describes (for example) the *steady-state* temperature distribution in a solid in which there are no heat sources – i.e. the temperature distribution after a long time has elapsed.

Laplace's equation also describes the gravitational potential in a region containing no matter or the electrostatic potential in a charge-free region. Further, it applies to the flow of an incompressible fluid with no sources, sinks or vortices; in this case u is the velocity potential, from which the velocity is given by $\boldsymbol{v} = \nabla u$.

10.1.4 Poisson's equation

Poisson's equation,

$$\nabla^2 u = \rho(\mathbf{r}), \tag{10.6}$$

describes the same physical situations as Laplace's equation, but in regions containing matter, charges or sources of heat or fluid. The function $\rho(\mathbf{r})$ is called the source density and in physical applications usually contains some multiplicative physical constants. For example, if u is the electrostatic potential in some region of space, in which case ρ is the density of electric charge, then $\nabla^2 u = -\rho(\mathbf{r})/\epsilon_0$, where ϵ_0 is the permittivity of free space. Alternatively, u might represent the gravitational potential in some region where the matter density is given by ρ; then $\nabla^2 u = 4\pi G\rho(\mathbf{r})$, where G is the gravitational constant.

10.1.5 Schrödinger's equation

The Schrödinger equation

$$-\frac{\hbar^2}{2m}\nabla^2 u + V(\mathbf{r})u = i\hbar\frac{\partial u}{\partial t} \tag{10.7}$$

describes the quantum mechanical wavefunction $u(\mathbf{r}, t)$ of a non-relativistic particle of mass m; $\hbar$ is Planck's constant divided by 2π. Like the diffusion equation it is second order in the three spatial variables and first order in time.

10.2 General form of solution

Before turning to the methods by which we may hope to solve PDEs such as those listed in the previous section, it is instructive, as for ODEs in Chapter 6, to study how PDEs may be formed from a set of possible solutions. Such a study can provide an indication of how equations obtained not from possible solutions but from physical arguments might be solved.

For definiteness let us suppose we have a set of functions involving two independent variables x and y. Without further specification this is of course a very wide set of functions, and we could not expect to find a useful equation that they all satisfy. However, let us consider a type of function $u_i(x, y)$ in which x and y appear in a particular way, such that u_i can be written as a function (however complicated) *of a single variable* p, itself a simple function of x and y.

We can illustrate this by considering the three functions

$$u_1(x, y) = x^4 + 4(x^2 y + y^2 + 1),$$

$$u_2(x, y) = \sin x^2 \cos 2y + \cos x^2 \sin 2y,$$

$$u_3(x, y) = \frac{x^2 + 2y + 2}{3x^2 + 6y + 5}.$$

These are all fairly complicated functions of x and y and a single differential equation of which each one is a solution is not obvious. However, if we observe that in fact each can be expressed as a function of the variable $p = x^2 + 2y$ alone (with no other x or y involved) then a great simplification takes place. Written in terms of p the above equations become

$$u_1(x, y) = (x^2 + 2y)^2 + 4 = p^2 + 4 = f_1(p),$$

$$u_2(x, y) = \sin(x^2 + 2y) = \sin p = f_2(p),$$

$$u_3(x, y) = \frac{(x^2 + 2y) + 2}{3(x^2 + 2y) + 5} = \frac{p + 2}{3p + 5} = f_3(p).$$

Let us now form, for each u_i, the partial derivatives $\partial u_i/\partial x$ and $\partial u_i/\partial y$. In each case these are (writing both the form for general p and the one appropriate to our particular case, $p = x^2 + 2y$)

$$\frac{\partial u_i}{\partial x} = \frac{d f_i(p)}{dp} \frac{\partial p}{\partial x} = 2x f_i',$$

$$\frac{\partial u_i}{\partial y} = \frac{d f_i(p)}{dp} \frac{\partial p}{\partial y} = 2 f_i',$$

for $i = 1, 2, 3$. All reference to the form of f_i can be eliminated from these equations by cross-multiplication, obtaining

$$\frac{\partial p}{\partial y} \frac{\partial u_i}{\partial x} = \frac{\partial p}{\partial x} \frac{\partial u_i}{\partial y},$$

or, for our specific form, $p = x^2 + 2y$,

$$\frac{\partial u_i}{\partial x} = x\frac{\partial u_i}{\partial y}. \tag{10.8}$$

It is thus apparent that not only are the three functions u_1, u_2, u_3 solutions of the PDE (10.8) but so also is *any arbitrary function* $f(p)$ of which the argument p has the form $x^2 + 2y$.

10.3 General and particular solutions

In the last section we found that the first-order PDE (10.8) has as a solution *any* function of the variable $x^2 + 2y$. This points the way for the solution of PDEs of other orders, as follows. It is *not* generally true that an nth-order PDE can always be considered as resulting from the elimination of n arbitrary *functions* from its solution (as opposed to the elimination of n arbitrary *constants* for an nth-order ODE, see Subsection 6.1.1). However, given specific PDEs we can try to solve them by seeking combinations of variables in terms of which the solutions may be expressed as arbitrary functions. Where this is possible we may expect n combinations to be involved in the solution.

Naturally, the exact functional form of the solution for any particular situation must be determined by some set of boundary conditions. For instance, if the PDE contains two independent variables x and y then for complete determination of its solution the boundary conditions will take a form equivalent to specifying $u(x, y)$ along a suitable continuum of points in the xy-plane (usually along a line).

We now discuss the general and particular solutions of first- and second-order PDEs. In order to simplify the algebra, we will restrict our discussion to equations containing just two independent variables x and y. Nevertheless, the method presented below may be extended to equations containing several independent variables.

10.3.1 First-order equations

Although most of the PDEs encountered in physical contexts are second order (i.e. they contain $\partial^2 u/\partial x^2$ or $\partial^2 u/\partial x \partial y$, etc.), we now discuss first-order equations to illustrate the general considerations involved in the form of the solution and in satisfying any boundary conditions on the solution.

The most general first-order linear PDE (containing two independent variables) is of the form

$$A(x, y)\frac{\partial u}{\partial x} + B(x, y)\frac{\partial u}{\partial y} + C(x, y)u = R(x, y), \tag{10.9}$$

where $A(x, y)$, $B(x, y)$, $C(x, y)$ and $R(x, y)$ are given functions. Clearly, if either $A(x, y)$ or $B(x, y)$ is zero then the PDE may be solved straightforwardly as a first-order linear ODE, the only modification being that the arbitrary constant of integration becomes an *arbitrary function* of x or y, respectively. As a simple example, consider the following.

Example Find the general solution $u(x, y)$ of

$$x\frac{\partial u}{\partial x} + 3u = x^2.$$

Dividing through by x we obtain

$$\frac{\partial u}{\partial x} + \frac{3u}{x} = x,$$

which is a linear equation with integrating factor

$$\exp\left(\int \frac{3}{x}\,dx\right) = \exp(3\ln x) = x^3.$$

Multiplying through by this factor we find

$$\frac{\partial}{\partial x}(x^3 u) = x^4,$$

which, on integrating with respect to x, gives

$$x^3 u = \frac{x^5}{5} + f(y),$$

where $f(y)$ is an *arbitrary function* of y. Dividing through by x^3, we obtain

$$u(x, y) = \frac{x^2}{5} + \frac{f(y)}{x^3}$$

as the final solution.[2] ◀

When the PDE contains partial derivatives with respect to both independent variables then, of course, we cannot employ the above procedure but must seek an alternative method. Let us for the moment restrict our attention to the special case in which $C(x, y) = R(x, y) = 0$ and, following the discussion of the previous section, look for solutions of the form $u(x, y) = f(p)$ where p is some, at present unknown, combination of x and y. We then have

$$\frac{\partial u}{\partial x} = \frac{df(p)}{dp}\frac{\partial p}{\partial x},$$
$$\frac{\partial u}{\partial y} = \frac{df(p)}{dp}\frac{\partial p}{\partial y},$$

which, when substituted into the PDE (10.9), give

$$\left[A(x, y)\frac{\partial p}{\partial x} + B(x, y)\frac{\partial p}{\partial y}\right]\frac{df(p)}{dp} = 0.$$

2 Substitute this answer into the original equation and verify that the latter is satisfied for any $f(y)$.

This removes all reference to the actual form of the function $f(p)$ if, for non-trivial p, we have

$$A(x, y)\frac{\partial p}{\partial x} + B(x, y)\frac{\partial p}{\partial y} = 0. \tag{10.10}$$

Let us now consider the necessary condition for $f(p)$ to remain constant as x and y vary; this is that p itself remains constant. Thus for f to remain constant implies that x and y must vary in such a way that

$$dp = \frac{\partial p}{\partial x}\,dx + \frac{\partial p}{\partial y}\,dy = 0. \tag{10.11}$$

The forms of (10.10) and (10.11) are very alike and become the same if we require that

$$\frac{dx}{A(x, y)} = \frac{dy}{B(x, y)}. \tag{10.12}$$

By integrating this expression the form of p can be found. This next example illustrates the point.

Example For

$$x\frac{\partial u}{\partial x} - 2y\frac{\partial u}{\partial y} = 0, \tag{10.13}$$

find (i) the solution that takes the value $2y + 1$ on the line $x = 1$, and (ii) a solution that has the value 4 at the point $(1, 1)$.

If we seek a solution of the form $u(x, y) = f(p)$, we deduce from (10.12) that $u(x, y)$ will be constant along lines of (x, y) that satisfy

$$\frac{dx}{x} = \frac{dy}{-2y},$$

which on integrating gives $x = cy^{-1/2}$. Identifying the constant of integration c with $p^{1/2}$ (to avoid fractional powers), we conclude that $p = x^2y$. Thus the general solution of the PDE (10.13) is

$$u(x, y) = f(x^2y),$$

where f is an arbitrary function.

We must now find the particular solutions that obey each of the imposed boundary conditions. For boundary condition (i) a little thought shows that the particular solution required is

$$u(x, y) = 2(x^2y) + 1 = 2x^2y + 1. \tag{10.14}$$

For boundary condition (ii) some obviously acceptable solutions are

$$u(x, y) = x^2y + 3,$$

$$u(x, y) = 4x^2y,$$

$$u(x, y) = 4.$$

Each is a valid solution [the freedom of choice of form arises from the fact that u is specified at only one point $(1, 1)$, and not along a continuum (say), as in boundary condition (i)]. All three are particular examples of the general solution, which may be written, for example, as

$$u(x, y) = x^2y + 3 + g(x^2y),$$

where $g = g(x^2y) = g(p)$ is an arbitrary function subject only to $g(1) = 0$. For this example, the forms of g corresponding to the particular solutions listed above are $g(p) = 0$, $g(p) = 3p - 3$, $g(p) = 1 - p$. ◀

As mentioned above, in order to find a solution of the form $u(x, y) = f(p)$ we require that the original PDE contains no term in u, but only terms containing its partial derivatives. If a term in u is present, so that $C(x, y) \neq 0$ in (10.9), then the procedure needs some modification, since we cannot simply divide out the dependence on $f(p)$ to obtain (10.10). In such cases we look instead for a solution of the form $u(x, y) = h(x, y)f(p)$. We illustrate this method in the following example.

Example Find the general solution of

$$x\frac{\partial u}{\partial x} + 2\frac{\partial u}{\partial y} - 2u = 0. \qquad (10.15)$$

We seek a solution of the form $u(x, y) = h(x, y)f(p)$, with the consequence that

$$\frac{\partial u}{\partial x} = \frac{\partial h}{\partial x}f(p) + h\frac{df(p)}{dp}\frac{\partial p}{\partial x},$$

$$\frac{\partial u}{\partial y} = \frac{\partial h}{\partial y}f(p) + h\frac{df(p)}{dp}\frac{\partial p}{\partial y}.$$

Substituting these expressions into the PDE (10.15) and rearranging, we obtain

$$\left(x\frac{\partial h}{\partial x} + 2\frac{\partial h}{\partial y} - 2h\right)f(p) + \left(x\frac{\partial p}{\partial x} + 2\frac{\partial p}{\partial y}\right)h\frac{df(p)}{dp} = 0.$$

The first factor in parentheses is just the original PDE with u replaced by h. Therefore, if h is *any* solution of the PDE, *however simple*, this term will vanish, to leave

$$\left(x\frac{\partial p}{\partial x} + 2\frac{\partial p}{\partial y}\right)h\frac{df(p)}{dp} = 0,$$

from which, as in the previous case, we obtain

$$x\frac{\partial p}{\partial x} + 2\frac{\partial p}{\partial y} = 0.$$

From (10.11) and (10.12) we see that $u(x, y)$ will be constant along lines of (x, y) that satisfy

$$\frac{dx}{x} = \frac{dy}{2},$$

which integrates to give $x = c\exp(y/2)$. Identifying the constant of integration c with p we find $p = x\exp(-y/2)$. Thus the general solution of (10.15) is

$$u(x, y) = h(x, y)f(x\exp(-\tfrac{1}{2}y)),$$

where $f(p)$ is any arbitrary function of p and $h(x, y)$ is any solution of (10.15).

If we take, for example, $h(x, y) = \exp y$, which clearly satisfies (10.15), then the general solution is

$$u(x, y) = (\exp y) f(x \exp(-\tfrac{1}{2}y)).$$

Alternatively, $h(x, y) = x^2$ also satisfies (10.15) and so the general solution to the equation can also be written

$$u(x, y) = x^2 g(x \exp(-\tfrac{1}{2}y)),$$

where g is an arbitrary function of p; clearly $g(p) = f(p)/p^2$. ◀

10.3.2 Inhomogeneous equations and problems

Let us discuss in a more general form the particular solutions of (10.13) found in the second example of the previous subsection. It is clear that, so far as this equation is concerned, if $u(x, y)$ is a solution then so is any multiple of $u(x, y)$ or any linear sum of separate solutions $u_1(x, y) + u_2(x, y)$. However, when it comes to fitting the boundary conditions this is not so.

For example, although $u(x, y)$ in (10.14) satisfies the PDE and the boundary condition $u(1, y) = 2y + 1$, the function $u_1(x, y) = 4u(x, y) = 8xy + 4$, whilst satisfying the PDE, takes the value $8y + 4$ on the line $x = 1$ and so does not satisfy the required boundary condition. Likewise the function $u_2(x, y) = u(x, y) + f_1(x^2 y)$, for arbitrary f_1, satisfies (10.13) but takes the value $u_2(1, y) = 2y + 1 + f_1(y)$ on the line $x = 1$, and so is not of the required form unless f_1 is identically zero.

Thus we see that when treating the superposition of solutions of PDEs two considerations arise, one concerning the equation itself and the other connected to the boundary conditions. The *equation* is said to be homogeneous if the fact that $u(x, y)$ is a solution implies that $\lambda u(x, y)$, for any constant λ, is also a solution. However, the *problem* is said to be homogeneous if, in addition, the boundary conditions are such that if they are satisfied by $u(x, y)$ then they are also satisfied by $\lambda u(x, y)$. The last requirement itself is referred to as that of *homogeneous boundary conditions*.

For example, the PDE (10.13) is homogeneous but the general first-order equation (10.9) would not be homogeneous unless $R(x, y) = 0$. Furthermore, the boundary condition (i) imposed on the solution of (10.13) in the previous subsection is not homogeneous though, in this case, the boundary condition

$$u(x, y) = 0 \quad \text{on the line } y = 4x^{-2}$$

would be, since $u(x, y) = \lambda(x^2 y - 4)$ satisfies this condition for any λ and, being a function of $x^2 y$, satisfies (10.13).

The reason for discussing the homogeneity of PDEs and their boundary conditions is that in linear PDEs there is a close parallel to the complementary-function and particular-integral property of ODEs. The general solution of an inhomogeneous problem can be written as the sum of *any* particular solution of the problem and the general solution of the corresponding homogeneous problem (as for ODEs, we require that the particular solution is not already contained in the general solution of the homogeneous problem).

Thus, for example, the general solution of

$$\frac{\partial u}{\partial x} - x\frac{\partial u}{\partial y} + au = f(x, y), \tag{10.16}$$

subject to, say, the boundary condition $u(0, y) = g(y)$, is given by

$$u(x, y) = v(x, y) + w(x, y),$$

where $v(x, y)$ is any solution (however simple) of (10.16) such that $v(0, y) = g(y)$ and $w(x, y)$ is the general solution of

$$\frac{\partial w}{\partial x} - x\frac{\partial w}{\partial y} + aw = 0, \tag{10.17}$$

with $w(0, y) = 0$. If the boundary conditions are sufficiently specified then the only possible solution of (10.17) will be $w(x, y) \equiv 0$ and $v(x, y)$ will be the complete solution by itself.

Alternatively, we may begin by finding the general solution of the inhomogeneous equation (10.16) *without* regard for any boundary conditions; it is just the sum of the general solution to the homogeneous equation and a particular integral of (10.16), both without reference to the boundary conditions. The boundary conditions can then be used to find the appropriate particular solution from the general solution.

We will not discuss at length general methods of obtaining particular integrals of PDEs but merely note that some of those methods available for ordinary differential equations can be suitably extended.[3]

Example Find the general solution of

$$y\frac{\partial u}{\partial x} - x\frac{\partial u}{\partial y} = 3x. \tag{10.18}$$

Hence find the most general particular solution (i) which satisfies $u(x, 0) = x^2$, and (ii) which has the value $u(x, y) = 2$ at the point $(1, 0)$.

This equation is inhomogeneous, and so let us first find the general solution of (10.18) without regard for any boundary conditions. We begin by looking for the solution of the corresponding homogeneous equation [(10.18) but with the RHS equal to zero] of the form $u(x, y) = f(p)$. Following the same procedure as that used in the solution of (10.13) we find that $u(x, y)$ will be constant along lines of (x, y) that satisfy

$$\frac{dx}{y} = \frac{dy}{-x} \quad \Rightarrow \quad \frac{x^2}{2} + \frac{y^2}{2} = c.$$

Identifying the constant of integration c with $p/2$, we find that the general solution of the homogeneous equation is $u(x, y) = f(x^2 + y^2)$ for arbitrary function f. Now by inspection a particular

3 See for example H. T. H. Piaggio, *An Elementary Treatise on Differential Equations and their Applications* (London: G. Bell and Sons, Ltd, 1954), pp. 175 ff.

integral of (10.18) is $u(x, y) = -3y$, and so the general solution to (10.18) is

$$u(x, y) = f(x^2 + y^2) - 3y.$$

Boundary condition (i) requires $u(x, 0) = f(x^2) = x^2$, i.e. $f(z) = z$, and so the particular solution in this case is

$$u(x, y) = x^2 + y^2 - 3y.$$

Similarly, boundary condition (ii) requires $u(1, 0) = f(1) = 2$. One possibility is $f(z) = 2z$, and if we make this choice, then one way of writing the most general particular solution is

$$u(x, y) = 2x^2 + 2y^2 - 3y + g(x^2 + y^2),$$

where g is any arbitrary function for which $g(1) = 0$. Alternatively, a simpler choice would be $f(z) = 2$, leading to

$$u(x, y) = 2 - 3y + h(x^2 + y^2),$$

where, this time, $h(1) = 0$. Clearly, if the two solutions are to represent the same explicit solution, we must have that $h(z) = g(z) + 2(z - 1)$, but, for the most general solution satisfying this one-point boundary condition, either form will do. ◀

Although we have discussed the solution of inhomogeneous problems only for first-order equations, the general considerations hold true for linear PDEs of higher order.

10.3.3 Second-order equations

As noted in Section 10.1, second-order linear PDEs are of great importance in describing the behavior of many physical systems. As in our discussion of first-order equations, for the moment we will restrict our discussion to equations with just two independent variables; extensions to a greater number of independent variables are straightforward.

The most general second-order linear PDE (containing two independent variables) has the form

$$A\frac{\partial^2 u}{\partial x^2} + B\frac{\partial^2 u}{\partial x \partial y} + C\frac{\partial^2 u}{\partial y^2} + D\frac{\partial u}{\partial x} + E\frac{\partial u}{\partial y} + Fu = R(x, y), \tag{10.19}$$

where $A, B, \ldots, F$ and $R(x, y)$ are given functions of x and y. Because of the nature of the solutions to such equations, they are usually divided into three classes, a division of which we will make further use in Section 10.6. The equation (10.19) is called *hyperbolic* if $B^2 > 4AC$, *parabolic* if $B^2 = 4AC$ and *elliptic* if $B^2 < 4AC$. Clearly, if A, B and C are functions of x and y (rather than just constants) then the equation might be of different types in different parts of the xy-plane.

Equation (10.19) obviously represents a very large class of PDEs, and it is usually impossible to find closed-form solutions to most of these equations. Therefore, for the moment we shall consider only homogeneous equations, with $R(x, y) = 0$, and make the further (greatly simplifying) restriction that, throughout the remainder of this section, $A, B, \ldots, F$ are not functions of x and y but merely constants.

We now tackle the problem of solving some types of second-order PDE with constant coefficients by seeking solutions that are arbitrary functions of particular combinations of independent variables, just as we did for first-order equations.

Following the discussion of the previous section, we can hope to find such solutions only if all the terms of the equation involve the same total number of differentiations, i.e. all terms are of the same order, although the number of differentiations with respect to the individual independent variables may be different. This means that in (10.19) we require the constants D, E and F to be identically zero (we have, of course, already assumed that $R(x, y)$ is zero), so that we are now considering only equations of the form

$$A\frac{\partial^2 u}{\partial x^2} + B\frac{\partial^2 u}{\partial x \partial y} + C\frac{\partial^2 u}{\partial y^2} = 0, \tag{10.20}$$

where A, B and C are constants. We note that both the one-dimensional wave equation,

$$\frac{\partial^2 u}{\partial x^2} - \frac{1}{c^2}\frac{\partial^2 u}{\partial t^2} = 0,$$

and the two-dimensional Laplace equation,

$$\frac{\partial^2 u}{\partial x^2} + \frac{\partial^2 u}{\partial y^2} = 0,$$

are of this form, but that the diffusion equation,

$$\kappa\frac{\partial^2 u}{\partial x^2} - \frac{\partial u}{\partial t} = 0,$$

is not, since it contains a first-order derivative.

Since all the terms in (10.20) involve two differentiations, by assuming a solution of the form $u(x, y) = f(p)$, where p is some unknown function of x and y (or t), we may be able to obtain a common factor $d^2 f(p)/dp^2$ as the only appearance of f on the LHS. Then, because of the zero RHS, all reference to the form of f can be canceled out.

We can gain some guidance on suitable forms for the combination $p = p(x, y)$ by considering $\partial u/\partial x$ when u is given by $u(x, y) = f(p)$, for then

$$\frac{\partial u}{\partial x} = \frac{d f(p)}{dp}\frac{\partial p}{\partial x}.$$

Clearly differentiation of this equation with respect to x (or y) will not lead to a single term on the RHS, containing f only as $d^2 f(p)/dp^2$, unless the factor $\partial p/\partial x$ is a constant so that $\partial^2 p/\partial x^2$ and $\partial^2 p/\partial x \partial y$ are necessarily zero. This shows that p must be a linear function of x. In an exactly similar way p must also be a linear function of y, i.e. $p = ax + by$.[4]

If we assume a solution to (10.20) of the form $u(x, y) = f(ax + by)$, and evaluate the terms ready for substitution into (10.20), we obtain

$$\frac{\partial u}{\partial x} = a\frac{d f(p)}{dp}, \qquad \frac{\partial u}{\partial y} = b\frac{d f(p)}{dp},$$

$$\frac{\partial^2 u}{\partial x^2} = a^2\frac{d^2 f(p)}{dp^2}, \qquad \frac{\partial^2 u}{\partial x \partial y} = ab\frac{d^2 f(p)}{dp^2}, \qquad \frac{\partial^2 u}{\partial y^2} = b^2\frac{d^2 f(p)}{dp^2},$$

4 It might seem that a more general form would be $q = p + c = ax + by + c$, where c is a constant. However, any arbitrary function $g(q)$ can always be written as $g(p + c) \equiv f(p)$ for some suitable function f. Hence it is sufficient to take $p = ax + by$ as the argument of an arbitrary function.

which on substitution give

$$\left(Aa^2 + Bab + Cb^2\right)\frac{d^2 f(p)}{dp^2} = 0. \qquad (10.21)$$

This is the form we have been seeking, since now a solution independent of the form of f can be obtained if we require that a and b satisfy

$$Aa^2 + Bab + Cb^2 = 0.$$

From this quadratic, two values for the ratio of the two constants a and b are obtained,

$$b/a = [-B \pm (B^2 - 4AC)^{1/2}]/2C.$$

If we denote these two ratios by λ_1 and λ_2 then *any* functions of the two variables

$$p_1 = x + \lambda_1 y, \qquad p_2 = x + \lambda_2 y$$

will be solutions of the original equation (10.20). The omission of the constant factor a from p_1 and p_2 is of no consequence since this can always be absorbed into the particular form of any chosen function; only the *relative* weighting of x and y in p is important.

Since p_1 and p_2 are in general different, we can thus write the general solution of (10.20) as

$$u(x, y) = f(x + \lambda_1 y) + g(x + \lambda_2 y), \qquad (10.22)$$

where f and g are arbitrary functions.

Finally, we note that the alternative solution $d^2 f(p)/dp^2 = 0$ to (10.21) leads only to the trivial solution $u(x, y) = kx + ly + m$, for which all second derivatives are individually zero.

As the next worked example we solve the one-dimensional wave equation.

Example Find the general solution of the one-dimensional wave equation

$$\frac{\partial^2 u}{\partial x^2} - \frac{1}{c^2}\frac{\partial^2 u}{\partial t^2} = 0.$$

The wave equation[5] has the form of (10.20) with $A = 1$, $B = 0$ and $C = -1/c^2$, and so the values of λ_1 and λ_2 are the solutions of

$$1 - \frac{\lambda^2}{c^2} = 0,$$

namely $\lambda_1 = -c$ and $\lambda_2 = c$. This means that arbitrary functions of the quantities

$$p_1 = x - ct, \qquad p_2 = x + ct$$

will be satisfactory solutions of the equation and that the general solution will be

$$u(x, t) = f(x - ct) + g(x + ct), \qquad (10.23)$$

where f and g are arbitrary functions. This solution is discussed in greater detail in Section 10.4. ◀

5 Is this equation hyperbolic, parabolic or elliptic?

The method used to obtain the general solution of the wave equation may also be applied straightforwardly to Laplace's equation.

Example Find the general solution of the two-dimensional Laplace equation

$$\frac{\partial^2 u}{\partial x^2} + \frac{\partial^2 u}{\partial y^2} = 0. \tag{10.24}$$

Following the established procedure, we look for a solution that is a function $f(p)$ of $p = x + \lambda y$, where from (10.24) λ satisfies

$$1 + \lambda^2 = 0.$$

This requires that $\lambda = \pm i$, and satisfactory variables p are $p = x \pm iy$. The general solution required is therefore, in terms of arbitrary functions f and g,

$$u(x, y) = f(x + iy) + g(x - iy).$$

Thus if f and g were arbitrarily chosen as, say, $f(p) = 3 + p$ and $g(p) = p^2$, then $u(x, y) = 3 + x + iy + (x - iy)^2 = 3 + x^2 - y^2 + x + i(y - 2xy)$. It should be remembered that, although f and g are arbitrary functions, this does not mean that u is an arbitrary function of x and y. For example, $u(x, y) = x(x + y)$ could not be a solution because, when it is expressed in terms of $p_1 = x + iy$ and $p_2 = x - iy$, it takes the form

$$u(x, y) = \tfrac{1}{4}(p_1 + p_2)[p_1 + p_2 - i(p_1 - p_2)]$$

and this *cannot* be manipulated into the form $f(p_1) + g(p_2)$.[6] ◀

It will be apparent from the last two examples that the nature of the appropriate linear combination of x and y depends upon whether $B^2 > 4AC$ or $B^2 < 4AC$. This is exactly the same criterion as determines whether the PDE is hyperbolic or elliptic. Hence as a general result, hyperbolic and elliptic equations of the form (10.20), given the restriction that the constants A, B and C are real, have as solutions functions whose arguments have the form $x + \alpha y$ and $x + i\beta y$ respectively, where α and β themselves are real.

The one case not covered by this result is that in which $B^2 = 4AC$, i.e. a parabolic equation. In this case λ_1 and λ_2 are not different and only one suitable combination of x and y results, namely

$$u(x, y) = f(x - (B/2C)y).$$

To find the second part of the general solution we try, in analogy with the corresponding situation for ordinary differential equations, a solution of the form

$$u(x, y) = h(x, y)g(x - (B/2C)y).$$

Substituting this into (10.20) and using $A = B^2/4C$ results in

$$\left(A\frac{\partial^2 h}{\partial x^2} + B\frac{\partial^2 h}{\partial x \partial y} + C\frac{\partial^2 h}{\partial y^2}\right) g = 0.$$

6 Show that this conclusion can be reached much more simply by direct substitution in this case.

Therefore we require $h(x, y)$ to be any solution of the original PDE. There are several simple solutions of this equation, but as only one is required we take the simplest non-trivial one, $h(x, y) = x$, to give the general solution of the parabolic equation

$$u(x, y) = f(x - (B/2C)y) + xg(x - (B/2C)y). \tag{10.25}$$

We could, of course, have taken $h(x, y) = y$, but this only leads to a solution that is already represented by (10.25).[7] As an example of a parabolic equation, consider the following.

Example Solve

$$\frac{\partial^2 u}{\partial x^2} + 2\frac{\partial^2 u}{\partial x \partial y} + \frac{\partial^2 u}{\partial y^2} = 0,$$

subject to the boundary conditions $u(0, y) = 0$ and $u(x, 1) = x^2$.

From our general result, functions of $p = x + \lambda y$ will be solutions provided

$$1 + 2\lambda + \lambda^2 = 0,$$

i.e. $\lambda = -1$ (twice) and the equation is parabolic. The general solution is therefore

$$u(x, y) = f(x - y) + xg(x - y).$$

The boundary condition $u(0, y) = 0$ implies $f(p) \equiv 0$, and then $u(x, 1) = x^2$ yields

$$xg(x - 1) = x^2,$$

which gives $g(p) = p + 1$. Therefore the particular solution required is

$$u(x, y) = x(p + 1) = x(x - y + 1).$$

Note that since values are given along two boundaries, $x = 0$ and $y = 1$, the solution is completely determined and it contains no arbitrary functions. ◀

To reinforce the material discussed above we will now give alternative derivations of the general solutions (10.22) and (10.25) by expressing the original PDE in terms of new variables before solving it. The actual solution will then become almost trivial; but, of course, it will be recognized that suitable new variables could hardly have been guessed if it were not for the work already done. This does not detract from the validity of the derivation to be described, only from the likelihood that it would be discovered by inspection.

We start again with (10.20) and change to new variables

$$\zeta = x + \lambda_1 y, \qquad \eta = x + \lambda_2 y.$$

7 Prove this by writing $q = x - (B/2C)y$ and showing that (10.25) can be rewritten as $u(x, y) = f_1(q) + yg_1(q)$, where $f_1(q) = f(q) + qg(q)$ and $g_1(q) = (B/2C)g(q)$.

With this change of variables, we have from the chain rule that

$$\frac{\partial}{\partial x} = \frac{\partial}{\partial \zeta} + \frac{\partial}{\partial \eta},$$
$$\frac{\partial}{\partial y} = \lambda_1 \frac{\partial}{\partial \zeta} + \lambda_2 \frac{\partial}{\partial \eta}.$$

Using these and the fact that

$$A + B\lambda_i + C\lambda_i^2 = 0 \quad \text{for } i = 1, 2,$$

our initial equation,

$$A\frac{\partial^2 u}{\partial x^2} + B\frac{\partial^2 u}{\partial x \partial y} + C\frac{\partial^2 u}{\partial y^2} = 0,$$

becomes

$$[2A + B(\lambda_1 + \lambda_2) + 2C\lambda_1\lambda_2]\frac{\partial^2 u}{\partial \zeta \partial \eta} = 0.$$

Then, providing the factor in brackets does not vanish, for which the required condition is $B^2 \neq 4AC$,[8] we obtain

$$\frac{\partial^2 u}{\partial \zeta \partial \eta} = 0,$$

which has the successive integrals

$$\frac{\partial u}{\partial \eta} = F(\eta), \quad u(\zeta, \eta) = f(\eta) + g(\zeta).$$

This solution is just the same as (10.22),

$$u(x, y) = f(x + \lambda_2 y) + g(x + \lambda_1 y).$$

If the equation *is* parabolic (i.e. $B^2 = 4AC$), we use an alternative set of new variables,

$$\zeta = x + \lambda y, \qquad \eta = x,$$

and then, recalling that $\lambda = -(B/2C)$, we can reduce (10.20) to

$$A\frac{\partial^2 u}{\partial \eta^2} = 0.$$

Two straightforward integrations give as the general solution

$$u(\zeta, \eta) = \eta g(\zeta) + f(\zeta),$$

which in terms of x and y has exactly the form of (10.25),

$$u(x, y) = xg(x + \lambda y) + f(x + \lambda y).$$

8 Show that this is the case.

Finally, as hinted at in Subsection 10.3.2 with reference to first-order linear PDEs, some of the methods used to find particular integrals of linear ODEs can be suitably modified to find particular integrals of PDEs of higher order. In simple cases, however, an appropriate solution may often be found by inspection.

Example Find the general solution of

$$\frac{\partial^2 u}{\partial x^2} + \frac{\partial^2 u}{\partial y^2} = 6(x + y).$$

Following our previous methods and results, the complementary function is

$$u(x, y) = f(x + iy) + g(x - iy),$$

and only a particular integral remains to be found. By inspection a particular integral of the equation is $u(x, y) = x^3 + y^3$, and so the general solution,

$$u(x, y) = f(x + iy) + g(x - iy) + x^3 + y^3,$$

can be found by combining the two. ◀

10.4 The wave equation

We have already found that the general solution of the one-dimensional wave equation is

$$u(x, t) = f(x - ct) + g(x + ct), \tag{10.26}$$

where f and g are arbitrary functions. However, the equation is of such general importance that further discussion will not be out of place.

Let us imagine that $u(x, t) = f(x - ct)$ represents the displacement of a string at time t and position x. It is clear that all positions x and times t for which $x - ct =$ constant will have the same instantaneous displacement. But $x - ct =$ constant is exactly the relation between the time and position of an observer traveling with speed c along the positive x-direction. Consequently this moving observer sees a constant displacement of the string, whereas to a stationary observer, the initial profile $u(x, 0)$ moves with speed c along the x-axis as if it were a rigid system. Thus $f(x - ct)$ represents a wave form of constant shape traveling along the positive x-axis with speed c, the actual form of the wave depending upon the function f. Similarly, the term $g(x + ct)$ is a constant wave form traveling with speed c in the negative x-direction. The general solution (10.23) represents a superposition of these.

If the functions f and g are the same then the complete solution (10.23) represents identical progressive waves going in opposite directions. This may result in a wave pattern whose profile does not progress, described as a *standing wave*. As a simple example, suppose both $f(p)$ and $g(p)$ have the form[9]

$$f(p) = g(p) = A\cos(kp + \epsilon).$$

9 In the usual notation, k is the wave number ($= 2\pi$/wavelength) and $kc = \omega$, the angular frequency of the wave.

Then (10.23) can be written as

$$u(x,t) = A[\cos(kx - kct + \epsilon) + \cos(kx + kct + \epsilon)]$$
$$= 2A\cos(kct)\cos(kx + \epsilon).$$

The important thing to notice is that the shape of the wave pattern, given by the factor involving x, is the same at all times but that its amplitude $2A\cos(kct)$ depends upon time. At some points x that satisfy

$$\cos(kx + \epsilon) = 0$$

there is no displacement at any time; such points are called *nodes*.

So far we have not imposed any boundary conditions on the solution (10.26). The problem of finding a solution to the wave equation that satisfies given boundary conditions is normally treated using the method of separation of variables discussed in the next chapter. Nevertheless, we now consider *D'Alembert's solution* $u(x,t)$ of the wave equation subject to initial conditions (boundary conditions) in the following general form:

$$\text{initial displacement, } u(x,0) = \phi(x); \quad \text{initial velocity, } \frac{\partial u(x,0)}{\partial t} = \psi(x).$$

The functions $\phi(x)$ and $\psi(x)$ are given and describe the displacement and velocity of each part of the string at the (arbitrary) time $t = 0$.

It is clear that what we need are the particular forms of the functions f and g in (10.26) that lead to the required values at $t = 0$. This means that

$$\phi(x) = u(x,0) = f(x-0) + g(x+0), \tag{10.27}$$

$$\psi(x) = \frac{\partial u(x,0)}{\partial t} = -cf'(x-0) + cg'(x+0), \tag{10.28}$$

where it should be noted that $f'(x-0)$ stands for $df(p)/dp$ evaluated, after the differentiation, at $p = x - c \times 0$; likewise for $g'(x+0)$.

Looking on the above two left-hand sides as functions of $p = x \pm ct$, but everywhere evaluated at $t = 0$, we may integrate (10.28) between an arbitrary (and irrelevant) lower limit p_0 and an indefinite upper limit p to obtain

$$\frac{1}{c}\int_{p_0}^{p} \psi(q)\,dq + K = -f(p) + g(p),$$

the constant of integration K depending on p_0. Comparing this equation with (10.27), with x replaced by p, we can establish the forms of the functions f and g as

$$f(p) = \frac{\phi(p)}{2} - \frac{1}{2c}\int_{p_0}^{p} \psi(q)\,dq - \frac{K}{2}, \tag{10.29}$$

$$g(p) = \frac{\phi(p)}{2} + \frac{1}{2c}\int_{p_0}^{p} \psi(q)\,dq + \frac{K}{2}. \tag{10.30}$$

Adding (10.29) with $p = x - ct$ to (10.30) with $p = x + ct$ gives as the solution to the original problem

$$u(x,t) = \frac{1}{2}[\phi(x-ct) + \phi(x+ct)] + \frac{1}{2c}\int_{x-ct}^{x+ct} \psi(q)\,dq, \tag{10.31}$$

in which we notice that all dependence on p_0 has disappeared.

Each of the terms in (10.31) has a fairly straightforward physical interpretation. In each case the factor $1/2$ represents the fact that only half a displacement profile that starts at any particular point on the string travels towards any other position x, the other half traveling away from it. The first term $\frac{1}{2}\phi(x-ct)$ arises from the initial displacement at a distance ct to the left of x; this travels forward arriving at x at time t. Similarly, the second contribution is due to the initial displacement at a distance ct to the right of x. The interpretation of the final term is a little less obvious. It can be viewed as representing the accumulated transverse displacement at position x due to the passage past x of all parts of the initial motion whose effects can reach x within a time t, both backward and forward traveling.

The extension to the three-dimensional wave equation of solutions of the type we have so far encountered presents no serious difficulty. In Cartesian coordinates the three-dimensional wave equation is

$$\frac{\partial^2 u}{\partial x^2} + \frac{\partial^2 u}{\partial y^2} + \frac{\partial^2 u}{\partial z^2} - \frac{1}{c^2}\frac{\partial^2 u}{\partial t^2} = 0. \tag{10.32}$$

In close analogy with the one-dimensional case we try solutions that are functions of linear combinations of all four variables,

$$p = lx + my + nz + \mu t.$$

It is clear that a solution $u(x, y, z, t) = f(p)$ will be acceptable provided that

$$\left(l^2 + m^2 + n^2 - \frac{\mu^2}{c^2}\right)\frac{d^2 f(p)}{dp^2} = 0.$$

Thus, as in the one-dimensional case, f can be arbitrary provided that

$$l^2 + m^2 + n^2 = \mu^2/c^2.$$

Using an obvious normalization, we take $\mu = \pm c$ and l, m, n as three numbers such that

$$l^2 + m^2 + n^2 = 1.$$

In other words (l, m, n) are the Cartesian components of a unit vector $\hat{\mathbf{n}}$ that points along the direction of propagation of the wave. The quantity p can be written in terms of vectors as the scalar expression $p = \hat{\mathbf{n}} \cdot \mathbf{r} \pm ct$, and the general solution of (10.32) is then

$$u(x, y, z, t) = u(\mathbf{r}, t) = f(\hat{\mathbf{n}} \cdot \mathbf{r} - ct) + g(\hat{\mathbf{n}} \cdot \mathbf{r} + ct), \tag{10.33}$$

where $\hat{\mathbf{n}}$ is *any* unit vector. It would perhaps be more transparent to write $\hat{\mathbf{n}}$ explicitly as one of the arguments of u.

10.5 The diffusion equation

One important class of second-order PDEs, which we have not yet considered in detail, is that in which the second derivative with respect to one variable appears, but only the first derivative with respect to another (usually time). This is exemplified by the one-dimensional diffusion equation

$$\kappa \frac{\partial^2 u(x,t)}{\partial x^2} = \frac{\partial u}{\partial t}, \tag{10.34}$$

in which κ is a constant with the dimensions length^2 $\times$ time^{-1}. The physical constants that go to make up κ in a particular case depend upon the nature of the process (e.g. solute diffusion, heat flow, etc.) and the material being described.

With (10.34) we cannot hope to repeat successfully the method of Subsection 10.3.3, since now $u(x,t)$ is differentiated a different number of times on the two sides of the equation; any attempted solution in the form $u(x,t) = f(p)$ with $p = ax + bt$ will lead only to an equation in which the form of f cannot be canceled out. Clearly we must try other methods.

Solutions may be obtained by using the standard method of separation of variables discussed in the next chapter. Alternatively, a simple solution is also given if both sides of (10.34), as it stands, are separately set equal to a constant α (say), so that

$$\frac{\partial^2 u}{\partial x^2} = \frac{\alpha}{\kappa}, \qquad \frac{\partial u}{\partial t} = \alpha.$$

These equations have the general solutions

$$u(x,t) = \frac{\alpha}{2\kappa}x^2 + xg(t) + h(t) \quad \text{and} \quad u(x,t) = \alpha t + m(x)$$

respectively and may be made compatible with each other if $g(t)$ is taken as constant, $g(t) = g$ (where g could be zero), $h(t) = \alpha t$ and $m(x) = (\alpha/2\kappa)x^2 + gx$. An acceptable solution is thus

$$u(x,t) = \frac{\alpha}{2\kappa}x^2 + gx + \alpha t + \text{constant}. \tag{10.35}$$

Let us now return to seeking solutions of equations by combining the independent variables in particular ways. Having seen that a linear combination of x and t will be of no value, we must search for other possible combinations. It has been noted already that κ has the dimensions length^2 $\times$ time^{-1} and so the combination of variables

$$\eta = \frac{x^2}{\kappa t}$$

will be dimensionless. Let us see if we can satisfy (10.34) with a solution of the form $u(x,t) = f(\eta)$. Evaluating the necessary derivatives we have

$$\frac{\partial u}{\partial x} = \frac{d f(\eta)}{d\eta}\frac{\partial \eta}{\partial x} = \frac{2x}{\kappa t}\frac{d f(\eta)}{d\eta},$$

$$\frac{\partial^2 u}{\partial x^2} = \frac{2}{\kappa t}\frac{d f(\eta)}{d\eta} + \left(\frac{2x}{\kappa t}\right)^2 \frac{d^2 f(\eta)}{d\eta^2},$$

$$\frac{\partial u}{\partial t} = -\frac{x^2}{\kappa t^2}\frac{d f(\eta)}{d\eta}.$$

Substituting these expressions into (10.34) we find that the new equation can be written entirely in terms of η,

$$4\eta \frac{d^2 f(\eta)}{d\eta^2} + (2+\eta)\frac{d f(\eta)}{d\eta} = 0.$$

This is a straightforward ODE, which can be solved as follows. Writing $f'(\eta) = df(\eta)/d\eta$, etc., we have

$$\frac{f''(\eta)}{f'(\eta)} = -\frac{1}{2\eta} - \frac{1}{4}$$

$$\Rightarrow \quad \ln[\eta^{1/2} f'(\eta)] = -\frac{\eta}{4} + c$$

$$\Rightarrow \quad f'(\eta) = \frac{A}{\eta^{1/2}} \exp\left(\frac{-\eta}{4}\right)$$

$$\Rightarrow \quad f(\eta) = A\int_{\eta_0}^{\eta} \mu^{-1/2} \exp\left(\frac{-\mu}{4}\right) d\mu.$$

If we now write this in terms of a slightly different variable

$$\zeta = \frac{\eta^{1/2}}{2} = \frac{x}{2(\kappa t)^{1/2}},$$

then $d\zeta = \frac{1}{4}\eta^{-1/2}\, d\eta$, and the solution to (10.34) is given by

$$u(x,t) = f(\eta) = g(\zeta) = B\int_{\zeta_0}^{\zeta} \exp(-\nu^2)\, d\nu. \tag{10.36}$$

Here B is a constant and it should be noticed that x and t appear on the RHS only in the indefinite upper limit ζ, and then only in the combination $xt^{-1/2}$. If ζ_0 is chosen as zero then $u(x,t)$ is, to within a constant factor,[10] the error function $\text{erf}[x/2(\kappa t)^{1/2}]$, which is tabulated in many reference books. Only non-negative values of x and t are to be considered here, and so $\zeta \geq \zeta_0$.

Let us try to determine what kind of (say) temperature distribution and flow this represents. For definiteness we take $\zeta_0 = 0$. Firstly, since $u(x,t)$ in (10.36) depends only

10 Take $B = 2\pi^{-1/2}$ to give the usual error function normalized in such a way that $\text{erf}(\infty) = 1$. See Subsection 9.10.3.

upon the product $xt^{-1/2}$, it is clear that all points x at times t such that $xt^{-1/2}$ has the same value have the same temperature. Put another way, at any specific time t the region having a particular temperature has moved along the positive x-axis a distance proportional to the square root of t. This is a typical *diffusion* process.

Notice that, on the one hand, at $t = 0$ the variable $\zeta \to \infty$ and u becomes quite independent of x (except perhaps at $x = 0$); the solution then represents a uniform spatial temperature distribution. On the other hand, at $x = 0$ we have that $u(x, t)$ is identically zero for all t. Our next worked example shows a solution of this type in action.

Example An infrared laser delivers a pulse of (heat) energy E to a point P on a large insulated sheet of thickness b, thermal conductivity k, specific heat s and density ρ. The sheet is initially at a uniform temperature. If $u(r, t)$ is the excess temperature a time t later, at a point that is a distance r ($\gg b$) from P, then show that a suitable expression for u is

$$u(r, t) = \frac{\alpha}{t}\exp\left(-\frac{r^2}{2\beta t}\right), \tag{10.37}$$

where α and β are constants. (Note that we use r instead of ρ to denote the radial coordinate in plane polars so as to avoid confusion with the density.)

Further, (i) show that $\beta = 2k/(s\rho)$; (ii) demonstrate that the excess heat energy in the sheet is independent of t, and hence evaluate α; and (iii) prove that the total heat flow past any circle of radius r is E.

The equation to be solved is the heat diffusion equation

$$k\nabla^2 u(\mathbf{r}, t) = s\rho\frac{\partial u(\mathbf{r}, t)}{\partial t}.$$

Since we only require the solution for $r \gg b$ we can treat the problem as two-dimensional with obvious circular symmetry. Thus only the r-derivative term in the expression for $\nabla^2 u$ is non-zero, giving

$$\frac{k}{r}\frac{\partial}{\partial r}\left(r\frac{\partial u}{\partial r}\right) = s\rho\frac{\partial u}{\partial t}, \tag{10.38}$$

where now $u(\mathbf{r}, t) = u(r, t)$.

(i) Substituting the given expression (10.37) into (10.38) we obtain

$$\frac{2k\alpha}{\beta t^2}\left(\frac{r^2}{2\beta t} - 1\right)\exp\left(-\frac{r^2}{2\beta t}\right) = \frac{s\rho\alpha}{t^2}\left(\frac{r^2}{2\beta t} - 1\right)\exp\left(-\frac{r^2}{2\beta t}\right),$$

from which we find that (10.37) is a solution, provided $\beta = 2k/(s\rho)$.

(ii) The excess heat in the system at any time t is

$$\begin{aligned} b\rho s\int_0^\infty u(r, t)2\pi r\,dr &= 2\pi b\rho s\alpha\int_0^\infty \frac{r}{t}\exp\left(-\frac{r^2}{2\beta t}\right)dr \\ &= 2\pi b\rho s\alpha\beta. \end{aligned}$$

The excess heat is therefore independent of t and so must be equal to the total heat input E, implying that

$$\alpha = \frac{E}{2\pi b\rho s\beta} = \frac{E}{4\pi bk}.$$

(iii) The total heat flow past a circle of radius r is

$$-2\pi rbk\int_0^\infty \frac{\partial u(r,t)}{\partial r}dt = -2\pi rbk\int_0^\infty \frac{E}{4\pi bkt}\left(\frac{-r}{\beta t}\right)\exp\left(-\frac{r^2}{2\beta t}\right)dt$$

$$= E\left[\exp\left(-\frac{r^2}{2\beta t}\right)\right]_0^\infty = E \quad \text{for all } r.$$

As we would expect, all the heat energy E deposited by the laser will eventually flow past a circle of any given radius r. ◀

10.6 Boundary conditions and the uniqueness of solutions

So far in this chapter we have discussed how to find general solutions to various types of first- and second-order linear PDE. Moreover, given a set of boundary conditions we have shown how to find the particular solution (or class of solutions) that satisfies them. For first-order equations, for example, we found that if the value of $u(x, y)$ is specified along some curve in the xy-plane then the solution to the PDE is in general unique, but that if $u(x, y)$ is specified at only a single point then the solution is not unique, because there exists a whole class of particular solutions that satisfy the boundary condition. For second-order equations, boundary values that are given only on a finite length of curve generally limit the region in the xy-plane in which valid solutions can be obtained.

The general topic of the types of boundary condition that cause a PDE to have a unique solution, a class of solutions, or even no solution at all, is a complex one and beyond the scope of the treatment given in this book.[11] We will, however, summarize the main results for the types of PDEs that predominate in physics and engineering; these tend to be second-order equations.

For second-order equations we might expect that relevant boundary conditions would involve specifying u, or some of its first derivatives, or both, along a suitable set of boundaries bordering or enclosing the region over which a solution is sought. Three common types of boundary condition occur and are associated with the names of Dirichlet, Neumann and Cauchy. They are as follows.

(i) *Dirichlet*: The value of u is specified at each point of the boundary.
(ii) *Neumann*: The value of $\partial u/\partial n$, the *normal derivative* of u, is specified at each point of the boundary. Note that $\partial u/\partial n = \nabla u \cdot \hat{\mathbf{n}}$, where $\hat{\mathbf{n}}$ is the (outward) normal to the boundary at each point.
(iii) *Cauchy*: Both u and $\partial u/\partial n$ are specified at each point of the boundary.

It can be shown that the type of boundary conditions needed is very closely related to the nature (hyperbolic, parabolic or elliptic) of the PDE, but that complications can arise in some cases. The general considerations involved in deciding exactly which boundary conditions are appropriate for a particular problem are complex, and we do not discuss

11 For reference: the determining factors are known as the *characteristic curves* (or just *characteristics*) of the PDE.

Table 10.1 *The appropriate boundary conditions for different types of partial differential equation*

Equation type	Boundary	Conditions
Hyperbolic	open	Cauchy
Parabolic	open	Dirichlet or Neumann
Elliptic	closed	Dirichlet or Neumann

them here.[12] We merely note that whether the various types of boundary condition are appropriate (in that they give a solution that is unique, sometimes to within a constant, and is well defined) depends not only upon the type of second-order equation under consideration but also on whether the solution region is bounded by a closed or an open curve (or a surface if there are more than two independent variables). Note that part of a closed boundary may be at infinity if conditions are imposed on u or $\partial u/\partial n$ there.

It may be shown that the appropriate boundary-condition and equation-type pairings for second-order equations are as given in Table 10.1. For example, Laplace's equation $\nabla^2 u = 0$ is elliptic and thus requires either Dirichlet or Neumann boundary conditions on a closed boundary which, as we have already noted, may be at infinity if the behavior of u is specified there (most often u or $\partial u/\partial n \to 0$ at infinity).

10.6.1 Uniqueness of solutions

Although we have merely stated the appropriate boundary types and conditions for which, in the general case, a PDE has a unique, well-defined solution, sometimes to within an additive constant, it is often important to be able to prove that a unique solution is obtained.

As an important example, let us consider Poisson's equation in three dimensions,

$$\nabla^2 u(\mathbf{r}) = \rho(\mathbf{r}), \tag{10.39}$$

with either Dirichlet or Neumann conditions on a closed boundary appropriate to such an elliptic equation; for brevity, in (10.39), we have absorbed any physical constants into ρ. We aim to show that, to within an unimportant constant, the solution of (10.39) is *unique* if either the potential u or its normal derivative $\partial u/\partial n$ is specified on all surfaces bounding a given region of space (including, if necessary, a hypothetical spherical surface of indefinitely large radius on which u or $\partial u/\partial n$ is prescribed to have an arbitrarily small value). Stated more formally this is as follows.

Uniqueness theorem. *If u is real and its first and second partial derivatives are continuous in a region V and on its boundary S, and $\nabla^2 u = \rho$ in V and either $u = f$ or $\partial u/\partial n = g$ on S, where ρ, f and g are prescribed functions, then u is unique (at least to within an additive constant).*

We now prove this statement using a method based on proof by contradiction.

12 For a discussion the reader is referred to, for example, P. M. Morse and H. Feshbach, *Methods of Theoretical Physics, Part I* (New York: McGraw-Hill, 1953), chap. 6.

Example Prove the uniqueness theorem for Poisson's equation.

Let us suppose, on the contrary, that the solution is not unique and that two solutions $u_1(\mathbf{r})$ and $u_2(\mathbf{r})$ both satisfy the conditions given above. Denote their difference by the function $w = u_1 - u_2$. We then have

$$\nabla^2 w = \nabla^2 u_1 - \nabla^2 u_2 = \rho - \rho = 0,$$

so that w satisfies Laplace's equation in V. Furthermore, since either $u_1 = f = u_2$ or $\partial u_1/\partial n = g = \partial u_2/\partial n$ on S, we must have either $w = 0$ or $\partial w/\partial n = 0$ on S.

If we now use Green's first theorem, (3.19), for the case where both scalar functions are taken as w we have

$$\int_V \left[w\nabla^2 w + (\nabla w)\cdot(\nabla w)\right] dV = \int_S w\frac{\partial w}{\partial n}\, dS.$$

However, either condition, $w = 0$ or $\partial w/\partial n = 0$, makes the RHS vanish whilst the first term on the LHS vanishes since $\nabla^2 w = 0$ in V. Thus we are left with

$$\int_V |\nabla w|^2\, dV = 0.$$

Since $|\nabla w|^2$ can never be negative, this can only be satisfied if

$$\nabla w = \mathbf{0},$$

i.e. if w, and hence $u_1 - u_2$, is a constant in V.

If Dirichlet conditions are given then $u_1 \equiv u_2$ on (some part of) S and hence $u_1 = u_2$ everywhere in V. For Neumann conditions, however, u_1 and u_2 can differ throughout V by an arbitrary (but unimportant) constant. ◀

The importance of this uniqueness theorem lies in the fact that if a solution to Poisson's (or Laplace's) equation that fits the given set of Dirichlet or Neumann conditions can be found by any means whatever, then that solution is the correct one, since only one exists. This result is the mathematical justification for the *method of images*, which is discussed more fully in the next chapter.

We also note that often the same general method, used in the above example for proving the uniqueness theorem for Poisson's equation, can be employed to prove the uniqueness (or otherwise) of solutions to other equations and boundary conditions.

SUMMARY

1. *General forms of equations and solutions*
   P, Q, and R are functions of x and y; A, B and C are constants

Name	General form	Typical "combination of variables" solution
1st-order	$P\frac{\partial u}{\partial x} + Q\frac{\partial u}{\partial y} = 0$	Any function $f(p)$ where $p = \int \frac{dx}{P} - \int \frac{dy}{Q}$

(*cont.*)

(*cont.*)

Name	General form	Typical "combination of variables" solution
1st-order	$P\frac{\partial u}{\partial x} + Q\frac{\partial u}{\partial y} + Ru = 0$	$h(x, y)f(p)$ where p is as above and $h(x, y)$ is *any* solution of the given equation
2nd-order	$A\frac{\partial^2 u}{\partial x^2} + B\frac{\partial^2 u}{\partial x \partial y} + C\frac{\partial^2 u}{\partial y^2} = 0$	$f(x + \lambda_1 y) + g(x + \lambda_2 y)$, where the λ_i satisfy $A + B\lambda + C\lambda^2 = 0$
2nd-order with $B^2 = 4AC$	$A\frac{\partial^2 u}{\partial x^2} + B\frac{\partial^2 u}{\partial x \partial y} + C\frac{\partial^2 u}{\partial y^2} = 0$	$f(x + \lambda_1 y) + xg(x + \lambda_1 y)$
Laplace	$\nabla^2 u = 0$	2D: $f(x + iy) + g(x - iy)$
Poisson	$\nabla^2 u = \rho(\mathbf{r})$	2D: $f(x + iy) + g(x - iy) + h(x, y)$, where $h(x, y)$ is *any* solution of the given equation
Wave	$\nabla^2 u = \frac{1}{c^2}\frac{\partial^2 u}{\partial t^2}$	1D: $f(x - ct) + g(x + ct)$ 3D: $f(\hat{\mathbf{n}} \cdot \mathbf{r} - ct) + g(\hat{\mathbf{n}} \cdot \mathbf{r} + ct)$
Diffusion	$\kappa \nabla^2 u = \frac{\partial u}{\partial t}$	1D: $\frac{\alpha}{2\kappa}x^2 + gx + \alpha t + c$ 1D: $\propto [\text{erf}(\zeta) - \text{erf}(\zeta_0)]$ where $\zeta = x/2(\kappa t)^{1/2}$

2. *Satisfying boundary conditions, which may be at* ∞
   - The solutions of Poisson's, Laplace's and the Klein–Gordon equations with u or $\partial u/\partial n$ specified on a closed boundary are unique, at least to within an additive constant.
   - The solution to an inhomogeneous problem = the general solution of the homogeneous problem + any particular solution satisfying the boundary conditions that is not already contained in the general solution.
   - Required boundary conditions according to equation type:

Equation type	Boundary	Specification	Examples
Elliptic ($B^2 < 4AC$)	closed	u or $\partial u/\partial n$	Laplace
Parabolic ($B^2 = 4AC$)	open	u or $\partial u/\partial n$	Diffusion, Schrödinger
Hyperbolic ($B^2 > 4AC$)	open	u and $\partial u/\partial n$	Wave

PROBLEMS

10.1. Determine whether the following can be written as functions of $p = x^2 + 2y$ only, and hence whether they are solutions of (10.8):

(a) $x^2(x^2-4)+4y(x^2-2)+4(y^2-1)$;
(b) $x^4+2x^2y+y^2$;
(c) $[x^4+4x^2y+4y^2+4]/[2x^4+x^2(8y+1)+8y^2+2y]$.

10.2. Find partial differential equations satisfied by the following functions $u(x,y)$ for all arbitrary functions f and all arbitrary constants a and b:
(a) $u(x,y)=f(x^2-y^2)$;
(b) $u(x,y)=(x-a)^2+(y-b)^2$;
(c) $u(x,y)=y^n f(y/x)$;
(d) $u(x,y)=f(x+ay)$.

10.3. Solve the following partial differential equations for $u(x,y)$ with the boundary conditions given:
(a) $x\dfrac{\partial u}{\partial x}+xy=u, \quad u=2y$ on the line $x=1$;
(b) $1+x\dfrac{\partial u}{\partial y}=xu, \quad u(x,0)=x$.

10.4. Find the most general solutions $u(x,y)$ of the following equations, consistent with the boundary conditions stated:
(a) $y\dfrac{\partial u}{\partial x}-x\dfrac{\partial u}{\partial y}=0, \quad u(x,0)=1+\sin x$;
(b) $i\dfrac{\partial u}{\partial x}=3\dfrac{\partial u}{\partial y}, \quad u=(4+3i)x^2$ on the line $x=y$;
(c) $\sin x\sin y\dfrac{\partial u}{\partial x}+\cos x\cos y\dfrac{\partial u}{\partial y}=0, \quad u=\cos 2y$ on $x+y=\pi/2$;
(d) $\dfrac{\partial u}{\partial x}+2x\dfrac{\partial u}{\partial y}=0, \quad u=2$ on the parabola $y=x^2$.

10.5. Find solutions of

$$\frac{1}{x}\frac{\partial u}{\partial x}+\frac{1}{y}\frac{\partial u}{\partial y}=0$$

for which (a) $u(0,y)=y$ and (b) $u(1,1)=1$.

10.6. Find the most general solutions $u(x,y)$ of the following equations consistent with the boundary conditions stated:
(a) $y\dfrac{\partial u}{\partial x}-x\dfrac{\partial u}{\partial y}=3x, \quad u=x^2$ on the line $y=0$;
(b) $y\dfrac{\partial u}{\partial x}-x\dfrac{\partial u}{\partial y}=3x, \quad u(1,0)=2$;
(c) $y^2\dfrac{\partial u}{\partial x}+x^2\dfrac{\partial u}{\partial y}=x^2y^2(x^3+y^3)$, no boundary conditions.

10.7. Solve

$$\sin x\frac{\partial u}{\partial x}+\cos x\frac{\partial u}{\partial y}=\cos x$$

subject to (a) $u(\pi/2,y)=0$ and (b) $u(\pi/2,y)=y(y+1)$.

10.8. A function $u(x, y)$ satisfies

$$2\frac{\partial u}{\partial x} + 3\frac{\partial u}{\partial y} = 10,$$

and takes the value 3 on the line $y = 4x$. Evaluate $u(2, 4)$.

10.9. If $u(x, y)$ satisfies

$$\frac{\partial^2 u}{\partial x^2} - 3\frac{\partial^2 u}{\partial x \partial y} + 2\frac{\partial^2 u}{\partial y^2} = 0$$

and $u = -x^2$ and $\partial u/\partial y = 0$ for $y = 0$ and all x, find the value of $u(0, 1)$.

10.10. Consider the partial differential equation

$$\frac{\partial^2 u}{\partial x^2} - 3\frac{\partial^2 u}{\partial x \partial y} + 2\frac{\partial^2 u}{\partial y^2} = 0. \qquad (*)$$

Find the function $u(x, y)$ that satisfies $(*)$ and the boundary condition $u = \partial u/\partial y = 1$ when $y = 0$ for all x. Evaluate $u(0, 1)$.

10.11. In those cases in which it is possible to do so, evaluate $u(2, 2)$, where $u(x, y)$ is the solution of

$$2y\frac{\partial u}{\partial x} - x\frac{\partial u}{\partial y} = xy(2y^2 - x^2)$$

that satisfies the (separate) boundary conditions given below.
(a) $u(x, 1) = x^2$.
(b) $u(1, \sqrt{10}) = 5$.
(c) $u(\sqrt{10}, 1) = 5$.

10.12. Solve

$$6\frac{\partial^2 u}{\partial x^2} - 5\frac{\partial^2 u}{\partial x \partial y} + \frac{\partial^2 u}{\partial y^2} = 14,$$

subject to $u = 2x + 1$ and $\partial u/\partial y = 4 - 6x$, both on the line $y = 0$.

10.13. Find the most general solution of $\dfrac{\partial^2 u}{\partial x^2} + \dfrac{\partial^2 u}{\partial y^2} = x^2 y^2$.

10.14. Solve

$$\frac{\partial^2 u}{\partial x \partial y} + 3\frac{\partial^2 u}{\partial y^2} = x(2y + 3x).$$

10.15. The non-relativistic Schrödinger equation (10.7) is similar to the diffusion equation in having different orders of derivatives in its various terms; this precludes solutions that are arbitrary functions of particular linear combinations of variables. However, since exponential functions do not change their forms under differentiation, solutions in the form of exponential functions of combinations of the variables may still be possible.

Consider the Schrödinger equation for the case of a constant potential, i.e. for a free particle, and show that it has solutions of the form $A\exp(lx + my + nz + \lambda t)$, where the only requirement is that

$$-\frac{\hbar^2}{2m}\left(l^2 + m^2 + n^2\right) = i\hbar\lambda.$$

In particular, identify the equation and wavefunction obtained by taking λ as $-iE/\hbar$, and l, m and n as $ip_x/\hbar$, $ip_y/\hbar$ and $ip_z/\hbar$, respectively, where E is the energy and $\mathbf{p}$ the momentum of the particle; these identifications are essentially the content of the de Broglie and Einstein relationships.

10.16. An infinitely long string on which waves travel at speed c has an initial displacement

$$y(x) = \begin{cases} \sin(\pi x/a), & -a \le x \le a, \\ 0, & |x| > a. \end{cases}$$

It is released from rest at time $t = 0$, and its subsequent displacement is described by $y(x, t)$.

By expressing the initial displacement as one explicit function incorporating Heaviside step functions, find an expression for $y(x, t)$ at a general time $t > 0$. In particular, determine the displacement as a function of time (a) at $x = 0$, (b) at $x = a$, and (c) at $x = a/2$.

10.17. An incompressible fluid of density ρ and negligible viscosity flows with velocity v along a thin, straight, perfectly light and flexible tube, of cross-section A which is held under tension T. Assume that small transverse displacements u of the tube are governed by

$$\frac{\partial^2 u}{\partial t^2} + 2v\frac{\partial^2 u}{\partial x \partial t} + \left(v^2 - \frac{T}{\rho A}\right)\frac{\partial^2 u}{\partial x^2} = 0.$$

(a) Show that the general solution consists of a superposition of two waveforms traveling with different speeds.

(b) The tube initially has a small transverse displacement $u = a\cos kx$ and is suddenly released from rest. Find its subsequent motion.

10.18. Like the Schrödinger equation, the equation describing the transverse vibrations of a rod,

$$a^4\frac{\partial^4 u}{\partial x^4} + \frac{\partial^2 u}{\partial t^2} = 0,$$

has different orders of derivatives in its various terms. Show, however, that it has solutions of exponential form, $u(x, t) = A\exp(\lambda x + i\omega t)$, provided that the relation $a^4\lambda^4 = \omega^2$ is satisfied.

Use a linear combination of such allowed solutions, expressed as the sum of sinusoids and hyperbolic sinusoids of λx, to describe the transverse vibrations of a rod of length L clamped at both ends. At a clamped point both u and $\partial u/\partial x$ must vanish; show that this implies that $\cos(\lambda L)\cosh(\lambda L) = 1$, thus determining the frequencies ω at which the rod can vibrate.

10.19. In an electrical cable of resistance R and capacitance C, each per unit length, voltage signals obey the equation $\partial^2 V/\partial x^2 = RC\partial V/\partial t$. This has solutions of the form given in (10.36) and also of the form $V = Ax + D$.

(a) Find a combination of these that represents the situation after a steady voltage V_0 is applied at $x = 0$ at time $t = 0$.

(b) Obtain a solution describing the propagation of the voltage signal resulting from the application of the signal $V = V_0$ for $0 < t < T$, $V = 0$ otherwise, to the end $x = 0$ of an infinite cable.

(c) Show that for $t \gg T$ the maximum signal occurs at a value of x proportional to $t^{1/2}$ and has a magnitude proportional to t^{-1}.

10.20. A sheet of material of thickness w, specific heat capacity c and thermal conductivity k is isolated in a vacuum, but its two sides are exposed to fluxes of radiant heat of strengths J_1 and J_2. Ignoring short-term transients, show that the temperature difference between its two surfaces is steady at $(J_2 - J_1)w/2k$, whilst their average temperature increases at a rate $(J_2 + J_1)/cw$.

10.21. Consider each of the following situations in a qualitative way and determine the equation type, the nature of the boundary curve and the type of boundary conditions involved:

(a) a conducting bar given an initial temperature distribution and then thermally isolated;

(b) two long conducting concentric cylinders, on each of which the voltage distribution is specified;

(c) two long conducting concentric cylinders, on each of which the charge distribution is specified;

(d) a semi-infinite string, the end of which is made to move in a prescribed way.

10.22. The daily and annual variations of temperature at the surface of the earth may be represented by sine-wave oscillations, with equal amplitudes and periods of 1 day and 365 days respectively. Assume that for (angular) frequency ω the temperature at depth x in the earth is given by $u(x, t) = A\sin(\omega t + \mu x)\exp(-\lambda x)$, where λ and μ are constants.

(a) Use the diffusion equation to find the values of λ and μ.

(b) Find the ratio of the depths below the surface at which the two amplitudes have dropped to $1/20$ of their surface values.

(c) At what time of year is the soil coldest at the greater of these depths, assuming that the smoothed annual variation in temperature at the surface has a minimum on February 1st?

10.23. The Klein–Gordon equation (which is satisfied by the quantum-mechanical wavefunction $\Phi(\mathbf{r})$ of a relativistic spinless particle of non-zero mass m) is

$$\nabla^2\Phi - m^2\Phi = 0.$$

Show that the solution for the scalar field $\Phi(\mathbf{r})$ in any volume V bounded by a surface S is unique if either Dirichlet or Neumann boundary conditions are specified on S.

HINTS AND ANSWERS

10.1. (a) Yes, $p^2 - 4p - 4$; (b) no, $(p - y)^2$; (c) yes, $(p^2 + 4)/(2p^2 + p)$.

10.3. Each equation is effectively an ordinary differential equation, but with a function of the non-integrated variable as the constant of integration;
(a) $u = xy(2 - \ln x)$; (b) $u = x^{-1}(1 - e^y) + xe^y$.

10.5. (a) $(y^2 - x^2)^{1/2}$; (b) $1 + f(y^2 - x^2)$ where $f(0) = 0$.

10.7. $u = y + f(y - \ln(\sin x))$; (a) $u = \ln(\sin x)$; (b) $u = y + [y - \ln(\sin x)]^2$.

10.9. General solution is $u(x, y) = f(x + y) + g(x + y/2)$. Show that $2p = -g'(p)/2$, and hence $g(p) = k - 2p^2$, whilst $f(p) = p^2 - k$, leading to $u(x, y) = -x^2 + y^2/2$; $u(0, 1) = 1/2$.

10.11. $p = x^2 + 2y^2$; $u(x, y) = f(p) + x^2y^2/2$.
(a) $u(x, y) = (x^2 + 2y^2 + x^2y^2 - 2)/2$; $u(2, 2) = 13$.
(b) The solution is only specified on $p = 21$, and so $u(2, 2)$ is undetermined.
(c) The solution is specified on $p = 12$, and so $u(2, 2) = 5 + \frac{1}{2}(4)(4) = 13$.

10.13. $u(x, y) = f(x + iy) + g(x - iy) + (1/12)x^4(y^2 - (1/15)x^2)$. In the last term, x and y may be interchanged. There are (infinitely) many other possibilities for the specific PI, e.g. $[\,15x^2y^2(x^2 + y^2) - (x^6 + y^6)\,]/360$.

10.15. $E = p^2/(2m)$, the relationship between energy and momentum for a non-relativistic particle; $u(\mathbf{r}, t) = A\exp[i(\mathbf{p}\cdot\mathbf{r} - Et)/\hbar]$, a plane wave of wave number $\mathbf{k} = \mathbf{p}/\hbar$ and angular frequency $\omega = E/\hbar$ traveling in the direction $\mathbf{p}/p$.

10.17. (a) $c = v \pm \alpha$ where $\alpha^2 = T/\rho A$;
(b) $u(x, t) = a\cos[k(x - vt)]\cos(k\alpha t) - (va/\alpha)\sin[k(x - vt)]\sin(k\alpha t)$.

10.19. (a) $V_0\left[1 - (2/\sqrt{\pi})\int^{\frac{1}{2}x(CR/t)^{1/2}} \exp(-\nu^2)\,d\nu\right]$;
(b) consider the input as equivalent to V_0 applied at $t = 0$ and continued and $-V_0$ applied at $t = T$ and continued;

$$V(x, t) = \frac{2V_0}{\sqrt{\pi}}\int_{\frac{1}{2}x(CR/t)^{1/2}}^{\frac{1}{2}x[CR/(t-T)]^{1/2}} \exp\left(-\nu^2\right)\,d\nu;$$

(c) for $t \gg T$, maximum at $x = [2t/(CR)]^{1/2}$ with value $\dfrac{V_0T\exp(-\frac{1}{2})}{(2\pi)^{1/2}t}$.

10.21. (a) Parabolic, open, Dirichlet $u(x, 0)$ given, Neumann $\partial u/\partial x = 0$ at $x = \pm L/2$ for all t;
(b) elliptic, closed, Dirichlet;
(c) elliptic, closed, Neumann $\partial u/\partial n = \sigma/\epsilon_0$;
(d) hyperbolic, open, Cauchy.

10.23. Follow an argument similar to that in Section 10.6 and argue that the additional term $\int m^2|w|^2\, dV$ must be zero, and hence that $w = 0$ everywhere.

11 Solution methods for PDEs

In the previous chapter we demonstrated the methods by which general solutions of some PDEs may be obtained in terms of arbitrary functions. In particular, solutions containing the independent variables in definite combinations were sought, thus reducing the effective number of them.

In the present chapter we begin by taking the opposite approach, namely that of trying to keep the independent variables as separate as possible; the aim is to reduce the partial differential equation to a set of ordinary differential equations, each of which contains only one of the independent variables. We then consider integral transform methods by which one of the independent variables may be eliminated, at least from differential coefficients. Finally, we discuss the use of Green's functions in solving inhomogeneous problems.

11.1 Separation of variables: the general method

Suppose we seek a solution $u(x, y, z, t)$ to some PDE (expressed in Cartesian coordinates). Let us attempt to obtain one that has the product form[1]

$$u(x, y, z, t) = X(x)Y(y)Z(z)T(t). \tag{11.1}$$

A solution that has this form is said to be *separable* in x, y, z and t, and seeking solutions of this form is called the method of *separation of variables*.

As simple examples we may observe that, of the functions

$$\text{(i) } xyz^2 \sin bt, \qquad \text{(ii) } xy + zt, \qquad \text{(iii) } (x^2 + y^2)z \cos \omega t,$$

(i) is completely separable, (ii) is inseparable in that no single variable can be separated out from it and written as a multiplicative factor, whilst (iii) is separable in z and t but not in x and y.

When seeking PDE solutions of the form (11.1), we are requiring not that there is no connection at all between the functions X, Y, Z and T (for example, certain parameters may appear in two or more of them), but only that X does not depend upon y, z, t, that Y does not depend on x, z, t, and so on.

1 It should be noted that the conventional use here of upper-case (capital) letters to denote the functions of the corresponding lower-case variable is intended to enable an easy correspondence between a function and its argument to be made.

For a general PDE it is likely that a separable solution is impossible, but certainly some common and important equations do have useful solutions of this form, and we will illustrate the method of solution by studying the three-dimensional wave equation

$$\nabla^2 u(\mathbf{r}) = \frac{1}{c^2}\frac{\partial^2 u(\mathbf{r})}{\partial t^2}. \tag{11.2}$$

We will work in Cartesian coordinates for the present and assume a solution of the form (11.1); the solutions in alternative coordinate systems, e.g. spherical or cylindrical polars, are considered in Section 11.3.

Expressed in Cartesian coordinates (11.2) takes the form

$$\frac{\partial^2 u}{\partial x^2} + \frac{\partial^2 u}{\partial y^2} + \frac{\partial^2 u}{\partial z^2} = \frac{1}{c^2}\frac{\partial^2 u}{\partial t^2}; \tag{11.3}$$

substituting (11.1) gives

$$\frac{d^2X}{dx^2}YZT + X\frac{d^2Y}{dy^2}ZT + XY\frac{d^2Z}{dz^2}T = \frac{1}{c^2}XYZ\frac{d^2T}{dt^2},$$

which can also be written as

$$X''YZT + XY''ZT + XYZ''T = \frac{1}{c^2}XYZT'', \tag{11.4}$$

where in each case the primes refer to the *ordinary* derivative with respect to the independent variable upon which the function depends. This emphasizes the fact that each of the functions X, Y, Z and T has only one independent variable and thus its only derivative is its total derivative. For the same reason, in each term in (11.4) three of the four functions are unaltered by the partial differentiation and behave exactly as constant multipliers.

If we now divide (11.4) throughout by $u = XYZT$ we obtain

$$\frac{X''}{X} + \frac{Y''}{Y} + \frac{Z''}{Z} = \frac{1}{c^2}\frac{T''}{T}. \tag{11.5}$$

This form shows the particular characteristic that is the basis of the method of separation of variables, namely that of the four terms the first is a function of x only, the second of y only, the third of z only and the RHS a function of t only and yet there is an equation connecting them. This can only be so for all x, y, z and t if *each* of the terms does not in fact, despite appearances, depend upon the corresponding independent variable but *is equal to a constant*, the four constants being such that (11.5) is satisfied.

Since there is only one equation to be satisfied and four constants involved, there is considerable freedom in the values they may take. For the purposes of our illustrative example let us make the choice of $-l^2$, $-m^2$, $-n^2$, for the first three constants. The constant associated with $c^{-2}T''/T$ must then have the value $-\mu^2 = -(l^2 + m^2 + n^2)$.

Having recognized that each term of (11.5) is individually equal to a constant (or parameter), we can now replace (11.5) by four separate ordinary differential equations:

$$\frac{X''}{X} = -l^2, \quad \frac{Y''}{Y} = -m^2, \quad \frac{Z''}{Z} = -n^2, \quad \frac{1}{c^2}\frac{T''}{T} = -\mu^2. \tag{11.6}$$

The important point to notice is not the simplicity of the equations (11.6) (the corresponding ones for a general PDE are usually far from simple) but that, by the device of assuming a separable solution, a *partial* differential equation (11.3), containing derivatives with respect to the four independent variables all in one equation, has been reduced to four *separate ordinary* differential equations (11.6). The ordinary equations are connected through four constant parameters that satisfy an algebraic relation. These constants are called *separation constants*.

The general solutions of the equations (11.6) can be deduced straightforwardly and are

$$\begin{aligned} X(x) &= A\exp(ilx) + B\exp(-ilx), \\ Y(y) &= C\exp(imy) + D\exp(-imy), \\ Z(z) &= E\exp(inz) + F\exp(-inz), \\ T(t) &= G\exp(ic\mu t) + H\exp(-ic\mu t), \end{aligned} \tag{11.7}$$

where $A, B, \ldots, H$ are constants, which may be determined if boundary conditions are imposed on the solution. Depending on the geometry of the problem and any boundary conditions, it is sometimes more appropriate to write the solutions (11.7) in the alternative form

$$\begin{aligned} X(x) &= A'\cos lx + B'\sin lx, \\ Y(y) &= C'\cos my + D'\sin my, \\ Z(z) &= E'\cos nz + F'\sin nz, \\ T(t) &= G'\cos(c\mu t) + H'\sin(c\mu t), \end{aligned} \tag{11.8}$$

for some different set of constants $A', B', \ldots, H'$. Clearly the choice of how best to represent the solution depends on the problem being considered.

As an example, suppose that we take as particular solutions the four functions

$$X(x) = \exp(ilx), \qquad Y(y) = \exp(imy),$$

$$Z(z) = \exp(inz), \qquad T(t) = \exp(-ic\mu t).$$

This gives a particular solution of the original PDE (11.3)

$$\begin{aligned} u(x, y, z, t) &= \exp(ilx)\exp(imy)\exp(inz)\exp(-ic\mu t) \\ &= \exp[i(lx + my + nz - c\mu t)], \end{aligned}$$

which is a special case of the solution (10.33) obtained in the previous chapter and represents a plane wave of unit amplitude propagating in a direction given by the vector with components l, m, n in a Cartesian coordinate system. In the conventional notation of wave theory, l, m and n are the components of the wave-number vector $\mathbf{k}$, whose magnitude is given by $\mu = k = 2\pi/\lambda$, where λ is the wavelength of the wave; $c\mu$ is the angular frequency ω of the wave. This gives the equation in the form

$$\begin{aligned} u(x, y, z, t) &= \exp[i(k_x x + k_y y + k_z z - \omega t)] \\ &= \exp[i(\mathbf{k}\cdot\mathbf{r} - \omega t)], \end{aligned}$$

and makes the exponent dimensionless.

The method of separation of variables can be applied to many commonly occurring PDEs encountered in physical applications. As a further example we consider the diffusion equation.

Example Use the method of separation of variables to obtain for the one-dimensional diffusion equation

$$\kappa \frac{\partial^2 u}{\partial x^2} = \frac{\partial u}{\partial t}, \tag{11.9}$$

a solution that tends to zero as $t \to \infty$ for all x.

Here we have only two independent variables x and t and we therefore assume a solution of the form

$$u(x, t) = X(x)T(t).$$

Substituting this expression into (11.9) and dividing through by $u = XT$ (and also by κ) we obtain

$$\frac{X''}{X} = \frac{T'}{\kappa T}.$$

Now, arguing exactly as above that the LHS is a function of x only and the RHS is a function of t only, we conclude that each side must equal a constant, which, anticipating the result and noting the imposed boundary condition, we will take as $-\lambda^2$. This gives us two ordinary equations,

$$X'' + \lambda^2 X = 0, \tag{11.10}$$

$$T' + \lambda^2 \kappa T = 0, \tag{11.11}$$

which have the solutions

$$X(x) = A\cos\lambda x + B\sin\lambda x,$$

$$T(t) = C\exp(-\lambda^2\kappa t).$$

Combining these to give the assumed solution $u = XT$ yields (absorbing the constant C into A and B)

$$u(x, t) = (A\cos\lambda x + B\sin\lambda x)\exp(-\lambda^2\kappa t). \tag{11.12}$$

In order to satisfy the boundary condition $u \to 0$ as $t \to \infty$, $\lambda^2\kappa$ must be > 0. Since κ is real and > 0, this implies that λ is a real non-zero number and that the solution is sinusoidal in x and is not a disguised hyperbolic function; this was our reason for choosing the separation constant as $-\lambda^2$. Mathematically, other solutions are possible, but they are unlikely to have physical applications.[2] ◄

As a final example we consider Laplace's equation in Cartesian coordinates; this may be treated in a similar manner.

2 Find a solution that has the property that the point at which the solution has any given amplitude moves along the x-axis with a constant speed. What is that speed?

Example Use the method of separation of variables to obtain a solution for the two-dimensional Laplace equation,

$$\frac{\partial^2 u}{\partial x^2} + \frac{\partial^2 u}{\partial y^2} = 0. \tag{11.13}$$

If we assume a solution of the form $u(x, y) = X(x)Y(y)$ then, following the above method, and taking the separation constant as λ^2, we find

$$X'' = \lambda^2 X, \qquad Y'' = -\lambda^2 Y.$$

Taking λ^2 as > 0, the general solution becomes

$$u(x, y) = (A \cosh \lambda x + B \sinh \lambda x)(C \cos \lambda y + D \sin \lambda y). \tag{11.14}$$

An alternative form, in which the exponentials are written explicitly, may be useful for other geometries or boundary conditions:

$$u(x, y) = [A \exp \lambda x + B \exp(-\lambda x)](C \cos \lambda y + D \sin \lambda y), \tag{11.15}$$

with different constants A and B.

If $\lambda^2 < 0$ then the roles of x and y interchange. The particular combination of sinusoidal and hyperbolic functions and the values of λ allowed will be determined by the geometrical properties of any specific problem, together with any prescribed or necessary boundary conditions. ◀

We note here that a particular case of the solution (11.14) links up with the "combination" result $u(x, y) = f(x + iy)$ of the previous chapter (equations (10.24) and following), namely that if $A = B$ and $D = iC$ then the solution is the same as $f(p) = AC \exp \lambda p$ with $p = x + iy$.

11.2 Superposition of separated solutions

It will be noticed in the previous two examples that there is considerable freedom in the values of the separation constant λ, the only essential requirement being that λ has the *same* value in both parts of the solution, i.e. the part depending on x and the part depending on y (or t). This is a general feature for solutions in separated form, which, if the original PDE has n independent variables, will contain $n - 1$ separation constants. All that is required in general is that we associate the correct function of one independent variable with the appropriate functions of the others, the correct function being the one with the same set of values for the separation constants.

If the original PDE is linear (as are the Laplace, Schrödinger, diffusion and wave equations) then mathematically acceptable solutions can be formed by superposing solutions corresponding to different allowed values of the separation constants. To take a two-variable example: if

$$u_{\lambda_j}(x, y) = X_{\lambda_j}(x)Y_{\lambda_j}(y)$$

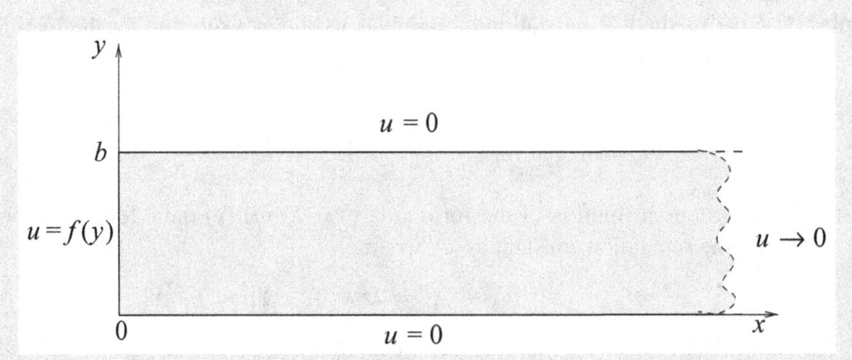

Figure 11.1 A semi-infinite metal plate whose edges are kept at fixed temperatures.

is a solution of a linear PDE obtained by giving the separation constant the value λ_j, then the superposition

$$u(x, y) = a_1 X_{\lambda_1}(x)Y_{\lambda_1}(y) + a_2 X_{\lambda_2}(x)Y_{\lambda_2}(y) + \cdots = \sum_i a_i X_{\lambda_i}(x)Y_{\lambda_i}(y) \qquad (11.16)$$

is also a solution for any constants a_i, provided that the λ_i are the allowed values of the separation constant λ given the imposed boundary conditions. Note that if the boundary conditions allow any of the separation constants to be zero then the form of the general solution is normally different and must be deduced by returning to the separated ordinary differential equations. We will encounter this behavior in Section 11.3.

The value of the superposition approach is that a boundary condition, say that $u(x, y)$ takes a particular form $f(x)$ when $y = 0$, might be met by choosing the constants a_i such that

$$f(x) = \sum_i a_i X_{\lambda_i}(x)Y_{\lambda_i}(0).$$

In general, this will be possible provided that the functions $X_{\lambda_i}(x)$ form a complete set – as do the sinusoidal functions of Fourier series or the spherical harmonics discussed in Subsection 9.3. Our first example concerns the temperature distribution throughout a thermally conducting plate.

Example A semi-infinite rectangular metal plate occupies the region $0 \le x \le \infty$ and $0 \le y \le b$ in the xy-plane. The temperature at the far end of the plate and along its two long sides is fixed at 0 °C. If the temperature of the plate at $x = 0$ is also fixed and is given by $f(y)$, find the steady-state temperature distribution $u(x,y)$ of the plate. Hence find the temperature distribution if $f(y) = u_0$, where u_0 is a constant.

The physical situation is illustrated in Figure 11.1. With the notation we have used several times before, the two-dimensional heat diffusion equation satisfied by the temperature $u(x, y, t)$ is

$$\kappa\left(\frac{\partial^2 u}{\partial x^2} + \frac{\partial^2 u}{\partial y^2}\right) = \frac{\partial u}{\partial t},$$

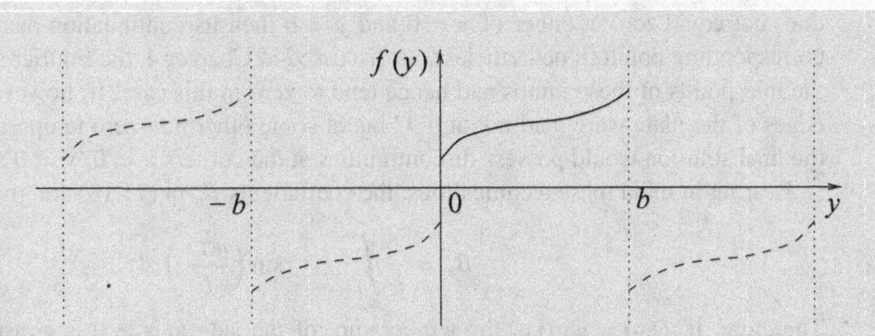

Figure 11.2 The continuation of $f(y)$ for a Fourier sine series.

with $\kappa = k/(s\rho)$. In this case, however, we are asked to find the steady-state temperature, which corresponds to $\partial u/\partial t = 0$, and so we are led to consider the (two-dimensional) Laplace equation

$$\frac{\partial^2 u}{\partial x^2} + \frac{\partial^2 u}{\partial y^2} = 0.$$

We saw that assuming a separable solution of the form $u(x, y) = X(x)Y(y)$ led to solutions such as (11.14) or (11.15), or equivalent forms with x and y interchanged. In the current problem we have to satisfy the boundary conditions $u(x, 0) = 0 = u(x, b)$ and so a solution that is sinusoidal in y seems appropriate. Furthermore, since we require $u(\infty, y) = 0$ it is best to write the x-dependence of the solution explicitly in terms of exponentials rather than of hyperbolic functions. We therefore write the separable solution in the form (11.15) as

$$u(x, y) = [A \exp \lambda x + B \exp(-\lambda x)](C \cos \lambda y + D \sin \lambda y).$$

Applying the boundary conditions, we see firstly that $u(\infty, y) = 0$ implies $A = 0$ if we take $\lambda > 0$. Secondly, since $u(x, 0) = 0$ we may set $C = 0$, which, if we absorb the constant D into B, leaves us with

$$u(x, y) = B \exp(-\lambda x) \sin \lambda y.$$

But, using the condition $u(x, b) = 0$, we require $\sin \lambda b = 0$ and so λ must be equal to $n\pi/b$, where n is any positive integer.

Using the principle of superposition (11.16), the general solution satisfying the given boundary conditions can therefore be written

$$u(x, y) = \sum_{n=1}^{\infty} B_n \exp(-n\pi x/b) \sin(n\pi y/b), \tag{11.17}$$

for some constants B_n. Notice that in the sum in (11.17) we have omitted negative values of n since they would lead to exponential terms that diverge as $x \to \infty$. The $n = 0$ term is also omitted since it is identically zero. Using the remaining boundary condition $u(0, y) = f(y)$ we see that the constants B_n must satisfy

$$f(y) = \sum_{n=1}^{\infty} B_n \sin(n\pi y/b). \tag{11.18}$$

This is clearly a Fourier sine series expansion of $f(y)$ (see Chapter 4). For (11.18) to hold, however, the continuation of $f(y)$ outside the region $0 \leq y \leq b$ must be an odd periodic function with period $2b$ (see Figure 11.2). We also see from Figure 11.2 that if the original function $f(y)$

does not equal zero at either of $y = 0$ and $y = b$ then its continuation has a discontinuity at the corresponding point(s); nevertheless, as discussed in Chapter 4, the Fourier series will converge to the mid-points of these jumps and hence tend to zero in this case. If, however, the top and bottom edges of the plate were held not at 0 °C but at some other non-zero temperature, then, in general, the final solution would possess discontinuities at the corners $x = 0, y = 0$ and $x = 0, y = b$.

Bearing in mind these technicalities, the coefficients B_n in (11.18) are given by

$$B_n = \frac{2}{b}\int_0^b f(y)\sin\left(\frac{n\pi y}{b}\right)dy. \tag{11.19}$$

Therefore, if $f(y) = u_0$ (i.e. the temperature of the side at $x = 0$ is constant along its length), (11.19) becomes

$$\begin{aligned} B_n &= \frac{2}{b}\int_0^b u_0 \sin\left(\frac{n\pi y}{b}\right)dy \\ &= \left[-\frac{2u_0}{b}\frac{b}{n\pi}\cos\left(\frac{n\pi y}{b}\right)\right]_0^b \\ &= -\frac{2u_0}{n\pi}[(-1)^n - 1] = \begin{cases} 4u_0/n\pi & \text{for } n \text{ odd,} \\ 0 & \text{for } n \text{ even.} \end{cases} \end{aligned}$$

Therefore the required solution is

$$u(x, y) = \sum_{n \text{ odd}} \frac{4u_0}{n\pi}\exp\left(-\frac{n\pi x}{b}\right)\sin\left(\frac{n\pi y}{b}\right).$$

As expected, for $x = 0$ this becomes the Fourier sine series for a square-wave of amplitude u_0 and width b.[3] ◄

In the above example the boundary conditions meant that one term in each part of the separable solution could be immediately discarded, making the problem much easier to solve. Sometimes, as we now illustrate, a little ingenuity is required in writing the separable solution in such a way that certain parts can be neglected immediately.

Example Suppose that the semi-infinite rectangular metal plate in the previous example is replaced by one that in the x-direction has finite length a. The temperature of the right-hand edge is fixed at 0 °C and all other boundary conditions remain as before. Find the steady-state temperature in the plate.

As in the previous example, the boundary conditions $u(x, 0) = 0 = u(x, b)$ suggest a solution that is sinusoidal in y. In this case, however, we require $u = 0$ on $x = a$ (rather than at infinity) and so a solution in which the x-dependence is written in terms of hyperbolic functions, such as (11.14), rather than exponentials is more appropriate. Moreover, since the constants in front of the hyperbolic functions are, at this stage, arbitrary, we may write the separable solution in the most convenient way that ensures that the condition $u(a, y) = 0$ is straightforwardly satisfied. We therefore write

$$u(x, y) = [A\cosh\lambda(a - x) + B\sinh\lambda(a - x)](C\cos\lambda y + D\sin\lambda y).$$

3 Find a closed form solution for the temperature throughout the plate if the temperature distribution at the $x = 0$ end were $u(0, y) = T_0\sin(3\pi y/b)$.

Now the condition $u(a, y) = 0$ is easily satisfied by setting $A = 0$. As before the conditions $u(x, 0) = 0 = u(x, b)$ imply $C = 0$ and $\lambda = n\pi/b$ for integer n. Superposing the solutions for different n we then obtain

$$u(x, y) = \sum_{n=1}^{\infty} B_n \sinh[n\pi(a - x)/b] \sin(n\pi y/b), \tag{11.20}$$

for some constants B_n. We have omitted negative values of n in the sum (11.20) since the relevant terms are already included in those obtained for positive n. Again the $n = 0$ term is identically zero. Using the final boundary condition $u(0, y) = f(y)$ as above we find that the constants B_n must satisfy

$$f(y) = \sum_{n=1}^{\infty} B_n \sinh(n\pi a/b) \sin(n\pi y/b),$$

and, remembering the caveats discussed in the previous example, the B_n are therefore given by

$$B_n = \frac{2}{b \sinh(n\pi a/b)} \int_0^b f(y) \sin(n\pi y/b)\, dy. \tag{11.21}$$

For the case where $f(y) = u_0$, following the working of the previous example gives (11.21) as

$$B_n = \frac{4u_0}{n\pi \sinh(n\pi a/b)} \quad \text{for } n \text{ odd}, \qquad B_n = 0 \quad \text{for } n \text{ even}. \tag{11.22}$$

The required solution is thus

$$u(x, y) = \sum_{n \text{ odd}} \frac{4u_0}{n\pi \sinh(n\pi a/b)} \sinh[n\pi(a - x)/b] \sin(n\pi y/b).$$

We note that, as required, in the limit $a \to \infty$ this solution tends to the solution of the previous example. ◀

Often the principle of superposition can be used to write the solution to problems with more complicated boundary conditions as the sum of solutions to problems that each satisfy only some part of the boundary condition but when added together satisfy all the conditions. Such is the case in the following worked example.

Example Find the steady-state temperature in the (finite) rectangular plate of the previous example, subject to the boundary conditions $u(x, b) = 0$, $u(a, y) = 0$ and $u(0, y) = f(y)$ as before, but now, in addition, $u(x, 0) = g(x)$.

Figure 11.3(c) shows the imposed boundary conditions for the metal plate. Although we could find a solution to this problem using the methods presented above, we can arrive at the answer almost immediately by using the principle of superposition and the result of the previous example.

Let us suppose the required solution $u(x, y)$ is made up of two parts:

$$u(x, y) = v(x, y) + w(x, y),$$

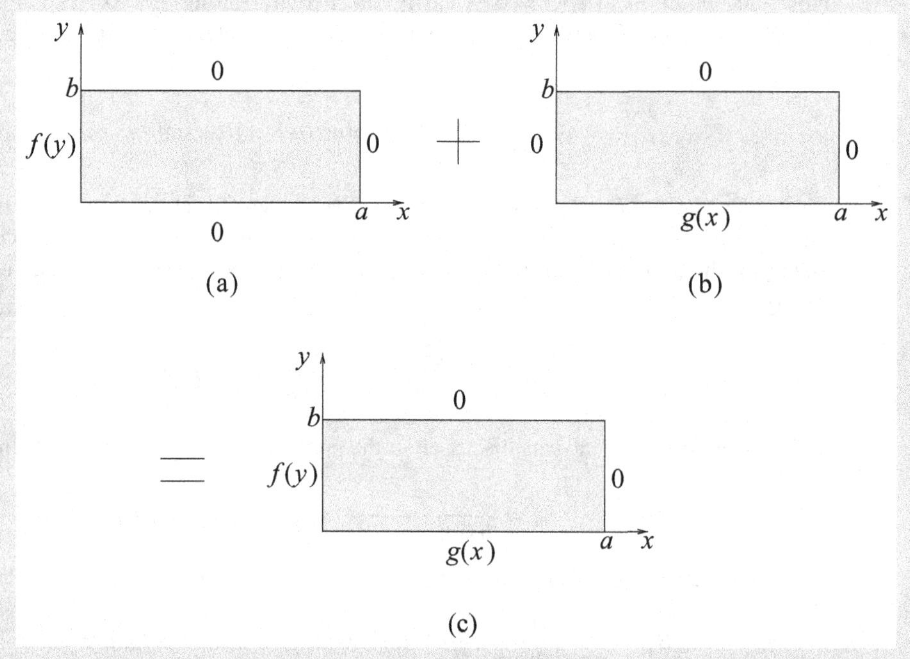

Figure 11.3 Superposition of boundary conditions for a metal plate.

where $v(x, y)$ is the solution satisfying the boundary conditions shown in Figure 11.3(a), whilst $w(x, y)$ is the solution satisfying the boundary conditions in Figure 11.3(b). It is clear that $v(x, y)$ is simply given by the solution to the previous example,

$$v(x, y) = \sum_{n \text{ odd}} B_n \sinh\left[\frac{n\pi(a-x)}{b}\right] \sin\left(\frac{n\pi y}{b}\right),$$

where B_n is given by (11.21). Moreover, by symmetry, $w(x, y)$ must be of the same form as $v(x, y)$ but with x and a interchanged with y and b, respectively, and with $f(y)$ in (11.21) replaced by $g(x)$. Therefore the required solution can be written down immediately without further calculation as

$$u(x, y) = \sum_{n \text{ odd}} B_n \sinh\left[\frac{n\pi(a-x)}{b}\right] \sin\left(\frac{n\pi y}{b}\right) + \sum_{n \text{ odd}} C_n \sinh\left[\frac{n\pi(b-y)}{a}\right] \sin\left(\frac{n\pi x}{a}\right),$$

the B_n being given by (11.21) and the C_n by

$$C_n = \frac{2}{a\sinh(n\pi b/a)} \int_0^a g(x)\sin(n\pi x/a)\,dx.$$

Clearly, this method may be extended to cases in which three or four sides of the plate have non-zero boundary conditions. ◀

As a final example of the usefulness of the principle of superposition we now consider a problem that illustrates how to deal with inhomogeneous boundary conditions by a suitable change of variables.

Example A bar of length L is initially at a temperature of 0 °C. One end of the bar ($x = 0$) is held at 0 °C and the other is supplied with heat at a constant rate per unit area of H. Find the temperature distribution within the bar after a time t.

With our usual notation, the heat diffusion equation satisfied by the temperature $u(x, t)$ is

$$\kappa \frac{\partial^2 u}{\partial x^2} = \frac{\partial u}{\partial t},$$

with $\kappa = k/(s\rho)$, where k is the thermal conductivity of the bar, s is its specific heat capacity and ρ is its density.

The boundary conditions can be written as

$$u(x, 0) = 0, \qquad u(0, t) = 0, \qquad \frac{\partial u(L, t)}{\partial x} = \frac{H}{k},$$

the last of which is inhomogeneous. In general, inhomogeneous boundary conditions can cause difficulties and it is usual to attempt a transformation of the problem into an equivalent homogeneous one. To this end, let us assume that the solution to our problem takes the form

$$u(x, t) = v(x, t) + w(x),$$

where the function $w(x)$ is to be suitably determined. In terms of v and w the problem becomes

$$\kappa \left(\frac{\partial^2 v}{\partial x^2} + \frac{d^2 w}{dx^2} \right) = \frac{\partial v}{\partial t},$$

$$v(x, 0) + w(x) = 0,$$

$$v(0, t) + w(0) = 0,$$

$$\frac{\partial v(L, t)}{\partial x} + \frac{dw(L)}{dx} = \frac{H}{k}.$$

There are several ways of choosing $w(x)$ so as to make the new problem straightforward. Using some physical insight, however, it is clear that ultimately (at $t = \infty$), when all transients have died away, the end $x = L$ will attain a temperature u_0 such that $ku_0/L = H$ and there will be a constant temperature gradient given by $u(x, \infty) = u_0 x/L$. We therefore choose

$$w(x) = \frac{Hx}{k}.$$

Since the second derivative of $w(x)$ is zero, v by itself satisfies the diffusion equation and the boundary conditions on v are now given by

$$v(x, 0) = -\frac{Hx}{k}, \qquad v(0, t) = 0, \qquad \frac{\partial v(L, t)}{\partial x} = 0.$$

These are homogeneous in x, i.e. at each of the two x-boundaries, 0 and L, if v satisfies the boundary condition, so does μv for any constant μ.

From (11.12) a separated solution for the one-dimensional diffusion equation is

$$v(x, t) = (A \cos \lambda x + B \sin \lambda x) \exp(-\lambda^2 \kappa t),$$

corresponding to a separation constant $-\lambda^2$. If we restrict λ to be real then all these solutions are transient ones decaying to zero as $t \to \infty$. These are just what is required to add to $w(x)$ to give the correct solution as $t \to \infty$. In order to satisfy $v(0, t) = 0$, however, we require $A = 0$.

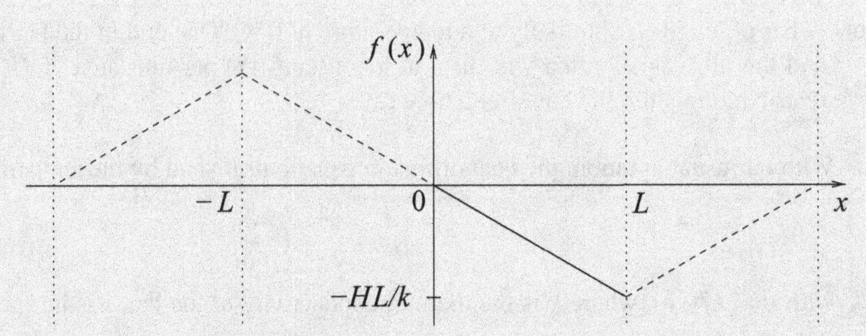

Figure 11.4 The appropriate continuation for a Fourier series containing only sine terms.

Furthermore, since

$$\frac{\partial v}{\partial x} = B \exp(-\lambda^2 \kappa t)\lambda \cos \lambda x,$$

in order to satisfy $\partial v(L, t)/\partial x = 0$ we require $\cos \lambda L = 0$, and so λ is restricted to the values

$$\lambda = \frac{n\pi}{2L},$$

where n is an odd non-negative integer, i.e. $n = 1, 3, 5, \ldots$.

Thus, to satisfy the boundary condition $v(x, 0) = -Hx/k$, we must have

$$\sum_{n \text{ odd}} B_n \sin\left(\frac{n\pi x}{2L}\right) = -\frac{Hx}{k},$$

in the range $x = 0$ to $x = L$. In this case we must be more careful about the continuation of the function $-Hx/k$, for which the Fourier sine series is required. We want a series that is odd in x (sine terms only) and continuous as $x = 0$ and $x = L$ (no discontinuities, since the series must converge at the end-points). This leads to a continuation of the function as shown in Figure 11.4, with a period of $L' = 4L$. Following the discussion of Section 4.3, since this continuation is odd about $x = 0$ and even about $x = L'/4 = L$ it can indeed be expressed as a Fourier sine series containing only odd-numbered terms.

The corresponding Fourier series coefficients are found to be

$$B_n = \frac{-8HL}{k\pi^2} \frac{(-1)^{(n-1)/2}}{n^2} \quad \text{for } n \text{ odd},$$

and thus the final formula for $u(x, t)$ is

$$u(x, t) = \frac{Hx}{k} - \frac{8HL}{k\pi^2} \sum_{n \text{ odd}} \frac{(-1)^{(n-1)/2}}{n^2} \sin\left(\frac{n\pi x}{2L}\right) \exp\left(-\frac{kn^2\pi^2 t}{4L^2 s\rho}\right),$$

giving the temperature for all positions $0 \le x \le L$ and for all times $t \ge 0$.[4] ◀

We note that in all the above examples the boundary conditions restricted the separation constant(s) to an infinite number of *discrete* values, usually integers. If, however, the boundary conditions allow the separation constant(s) λ to take a *continuum* of values then

4 Note the very large damping of the higher frequency components after a time $t \sim L^2 s\rho/k$.

the summation in (11.16) is replaced by an integral over λ. This is discussed further in connection with integral transform methods in Section 11.4.

11.3 Separation of variables in polar coordinates

So far we have considered the solution of PDEs only in Cartesian coordinates, but many systems in two and three dimensions are more naturally expressed in some form of polar coordinates, in which full advantage can be taken of any inherent symmetries. For example, the potential associated with an isolated point charge has a very simple expression, $q/(4\pi\epsilon_0 r)$, when polar coordinates are used, but involves all three coordinates and square roots when Cartesians are employed. For these reasons we now turn to the separation of variables in plane polar, cylindrical polar and spherical polar coordinates.

Most of the PDEs we have considered so far have involved the operator ∇^2, e.g. the wave equation, the diffusion equation, Schrödinger's equation and Poisson's equation (and of course Laplace's equation). It is therefore appropriate that we recall the expressions for ∇^2 when expressed in polar coordinate systems. From Chapter 2, in plane polars, cylindrical polars and spherical polars, respectively, we have

$$\nabla^2 = \frac{1}{\rho}\frac{\partial}{\partial \rho}\left(\rho\frac{\partial}{\partial \rho}\right) + \frac{1}{\rho^2}\frac{\partial^2}{\partial \phi^2}, \tag{11.23}$$

$$\nabla^2 = \frac{1}{\rho}\frac{\partial}{\partial \rho}\left(\rho\frac{\partial}{\partial \rho}\right) + \frac{1}{\rho^2}\frac{\partial^2}{\partial \phi^2} + \frac{\partial^2}{\partial z^2}, \tag{11.24}$$

$$\nabla^2 = \frac{1}{r^2}\frac{\partial}{\partial r}\left(r^2\frac{\partial}{\partial r}\right) + \frac{1}{r^2 \sin\theta}\frac{\partial}{\partial \theta}\left(\sin\theta\frac{\partial}{\partial \theta}\right) + \frac{1}{r^2\sin^2\theta}\frac{\partial^2}{\partial \phi^2}. \tag{11.25}$$

Of course the first of these may be obtained from the second by taking z to be identically zero.

11.3.1 Laplace's equation in polar coordinates

The simplest of the equations containing ∇^2 is Laplace's equation,

$$\nabla^2 u(\mathbf{r}) = 0. \tag{11.26}$$

Since it contains most of the essential features of the other more complicated equations, we will consider its solution first.

Laplace's equation in plane polars

Suppose that we need to find a solution of (11.26) that has a prescribed behavior on the circle $\rho = a$ (e.g. if we are finding the shape taken up by a circular drum skin when its rim is slightly deformed from being planar). Then we may seek solutions of (11.26) that are separable in ρ and ϕ (measured from some arbitrary radius as $\phi = 0$) and hope to accommodate the boundary condition by examining the solution for $\rho = a$.

Thus, writing $u(\rho, \phi) = P(\rho)\Phi(\phi)$ and using expression (11.23), Laplace's equation (11.26) becomes

$$\frac{\Phi}{\rho}\frac{\partial}{\partial \rho}\left(\rho\frac{\partial P}{\partial \rho}\right) + \frac{P}{\rho^2}\frac{\partial^2 \Phi}{\partial \phi^2} = 0.$$

Now, employing the same device as previously, that of dividing through by $u = P\Phi$ and multiplying through by ρ^2, results in the separated equation

$$\frac{\rho}{P}\frac{\partial}{\partial \rho}\left(\rho\frac{\partial P}{\partial \rho}\right) + \frac{1}{\Phi}\frac{\partial^2 \Phi}{\partial \phi^2} = 0.$$

Following our earlier argument, since the first term on the RHS is a function of ρ only, whilst the second term depends only on ϕ, we obtain the two *ordinary* equations

$$\frac{\rho}{P}\frac{d}{d\rho}\left(\rho\frac{dP}{d\rho}\right) = n^2, \tag{11.27}$$

$$\frac{1}{\Phi}\frac{d^2\Phi}{d\phi^2} = -n^2, \tag{11.28}$$

where we have taken the separation constant to have the form n^2 for later convenience; for the present, n is a general (complex) number.

Let us first consider the case in which $n \neq 0$. The second equation, (11.28), then has the general solution

$$\Phi(\phi) = A\exp(in\phi) + B\exp(-in\phi). \tag{11.29}$$

Equation (11.27), on the other hand, is the homogeneous equation

$$\rho^2 P'' + \rho P' - n^2 P = 0,$$

which must be solved either by trying a power solution in ρ or by making the substitution $\rho = \exp t$ as described in Subsection 6.5.1 and so reducing it to an equation with constant coefficients. Carrying out this procedure we find

$$P(\rho) = C\rho^n + D\rho^{-n}. \tag{11.30}$$

Returning to the solution (11.29) of the azimuthal equation (11.28), we can see that if Φ, and hence u, is to be single-valued and so not change when ϕ increases by 2π then n must be an integer. Mathematically, other values of n are permissible, but for the description of real physical situations it is clear that this limitation must be imposed. Having thus restricted the possible values of n in one part of the solution, the same limitations must be carried over into the radial part, (11.30). Thus we may write a particular solution of the two-dimensional Laplace equation as

$$u(\rho, \phi) = (A\cos n\phi + B\sin n\phi)(C\rho^n + D\rho^{-n}),$$

where A, B, C, D are arbitrary constants and n is any integer.

We have not yet, however, considered the solution when $n = 0$. In this case, the solutions of the separated ordinary equations (11.28) and (11.27), respectively, are easily shown to be

$$\Phi(\phi) = A\phi + B,$$
$$P(\rho) = C \ln \rho + D.$$

But, in order that $u = P\Phi$ is single-valued, we require $A = 0$, and so the solution for $n = 0$ is simply (absorbing B into C and D)

$$u(\rho, \phi) = C \ln \rho + D.$$

Superposing the solutions for the different allowed values of n, we can write the general solution to Laplace's equation in plane polars as

$$u(\rho, \phi) = (C_0 \ln \rho + D_0) + \sum_{n=1}^{\infty} (A_n \cos n\phi + B_n \sin n\phi)(C_n \rho^n + D_n \rho^{-n}), \qquad (11.31)$$

where n can take only integer values. Negative values of n have been omitted from the sum since they are already included in the terms obtained for positive n. We note that, since $\ln \rho$ is singular at $\rho = 0$, whenever we solve Laplace's equation in a region containing the origin, C_0 must be identically zero.

Example A circular drum skin has a supporting rim at $\rho = a$. If the rim is twisted so that it is displaced vertically by a small amount $\epsilon(\sin \phi + 2 \sin 2\phi)$, where ϕ is the azimuthal angle with respect to a given radius, find the resulting displacement $u(\rho, \phi)$ over the entire drum skin.

The transverse displacement of a circular drum skin is usually described by the two-dimensional wave equation. In this case, however, there is no time dependence and so $u(\rho, \phi)$ solves the two-dimensional Laplace equation, subject to the imposed boundary condition.

Referring to (11.31), since we wish to find a solution that is finite everywhere inside $\rho = a$, we require $C_0 = 0$ and $D_n = 0$ for all $n > 0$. Now the boundary condition at the rim requires

$$u(a, \phi) = D_0 + \sum_{n=1}^{\infty} C_n a^n (A_n \cos n\phi + B_n \sin n\phi) = \epsilon(\sin \phi + 2 \sin 2\phi).$$

Equating coefficients in the two expressions[5] shows that we require $D_0 = 0$ and $A_n = 0$ for all n. Furthermore, we must have $C_1 B_1 a = \epsilon$, $C_2 B_2 a^2 = 2\epsilon$ and $B_n = 0$ for $n > 2$. Hence the appropriate shape for the drum skin is

$$u(\rho, \phi) = \frac{\epsilon\rho}{a} \sin \phi + \frac{2\epsilon\rho^2}{a^2} \sin 2\phi = \frac{\epsilon\rho}{a} \left(\sin \phi + \frac{2\rho}{a} \sin 2\phi \right).$$

This form is valid over the whole skin, and not just on the rim. ◀

5 Here we have simply equated the coefficients of each sinusoid separately. If necessary, this could be formally justified by multiplying the equation through by $\sin m\phi$, integrating from 0 to 2π, and using the mutual orthogonality of the sinusoids.

Laplace's equation in cylindrical polars

Passing to three dimensions, we now consider the solution of Laplace's equation in cylindrical polar coordinates,

$$\frac{1}{\rho}\frac{\partial}{\partial \rho}\left(\rho\frac{\partial u}{\partial \rho}\right)+\frac{1}{\rho^2}\frac{\partial^2 u}{\partial \phi^2}+\frac{\partial^2 u}{\partial z^2}=0. \tag{11.32}$$

We note here that, even when considering a cylindrical physical system, if there is no dependence of the physical variables on z (i.e. along the length of the cylinder) then the problem may be treated using two-dimensional plane polars, as discussed above.

For the more general case, however, we proceed as previously by trying a solution of the form

$$u(\rho,\phi,z)=P(\rho)\Phi(\phi)Z(z),$$

which, on substitution into (11.32) and division through by $u = P\Phi Z$, gives

$$\frac{1}{P\rho}\frac{d}{d\rho}\left(\rho\frac{dP}{d\rho}\right)+\frac{1}{\Phi\rho^2}\frac{d^2\Phi}{d\phi^2}+\frac{1}{Z}\frac{d^2Z}{dz^2}=0.$$

The last term depends only on z, and the first and second (taken together) depend only on ρ and ϕ. Taking the separation constant to be k^2, we find

$$\frac{1}{Z}\frac{d^2Z}{dz^2}=k^2,$$

$$\frac{1}{P\rho}\frac{d}{d\rho}\left(\rho\frac{dP}{d\rho}\right)+\frac{1}{\Phi\rho^2}\frac{d^2\Phi}{d\phi^2}+k^2=0.$$

The first of these equations has the straightforward solution

$$Z(z)=E\exp(-kz)+F\exp kz.$$

Multiplying the second equation through by ρ^2, we obtain

$$\frac{\rho}{P}\frac{d}{d\rho}\left(\rho\frac{dP}{d\rho}\right)+\frac{1}{\Phi}\frac{d^2\Phi}{d\phi^2}+k^2\rho^2=0,$$

in which the second term depends only on Φ and the other terms depend only on ρ. Taking the second separation constant to be m^2, we find

$$\frac{1}{\Phi}\frac{d^2\Phi}{d\phi^2}=-m^2, \tag{11.33}$$

$$\rho\frac{d}{d\rho}\left(\rho\frac{dP}{d\rho}\right)+(k^2\rho^2-m^2)P=0. \tag{11.34}$$

The equation in the azimuthal angle ϕ has the very familiar solution

$$\Phi(\phi)=C\cos m\phi+D\sin m\phi.$$

As in the two-dimensional case, single-valuedness of u requires that m is an integer. However, in the particular case $m = 0$ the solution is

$$\Phi(\phi)=C\phi+D.$$

This form is appropriate to a solution with axial symmetry ($C = 0$) or one that is multi-valued, but manageably so, such as the magnetic scalar potential associated with a current I (in which case $C = I/(2\pi)$ and D is arbitrary).

Finally, the ρ-equation (11.34) may be transformed into Bessel's equation of order m by writing $\mu = k\rho$. This has the solution

$$P(\rho) = AJ_m(k\rho) + BY_m(k\rho).$$

The properties of these functions were investigated in Chapter 9 and will not be pursued here. We merely note that $Y_m(k\rho)$ is singular at $\rho = 0$, and so, when seeking solutions to Laplace's equation in cylindrical coordinates within some region containing the $\rho = 0$ axis, we require $B = 0$.

The complete separated-variable solution in cylindrical polars of Laplace's equation $\nabla^2 u = 0$ is thus given by

$$u(\rho, \phi, z) = [AJ_m(k\rho) + BY_m(k\rho)][C\cos m\phi + D\sin m\phi][E\exp(-kz) + F\exp kz]. \tag{11.35}$$

Of course we may use the principle of superposition to build up more general solutions by adding together solutions of the form (11.35) for all allowed values of the separation constants k and m. The following example has a solution that involves only one value for m but an infinite number of k-values.

Example A semi-infinite solid cylinder of radius a has its curved surface held at 0 °C and its base held at a temperature T_0. Find the steady-state temperature distribution in the cylinder.

The physical situation is shown in Figure 11.5. The steady-state temperature distribution $u(\rho, \phi, z)$ must satisfy Laplace's equation subject to the imposed boundary conditions. Let us take the cylinder to have its base in the $z = 0$ plane and to extend along the positive z-axis. From (11.35), in order that u is finite everywhere in the cylinder we immediately require $B = 0$ and $F = 0$. Furthermore, since the boundary conditions, and hence the temperature distribution, are axially symmetric (no ϕ-dependence), only $m = 0$ solutions are acceptable. So, the general solution must be a superposition of solutions of the form $J_0(k\rho)\exp(-kz)$ for all allowed values of the separation constant k.

The boundary condition $u(a, \phi, z) = 0$ restricts the allowed values of k, since we must have $J_0(ka) = 0$. The zeros of Bessel functions are given in most books of mathematical tables, and we find that, to two decimal places,

$$J_0(x) = 0 \quad \text{for } x = 2.40,\ 5.52,\ 8.65,\ \ldots.$$

Writing the allowed values of k as k_n for $n = 1, 2, 3, \ldots$ (so, for example, $k_1 = 2.40/a$), the required solution takes the form

$$u(\rho, \phi, z) = \sum_{n=1}^{\infty} A_n J_0(k_n\rho)\exp(-k_n z).$$

By imposing the remaining boundary condition $u(\rho, \phi, 0) = T_0$, the coefficients A_n can be found in a similar way to Fourier coefficients but this time by exploiting the orthogonality of the Bessel

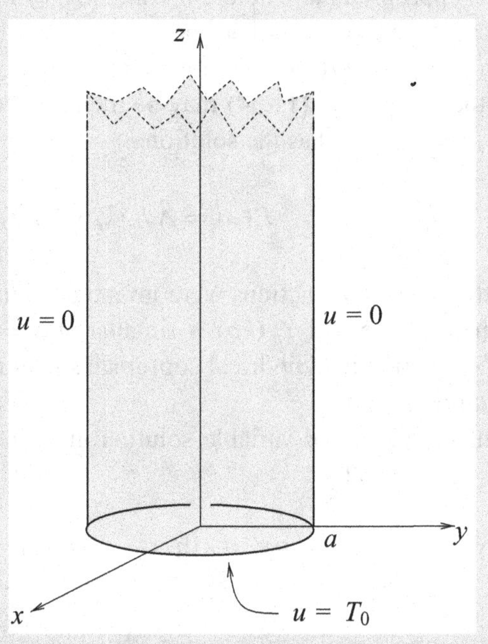

Figure 11.5 A uniform metal cylinder whose curved surface is kept at 0 °C and whose base is held at a temperature T_0.

functions, as discussed in Chapter 9. From this boundary condition we require

$$u(\rho, \phi, 0) = \sum_{n=1}^{\infty} A_n J_0(k_n \rho) = T_0.$$

If we multiply this expression by $\rho J_0(k_r \rho)$ and integrate from $\rho = 0$ to $\rho = a$, using the orthogonality of the Bessel functions $J_0(k_n \rho)$, then we find that the coefficients are given by (9.88) as

$$A_n = \frac{2T_0}{a^2 J_1^2(k_n a)} \int_0^a J_0(k_n \rho) \rho \, d\rho. \tag{11.36}$$

The integral on the RHS can be evaluated using the recurrence relation (9.89),

$$\frac{d}{dz}[z J_1(z)] = z J_0(z),$$

which on setting $z = k_n \rho$ yields

$$\frac{1}{k_n} \frac{d}{d\rho}[k_n \rho J_1(k_n \rho)] = k_n \rho J_0(k_n \rho).$$

Therefore the integral in (11.36) is given by

$$\int_0^a J_0(k_n \rho) \rho \, d\rho = \left[\frac{1}{k_n} \rho J_1(k_n \rho) \right]_0^a = \frac{1}{k_n} a J_1(k_n a),$$

and the coefficients A_n may be expressed as

$$A_n = \frac{2T_0}{a^2 J_1^2(k_n a)} \left[\frac{a J_1(k_n a)}{k_n} \right] = \frac{2T_0}{k_n a J_1(k_n a)}.$$

The infinite sum

$$u(\rho, \phi, z) = \sum_{n=1}^{\infty} \frac{2T_0}{k_n a J_1(k_n a)} J_0(k_n \rho) \exp(-k_n z)$$

now gives the steady-state temperature at position (ρ, ϕ, z) in the cylinder. ◀

We note that if, in the above example, the base of the cylinder were not kept at a uniform temperature T_0, but instead had some fixed temperature distribution $T(\rho, \phi)$, then the solution of the problem would become more complicated. In such a case, the required temperature distribution $u(\rho, \phi, z)$ is in general *not* axially symmetric, and so the separation constant m is not restricted to be zero but may take any integer value. The solution will then take the form

$$u(\rho, \phi, z) = \sum_{m=0}^{\infty} \sum_{n=1}^{\infty} J_m(k_{nm}\rho)(C_{nm} \cos m\phi + D_{nm} \sin m\phi) \exp(-k_{nm} z),$$

where the separation constants k_{nm} are such that $J_m(k_{nm}a) = 0$, i.e. $k_{nm}a$ is the nth zero of the mth-order Bessel function. At the base of the cylinder we would then require

$$u(\rho, \phi, 0) = \sum_{m=0}^{\infty} \sum_{n=1}^{\infty} J_m(k_{nm}\rho)(C_{nm} \cos m\phi + D_{nm} \sin m\phi) = T(\rho, \phi). \qquad (11.37)$$

The coefficients C_{nm} could be found by multiplying (11.37) by $J_q(k_{rq}\rho) \cos q\phi$, integrating with respect to ρ and ϕ over the base of the cylinder and exploiting the orthogonality of the Bessel functions and of the trigonometric functions. The D_{nm} could be found in a similar way by multiplying (11.37) by $J_q(k_{rq}\rho) \sin q\phi$.

Laplace's equation in spherical polars

We now come to an equation that is very widely applicable in physical science, namely $\nabla^2 u = 0$ in spherical polar coordinates:

$$\frac{1}{r^2}\frac{\partial}{\partial r}\left(r^2 \frac{\partial u}{\partial r}\right) + \frac{1}{r^2 \sin\theta}\frac{\partial}{\partial \theta}\left(\sin\theta \frac{\partial u}{\partial \theta}\right) + \frac{1}{r^2 \sin^2\theta}\frac{\partial^2 u}{\partial \phi^2} = 0. \qquad (11.38)$$

Our method of procedure will be as before; we try a solution of the form

$$u(r, \theta, \phi) = R(r)\Theta(\theta)\Phi(\phi).$$

Substituting this in (11.38), dividing through by $u = R\Theta\Phi$ and multiplying by r^2, we obtain

$$\frac{1}{R}\frac{d}{dr}\left(r^2 \frac{dR}{dr}\right) + \frac{1}{\Theta \sin\theta}\frac{d}{d\theta}\left(\sin\theta \frac{d\Theta}{d\theta}\right) + \frac{1}{\Phi \sin^2\theta}\frac{d^2\Phi}{d\phi^2} = 0. \qquad (11.39)$$

The first term depends only on r and the second and third terms (taken together) depend only on θ and ϕ. Thus (11.39) is equivalent to the two equations

$$\frac{1}{R}\frac{d}{dr}\left(r^2\frac{dR}{dr}\right) = \lambda, \tag{11.40}$$

$$\frac{1}{\Theta \sin\theta}\frac{d}{d\theta}\left(\sin\theta\frac{d\Theta}{d\theta}\right) + \frac{1}{\Phi\sin^2\theta}\frac{d^2\Phi}{d\phi^2} = -\lambda. \tag{11.41}$$

Equation (11.40) is a homogeneous equation,

$$r^2\frac{d^2R}{dr^2} + 2r\frac{dR}{dr} - \lambda R = 0,$$

which can be reduced, by the substitution $r = \exp t$ (and writing $R(r) = S(t)$), to

$$\frac{d^2S}{dt^2} + \frac{dS}{dt} - \lambda S = 0.$$

This has the straightforward solution

$$S(t) = A\exp\lambda_1 t + B\exp\lambda_2 t,$$

and so the solution to the radial equation is

$$R(r) = Ar^{\lambda_1} + Br^{\lambda_2},$$

where $\lambda_1 + \lambda_2 = -1$ and $\lambda_1\lambda_2 = -\lambda$. We can thus take λ_1 and λ_2 as given by ℓ and $-(\ell+1)$; λ then has the form $\ell(\ell+1)$. (It should be noted that at this stage nothing has been either assumed or proved about whether ℓ is an integer.)

Hence we have obtained some information about the first factor in the separated-variable solution, which will now have the form

$$u(r,\theta,\phi) = \left[Ar^{\ell} + Br^{-(\ell+1)}\right]\Theta(\theta)\Phi(\phi), \tag{11.42}$$

where Θ and Φ must satisfy (11.41) with $\lambda = \ell(\ell+1)$.

The next step is to take (11.41) further. Multiplying through by $\sin^2\theta$ and substituting for λ, it too takes a separated form:

$$\left[\frac{\sin\theta}{\Theta}\frac{d}{d\theta}\left(\sin\theta\frac{d\Theta}{d\theta}\right) + \ell(\ell+1)\sin^2\theta\right] + \frac{1}{\Phi}\frac{d^2\Phi}{d\phi^2} = 0. \tag{11.43}$$

Taking the separation constant as m^2, the equation in the azimuthal angle ϕ has the same solution as in cylindrical polars, namely

$$\Phi(\phi) = C\cos m\phi + D\sin m\phi.$$

As before, single-valuedness of u requires that m is an integer; for $m = 0$ we again have $\Phi(\phi) = C\phi + D$.

Having settled the form of $\Phi(\phi)$, we are left only with the equation satisfied by $\Theta(\theta)$, which is

$$\frac{\sin\theta}{\Theta}\frac{d}{d\theta}\left(\sin\theta\frac{d\Theta}{d\theta}\right) + \ell(\ell+1)\sin^2\theta = m^2. \tag{11.44}$$

A change of independent variable from θ to $\mu = \cos\theta$ will reduce this to a form for which solutions are known, and of which some study has been made in Chapter 9. Putting

$$\mu = \cos\theta, \qquad \frac{d\mu}{d\theta} = -\sin\theta, \qquad \frac{d}{d\theta} = -(1-\mu^2)^{1/2}\frac{d}{d\mu},$$

the equation for $M(\mu) \equiv \Theta(\theta)$ reads

$$\frac{d}{d\mu}\left[(1-\mu^2)\frac{dM}{d\mu}\right] + \left[\ell(\ell+1) - \frac{m^2}{1-\mu^2}\right]M = 0. \tag{11.45}$$

This equation is the *associated Legendre equation*, which was mentioned in Section 9.2 in the context of Sturm–Liouville equations.

We recall that for the case $m = 0$, (11.45) reduces to Legendre's equation, which was studied at length in Chapter 9, and has the solution

$$M(\mu) = EP_\ell(\mu) + FQ_\ell(\mu). \tag{11.46}$$

We have not solved (11.45) explicitly for general m, but the solutions were given in Section 9.2 and are the associated Legendre functions $P_\ell^m(\mu)$ and $Q_\ell^m(\mu)$, where

$$P_\ell^m(\mu) = (1-\mu^2)^{|m|/2}\frac{d^{|m|}}{d\mu^{|m|}}P_\ell(\mu), \tag{11.47}$$

and similarly for $Q_\ell^m(\mu)$. We then have

$$M(\mu) = EP_\ell^m(\mu) + FQ_\ell^m(\mu); \tag{11.48}$$

here m must be an integer, $0 \le |m| \le \ell$. We note that if we require solutions to Laplace's equation that are finite when $\mu = \cos\theta = \pm 1$ (i.e. on the polar axis where $\theta = 0, \pi$), then we must have $F = 0$ in (11.46) and (11.48) since $Q_\ell^m(\mu)$ diverges at $\mu = \pm 1$.

It will be remembered that one of the important conditions for obtaining finite polynomial solutions of Legendre's equation is that ℓ is an integer ≥ 0. This condition therefore applies also to the solutions (11.46) and (11.48) and is reflected back into the radial part of the general solution given in (11.42).

Now that the solutions of each of the three ordinary differential equations governing R, Θ and Φ have been obtained, we may assemble a complete separated-variable solution of Laplace's equation in spherical polars. It is

$$u(r,\theta,\phi) = (Ar^\ell + Br^{-(\ell+1)})(C\cos m\phi + D\sin m\phi)[EP_\ell^m(\cos\theta) + FQ_\ell^m(\cos\theta)], \tag{11.49}$$

where the three bracketed factors are connected only through the *integer* parameters ℓ and m, $0 \le |m| \le \ell$. As before, a general solution may be obtained by superposing solutions of this form for the allowed values of the separation constants ℓ and m. As mentioned above, if the solution is required to be finite on the polar axis then $F = 0$ for all ℓ and m. In our first illustrative example, not only must $F = 0$, but only $m = 0$ terms are allowed.

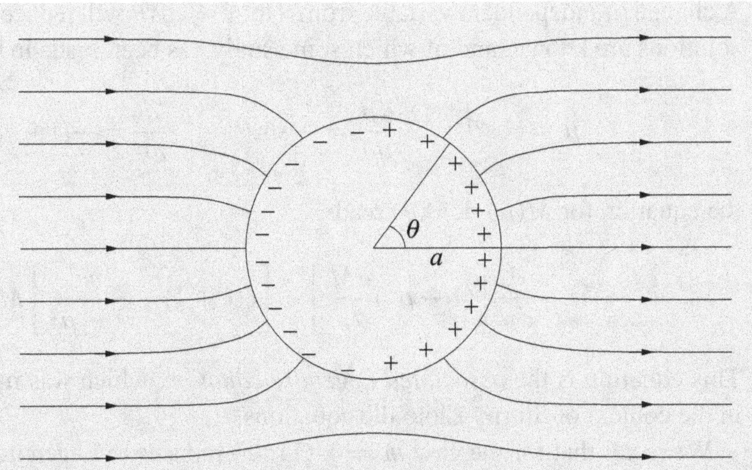

Figure 11.6 Induced charge and field lines associated with a conducting sphere placed in an initially uniform electrostatic field.

Example An uncharged conducting sphere of radius a is placed at the origin in an initially uniform electrostatic field E. Show that it behaves as an electric dipole.

The uniform field, taken in the direction of the polar axis, has an electrostatic potential

$$u = -Ez = -Er\cos\theta,$$

where u is arbitrarily taken as zero at $z = 0$. This satisfies Laplace's equation $\nabla^2 u = 0$, as must the potential v when the sphere is present; for large r the asymptotic form of v must still be $-Er\cos\theta$.

Since the problem is clearly axially symmetric, we have immediately that $m = 0$, and since we require v to be finite on the polar axis we must have $F = 0$ in (11.49). Therefore the solution must be of the form

$$v(r, \theta, \phi) = \sum_{\ell=0}^{\infty}(A_\ell r^\ell + B_\ell r^{-(\ell+1)})P_\ell(\cos\theta).$$

Now the $\cos\theta$-dependence of v for large r indicates that the (θ, ϕ)-dependence of $v(r, \theta, \phi)$ is given by $P_1^0(\cos\theta) = \cos\theta$. Thus the r-dependence of v must also correspond to an $\ell = 1$ solution, and the most general such solution (outside the sphere, i.e. for $r \geq a$) is

$$v(r, \theta, \phi) = (A_1 r + B_1 r^{-2})P_1(\cos\theta).$$

The asymptotic form of v for large r immediately gives $A_1 = -E$ and so yields the solution

$$v(r, \theta, \phi) = \left(-Er + \frac{B_1}{r^2}\right)\cos\theta.$$

Since the sphere is conducting, it is an equipotential region and so v must not depend on θ for $r = a$. This can only be the case if $B_1/a^2 = Ea$, thus fixing B_1. The final solution is therefore

$$v(r, \theta, \phi) = -Er\left(1 - \frac{a^3}{r^3}\right)\cos\theta.$$

Since a dipole of moment p gives rise to a potential $p/(4\pi\epsilon_0 r^2)$, this result shows that the sphere behaves as a dipole of moment $4\pi\epsilon_0 a^3 E$. The physical origin of this effect is the charge distribution induced on its surface by the external field; see Figure 11.6. ◀

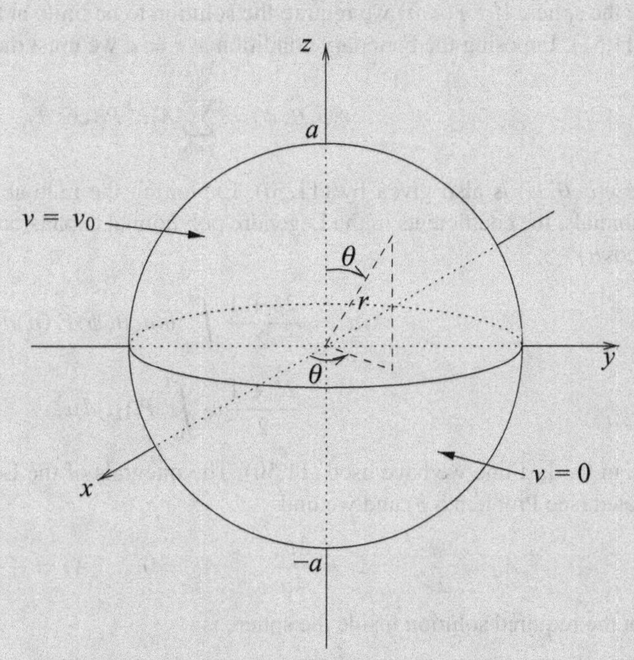

Figure 11.7 A hollow split conducting sphere with its top half charged to a potential v_0 and its bottom half at zero potential.

Often the boundary conditions are not so easily met, and it is necessary to use the mutual orthogonality of the associated Legendre functions (and the trigonometric functions) to obtain the coefficients in the general solution.

Example A hollow split conducting sphere of radius a is placed at the origin. If one half of its surface is charged to a potential v_0 and the other half is kept at zero potential, find the potential v inside and outside the sphere.

Let us choose the top hemisphere to be charged to v_0 and the bottom hemisphere to be at zero potential, with the plane in which the two hemispheres meet perpendicular to the polar axis; this is shown in Figure 11.7. The boundary condition then becomes

$$v(a, \theta, \phi) = \begin{cases} v_0 & \text{for } 0 < \theta < \pi/2 \quad (0 < \cos\theta < 1), \\ 0 & \text{for } \pi/2 < \theta < \pi \quad (-1 < \cos\theta < 0). \end{cases} \tag{11.50}$$

The problem is clearly axially symmetric and so we may set $m = 0$. Also, we require the solution to be finite on the polar axis and so it cannot contain $Q_\ell(\cos\theta)$. Therefore the general form of the solution to (11.38) is

$$v(r, \theta, \phi) = \sum_{\ell=0}^{\infty} (A_\ell r^\ell + B_\ell r^{-(\ell+1)}) P_\ell(\cos\theta). \tag{11.51}$$

Inside the sphere (for $r < a$) we require the solution to be finite at the origin and so $B_\ell = 0$ for all ℓ in (11.51). Imposing the boundary condition at $r = a$ we must then have

$$v(a, \theta, \phi) = \sum_{\ell=0}^{\infty} A_\ell a^\ell P_\ell(\cos\theta),$$

where $v(a, \theta, \phi)$ is also given by (11.50). Exploiting the mutual orthogonality of the Legendre polynomials, the coefficients in the Legendre polynomial expansion are given by (9.14) as (writing $\mu = \cos\theta$)

$$\begin{aligned} A_\ell a^\ell &= \frac{2\ell+1}{2} \int_{-1}^{1} v(a, \theta, \phi) P_\ell(\mu) d\mu \\ &= \frac{2\ell+1}{2} v_0 \int_0^1 P_\ell(\mu) d\mu, \end{aligned}$$

where in the last line we have used (11.50). The integrals of the Legendre polynomials are easily evaluated (see Problem 8.3) and we find

$$A_0 = \frac{v_0}{2}, \qquad A_1 = \frac{3v_0}{4a}, \qquad A_2 = 0, \qquad A_3 = -\frac{7v_0}{16a^3}, \qquad \ldots,$$

so that the required solution inside the sphere is

$$v(r, \theta, \phi) = \frac{v_0}{2}\left[1 + \frac{3r}{2a} P_1(\cos\theta) - \frac{7r^3}{8a^3} P_3(\cos\theta) + \cdots\right].$$

Outside the sphere (for $r > a$) we require the solution to be bounded as r tends to infinity and so in (11.51) we must have $A_\ell = 0$ for all ℓ. In this case, by imposing the boundary condition at $r = a$ we require

$$v(a, \theta, \phi) = \sum_{\ell=0}^{\infty} B_\ell a^{-(\ell+1)} P_\ell(\cos\theta),$$

where $v(a, \theta, \phi)$ is given by (11.50). Following the above argument the coefficients in the expansion are given by

$$B_\ell a^{-(\ell+1)} = \frac{2\ell+1}{2} v_0 \int_0^1 P_\ell(\mu) d\mu,$$

so that the required solution outside the sphere is

$$v(r, \theta, \phi) = \frac{v_0 a}{2r}\left[1 + \frac{3a}{2r} P_1(\cos\theta) - \frac{7a^3}{8r^3} P_3(\cos\theta) + \cdots\right].$$

We note that for $r \gg a$ the sphere behaves as a spherical capacitor of radius a and capacitance $4\pi\epsilon_0 a$, charged to a potential of $v_0/2$. ◀

In the above example, on the equator of the sphere (i.e. at $r = a$ and $\theta = \pi/2$) the potential is given, in both results, by[6]

$$v(a, \pi/2, \phi) = v_0/2,$$

6 Recall that for odd ℓ, $P_\ell(x)$ contains only odd powers of x and therefore no constant term, and so $P_\ell(\cos\pi/2) = P_\ell(0) = 0$.

i.e. mid-way between the potentials of the top and bottom hemispheres. This is so because a Legendre polynomial expansion of a function behaves in the same way as a Fourier series expansion, in that it converges to the average of the two values at any discontinuities present in the original function.

If the potential on the surface of the sphere had been given as a function of θ *and* ϕ, then we would have had to consider a double series summed over ℓ and m (for $-\ell \leq m \leq \ell$), since, in general, the solution would not have been axially symmetric.

Finally, we note in general that, when obtaining solutions of Laplace's equation in spherical polar coordinates, one finds that, for solutions that are finite on the polar axis, the angular part of the solution is given by

$$\Theta(\theta)\Phi(\phi) = P_\ell^m(\cos\theta)(C\cos m\phi + D\sin m\phi),$$

where ℓ and m are integers with $-\ell \leq m \leq \ell$. This general form is sufficiently common that particular functions of θ and ϕ called *spherical harmonics* are defined and tabulated (see Section 9.3).

11.3.2 Other equations in polar coordinates

The development of the solutions of $\nabla^2 u = 0$ carried out in the previous subsection can be employed to solve other equations in which the ∇^2 operator appears. Since we have discussed the general method in some depth already, only an outline of the solutions will be given here.

Let us first consider the wave equation

$$\nabla^2 u = \frac{1}{c^2}\frac{\partial^2 u}{\partial t^2}, \tag{11.52}$$

and look for a separated solution of the form $u = F(\mathbf{r})T(t)$, so that initially we are separating only the spatial and time dependences. Substituting this form into (11.52) and taking the separation constant as k^2 we obtain

$$\nabla^2 F + k^2 F = 0, \qquad \frac{d^2T}{dt^2} + k^2c^2T = 0. \tag{11.53}$$

The second equation has the simple solution

$$T(t) = A\exp(i\omega t) + B\exp(-i\omega t), \tag{11.54}$$

where $\omega = kc$; this may also be expressed in terms of sines and cosines, of course. The first equation in (11.53) is referred to as *Helmholtz's equation*; we discuss it below.

We may treat the diffusion equation

$$\kappa\nabla^2 u = \frac{\partial u}{\partial t}$$

in a similar way. Separating the spatial and time dependences by assuming a solution of the form $u = F(\mathbf{r})T(t)$, and taking the separation constant as k^2, we find

$$\nabla^2 F + k^2 F = 0, \qquad \frac{dT}{dt} + k^2\kappa T = 0.$$

Just as in the case of the wave equation, the spatial part of the solution satisfies Helmholtz's equation. It only remains to consider the time dependence, which has the simple solution

$$T(t) = A \exp(-k^2 \kappa t).$$

Helmholtz's equation is clearly of central importance in the solutions of the wave and diffusion equations. It can be solved in polar coordinates in much the same way as Laplace's equation, and indeed reduces to Laplace's equation when $k = 0$. Therefore, we will merely sketch the method of its solution in each of the three polar coordinate systems.

Helmholtz's equation in plane polars

In two-dimensional plane polar coordinates, Helmholtz's equation takes the form

$$\frac{1}{\rho}\frac{\partial}{\partial \rho}\left(\rho \frac{\partial F}{\partial \rho}\right) + \frac{1}{\rho^2}\frac{\partial^2 F}{\partial \phi^2} + k^2 F = 0.$$

If we try a separated solution of the form $F(\mathbf{r}) = P(\rho)\Phi(\phi)$, and take the separation constant as m^2, we find

$$\frac{d^2\Phi}{d\phi^2} + m^2\phi = 0,$$

$$\frac{d^2 P}{d\rho^2} + \frac{1}{\rho}\frac{dP}{d\rho} + \left(k^2 - \frac{m^2}{\rho^2}\right) P = 0.$$

As for Laplace's equation, the angular part has the familiar solution (if $m \neq 0$)

$$\Phi(\phi) = A \cos m\phi + B \sin m\phi,$$

or an equivalent form in terms of complex exponentials. The radial equation differs from that found in the solution of Laplace's equation, but by making the substitution $\mu = k\rho$ it is easily transformed into Bessel's equation of order m (discussed in Chapter 9), and has the solution

$$P(\rho) = C J_m(k\rho) + D Y_m(k\rho),$$

where Y_m is a Bessel function of the second kind, which is infinite at the origin and is not to be confused with a spherical harmonic (these are written with a superscript as well as a subscript).

Putting the two parts of the solution together we have

$$F(\rho, \phi) = [A \cos m\phi + B \sin m\phi][C J_m(k\rho) + D Y_m(k\rho)]. \tag{11.55}$$

Clearly, for solutions of Helmholtz's equation that are required to be finite at the origin, as in the following worked example, we must set $D = 0$.

Example Find the four lowest frequency modes of oscillation of a circular drum skin of radius a whose circumference is held fixed in a plane.

The transverse displacement $u(\mathbf{r}, t)$ of the drum skin satisfies the two-dimensional wave equation

$$\nabla^2 u = \frac{1}{c^2}\frac{\partial^2 u}{\partial t^2},$$

with $c^2 = T/\sigma$, where T is the tension of the drum skin and σ is its mass per unit area. From (11.54) and (11.55) a separated solution of this equation, in plane polar coordinates, that is finite at the origin is

$$u(\rho, \phi, t) = J_m(k\rho)(A\cos m\phi + B\sin m\phi)\exp(\pm i\omega t),$$

where $\omega = kc$. Since we require the solution to be single-valued we must have m as an integer. Furthermore, if the drum skin is clamped at its outer edge $\rho = a$ then we also require $u(a, \phi, t) = 0$. Thus we need

$$J_m(ka) = 0,$$

which in turn restricts the allowed values of k. The zeros of Bessel functions can be obtained from most books of tables; the first few are

$$J_0(x) = 0 \quad \text{for } x \approx 2.40,\ 5.52,\ 8.65, \ldots,$$

$$J_1(x) = 0 \quad \text{for } x \approx 3.83,\ 7.02,\ 10.17, \ldots,$$

$$J_2(x) = 0 \quad \text{for } x \approx 5.14,\ 8.42,\ 11.62 \ldots.$$

The smallest value of x for which any of the Bessel functions is zero is $x \approx 2.40$, which occurs for $J_0(x)$. Thus the lowest-frequency mode has $k = 2.40/a$ and angular frequency $\omega = 2.40c/a$. Since $m = 0$ for this mode, the shape of the drum skin is

$$u \propto J_0\left(2.40\frac{\rho}{a}\right);$$

this is illustrated in the top left-hand corner of Figure 11.8. In this figure dashed lines indicate those interior parts of the drum skin that do not move during the motion; for the lowest mode there are no such parts.

Continuing in the same way, the next three modes are given by

$$\omega = 3.83\frac{c}{a}, \qquad u \propto J_1\left(3.83\frac{\rho}{a}\right)\cos\phi, \qquad J_1\left(3.83\frac{\rho}{a}\right)\sin\phi;$$

$$\omega = 5.14\frac{c}{a}, \qquad u \propto J_2\left(5.14\frac{\rho}{a}\right)\cos 2\phi, \qquad J_2\left(5.14\frac{\rho}{a}\right)\sin 2\phi;$$

$$\omega = 5.52\frac{c}{a}, \qquad u \propto J_0\left(5.52\frac{\rho}{a}\right).$$

These modes are also shown in Figure 11.8. We note that the second and third frequencies have *two* corresponding modes of oscillation; these frequencies are therefore two-fold degenerate. ◀

Helmholtz's equation in cylindrical polars

Generalizing the above method to three-dimensional cylindrical polars is straightforward, and following a similar procedure to that used for Laplace's equation we find the separated

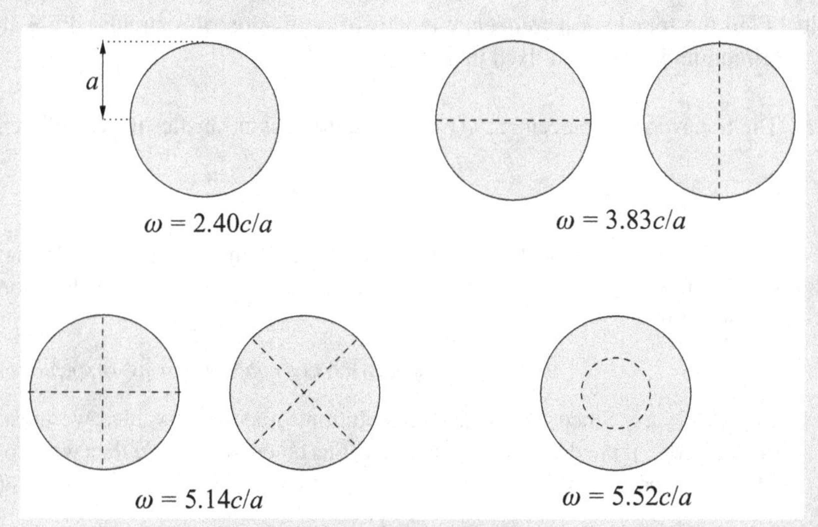

Figure 11.8 The modes of oscillation with the four lowest frequencies for a circular drum skin of radius a. The dashed lines indicate the nodes, where the displacement of the drum skin is always zero.

solution of Helmholtz's equation takes the form

$$F(\rho,\phi,z) = \left[AJ_m\left(\sqrt{k^2-\alpha^2}\,\rho\right) + BY_m\left(\sqrt{k^2-\alpha^2}\,\rho\right)\right]$$
$$\times (C\cos m\phi + D\sin m\phi)[E\exp(i\alpha z) + F\exp(-i\alpha z)],$$

where α and m are separation constants. We note that the angular part of the solution is the same as for Laplace's equation in cylindrical polars.

Helmholtz's equation in spherical polars

In spherical polars, we again find that the angular parts of the solution $\Theta(\theta)\Phi(\phi)$ are identical to those of Laplace's equation in this coordinate system, i.e. they are the spherical harmonics $Y_\ell^m(\theta,\phi)$, and so we will not discuss them further.

The radial equation in this case is given by

$$r^2R'' + 2rR' + [k^2r^2 - \ell(\ell+1)]R = 0, \tag{11.56}$$

which has an additional term k^2r^2R compared with the radial equation for the Laplace solution. The equation (11.56) looks very much like Bessel's equation. In fact, by writing $R(r) = r^{-1/2}S(r)$ and making the change of variable $\mu = kr$, it can be reduced to Bessel's equation of order $\ell + \frac{1}{2}$, which has as its solutions $S(\mu) = J_{\ell+1/2}(\mu)$ and $Y_{\ell+1/2}(\mu)$ (see Section 9.6). The separated solution to Helmholtz's equation in spherical polars is thus

$$F(r,\theta,\phi) = r^{-1/2}[AJ_{\ell+1/2}(kr) + BY_{\ell+1/2}(kr)](C\cos m\phi + D\sin m\phi)$$
$$\times[EP_\ell^m(\cos\theta) + FQ_\ell^m(\cos\theta)]. \tag{11.57}$$

For solutions that are finite at the origin we require $B = 0$, and for solutions that are finite on the polar axis we require $F = 0$. It is worth mentioning that the solutions proportional to $r^{-1/2}J_{\ell+1/2}(kr)$ and $r^{-1/2}Y_{\ell+1/2}(kr)$ when suitably normalized are called *spherical Bessel functions* of the first and second kind, respectively, and are denoted by $j_\ell(kr)$ and $n_\ell(\mu)$ (see Section 9.6).

As mentioned at the beginning of this subsection, the separated solution of the wave equation in spherical polars is the product of the time-dependent part (11.54) and a spatial part (11.57). It will be noticed that, although this solution corresponds to a solution of definite frequency $\omega = kc$, the zeros of the radial function $j_\ell(kr)$ are not equally spaced in r, except for the case $\ell = 0$ involving $j_0(kr)$, and so there is no precise wavelength associated with the solution.

To conclude this subsection, let us mention briefly the Schrödinger equation for the electron in a hydrogen atom, the nucleus of which is taken at the origin and is assumed massive compared with the electron. Under these circumstances the Schrödinger equation is

$$-\frac{\hbar^2}{2m}\nabla^2 u - \frac{e^2}{4\pi\epsilon_0}\frac{u}{r} = i\hbar\frac{\partial u}{\partial t}.$$

For a "stationary-state" solution, for which the energy is a constant E and the time-dependent factor T in u is given by $T(t) = A\exp(-iEt/\hbar)$, the above equation is similar to, but not quite the same as, the Helmholtz equation.[7] However, as with the wave equation, the angular parts of the solution are identical to those for Laplace's equation and are expressed in terms of spherical harmonics.

The important point to note is that for *any* equation involving ∇^2, provided θ and ϕ do not appear in the equation other than as part of ∇^2, a separated-variable solution in spherical polars will always lead to spherical harmonic solutions. This is the case for the Schrödinger equation describing an atomic electron in *any* central potential $V(r)$.

11.3.3 Solution by expansion

It is sometimes possible to use the uniqueness theorem discussed in the previous chapter, together with the results of the last few subsections, in which Laplace's equation (and other equations) were considered in polar coordinates, to obtain solutions of such equations appropriate to particular physical situations.

We will illustrate the method for Laplace's equation in spherical polars and first assume that the required solution of $\nabla^2 u = 0$ can be written as a superposition in the normal way:

$$u(r,\theta,\phi) = \sum_{\ell=0}^{\infty}\sum_{m=-\ell}^{\ell}(Ar^\ell + Br^{-(\ell+1)})P_\ell^m(\cos\theta)(C\cos m\phi + D\sin m\phi). \quad (11.58)$$

Here, all the constants A, B, C, D may depend upon ℓ and m, and we have assumed that the required solution is finite on the polar axis. As usual, boundary conditions of a physical nature will then fix or eliminate some of the constants; for example, u finite at the origin implies all $B = 0$, or axial symmetry implies that only $m = 0$ terms are present.

7 For the solution by series of the r-equation in this case the reader may consult, for example, L. Schiff, *Quantum Mechanics* (New York: McGraw-Hill, 1955), p. 82.

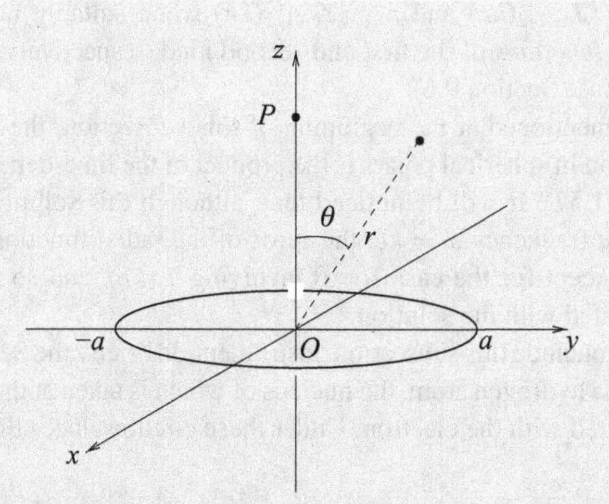

Figure 11.9 The polar axis Oz is taken as normal to the plane of the ring of matter and passing through its center.

The essence of the method is then to find the remaining constants by determining u at values of r, θ, ϕ for which it can be evaluated *by other means*, e.g. by direct calculation on an axis of symmetry. Once the remaining constants have been fixed by these special considerations to have particular values, the uniqueness theorem can be invoked to establish that they must have these values in general. Our worked example is taken from the area of gravitation.

Example Calculate the gravitational potential at a general point in space due to a uniform ring of matter of radius a and total mass M.

Everywhere except on the ring the potential $u(\mathbf{r})$ satisfies the Laplace equation, and so if we use polar coordinates with the normal to the ring as polar axis, as in Figure 11.9, a solution of the form (11.58) can be assumed.

We expect the potential $u(r, \theta, \phi)$ to tend to zero as $r \to \infty$, and also to be finite at $r = 0$. At first sight this might seem to imply that all A and B, and hence u, must be identically zero, an unacceptable result. In fact, what it means is that different expressions must apply to different regions of space. On the ring itself we no longer have $\nabla^2 u = 0$ and so it is not surprising that the form of the expression for u changes there. Let us therefore take two separate regions.

In the region $r > a$

(i) we must have $u \to 0$ as $r \to \infty$, implying that all $A = 0$, *and*
(ii) the system is axially symmetric and so only $m = 0$ terms appear.

With these restrictions we can write as a trial form

$$u(r, \theta, \phi) = \sum_{\ell=0}^{\infty} B_\ell r^{-(\ell+1)} P_\ell^0(\cos\theta). \tag{11.59}$$

The constants B_ℓ are still to be determined; this we do by calculating *directly* the potential where this can be done simply – in this case, on the polar axis.

Considering a point P on the polar axis at a distance z ($> a$) from the plane of the ring (taken as $\theta = \pi/2$), all parts of the ring are at a distance $(z^2 + a^2)^{1/2}$ from it. The potential at P is thus straightforwardly

$$u(z, 0, \phi) = -\frac{GM}{(z^2 + a^2)^{1/2}}, \tag{11.60}$$

where G is the gravitational constant. This must be the same as (11.59) for the particular values $r = z$, $\theta = 0$, and ϕ undefined. Since $P_\ell^0(\cos\theta) = P_\ell(\cos\theta)$ with $P_\ell(1) = 1$, putting $r = z$ in (11.59) gives

$$u(z, 0, \phi) = \sum_{\ell=0}^{\infty} \frac{B_\ell}{z^{\ell+1}}. \tag{11.61}$$

However, using the binomial theorem to expand (11.60) for $z > a$ (since it applies to this region of space) we obtain

$$u(z, 0, \phi) = -\frac{GM}{z}\left[1 - \frac{1}{2}\left(\frac{a}{z}\right)^2 + \frac{3}{8}\left(\frac{a}{z}\right)^4 - \cdots\right],$$

which on comparison with (11.61) gives[8]

$$\begin{aligned} B_0 &= -GM, \\ B_{2\ell} &= -\frac{GMa^{2\ell}(-1)^\ell(2\ell-1)!!}{2^\ell \ell!} \quad \text{for } \ell \geq 1, \\ B_{2\ell+1} &= 0. \end{aligned} \tag{11.62}$$

We now conclude the argument by saying that if a solution for a general point (r, θ, ϕ) exists at all, which of course we very much expect on physical grounds, then it must be (11.59) with the B_ℓ given by (11.62). This is so, because, thus defined, it is a solution of Laplace's equation with no arbitrary constants that satisfies all the boundary conditions; the uniqueness theorem states that there is only one such function. The expression for the potential in the region $r > a$ is therefore

$$u(r, \theta, \phi) = -\frac{GM}{r}\left[1 + \sum_{\ell=1}^{\infty} \frac{(-1)^\ell(2\ell-1)!!}{2^\ell \ell!}\left(\frac{a}{r}\right)^{2\ell} P_{2\ell}(\cos\theta)\right].$$

The expression for $r < a$ can be found in a similar way. The finiteness of u at $r = 0$ and the axial symmetry give

$$u(r, \theta, \phi) = \sum_{\ell=0}^{\infty} A_\ell r^\ell P_\ell^0(\cos\theta).$$

Comparing this expression for $r = z$, $\theta = 0$ with the $z < a$ expansion of (11.60), which is valid for any z, establishes $A_{2\ell+1} = 0$, $A_0 = -GM/a$ and

$$A_{2\ell} = -\frac{GM}{a^{2\ell+1}} \frac{(-1)^\ell(2\ell-1)!!}{2^\ell \ell!},$$

8 $(2\ell-1)!! = 1 \times 3 \times \cdots \times (2\ell-1)$. It can also be written as $(2\ell)!/[2^\ell \ell!]$.

so that the final expression valid, and convergent, for $r < a$ is thus

$$u(r, \theta, \phi) = -\frac{GM}{a}\left[1 + \sum_{\ell=1}^{\infty} \frac{(-1)^\ell (2\ell - 1)!!}{2^\ell \ell!}\left(\frac{r}{a}\right)^{2\ell} P_{2\ell}(\cos\theta)\right].$$

It is easy to check that the solution obtained has the expected physical values[9] for large r and for $r = 0$ and is continuous at $r = a$. ◀

11.3.4 Separation of variables for inhomogeneous equations

So far our discussion of the method of separation of variables has been limited to the solution of homogeneous equations such as the Laplace equation and the wave equation. The solutions of inhomogeneous PDEs are usually obtained using the Green's function methods to be discussed later in Section 11.5. However, as a final illustration of the usefulness of the separation of variables, we now consider its application to the solution of inhomogeneous equations.

Because of the added complexity in dealing with inhomogeneous equations, we shall restrict our discussion to the solution of Poisson's equation,

$$\nabla^2 u = \rho(\mathbf{r}), \tag{11.63}$$

in spherical polar coordinates, although the general method can accommodate other coordinate systems and equations. In physical problems the RHS of (11.63) usually contains some multiplicative constant(s). If u is the electrostatic potential in some region of space in which ρ is the density of electric charge then $\nabla^2 u = -\rho(\mathbf{r})/\epsilon_0$. Alternatively, u might represent the gravitational potential in some region where the matter density is given by ρ, so that $\nabla^2 u = 4\pi G\rho(\mathbf{r})$.

We will simplify our discussion by assuming that the required solution u is finite on the polar axis and also that the system possesses axial symmetry about that axis – in which case ρ does not depend on the azimuthal angle ϕ. The key to the method is then to assume a separated form for both the solution u *and* the density term ρ.

From the discussion of Laplace's equation, for systems with axial symmetry only $m = 0$ terms appear, and so the angular part of the solution can be expressed in terms of Legendre polynomials $P_\ell(\cos\theta)$. Since these functions form an orthogonal set let us expand both u and ρ in terms of them:

$$u = \sum_{\ell=0}^{\infty} R_\ell(r) P_\ell(\cos\theta), \tag{11.64}$$

$$\rho = \sum_{\ell=0}^{\infty} F_\ell(r) P_\ell(\cos\theta), \tag{11.65}$$

where the coefficients $R_\ell(r)$ and $F_\ell(r)$ in the Legendre polynomial expansions are functions of r. Since in any particular problem ρ is given, we can find the coefficients $F_\ell(r)$ in the expansion in the usual way (see Subsection 9.1.2). It then only remains to find the coefficients $R_\ell(r)$ in the expansion of the solution u.

9 Without looking at the calculated values, what should they be?

Writing ∇^2 in spherical polars and substituting (11.64) and (11.65) into (11.63) we obtain

$$\sum_{\ell=0}^{\infty}\left[\frac{P_\ell(\cos\theta)}{r^2}\frac{d}{dr}\left(r^2\frac{dR_\ell}{dr}\right)+\frac{R_\ell}{r^2\sin\theta}\frac{d}{d\theta}\left(\sin\theta\frac{dP_\ell(\cos\theta)}{d\theta}\right)\right]=\sum_{\ell=0}^{\infty}F_\ell(r)P_\ell(\cos\theta). \tag{11.66}$$

However, if, in equation (11.44) of our discussion of the angular part of the solution to Laplace's equation, we set $m = 0$, then we conclude that

$$\frac{1}{\sin\theta}\frac{d}{d\theta}\left(\sin\theta\frac{dP_\ell(\cos\theta)}{d\theta}\right)=-\ell(\ell+1)P_\ell(\cos\theta).$$

Substituting this into (11.66), we find that the LHS is greatly simplified and we obtain

$$\sum_{\ell=0}^{\infty}\left[\frac{1}{r^2}\frac{d}{dr}\left(r^2\frac{dR_\ell}{dr}\right)-\frac{\ell(\ell+1)R_\ell}{r^2}\right]P_\ell(\cos\theta)=\sum_{\ell=0}^{\infty}F_\ell(r)P_\ell(\cos\theta).$$

This relation is most easily satisfied by equating terms on both sides for each value of ℓ separately, so that for $\ell = 0, 1, 2, \ldots$ we have

$$\frac{1}{r^2}\frac{d}{dr}\left(r^2\frac{dR_\ell}{dr}\right)-\frac{\ell(\ell+1)R_\ell}{r^2}=F_\ell(r). \tag{11.67}$$

This is an ODE in which $F_\ell(r)$ is given, and it can therefore be solved for $R_\ell(r)$. The solution to Poisson's equation, u, is then obtained by making the superposition (11.64), as in the following worked example.

Example In a certain system, the electric charge density ρ is distributed as follows:

$$\rho=\begin{cases}Ar\cos\theta & \text{for } 0\le r<a,\\ 0 & \text{for } r\ge a.\end{cases}$$

Find the electrostatic potential inside and outside the charge distribution, given that both the potential and its radial derivative are continuous everywhere.

The electrostatic potential u satisfies

$$\nabla^2 u=\begin{cases}-(A/\epsilon_0)r\cos\theta & \text{for } 0\le r<a,\\ 0 & \text{for } r\ge a.\end{cases}$$

For $r < a$ the RHS can be written $-(A/\epsilon_0)rP_1(\cos\theta)$, and the coefficients in (11.65) are simply $F_1(r) = -(Ar/\epsilon_0)$ and $F_\ell(r) = 0$ for $\ell \neq 1$. Therefore we need only calculate $R_1(r)$, which satisfies (11.67) for $\ell = 1$:

$$\frac{1}{r^2}\frac{d}{dr}\left(r^2\frac{dR_1}{dr}\right)-\frac{2R_1}{r^2}=-\frac{Ar}{\epsilon_0}.$$

This can be rearranged to give

$$r^2R_1''+2rR_1'-2R_1=-\frac{Ar^3}{\epsilon_0},$$

where the prime denotes differentiation with respect to r. The LHS is homogeneous and the equation can be reduced by the substitution $r = \exp t$, and writing $R_1(r) = S(t)$, to

$$\ddot{S} + \dot{S} - 2S = -\frac{A}{\epsilon_0}\exp 3t, \qquad (11.68)$$

where the dots indicate differentiation with respect to t.

This is an inhomogeneous second-order ODE with constant coefficients and can be straightforwardly solved by the methods of Subsection 6.5.1 to give

$$S(t) = c_1 \exp t + c_2 \exp(-2t) - \frac{A}{10\epsilon_0}\exp 3t.$$

Recalling that $r = \exp t$ we find[10]

$$R_1(r) = c_1 r + c_2 r^{-2} - \frac{A}{10\epsilon_0} r^3.$$

Since we are interested in the region $r < a$ we must have $c_2 = 0$ for the solution to remain finite. Thus inside the charge distribution the electrostatic potential has the form

$$u_1(r, \theta, \phi) = \left(c_1 r - \frac{A}{10\epsilon_0} r^3\right) P_1(\cos\theta). \qquad (11.69)$$

Outside the charge distribution (for $r \geq a$), however, the electrostatic potential obeys Laplace's equation, $\nabla^2 u = 0$, and so given the symmetry of the problem and the requirement that $u \to \infty$ as $r \to \infty$ the solution must take the form

$$u_2(r, \theta, \phi) = \sum_{\ell=0}^{\infty} \frac{B_\ell}{r^{\ell+1}} P_\ell(\cos\theta). \qquad (11.70)$$

We can now use the boundary conditions at $r = a$ to fix the constants in (11.69) and (11.70). The requirement of continuity of the potential and its radial derivative at $r = a$ imply that

$$u_1(a, \theta, \phi) = u_2(a, \theta, \phi),$$

$$\frac{\partial u_1}{\partial r}(a, \theta, \phi) = \frac{\partial u_2}{\partial r}(a, \theta, \phi).$$

Since u_1 contains only an $\ell = 1$ term, the first continuity requirement means that $B_\ell = 0$ for $\ell \neq 1$. Carrying out the necessary differentiations and setting $r = a$ in (11.69) and (11.70) we obtain the simultaneous equations

$$c_1 a - \frac{A}{10\epsilon_0} a^3 = \frac{B_1}{a^2},$$

$$c_1 - \frac{3A}{10\epsilon_0} a^2 = -\frac{2B_1}{a^3},$$

which may be solved to give $c_1 = Aa^2/(6\epsilon_0)$ and $B_1 = Aa^5/(15\epsilon_0)$. Since $P_1(\cos\theta) = \cos\theta$, the electrostatic potentials inside and outside the charge distribution are given, respectively, by

$$u_1(r, \theta, \phi) = \frac{A}{\epsilon_0}\left(\frac{a^2 r}{6} - \frac{r^3}{10}\right)\cos\theta, \qquad u_2(r, \theta, \phi) = \frac{Aa^5}{15\epsilon_0}\frac{\cos\theta}{r^2}.$$

It should be noted that the calculation would have been very much longer if the given charge distribution had not had an angular variation that was the same as that of a single Legendre polynomial. ◀

10 Since the original equation for $R_1(r)$ is homogeneous, this result can be obtained more directly by trying Br^n for the CF and Cr^3 for the PI. Confirm the given result using this approach.

11.4 Integral transform methods

In the method of separation of variables our aim was to keep the independent variables in a PDE as separate as possible. We now discuss the use of integral transforms in solving PDEs, a method by which one of the independent variables can be eliminated from the differential coefficients. It will be assumed that the reader is familiar with Laplace and Fourier transforms and their properties, as discussed in Chapter 5.

The method consists simply of transforming the PDE into one containing derivatives with respect to a smaller number of variables. Thus, if the original equation has just two independent variables, it may be possible to reduce the PDE into a soluble ODE. The solution obtained can then (where possible) be transformed back to give the solution of the original PDE. As we shall see, boundary conditions can usually be incorporated in a natural way.

Which sort of transform to use, and the choice of the variable(s) with respect to which the transform is to be taken, is a matter of experience; we illustrate this in the example below. In practice, transforms can be taken with respect to each variable in turn, and the transformation that affords the greatest simplification can be pursued further. Our worked examples are based on the diffusion equation.

Example A semi-infinite tube of constant cross-section contains initially pure water. At time $t = 0$, one end of the tube is put into contact with a salt solution and maintained at a concentration u_0. Find the total amount of salt that has diffused into the tube after time t, if the diffusion constant is κ.

The concentration $u(x, t)$ at time t and distance x from the end of the tube satisfies the diffusion equation

$$\kappa \frac{\partial^2 u}{\partial x^2} = \frac{\partial u}{\partial t}, \qquad (11.71)$$

which has to be solved subject to the boundary conditions $u(0, t) = u_0$ for all t and $u(x, 0) = 0$ for all $x > 0$.

Since we are interested only in $t > 0$, the use of the Laplace transform is suggested. Furthermore, it will be recalled from Chapter 5 that one of the major virtues of Laplace transformations is the possibility they afford of replacing derivatives of functions by simple multiplication by a scalar. If the derivative with respect to time were so removed, equation (11.71) would contain only differentiation with respect to a single variable. Let us therefore take the Laplace transform of (11.71) with respect to t:

$$\int_0^\infty \kappa \frac{\partial^2 u}{\partial x^2} \exp(-st)\, dt = \int_0^\infty \frac{\partial u}{\partial t} \exp(-st)\, dt.$$

On the LHS the (double) differentiation is with respect to x, whereas the integration is with respect to the independent variable t. Therefore the derivative can be taken outside the integral. Denoting the Laplace transform of $u(x, t)$ by $\bar{u}(x, s)$ and using result (5.51) to rewrite the transform of the derivative on the RHS (or by integrating directly by parts), we obtain

$$\kappa \frac{\partial^2 \bar{u}}{\partial x^2} = s\bar{u}(x, s) - u(x, 0).$$

But from the boundary condition $u(x, 0) = 0$ the last term on the RHS vanishes, and the solution to the resulting second-order differential equation with x as the independent variable is immediate:

$$\bar{u}(x, s) = A \exp\left(\sqrt{\frac{s}{\kappa}}\, x\right) + B \exp\left(-\sqrt{\frac{s}{\kappa}}\, x\right),$$

where the constants A and B may depend on s.

We require $u(x, t) \to 0$ as $x \to \infty$ and so we must also have $\bar{u}(\infty, s) = 0$; consequently we require that $A = 0$. The value of B is determined by the need for $u(0, t) = u_0$ and hence that

$$\bar{u}(0, s) = \int_0^\infty u_0 \exp(-st)\, dt = \frac{u_0}{s}.$$

We thus conclude that $B = u_0/s$ and that the appropriate expression for the Laplace transform of $u(x, t)$ is

$$\bar{u}(x, s) = \frac{u_0}{s} \exp\left(-\sqrt{\frac{s}{\kappa}}\, x\right). \tag{11.72}$$

To obtain $u(x, t)$ from this result requires the inversion of this transform – a task that is generally difficult and requires a contour integration. This is discussed in Chapter 14, but for completeness we note here that the solution is

$$u(x, t) = u_0 \left[1 - \operatorname{erf}\left(\frac{x}{\sqrt{4\kappa t}}\right)\right],$$

where $\operatorname{erf}(x)$ is the error function discussed in Subsection 9.10.3. (The more complete sets of mathematical tables list this inverse Laplace transform.)

In the present problem, however, an alternative method is available. Let $w(t)$ be the amount of salt that has diffused into the tube in time t; then

$$w(t) = \int_0^\infty u(x, t)\, dx,$$

and its transform is given by

$$\begin{aligned}
\bar{w}(s) &= \int_0^\infty dt\, \exp(-st) \int_0^\infty u(x, t)\, dx \\
&= \int_0^\infty dx \int_0^\infty u(x, t) \exp(-st)\, dt \\
&= \int_0^\infty \bar{u}(x, s)\, dx.
\end{aligned}$$

Substituting for $\bar{u}(x, s)$ from (11.72) into the last integral and integrating,[11] we obtain

$$\bar{w}(s) = u_0 \kappa^{1/2} s^{-3/2}.$$

This expression is much simpler to invert, and referring to the table of standard Laplace transforms (Table 5.1) we find

$$w(t) = 2(\kappa/\pi)^{1/2} u_0 t^{1/2},$$

which is thus the total amount of salt in the tube at time t. ◀

11 Note that the integration is with respect to x, and so the awkward s-dependence of the integrand is not a problem.

The above example shows that in some circumstances the use of a Laplace transformation can greatly simplify the solution of a PDE. However, it will have been observed that (as with ODEs) the easy elimination of some derivatives is usually paid for by the introduction of a difficult inverse transformation. This problem, although still present, is less severe for Fourier transformations, as we now illustrate.

Example An infinite metal bar has an initial temperature distribution $f(x)$ along its length. Find the temperature distribution at a later time t.

We are interested in values of x from $-\infty$ to ∞, which suggests Fourier transformation with respect to x. Assuming that the solution obeys the boundary conditions $u(x,t) \to 0$ and $\partial u/\partial x \to 0$ as $|x| \to \infty$, we may Fourier-transform the one-dimensional diffusion equation (11.71) to obtain

$$\frac{\kappa}{\sqrt{2\pi}} \int_{-\infty}^{\infty} \frac{\partial^2 u(x,t)}{\partial x^2} \exp(-ikx)\, dx = \frac{1}{\sqrt{2\pi}} \frac{\partial}{\partial t} \int_{-\infty}^{\infty} u(x,t) \exp(-ikx)\, dx,$$

where on the RHS we have taken the partial derivative with respect to t outside the integral. Denoting the Fourier transform of $u(x,t)$ by $\widetilde{u}(k,t)$, and using equation (5.28) to rewrite the Fourier transform of the second derivative on the LHS, we then have

$$-\kappa k^2 \widetilde{u}(k,t) = \frac{\partial \widetilde{u}(k,t)}{\partial t}.$$

This first-order equation has the simple solution

$$\widetilde{u}(k,t) = \widetilde{u}(k,0) \exp(-\kappa k^2 t),$$

where the initial conditions give

$$\begin{aligned} \widetilde{u}(k,0) &= \frac{1}{\sqrt{2\pi}} \int_{-\infty}^{\infty} u(x,0) \exp(-ikx)\, dx \\ &= \frac{1}{\sqrt{2\pi}} \int_{-\infty}^{\infty} f(x) \exp(-ikx)\, dx = \widetilde{f}(k). \end{aligned}$$

Thus we may write the Fourier transform of the solution as

$$\widetilde{u}(k,t) = \widetilde{f}(k) \exp(-\kappa k^2 t) = \sqrt{2\pi}\, \widetilde{f}(k) \widetilde{G}(k,t), \tag{11.73}$$

where we have defined the function $\widetilde{G}(k,t) = (\sqrt{2\pi})^{-1} \exp(-\kappa k^2 t)$. Since $\widetilde{u}(k,t)$ can be written as the product of two Fourier transforms, we can use the convolution theorem, Subsection 5.3.2, to write the solution as

$$u(x,t) = \int_{-\infty}^{\infty} G(x-x',t) f(x')\, dx',$$

where $G(x,t)$ is the Green's function for this problem (see Subsection 6.5.5). This function is the inverse Fourier transform of $\widetilde{G}(k,t)$ and is thus given by

$$\begin{aligned} G(x,t) &= \frac{1}{2\pi} \int_{-\infty}^{\infty} \exp(-\kappa k^2 t) \exp(ikx)\, dk \\ &= \frac{1}{2\pi} \int_{-\infty}^{\infty} \exp\left[-\kappa t \left(k^2 - \frac{ix}{\kappa t} k\right)\right] dk. \end{aligned}$$

Completing the square in the integrand we find

$$\begin{aligned} G(x,t) &= \frac{1}{2\pi}\exp\left(-\frac{x^2}{4\kappa t}\right)\int_{-\infty}^{\infty}\exp\left[-\kappa t\left(k-\frac{ix}{2\kappa t}\right)^2\right]dk \\ &= \frac{1}{2\pi}\exp\left(-\frac{x^2}{4\kappa t}\right)\int_{-\infty}^{\infty}\exp\left(-\kappa t k'^2\right)dk' \\ &= \frac{1}{\sqrt{4\pi\kappa t}}\exp\left(-\frac{x^2}{4\kappa t}\right), \end{aligned}$$

where in the second line we have made the substitution $k' = k - ix/(2\kappa t)$, and in the last line we have used the standard result for the integral of a Gaussian. (Strictly speaking the change of variable from k to k' shifts the path of integration off the real axis, since k' is complex for real k, and so results in a complex integral, as will be discussed in Chapter 14. Nevertheless, in this case the path of integration can be shifted back to the real axis without affecting the value of the integral.)

Thus the temperature in the bar at a later time t is given by

$$u(x,t) = \frac{1}{\sqrt{4\pi\kappa t}}\int_{-\infty}^{\infty}\exp\left[-\frac{(x-x')^2}{4\kappa t}\right]f(x')\,dx', \qquad (11.74)$$

which may be evaluated (numerically if necessary) when the form of $f(x)$ is given. ◀

As we might expect from our discussion of Green's functions in Chapter 6, we see from (11.74) that, if the initial temperature distribution is $f(x) = \delta(x-a)$, i.e. a "point" source at $x = a$, then the temperature distribution at later times is simply given by

$$u(x,t) = G(x-a,t) = \frac{1}{\sqrt{4\pi\kappa t}}\exp\left[-\frac{(x-a)^2}{4\kappa t}\right].$$

The temperature at several later times is illustrated in Figure 11.10, which shows that the heat diffuses out from its initial position; the width of the Gaussian increases as $\sqrt{t}$, a dependence on time which is characteristic of diffusion processes.

The reader may have noticed that in both examples using integral transforms the solutions have been obtained in closed form – albeit in one case in the form of an integral. This differs from the infinite series solutions usually obtained via the separation of variables. It should be noted that this behavior is a result of the infinite range in x rather than of the transform method itself. In fact the method of separation of variables would yield the same solutions, since in the infinite-range case the separation constant is not restricted to take on an infinite set of discrete values but may have any real value, with the result that the sum over λ becomes an integral, as mentioned at the end of Section 11.2. Our final worked example in this section proves this assertion.

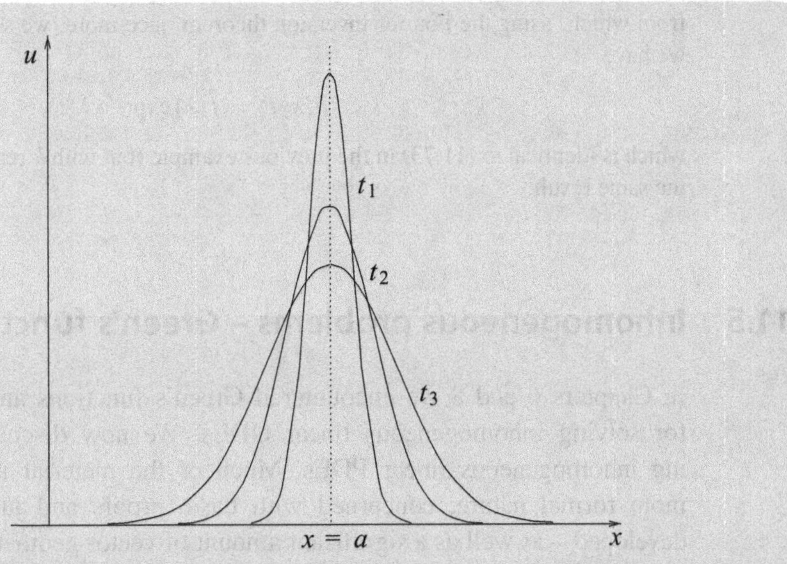

Figure 11.10 Diffusion of heat from a point source in a metal bar: the curves show the temperature u at position x for various times $t_1 < t_2 < t_3$. The area under the curves remains constant, since the total heat energy is conserved.

Example An infinite metal bar has an initial temperature distribution $f(x)$ along its length. Find the temperature distribution at a later time t using the method of separation of variables.

This is the same problem as in the previous example, but we now seek a solution by separating variables. From (11.12) a separated solution for the one-dimensional diffusion equation is given by

$$u(x,t) = [A\exp(i\lambda x) + B\exp(-i\lambda x)]\exp(-\kappa\lambda^2 t),$$

where $-\lambda^2$ is the separation constant. Since the bar is infinite we do not require the solution to take a given form at any finite value of x (for instance at $x = 0$) and so there is no restriction on λ other than its being real. Therefore instead of the superposition of such solutions in the form of a sum over allowed values of λ we have an integral over all λ,

$$u(x,t) = \frac{1}{\sqrt{2\pi}}\int_{-\infty}^{\infty} A(\lambda)\exp(-\kappa\lambda^2 t)\exp(i\lambda x)\,d\lambda, \tag{11.75}$$

where in taking λ from $-\infty$ to ∞ we need include only one of the complex exponentials; we have taken a factor $1/\sqrt{2\pi}$ out of $A(\lambda)$ for convenience. We can see from (11.75) that the expression for $u(x,t)$ has the form of an inverse Fourier transform (where λ is the transform variable). Therefore, Fourier-transforming both sides and using the Fourier inversion theorem, we find

$$\tilde{u}(\lambda,t) = A(\lambda)\exp(-\kappa\lambda^2 t).$$

Now, the initial boundary condition requires

$$u(x,0) = \frac{1}{\sqrt{2\pi}}\int_{-\infty}^{\infty} A(\lambda)\exp(i\lambda x)\,d\lambda = f(x),$$

from which, using the Fourier inversion theorem once more, we see that $A(\lambda) = \tilde{f}(\lambda)$. Therefore we have

$$\tilde{u}(\lambda, t) = \tilde{f}(\lambda) \exp(-\kappa\lambda^2 t),$$

which is identical to (11.73) in the previous example (but with k replaced by λ), and hence leads to the same result. ◀

11.5 Inhomogeneous problems – Green's functions

In Chapters 6 and 8 we encountered Green's functions and found them a useful tool for solving inhomogeneous linear ODEs. We now discuss their usefulness in solving inhomogeneous linear PDEs. Much of the material in this final section is of a more formal nature, concerned with basic proofs and justifications of the methods developed – as well as a significant amount of vector geometry – and it could be omitted on a first reading.

For the sake of brevity we shall again denote a linear PDE by

$$\mathcal{L}u(\mathbf{r}) = \rho(\mathbf{r}), \tag{11.76}$$

where $\mathcal{L}$ is a linear partial differential operator. For example, in Laplace's equation we have $\mathcal{L} = \nabla^2$, whereas for Helmholtz's equation $\mathcal{L} = \nabla^2 + k^2$. Note that we have not specified the dimensionality of the problem, and (11.76) may, for example, represent Poisson's equation in two or three (or more) dimensions. The reader will also notice that for the sake of simplicity we have not included any time dependence in (11.76). Nevertheless, the following discussion can be generalized to include it.

As we discussed in Subsection 10.3.2, a problem is inhomogeneous if the fact that $u(\mathbf{r})$ is a solution does *not* imply that any constant multiple $\lambda u(\mathbf{r})$ is also a solution. This inhomogeneity may derive from either the PDE itself or from the boundary conditions imposed on the solution.

In our discussion of Green's function solutions of inhomogeneous ODEs (see Subsection 6.5.5) we dealt with inhomogeneous boundary conditions by making a suitable change of variable such that in the new variable the boundary conditions were homogeneous. In an analogous way, as illustrated in the final example of Section 11.2, it is usually possible to make a change of variables in PDEs to eliminate inhomogeneity in the boundary conditions, but at the expense of introducing it into the equation. Therefore let us assume for the moment that the boundary conditions imposed on the solution $u(\mathbf{r})$ of (11.76) are homogeneous. This most commonly means that if we seek a solution to (11.76) in some region V then on the surface S that bounds V the solution obeys the conditions $u(\mathbf{r}) = 0$ or $\partial u/\partial n = 0$, where $\partial u/\partial n$ is the normal derivative of u at the surface S.

We shall discuss the extension of the Green's function method to the direct solution of problems with inhomogeneous boundary conditions in Subsection 11.5.2, but we first highlight how the Green's function approach to solving ODEs can be simply extended to PDEs for homogeneous boundary conditions.

11.5.1 Similarities to Green's functions for ODEs

As in the discussion of ODEs in Chapter 6, we may consider the Green's function for a system described by a PDE as the response of the system to a "unit impulse" or "point source". Thus if we seek a solution to (11.76) that satisfies some homogeneous boundary conditions on $u(\mathbf{r})$ then the Green's function $G(\mathbf{r}, \mathbf{r}_0)$ for the problem is a solution of

$$\mathcal{L}G(\mathbf{r}, \mathbf{r}_0) = \delta(\mathbf{r} - \mathbf{r}_0), \tag{11.77}$$

where $\mathbf{r}_0$ lies in V. The Green's function $G(\mathbf{r}, \mathbf{r}_0)$ must also satisfy the imposed (homogeneous) boundary conditions.

It is understood that in (11.77) the $\mathcal{L}$ operator expresses differentiation with respect to $\mathbf{r}$ as opposed to $\mathbf{r}_0$. Also, $\delta(\mathbf{r} - \mathbf{r}_0)$ is the Dirac delta function (see Chapter 5) of dimension appropriate to the problem; it may be thought of as representing a unit-strength point source at $\mathbf{r} = \mathbf{r}_0$.

Following an analogous argument to that given in Subsection 6.5.5 for ODEs, if the boundary conditions on $u(\mathbf{r})$ are homogeneous then a solution to (11.76) that satisfies the imposed boundary conditions is given by

$$u(\mathbf{r}) = \int G(\mathbf{r}, \mathbf{r}_0)\rho(\mathbf{r}_0)\, dV(\mathbf{r}_0), \tag{11.78}$$

where the integral on $\mathbf{r}_0$ is over some appropriate "volume". In two or more dimensions, however, the task of finding directly a solution to (11.77) that satisfies the imposed boundary conditions on S can be a difficult one, and we return to this in the next subsection.

An alternative approach is to follow a similar argument to that presented in Chapter 8 for ODEs and so to construct the Green's function for (11.76) as a superposition of eigenfunctions of the operator $\mathcal{L}$, provided $\mathcal{L}$ is Hermitian. By analogy with an ordinary differential operator, a partial differential operator is Hermitian if it satisfies

$$\int_V v^*(\mathbf{r})\mathcal{L}w(\mathbf{r})\, dV = \left[\int_V w^*(\mathbf{r})\mathcal{L}v(\mathbf{r})\, dV\right]^*,$$

where the asterisk denotes complex conjugation and v and w are arbitrary functions obeying the imposed (homogeneous) boundary condition on the solution of $\mathcal{L}u(\mathbf{r}) = 0$.

The eigenfunctions $u_n(\mathbf{r})$, $n = 0, 1, 2, \ldots$, of $\mathcal{L}$ satisfy

$$\mathcal{L}u_n(\mathbf{r}) = \lambda_n u_n(\mathbf{r}),$$

where λ_n are the corresponding eigenvalues, which are all real for an Hermitian operator $\mathcal{L}$. Furthermore, each eigenfunction must obey any imposed (homogeneous) boundary conditions. Using an argument analogous to that given in Chapter 8, the Green's function for the problem is given by

$$G(\mathbf{r}, \mathbf{r}_0) = \sum_{n=0}^{\infty} \frac{u_n(\mathbf{r})u_n^*(\mathbf{r}_0)}{\lambda_n}. \tag{11.79}$$

From (11.79) we see immediately that the Green's function (irrespective of how it is found) enjoys the property

$$G(\mathbf{r}, \mathbf{r}_0) = G^*(\mathbf{r}_0, \mathbf{r}).$$

Thus, if the Green's function is real then it is symmetric in its two arguments.

Once the Green's function has been obtained, the solution to (11.76) is again given by (11.78). For PDEs this approach can become very cumbersome, however, and so we shall not pursue it further here.

11.5.2 General boundary-value problems

As mentioned above, often inhomogeneous boundary conditions can be dealt with by making an appropriate change of variables, such that the boundary conditions in the new variables are homogeneous although the equation itself is generally inhomogeneous. In this section, however, we extend the use of Green's functions to problems with inhomogeneous boundary conditions (and equations). This provides a more consistent and intuitive approach to the solution of such *boundary-value problems*.

For definiteness we shall consider Poisson's equation

$$\nabla^2 u(\mathbf{r}) = \rho(\mathbf{r}), \tag{11.80}$$

but the material of this section may be extended to other linear PDEs of the form (11.76). Clearly, Poisson's equation reduces to Laplace's equation for $\rho(\mathbf{r}) = 0$ and so our discussion is equally applicable to this case.

We wish to solve (11.80) in some region V bounded by a surface S, which may consist of several disconnected parts. As stated above, we shall allow the possibility that the boundary conditions on the solution $u(\mathbf{r})$ may be inhomogeneous on S, although, as we shall see, this method reduces to those discussed above in the special case that the boundary conditions are in fact homogeneous.

The two common types of inhomogeneous boundary condition for Poisson's equation are (as discussed in Section 10.6):

(i) Dirichlet conditions, in which $u(\mathbf{r})$ is specified on S, and
(ii) Neumann conditions, in which $\partial u/\partial n$ is specified on S.

Generally, specifying *both* Dirichlet *and* Neumann conditions on S overdetermines the problem and leads to there being no solution.

The specification of the surface S requires some further comment, since S may have several disconnected parts. If we wish to solve Poisson's equation inside some closed surface S then the situation is straightforward and is shown in Figure 11.11(a). If, however, we wish to solve Poisson's equation in the gap between two closed surfaces (for example in the gap between two concentric conducting cylinders) then the volume V is bounded by a surface S that has two disconnected parts S_1 and S_2, as shown in Figure 11.11(b); the direction of the normal to the surface is always taken as pointing *out* of the volume V. A similar situation arises when we wish to solve Poisson's equation *outside* some closed surface S_1. In this case the volume V is infinite but is treated formally by taking the surface S_2 as a large sphere of radius R and letting R tend to infinity.

In order to solve (11.80) subject to either Dirichlet or Neumann boundary conditions on S, we will remind ourselves of Green's second theorem, equation (3.20), which states that, for two scalar functions $\phi(\mathbf{r})$ and $\psi(\mathbf{r})$ defined in some volume V bounded by a

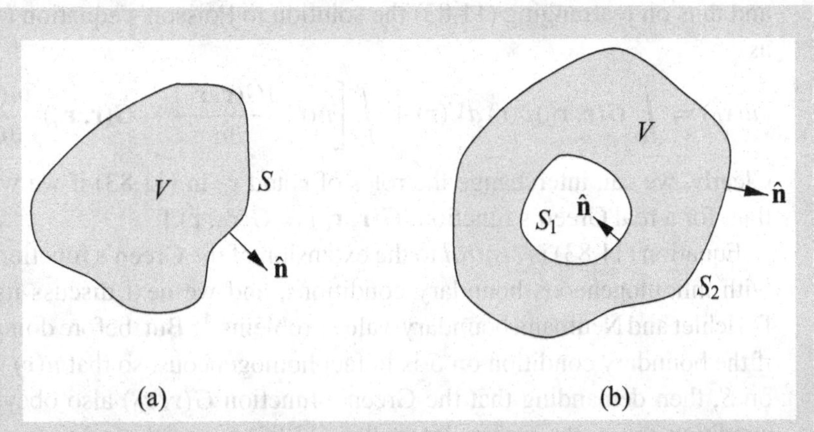

Figure 11.11 Surfaces used for solving Poisson's equation in different regions V.

surface S,

$$\int_V (\phi \nabla^2 \psi - \psi \nabla^2 \phi)\, dV = \int_S (\phi \nabla \psi - \psi \nabla \phi) \cdot \hat{\mathbf{n}}\, dS, \qquad (11.81)$$

where on the RHS it is common to write, for example, $\nabla \psi \cdot \hat{\mathbf{n}}\, dS$ as $(\partial \psi / \partial n)\, dS$. The expression $\partial \psi / \partial n$ stands for $\nabla \psi \cdot \hat{\mathbf{n}}$, the rate of change of ψ in the direction of the unit outward normal $\hat{\mathbf{n}}$ to the surface S.

The Green's function for Poisson's equation (11.80) must satisfy

$$\nabla^2 G(\mathbf{r}, \mathbf{r}_0) = \delta(\mathbf{r} - \mathbf{r}_0), \qquad (11.82)$$

where $\mathbf{r}_0$ lies in V. (As mentioned above, we may think of $G(\mathbf{r}, \mathbf{r}_0)$ as the solution to Poisson's equation for a unit-strength point source located at $\mathbf{r} = \mathbf{r}_0$.) Let us for the moment impose no boundary conditions on $G(\mathbf{r}, \mathbf{r}_0)$.

If we now let $\phi = u(\mathbf{r})$ and $\psi = G(\mathbf{r}, \mathbf{r}_0)$ in Green's theorem (11.81) then we obtain

$$\int_V \left[u(\mathbf{r}) \nabla^2 G(\mathbf{r}, \mathbf{r}_0) - G(\mathbf{r}, \mathbf{r}_0)\, \nabla^2 u(\mathbf{r}) \right] dV(\mathbf{r})$$

$$= \int_S \left[u(\mathbf{r}) \frac{\partial G(\mathbf{r}, \mathbf{r}_0)}{\partial n} - G(\mathbf{r}, \mathbf{r}_0) \frac{\partial u(\mathbf{r})}{\partial n} \right] dS(\mathbf{r}),$$

where we have made explicit that the volume and surface integrals are with respect to $\mathbf{r}$. Using (11.80) and (11.82) the LHS can be simplified to give

$$\int_V [u(\mathbf{r}) \delta(\mathbf{r} - \mathbf{r}_0) - G(\mathbf{r}, \mathbf{r}_0) \rho(\mathbf{r})]\, dV(\mathbf{r})$$

$$= \int_S \left[u(\mathbf{r}) \frac{\partial G(\mathbf{r}, \mathbf{r}_0)}{\partial n} - G(\mathbf{r}, \mathbf{r}_0) \frac{\partial u(\mathbf{r})}{\partial n} \right] dS(\mathbf{r}).$$

Since $\mathbf{r}_0$ lies within the volume V,

$$\int_V u(\mathbf{r}) \delta(\mathbf{r} - \mathbf{r}_0)\, dV(\mathbf{r}) = u(\mathbf{r}_0),$$

and thus on rearranging (11.83) the solution to Poisson's equation (11.80) can be written as

$$u(\mathbf{r}_0) = \int_V G(\mathbf{r}, \mathbf{r}_0)\rho(\mathbf{r})\, dV(\mathbf{r}) + \int_S \left[u(\mathbf{r}) \frac{\partial G(\mathbf{r}, \mathbf{r}_0)}{\partial n} - G(\mathbf{r}, \mathbf{r}_0) \frac{\partial u(\mathbf{r})}{\partial n} \right] dS(\mathbf{r}). \quad (11.83)$$

Clearly, we can interchange the roles of $\mathbf{r}$ and $\mathbf{r}_0$ in (11.83) if we wish. [Remember also that, for a real Green's function, $G(\mathbf{r}, \mathbf{r}_0) = G(\mathbf{r}_0, \mathbf{r})$.]

Equation (11.83) is *central* to the extension of the Green's function method to problems with inhomogeneous boundary conditions, and we next discuss its application to both Dirichlet and Neumann boundary-value problems.[12] But, before doing so, we also note that if the boundary condition on S is in fact homogeneous, so that $u(\mathbf{r}) = 0$ or $\partial u(\mathbf{r})/\partial n = 0$ on S, then demanding that the Green's function $G(\mathbf{r}, \mathbf{r}_0)$ also obeys the same boundary condition causes the surface integral in (11.83) to vanish, and we are left with the familiar form of solution given in (11.78).

11.5.3 Dirichlet problems

In a Dirichlet problem we require the solution $u(\mathbf{r})$ of Poisson's equation (11.80) to take specific values on some surface S that bounds V, i.e. we require that $u(\mathbf{r}) = f(\mathbf{r})$ on S where f is a given function.

If we seek a Green's function $G(\mathbf{r}, \mathbf{r}_0)$ for this problem it must clearly satisfy (11.82), but we are free to choose the boundary conditions satisfied by $G(\mathbf{r}, \mathbf{r}_0)$ in such a way as to make the solution (11.83) as simple as possible. From (11.83), we see that by choosing

$$G(\mathbf{r}, \mathbf{r}_0) = 0 \quad \text{for } \mathbf{r} \text{ on } S \quad (11.84)$$

the second term in the surface integral vanishes. Since $u(\mathbf{r}) = f(\mathbf{r})$ on S, (11.83) then becomes

$$u(\mathbf{r}_0) = \int_V G(\mathbf{r}, \mathbf{r}_0)\rho(\mathbf{r})\, dV(\mathbf{r}) + \int_S f(\mathbf{r}) \frac{\partial G(\mathbf{r}, \mathbf{r}_0)}{\partial n}\, dS(\mathbf{r}). \quad (11.85)$$

Thus we wish to find the *Dirichlet Green's function* that

(i) satisfies (11.82) and hence is singular at $\mathbf{r} = \mathbf{r}_0$, and
(ii) obeys the boundary condition $G(\mathbf{r}, \mathbf{r}_0) = 0$ for $\mathbf{r}$ on S.

In general, it is difficult to obtain this function directly, and so it is useful to separate these two requirements. We therefore look for a solution of the form

$$G(\mathbf{r}, \mathbf{r}_0) = F(\mathbf{r}, \mathbf{r}_0) + H(\mathbf{r}, \mathbf{r}_0),$$

where $F(\mathbf{r}, \mathbf{r}_0)$ satisfies (11.82) and has the required singular character at $\mathbf{r} = \mathbf{r}_0$ but does not necessarily obey the boundary condition on S, whilst $H(\mathbf{r}, \mathbf{r}_0)$ satisfies the

12 The equation looks formidable, but it can be broken down into three parts representing different physical contributions to the potential at $\mathbf{r}_0$: the first term gives that due to the element of charge $\rho\, dV$ positioned at $\mathbf{r}$; the second gives the effect of the (prescribed) potential at the bounding surface, propagated (weighted) by the derivative of the Green's function at that surface; and the third gives the effect of the (prescribed) derivative of the potential at the bounding surface, propagated (weighted) by the Green's function at that surface. The minus sign in the final term can be thought of as arising from the fact that $\mathbf{r}_0$ is inside the surface, but the normal derivative is in the direction of the outward normal.

corresponding homogeneous equation (i.e. Laplace's equation) inside V but is adjusted in such a way that the sum $G(\mathbf{r}, \mathbf{r}_0)$ equals zero on S. The Green's function $G(\mathbf{r}, \mathbf{r}_0)$ is still a solution of (11.82) since

$$\nabla^2 G(\mathbf{r}, \mathbf{r}_0) = \nabla^2 F(\mathbf{r}, \mathbf{r}_0) + \nabla^2 H(\mathbf{r}, \mathbf{r}_0) = \nabla^2 F(\mathbf{r}, \mathbf{r}_0) + 0 = \delta(\mathbf{r} - \mathbf{r}_0).$$

The function $F(\mathbf{r}, \mathbf{r}_0)$ is called the *fundamental solution* and will clearly take different forms depending on the dimensionality of the problem. Let us first consider the fundamental solution to (11.82) in three dimensions.

Example Find the fundamental solution to Poisson's equation in three dimensions that tends to zero as $|\mathbf{r}| \to \infty$.

We wish to solve

$$\nabla^2 F(\mathbf{r}, \mathbf{r}_0) = \delta(\mathbf{r} - \mathbf{r}_0) \tag{11.86}$$

in three dimensions, subject to the boundary condition $F(\mathbf{r}, \mathbf{r}_0) \to 0$ as $|\mathbf{r}| \to \infty$. Since the problem is spherically symmetric about $\mathbf{r}_0$, let us consider a large sphere S of radius R centered on $\mathbf{r}_0$, and integrate (11.86) over the enclosed volume V. We then obtain

$$\int_V \nabla^2 F(\mathbf{r}, \mathbf{r}_0)\, dV = \int_V \delta(\mathbf{r} - \mathbf{r}_0)\, dV = 1, \tag{11.87}$$

since V encloses the point $\mathbf{r}_0$. However, using the divergence theorem,

$$\int_V \nabla^2 F(\mathbf{r}, \mathbf{r}_0)\, dV = \int_S \nabla F(\mathbf{r}, \mathbf{r}_0) \cdot \hat{\mathbf{n}}\, dS, \tag{11.88}$$

where $\hat{\mathbf{n}}$ is the unit (outward) normal to the large sphere S at any point.

Since the problem is spherically symmetric about $\mathbf{r}_0$, we expect that

$$F(\mathbf{r}, \mathbf{r}_0) = F(|\mathbf{r} - \mathbf{r}_0|) = F(r),$$

i.e. that F has the same value everywhere on S. Thus, evaluating the surface integral in (11.88) and equating it to unity from (11.87), we have[13]

$$4\pi r^2 \frac{dF}{dr}\bigg|_{r=R} = 1.$$

Integrating this expression we obtain

$$F(r) = -\frac{1}{4\pi r} + \text{constant},$$

but, since we require $F(\mathbf{r}, \mathbf{r}_0) \to 0$ as $|\mathbf{r}| \to \infty$, the constant must be zero. The fundamental solution in three dimensions is consequently given by

$$F(\mathbf{r}, \mathbf{r}_0) = -\frac{1}{4\pi |\mathbf{r} - \mathbf{r}_0|}. \tag{11.89}$$

This is clearly also the full Green's function for Poisson's equation subject to the boundary condition $u(\mathbf{r}) \to 0$ as $|\mathbf{r}| \to \infty$. ◀

13 A vertical bar to the right of an expression is a common alternative to enclosing the expression in square brackets; as usual, the subscript shows the value of the variable at which the expression is to be evaluated.

Using (11.89) we can write down the solution of Poisson's equation to find, for example, the electrostatic potential $u(\mathbf{r})$ due to some distribution of electric charge $\rho(\mathbf{r})$. The electrostatic potential satisfies

$$\nabla^2 u(\mathbf{r}) = -\frac{\rho}{\epsilon_0},$$

where $u(\mathbf{r}) \to 0$ as $|\mathbf{r}| \to \infty$. Since the boundary condition on the surface at infinity is homogeneous the surface integral in (11.85) vanishes, and using (11.89) in the one remaining term we recover the familiar solution

$$u(\mathbf{r}_0) = \int \frac{\rho(\mathbf{r})}{4\pi\epsilon_0 |\mathbf{r} - \mathbf{r}_0|} \, dV(\mathbf{r}), \tag{11.90}$$

where the volume integral is over all space.

We can develop an analogous theory in two dimensions. As before the fundamental solution satisfies

$$\nabla^2 F(\mathbf{r}, \mathbf{r}_0) = \delta(\mathbf{r} - \mathbf{r}_0), \tag{11.91}$$

where $\delta(\mathbf{r} - \mathbf{r}_0)$ is now the two-dimensional delta function. Following an analogous method to that used in the previous example, we find the fundamental solution in two dimensions to be given by

$$F(\mathbf{r}, \mathbf{r}_0) = \frac{1}{2\pi} \ln |\mathbf{r} - \mathbf{r}_0| + \text{constant}. \tag{11.92}$$

From the form of the solution we see that in two dimensions we cannot apply the condition $F(\mathbf{r}, \mathbf{r}_0) \to 0$ as $|\mathbf{r}| \to \infty$, and in this case the constant does not necessarily vanish.

We now return to the task of constructing the full Dirichlet Green's function. To do so we wish to add to the fundamental solution a solution of the homogeneous equation (in this case Laplace's equation) such that $G(\mathbf{r}, \mathbf{r}_0) = 0$ on S, as required by (11.85) and its attendant conditions. The appropriate Green's function is constructed by adding to the fundamental solution "copies" of itself that represent "image" sources at different locations *outside* V. Hence this approach is called the *method of images*.

In summary, if we wish to solve Poisson's equation in some region V subject to Dirichlet boundary conditions on its surface S then the procedure and argument are as follows.

(i) To the single source $\delta(\mathbf{r} - \mathbf{r}_0)$ inside V add image sources *outside* V

$$\sum_{n=1}^{N} q_n \delta(\mathbf{r} - \mathbf{r}_n) \quad \text{with } \mathbf{r}_n \text{ outside } V,$$

where the positions $\mathbf{r}_n$ and the strengths q_n of the image sources are to be determined as described in step (iii) below.

(ii) Since all the image sources lie outside V, the fundamental solution corresponding to each source satisfies Laplace's equation *inside* V. Thus we may add the fundamental solutions $F(\mathbf{r}, \mathbf{r}_n)$ corresponding to each image source to that corresponding to the single source inside V, obtaining the Green's function

$$G(\mathbf{r}, \mathbf{r}_0) = F(\mathbf{r}, \mathbf{r}_0) + \sum_{n=1}^{N} q_n F(\mathbf{r}, \mathbf{r}_n).$$

(iii) Now adjust the positions $\mathbf{r}_n$ and strengths q_n of the image sources so that the required boundary conditions are satisfied on S. For a Dirichlet Green's function we require $G(\mathbf{r}, \mathbf{r}_0) = 0$ for $\mathbf{r}$ on S.

(iv) The solution to Poisson's equation subject to the Dirichlet boundary condition $u(\mathbf{r}) = f(\mathbf{r})$ on S is then given by (11.85).

In general it is very difficult to find the correct positions and strengths for the images, i.e. to make them such that the boundary conditions on S are satisfied. Nevertheless, it is possible to do so for certain problems that have simple geometry. In particular, for problems in which the boundary S consists of straight lines (in two dimensions) or planes (in three dimensions), positions of the image points can be deduced simply by imagining the boundary lines or planes to be mirrors in which the single source in V (at $\mathbf{r}_0$) is reflected.

Example Solve Laplace's equation $\nabla^2 u = 0$ in three dimensions in the half-space $z > 0$, given that $u(\mathbf{r}) = f(\mathbf{r})$ on the plane $z = 0$.

The surface S bounding V consists of the xy-plane and the surface at infinity.[14] Therefore, the Dirichlet Green's function for this problem must satisfy $G(\mathbf{r}, \mathbf{r}_0) = 0$ on $z = 0$ and $G(\mathbf{r}, \mathbf{r}_0) \to 0$ as $|\mathbf{r}| \to \infty$. Thus it is clear in this case that we require one image source at the position $\mathbf{r}_1$ that is the reflection of $\mathbf{r}_0$ in the plane $z = 0$, as shown in Figure 11.12. This means that $\mathbf{r}_1$ lies in $z < 0$, i.e. as required, outside the region in which we wish to obtain a solution. It is also clear that, to make the net potential zero on the plane, the strength of this image should be -1.

Therefore by adding the fundamental solutions corresponding to the original source and its image we obtain the Green's function

$$G(\mathbf{r}, \mathbf{r}_0) = -\frac{1}{4\pi|\mathbf{r} - \mathbf{r}_0|} + \frac{1}{4\pi|\mathbf{r} - \mathbf{r}_1|}, \tag{11.93}$$

where $\mathbf{r}_1$ is the reflection of $\mathbf{r}_0$ in the plane $z = 0$, i.e. if $\mathbf{r}_0 = (x_0, y_0, z_0)$ then $\mathbf{r}_1 = (x_0, y_0, -z_0)$. Clearly $G(\mathbf{r}, \mathbf{r}_0) \to 0$ as $|\mathbf{r}| \to \infty$ as required. Also, by design, $G(\mathbf{r}, \mathbf{r}_0) = 0$ on $z = 0$, and so (11.93) is the desired Dirichlet Green's function.

The solution to Laplace's equation is then given by (11.85) with $\rho(\mathbf{r}) = 0$,

$$u(\mathbf{r}_0) = \int_S f(\mathbf{r}) \frac{\partial G(\mathbf{r}, \mathbf{r}_0)}{\partial n}\, dS(\mathbf{r}). \tag{11.94}$$

Clearly the surface at infinity makes no contribution to this integral. The outward-pointing unit vector normal to the xy-plane is simply $\hat{\mathbf{n}} = -\mathbf{k}$ (where $\mathbf{k}$ is the unit vector in the z-direction), and so

$$\frac{\partial G(\mathbf{r}, \mathbf{r}_0)}{\partial n} = -\frac{\partial G(\mathbf{r}, \mathbf{r}_0)}{\partial z} = -\mathbf{k} \cdot \nabla G(\mathbf{r}, \mathbf{r}_0).$$

We may evaluate this normal derivative by writing the Green's function (11.93) explicitly in terms of x, y and z (and x_0, y_0 and z_0) and calculating the partial derivative with respect to z directly. It

14 Most easily thought of in this case as a hemisphere of very large radius, centered on the origin and lying in the region $z > 0$.

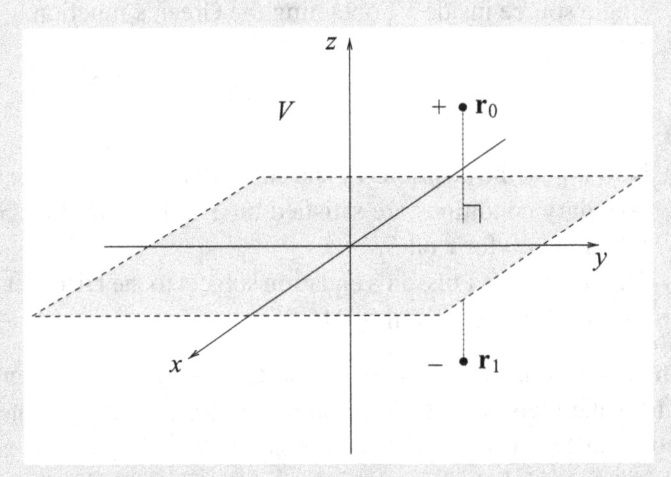

Figure 11.12 The arrangement of images for solving Laplace's equation in the half-space $z > 0$.

is usually quicker, however, to use the fact that[15]

$$\nabla|\mathbf{r}-\mathbf{r}_0| = \frac{\mathbf{r}-\mathbf{r}_0}{|\mathbf{r}-\mathbf{r}_0|}; \tag{11.95}$$

thus

$$\nabla G(\mathbf{r},\mathbf{r}_0) = \frac{\mathbf{r}-\mathbf{r}_0}{4\pi|\mathbf{r}-\mathbf{r}_0|^3} - \frac{\mathbf{r}-\mathbf{r}_1}{4\pi|\mathbf{r}-\mathbf{r}_1|^3}.$$

Since $\mathbf{r}_0 = (x_0, y_0, z_0)$ and $\mathbf{r}_1 = (x_0, y_0, -z_0)$ the normal derivative is given by

$$-\frac{\partial G(\mathbf{r},\mathbf{r}_0)}{\partial z} = -\mathbf{k}\cdot\nabla G(\mathbf{r},\mathbf{r}_0)$$

$$= -\frac{z-z_0}{4\pi|\mathbf{r}-\mathbf{r}_0|^3} + \frac{z+z_0}{4\pi|\mathbf{r}-\mathbf{r}_1|^3}.$$

Therefore on the surface $z = 0$, and writing out the dependence on x, y and z explicitly, we have

$$-\left.\frac{\partial G(\mathbf{r},\mathbf{r}_0)}{\partial z}\right|_{z=0} = \frac{2z_0}{4\pi[(x-x_0)^2+(y-y_0)^2+z_0^2]^{3/2}}.$$

Inserting this expression into (11.94) we obtain the solution,

$$u(x_0, y_0, z_0) = \frac{z_0}{2\pi}\int_{-\infty}^{\infty}\int_{-\infty}^{\infty}\frac{f(x,y)}{[(x-x_0)^2+(y-y_0)^2+z_0^2]^{3/2}}\,dx\,dy,$$

in the form of an integral that can be evaluated numerically, if necessary, once an explicit form for $f(x, y)$ is given.[16] ◀

15 Since $|\mathbf{r}-\mathbf{r}_0|^2 = (\mathbf{r}-\mathbf{r}_0)\cdot(\mathbf{r}-\mathbf{r}_0)$ we have $\nabla|\mathbf{r}-\mathbf{r}_0|^2 = 2(\mathbf{r}-\mathbf{r}_0)$, from which we obtain

$$\nabla(|\mathbf{r}-\mathbf{r}_0|^2)^{1/2} = \frac{1}{2}\frac{2(\mathbf{r}-\mathbf{r}_0)}{(|\mathbf{r}-\mathbf{r}_0|^2)^{1/2}} = \frac{\mathbf{r}-\mathbf{r}_0}{|\mathbf{r}-\mathbf{r}_0|}.$$

Note that this result holds in two *and* three dimensions.

16 Complete this calculation to find u on the z-axis if $f(x, y) = V_0$ for $x^2 + y^2 < a^2$, but is zero elsewhere on the $z = 0$ plane. Deduce the potential at a distance z_0 from an infinite plane raised to a potential V_0.

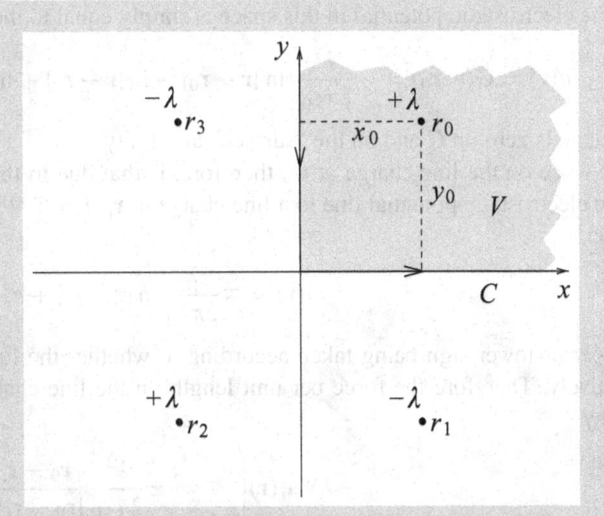

Figure 11.13 The arrangement of images for finding the force on a line charge situated in the (two-dimensional) quarter-space $x > 0$, $y > 0$, when the planes $x = 0$ and $y = 0$ are earthed.

An analogous procedure may be applied in two-dimensional problems. For example, in solving Poisson's equation in two dimensions in the half-space $x > 0$ we again require just one image charge, of strength $q_1 = -1$, at a position $\mathbf{r}_1$ that is the reflection of $\mathbf{r}_0$ in the line $x = 0$. Since we require $G(\mathbf{r}, \mathbf{r}_0) = 0$ when $\mathbf{r}$ lies on $x = 0$, the constant in (11.92) must equal zero, and so the Dirichlet Green's function is

$$G(\mathbf{r}, \mathbf{r}_0) = \frac{1}{2\pi}\left(\ln|\mathbf{r} - \mathbf{r}_0| - \ln|\mathbf{r} - \mathbf{r}_1|\right).$$

Clearly $G(\mathbf{r}, \mathbf{r}_0)$ tends to zero as $|\mathbf{r}| \to \infty$. If, however, we wish to solve the two-dimensional Poisson equation in the quarter-space $x > 0$, $y > 0$, then more image points are required, as the next example shows.

Example A line charge in the z-direction of charge density λ is placed at some position $\mathbf{r}_0$ in the quarter-space $x > 0$, $y > 0$. Calculate the force per unit length on the line charge due to the presence of thin earthed plates along $x = 0$ and $y = 0$.

Here we wish to solve Poisson's equation,

$$\nabla^2 u = -\frac{\lambda}{\epsilon_0}\delta(\mathbf{r} - \mathbf{r}_0),$$

in the quarter-space $x > 0$, $y > 0$. It is clear that we require three image line charges with positions and strengths as shown in Figure 11.13 (all of which lie outside the region in which we seek a solution). The boundary condition that the electrostatic potential u is zero on $x = 0$ and $y = 0$ (shown as the "curve" C in Figure 11.13) is then automatically satisfied, and so this system of image charges is directly equivalent, so far as the quarter-space $x > 0$, $y > 0$ is concerned, to the original situation of a single line charge in the presence of the earthed plates along $x = 0$ and $y = 0$.

Thus the electrostatic potential in this space is simply equal to the Dirichlet Green's function

$$u(\mathbf{r}) = G(\mathbf{r}, \mathbf{r}_0) = -\frac{\lambda}{2\pi\epsilon_0}(\ln|\mathbf{r}-\mathbf{r}_0| - \ln|\mathbf{r}-\mathbf{r}_1| + \ln|\mathbf{r}-\mathbf{r}_2| - \ln|\mathbf{r}-\mathbf{r}_3|),$$

which equals zero on C and on the "surface" at infinity.

The force on the line charge at $\mathbf{r}_0$, therefore, is that due to the three line charges at $\mathbf{r}_1$, $\mathbf{r}_2$ and $\mathbf{r}_3$. The electrostatic potential due to a line charge at $\mathbf{r}_i$, $i = 1, 2$ or 3, is given by the fundamental solution

$$u_i(\mathbf{r}) = \mp\frac{\lambda}{2\pi\epsilon_0}\ln|\mathbf{r}-\mathbf{r}_i| + c,$$

the upper or lower sign being taken according to whether the line charge is positive or negative, respectively. Therefore the force per unit length on the line charge at $\mathbf{r}_0$, due to the one at $\mathbf{r}_i$, is given by

$$-\lambda\nabla u_i(\mathbf{r})\bigg|_{\mathbf{r}=\mathbf{r}_0} = \pm\frac{\lambda^2}{2\pi\epsilon_0}\frac{\mathbf{r}_0-\mathbf{r}_i}{|\mathbf{r}_0-\mathbf{r}_i|^2}.$$

Adding the contributions from the three image charges shown in Figure 11.13, the total force experienced by the line charge at $\mathbf{r}_0$ is given by

$$\mathbf{F} = \frac{\lambda^2}{2\pi\epsilon_0}\left(-\frac{\mathbf{r}_0-\mathbf{r}_1}{|\mathbf{r}_0-\mathbf{r}_1|^2} + \frac{\mathbf{r}_0-\mathbf{r}_2}{|\mathbf{r}_0-\mathbf{r}_2|^2} - \frac{\mathbf{r}_0-\mathbf{r}_3}{|\mathbf{r}_0-\mathbf{r}_3|^2}\right),$$

where, from the figure, $\mathbf{r}_0 - \mathbf{r}_1 = 2y_0\mathbf{j}$, $\mathbf{r}_0 - \mathbf{r}_2 = 2x_0\mathbf{i} + 2y_0\mathbf{j}$ and $\mathbf{r}_0 - \mathbf{r}_3 = 2x_0\mathbf{i}$. Thus, in terms of x_0 and y_0, the total force on the line charge due to the charge induced on the plates is given by

$$\begin{aligned}\mathbf{F} &= \frac{\lambda^2}{2\pi\epsilon_0}\left(-\frac{1}{2y_0}\mathbf{j} + \frac{2x_0\,\mathbf{i} + 2y_0\,\mathbf{j}}{4x_0^2 + 4y_0^2} - \frac{1}{2x_0}\mathbf{i}\right)\\ &= -\frac{\lambda^2}{4\pi\epsilon_0(x_0^2+y_0^2)}\left(\frac{y_0^2}{x_0}\mathbf{i} + \frac{x_0^2}{y_0}\mathbf{j}\right).\end{aligned}$$

The reader should attempt to interpret this more general result in the special cases when $x_0 \gg y_0$, or vice-versa. ◀

Further generalizations are possible. For instance, solving Poisson's equation in the two-dimensional strip $-\infty < x < \infty$, $0 < y < b$ requires an infinite series of image points.

So far we have considered problems in which the boundary S consists of straight lines (in two dimensions) or planes (in three dimensions), in which simple reflections of the source at $\mathbf{r}_0$ in these boundaries fix the positions of the image points. For more complicated (curved) boundaries this is no longer possible, and finding the appropriate position(s) and strength(s) of the image source(s) requires further work.

Example Use the method of images to find the Dirichlet Green's function for solving Poisson's equation outside a sphere of radius a centered at the origin.

We need to find a solution of Poisson's equation valid outside the sphere of radius a. Since an image point $\mathbf{r}_1$ cannot lie in this region, it must be located within the sphere. The Green's function for this

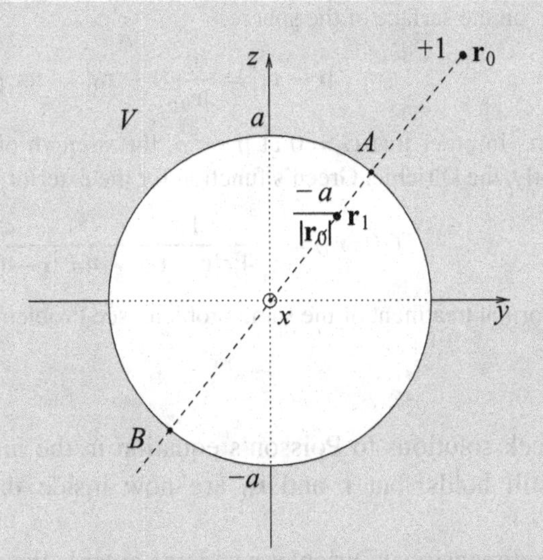

Figure 11.14 The arrangement of images for solving Poisson's equation outside a sphere of radius a centered at the origin. For a charge $+1$ at $\mathbf{r}_0$, the image point $\mathbf{r}_1$ is given by $(a/|\mathbf{r}_0|)^2\mathbf{r}_0$ and the strength of the image charge is $-a/|\mathbf{r}_0|$.

problem is therefore

$$G(\mathbf{r}, \mathbf{r}_0) = -\frac{1}{4\pi|\mathbf{r} - \mathbf{r}_0|} - \frac{q}{4\pi|\mathbf{r} - \mathbf{r}_1|},$$

where $|\mathbf{r}_0| > a$, $|\mathbf{r}_1| < a$ and q is the strength of the image which we have yet to determine. Clearly, $G(\mathbf{r}, \mathbf{r}_0) \to 0$ on the surface at infinity.

By symmetry we expect the image point $\mathbf{r}_1$ to lie on the same radial line as the original source, $\mathbf{r}_0$, as shown in Figure 11.14, and so $\mathbf{r}_1 = k\mathbf{r}_0$ where $k < 1$. However, for a Dirichlet Green's function we require $G(\mathbf{r} - \mathbf{r}_0) = 0$ on $|\mathbf{r}| = a$, and the form of the Green's function suggests that we need

$$|\mathbf{r} - \mathbf{r}_0| \propto |\mathbf{r} - \mathbf{r}_1| \quad \text{for all } |\mathbf{r}| = a. \tag{11.96}$$

Referring to Figure 11.14, if this relationship is to hold over the whole surface of the sphere, then it must certainly hold for the points A and B. We thus require

$$\frac{|\mathbf{r}_0| - a}{a - |\mathbf{r}_1|} = \frac{|\mathbf{r}_0| + a}{a + |\mathbf{r}_1|},$$

which reduces to $|\mathbf{r}_1| = a^2/|\mathbf{r}_0|$. Therefore the image point must be located at the position

$$\mathbf{r}_1 = \frac{a^2}{|\mathbf{r}_0|^2}\mathbf{r}_0.$$

It may now be checked that, for this location of the image point, (11.96) is satisfied over the whole sphere, and not just on the radial line through $\mathbf{r}_0$. Using the geometrical result

$$\begin{aligned}|\mathbf{r} - \mathbf{r}_1|^2 &= |\mathbf{r}|^2 - \frac{2a^2}{|\mathbf{r}_0|^2}\mathbf{r}\cdot\mathbf{r}_0 + \frac{a^4}{|\mathbf{r}_0|^2} \\ &= \frac{a^2}{|\mathbf{r}_0|^2}\left(|\mathbf{r}_0|^2 - 2\mathbf{r}\cdot\mathbf{r}_0 + a^2\right) \quad \text{for } |\mathbf{r}| = a,\end{aligned} \tag{11.97}$$

we see that, on the surface of the sphere,

$$|\mathbf{r}-\mathbf{r}_1| = \frac{a}{|\mathbf{r}_0|}|\mathbf{r}-\mathbf{r}_0| \qquad \text{for } |\mathbf{r}| = a. \tag{11.98}$$

Therefore, in order that $G = 0$ at $|\mathbf{r}| = a$, the strength of the image charge must be $-a/|\mathbf{r}_0|$. Consequently, the Dirichlet Green's function for the exterior of the sphere is

$$G(\mathbf{r}, \mathbf{r}_0) = -\frac{1}{4\pi|\mathbf{r}-\mathbf{r}_0|} + \frac{a/|\mathbf{r}_0|}{4\pi\left|\mathbf{r}-(a^2/|\mathbf{r}_0|^2)\mathbf{r}_0\right|}.$$

For a less formal treatment of the same problem, see Problem 11.22. ◀

If we seek solutions to Poisson's equation in the *interior* of a sphere then the above analysis still holds, but $\mathbf{r}$ and $\mathbf{r}_0$ are now inside the sphere and the image $\mathbf{r}_1$ lies outside it.

For two-dimensional Dirichlet problems outside the circle $|\mathbf{r}| = a$, we are led by arguments similar to those employed previously to use the same image point as in the three-dimensional case, namely

$$\mathbf{r}_1 = \frac{a^2}{|\mathbf{r}_0|^2}\mathbf{r}_0. \tag{11.99}$$

As illustrated below, however, it is usually necessary to take the image strength as -1 in two-dimensional problems.

Example Solve Laplace's equation in the two-dimensional region $|\mathbf{r}| \leq a$, subject to the boundary condition $u = f(\phi)$ on $|\mathbf{r}| = a$.

In this case we wish to find the Dirichlet Green's function in the interior of a disc of radius a, so the image charge must lie outside the disc. Taking the strength of the image to be -1, we have

$$G(\mathbf{r}, \mathbf{r}_0) = \frac{1}{2\pi}\ln|\mathbf{r}-\mathbf{r}_0| - \frac{1}{2\pi}\ln|\mathbf{r}-\mathbf{r}_1| + c, \tag{11.100}$$

where $\mathbf{r}_1 = (a^2/|\mathbf{r}_0|^2)\mathbf{r}_0$ lies outside the disc, and c is a constant that includes the strength of the image charge and does not necessarily equal zero.

Since we require $G(\mathbf{r}, \mathbf{r}_0) = 0$ when $|\mathbf{r}| = a$, the value of the constant c is determined by (11.98) as $-\ln(\mathbf{r}_0/a)$,[17] and the Dirichlet Green's function for this problem is given by

$$G(\mathbf{r}, \mathbf{r}_0) = \frac{1}{2\pi}\left(\ln|\mathbf{r}-\mathbf{r}_0| - \ln\left|\mathbf{r} - \frac{a^2}{|\mathbf{r}_0|^2}\mathbf{r}_0\right| - \ln\frac{|\mathbf{r}_0|}{a}\right). \tag{11.101}$$

17 To confirm that its value is as given, take $\mathbf{r}$ as the specific point $\mathbf{r} = a\mathbf{r}_0/|\mathbf{r}_0|$ on the disc. For this point, all the vectors appearing in (11.100) are collinear and the magnitudes of their differences are easily determined.

Using plane polar coordinates, the solution to the boundary-value problem can be written as a line integral around the circle $\rho = a$:

$$u(\mathbf{r}_0) = \int_C f(\mathbf{r}) \frac{\partial G(\mathbf{r}, \mathbf{r}_0)}{\partial n} \, dl$$
$$= \int_0^{2\pi} f(\mathbf{r}) \left. \frac{\partial G(\mathbf{r}, \mathbf{r}_0)}{\partial \rho} \right|_{\rho = a} a \, d\phi. \tag{11.102}$$

The normal derivative of the Green's function (11.101) is given by

$$\frac{\partial G(\mathbf{r}, \mathbf{r}_0)}{\partial \rho} = \frac{\mathbf{r}}{|\mathbf{r}|} \cdot \nabla G(\mathbf{r}, \mathbf{r}_0)$$
$$= \frac{\mathbf{r}}{2\pi |\mathbf{r}|} \cdot \left(\frac{\mathbf{r} - \mathbf{r}_0}{|\mathbf{r} - \mathbf{r}_0|^2} - \frac{\mathbf{r} - \mathbf{r}_1}{|\mathbf{r} - \mathbf{r}_1|^2} \right). \tag{11.103}$$

Using the fact that $\mathbf{r}_1 = (a^2/|\mathbf{r}_0|^2)\mathbf{r}_0$ and the geometrical result (11.98), we find that

$$\left. \frac{\partial G(\mathbf{r}, \mathbf{r}_0)}{\partial \rho} \right|_{\rho = a} = \frac{a^2 - |\mathbf{r}_0|^2}{2\pi a |\mathbf{r} - \mathbf{r}_0|^2}.$$

In plane polar coordinates, $\mathbf{r} = \rho \cos\phi \, \mathbf{i} + \rho \sin\phi \, \mathbf{j}$ and $\mathbf{r}_0 = \rho_0 \cos\phi_0 \, \mathbf{i} + \rho_0 \sin\phi_0 \, \mathbf{j}$, and so

$$\left. \frac{\partial G(\mathbf{r}, \mathbf{r}_0)}{\partial \rho} \right|_{\rho = a} = \left(\frac{1}{2\pi a} \right) \frac{a^2 - \rho_0^2}{a^2 + \rho_0^2 - 2a\rho_0 \cos(\phi - \phi_0)}.$$

On substituting into (11.102), we obtain

$$u(\rho_0, \phi_0) = \frac{1}{2\pi} \int_0^{2\pi} \frac{(a^2 - \rho_0^2) f(\phi) \, d\phi}{a^2 + \rho_0^2 - 2a\rho_0 \cos(\phi - \phi_0)}, \tag{11.104}$$

which is the solution to the problem. ◀

11.5.4 Neumann problems

In a Neumann problem we require the normal derivative of the solution of Poisson's equation to take on specific values on some surface S that bounds V, i.e. we require $\partial u(\mathbf{r})/\partial n = f(\mathbf{r})$ on S, where f is a given function. As we shall see, much of our discussion of Dirichlet problems can be immediately taken over into the solution of Neumann problems.

As we proved in Section 10.6 of the previous chapter, specifying Neumann boundary conditions determines the relevant solution of Poisson's equation to within an (unimportant) additive constant. However, unlike Dirichlet conditions, Neumann conditions impose a self-consistency requirement. In order for a solution u to exist, it is necessary that the following consistency condition holds:

$$\int_S f \, dS = \int_S \nabla u \cdot \hat{\mathbf{n}} \, dS = \int_V \nabla^2 u \, dV = \int_V \rho \, dV, \tag{11.105}$$

where we have used the divergence theorem to convert the surface integral into a volume integral. As a physical example, the integral of the normal component of an electric field over a surface bounding a given volume cannot be chosen arbitrarily when the charge inside the volume has already been specified (Gauss's theorem).

Let us again consider (11.83), which is central to our discussion of Green's functions in inhomogeneous problems. It reads

$$u(\mathbf{r}_0) = \int_V G(\mathbf{r}, \mathbf{r}_0)\rho(\mathbf{r})\, dV(\mathbf{r}) + \int_S \left[u(\mathbf{r}) \frac{\partial G(\mathbf{r}, \mathbf{r}_0)}{\partial n} - G(\mathbf{r}, \mathbf{r}_0) \frac{\partial u(\mathbf{r})}{\partial n} \right] dS(\mathbf{r}).$$

As always, the Green's function must obey

$$\nabla^2 G(\mathbf{r}, \mathbf{r}_0) = \delta(\mathbf{r} - \mathbf{r}_0),$$

where $\mathbf{r}_0$ lies in V. In the solution of Dirichlet problems in the previous subsection, we chose the Green's function to obey the boundary condition $G(\mathbf{r}, \mathbf{r}_0) = 0$ on S and, in a similar way, we might wish to choose $\partial G(\mathbf{r}, \mathbf{r}_0)/\partial n = 0$ in the solution of Neumann problems. However, in general this is *not* permitted since the Green's function must obey the consistency condition

$$\int_S \frac{\partial G(\mathbf{r}, \mathbf{r}_0)}{\partial n}\, dS = \int_S \nabla G(\mathbf{r}, \mathbf{r}_0) \cdot \hat{\mathbf{n}}\, dS = \int_V \nabla^2 G(\mathbf{r}, \mathbf{r}_0)\, dV = 1.$$

The simplest permitted boundary condition is therefore

$$\frac{\partial G(\mathbf{r}, \mathbf{r}_0)}{\partial n} = \frac{1}{A} \qquad \text{for } \mathbf{r} \text{ on } S,$$

where A is the area of the surface S; this defines a *Neumann Green's function*.

If we require $\partial u(\mathbf{r})/\partial n = f(\mathbf{r})$ on S, the solution to Poisson's equation is given by

$$\begin{aligned} u(\mathbf{r}_0) &= \int_V G(\mathbf{r}, \mathbf{r}_0)\rho(\mathbf{r})\, dV(\mathbf{r}) + \frac{1}{A}\int_S u(\mathbf{r})\, dS(\mathbf{r}) - \int_S G(\mathbf{r}, \mathbf{r}_0) f(\mathbf{r})\, dS(\mathbf{r}) \\ &= \int_V G(\mathbf{r}, \mathbf{r}_0)\rho(\mathbf{r})\, dV(\mathbf{r}) + \langle u(\mathbf{r})\rangle_S - \int_S G(\mathbf{r}, \mathbf{r}_0) f(\mathbf{r})\, dS(\mathbf{r}), \end{aligned} \qquad (11.106)$$

where $\langle u(\mathbf{r})\rangle_S$ is the average of u over the surface S and is a freely specifiable constant. For Neumann problems in which the volume V is bounded by a surface S at infinity, we do not need the $\langle u(\mathbf{r})\rangle_S$ term. For example, if we wish to solve a Neumann problem outside the unit sphere centered at the origin then $r > a$ is the region V throughout which we require the solution; this region may be considered as being bounded by two disconnected surfaces, the surface of the sphere and a surface at infinity. By requiring that $u(\mathbf{r}) \to 0$ as $|\mathbf{r}| \to \infty$, the term $\langle u(\mathbf{r})\rangle_S$ becomes zero.

As mentioned above, much of our discussion of Dirichlet problems can be taken over into the solution of Neumann problems. In particular, we may use the method of images to find the appropriate Neumann Green's function.

Example Solve Laplace's equation in the two-dimensional region $|\mathbf{r}| \leq a$ subject to the boundary condition $\partial u/\partial n = f(\phi)$ on $|\mathbf{r}| = a$, with $\int_0^{2\pi} f(\phi)\,d\phi = 0$ as required by the consistency condition (11.105).

Let us assume, as in Dirichlet problems with this geometry, that a single image charge is placed outside the circle at

$$\mathbf{r}_1 = \frac{a^2}{|\mathbf{r}_0|^2}\mathbf{r}_0,$$

where $\mathbf{r}_0$ is the position of the source inside the circle [see equation (11.99)]. Then, from (11.98), we have the useful geometrical result

$$|\mathbf{r} - \mathbf{r}_1| = \frac{a}{|\mathbf{r}_0|}|\mathbf{r} - \mathbf{r}_0| \quad \text{for } |\mathbf{r}| = a. \tag{11.107}$$

Leaving the strength q of the image as a parameter, the Green's function has the form

$$G(\mathbf{r}, \mathbf{r}_0) = \frac{1}{2\pi}\left(\ln|\mathbf{r} - \mathbf{r}_0| + q\ln|\mathbf{r} - \mathbf{r}_1| + c\right). \tag{11.108}$$

Using plane polar coordinates, the radial (i.e. normal) derivative of this function is given by

$$\begin{aligned}\frac{\partial G(\mathbf{r}, \mathbf{r}_0)}{\partial \rho} &= \frac{\mathbf{r}}{|\mathbf{r}|}\cdot\nabla G(\mathbf{r}, \mathbf{r}_0)\\ &= \frac{\mathbf{r}}{2\pi|\mathbf{r}|}\cdot\left[\frac{\mathbf{r} - \mathbf{r}_0}{|\mathbf{r} - \mathbf{r}_0|^2} + \frac{q(\mathbf{r} - \mathbf{r}_1)}{|\mathbf{r} - \mathbf{r}_1|^2}\right].\end{aligned}$$

Using (11.107) to write $|\mathbf{r} - \mathbf{r}_1|$ in terms of $|\mathbf{r} - \mathbf{r}_0|$, on the perimeter of the circle $\rho = a$ the radial derivative takes the form

$$\begin{aligned}\left.\frac{\partial G(\mathbf{r}, \mathbf{r}_0)}{\partial \rho}\right|_{\rho=a} &= \frac{1}{2\pi|\mathbf{r}|}\left[\frac{|\mathbf{r}|^2 - \mathbf{r}\cdot\mathbf{r}_0}{|\mathbf{r} - \mathbf{r}_0|^2} + \frac{q|\mathbf{r}|^2 - q(a^2/|\mathbf{r}_0|^2)\mathbf{r}\cdot\mathbf{r}_0}{(a^2/|\mathbf{r}_0|^2)|\mathbf{r} - \mathbf{r}_0|^2}\right]\\ &= \frac{1}{2\pi a}\frac{1}{|\mathbf{r} - \mathbf{r}_0|^2}\left[|\mathbf{r}|^2 + q|\mathbf{r}_0|^2 - (1+q)\mathbf{r}\cdot\mathbf{r}_0\right],\end{aligned}$$

where we have set $|\mathbf{r}|^2 = a^2$ in the second term on the RHS, but not in the first. If we take $q = 1$, the radial derivative simplifies to

$$\left.\frac{\partial G(\mathbf{r}, \mathbf{r}_0)}{\partial \rho}\right|_{\rho=a} = \frac{1}{2\pi a},$$

or $1/L$, where L is the circumference, and so (11.108) with $q = 1$ is the required Neumann Green's function.

Since $\rho(\mathbf{r}) = 0$, the solution to our boundary-value problem is now given by (11.106) as

$$u(\mathbf{r}_0) = \langle u(\mathbf{r})\rangle_C - \int_C G(\mathbf{r}, \mathbf{r}_0)f(\mathbf{r})\,dl(\mathbf{r}),$$

where the integral is around the circumference of the circle C. In plane polar coordinates $\mathbf{r} = \rho\cos\phi\,\mathbf{i} + \rho\sin\phi\,\mathbf{j}$ and $\mathbf{r}_0 = \rho_0\cos\phi_0\,\mathbf{i} + \rho_0\sin\phi_0\,\mathbf{j}$, and again using (11.107) we find that on C the Green's function is given by

$$G(\mathbf{r},\mathbf{r}_0)|_{\rho=a} = \frac{1}{2\pi}\left[\ln|\mathbf{r}-\mathbf{r}_0| + \ln\left(\frac{a}{|\mathbf{r}_0|}|\mathbf{r}-\mathbf{r}_0|\right) + c\right]$$

$$= \frac{1}{2\pi}\left(\ln|\mathbf{r}-\mathbf{r}_0|^2 + \ln\frac{a}{|\mathbf{r}_0|} + c\right)$$

$$= \frac{1}{2\pi}\left\{\ln\left[a^2+\rho_0^2-2a\rho_0\cos(\phi-\phi_0)\right] + \ln\frac{a}{\rho_0} + c\right\}. \qquad (11.109)$$

Since $dl = a\,d\phi$ on C, the solution to the problem is given by

$$u(\rho_0,\phi_0) = \langle u\rangle_C - \frac{a}{2\pi}\int_0^{2\pi} f(\phi)\ln[a^2+\rho_0^2-2a\rho_0\cos(\phi-\phi_0)]\,d\phi.$$

The contributions of the final two terms in the Green's function (11.109) vanish because $\int_0^{2\pi} f(\phi)\,d\phi = 0$. The average value of u around the circumference, $\langle u\rangle_C$, is a freely specifiable constant as we would expect for a Neumann problem. This result should be compared with the result (11.104) for the corresponding Dirichlet problem, but it should be remembered that in the one case $f(\phi)$ is a potential, and in the other the gradient of a potential. ◀

SUMMARY

1. *The method of separation of variables*
   To solve a PDE for $u(x_i)$ in which there are n independent variables $x_1, x_2, \ldots, x_n$, assume a solution of the form $u(x_i) = X_1(x_1)X_2(x_2)\cdots X_n(x_n)$, substitute it, and arrange the resulting equation in the form

$$\sum_{i=1}^{n} \frac{1}{X_i(x_i)}\mathcal{L}_i X_i(x_i) = 0,$$

   where $\mathcal{L}_k$ is a differential operator that depends only on x_k (and does not involve any other of the x_i).

   Then, for each i, set $\dfrac{1}{X_i}\mathcal{L}_i X_i = \lambda_i$, where the λ_i are a set of *separation constants* satisfying $\sum_{i=1}^{n}\lambda_i = 0$, but are otherwise arbitrary; if any of the $\lambda_i = 0$, then special (usually simplifying) conditions apply. The equations $\mathcal{L}_i X_i = \lambda_i X_i$ are now a set of ODEs that can be solved subject to any given boundary conditions, including ones that require only the finiteness of the $y_i(x)$.

2. *The most general solutions of* $\nabla^2 u = 0$
   In all cases, the summations indicated are carried out only over those values of the summation index that make the corresponding functions satisfy the boundary

conditions. For example, in 3D cylindrical polars, the sum over k is typically limited to those values of k for which $J_m(ka) = 0$ for some given a.

- 2D Cartesian:
$$u(x,y) = \sum_{\lambda} \left(A_\lambda e^{\lambda x} + B_\lambda e^{-\lambda x}\right)(C_\lambda \cos \lambda y + D_\lambda \sin \lambda y).$$
- 2D plane polar:
$$u(\rho,\phi) = C_0 \ln \rho + D_0 + \sum_{n=1}^{\infty}(C_n \rho^n + D_n \rho^{-n})(A_n \cos n\phi + B_n \sin n\phi).$$
- 3D cylindrical polar:
$$\begin{aligned} u(\rho,\phi,z) = \sum_{k,m} &[A_{km} J_m(k\rho) + B_{km} Y_m(k\rho)] \\ &\times [C_{km} \cos m\phi + D_{km} \sin m\phi][E_{km} e^{-kz} + F_{km} e^{kz}]. \end{aligned}$$
- 3D spherical polar:
$$\begin{aligned} u(r,\theta,\phi) = \sum_{\ell,m} &(A_{\ell m} r^{\ell} + B_{\ell m} r^{-(\ell+1)}) \\ &\times [E_{\ell m} P_\ell^m(\cos\theta) + F_{\ell m} Q_\ell^m(\cos\theta)] \\ &\times (C_{\ell m} \cos m\phi + D_{\ell m} \sin m\phi), \end{aligned}$$
with $\ell \geq 0$ and $-\ell \leq m \leq \ell$.

3. *Typical solutions of $\nabla^2 = 0$ for regions that include/exclude the origin and that have $\nabla u \to 0$ at infinity, contain no singularities, and (where relevant) have axial symmetry*

Coordinates	Include the origin	Exclude the origin
2D Cartesian	$\sum_{\lambda\geq 0}(A_\lambda \cos \lambda y + B_\lambda \sin \lambda y)e^{-\lambda x}$	the same
	$\sum_{\lambda\geq 0}(A_\lambda \cos \lambda x + B_\lambda \sin \lambda x)e^{-\lambda y}$	the same
2D polar	$\sum_{n=0}^{\infty}(A_n \cos n\theta + B_n \sin n\theta)\rho^n$	$C_0 \ln \rho + \sum_{n=0}^{\infty}(A_n \cos n\theta + B_n \sin n\theta)\rho^{-n}$
3D cylindrical polar	$\sum_k A_k J_0(k\rho)e^{-kz}$	$\sum_k [A_k J_0(k\rho) + B_k Y_0(k\rho)]e^{-kz}$
3D spherical polar	$\sum_{\ell=0}^{\infty} A_\ell r^\ell P_\ell(\cos\theta)$	$\sum_{\ell=0}^{\infty} \frac{A_\ell}{r^{\ell+1}} P_\ell(\cos\theta)$

The expansion coefficients can sometimes be found by calculating the solution directly for a particular subset of the independent variables, and then invoking the uniqueness theorem to find them for the full range of values of the variables.

4. *Some solutions, without singularities, of other equations*
   - 3D Wave equation $\nabla^2 u(\mathbf{r}) - \dfrac{1}{c^2}\dfrac{\partial^2 u(\mathbf{r})}{\partial t^2} = 0$

$$u(x, y, z, t) = \sum_{\ell,m,n} A_{\ell m n} e^{\pm i\ell x} e^{\pm imy} e^{\pm inz} e^{\pm ic\mu t}$$

   where $\ell^2 + m^2 + n^2 = \mu^2$.
   - 1D Diffusion equation $\kappa \nabla^2 u = \dfrac{\partial u}{\partial t}$

$$u(x, t) = \sum_{\lambda} (a_\lambda \cos \lambda x + B_\lambda \sin \lambda x) e^{-\lambda^2 \kappa t}.$$

   - Helmholtz equation $\nabla^2 u + k^2 u = 0$:
     (i) Plane polars

$$u(\rho, \phi) = \sum_m (A_m \cos m\phi + B_m \sin m\phi) J_m(k\rho).$$

     (ii) Cylindrical polars

$$u(\rho, \phi, z) = \sum_{m,\alpha} J_m(\sqrt{k^2 - \alpha^2}\rho)(A_{m\alpha} \cos m\phi + B_{m\alpha} \sin m\phi) \times (C_{m\alpha} e^{i\alpha z} + D_{m\alpha} e^{-i\alpha z}).$$

     (iii) Spherical polars

$$u(r, \theta, \phi) = \frac{1}{\sqrt{r}} \sum_{\ell, m} J_{\ell+1/2}(kr) P_\ell^m(\cos\theta)(A_{\ell m} \cos m\phi + B_{\ell m} \sin m\phi).$$

5. *Inhomogeneous PDEs*
   An inhomogeneous equation can sometimes be tackled by expanding both the solution and the RHS of the PDE as a superposition of the orthogonal functions (of one of the independent variables) that form the solutions of the corresponding homogeneous equation. The multipliers (functions of the other independent variables) of the functions that appear in the expansion can then separately be set equal on the two sides of the equation. This procedure yields differential equations satisfied by those multipliers.

6. *Green's functions* $G(\mathbf{r}', \mathbf{r})$
   With $\mathcal{L}u(\mathbf{r}) = \rho(\mathbf{r})$ as the equation to be solved, if $G(\mathbf{r}', \mathbf{r})$ satisfies $\mathcal{L}G(\mathbf{r}', \mathbf{r}) = \delta(\mathbf{r}' - \mathbf{r})$, as well as the imposed homogeneous boundary conditions, then the solution is

$$u(\mathbf{r}) = \int G(\mathbf{r}', \mathbf{r}) \rho(\mathbf{r}')\, dV(\mathbf{r}').$$

For inhomogeneous boundary conditions, with $u(\mathbf{r}')$ and/or $\partial u(\mathbf{r}')/\partial n$ given on the boundary S, this generalizes to

$$u(\mathbf{r}) = \int_V G(\mathbf{r}',\mathbf{r})\rho(\mathbf{r}')\,dV(\mathbf{r}') + \int_S \left[u(\mathbf{r}')\frac{\partial G(\mathbf{r}',\mathbf{r})}{\partial n} - G(\mathbf{r}',\mathbf{r})\frac{\partial u(\mathbf{r}')}{\partial n}\right] dS(\mathbf{r}').$$

- For Dirichlet boundary conditions, $u(\mathbf{r}') = f(\mathbf{r}')$ on S, make $G(\mathbf{r}',\mathbf{r}) = 0$ for $\mathbf{r}'$ on S. Then

$$u(\mathbf{r}) = \int_V G(\mathbf{r}',\mathbf{r})\rho(\mathbf{r}')\,dV(\mathbf{r}') + \int_S f(\mathbf{r}')\frac{\partial G(\mathbf{r}',\mathbf{r})}{\partial n}\,dS(\mathbf{r}').$$

- For Neumann boundary conditions, $\partial u(\mathbf{r}')/\partial n = g(\mathbf{r}')$ on S, make $\partial G(\mathbf{r}',\mathbf{r})/\partial n = 1/A$ for $\mathbf{r}'$ on S, where A is the area of S. Then

$$u(\mathbf{r}) = \int_V G(\mathbf{r}',\mathbf{r})\rho(\mathbf{r}')\,dV(\mathbf{r}') + \langle u(\mathbf{r}')\rangle_S - \int_S G(\mathbf{r}',\mathbf{r})g(\mathbf{r}')\,dS(\mathbf{r}'),$$

where the constant $\langle u(\mathbf{r}')\rangle_S$ is the average value of u over S.
- One possible method of constructing $G(\mathbf{r}',\mathbf{r})$ is the *method of images*. See pp. 466 and following, for the steps in the method and several worked examples.
- The fundamental solutions of Poisson's equation needed for image construction are

$$\text{in 3D, } F(\mathbf{r}',\mathbf{r}) = -\frac{1}{4\pi|\mathbf{r}'-\mathbf{r}|}; \quad \text{in 2D, } F(\mathbf{r}',\mathbf{r}) = \frac{1}{2\pi}\ln|\mathbf{r}'-\mathbf{r}| + c.$$

- If $\mathcal{L}$ is Hermitian and has eigenfunctions $u_n(\mathbf{r})$ satisfying $\mathcal{L}u_n(\mathbf{r}) = \lambda_n u_n(\mathbf{r})$, then a formal, but not very useful, construction for $G(\mathbf{r}',\mathbf{r})$ is

$$G(\mathbf{r}',\mathbf{r}) = \sum_{n=0}^{\infty} \frac{u_n(\mathbf{r}')\,u_n^*(\mathbf{r})}{\lambda_n}.$$

7. *Transform methods*
By taking the Laplace or Fourier transform of a PDE, all functions of one of the independent variables, and all derivatives with respect to it, can be removed, replacing them with algebraic functions of the transform variable so introduced. This reduces the number of independent variables involved in differential operators by one, and, in some cases, reduces the PDE to an ODE.

PROBLEMS

11.1. Solve the following first-order partial differential equations by separating the variables:

$$\text{(a)}\ \frac{\partial u}{\partial x} - x\frac{\partial u}{\partial y} = 0; \qquad \text{(b)}\ x\frac{\partial u}{\partial x} - 2y\frac{\partial u}{\partial y} = 0.$$

11.2. A cube, made of material whose conductivity is k, has as its six faces the planes $x = \pm a$, $y = \pm a$ and $z = \pm a$, and contains no internal heat sources. Verify that the temperature distribution

$$u(x, y, z, t) = A \cos \frac{\pi x}{a} \sin \frac{\pi z}{a} \exp\left(-\frac{2\kappa\pi^2 t}{a^2}\right)$$

obeys the appropriate diffusion equation. Across which faces is there heat flow? What is the direction and rate of heat flow at the point $(3a/4, a/4, a)$ at time $t = a^2/(\kappa\pi^2)$?

11.3. The wave equation describing the transverse vibrations of a stretched membrane under tension T and having a uniform surface density ρ is

$$T\left(\frac{\partial^2 u}{\partial x^2} + \frac{\partial^2 u}{\partial y^2}\right) = \rho \frac{\partial^2 u}{\partial t^2}.$$

Find a separable solution appropriate to a membrane stretched on a frame of length a and width b, showing that the natural angular frequencies of such a membrane are given by

$$\omega^2 = \frac{\pi^2 T}{\rho}\left(\frac{n^2}{a^2} + \frac{m^2}{b^2}\right),$$

where n and m are any positive integers.

11.4. Schrödinger's equation for a non-relativistic particle in a constant potential region can be taken as

$$-\frac{\hbar^2}{2m}\left(\frac{\partial^2 u}{\partial x^2} + \frac{\partial^2 u}{\partial y^2} + \frac{\partial^2 u}{\partial z^2}\right) = i\hbar \frac{\partial u}{\partial t}.$$

(a) Find a solution, separable in the four independent variables, that can be written in the form of a plane wave,

$$\psi(x, y, z, t) = A \exp[i(\mathbf{k} \cdot \mathbf{r} - \omega t)].$$

Using the relationships associated with de Broglie ($\mathbf{p} = \hbar\mathbf{k}$) and Einstein ($E = \hbar\omega$), show that the separation constants must be such that

$$p_x^2 + p_y^2 + p_z^2 = 2mE.$$

(b) Obtain a different separable solution describing a particle confined to a box of side a (ψ must vanish at the walls of the box). Show that the energy of the particle can only take the quantized values

$$E = \frac{\hbar^2\pi^2}{2ma^2}(n_x^2 + n_y^2 + n_z^2),$$

where n_x, n_y, n_z are integers.

11.5. Denoting the three terms of ∇^2 in spherical polars by ∇_r^2, ∇_θ^2, ∇_ϕ^2 in an obvious way, evaluate $\nabla_r^2 u$, etc. for the two functions given below and verify that, in each case, although the individual terms are not necessarily zero their sum $\nabla^2 u$ is zero. Identify the corresponding values of ℓ and m.

(a) $u(r,\theta,\phi) = \left(Ar^2 + \dfrac{B}{r^3}\right)\dfrac{3\cos^2\theta - 1}{2}$.

(b) $u(r,\theta,\phi) = \left(Ar + \dfrac{B}{r^2}\right)\sin\theta\exp i\phi$.

11.6. Prove that the expression given in equation (11.47) for the associated Legendre function $P_\ell^m(\mu)$ satisfies the appropriate equation, (11.45), as follows.
(a) Evaluate $dP_\ell^m(\mu)/d\mu$ and $d^2P_\ell^m(\mu)/d\mu^2$, using the forms given in (11.47), and substitute them into (11.45).
(b) Differentiate Legendre's equation m times using Leibnitz' theorem.
(c) Show that the equations obtained in (a) and (b) are multiples of each other, and hence that the validity of (b) implies that of (a).

11.7. Continue the analysis of Problem 2.16, concerned with the flow of a very viscous fluid past a sphere, to find the full expression for the stream function $\psi(r,\theta)$. At the surface of the sphere $r = a$, the velocity field $\mathbf{u} = \mathbf{0}$, whilst far from the sphere $\psi \simeq (Ur^2\sin^2\theta)/2$.

Show that $f(r)$ can be expressed as a superposition of powers of r, and determine which powers give acceptable solutions. Hence show that

$$\psi(r,\theta) = \frac{U}{4}\left(2r^2 - 3ar + \frac{a^3}{r}\right)\sin^2\theta.$$

11.8. The motion of a very viscous fluid in the two-dimensional (wedge) region $-\alpha < \phi < \alpha$ can be described, in (ρ,ϕ), coordinates by the (biharmonic) equation

$$\nabla^2\nabla^2\psi \equiv \nabla^4\psi = 0,$$

together with the boundary conditions $\partial\psi/\partial\phi = 0$ at $\phi = \pm\alpha$, which represent the fact that there is no radial fluid velocity close to either of the bounding walls because of the viscosity, and $\partial\psi/\partial\rho = \pm\rho$ at $\phi = \pm\alpha$, which impose the condition that azimuthal flow increases linearly with r along any radial line. Assuming a solution in separated-variable form, show that the full expression for ψ is

$$\psi(\rho,\phi) = \frac{\rho^2}{2}\frac{\sin 2\phi - 2\phi\cos 2\alpha}{\sin 2\alpha - 2\alpha\cos 2\alpha}.$$

11.9. A circular disc of radius a is heated in such a way that its perimeter $\rho = a$ has a steady temperature distribution $A + B\cos^2\phi$, where ρ and ϕ are plane polar coordinates and A and B are constants. Find the temperature $T(\rho,\phi)$ everywhere in the region $\rho < a$.

11.10. Consider possible solutions of Laplace's equation inside a circular domain as follows.

(a) Find the solution in plane polar coordinates $\rho,\ \phi$ that takes the value $+1$ for $0 < \phi < \pi$ and the value -1 for $-\pi < \phi < 0$, when $\rho = a$.

(b) For a point (x, y) on or inside the circle $x^2 + y^2 = a^2$, identify the angles α and β defined by

$$\alpha = \tan^{-1}\frac{y}{a+x} \qquad \text{and} \qquad \beta = \tan^{-1}\frac{y}{a-x}.$$

Show that $u(x, y) = (2/\pi)(\alpha + \beta)$ is a solution of Laplace's equation that satisfies the boundary conditions given in (a).

(c) Deduce a Fourier series expansion for the function

$$\tan^{-1}\frac{\sin\phi}{1+\cos\phi} + \tan^{-1}\frac{\sin\phi}{1-\cos\phi}.$$

11.11. The free transverse vibrations of a thick rod satisfy the equation

$$a^4\frac{\partial^4 u}{\partial x^4} + \frac{\partial^2 u}{\partial t^2} = 0.$$

Obtain a solution in separated-variable form and, for a rod clamped at one end, $x = 0$, and free at the other, $x = L$, show that the angular frequency of vibration ω satisfies

$$\cosh\left(\frac{\omega^{1/2}L}{a}\right) = -\sec\left(\frac{\omega^{1/2}L}{a}\right).$$

[At a clamped end both u and $\partial u/\partial x$ vanish, whilst at a free end, where there is no bending moment, $\partial^2 u/\partial x^2$ and $\partial^3 u/\partial x^3$ are both zero.]

11.12. A membrane is stretched between two concentric rings of radii a and b $(b > a)$. If the smaller ring is transversely distorted from the planar configuration by an amount $c|\phi|$, $-\pi \le \phi \le \pi$, show that the membrane then has a shape given by

$$u(\rho, \phi) = \frac{c\pi}{2}\frac{\ln(b/\rho)}{\ln(b/a)} - \frac{4c}{\pi}\sum_{m\ \text{odd}}\frac{a^m}{m^2(b^{2m} - a^{2m})}\left(\frac{b^{2m}}{\rho^m} - \rho^m\right)\cos m\phi.$$

11.13. A string of length L, fixed at its two ends, is plucked at its mid-point by an amount A and then released. Prove that the subsequent displacement is given by

$$u(x,t) = \sum_{n=0}^{\infty}\frac{8A\,(-1)^n}{\pi^2(2n+1)^2}\sin\left[\frac{(2n+1)\pi x}{L}\right]\cos\left[\frac{(2n+1)\pi ct}{L}\right],$$

where, in the usual notation, $c^2 = T/\rho$.

Find the total kinetic energy of the string when it passes through its unplucked position, by calculating it in each mode (each n) and summing, using the result

$$\sum_0^{\infty}\frac{1}{(2n+1)^2} = \frac{\pi^2}{8}.$$

Confirm that the total energy is equal to the work done in plucking the string initially.

11.14. A conducting spherical shell of radius a is cut round its equator and the two halves connected to voltages of $+V$ and $-V$. Show that an expression for the potential at the point (r, θ, ϕ) anywhere inside the two hemispheres is

$$u(r, \theta, \phi) = V \sum_{n=0}^{\infty} \frac{(-1)^n (2n)!(4n+3)}{2^{2n+1} n!(n+1)!} \left(\frac{r}{a}\right)^{2n+1} P_{2n+1}(\cos\theta).$$

11.15. Prove that the potential for $\rho < a$ associated with a vertical split cylinder of radius a, the two halves of which ($\cos\phi > 0$ and $\cos\phi < 0$) are maintained at equal and opposite potentials $\pm V$, is given by

$$u(\rho, \phi) = \frac{4V}{\pi} \sum_{n=0}^{\infty} \frac{(-1)^n}{2n+1} \left(\frac{\rho}{a}\right)^{2n+1} \cos(2n+1)\phi.$$

[This is the cylindrical polar analogue of the previous question.]

11.16. A slice of biological material of thickness L is placed into a solution of a radioactive isotope of constant concentration C_0 at time $t = 0$. For a later time t find the concentration of radioactive ions at a depth x inside one of its surfaces if the diffusion constant is κ.

11.17. Two identical copper bars are each of length a. Initially, one is at 0 °C and the other is at 100 °C; they are then joined together end to end and thermally isolated. Obtain in the form of a Fourier series an expression $u(x, t)$ for the temperature at any point a distance x from the join at a later time t. Bear in mind the heat flow conditions at the free ends of the bars.

Taking $a = 0.5$ m estimate the time it takes for one of the free ends to attain a temperature of 55 °C. The thermal conductivity of copper is $3.8 \times 10^2\ \text{J m}^{-1}\,\text{K}^{-1}\,\text{s}^{-1}$, and its specific heat capacity is $3.4 \times 10^6\ \text{J m}^{-3}\,\text{K}^{-1}$.

11.18. A sphere of radius a and thermal conductivity k_1 is surrounded by an infinite medium of conductivity k_2 in which far away the temperature tends to T_∞. A distribution of heat sources $q(\theta)$ embedded in the sphere's surface establish steady temperature fields $T_1(r, \theta)$ inside the sphere and $T_2(r, \theta)$ outside it. It can be shown, by considering the heat flow through a small volume that includes part of the sphere's surface, that

$$k_1 \frac{\partial T_1}{\partial r} - k_2 \frac{\partial T_2}{\partial r} = q(\theta) \quad \text{on} \quad r = a.$$

Given that

$$q(\theta) = \frac{1}{a} \sum_{n=0}^{\infty} q_n P_n(\cos\theta),$$

find complete expressions for $T_1(r, \theta)$ and $T_2(r, \theta)$. What is the temperature at the center of the sphere?

11.19. Using result (11.74) from the worked example in the text, find the general expression for the temperature $u(x, t)$ in the bar, given that the temperature distribution at time $t = 0$ is $u(x, 0) = \exp(-x^2/a^2)$.

11.20. Working in *spherical* polar coordinates $\mathbf{r} = (r, \theta, \phi)$, but for a system that has azimuthal symmetry around the polar axis, consider the following gravitational problem.

(a) Show that the gravitational potential due to a uniform disc of radius a and mass M, centered at the origin, is given for $r < a$ by

$$-\frac{2GM}{a}\left[1 - \frac{r}{a}P_1(\cos\theta) + \frac{1}{2}\left(\frac{r}{a}\right)^2 P_2(\cos\theta) - \frac{1}{8}\left(\frac{r}{a}\right)^4 P_4(\cos\theta) + \cdots\right],$$

and for $r > a$ by

$$-\frac{GM}{r}\left[1 - \frac{1}{4}\left(\frac{a}{r}\right)^2 P_2(\cos\theta) + \frac{1}{8}\left(\frac{a}{r}\right)^4 P_4(\cos\theta) - \cdots\right],$$

where the polar axis is normal to the plane of the disc.

(b) Reconcile the presence of a term $P_1(\cos\theta)$, which is odd under $\theta \to \pi - \theta$, with the symmetry with respect to the plane of the disc of the physical system.

(c) Deduce that the gravitational field near an infinite sheet of matter of constant density ρ per unit area is $2\pi G\rho$.

11.21. In the region $-\infty < x, y < \infty$ and $-t \le z \le t$, a charge-density wave $\rho(\mathbf{r}) = A\cos qx$, in the x-direction, is represented by

$$\rho(\mathbf{r}) = \frac{e^{iqx}}{\sqrt{2\pi}}\int_{-\infty}^{\infty} \tilde{\rho}(\alpha)e^{i\alpha z}\,d\alpha.$$

The resulting potential is represented by

$$V(\mathbf{r}) = \frac{e^{iqx}}{\sqrt{2\pi}}\int_{-\infty}^{\infty} \tilde{V}(\alpha)e^{i\alpha z}\,d\alpha.$$

Determine the relationship between $\tilde{V}(\alpha)$ and $\tilde{\rho}(\alpha)$, and hence show that the potential at the point $(0, 0, 0)$ is

$$\frac{A}{\pi\epsilon_0}\int_{-\infty}^{\infty} \frac{\sin kt}{k(k^2 + q^2)}\,dk.$$

11.22. Point charges q and $-qa/b$ (with $a < b$) are placed, respectively, at a point P, a distance b from the origin O, and a point Q between O and P, a distance a^2/b from O. Show, by considering similar triangles QOS and SOP, where S is any point on the surface of the sphere centered at O and of radius a, that the net potential anywhere on the sphere due to the two charges is zero.

Use this result (backed up by the uniqueness theorem) to find the force with which a point charge q placed a distance b from the center of a spherical conductor of radius a ($< b$) is attracted to the sphere (i) if the sphere is earthed, and (ii) if the sphere is uncharged and insulated.

11.23. Find the Green's function $G(\mathbf{r}, \mathbf{r}_0)$ in the half-space $z > 0$ for the solution of $\nabla^2\Phi = 0$ with Φ specified in cylindrical polar coordinates (ρ, ϕ, z) on the plane $z = 0$ by

$$\Phi(\rho, \phi, z) = \begin{cases} 1 & \text{for } \rho \leq 1, \\ 1/\rho & \text{for } \rho > 1. \end{cases}$$

Determine the variation of $\Phi(0, 0, z)$ along the z-axis.

11.24. Electrostatic charge is distributed in a sphere of radius R centered on the origin. Determine the form of the resultant potential $\phi(\mathbf{r})$ at distances much greater than R, as follows.
(a) Express in the form of an integral over all space the solution of

$$\nabla^2\phi = -\frac{\rho(\mathbf{r})}{\epsilon_0}.$$

(b) Show that, for $r \gg r'$,

$$|\mathbf{r} - \mathbf{r}'| = r - \frac{\mathbf{r}\cdot\mathbf{r}'}{r} + \mathrm{O}\left(\frac{1}{r}\right).$$

(c) Use results (a) and (b) to show that $\phi(\mathbf{r})$ has the form

$$\phi(\mathbf{r}) = \frac{M}{r} + \frac{\mathbf{d}\cdot\mathbf{r}}{r^3} + \mathrm{O}\left(\frac{1}{r^3}\right).$$

Find expressions for M and $\mathbf{d}$, and identify them physically.

11.25. Find, in the form of an infinite series, the Green's function of the ∇^2 operator for the Dirichlet problem in the region $-\infty < x < \infty$, $-\infty < y < \infty$, $-c \leq z \leq c$.

11.26. Find the Green's function for the three-dimensional Neumann problem

$$\nabla^2\phi = 0 \quad \text{for } z > 0 \quad \text{and} \quad \frac{\partial\phi}{\partial z} = f(x, y) \quad \text{on } z = 0.$$

Determine $\phi(x, y, z)$ if

$$f(x, y) = \begin{cases} \delta(y) & \text{for } |x| < a, \\ 0 & \text{for } |x| \geq a. \end{cases}$$

11.27. Determine the Green's function for the Klein–Gordon equation in a half-space as follows.

(a) By applying the divergence theorem to the volume integral

$$\int_V \left[\phi(\nabla^2 - m^2)\psi - \psi(\nabla^2 - m^2)\phi\right] dV,$$

obtain a Green's function expression, as the sum of a volume integral and a surface integral, for the function $\phi(\mathbf{r}')$ that satisfies

$$\nabla^2\phi - m^2\phi = \rho$$

in V and takes the specified form $\phi = f$ on S, the boundary of V. The Green's function, $G(\mathbf{r}, \mathbf{r}')$, to be used satisfies

$$\nabla^2 G - m^2 G = \delta(\mathbf{r} - \mathbf{r}')$$

and vanishes when $\mathbf{r}$ is on S.

(b) When V is all space, $G(\mathbf{r}, \mathbf{r}')$ can be written as $G(t) = g(t)/t$, where $t = |\mathbf{r} - \mathbf{r}'|$ and $g(t)$ is bounded as $t \to \infty$. Find the form of $G(t)$.

(c) Find $\phi(\mathbf{r})$ in the half-space $x > 0$ if $\rho(\mathbf{r}) = \delta(\mathbf{r} - \mathbf{r}_1)$ and $\phi = 0$ both on $x = 0$ and as $r \to \infty$.

HINTS AND ANSWERS

11.1. (a) $C \exp[\lambda(x^2 + 2y)]$; (b) $C(x^2 y)^\lambda$.

11.3. $u(x, y, t) = \sin(n\pi x/a)\sin(m\pi y/b)(A \sin\omega t + B\cos\omega t)$.

11.5. (a) $6u/r^2, -6u/r^2, 0, \ell = 2$ (or -3), $m = 0$;
(b) $2u/r^2, (\cot^2\theta - 1)u/r^2; -u/(r^2\sin^2\theta), \ell = 1$ (or -2), $m = \pm 1$.

11.7. Solutions of the form r^ℓ give ℓ as $-1, 1, 2, 4$. Because of the asymptotic form of ψ, an r^4 term cannot be present. The coefficients of the three remaining terms are determined by the two boundary conditions $\mathbf{u} = \mathbf{0}$ on the sphere and the form of ψ for large r.

11.9. Express $\cos^2\phi$ in terms of $\cos 2\phi$; $T(\rho, \phi) = A + B/2 + (B\rho^2/2a^2)\cos 2\phi$.

11.11. $(A\cos mx + B\sin mx + C\cosh mx + D\sinh mx)\cos(\omega t + \epsilon)$, with $m^4 a^4 = \omega^2$.

11.13. $E_n = 16\rho A^2 c^2/[(2n+1)^2\pi^2 L]$; $E = 2\rho c^2 A^2/L = \int_0^A [2Tv/(\frac{1}{2}L)]\,dv$.

11.15. Note that the boundary value function is a square wave that is *symmetric* in ϕ.

11.17. Since there is no heat flow at $x = \pm a$, use a series of period $4a$, $u(x, 0) = 100$ for $0 < x \le 2a$, $u(x, 0) = 0$ for $-2a \le x < 0$.

$$u(x,t) = 50 + \frac{200}{\pi}\sum_{n=0}^{\infty}\frac{1}{2n+1}\sin\left[\frac{(2n+1)\pi x}{2a}\right]\exp\left[-\frac{k(2n+1)^2\pi^2 t}{4a^2 s}\right].$$

Taking only the $n = 0$ term gives $t \approx 2300$ s.

11.19. $u(x, t) = [a/(a^2 + 4\kappa t)^{1/2}]\exp[-x^2/(a^2 + 4\kappa t)]$.

11.21. Fourier-transform Poisson's equation to show that $\tilde{\rho}(\alpha) = \epsilon_0(\alpha^2 + q^2)\tilde{V}(\alpha)$.

11.23. Follow the worked example that includes result (11.94). For part of the explicit integration, substitute $\rho = z\tan\alpha$.

$$\Phi(0,0,z) = \frac{z(1+z^2)^{1/2} - z^2 + (1+z^2)^{1/2} - 1}{z(1+z^2)^{1/2}}.$$

11.25. The terms in $G(\mathbf{r}, \mathbf{r}_0)$ that are additional to the fundamental solution are

$$\frac{1}{4\pi}\sum_{n=2}^{\infty}(-1)^n\left\{\left[(x-x_0)^2 + (y-y_0)^2 + (z+(-1)^n z_0 - nc)^2\right]^{-1/2} + \left[(x-x_0)^2 + (y-y_0)^2 + (z+(-1)^n z_0 + nc)^2\right]^{-1/2}\right\}.$$

11.27. (a) As given in equation (11.85), but with $\mathbf{r}_0$ replaced by $\mathbf{r}'$.

(b) Move the origin to $\mathbf{r}'$ and integrate the defining Green's equation to obtain

$$4\pi t^2\frac{dG}{dt} - m^2\int_0^t G(t')4\pi t'^2\,dt' = 1,$$

leading to $G(t) = [-1/(4\pi t)]e^{-mt}$.

(c) $\phi(\mathbf{r}) = [-1/(4\pi)](p^{-1}e^{-mp} - q^{-1}e^{-mq})$, where $p = |\mathbf{r} - \mathbf{r}_1|$ and $q = |\mathbf{r} - \mathbf{r}_2|$ with $\mathbf{r}_1 = (x_1, y_1, z_1)$ and $\mathbf{r}_2 = (-x_1, y_1, z_1)$.

12 Calculus of variations

How to find stationary values of functions of a single variable $f(x)$, of several variables $f(x, y, \ldots)$ and of constrained variables, where $x, y, \ldots$ are subject to the n constraints $g_i(x, y, \ldots) = 0$, $i = 1, 2, \ldots, n$ will be known to the reader and is summarized in Sections A.3 and A.7 of Appendix A. In all those cases the forms of the functions f and g_i were known, and the problem was one of finding the appropriate values of the variables x, y, etc.

We now turn to a different kind of problem in which we are interested in bringing about a particular condition for a given expression (usually maximizing or minimizing it) by varying the *functions* on which the expression depends. For instance, we might want to know in what shape a fixed length of rope should be arranged so as to enclose the largest possible area, or in what shape it will hang when suspended under gravity from two fixed points. In each case we are concerned with a general maximization or minimization criterion by which the function $y(x)$ that satisfies the given problem may be found.

The calculus of variations provides a method for finding the function $y(x)$. The problem must first be expressed in a mathematical form, and the form most commonly applicable to such problems is an *integral*. In each of the above questions, the quantity that has to be maximized or minimized by an appropriate choice of the function $y(x)$ may be expressed as an integral involving $y(x)$ and the variables describing the geometry of the situation.

In our example of the rope hanging from two fixed points, we need to find the shape function $y(x)$ that minimizes the gravitational potential energy of the rope. Each elementary piece of the rope has a gravitational potential energy proportional both to its vertical height above an arbitrary zero level and to the length of the piece. Therefore the total potential energy is given by an integral for the whole rope of such elementary contributions. The particular function $y(x)$ for which the value of this integral is a minimum will give the shape assumed by the hanging rope.

So in general we are led by this type of question to study the value of an integral whose integrand has a specified form in terms of an unknown function and its derivatives, and to study how that value changes when the form of the function is varied. Specifically, we aim to find the function that makes the integral *stationary*, i.e. the function that makes the value of the integral a local maximum or minimum. Note that, unless it is stated otherwise, y' is used to denote dy/dx throughout this chapter. We

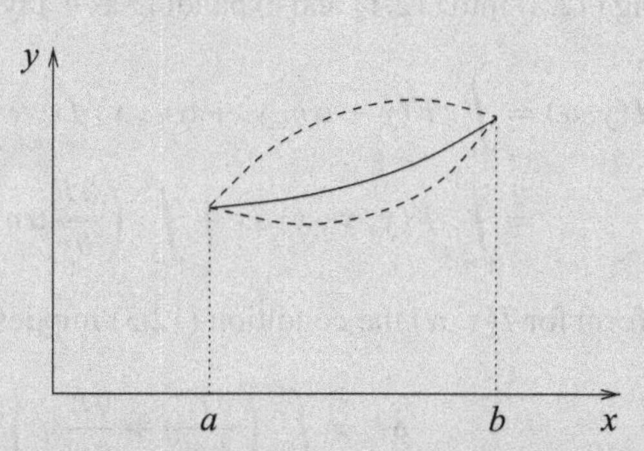

Figure 12.1 Possible paths for the integral (12.1). The solid line is the curve along which the integral is assumed stationary. The broken curves represent small variations from this path.

also assume that all the functions we need to deal with are sufficiently smooth and differentiable.

12.1 The Euler–Lagrange equation

Let us consider the integral

$$I = \int_a^b F(y, y', x)\, dx, \tag{12.1}$$

where a, b and the form of the function F are fixed by given considerations, e.g. the physics of the problem, but the curve $y(x)$ is to be chosen so as to make stationary the value of I, which is clearly a function, or more accurately a *functional*, of this curve, i.e. $I = I[\,y(x)]$. Referring to Figure 12.1, we wish to find the function $y(x)$ (given, say, by the solid line) such that first-order small changes in it (for example the two broken lines) will make only second-order changes in the value of I.

Writing this in a more mathematical form, let us suppose that $y(x)$ is the function required to make I stationary and consider making the replacement

$$y(x) \to y(x) + \alpha\eta(x), \tag{12.2}$$

where the parameter α is small and $\eta(x)$ is an arbitrary function with sufficiently amenable mathematical properties. For the value of I to be stationary with respect to these variations, we require

$$\left.\frac{dI}{d\alpha}\right|_{\alpha=0} = 0 \quad \text{for all } \eta(x). \tag{12.3}$$

Substituting (12.2) into (12.1) and expanding as a Taylor series in α we obtain

$$\begin{aligned} I(y,\alpha) &= \int_a^b F(y+\alpha\eta,\, y'+\alpha\eta',\, x)\,dx \\ &= \int_a^b F(y,y',x)\,dx + \int_a^b \left(\frac{\partial F}{\partial y}\alpha\eta + \frac{\partial F}{\partial y'}\alpha\eta'\right)dx + \mathrm{O}(\alpha^2). \end{aligned}$$

With this form for $I(y,\alpha)$ the condition (12.3) implies that for all $\eta(x)$ we require

$$\delta I = \int_a^b \left(\frac{\partial F}{\partial y}\eta + \frac{\partial F}{\partial y'}\eta'\right)dx = 0,$$

where δI denotes the first-order variation in the value of I due to the variation (12.2) in the function $y(x)$. Integrating the second term by parts this becomes

$$\left[\eta\frac{\partial F}{\partial y'}\right]_a^b + \int_a^b \left[\frac{\partial F}{\partial y} - \frac{d}{dx}\left(\frac{\partial F}{\partial y'}\right)\right]\eta(x)\,dx = 0. \tag{12.4}$$

In order to simplify the result we will assume that the end-points are fixed, i.e. that not only are a and b given, but so are $y(a)$ and $y(b)$. This restriction means that we require $\eta(a) = \eta(b) = 0$, in which case the first term on the LHS of (12.4) equals zero at both end-points. Since (12.4) must be satisfied for arbitrary $\eta(x)$, it is easy to see that we require

$$\frac{\partial F}{\partial y} = \frac{d}{dx}\left(\frac{\partial F}{\partial y'}\right). \tag{12.5}$$

This is known as the *Euler–Lagrange* (EL) equation, and is a differential equation for $y(x)$, since the function F is known.

12.2 Special cases

The next general step is to solve for $y = y(x)$ the differential equation that the Euler–Lagrange equation has generated in any particular case; different forms of F will generate different differential equations for y. If F falls into one of several special classes, a first integral of the EL equation can be obtained in a general form; this saves the labour of the first stage of integration and reduces the order of the differential equation by one. We now consider these special classes.

12.2.1 *F* does not contain *y* explicitly

In this case $\partial F/\partial y = 0$, and (12.5) can be integrated immediately giving

$$\frac{\partial F}{\partial y'} = \text{constant}. \tag{12.6}$$

This simplest of cases, we illustrate with what is probably the simplest of worked examples – certainly the least surprising of results.

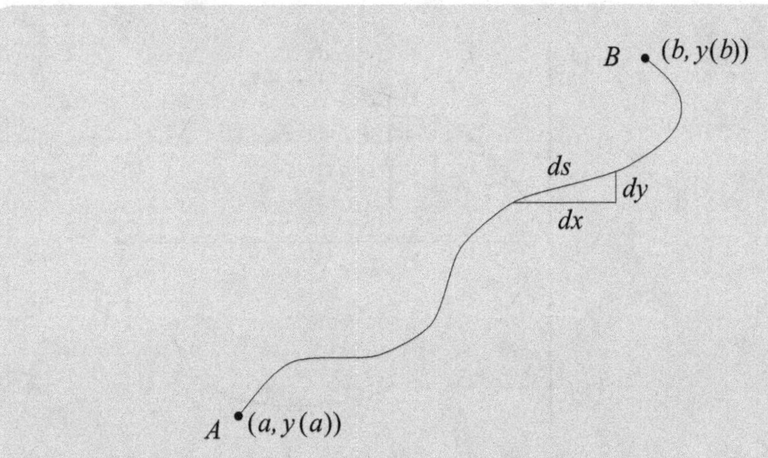

Figure 12.2 An arbitrary path between two fixed points.

Example Show that the shortest curve joining two points is a straight line and find the length of the line.

Let the two points be labeled A and B and have coordinates $(a, y(a))$ and $(b, y(b))$ respectively (see Figure 12.2). Whatever the shape of the curve joining A to B, the length of an element of path ds is given by

$$ds = \left[(dx)^2 + (dy)^2\right]^{1/2} = (1 + y'^2)^{1/2} dx,$$

and hence the total path length along the curve is given by

$$L = \int_a^b (1 + y'^2)^{1/2}\, dx. \tag{12.7}$$

We must now apply the results of the previous section to determine that path which makes L stationary (clearly a minimum in this case). Since the integral does not contain y (or indeed x) explicitly, we may use (12.6) to obtain

$$k = \frac{\partial F}{\partial y'} = \frac{y'}{(1 + y'^2)^{1/2}},$$

where k is a constant. This is easily rearranged and integrated to give

$$y = \frac{k}{(1 - k^2)^{1/2}} x + c,$$

which, as expected, is the equation of a straight line in the form $y = mx + c$, with $m = k/(1 - k^2)^{1/2}$; this answers the first part of the question.

The value of m (or k) can be found by demanding that the straight line passes through the points A and B and is given by $m = [\,y(b) - y(a)]/(b - a)$. Substituting the equation of the straight line into (12.7) we find that the total path length is given by

$$L^2 = [\,y(b) - y(a)]^2 + (b - a)^2,$$

again as expected.[1] ◀

1 The reader should carry out this calculation for themselves in order to gain practice at using the minimizing curve once it has been found.

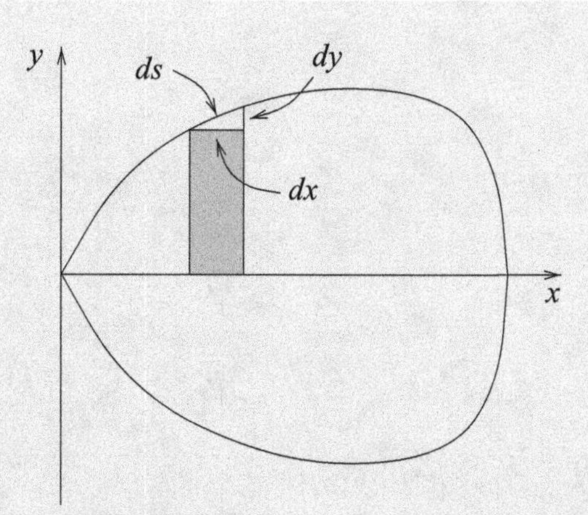

Figure 12.3 A convex closed curve that is symmetrical about the x-axis.

12.2.2 *F* does not contain *x* explicitly

In this case, we start from the differentiation of a product,

$$\frac{d}{dx}\left(y'\frac{\partial F}{\partial y'}\right) = y'\frac{d}{dx}\left(\frac{\partial F}{\partial y'}\right) + y''\frac{\partial F}{\partial y'}$$

and use the EL equation (12.5), multiplied through by y', to replace the first term on the RHS, thus obtaining, after a little rearrangement,

$$y'\frac{\partial F}{\partial y} + y''\frac{\partial F}{\partial y'} = \frac{d}{dx}\left(y'\frac{\partial F}{\partial y'}\right).$$

But since F is a function of y and y' only, and not explicitly of x, the LHS of this equation is just the total derivative of F, namely dF/dx. Hence, integrating we obtain

$$F - y'\frac{\partial F}{\partial y'} = \text{constant}. \tag{12.8}$$

This, therefore, is the form of the first integral of the EL equation for any case in which the variable of integration does not appear explicitly in the integrand of (12.1). Subsequent analysis can start from here, rather than from the EL equation itself, as we now demonstrate.

Example Find the closed convex curve of length l that encloses the greatest possible area.

Without any loss of generality we can assume that the curve passes through the origin and can further suppose that it is symmetric with respect to the x-axis; this assumption is not essential. Using the distance s along the curve, measured from the origin, as the independent variable and y as the dependent one, we have the boundary conditions $y(0) = y(l/2) = 0$. The element of area shown in Figure 12.3 is then given by

$$dA = y\,dx = y\left[(ds)^2 - (dy)^2\right]^{1/2},$$

and the total area by

$$A = 2\int_0^{l/2} y(1 - y'^2)^{1/2}\, ds; \tag{12.9}$$

here y' stands for dy/ds rather than dy/dx. Since the integrand does not contain s explicitly, we can use (12.8) to obtain a first integral of the EL equation for y, namely

$$y(1 - y'^2)^{1/2} + yy'^2(1 - y'^2)^{-1/2} = k,$$

where k is a constant. On rearranging this gives

$$ky' = \pm(k^2 - y^2)^{1/2},$$

which, using $y(0) = 0$, integrates to

$$y/k = \sin(s/k). \tag{12.10}$$

The other end-point, $y(l/2) = 0$, fixes the value of k as $l/(2\pi)$ to yield

$$y = \frac{l}{2\pi} \sin \frac{2\pi s}{l}.$$

From this we obtain $dy = \cos(2\pi s/l)\, ds$ and since $(ds)^2 = (dx)^2 + (dy)^2$ we also find that $dx = \pm \sin(2\pi s/l)\, ds$. This in turn can be integrated and, using $x(0) = 0$, gives x in terms of s as

$$x - \frac{l}{2\pi} = -\frac{l}{2\pi} \cos \frac{2\pi s}{l}.$$

We thus obtain the expected result that x and y lie on the circle of radius $l/(2\pi)$ given by

$$\left(x - \frac{l}{2\pi}\right)^2 + y^2 = \frac{l^2}{4\pi^2}.$$

Substituting the solution (12.10) into the expression for the total area (12.9), it is easily verified that $A = l^2/(4\pi)$. A much quicker derivation of this result is possible using plane polar coordinates. ◀

The previous two examples have been carried out in some detail, even though the answers are more easily obtained in other ways, expressly so that the method is transparent and the way in which it works can be filled in mentally at almost every step. The next example, however, does not have such an intuitively obvious solution.

Example Two rings, each of radius a, are placed parallel with their centers $2b$ apart and on a common normal. An open-ended axially symmetric soap film is formed between them (see Figure 12.4). Find the shape assumed by the film.

Creating the soap film requires an energy γ per unit area (numerically equal to the surface tension of the soap solution). So the stable shape of the soap film, i.e. the one that minimizes the energy, will also be the one that minimizes the surface area (neglecting gravitational effects).

It is obvious that any convex surface, shaped such as that shown by the broken line in Figure 12.4(a), cannot be a minimum but it is not clear whether some shape intermediate between the cylinder shown by solid lines in (a), with area $4\pi ab$ (or twice this for the double surface of the film), and the form shown in (b), with area approximately $2\pi a^2$, will produce a lower total area than both of these extremes. If there is such a shape [e.g. that in Figure 12.4(c)], then it will be that which is the best compromise between two requirements, the need to minimize the ring-to-ring

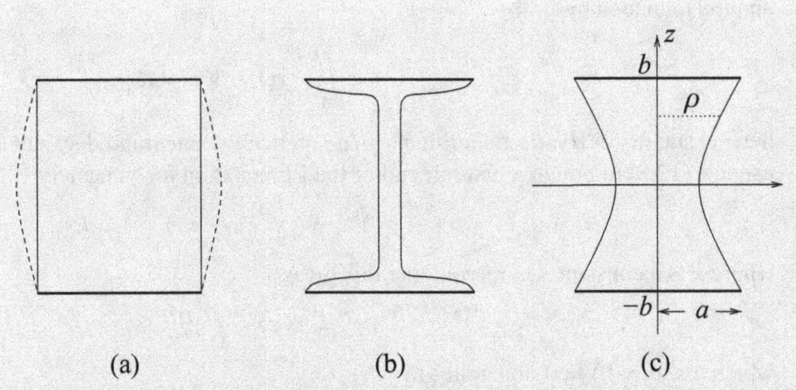

Figure 12.4 Possible soap films between two parallel circular rings.

distance measured on the film surface (a) and the need to minimize the average waist measurement of the surface (b).

We take cylindrical polar coordinates as in Figure 12.4(c) and let the radius of the soap film at height z be $\rho(z)$ with $\rho(\pm b) = a$. Counting only one side of the film, the element of surface area between z and $z + dz$ is

$$dS = 2\pi\rho\left[(dz)^2 + (d\rho)^2\right]^{1/2},$$

so the total surface area is given by

$$S = 2\pi \int_{-b}^{b} \rho(1 + \rho'^2)^{1/2}\, dz. \tag{12.11}$$

Since the integrand does not contain z explicitly, we can use (12.8) to obtain an equation for ρ that minimizes S, i.e.

$$\rho(1 + \rho'^2)^{1/2} - \rho\rho'^2(1 + \rho'^2)^{-1/2} = k,$$

where k is a constant. Multiplying through by $(1 + \rho'^2)^{1/2}$, rearranging to find an explicit expression for ρ' and integrating we find

$$\cosh^{-1}\frac{\rho}{k} = \frac{z}{k} + c,$$

where c is the constant of integration. Using the boundary conditions $\rho(\pm b) = a$, we require $c = 0$ and k such that $a/k = \cosh b/k$ (if b/a is too large, no such k can be found). Thus the curve that minimizes the surface area is

$$\rho/k = \cosh(z/k),$$

and in profile the soap film is a catenary (see Section 12.4) with the minimum distance from the axis equal to k. ◀

12.3 Some extensions

It is quite possible to relax many of the restrictions we have imposed so far. For example, we can allow end-points that are constrained to lie on given curves rather than being fixed, or we can consider problems with several dependent and/or independent variables or higher-order derivatives of the dependent variable. The first extension, whilst perhaps the most interesting, is beyond the scope of this present treatment, but the results for the others are summarized in the next three subsections.

12.3.1 Several dependent variables

Here we have $F = F(y_1, y_1', y_2, y_2', \ldots, y_n, y_n', x)$ where each $y_i = y_i(x)$. The analysis in this case proceeds as before, leading to n separate but simultaneous equations for the $y_i(x)$,

$$\frac{\partial F}{\partial y_i} = \frac{d}{dx}\left(\frac{\partial F}{\partial y_i'}\right), \qquad i = 1, 2, \ldots, n. \tag{12.12}$$

If x does not appear explicitly in the integrand, then a first integral of the above set of Euler–Lagrange equations is

$$F - \sum_{i=1}^{n} y_i' \frac{\partial F}{\partial y_i'} = \text{constant}. \tag{12.13}$$

This is the analogue of the corresponding result for a single dependent variable.

12.3.2 Several independent variables

With n independent variables, we need to extremize multiple integrals of the form

$$I = \int\int\cdots\int F\left(y, \frac{\partial y}{\partial x_1}, \frac{\partial y}{\partial x_2}, \ldots, \frac{\partial y}{\partial x_n}, x_1, x_2, \ldots, x_n\right) dx_1\, dx_2 \cdots dx_n.$$

Using the same kind of analysis as before, we find that the extremizing function $y = y(x_1, x_2, \ldots, x_n)$ must satisfy

$$\frac{\partial F}{\partial y} = \sum_{i=1}^{n} \frac{\partial}{\partial x_i}\left(\frac{\partial F}{\partial y_{x_i}}\right), \tag{12.14}$$

where y_{x_i} stands for $\partial y/\partial x_i$.

12.3.3 Higher-order derivatives

If in (12.1) $F = F(y, y', y'', \ldots, y^{(n)}, x)$ then using the same method as before and performing repeated integration by parts, it can be shown that the required extremizing function $y(x)$ satisfies

$$\frac{\partial F}{\partial y} - \frac{d}{dx}\left(\frac{\partial F}{\partial y'}\right) + \frac{d^2}{dx^2}\left(\frac{\partial F}{\partial y''}\right) - \cdots + (-1)^n \frac{d^n}{dx^n}\left(\frac{\partial F}{\partial y^{(n)}}\right) = 0, \tag{12.15}$$

provided that $y = y' = \cdots = y^{(n-1)} = 0$ at both end-points. If y, or any of its derivatives, are not zero at the end-points then corresponding contributions will appear on the RHS of (12.15).

12.4 Constrained variation

The problem of finding the stationary values of a function $f(x, y)$ subject to the constraint $g(x, y) =$ constant is solved by means of Lagrange's undetermined multipliers; the corresponding problem in the calculus of variations is solved by an analogous method.

Suppose that we wish to find the stationary values of

$$I = \int_a^b F(y, y', x)\,dx,$$

subject to the constraint that the value of

$$J = \int_a^b G(y, y', x)\,dx$$

is held constant. Following the method of Lagrange undetermined multipliers let us define a new functional

$$K = I + \lambda J = \int_a^b (F + \lambda G)\,dx,$$

and find its *unconstrained* stationary values. Repeating the analysis of Section 12.1 we find that we require

$$\frac{\partial F}{\partial y} - \frac{d}{dx}\left(\frac{\partial F}{\partial y'}\right) + \lambda\left[\frac{\partial G}{\partial y} - \frac{d}{dx}\left(\frac{\partial G}{\partial y'}\right)\right] = 0,$$

which, together with the original constraint $J =$ constant, will yield the required solution $y(x)$.

This method is easily generalized to cases with more than one constraint by the introduction of more Lagrange multipliers. If we wish to find the stationary values of an integral I subject to the multiple constraints that the values of the integrals J_i be held constant for $i = 1, 2, \ldots, n$, then we simply find the unconstrained stationary values of the new integral

$$K = I + \sum_1^n \lambda_i J_i.$$

Our worked example has only one such constraint.

Example Find the shape assumed by a uniform rope when suspended by its ends from two points at equal heights.

We will solve this problem using x (see Figure 12.5) as the independent variable. Let the rope of length $2L$ be suspended between the points $x = \pm a$, $y = 0$ $(L > a)$ and have uniform linear density ρ. We then need to find the stationary value of the rope's gravitational potential energy,

$$I = -\rho g \int y\,ds = -\rho g \int_{-a}^{a} y(1 + y'^2)^{1/2}\,dx,$$

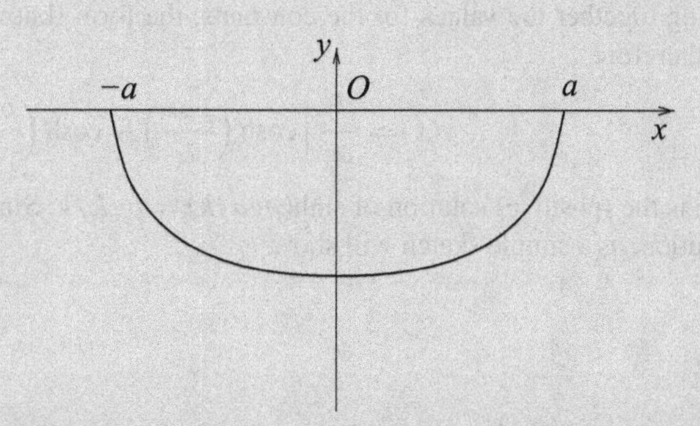

Figure 12.5 A uniform rope with fixed end-points suspended under gravity.

with respect to small changes in the form of the rope but subject to the constraint that the total length of the rope remains constant, i.e.

$$J = \int ds = \int_{-a}^{a} (1 + y'^2)^{1/2}\, dx = 2L.$$

We thus define a new integral (omitting the factor -1 from I for brevity)

$$K = I + \lambda J = \int_{-a}^{a} (\rho g y + \lambda)(1 + y'^2)^{1/2}\, dx$$

and find its stationary values. Since the integrand does not contain the independent variable x explicitly, we can use (12.8) to find the first integral:

$$(\rho g y + \lambda)(1 + y'^2)^{1/2} - (\rho g y + \lambda)(1 + y'^2)^{-1/2} y'^2 = k,$$

where k is a constant; this reduces to

$$y'^2 = \left(\frac{\rho g y + \lambda}{k}\right)^2 - 1.$$

Making the substitution $\rho g y + \lambda = k \cosh z$, this can be integrated easily to give

$$\frac{k}{\rho g} \cosh^{-1}\left(\frac{\rho g y + \lambda}{k}\right) = x + c,$$

where c is the constant of integration.

We now have three unknowns, λ, k and c, that must be evaluated using the two end conditions $y(\pm a) = 0$ and the constraint $J = 2L$. The end conditions give

$$\cosh \frac{\rho g(a + c)}{k} = \frac{\lambda}{k} = \cosh \frac{\rho g(-a + c)}{k},$$

and since $a \neq 0$, these imply $c = 0$ and $\lambda/k = \cosh(\rho g a/k)$. Putting $c = 0$ into the constraint, in which $y' = \sinh(\rho g x/k)$, we obtain

$$\begin{aligned} 2L &= \int_{-a}^{a} \left[1 + \sinh^2\left(\frac{\rho g x}{k}\right)\right]^{1/2} dx \\ &= \frac{2k}{\rho g} \sinh\left(\frac{\rho g a}{k}\right). \end{aligned}$$

Collecting together the values for the constants, the form (known as a catenary)[2] adopted by the rope is therefore

$$y(x) = \frac{k}{\rho g}\left[\cosh\left(\frac{\rho g x}{k}\right) - \cosh\left(\frac{\rho g a}{k}\right)\right],$$

where k is the (positive) solution of $\sinh(\rho g a/k) = \rho g L/k$. Since $L > a$, this equation is assured of a solution, as a simple sketch will show. ◀

12.5 Physical variational principles

Many results in both classical and quantum physics can be expressed as variational principles, and it is often when expressed in this form that their physical meaning is most clearly understood. Moreover, once a physical phenomenon has been written as a variational principle, we can use all the results derived in this chapter to investigate its behavior. It is usually possible to identify conserved quantities, or symmetries of the system of interest, that otherwise might be found only with considerable effort. From the wide range of physical variational principles we will select two examples from familiar areas of classical physics, namely geometric optics and mechanics.

12.5.1 Fermat's principle in optics

Fermat's principle in geometrical optics states that a ray of light traveling in a region of variable refractive index follows a path such that the total optical path length (physical length × refractive index) is stationary.

Example From Fermat's principle deduce Snell's law of refraction at an interface.

Let the interface be at $y =$ constant (see Figure 12.6) and let it separate two regions with refractive indices n_1 and n_2 respectively. On a ray the element of physical path length is $ds = (1 + y'^2)^{1/2}dx$, and so for a ray that passes through the points A and B, the total optical path length is

$$P = \int_A^B n(y)(1 + y'^2)^{1/2}\,dx.$$

Since the integrand does not contain the independent variable x explicitly, we use (12.8) to obtain a first integral, which, after some rearrangement, reads

$$n(y)\left(1 + y'^2\right)^{-1/2} = k,$$

where k is a constant.[3] Recalling that y' is the tangent of the angle ϕ between the instantaneous direction of the ray and the x-axis, this *general* result, which is not dependent on the configuration

2 Note that this is the same curve as appeared in the worked example on soap films on p. 493.
3 Obtain this result for yourself.

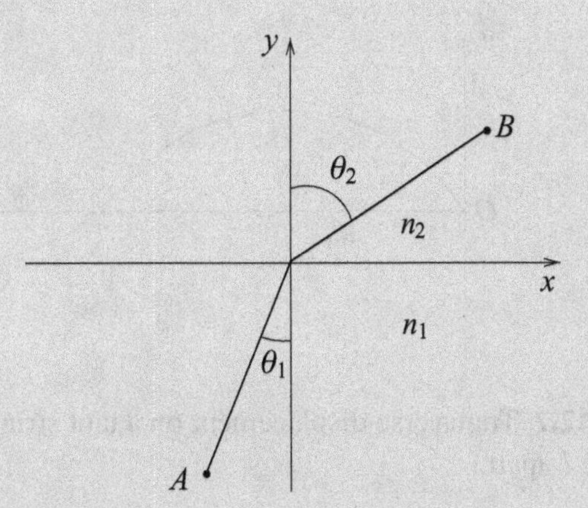

Figure 12.6 Path of a light ray at the plane interface between media with refractive indices n_1 and n_2, where $n_2 < n_1$.

presently under consideration, can be put in the form

$$n \cos\phi = \text{constant}$$

along a ray, even though n and ϕ vary individually.

For our particular configuration n is constant in each medium and therefore so is y'. Thus the rays travel in straight lines in each medium (as anticipated in Figure 12.6, but not assumed in our analysis), and since k is constant along the *whole* path we have $n_1 \cos\phi_1 = n_2 \cos\phi_2$, or, in terms of the conventional angles in the figure,

$$n_1 \sin\theta_1 = n_2 \sin\theta_2.$$

This is Snell's law for refraction at a general interface. When the second medium is air, n_2 is often approximated by unity and the law stated as $\sin\theta_2/\sin\theta_1 = \mu$ where μ is the refractive index of the first medium. ◀

12.5.2 Hamilton's principle in mechanics

Consider a mechanical system whose configuration can be uniquely defined by a number of coordinates q_i (usually distances and angles) together with time t and which experiences only forces derivable from a potential. Hamilton's principle states that in moving from one configuration at time t_0 to another at time t_1 the motion of such a system is such as to make

$$\mathcal{L} = \int_{t_0}^{t_1} L(q_1, q_2 \ldots, q_n, \dot{q}_1, \dot{q}_2, \ldots, \dot{q}_n, t)\, dt \qquad (12.16)$$

stationary. The *Lagrangian* L is defined, in terms of the kinetic energy T and the potential energy V (with respect to some reference situation), by $L = T - V$. Here V is a function of the q_i only, not of the $\dot{q}_i$. Applying the EL equation to $\mathcal{L}$ we obtain *Lagrange's equations*,

$$\frac{\partial L}{\partial q_i} = \frac{d}{dt}\left(\frac{\partial L}{\partial \dot{q}_i}\right), \qquad i = 1, 2, \ldots, n.$$

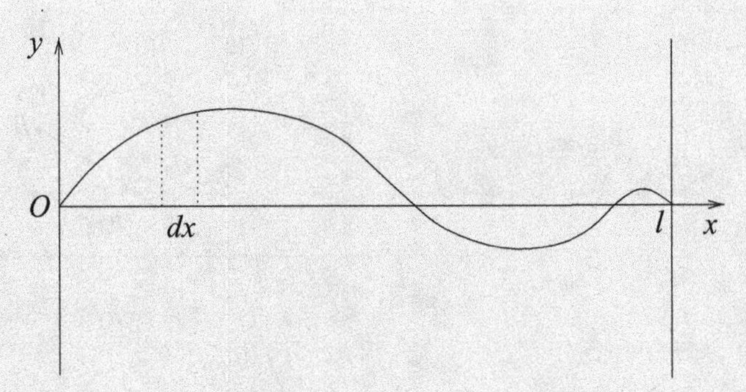

Figure 12.7 Transverse displacement on a taut string that is fixed at two points a distance l apart.

These equations apply to a wide range of systems meeting the requirements set out above and form the starting point for analyzing the motion of a general mechanical system once a set of coordinates sufficient to define any possible configuration has been established.

Example Using Hamilton's principle derive the wave equation for small transverse oscillations of a taut string.

In this example we are in fact considering a generalization of (12.16) to a case involving one isolated independent coordinate t, together with a *continuum* in which the q_i become the continuous variable x. The expressions for T and V therefore become integrals over x rather than sums over the label i.

If ρ and τ are the local density and tension of the string, both of which may depend on x, then, referring to Figure 12.7, the kinetic and potential energies of the string are given by

$$T = \int_0^l \frac{\rho}{2}\left(\frac{\partial y}{\partial t}\right)^2 dx, \qquad V = \int_0^l \frac{\tau}{2}\left(\frac{\partial y}{\partial x}\right)^2 dx$$

and (12.16) becomes

$$\mathcal{L} = \frac{1}{2}\int_{t_0}^{t_1} dt \int_0^l \left[\rho\left(\frac{\partial y}{\partial t}\right)^2 - \tau\left(\frac{\partial y}{\partial x}\right)^2\right] dx.$$

Using (12.14) and the fact that y does not appear explicitly, we obtain

$$\frac{\partial}{\partial t}\left(\rho\frac{\partial y}{\partial t}\right) - \frac{\partial}{\partial x}\left(\tau\frac{\partial y}{\partial x}\right) = 0.$$

If, in addition, ρ and τ do not depend on x or t then the Euler–Lagrange equation simplifies to

$$\frac{\partial^2 y}{\partial x^2} = \frac{1}{c^2}\frac{\partial^2 y}{\partial t^2},$$

where $c^2 = \tau/\rho$. This is the wave equation for small transverse oscillations of a taut uniform string. ◀

12.6 General eigenvalue problems

We have seen in this chapter that the problem of finding a curve that makes the value of a given integral stationary when the integral is taken along the curve results, in each case, in a differential equation for the curve. It is not a great extension to ask whether this may be used to solve differential equations, by setting up a suitable variational problem and then seeking ways other than the Euler equation of finding or estimating stationary solutions.

We shall be concerned with differential equations of the form $\mathcal{L}y = \lambda\rho(x)y$, where the differential operator $\mathcal{L}$ is self-adjoint, so that $\mathcal{L} = \mathcal{L}^\dagger$ (with appropriate boundary conditions on the solution y) and $\rho(x)$ is some weight function, as discussed in Chapter 8. In particular, we will concentrate on the Sturm–Liouville equation as an explicit example, but much of what follows can be applied to other equations of this type.

We have already discussed the solution of equations of the Sturm–Liouville type in Chapter 8 and the same notation will be used here. In this section, however, we will adopt a variational approach to estimating the eigenvalues of such equations.

Suppose we search for stationary values of the integral

$$I = \int_a^b \left[p(x)y'^2(x) - q(x)y^2(x) \right] dx, \tag{12.17}$$

with $y(a) = y(b) = 0$ and p and q any sufficiently smooth and differentiable functions of x. However, in addition we impose a normalization condition

$$J = \int_a^b \rho(x)y^2(x)\, dx = \text{constant}. \tag{12.18}$$

Here $\rho(x)$ is a positive weight function defined in the interval $a \le x \le b$, but which may in particular cases be a constant.

Then, as in Section 12.4, we use undetermined Lagrange multipliers,[4] and consider $K = I - \lambda J$ given by

$$K = \int_a^b \left[py'^2 - (q + \lambda\rho)y^2 \right] dx.$$

On application of the EL equation (12.5) this yields

$$\frac{d}{dx}\left(p\frac{dy}{dx} \right) + qy + \lambda\rho y = 0, \tag{12.19}$$

which is exactly the Sturm–Liouville equation (8.34), with eigenvalue λ.[5] Now, since both I and J are quadratic in y and its derivative, finding stationary values of K is equivalent to finding stationary values of I/J. This may also be shown by considering the functional

4 We use $-\lambda$, rather than λ, so that the final equation (12.19) appears in the conventional Sturm–Liouville form.

5 In other words, the Sturm–Liouville equation *is* the Euler–Lagrange equation for the integral (12.17) when it is subjected to the constraint (12.18).

$\Lambda = I/J$, for which

$$\begin{aligned}\delta\Lambda &= (\delta I/J) - (I/J^2)\,\delta J \\ &= (\delta I - \Lambda\delta J)/J \\ &= \delta K/J.\end{aligned}$$

Hence, extremizing Λ is equivalent to extremizing K. Thus we have the important result that *finding functions y that make I/J stationary is equivalent to finding functions y that are solutions of the Sturm–Liouville equation; the resulting value of I/J equals the corresponding eigenvalue of the equation.*

Of course this does not tell us how to find such a function y and, naturally, to have to do this by solving (12.19) directly defeats the purpose of the exercise. We will see in the next section how some progress can be made. It is worth recalling that the functions $p(x)$, $q(x)$ and $\rho(x)$ can have many different forms, and so (12.19) represents quite a wide variety of equations.

We now recall some properties of the solutions of the Sturm–Liouville equation. The eigenvalues λ_i of (12.19) are real and will be assumed non-degenerate (for simplicity). We also assume that the corresponding eigenfunctions have been made real, so that normalized eigenfunctions $y_i(x)$ satisfy the orthogonality relation [as in (8.24)]

$$\int_a^b y_i y_j \rho\, dx = \delta_{ij}. \tag{12.20}$$

Further, we take the boundary condition in the form

$$\left[y_i p y_j'\right]_{x=a}^{x=b} = 0; \tag{12.21}$$

this can be satisfied by $y(a) = y(b) = 0$, but also by many other sets of boundary conditions.

Example Show that

$$\int_a^b \left(y_j' p y_i' - y_j q y_i\right) dx = \lambda_i \delta_{ij}. \tag{12.22}$$

Let y_i be an eigenfunction of (12.19), corresponding to a particular eigenvalue λ_i, so that

$$(py_i')' + (q + \lambda_i \rho) y_i = 0.$$

Multiplying this through by y_j and integrating from a to b (the first term by parts) we obtain

$$\Big[y_j\,(py_i')\Big]_a^b - \int_a^b y_j'(py_i')\,dx + \int_a^b y_j(q + \lambda_i\rho)y_i\,dx = 0. \tag{12.23}$$

The first term vanishes by virtue of (12.21), and on rearranging the other terms and using (12.20), we find the result (12.22). ◂

We see at once that, if the function $y(x)$ minimizes I/J, i.e. satisfies the Sturm–Liouville equation, then putting $y_i = y_j = y$ in (12.20) and (12.22) yields J and I respectively on the left-hand sides; thus, as mentioned above, the minimized value of I/J is just the eigenvalue λ, introduced originally as the undetermined multiplier. The formal proof of this is as follows.

Example For a function y satisfying the Sturm–Liouville equation verify that, provided (12.21) is satisfied, $\lambda = I/J$.

Firstly, we multiply (12.19) through by y to give

$$y(py')' + qy^2 + \lambda \rho y^2 = 0.$$

Now integrating this expression by parts we have

$$\left[ypy'\right]_a^b - \int_a^b \left(py'^2 - qy^2\right) dx + \lambda \int_a^b \rho y^2 \, dx = 0.$$

The first term on the LHS is zero, the second is simply $-I$ and the third is λJ. Thus $\lambda = I/J$. ◀

12.7 Estimation of eigenvalues and eigenfunctions

Since the eigenvalues λ_i of the Sturm–Liouville equation are the stationary values of I/J (see above), it follows that any evaluation of I/J must yield a value that lies between the lowest and highest eigenvalues of the corresponding Sturm–Liouville equation, i.e.

$$\lambda_{\text{min}} \leq \frac{I}{J} \leq \lambda_{\text{max}},$$

where, depending on the equation under consideration, either $\lambda_{\text{min}} = -\infty$ and λ_{max} is finite, or $\lambda_{\text{max}} = \infty$ and λ_{min} is finite. Notice that here we have departed from direct consideration of the minimizing problem and made a statement about a calculation in which no actual minimization is necessary.

Thus, as an example, for an equation with a finite lowest eigenvalue λ_0 any evaluation of I/J provides an upper bound on λ_0. Further, we will now show that the estimate λ obtained is a better estimate of λ_0 than the estimated (guessed) function y is of y_0, the true eigenfunction corresponding to λ_0. The sense in which "better" is used here will be clear from the final result.

Firstly, we expand the estimated or *trial function* y in terms of the complete set y_i:

$$y = y_0 + c_1 y_1 + c_2 y_2 + \cdots,$$

where, if a good trial function has been guessed, the c_i will be small. Using (12.20) we have immediately that $J = 1 + \sum_i |c_i|^2$. The other required integral is

$$I = \int_a^b \left[p\left(y_0' + \sum_i c_i y_i'\right)^2 - q\left(y_0 + \sum_i c_i y_i\right)^2 \right] dx.$$

On multiplying out the squared terms, all the cross terms vanish because of (12.22) to leave

$$\begin{aligned}\lambda &= \frac{I}{J} \\ &= \frac{\lambda_0 + \sum_i |c_i|^2\lambda_i}{1+\sum_j |c_j|^2} \\ &= \lambda_0 + \sum_i |c_i|^2(\lambda_i - \lambda_0) + O(c^4).\end{aligned}$$

Hence λ differs from λ_0 by a term second order in the c_i, even though y differed from y_0 by a term first order in the c_i; this is what we aimed to show. We notice incidentally that, since $\lambda_0 < \lambda_i$ for all i, λ is shown to be necessarily $\geq \lambda_0$, with equality only if all $c_i = 0$, i.e. if $y \equiv y_0$.

The method can be extended to the second and higher eigenvalues by imposing, in addition to the original constraints and boundary conditions, a restriction of the trial functions to only those that are orthogonal to the eigenfunctions corresponding to lower eigenvalues. (Of course, this requires complete or nearly complete knowledge of these latter eigenfunctions.) An example is given at the end of this chapter (Problem 12.23).

We now illustrate the method we have discussed by considering a simple example, one for which, as on previous occasions, the answer is obvious.

Example Estimate the lowest eigenvalue of the equation

$$-\frac{d^2y}{dx^2} = \lambda y, \qquad 0 \leq x \leq 1, \tag{12.24}$$

with boundary conditions

$$y(0) = 0, \qquad y'(1) = 0. \tag{12.25}$$

We need to find the lowest value λ_0 of λ for which (12.24) has a solution $y(x)$ that satisfies (12.25). The exact answer is of course $y = A\sin(x\pi/2)$ and $\lambda_0 = \pi^2/4 \approx 2.47$.

Firstly we note that the Sturm–Liouville equation reduces to (12.24) if we take $p(x) = 1$, $q(x) = 0$ and $\rho(x) = 1$ and that the boundary conditions satisfy (12.21). Thus we are able to apply the previous theory.

We will use three trial functions so that the effect on the estimate of λ_0 of making better or worse "guesses" can be seen. One further preliminary remark is relevant, namely that the estimate is independent of any constant multiplicative factor in the function used. This is easily verified by looking at the form of I/J. We normalize each trial function so that $y(1) = 1$, purely in order to facilitate comparison of the various function shapes.

Figure 12.8 illustrates the trial functions used, curve (a) being the exact solution $y = \sin(\pi x/2)$. The other curves are (b) $y(x) = 2x - x^2$, (c) $y(x) = x^3 - 3x^2 + 3x$, and (d) $y(x) = \sin^2(\pi x/2)$. The choice of trial function is governed by the following considerations:

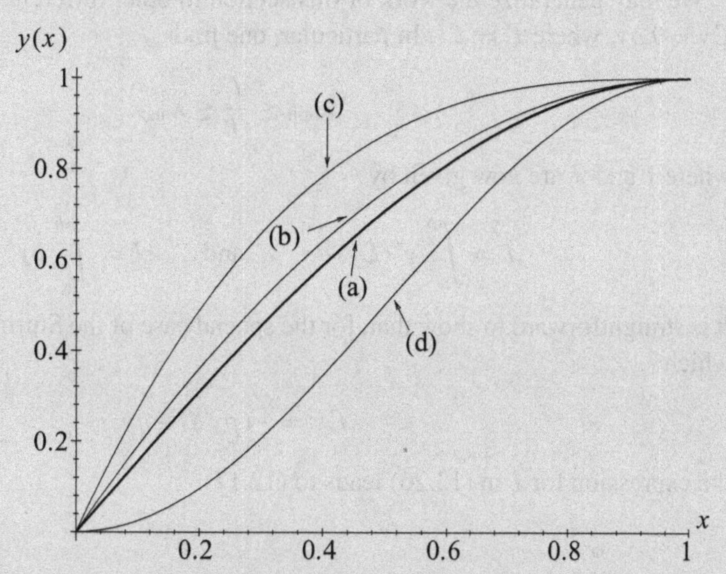

Figure 12.8 Trial solutions used to estimate the lowest eigenvalue λ of $-y'' = \lambda y$ with $y(0) = y'(1) = 0$. They are: (a) $y = \sin(\pi x/2)$, the exact result; (b) $y = 2x - x^2$; (c) $y = x^3 - 3x^2 + 3x$; (d) $y = \sin^2(\pi x/2)$.

(i) the boundary conditions (12.25) *must* be satisfied;
(ii) a "good" trial function ought to mimic the correct solution as far as possible, but it may not be easy to guess even the general shape of the correct solution in some cases;
(iii) the evaluation of I/J should be as simple as possible.

It is easily verified that functions (b), (c) and (d) all satisfy (12.25) but, so far as mimicking the correct solution is concerned, we would expect from the figure that (b) would be superior to the other two. The three evaluations are straightforward, using (12.17) and (12.18) and remembering that $q(x) = 0$ and so there is only one term in each numerator:

$$\lambda_b = \frac{\int_0^1 (2 - 2x)^2\, dx}{\int_0^1 (2x - x^2)^2\, dx} = \frac{4/3}{8/15} = 2.50,$$

$$\lambda_c = \frac{\int_0^1 (3x^2 - 6x + 3)^2\, dx}{\int_0^1 (x^3 - 3x^2 + 3x)^2\, dx} = \frac{9/5}{9/14} = 2.80,$$

$$\lambda_d = \frac{\int_0^1 (\pi^2/4)\sin^2(\pi x)\, dx}{\int_0^1 \sin^4(\pi x/2)\, dx} = \frac{\pi^2/8}{3/8} = 3.29.$$

We expected all evaluations to yield estimates greater than the lowest eigenvalue, 2.47, and this is indeed so.[6] From these trials alone we are able to say (only) that $\lambda_0 \leq 2.50$. As expected, the best approximation (b) to the true eigenfunction yields the lowest, and therefore the best, upper bound on λ_0. ◀

6 Show that using the exact solution as a trial function *does* give the true lowest eigenvalue.

We may generalize the work of this section to other differential equations of the form $\mathcal{L}y = \lambda\rho y$, where $\mathcal{L} = \mathcal{L}^\dagger$. In particular, one finds

$$\lambda_{\min} \leq \frac{I}{J} \leq \lambda_{\max},$$

where I and J are now given by

$$I = \int_a^b y^*(\mathcal{L}y)\,dx \qquad \text{and} \qquad J = \int_a^b \rho y^* y\,dx. \tag{12.26}$$

It is straightforward to show that, for the special case of the Sturm–Liouville equation, for which

$$\mathcal{L}y = -(py')' - qy,$$

the expression for I in (12.26) leads to (12.17).

12.8 Adjustment of parameters

Instead of trying to estimate the lowest eigenvalue λ_0 by selecting a large number of different trial functions, we may also use trial functions that include one or more parameters which themselves may be adjusted to give the lowest value to $\lambda = I/J$ and hence the best estimate of λ_0. The justification for this method comes from the knowledge that no matter what form of function is chosen, nor what values are assigned to the parameters, provided the boundary conditions are satisfied, λ can never be less than the required λ_0.

To illustrate this method an example from quantum mechanics will be used. The time-independent Schrödinger equation is formally written as the eigenvalue equation $H\psi = E\psi$, where H is a linear operator, ψ the wavefunction describing a quantum mechanical system and E the energy of the system. The energy operator H is called the Hamiltonian and for a particle of mass m moving in a one-dimensional harmonic oscillator potential is given by

$$H = -\frac{\hbar^2}{2m}\frac{d^2}{dx^2} + \frac{kx^2}{2}, \tag{12.27}$$

where $\hbar$ is Planck's constant divided by 2π.

Example Estimate the ground-state energy of a quantum harmonic oscillator.

Using (12.27) in $H\psi = E\psi$, the Schrödinger equation is

$$-\frac{\hbar^2}{2m}\frac{d^2\psi}{dx^2} + \frac{kx^2}{2}\psi = E\psi, \qquad -\infty < x < \infty. \tag{12.28}$$

The boundary conditions are that ψ should vanish as $x \to \pm\infty$. Equation (12.28) is a form of the Sturm–Liouville equation in which $p = \hbar^2/(2m)$, $q = -kx^2/2$, $\rho = 1$ and $\lambda = E$; it can be solved by the methods developed previously, e.g. by writing the eigenfunction ψ as a power series in x.

However, our purpose here is to illustrate variational methods and so we take as a trial wavefunction $\psi = \exp(-\alpha x^2)$, where α is a positive parameter whose value we will choose later. This function certainly $\to 0$ as $x \to \pm\infty$ and is convenient for calculations. Whether it approximates the true wave function is unknown, but if it does not our estimate will still be valid, although the upper bound will be a poor one.

With $y = \exp(-\alpha x^2)$ and therefore $y' = -2\alpha x \exp(-\alpha x^2)$, the required estimate is

$$E = \lambda = \frac{\int_{-\infty}^{\infty} [(\hbar^2/2m)4\alpha^2 x^2 + (k/2)x^2]e^{-2\alpha x^2}\,dx}{\int_{-\infty}^{\infty} e^{-2\alpha x^2}\,dx} = \frac{\hbar^2 \alpha}{2m} + \frac{k}{8\alpha}. \tag{12.29}$$

This evaluation is easily carried out using the reduction formula

$$I_n = \frac{n-1}{4\alpha} I_{n-2}, \quad \text{for integrals of the form} \quad I_n = \int_{-\infty}^{\infty} x^n e^{-2\alpha x^2}\,dx. \tag{12.30}$$

So, we have obtained the estimate (12.29), involving the parameter α, for the oscillator's ground-state energy, i.e. the lowest eigenvalue of H. In line with our previous discussion we now minimize λ with respect to α. Putting $d\lambda/d\alpha = 0$ (clearly a minimum), yields $\alpha = (km)^{1/2}/(2\hbar)$, which in turn gives as the minimum value for λ

$$\lambda = E = \frac{\hbar}{2}\left(\frac{k}{m}\right)^{1/2} = \frac{\hbar\omega}{2}, \tag{12.31}$$

where we have put $(k/m)^{1/2}$ equal to the classical angular frequency ω.

The method thus leads to the conclusion that the ground-state energy E_0 is $\le \frac{1}{2}\hbar\omega$. In fact, as is well known, the equality sign holds, $\frac{1}{2}\hbar\omega$ being just the zero-point energy of a quantum mechanical oscillator. Our estimate gives the exact value because $\psi(x) = \exp(-\alpha x^2)$ is the correct functional form for the ground-state wavefunction and the particular value of α that we have found is that needed to make ψ an eigenfunction of H with eigenvalue $\frac{1}{2}\hbar\omega$. ◀

An alternative but equivalent approach to this is developed in the problems that follow, as is an extension of this particular problem to estimating the second-lowest eigenvalue (see Problem 12.23).

SUMMARY

1. *Euler–Lagrange equations*

With $I = \int_a^b F(y, y', x)\,dx$ as the basic integral to be made stationary with respect to variations in $y(x)$, the general equation to be satisfied by $y(x)$ is the corresponding Euler–Lagrange equation as given below.

Form of F	Euler–Lagrange equation
$F(y, y', x)$	$\dfrac{\partial F}{\partial y} = \dfrac{d}{dx}\left(\dfrac{\partial F}{\partial y'}\right)$
$F(y', x)$	$0 = \dfrac{d}{dx}\left(\dfrac{\partial F}{\partial y'}\right)$

(cont.)

(cont.)

Form of F	Euler–Lagrange equation
$F(y, y')$	$\dfrac{\partial F}{\partial y} = \dfrac{d}{dx}\left(\dfrac{\partial F}{\partial y'}\right)$
$F(y_i, y_i', x)$	$\dfrac{\partial F}{\partial y_i} = \dfrac{d}{dx}\left(\dfrac{\partial F}{\partial y_i'}\right)$ for each $i = 1, 2, \ldots, n$
$F(y, \dfrac{\partial y}{\partial x_j}, x_j)$	$\dfrac{\partial F}{\partial y} = \sum_{j=1}^{n} \dfrac{\partial}{\partial x_j}\left(\dfrac{\partial F}{\partial y_{x_j}}\right)$ where $j = 1, 2, \ldots, n$
$F(y, y', \ldots, y^{(n)}, x)$	$\dfrac{\partial F}{\partial y} = \dfrac{d}{dx}\left(\dfrac{\partial F}{\partial y'}\right) - \cdots + (-1)^{n+1}\dfrac{d^n}{dx^n}\left(\dfrac{\partial F}{\partial y^{(n)}}\right)$

2. *Particular first integrals*
   - If $F = F(y', x)$, then $\dfrac{\partial F}{\partial y'} = c$ is a first integral.
   - If $F = F(y, y')$, then $F - y'\dfrac{\partial F}{\partial y'} = c$ is a first integral.
   - If $F = F(y_i, y_i')$, then $F - \sum_{i=1}^{n} y_i' \dfrac{\partial F}{\partial y_i'} = c$ is a first integral.

3. *Constrained variation*
   - If $I = \int F(y, y', x)\,dx$ is to be stationary with respect to variations in y subject to n constraints $J_i = \int G_i(y, y', x)\,dx$, replace F by $F_1 = F + \sum_{i=1}^{n} \lambda_i G_i$. Then solve the set of $n+1$ equations comprising the EL equation for F_1 and the constraints, and so determine the required $y(x)$ and, if needed, the n values of λ_i.
   - The functions $y = y(x)$ that make $I = \int_a^b [p(x)y'^2 - q(x)y^2]\,dx$ stationary, subject to $y(a) = y(b) = 0$ and constraint $J = \int_a^b \rho(x)y^2\,dx$, are solutions of the Sturm–Liouville equation
   $$\frac{d}{dx}\left(p\frac{dy}{dx}\right) + qy + \lambda\rho y = 0,$$
   with $I/J = \lambda$.
   - For all $y(x)$ that satisfy the boundary conditions, I/J *always* lies between the lowest and highest eigenvalues λ of the corresponding SL equation. In particular, for *any* such choice of $y(x)$, $I/J \geq \lambda_0$, the lowest eigenvalue.
   - Trial functions aimed at putting an upper bound on λ_0 may contain any number of parameters, the values of which may be adjusted to minimize the upper bound.

PROBLEMS

12.1. A surface of revolution, whose equation in cylindrical polar coordinates is $\rho = \rho(z)$, is bounded by the circles $\rho = a$, $z = \pm c$ $(a > c)$. Show that the function that makes the surface integral $I = \int \rho^{-1/2}\, dS$ stationary with respect to small variations is given by $\rho(z) = k + z^2/(4k)$, where $k = [a \pm (a^2 - c^2)^{1/2}]/2$.

12.2. Show that the lowest value of the integral

$$\int_A^B \frac{(1 + y'^2)^{1/2}}{y}\, dx,$$

where A is $(-1, 1)$ and B is $(1, 1)$, is $2\ln(1 + \sqrt{2})$. Assume that the Euler–Lagrange equation gives a minimizing curve.

12.3. The refractive index n of a medium is a function only of the distance r from a fixed point O. Prove that the equation of a light ray, assumed to lie in a plane through O, traveling in the medium satisfies (in plane polar coordinates)

$$\frac{1}{r^2}\left(\frac{dr}{d\phi}\right)^2 = \frac{r^2\, n^2(r)}{a^2\, n^2(a)} - 1,$$

where a is the distance of the ray from O at the point at which $dr/d\phi = 0$. If $n = [1 + (\alpha^2/r^2)]^{1/2}$ and the ray starts and ends far from O, find its deviation (the angle through which the ray is turned), if its minimum distance from O is a.

12.4. The Lagrangian for a π-meson is given by

$$L(\mathbf{x}, t) = \tfrac{1}{2}(\dot{\phi}^2 - |\nabla\phi|^2 - \mu^2\phi^2),$$

where μ is the meson mass and $\phi(\mathbf{x}, t)$ is its wavefunction. Assuming Hamilton's principle, find the wave equation satisfied by ϕ.

12.5. Prove the following results about general systems.

(a) For a system described in terms of coordinates q_i and t, show that if t does not appear explicitly in the expressions for x, y and z ($x = x(q_i, t)$, etc.) then the kinetic energy T is a homogeneous quadratic function of the $\dot{q}_i$ (it may also involve the q_i). Deduce that $\sum_i \dot{q}_i(\partial T/\partial \dot{q}_i) = 2T$.

(b) Assuming that the forces acting on the system are derivable from a potential V, show, by expressing dT/dt in terms of q_i and $\dot{q}_i$, that $d(T + V)/dt = 0$.

12.6. For a system specified by the coordinates q and t, show that the equation of motion is unchanged if the Lagrangian $L(q, \dot{q}, t)$ is replaced by

$$L_1 = L + \frac{d\phi(q, t)}{dt},$$

where ϕ is an arbitrary function. Deduce that the equation of motion of a particle that moves in one dimension subject to a force $-dV(x)/dx$ (x being measured from a point O) is unchanged if O is forced to move with a constant velocity v (x still being measured from O).

12.7. In cylindrical polar coordinates, the curve $(\rho(\theta), \theta, \alpha\rho(\theta))$ lies on the surface of the cone $z = \alpha\rho$. Show that geodesics (curves of minimum length joining two points) on the cone satisfy

$$\rho^4 = c^2[\beta^2\rho'^2 + \rho^2],$$

where c is an arbitrary constant, but β has to have a particular value. Determine the form of $\rho(\theta)$ and hence find the equation of the shortest path on the cone between the points $(R, -\theta_0, \alpha R)$ and $(R, \theta_0, \alpha R)$.

[You will find it useful to determine the form of the derivative of $\cos^{-1}(u^{-1})$.]

12.8. Derive the differential equations for the plane-polar coordinates, r and ϕ, of a particle of unit mass moving in a field of potential $V(r)$. Find the form of V if the path of the particle is given by $r = a \sin\phi$.

12.9. You are provided with a line of length $\pi a/2$ and negligible mass and some lead shot of total mass M. Use a variational method to determine how the lead shot must be distributed along the line if the loaded line is to hang in a circular arc of radius a when its ends are attached to two points at the same height. Measure the distance s along the line from its center.

12.10. A general result is that light travels through a variable medium by the path that minimizes the travel time (this is an alternative formulation of Fermat's principle).

Light travels in the vertical xz-plane through a slab of material which lies between the planes $z = z_0$ and $z = 2z_0$, and in which the speed of light $v(z) = c_0 z/z_0$. Show that the ray paths are arcs of circles. Deduce that, if a ray enters the material at $(0, z_0)$ at an angle to the vertical, $\pi/2 - \theta$, of more than $30°$, then it does not reach the far side of the slab.

12.11. With respect to a particular cylindrical polar coordinate system (ρ, ϕ, z), the speed of light $v(\rho, \phi)$ is independent of z. If the path of the light is parameterized as $\rho = \rho(z)$, $\phi = \phi(z)$, use the alternative formulation of Fermat's principle, given in the previous question, to show that

$$v^2(\rho'^2 + \rho^2\phi'^2 + 1)$$

is constant along the path.

For the particular case when $v = v(\rho) = b(a^2 + \rho^2)^{1/2}$, show that the two Euler–Lagrange equations have a common solution in which the light travels along a helical path given by $\phi = Az + B$, $\rho = C$, provided that A has a particular value.

12.12. Use result (12.22) to evaluate

$$J = \int_{-1}^{1} (1 - x^2) P_m'(x) P_n'(x)\, dx,$$

where $P_m(x)$ is a Legendre polynomial of order m.

12.13. The Schwarzchild metric for the static field of a non-rotating spherically symmetric black hole of mass M is given by

$$(ds)^2 = c^2 \left(1 - \frac{2GM}{c^2 r}\right)(dt)^2 - \frac{(dr)^2}{1 - 2GM/(c^2 r)} - r^2\,(d\theta)^2 - r^2 \sin^2\theta\,(d\phi)^2.$$

Considering only motion confined to the plane $\theta = \pi/2$, and assuming that the path of a small test particle is such as to make $\int ds$ stationary, find two first integrals of the equations of motion. From their Newtonian limits, in which GM/r, $\dot{r}^2$ and $r^2\dot{\phi}^2$ are all $\ll c^2$, identify the constants of integration.

12.14. Show that $y'' - xy + \lambda x^2 y = 0$ has a solution for which $y(0) = y(1) = 0$ and $\lambda \leq 147/4$.

12.15. Determine the minimum value that the integral

$$J = \int_0^1 [x^4 (y'')^2 + 4x^2 (y')^2]\, dx$$

can have, given that y is not singular at $x = 0$ and that $y(1) = y'(1) = 1$. Assume that the Euler–Lagrange equation gives the lower limit.

12.16. Estimate the lowest eigenvalue, λ_0, of the equation

$$\frac{d^2 y}{dx^2} - x^2 y + \lambda y = 0, \qquad y(-1) = y(1) = 0,$$

using a quadratic trial function.

12.17. Find an appropriate, but simple, trial function and use it to estimate the lowest eigenvalue λ_0 of Stokes' equation,

$$\frac{d^2 y}{dx^2} + \lambda x y = 0, \qquad \text{with } y(0) = y(\pi) = 0.$$

Explain why your estimate must be strictly greater than λ_0.

12.18. Consider the problem of finding the lowest eigenvalue, λ_0, of the equation

$$(1 + x^2)\frac{d^2 y}{dx^2} + 2x \frac{dy}{dx} + \lambda y = 0, \qquad y(\pm 1) = 0.$$

(a) Recast the problem in variational form, and derive an approximation λ_1 to λ_0 by using the trial function $y_1(x) = 1 - x^2$.
(b) Show that an improved estimate λ_2 is obtained by using $y_2(x) = \cos(\pi x/2)$.
(c) Prove that the estimate $\lambda(\gamma)$ obtained by taking $y_1(x) + \gamma y_2(x)$ as the trial function is

$$\lambda(\gamma) = \frac{64/15 + 64\gamma/\pi - 384\gamma/\pi^3 + (\pi^2/3 + 1/2)\gamma^2}{16/15 + 64\gamma/\pi^3 + \gamma^2}.$$

Investigate $\lambda(\gamma)$ numerically as γ is varied, or, more simply, show that $\lambda(-1.80) = 3.668$, an improvement on both λ_1 and λ_2.

12.19. A drum skin is stretched across a fixed circular rim of radius a. Small transverse vibrations of the skin have an amplitude $z(\rho, \phi, t)$ that satisfies

$$\nabla^2 z = \frac{1}{c^2}\frac{\partial^2 z}{\partial t^2}$$

in plane polar coordinates. For a normal mode independent of azimuth, $z = Z(\rho)\cos\omega t$, find the differential equation satisfied by $Z(\rho)$. By using a trial function of the form $a^\nu - \rho^\nu$, with adjustable parameter ν, obtain an estimate for the lowest normal mode frequency.

[The exact answer is $(5.78)^{1/2}c/a$.]

12.20. This is an alternative approach to the example in Section 12.8. Using the notation of that section, the expectation value of the energy of the state ψ is given by $\int \psi^* H\psi\, dv$. Denote the eigenfunctions of H by ψ_i, so that $H\psi_i = E_i\psi_i$, and, since H is self-adjoint (Hermitian), $\int \psi_j^*\psi_i\, dv = \delta_{ij}$.
(a) By writing any function ψ as $\sum c_j\psi_j$ and following an argument similar to that in Section 12.7, show that

$$E = \frac{\int \psi^* H\psi\, dv}{\int \psi^*\psi\, dv} \geq E_0,$$

the energy of the lowest state. This is the Rayleigh–Ritz principle.
(b) Using the same trial function as in Section 12.8, $\psi = \exp(-\alpha x^2)$, show that the same result is obtained.

12.21. For the boundary conditions given below, obtain a functional $\Lambda(y)$ whose stationary values give the eigenvalues of the equation

$$(1+x)\frac{d^2y}{dx^2} + (2+x)\frac{dy}{dx} + \lambda y = 0, \qquad y(0) = 0,\ y'(2) = 0.$$

Derive an approximation to the lowest eigenvalue λ_0 using the trial function $y(x) = xe^{-x/2}$. For what value(s) of γ would

$$y(x) = xe^{-x/2} + \beta\sin\gamma x$$

be a suitable trial function for attempting to obtain an improved estimate of λ_0?

12.22. The Hamiltonian H for the hydrogen atom is

$$-\frac{\hbar^2}{2m}\nabla^2 - \frac{q^2}{4\pi\epsilon_0 r}.$$

For a spherically symmetric state, as may be assumed for the ground state, the only relevant part of ∇^2 is that involving differentiation with respect to r.

(a) Define the integrals J_n by

$$J_n = \int_0^\infty r^n e^{-2\beta r}\, dr$$

and show that, for a trial wavefunction of the form $\exp(-\beta r)$ with $\beta > 0$, $\int \psi^* H\psi\, dv$ and $\int \psi^*\psi\, dv$ (see Problem 12.20(a)) can be expressed as $aJ_1 - bJ_2$ and cJ_2 respectively, where a, b and c are factors which you should determine.

(b) Show that the estimate of E is minimized when $\beta = mq^2/(4\pi\epsilon_0\hbar^2)$.

(c) Hence find an upper limit for the ground-state energy of the hydrogen atom. In fact, $\exp(-\beta r)$ is the correct form for the wavefunction and the limit gives the actual value.

12.23. This is an extension to Section 12.8 and Problem 12.20. With the ground-state (i.e. the lowest-energy) wavefunction as $\exp(-\alpha x^2)$, take as a trial function the orthogonal wave function $x^{2n+1}\exp(-\alpha x^2)$, using the integer n as a variable parameter. Use either Sturm–Liouville theory or the Rayleigh–Ritz principle to show that the energy of the second lowest state of a quantum harmonic oscillator is $\leq 3\hbar\omega/2$.

12.24. A particle of mass m moves in a one-dimensional potential well of the form

$$V(x) = -\mu\frac{\hbar^2\alpha^2}{m}\operatorname{sech}^2\alpha x,$$

where μ and α are positive constants. As in Problem 12.22, the expectation value $\langle E\rangle$ of the energy of the system is $\int \psi^* H\psi\, dx$, where the self-adjoint operator H is given by $-(\hbar^2/2m)d^2/dx^2 + V(x)$. Using trial wavefunctions of the form $y = A\operatorname{sech}\beta x$, show the following:

(a) for $\mu = 1$, there is an exact eigenfunction of H, with a corresponding $\langle E\rangle$ of half of the maximum depth of the well;

(b) for $\mu = 6$, the "binding energy" of the ground state is at least $10\hbar^2\alpha^2/(3m)$.

[You will find it useful to note that for $u, v \geq 0$, $\operatorname{sech} u \operatorname{sech} v \geq \operatorname{sech}(u+v)$.]

12.25. The upper and lower surfaces of a film of liquid, which has surface energy per unit area (surface tension) γ and density ρ, have equations $z = p(x)$ and $z = q(x)$, respectively. The film has a given volume V (per unit depth in the y-direction) and lies in the region $-L < x < L$, with $p(0) = q(0) = p(L) = q(L) = 0$. The total energy (per unit depth) of the film consists of its surface energy and its gravitational energy, and is expressed by

$$E = \frac{1}{2}\rho g \int_{-L}^{L} (p^2 - q^2)\, dx + \gamma \int_{-L}^{L} \left[(1 + p'^2)^{1/2} + (1 + q'^2)^{1/2}\right] dx.$$

(a) Express V in terms of p and q.
(b) Show that, if the total energy is minimized, p and q must satisfy

$$\frac{p'^2}{(1 + p'^2)^{1/2}} - \frac{q'^2}{(1 + q'^2)^{1/2}} = \text{constant}.$$

(c) As an approximate solution, consider the equations

$$p = a(L - |x|), \qquad q = b(L - |x|),$$

where a and b are sufficiently small that a^3 and b^3 can be neglected compared with unity. Find the values of a and b that minimize E.

12.26. The Sturm–Liouville equation can be extended to two independent variables, x and z, with little modification. In equation (12.17), y'^2 is replaced by $(\nabla y)^2$ and the integrals of the various functions of $y(x, z)$ become two-dimensional, i.e. the infinitesimal is $dx\, dz$.

The vibrations of a trampoline 4 units long and 1 unit wide satisfy the equation

$$\nabla^2 y + k^2 y = 0.$$

By taking the simplest possible permissible polynomial as a trial function, show that the lowest mode of vibration has $k^2 \leq 10.63$ and, by direct solution, that the actual value is 10.49.

HINTS AND ANSWERS

12.1. Note that the integrand, $2\pi\rho^{1/2}(1 + \rho'^2)^{1/2}$, does not contain z explicitly.

12.3. $I = \int n(r)[r^2 + (dr/d\phi)^2]^{1/2}\, d\phi$. Take axes such that $\phi = 0$ when $r = \infty$. If $\beta = (\pi - \text{deviation angle})/2$ then $\beta = \phi$ at $r = a$, and the equation reduces to

$$\frac{\beta}{(a^2 + \alpha^2)^{1/2}} = \int_{-\infty}^{\infty} \frac{dr}{r(r^2 - a^2)^{1/2}},$$

which can be evaluated by putting $r = a(y + y^{-1})/2$, or successively $r = a\cosh\psi$, $y = \exp\psi$ to yield a deviation of $\pi[(a^2 + \alpha^2)^{1/2} - a]/a$.

12.5. (a) $\partial x/\partial t = 0$ and so $\dot{x} = \sum_i \dot{q}_i\, \partial x/\partial q_i$; (b) use

$$\sum_i \dot{q}_i \frac{d}{dt}\left(\frac{\partial T}{\partial \dot{q}_i}\right) = \frac{d}{dt}(2T) - \sum_i \ddot{q}_i \frac{\partial T}{\partial \dot{q}_i}.$$

12.7. Use result (12.8); $\beta^2 = 1 + \alpha^2$.
Put $\rho = uc$ to obtain $d\theta/du = \beta/[u(u^2 - 1)^{1/2}]$. Remember that $\cos^{-1}$ is a multivalued function; $\rho(\theta) = [R\cos(\theta_0/\beta)]/[\cos(\theta/\beta)]$.

12.9. $-\lambda y'(1 - y'^2)^{-1/2} = 2gP(s)$, $y = y(s)$, $P(s) = \int_0^s \rho(s')\,ds'$. The solution $y = -a\cos(s/a)$ and $2P(\pi a/4) = M$ give $\lambda = -gM$. The required $\rho(s)$ is given by $[M/(2a)]\sec^2(s/a)$.

12.11. Note that the ϕ EL equation is automatically satisfied if $v \neq v(\phi)$. $A = 1/a$.

12.13. Denoting $(ds)^2/(dt)^2$ by f^2, the Euler–Lagrange equation for ϕ gives $r^2\dot{\phi} = Af$, where A corresponds to the angular momentum of the particle. Use result (12.13) to obtain $c^2 - (2GM/r) = Bf$, where to first order in small quantities

$$cB = c^2 - \frac{GM}{r} + \frac{1}{2}(\dot{r}^2 + r^2\dot{\phi}^2),$$

which reads "total energy = rest mass + gravitational energy + radial and azimuthal kinetic energy".

12.15. Convert the equation to the usual form, by writing $y'(x) = u(x)$, and obtain $x^2u'' + 4xu' - 4u = 0$ with general solution $Ax^{-4} + Bx$. Integrating a second time and using the boundary conditions gives $y(x) = (1 + x^2)/2$ and $J = 1$.

12.17. Using $y = \sin x$ as a trial function shows that $\lambda_0 \leq 2/\pi$. The estimate must be $> \lambda_0$ since the trial function does not satisfy the original equation.

12.19. $Z'' + \rho^{-1}Z' + (\omega/c)^2 Z = 0$, with $Z(a) = 0$ and $Z'(0) = 0$; this is an SL equation with $p = \rho$, $q = 0$ and weight function ρ/c^2. Estimate of $\omega^2 = [c^2\nu/(2a^2)][0.5 - 2(\nu + 2)^{-1} + (2\nu + 2)^{-1}]^{-1}$, which minimizes to $c^2(2 + \sqrt{2})^2/(2a^2) = 5.83c^2/a^2$ when $\nu = \sqrt{2}$.

12.21. Note that the original equation is not self-adjoint; it needs an integrating factor of e^x. $\Lambda(y) = [\int_0^2 (1 + x)e^x y'^2\,dx]/[\int_0^2 e^x y^2\,dx]$; $\lambda_0 \leq 3/8$. Since $y'(2)$ must equal 0, $\gamma = (\pi/2)(n + \frac{1}{2})$ for some integer n.

12.23. $E_1 \leq (\hbar\omega/2)(8n^2 + 12n + 3)/(4n + 1)$, which has a minimum value $3\hbar\omega/2$ when integer $n = 0$.

12.25. (a) $V = \int_{-L}^{L} (p - q)\,dx$. (c) Use $V = (a - b)L^2$ to eliminate b from the expression for E; now the minimization is with respect to a alone. The values for a and b are $\pm V/(2L^2) - V\rho g/(6\gamma)$.

13 Integral equations

It is not unusual in the analysis of a physical system to encounter an equation in which an unknown but required function $y(x)$, say, appears under an integral sign. Such an equation is called an *integral equation*, and in this chapter we discuss several methods for solving the more straightforward examples of such equations.

Before embarking on our discussion of methods for solving various integral equations, we begin with a warning that many of the integral equations met in practice cannot be solved by the elementary methods presented here but must instead be solved numerically, usually on a computer. Nevertheless, the regular occurrence of several simple types of integral equation that may be solved analytically is sufficient reason to explore these equations more fully.

We begin this chapter by discussing how a differential equation can be transformed into an integral equation and by considering the most common types of linear integral equation. After introducing the operator notation and considering the existence of solutions for various types of equation, we go on to discuss elementary methods of obtaining closed-form solutions of simple integral equations. We then consider the solution of integral equations in terms of infinite series and conclude by discussing the properties of integral equations with Hermitian kernels, i.e. those in which the integrands have particular symmetry properties.

13.1 Obtaining an integral equation from a differential equation

Integral equations occur in many situations, partly because we may always rewrite a differential equation as an integral equation. It is sometimes advantageous to make this transformation, since questions concerning the existence of a solution are more easily answered for integral equations (see Section 13.3), and, furthermore, an integral equation can incorporate automatically any boundary conditions on the solution.

We shall illustrate the principles involved by considering the differential equation

$$y''(x) = f(x, y), \tag{13.1}$$

where $f(x, y)$ can be any function of x and y but not of $y'(x)$. Equation (13.1) thus represents a large class of linear and non-linear second-order differential equations.

We can convert (13.1) into the corresponding integral equation by first integrating with respect to x to obtain

$$y'(x) = \int_0^x f(z, y(z))\,dz + c_1.$$

Integrating once more, we find

$$y(x) = \int_0^x du \int_0^u f(z, y(z))\,dz + c_1 x + c_2.$$

Provided we do not change the region in the uz-plane over which the double integral is taken, we can reverse the order of the two integrations. Changing the integration limits appropriately,[1] we find

$$y(x) = \int_0^x f(z, y(z))\,dz \int_z^x du + c_1 x + c_2 \qquad (13.2)$$

$$= \int_0^x (x - z) f(z, y(z))\,dz + c_1 x + c_2; \qquad (13.3)$$

this is a non-linear (for general $f(x, y)$) *Volterra* integral equation.

It is straightforward to incorporate any boundary conditions on the solution $y(x)$ by fixing the constants c_1 and c_2 in (13.3). For example, we might have the one-point boundary condition $y(0) = a$ and $y'(0) = b$, for which it is clear that we must set $c_1 = b$ and $c_2 = a$.

13.2 Types of integral equation

From (13.3), we can see that even a relatively simple differential equation such as (13.1) can lead to a corresponding integral equation that is non-linear. In this chapter, however, we will restrict our attention to *linear* integral equations, which have the general form

$$g(x)y(x) = f(x) + \lambda \int_a^b K(x, z)y(z)\,dz. \qquad (13.4)$$

In (13.4), $y(x)$ is the unknown function, while the functions $f(x)$, $g(x)$ and $K(x, z)$ are assumed known. $K(x, z)$ is called the *kernel* of the integral equation. The integration limits a and b are also assumed known, and may be constants or functions of x, and λ is a known constant or parameter.

In fact, we will be concerned only with various special cases of (13.4), which are known by particular names. Firstly, if $g(x) = 0$ then the unknown function $y(x)$ appears only under the integral sign, and (13.4) is called a linear integral equation *of the first kind*. Alternatively, if $g(x) = 1$, so that $y(x)$ appears twice, once inside the integral and once outside, then (13.4) is called a linear integral equation *of the second kind*. In either case, if $f(x) = 0$ the equation is called *homogeneous*, otherwise it is *inhomogeneous*.

We can distinguish further between different types of integral equation by the form of the integration limits a and b. If these limits are fixed constants then the equation is

1 Identify the relevant area in the uz-plane and verify that the two prescriptions for it are consistent.

called a *Fredholm* equation. If, however, the upper limit $b = x$ (i.e. it is variable) then the equation is called a *Volterra* equation; such an equation is analogous to one with fixed limits but for which the kernel $K(x, z) = 0$ for $z > x$. Finally, we note that any equation for which either (or both) of the integration limits is infinite, or for which $K(x, z)$ becomes infinite in the range of integration, is called a *singular* integral equation.

13.3 Operator notation and the existence of solutions

There is a close correspondence between linear integral equations and the matrix equations discussed in Chapter 1. However, the former involve linear, integral relations between functions in an infinite-dimensional function space (see Chapter 8), whereas the latter specify linear relations among vectors in a finite-dimensional vector space.

Since we are restricting our attention to linear integral equations, it will be convenient to introduce the linear integral operator $\mathcal{K}$, whose action on an arbitrary function y is given by

$$\mathcal{K}y = \int_a^b K(x, z)y(z)\, dz. \tag{13.5}$$

This is analogous to the introduction in Chapters 7 and 8 of the notation $\mathcal{L}$ to describe a linear differential operator. Furthermore, we may define the Hermitian conjugate $\mathcal{K}^\dagger$ by

$$\mathcal{K}^\dagger y = \int_a^b K^*(z, x)y(z)\, dz,$$

where the asterisk denotes complex conjugation and we have reversed the order of the arguments in the kernel.[2]

It is clear from (13.5) that $\mathcal{K}$ is indeed linear. Moreover, since $\mathcal{K}$ operates on the infinite-dimensional space of (reasonable) functions, we may make an obvious analogy with matrix equations and consider the action of $\mathcal{K}$ on a function f as that of a matrix on a column vector (both of infinite dimension).

When written in operator form, the integral equations discussed in the previous section resemble equations familiar from linear algebra. For example, the inhomogeneous Fredholm equation of the first kind may be written as

$$0 = f + \lambda\mathcal{K}y,$$

which has the unique solution $y = -\mathcal{K}^{-1}f/\lambda$, provided that $f \neq 0$ and the inverse operator $\mathcal{K}^{-1}$ exists.

Similarly, we may write the corresponding Fredholm equation of the second kind as

$$y = f + \lambda\mathcal{K}y. \tag{13.6}$$

In the homogeneous case, where $f = 0$, this reduces to $y = \lambda\mathcal{K}y$, which is reminiscent of an eigenvalue problem in linear algebra (except that λ appears on the other side of

2 If $\mathcal{K}y = \int_0^x (x^2 - z^2)e^{ikz}y(z)\, dz$, what is the explicit integral representation of $\mathcal{K}^\dagger y$?

the equation) and, similarly, only has solutions for at most a countably infinite set of *eigenvalues* λ_i. The corresponding solutions y_i are called the eigenfunctions.

In the inhomogeneous case ($f \neq 0$), the solution to (13.6) can be written symbolically as

$$y = (1 - \lambda\mathcal{K})^{-1} f,$$

again provided that the inverse operator exists. It may be shown that, in general, (13.6) does possess a unique solution if $\lambda \neq \lambda_i$, i.e. when λ does not equal one of the eigenvalues of the corresponding homogeneous equation.

When λ does equal one of these eigenvalues, (13.6) may have either many solutions or no solution at all, depending on the form of f. If the function f is orthogonal to *every* eigenfunction g of the equation

$$g = \lambda^* \mathcal{K}^\dagger g \tag{13.7}$$

that belongs to the eigenvalue λ^*, i.e.

$$\langle g|f \rangle = \int_a^b g^*(x) f(x)\, dx = 0$$

for every function g obeying (13.7), then it can be shown that (13.6) has many solutions. Otherwise the equation has no solution. These statements are discussed further in Section 13.7, for the special case of integral equations with Hermitian kernels, i.e. those for which $\mathcal{K} = \mathcal{K}^\dagger$.

13.4 Closed-form solutions

In certain very special cases, it may be possible to obtain a closed-form solution of an integral equation. The reader should realize, however, when faced with an integral equation, that in general it will not be soluble by the simple methods presented in this section but must instead be solved using (numerical) iterative methods, such as those outlined in Section 13.5.

13.4.1 Separable kernels

The most straightforward integral equations to solve are Fredholm equations with *separable* (or *degenerate*) kernels. A kernel is separable if it has the form

$$K(x, z) = \sum_{i=1}^{n} \phi_i(x)\psi_i(z), \tag{13.8}$$

where $\phi_i(x)$ and $\psi_i(z)$ are respectively functions of x only and of z only and the number of terms in the sum, n, is finite.[3]

3 Which of the following forms for $K(x, z)$ would be separable kernels: (a) $(\sinh x)(\sin z)$, (b) $\sin(xz)$, (c) $\sinh(x + z)$, (d) $\sinh(x/z)$, (e) $(\sin x)/(\sinh z)$, (f) $\sin(x \sin z)$?

Let us consider the solution of the (inhomogeneous) Fredholm equation of the second kind,

$$y(x) = f(x) + \lambda \int_a^b K(x,z)y(z)\,dz, \tag{13.9}$$

which has a separable kernel of the form (13.8). Writing the kernel in its separated form, the functions $\phi_i(x)$ may be taken outside the integral over z to obtain

$$y(x) = f(x) + \lambda \sum_{i=1}^{n} \phi_i(x) \int_a^b \psi_i(z)y(z)\,dz.$$

Since the integration limits a and b are constant for a Fredholm equation, the integral over z in each term of the sum is just a constant. Denoting these constants by

$$c_i = \int_a^b \psi_i(z)y(z)\,dz, \tag{13.10}$$

the solution to (13.9) is found to be

$$y(x) = f(x) + \lambda \sum_{i=1}^{n} c_i\phi_i(x), \tag{13.11}$$

where the constants c_i can be evaluated by substituting (13.11) into (13.10).

Example Solve the integral equation

$$y(x) = x + \lambda \int_0^1 (xz + z^2)y(z)\,dz. \tag{13.12}$$

The kernel for this equation is $K(x,z) = xz + z^2$, which is clearly separable, and using the notation in (13.8) we have $\phi_1(x) = x$, $\phi_2(x) = 1$, $\psi_1(z) = z$ and $\psi_2(z) = z^2$. From (13.11) the solution to (13.12) has the form

$$y(x) = x + \lambda(c_1x + c_2),$$

where the constants c_1 and c_2 are given by (13.10) as

$$c_1 = \int_0^1 z[z + \lambda(c_1z + c_2)]\,dz = \tfrac{1}{3} + \tfrac{1}{3}\lambda c_1 + \tfrac{1}{2}\lambda c_2,$$

$$c_2 = \int_0^1 z^2[z + \lambda(c_1z + c_2)]\,dz = \tfrac{1}{4} + \tfrac{1}{4}\lambda c_1 + \tfrac{1}{3}\lambda c_2.$$

These two simultaneous linear equations may be straightforwardly solved for c_1 and c_2 to give

$$c_1 = \frac{24+\lambda}{72 - 48\lambda - \lambda^2} \quad \text{and} \quad c_2 = \frac{18}{72 - 48\lambda - \lambda^2}.$$

Resubstitution of these values shows that

$$y(x) = \frac{(72 - 24\lambda)x + 18\lambda}{72 - 48\lambda - \lambda^2}$$

is the solution to the original integral equation. ◀

In the above example, we see that (13.12) has a (finite) unique solution provided that λ is not equal to either root of the quadratic in the denominator of $y(x)$. The roots of this quadratic are in fact the *eigenvalues* of the corresponding homogeneous equation, as mentioned in the previous section. In general, if the separable kernel contains n terms, as in (13.8), there will be n such eigenvalues, although they may not all be different.

Kernels consisting of trigonometric (or hyperbolic) functions of sums or differences of x and z are often separable.

Example Find the eigenvalues and corresponding eigenfunctions of the homogeneous Fredholm equation

$$y(x) = \lambda \int_0^{\pi} \sin(x+z)\, y(z)\, dz. \tag{13.13}$$

The kernel of this integral equation can be written in separated form as

$$K(x,z) = \sin(x+z) = \sin x \cos z + \cos x \sin z,$$

so, comparing with (13.8), we have $\phi_1(x) = \sin x$, $\phi_2(x) = \cos x$, $\psi_1(z) = \cos z$ and $\psi_2(z) = \sin z$.

Thus, from (13.11), the solution to (13.13) has the form

$$y(x) = \lambda(c_1 \sin x + c_2 \cos x),$$

where the constants c_1 and c_2 are given by

$$c_1 = \lambda \int_0^{\pi} \cos z\,(c_1 \sin z + c_2 \cos z)\, dz \;=\; \frac{\lambda\pi}{2} c_2, \tag{13.14}$$

$$c_2 = \lambda \int_0^{\pi} \sin z\,(c_1 \sin z + c_2 \cos z)\, dz \;=\; \frac{\lambda\pi}{2} c_1. \tag{13.15}$$

Combining these two equations we find $c_1 = (\lambda\pi/2)^2 c_1$, and, assuming that $c_1 \neq 0$, this gives $\lambda = \pm 2/\pi$, the two eigenvalues of the integral equation (13.13).

By substituting each of the eigenvalues back into (13.14) and (13.15), we find that the eigenfunctions corresponding to the eigenvalues $\lambda_1 = 2/\pi$ and $\lambda_2 = -2/\pi$ are given respectively by

$$y_1(x) = A(\sin x + \cos x) \qquad \text{and} \qquad y_2(x) = B(\sin x - \cos x), \tag{13.16}$$

where A and B are arbitrary constants. ◀

13.4.2 Integral transform methods

If the kernel of an integral equation can be written as a function of the difference $x - z$ of its two arguments, then it is called a *displacement* kernel. An integral equation having such a kernel, and which also has the integration limits $-\infty$ to ∞, may be solved by the use of Fourier transforms (Chapter 5).

If we consider the following integral equation with a displacement kernel,

$$y(x) = f(x) + \lambda \int_{-\infty}^{\infty} K(x-z) y(z)\, dz, \tag{13.17}$$

the integral over z clearly takes the form of a convolution (see Chapter 5). Therefore, Fourier-transforming (13.17) and using the convolution theorem, we obtain

$$\tilde{y}(k) = \tilde{f}(k) + \sqrt{2\pi}\lambda\tilde{K}(k)\tilde{y}(k),$$

which may be rearranged to give

$$\tilde{y}(k) = \frac{\tilde{f}(k)}{1 - \sqrt{2\pi}\lambda\tilde{K}(k)}. \tag{13.18}$$

Taking the inverse Fourier transform, the solution to (13.17) is given by

$$y(x) = \frac{1}{\sqrt{2\pi}} \int_{-\infty}^{\infty} \frac{\tilde{f}(k)\exp(ikx)}{1 - \sqrt{2\pi}\lambda\tilde{K}(k)}\, dk.$$

If we can evaluate this inverse Fourier transformation then the solution may be found explicitly; otherwise it must be left in the form of an integral.

Example Find the Fourier transform of the function

$$g(x) = \begin{cases} 1 & \text{if } |x| \le a, \\ 0 & \text{if } |x| > a. \end{cases}$$

Hence find an explicit expression for the solution of the integral equation

$$y(x) = f(x) + \lambda \int_{-\infty}^{\infty} \frac{\sin(x-z)}{x-z} y(z)\, dz. \tag{13.19}$$

Find the solution for the special case $f(x) = (\sin x)/x$.

The Fourier transform of $g(x)$ is given directly by

$$\tilde{g}(k) = \frac{1}{\sqrt{2\pi}} \int_{-a}^{a} \exp(-ikx)\, dx = \left[\frac{1}{\sqrt{2\pi}} \frac{\exp(-ikx)}{(-ik)}\right]_{-a}^{a} = \sqrt{\frac{2}{\pi}} \frac{\sin ka}{k}. \tag{13.20}$$

The kernel of the integral equation (13.19) is $K(x-z) = [\sin(x-z)]/(x-z)$. Using (13.20), it is straightforward to show that the Fourier transform of the kernel is

$$\tilde{K}(k) = \begin{cases} \sqrt{\pi/2} & \text{if } |k| \le 1, \\ 0 & \text{if } |k| > 1. \end{cases} \tag{13.21}$$

Thus, using (13.18), we find the Fourier transform of the solution to be

$$\tilde{y}(k) = \begin{cases} \tilde{f}(k)/(1 - \pi\lambda) & \text{if } |k| \le 1, \\ \tilde{f}(k) & \text{if } |k| > 1. \end{cases} \tag{13.22}$$

Inverse Fourier-transforming, and writing the result in a slightly more convenient form, the solution to (13.19) is given by

$$\begin{aligned} y(x) &= f(x) + \left(\frac{1}{1-\pi\lambda} - 1\right) \frac{1}{\sqrt{2\pi}} \int_{-1}^{1} \tilde{f}(k)\exp(ikx)\, dk \\ &= f(x) + \frac{\pi\lambda}{1-\pi\lambda} \frac{1}{\sqrt{2\pi}} \int_{-1}^{1} \tilde{f}(k)\exp(ikx)\, dk. \end{aligned} \tag{13.23}$$

It is clear from (13.22) that when $\lambda = 1/\pi$, which is the only eigenvalue of the corresponding homogeneous equation to (13.19), the solution becomes infinite, as we would expect.

For the special case $f(x) = (\sin x)/x$, the Fourier transform $\tilde{f}(k)$ is identical to that in (13.21), and the solution (13.23) becomes

$$\begin{aligned} y(x) &= \frac{\sin x}{x} + \left(\frac{\pi\lambda}{1-\pi\lambda}\right)\frac{1}{\sqrt{2\pi}}\int_{-1}^{1}\sqrt{\frac{\pi}{2}}\exp(ikx)\,dk \\ &= \frac{\sin x}{x} + \left(\frac{\pi\lambda}{1-\pi\lambda}\right)\frac{1}{2}\left[\frac{\exp(ikx)}{ix}\right]_{k=-1}^{k=1} \\ &= \frac{\sin x}{x} + \left(\frac{\pi\lambda}{1-\pi\lambda}\right)\frac{\sin x}{x} = \left(\frac{1}{1-\pi\lambda}\right)\frac{\sin x}{x}. \end{aligned}$$

In this particular case the solution $y(x)$ is simply a multiple of $f(x)$. Of course, this is not normally the case, and only comes about because of the obvious relationship between $f(x)$ and the kernel of the integral. ◀

If the integral equation (13.17) had integration limits 0 and x (so making it a Volterra equation), then its solution could be found in a similar way, but by using the convolution theorem for Laplace transforms (see Chapter 5) rather than that for Fourier transforms. We would find

$$\bar{y}(s) = \frac{\bar{f}(s)}{1-\lambda\bar{K}(s)},$$

where s is the Laplace transform variable. Often one is able to use the dictionary of Laplace transforms given in Table 5.1 to invert this equation and find the solution $y(x)$. In general, however, the evaluation of inverse Laplace transform integrals is difficult, since (in principle) it requires a contour integration; see Chapter 14.

As a final example of the use of integral transforms in solving integral equations, we mention equations that have integration limits $-\infty$ and ∞ and a kernel of the form

$$K(x, z) = \exp(-ixz).$$

Consider, for example, the inhomogeneous Fredholm equation

$$y(x) = f(x) + \lambda\int_{-\infty}^{\infty}\exp(-ixz)\,y(z)\,dz. \tag{13.24}$$

The integral over z is clearly just (a multiple of) the Fourier transform of $y(z)$, so we can write

$$y(x) = f(x) + \sqrt{2\pi}\lambda\tilde{y}(x). \tag{13.25}$$

If we now take the Fourier transform of (13.25) but continue to denote the independent variable by x (i.e. rather than k, for example), we obtain

$$\tilde{y}(x) = \tilde{f}(x) + \sqrt{2\pi}\lambda y(-x). \tag{13.26}$$

Substituting (13.26) into (13.25) we find

$$y(x) = f(x) + \sqrt{2\pi}\lambda\left[\tilde{f}(x) + \sqrt{2\pi}\lambda y(-x)\right],$$

but on making the change $x \to -x$ and substituting back in for $y(-x)$, this gives

$$y(x) = f(x) + \sqrt{2\pi}\lambda \tilde{f}(x) + 2\pi\lambda^2\left[f(-x) + \sqrt{2\pi}\lambda \tilde{f}(-x) + 2\pi\lambda^2 y(x)\right].$$

Thus the solution to (13.24) is given by

$$y(x) = \frac{1}{1-(2\pi)^2\lambda^4}\left[f(x) + (2\pi)^{1/2}\lambda \tilde{f}(x) + 2\pi\lambda^2 f(-x) + (2\pi)^{3/2}\lambda^3 \tilde{f}(-x)\right]. \tag{13.27}$$

Clearly, (13.24) possesses a unique solution provided $\lambda \neq \pm 1/\sqrt{2\pi}$ or $\pm i/\sqrt{2\pi}$; these are easily shown to be the eigenvalues of the corresponding homogeneous equation (i.e. one in which $f(x) \equiv 0$).

Example Solve the integral equation

$$y(x) = \exp\left(-\frac{x^2}{2}\right) + \lambda \int_{-\infty}^{\infty} \exp(-ixz)\, y(z)\, dz, \tag{13.28}$$

where λ is a real constant. Show that the solution is unique unless λ has one of two particular values. Does a solution exist for either of these two values of λ?

Following the argument given above, the solution to (13.28) is given by (13.27) with $f(x) = \exp(-x^2/2)$. In order to write the solution explicitly, however, we must calculate the Fourier transform of $f(x)$. Using equation (5.7), we find $\tilde{f}(k) = \exp(-k^2/2)$, from which we note that $f(x)$ has the special property that its functional form is identical to that of its Fourier transform. Thus, the solution to (13.28) is given by

$$y(x) = \frac{1}{1-(2\pi)^2\lambda^4}\left[1 + (2\pi)^{1/2}\lambda + 2\pi\lambda^2 + (2\pi)^{3/2}\lambda^3\right]\exp\left(-\frac{x^2}{2}\right). \tag{13.29}$$

Since λ is restricted to be real, the solution to (13.28) will be unique unless $\lambda = \pm 1/\sqrt{2\pi}$, at which points (13.29) becomes infinite. In order to find whether solutions exist for either of these values of λ we must return to equations (13.25) and (13.26).

Let us first consider the case $\lambda = +1/\sqrt{2\pi}$. Putting this value into (13.25) and (13.26), we obtain

$$y(x) = f(x) + \tilde{y}(x), \tag{13.30}$$

$$\tilde{y}(x) = \tilde{f}(x) + y(-x). \tag{13.31}$$

Substituting (13.31) into (13.30) we find

$$y(x) = f(x) + \tilde{f}(x) + y(-x),$$

but on changing x to $-x$ and substituting back in for $y(-x)$, this gives

$$y(x) = f(x) + \tilde{f}(x) + f(-x) + \tilde{f}(-x) + y(x).$$

Thus, in order for a solution to exist, we require that the function $f(x)$ obeys

$$f(x) + \tilde{f}(x) + f(-x) + \tilde{f}(-x) = 0.$$

This is satisfied if $f(x) = -\tilde{f}(x)$, i.e. if the functional form of $f(x)$ is minus the form of its Fourier transform. We may repeat this analysis for the case $\lambda = -1/\sqrt{2\pi}$, and, in a similar way, we find that this time we require $f(x) = \tilde{f}(x)$.

In our case $f(x) = \exp(-x^2/2)$, for which, as we mentioned above, $f(x) = \tilde{f}(x)$. Therefore, (13.28) possesses no solution when $\lambda = +1/\sqrt{2\pi}$ but has many solutions when $\lambda = -1/\sqrt{2\pi}$. ◀

A similar approach to the above may be taken to solve equations with kernels of the form $K(x, y) = \cos xy$ or $\sin xy$, either by considering the integral over y in each case as the real or imaginary part of the corresponding Fourier transform or by using Fourier cosine or sine transforms directly.

13.4.3 Differentiation

A closed-form solution to a Volterra equation can sometimes be obtained by differentiating the equation to obtain the corresponding differential equation; this may be easier to solve than the original integral equation.[4]

Example Solve the integral equation

$$y(x) = x - \int_0^x xz^2 y(z)\,dz. \tag{13.32}$$

Dividing through by x, we obtain

$$\frac{y(x)}{x} = 1 - \int_0^x z^2 y(z)\,dz,$$

which may be differentiated with respect to x to give

$$\frac{d}{dx}\left[\frac{y(x)}{x}\right] = -x^2 y(x) = -x^3\left[\frac{y(x)}{x}\right].$$

This equation may be integrated straightforwardly, and we find

$$\ln\left[\frac{y(x)}{x}\right] = -\frac{x^4}{4} + c,$$

where c is a constant of integration. Thus the solution to (13.32) has the form

$$y(x) = Ax\exp\left(-\frac{x^4}{4}\right), \tag{13.33}$$

where A is an arbitrary constant.[5]

Since the original integral equation (13.32) contains no arbitrary constants, neither should its solution. We may calculate the value of the constant, A, by substituting the solution (13.33) back into (13.32), from which we find $A = 1$. ◀

4 Note that this general approach cannot be applied to integral equations involving fixed integration limits, e.g. Fredholm equations.

5 Show that if the original equation is not divided through by x before differentiation, the second-order equation $y'' + x^3 y' + 4x^2 y = 0$ is obtained for $y(x)$. Verify that the solution actually obtained does indeed satisfy this differential equation.

13.5 Neumann series

As mentioned above, most integral equations met in practice will not be of the simple forms discussed in the previous section and so, in general, it is not possible to find closed-form solutions. In such cases, we might try to obtain a solution in the form of an infinite series, as we did for differential equations (see Chapter 7).

Let us consider the equation

$$y(x) = f(x) + \lambda \int_a^b K(x,z) y(z)\, dz, \tag{13.34}$$

where either both integration limits are constants (for a Fredholm equation) or the upper limit is variable (for a Volterra equation). Clearly, if λ were small then a crude (but reasonable) approximation to the solution would be

$$y(x) \approx y_0(x) = f(x),$$

where $y_0(x)$ stands for our "zeroth-order" approximation to the solution (and is not to be confused with an eigenfunction).

Substituting this crude guess under the integral sign in the original equation, we obtain what should be a better approximation:

$$y_1(x) = f(x) + \lambda \int_a^b K(x,z) y_0(z)\, dz = f(x) + \lambda \int_a^b K(x,z) f(z)\, dz,$$

which is first order in λ. Repeating the procedure once more results in the second-order approximation

$$\begin{aligned} y_2(x) &= f(x) + \lambda \int_a^b K(x,z) y_1(z)\, dz \\ &= f(x) + \lambda \int_a^b K(x,z_1) f(z_1)\, dz_1 + \lambda^2 \int_a^b dz_1 \int_a^b K(x,z_1) K(z_1,z_2) f(z_2)\, dz_2. \end{aligned}$$

It is clear that we may continue this process to obtain progressively higher-order approximations to the solution. Introducing the functions

$$\begin{aligned} K_1(x,z) &= K(x,z), \\ K_2(x,z) &= \int_a^b K(x,z_1) K(z_1,z)\, dz_1, \\ K_3(x,z) &= \int_a^b dz_1 \int_a^b K(x,z_1) K(z_1,z_2) K(z_2,z)\, dz_2, \end{aligned}$$

and so on, which obey the recurrence relation

$$K_n(x,z) = \int_a^b K(x,z_1) K_{n-1}(z_1,z)\, dz_1,$$

we may write the nth-order approximation as

$$y_n(x) = f(x) + \sum_{m=1}^{n} \lambda^m \int_a^b K_m(x, z) f(z)\, dz. \tag{13.35}$$

The solution to the original integral equation is then given by $y(x) = \lim_{n\to\infty} y_n(x)$, *provided the infinite series converges*. Using (13.35), this solution may be written as

$$y(x) = f(x) + \lambda \int_a^b R(x, z; \lambda) f(z)\, dz, \tag{13.36}$$

where the *resolvent kernel* $R(x, z; \lambda)$ is given by

$$R(x, z; \lambda) = \sum_{m=0}^{\infty} \lambda^m K_{m+1}(x, z). \tag{13.37}$$

Clearly, the resolvent kernel, and hence the series solution, will converge provided λ is sufficiently small. In fact, it may be shown that the series converges in some domain of $|\lambda|$ provided the original kernel $K(x, z)$ is bounded in such a way that

$$|\lambda|^2 \int_a^b dx \int_a^b |K(x, z)|^2\, dz < 1. \tag{13.38}$$

Example Use the Neumann series method to solve the integral equation

$$y(x) = x + \lambda \int_0^1 xzy(z)\, dz. \tag{13.39}$$

Following the method outlined above, we begin with the crude approximation $y(x) \approx y_0(x) = x$. Substituting this under the integral sign in (13.39), we obtain the next approximation

$$y_1(x) = x + \lambda \int_0^1 xzy_0(z)\, dz = x + \lambda \int_0^1 xz^2 dz = x + \frac{\lambda x}{3}.$$

Repeating the procedure once more, we obtain

$$\begin{aligned} y_2(x) &= x + \lambda \int_0^1 xzy_1(z)\, dz \\ &= x + \lambda \int_0^1 xz\left(z + \frac{\lambda z}{3}\right) dz = x + \left(\frac{\lambda}{3} + \frac{\lambda^2}{9}\right) x. \end{aligned}$$

For this simple example, it is easy to see that by continuing this process the solution to (13.39) is obtained as

$$y(x) = x + \left[\frac{\lambda}{3} + \left(\frac{\lambda}{3}\right)^2 + \left(\frac{\lambda}{3}\right)^3 + \cdots\right] x.$$

Clearly the expression in brackets is an infinite geometric series with first term $\lambda/3$ and common ratio $\lambda/3$. Thus, *provided* $|\lambda| < 3$, this infinite series converges to the value $\lambda/(3 - \lambda)$, and the solution to (13.39) is

$$y(x) = x + \frac{\lambda x}{3-\lambda} = \frac{3x}{3-\lambda}. \tag{13.40}$$

Finally, we note that the requirement that $|\lambda| < 3$ may also be derived very easily from the condition (13.38).[6] ◀

13.6 Fredholm theory

In the previous section, we found that a solution to the integral equation (13.34) can be obtained as a Neumann series of the form (13.36), where the resolvent kernel $R(x, z; \lambda)$ is written as an infinite power series in λ. This solution is valid provided the infinite series converges.

A related, but more elegant, approach to the solution of integral equations using infinite series was found by Fredholm. We will not reproduce Fredholm's analysis here, but merely state the results we need. Essentially, *Fredholm theory* provides a formula for the resolvent kernel $R(x, z; \lambda)$ in (13.36) in terms of the ratio of two infinite series:

$$R(x, z; \lambda) = \frac{D(x, z; \lambda)}{d(\lambda)}. \tag{13.41}$$

The numerator and denominator in (13.41) are given by

$$D(x, z; \lambda) = \sum_{n=0}^{\infty} \frac{(-1)^n}{n!} D_n(x, z)\lambda^n, \tag{13.42}$$

$$d(\lambda) = \sum_{n=0}^{\infty} \frac{(-1)^n}{n!} d_n \lambda^n, \tag{13.43}$$

where the functions $D_n(x, z)$ and the constants d_n are found from recurrence relations as follows. We start with

$$D_0(x, z) = K(x, z) \qquad \text{and} \qquad d_0 = 1, \tag{13.44}$$

where $K(x, z)$ is the kernel of the original integral equation (13.34). The higher-order coefficients of λ in (13.43) and (13.42) are then obtained from the two recurrence relations

$$d_n = \int_a^b D_{n-1}(x, x)\, dx, \tag{13.45}$$

$$D_n(x, z) = K(x, z)d_n - n \int_a^b K(x, z_1) D_{n-1}(z_1, z)\, dz_1. \tag{13.46}$$

Although the formulae for the resolvent kernel appear complicated, they are often simple to apply. Moreover, for the Fredholm solution the power series (13.42) and (13.43) are

6 It will be clear that this upper bound for $|\lambda|$ plays much the same role as does the radius of convergence in the study of power series.

both guaranteed to converge for all values of λ, unlike Neumann series, which converge only if the condition (13.38) is satisfied. Thus the Fredholm method leads to a unique, non-singular solution, provided that $d(\lambda) \neq 0$. In fact, as we might suspect, the solutions of $d(\lambda) = 0$ give the eigenvalues of the homogeneous equation corresponding to (13.34), i.e. with $f(x) \equiv 0$.

Example Use Fredholm theory to solve the integral equation (13.39).

Using (13.36) and (13.41), the solution to (13.39) can be written in the form

$$y(x) = x + \lambda \int_0^1 R(x, z; \lambda) z \, dz = x + \lambda \int_0^1 \frac{D(x, z; \lambda)}{d(\lambda)} z \, dz. \tag{13.47}$$

In order to find the form of the resolvent kernel $R(x, z; \lambda)$, we begin by setting

$$D_0(x, z) = K(x, z) = xz \qquad \text{and} \qquad d_0 = 1$$

and use the recurrence relations (13.45) and (13.46) to obtain

$$d_1 = \int_0^1 D_0(x, x) \, dx = \int_0^1 x^2 \, dx = \frac{1}{3},$$

$$D_1(x, z) = \frac{xz}{3} - \int_0^1 xz_1^2 z \, dz_1 = \frac{xz}{3} - xz \left[\frac{z_1^3}{3} \right]_0^1 = 0.$$

Applying the recurrence relations again we find that $d_n = 0$ and $D_n(x, z) = 0$ for $n > 1$. Thus, from (13.42) and (13.43), the numerator and denominator of the resolvent are given, respectively, by

$$D(x, z; \lambda) = xz \qquad \text{and} \qquad d(\lambda) = 1 - \frac{\lambda}{3}.$$

Substituting these expressions into (13.47), we find that the solution to (13.39) is given by

$$y(x) = x + \lambda \int_0^1 \frac{xz^2}{1 - \lambda/3} \, dz$$

$$= x + \lambda \left[\frac{x}{1 - \lambda/3} \frac{z^3}{3} \right]_0^1 = x + \frac{\lambda x}{3 - \lambda} = \frac{3x}{3 - \lambda},$$

which, as expected, is the same as the solution (13.40) found by constructing a Neumann series. ◀

13.7 Schmidt–Hilbert theory

The Schmidt–Hilbert (SH) theory of integral equations may be considered as analogous to the Sturm–Liouville theory of differential equations, discussed in Chapter 8, and is concerned with the properties of integral equations with *Hermitian* kernels. An Hermitian kernel enjoys the property

$$K(x, z) = K^*(z, x), \tag{13.48}$$

and it is clear that a special case of (13.48) occurs for a real kernel that is also symmetric with respect to its two arguments.[7]

Let us begin by considering the homogeneous integral equation

$$y = \lambda \mathcal{K} y,$$

where the integral operator $\mathcal{K}$ has an Hermitian kernel. As discussed in Section 13.3, in general this equation will have solutions only for $\lambda = \lambda_i$, where the λ_i are the eigenvalues of the integral equation, the corresponding solutions y_i being the eigenfunctions of the equation.

By following similar arguments to those presented in Chapter 8 for SL theory, it may be shown that the eigenvalues λ_i of an Hermitian kernel are real and that the corresponding eigenfunctions y_i belonging to different eigenvalues are orthogonal and form a complete set. If the eigenfunctions are suitably normalized, we have

$$\langle y_i | y_j \rangle = \int_a^b y_i^*(x) y_j(x)\, dx = \delta_{ij}. \tag{13.49}$$

If an eigenvalue is degenerate then the eigenfunctions corresponding to that eigenvalue can be made orthogonal by the Gram–Schmidt procedure, in a similar way to that discussed in Chapter 8 in the context of SL theory.

Like SL theory, SH theory does not provide a method of obtaining the eigenvalues and eigenfunctions of any particular homogeneous integral equation with an Hermitian kernel; for this we have to turn to the methods discussed in the previous sections of this chapter. Rather, SH theory is concerned with the general properties of the solutions to such equations. Where SH theory becomes applicable, however, is in the solution of inhomogeneous integral equations with Hermitian kernels for which the eigenvalues and eigenfunctions of the corresponding homogeneous equation are already known.

Let us consider the inhomogeneous equation

$$y = f + \lambda \mathcal{K} y, \tag{13.50}$$

where $\mathcal{K} = \mathcal{K}^\dagger$ and for which we know the eigenvalues λ_i and normalized eigenfunctions y_i of the corresponding homogeneous problem. The function f may or may not be expressible solely in terms of the eigenfunctions y_i, and to accommodate this situation we write the unknown solution y as $y = f + \sum_i a_i y_i$, where the a_i are expansion coefficients to be determined.

Substituting this into (13.50), we obtain

$$f + \sum_i a_i y_i = f + \lambda \sum_i \frac{a_i y_i}{\lambda_i} + \lambda \mathcal{K} f, \tag{13.51}$$

7 Which of the kernels listed in footnote 3 are Hermitian?

where we have used the fact that $y_i = \lambda_i \mathcal{K} y_i$. Forming the inner product of both sides of (13.51) with y_j, we find

$$\sum_i a_i \langle y_j | y_i \rangle = \lambda \sum_i \frac{a_i}{\lambda_i} \langle y_j | y_i \rangle + \lambda \langle y_j | \mathcal{K} f \rangle. \tag{13.52}$$

Since the eigenfunctions are orthonormal and $\mathcal{K}$ is an Hermitian operator, we have that both $\langle y_j | y_i \rangle = \delta_{ij}$ and $\langle y_j | \mathcal{K} f \rangle = \langle \mathcal{K} y_j | f \rangle = \lambda_j^{-1} \langle y_j | f \rangle$. Thus the coefficients a_j are given by

$$a_j = \frac{\lambda \lambda_j^{-1} \langle y_j | f \rangle}{1 - \lambda \lambda_j^{-1}} = \frac{\lambda \langle y_j | f \rangle}{\lambda_j - \lambda}, \tag{13.53}$$

and the solution is

$$y = f + \sum_i a_i y_i = f + \lambda \sum_i \frac{\langle y_i | f \rangle}{\lambda_i - \lambda} y_i. \tag{13.54}$$

This also shows, incidentally, that a formal representation for the resolvent kernel is[8]

$$R(x, z; \lambda) = \sum_i \frac{y_i(x) y_i^*(z)}{\lambda_i - \lambda}. \tag{13.55}$$

If f *can* be expressed as a linear superposition of the y_i, i.e. $f = \sum_i b_i y_i$, then $b_i = \langle y_i | f \rangle$ and the solution can be written more briefly as

$$y = \sum_i \frac{b_i}{1 - \lambda \lambda_i^{-1}} y_i. \tag{13.56}$$

We see from (13.54) that the inhomogeneous equation (13.50) has a unique solution provided $\lambda \neq \lambda_i$, i.e. when λ is not equal to one of the eigenvalues of the corresponding homogeneous equation. However, if λ does equal one of the eigenvalues λ_j then, in general, the coefficients a_j become singular and no (finite) solution exists.

Returning to (13.53), we notice that even if $\lambda = \lambda_j$ a non-singular solution to the integral equation is still possible provided that the function f is orthogonal to every eigenfunction corresponding to the eigenvalue λ_j, i.e.

$$\langle y_j | f \rangle = \int_a^b y_j^*(x) f(x)\, dx = 0.$$

The following worked example illustrates the case in which f can be expressed in terms of the y_i. One in which it cannot is considered in Problem 13.14.

8 That is, much like a Green's function.

Example Use Schmidt–Hilbert theory to solve the integral equation

$$y(x) = \sin(x+\alpha) + \lambda \int_0^{\pi} \sin(x+z)y(z)\,dz. \tag{13.57}$$

It is clear that the kernel $K(x,z) = \sin(x+z)$ is real and symmetric in x and z and is thus Hermitian. In order to solve this inhomogeneous equation using SH theory, however, we must first find the eigenvalues and eigenfunctions of the corresponding homogeneous equation.

In fact, we have considered the solution of the corresponding homogeneous equation (13.13) already, in Subsection 13.4.1, where we found that it has two eigenvalues $\lambda_1 = 2/\pi$ and $\lambda_2 = -2/\pi$, with eigenfunctions given by (13.16). The normalized eigenfunctions are

$$y_1(x) = \frac{1}{\sqrt{\pi}}(\sin x + \cos x) \quad \text{and} \quad y_2(x) = \frac{1}{\sqrt{\pi}}(\sin x - \cos x) \tag{13.58}$$

and are easily shown to obey the orthonormality condition (13.49).

Using (13.54), the solution to the inhomogeneous equation (13.57) has the form

$$y(x) = a_1 y_1(x) + a_2 y_2(x), \tag{13.59}$$

where the coefficients a_1 and a_2 are given by (13.53) with $f(x) = \sin(x+\alpha)$. Therefore, using (13.58),

$$a_1 = \frac{1}{1-\pi\lambda/2}\int_0^{\pi} \frac{1}{\sqrt{\pi}}(\sin z + \cos z)\sin(z+\alpha)\,dz = \frac{\sqrt{\pi}}{2-\pi\lambda}(\cos\alpha + \sin\alpha),$$

$$a_2 = \frac{1}{1+\pi\lambda/2}\int_0^{\pi} \frac{1}{\sqrt{\pi}}(\sin z - \cos z)\sin(z+\alpha)\,dz = \frac{\sqrt{\pi}}{2+\pi\lambda}(\cos\alpha - \sin\alpha).$$

Substituting these expressions for a_1 and a_2 into (13.59) and simplifying, we find that the solution to (13.57) is given by

$$y(x) = \frac{1}{1-(\pi\lambda/2)^2}\left[\sin(x+\alpha) + (\pi\lambda/2)\cos(x-\alpha)\right].$$

◀

SUMMARY

1. *Equation types*

- Linear of first kind: $0 = f(x) + \lambda \int_a^b K(x,z)y(z)\,dz$.
- Linear of second kind: $y(x) = f(x) + \lambda \int_a^b K(x,z)y(z)\,dz$.
- Fredholm: b fixed. Volterra: $b = x$. Singular: $a = \infty$ or $b = \infty$ or $K(x,z) = \infty$ for some z in $a \le z \le b$.

2. *Operator notation*
   - $\mathcal{K}y = \int_a^b K(x,z)y(z)\,dz$ defines linear operator $\mathcal{K}$.
   - $\mathcal{K}^\dagger y = \int_a^b K^*(z,x)y(z)\,dz$ defines Hermitian conjugate $\mathcal{K}^\dagger$.
   - $y_i = \lambda_i \mathcal{K} y_i$ gives the eigenvalues λ_i and eigenfunctions $y_i(x)$ of $\mathcal{K}$.
   - $y = f + \lambda \mathcal{K} y$ has a unique solution provided $\lambda \neq \lambda_i$ for any i.
   - Schmidt–Hilbert theory for Hermitian kernels ($\mathcal{K}^\dagger = \mathcal{K}$) and orthonormal eigenfunctions, $\langle y_i | y_j \rangle = \delta_{ij}$:

     (a) $y = f + \lambda \mathcal{K} y$ has solution $y = f + \lambda \sum_i \dfrac{\langle y_i | f \rangle}{\lambda_i - \lambda} y_i$.

     (b) For a case in which $f = \sum_i b_i y_i$ the solution is $y = \sum_i \dfrac{b_i}{1 - \lambda \lambda_i^{-1}} y_i$.

3. *Closed form solutions*

   (a) Separable kernel: $K(x,z) = \sum_{i=1}^{n} \phi_i(x)\psi_i(z)$.

   The solution is $y(x) = f(x) + \lambda \sum_{i=1}^{n} c_i \phi_i(x)$, with the c_i found by substitution and solving the resultant simultaneous algebraic equations, provided $\lambda \neq$ any of the n eigenvalues λ_i.

   (b) Displacement kernel: $K(x,z) = K(x - z)$.

   (i) If $a = -\infty$ and $b = +\infty$, then

   $$y(x) = \frac{1}{\sqrt{2\pi}} \int_{-\infty}^{\infty} \frac{\tilde{f}(k)\exp(ikx)}{1 - \sqrt{2\pi}\lambda \tilde{K}(k)}\,dk,$$

   where $\tilde{g}(k)$ is the Fourier transform of $g(z)$.

   (ii) If a is finite and $b = x$, then

   $$\bar{y}(s) = \frac{\bar{f}(s)}{1 - \lambda \bar{K}(s)}$$

   gives the Laplace transform of the required solution.

   (c) Kernels of the forms $\exp(-ixz)$, $\cos(xz)$ and $\sin(xz)$:
   Substitution and resubstitution for y and $\tilde{y}$ gives an explicit expression for $y(x)$ in terms of f and $\tilde{f}$.

4. *Iteration methods*

   (a) The Neumann series solution for $y = f + \lambda \mathcal{K} y$ is

   $$y(x) = f(x) + \lambda \int_a^b R(x,z;\lambda) f(z)\,dz,$$

where the resolvent kernel $R(x, z : \lambda)$ is given by

$$R(x, z; \lambda) = \sum_{m=0}^{\infty} \lambda^m K_{m+1}(x, z),$$

with $K_1(x, z) = K(x, z)$ and $K_n(x, z) = \int_a^b K(x, z_1)K_{n-1}(z_1, z)\, dz_1$. There is a range restriction on λ.

(b) Fredholm theory computes the resolvent kernel as

$$R(x, z : \lambda) = \left[\sum_{n=0}^{\infty} \frac{(-1)^n}{n!} D_n(x, z)\lambda^n\right] \div \left[\sum_{n=0}^{\infty} \frac{(-1)^n}{n!} d_n \lambda^n\right],$$

where $D_0(x, z) = K(x, z)$ and $d_0 = 1$. The D_n and d_n are found alternately from

$$d_n = \int_a^b D_{n-1}(x, x)\, dx,$$

$$D_n(x, z) = K(x, z)d_n - n \int_a^b K(x, z_1)D_{n-1}(z_1, z)\, dz.$$

PROBLEMS

13.1. Solve the integral equation

$$\int_0^{\infty} \cos(xv)y(v)\, dv = \exp(-x^2/2)$$

for the function $y = y(x)$ for $x > 0$. Note that for $x < 0$, $y(x)$ can be chosen as is most convenient.

13.2. Solve

$$\int_0^{\infty} f(t)\exp(-st)\, dt = \frac{a}{a^2 + s^2}.$$

13.3. Convert

$$f(x) = \exp x + \int_0^x (x - y)f(y)\, dy$$

into a differential equation, and hence show that its solution is

$$(\alpha + \beta x)\exp x + \gamma\, \exp(-x),$$

where α, β and γ are constants that should be determined.

13.4. Use the fact that its kernel is separable, to solve for $y(x)$ the integral equation

$$y(x) = A\cos(x+a) + \lambda \int_0^{\pi} \sin(x+z)y(z)\,dz.$$

[This equation is an inhomogeneous extension of the homogeneous Fredholm equation (13.13), and is similar to equation (13.57).]

13.5. Solve for $\phi(x)$ the integral equation

$$\phi(x) = f(x) + \lambda \int_0^1 \left[\left(\frac{x}{y}\right)^n + \left(\frac{y}{x}\right)^n\right]\phi(y)\,dy,$$

where $f(x)$ is bounded for $0 < x < 1$ and $-\frac{1}{2} < n < \frac{1}{2}$, expressing your answer in terms of the quantities $F_m = \int_0^1 f(y)y^m\,dy$.

(a) Give the explicit solution when $\lambda = 1$.

(b) For what values of λ are there no solutions unless $F_{\pm n}$ are in a particular ratio? What is this ratio?

13.6. Consider the inhomogeneous integral equation

$$f(x) = g(x) + \lambda \int_a^b K(x,y)f(y)\,dy,$$

for which the kernel $K(x, y)$ is real, symmetric and continuous in $a \le x \le b$, $a \le y \le b$.

(a) If λ is one of the eigenvalues λ_i of the homogeneous equation

$$f_i(x) = \lambda_i \int_a^b K(x,y)f_i(y)\,dy,$$

prove that the inhomogeneous equation can only have a non-trivial solution if $g(x)$ is orthogonal to the corresponding eigenfunction $f_i(x)$.

(b) Show that the only values of λ for which

$$f(x) = \lambda \int_0^1 xy(x+y)f(y)\,dy$$

has a non-trivial solution are the roots of the equation

$$\lambda^2 + 120\lambda - 240 = 0.$$

(c) Solve

$$f(x) = \mu x^2 + \int_0^1 2xy(x+y)f(y)\,dy.$$

13.7. The kernel of the integral equation

$$\psi(x) = \lambda \int_a^b K(x,y)\psi(y)\,dy$$

has the form

$$K(x, y) = \sum_{n=0}^{\infty} h_n(x) g_n(y),$$

where the $h_n(x)$ form a complete orthonormal set of functions over the interval $[a, b]$.

(a) Show that the eigenvalues λ_i are given by

$$|\mathsf{M} - \lambda^{-1}\mathsf{I}| = 0,$$

where M is the matrix with elements

$$M_{kj} = \int_a^b g_k(u) h_j(u)\, du.$$

If the corresponding solutions are $\psi^{(i)}(x) = \sum_{n=0}^{\infty} a_n^{(i)} h_n(x)$, find an expression for $a_n^{(i)}$.

(b) Obtain the eigenvalues and eigenfunctions over the interval $[0, 2\pi]$ if

$$K(x, y) = \sum_{n=1}^{\infty} \frac{1}{n} \cos nx \cos ny.$$

13.8. By taking its Laplace transform, and that of $x^n e^{-ax}$, obtain the explicit solution of

$$f(x) = e^{-x}\left[x + \int_0^x (x - u)e^u f(u)\, du\right].$$

Verify your answer by substitution.

13.9. For $f(t) = \exp(-t^2/2)$, use the relationships of the Fourier transforms of $f'(t)$ and $tf(t)$ to that of $f(t)$ itself to find a simple differential equation satisfied by $\tilde{f}(\omega)$, the Fourier transform of $f(t)$, and hence determine $\tilde{f}(\omega)$ to within a constant. Use this result to solve for $h(t)$ the integral equation

$$\int_{-\infty}^{\infty} e^{-t(t-2x)/2} h(t)\, dt = e^{3x^2/8}.$$

13.10. Show that the equation

$$f(x) = x^{-1/3} + \lambda \int_0^{\infty} f(y) \exp(-xy)\, dy$$

has a solution of the form $Ax^{\alpha} + Bx^{\beta}$. Determine the values of α and β, and show that those of A and B are

$$\frac{1}{1 - \lambda^2 \Gamma(\frac{1}{3})\Gamma(\frac{2}{3})} \quad \text{and} \quad \frac{\lambda \Gamma(\frac{2}{3})}{1 - \lambda^2 \Gamma(\frac{1}{3})\Gamma(\frac{2}{3})},$$

where $\Gamma(z)$ is the gamma function.

13.11. At an international "peace" conference a large number of delegates are seated around a circular table with each delegation sitting near its allies and diametrically opposite the delegation most bitterly opposed to it. The position of a delegate is denoted by θ, with $0 \leq \theta \leq 2\pi$. The fury $f(\theta)$ felt by the delegate at θ is the sum of his own natural hostility $h(\theta)$ and the influences on him of each of the other delegates; a delegate at position ϕ contributes an amount $K(\theta - \phi)f(\phi)$. Thus

$$f(\theta) = h(\theta) + \int_0^{2\pi} K(\theta - \phi) f(\phi)\, d\phi.$$

Show that if $K(\psi)$ takes the form $K(\psi) = k_0 + k_1 \cos\psi$ then

$$f(\theta) = h(\theta) + p + q\cos\theta + r\sin\theta$$

and evaluate p, q and r. A positive value for k_1 implies that delegates tend to placate their opponents but upset their allies, whilst negative values imply that they calm their allies but infuriate their opponents. A walkout will occur if $f(\theta)$ exceeds a certain threshold value for some θ. Is this more likely to happen for positive or for negative values of k_1?

13.12. By considering functions of the form $h(x) = \int_0^x (x - y) f(y)\, dy$, show that the solution $f(x)$ of the integral equation

$$f(x) = x + \tfrac{1}{2}\int_0^1 |x - y| f(y)\, dy$$

satisfies the equation $f''(x) = f(x)$.

By examining the special cases $x = 0$ and $x = 1$, show that

$$f(x) = \frac{2}{(e+3)(e+1)}[(e+2)e^x - ee^{-x}].$$

13.13. The operator $\mathcal{M}$ is defined by

$$\mathcal{M}f(x) \equiv \int_{-\infty}^{\infty} K(x, y) f(y)\, dy,$$

where $K(x, y) = 1$ inside the square $|x| < a$, $|y| < a$ and $K(x, y) = 0$ elsewhere. Consider the possible eigenvalues of $\mathcal{M}$ and the eigenfunctions that correspond to them; show that the only possible eigenvalues are 0 and $2a$ and determine the corresponding eigenfunctions. Hence find the general solution of

$$f(x) = g(x) + \lambda \int_{-\infty}^{\infty} K(x, y) f(y)\, dy.$$

13.14. For the integral equation

$$y(x) = x^{-3} + \lambda \int_a^b x^2 z^2 y(z)\, dz,$$

show that the resolvent kernel is $5x^2z^2/[5 - \lambda(b^5 - a^5)]$ and hence solve the equation. For what range of λ is the solution valid?

13.15. Use Fredholm theory to show that, for the kernel

$$K(x, z) = (x + z)\exp(x - z)$$

over the interval [0, 1], the resolvent kernel is

$$R(x, z; \lambda) = \frac{\exp(x - z)[(x + z) - \lambda(\frac{1}{2}x + \frac{1}{2}z - xz - \frac{1}{3})]}{1 - \lambda - \frac{1}{12}\lambda^2},$$

and hence solve

$$y(x) = x^2 + 2\int_0^1 (x + z)\exp(x - z)\, y(z)\, dz,$$

expressing your answer in terms of I_n, where $I_n = \int_0^1 u^n \exp(-u)\, du$.

13.16. This exercise shows that following formal theory is not necessarily the best way to get practical results!

(a) Determine the eigenvalues $\lambda_\pm$ of the kernel $K(x, z) = (xz)^{1/2}(x^{1/2} + z^{1/2})$ and show that the corresponding eigenfunctions have the forms

$$y_\pm(x) = A_\pm(\sqrt{2}x^{1/2} \pm \sqrt{3}x),$$

where $A_\pm^2 = 5/(10 \pm 4\sqrt{6})$.

(b) Use Schmidt–Hilbert theory to solve

$$y(x) = 1 + \tfrac{5}{2}\int_0^1 K(x, z)y(z)\, dz.$$

(c) As will have been apparent, the algebra involved in the formal method used in (b) is long and error-prone, and it is in fact much more straightforward to use a trial function $1 + \alpha x^{1/2} + \beta x$. Check your answer by doing so.

HINTS AND ANSWERS

13.1. Define $y(-x) = y(x)$ and use the cosine Fourier transform inversion theorem; $y(x) = (2/\pi)^{1/2}\exp(-x^2/2)$.

13.3. $f''(x) - f(x) = \exp x$; $\alpha = 3/4$, $\beta = 1/2$, $\gamma = 1/4$.

13.5. (a) $\phi(x) = f(x) - (1 + 2n)F_n x^n - (1 - 2n)F_{-n}x^{-n}$. (b) There are no solutions for $\lambda = [1 \pm (1 - 4n^2)^{-1/2}]^{-1}$ unless $F_{\pm n} = 0$ or $F_n/F_{-n} = \mp[(1 - 2n)/(1 + 2n)]^{1/2}$.

13.7. (a) $a_n^{(i)} = \int_a^b h_n(x)\psi^{(i)}(x)\, dx$; (b) use $(1/\sqrt{\pi})\cos nx$ and $(1/\sqrt{\pi})\sin nx$; M is diagonal; eigenvalues $\lambda_k = k/\pi$ with eigenfunctions $\psi^{(k)}(x) = (1/\sqrt{\pi})\cos kx$.

13.9. $d\tilde{f}/d\omega = -\omega\tilde{f}$, leading to $\tilde{f}(\omega) = Ae^{-\omega^2/2}$. Rearrange the integral as a convolution and deduce that $\tilde{h}(\omega) = Be^{-3\omega^2/2}$; $h(t) = Ce^{-t^2/6}$, where resubstitution and Gaussian normalization show that $C = \sqrt{2/(3\pi)}$.

13.11. $p = k_0 H/(1 - 2\pi k_0)$, $q = k_1 H_c/(1 - \pi k_1)$ and $r = k_1 H_s/(1 - \pi k_1)$, where $H = \int_0^{2\pi} h(z)\,dz$, $H_c = \int_0^{2\pi} h(z)\cos z\,dz$, and $H_s = \int_0^{2\pi} h(z)\sin z\,dz$. Positive values of $k_1 (\approx \pi^{-1})$ are most likely to cause a conference breakdown.

13.13. For eigenvalue $0 : f(x) = 0$ for $|x| < a$ or $f(x)$ is such that $\int_{-a}^{a} f(y)dy = 0$. For eigenvalue $2a : f(x) = \mu S(x, a)$ with μ a constant and $S(x, a) \equiv [H(a + x) - H(x - a)]$, where $H(z)$ is the Heaviside step function. Take $f(x) = g(x) + cGS(x, a)$, where $G = \int_{-a}^{a} g(z)\,dz$. Show that $c = \lambda/(1 - 2a\lambda)$.

13.15. $y(x) = x^2 - (3I_3 x + I_2)\exp x$.

14 Complex variables

Throughout this book references have been made to results derived from the theory of complex variables. This theory thus becomes an integral part of the mathematics appropriate to physical applications. Indeed, so numerous and widespread are these applications that the whole of the next chapter is devoted to a systematic presentation of some of the more important ones and a summary of some of the others. This current chapter develops the general theory on which these applications are based. The difficulty with it, from the point of view of a book such as the present one, is that the underlying basis has a distinctly pure mathematics flavor.

Thus, to adopt a comprehensive rigorous approach would involve a large amount of groundwork in analysis, for example formulating precise definitions of continuity and differentiability, developing the theory of sets and making a detailed study of boundedness. Instead, we will be selective and pursue only those parts of the formal theory that are needed to establish the results used in the next chapter and elsewhere in this book.

In this spirit, the proofs that have been adopted for some of the standard results of complex variable theory have been chosen with an eye to simplicity rather than sophistication. This means that in some cases the imposed conditions are more stringent than would be strictly necessary if more sophisticated proofs were used; where this happens the less restrictive results are usually stated as well. The reader who is interested in a fuller treatment should consult one of the many excellent textbooks on this fascinating subject.[1]

One further concession to "hand-waving" has been made in the interests of keeping the treatment to a moderate length. In several places phrases such as "can be made as small as we like" are used, rather than a careful treatment in terms of "given $\epsilon > 0$, there exists a $\delta > 0$ such that". In the authors' experience, some students are more at ease with the former type of statement, despite its lack of precision, whilst others, those who would contemplate only the latter, are usually well able to supply it for themselves.

1 For example, K. Knopp, *Theory of Functions, Part I* (New York: Dover, 1945); E. G. Phillips, *Functions of a Complex Variable with Applications*, 7th edn (Edinburgh: Oliver and Boyd, 1951); E. C. Titchmarsh, *The Theory of Functions* (Oxford: Oxford University Press, 1952).

14.1 Functions of a complex variable

The quantity $f(z)$ is said to be a function of the complex variable z if to every value of z in a certain domain R (a region of the Argand diagram) there corresponds one or more values of $f(z)$. Stated like this $f(z)$ could be any function consisting of a real and an imaginary part, each of which is, in general, itself a function of x and y. If we denote the real and imaginary parts of $f(z)$ by u and v, respectively, then

$$f(z) = u(x, y) + iv(x, y).$$

In this chapter, however, we will be primarily concerned with functions that are single-valued, so that to each value of z there corresponds just one value of $f(z)$, and are differentiable in a particular sense, which we now discuss.

A function $f(z)$ that is single-valued in some domain R is *differentiable* at the point z in R if the *derivative*

$$f'(z) = \lim_{\Delta z \to 0} \left[\frac{f(z + \Delta z) - f(z)}{\Delta z} \right] \tag{14.1}$$

exists and is unique, in that its value does not depend upon the direction in the Argand diagram from which Δz tends to zero. Our first two worked examples show one function that is differentiable in this sense, and another that is not.

Example Show that the function $f(z) = x^2 - y^2 + i2xy$ is differentiable for all values of z.

Considering the definition (14.1), and taking $\Delta z = \Delta x + i\Delta y$, we have

$$\begin{aligned}
\frac{f(z + \Delta z) - f(z)}{\Delta z}
&= \frac{(x + \Delta x)^2 - (y + \Delta y)^2 + 2i(x + \Delta x)(y + \Delta y) - x^2 + y^2 - 2ixy}{\Delta x + i\Delta y} \\
&= \frac{2x\Delta x + (\Delta x)^2 - 2y\Delta y - (\Delta y)^2 + 2i(x\Delta y + y\Delta x + \Delta x\Delta y)}{\Delta x + i\Delta y} \\
&= 2x + i2y + \frac{(\Delta x)^2 - (\Delta y)^2 + 2i\Delta x\Delta y}{\Delta x + i\Delta y}.
\end{aligned}$$

Now, in whatever way Δx and Δy are allowed to tend to zero (e.g. taking $\Delta y = 0$ and letting $\Delta x \to 0$ or vice versa, or letting both tend to zero in a particular non-zero ratio), the last term on the RHS will tend to zero and the unique limit $2x + i2y$ will be obtained. Since z was arbitrary, $f(z)$ with $u = x^2 - y^2$ and $v = 2xy$ is differentiable at all points in the (finite) complex plane. ◄

We note that the above working can be considerably reduced by recognising that, since $z = x + iy$, we can write $f(z)$ as

$$f(z) = x^2 - y^2 + 2ixy = (x + iy)^2 = z^2.$$

We then find that

$$f'(z) = \lim_{\Delta z \to 0} \left[\frac{(z + \Delta z)^2 - z^2}{\Delta z} \right] = \lim_{\Delta z \to 0} \left[\frac{2z\Delta z + (\Delta z)^2}{\Delta z} \right]$$
$$= \lim_{\Delta z \to 0} (2z + \Delta z) = 2z,$$

from which we see immediately that the limit both exists and is independent of the way in which $\Delta z \to 0$. Thus we have verified that $f(z) = z^2$ is differentiable for all (finite) z. We also note that the derivative is analogous to that found for real variables.

Although the definition of a differentiable function clearly includes a wide class of functions, the concept of differentiability is restrictive and, indeed, some functions are not differentiable at any point in the complex plane.

Example Show that the function $f(z) = 2y + ix$ is not differentiable anywhere in the complex plane.

In this case $f(z)$ cannot be written simply in terms of z, and so we must consider the limit (14.1) in terms of x and y explicitly. Following the same procedure as in the previous example we find

$$\frac{f(z + \Delta z) - f(z)}{\Delta z} = \frac{2y + 2\Delta y + ix + i\Delta x - 2y - ix}{\Delta x + i\Delta y}$$
$$= \frac{2\Delta y + i\Delta x}{\Delta x + i\Delta y}.$$

In this case the limit will clearly depend on the direction from which $\Delta z \to 0$. Suppose $\Delta z \to 0$ along a line through z of slope m, so that $\Delta y = m\Delta x$, then

$$\lim_{\Delta z \to 0} \left[\frac{f(z + \Delta z) - f(z)}{\Delta z} \right] = \lim_{\Delta x,\, \Delta y \to 0} \left[\frac{2\Delta y + i\Delta x}{\Delta x + i\Delta y} \right] = \frac{2m + i}{1 + im}.$$

This limit is dependent on m and hence on the direction from which $\Delta z \to 0$. Since this conclusion is independent of the value of z, and hence true for all z, $f(z) = 2y + ix$ is nowhere differentiable. ◀

A function that is single-valued and differentiable at all points of a domain R is said to be *analytic* (or *regular*) in R. A function may be analytic in a domain except at a finite number of points (or an infinite number if the domain is infinite); in this case it is said to be analytic except at these points, which are called the *singularities* of $f(z)$. In our treatment we will not consider cases in which an infinite number of singularities occur in a finite domain.

Example Show that the function $f(z) = 1/(1 - z)$ is analytic everywhere except at $z = 1$.

Since $f(z)$ is given explicitly as a function of z, evaluation of the limit (14.1) is somewhat easier. We find

$$\begin{aligned} f'(z) &= \lim_{\Delta z \to 0} \left[\frac{f(z + \Delta z) - f(z)}{\Delta z} \right] \\ &= \lim_{\Delta z \to 0} \left[\frac{1}{\Delta z} \left(\frac{1}{1 - z - \Delta z} - \frac{1}{1 - z} \right) \right] \\ &= \lim_{\Delta z \to 0} \left[\frac{1}{(1 - z - \Delta z)(1 - z)} \right] = \frac{1}{(1 - z)^2}, \end{aligned}$$

independently of the way in which $\Delta z \to 0$, provided $z \neq 1$. Hence $f(z)$ is analytic everywhere except at the singularity $z = 1$. ◀

14.2 The Cauchy–Riemann relations

From examining the previous examples, it is apparent that for a function $f(z)$ to be differentiable and hence analytic there must be some particular connection between its real and imaginary parts u and v.

By considering a general function we next establish what this connection must be. If the limit

$$L = \lim_{\Delta z \to 0} \left[\frac{f(z + \Delta z) - f(z)}{\Delta z} \right] \tag{14.2}$$

is to exist and be unique, in the way required for differentiability, then any two specific ways of letting $\Delta z \to 0$ must produce the same limit. In particular, moving parallel to the real axis and moving parallel to the imaginary axis must do so. This is certainly a necessary condition, although it may not be sufficient.

If we let $f(z) = u(x, y) + iv(x, y)$ and $\Delta z = \Delta x + i\Delta y$ then we have

$$f(z + \Delta z) = u(x + \Delta x,\ y + \Delta y) + iv(x + \Delta x,\ y + \Delta y),$$

and the limit (14.2) is given by

$$L = \lim_{\Delta x,\, \Delta y \to 0} \left[\frac{u(x + \Delta x, y + \Delta y) + iv(x + \Delta x, y + \Delta y) - u(x, y) - iv(x, y)}{\Delta x + i\Delta y} \right].$$

If we first suppose that Δz is purely real, so that $\Delta y = 0$, we obtain

$$L = \lim_{\Delta x \to 0} \left[\frac{u(x + \Delta x, y) - u(x, y)}{\Delta x} + i\frac{v(x + \Delta x, y) - v(x, y)}{\Delta x} \right] = \frac{\partial u}{\partial x} + i\frac{\partial v}{\partial x}, \tag{14.3}$$

provided each limit exists at the point z. Similarly, if Δz is taken as purely imaginary, so that $\Delta x = 0$, we find

$$L = \lim_{\Delta y \to 0} \left[\frac{u(x, y + \Delta y) - u(x, y)}{i \Delta y} + i \frac{v(x, y + \Delta y) - v(x, y)}{i \Delta y} \right] = \frac{1}{i} \frac{\partial u}{\partial y} + \frac{\partial v}{\partial y}. \tag{14.4}$$

For f to be differentiable at the point z, expressions (14.3) and (14.4) must be identical. It follows from equating real and imaginary parts that *necessary* conditions for this are

$$\frac{\partial u}{\partial x} = \frac{\partial v}{\partial y} \quad \text{and} \quad \frac{\partial v}{\partial x} = -\frac{\partial u}{\partial y}. \tag{14.5}$$

These two equations are known as the *Cauchy–Riemann relations*.

We can now see why for the earlier examples (i) $f(z) = x^2 - y^2 + i2xy$ might be differentiable and (ii) $f(z) = 2y + ix$ could not be.

(i) $u = x^2 - y^2$, $v = 2xy$:

$$\frac{\partial u}{\partial x} = 2x = \frac{\partial v}{\partial y} \quad \text{and} \quad \frac{\partial v}{\partial x} = 2y = -\frac{\partial u}{\partial y},$$

(ii) $u = 2y$, $v = x$:

$$\frac{\partial u}{\partial x} = 0 = \frac{\partial v}{\partial y} \quad \text{but} \quad \frac{\partial v}{\partial x} = 1 \neq -2 = -\frac{\partial u}{\partial y}.$$

It is apparent that for $f(z)$ to be analytic something more than the existence of the partial derivatives of u and v with respect to x and y is required; this something is that they satisfy the Cauchy–Riemann relations.[2]

We may enquire also as to the *sufficient* conditions for $f(z)$ to be analytic in R. It can be shown[3] that a sufficient condition is that the four partial derivatives exist, *are continuous* and satisfy the Cauchy–Riemann relations. It is the additional requirement of continuity that makes the difference between the necessary conditions and the sufficient conditions. We now consider a function that is analytic in some parts of the complex plane, but not in others.

Example In which domain(s) of the complex plane is $f(z) = |x| - i|y|$ an analytic function?

Writing $f = u + iv$ it is clear that both $\partial u/\partial y$ and $\partial v/\partial x$ are zero in all four quadrants and hence that the second Cauchy–Riemann relation in (14.5) is satisfied everywhere.

Turning to the first Cauchy–Riemann relation, in the first quadrant ($x > 0$, $y > 0$) we have $f(z) = x - iy$ so that

$$\frac{\partial u}{\partial x} = 1, \qquad \frac{\partial v}{\partial y} = -1,$$

which clearly violates the first relation in (14.5). Thus $f(z)$ is not analytic in the first quadrant.

2 Are the functions $x \pm iy$ analytic?
3 See, for example, any of the references given on p. 540.

Following a similar argument for the other quadrants, we find

$$\frac{\partial u}{\partial x} = -1 \quad \text{or} \quad +1 \qquad \text{for } x < 0 \text{ and } x > 0, \text{ respectively,}$$

$$\frac{\partial v}{\partial y} = -1 \quad \text{or} \quad +1 \qquad \text{for } y > 0 \text{ and } y < 0, \text{ respectively.}$$

Therefore $\partial u/\partial x$ and $\partial v/\partial y$ are equal, and hence $f(z)$ is analytic only in the second and fourth quadrants. ◀

Since x and y are related to z and its complex conjugate z^* by

$$x = \frac{1}{2}(z + z^*) \qquad \text{and} \qquad y = \frac{1}{2i}(z - z^*), \tag{14.6}$$

we may formally regard any function $f = u + iv$ as a function of z and z^*, rather than x and y. If we do this and examine $\partial f/\partial z^*$ we obtain

$$\begin{aligned}
\frac{\partial f}{\partial z^*} &= \frac{\partial f}{\partial x}\frac{\partial x}{\partial z^*} + \frac{\partial f}{\partial y}\frac{\partial y}{\partial z^*} \\
&= \left(\frac{\partial u}{\partial x} + i\frac{\partial v}{\partial x}\right)\left(\frac{1}{2}\right) + \left(\frac{\partial u}{\partial y} + i\frac{\partial v}{\partial y}\right)\left(-\frac{1}{2i}\right) \\
&= \frac{1}{2}\left(\frac{\partial u}{\partial x} - \frac{\partial v}{\partial y}\right) + \frac{i}{2}\left(\frac{\partial v}{\partial x} + \frac{\partial u}{\partial y}\right).
\end{aligned} \tag{14.7}$$

Now, if f is analytic then the Cauchy–Riemann relations (14.5) must be satisfied, and these immediately give that $\partial f/\partial z^*$ is identically zero. Thus we conclude that if f is analytic then f cannot be a function of z^* and any expression representing an analytic function of z can contain x and y only in the combination $x + iy$, *not* in the combination $x - iy$.[4]

We conclude this section by discussing some properties of analytic functions that are of great practical importance in theoretical physics. These can be obtained simply from the requirement that the Cauchy–Riemann relations must be satisfied by the real and imaginary parts of an analytic function.

The most important of these results can be obtained by differentiating the first Cauchy–Riemann relation with respect to one independent variable, and the second with respect to the other independent variable, to obtain the two chains of equalities

$$\frac{\partial}{\partial x}\left(\frac{\partial u}{\partial x}\right) = \frac{\partial}{\partial x}\left(\frac{\partial v}{\partial y}\right) = \frac{\partial}{\partial y}\left(\frac{\partial v}{\partial x}\right) = -\frac{\partial}{\partial y}\left(\frac{\partial u}{\partial y}\right),$$

$$\frac{\partial}{\partial x}\left(\frac{\partial v}{\partial x}\right) = -\frac{\partial}{\partial x}\left(\frac{\partial u}{\partial y}\right) = -\frac{\partial}{\partial y}\left(\frac{\partial u}{\partial x}\right) = -\frac{\partial}{\partial y}\left(\frac{\partial v}{\partial y}\right).$$

4 Use this criterion to decide which of the following functions are analytic: (a) $|z|^2$, (b) $(x - iy)/(x^2 + y^2)$, (c) $\cos x \cosh y \pm i \sin x \sinh y$. Confirm your answers using the C–R relations.

Thus both u and v are *separately* solutions of Laplace's equation in two dimensions, i.e.

$$\frac{\partial^2 u}{\partial x^2} + \frac{\partial^2 u}{\partial y^2} = 0 \qquad \text{and} \qquad \frac{\partial^2 v}{\partial x^2} + \frac{\partial^2 v}{\partial y^2} = 0. \tag{14.8}$$

We will make significant use of this result[5] in the next chapter.

A further useful result concerns the two families of curves $u(x, y) =$ constant and $v(x, y) =$ constant, where u and v are the real and imaginary parts of any analytic function $f = u + iv$. As discussed in Chapter 2, the vector normal to the curve $u(x, y) =$ constant is given by

$$\nabla u = \frac{\partial u}{\partial x}\mathbf{i} + \frac{\partial u}{\partial y}\mathbf{j}, \tag{14.9}$$

where $\mathbf{i}$ and $\mathbf{j}$ are the unit vectors along the x- and y-axes, respectively. A similar expression exists for ∇v, the normal to the curve $v(x, y) =$ constant. Taking the scalar product of these two normal vectors, we obtain

$$\begin{aligned}\nabla u \cdot \nabla v &= \frac{\partial u}{\partial x}\frac{\partial v}{\partial x} + \frac{\partial u}{\partial y}\frac{\partial v}{\partial y} \\ &= -\frac{\partial u}{\partial x}\frac{\partial u}{\partial y} + \frac{\partial u}{\partial y}\frac{\partial u}{\partial x} = 0,\end{aligned}$$

where in the last line we have used the Cauchy–Riemann relations to rewrite the partial derivatives of v as partial derivatives of u. Since the scalar product of the normal vectors is zero, they must be orthogonal, and the curves $u(x, y) =$ constant and $v(x, y) =$ constant must therefore intersect at *right angles*.

A further property of analytic functions is proved in the following worked example.

Example Use the Cauchy–Riemann relations to show that, for any analytic function $f = u + iv$, the relation $|\nabla u| = |\nabla v|$ must hold.

From (14.9) we have

$$|\nabla u|^2 = \nabla u \cdot \nabla u = \left(\frac{\partial u}{\partial x}\right)^2 + \left(\frac{\partial u}{\partial y}\right)^2.$$

Using the Cauchy–Riemann relations to write the partial derivatives of u in terms of those of v, we obtain

$$|\nabla u|^2 = \left(\frac{\partial v}{\partial y}\right)^2 + \left(\frac{\partial v}{\partial x}\right)^2 = |\nabla v|^2,$$

from which the result $|\nabla u| = |\nabla v|$ follows immediately. ◀

5 Confirm by direct calculation that this is so for the function $f(z) = z^3$.

14.3 Power series in a complex variable

The theory of power series in a real variable can be extended to a series such as

$$f(z) = \sum_{n=0}^{\infty} a_n z^n, \tag{14.10}$$

where z is a complex variable and the a_n are, in general, complex. We next study the additional considerations that arise, and the properties of these more complex series.

Expression (14.10) is a power series about the origin and may be used for general discussion, since a power series about any other point z_0 can be obtained by a change of variable from z to $z - z_0$. If z were written in its modulus and argument form, $z = r \exp i\theta$, expression (14.10) would become

$$f(z) = \sum_{n=0}^{\infty} a_n r^n \exp(in\theta). \tag{14.11}$$

This series is absolutely convergent if

$$\sum_{n=0}^{\infty} |a_n| r^n, \tag{14.12}$$

which is a series of positive real terms, is convergent. Thus tests for the absolute convergence of real series can be used in the present context, and of these the most appropriate form is based on the Cauchy root test. With the *radius of convergence* R defined by

$$\frac{1}{R} = \lim_{n\to\infty} |a_n|^{1/n}, \tag{14.13}$$

the series (14.10) is absolutely convergent if $|z| < R$ and divergent if $|z| > R$. If $|z| = R$ then no particular conclusion may be drawn, and this case must be considered separately.

A circle of radius R centered on the origin is called the *circle of convergence* of the series $\sum a_n z^n$. The cases $R = 0$ and $R = \infty$ correspond, respectively, to convergence at the origin only and convergence everywhere. For R finite the convergence occurs in a restricted part of the z-plane (the Argand diagram). For a power series about a general point z_0, the circle of convergence is, of course, centered on that point.

Example Find the parts of the z-plane for which the following series are convergent:

$$\text{(i)} \sum_{n=0}^{\infty} \frac{z^n}{n!}, \qquad \text{(ii)} \sum_{n=0}^{\infty} n! z^n, \qquad \text{(iii)} \sum_{n=1}^{\infty} \frac{z^n}{n}.$$

(i) Since $(n!)^{1/n}$ behaves like n as $n \to \infty$ we find $\lim(1/n!)^{1/n} = 0$. Hence $R = \infty$ and the series is convergent for all z. (ii) Correspondingly, $\lim(n!)^{1/n} = \infty$. Thus $R = 0$ and the series converges only at $z = 0$. (iii) As $n \to \infty$, $(n)^{1/n}$ has a lower limit of 1 and hence $\lim(1/n)^{1/n} = 1/1 = 1$. Thus the series is absolutely convergent if the condition $|z| < 1$ is satisfied. ◀

Case (iii) in the above example provides a good illustration of the fact that on its circle of convergence a power series may or may not converge. For this particular series, the circle of convergence is $|z| = 1$, so let us consider the convergence of the series at two different points on this circle. Taking $z = 1$, the series becomes

$$\sum_{n=1}^{\infty} \frac{1}{n} = 1 + \frac{1}{2} + \frac{1}{3} + \frac{1}{4} + \cdots,$$

which is easily shown to diverge.[6] However, when $z = -1$ the series is given by

$$\sum_{n=1}^{\infty} \frac{(-1)^n}{n} = -1 + \frac{1}{2} - \frac{1}{3} + \frac{1}{4} - \cdots;$$

this is an alternating series whose terms decrease in magnitude and which therefore converges.

The ratio test used for ordinary power series may also be employed to investigate the absolute convergence of a complex power series. A series is absolutely convergent if

$$\lim_{n\to\infty} \frac{|a_{n+1}||z|^{n+1}}{|a_n||z|^n} = \lim_{n\to\infty} \frac{|a_{n+1}||z|}{|a_n|} < 1 \qquad (14.14)$$

and hence the radius of convergence R of the series is given by

$$\frac{1}{R} = \lim_{n\to\infty} \frac{|a_{n+1}|}{|a_n|}.$$

For instance, in case (i) of the previous example, we have

$$\frac{1}{R} = \lim_{n\to\infty} \frac{n!}{(n+1)!} = \lim_{n\to\infty} \frac{1}{n+1} = 0.$$

Thus the series is absolutely convergent for all (finite) z, confirming the previous result.[7]

Before turning to particular power series, we conclude this section by stating the important result[8] that *the power series* $\sum_0^\infty a_n z^n$ *has a sum that is an analytic function of* z *inside its circle of convergence.*

As a corollary to the above theorem, it may further be shown that if $f(z) = \sum a_n z^n$ then, inside the circle of convergence of the series,

$$f'(z) = \sum_{n=0}^{\infty} n a_n z^{n-1}.$$

Repeated application of this result demonstrates that any power series can be differentiated any number of times inside its circle of convergence.

6 By, for example, forming groups of enough successive terms that the sum of terms in each group $\geq \frac{1}{2}$. As many such groups can be made as is needed for the partial sum of the series to exceed any given finite number.

7 What is the radius of convergence of a power series whose nth term is $(-1)^n z^{2n}/2^n$?

8 For a proof see, for example, K. F. Riley, *Mathematical Methods for the Physical Sciences* (Cambridge: Cambridge University Press, 1974), p. 446.

14.4 Some elementary functions

In the example at the end of the previous section it was shown that the function $\exp z$ *defined* by

$$\exp z = \sum_{n=0}^{\infty} \frac{z^n}{n!} \qquad (14.15)$$

is convergent for all z of finite modulus and is thus, by the discussion of the previous section, an analytic function over the whole z-plane.[9] Like its real-variable counterpart it is called the *exponential function*; also like its real counterpart it is equal to its own derivative.

The multiplication of two exponential functions results in a further exponential function, in accordance with the corresponding result for real variables, as we now prove.

Example Show that $\exp z_1 \exp z_2 = \exp(z_1 + z_2)$.

From the series expansion (14.15) of $\exp z_1$ and a similar expansion for $\exp z_2$, it is clear that the coefficient of $z_1^r z_2^s$ in the corresponding series expansion of $\exp z_1 \exp z_2$ is simply $1/(r!s!)$.

But, from (14.15) we also have

$$\exp(z_1 + z_2) = \sum_{n=0}^{\infty} \frac{(z_1 + z_2)^n}{n!}.$$

In order to find the coefficient of $z_1^r z_2^s$ in this expansion, we clearly have to consider the term in which $n = r + s$, namely

$$\frac{(z_1 + z_2)^{r+s}}{(r+s)!} = \frac{1}{(r+s)!}\left({}^{r+s}C_0 z_1^{r+s} + \cdots + {}^{r+s}C_s z_1^r z_2^s + \cdots + {}^{r+s}C_{r+s} z_2^{r+s}\right).$$

The coefficient of $z_1^r z_2^s$ in this is given by

$${}^{r+s}C_s \frac{1}{(r+s)!} = \frac{(r+s)!}{s!r!}\frac{1}{(r+s)!} = \frac{1}{r!s!}.$$

Thus, since the corresponding coefficients on the two sides are equal, and all the series involved are absolutely convergent for all z, we can conclude that $\exp z_1 \exp z_2 = \exp(z_1 + z_2)$. ◀

As an extension of (14.15) we may also define the complex exponent of a real number $a > 0$ by the equation

$$a^z = \exp(z \ln a), \qquad (14.16)$$

where $\ln a$ is the natural logarithm of a. The particular case $a = e$ and the fact that $\ln e = 1$ enable us to write $\exp z$ interchangeably with e^z. If z is real then the definition agrees with the familiar one.

The result for $z = iy$ and $a = e$,

$$\exp iy = \cos y + i \sin y, \qquad (14.17)$$

9 Functions that are analytic in the *whole* z-plane are usually called *integral* or *entire* functions.

reproduces Euler's equation. Its immediate extension is

$$\exp z = (\exp x)(\cos y + i \sin y). \qquad (14.18)$$

As z varies over the complex plane, the modulus of $\exp z$ takes all real positive values, except that of 0. However, two values of z that differ by $2\pi k i$, for any integer k, produce the same value of $\exp z$, as given by (14.18), and so $\exp z$ is periodic with period $2\pi i$.[10] If we denote $\exp z$ by t, then the strip $-\pi < y \leq \pi$ in the z-plane corresponds to the whole of the t-plane, except for the point $t = 0$.

The sine, cosine, sinh and cosh functions of a complex variable are defined from the exponential function exactly as are those for real variables. The functions derived from them (e.g. tan and tanh), the identities they satisfy and their derivative properties are also just as for real variables. In view of this we will not give them further attention here.

The inverse function of $\exp z$ is given by w, the solution of

$$\exp w = z. \qquad (14.19)$$

By virtue of the discussion following (14.18), w is not uniquely defined and is indeterminate to the extent of any integer multiple of $2\pi i$. If we express z as

$$z = r \exp i\theta,$$

where r is the (real) modulus of z and θ is its argument ($-\pi < \theta \leq \pi$), then multiplying z by $\exp(2ik\pi)$, where k is an integer, will result in the same complex number z. Thus we may write

$$z = r \exp[i(\theta + 2k\pi)],$$

where k is an integer. If we denote w in (14.19) by

$$w = \operatorname{Ln} z = \ln r + i(\theta + 2k\pi), \qquad (14.20)$$

where $\ln r$ is the natural logarithm (to base e) of the real positive quantity r, then $\operatorname{Ln} z$ is an infinitely multivalued function of z. Its *principal value*, denoted by $\ln z$, is obtained by taking $k = 0$ so that its argument lies in the range $-\pi$ to π. Thus

$$\ln z = \ln r + i\theta, \qquad \text{with } -\pi < \theta \leq \pi. \qquad (14.21)$$

Now that the logarithm of a complex variable has been defined, definition (14.16) of a general power can be extended to cases other than those in which a is real and positive. If t ($\neq 0$) and z are both complex, then the zth power of t is defined by

$$t^z = \exp(z \operatorname{Ln}\, t). \qquad (14.22)$$

Since $\operatorname{Ln} t$ is multivalued, so is this definition. Its principal value is obtained by giving $\operatorname{Ln} t$ its principal value, $\ln t$.[11]

If t ($\neq 0$) is complex but z is real and equal to $1/n$, then (14.22) provides a definition of the nth root of t. Because of the multivaluedness of $\operatorname{Ln} t$, there will be more than one nth root of any given t.

10 Note, the period is $2\pi i$, *not* 2π.

11 Are the following real, imaginary or complex? (a) i, (b) i^i, (c) $(i^i)^i$.

Example Show that there are exactly n distinct nth roots of t.

From (14.22) the nth roots of t are given by

$$t^{1/n} = \exp\left(\frac{1}{n}\text{Ln } t\right).$$

On the RHS let us write t as follows:

$$t = r\exp[i(\theta + 2k\pi)],$$

where k is an integer. We then obtain

$$t^{1/n} = \exp\left[\frac{1}{n}\ln r + i\frac{(\theta + 2k\pi)}{n}\right]$$
$$= r^{1/n}\exp\left[i\frac{(\theta + 2k\pi)}{n}\right],$$

where $k = 0, 1, \ldots, n-1$; for other values of k we simply recover the roots already found. Thus t has n distinct nth roots. We note that they all have the same modulus and would be equally spaced around a circle in the Argand diagram. ◀

14.5 Multivalued functions and branch cuts

In the definition of an analytic function, one of the conditions imposed was that the function is single-valued. However, as shown in the previous section, the logarithmic function, a complex power and a complex root are all multivalued. Nevertheless, it happens that the properties of analytic functions can still be applied to these and other multivalued functions of a complex variable provided that suitable care is taken. This care amounts to identifying the *branch points* of the multivalued function $f(z)$ in question. A branch point is defined by the property that if z is varied in such a way that its path in the Argand diagram forms a closed curve that encloses a branch point, then, in general, $f(z)$ will not return to its original value.

As a particular illustrative example, let us consider the multivalued function $f(z) = z^{1/2}$ and express z as $z = r\exp i\theta$. From Figure 14.1(a), it is clear that, as the point z traverses any closed contour C that does not enclose the origin, θ will return to its original value after one complete circuit. However, for any closed contour C' that does enclose the origin, after one circuit $\theta \to \theta + 2\pi$ (see Figure 14.1(b)). Thus, for the function $f(z) = z^{1/2}$, after one circuit

$$r^{1/2}\exp(i\theta/2) \to r^{1/2}\exp[i(\theta + 2\pi)/2] = -r^{1/2}\exp(i\theta/2).$$

In other words, the value of $f(z)$ changes around any closed loop enclosing the origin; in this case $f(z) \to -f(z)$. Thus $z = 0$ is a branch point of the function $f(z) = z^{1/2}$.

We note in this case that if any closed contour enclosing the origin is traversed *twice* then $f(z) = z^{1/2}$ returns to its original value. The number of loops around a branch point required for any given function $f(z)$ to return to its original value depends on the function in question, and for some functions (e.g. Ln z, which also has a branch point at the origin) the original value is never recovered.

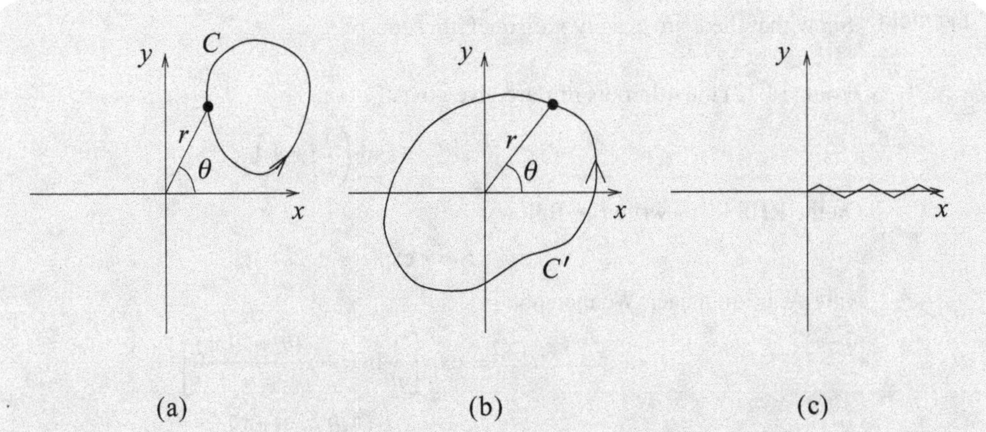

Figure 14.1 (a) A closed contour not enclosing the origin; (b) a closed contour enclosing the origin; (c) a possible branch cut for $f(z) = z^{1/2}$.

In order that $f(z)$ may be treated as single-valued, we may define a *branch cut* in the Argand diagram. A branch cut is a line (or curve) in the complex plane and may be regarded as an artificial barrier that we must not cross. Branch cuts are positioned in such a way that we are prevented from making a complete circuit around any one branch point, and so the function in question remains single-valued.

For the function $f(z) = z^{1/2}$, we may take as a branch cut any curve starting at the origin $z = 0$ and extending out to $|z| = \infty$ in any direction, since all such curves would equally well prevent us from making a closed loop around the branch point at the origin. It is usual, however, to take the cut along either the real or the imaginary axis. For example, in Figure 14.1(c), we take the cut as the positive real axis. By agreeing not to cross this cut, we restrict θ to lie in the range $0 \leq \theta < 2\pi$, and so keep $f(z)$ single-valued.

These ideas can be extended to functions with more than one branch point, as in the following example.

Example Find the branch points of $f(z) = \sqrt{z^2 + 1}$, and hence sketch suitable arrangements of branch cuts.

We begin by writing $f(z)$ as

$$f(z) = \sqrt{z^2 + 1} = \sqrt{(z - i)(z + i)}.$$

As shown above, the function $g(z) = z^{1/2}$ has a branch point at $z = 0$. Thus we might expect $f(z)$ to have branch points at values of z that make the expression under the square root equal to zero, i.e. at $z = i$ and $z = -i$.

As shown in Figure 14.2(a), we use the notation

$$z - i = r_1 \exp i\theta_1 \qquad \text{and} \qquad z + i = r_2 \exp i\theta_2.$$

We can therefore write $f(z)$ as

$$f(z) = \sqrt{r_1 r_2} \exp(i\theta_1/2) \exp(i\theta_2/2) = \sqrt{r_1 r_2} \exp\left[i(\theta_1 + \theta_2)/2\right].$$

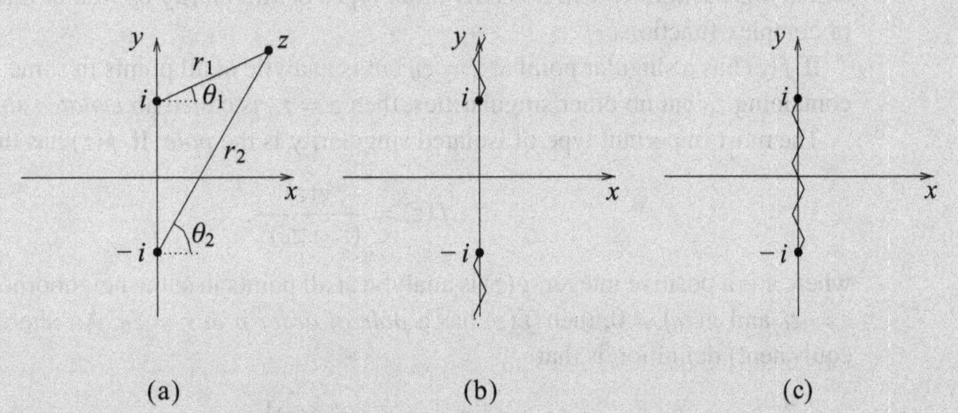

Figure 14.2 (a) Coordinates used in the analysis of the branch points of $f(z) = (z^2 + 1)^{1/2}$; (b) one possible arrangement of branch cuts; (c) another possible branch cut, which is finite.

Let us now consider how $f(z)$ changes as we make one complete circuit around various closed loops C in the Argand diagram. If C encloses

(i) neither branch point, then $\theta_1 \to \theta_1, \theta_2 \to \theta_2$ and so $f(z) \to f(z)$;

(ii) $z = i$ but not $z = -i$, then $\theta_1 \to \theta_1 + 2\pi, \theta_2 \to \theta_2$ and so $f(z) \to -f(z)$;

(iii) $z = -i$ but not $z = i$, then $\theta_1 \to \theta_1, \theta_2 \to \theta_2 + 2\pi$ and so $f(z) \to -f(z)$;

(iv) both branch points, then $\theta_1 \to \theta_1 + 2\pi, \theta_2 \to \theta_2 + 2\pi$ and so $f(z) \to f(z)$.

Thus, as expected, $f(z)$ changes value around loops containing either $z = i$ or $z = -i$ (but not both). We must therefore choose branch cuts that prevent us from making a complete loop around either branch point; one suitable choice is shown in Figure 14.2(b).

For this $f(z)$, however, we have noted that after traversing a loop containing *both* branch points the function returns to its original value. Thus we may choose an alternative, *finite*, branch cut that allows this possibility but still prevents us from making a complete loop around just one of the points. A suitable cut is shown in Figure 14.2(c). ◀

Before we leave this section, it should be stated that the branch cuts are more of an aid to the user than a mathematical reality; their main purpose is to avoid any inadvertent change in the value of a function that might result from a careless choice of contour. In some more sophisticated applications the difference in the values of a function on the two sides of a cut does have mathematical or physical significance, but this is more directly related to properties of the branch point rather than to those of the cut.

14.6 Singularities and zeros of complex functions

A singular point of a complex function $f(z)$ is any point in the Argand diagram at which $f(z)$ fails to be analytic. We have already met one sort of singularity, the branch point, and in this section we will consider other types of singularity as well as discuss the zeros of complex functions.

If $f(z)$ has a singular point at $z = z_0$ but is analytic at all points in some neighborhood containing z_0 but no other singularities, then $z = z_0$ is called an *isolated singularity*.[12]

The most important type of isolated singularity is the *pole*. If $f(z)$ has the form

$$f(z) = \frac{g(z)}{(z - z_0)^n}, \qquad (14.23)$$

where n is a positive integer, $g(z)$ is analytic at all points in some neighborhood containing $z = z_0$ and $g(z_0) \neq 0$, then $f(z)$ has a *pole of order n* at $z = z_0$. An alternative (though equivalent) definition is that

$$\lim_{z \to z_0} \left[(z - z_0)^n f(z) \right] = a, \qquad (14.24)$$

where a is a finite, non-zero complex number. We note that if the above limit is equal to zero then $z = z_0$ is a pole of order less than n, or $f(z)$ is analytic there; if the limit is infinite then the pole is of an order greater than n. It may also be shown that if $f(z)$ has a pole at $z = z_0$, then $|f(z)| \to \infty$ as $z \to z_0$ from any direction in the Argand diagram.[13] If no finite value of n can be found such that (14.24) is satisfied, then $z = z_0$ is called an *essential singularity*.

As illustrative examples, consider the following.

Example Find the singularities of the functions

$$\text{(i) } f(z) = \frac{1}{1-z} - \frac{1}{1+z}, \qquad \text{(ii) } f(z) = \tanh z.$$

(i) If we write $f(z)$ as

$$f(z) = \frac{1}{1-z} - \frac{1}{1+z} = \frac{2z}{(1-z)(1+z)},$$

we see immediately from either (14.23) or (14.24) that $f(z)$ has poles of order 1 (or *simple poles*) at $z = 1$ and $z = -1$.

(ii) In this case we write

$$f(z) = \tanh z = \frac{\sinh z}{\cosh z} = \frac{\exp z - \exp(-z)}{\exp z + \exp(-z)}.$$

Thus $f(z)$ has a singularity when $\exp z = -\exp(-z)$ or, equivalently, when

$$\exp z = \exp[i(2n+1)\pi] \exp(-z),$$

where n is any integer. Equating the arguments of the exponentials we find $z = (n + \frac{1}{2})\pi i$, for integer n.

12 Clearly, branch points are not isolated singularities.

13 Although perhaps intuitively obvious, this result really requires formal demonstration by analysis.

Furthermore, using l'Hôpital's rule we have

$$\lim_{z\to(n+\frac{1}{2})\pi i}\left\{\frac{[z-(n+\frac{1}{2})\pi i]\sinh z}{\cosh z}\right\}$$

$$= \lim_{z\to(n+\frac{1}{2})\pi i}\left\{\frac{[z-(n+\frac{1}{2})\pi i]\cosh z+\sinh z}{\sinh z}\right\}=1.$$

Therefore, from (14.24), each singularity is a simple pole. ◀

Another type of singularity exists at points for which the value of $f(z)$ takes an indeterminate form such as $0/0$ but $\lim_{z\to z_0} f(z)$ exists and is independent of the direction from which z_0 is approached. Such points are called *removable singularities*.

Example Show that $f(z) = (\sin z)/z$ has a removable singularity at $z = 0$.

It is clear that $f(z)$ takes the indeterminate form $0/0$ at $z = 0$. However, by expanding $\sin z$ as a power series in z, we find

$$f(z) = \frac{1}{z}\left(z - \frac{z^3}{3!} + \frac{z^5}{5!} - \cdots\right) = 1 - \frac{z^2}{3!} + \frac{z^4}{5!} - \cdots.$$

Thus $\lim_{z\to 0} f(z) = 1$ independently of the way in which $z \to 0$, and so $f(z)$ has a removable singularity at $z = 0$. ◀

The term "removable singularity", whilst in common use, is somewhat misleading since the function is, in fact, analytic at the point and there is no singularity there.[14]

An expression common in mathematics, but which we have so far avoided using explicitly in this chapter, is "z tends to infinity". For a real variable such as $|z|$ or R, "tending to infinity" has a reasonably well-defined meaning. For a complex variable needing a two-dimensional plane to represent it, the meaning is not intrinsically well defined. However, it is convenient to have a unique meaning and this is provided by the following *definition*: the behavior of $f(z)$ *at infinity* is given by that of $f(1/\xi)$ at $\xi = 0$, where $\xi = 1/z$.

Example Find the behavior at infinity of (i) $f(z) = a + bz^{-2}$, (ii) $f(z) = z(1 + z^2)$ and (iii) $f(z) = \exp z$.

(i) $f(z) = a + bz^{-2}$: on putting $z = 1/\xi$, $f(1/\xi) = a + b\xi^2$, which is analytic at $\xi = 0$; thus f is analytic at $z = \infty$.
(ii) $f(z) = z(1 + z^2)$: $f(1/\xi) = 1/\xi + 1/\xi^3$; thus f has a pole of order 3 at $z = \infty$.
(iii) $f(z) = \exp z$: $f(1/\xi) = \sum_0^\infty (n!)^{-1}\xi^{-n}$; thus f has an essential singularity at $z = \infty$. ◀

We conclude this section by briefly mentioning the *zeros* of a complex function. As the name suggests, if $f(z_0) = 0$ then $z = z_0$ is called a zero of the function $f(z)$. Zeros are

14 What type of singularities, if any, do the functions $(\sin z \pm z\cos z)/z^3$ possess at the origin?

classified in a similar way to poles, in that if

$$f(z) = (z - z_0)^n g(z),$$

where n is a positive integer and $g(z_0) \neq 0$, then $z = z_0$ is called a *zero of order n* of $f(z)$. If $n = 1$ then $z = z_0$ is called a *simple zero*. It may further be shown that if $z = z_0$ is a zero of order n of $f(z)$ then it is also a pole of order n of the function $1/f(z)$.

We will return in Section 14.11 to the classification of zeros and poles in terms of their series expansions.

14.7 Conformal transformations

We now turn our attention to the subject of transformations, by which we mean a change of coordinates from the complex variable $z = x + iy$ to another, say $w = r + is$, by means of a prescribed formula:

$$w = g(z) = r(x, y) + is(x, y).$$

Under such a transformation, or *mapping*, the Argand diagram for the z-variable is transformed into one for the w-variable, although the complete z-plane might be mapped onto only a part of the w-plane, or onto the whole of the w-plane, or onto some or all of the w-plane covered more than once.

We shall consider only those mappings for which w and z are related by a function $w = g(z)$ and its inverse $z = h(w)$ with both functions analytic, except possibly at a few isolated points; such mappings are called *conformal*. Their important properties are that, except at points at which $g'(z)$, and hence $h'(w)$, is zero or infinite:

(i) continuous lines in the z-plane transform into continuous lines in the w-plane;
(ii) the angle between two intersecting curves in the z-plane equals the angle between the corresponding curves in the w-plane;
(iii) the magnification, as between the z-plane and the w-plane, of a small line element in the neighborhood of any particular point is independent of the direction of the element;
(iv) any analytic function of z transforms to an analytic function of w and vice versa.

Result (i) is immediate, and results (ii) and (iii) can be justified by the following argument. Let two curves C_1 and C_2 pass through the point z_0 in the z-plane and let z_1 and z_2 be two points on their respective tangents at z_0, each a distance ρ from z_0. The same prescription with w replacing z describes the transformed situation; however, the transformed tangents may not be straight lines and the distances of w_1 and w_2 from w_0 have not yet been shown to be equal. This situation is illustrated in Figure 14.3.

In the z-plane z_1 and z_2 are given by

$$z_1 - z_0 = \rho \exp i\theta_1 \qquad \text{and} \qquad z_2 - z_0 = \rho \exp i\theta_2.$$

The corresponding descriptions in the w-plane are

$$w_1 - w_0 = \rho_1 \exp i\phi_1 \qquad \text{and} \qquad w_2 - w_0 = \rho_2 \exp i\phi_2.$$

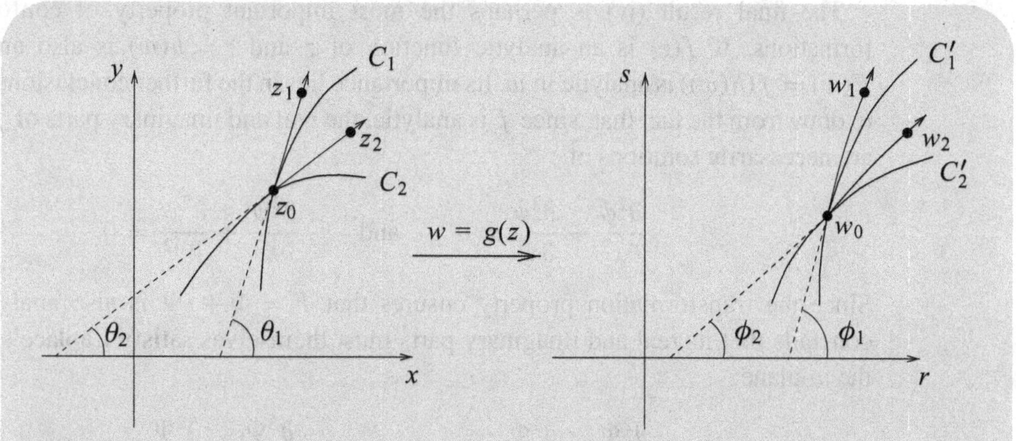

Figure 14.3 Two curves C_1 and C_2 in the z-plane, which are mapped onto C_1' and C_2' in the w-plane.

The angles θ_i and ϕ_i are clear from Figure 14.3. The transformed angles ϕ_i are those made with the r-axis by the tangents to the transformed curves at their point of intersection. Since any finite-length tangent may be curved, w_i is more strictly given by $w_i - w_0 = \rho_i \exp i(\phi_i + \delta\phi_i)$, where $\delta\phi_i \to 0$ as $\rho_i \to 0$, i.e. as $\rho \to 0$.

Now since $w = g(z)$, where g is analytic, we have

$$\lim_{z_1 \to z_0} \left(\frac{w_1 - w_0}{z_1 - z_0} \right) = \lim_{z_2 \to z_0} \left(\frac{w_2 - w_0}{z_2 - z_0} \right) = \left. \frac{dg}{dz} \right|_{z=z_0},$$

which may be written as

$$\lim_{\rho \to 0} \left\{ \frac{\rho_1}{\rho} \exp[i(\phi_1 + \delta\phi_1 - \theta_1)] \right\} = \lim_{\rho \to 0} \left\{ \frac{\rho_2}{\rho} \exp[i(\phi_2 + \delta\phi_2 - \theta_2)] \right\} = g'(z_0). \tag{14.25}$$

Comparing magnitudes and phases (i.e. arguments) in the equalities (14.25) gives the stated results (ii) and (iii) and adds quantitative information to them, namely that for *small* line elements

$$\frac{\rho_1}{\rho} \approx \frac{\rho_2}{\rho} \approx |g'(z_0)|, \tag{14.26}$$

$$\phi_1 - \theta_1 \approx \phi_2 - \theta_2 \approx \arg g'(z_0). \tag{14.27}$$

For strict comparison with result (ii), (14.27) must be written as $\theta_1 - \theta_2 = \phi_1 - \phi_2$, with an ordinary equality sign, since the angles are only defined in the limit $\rho \to 0$ when (14.27) becomes a true identity. We also see from (14.26) that the linear magnification factor is $|g'(z_0)|$; similarly, small areas are magnified by $|g'(z_0)|^2$.

Since in the neighborhoods of corresponding points in a transformation angles are preserved and magnifications are independent of direction, it follows that small plane figures are transformed into figures of the same shape, but, in general, ones that are magnified and rotated (though not distorted). However, we also note that at points where $g'(z) = 0$, the angle $\arg g'(z)$ through which line elements are rotated is undefined; these are called *critical points* of the transformation.

The final result (iv) is perhaps the most important property of conformal transformations. If $f(z)$ is an analytic function of z and $z = h(w)$ is also analytic, then $F(w) = f(h(w))$ is analytic in w. Its importance lies in the further conclusions it allows us to draw from the fact that, since f is analytic, the real and imaginary parts of $f = \phi + i\psi$ are necessarily solutions of

$$\frac{\partial^2 \phi}{\partial x^2} + \frac{\partial^2 \phi}{\partial y^2} = 0 \quad \text{and} \quad \frac{\partial^2 \psi}{\partial x^2} + \frac{\partial^2 \psi}{\partial y^2} = 0. \tag{14.28}$$

Since the transformation property ensures that $F = \Phi + i\Psi$ is also analytic, we can conclude that its real and imaginary parts must themselves satisfy Laplace's equation in the w-plane:

$$\frac{\partial^2 \Phi}{\partial r^2} + \frac{\partial^2 \Phi}{\partial s^2} = 0 \quad \text{and} \quad \frac{\partial^2 \Psi}{\partial r^2} + \frac{\partial^2 \Psi}{\partial s^2} = 0. \tag{14.29}$$

Further, suppose that (say) Re $f(z) = \phi$ is constant over a boundary C in the z-plane; then Re $F(w) = \Phi$ is constant over C in the z-plane. But this is the same as saying that Re $F(w)$ is constant over the boundary C' in the w-plane, C' being the curve into which C is transformed by the conformal transformation $w = g(z)$. This result is exploited extensively in the next chapter to solve Laplace's equation for a variety of two-dimensional geometries.

Examples of useful conformal transformations are numerous. For instance, $w = z + b$, $w = (\exp i\phi)z$ and $w = az$ correspond, respectively, to a translation by b, a rotation through an angle ϕ and a stretching (or contraction) in the radial direction (for a real). These three examples can be combined into the general linear transformation $w = az + b$, where, in general, a and b are complex. Another example is the inversion mapping $w = 1/z$, which maps the interior of the unit circle to the exterior and vice versa. Other, more complicated, examples also exist, and the next worked example illustrates one that combines three of the above elements.

Example Show that if the point z_0 lies in the upper half of the z-plane then the transformation

$$w = (\exp i\phi)\frac{z - z_0}{z - z_0^*}$$

maps the upper half of the z-plane into the interior of the unit circle in the w-plane. Hence find a similar transformation that maps the point $z = i$ onto $w = 0$ and the points $x = \pm\infty$ onto $w = 1$.

Taking the modulus of w, we have

$$|w| = \left|(\exp i\phi)\frac{z - z_0}{z - z_0^*}\right| = \left|\frac{z - z_0}{z - z_0^*}\right|.$$

However, since the complex conjugate z_0^* is the reflection of z_0 in the real axis, if z and z_0 both lie in the upper half of the z-plane then $|z - z_0| \leq |z - z_0^*|$; thus $|w| \leq 1$, as required. We also note that (i) the equality holds only when z lies on the real axis, and so this axis is mapped onto the boundary of the unit circle in the w-plane; (ii) the point z_0 is mapped onto $w = 0$, the origin of the w-plane.

By fixing the images of two points in the z-plane, the constants z_0 and ϕ can also be fixed. Since we require the point $z = i$ to be mapped onto $w = 0$, we have immediately $z_0 = i$. By further requiring $z = \pm\infty$ to be mapped onto $w = 1$, we find $1 = w = \exp i\phi$ and so $\phi = 0$. The required

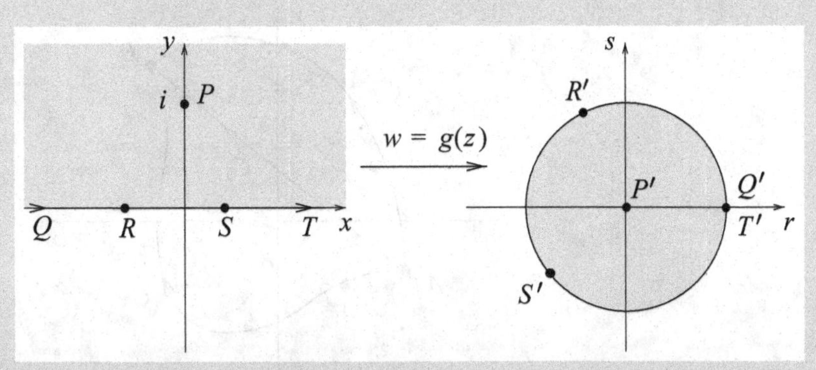

Figure 14.4 Transforming the upper half of the z-plane into the interior of the unit circle in the w-plane, in such a way that $z = i$ is mapped onto $w = 0$ and the points $x = \pm\infty$ are mapped onto $w = 1$.

transformation is therefore

$$w = \frac{z - i}{z + i}.$$

Figure 14.4 illustrates its effect.[15] ◀

14.8 Complex integrals

Corresponding to integration with respect to a real variable, it is possible to define integration with respect to a complex variable between two complex limits. Since the z-plane is two-dimensional there is clearly greater freedom and hence ambiguity in what is meant by a complex integral. If a complex function $f(z)$ is single-valued and continuous in some region R in the complex plane, then we can define the complex integral of $f(z)$ between two points A and B along some curve in R; its value will depend, in general, upon the path taken between A and B (see Figure 14.5). However, we will find that for some paths that are different but bear a particular relationship to each other the value of the integral does *not* depend upon which of the paths is adopted.

Let a particular path C be described by a continuous (real) parameter t ($\alpha \le t \le \beta$) that gives successive positions on C by means of the equations

$$x = x(t), \qquad y = y(t), \tag{14.30}$$

with $t = \alpha$ and $t = \beta$ corresponding to the points A and B, respectively. Then the integral along path C of a continuous function $f(z)$ is written

$$\int_C f(z)\,dz \tag{14.31}$$

15 Which point in the z-plane maps onto $w = -i$? Determine the form of the inverse mapping $z = h(w)$.

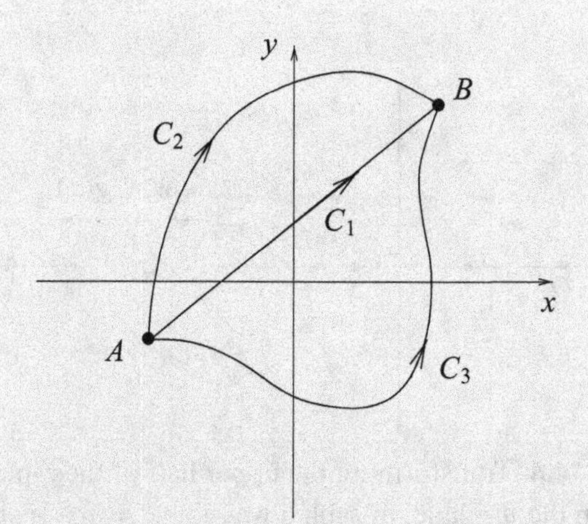

Figure 14.5 Some alternative paths for the integral of a function $f(z)$ between A and B.

and can be given explicitly as a sum of real integrals as follows:

$$\begin{aligned}\int_C f(z)\,dz &= \int_C (u+iv)(dx+idy) \\ &= \int_C u\,dx - \int_C v\,dy + i\int_C u\,dy + i\int_C v\,dx \\ &= \int_\alpha^\beta u\frac{dx}{dt}\,dt - \int_\alpha^\beta v\frac{dy}{dt}\,dt + i\int_\alpha^\beta u\frac{dy}{dt}\,dt + i\int_\alpha^\beta v\frac{dx}{dt}\,dt.\end{aligned} \tag{14.32}$$

The question of when such an integral exists will not be pursued, except to state that a sufficient condition is that dx/dt and dy/dt are continuous.

Example Evaluate the complex integral of $f(z) = z^{-1}$ along the circle $|z| = R$, starting and finishing at $z = R$.

The path C_1 is parameterized as follows (Figure 14.6(a)):

$$z(t) = R\cos t + iR\sin t, \qquad 0 \le t \le 2\pi,$$

whilst $f(z)$ is given by

$$f(z) = \frac{1}{x+iy} = \frac{x-iy}{x^2+y^2}.$$

Thus the real and imaginary parts of $f(z)$ are

$$u = \frac{x}{x^2+y^2} = \frac{R\cos t}{R^2} \qquad \text{and} \qquad v = \frac{-y}{x^2+y^2} = -\frac{R\sin t}{R^2}.$$

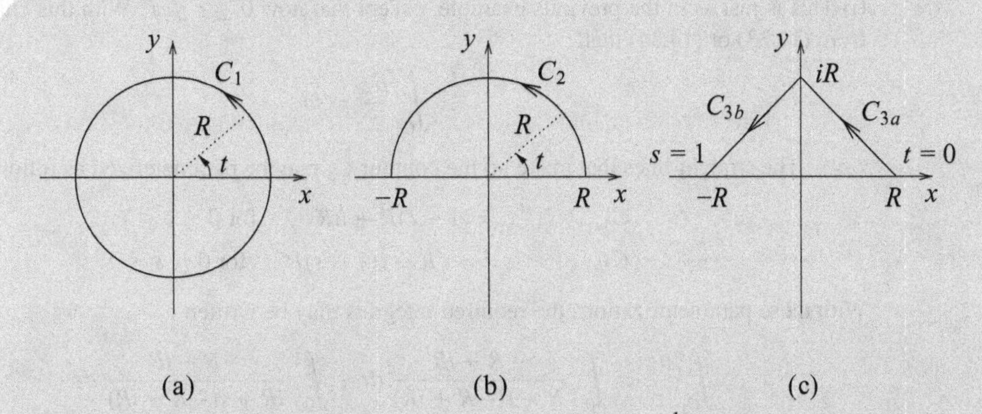

Figure 14.6 Different paths for an integral of $f(z) = z^{-1}$. See the text for details.

Hence, using expression (14.32),

$$\begin{aligned}\int_{C_1} \frac{1}{z}\,dz &= \int_0^{2\pi} \frac{\cos t}{R}(-R\sin t)\,dt - \int_0^{2\pi} \left(\frac{-\sin t}{R}\right) R\cos t\,dt \\ &\quad + i\int_0^{2\pi} \frac{\cos t}{R} R\cos t\,dt + i\int_0^{2\pi} \left(\frac{-\sin t}{R}\right)(-R\sin t)\,dt \qquad (14.33) \\ &= 0 + 0 + i\pi + i\pi = 2\pi i.\end{aligned}$$

This very important result will be used many times later, and the following should be carefully noted: (i) its value, (ii) that this value is independent of R. ◀

With a bit of experience, the reader may be able to evaluate integrals like the LHS of (14.33) directly without having to write them as four separate real integrals. In the present case,

$$\int_{C_1} \frac{dz}{z} = \int_0^{2\pi} \frac{-R\sin t + iR\cos t}{R\cos t + iR\sin t}\,dt = \int_0^{2\pi} i\,dt = 2\pi i. \qquad (14.34)$$

In the above example the contour was closed, and so it began and ended at the same point in the Argand diagram. We can evaluate complex integrals along open paths in a similar way.

Example Evaluate the complex integral of $f(z) = z^{-1}$ along the following paths (see Figure 14.6):

(i) the contour C_2 consisting of the semi-circle $|z| = R$ in the half-plane $y \geq 0$,

(ii) the contour C_3 made up of the two straight lines C_{3a} and C_{3b}.

(i) This is just as in the previous example, except that now $0 \le t \le \pi$. With this change, we have from (14.33) or (14.34) that

$$\int_{C_2} \frac{dz}{z} = \pi i. \tag{14.35}$$

(ii) The straight lines that make up the contour C_3 may be parameterized as follows:

$$C_{3a}, \qquad z = (1-t)R + itR \qquad \text{for } 0 \le t \le 1;$$

$$C_{3b}, \qquad z = -sR + i(1-s)R \qquad \text{for } 0 \le s \le 1.$$

With these parameterizations the required integrals may be written

$$\int_{C_3} \frac{dz}{z} = \int_0^1 \frac{-R+iR}{R+t(-R+iR)}\,dt + \int_0^1 \frac{-R-iR}{iR+s(-R-iR)}\,ds. \tag{14.36}$$

If we could take over from real-variable theory that, for real t, $\int (a+bt)^{-1}\,dt = b^{-1}\ln(a+bt)$ even if a and b are complex, then these integrals could be evaluated immediately. However, to do this would be presuming to some extent what we wish to show, and so the evaluation must be made in terms of entirely real integrals. For example, the first is given by

$$\begin{aligned}
\int_0^1 \frac{-R+iR}{R(1-t)+itR}\,dt &= \int_0^1 \frac{(-1+i)(1-t-it)}{(1-t)^2+t^2}\,dt \\
&= \int_0^1 \frac{2t-1}{1-2t+2t^2}\,dt + i\int_0^1 \frac{1}{1-2t+2t^2}\,dt \\
&= \frac{1}{2}\left[\ln(1-2t+2t^2)\right]_0^1 + \frac{i}{2}\left[2\tan^{-1}\left(\frac{t-\frac{1}{2}}{\frac{1}{2}}\right)\right]_0^1 \\
&= 0 + \frac{i}{2}\left[\frac{\pi}{2} - \left(-\frac{\pi}{2}\right)\right] = \frac{\pi i}{2}.
\end{aligned}$$

The second integral on the RHS of (14.36) can also be shown to have the value $\pi i/2$. Thus

$$\int_{C_3} \frac{dz}{z} = \pi i$$

gives the value of the integral for the full path C_3. ◀

Considering the results of the preceding two examples, which have common integrands and limits, some interesting observations are possible. Firstly, the two integrals from $z = R$ to $z = -R$, along C_2 and C_3, respectively, have the same value, even though the paths taken are different. It also follows that if we took a closed path C_4, given by C_2 from R to $-R$ and C_3 traversed backwards from $-R$ to R, then the integral round C_4 of z^{-1} would be zero (both parts contributing equal and opposite amounts). This is to be compared with result (14.34), in which closed path C_1, beginning and ending at the same place as C_4, yields a value $2\pi i$.

It is not true, however, that the integrals along the paths C_2 and C_3 are equal for any function $f(z)$, or, indeed, that their values are independent of R in general. The following example illustrates this.

Example Evaluate the complex integral of $f(z) = \text{Re}\, z$ along the paths C_1, C_2 and C_3 shown in Figure 14.6.

(i) If we take $f(z) = \text{Re}\, z$ and the contour C_1 then

$$\int_{C_1} \text{Re}\, z\, dz = \int_0^{2\pi} R\cos t(-R\sin t + iR\cos t)\, dt = i\pi R^2.$$

(ii) Using C_2 as the contour,

$$\int_{C_2} \text{Re}\, z\, dz = \int_0^{\pi} R\cos t(-R\sin t + iR\cos t)\, dt = \tfrac{1}{2}i\pi R^2.$$

(iii) Finally the integral along $C_3 = C_{3a} + C_{3b}$ is given by

$$\begin{aligned}\int_{C_3} \text{Re}\, z\, dz &= \int_0^1 (1-t)R(-R+iR)\, dt + \int_0^1 (-sR)(-R-iR)\, ds \\ &= \tfrac{1}{2}R^2(-1+i) + \tfrac{1}{2}R^2(1+i) = iR^2.\end{aligned}$$

We note that all three results are different, and that each is a function of R. ◀

The results of this section demonstrate that the value of an integral between the same two points may depend upon the path that is taken between them but, at the same time, suggest that, under some circumstances, the value is independent of the path. The general situation is summarized in the result of the next section, namely Cauchy's theorem, which is the cornerstone of the integral calculus of complex variables.

Before passing on to Cauchy's theorem, however, we note an important result concerning complex integrals that will be of some use later. Let us consider the integral of a function $f(z)$ along some path C. If M is an upper bound on the value of $|f(z)|$ on the path, i.e. $|f(z)| \leq M$ on C, and L is the length of the path C, then

$$\left|\int_C f(z)\, dz\right| \leq \int_c |f(z)||dz| \leq M\int_C dl = ML. \qquad (14.37)$$

It is straightforward to verify that this result does indeed hold for the complex integrals considered earlier in this section.

14.9 Cauchy's theorem

Cauchy's theorem states that if $f(z)$ is an analytic function, and $f'(z)$ is continuous at each point within and on a closed contour C, then

$$\oint_C f(z)\, dz = 0. \qquad (14.38)$$

In this statement and from now on we denote an integral around a closed contour by $\oint_C$.

To prove this theorem we will need the two-dimensional form of the divergence theorem, known as Green's theorem in a plane (see Section 3.3). This says that if p and q are two functions with continuous first derivatives within and on a closed contour C (bounding a domain R) in the xy-plane, then

$$\iint_R \left(\frac{\partial p}{\partial x} + \frac{\partial q}{\partial y} \right) dxdy = \oint_C (p\,dy - q\,dx). \tag{14.39}$$

With $f(z) = u + iv$ and $dz = dx + i\,dy$, this can be applied to

$$I = \oint_C f(z)\,dz = \oint_C (u\,dx - v\,dy) + i \oint_C (v\,dx + u\,dy)$$

to give

$$I = \iint_R \left[\frac{\partial(-u)}{\partial y} + \frac{\partial(-v)}{\partial x} \right] dx\,dy + i \iint_R \left[\frac{\partial(-v)}{\partial y} + \frac{\partial u}{\partial x} \right] dx\,dy. \tag{14.40}$$

Now, recalling that $f(z)$ is analytic and therefore that the Cauchy–Riemann relations (14.5) apply, we see that each integrand is identically zero and thus I is also zero; this proves Cauchy's theorem.

In fact, the conditions of the above proof are more stringent than they need be. The continuity of $f'(z)$ is not necessary for the proof of Cauchy's theorem, analyticity of $f(z)$ within and on C being sufficient. However, the proof then becomes more complicated and is too long to be given here.[16]

The connection between Cauchy's theorem and the zero value of the integral of z^{-1} around the composite path C_4 discussed towards the end of the previous section is now apparent: the function z^{-1} is analytic in the two regions of the z-plane enclosed by contours (C_2 and C_{3a}) and (C_2 and C_{3b}). An immediate consequence of Cauchy's theorem is contained in the next worked example.

Example Suppose two points A and B in the complex plane are joined by two different paths C_1 and C_2. Show that if $f(z)$ is an analytic function on each path and in the region enclosed by the two paths, then the integral of $f(z)$ is the same along C_1 and C_2.

The situation is shown in Figure 14.7. Since $f(z)$ is analytic in R, it follows from Cauchy's theorem that we have

$$\int_{C_1} f(z)\,dz - \int_{C_2} f(z)\,dz = \oint_{C_1 - C_2} f(z)\,dz = 0,$$

since $C_1 - C_2$ forms a closed contour enclosing R. Thus we immediately obtain

$$\int_{C_1} f(z)\,dz = \int_{C_2} f(z)\,dz,$$

and so the values of the integrals along C_1 and C_2 are equal. ◀

An important application of Cauchy's theorem is the proof that, in some cases, it is possible to deform a closed contour C into another contour γ in such a way that the integrals of a function $f(z)$ around each of the contours have the same value.

16 The reader may refer to almost any book that is devoted to complex variables and the theory of functions.

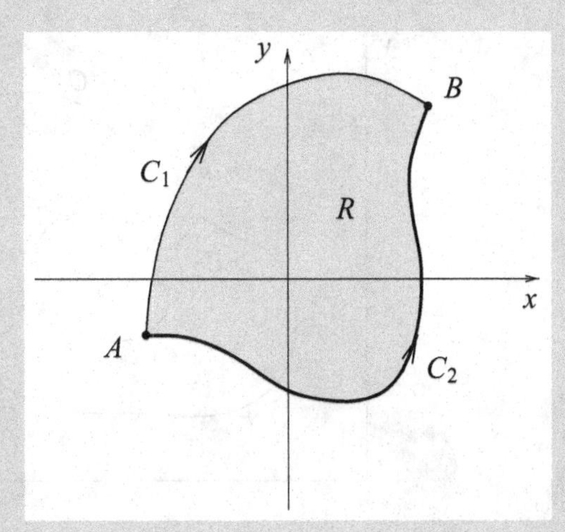

Figure 14.7 Two paths C_1 and C_2 enclosing a region R.

Example Consider two closed contours C and γ in the Argand diagram, γ being sufficiently small that it lies completely within C. Show that if the function $f(z)$ is analytic in the region between the two contours then

$$\oint_C f(z)\,dz = \oint_\gamma f(z)\,dz. \tag{14.41}$$

To prove this result we consider a contour as shown in Figure 14.8. The two close parallel lines C_1 and C_2 join γ and C, which are "cut" to accommodate them. The new contour Γ so formed consists of C, C_1, γ and C_2.

Within the area bounded by Γ, the function $f(z)$ is analytic, and therefore, by Cauchy's theorem (14.38),

$$\oint_\Gamma f(z)\,dz = 0. \tag{14.42}$$

Now the parts C_1 and C_2 of Γ are traversed in opposite directions, and in the limit lie on top of each other, and so their contributions to (14.42) cancel. Thus

$$\oint_C f(z)\,dz + \oint_\gamma f(z)\,dz = 0. \tag{14.43}$$

The sense of the integral round γ is opposite to the conventional (anticlockwise) one, and so by traversing γ in the usual sense, we establish the result (14.41). ◀

A sort of converse of Cauchy's theorem is known as *Morera's theorem*, which states that if $f(z)$ is a continuous function of z in a closed domain R bounded by a curve C and, further, $\oint_C f(z)\,dz = 0$, then $f(z)$ is analytic in R.

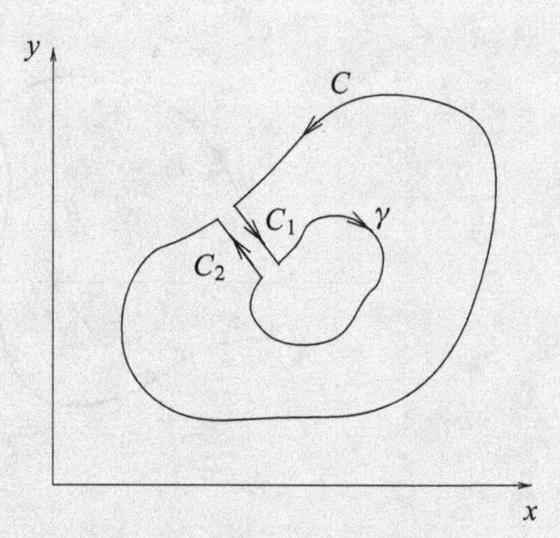

Figure 14.8 The contour used to prove the result (14.41).

14.10 Cauchy's integral formula

Another very important theorem in the theory of complex variables is *Cauchy's integral formula*, which states that if $f(z)$ is analytic within and on a closed contour C and z_0 is a point within C then

$$f(z_0) = \frac{1}{2\pi i} \oint_C \frac{f(z)}{z - z_0}\, dz. \tag{14.44}$$

This formula is saying that the value of an analytic function anywhere inside a closed contour is uniquely determined by its values on the contour[17] and that the specific expression (14.44) can be given for the value at the interior point.

We may prove Cauchy's integral formula by using (14.41) and taking γ to be a circle centered on the point $z = z_0$, of small enough radius ρ that it all lies inside C. Then, since $f(z)$ is analytic inside C, the integrand $f(z)/(z - z_0)$ is analytic in the space between C and γ. Thus, from (14.41), the integral around γ has the same value as that around C.

We then use the fact that any point z on γ is given by $z = z_0 + \rho \exp i\theta$ (and so $dz = i\rho \exp i\theta\, d\theta$). Thus the value of the integral around γ is given by

$$\begin{aligned} I = \oint_\gamma \frac{f(z)}{z - z_0}\, dz &= \int_0^{2\pi} \frac{f(z_0 + \rho \exp i\theta)}{\rho \exp i\theta} i\rho \exp i\theta\, d\theta \\ &= i \int_0^{2\pi} f(z_0 + \rho \exp i\theta)\, d\theta. \end{aligned}$$

17 The similarity between this and the uniqueness theorem for the Laplace equation with Dirichlet boundary conditions (see Chapter 10) is apparent.

If the radius of the circle γ is now shrunk to zero, i.e. $\rho \to 0$, then $I \to 2\pi i f(z_0)$, thus establishing the result (14.44).

An extension to Cauchy's integral formula can be made, yielding an integral expression for $f'(z_0)$:

$$f'(z_0) = \frac{1}{2\pi i}\int_C \frac{f(z)}{(z-z_0)^2}\,dz, \tag{14.45}$$

under the same conditions as previously stated. We now prove this.

Example Prove Cauchy's integral formula for $f'(z_0)$ as given in (14.45).

To show this, we use the definition of a derivative and (14.44) itself to evaluate

$$\begin{aligned} f'(z_0) &= \lim_{h\to 0}\frac{f(z_0+h)-f(z_0)}{h} \\ &= \lim_{h\to 0}\left[\frac{1}{2\pi i}\oint_C \frac{f(z)}{h}\left(\frac{1}{z-z_0-h}-\frac{1}{z-z_0}\right)dz\right] \\ &= \lim_{h\to 0}\left[\frac{1}{2\pi i}\oint_C \frac{f(z)}{(z-z_0-h)(z-z_0)}\,dz\right] \\ &= \frac{1}{2\pi i}\oint_C \frac{f(z)}{(z-z_0)^2}\,dz, \end{aligned}$$

which establishes result (14.45). ◀

Further, it may be proved by induction that the nth derivative of $f(z)$ is also given by a Cauchy integral,

$$f^{(n)}(z_0) = \frac{n!}{2\pi i}\oint_C \frac{f(z)\,dz}{(z-z_0)^{n+1}}. \tag{14.46}$$

Thus, if the value of the analytic function is known on C then not only may the value of the function at any interior point be calculated, but also the values of *all* its derivatives.

The observant reader will notice that (14.46) may also be obtained by the formal device of differentiating under the integral sign with respect to z_0 in Cauchy's integral formula (14.44):

$$\begin{aligned} f^{(n)}(z_0) &= \frac{1}{2\pi i}\oint_C \frac{\partial^n}{\partial z_0^n}\left[\frac{f(z)}{(z-z_0)}\right]dz \\ &= \frac{n!}{2\pi i}\oint_C \frac{f(z)\,dz}{(z-z_0)^{n+1}}. \end{aligned}$$

Yet another contribution to complex variable theory due to Cauchy is the following inequality.

Example Suppose that $f(z)$ is analytic inside and on a circle C of radius R centered on the point $z = z_0$. If $|f(z)| \leq M$ on the circle, where M is some constant, show that

$$\left|f^{(n)}(z_0)\right| \leq \frac{Mn!}{R^n}. \tag{14.47}$$

From (14.46) we have

$$\left|f^{(n)}(z_0)\right| = \frac{n!}{2\pi}\left|\oint_C \frac{f(z)\,dz}{(z-z_0)^{n+1}}\right|,$$

and on using (14.37) this becomes

$$\left|f^{(n)}(z_0)\right| \leq \frac{n!}{2\pi}\frac{M}{R^{n+1}}2\pi R = \frac{Mn!}{R^n}.$$

This result is known as *Cauchy's inequality*. ◀

We may use Cauchy's inequality to prove *Liouville's theorem*, which states that if $f(z)$ is analytic and bounded for all z then f is a constant. Setting $n = 1$ in (14.47) and letting $R \to \infty$, we find $|f'(z_0)| = 0$ and hence $f'(z_0) = 0$. Since $f(z)$ is analytic for all z, we may take z_0 as any point in the z-plane and thus $f'(z) = 0$ for all z; this implies $f(z) =$ constant. Liouville's theorem may be used in turn to prove the *fundamental theorem of algebra* (see Problem 14.9).

14.11 Taylor and Laurent series

Following on from (14.46), we may establish *Taylor's theorem* for functions of a complex variable. If $f(z)$ is analytic inside and on a circle C of radius R centered on the point $z = z_0$, and z is a point inside C, then

$$f(z) = \sum_{n=0}^{\infty} a_n(z-z_0)^n, \tag{14.48}$$

where a_n is given by $f^{(n)}(z_0)/n!$. The Taylor expansion is valid inside the region of analyticity and, for any particular z_0, can be shown to be unique.

To prove Taylor's theorem (14.48), we note that, since $f(z)$ is analytic inside and on C, we may use Cauchy's formula to write $f(z)$ as

$$f(z) = \frac{1}{2\pi i}\oint_C \frac{f(\xi)}{\xi - z}\,d\xi, \tag{14.49}$$

where ξ lies on C. Now we may expand the factor $(\xi - z)^{-1}$ as a geometric series in $(z - z_0)/(\xi - z_0)$,

$$\frac{1}{\xi - z} = \frac{1}{\xi - z_0 + z_0 - z} = \frac{1}{\xi - z_0}\left(1 - \frac{z - z_0}{\xi - z_0}\right)^{-1} = \frac{1}{\xi - z_0}\sum_{n=0}^{\infty}\left(\frac{z - z_0}{\xi - z_0}\right)^n,$$

so (14.49) becomes

$$f(z) = \frac{1}{2\pi i}\oint_C \frac{f(\xi)}{\xi - z_0}\sum_{n=0}^{\infty}\left(\frac{z - z_0}{\xi - z_0}\right)^n d\xi$$

$$= \frac{1}{2\pi i}\sum_{n=0}^{\infty}(z - z_0)^n \oint_C \frac{f(\xi)}{(\xi - z_0)^{n+1}}\,d\xi$$

$$= \frac{1}{2\pi i}\sum_{n=0}^{\infty}(z - z_0)^n \frac{2\pi i f^{(n)}(z_0)}{n!}, \qquad (14.50)$$

where we have used Cauchy's integral formula (14.46) for the derivatives of $f(z)$. Canceling the factors of $2\pi i$, we thus establish the result (14.48) with $a_n = f^{(n)}(z_0)/n!$.

Being analytic places considerable constraints on the variation a function may show in the complex plane; all of its derivatives at any one point are fixed by its behavior in a surrounding infinitesimally small neighborhood, but the corresponding power series expansion then determines its behavior over a finite area extending well beyond that neighborhood. One consequence of this is shown in the following worked example.

Example Show that if $f(z)$ and $g(z)$ are analytic in some region R, and $f(z) = g(z)$ within some subregion S of R, then $f(z) = g(z)$ throughout R.

It is simpler to consider the (analytic) function $h(z) = f(z) - g(z)$, and to show that because $h(z) = 0$ in S it follows that $h(z) = 0$ throughout R.

If we choose a point $z = z_0$ in S, then we can expand $h(z)$ in a Taylor series about z_0,

$$h(z) = h(z_0) + h'(z_0)(z - z_0) + \tfrac{1}{2}h''(z_0)(z - z_0)^2 + \cdots,$$

which will converge inside some circle C that extends at least as far as the nearest part of the boundary of R, since $h(z)$ is analytic in R. But since z_0 lies in S, we have

$$h(z_0) = h'(z_0) = h''(z_0) = \cdots = 0,$$

and so $h(z) = 0$ inside C. We may now expand about a new point, which can lie anywhere within C, and repeat the process. By continuing this procedure we may show that $h(z) = 0$ throughout R.

This result is called the *identity theorem* and, in fact, the equality of $f(z)$ and $g(z)$ throughout R follows from their equality along any curve of non-zero length in R, or even at a countably infinite number of points in R. ◀

So far we have assumed that $f(z)$ is analytic inside and on the (circular) contour C. If, however, $f(z)$ has a singularity inside C at the point $z = z_0$, then it cannot be expanded in a Taylor series. Nevertheless, suppose that $f(z)$ has a pole of order p at $z = z_0$ but is analytic at every other point inside and on C. Then the function $g(z) = (z - z_0)^p f(z)$ is analytic at $z = z_0$, and so may be expanded as a Taylor series about $z = z_0$:

$$g(z) = \sum_{n=0}^{\infty} b_n (z - z_0)^n. \qquad (14.51)$$

Thus, for all z inside C, $f(z)$ will have a power series representation of the form

$$f(z) = \frac{a_{-p}}{(z-z_0)^p} + \cdots + \frac{a_{-1}}{z-z_0} + a_0 + a_1(z-z_0) + a_2(z-z_0)^2 + \cdots, \qquad (14.52)$$

with $a_{-p} \neq 0$. Such a series, which is an extension of the Taylor expansion, is called a *Laurent series*. By comparing the coefficients in (14.51) and (14.52), we see that $a_n = b_{n+p}$. Now, the coefficients b_n in the Taylor expansion of $g(z)$ are seen from (14.50) to be given by

$$b_n = \frac{g^{(n)}(z_0)}{n!} = \frac{1}{2\pi i}\oint \frac{g(z)}{(z-z_0)^{n+1}}\,dz,$$

and so for the coefficients a_n in (14.52) we have

$$a_n = \frac{1}{2\pi i}\oint \frac{g(z)}{(z-z_0)^{n+1+p}}\,dz = \frac{1}{2\pi i}\oint \frac{f(z)}{(z-z_0)^{n+1}}\,dz,$$

an expression that is valid for both positive and negative n.

The terms in the Laurent series with $n \geq 0$ are collectively called the *analytic part*, whilst the remainder of the series, consisting of terms in inverse powers of $z - z_0$, is called the *principal part*. Depending on the nature of the point $z = z_0$, the principal part may contain an infinite number of terms, so that

$$f(z) = \sum_{n=-\infty}^{+\infty} a_n(z-z_0)^n. \qquad (14.53)$$

In this case we would expect the principal part to converge only for $|(z-z_0)^{-1}|$ less than some constant, i.e. *outside* some circle centered on z_0. However, the analytic part will converge *inside* some (different) circle also centered on z_0. If the latter circle has the greater radius then the Laurent series will converge in the region R *between* the two circles (see Figure 14.9); otherwise it does not converge at all.

In fact, it may be shown that any function $f(z)$ that is analytic in a region R between two such circles C_1 and C_2 centered on $z = z_0$ can be expressed as a Laurent series about z_0 that converges in R. We note that, depending on the nature of the point $z = z_0$, the inner circle may be a point (when the principal part contains only a finite number of terms) and the outer circle may have an infinite radius.

We may use the Laurent series of a function $f(z)$ about any point $z = z_0$ to classify the nature of that point. If $f(z)$ is actually analytic at $z = z_0$, then in (14.53) all a_n for $n < 0$ must be zero. It may happen that not only are all a_n zero for $n < 0$ but $a_0, a_1, \ldots, a_{m-1}$ are all zero as well. In this case, the first non-vanishing term in (14.53) is $a_m(z-z_0)^m$, with $m > 0$, and $f(z)$ is then said to have a *zero of order* m at $z = z_0$.

If $f(z)$ is not analytic at $z = z_0$, then two cases arise, as discussed above (p is here taken as positive):

(i) it is possible to find an integer p such that $a_{-p} \neq 0$ but $a_{-p-k} = 0$ for all integers $k > 0$;

(ii) it is not possible to find such a lowest value of $-p$.

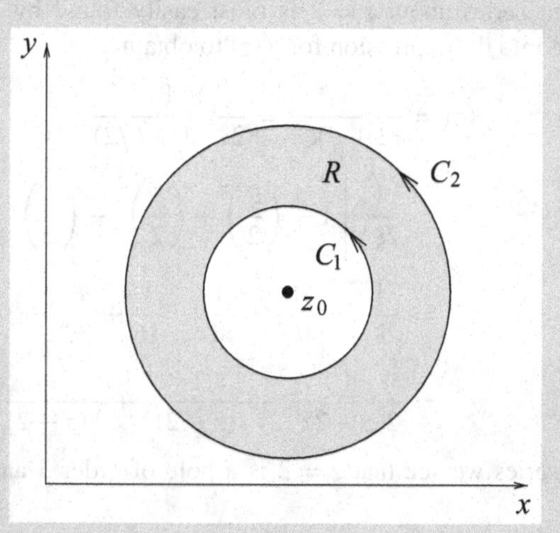

Figure 14.9 The region of convergence R for a Laurent series of $f(z)$ about a point $z = z_0$ where $f(z)$ has a singularity.

In case (i), $f(z)$ is of the form (14.52) and is described as having a *pole of order* p at $z = z_0$; the value of a_{-1} (not a_{-p}) is called the *residue* of $f(z)$ at the pole $z = z_0$, and will play an important part in later applications.

For case (ii), in which the negatively decreasing powers of $z - z_0$ do not terminate, $f(z)$ is said to have an *essential singularity*. These definitions should be compared with those given in Section 14.6.

Example Find the Laurent series of

$$f(z) = \frac{1}{z(z-2)^3}$$

about the singularities $z = 0$ and $z = 2$ (separately). Hence verify that $z = 0$ is a pole of order 1 and $z = 2$ is a pole of order 3, and find the residue of $f(z)$ at each pole.

To obtain the Laurent series about $z = 0$, we make the factor in parentheses in the denominator take the form $(1 - \alpha z)$, where α is some constant, and thus obtain

$$\begin{aligned} f(z) &= -\frac{1}{8z(1-z/2)^3} \\ &= -\frac{1}{8z}\left[1 + (-3)\left(-\frac{z}{2}\right) + \frac{(-3)(-4)}{2!}\left(-\frac{z}{2}\right)^2 + \frac{(-3)(-4)(-5)}{3!}\left(-\frac{z}{2}\right)^3 + \cdots\right] \\ &= -\frac{1}{8z} - \frac{3}{16} - \frac{3z}{16} - \frac{5z^2}{32} - \cdots. \end{aligned}$$

Since the lowest power of z is -1, the point $z = 0$ is a pole of order 1. The residue of $f(z)$ at $z = 0$ is simply the coefficient of z^{-1} in the Laurent expansion about that point and is equal to $-1/8$.

The Laurent series about $z = 2$ is most easily found by letting $z = 2 + \xi$ (or $z - 2 = \xi$) and substituting into the expression for $f(z)$ to obtain

$$\begin{aligned} f(z) &= \frac{1}{(2+\xi)\xi^3} = \frac{1}{2\xi^3(1+\xi/2)} \\ &= \frac{1}{2\xi^3}\left[1 - \left(\frac{\xi}{2}\right) + \left(\frac{\xi}{2}\right)^2 - \left(\frac{\xi}{2}\right)^3 + \left(\frac{\xi}{2}\right)^4 - \cdots\right] \\ &= \frac{1}{2\xi^3} - \frac{1}{4\xi^2} + \frac{1}{8\xi} - \frac{1}{16} + \frac{\xi}{32} - \cdots \\ &= \frac{1}{2(z-2)^3} - \frac{1}{4(z-2)^2} + \frac{1}{8(z-2)} - \frac{1}{16} + \frac{z-2}{32} - \cdots. \end{aligned}$$

From this series we see that $z = 2$ is a pole of order 3 and that the residue of $f(z)$ at $z = 2$ is 1/8. ◀

As we shall see in the next few sections, finding the residue of a function at a singularity is of crucial importance in the evaluation of complex integrals. Specifically, formulae exist for calculating the residue of a function at a particular (singular) point $z = z_0$ without having to expand the function explicitly as a Laurent series about z_0 and identify the coefficient of $(z - z_0)^{-1}$. The type of formula generally depends on the nature of the singularity at which the residue is required.

Example Suppose that $f(z)$ has a pole of order m at the point $z = z_0$. By considering the Laurent series of $f(z)$ about z_0, derive a general expression for the residue $R(z_0)$ of $f(z)$ at $z = z_0$. Hence evaluate the residue of the function

$$f(z) = \frac{\exp iz}{(z^2+1)^2}$$

at the point $z = i$.

If $f(z)$ has a pole of order m at $z = z_0$, then its Laurent series about this point has the form

$$f(z) = \frac{a_{-m}}{(z-z_0)^m} + \cdots + \frac{a_{-1}}{(z-z_0)} + a_0 + a_1(z-z_0) + a_2(z-z_0)^2 + \cdots,$$

which, on multiplying both sides of the equation by $(z - z_0)^m$, gives

$$(z-z_0)^m f(z) = a_{-m} + a_{-m+1}(z-z_0) + \cdots + a_{-1}(z-z_0)^{m-1} + \cdots.$$

Differentiating both sides $m - 1$ times, we obtain

$$\frac{d^{m-1}}{dz^{m-1}}[(z-z_0)^m f(z)] = (m-1)!\, a_{-1} + \sum_{n=1}^{\infty} b_n(z-z_0)^n,$$

for some coefficients b_n. In the limit $z \to z_0$, however, the terms in the sum disappear, and after rearranging we obtain the formula

$$R(z_0) = a_{-1} = \lim_{z\to z_0}\left\{\frac{1}{(m-1)!}\frac{d^{m-1}}{dz^{m-1}}[(z-z_0)^m f(z)]\right\}, \qquad (14.54)$$

which gives the value of the residue of $f(z)$ at the point $z = z_0$.

If we now consider the function

$$f(z) = \frac{\exp iz}{(z^2+1)^2} = \frac{\exp iz}{(z+i)^2(z-i)^2},$$

we see immediately that it has poles of order 2 (*double* poles) at $z = i$ and $z = -i$. To calculate the residue at (for example) $z = i$, we may apply the formula (14.54) with $m = 2$. Performing the required differentiation, we obtain

$$\begin{aligned}\frac{d}{dz}[(z-i)^2 f(z)] &= \frac{d}{dz}\left[\frac{\exp iz}{(z+i)^2}\right] \\ &= \frac{1}{(z+i)^4}[(z+i)^2 i \exp iz - 2(\exp iz)(z+i)].\end{aligned}$$

Setting $z = i$, we find

$$R(i) = \frac{1}{1!}\frac{1}{16}\left(-4ie^{-1} - 4ie^{-1}\right) = -\frac{i}{2e}$$

as the value of the residue at this pole. ◀

An important special case of (14.54) occurs when $f(z)$ has a *simple pole* (a pole of order 1) at $z = z_0$. Then the residue at z_0 is given by

$$R(z_0) = \lim_{z \to z_0} [(z - z_0) f(z)]. \tag{14.55}$$

If $f(z)$ has a simple pole at $z = z_0$ and, as is often the case, has the form $g(z)/h(z)$, where $g(z)$ is analytic and non-zero at z_0 and $h(z_0) = 0$, then (14.55) becomes

$$\begin{aligned}R(z_0) &= \lim_{z \to z_0} \frac{(z-z_0)g(z)}{h(z)} = g(z_0) \lim_{z \to z_0} \frac{(z-z_0)}{h(z)} \\ &= g(z_0) \lim_{z \to z_0} \frac{1}{h'(z)} = \frac{g(z_0)}{h'(z_0)},\end{aligned} \tag{14.56}$$

where we have used l'Hôpital's rule. This result often provides the simplest way of determining the residue at a simple pole.

14.12 Residue theorem

Having seen from Cauchy's theorem that the value of an integral round a closed contour C is zero if the integrand is analytic inside the contour, it is natural to ask what value it takes when the integrand is not analytic inside C. The answer to this is contained in the residue theorem, which we now discuss.

Suppose the function $f(z)$ has a pole of order m at the point $z = z_0$, and so can be written as a Laurent series about z_0 of the form

$$f(z) = \sum_{n=-m}^{\infty} a_n (z - z_0)^n. \tag{14.57}$$

Now consider the integral I of $f(z)$ around a closed contour C that encloses $z = z_0$, but no other singular points. Using Cauchy's theorem, this integral has the same value as the integral around a circle γ of radius ρ centered on $z = z_0$, since $f(z)$ is analytic in the region between C and γ. On the circle we have $z = z_0 + \rho \exp i\theta$ (and $dz = i\rho \exp i\theta \, d\theta$), and so

$$\begin{aligned} I &= \oint_\gamma f(z)\,dz \\ &= \sum_{n=-m}^{\infty} a_n \oint (z - z_0)^n\,dz \\ &= \sum_{n=-m}^{\infty} a_n \int_0^{2\pi} i\rho^{n+1} \exp[i(n+1)\theta]\,d\theta. \end{aligned}$$

For every term in the series with $n \neq -1$, we have

$$\int_0^{2\pi} i\rho^{n+1} \exp[i(n+1)\theta]\,d\theta = \left[\frac{i\rho^{n+1} \exp[i(n+1)\theta]}{i(n+1)}\right]_0^{2\pi} = 0,$$

but for the $n = -1$ term we obtain

$$\int_0^{2\pi} i\,d\theta = 2\pi i.$$

Therefore only the term in $(z - z_0)^{-1}$ contributes to the value of the integral around γ (and therefore C), and I takes the value

$$I = \oint_C f(z)\,dz = 2\pi i a_{-1}. \tag{14.58}$$

Thus the integral around any closed contour containing a single pole of general order m (or, by extension, an essential singularity) is equal to $2\pi i$ times the residue of $f(z)$ at $z = z_0$.

If we extend the above argument to the case where $f(z)$ is continuous within and on a closed contour C and analytic, except for a finite number of poles, within C, then we arrive at the *residue theorem*

$$\oint_C f(z)\,dz = 2\pi i \sum_j R_j, \tag{14.59}$$

where $\sum_j R_j$ is the sum of the residues of $f(z)$ at its poles within C.

The method of proof is indicated by Figure 14.10, in which (a) shows the original contour C referred to in (14.59) and (b) shows a contour C' giving the same value to the integral, because f is analytic between C and C'. Now the contribution to the C' integral from the polygon (a triangle for the case illustrated) joining the small circles is zero, since f is also analytic inside C'. Hence the whole value of the integral comes from the circles and, by result (14.58), each of these contributes $2\pi i$ times the residue at the pole it encloses. All the circles are traversed in their positive sense if C is thus traversed and so the residue theorem follows. Formally, Cauchy's theorem (14.38) is a particular case of (14.59) in which C encloses no poles.

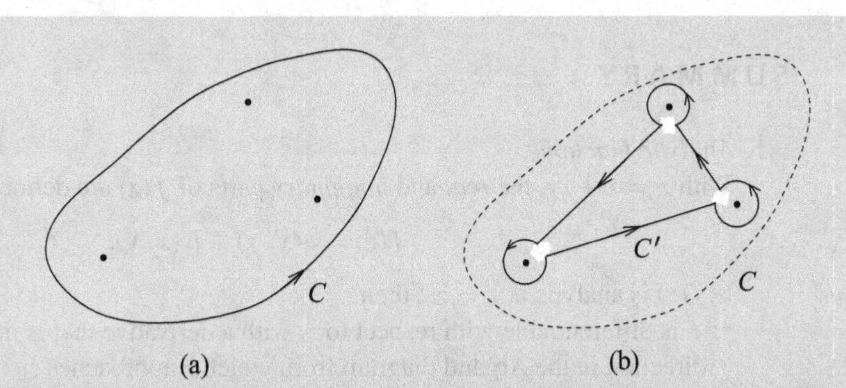

Figure 14.10 The contours used to prove the residue theorem: (a) the original contour; (b) the contracted contour encircling each of the poles.

Finally we prove another important result, for later use. Suppose that $f(z)$ has a simple pole at $z = z_0$ and so may be expanded as the Laurent series

$$f(z) = \phi(z) + a_{-1}(z - z_0)^{-1},$$

where $\phi(z)$ is analytic within some neighborhood surrounding z_0. We wish to find an expression for the integral I of $f(z)$ along an *open* contour C, which is the arc of a circle of radius ρ centered on $z = z_0$ given by

$$|z - z_0| = \rho, \qquad \theta_1 \leq \arg(z - z_0) \leq \theta_2, \tag{14.60}$$

where ρ is chosen small enough that no singularity of f, other than $z = z_0$, lies within the circle. Then I is given by

$$I = \int_C f(z)\,dz = \int_C \phi(z)\,dz + a_{-1} \int_C (z - z_0)^{-1}\,dz.$$

If the radius of the arc C is now allowed to tend to zero, then the first integral tends to zero, since the path becomes of zero length and ϕ is analytic and therefore continuous along it. On C, $z = \rho e^{i\theta}$ and hence the required expression for I is

$$I = \lim_{\rho \to 0} \int_C f(z)\,dz = \lim_{\rho \to 0} \left(a_{-1} \int_{\theta_1}^{\theta_2} \frac{1}{\rho e^{i\theta}} i\rho e^{i\theta}\,d\theta \right) = ia_{-1}(\theta_2 - \theta_1). \tag{14.61}$$

We note that result (14.58) is a special case of (14.61) in which θ_2 is equal to $\theta_1 + 2\pi$.

In the next chapter, some ways in which the properties of a function of a complex variable can be used to solve practical problems in the physical sciences are demonstrated; other methods also based on complex variable theory can be no more than mentioned. Some of the methods depend upon the differential properties of such functions, in particular the property that each part is a solution of Laplace's equation, whilst others are based on contour integration and the residue theorem. Amongst the latter are the evaluation of certain types of definite integrals, the summation of infinite series, and the determination of inverse Laplace transforms.

SUMMARY

1. *Analytic functions*

   With $z = x + iy$, the *real* and *imaginary* parts of $f(z)$ are defined by

$$f(z) = u(x, y) + iv(x, y).$$

   If $f(z)$ is analytic at $z = z_0$, then

   - f is differentiable with respect to z, with a derivative that is independent of the direction in the Argand diagram from which z approaches z_0.
   - The Cauchy–Riemann relations

$$\frac{\partial u}{\partial x} = \frac{\partial v}{\partial y} \quad \text{and} \quad \frac{\partial u}{\partial y} = -\frac{\partial v}{\partial x}$$

     are satisfied.
   - With x and y written as $x = \frac{1}{2}(z + z^*)$ and $y = -\frac{1}{2}i(z - z^*)$, the function $u(x, y) + iv(x, y)$ is a function of z only (and not of z^*).
   - Both u and v are separately solutions of Laplace's equation, i.e. $\nabla^2 u = 0$ and $\nabla^2 v = 0$.
   - ∇u and ∇v are orthogonal vectors, and $|\nabla u| = |\nabla v|$.

2. *Power series and other functions of z*

   - The radius of convergence R of the power series $f(z) = \sum_{n=0}^{\infty} a_n z^n$ is given by $R^{-1} = \lim_{n\to\infty} |a_n|^{1/n}$. Terms for which $a_n = 0$ are to be ignored when establishing the existence and value of the limit.
   - Any power series can be differentiated any number of times inside its circle of convergence, with

$$f^{(r)}(z) = \sum_{n=0}^{\infty} n(n-1)\cdots(n-r+1)a_n z^{n-r}.$$

   - With $z = re^{i\theta}$, $\text{Ln}\, z = \ln r + i(\theta + 2\pi k)$ and $\ln z = \ln r + i\theta$ with $-\pi < \theta \le \pi$.
   - With t and z both complex, $t^z = \exp(z\text{Ln}\, t)$, a multivalued function.
   - Multivalued functions, such as logarithms, complex powers and real or complex roots, can effectively be made single-valued by the judicious use of branch cuts.

3. *Singularities of complex functions at $z = z_0$*

Singularity type	Criterion
Removable	$f(z_0) = \dfrac{0}{0}$, but $\lim_{z\to z_0} f(z)$ exists
Pole of order n, with n a positive integer	There is a positive integer n such that $\lim_{z\to z_0}[(z-z_0)^n f(z)] = a \neq 0$
Essential	There is no positive integer n such that $\lim_{z\to z_0}[(z-z_0)^n f(z)] = a \neq 0$
At infinity	$f(z)$ has a singularity at infinity if $f(\xi)$, where $\xi = 1/z$, has a singularity at $\xi = 0$

4. *Conformal transformations*
A mapping $w = g(z) = r(x, y) + is(x, y)$ and its inverse $z = h(w)$ are conformal if both g and h are analytic (except possibly at a finite number of isolated points).
The *critical points* of the mapping are those at which either or both of $g'(z)$ or $h'(w)$ is zero or infinite. Except at critical points, a conformal transformation has the following properties.

Feature in z-plane	Feature in w-plane
Continuous line	Continuous line
Angle between two curves	Corresponding angle has the same value
Infinitesimal line segment	The magnification, $\lvert g'(z)\rvert$, and rotation angle, $\arg g'(z)$, are independent of the line's direction
Small plane figures	Though magnified and rotated, the shape is unchanged
$f(z) = u + iv$ is analytic	The transformed function $F(g(z)) = F(w) = U + iV$ is analytic
Any 2D solution of $\nabla^2\phi = 0$	Another 2D solution of $\nabla^2\phi = 0$
u (or v) is constant over a boundary C	U (or V) is constant, with the same value, over the transformed boundary C'

5. *Complex integrals*
   - The integral $\int_A^B f(z)\,dz$ may have different values for two different paths, C_1 and C_2, between A and B. The two values will be equal if the closed path $C_1 + (-C_2)$ encloses no singularities of $f(z)$.
   - Cauchy's theorem: $\oint_C f(z)\,dz = 0$ if $f(z)$ is analytic and $f'(z)$ continuous, within and on the closed contour C.
   - Cauchy's integral formulae: with $f(z)$ analytic in and on C and z_0 within C,

$$f(z_0) = \frac{1}{2\pi i}\oint_C \frac{f(z)}{z-z_0}\,dz \quad \text{and} \quad f^{(n)}(z_0) = \frac{n!}{2\pi i}\oint_C \frac{f(z)}{(z-z_0)^{n+1}}\,dz.$$

6. *Taylor and Laurent series*

- Taylor's theorem: $f(z) = \sum_{n=0}^{\infty} a_n(z - z_0)^n$ with $a_n = f^{(n)}(z_0)/n!$ and $f^{(n)}(z_0)$ as given by Cauchy's formula. If $a_0 = a_1 = \cdots = a_{m-1} = 0$ but $a_m \neq 0$, then $f(z)$ has a zero of order m at $z = z_0$.
- Two functions that are analytic in a region R and are equal along any curve of finite length in R, are equal throughout R.
- A function $f(z)$ that has a pole of order p at $z = z_0$ has a Laurent expansion, in which $a_{-p} \neq 0$, of the form

$$f(z) = \frac{a_{-p}}{(z - z_0)^p} + \cdots + \frac{a_{-1}}{z - z_0} + a_0 + a_1(z - z_0) + a_2(z - z_0)^2 + \cdots .$$

- The *residue* $R(z_0)$ at the pole is the value of a_{-1}. (Warning: *not* the value of a_{-p}.)

7. *Calculating the residue of $f(z)$ at a pole located at $z = z_0$*

Pole type	How to calculate the residue
Simple pole ($p = 1$)	$\lim_{z \to z_0} \{(z - z_0) f(z)\}$
$f(z) = \dfrac{g(z)}{h(z)}$	If $h(z)$ has a simple ($m = 1$) zero at $z = z_0$ whilst $g(z)$ is analytic and $g(z_0) \neq 0$, then the residue is $g(z_0)/h'(z_0)$
General pole of order p	$\lim_{z \to z_0} \left\{ \dfrac{1}{(p-1)!} \dfrac{d^{p-1}}{dz^{p-1}} [(z - z_0)^p f(z)] \right\}$
General	Set $z = z_0 + \xi$, use the binomial, Maclaurin and/or other expansions and determine the coefficient of ξ^{-1}. Other coefficients can be ignored

- Residue theorem: $\oint_C f(z)\,dz = 2\pi i \sum_j R_j$, where R_j is the residue at the jth pole of $f(z)$ that lies within C.

PROBLEMS

14.1. Find an analytic function of $z = x + iy$ whose imaginary part is

$$(y \cos y + x \sin y) \exp x.$$

14.2. Find a function $f(z)$, analytic in a suitable part of the Argand diagram, for which

$$\text{Re } f = \frac{\sin 2x}{\cosh 2y - \cos 2x}.$$

Where are the singularities of $f(z)$?

14.3. Find the radii of convergence of the following Taylor series:

(a) $\displaystyle\sum_{n=2}^{\infty} \frac{z^n}{\ln n}$, (b) $\displaystyle\sum_{n=1}^{\infty} \frac{n!z^n}{n^n}$,

(c) $\displaystyle\sum_{n=1}^{\infty} z^n n^{\ln n}$, (d) $\displaystyle\sum_{n=1}^{\infty} \left(\frac{n+p}{n}\right)^{n^2} z^n$, with p real.

14.4. Find the Taylor series expansion about the origin of the function $f(z)$ defined by

$$f(z) = \sum_{r=1}^{\infty} (-1)^{r+1} \sin\left(\frac{pz}{r}\right),$$

where p is a constant. Hence verify that $f(z)$ is a convergent series for all z.

14.5. Determine the types of singularities (if any) possessed by the following functions at $z = 0$ and $z = \infty$:

(a) $(z-2)^{-1}$, (b) $(1+z^3)/z^2$, (c) $\sinh(1/z)$,
(d) e^z/z^3, (e) $z^{1/2}/(1+z^2)^{1/2}$.

14.6. Identify the zeros, poles and essential singularities of the following functions:

(a) $\tan z$, (b) $[(z-2)/z^2]\sin[1/(1-z)]$, (c) $\exp(1/z)$,
(d) $\tan(1/z)$, (e) $z^{2/3}$.

14.7. Find the real and imaginary parts of the functions (i) z^2, (ii) e^z, and (iii) $\cosh \pi z$. By considering the values taken by these parts on the boundaries of the region $0 \le x, y \le 1$, determine the solution of Laplace's equation in that region that satisfies the boundary conditions

$$\phi(x,0) = 0, \qquad \phi(0,y) = 0,$$
$$\phi(x,1) = x, \qquad \phi(1,y) = y + \sin \pi y.$$

14.8. Show that $\arg(\tanh z)$ is given by $\tan^{-1}(\sin 2y/\sinh 2x)$. Identify $\arg(\tanh z)$ as the imaginary part of a suitable function of a complex variable and hence deduce that, for $x > 0$, $\tan^{-1}(\sin 2y/\sinh 2x)$ is a solution of Laplace's equation. By means of an appropriate scaling, find an explicit solution of Laplace's equation that takes the value 0 on the lines $y = 0$ and $y = a$ (for $x > 0$) and the value A on the line $x = 0$ (for $0 < y < a$).

14.9. The *fundamental theorem of algebra* states that, for a complex polynomial $p_n(z)$ of degree n, the equation $p_n(z) = 0$ has precisely n complex roots. By applying Liouville's theorem (see the end of Section 14.10) to $f(z) = 1/p_n(z)$, prove that $p_n(z) = 0$ has at least one complex root. Factor out that root to obtain $p_{n-1}(z)$ and, by repeating the process, prove the above theorem.

14.10. Show that, if a is a positive real constant, the function $\exp(iaz^2)$ is analytic and $\to 0$ as $|z| \to \infty$ for $0 < \arg z \leq \pi/4$. By applying Cauchy's theorem to a suitable contour prove that

$$\int_0^\infty \cos(ax^2)\, dx = \sqrt{\frac{\pi}{8a}}.$$

14.11. The function

$$f(z) = (1 - z^2)^{1/2}$$

of the complex variable z is defined to be real and positive on the real axis in the range $-1 < x < 1$. Using cuts running along the real axis for $1 < x < +\infty$ and $-\infty < x < -1$, show how $f(z)$ is made single-valued and evaluate it on the upper and lower sides of both cuts.

Use these results and a suitable contour in the complex z-plane to evaluate the integral

$$I = \int_1^\infty \frac{dx}{x(x^2 - 1)^{1/2}}.$$

Confirm your answer by making the substitution $x = \sec\theta$.

14.12. By applying the residue theorem around a wedge-shaped contour of angle $2\pi/n$, with one side along the real axis, prove that the integral

$$\int_0^\infty \frac{dx}{1 + x^n},$$

where n is real and ≥ 2, has the value $(\pi/n)\text{cosec}(\pi/n)$.

14.13. The following is an alternative (and roundabout!) way of evaluating the Gaussian integral.

(a) Prove that the integral of $[\exp(i\pi z^2)]\text{cosec}\pi z$ around the parallelogram with corners $\pm 1/2 \pm R\exp(i\pi/4)$ has the value $2i$.

(b) Show that the parts of the contour parallel to the real axis do not contribute when $R \to \infty$.

(c) Evaluate the integrals along the other two sides by putting $z' = r\exp(i\pi/4)$ and working in terms of $z' + \frac{1}{2}$ and $z' - \frac{1}{2}$. Hence, by letting $R \to \infty$ show that

$$\int_{-\infty}^\infty e^{-\pi r^2}\, dr = 1.$$

HINTS AND ANSWERS

14.1. $\partial u/\partial y = -(\exp x)(y\cos y + x\sin y + \sin y)$; $z\exp z$.

14.3. (a) 1; (b) e; (c) 1; (d) e^{-p}.

14.5. (a) Analytic, analytic; (b) double pole, single pole; (c) essential singularity, analytic; (d) triple pole, essential singularity; (e) branch point, branch point.

14.7. (i) $x^2 - y^2$, $2xy$; (ii) $e^x \cos y$, $e^x \sin y$; (iii) $\cosh \pi x \cos \pi y$, $\sinh \pi x \sin \pi y$; $\phi(x, y) = xy + (\sinh \pi x \sin \pi y)/\sinh \pi$.

14.9. Assume that $p_r(x)$ $(r = n, n-1, \ldots, 1)$ has no roots and then argue by the method of contradiction.

14.11. With $0 \le \theta_1 < 2\pi$ and $-\pi < \theta_2 \le \pi$, $f(z) = (r_1 r_2)^{1/2} \exp[\,i(\theta_1 + \theta_2 - \pi)\,]$. The four values are $\pm i(x^2 - 1)^{1/2}$, with the plus sign corresponding to points near the cut that lie in the second and fourth quadrants. $I = \pi/2$.

14.13. (a) The only pole is at the origin with residue π^{-1};
(b) each is $O[\,\exp(-\pi R^2 \mp \sqrt{2}\pi R)\,]$;
(c) the sum of the integrals is $2i \int_{-R}^{R} \exp(-\pi r^2)\, dr$.

15 Applications of complex variables

In Chapter 14, we developed the basic theory of the functions of a complex variable, $z = x + iy$, studied their analyticity (differentiability) properties and derived a number of results concerned with values of contour integrals in the complex plane. In this current chapter we will show how some of those results and properties can be exploited to tackle problems arising directly from physical situations or from apparently unrelated parts of mathematics.

In the former category will be the use of the differential properties of the real and imaginary parts of a function of a complex variable to solve problems involving Laplace's equation in two dimensions, whilst an example of the latter might be the summation of certain types of infinite series. Other applications, such as the Bromwich inversion formula for Laplace transforms, appear as mathematical problems that have their origins in physical applications; the Bromwich inversion enables us to extract the spatial or temporal response of a system to an initial input from the representation of that response in "frequency space" – or, more correctly, imaginary frequency space.

Some other topics that could have been considered, had space permitted, are the location of the (complex) zeros of a polynomial, the approximate evaluation of certain types of contour integrals using the methods of steepest descent and stationary phase, and the so-called "phase-integral" solutions to some differential equations. However, for these and many more, the brief outlines given in the final section of this chapter will have to suffice.

15.1 Complex potentials

Towards the end of Section 14.2 of the previous chapter it was shown that the real and imaginary parts of an analytic function of z are separately solutions of Laplace's equation in two dimensions. Analytic functions thus offer a possible way of solving some two-dimensional physical problems describable by a potential satisfying $\nabla^2\phi = 0$. The general method is known as that of *complex potentials*.

We also found that if $f = u + iv$ is an analytic function of z then any curve $u =$ constant intersects any curve $v =$ constant at right angles. In the context of solutions of Laplace's equation, this result implies that the real and imaginary parts of $f(z)$ have an additional connection between them, for if the set of contours on which one of them is a constant represents the equipotentials of a system then the contours on which the other is constant, being orthogonal to each of the first set, must represent the *corresponding* field lines or

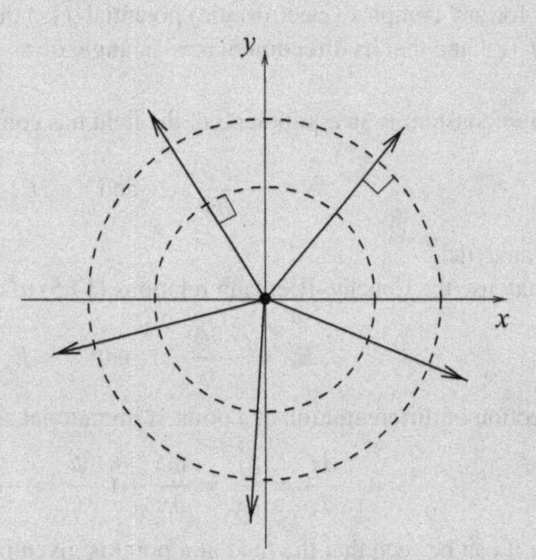

Figure 15.1 The equipotentials (dashed circles) and field lines (solid lines) for a line charge perpendicular to the z-plane.

stream lines, depending on the context. The analytic function f is the complex potential. It is conventional to use ϕ and ψ (rather than u and v) to denote the real and imaginary parts of a complex potential, so that $f = \phi + i\psi$.

As an example, consider the function

$$f(z) = \frac{-q}{2\pi\epsilon_0} \ln z \tag{15.1}$$

in connection with the physical situation of a line charge of strength q per unit length passing through the origin, perpendicular to the z-plane (Figure 15.1). Its real and imaginary parts are

$$\phi = \frac{-q}{2\pi\epsilon_0} \ln |z|, \qquad \psi = \frac{-q}{2\pi\epsilon_0} \arg z. \tag{15.2}$$

The contours in the z-plane of $\phi =$ constant are concentric circles and those of $\psi =$ constant are radial lines. As expected, these are orthogonal sets, but in addition they are, respectively, the equipotentials and electric field lines appropriate to the field produced by the line charge. The minus sign is needed in (15.1) because the value of ϕ must decrease with increasing distance from the origin.

Suppose, for a general complex potential, we make the choice that the real part ϕ of the analytic function f gives the conventional potential function; ψ could equally well be selected. Then we may consider how the direction and magnitude of the field are related to f. This we now do.

Example Show that for any complex (electrostatic) potential $f(z)$ the strength of the electric field is given by $E = |f'(z)|$ and that its direction makes an angle of $\pi - \arg[f'(z)]$ with the x-axis.

Because $\phi =$ constant is an equipotential, the field has components

$$E_x = -\frac{\partial \phi}{\partial x} \quad \text{and} \quad E_y = -\frac{\partial \phi}{\partial y}. \tag{15.3}$$

Since f is analytic:

(i) we may use the Cauchy–Riemann relations (14.5) to change the second of these, obtaining

$$E_x = -\frac{\partial \phi}{\partial x} \quad \text{and} \quad E_y = \frac{\partial \psi}{\partial x}; \tag{15.4}$$

(ii) the direction of differentiation at a point is immaterial and so

$$\frac{df}{dz} = \frac{\partial f}{\partial x} = \frac{\partial \phi}{\partial x} + i\frac{\partial \psi}{\partial x} = -E_x + iE_y. \tag{15.5}$$

From these it can be seen that the field at a point is given in magnitude by $E = |f'(z)|$ and that it makes an angle with the x-axis given by $\pi - \arg[f'(z)]$. ◀

It will be apparent from the above that much of physical interest can be calculated by working directly in terms of f and z. In particular, the electric field vector $\mathbf{E}$ may be represented, using (15.5) above, by the quantity[1]

$$\mathcal{E} = E_x + iE_y = -[f'(z)]^*.$$

Complex potentials can be used in two-dimensional fluid mechanics problems in a similar way. If the flow is stationary (i.e. the velocity of the fluid does not depend on time) and irrotational, and the fluid is both incompressible and non-viscous, then the velocity of the fluid can be described by $\mathbf{V} = \nabla\phi$, where ϕ is the velocity potential and satisfies $\nabla^2\phi = 0$. If, for a complex potential $f = \phi + i\psi$, the real part ϕ is taken to represent the velocity potential then the curves $\psi =$ constant will be the streamlines of the flow. In a direct parallel with the electric field, the velocity may be represented in terms of the complex potential by

$$\mathcal{V} = V_x + iV_y = [f'(z)]^*,$$

the difference of a minus sign reflecting the corresponding difference between the definitions of $\mathbf{E}$ and $\mathbf{V}$. The speed of the flow is equal to $|f'(z)|$. Points where $f'(z) = 0$, and thus the velocity is zero, are called *stagnation points* of the flow.

Analogously to the electrostatic case, a line *source* of fluid at $z = z_0$, perpendicular to the z-plane (i.e. a point from which fluid is emerging at a constant rate), is described by the complex potential

$$f(z) = k\ln(z - z_0),$$

1 Apply this result to the line charge discussed above to obtain the associated electric field. Confirm that its strength and direction are as expected.

where k is the strength of the source. A sink is similarly represented, but with k replaced by $-k$. Other simple examples are as follows.

(i) The flow of a fluid at a constant speed V_0 and at an angle α to the x-axis is described by $f(z) = V_0(\exp i\alpha)z$.

(ii) Vortex flow, in which fluid flows azimuthally in an anticlockwise direction around some point z_0, the speed of the flow being inversely proportional to the distance from z_0, is described by $f(z) = -ik\ln(z - z_0)$, where k is the strength of the vortex. For a clockwise vortex k is replaced by $-k$.

Our next worked example is taken from the area of fluid flow.

Example Verify that the complex potential

$$f(z) = V_0\left(z + \frac{a^2}{z}\right)$$

is appropriate to a circular cylinder of radius a placed so that it is perpendicular to a uniform fluid flow of speed V_0 parallel to the x-axis.

Firstly, since $f(z)$ is analytic except at $z = 0$, both its real and imaginary parts satisfy Laplace's equation in the region exterior to the cylinder. Also $f(z) \to V_0 z$ as $z \to \infty$, so that $\operatorname{Re} f(z) \to V_0 x$, which is appropriate to a uniform flow of speed V_0 in the x-direction far from the cylinder.

Writing $z = r\exp i\theta$ and using de Moivre's theorem we have

$$\begin{aligned} f(z) &= V_0\left[r\exp i\theta + \frac{a^2}{r}\exp(-i\theta)\right] \\ &= V_0\left(r + \frac{a^2}{r}\right)\cos\theta + iV_0\left(r - \frac{a^2}{r}\right)\sin\theta. \end{aligned}$$

Thus we see that the streamlines of the flow described by $f(z)$ are given by

$$\psi = V_0\left(r - \frac{a^2}{r}\right)\sin\theta = \text{constant}.$$

In particular, $\psi = 0$ on $r = a$, independently of the value of θ, and so $r = a$ must be a streamline. Since there can be no flow of fluid across streamlines, $r = a$ must correspond to a boundary along which the fluid flows tangentially. Thus $f(z)$ is a solution of Laplace's equation that satisfies all the required physical boundary conditions, and so, by the uniqueness theorem, it is the appropriate complex potential.[2] ◀

By a similar argument, the complex potential $f(z) = -E(z - a^2/z)$ (note the minus signs) is appropriate to a conducting circular cylinder of radius a placed perpendicular to a uniform electric field $\mathbf{E}$ in the x-direction.

The real and imaginary parts of a complex potential $f = \phi + i\psi$ have another interesting relationship in the context of Laplace's equation in electrostatics or fluid mechanics. Let us choose ϕ as the conventional potential, so that ψ represents the stream function (or

2 Find, for a fixed value of r, the variation with θ of the flow speed, showing that it lies between $V_0(r^2 - a^2)/r^2$ and $V_0(r^2 + a^2)/r^2$.

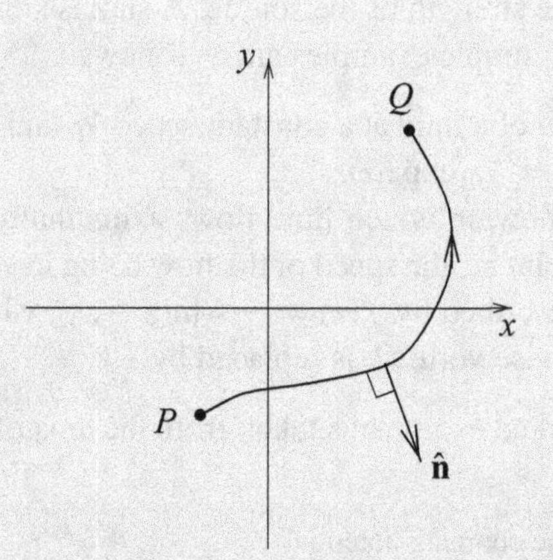

Figure 15.2 A curve joining the points P and Q. Also shown is $\hat{\mathbf{n}}$, the unit vector normal to the curve.

electric field, depending on the application), and consider the difference in the values of ψ at any two points P and Q connected by some path C, as shown in Figure 15.2. This difference is given by

$$\psi(Q) - \psi(P) = \int_P^Q d\psi = \int_P^Q \left(\frac{\partial \psi}{\partial x}\, dx + \frac{\partial \psi}{\partial y}\, dy \right),$$

which, on using the Cauchy–Riemann relations, becomes

$$\begin{aligned} \psi(Q) - \psi(P) &= \int_P^Q \left(-\frac{\partial \phi}{\partial y}\, dx + \frac{\partial \phi}{\partial x}\, dy \right) \\ &= \int_P^Q \nabla\phi \cdot \hat{\mathbf{n}}\, ds \;=\; \int_P^Q \frac{\partial \phi}{\partial n}\, ds, \end{aligned}$$

where $\hat{\mathbf{n}}$ is the vector unit normal[3] to the path C and s is the arc-length along the path; the last equality is written in terms of the normal derivative $\partial\phi/\partial n \equiv \nabla\phi \cdot \hat{\mathbf{n}}$.

Now suppose that in an electrostatics application, the path C is the surface of a conductor; then

$$\frac{\partial \phi}{\partial n} = -\frac{\sigma}{\epsilon_0},$$

where σ is the surface charge density per unit length normal to the xy-plane. Therefore $-\epsilon_0[\psi(Q) - \psi(P)]$, being the integral between P and Q of this charge density, is equal to the total charge per unit length normal to the xy-plane on the surface of the conductor between the points P and Q. Similarly, in fluid mechanics applications, if the density of the fluid is ρ and its velocity is $\mathbf{V}$ then

$$\rho[\psi(Q) - \psi(P)] = \rho \int_P^Q \nabla\phi \cdot \hat{\mathbf{n}}\, ds = \rho \int_P^Q \mathbf{V} \cdot \hat{\mathbf{n}}\, ds$$

3 $\hat{\mathbf{n}}$ has components $(ds)^{-1}(dy, -dx)$.

is equal to the mass flux between P and Q per unit length perpendicular to the xy-plane. The next example applies this to the electrostatic case.

Example A conducting circular cylinder of radius a is placed with its center line passing through the origin and perpendicular to a uniform electric field $\mathbf{E}$ in the x-direction. Find the charge per unit length induced on the half of the cylinder that lies in the region $x < 0$.

As mentioned immediately following the previous example, the appropriate complex potential for this problem is $f(z) = -E(z - a^2/z)$. Writing $z = r \exp i\theta$ this becomes

$$f(z) = -E\left[r \exp i\theta - \frac{a^2}{r}\exp(-i\theta)\right]$$
$$= -E\left(r - \frac{a^2}{r}\right)\cos\theta - iE\left(r + \frac{a^2}{r}\right)\sin\theta,$$

so that on $r = a$ the imaginary part of f is given by

$$\psi = -2Ea\sin\theta.$$

Therefore the induced charge q per unit length on the left half of the cylinder, between $\theta = \pi/2$ and $\theta = 3\pi/2$, is given by

$$q = 2\epsilon_0 Ea[\sin(3\pi/2) - \sin(\pi/2)] = -4\epsilon_0 Ea.$$

It should be remembered that ψ measures the integral of the charge density, and not the density itself; the $\sin\theta$ dependence of ψ implies a $\cos\theta$ dependence of the induced charge density, in line with the intuitive result that the density is maximal on the $\theta = 0$ part of the cylinder. ◀

15.2 Applications of conformal transformations

In Section 14.7 of the previous chapter it was shown that, under a conformal transformation $w = g(z)$ from $z = x + iy$ to a new variable $w = r + is$, if a solution of Laplace's equation in some region R of the xy-plane can be found as the real or imaginary part of an analytic function[4] of z, then the same expression put in terms of r and s will be a solution of Laplace's equation in the corresponding region R' of the w-plane, and vice versa. In addition, if the solution is constant over the boundary C of the region R in the xy-plane, then the solution in the w-plane will take the same constant value over the corresponding curve C' that bounds R'.

Thus, from any two-dimensional solution of Laplace's equation for a particular geometry, typified by those discussed in the previous section, further solutions for other geometries can be obtained by making conformal transformations. From the physical point of view the given geometry is usually complicated, and so the solution is sought by transforming to a simpler one. However, first working from simpler to more complicated situations

4 In fact, the original solution in the xy-plane need not be given explicitly as the real or imaginary part of an analytic function. Any solution of $\nabla^2\phi = 0$ in the xy-plane is carried over into another solution of $\nabla^2\phi = 0$ in the new variables by a conformal transformation, and vice versa.

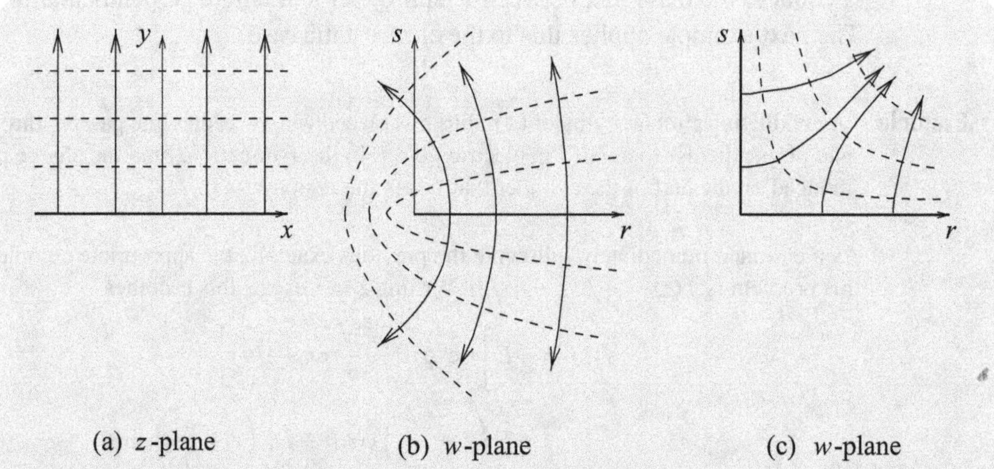

Figure 15.3 The equipotential lines (broken) and field lines (solid) (a) for an infinite charged conducting plane at $y = 0$, where $z = x + iy$, and after the transformations (b) $w = z^2$ and (c) $w = z^{1/2}$ of the situation shown in (a).

can provide useful experience and make it more likely that the reverse procedure can be tackled successfully; our first worked example takes this approach.

Example Find the complex electrostatic potential associated with an infinite charged conducting plate $y = 0$, and thus obtain those associated with

(i) a semi-infinite charged conducting plate ($r > 0$, $s = 0$);
(ii) the inside of a right-angled charged conducting wedge ($r > 0$, $s = 0$ and $r = 0$, $s > 0$).

Figure 15.3(a) shows the equipotentials (broken lines) and field lines (solid lines) for the infinite charged conducting plane $y = 0$. Suppose that we elect to make the real part of the complex potential coincide with the conventional electrostatic potential. If the plate is charged to a potential V then clearly

$$\phi(x, y) = V - ky, \tag{15.6}$$

where k is related to the charge density σ by $k = \sigma/\epsilon_0$, since physically the electric field $\mathbf{E}$ has components $(0, \sigma/\epsilon_0)$ and $\mathbf{E} = -\nabla\phi$.

Thus what is needed is an analytic function of z, of which the real part is $V - ky$. This can be obtained by inspection, but we may proceed formally and use the Cauchy–Riemann relations to obtain the imaginary part $\psi(x, y)$ as follows:

$$\frac{\partial \psi}{\partial y} = \frac{\partial \phi}{\partial x} = 0 \quad \text{and} \quad \frac{\partial \psi}{\partial x} = -\frac{\partial \phi}{\partial y} = k.$$

Hence $\psi = kx + c$ and, absorbing c into V, the required complex potential is

$$f(z) = V - ky + ikx = V + ikz. \tag{15.7}$$

(i) Now consider the transformation

$$w = g(z) = z^2. \tag{15.8}$$

This satisfies the criteria for a conformal mapping (except at $z = 0$) and carries the upper half of the z-plane into the entire w-plane; the equipotential plane $y = 0$ goes into the half-plane $r > 0$, $s = 0$.

By the general results proved, $f(z)$, when expressed in terms of r and s, will give a complex potential whose real part will be constant on the half-plane in question; we deduce that

$$F(w) = f(z) = V + ikz = V + ikw^{1/2} \tag{15.9}$$

is the required potential. Expressed in terms of r, s and $\rho = (r^2 + s^2)^{1/2}$, $w^{1/2}$ is given by[5]

$$w^{1/2} = \left(\frac{\rho + r}{2}\right)^{1/2} + i\left(\frac{\rho - r}{2}\right)^{1/2}, \tag{15.10}$$

from which it follows that the electrostatic potential, expressed in terms of r and s, is

$$\Phi(r, s) = \text{Re}\, F(w) = V - \frac{k}{\sqrt{2}}\left[(r^2 + s^2)^{1/2} - r\right]^{1/2}. \tag{15.11}$$

The corresponding equipotentials and field lines are shown in Figure 15.3(b). Using results (15.3)–(15.5), the magnitude of the electric field is

$$|\mathbf{E}| = |F'(w)| = |\tfrac{1}{2}ikw^{-1/2}| = \tfrac{1}{2}k(r^2 + s^2)^{-1/4}.$$

(ii) A transformation "converse" to that used in (i) is

$$w = g(z) = z^{1/2},$$

and has the effect of mapping the upper half of the z-plane into the first quadrant of the w-plane and the conducting plane $y = 0$ into the wedge $r > 0$, $s = 0$ and $r = 0$, $s > 0$.

The complex potential now becomes

$$\begin{aligned} F(w) &= V + ikw^2 \\ &= V + ik[(r^2 - s^2) + 2irs], \end{aligned} \tag{15.12}$$

showing that the electrostatic potential is $V - 2krs$ and that the electric field has components

$$\mathbf{E} = (2ks, 2kr). \tag{15.13}$$

Figure 15.3(c) indicates the approximate equipotentials and field lines. (Note that, in both transformations, $g'(z)$ is either 0 or ∞ at the origin, and so neither transformation is conformal there. Consequently there is no violation of result (ii), given at the start of Section 14.7, concerning the angles between intersecting lines.) ◀

The *method of images*, discussed in Section 11.5, can be used in conjunction with conformal transformations to solve some problems involving Laplace's equation in two dimensions, as we now demonstrate.

5 Confirm that this is so by computing the square (*not* the modulus squared) of the given expression, showing that it is equal to $r + is$.

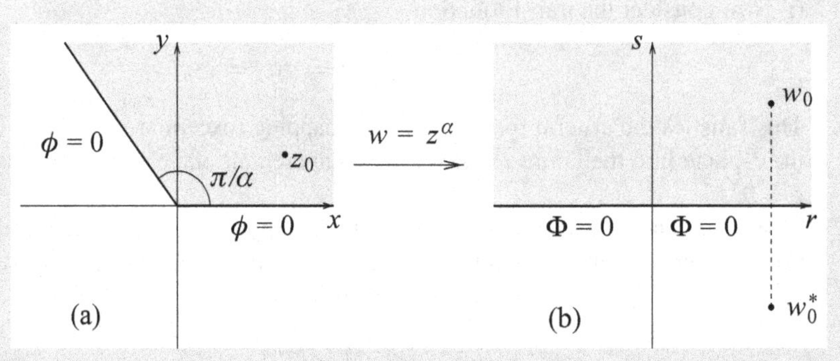

Figure 15.4 (a) An infinite conducting wedge with interior angle π/α and a line charge at $z = z_0$; (b) after the transformation $w = z^\alpha$, with an additional image charge placed at $w = w_0^*$.

Example A wedge of angle π/α with its vertex at $z = 0$ is formed by two semi-infinite conducting plates, as shown in Figure 15.4(a). A line charge of strength q per unit length is positioned at $z = z_0$, perpendicular to the z-plane. By considering the transformation $w = z^\alpha$, find the complex electrostatic potential for this situation.

Let us consider the action of the transformation $w = z^\alpha$ on the lines defining the positions of the conducting plates. The plate that lies along the positive x-axis is mapped onto the positive r-axis in the w-plane, whereas the plate that lies along the direction $\exp(i\pi/\alpha)$ is mapped into the negative r-axis, as shown in Figure 15.4(b). Similarly the line charge at z_0 is mapped onto the point $w_0 = z_0^\alpha$.

From Figure 15.4(b), we see that in the w-plane the problem can be solved by introducing a second line charge of opposite sign (known as an *image source*) at the point w_0^*, so that the potential $\Phi = 0$ along the r-axis. The complex potential for such an arrangement is simply

$$F(w) = -\frac{q}{2\pi\epsilon_0}\ln(w - w_0) + \frac{q}{2\pi\epsilon_0}\ln(w - w_0^*).$$

Substituting $w = z^\alpha$ into the above shows that the required complex potential in the original z-plane is

$$f(z) = \frac{q}{2\pi\epsilon_0}\ln\left(\frac{z^\alpha - z_0^{*\alpha}}{z^\alpha - z_0^\alpha}\right).$$

Note that the appearance of a complex conjugate in the final expression is not in conflict with the general requirement that the complex potential be analytic. It is z^* that must not appear; here, $z_0^{*\alpha}$ is no more than a parameter of the problem. ◀

15.3 Definite integrals using contour integration

We now turn our attention to the most direct application of contour integration and the residue theorem, namely that of evaluating various types of definite integral. The different sorts are largely characterized by the type of contour employed, but in each case little preamble is needed, since a series of worked examples provides the best explanation.

15.3.1 Integrals of sinusoidal functions

Suppose that an integral of the form

$$\int_0^{2\pi} F(\cos\theta, \sin\theta)\, d\theta \tag{15.14}$$

is to be evaluated. It can be made into a contour integral around the unit circle C by writing $z = \exp i\theta$, and making the substitutions

$$\cos\theta = \tfrac{1}{2}(z + z^{-1}), \qquad \sin\theta = -\tfrac{1}{2}i(z - z^{-1}), \qquad d\theta = -iz^{-1}\, dz. \tag{15.15}$$

This contour integral can then be evaluated using the residue theorem, provided the transformed integrand has only a finite number of poles inside the unit circle and none on it. Our first worked example in this section has these properties.

Example Evaluate

$$I = \int_0^{2\pi} \frac{\cos 2\theta}{a^2 + b^2 - 2ab\cos\theta}\, d\theta, \qquad b > a > 0. \tag{15.16}$$

By de Moivre's theorem,

$$\cos n\theta = \tfrac{1}{2}(z^n + z^{-n}). \tag{15.17}$$

Using $n = 2$ in (15.17) and straightforward substitution for the other functions of θ in (15.16) gives[6]

$$I = \frac{i}{2ab}\oint_C \frac{z^4 + 1}{z^2(z - a/b)(z - b/a)}\, dz.$$

Thus there are two poles inside C, a double pole at $z = 0$ and a simple pole at $z = a/b$ (recall that $b > a$ and so the pole at $z = b/a$ is outside the unit circle).

We could find the residue of the integrand at $z = 0$ by expanding the integrand as a Laurent series in z and identifying the coefficient of z^{-1}. Alternatively, we may use the formula (14.54) with $m = 2$. Choosing the latter method and denoting the integrand, excluding the factor $i/2ab$, by $f(z)$, we have

$$\begin{aligned}\frac{d}{dz}[z^2 f(z)] &= \frac{d}{dz}\left[\frac{z^4 + 1}{(z - a/b)(z - b/a)}\right] \\ &= \frac{(z - a/b)(z - b/a)4z^3 - (z^4 + 1)[(z - a/b) + (z - b/a)]}{(z - a/b)^2(z - b/a)^2}.\end{aligned}$$

6 For practice, make these substitutions and confirm the final form of the integral.

Now setting $z = 0$ and applying (14.54), we find

$$R(0) = \frac{a}{b} + \frac{b}{a}.$$

For the simple pole at $z = a/b$, equation (14.55) gives the residue as

$$R(a/b) = \lim_{z \to (a/b)} [(z - a/b) f(z)] = \frac{(a/b)^4 + 1}{(a/b)^2 (a/b - b/a)}$$

$$= -\frac{a^4 + b^4}{ab(b^2 - a^2)}.$$

The residue theorem now gives

$$I = 2\pi i \times \frac{i}{2ab} \left[\frac{a^2 + b^2}{ab} - \frac{a^4 + b^4}{ab(b^2 - a^2)} \right] = \frac{2\pi a^2}{b^2 (b^2 - a^2)}$$

as the final result.[7] ◀

It is clear that the integral of any other specified function of sinusoids can, in principle, be tackled in the same way using only (15.15), (15.17), and the latter's $\sin n\theta$ counterpart, provided the integral is over the range $0 \le \theta \le 2\pi$. Integrals over other ranges can sometimes be cast in this form by the use of symmetry or a change of integration variable.

15.3.2 Some infinite integrals

We next consider the evaluation of an integral of the form

$$\int_{-\infty}^{\infty} f(x)\, dx,$$

where $f(z)$ has the following properties:

(i) $f(z)$ is analytic in the upper half-plane, $\mathrm{Im}\, z \ge 0$, except for a finite number of poles, none of which is on the real axis;
(ii) on a semi-circle Γ of radius R (Figure 15.5), R times the maximum of $|f|$ on Γ tends to zero as $R \to \infty$ (a sufficient condition is that $zf(z) \to 0$ as $|z| \to \infty$);
(iii) $\int_{-\infty}^{0} f(x)\, dx$ and $\int_{0}^{\infty} f(x)\, dx$ both exist.

Since

$$\left| \int_\Gamma f(z)\, dz \right| \le 2\pi R \times (\text{maximum of } |f| \text{ on } \Gamma),$$

condition (ii) ensures that the integral along Γ tends to zero as $R \to \infty$, after which it is obvious from the residue theorem that the required integral is given by

$$\int_{-\infty}^{\infty} f(x)\, dx = 2\pi i \times (\text{sum of the residues at poles with } \mathrm{Im}\, z > 0). \tag{15.18}$$

The next example illustrates this.

7 Explain, in terms of the behavior of the integrand, why I is large if b is only a little greater than a.

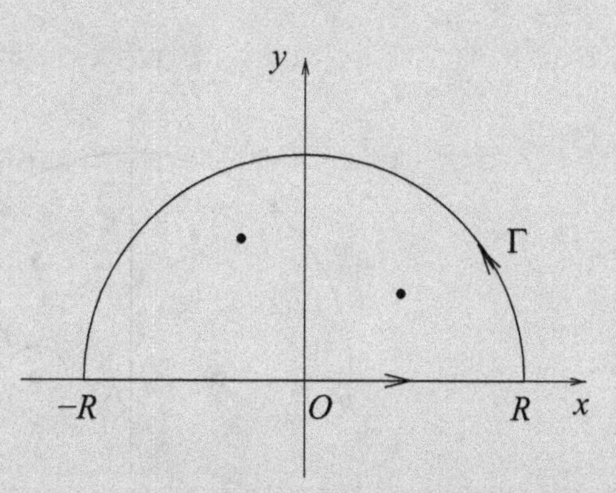

Figure 15.5 A semi-circular contour in the upper half-plane.

Example Evaluate

$$I = \int_0^\infty \frac{dx}{(x^2+a^2)^4}, \qquad \text{where } a \text{ is real.}$$

The complex function $(z^2+a^2)^{-4}$ has poles of order 4 at $z = \pm ai$, of which only $z = ai$ is in the upper half-plane. Conditions (ii) and (iii) are clearly satisfied. For higher-order poles, formula (14.54) for evaluating residues can be tiresome to apply. So, instead, we put $z = ai + \xi$ and expand for small ξ to obtain[8]

$$\frac{1}{(z^2+a^2)^4} = \frac{1}{(2ai\xi+\xi^2)^4} = \frac{1}{(2ai\xi)^4}\left[1 - \frac{i\xi}{2a}\right]^{-4}.$$

The coefficient of ξ^{-1} is determined by that of the ξ^3 term in the binomial expansion of the expression enclosed in square brackets. It, and hence the residue, is thus

$$\frac{1}{(2a)^4}\frac{(-4)(-5)(-6)}{3!}\left(\frac{-i}{2a}\right)^3 = \frac{-5i}{32a^7},$$

and so by the residue theorem

$$\int_{-\infty}^{\infty} \frac{dx}{(x^2+a^2)^4} = \frac{10\pi}{32a^7}.$$

From the symmetry of the integrand about $x = 0$, I is one-half of this value and so $I = 5\pi/(32a^7)$.[9] ◀

Condition (i) of the previous method required there to be no poles of the integrand on the real axis, but in fact simple poles on the real axis can be accommodated by indenting the contour as shown in Figure 15.6. The indentation at the pole $z = z_0$ is in the form of a

8 This illustrates another useful technique for determining residues.

9 Can you find a substitution that will transform the original integral into one that can be evaluated by means of the technique of the previous section?

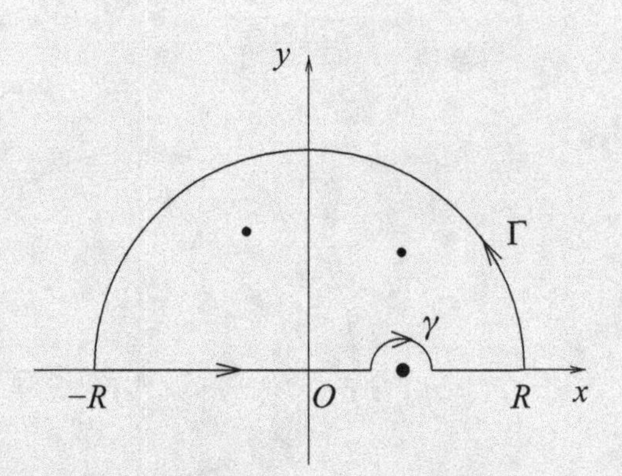

Figure 15.6 An indented contour used when the integrand has a simple pole on the real axis.

semi-circle γ of radius ρ in the upper half-plane, thus excluding the pole from the interior of the contour.

What is then obtained from a contour integration, apart from the contributions for Γ and γ, is called the *principal value of the integral,* defined as $\rho \to 0$ by

$$P \int_{-R}^{R} f(x)\,dx \equiv \int_{-R}^{z_0-\rho} f(x)\,dx + \int_{z_0+\rho}^{R} f(x)\,dx.$$

The remainder of the calculation goes through as before, but the contribution from the semi-circle, γ, must be included. Result (14.61) of Section 14.12 shows that since only a simple pole is involved its contribution is

$$-ia_{-1}\pi, \tag{15.19}$$

where a_{-1} is the residue at the pole and the minus sign arises because γ is traversed in the clockwise (negative) sense.

We defer giving an example of an indented contour until we have established *Jordan's lemma*; we will then work through an example illustrating both. Jordan's lemma enables infinite integrals involving sinusoidal functions to be evaluated.

For a function $f(z)$ of a complex variable z, if

(i) *$f(z)$ is analytic in the upper half-plane except for a finite number of poles in* $\text{Im}\, z > 0$,
(ii) *the maximum of $|f(z)| \to 0$ as $|z| \to \infty$ in the upper half-plane,*
(iii) *$m > 0$,*

then

$$I_\Gamma = \int_\Gamma e^{imz} f(z)\,dz \to 0 \quad \text{as } R \to \infty, \tag{15.20}$$

where Γ is the same semi-circular contour as in Figure 15.5.

Note that this condition (ii) is less stringent than the earlier condition (ii) (see the start of this section), since we now only require $M(R) \to 0$ and not $RM(R) \to 0$, where M is the maximum[10] of $|f(z)|$ on $|z| = R$.

10 More strictly, the least upper bound.

The proof of the lemma is straightforward once it has been observed that, for $0 \leq \theta \leq \pi/2$,

$$1 \geq \frac{\sin\theta}{\theta} \geq \frac{2}{\pi}. \qquad (15.21)$$

Then, since on Γ we have $|\exp(imz)| = |\exp(-mR\sin\theta)|$,

$$I_\Gamma \leq \int_\Gamma |e^{imz} f(z)|\,|dz| \leq MR\int_0^\pi e^{-mR\sin\theta}\,d\theta = 2MR\int_0^{\pi/2} e^{-mR\sin\theta}\,d\theta.$$

Thus, using (15.21),

$$I_\Gamma \leq 2MR\int_0^{\pi/2} e^{-mR(2\theta/\pi)}\,d\theta = \frac{\pi M}{m}\left(1 - e^{-mR}\right) < \frac{\pi M}{m};$$

hence, as $R \to \infty$, I_Γ tends to zero since M tends to zero.

As noted earlier, our next example involves both Jordan's lemma and the notion of a principal value.

Example Find the principal value of

$$\int_{-\infty}^{\infty} \frac{\cos mx}{x-a}\,dx, \qquad \text{for } a \text{ real, } m > 0.$$

Consider the function $(z-a)^{-1}\exp(imz)$; although it has no poles in the upper half-plane it does have a simple pole at $z = a$, and further $|(z-a)^{-1}| \to 0$ as $|z| \to \infty$. We will use a contour like that shown in Figure 15.6 and apply the residue theorem. Symbolically,

$$\int_{-R}^{a-\rho} + \int_\gamma + \int_{a+\rho}^{R} + \int_\Gamma = 0. \qquad (15.22)$$

Now as $R \to \infty$ and $\rho \to 0$ we have $\int_\Gamma \to 0$, by Jordan's lemma, and from (15.18) and (15.19) we obtain

$$P\int_{-\infty}^{\infty} \frac{e^{imx}}{x-a}\,dx - i\pi a_{-1} = 0, \qquad (15.23)$$

where a_{-1} is the residue of $(z-a)^{-1}\exp(imz)$ at $z = a$, which is $\exp(ima)$. Then taking the real and imaginary parts of (15.23) gives

$$P\int_{-\infty}^{\infty} \frac{\cos mx}{x-a}\,dx = -\pi\sin ma, \qquad \text{as required,}$$

$$P\int_{-\infty}^{\infty} \frac{\sin mx}{x-a}\,dx = \pi\cos ma, \qquad \text{as a bonus.}$$

For the particular case $a = 0$, the cosine integral gives the expected zero result for an odd integrand, whilst the sine integral, after a change of variable $mx \to x$, gives the normalization of the sinc function: $\int_{-\infty}^{\infty}(\sin x)/x\,dx = \pi$. ◀

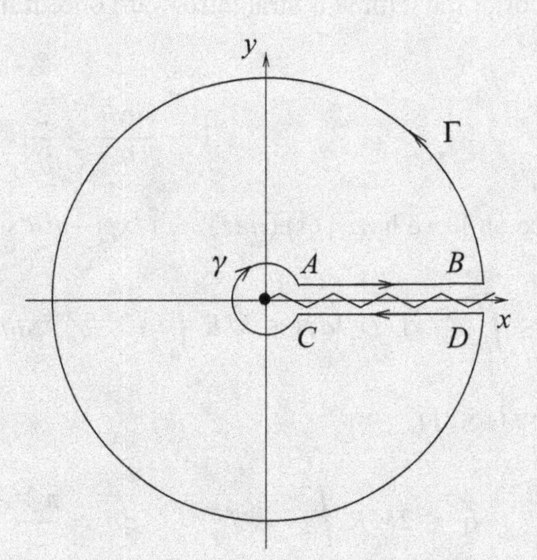

Figure 15.7 A typical cut-plane contour for use with multivalued functions that have a single branch point located at the origin.

15.3.3 Integrals of multivalued functions

In the previous chapter we discussed briefly some of the properties and difficulties associated with certain multivalued functions such as $z^{1/2}$ or $\mathrm{Ln}\, z$. It was mentioned that one method of managing such functions is by means of a "cut plane". A similar technique can be used with advantage to evaluate some kinds of infinite integral involving real functions for which the corresponding complex functions are multivalued. A typical contour employed for functions with a single branch point located at the origin is shown in Figure 15.7. Here Γ is a large circle of radius R and γ is a small one of radius ρ, both centered on the origin. Eventually we will let $R \to \infty$ and $\rho \to 0$.

The success of the method is due to the fact that because the integrand is multivalued, its values along the two lines AB and CD joining $z = \rho$ to $z = R$ are *not* equal and opposite, although both are related to the corresponding real integral. Again an example provides the best explanation.

Example Evaluate

$$I = \int_0^\infty \frac{dx}{(x+a)^3 x^{1/2}}, \qquad a > 0.$$

We consider the integrand $f(z) = (z+a)^{-3} z^{-1/2}$ and note that $|zf(z)| \to 0$ on the two circles as $\rho \to 0$ and $R \to \infty$.[11] Thus, in the limits, the two circles make no contribution to the contour integral.

11 On the smaller circle $|zf(z)|$ behaves like $a^{-3}\rho^{1/2}$, whilst on the larger one it $\sim R^{-5/2}$.

The only pole of the integrand inside the contour is at $z = -a$ (and is of order 3). To determine its residue we put $z = -a + \xi$ and expand (noting that $(-a)^{1/2}$ equals $a^{1/2}\exp(i\pi/2) = ia^{1/2}$):

$$\frac{1}{(z+a)^3 z^{1/2}} = \frac{1}{\xi^3 i a^{1/2}(1-\xi/a)^{1/2}}$$
$$= \frac{1}{i\xi^3 a^{1/2}}\left(1 + \frac{1}{2}\frac{\xi}{a} + \frac{3}{8}\frac{\xi^2}{a^2} + \cdots\right).$$

The residue is thus $-3i/(8a^{5/2})$.

The residue theorem (14.59) now gives

$$\int_{AB} + \int_{\Gamma} + \int_{DC} + \int_{\gamma} = 2\pi i\left(\frac{-3i}{8a^{5/2}}\right).$$

We have seen that $\int_{\Gamma}$ and $\int_{\gamma}$ vanish, and if we denote z by x along the line AB then it has the value $z = x\exp 2\pi i$ along the line DC (note that $\exp 2\pi i$ must not be set equal to 1 until after the substitution for z has been made in $\int_{DC}$). Substituting these expressions,

$$\int_0^{\infty} \frac{dx}{(x+a)^3 x^{1/2}} + \int_{\infty}^{0} \frac{dx}{[x\exp 2\pi i + a]^3 x^{1/2}\exp(\frac{1}{2}2\pi i)} = \frac{3\pi}{4a^{5/2}}.$$

Thus, on setting $e^{2\pi i} = 1$ and reversing the limits in the second integral, we have

$$\left(1 - \frac{1}{\exp \pi i}\right)\int_0^{\infty} \frac{dx}{(x+a)^3 x^{1/2}} = \frac{3\pi}{4a^{5/2}}$$

and the result,

$$I = \frac{1}{2} \times \frac{3\pi}{4a^{5/2}},$$

follows directly. ◀

Several other examples of integrals of multivalued functions around a variety of contours are included in the problems at the end of this chapter.

15.4 Summation of series

We now turn to an application of contour integration which at first sight might seem to lie in an unrelated area of mathematics, namely the summation of infinite series. Sometimes a real infinite series with index n, say, can be summed with the help of a suitable complex function that has poles on the real axis at the various positions $z = n$ with the corresponding residues at those poles equal to the values of the terms of the series. A worked example provides the best explanation of how the technique is applied; other examples will be found in the problems.

Example By considering

$$\oint_C \frac{\pi \cot \pi z}{(a+z)^2}\,dz,$$

where a is not an integer and C is a circle of large radius, evaluate

$$\sum_{n=-\infty}^{\infty} \frac{1}{(a+n)^2}.$$

The integrand has (i) simple poles at $z =$ integer n, for $-\infty < n < \infty$, due to the factor $\cot \pi z$ and (ii) a double pole at $z = -a$.

(i) To find the residue of $\cot \pi z$, put $z = n + \xi$ for small ξ. We write $\cot \pi z$ as $\cos \pi z / \sin \pi z$ and Taylor expand both numerator and denominator to first order in $\pi\xi$, remembering that $\sin n\pi = 0$:

$$\cot \pi z = \frac{\cos(n\pi + \xi\pi)}{\sin(n\pi + \xi\pi)} \approx \frac{\cos n\pi + 0\xi\pi}{0 + (\cos n\pi)\xi\pi} = \frac{1}{\xi\pi} \quad \text{for all } n.$$

The residue of the integrand at $z = n$ is thus $\pi(a+n)^{-2}\pi^{-1}$.

(ii) Putting $z = -a + \xi$ for small ξ and determining the coefficient of ξ^{-1} gives[12]

$$\begin{aligned}\frac{\pi \cot \pi z}{(a+z)^2} &= \frac{\pi}{\xi^2}\cot(-a\pi + \xi\pi)\\ &= \frac{\pi}{\xi^2}\left\{\cot(-a\pi) + \xi\left[\frac{d}{dz}(\cot \pi z)\right]_{z=-a} + \cdots\right\},\end{aligned}$$

and so the residue at the double pole $z = -a$ is given by

$$\pi[-\pi \operatorname{cosec}^2 \pi z]_{z=-a} = -\pi^2 \operatorname{cosec}^2 \pi a.$$

Collecting together these results to express the residue theorem gives

$$I = \oint_C \frac{\pi \cot \pi z}{(a+z)^2}\,dz = 2\pi i\left[\sum_{n=-N}^{N} \frac{1}{(a+n)^2} - \pi^2 \operatorname{cosec}^2 \pi a\right], \tag{15.24}$$

where N equals the integer part of R. But as the radius R of C tends to ∞, $\cot \pi z \to \mp i$ (depending on whether Im z is greater or less than zero, respectively[13]). Thus, for some finite constant k,

$$|I| < \left| k \oint_C \frac{dz}{(a+z)^2} \right|,$$

which tends to 0 as $R \to \infty$. Thus $I \to 0$ as R (and hence N) $\to \infty$, and (15.24) establishes the result

$$\sum_{n=-\infty}^{\infty} \frac{1}{(a+n)^2} = \frac{\pi^2}{\sin^2 \pi a}$$

for the sum of the infinite series. ◀

12 This again illustrates one of the techniques for determining residues.

13 Prove this by writing $\cot \pi z$ in terms of exponential functions and then putting $z = x + iy$.

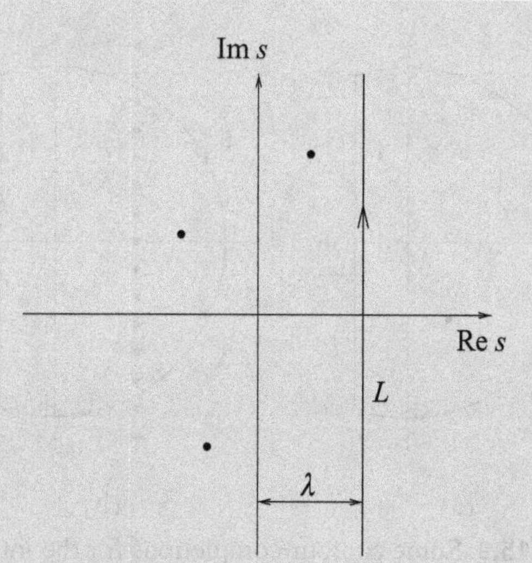

Figure 15.8 The integration path of the inverse Laplace transform is along the infinite line L. The quantity λ must be positive and large enough for all poles of the integrand to lie to the left of L.

Series with alternating signs in the terms, i.e. $(-1)^n$, can also be attempted in this way but using $\operatorname{cosec} \pi z$ rather than $\cot \pi z$, since the former has residue $(-1)^n \pi^{-1}$ at $z = n$ (see Problem 15.17).

15.5 Inverse Laplace transform

As a further example of the use of contour integration we now discuss a method whereby the process of Laplace transformation, discussed in Chapter 5, can be inverted.

It will be recalled that the Laplace transform $\bar{f}(s)$ of a function $f(x)$, $x \geq 0$, is given by

$$\bar{f}(s) = \int_0^\infty e^{-sx} f(x)\, dx, \qquad \text{Re } s > s_0. \tag{15.25}$$

In Chapter 5, functions $f(x)$ were deduced from the transforms by means of a prepared dictionary. However, an explicit formula for an unknown inverse does exist and may be written in the form of an integral. It is known as the *Bromwich integral* and is given by

$$f(x) = \frac{1}{2\pi i} \int_{\lambda - i\infty}^{\lambda + i\infty} e^{sx} \bar{f}(s)\, ds, \qquad \lambda > 0, \tag{15.26}$$

where s is treated as a complex variable and the integration is along the line L indicated in Figure 15.8. The position of the line is dictated by the requirements that λ is positive and that all singularities of $\bar{f}(s)$ lie to the left of the line.

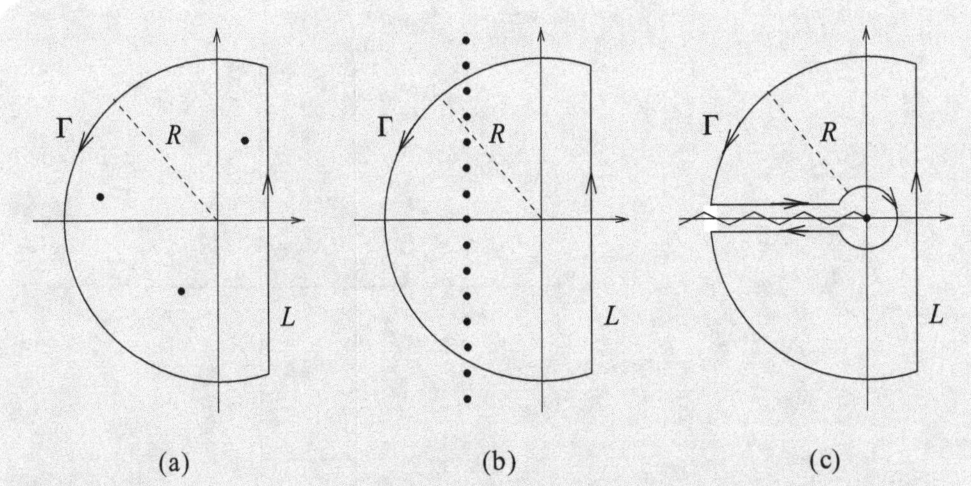

Figure 15.9 Some contour completions for the integration path L of the inverse Laplace transform. For details of when each is appropriate see the main text.

That (15.26) really is the unique inverse of (15.25) is difficult to show for general functions and transforms, but the following verification should at least make it plausible:

$$
\begin{aligned}
f(x) &= \frac{1}{2\pi i}\int_{\lambda-i\infty}^{\lambda+i\infty} ds\, e^{sx}\int_0^\infty e^{-su} f(u)\, du, \qquad \text{Re}(s) > 0, \ \text{ i.e. } \lambda > 0, \\
&= \frac{1}{2\pi i}\int_0^\infty du\, f(u)\int_{\lambda-i\infty}^{\lambda+i\infty} e^{s(x-u)}\, ds \\
&= \frac{1}{2\pi i}\int_0^\infty du\, f(u)\int_{-\infty}^{\infty} e^{\lambda(x-u)} e^{ip(x-u)} i\, dp, \qquad \text{putting } s = \lambda + ip, \\
&= \frac{1}{2\pi}\int_0^\infty f(u) e^{\lambda(x-u)} 2\pi\, \delta(x-u)\, du \\
&= \begin{cases} f(x) & x \geq 0, \\ 0 & x < 0. \end{cases}
\end{aligned} \tag{15.27}
$$

Our main purpose here is to demonstrate the use of contour integration. To employ it in the evaluation of the line integral (15.26), the path L must be made part of a closed contour in such a way that the contribution from the completion either vanishes or is simply calculable.

A typical completion is shown in Figure 15.9(a) and would be appropriate if $\bar{f}(s)$ had a finite number of poles. For more complicated cases, in which $\bar{f}(s)$ has an infinite sequence of poles but all to the left of L as in Figure 15.9(b), a sequence of circular-arc completions that pass between the poles must be used and $f(x)$ is obtained as a series. If $\bar{f}(s)$ is a multivalued function then a cut plane is needed and a contour such as that shown in Figure 15.9(c) might be appropriate.

We consider here only the simple case in which the contour in Figure 15.9(a) is used; we refer the reader to the problems at the end of the chapter for others. Ideally, we would like the contribution to the integral from the circular arc Γ to tend to zero as its radius $R \to \infty$. Using a modified version of Jordan's lemma, it may be shown that this is indeed the case if there exist constants $M > 0$ and $\alpha > 0$ such that on Γ

$$|\bar{f}(s)| \leq \frac{M}{R^\alpha}.$$

Moreover, this condition always holds when $\bar{f}(s)$ has the form

$$\bar{f}(s) = \frac{P(s)}{Q(s)},$$

where $P(s)$ and $Q(s)$ are polynomials and the degree of $Q(s)$ is greater than that of $P(s)$.

When the contribution from the part-circle Γ tends to zero as $R \to \infty$, we have from the residue theorem that the inverse Laplace transform (15.26) is given simply by[14]

$$f(t) = \sum \left(\text{residues of } \bar{f}(s)e^{sx} \text{ at all of its poles}\right). \tag{15.28}$$

The integrand in the next example is the ratio of two polynomials and has two such poles.

Example Find the function $f(x)$ whose Laplace transform is

$$\bar{f}(s) = \frac{s}{s^2 - k^2},$$

where k is a constant.

It is clear that $\bar{f}(s)$ is of the form required for the integral over the circular arc Γ to tend to zero as $R \to \infty$, and so we may use the result (15.28) directly. Now

$$\bar{f}(s)e^{sx} = \frac{se^{sx}}{(s-k)(s+k)},$$

and thus has simple poles at $s = k$ and $s = -k$. Using (14.55) the residues at each pole can be easily calculated as

$$R(k) = \frac{ke^{kx}}{2k} \quad \text{and} \quad R(-k) = \frac{ke^{-kx}}{2k}.$$

Thus the inverse Laplace transform is given by

$$f(x) = \tfrac{1}{2}\left(e^{kx} + e^{-kx}\right) = \cosh kx.$$

This result may be checked by computing the forward transform of $\cosh kx$. ◀

Sometimes a little more care is required when deciding in which half-plane to close the contour C, as in the following example.

14 The factors of $(2\pi i)^{-1}$ from the Bromwich integral and $2\pi i$ from the residue theorem having canceled.

Example Find the function $f(x)$ whose Laplace transform is

$$\bar{f}(s) = \frac{1}{s}(e^{-as} - e^{-bs}),$$

where a and b are fixed and positive, with $b > a$.

From (15.26) we have the integral

$$f(x) = \frac{1}{2\pi i} \int_{\lambda - i\infty}^{\lambda + i\infty} \frac{e^{(x-a)s} - e^{(x-b)s}}{s} \, ds. \tag{15.29}$$

Now, despite appearances to the contrary, the integrand has no poles, as may be confirmed by expanding the exponentials as Taylor series about $s = 0$. Depending on the value of x, several cases arise.

(i) For $x < a$ both exponentials in the integrand will tend to zero as $\text{Re}\, s \to \infty$. Thus we may close L with a circular arc Γ in the *right* half-plane (λ can be as small as desired), and we observe that $s \times$ integrand tends to zero everywhere on Γ as $R \to \infty$. With no poles enclosed and no contribution from Γ, the integral along L must also be zero. Thus

$$f(x) = 0 \qquad \text{for } x < a. \tag{15.30}$$

(ii) For $x > b$ the exponentials in the integrand will tend to zero as $\text{Re}\, s \to -\infty$, and so we may close L in the left half-plane, as in Figure 15.9(a). Again the integral around Γ vanishes for infinite R, and so, by the residue theorem,

$$f(x) = 0 \qquad \text{for } x > b. \tag{15.31}$$

(iii) For $a < x < b$ the two parts of the integrand behave in different ways and have to be treated separately:

$$I_1 - I_2 \equiv \frac{1}{2\pi i} \int_L \frac{e^{(x-a)s}}{s} \, ds - \frac{1}{2\pi i} \int_L \frac{e^{(x-b)s}}{s} \, ds.$$

The integrand of I_1 then vanishes in the far left-hand half-plane, but does now have a (simple) pole at $s = 0$. Closing L in the left half-plane, and using the residue theorem, we obtain

$$I_1 = \text{residue at } s = 0 \text{ of } s^{-1} e^{(x-a)s} = 1. \tag{15.32}$$

The integrand of I_2, however, vanishes in the far right-hand half-plane (and also has a simple pole at $s = 0$) and is evaluated by a circular-arc completion in that half-plane. Since the only pole is at $s = 0$ and λ is strictly > 0, such a contour encloses no poles and leads to $I_2 = 0$.

Thus, collecting together results (15.30)–(15.32) we obtain

$$f(x) = \begin{cases} 0 & \text{for } x < a, \\ 1 & \text{for } a < x < b, \\ 0 & \text{for } x > b, \end{cases}$$

as shown in Figure 15.10. ◀

15.6 Some more advanced applications

As indicated at the start of this chapter, there are many more applications of complex numbers and complex variable theory than we are able to deal with in the present volume.

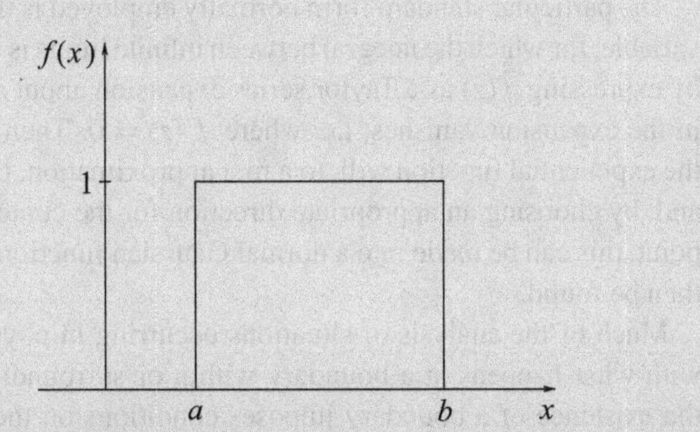

Figure 15.10 The result of the Laplace inversion of $\bar{f}(s) = s^{-1}(e^{-as} - e^{-bs})$ with $b > a$.

Brief summaries of some of them are given below, but the reader will need to look elsewhere for the detail needed to apply them to specific problems.[15]

Nearly all of the applications we must omit involve complex variables, but one use that is made of complex numbers is for the simplification of the analysis of a.c. electrical circuits, or their mechanical analogues. By assigning complex resistances (impedances) to the reactive components of such a circuit, the system can be treated as if it were a d.c. one. The method is explained in the first problem below, and will not be described further here.

The same general area of physics and engineering provides an example of the use of complex variables. The residue theorem, relating the value of a closed contour integral to the sum of the residues at the poles enclosed by the contour, was established in the previous chapter and used in the current one. One important practical use of an extension to the theorem is that of locating the zeros of functions of a complex variable. The location of such zeros has relevance to the study of electrical networks and general oscillations since the complex zeros of certain functions (usually polynomials) give the system parameters (usually frequencies) at which system instabilities occur. The basis of the method is a theorem, known as *the principle of the argument* that relates the change in the argument of a function as it traverses a closed contour in the complex plane to the number of zeros enclosed by the contour.

A second application of complex variable theory is a method of finding approximations to certain types of infinite integrals that cannot be found in closed form. The class of integrals that can be tackled by this method is that containing integrands that are, or can be, represented by exponential functions of the general form $g(z)\exp[f(z)]$. The exponents $f(z)$ may be complex, and so integrals of sinusoids can be handled, as well as those with more obvious exponential properties. The analyticity properties of the functions of a complex variable are used to move the integration path to a part of the complex plane where a general integrand can be well approximated by a standard form; the standard form is then integrated explicitly.

15 Some of them are considered in detail in K. F. Riley, M. P. Hobson and S. J. Bence, *Mathematical Methods for Physics and Engineering*, 3rd edn (Cambridge: Cambridge University Press, 2006), Chapter 25.

The particular standard form normally employed is that of a Gaussian function of a real variable, for which the integral between infinite limits is well known. This form is generated by expressing $f(z)$ as a Taylor series expansion about a point z_0, at which the linear term in the expansion vanishes, i.e. where $f'(z) = 0$. Then, apart from a constant multiplier, the exponential function will, to a first approximation, behave like $\exp[\,\frac{1}{2}f''(z_0)(z-z_0)^2\,]$ and, by choosing an appropriate direction for the contour to take as it passes through the point, this can be made into a normal Gaussian function of a real variable. Its integral may then be found.

Much of the analysis of situations occurring in physics and engineering is concerned with what happens at a boundary within or surrounding a physical system. Sometimes the existence of a boundary imposes conditions on the behavior of variables describing the state of the system; obvious examples include the zero displacement at its end-points of an anchored vibrating string and the zero potential contour that must coincide with a grounded electrical conductor.

More subtle are the effects at internal boundaries, where the same non-vanishing variable has to describe the situation on either side of the boundary but its behavior is quantitatively, or even *qualitatively*, different in the two regions. This type of behavior is exhibited by solutions of the Stokes' differential equation. As well as solutions written as series in the usual way (see Chapter 7), it possesses others that can be expressed as complex contour integrals; hence its connection with the current chapter.

The Stokes' equation can be written in the form

$$\frac{d^2y}{dz^2} = zy \tag{15.33}$$

where z is the complex independent variable. It will be immediately apparent that, even for z restricted to be real and denoted by x, the behavior of the solutions to (15.33) will change markedly as x passes through $x = 0$. For positive x they will have similar characteristics to the solutions of $y'' = k^2y$, where k is real; these have monotonic exponential forms, either increasing or decreasing. On the other hand, when x is negative the solutions will be similar to those of $y'' + k^2y = 0$, i.e. oscillatory functions of x. This is just the sort of behavior shown by the wavefunction describing light diffracted by a sharp edge or by the quantum wavefunction describing a particle near to the boundary of a region which it is classically forbidden to enter on energy grounds. Other examples could be taken from the propagation of electromagnetic radiation in an ion plasma or wave-guide.

An integral of the form

$$y(z) = \int_a^b f(t)\exp(zt)\,dt,$$

in which z is a parameter so far as the integration is concerned but the independent variable so far as $y(z)$ is concerned, is a solution of (15.33) if $f(t)$ is appropriately chosen. Such considerations lead to what are known as *Airy integrals*; these provide physically acceptable solutions to the Stokes' equation.

As a final example of important applications of complex variable theory, we can mention the so-called WKB approach to wave propagation in a non-uniform medium; this approach aims to treat the phase of any solution to the corresponding wave equation with some accuracy, whilst allowing its amplitude to vary with position. The Stokes' equation is one equation to which the method can be applied. Physical applications taken from atomic and nuclear physics include the energy levels in a quantum well, scattering from atomic nuclei and some aspects of quantum tunneling.

SUMMARY

1. *Definite integrals*
   The residue theorem can be used to find the integral of a complex function taken round a closed contour, some or all of which coincides with the range for a required integral I. If the integral along the non-coincident parts of the contour can be evaluated (or shown to be zero), the value of I can be deduced. How to select an appropriate contour for some possible applications is shown below; in each case the final steps are to locate the poles of the integral that lie within the closed contour, calculate their residues, and then apply the residue theorem.

Integral type	Constructing the contour
Sinusoidal functions $\int_0^{2\pi} f(\theta)\,d\theta$	Write $z = e^{i\theta}$, $\cos n\theta = \frac{1}{2}(z^n + z^{-n})$, $\sin\theta = -\frac{1}{2}i(z^n - z^{-n})$ and $d\theta = -iz^{-1}\,dz$. Integrate the resulting integrand around the unit circle.
Infinite integrals $\int_{-\infty}^{\infty} f(x)\,dx$	Construct a closed contour C using the real axis and a semi-circle in either the upper or lower half-plane (whichever can be shown to give zero contribution to $\oint_C f(z)\,dz$). For improper integrals, make contour indentations to deliberately include or exclude any simple poles on the real axis; each indentation contributes $\pm\pi i \times$ the residue at the corresponding pole.
Multivalued functions $\int_0^{\infty} f(x)\,dx$	Use a cut plane with the cut running, typically, along either the positive or negative real axis joining a small circle γ enclosing the origin and a large circle Γ at infinity. At corresponding points on the two branches of the cut, where $z = \pm xe^{i0}$ and $z = \pm xe^{2\pi i}$, the values of the integrand will differ and, although related, the two integrals along the branches will *not* cancel each other.
Series summation $\sum_{n=-\infty}^{\infty} f(n)$	Evaluate the integral $\oint_C f(z)\pi\cot\pi z\,dz$ around an infinite contour that encloses the real axis. The residues to be included are $f(n)$, at the poles $z = n$ of the $\pi\cot\pi z$ factor, and those at any poles of $f(z)$ lying inside C. For alternating sign series, use $\pi\operatorname{cosec}\pi z$ rather than $\pi\cot\pi z$.
Inverse Laplace transforms $\int_{\lambda-i\infty}^{\lambda+i\infty} e^{sx}\bar{f}(s)\,ds$	Make the line of the Bromwich integral part of a closed contour C, typically D or ᗡ in shape, chosen so that the integral around the semi-circle at infinity is zero. Different contours may be needed for different parts of the x-range.

2. *Complex potentials in 2D*
The complex potential in the xy-plane is $f(z)$ with $z = x + iy$; with $f(z) = \phi + i\psi$, take ϕ as the conventional physical potential, i.e. $\nabla^2\phi = 0$.
   - The curves $\psi =$ constant are the field lines (or stream lines) of the system.
   - The electric field is represented in both magnitude and direction by $\mathcal{E} = E_x + iE_y = -[f'(z)]^*$. Similarly, for the 2D flow of an incompressible inviscid fluid, the velocity vector is $\mathcal{V} = V_x + iV_y = [f'(z)]^*$.
   - If P and Q are two points on a curve along which ϕ is constant (e.g. the surface of a conductor), then $-\epsilon_0[\psi(Q) - \psi(P)]$ is the total charge (per unit length perpendicular to the xy-plane) lying between P and Q. Similarly, $\rho[\psi(Q) - \psi(P)]$ is the total mass flux crossing the curve between P and Q in the fluid-flow case.
   - Conformal transformations can sometimes be used to reduce the problem of solving $\nabla^2 u = 0$ for complicated 2D geometry to the same problem but with a simpler geometry. Conformal transformations can be combined with the method of images and appeal to the uniqueness theorem.

PROBLEMS

15.1. In the method of complex impedances for a.c. circuits, an inductance L is represented by a complex impedance $Z_L = i\omega L$ and a capacitance C by $Z_C = 1/(i\omega C)$. Kirchhoff's circuit laws,

$$\sum_i I_i = 0 \text{ at a node and } \sum_i Z_i I_i = \sum_j V_j \text{ around any closed loop,}$$

are then applied as if the circuit were a d.c. one.

Apply this method to the a.c. bridge connected as in Figure 15.11 to show that if the resistance R is chosen as $R = (L/C)^{1/2}$ then the amplitude of the current, I_R, through it is independent of the angular frequency ω of the applied a.c. voltage $V_0\, e^{i\omega t}$.

Determine how the phase of I_R, relative to that of the voltage source, varies with the angular frequency ω.

15.2. A long straight fence made of conducting wire mesh separates two fields and stands one meter high. Sometimes, on fine days, there is a vertical electric field over flat open countryside. Well away from the fence the strength of the field is E_0. By considering the effect of the transformation $w = (1 - z^2)^{1/2}$ on the real and imaginary z-axes, find the strengths of the field (a) at a point one meter directly above the fence, (b) at ground level one meter to the side of the fence, and (c) at a point that is level with the top of the fence but one meter to the side of it. What is the direction of the field in case (c)?

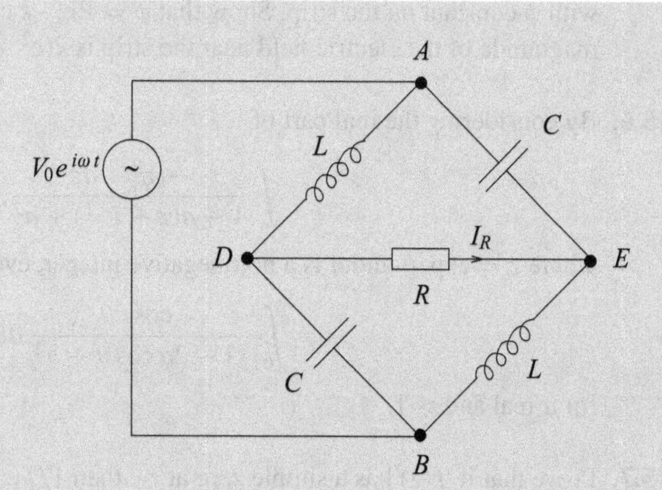

Figure 15.11 The inductor–capacitor–resistor network for Problem 15.1.

15.3. For the function

$$f(z) = \ln\left(\frac{z+c}{z-c}\right),$$

where c is real, show that the real part u of f is constant on a circle of radius $c \operatorname{cosech} u$ centered on the point $z = c \coth u$. Use this result to show that the electrical capacitance per unit length of two parallel cylinders of radii a, placed with their axes $2d$ apart, is proportional to $[\cosh^{-1}(d/a)]^{-1}$.

15.4. Find a complex potential in the z-plane appropriate to a physical situation in which the half-plane $x > 0$, $y = 0$ has zero potential and the half-plane $x < 0$, $y = 0$ has potential V.

By making the transformation $w = a(z + z^{-1})/2$, with a real and positive, find the electrostatic potential associated with the half-plane $r > a$, $s = 0$ and the half-plane $r < -a$, $s = 0$ at potentials 0 and V, respectively.

15.5. By considering in turn the transformations

$$z = \tfrac{1}{2}c(w + w^{-1}) \quad \text{and} \quad w = \exp \zeta,$$

where $z = x + iy$, $w = r \exp i\theta$, $\zeta = \xi + i\eta$ and c is a real positive constant, show that $z = c \cosh \zeta$ maps the strip $\xi \geq 0$, $0 \leq \eta \leq 2\pi$, onto the whole z-plane. Which curves in the z-plane correspond to the lines $\xi =$ constant and $\eta =$ constant? Identify those corresponding to $\xi = 0$, $\eta = 0$ and $\eta = 2\pi$.

The electric potential ϕ of a charged conducting strip $-c \leq x \leq c$, $y = 0$, satisfies

$$\phi \sim -k \ln(x^2 + y^2)^{1/2} \quad \text{for large values of } (x^2 + y^2)^{1/2},$$

with ϕ constant on the strip. Show that $\phi = \text{Re}[-k\cosh^{-1}(z/c)]$ and that the magnitude of the electric field near the strip is $k(c^2 - x^2)^{-1/2}$.

15.6. By considering the real part of

$$\int \frac{-iz^{n-1}\,dz}{1 - a(z + z^{-1}) + a^2},$$

where $z = \exp i\theta$ and n is a non-negative integer, evaluate

$$\int_0^{\pi} \frac{\cos n\theta}{1 - 2a\cos\theta + a^2}\,d\theta$$

for a real and > 1.

15.7. Prove that if $f(z)$ has a simple zero at z_0, then $1/f(z)$ has residue $1/f'(z_0)$ there. Hence evaluate

$$\int_{-\pi}^{\pi} \frac{\sin\theta}{a - \sin\theta}\,d\theta,$$

where a is real and > 1.

15.8. Prove that, for $\alpha > 0$, the integral

$$\int_0^{\infty} \frac{t\sin\alpha t}{1 + t^2}\,dt$$

has the value $(\pi/2)\exp(-\alpha)$.

15.9. Prove that

$$\int_0^{\infty} \frac{\cos mx}{4x^4 + 5x^2 + 1}\,dx = \frac{\pi}{6}\left(4e^{-m/2} - e^{-m}\right) \qquad \text{for } m > 0.$$

15.10. Show that the principal value of the integral

$$\int_{-\infty}^{\infty} \frac{\cos(x/a)}{x^2 - a^2}\,dx$$

is $-(\pi/a)\sin 1$.

15.11. Using a suitable cut plane, prove that if α is real and $0 < \alpha < 1$ then

$$\int_0^{\infty} \frac{x^{-\alpha}}{1 + x}\,dx$$

has the value $\pi\,\text{cosec}\,\pi\alpha$.

15.12. Show that

$$\int_0^{\infty} \frac{\ln x}{x^{3/4}(1 + x)}\,dx = -\sqrt{2}\pi^2.$$

15.13. By integrating a suitable function around a large semi-circle in the upper half-plane and a small semi-circle centered on the origin, determine the value of

$$I = \int_0^\infty \frac{(\ln x)^2}{1+x^2}\,dx$$

and deduce, as a by-product of your calculation, that

$$\int_0^\infty \frac{\ln x}{1+x^2}\,dx = 0.$$

15.14. The equation of an ellipse in plane polar coordinates r, θ, with one of its foci at the origin, is

$$\frac{l}{r} = 1 - \epsilon\cos\theta,$$

where l is a length (that of the latus rectum) and ϵ $(0 < \epsilon < 1)$ is the eccentricity of the ellipse. Express the area of the ellipse as an integral around the unit circle in the complex plane, and show that the only singularity of the integrand inside the circle is a double pole at $z_0 = \epsilon^{-1} - (\epsilon^{-2} - 1)^{1/2}$.

By setting $z = z_0 + \xi$ and expanding the integrand in powers of ξ, find the residue at z_0 and hence show that the area is equal to $\pi l^2(1-\epsilon^2)^{-3/2}$.

[In terms of the semi-axes a and b of the ellipse, $l = b^2/a$ and $\epsilon^2 = (a^2 - b^2)/a^2$.]

15.15. Prove that

$$\sum_{-\infty}^{\infty} \frac{1}{n^2 + \frac{3}{4}n + \frac{1}{8}} = 4\pi.$$

Carry out the summation numerically, say between -4 and 4, and note how much of the sum comes from values near the poles of the contour integration.

15.16. This problem illustrates a method of summing some infinite series.

(a) Determine the residues at all the poles of the function

$$f(z) = \frac{\pi\cot\pi z}{a^2 + z^2},$$

where a is a positive real constant.

(b) By evaluating, in two different ways, the integral I of $f(z)$ along the straight line joining $-\infty - ia/2$ and $+\infty - ia/2$, show that

$$\sum_{n=1}^{\infty} \frac{1}{a^2 + n^2} = \frac{\pi\coth\pi a}{2a} - \frac{1}{2a^2}.$$

(c) Deduce the value of $\sum_1^\infty n^{-2}$.

15.17. By considering the integral of

$$\left(\frac{\sin \alpha z}{\alpha z}\right)^2 \frac{\pi}{\sin \pi z}, \qquad \alpha < \frac{\pi}{2},$$

around a circle of large radius, prove that

$$\sum_{m=1}^{\infty}(-1)^{m-1}\frac{\sin^2 m\alpha}{(m\alpha)^2} = \frac{1}{2}.$$

15.18. Use the Bromwich inversion, and contours similar to that shown in Figure 15.9(a), to find the functions of which the following are the Laplace transforms:
(a) $s(s^2+b^2)^{-1}$;
(b) $n!(s-a)^{-(n+1)}$, with n a positive integer and $s > a$;
(c) $a(s^2-a^2)^{-1}$, with $s > |a|$.
Compare your answers with those given in a table of standard Laplace transforms.

15.19. Find the function $f(t)$ whose Laplace transform is

$$\bar{f}(s) = \frac{e^{-s} - 1 + s}{s^2}.$$

15.20. A function $f(t)$ has the Laplace transform

$$F(s) = \frac{1}{2i}\ln\left(\frac{s+i}{s-i}\right),$$

the complex logarithm being defined by a finite branch cut running along the imaginary axis from $-i$ to i.
(a) Convince yourself that, for $t > 0$, $f(t)$ can be expressed as a closed contour integral that encloses only the branch cut.
(b) Calculate $F(s)$ on either side of the branch cut, evaluate the integral and hence determine $f(t)$.
(c) Confirm that the derivative with respect to s of the Laplace transform integral of your answer is the same as that given by dF/ds.

15.21. Use the contour in Figure 15.9(c) to show that the function with Laplace transform $s^{-1/2}$ is $(\pi x)^{-1/2}$.

[For an integrand of the form $r^{-1/2}\exp(-rx)$ change variable to $t = r^{1/2}$.]

HINTS AND ANSWERS

15.1. Apply Kirchhoff's laws to three independent loops, say *ADBA*, *ADEA* and *DBED*. Eliminate other currents from the equations to obtain

$I_R = \omega_0 C V_0[\,(\omega_0^2 - \omega^2 - 2i\omega\omega_0)/(\omega_0^2 + \omega^2)\,]$, where $\omega_0^2 = (LC)^{-1}$;
$|I_R| = \omega_0 C V_0$; the phase of I_R is $\tan^{-1}[\,(-2\omega\omega_0)/(\omega_0^2 - \omega^2)\,]$.

15.3. Set $c \coth u_1 = -d$, $c \coth u_2 = +d$, $|c\,\text{cosech} u| = a$ and note that the capacitance is proportional to $(u_2 - u_1)^{-1}$.

15.5. $\xi =$ constant, ellipses $x^2(a+1)^{-2} + y^2(a-1)^{-2} = c^2/(4a^2)$; $\eta =$ constant, hyperbolae $x^2(\cos\alpha)^{-2} - y^2(\sin\alpha)^{-2} = c^2$. The curves are the cuts $-c \le x \le c$, $y = 0$ and $|x| \ge c$, $y = 0$. The curves for $\eta = 2\pi$ are the same as those for $\eta = 0$.

15.7. The only pole inside the unit circle is at $z = ia - i(a^2-1)^{1/2}$; the residue is given by $-(i/2)(a^2-1)^{-1/2}$; the integral has value $2\pi[a(a^2-1)^{-1/2} - 1]$.

15.9. Factorize the denominator, showing that the relevant simple poles are at $i/2$ and i.

15.11. Use a contour like that shown in Figure 15.7.

15.13. Note that $\rho \ln^n \rho \to 0$ as $\rho \to 0$ for all n. When z is on the negative real axis, $(\ln z)^2$ contains three terms; one of the corresponding integrals is a standard form. The residue at $z = i$ is $i\pi^2/8$; $I = \pi^3/8$.

15.15. Evaluate

$$\int \frac{\pi \cot \pi z}{\left(\frac{1}{2}+z\right)\left(\frac{1}{4}+z\right)}\,dz$$

around a large circle centered on the origin; residue at $z = -1/2$ is 0; residue at $z = -1/4$ is $4\pi \cot(-\pi/4)$.

15.17. The behavior of the integrand for large $|z|$ is $|z|^{-2}\exp[(2\alpha - \pi)|z|]$. The residue at $z = \pm m$, for each integer m, is $\sin^2(m\alpha)(-1)^m/(m\alpha)^2$. The contour contributes nothing.
Required summation = [total sum − ($m = 0$ term)]/2.

15.19. Note that $\bar{f}(s)$ has no pole at $s = 0$. For $t < 0$ close the Bromwich contour in the right half-plane, and for $t > 1$ in the left half-plane. For $0 < t < 1$ the integrand has to be split into separate terms containing e^{-s} and $s - 1$ and the completions made in the right and left half-planes, respectively. The last of these completed contours now contains a second-order pole at $s = 0$. $f(t) = 1 - t$ for $0 < t < 1$, but is 0 otherwise.

15.21. $\int_\Gamma$ and $\int_\gamma$ tend to 0 as $R \to \infty$ and $\rho \to 0$. Put $s = r \exp i\pi$ and $s = r\exp(-i\pi)$ on the two sides of the cut and use $\int_0^\infty \exp(-t^2 x)\,dt = \frac{1}{2}(\pi/x)^{1/2}$. There are no poles inside the contour.

16 Probability

All scientists will know the importance of experiment and observation and, equally, be aware that the results of some experiments depend to a degree on chance. For example, in an experiment to measure the heights of a random sample of people, we would not be in the least surprised if all the heights were found to be different; but, if the experiment were repeated often enough, we would expect to find some sort of regularity in the results. Statistics, which is the subject of the following chapter, is concerned with the analysis of real experimental data of this sort. In this chapter, however, we discuss probability, which, to a pure mathematician, is an entirely theoretical subject based on axioms and deductions from them. Although this axiomatic approach to probability is important, and we discuss it briefly, a treatment more in keeping with its eventual applications in statistics is adopted here.

We first discuss the terminology required, with particular reference to the convenient graphical representation of experimental results as Venn diagrams. The concepts of random variables and distributions of random variables are then introduced. It is here that the connection with statistics is made; we assert that the results of many experiments are random variables and that those results have some sort of regularity, which is represented by a distribution. Precise definitions of a random variable and a distribution are then given, as are the defining equations for some important distributions. We also derive some useful quantities that characterize these distributions.

16.1 Venn diagrams

We call a single performance of an experiment a *trial* and each possible result an *outcome*. The *sample space* S of the experiment is then the set of all possible outcomes of an individual trial. For example, if we throw a six-sided die then there are six possible outcomes that together form the sample space of the experiment. At this stage we are not concerned with how likely a particular outcome might be (we will return to the probability of an outcome in due course) but rather will concentrate on the classification of possible outcomes. It is clear that some sample spaces are finite (e.g. the outcomes of throwing a die) whilst others are infinite (e.g. the outcomes of measuring people's heights). Most often, one is not interested in individual outcomes but in whether an outcome belongs to a given subset A (say) of the sample space S; these subsets are called *events*. For example, we might be interested in whether a person is taller or shorter than 180 cm, in which

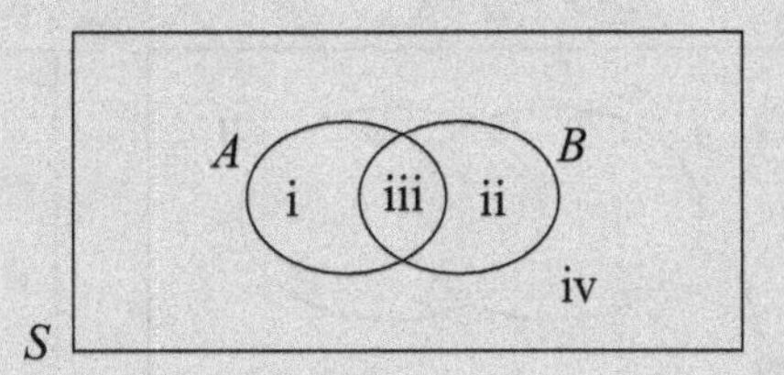

Figure 16.1 A Venn diagram.

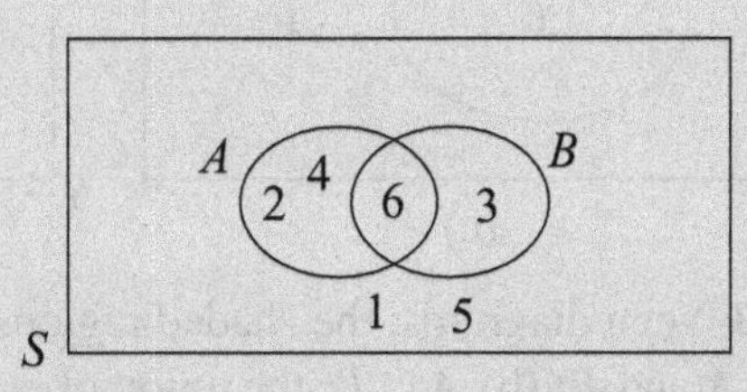

Figure 16.2 The Venn diagram for the outcomes of the die-throwing trials described in the worked example.

case we divide the sample space into just two events: namely, that the outcome (height measured) is (i) greater than 180 cm or (ii) less than 180 cm.

A common graphical representation of the outcomes of an experiment is the *Venn diagram*. A Venn diagram usually consists of a rectangle, the interior of which represents the sample space, together with one or more closed curves inside it. The interior of each closed curve then represents an event. Figure 16.1 shows a typical Venn diagram representing a sample space S and two events A and B. Every possible outcome is assigned to an appropriate region; in this example there are four regions to consider (marked i to iv in Figure 16.1):

(i) outcomes that belong to event A but not to event B;
(ii) outcomes that belong to event B but not to event A;
(iii) outcomes that belong to both event A and event B;
(iv) outcomes that belong to neither event A nor event B.

As a concrete example, consider the following.

Example A six-sided die is thrown. Let event A be "the number obtained is divisible by 2" and event B be "the number obtained is divisible by 3". Draw a Venn diagram to represent these events.

It is clear that the outcomes 2, 4, 6 belong to event A and that the outcomes 3, 6 belong to event B. Of these, 6 belongs to both A and B. The remaining outcomes, 1, 5, belong to neither A nor B. These observations are recorded schematically in the Venn diagram shown in Figure 16.2. ◀

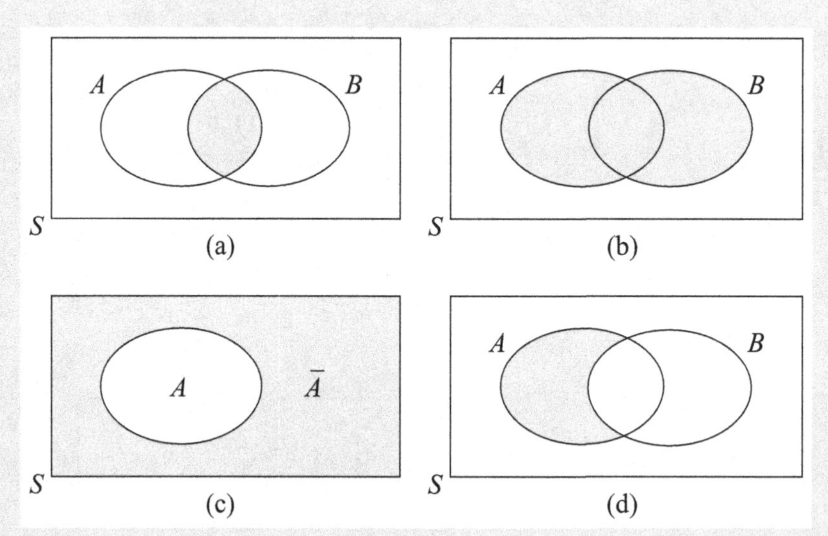

Figure 16.3 Venn diagrams: the shaded regions show (a) $A \cap B$, the intersection of two events A and B; (b) $A \cup B$, the union of events A and B; (c) the complement $\bar{A}$ of an event A; (d) $A - B$, those outcomes in A that do not belong to B.

In the above example, one outcome, 6, is divisible by both 2 and 3 and so belongs to both A and B. This outcome is placed in region iii of Figure 16.1, which is called the *intersection* of A and B and is denoted by $A \cap B$ [see Figure 16.3(a)]. If no events lie in the region of intersection then A and B are said to be *mutually exclusive* or *disjoint*. In this case, the Venn diagram is often (re-)drawn so that the closed curves representing the events A and B do not overlap, so as to make graphically explicit the fact that A and B are disjoint. It is not necessary, however, to draw the diagram in this way, since we may simply assign zero outcomes to the shaded region in Figure 16.3(a). An event that contains no outcomes is called the *empty event* and denoted by $\emptyset$.

The event comprising all the elements that belong to either A or B, or to both, is called the *union* of A and B and is denoted by $A \cup B$ [see Figure 16.3(b)]. In the previous example, $A \cup B = \{2, 3, 4, 6\}$. It is sometimes convenient to talk about those outcomes that do *not* belong to a particular event. The set of outcomes that do not belong to A is called the *complement* of A and is denoted by $\bar{A}$ [see Figure 16.3(c)]; this can also be written as $\bar{A} = S - A$. It is clear that $A \cup \bar{A} = S$ and $A \cap \bar{A} = \emptyset$.

The above notation can be extended in an obvious way, so that $A - B$ denotes the outcomes in A that do not belong to B. It is clear from Figure 16.3(d) that $A - B$ can also be written as $A \cap \bar{B}$. Finally, when *all* the outcomes in event B (say) also belong to event A, but A may contain, in addition, outcomes that do not belong to B, then B is called a *subset* of A, a situation that is denoted by $B \subset A$; alternatively, one may write $A \supset B$, which states that A *contains* B. In this case, the closed curve representing

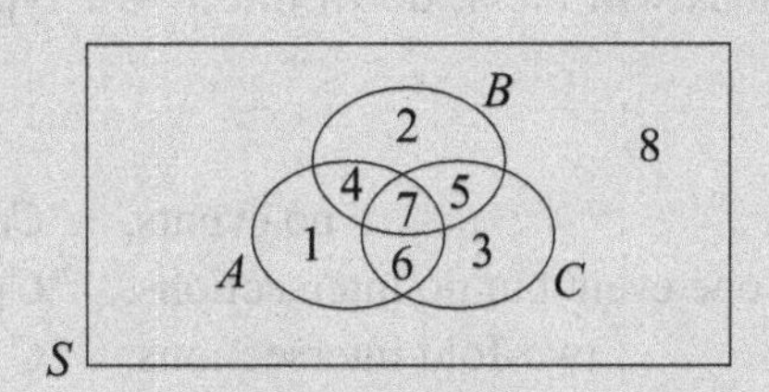

Figure 16.4 The general Venn diagram for three events is divided into eight regions.

the event B is often drawn lying completely within the closed curve representing the event A.[1]

The operations $\cup$ and $\cap$ are extended straightforwardly to more than two events. If there exist n events $A_1, A_2, \ldots, A_n$, in some sample space S, then the event consisting of all those outcomes that belong to *one or more* of the A_i is the *union* of $A_1, A_2, \ldots, A_n$ and is denoted by

$$A_1 \cup A_2 \cup \cdots \cup A_n. \tag{16.1}$$

Similarly, the event consisting of all the outcomes that belong to *every one* of the A_i is called the *intersection* of $A_1, A_2, \ldots, A_n$ and is denoted by

$$A_1 \cap A_2 \cap \cdots \cap A_n. \tag{16.2}$$

If, for *any* pair of values i, j with $i \neq j$,

$$A_i \cap A_j = \emptyset \tag{16.3}$$

then the events A_i and A_j are said to be *mutually exclusive* or *disjoint*.

Consider three events A, B and C with a Venn diagram such as is shown in Figure 16.4. It will be clear that, in general, the diagram will be divided into eight regions and they will be of four different types. Three regions correspond to a single event; three regions are each the intersection of exactly two events; one region is the three-fold intersection of all three events; and finally one region corresponds to none of the events. Let us now consider the numbers of different regions in a general n-event Venn diagram.

For one-event Venn diagrams there are two regions, for the two-event case there are four regions and, as we have just seen, for the three-event case there are eight. In the general n-event case there are 2^n regions, as is clear from the fact that any particular region R lies either inside or outside the closed curve of any particular event. With two choices (inside or outside) for each of n closed curves, there are 2^n different possible combinations with which to characterize R. Once n gets beyond three it becomes impossible to draw a simple two-dimensional Venn diagram, but this does not change the results.

1 What would be the meaning of $A - B$ if $B \supset A$?

The 2^n regions will break down into $n+1$ types, with the numbers of each type as follows:[2]

$$\begin{aligned}
\text{no events,} \quad & {}^nC_0 = 1; \\
\text{one event but no intersections,} \quad & {}^nC_1 = n; \\
\text{two-fold intersections,} \quad & {}^nC_2 = \tfrac{1}{2}n(n-1); \\
\text{three-fold intersections,} \quad & {}^nC_3 = \tfrac{1}{3!}n(n-1)(n-2); \\
& \vdots \\
\text{an n-fold intersection,} \quad & {}^nC_n = 1.
\end{aligned}$$

That this makes a total of 2^n can be checked by considering the binomial expansion

$$2^n = (1+1)^n = 1 + n + \tfrac{1}{2}n(n-1) + \cdots + 1.$$

Using Venn diagrams, it is straightforward to show that the operations $\cap$ and $\cup$ obey the following algebraic laws:

$$\begin{aligned}
&\text{commutativity,} && A \cap B = B \cap A, \quad A \cup B = B \cup A; \\
&\text{associativity,} && (A \cap B) \cap C = A \cap (B \cap C), \quad (A \cup B) \cup C = A \cup (B \cup C); \\
&\text{distributivity,} && A \cap (B \cup C) = (A \cap B) \cup (A \cap C), \\
& && A \cup (B \cap C) = (A \cup B) \cap (A \cup C); \\
&\text{idempotency,} && A \cap A = A, \quad A \cup A = A.
\end{aligned}$$

The following illustrates the operation of this algebra.

Example Show that (i) $A \cup (A \cap B) = A \cap (A \cup B) = A$, (ii) $(A - B) \cup (A \cap B) = A$.

(i) Using the distributivity and idempotency laws above, we see that

$$A \cup (A \cap B) = (A \cup A) \cap (A \cup B) = A \cap (A \cup B).$$

By sketching a Venn diagram[3] it is immediately clear that both expressions are equal to A. Nevertheless, we here proceed in a more formal manner in order to deduce this result algebraically. Let us begin by writing

$$X = A \cup (A \cap B) = A \cap (A \cup B), \tag{16.4}$$

from which we want to deduce a simpler expression for the event X. Using the first equality in (16.4) and the algebraic laws for $\cap$ and $\cup$, we may write

$$\begin{aligned}
A \cap X &= A \cap [A \cup (A \cap B)] \\
&= (A \cap A) \cup [A \cap (A \cap B)] \\
&= A \cup (A \cap B) = X.
\end{aligned}$$

2 The symbols ${}^nC_i$, for $i = 0, 1, 2, \ldots, n$, are a convenient notation for combinations; their general definitions are given in Section A.1.

3 Make this sketch and confirm both of the stated results.

Since $A \cap X = X$ we must have $X \subset A$. Now, using the second equality in (16.4) in a similar way, we find

$$\begin{aligned} A \cup X &= A \cup [A \cap (A \cup B)] \\ &= (A \cup A) \cap [A \cup (A \cup B)] \\ &= A \cap (A \cup B) = X, \end{aligned}$$

from which we deduce that $A \subset X$. Thus, since $X \subset A$ and $A \subset X$, we must conclude that $X = A$.

(ii) Since we do not know how to deal with compound expressions containing a minus sign, we begin by writing $A - B = A \cap \bar{B}$ as mentioned above. Then, using the distributivity law, we obtain

$$\begin{aligned} (A - B) \cup (A \cap B) &= (A \cap \bar{B}) \cup (A \cap B) \\ &= A \cap (\bar{B} \cup B) \\ &= A \cap S = A. \end{aligned}$$

As noted earlier, both of these results can be proved trivially by drawing appropriate Venn diagrams. ◀

Further useful results may be derived from Venn diagrams. In particular, it is simple to show that the following rules hold:

(i) if $A \subset B$ then $\bar{A} \supset \bar{B}$;
(ii) $\overline{A \cup B} = \bar{A} \cap \bar{B}$;
(iii) $\overline{A \cap B} = \bar{A} \cup \bar{B}$.

Statements (ii) and (iii) are known jointly as *de Morgan's laws* and are sometimes useful in simplifying logical expressions, as in the following worked example.

Example There exist two events A and B such that

$$\overline{(X \cup A)} \cup \overline{(X \cup \bar{A})} = B.$$

Find an expression for the event X in terms of A and B.

We begin by taking the complement of both sides of the above expression: applying de Morgan's laws we obtain

$$\bar{B} = (X \cup A) \cap (X \cup \bar{A}).$$

We may then use the algebraic laws obeyed by $\cap$ and $\cup$ to yield

$$\bar{B} = X \cup (A \cap \bar{A}) = X \cup \emptyset = X.$$

Thus, we find that $X = \bar{B}$. ◀

16.2 Probability

In the previous section we discussed Venn diagrams, which are graphical representations of the possible outcomes of experiments. We did not, however, give any indication of how likely each outcome or event might be when any particular experiment is performed. Most experiments show some regularity. By this we mean that the relative frequency of an event is approximately the same on each occasion that a set of trials is performed. For example, if we throw a die N times then we expect that a six will occur approximately $N/6$ times (assuming, of course, that the die is not biased). The regularity of outcomes allows us to define the *probability*, $\Pr(A)$, as the expected relative frequency of event A in a large number of trials. More quantitatively, if an experiment has a total of n_S outcomes in the sample space S, and n_A of these outcomes correspond to the event A, then the probability that event A will occur is

$$\Pr(A) = \frac{n_A}{n_S}. \tag{16.5}$$

16.2.1 Axioms and theorems

From (16.5) we may deduce the following properties of the probability $\Pr(A)$.

(i) For any event A in a sample space S,

$$0 \leq \Pr(A) \leq 1. \tag{16.6}$$

If $\Pr(A) = 1$ then A is a certainty; if $\Pr(A) = 0$ then A is an impossibility.

(ii) For the entire sample space S we have

$$\Pr(S) = \frac{n_S}{n_S} = 1, \tag{16.7}$$

which simply states that we are certain to obtain one of the possible outcomes.

(iii) If A and B are two events in S then, from the Venn diagrams in Figure 16.3, we see that

$$n_{A\cup B} = n_A + n_B - n_{A\cap B}, \tag{16.8}$$

the final subtraction arising because the outcomes in the intersection of A and B are counted twice when the outcomes of A are added to those of B. Dividing both sides of (16.8) by n_S, we obtain the *addition rule* for probabilities

$$\Pr(A \cup B) = \Pr(A) + \Pr(B) - \Pr(A \cap B). \tag{16.9}$$

However, if A and B are *mutually exclusive* events ($A \cap B = \emptyset$) then $\Pr(A \cap B) = 0$ and we obtain the special case

$$\Pr(A \cup B) = \Pr(A) + \Pr(B). \tag{16.10}$$

(iv) If $\bar{A}$ is the complement of A then $\bar{A}$ and A are mutually exclusive events. Thus, from (16.7) and (16.10) we have

$$1 = \Pr(S) = \Pr(A \cup \bar{A}) = \Pr(A) + \Pr(\bar{A}),$$

from which we obtain the *complement law*

$$\Pr(\bar{A}) = 1 - \Pr(A). \tag{16.11}$$

This is particularly useful for problems in which evaluating the probability of the complement is easier than evaluating the probability of the event itself.

Our next worked example is a simple application of the addition rule.

Example Calculate the probability of drawing an ace or a spade from a pack of cards.

Let A be the event that an ace is drawn and B the event that a spade is drawn. It immediately follows that $\Pr(A) = \frac{4}{52} = \frac{1}{13}$ and $\Pr(B) = \frac{13}{52} = \frac{1}{4}$. The intersection of A and B consists of only the ace of spades and so $\Pr(A \cap B) = \frac{1}{52}$. Thus, from (16.9)

$$\Pr(A \cup B) = \tfrac{1}{13} + \tfrac{1}{4} - \tfrac{1}{52} = \tfrac{4}{13}.$$

In this case it is just as simple to recognize that there are 16 cards in the pack that satisfy the required condition (13 spades plus three other aces) and so the probability is $\frac{16}{52}$.[4] ◀

The above theorems can be extended to a greater number of events. For example, if $A_1, A_2, \ldots, A_n$ are mutually exclusive events then (16.10) becomes

$$\Pr(A_1 \cup A_2 \cup \cdots \cup A_n) = \Pr(A_1) + \Pr(A_2) + \cdots + \Pr(A_n). \qquad (16.12)$$

Furthermore, if $A_1, A_2, \ldots, A_n$ (whether mutually exclusive or not) *exhaust* S, i.e. are such that $A_1 \cup A_2 \cup \cdots \cup A_n = S$, then

$$\Pr(A_1 \cup A_2 \cup \cdots \cup A_n) = \Pr(S) = 1. \qquad (16.13)$$

The die described in the next and several later worked examples is so blatantly biased that it could never be used in any notionally fair game of chance, but it *is* convenient for mathematical illustration!

Example A biased six-sided die has probabilities $\frac{1}{2}p, p, p, p, p, 2p$ of showing 1, 2, 3, 4, 5, 6 respectively. Calculate p.

Given that the individual events are mutually exclusive, (16.12) can be applied to give

$$\Pr(1 \cup 2 \cup 3 \cup 4 \cup 5 \cup 6) = \tfrac{1}{2}p + p + p + p + p + 2p = \tfrac{13}{2}p.$$

The union of all possible outcomes on the LHS of this equation is clearly the sample space, S, and so

$$\Pr(S) = \tfrac{13}{2}p.$$

Now (16.7) requires that $\Pr(S) = 1$, and so $p = \frac{2}{13}$. ◀

When the possible outcomes of a trial correspond to more than two events, and those events are *not* mutually exclusive, the calculation of the probability of the union of a number of events is more complicated, and the generalization of the addition law (16.9)

4 Suppose that the card drawing were scored and the drawn card replaced in the pack, at random, before a subsequent draw. What would be the expected average score for N such drawings (a) with 1 point scored if the card is an ace or a spade, zero otherwise, and (b) with 1 point scored if the card is an ace and 1 point scored if the card is a spade, zero otherwise?

requires further work. Let us begin by considering the union of three events A_1, A_2 and A_3, which need not be mutually exclusive. We first define the event $B = A_2 \cup A_3$ and, using the addition law (16.9), we obtain

$$\Pr(A_1 \cup A_2 \cup A_3) = \Pr(A_1 \cup B) = \Pr(A_1) + \Pr(B) - \Pr(A_1 \cap B). \tag{16.14}$$

However, we may write $\Pr(A_1 \cap B)$ as

$$\begin{aligned}\Pr(A_1 \cap B) &= \Pr[A_1 \cap (A_2 \cup A_3)] \\ &= \Pr[(A_1 \cap A_2) \cup (A_1 \cap A_3)] \\ &= \Pr(A_1 \cap A_2) + \Pr(A_1 \cap A_3) - \Pr(A_1 \cap A_2 \cap A_3).\end{aligned}$$

Substituting this expression, and that for $\Pr(B)$ obtained from (16.9), into (16.14) we obtain the probability addition law for three general events,

$$\begin{aligned}\Pr(A_1 \cup A_2 \cup A_3) = \Pr(A_1) + \Pr(A_2) + \Pr(A_3) - \Pr(A_2 \cap A_3) - \Pr(A_1 \cap A_3) \\ - \Pr(A_1 \cap A_2) + \Pr(A_1 \cap A_2 \cap A_3),\end{aligned} \tag{16.15}$$

which we now apply to a card-drawing problem.

Example Calculate the probability of drawing from a pack of cards one that is an ace or is a spade or shows an even number (2, 4, 6, 8, 10).

If, as previously, A is the event that an ace is drawn, $\Pr(A) = \frac{4}{52}$. Similarly the event B, that a spade is drawn, has $\Pr(B) = \frac{13}{52}$. The further possibility C, that the card is even (but not a picture card) has $\Pr(C) = \frac{20}{52}$. The two-fold intersections have probabilities

$$\Pr(A \cap B) = \frac{1}{52}, \quad \Pr(A \cap C) = 0, \quad \Pr(B \cap C) = \frac{5}{52}.$$

There is no three-fold intersection as events A and C are mutually exclusive. Hence

$$\Pr(A \cup B \cup C) = \frac{1}{52}[(4 + 13 + 20) - (1 + 0 + 5) + (0)] = \frac{31}{52}.$$

The reader should identify the 31 cards involved. ◂

When the probabilities are combined to calculate the probability for the union of the n general events, the result, which may be proved by induction upon n, is

$$\begin{aligned}\Pr(A_1 \cup A_2 \cup \cdots \cup A_n) = \sum_i \Pr(A_i) - \sum_{i,j} \Pr(A_i \cap A_j) + \sum_{i,j,k} \Pr(A_i \cap A_j \cap A_k) \\ - \cdots + (-1)^{n+1} \Pr(A_1 \cap A_2 \cap \cdots \cap A_n).\end{aligned} \tag{16.16}$$

Each summation runs over all possible sets of subscripts, except those in which any two subscripts in a set are the same; each pair of unequal subscripts must be counted only once. The number of terms in the summation of probabilities of m-fold intersections of the n events is given by ${}^nC_m$ (as discussed in Section 16.1). Equation (16.9) is a special case of

(16.16) in which $n = 2$ and only the first two terms on the RHS survive. We now illustrate this result with a worked example that has $n = 4$ and includes a four-fold intersection.

Example Find the probability of drawing from a pack a card that has at least one of the following properties:
A, it is an ace;
B, it is a spade;
C, it is a black honor card (ace, king, queen, jack or 10);
D, it is a black ace.

Measuring all probabilities in units of $\frac{1}{52}$, the single-event probabilities are

$$\Pr(A) = 4, \qquad \Pr(B) = 13, \qquad \Pr(C) = 10, \qquad \Pr(D) = 2.$$

The two-fold intersection probabilities, measured in the same units, are

$$\Pr(A \cap B) = 1, \qquad \Pr(A \cap C) = 2, \qquad \Pr(A \cap D) = 2,$$

$$\Pr(B \cap C) = 5, \qquad \Pr(B \cap D) = 1, \qquad \Pr(C \cap D) = 2.$$

The three-fold intersections have probabilities

$$\Pr(A \cap B \cap C) = 1, \quad \Pr(A \cap B \cap D) = 1, \quad \Pr(A \cap C \cap D) = 2, \quad \Pr(B \cap C \cap D) = 1.$$

Finally, the four-fold intersection, requiring all four conditions to hold, is satisfied only by the ace of spades, and hence (again in units of $\frac{1}{52}$)

$$\Pr(A \cap B \cap C \cap D) = 1.$$

Substituting in (16.16) gives

$$P = \frac{1}{52}[(4 + 13 + 10 + 2) - (1 + 2 + 2 + 5 + 1 + 2) + (1 + 1 + 2 + 1) - (1)] = \frac{20}{52}$$

for the probability that the drawn card has at least one of the listed properties. ◀

We conclude this section on basic theorems by deriving a useful general expression for the probability $\Pr(A \cap B)$ that two events A and B both occur in the case where A (say) is the union of a set of n *mutually exclusive* events A_i. In this case

$$A \cap B = (A_1 \cap B) \cup \cdots \cup (A_n \cap B),$$

where the events $A_i \cap B$ are also mutually exclusive. Thus, from the addition law (16.12) for mutually exclusive events, we find

$$\Pr(A \cap B) = \sum_i \Pr(A_i \cap B). \tag{16.17}$$

Moreover, in the special case where the events A_i *exhaust* the sample space S, we have $A \cap B = S \cap B = B$, and we obtain the *total probability law*

$$\Pr(B) = \sum_i \Pr(A_i \cap B). \tag{16.18}$$

16.2.2 Conditional probability

So far we have defined only probabilities of the form "what is the probability that event A happens?". In this section we turn to *conditional probability*, the probability that a particular event occurs *given* the occurrence of another, possibly related, event. For example, we may wish to know the probability of event B, drawing an ace from a pack of cards from which one card has already been removed, given that event A, the card already removed was itself an ace, has occurred.

We denote this probability by $\Pr(B|A)$ and may obtain a formula for it by considering the total probability $\Pr(A \cap B) = \Pr(B \cap A)$ that both A and B will occur. This may be written in two ways, i.e.

$$\begin{aligned}\Pr(A \cap B) &= \Pr(A)\Pr(B|A) \\ &= \Pr(B)\Pr(A|B).\end{aligned}$$

From this we obtain

$$\Pr(A|B) = \frac{\Pr(A \cap B)}{\Pr(B)} \tag{16.19}$$

and

$$\Pr(B|A) = \frac{\Pr(B \cap A)}{\Pr(A)}. \tag{16.20}$$

In terms of Venn diagrams, we may think of $\Pr(B|A)$ as the probability of B in the reduced sample space defined by A. Thus, if two events A and B are mutually exclusive then

$$\Pr(A|B) = 0 = \Pr(B|A). \tag{16.21}$$

When an experiment consists of drawing objects at random from a given set of objects, it is termed *sampling a population*. We need to distinguish between two different ways in which such a *sampling experiment* may be performed. After an object has been drawn at random from the set it may either be put aside or returned to the set before the next object is randomly drawn. The former is termed "sampling without replacement", the latter "sampling with replacement". The following demonstrates the difference.

Example Find the probability of drawing two aces at random from a pack of cards (i) when the first card drawn is replaced at random into the pack before the second card is drawn, and (ii) when the first card is put aside after being drawn.

Let A be the event that the first card is an ace, and B the event that the second card is an ace. Now

$$\Pr(A \cap B) = \Pr(A)\Pr(B|A),$$

and for both (i) and (ii) we know that $\Pr(A) = \frac{4}{52} = \frac{1}{13}$.

(i) If the first card is replaced in the pack before the next is drawn then $\Pr(B|A) = \Pr(B) = \frac{4}{52} = \frac{1}{13}$, since A and B are independent events. We then have

$$\Pr(A \cap B) = \Pr(A)\Pr(B) = \frac{1}{13} \times \frac{1}{13} = \frac{1}{169}.$$

(ii) If the first card is put aside and the second then drawn, A and B are not independent and $\Pr(B|A) = \frac{3}{51}$. We then have

$$\Pr(A \cap B) = \Pr(A)\Pr(B|A) = \frac{1}{13} \times \frac{3}{51} = \frac{1}{221}$$

for the combined probability of the two events.[5] ◀

Two events A and B are *statistically independent* if $\Pr(A|B) = \Pr(A)$ [or equivalently if $\Pr(B|A) = \Pr(B)$]. In words, the probability of A given B is then the same as the probability of A regardless of whether B occurs. For example, if we throw a coin and a die at the same time, we would normally expect that the probability of throwing a six was independent of whether a head was thrown. If A and B are statistically independent then it follows that

$$\Pr(A \cap B) = \Pr(A)\Pr(B). \tag{16.22}$$

In fact, on the basis of intuition and experience, (16.22) may be regarded as the *definition* of the statistical independence of two events.

The idea of statistical independence is easily extended to an arbitrary number of events $A_1, A_2, \ldots, A_n$. The events are said to be (mutually) independent if

$$\begin{aligned}
\Pr(A_i \cap A_j) &= \Pr(A_i)\Pr(A_j), \\
\Pr(A_i \cap A_j \cap A_k) &= \Pr(A_i)\Pr(A_j)\Pr(A_k), \\
&\;\;\vdots \\
\Pr(A_1 \cap A_2 \cap \cdots \cap A_n) &= \Pr(A_1)\Pr(A_2)\cdots\Pr(A_n),
\end{aligned}$$

for all combinations of indices i, j and k for which no two indices are the same. Even if all n events are not mutually independent, any two events for which $\Pr(A_i \cap A_j) = \Pr(A_i)\Pr(A_j)$ are said to be *pairwise independent*.

We now derive two results that often prove useful when working with conditional probabilities. Let us suppose that an event A is the union of n *mutually exclusive* events A_i. If B is some other event then from (16.17) we have

$$\Pr(A \cap B) = \sum_i \Pr(A_i \cap B).$$

Dividing both sides of this equation by $\Pr(B)$, and using (16.19), we obtain

$$\Pr(A|B) = \sum_i \Pr(A_i|B), \tag{16.23}$$

which is the *addition law for conditional probabilities*.

Furthermore, if the set of mutually exclusive events A_i exhausts the sample space S then, from the *total probability law* (16.18), the probability $\Pr(B)$ of some event B in

5 Would the answer be different if the first card had been replaced but the requirement had been for two different aces?

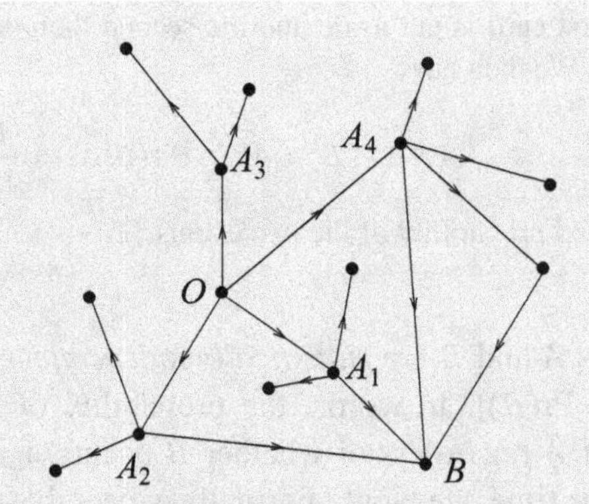

Figure 16.5 A collection of traffic islands connected by one-way roads.

S can be written as

$$\Pr(B) = \sum_i \Pr(A_i) \Pr(B|A_i). \tag{16.24}$$

The following very artificial car-routing problem illustrates this point.

Example A collection of traffic islands connected by a system of one-way roads is shown in Figure 16.5. At any given island a car driver chooses a direction at random from those available. What is the probability that a driver starting at O will arrive at B?

In order to leave O the driver must pass through one of A_1, A_2, A_3 or A_4, which thus form a complete set of mutually exclusive events. Since at each island (including O) the driver chooses a direction at random from those available, we have that $\Pr(A_i) = \frac{1}{4}$ for $i = 1, 2, 3, 4$. From Figure 16.5, we see also that

$$\Pr(B|A_1) = \tfrac{1}{3}, \quad \Pr(B|A_2) = \tfrac{1}{3}, \quad \Pr(B|A_3) = 0, \quad \Pr(B|A_4) = \tfrac{2}{4} = \tfrac{1}{2}.$$

Now the total probability law, (16.24), gives

$$\Pr(B) = \sum_i \Pr(A_i) \Pr(B|A_i) = \tfrac{1}{4}\left(\tfrac{1}{3} + \tfrac{1}{3} + 0 + \tfrac{1}{2}\right) = \tfrac{7}{24}$$

as the probability of arriving at B. ◀

Finally, we note that the concept of conditional probability may be straightforwardly extended to several compound events. For example, in the case of three events A, B, C, we may write $\Pr(A \cap B \cap C)$ in several ways, e.g.

$$\begin{aligned}\Pr(A \cap B \cap C) &= \Pr(C) \Pr(A \cap B|C) \\ &= \Pr(B \cap C) \Pr(A|B \cap C) \\ &= \Pr(C) \Pr(B|C) \Pr(A|B \cap C).\end{aligned}$$

The result of the following example is sometimes useful when a set of mutually exclusive events that exhaust the sample space can be identified.

Example Suppose $\{A_i\}$ is a set of mutually exclusive events that exhausts the sample space S. If B and C are two other events in S, show that

$$\Pr(B|C) = \sum_i \Pr(A_i|C)\Pr(B|A_i \cap C).$$

Using (16.19) and (16.17), we may write

$$\Pr(C)\Pr(B|C) = \Pr(B \cap C) = \sum_i \Pr(A_i \cap B \cap C). \qquad (16.25)$$

Each term in the sum on the RHS can be expanded as an appropriate product of conditional probabilities,

$$\Pr(A_i \cap B \cap C) = \Pr(C)\Pr(A_i|C)\Pr(B|A_i \cap C).$$

Substituting this form into (16.25) and dividing through by $\Pr(C)$ gives the required result. ◀

16.2.3 Bayes' theorem

In the previous section we saw that the probability that both an event A and a related event B will occur can be written either as $\Pr(A)\Pr(B|A)$ or $\Pr(B)\Pr(A|B)$. Hence

$$\Pr(A)\Pr(B|A) = \Pr(B)\Pr(A|B),$$

from which we obtain *Bayes' theorem*,

$$\Pr(A|B) = \frac{\Pr(A)}{\Pr(B)}\Pr(B|A). \qquad (16.26)$$

This theorem clearly shows that $\Pr(B|A) \neq \Pr(A|B)$, unless $\Pr(A) = \Pr(B)$. It is sometimes useful to rewrite $\Pr(B)$, if it is not known directly, as

$$\Pr(B) = \Pr(A)\Pr(B|A) + \Pr(\bar{A})\Pr(B|\bar{A})$$

so that Bayes' theorem becomes

$$\Pr(A|B) = \frac{\Pr(A)\Pr(B|A)}{\Pr(A)\Pr(B|A) + \Pr(\bar{A})\Pr(B|\bar{A})}. \qquad (16.27)$$

The next example illustrates its use.

Example Suppose that the blood test for some disease is reliable in the following sense: for people who are infected with the disease the test produces a positive result in 99.99% of cases; for people not infected a positive test result is obtained in only 0.02% of cases. Furthermore, assume that in the general population one person in 10 000 people is infected. A person is selected at random and found to test positive for the disease. What is the probability that the individual is actually infected?

Let A be the event that the individual is infected and B be the event that the individual tests positive for the disease. Using Bayes' theorem the probability that a person who tests positive is actually infected is

$$\Pr(A|B) = \frac{\Pr(A)\Pr(B|A)}{\Pr(A)\Pr(B|A) + \Pr(\bar{A})\Pr(B|\bar{A})}.$$

Now $\Pr(A) = 1/10\,000 = 1 - \Pr(\bar{A})$, and we are told that $\Pr(B|A) = 9999/10\,000$ and $\Pr(B|\bar{A}) = 2/10\,000$. Thus we obtain

$$\Pr(A|B) = \frac{1/10\,000 \times 9999/10\,000}{(1/10\,000 \times 9999/10\,000) + (9999/10\,000 \times 2/10\,000)} = \frac{1}{3}.$$

Thus, there is only a one in three chance that a person chosen at random, who tests positive for the disease, is actually infected.

At first glance, this answer may seem a little surprising, but the reason for the counter-intuitive result is that the 0.02% chance of an erroneous positive test for an uninfected person is of the same order of magnitude as the 0.01% chance of selecting somebody who is actually infected. ◀

We note that (16.27) may be written in a more general form if S is not simply divided into A and $\bar{A}$ but, rather, into *any* set of mutually exclusive events A_i that exhaust S. Using the total probability law (16.24), we may then write

$$\Pr(B) = \sum_i \Pr(A_i)\Pr(B|A_i),$$

so that Bayes' theorem takes the form

$$\Pr(A|B) = \frac{\Pr(A)\Pr(B|A)}{\sum_i \Pr(A_i)\Pr(B|A_i)}, \tag{16.28}$$

where the event A need not coincide with any of the A_i.

As a final point, we comment that sometimes we are concerned only with the *relative* probabilities of two events A and C (say), given the occurrence of some other event B. From (16.26) we then obtain a different form of Bayes' theorem,

$$\frac{\Pr(A|B)}{\Pr(C|B)} = \frac{\Pr(A)\Pr(B|A)}{\Pr(C)\Pr(B|C)}, \tag{16.29}$$

which does not contain $\Pr(B)$ at all.

16.3 Permutations and combinations

In equation (16.5) we defined the probability of an event A in a sample space S as

$$\Pr(A) = \frac{n_A}{n_S},$$

where n_A is the number of outcomes belonging to event A and n_S is the total number of possible outcomes. It is therefore necessary to be able to count the number of possible outcomes in various common situations.

16.3.1 Permutations

Let us first consider a set of n objects that are all different. We may ask in how many ways these n objects may be arranged, i.e. how many *permutations* of these objects exist. This is straightforward to deduce, as follows: the object in the first position may be chosen in n different ways, that in the second position in $n-1$ ways, and so on until the final object is positioned. The number of possible arrangements is therefore

$$n(n-1)(n-2)\cdots(1) = n! \tag{16.30}$$

Generalizing (16.30) slightly, let us suppose we choose only k $(< n)$ objects from n. The number of possible permutations of these k objects selected from n is given by

$$\underbrace{n(n-1)(n-2)\cdots(n-k+1)}_{k \text{ factors}} = \frac{n!}{(n-k)!} \equiv {}^nP_k. \tag{16.31}$$

In calculating the number of permutations of the various objects we have so far assumed that the objects are sampled *without replacement* – i.e. once an object has been drawn from the set it is put aside. As mentioned previously, however, we may instead replace each object before the next is chosen. The number of permutations of k objects from n *with replacement* may be calculated very easily since the first object can be chosen in n different ways, as can the second, the third, etc. Therefore the number of permutations is simply n^k. This may also be viewed as the number of permutations of k objects from n where repetitions are allowed, i.e. each object may be used as often as one likes.

The following worked example requires both sorts of calculation and produces a mildly surprising result.

Example Find the probability that in a group of k people at least two have the same birthday (ignoring 29 February).

This is a situation in which it is easier to first calculate the probability p that no two people share a birthday, and then to calculate the required probability that at least two do, as $q = 1 - p$.

Firstly, we imagine each of the k people in turn pointing to their birthday on a year planner. Thus, we are sampling the 365 days of the year "with replacement" and so the total number of possible outcomes is $(365)^k$.

Now (for the moment) we assume that no two people share a birthday and imagine the process being repeated, except that as each person points out their birthday it is crossed off the planner. In this case, we are sampling the days of the year "without replacement", and so the possible number of outcomes for which all the birthdays are different is, as in (16.31),

$$^{365}P_k = \frac{365!}{(365-k)!}.$$

Hence the probability that all the birthdays are different is

$$p = \frac{365!}{(365-k)!\,365^k}.$$

Now using the complement rule (16.11), the probability q that two or more people have the same birthday is simply

$$q = 1 - p = 1 - \frac{365!}{(365-k)!\,365^k}.$$

This expression may be conveniently evaluated using Stirling's approximation for $n!$ when n is large, namely

$$n! \sim \sqrt{2\pi n}\left(\frac{n}{e}\right)^n,$$

to give[6]

$$q \approx 1 - e^{-k}\left(\frac{365}{365-k}\right)^{365-k+0.5}.$$

It is interesting to note that if $k = 23$ the probability is a little greater than a half that at least two people have the same birthday, and if $k = 50$ the probability rises to 0.970. This can prove a good bet at a party of non-mathematicians! ◀

So far we have assumed that all n objects are different (or *distinguishable*). Let us now consider n objects of which n_1 are identical and of type 1, n_2 are identical and of type 2, ..., n_m are identical and of type m (clearly $n = n_1 + n_2 + \cdots + n_m$). From (16.30) the number of permutations of these n objects is again $n!$. However, the number of *distinguishable* permutations is only

$$\frac{n!}{n_1!n_2!\cdots n_m!}, \tag{16.32}$$

since the ith group of identical objects can be rearranged in $n_i!$ ways without changing the distinguishable permutation.

Example A set of snooker balls consists of a white, a yellow, a green, a brown, a blue, a pink, a black and 15 reds. How many distinguishable permutations of the balls are there?

In total there are 22 balls, the 15 reds being indistinguishable. Thus from (16.32) the number of distinguishable permutations is

$$\frac{22!}{(1!)(1!)(1!)(1!)(1!)(1!)(1!)(15!)} = \frac{22!}{15!} = 859\,541\,760.$$

6 Carry through this calculation.

Although a factor of (1!) does not change a computed value, it is advisable to write and account for each one in a calculation such as this. That the sum of the "arguments" of the factorial functions appearing in the denominator equals the "argument" of the factorial function in the numerator gives a useful check that nothing has been overlooked. ◀

16.3.2 Combinations

We now consider the number of *combinations* of various objects when their order is immaterial. Assuming all the objects to be distinguishable, from (16.31) we see that the number of permutations of k objects chosen from n is ${}^nP_k = n!/(n-k)!$. Now, since we are no longer concerned with the order of the chosen objects, which can be internally arranged in $k!$ different ways, the number of combinations of k objects from n is

$$\frac{n!}{(n-k)!k!} \equiv {}^nC_k \equiv \binom{n}{k} \qquad \text{for } 0 \le k \le n, \tag{16.33}$$

where ${}^nC_k$ is called the *binomial coefficient* since it also appears in the binomial expansion for positive integer n, namely

$$(a+b)^n = \sum_{k=0}^{n} {}^nC_k a^k b^{n-k}. \tag{16.34}$$

Result (16.33) is used several times in the following card-dealing calculation.

Example A hand of 13 playing cards is dealt from a well-shuffled pack of 52. What is the probability that the hand contains two aces?

Since the order of the cards in the hand is immaterial, the total number of distinct hands is simply equal to the number of combinations of 13 objects drawn from 52, i.e. ${}^{52}C_{13}$. However, the number of hands containing two aces is equal to the number of ways, ${}^4C_2$, in which the two aces can be drawn from the four available, multiplied by the number of ways, ${}^{48}C_{11}$, in which the remaining 11 cards in the hand can be drawn from the 48 cards that are not aces. Thus the required probability is given by

$$\begin{aligned}\frac{{}^4C_2\,{}^{48}C_{11}}{{}^{52}C_{13}} &= \frac{4!}{2!2!}\,\frac{48!}{11!37!}\,\frac{13!39!}{52!}\\ &= \frac{(3)(4)}{2}\,\frac{(12)(13)(38)(39)}{(49)(50)(51)(52)} = 0.213.\end{aligned}$$

This calculation is easily adapted for any number of aces.[7] ◀

Another useful result that may be derived using the binomial coefficients is the number of ways in which n distinguishable objects can be divided into m piles, with n_i objects in the ith pile, $i = 1, 2, \ldots, m$ (the ordering of objects within each pile being unimportant). This may be straightforwardly calculated as follows. We may choose the n_1 objects in the first pile from the original n objects in ${}^nC_{n_1}$ ways. The n_2 objects in the second pile can

7 Is it more likely that a particular hand contains exactly one king or that it contains no kings?

then be chosen from the $n - n_1$ remaining objects in ${}^{n-n_1}C_{n_2}$ ways, etc. We may continue in this fashion until we reach the $(m-1)$th pile, which may be formed in ${}^{n-n_1-\cdots-n_{m-2}}C_{n_{m-1}}$ ways. The remaining objects then form the mth pile and so can only be "chosen" in one way. Thus the total number of ways of dividing the original n objects into m piles is given by the product

$$\begin{aligned} N &= {}^nC_{n_1}\,{}^{n-n_1}C_{n_2}\cdots{}^{n-n_1-\cdots-n_{m-2}}C_{n_{m-1}} \\ &= \frac{n!}{n_1!(n-n_1)!}\frac{(n-n_1)!}{n_2!(n-n_1-n_2)!}\cdots\frac{(n-n_1-n_2-\cdots-n_{m-2})!}{n_{m-1}!(n-n_1-n_2-\cdots-n_{m-2}-n_{m-1})!} \\ &= \frac{n!}{n_1!(n-n_1)!}\frac{(n-n_1)!}{n_2!(n-n_1-n_2)!}\cdots\frac{(n-n_1-n_2-\cdots-n_{m-2})!}{n_{m-1}!n_m!} \\ &= \frac{n!}{n_1!n_2!\cdots n_m!}. \end{aligned} \qquad (16.35)$$

These numbers are called *multinomial coefficients* since (16.35) is the coefficient of $x_1^{n_1}x_2^{n_2}\cdots x_m^{n_m}$ in the multinomial expansion of $(x_1+x_2+\cdots+x_m)^n$, i.e. for positive integer n

$$(x_1+x_2+\cdots+x_m)^n = \sum_{\substack{n_1,n_2,\ldots,n_m \\ n_1+n_2+\cdots+n_m=n}} \frac{n!}{n_1!n_2!\cdots n_m!}x_1^{n_1}x_2^{n_2}\cdots x_m^{n_m}.$$

For the case $m=2$, $n_1=k$, $n_2=n-k$, (16.35) reduces to the binomial coefficient ${}^nC_k$. Furthermore, we note that the multinomial coefficient (16.35) is identical to the expression (16.32) for the number of distinguishable permutations of n objects, n_i of which are identical and of type i (for $i=1,2,\ldots,m$ and $n_1+n_2+\cdots+n_m=n$). A few moments' thought should convince the reader that the two expressions (16.35) and (16.32) must be identical.

The following example applies these ideas to a further card-dealing problem.

Example In the card game of bridge, each of four players is dealt 13 cards from a full pack of 52. What is the probability that each player is dealt an ace?

From (16.35), the total number of distinct bridge deals is 52!/(13!13!13!13!). However, the number of ways in which the four aces can be distributed with one in each hand is $4!/(1!1!1!1!) = 4!$; the remaining 48 cards can then be dealt out in 48!/(12!12!12!12!) ways. Thus the probability that each player receives an ace is

$$4!\frac{48!}{(12!)^4}\frac{(13!)^4}{52!} = \frac{24(13)^4}{(49)(50)(51)(52)} = 0.105.$$

Note that result (16.35) has been applied three times in this calculation. ◀

As in the case of permutations we might ask how many combinations of k objects can be chosen from n *with replacement* (repetition). To calculate this, we may imagine the n (distinguishable) objects set out on a table. Each combination of k objects can then be made by pointing to k of the n objects in turn (with repetitions allowed). These k equivalent selections distributed amongst n different but re-choosable objects are strictly analogous to the placing of k indistinguishable "balls" in n different boxes with no restriction on the number of balls in each box. A particular selection in the case $k = 7$, $n = 5$ may be symbolized as

$$\text{xxx}|\quad|\text{x}|\text{xx}|\text{x}.$$

This denotes three balls in the first box, none in the second, one in the third, two in the fourth and one in the fifth. We therefore need only consider the number of (distinguishable) ways in which k crosses and $n-1$ vertical lines can be arranged, i.e. the number of permutations of $k+n-1$ objects of which k are identical crosses and $n-1$ are identical lines. This is given by (16.33) as

$$\frac{(k+n-1)!}{k!(n-1)!} = {}^{n+k-1}C_k. \tag{16.36}$$

We note that this expression also occurs in the binomial expansion for negative integer powers. If n is a positive integer, then

$$(a+b)^{-n} = \sum_{k=0}^{\infty}(-1)^k\, {}^{n+k-1}C_k a^{-n-k} b^k,$$

where a is taken to be larger than b in magnitude.

As our first direct application of the notions of probability to the physical sciences, consider the following set of calculations.

Example A system contains a number N of (non-interacting) particles, each of which can be in any of the quantum states of the system. The structure of the set of quantum states is such that there exist R energy levels with corresponding energies E_i and degeneracies g_i (i.e. the ith energy level contains g_i quantum states). Find the numbers of distinct ways in which the particles can be distributed among the quantum states of the system such that the ith energy level contains n_i particles, for $i = 1, 2, \ldots, R$, in the cases where the particles are

(i) distinguishable with no restriction on the number in each state;
(ii) indistinguishable with no restriction on the number in each state;
(iii) indistinguishable with a maximum of one particle in each state;
(iv) distinguishable with a maximum of one particle in each state.

It is easiest to solve this problem in two stages. Let us first consider distributing the N particles among the R energy levels, *without* regard for the individual degenerate quantum states that comprise each level. If the particles are *distinguishable* then the number of distinct arrangements with n_i particles in the ith level, $i = 1, 2, \ldots, R$, is given by (16.35) as

$$\frac{N!}{n_1!n_2!\cdots n_R!}.$$

If, however, the particles are *indistinguishable* then clearly there exists only one distinct arrangement having n_i particles in the ith level, $i = 1, 2, \ldots, R$.

If we now suppose that there exist w_i ways in which the n_i particles in the ith energy level can be distributed among the g_i degenerate states, then it follows that the number of distinct ways in which the N particles can be distributed among all R quantum states of the system, with n_i particles in the ith level, is given by

$$W\{n_i\} = \begin{cases} \dfrac{N!}{n_1!n_2!\cdots n_R!}\displaystyle\prod_{i=1}^{R} w_i & \text{for distinguishable particles,} \\ \displaystyle\prod_{i=1}^{R} w_i & \text{for indistinguishable particles.} \end{cases} \qquad (16.37)$$

It therefore remains only for us to find the appropriate expression for w_i in each of the cases (i)–(iv) above.

Case (i). If there is no restriction on the number of particles in each quantum state, then in the ith energy level each particle can reside in any of the g_i degenerate quantum states. Thus, if the particles are distinguishable, the number of distinct arrangements is simply $w_i = g_i^{n_i}$. Then, from (16.37),

$$W\{n_i\} = \frac{N!}{n_1!n_2!\cdots n_R!}\prod_{i=1}^{R} g_i^{n_i} = N!\prod_{i=1}^{R}\frac{g_i^{n_i}}{n_i!}.$$

Such a system of particles (for example, atoms or molecules in a classical gas) is said to obey Maxwell–Boltzmann statistics.

Case (ii). If the particles are indistinguishable and there is no restriction on the number in each state then, from (16.36), the number of distinct arrangements of the n_i particles among the g_i states in the ith energy level is

$$w_i = \frac{(n_i + g_i - 1)!}{n_i!(g_i - 1)!}.$$

Substituting this expression into the appropriate part of (16.37), we obtain

$$W\{n_i\} = \prod_{i=1}^{R}\frac{(n_i + g_i - 1)!}{n_i!(g_i - 1)!}.$$

Such a system of particles (for example, a gas of photons) is said to obey Bose–Einstein statistics.

Case (iii). If a maximum of one particle can reside in each of the g_i degenerate quantum states in the ith energy level then the number of particles in each state is either 0 or 1. Since the particles are indistinguishable, w_i is equal to the number of distinct arrangements in which n_i states are occupied and $g_i - n_i$ states are unoccupied; this is given by

$$w_i = {}^{g_i}C_{n_i} = \frac{g_i!}{n_i!(g_i - n_i)!}.$$

Thus, from (16.37), we have

$$W\{n_i\} = \prod_{i=1}^{R}\frac{g_i!}{n_i!(g_i - n_i)!}.$$

Such a system is said to obey Fermi–Dirac statistics, and an example is provided by an electron gas.

Case (iv). Again, the number of particles in each state is either 0 or 1. If the particles are distinguishable, however, each arrangement identified in case (iii) can be reordered in $n_i!$ different ways, so that

$$w_i = {}^{g_i}P_{n_i} = \frac{g_i!}{(g_i - n_i)!}.$$

Substituting this expression into the first expression given in (16.37) yields

$$W\{n_i\} = N!\prod_{i=1}^{R}\frac{g_i!}{n_i!(g_i-n_i)!}.$$

Such a system of particles has the names of no famous scientists attached to it, since it appears that it never occurs in nature. ◀

16.4 Random variables and distributions

Suppose an experiment has an outcome sample space S. A real variable X that is defined for all possible outcomes in S (so that a real number – not necessarily unique – is assigned to each possible outcome) is called a *random variable* (RV). The outcome of the experiment may already be a real number and hence a random variable, e.g. the number of heads obtained in 10 throws of a coin, or the sum of the values if two dice are thrown. However, more arbitrary assignments are possible, e.g. the assignment of a "quality" rating to each successive item produced by a manufacturing process. Furthermore, assuming that a probability can be assigned to all possible outcomes in a sample space S, it is possible to assign a *probability distribution* to any random variable. Random variables may be divided into two classes, discrete and continuous, and we now examine each of these in turn.

16.4.1 Discrete random variables

A random variable X that takes only discrete values $x_1, x_2, \ldots, x_n$, with probabilities $p_1, p_2, \ldots, p_n$, is called a discrete random variable. The number of values n for which X has a non-zero probability is finite or at most countably infinite. As mentioned above, an example of a discrete random variable is the number of heads obtained in 10 throws of a coin. If X is a discrete random variable, we can define a *probability function* (PF) $f(x)$ that assigns probabilities to all the distinct values that X can take, such that

$$f(x) = \Pr(X = x) = \begin{cases} p_i & \text{if } x = x_i, \\ 0 & \text{otherwise.} \end{cases} \tag{16.38}$$

A typical PF (see Figure 16.6) thus consists of spikes, at *valid values* of X, whose height at x corresponds to the probability that $X = x$. Since the probabilities must sum to unity, we require

$$\sum_{i=1}^{n} f(x_i) = 1. \tag{16.39}$$

We may also define the *cumulative probability function* (CPF) of X, $F(x)$, whose value gives the probability that $X \leq x$, so that

$$F(x) = \Pr(X \leq x) = \sum_{x_i \leq x} f(x_i). \tag{16.40}$$

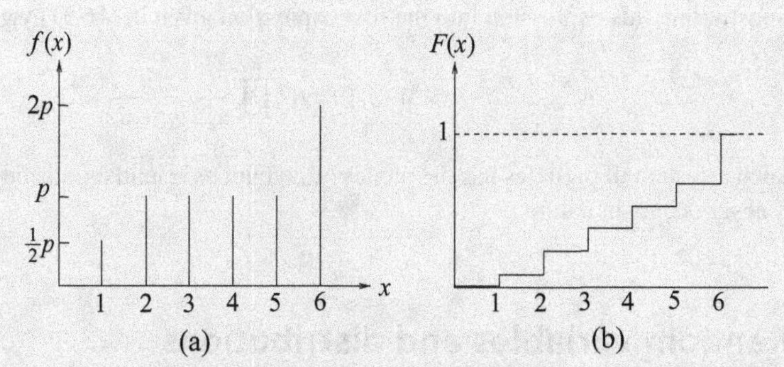

Figure 16.6 (a) A typical probability function for a discrete distribution, that for the biased die discussed earlier. Since the probabilities must sum to unity we require $p = 2/13$. (b) The cumulative probability function for the same discrete distribution. [Note that a different scale has been used for (b).]

Hence $F(x)$ is a step function that has upward jumps of p_i at $x = x_i$, $i = 1, 2, \ldots, n$, and is constant between possible values of X. We may also calculate the probability that X lies between two limits, l_1 and l_2 ($l_1 < l_2$); this is given by

$$\Pr(l_1 < X \le l_2) = \sum_{l_1 < x_i \le l_2} f(x_i) = F(l_2) - F(l_1), \qquad (16.41)$$

i.e. it is the sum of all the probabilities for which x_i lies within the relevant interval.

We now give an example of how a set of $f(x_1)$, denoted in the calculation by $f(i)$, might be calculated for a process whose outcome is a random variable.

Example A bag contains seven red balls and three white balls. Three balls are drawn at random and not replaced. Find the probability function for the number of red balls drawn.

Let X be the number of red balls drawn. Then

$$\Pr(X = 0) = f(0) = \frac{3}{10} \times \frac{2}{9} \times \frac{1}{8} = \frac{1}{120},$$

$$\Pr(X = 1) = f(1) = \frac{3}{10} \times \frac{2}{9} \times \frac{7}{8} \times 3 = \frac{7}{40},$$

$$\Pr(X = 2) = f(2) = \frac{3}{10} \times \frac{7}{9} \times \frac{6}{8} \times 3 = \frac{21}{40},$$

$$\Pr(X = 3) = f(3) = \frac{7}{10} \times \frac{6}{9} \times \frac{5}{8} = \frac{7}{24}.$$

The factors of 3 that appear for $X = 1$ and $X = 2$ should be noted and explained.[8] It should also be noted that, as expected, $\sum_{i=0}^{3} f(i) = 1$. ◀

8 For $X = 1$, say, consider each relevant scenario, *WWR*, *WRW* and *RWW*, and show that although the contributing probability ratios are not the same in each case, their products are.

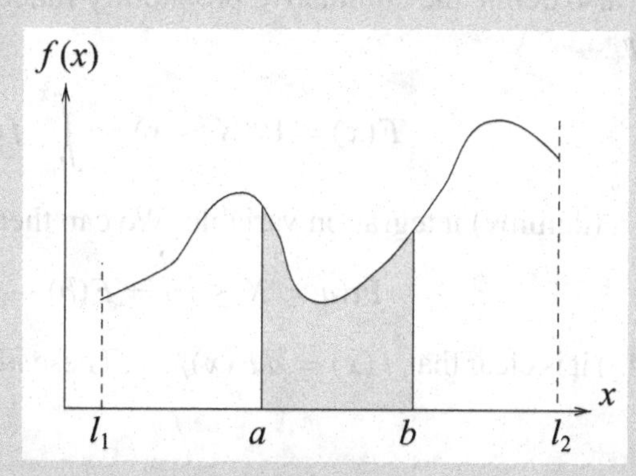

Figure 16.7 The probability density function for a continuous random variable X that can take values only between the limits l_1 and l_2. The shaded area under the curve gives $\Pr(a < X \leq b)$, whereas the total area under the curve, between the limits l_1 and l_2, is equal to unity.

16.4.2 Continuous random variables

A random variable X is said to have a *continuous* distribution if X is defined for a continuous range of values between given limits (often $-\infty$ to ∞). An example of a continuous random variable is the height of a person drawn from a population, which can take *any* value (within limits!). We can define the *probability density function* (PDF) $f(x)$ of a continuous random variable X such that

$$\Pr(x < X \leq x + dx) = f(x)\,dx,$$

i.e. $f(x)\,dx$ is the probability that X lies in the interval $x < X \leq x + dx$. Clearly $f(x)$ must be a real function that is everywhere ≥ 0. If X can take only values between the limits l_1 and l_2 then, in order for the sum of the probabilities of all possible outcomes to be equal to unity, we require

$$\int_{l_1}^{l_2} f(x)\,dx = 1.$$

Often X can take any value between $-\infty$ and ∞ and so

$$\int_{-\infty}^{\infty} f(x)\,dx = 1.$$

The probability that X lies in the interval $a < X \leq b$ is then given by

$$\Pr(a < X \leq b) = \int_a^b f(x)\,dx, \qquad (16.42)$$

i.e. $\Pr(a < X \leq b)$ is equal to the area under the curve of $f(x)$ between these limits (see Figure 16.7).

We may also define the cumulative probability function $F(x)$ for a continuous random variable by

$$F(x) = \Pr(X \leq x) = \int_{l_1}^{x} f(u)\,du, \tag{16.43}$$

where u is a (dummy) integration variable. We can then write

$$\Pr(a < X \leq b) = F(b) - F(a).$$

From (16.43) it is clear that $f(x) = dF(x)/dx$. These ideas are illustrated by the following example.

Example A random variable X has a PDF $f(x)$ given by Ae^{-x} in the interval $0 < x < \infty$ and zero elsewhere. Find the value of the constant A and hence calculate the probability that X lies in the interval $1 < X \leq 2$.

We require the integral of $f(x)$ between 0 and ∞ to equal unity. Evaluating this integral, we find

$$\int_0^\infty Ae^{-x}\,dx = \left[-Ae^{-x}\right]_0^\infty = A,$$

and hence $A = 1$. From (16.42), we then obtain

$$\Pr(1 < X \leq 2) = \int_1^2 f(x)\,dx = \int_1^2 e^{-x}\,dx = -e^{-2} - (-e^{-1}) = 0.23$$

as the probability that X lies in the interval $1 < X \leq 2$. ◀

It is worth mentioning here that a *discrete* RV can in fact be treated as continuous and assigned a corresponding probability density function. If X is a discrete RV that takes only the values $x_1, x_2, \ldots, x_n$ with probabilities $p_1, p_2, \ldots, p_n$ then we may describe X as a continuous RV with PDF

$$f(x) = \sum_{i=1}^{n} p_i \delta(x - x_i), \tag{16.44}$$

where $\delta(x)$ is the Dirac delta function discussed in Subsection 5.2. From (16.42) and the fundamental property of the delta function (5.12), we see that

$$\begin{aligned}\Pr(a < X \leq b) &= \int_a^b f(x)\,dx,\\ &= \sum_{i=1}^{n} p_i \int_a^b \delta(x - x_i)\,dx = \sum_i p_i,\end{aligned}$$

where the final sum extends over those values of i for which $a < x_i \leq b$.

16.4.3 Sets of random variables

It is common in practice to consider two or more random variables simultaneously. For example, one might be interested in both the height and weight of a person drawn at random from a population. In the general case, these variables may depend on one another and are described by *joint probability density functions*; these are discussed fully in Section 16.11. We simply note here that if we have (say) two random variables X and Y then by analogy with the single-variable case we define their joint probability density function $f(x, y)$ in such a way that, if X and Y are discrete RVs,

$$\Pr(X = x_i,\ Y = y_j) = f(x_i, y_j),$$

or, if X and Y are continuous RVs,

$$\Pr(x < X \le x + dx,\ y < Y \le y + dy) = f(x, y)\, dx\, dy.$$

In many circumstances, however, random variables do not depend on one another, i.e. they are *independent*. As an example, for a person drawn at random from a population, we might expect height and IQ to be independent random variables. Let us suppose that X and Y are two random variables with probability density functions $g(x)$ and $h(y)$ respectively. In mathematical terms, X and Y are independent RVs if their joint probability density function is given by $f(x, y) = g(x)h(y)$. Thus, for independent RVs, if X and Y are both discrete then

$$\Pr(X = x_i,\ Y = y_j) = g(x_i)h(y_j)$$

or, if X and Y are both continuous, then

$$\Pr(x < X \le x + dx,\ y < Y \le y + dy) = g(x)h(y)\, dx\, dy.$$

The important point in each case is that the RHS is simply the product of the individual probability density functions (compare with the expression for $\Pr(A \cap B)$ in (16.22) for statistically independent events A and B). By a simple extension, one may also consider the case where one of the random variables is discrete and the other continuous. The above discussion may also be trivially extended to any number of independent RVs X_i, $i = 1, 2, \ldots, N$. In the following worked example $N = 2$.

Example The independent random variables X and Y have the PDFs $g(x) = e^{-x}$ and $h(y) = 2e^{-2y}$ respectively. Calculate the probability that X lies in the interval $1 < X \le 2$ and Y lies in the interval $0 < Y \le 1$.

Since X and Y are independent RVs, the joint PDF is the product of the two separate PDFs and

$$\begin{aligned}\Pr(1 < X \le 2,\ 0 < Y \le 1) &= \int_1^2 g(x)\, dx \int_0^1 h(y)\, dy \\ &= \int_1^2 e^{-x}\, dx \int_0^1 2e^{-2y}\, dy \\ &= \left[-e^{-x}\right]_1^2 \times \left[-e^{-2y}\right]_0^1 = 0.23 \times 0.86 = 0.20\end{aligned}$$

gives the required probability that both variables lie in the specified ranges. ◀

16.5 Properties of distributions

For a single random variable X, the probability density function $f(x)$ contains all possible information about how the variable is distributed. However, for the purposes of comparison, it is conventional and useful to characterize $f(x)$ by certain of its properties. Most of these standard properties are defined in terms of *averages* or *expectation values*. In the most general case, the expectation value $E[g(X)]$ of any function $g(X)$ of the random variable X is defined as

$$E[g(X)] = \begin{cases} \sum_i g(x_i)f(x_i) & \text{for a discrete distribution,} \\ \int g(x)f(x)\,dx & \text{for a continuous distribution,} \end{cases} \qquad (16.45)$$

where the sum or integral is over all allowed values of X. It is assumed that the series is absolutely convergent or that the integral exists, as the case may be. From its definition it is straightforward to show that the expectation value has the following properties:

(i) if a is a constant then $E[a] = a$;
(ii) if a is a constant then $E[ag(X)] = aE[g(X)]$;
(iii) if $g(X) = s(X) + t(X)$ then $E[g(X)] = E[s(X)] + E[t(X)]$.

It should be noted that the expectation value is not a function of X but is instead a number that depends on the form of the probability density function $f(x)$ and the function $g(x)$. Most of the standard quantities used to characterize $f(x)$ are simply the expectation values of various functions of the random variable X. We now consider these standard quantities.

16.5.1 Mean

The property most commonly used to characterize a probability distribution is its *mean*, which is defined simply as the expectation value $E[X]$ of the variable X itself. Thus, the mean is given by

$$E[X] = \begin{cases} \sum_i x_i f(x_i) & \text{for a discrete distribution,} \\ \int xf(x)\,dx & \text{for a continuous distribution.} \end{cases} \qquad (16.46)$$

The alternative notations μ and $\langle x \rangle$ are also commonly used to denote the mean. If in (16.46) the series is not absolutely convergent, or the integral does not exist, we say that the distribution does not have a mean, but this is very rare in physical applications. Our worked example is taken from quantum physics.

Example The probability of finding a 1s electron in a hydrogen atom in a given infinitesimal volume dV is $\psi^*\psi\,dV$, where the quantum mechanical wavefunction ψ is given by

$$\psi = Ae^{-r/a_0}.$$

Find the value of the real constant A and thereby deduce the mean distance of the electron from the origin.

Let us consider the random variable R = "distance of the electron from the origin". Since the 1s orbital has no θ- or ϕ-dependence (it is spherically symmetric), we may consider the infinitesimal volume element dV as the spherical shell with inner radius r and outer radius $r + dr$. Thus, $dV = 4\pi r^2\, dr$ and the PDF of R is simply

$$\Pr(r < R \leq r + dr) \equiv f(r)\, dr = 4\pi r^2 A^2 e^{-2r/a_0}\, dr.$$

The value of A is found by requiring the total probability (i.e. the probability that the electron is *somewhere*) to be unity. Since R must lie between zero and infinity, we require that

$$A^2 \int_0^\infty e^{-2r/a_0} 4\pi r^2\, dr = 1.$$

Integrating by parts we find that we must have $A = 1/(\pi a_0^3)^{1/2}$. Now, using the definition of the mean (16.46), we find

$$E[R] = \int_0^\infty r f(r)\, dr = \frac{4}{a_0^3} \int_0^\infty r^3 e^{-2r/a_0}\, dr.$$

The integral on the RHS may also be integrated by parts and takes the value $3a_0^4/8$; consequently we find that $E[R] = 3a_0/2$.[9] ◀

16.5.2 Mode and median

Although the mean discussed in the last section is the most common measure of the "average" of a distribution, two other measures, which do not rely on the concept of expectation values, are frequently encountered.

The *mode* of a distribution is the value of the random variable X at which the probability (density) function $f(x)$ has its greatest value. If there is more than one value of X for which this is true then each value may equally be called the mode of the distribution.

The *median* M of a distribution is the value of the random variable X at which the cumulative probability function $F(x)$ takes the value $\frac{1}{2}$, i.e. $F(M) = \frac{1}{2}$. Related to the median are the lower and upper quartiles Q_l and Q_u of the PDF, which are defined such that

$$F(Q_\text{l}) = \tfrac{1}{4}, \qquad F(Q_\text{u}) = \tfrac{3}{4}.$$

Thus the median and lower and upper quartiles divide the PDF into four regions each containing one quarter of the probability. Smaller subdivisions are also possible, e.g. the nth percentile, P_n, of a PDF is defined by $F(P_n) = n/100$.

Example Find the mode of the PDF for the distance from the origin of the electron whose wavefunction was given in the previous example.

We found in the previous example that the PDF for the electron's distance from the origin was given by

$$f(r) = \frac{4r^2}{a_0^3} e^{-2r/a_0}. \qquad (16.47)$$

9 a_0 is known as the Bohr radius; it can be expressed in terms of fundamental constants as $(\epsilon_0 h^2)/(\pi m_e e^2)$ and has a value of 52.9×10^{-12} m.

As we need the value of r that makes $f(r)$ maximal, we differentiate $f(r)$ with respect to r and obtain

$$\frac{df}{dr} = \frac{8r}{a_0^3}\left(1 - \frac{r}{a_0}\right)e^{-2r/a_0}.$$

The derivative has zero value at $r = 0$, $r = a_0$, and $r = \infty$. Since $f(0) = f(\infty) = 0$ and $f(a_0)$ is clearly positive, the maximum must occur at $r = a_0$. Moreover, it is a global maximum (as opposed to just a local one). Thus the mode of $f(r)$ occurs at $r = a_0$. ◀

16.5.3 Variance and standard deviation

The *variance* of a distribution, $V[X]$, also written σ^2, is defined by

$$V[X] = E\left[(X-\mu)^2\right] = \begin{cases} \sum_j (x_j - \mu)^2 f(x_j) & \text{for a discrete distribution,} \\ \int (x-\mu)^2 f(x)\,dx & \text{for a continuous distribution.} \end{cases} \tag{16.48}$$

Here μ has been written for the expectation value $E[X]$ of X. As in the case of the mean, unless the series and the integral in (16.48) converge the distribution does not have a variance. From the definition (16.48) we may easily derive the following useful properties of $V[X]$. If a and b are constants then

(i) $V[a] = 0$,
(ii) $V[aX + b] = a^2 V[X]$.

The variance of a distribution is always positive; its positive square root is known as the *standard deviation* of the distribution and is often denoted by σ. Roughly speaking, σ measures the spread (about $x = \mu$) of the values that X can assume. For a distribution with a finite variance it can be shown that, for any positive constant c,

$$\Pr(|X - \mu| \geq c) \leq \frac{\sigma^2}{c^2}.$$

Thus, for *any* such distribution $f(x)$ we have, for example,

$$\Pr(|X - \mu| \geq 2\sigma) \leq \frac{1}{4} \quad \text{and} \quad \Pr(|X - \mu| \geq 3\sigma) \leq \frac{1}{9}.$$

We finish this section with a variance calculation for a continuous distribution.

Example Find the standard deviation of the PDF for the distance from the origin of the electron whose wavefunction was discussed in the previous two examples.

Inserting the expression (16.47) for the PDF $f(r)$ into (16.48), the variance of the random variable R is given by

$$V[R] = \int_0^\infty (r-\mu)^2 \frac{4r^2}{a_0^3} e^{-2r/a_0}\,dr = \frac{4}{a_0^3}\int_0^\infty (r^4 - 2r^3\mu + r^2\mu^2)e^{-2r/a_0}\,dr,$$

where the mean $\mu = E[R] = 3a_0/2$. Integrating each term in the integrand by parts we obtain

$$V[R] = 3a_0^2 - 3\mu a_0 + \mu^2 = \frac{3a_0^2}{4}.$$

Thus the standard deviation of the distribution is $\sigma = \sqrt{3}a_0/2$. ◀

16.5.4 Moments

The mean (or expectation) of X is sometimes called the *first moment* of X, since it is defined as the sum or integral of the probability density function multiplied by the first power of x. By a simple extension the kth moment of a distribution is defined by

$$\mu_k \equiv E[X^k] = \begin{cases} \sum_j x_j^k f(x_j) & \text{for a discrete distribution,} \\ \int x^k f(x)\,dx & \text{for a continuous distribution.} \end{cases} \tag{16.49}$$

For notational convenience, we have introduced the symbol μ_k to denote $E[X^k]$, the kth moment of the distribution. Clearly, the mean of the distribution is then denoted by μ_1, often abbreviated simply to μ, as in the previous subsection, as this rarely causes confusion.

A useful result that relates the second moment, the mean and the variance of a distribution is proved using the properties of the expectation operator:

$$\begin{aligned} V[X] &= E\left[(X-\mu)^2\right] \\ &= E\left[X^2 - 2\mu X + \mu^2\right] \\ &= E\left[X^2\right] - 2\mu E[X] + \mu^2 \\ &= E\left[X^2\right] - 2\mu^2 + \mu^2 \\ &= E\left[X^2\right] - \mu^2. \end{aligned} \tag{16.50}$$

In alternative notations, this result can be written

$$\langle (x-\mu)^2 \rangle = \langle x^2 \rangle - \langle x \rangle^2 \qquad \text{or} \qquad \sigma^2 = \mu_2 - \mu_1^2.$$

We now use the biased die introduced in a previous example to illustrate the numerical workings of the definitions we have made so far.

Example A biased die has probabilities $p/2$, p, p, p, p, $2p$ of showing 1, 2, 3, 4, 5, 6 respectively. Find (i) the mean, (ii) the second moment and (iii) the variance of this probability distribution.

As shown previously, by demanding that the sum of the probabilities equals unity we require $p = 2/13$. Now, using the definition of the mean (16.46) for a discrete distribution,

$$\begin{aligned} E[X] = \sum_j x_j f(x_j) &= 1 \times \tfrac{1}{2}p + 2 \times p + 3 \times p + 4 \times p + 5 \times p + 6 \times 2p \\ &= \frac{53}{2}p = \frac{53}{2} \times \frac{2}{13} = \frac{53}{13}. \end{aligned}$$

Similarly, using the definition of the second moment (16.49),

$$E[X^2] = \sum_j x_j^2 f(x_j) = 1^2 \times \tfrac{1}{2}p + 2^2 p + 3^2 p + 4^2 p + 5^2 p + 6^2 \times 2p$$
$$= \frac{253}{2} p = \frac{253}{13}.$$

Finally, using the definition of the variance (16.48), with $\mu = 53/13$, we obtain

$$V[X] = \sum_j (x_j - \mu)^2 f(x_j)$$
$$= (1-\mu)^2 \tfrac{1}{2} p + (2-\mu)^2 p + (3-\mu)^2 p + (4-\mu)^2 p + (5-\mu)^2 p + (6-\mu)^2 2p$$
$$= \left(\frac{3120}{169}\right) p = \frac{480}{169}.$$

It is easy to verify that $V[X] = E\left[X^2\right] - (E[X])^2$. ◀

In practice, to calculate the moments of a distribution it is often simpler to use the moment generating function discussed in Subsection 16.7.2. This is particularly true for higher-order moments, where direct evaluation of the sum or integral in (16.49) can be somewhat laborious.

16.6 Functions of random variables

Suppose X is some random variable for which the probability density function $f(x)$ is known. In many cases, we are more interested in a related random variable $Y = Y(X)$, where $Y(X)$ is some function of X. What is the probability density function $g(y)$ for the new random variable Y? We now discuss how to obtain this function, but restrict our discussion to continuous random variables and to cases in which $Y(X)$ has a single-valued inverse, i.e. only one x-value corresponds to any particular y-value. Discrete RVs and functions with multi-valued inverses can also be treated, but usually involve articulating all possible cases individually.

16.6.1 Continuous random variables

If X is a continuous RV, then so too is the new random variable $Y = Y(X)$. The probability that Y lies in the range y to $y + dy$ is given by

$$g(y)\,dy = \int_{dS} f(x)\,dx, \qquad (16.51)$$

where dS corresponds to all values of x for which Y lies in the range y to $y + dy$. We may write, to first order in dy, that

$$g(y)\,dy = \left| \int_{x(y)}^{x(y+dy)} f(x')\,dx' \right| = \int_{x(y)}^{x(y)+\left|\frac{dx}{dy}\right| dy} f(x')\,dx',$$

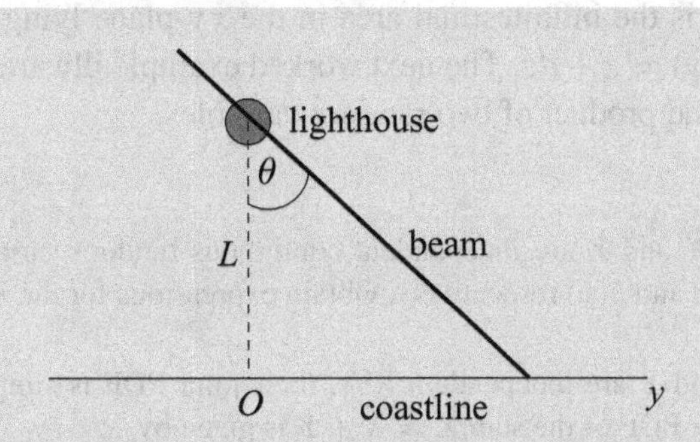

Figure 16.8 The illumination of a coastline by the beam from a lighthouse.

from which we obtain[10]

$$g(y) = f(x(y)) \left| \frac{dx}{dy} \right| . \qquad (16.52)$$

As an example, consider the following.

Example A lighthouse is situated at a distance L from a straight coastline, opposite a point O, and sends out a narrow continuous beam of light simultaneously in opposite directions. The beam rotates with constant angular velocity. If the random variable Y is the distance along the coastline, measured from O, of the spot that the light beam illuminates, find its probability density function.

The situation is illustrated in Figure 16.8. Since the light beam rotates at a constant angular velocity, θ is distributed uniformly between $-\pi/2$ and $\pi/2$, and so $f(\theta) = 1/\pi$. Now $y = L\tan\theta$, which possesses the single-valued inverse $\theta = \tan^{-1}(y/L)$, provided that θ lies between $-\pi/2$ and $\pi/2$. Since $dy/d\theta = L\sec^2\theta = L(1+\tan^2\theta) = L[1+(y/L)^2]$, from (16.52) we find

$$g(y) = \frac{1}{\pi}\left|\frac{d\theta}{dy}\right| = \frac{1}{\pi L[1+(y/L)^2]} \quad \text{for } -\infty < y < \infty.$$

A distribution of this form is called a *Cauchy distribution* and is discussed in Subsection 16.9.5. ◀

16.6.2 Functions of several random variables

We may extend our discussion further, to the case in which the new random variable is a function of *several* other random variables. For definiteness, let us consider the random variable $Z = Z(X, Y)$, which is a function of two other continuous RVs X and Y. Given that these variables are described by the joint probability density function $f(x, y)$, we wish to find the probability density function $p(z)$ of the variable Z.

As X and Y are both continuous RVs, $p(z)$ is found by requiring that

$$p(z)\,dz = \iint_{dS} f(x, y)\,dx\,dy, \qquad (16.53)$$

10 Show that, if a continuous random variable X has a probability density function $f(x)$ and the corresponding cumulative probability function is $F(x)$, then the random variable $Y = F(X)$ is uniformly distributed between 0 and 1.

where dS is the infinitesimal area in the xy-plane lying between the curves $Z(x, y) = z$ and $Z(x, y) = z + dz$. The next worked example illustrates this for the particular cases of the sum and product of two random variables.

Example Suppose X and Y are independent continuous random variables in the range $-\infty$ to ∞, with PDFs $g(x)$ and $h(y)$ respectively. Obtain expressions for the PDFs of $Z = X + Y$ and $W = XY$.

Since X and Y are independent RVs, their joint PDF is simply $f(x, y) = g(x)h(y)$. Thus, from (16.53), the PDF of the sum $Z = X + Y$ is given by

$$p(z)\,dz = \int_{-\infty}^{\infty} dx\; g(x) \int_{z-x}^{z+dz-x} dy\; h(y).$$

Now, since the second integral is over an infinitesimal range dz, its value is $h(z - x)\,dz$, and so

$$p(z)\,dz = \left(\int_{-\infty}^{\infty} g(x)h(z - x)\,dx \right) dz.$$

Thus $p(z)$ is the *convolution* of the PDFs of g and h (i.e. $p = g * h$, see Subsection 5.3.2).
In a similar way, the PDF of the product $W = XY$ is given by

$$q(w)\,dw = \int_{-\infty}^{\infty} dx\; g(x) \int_{w/|x|}^{(w+dw)/|x|} dy\; h(y)$$
$$= \left(\int_{-\infty}^{\infty} g(x)h(w/x)\frac{dx}{|x|} \right) dw.$$

The form of the argument of the function h should be noted in each case, as should the presence of $|x|^{-1}$ in the second case. ◀

16.6.3 Expectation values and variances

In some cases, one is interested only in the expectation value or the variance of the new variable Z rather than in its full probability density function. For definiteness, let us again consider the random variable $Z = Z(X, Y)$, which is a function of two RVs X and Y with a known joint distribution $f(x, y)$; the results we will obtain are readily generalized to more (or fewer) variables.

It is clear that $E[Z]$ and $V[Z]$ can be obtained, in principle, by first using the methods discussed above to obtain $p(z)$ and then evaluating the appropriate sums or integrals. The intermediate step of calculating $p(z)$ is not necessary, however, since it is straightforward to obtain expressions for $E[Z]$ and $V[Z]$ in terms of the variables X and Y. For example, if X and Y are continuous RVs then the expectation value of Z is given by

$$E[Z] = \int zp(z)\,dz = \iint Z(x, y)f(x, y)\,dx\,dy. \qquad (16.54)$$

An analogous result exists for discrete random variables.

Integrals of the form (16.54) are often difficult to evaluate. Nevertheless, we may use (16.54) to derive an important general result concerning expectation values. If X and Y are *any* two random variables and a and b are arbitrary constants then by letting $Z = aX + bY$

we find

$$E[aX + bY] = aE[X] + bE[Y].$$

Furthermore, we may use this result to obtain an *approximate* expression for the expectation value $E[Z(X, Y)]$ of any arbitrary function of X and Y. Letting $\mu_X = E[X]$ and $\mu_Y = E[Y]$, and provided $Z(X, Y)$ can be reasonably approximated by the linear terms of its Taylor expansion about the point (μ_X, μ_Y), we have

$$Z(X, Y) \approx Z(\mu_X, \mu_Y) + \left(\frac{\partial Z}{\partial X}\right)(X - \mu_X) + \left(\frac{\partial Z}{\partial Y}\right)(Y - \mu_Y), \tag{16.55}$$

where the partial derivatives are evaluated at $X = \mu_X$ and $Y = \mu_Y$. Taking the expectation values of both sides, we find

$$E[Z(X, Y)] \approx Z(\mu_X, \mu_Y) + \left(\frac{\partial Z}{\partial X}\right)(E[X] - \mu_X) + \left(\frac{\partial Z}{\partial Y}\right)(E[Y] - \mu_Y).$$

Now, since $E[X] = \mu_X$ and $E[Y] = \mu_Y$, this gives the approximate result $E[\,Z(X, Y)] \approx Z(\mu_X, \mu_Y)$.

By analogy with (16.54), the variance of $Z = Z(X, Y)$ is given by

$$V[Z] = \int (z - \mu_Z)^2 p(z)\, dz = \iint [Z(x, y) - \mu_Z]^2 f(x, y)\, dx\, dy, \tag{16.56}$$

where $\mu_Z = E[Z]$. We may use this expression to derive a second useful result. If X and Y are two *independent* random variables,[11] so that $f(x, y) = g(x)h(y)$, and a, b and c are constants then by setting $Z = aX + bY + c$ in (16.56) we obtain

$$V[aX + bY + c] = a^2 V[X] + b^2 V[Y]. \tag{16.57}$$

From (16.57) we also obtain the important special cases[12]

$$V[X + Y] = V[X - Y] = V[X] + V[Y].$$

Provided X and Y are indeed independent random variables, we may obtain an approximate expression for $V[\,Z(X, Y)]$, for any arbitrary function $Z(X, Y)$, in a similar manner to that used in approximating $E[\,Z(X, Y)]$ above. Taking the variance of both sides of (16.55), and using (16.57), we find

$$V[\,Z(X, Y)] \approx \left(\frac{\partial Z}{\partial X}\right)^2 V[X] + \left(\frac{\partial Z}{\partial Y}\right)^2 V[Y], \tag{16.58}$$

the partial derivatives being evaluated at $X = \mu_X$ and $Y = \mu_Y$.

11 If the RVs are not independent, then an additional term involving their covariance is needed in (16.57), as in equation (16.124).

12 That is, it does not matter whether X and Y are added or subtracted to form a new RV, the variance of the new RV is always formed by *adding* those of X and Y.

16.7 Generating functions

As we saw in Chapter 9, when dealing with particular sets of functions f_n, each member of the set being characterized by a different non-negative integer n, it is sometimes possible to summarize the whole set by a single function of a dummy variable (say t), called a generating function. The relationship between the generating function and the nth member f_n of the set is that if the generating function is expanded as a power series in t then f_n is the coefficient of t^n. For example, in the expansion of the generating function $G(z,t) = (1 - 2zt + t^2)^{-1/2}$, the coefficient of t^n is the nth Legendre polynomial $P_n(z)$, i.e.

$$G(z,t) = (1 - 2zt + t^2)^{-1/2} = \sum_{n=0}^{\infty} P_n(z)t^n.$$

We found that many useful properties of, and relationships between, the members of a set of functions could be established using the generating function and other functions obtained from it, e.g. its derivatives.

Similar ideas can be used in the area of probability theory, and two types of generating function can be usefully defined, one more generally applicable than the other. The more restricted of the two, applicable only to discrete integral distributions, is called a probability generating function; this is discussed in the next section. The second type, a moment generating function, can be used with both discrete and continuous distributions and is considered in Subsection 16.7.2.

16.7.1 Probability generating functions

As already indicated, probability generating functions are restricted in applicability to integer distributions, of which the most common (the binomial, the Poisson and the geometric) are considered in this and later subsections. In such distributions a random variable may take only non-negative integer values. The actual possible values may be finite or infinite in number, but, for formal purposes, all integers, 0, 1, 2, ... are considered possible. If only a finite number of integer values can occur in any particular case then those that cannot occur are included but are assigned zero probability.

If, as previously, the probability that the random variable X takes the value x_n is $f(x_n)$, then

$$\sum_n f(x_n) = 1.$$

In the present case, however, only non-negative integer values of x_n are possible, and we can, without ambiguity, write the probability that X takes the value n as f_n, with

$$\sum_{n=0}^{\infty} f_n = 1. \qquad (16.59)$$

We may now define the *probability generating function* $\Phi_X(t)$ by

$$\Phi_X(t) \equiv \sum_{n=0}^{\infty} f_n t^n. \tag{16.60}$$

It is immediately apparent that $\Phi_X(t) = E[t^X]$ and that, by virtue of (16.59), $\Phi_X(1) = 1$.

Probably the simplest example of a probability generating function (PGF) is provided by the random variable X defined by

$$X = \begin{cases} 1 & \text{if the outcome of a single trial is a "success"}, \\ 0 & \text{if the trial ends in "failure"}. \end{cases}$$

If the probability of success is p and that of failure q $(=1-p)$ then

$$\Phi_X(t) = qt^0 + pt^1 + 0 + 0 + \cdots = q + pt. \tag{16.61}$$

This type of random variable is discussed much more fully in Subsection 16.8.1. In a similar but slightly more complicated way, a Poisson-distributed integer variable with mean λ (see Subsection 16.8.5) has a PGF

$$\Phi_X(t) = \sum_{n=0}^{\infty} \frac{e^{-\lambda}\lambda^n}{n!} t^n = e^{-\lambda} e^{\lambda t}. \tag{16.62}$$

We note that, as required, $\Phi_X(1) = 1$ in both cases.

Useful results will be obtained from this kind of approach only if the summation (16.60) can be carried out explicitly in particular cases and the functions derived from $\Phi_X(t)$ can be shown to be related to meaningful parameters. Two such relationships can be obtained by differentiating (16.60) with respect to t. Taking the first derivative we find

$$\frac{d\Phi_X(t)}{dt} = \sum_{n=0}^{\infty} n f_n t^{n-1} \quad \Rightarrow \quad \Phi'_X(1) = \sum_{n=0}^{\infty} n f_n = E[X], \tag{16.63}$$

and differentiating once more we obtain

$$\frac{d^2\Phi_X(t)}{dt^2} = \sum_{n=0}^{\infty} n(n-1) f_n t^{n-2} \quad \Rightarrow \quad \Phi''_X(1) = \sum_{n=0}^{\infty} n(n-1) f_n = E[X(X-1)]. \tag{16.64}$$

Equation (16.63) shows that $\Phi'_X(1)$ gives the mean of X. Using both (16.64) and (16.50) allows us to write

$$\begin{aligned} \Phi''_X(1) + \Phi'_X(1) - \left[\Phi'_X(1)\right]^2 &= E[X(X-1)] + E[X] - (E[X])^2 \\ &= E\left[X^2\right] - E[X] + E[X] - (E[X])^2 \\ &= E\left[X^2\right] - (E[X])^2 \\ &= V[X], \end{aligned} \tag{16.65}$$

and so express the variance of X in terms of the derivatives of its probability generating function. The following example illustrates these ideas.

Example A random variable X is given by the number of trials needed to obtain a first success when the chance of success at each trial is constant and equal to p. Find the probability generating function for X and use it to determine the mean and variance of X.

Clearly, at least one trial is needed, and so $f_0 = 0$. If n (≥ 1) trials are needed for the first success, the first $n-1$ trials must have resulted in failure. Thus

$$\Pr(X = n) = q^{n-1}p, \qquad n \geq 1, \tag{16.66}$$

where $q = 1 - p$ is the probability of failure in each individual trial.

The corresponding probability generating function is thus

$$\begin{aligned}\Phi_X(t) &= \sum_{n=0}^{\infty} f_n t^n = \sum_{n=1}^{\infty} (q^{n-1}p)t^n \\ &= \frac{p}{q}\sum_{n=1}^{\infty}(qt)^n = \frac{p}{q} \times \frac{qt}{1-qt} = \frac{pt}{1-qt},\end{aligned} \tag{16.67}$$

where we have used the result for the sum of a geometric series, given in Section A.6, to obtain a closed-form expression for $\Phi_X(t)$. Again, as must be the case, $\Phi_X(1) = 1$.

To find the mean and variance of X we need to evaluate $\Phi'_X(1)$ and $\Phi''_X(1)$. Differentiating (16.67) gives

$$\Phi'_X(t) = \frac{p}{(1-qt)^2} \quad \Rightarrow \quad \Phi'_X(1) = \frac{p}{p^2} = \frac{1}{p},$$

$$\Phi''_X(t) = \frac{2pq}{(1-qt)^3} \quad \Rightarrow \quad \Phi''_X(1) = \frac{2pq}{p^3} = \frac{2q}{p^2}.$$

Thus, using (16.63) and (16.65),

$$\begin{aligned}E[X] &= \Phi'_X(1) = \frac{1}{p}, \\ V[X] &= \Phi''_X(1) + \Phi'_X(1) - [\Phi'_X(1)]^2 \\ &= \frac{2q}{p^2} + \frac{1}{p} - \frac{1}{p^2} = \frac{q}{p^2}.\end{aligned}$$

A distribution with probabilities of the general form (16.66) is known as a *geometric distribution* and is discussed in Subsection 16.8.3. This form of distribution is common in "waiting time" problems (Subsection 16.9.3). ◀

Sums of random variables

We now turn to considering the sum of two or more independent random variables, say X and Y, and denote by S_2 the random variable

$$S_2 = X + Y.$$

If $\Phi_{S_2}(t)$ is the PGF for S_2, the coefficient of t^n in its expansion is given by the probability that $X + Y = n$ and is thus equal to the sum of the probabilities that $X = r$ and $Y = n - r$ for all values of r in $0 \leq r \leq n$. Since such outcomes for different values of r are mutually

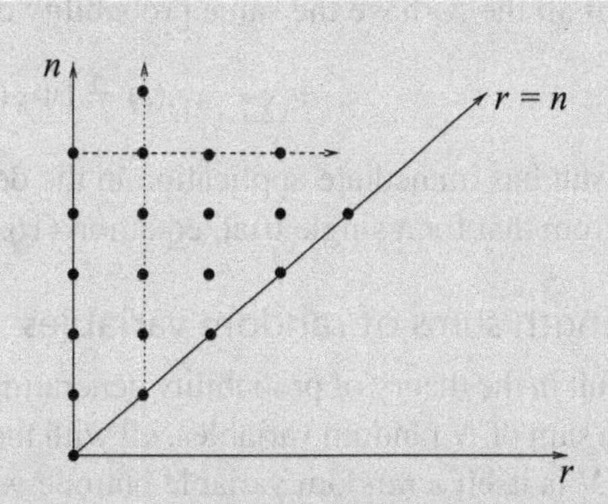

Figure 16.9 The pairs of values of n and r used in the evaluation of $\Phi_{X+Y}(t)$.

exclusive, we have

$$\Pr(X+Y=n)=\sum_{r=0}^{n}\Pr(X=r)\Pr(Y=n-r). \tag{16.68}$$

Multiplying both sides of (16.68) by t^n and summing over all values of n enables us to express this relationship in terms of probability generating functions as follows:

$$\Phi_{X+Y}(t)=\sum_{n=0}^{\infty}\Pr(X+Y=n)t^n=\sum_{n=0}^{\infty}\sum_{r=0}^{n}\Pr(X=r)t^r\Pr(Y=n-r)t^{n-r}$$

$$=\sum_{r=0}^{\infty}\sum_{n=r}^{\infty}\Pr(X=r)t^r\Pr(Y=n-r)t^{n-r}.$$

The change in summation order is justified by reference to Figure 16.9, which illustrates that the summations are over exactly the same pairs of values of n and r, but with the first (inner) summation over the points in a column rather than over the points in a row. Now, setting $n=r+s$ gives the final result,

$$\Phi_{X+Y}(t)=\sum_{r=0}^{\infty}\Pr(X=r)t^r\sum_{s=0}^{\infty}\Pr(Y=s)t^s$$

$$=\Phi_X(t)\Phi_Y(t), \tag{16.69}$$

i.e. the PGF of the sum of two independent random variables is equal to the product of their individual PGFs. The same result can be deduced in a less formal way by noting that if X and Y are independent then

$$E\left[t^{X+Y}\right]=E\left[t^X\right]E\left[t^Y\right].$$

Clearly result (16.69) can be extended to more than two random variables by writing $S_3=S_2+Z$, etc., to give

$$\Phi_{(\sum_{i=1}^n X_i)}(t)=\prod_{i=1}^{n}\Phi_{X_i}(t), \tag{16.70}$$

and, further, if all the X_i have the same probability distribution,

$$\Phi_{(\sum_{i=1}^{n} X_i)}(t) = [\Phi_X(t)]^n . \tag{16.71}$$

This latter result has immediate application in the deduction of the PGF for the binomial distribution from that for a single trial, equation (16.61).[13]

Variable-length sums of random variables

As a final result in the theory of probability generating functions we show how to calculate the PGF for a sum of N random variables, all with the same probability distribution, when the value of N is itself a random variable but one with a known probability distribution. In symbols, we wish to find the distribution of

$$S_N = X_1 + X_2 + \cdots + X_N, \tag{16.72}$$

where N is a random variable with $\Pr(N = n) = h_n$ and PGF $\chi_N(t) = \sum h_n t^n$.

The probability ξ_k that $S_N = k$ is given by a sum of conditional probabilities, namely[14]

$$\begin{aligned}
\xi_k &= \sum_{n=0}^{\infty} \Pr(N = n) \Pr(X_0 + X_1 + X_2 + \cdots + X_n = k) \\
&= \sum_{n=0}^{\infty} h_n \times \text{coefficient of } t^k \text{ in } [\Phi_X(t)]^n.
\end{aligned}$$

Multiplying both sides of this equation by t^k and summing over all k, we obtain an expression for the PGF $\Xi_S(t)$ of S_N:

$$\begin{aligned}
\Xi_S(t) = \sum_{k=0}^{\infty} \xi_k t^k &= \sum_{k=0}^{\infty} t^k \sum_{n=0}^{\infty} h_n \times \text{coefficient of } t^k \text{ in } [\Phi_X(t)]^n \\
&= \sum_{n=0}^{\infty} h_n \sum_{k=0}^{\infty} t^k \times \text{coefficient of } t^k \text{ in } [\Phi_X(t)]^n \\
&= \sum_{n=0}^{\infty} h_n [\Phi_X(t)]^n \\
&= \chi_N(\Phi_X(t)).
\end{aligned} \tag{16.73}$$

In words, the PGF of the sum S_N is given by the compound function $\chi_N(\Phi_X(t))$ obtained by substituting $\Phi_X(t)$ for t in the PGF for the number of terms N in the sum. We illustrate this with the following example.

13 Write down the PGF for the binomial distribution and confirm that it gives the correct values for the mean and variance of the distribution.

14 Formally $X_0 = 0$ has to be included, since $\Pr(N = 0)$ may be non-zero.

Example The probability distribution for the number of eggs in a clutch is Poisson distributed with mean λ, and the probability that each egg will hatch is p (and is independent of the size of the clutch). Use the results stated in (16.61) and (16.62) to show that the PGF (and hence the probability distribution) for the number of chicks that hatch corresponds to a Poisson distribution having mean λp.

The number of chicks that hatch is given by a sum of the form (16.72) in which $X_i = 1$ if the ith chick hatches and $X_i = 0$ if it does not. As given by (16.61), $\Phi_X(t)$ is thus $(1-p)+pt$. The value of N is given by a Poisson distribution with mean λ; thus, from (16.62), in the terminology of our previous discussion,

$$\chi_N(t) = e^{-\lambda}e^{\lambda t}.$$

We now substitute these forms into (16.73) to obtain

$$\begin{aligned}\Xi_S(t) &= \exp(-\lambda)\exp[\lambda\Phi_X(t)]\\ &= \exp(-\lambda)\exp\{\lambda[(1-p)+pt]\}\\ &= \exp(-\lambda p)\exp(\lambda pt).\end{aligned}$$

But this is exactly the PGF of a Poisson distribution with mean λp.

That this implies that the probability is Poisson distributed is intuitively obvious since, in the expansion of the PGF as a power series in t, every coefficient will be precisely that implied by such a distribution. A solution of the same problem by direct calculation appears in the answer to Problem 16.23. ◀

16.7.2 Moment generating functions

As we saw in Section 16.5 a probability function is often expressed in terms of its moments. This leads naturally to the second type of generating function, a *moment generating function*. For a random variable X, and a real number t, the moment generating function (MGF) is defined by

$$M_X(t) = E\left[e^{tX}\right] = \begin{cases}\sum_i e^{tx_i} f(x_i) & \text{for a discrete distribution,}\\ \int e^{tx} f(x)\,dx & \text{for a continuous distribution.}\end{cases} \qquad (16.74)$$

The MGF will exist for all values of t provided that X is bounded and always exists at the point $t = 0$ where $M(0) = E(1) = 1$.

It will be apparent that the PGF and the MGF for a random variable X are closely related. The former is the expectation of t^X whilst the latter is the expectation of e^{tX}:

$$\Phi_X(t) = E\left[t^X\right], \qquad M_X(t) = E\left[e^{tX}\right].$$

The MGF can thus be obtained from the PGF by replacing t by e^t, and vice versa. The MGF has more general applicability, however, since it can be used with both continuous and discrete distributions whilst the PGF is restricted to non-negative integer distributions.

As its name suggests, the MGF is particularly useful for obtaining the moments of a distribution, as is easily seen by noting that

$$E\left[e^{tX}\right] = E\left[1 + tX + \frac{t^2X^2}{2!} + \cdots\right]$$
$$= 1 + E[X]t + E\left[X^2\right]\frac{t^2}{2!} + \cdots.$$

Assuming that the MGF exists for all t around the point $t = 0$, we can deduce that the moments of a distribution are given in terms of its MGF by

$$E[X^n] = \left.\frac{d^n M_X(t)}{dt^n}\right|_{t=0}. \qquad (16.75)$$

Similarly, by substitution in (16.50), the variance of the distribution is given by

$$V[X] = M''_X(0) - \left[M'_X(0)\right]^2, \qquad (16.76)$$

where the prime denotes differentiation with respect to t. For probably the single most important continuous distribution, the Gaussian, these evaluations proceed as follows.

Example The MGF for the Gaussian distribution (see the end of Subsection 16.9.2) is given by

$$M_X(t) = \exp\left(\mu t + \tfrac{1}{2}\sigma^2 t^2\right).$$

Find the expectation and variance of this distribution.

Using (16.75),

$$M'_X(t) = \left(\mu + \sigma^2 t\right)\exp\left(\mu t + \tfrac{1}{2}\sigma^2 t^2\right) \quad \Rightarrow \quad E[X] = M'_X(0) = \mu,$$
$$M''_X(t) = \left[\sigma^2 + (\mu + \sigma^2 t)^2\right]\exp\left(\mu t + \tfrac{1}{2}\sigma^2 t^2\right) \quad \Rightarrow \quad M''_X(0) = \sigma^2 + \mu^2.$$

Thus, using (16.76),

$$V[X] = \sigma^2 + \mu^2 - \mu^2 = \sigma^2.$$

That the mean is found to be μ and the variance σ^2 justifies the use of these symbols in the Gaussian distribution. ◀

The moment generating function has several useful properties that follow from its definition and can be employed in simplifying calculations.

Scaling and shifting

If $Y = aX + b$, where a and b are arbitrary constants, then

$$M_Y(t) = E\left[e^{tY}\right] = E\left[e^{t(aX+b)}\right] = e^{bt}E\left[e^{atX}\right] = e^{bt}M_X(at). \qquad (16.77)$$

This result is sometimes useful for obtaining the moments of a distribution about its mean, rather than about $x = 0$. If the MFG of X is $M_X(t)$ then the variable $Y = X - \mu$ has the

MGF $M_Y(t) = e^{-\mu t} M_X(t)$, which generates the so-called *central* moments of X, i.e.

$$E[(X-\mu)^n] = E[Y^n] = M_Y^{(n)}(0) = \left(\frac{d^n}{dt^n} [e^{-\mu t} M_X(t)] \right)_{t=0}.$$

Clearly the second central moment is the same as the variance of the distribution.[15]

Sums of random variables

If $X_1, X_2, \ldots, X_N$ are independent random variables and $S_N = X_1 + X_2 + \cdots + X_N$ then

$$M_{S_N}(t) = E\left[e^{tS_N}\right] = E\left[e^{t(X_1+X_2+\cdots+X_N)}\right] = E\left[\prod_{i=1}^{N} e^{tX_i}\right].$$

Since the X_i are *independent*,

$$M_{S_N}(t) = \prod_{i=1}^{N} E\left[e^{tX_i}\right] = \prod_{i=1}^{N} M_{X_i}(t). \tag{16.78}$$

In words, the MGF of the sum of N independent random variables is the product of their individual MGFs. By combining (16.78) with (16.77), we obtain the more general result that the MGF of $S_N = c_1X_1 + c_2X_2 + \cdots + c_NX_N$ (where the c_i are constants) is given by

$$M_{S_N}(t) = \prod_{i=1}^{N} M_{X_i}(c_i t). \tag{16.79}$$

Variable-length sums of random variables

Although we will only quote it without proof, there is a prescription for the MGF of a variable-length sum of random variables analogous to that given for PGFs in (16.73).

If S_N is the sum of N independent random variables X_i ($i = 1, 2, \ldots, N$), all with the same probability distribution, and N itself is a random variable with probability generating function $\chi_N(t)$, then the MGF of S_N is related to that of the X_i by

$$M_{S_N}(t) = \chi_N(M_X(t)).$$

In words, the MGF of the sum S_N is given by the compound function $\chi_N(M_X(t))$ obtained by substituting $M_X(t)$ for t in the PGF for the number of terms N in the sum.

Uniqueness

If the MGF of the random variable X_1 is identical to that for X_2 then the probability distributions of X_1 and X_2 are identical. This is intuitively reasonable although a rigorous proof is complicated[16] and beyond the scope of this book.

15 Use Leibnitz' theorem to show that, for $n = 2$, the above formulation agrees with (16.76).

16 See, for example, P. A. Moran, *An Introduction to Probability Theory* (New York: Oxford Science Publications, 1984).

Table 16.1 *Some important discrete probability distributions*

Distribution	Probability law $f(x)$	MGF	$E[X]$	$V[X]$
binomial	${}^nC_x p^x q^{n-x}$	$(pe^t + q)^n$	np	npq
negative binomial	${}^{r+x-1}C_x p^r q^x$	$\left(\dfrac{p}{1-qe^t}\right)^r$	$\dfrac{rq}{p}$	$\dfrac{rq}{p^2}$
geometric	$q^{x-1}p$	$\dfrac{pe^t}{1-qe^t}$	$\dfrac{1}{p}$	$\dfrac{q}{p^2}$
hypergeometric	$\dfrac{(Np)!(Nq)!n!(N-n)!}{x!(Np-x)!(n-x)!(Nq-n+x)!N!}$		np	$\dfrac{N-n}{N-1}npq$
Poisson	$\dfrac{\lambda^x}{x!}e^{-\lambda}$	$e^{\lambda(e^t-1)}$	λ	λ

16.8 Important discrete distributions

Having discussed some general properties of distributions, we now consider the more important discrete distributions encountered in physical applications. These are discussed in detail below, and summarized for convenience in Table 16.1; we refer the reader to the relevant section below for an explanation of the symbols used.

16.8.1 The binomial distribution

Perhaps the most important discrete probability distribution is the *binomial distribution*. This distribution describes processes that consist of a number of independent identical *trials* with two possible outcomes, A and $B = \bar{A}$. We may call these outcomes "success" and "failure" respectively. If the probability of a success is $\Pr(A) = p$ then the probability of a failure is $\Pr(B) = q = 1 - p$. If we perform n trials then the discrete random variable

$$X = \text{number of times } A \text{ occurs}$$

can take the values $0, 1, 2, \ldots, n$; its distribution amongst these values is described by the *binomial distribution*.

We now calculate the probability that in n trials we obtain x successes (and so $n - x$ failures). One way of obtaining such a result is to have x successes followed by $n - x$ failures. Since the trials are assumed independent, the probability of this is

$$\underbrace{pp\cdots p}_{x \text{ times}} \times \underbrace{qq\cdots q}_{n-x \text{ times}} = p^x q^{n-x}.$$

This is, however, just one permutation of x successes and $n - x$ failures. The total number of permutations of n objects, of which x are identical and of type 1 and $n - x$ are identical and of type 2, is given by (16.33) as

$$\frac{n!}{x!(n-x)!} \equiv {}^nC_x.$$

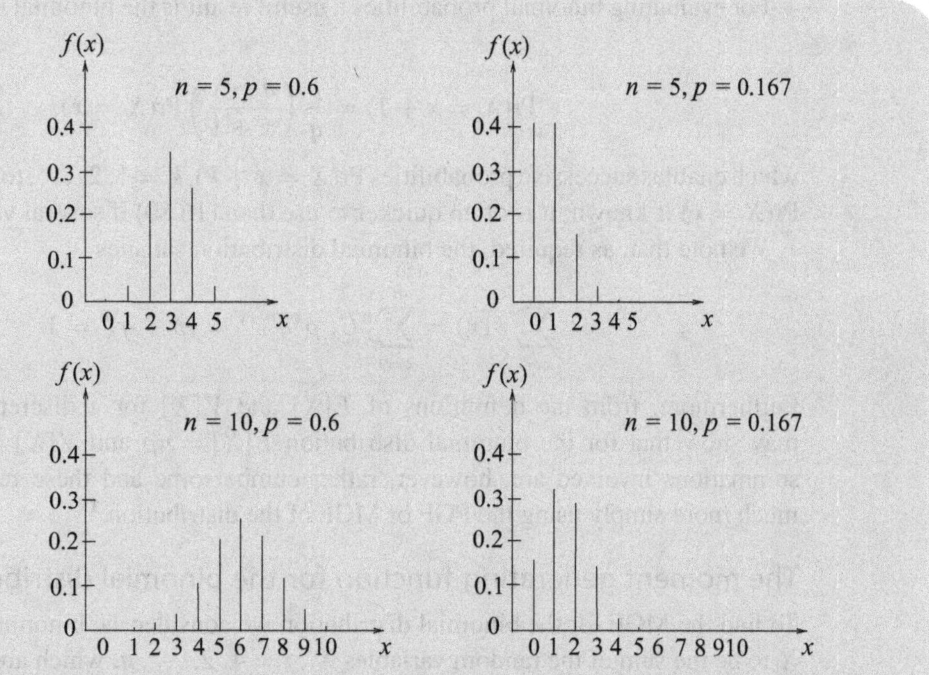

Figure 16.10 Some typical binomial distributions with various combinations of parameters n and p.

Therefore, the total probability of obtaining x successes from n trials is

$$f(x) = \Pr(X = x) = {}^nC_x\, p^x q^{n-x} = {}^nC_x\, p^x (1-p)^{n-x}, \tag{16.80}$$

which is the *binomial probability distribution formula*. When a random variable X follows the binomial distribution for n trials, with a probability of success p, we write $X \sim \text{Bin}(n, p)$; such a random variable X is often referred to as a binomial *variate*. Some typical binomial distributions are shown in Figure 16.10. As a simple example of a binomial calculation, consider the following.

Example If an unbiased single six-sided die is rolled five times, what is the probability that a six is thrown exactly three times?

Here the number of "trials" $n = 5$, and we are interested in the random variable

$$X = \text{number of sixes thrown.}$$

Since the probability of a "success" is $p = \frac{1}{6}$, equation (16.80) gives

$$\Pr(X = 3) = \frac{5!}{3!(5-3)!}\left(\frac{1}{6}\right)^3\left(\frac{5}{6}\right)^{(5-3)} = 0.032$$

as the probability of obtaining exactly three sixes in five throws. ◀

For evaluating binomial probabilities a useful result is the binomial recurrence formula

$$\Pr(X = x + 1) = \frac{p}{q}\left(\frac{n - x}{x + 1}\right)\Pr(X = x), \tag{16.81}$$

which enables successive probabilities $\Pr(X = x + k)$, $k = 1, 2, \ldots$, to be calculated once $\Pr(X = x)$ is known; it is often quicker to use than (16.80) if several values are needed.

We note that, as required, the binomial distribution satisfies

$$\sum_{x=0}^{n} f(x) = \sum_{x=0}^{n} {}^nC_x\, p^x q^{n-x} = (p + q)^n = 1.$$

Furthermore, from the definitions of $E[X]$ and $V[X]$ for a discrete distribution, we may show that for the binomial distribution $E[X] = np$ and $V[X] = npq$. The direct summations involved are, however, rather cumbersome and these results are obtained much more simply using the PGF or MGF of the distribution.[17]

The moment generating function for the binomial distribution

To find the MGF for the binomial distribution we consider the binomial random variable X to be the sum of the random variables X_i, $i = 1, 2, \ldots, n$, which are defined by

$$X_i = \begin{cases} 1 & \text{if a "success" occurs on the } i\text{th trial,} \\ 0 & \text{if a "failure" occurs on the } i\text{th trial.} \end{cases}$$

Thus

$$\begin{aligned} M_i(t) = E\left[e^{tX_i}\right] &= e^{0t} \times \Pr(X_i = 0) + e^{1t} \times \Pr(X_i = 1) \\ &= 1 \times q + e^t \times p \\ &= pe^t + q. \end{aligned}$$

From (16.78), it follows that the MGF for the binomial distribution is given by

$$M(t) = \prod_{i=1}^{n} M_i(t) = (pe^t + q)^n. \tag{16.82}$$

We can now use the moment generating function to derive the mean and variance of the binomial distribution. From (16.82)

$$M'(t) = npe^t(pe^t + q)^{n-1},$$

and from (16.75)

$$E[X] = M'(0) = np(p + q)^{n-1} = np,$$

where the last equality follows from $p + q = 1$.

Differentiating with respect to t once more gives

$$M''(t) = e^t(n - 1)np^2(pe^t + q)^{n-2} + e^t np(pe^t + q)^{n-1},$$

17 In the following section the MGF is used. The same results, using the PGF, were obtained in footnote 10.

and from (16.75)

$$E[X^2] = M''(0) = n^2p^2 - np^2 + np.$$

Thus, using (16.76)

$$V[X] = M''(0) - \left[M'(0)\right]^2 = n^2p^2 - np^2 + np - n^2p^2 = np(1-p) = npq.$$

Multiple binomial distributions

Suppose X and Y are two *independent* random variables, both of which are described by binomial distributions with a common probability of success p, but with (in general) different numbers of trials n_1 and n_2, so that $X \sim \text{Bin}(n_1, p)$ and $Y \sim \text{Bin}(n_2, p)$. Now consider the random variable $Z = X + Y$.

Since X and Y are independent random variables, the MGF $M_Z(t)$ of the new variable $Z = X + Y$ is given simply by the product of the individual MGFs $M_X(t)$ and $M_Y(t)$. Thus, we obtain

$$M_Z(t) = M_X(t)M_Y(t) = (pe^t + q)^{n_1}(pe^t + q)^{n_2} = (pe^t + q)^{n_1+n_2},$$

which we recognize as the MGF of $Z \sim \text{Bin}(n_1 + n_2, p)$. Hence Z is also described by a binomial distribution.[18]

This result may be extended to any number of binomial distributions. If X_i, $i = 1, 2, \ldots, N$, is distributed as $X_i \sim \text{Bin}(n_i, p)$ then $Z = X_1 + X_2 + \cdots + X_N$ is distributed as $Z \sim \text{Bin}(n_1 + n_2 + \cdots + n_N, p)$, as would be expected since the result of $\sum_i n_i$ trials cannot depend on how they are split up.

Unfortunately, no equivalent simple result exists for the probability distribution of the *difference* $Z = X - Y$ of two binomially distributed variables.

16.8.2 The multinomial distribution

The binomial distribution describes the probability of obtaining x "successes" from n independent trials, where each trial has only two possible outcomes. This may be generalized to the case where each trial has k possible outcomes with respective probabilities $p_1, p_2, \ldots, p_k$. If we consider the random variables X_i, $i = 1, 2, \ldots, n$, to be the number of outcomes of type i in n trials then we may calculate their joint probability function

$$f(x_1, x_2, \ldots, x_k) = \Pr(X_1 = x_1,\ X_2 = x_2,\ \ldots,\ X_k = x_k),$$

where we must have $\sum_{i=1}^{k} x_i = n$. In n trials the probability of obtaining x_1 outcomes of type 1, followed by x_2 outcomes of type 2, etc. is given by

$$p_1^{x_1} p_2^{x_2} \cdots p_k^{x_k}.$$

However, the number of distinguishable permutations of this result is

$$\frac{n!}{x_1!x_2!\cdots x_k!},$$

18 If $X \sim \text{Bin}(n, p_1)$ and $Y \sim \text{Bin}(n, p_2)$, find the mean and variance of $Z = X + Y$ using its MGF.

and thus

$$f(x_1, x_2, \ldots, x_k) = \frac{n!}{x_1!x_2!\cdots x_k!} p_1^{x_1} p_2^{x_2} \cdots p_k^{x_k}. \tag{16.83}$$

This is the *multinomial probability distribution.*

If $k = 2$ then the multinomial distribution reduces to the familiar binomial distribution. Although in this form the binomial distribution appears to be a function of two random variables, it must be remembered that, in fact, since $p_2 = 1 - p_1$ and $x_2 = n - x_1$, the distribution of X_1 is entirely determined by the parameters p_1 (or p_2) and n. That X_1 has a *binomial* distribution follows from recalling that it represents the number of objects of a particular type obtained from sampling with replacement, the very process that led to the original definition of the binomial distribution. In fact, any of the random variables X_i has a binomial distribution, i.e. the marginal distribution of each X_i is binomial with parameters n and p_i. It immediately follows that

$$E[X_i] = np_i \qquad \text{and} \qquad V[X_i]^2 = np_i(1 - p_i). \tag{16.84}$$

The following worked example applies the multinomial distribution several times.

Example At a village fête patrons were invited, for a 10 p entry fee, to pick without looking six tickets from a drum containing equal large numbers of red, blue and green tickets. If five or more of the tickets were of the same color a prize of 100 p was awarded. A consolation award of 40 p was made if two tickets of each color were picked. Was a good time had by all?

In this case, all types of outcome (red, blue and green) have the same probabilities. The probability of obtaining any given combination of tickets is given by the multinomial distribution with $n = 6$, $k = 3$ and $p_i = \frac{1}{3}, i = 1, 2, 3$.

(i) The probability of picking six tickets of the same color is given by

$$\text{Pr (six of the same color)} = 3 \times \frac{6!}{6!0!0!}\left(\frac{1}{3}\right)^6\left(\frac{1}{3}\right)^0\left(\frac{1}{3}\right)^0 = \frac{1}{243}.$$

The factor of 3 is present because there are three different colors.

(ii) The probability of picking five tickets of one color and one ticket of another color is

$$\text{Pr(five of one color; one of another)} = 3 \times 2 \times \frac{6!}{5!1!0!}\left(\frac{1}{3}\right)^5\left(\frac{1}{3}\right)^1\left(\frac{1}{3}\right)^0 = \frac{4}{81}.$$

The factors of 3 and 2 are included because there are three ways to choose the color of the five matching tickets, and then two ways to choose the color of the remaining ticket.

(iii) Finally, the probability of picking two tickets of each color is

$$\text{Pr (two of each color)} = \frac{6!}{2!2!2!}\left(\frac{1}{3}\right)^2\left(\frac{1}{3}\right)^2\left(\frac{1}{3}\right)^2 = \frac{10}{81}.$$

Thus the expected return to any patron was, in pence,

$$100\left(\frac{1}{243} + \frac{4}{81}\right) + \left(40 \times \frac{10}{81}\right) = 10.29.$$

A good time was had by all but the stall holder! ◀

16.8.3 The geometric and negative binomial distributions

A special case of the binomial distribution occurs when instead of the number of successes we consider the discrete random variable

$$X = \text{number of trials required to obtain the first success.}$$

The probability that x trials are required in order to obtain the first success, is simply the probability of obtaining $x - 1$ failures followed by one success. If the probability of a success on each trial is p, then for $x > 0$

$$f(x) = \Pr(X = x) = (1 - p)^{x-1}p = q^{x-1}p,$$

where $q = 1 - p$. This distribution is sometimes called the *geometric distribution*. The probability generating function for this distribution is given in (16.67). By replacing t by e^t in (16.67) we immediately obtain the MGF of the geometric distribution

$$M(t) = \frac{pe^t}{1 - qe^t},$$

from which its mean and variance are found to be

$$E[X] = \frac{1}{p}, \qquad V[X] = \frac{q}{p^2}.$$

Another distribution closely related to the binomial is the negative binomial distribution. This describes the probability distribution of the random variable

$$X = \text{number of failures before the } r\text{th success.}$$

One way of obtaining x failures before the rth success is to have $r - 1$ successes followed by x failures followed by the rth success, for which the probability is

$$\underbrace{pp\cdots p}_{r-1 \text{ times}} \times \underbrace{qq\cdots q}_{x \text{ times}} \times p = p^r q^x.$$

However, the first $r + x - 1$ factors constitute just one permutation of $r - 1$ successes and x failures. The total number of permutations of these $r + x - 1$ objects, of which $r - 1$ are identical and of type 1 and x are identical and of type 2, is ${}^{r+x-1}C_x$. Therefore, the total probability of obtaining x failures before the rth success is

$$f(x) = \Pr(X = x) = {}^{r+x-1}C_x p^r q^x,$$

which is called the *negative binomial distribution* (see the related discussion on p. 631). It is straightforward to show that the MGF of this distribution is

$$M(t) = \left(\frac{p}{1 - qe^t}\right)^r,$$

and that its mean and variance are given by

$$E[X] = \frac{rq}{p} \quad \text{and} \quad V[X] = \frac{rq}{p^2}.$$

16.8.4 The hypergeometric distribution

In Subsection 16.8.1 we saw that the probability of obtaining x successes in n *independent* trials was given by the binomial distribution. Suppose that these n "trials" actually consist of drawing at random n balls, from a set of N such balls of which M are red and the rest white. Let us consider the random variable X = number of red balls drawn.

On the one hand, if the balls are drawn *with replacement* then the trials are independent and the probability of drawing a red ball is $p = M/N$ each time. Therefore, the probability of drawing x red balls in n trials is given by the binomial distribution as

$$\Pr(X = x) = \frac{n!}{x!(n-x)!} p^x (1-p)^{n-x}.$$

On the other hand, if the balls are drawn *without replacement* the trials are not independent and the probability of drawing a red ball depends on how many red balls have already been drawn. We can, however, still derive a general formula for the probability of drawing x red balls in n trials, as follows.

The number of ways of drawing x red balls from M is ${}^M C_x$, and the number of ways of drawing $n - x$ white balls from $N - M$ is ${}^{N-M} C_{n-x}$. Therefore, the total number of ways to obtain x red balls in n trials is ${}^M C_x \, {}^{N-M} C_{n-x}$. However, the total number of ways of drawing n objects from N is simply ${}^N C_n$. Hence the probability of obtaining x red balls in n trials is

$$\begin{aligned}
\Pr(X = x) &= \frac{{}^M C_x \, {}^{N-M} C_{n-x}}{{}^N C_n} \\
&= \frac{M!}{x!(M-x)!} \frac{(N-M)!}{(n-x)!(N-M-n+x)!} \frac{n!(N-n)!}{N!}, \qquad (16.85) \\
&= \frac{(Np)!(Nq)!\, n!(N-n)!}{x!(Np-x)!(n-x)!(Nq-n+x)!\, N!}, \qquad (16.86)
\end{aligned}$$

where in the last line $p = M/N$ and $q = 1 - p$. This is called the *hypergeometric distribution.*

By performing the relevant summations directly, it may be shown that the hypergeometric distribution has mean

$$E[X] = n\frac{M}{N} = np$$

and variance[19]

$$V[X] = \frac{nM(N-M)(N-n)}{N^2(N-1)} = \frac{N-n}{N-1} npq.$$

An application of the hypergeometric distribution that is of interest (at least in the UK) far beyond the realms of physics and engineering, is the calculation of the odds against winning the National Lottery.

---

19 Note that for choosing just one ball, the variance is the same as that for a binomial trial, namely npq, but if all N are chosen there is no variance in the results – the number of red balls included will always be M.

Example In the UK National Lottery each participant chooses six different numbers between 1 and 49. In each weekly draw six numbered winning balls are subsequently drawn. Find the probabilities that a participant correctly predicts 0, 1, 2, 3, 4, 5, 6 winning numbers.

The probabilities are given by a hypergeometric distribution with N (the total number of balls) $= 49$, M (the number of winning balls drawn) $= 6$, and n (the number of numbers chosen by each participant) $= 6$. Thus, substituting in (16.85), we find

$$\Pr(0) = \frac{{}^6C_0\,{}^{43}C_6}{{}^{49}C_6} = \frac{1}{2.29}, \quad \Pr(1) = \frac{{}^6C_1\,{}^{43}C_5}{{}^{49}C_6} = \frac{1}{2.42},$$

$$\Pr(2) = \frac{{}^6C_2\,{}^{43}C_4}{{}^{49}C_6} = \frac{1}{7.55}, \quad \Pr(3) = \frac{{}^6C_3\,{}^{43}C_3}{{}^{49}C_6} = \frac{1}{56.6},$$

$$\Pr(4) = \frac{{}^6C_4\,{}^{43}C_2}{{}^{49}C_6} = \frac{1}{1032}, \quad \Pr(5) = \frac{{}^6C_5\,{}^{43}C_1}{{}^{49}C_6} = \frac{1}{54\,200},$$

$$\Pr(6) = \frac{{}^6C_6\,{}^{43}C_0}{{}^{49}C_6} = \frac{1}{13.98 \times 10^6}.$$

It can easily be seen that:

(i) $\sum_{i=0}^{6} \Pr(i) = 0.44 + 0.41 + 0.13 + 0.02 + O(10^{-3}) = 1$, as expected;

(ii) as the stake money is £1, the prize of £10 for three correct predictions, and a typical jackpot share of about £2M for six correct predictions are both less than 20% of a "fair return".

Perhaps a little surprising is the fact that the chance of no correct predictions is only a little larger than the chance of one correct – but neither gets you a prize! ◀

Note that if n, the number of trials (balls drawn), is small compared with each of N, M and $N - M$ then not replacing the balls is of little consequence, and we may approximate the hypergeometric distribution by the binomial distribution (with $p = M/N$); this is much easier to evaluate.

16.8.5 The Poisson distribution

We have seen that the binomial distribution describes the number of successful outcomes in a certain number of trials n. The Poisson distribution also describes the probability of obtaining a given number of successes but for situations in which the number of "trials" cannot be enumerated; rather it describes the situation in which discrete events occur in a continuum. Typical examples of discrete random variables X described by a Poisson distribution are the number of telephone calls received by a switchboard in a given interval, or the number of stars above a certain brightness in a particular area of the sky. Given a mean rate of occurrence λ of these events in the relevant interval or area, the Poisson distribution gives the probability $\Pr(X = x)$ that exactly x events will occur.

We may derive the form of the Poisson distribution as the limit of the binomial distribution when the number of trials $n \to \infty$ and the probability of "success" $p \to 0$, in such a way that $np = \lambda$ remains finite. Thus, in our example of a telephone switchboard, suppose we wish to find the probability that exactly x calls are received during some time interval, given that the mean number of calls in such an interval is λ.

Let us begin by dividing the time interval into a large number, n, of equal shorter intervals, in each of which the probability of receiving a call is p. As we let $n \to \infty$ then $p \to 0$, but since we require the mean number of calls in the interval to equal λ, we must have $np = \lambda$. The probability of x successes in n trials is given by the binomial formula as

$$\Pr(X = x) = \frac{n!}{x!(n-x)!} p^x (1-p)^{n-x}. \tag{16.87}$$

Now as $n \to \infty$, with x finite, the ratio of the n-dependent factorials in (16.87) behaves asymptotically as a power of n, i.e.

$$\lim_{n\to\infty} \frac{n!}{(n-x)!} = \lim_{n\to\infty} n(n-1)(n-2)\cdots(n-x+1) \sim n^x.$$

Also

$$\lim_{n\to\infty} \lim_{p\to 0} (1-p)^{n-x} = \lim_{p\to 0} \frac{(1-p)^{\lambda/p}}{(1-p)^x} = \frac{e^{-\lambda}}{1}.$$

Thus, using $\lambda = np$ and hence $n^x p^x = \lambda^x$, (16.87) tends to the *Poisson distribution*

$$f(x) = \Pr(X = x) = \frac{e^{-\lambda}\lambda^x}{x!}, \tag{16.88}$$

which gives the probability of obtaining exactly x calls in the given time interval. As we shall show below, λ is the mean of the distribution. Events following a Poisson distribution are usually said to occur randomly in time.

Alternatively, we may derive the Poisson distribution directly, without considering it as the limit of a binomial distribution. Let us again consider our example of a telephone switchboard. Suppose that the probability that x calls have been received in a time interval t is $P_x(t)$. If the average number of calls received in a unit time is λ then in a further small time interval Δt the probability of receiving a call is $\lambda\Delta t$, provided Δt is short enough that the probability of receiving two or more calls in this small interval is negligible. Similarly the probability of receiving no call during the same small interval is simply $1 - \lambda\Delta t$.

Thus, for $x > 0$, the probability of receiving exactly x calls in the total interval $t + \Delta t$ is given by

$$P_x(t + \Delta t) = P_x(t)(1 - \lambda\Delta t) + P_{x-1}(t)\lambda\Delta t.$$

Rearranging the equation, dividing through by Δt and letting $\Delta t \to 0$, we obtain the differential recurrence equation

$$\frac{dP_x(t)}{dt} = \lambda P_{x-1}(t) - \lambda P_x(t). \tag{16.89}$$

For $x = 0$ (i.e. no calls received), however, (16.89) simplifies to

$$\frac{dP_0(t)}{dt} = -\lambda P_0(t),$$

which may be integrated to give $P_0(t) = P_0(0)e^{-\lambda t}$. But since the probability $P_0(0)$ of receiving no calls in a zero time interval must equal unity, we have $P_0(t) = e^{-\lambda t}$. This expression for $P_0(t)$ may then be substituted back into (16.89) with $x = 1$ to obtain a differential equation for $P_1(t)$ that has the solution $P_1(t) = \lambda t e^{-\lambda t}$. We may repeat this process to obtain expressions for $P_2(t), P_3(t), \ldots, P_x(t)$, and we find[20]

$$P_x(t) = \frac{(\lambda t)^x}{x!} e^{-\lambda t}. \tag{16.90}$$

By setting $t = 1$ in (16.90), we again obtain the Poisson distribution (16.88) for obtaining exactly x calls in a unit time interval.

If a discrete random variable is described by a Poisson distribution of mean λ then we write $X \sim \text{Po}(\lambda)$. As it must be, the sum of the probabilities is unity:

$$\sum_{x=0}^{\infty} \Pr(X = x) = e^{-\lambda} \sum_{x=0}^{\infty} \frac{\lambda^x}{x!} = e^{-\lambda} e^{\lambda} = 1.$$

From (16.88) we may also derive the *Poisson recurrence formula*,

$$\Pr(X = x + 1) = \frac{\lambda}{x + 1} \Pr(X = x) \quad \text{for } x = 0, 1, 2, \ldots, \tag{16.91}$$

which enables successive probabilities to be calculated easily once one is known.

Example A person receives on average one e-mail message per half-hour interval. Assuming that the e-mails are received randomly in time, find the probabilities that in any particular hour 0, 1, 2, 3, 4, 5 messages are received.

Let X = number of e-mails received per hour. Clearly the mean number of e-mails per hour is two, and so X follows a Poisson distribution with $\lambda = 2$, i.e.

$$\Pr(X = x) = \frac{2^x}{x!} e^{-2}.$$

Thus $\Pr(X = 0) = e^{-2} = 0.135$, $\Pr(X = 1) = 2e^{-2} = 0.271$, $\Pr(X = 2) = 2^2 e^{-2}/2! = 0.271$, $\Pr(X = 3) = 2^3 e^{-2}/3! = 0.180$, $\Pr(X = 4) = 2^4 e^{-2}/4! = 0.090$, $\Pr(X = 5) = 2^5 e^{-2}/5! = 0.036$. These values are most conveniently found using the recurrence formula (16.91). ◀

The above example illustrates the point that a Poisson distribution typically rises and then falls. It either has a maximum when x is equal to the integer part of λ or, if λ happens to be an integer, has equal maximal values at $x = \lambda - 1$ and $x = \lambda$. The Poisson distribution always has a long "tail" towards higher values of X but the higher the value of the mean the more symmetric the distribution becomes. Typical Poisson distributions are shown in Figure 16.11. Using the definitions of mean and variance, we may show that, for the Poisson distribution, $E[X] = \lambda$ and $V[X] = \lambda$. Nevertheless, as in the case of the binomial distribution, performing the relevant summations directly is somewhat tiresome, and these results are much more easily proved using the moment generating function.

20 Verify, by substitution, that these solutions do satisfy (16.89).

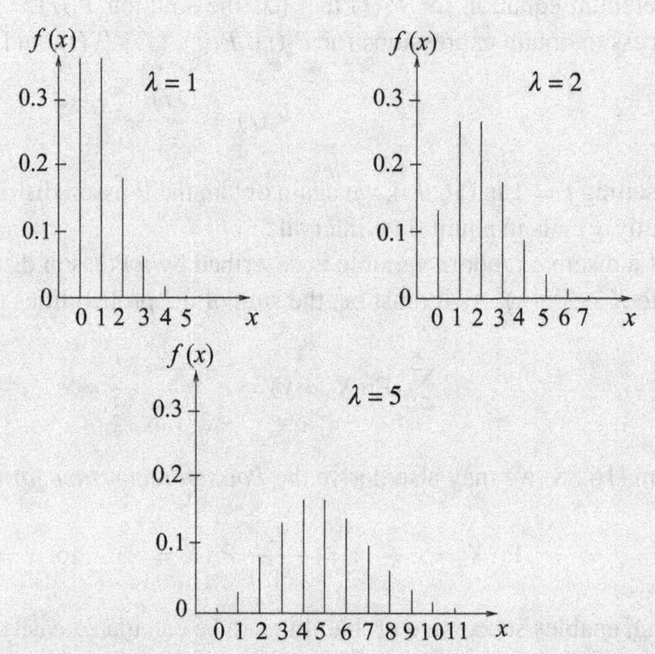

Figure 16.11 Three Poisson distributions for different values of the parameter λ.

The moment generating function for the Poisson distribution

The MGF of the Poisson distribution is given by

$$M_X(t) = E\left[e^{tX}\right] = \sum_{x=0}^{\infty} \frac{e^{tx}e^{-\lambda}\lambda^x}{x!} = e^{-\lambda}\sum_{x=0}^{\infty}\frac{(\lambda e^t)^x}{x!} = e^{-\lambda}e^{\lambda e^t} = e^{\lambda(e^t-1)}, \quad (16.92)$$

from which we obtain

$$M'_X(t) = \lambda e^t e^{\lambda(e^t-1)},$$
$$M''_X(t) = (\lambda^2 e^{2t} + \lambda e^t)e^{\lambda(e^t-1)}.$$

Thus, the mean and variance of the Poisson distribution are given by

$$E[X] = M'_X(0) = \lambda \qquad \text{and} \qquad V[X] = M''_X(0) - [M'_X(0)]^2 = \lambda.$$

The Poisson approximation to the binomial distribution

Earlier we derived the Poisson distribution as the limit of the binomial distribution when $n \to \infty$ and $p \to 0$ in such a way that $np = \lambda$ remains finite, where λ is the mean of the Poisson distribution. It is not surprising, therefore, that the Poisson distribution is a very good approximation to the binomial distribution for large n (≥ 50, say) and small p (≤ 0.1, say). Moreover, it is easier to calculate as it involves fewer factorials.

Example In a large batch of light bulbs, the probability that a bulb is defective is 0.5%. For a sample of 200 bulbs taken at random, find the approximate probabilities that 0, 1 and 2 of the bulbs respectively are defective.

Let the random variable $X =$ number of defective bulbs in a sample. This is distributed as $X \sim \text{Bin}(200, 0.005)$, implying that $\lambda = np = 1.0$. Since n is large and p small, we may approximate the distribution as $X \sim \text{Po}(1)$, giving

$$\Pr(X = x) \approx e^{-1}\frac{1^x}{x!},$$

from which we find $\Pr(X = 0) \approx 0.37$, $\Pr(X = 1) \approx 0.37$, $\Pr(X = 2) \approx 0.18$. For comparison, it may be noted that the exact values calculated from the binomial distribution are identical to those found here to two decimal places. ◀

Multiple Poisson distributions

Mirroring our discussion of multiple binomial distributions in Subsection 16.8.1, let us suppose X and Y are two *independent* random variables, both of which are described by Poisson distributions with (in general) different means, so that $X \sim \text{Po}(\lambda_1)$ and $Y \sim \text{Po}(\lambda_2)$. Now consider the random variable $Z = X + Y$. We may calculate the probability distribution of Z directly using

$$p(n) = \sum_{r=0}^{r=n} e^{-\lambda_1}\frac{\lambda_1^{n-r}}{(n-r)!}\, e^{-\lambda_2}\frac{\lambda_2^r}{r!},$$

but we may derive the result much more easily by using the moment generating function (or indeed the probability generating functions).

Since X and Y are independent RVs, the MGF for Z is simply the product of the individual MGFs for X and Y. Thus, from (16.92),

$$M_Z(t) = M_X(t)M_Y(t) = e^{\lambda_1(e^t-1)}e^{\lambda_2(e^t-1)} = e^{(\lambda_1+\lambda_2)(e^t-1)},$$

which we recognize as the MGF of $Z \sim \text{Po}(\lambda_1 + \lambda_2)$. Hence Z is also Poisson distributed and has mean $\lambda_1 + \lambda_2$; straightforward use of this result is made in the worked example below. Unfortunately, there is no such simple result for the *difference* $Z = X - Y$ of two independent Poisson variates. A closed-form expression for the PDF of this Z does exist, but it is a rather complicated combination of exponentials and a modified Bessel function.[21]

Example Two types of e-mail arrive independently and at random: external e-mails at a mean rate of one every five minutes and internal e-mails at a rate of two every five minutes. Calculate the probability of receiving two or more e-mails in any two-minute interval.

21 For a derivation see, for example, M. P. Hobson and A. N. Lasenby, *Monthly Notices of the Royal Astronomical Society*, **298**, 905 (1998).

Table 16.2 *Some important continuous probability distributions*

Distribution	Probability law $f(x)$	MGF	$E[X]$	$V[X]$
Gaussian	$\frac{1}{\sigma\sqrt{2\pi}}\exp\left[-\frac{(x-\mu)^2}{2\sigma^2}\right]$	$\exp\left(\mu t+\frac{1}{2}\sigma^2 t^2\right)$	μ	σ^2
exponential	$\lambda e^{-\lambda x}$	$\left(\frac{\lambda}{\lambda - t}\right)$	$\frac{1}{\lambda}$	$\frac{1}{\lambda^2}$
gamma	$\frac{\lambda}{\Gamma(r)}(\lambda x)^{r-1}e^{-\lambda x}$	$\left(\frac{\lambda}{\lambda - t}\right)^r$	$\frac{r}{\lambda}$	$\frac{r}{\lambda^2}$
chi-squared	$\frac{1}{2^{n/2}\Gamma(n/2)}x^{(n/2)-1}e^{-x/2}$	$\left(\frac{1}{1-2t}\right)^{n/2}$	n	$2n$
uniform	$\frac{1}{b-a}$	$\frac{e^{bt}-e^{at}}{(b-a)t}$	$\frac{a+b}{2}$	$\frac{(b-a)^2}{12}$

Let

$$X = \text{number of external e-mails per two-minute interval},$$
$$Y = \text{number of internal e-mails per two-minute interval}.$$

Since we expect on average one external e-mail and two internal e-mails every five minutes we have, when considering a two-minute interval, that $X \sim \text{Po}(0.4)$ and $Y \sim \text{Po}(0.8)$. Letting $Z = X + Y$ we have $Z \sim \text{Po}(0.4 + 0.8) = \text{Po}(1.2)$. Now

$$\Pr(Z \geq 2) = 1 - \Pr(Z < 2) = 1 - \Pr(Z = 0) - \Pr(Z = 1)$$

and

$$\Pr(Z = 0) = e^{-1.2} = 0.301,$$
$$\Pr(Z = 1) = e^{-1.2}\frac{1.2}{1} = 0.361.$$

Hence $\Pr(Z \geq 2) = 1 - 0.301 - 0.361 = 0.338$ gives the probability of receiving at least two e-mails within the two-minute interval. ◀

The above result can be extended, of course, to any number of Poisson processes, so that if $X_i = \text{Po}(\lambda_i)$, $i = 1, 2, \ldots, n$ then the random variable $Z = X_1 + X_2 + \cdots + X_n$ is distributed as $Z \sim \text{Po}(\lambda_1 + \lambda_2 + \cdots + \lambda_n)$.

16.9 Important continuous distributions

Having discussed the most commonly encountered discrete probability distributions, we now consider some of the more important continuous probability distributions. These are summarized for convenience in Table 16.2; we refer the reader to the relevant subsection below for an explanation of the symbols used.

16.9.1 The uniform distribution

Firstly we mention the very simple, but common, *uniform distribution*, which describes a continuous random variable that has a constant PDF over its allowed range of values. If the limits on X are a and b then

$$f(x) = \begin{cases} 1/(b-a) & \text{for } a \leq x \leq b, \\ 0 & \text{otherwise.} \end{cases}$$

The MGF of the uniform distribution is found to be

$$M(t) = \frac{e^{bt} - e^{at}}{(b-a)t},$$

and its mean and variance are given by

$$E[X] = \frac{a+b}{2}, \qquad V[X] = \frac{(b-a)^2}{12}.$$

16.9.2 The Gaussian distribution

By far the most important continuous probability distribution is the *Gaussian* or *normal* distribution. The reason for its importance is that a great many random variables of interest, in all areas of the physical sciences and beyond, are described either exactly or approximately by a Gaussian distribution. Moreover, the Gaussian distribution can be used to approximate other, more complicated, probability distributions.

The probability density function for a Gaussian distribution of a random variable X, with mean $E[X] = \mu$ and variance $V[X] = \sigma^2$, takes the form

$$f(x) = \frac{1}{\sigma\sqrt{2\pi}} \exp\left[-\frac{1}{2}\left(\frac{x-\mu}{\sigma}\right)^2\right]. \qquad (16.93)$$

The factor $1/\sqrt{2\pi}$ arises from the normalization of the distribution,

$$\int_{-\infty}^{\infty} f(x)dx = 1.$$

The Gaussian distribution is symmetric about the point $x = \mu$ and has the characteristic "bell" shape shown in Figure 16.12. The width of the curve is described by the standard deviation σ: if σ is large then the curve is broad, and if σ is small then the curve is narrow (see the figure). At $x = \mu \pm \sigma$, $f(x)$ falls to $e^{-1/2} \approx 0.61$ of its peak value; these points are points of inflection, where $d^2f/dx^2 = 0$. When a random variable X follows a Gaussian distribution with mean μ and variance σ^2, we write $X \sim N(\mu, \sigma^2)$.

The effects of changing μ and σ are only to shift the curve along the x-axis or to broaden or narrow it, respectively. Thus all Gaussians are equivalent in that a change of origin and scale can reduce them to a standard form. We therefore consider the random variable $Z = (X - \mu)/\sigma$, for which the PDF takes the form

$$\phi(z) = \frac{1}{\sqrt{2\pi}} \exp\left(-\frac{z^2}{2}\right), \qquad (16.94)$$

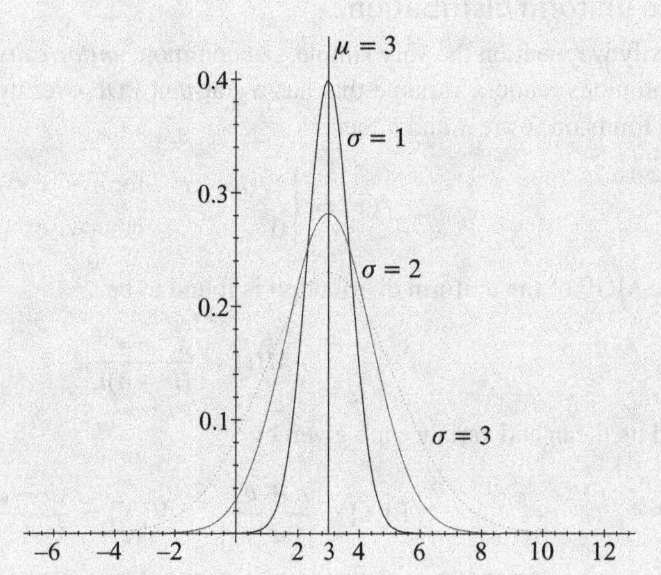

Figure 16.12 The Gaussian or normal distribution for mean $\mu = 3$ and various values of the standard deviation σ.

which is called the *standard Gaussian distribution* and has mean $\mu = 0$ and variance $\sigma^2 = 1$. The random variable Z is called the *standard variable*.

From (16.93) we can define the cumulative probability function for a Gaussian distribution as

$$F(x) = \Pr(X < x) = \frac{1}{\sigma\sqrt{2\pi}} \int_{-\infty}^{x} \exp\left[-\frac{1}{2}\left(\frac{u-\mu}{\sigma}\right)^2\right] du, \tag{16.95}$$

where u is a (dummy) integration variable. Unfortunately, this (indefinite) integral cannot be evaluated analytically. It is therefore standard practice to tabulate values of the cumulative probability function for the standard Gaussian distribution (see Figure 16.13), i.e.

$$\Phi(z) = \Pr(Z < z) = \frac{1}{\sqrt{2\pi}} \int_{-\infty}^{z} \exp\left(-\frac{u^2}{2}\right) du. \tag{16.96}$$

It is usual only to tabulate $\Phi(z)$ for $z > 0$, since it can easily be seen, from Figure 16.13 and the symmetry of the Gaussian distribution, that $\Phi(-z) = 1 - \Phi(z)$; see Table 16.3. Using such a table it is then straightforward to evaluate the probability that Z lies in a given range of z-values. For example, for a and b constant,

$$\begin{aligned} \Pr(Z < a) &= \Phi(a), \\ \Pr(Z > a) &= 1 - \Phi(a), \\ \Pr(a < Z \leq b) &= \Phi(b) - \Phi(a). \end{aligned}$$

Remembering that $Z = (X - \mu)/\sigma$ and comparing (16.95) and (16.96), we see that

$$F(x) = \Phi\left(\frac{x-\mu}{\sigma}\right),$$

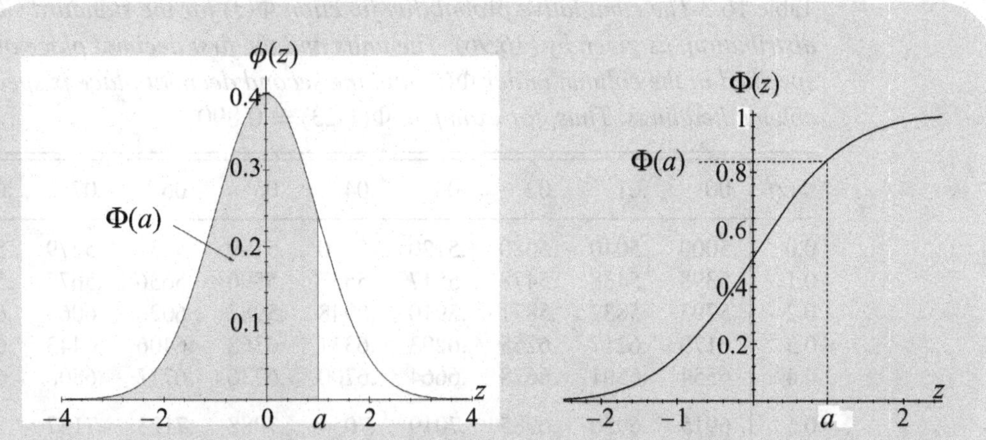

Figure 16.13 On the left, the standard Gaussian distribution $\phi(z)$; the shaded area gives $\Pr(Z < a) = \Phi(a)$. On the right, the cumulative probability function $\Phi(z)$ for a standard Gaussian distribution $\phi(z)$.

and so we may also calculate the probability that the original random variable X lies in a given x-range. For example,

$$\Pr(a < X \leq b) = \frac{1}{\sigma\sqrt{2\pi}} \int_a^b \exp\left[-\frac{1}{2}\left(\frac{u-\mu}{\sigma}\right)^2\right] du \tag{16.97}$$

$$= F(b) - F(a) \tag{16.98}$$

$$= \Phi\left(\frac{b-\mu}{\sigma}\right) - \Phi\left(\frac{a-\mu}{\sigma}\right). \tag{16.99}$$

We now derive some generally useful standard indicators of how close to its mean any individual RV value is likely to be.

Example If X is described by a Gaussian distribution of mean μ and variance σ^2, calculate the probabilities that X lies within 1σ, 2σ and 3σ of the mean.

From (16.99)

$$\Pr(\mu - n\sigma < X \leq \mu + n\sigma) = \Phi(n) - \Phi(-n) = \Phi(n) - [1 - \Phi(n)],$$

and so from Table 16.3

$$\Pr(\mu - \sigma < X \leq \mu + \sigma) = 2\Phi(1) - 1 = 0.6826 \approx 68.3\%,$$
$$\Pr(\mu - 2\sigma < X \leq \mu + 2\sigma) = 2\Phi(2) - 1 = 0.9544 \approx 95.4\%,$$
$$\Pr(\mu - 3\sigma < X \leq \mu + 3\sigma) = 2\Phi(3) - 1 = 0.9974 \approx 99.7\%.$$

Thus we expect X to be distributed in such a way that about two-thirds of the values will lie between $\mu - \sigma$ and $\mu + \sigma$, 95% will lie within 2σ of the mean and 99.7% will lie within 3σ of the mean. These limits are called the one-, two- and three-sigma limits respectively; it is particularly important to note that they are independent of the actual values of the mean and variance. ◀

Table 16.3 *The cumulative probability function* $\Phi(z)$ *for the standard Gaussian distribution, as given by (16.96). The units and the first decimal place of* z *are specified in the column under* $\Phi(z)$ *and the second decimal place is specified by the column headings. Thus, for example,* $\Phi(1.23) = 0.8907$

$\Phi(z)$	.00	.01	.02	.03	.04	.05	.06	.07	.08	.09
0.0	.5000	.5040	.5080	.5120	.5160	.5199	.5239	.5279	.5319	.5359
0.1	.5398	.5438	.5478	.5517	.5557	.5596	.5636	.5675	.5714	.5753
0.2	.5793	.5832	.5871	.5910	.5948	.5987	.6026	.6064	.6103	.6141
0.3	.6179	.6217	.6255	.6293	.6331	.6368	.6406	.6443	.6480	.6517
0.4	.6554	.6591	.6628	.6664	.6700	.6736	.6772	.6808	.6844	.6879
0.5	.6915	.6950	.6985	.7019	.7054	.7088	.7123	.7157	.7190	.7224
0.6	.7257	.7291	.7324	.7357	.7389	.7422	.7454	.7486	.7517	.7549
0.7	.7580	.7611	.7642	.7673	.7704	.7734	.7764	.7794	.7823	.7852
0.8	.7881	.7910	.7939	.7967	.7995	.8023	.8051	.8078	.8106	.8133
0.9	.8159	.8186	.8212	.8238	.8264	.8289	.8315	.8340	.8365	.8389
1.0	.8413	.8438	.8461	.8485	.8508	.8531	.8554	.8577	.8599	.8621
1.1	.8643	.8665	.8686	.8708	.8729	.8749	.8770	.8790	.8810	.8830
1.2	.8849	.8869	.8888	.8907	.8925	.8944	.8962	.8980	.8997	.9015
1.3	.9032	.9049	.9066	.9082	.9099	.9115	.9131	.9147	.9162	.9177
1.4	.9192	.9207	.9222	.9236	.9251	.9265	.9279	.9292	.9306	.9319
1.5	.9332	.9345	.9357	.9370	.9382	.9394	.9406	.9418	.9429	.9441
1.6	.9452	.9463	.9474	.9484	.9495	.9505	.9515	.9525	.9535	.9545
1.7	.9554	.9564	.9573	.9582	.9591	.9599	.9608	.9616	.9625	.9633
1.8	.9641	.9649	.9656	.9664	.9671	.9678	.9686	.9693	.9699	.9706
1.9	.9713	.9719	.9726	.9732	.9738	.9744	.9750	.9756	.9761	.9767
2.0	.9772	.9778	.9783	.9788	.9793	.9798	.9803	.9808	.9812	.9817
2.1	.9821	.9826	.9830	.9834	.9838	.9842	.9846	.9850	.9854	.9857
2.2	.9861	.9864	.9868	.9871	.9875	.9878	.9881	.9884	.9887	.9890
2.3	.9893	.9896	.9898	.9901	.9904	.9906	.9909	.9911	.9913	.9916
2.4	.9918	.9920	.9922	.9925	.9927	.9929	.9931	.9932	.9934	.9936
2.5	.9938	.9940	.9941	.9943	.9945	.9946	.9948	.9949	.9951	.9952
2.6	.9953	.9955	.9956	.9957	.9959	.9960	.9961	.9962	.9963	.9964
2.7	.9965	.9966	.9967	.9968	.9969	.9970	.9971	.9972	.9973	.9974
2.8	.9974	.9975	.9976	.9977	.9977	.9978	.9979	.9979	.9980	.9981
2.9	.9981	.9982	.9982	.9983	.9984	.9984	.9985	.9985	.9986	.9986
3.0	.9987	.9987	.9987	.9988	.9988	.9989	.9989	.9989	.9990	.9990
3.1	.9990	.9991	.9991	.9991	.9992	.9992	.9992	.9992	.9993	.9993
3.2	.9993	.9993	.9994	.9994	.9994	.9994	.9994	.9995	.9995	.9995
3.3	.9995	.9995	.9995	.9996	.9996	.9996	.9996	.9996	.9996	.9997
3.4	.9997	.9997	.9997	.9997	.9997	.9997	.9997	.9997	.9997	.9998

There are many other ways in which the Gaussian distribution may be used. We now illustrate some of the uses in more complicated examples.

Example Sawmill A produces boards whose distribution of lengths is well approximated by a Gaussian with mean 209.4 cm and standard deviation 5.0 cm. A board is accepted if it is longer than 200 cm but is rejected otherwise. Show that 3% of boards are rejected.

Sawmill B produces boards of the same standard deviation but of mean length 210.1 cm. Find the proportion of boards rejected if they are drawn at random from the outputs of A and B in the ratio 3 : 1.

Let X = length of boards from A, so that $X \sim N(209.4,\ (5.0)^2)$ and

$$\Pr(X < 200) = \Phi\left(\frac{200-\mu}{\sigma}\right) = \Phi\left(\frac{200-209.4}{5.0}\right) = \Phi(-1.88).$$

But, since $\Phi(-z) = 1 - \Phi(z)$ we have, using Table 16.3,

$$\Pr(X < 200) = 1 - \Phi(1.88) = 1 - 0.9699 = 0.0301,$$

i.e. 3.0% of boards are rejected.

Now let Y = length of boards from B, so that $Y \sim N(210.1,\ (5.0)^2)$ and

$$\begin{aligned}\Pr(Y < 200) &= \Phi\left(\frac{200-210.1}{5.0}\right) = \Phi(-2.02)\\ &= 1 - \Phi(2.02)\\ &= 1 - 0.9783 = 0.0217.\end{aligned}$$

Therefore, when taken alone, only 2.2% of boards from B are rejected. If, however, boards are drawn at random from A and B in the ratio 3 : 1 then

$$\tfrac{1}{4}(3 \times 0.030 + 1 \times 0.022) = 0.028 = 2.8$$

gives the percentage of boards rejected. ◀

We may sometimes work backwards to derive the mean and standard deviation of a population that is known to be Gaussian distributed.

Example The time taken for a computer "packet" to travel from Cambridge UK to Cambridge MA is Gaussian distributed. 6.8% of the packets take over 200 ms to make the journey, and 3.0% take under 140 ms. Find the mean and standard deviation of the distribution.

Let X = journey time in ms; we are told that $X \sim N(\mu, \sigma^2)$ where μ and σ are unknown. Since 6.8% of journey times are longer than 200 ms,

$$\Pr(X > 200) = 1 - \Phi\left(\frac{200-\mu}{\sigma}\right) = 0.068,$$

from which we find

$$\Phi\left(\frac{200-\mu}{\sigma}\right) = 1 - 0.068 = 0.932.$$

Using Table 16.3, we have therefore

$$\frac{200-\mu}{\sigma} = 1.49. \tag{16.100}$$

Also, 3.0% of journey times are under 140 ms, so

$$\Pr(X < 140) = \Phi\left(\frac{140-\mu}{\sigma}\right) = 0.030.$$

Now using $\Phi(-z) = 1 - \Phi(z)$ gives

$$\Phi\left(\frac{\mu-140}{\sigma}\right) = 1 - 0.030 = 0.970.$$

Using Table 16.3 again, we find

$$\frac{\mu-140}{\sigma} = 1.88. \tag{16.101}$$

Solving the simultaneous equations (16.100) and (16.101) gives $\mu = 173.5$ and $\sigma = 17.8$ for the mean and standard deviation, respectively. ◀

The moment generating function for the Gaussian distribution

Using the definition of the MGF (16.74),

$$\begin{aligned} M_X(t) = E\left[e^{tX}\right] &= \int_{-\infty}^{\infty} \frac{1}{\sigma\sqrt{2\pi}} \exp\left[tx - \frac{(x-\mu)^2}{2\sigma^2}\right] dx \\ &= c \exp\left(\mu t + \tfrac{1}{2}\sigma^2 t^2\right), \end{aligned}$$

where the final equality is established by completing the square in the argument of the exponential, as follows,

$$\begin{aligned} tx - \frac{(x-\mu)^2}{2\sigma^2} &= \frac{2\sigma^2 tx - x^2 + 2x\mu - \mu^2}{2\sigma^2} \\ &= \frac{-[x - \sigma^2 t - \mu]^2 + (\sigma^2 t + \mu)^2 - \mu^2}{2\sigma^2} \\ &= -\frac{[x - (\sigma^2 t + \mu)]^2}{2\sigma^2} + \frac{\sigma^4 t^2 + 2\mu\sigma^2 t}{2\sigma^2}, \end{aligned}$$

and then writing

$$c = \int_{-\infty}^{\infty} \frac{1}{\sigma\sqrt{2\pi}} \exp\left\{-\frac{[x - (\mu + \sigma^2 t)]^2}{2\sigma^2}\right\} dx.$$

However, the final integral is simply the normalization integral for the Gaussian distribution, and so $c = 1$ and the MGF is given by

$$M_X(t) = \exp\left(\mu t + \tfrac{1}{2}\sigma^2 t^2\right). \tag{16.102}$$

We showed in Subsection 16.7.2 that this MGF leads to $E[X] = \mu$ and $V[X] = \sigma^2$, as required.

Gaussian approximation to the binomial distribution

We may consider the Gaussian distribution as the limit of the binomial distribution when the number of trials $n \to \infty$ but the probability of a success p remains finite, so that $np \to \infty$ also. (This contrasts with the Poisson distribution, which corresponds to the limit $n \to \infty$ and $p \to 0$ with $np = \lambda$ remaining finite.) In other words, a Gaussian distribution results when an experiment with a finite probability of success is repeated a large number of times. We now show how this Gaussian limit arises.

The binomial probability function gives the probability of x successes in n trials as

$$f(x) = \frac{n!}{x!(n-x)!} p^x (1-p)^{n-x}.$$

Taking the limit as $n \to \infty$ (and $x \to \infty$) we may approximate the factorials by Stirling's approximation

$$n! \sim \sqrt{2\pi n} \left(\frac{n}{e}\right)^n$$

and obtain

$$\begin{aligned} f(x) &\approx \frac{1}{\sqrt{2\pi n}} \left(\frac{x}{n}\right)^{-x-1/2} \left(\frac{n-x}{n}\right)^{-n+x-1/2} p^x (1-p)^{n-x} \\ &= \frac{1}{\sqrt{2\pi n}} \exp\left[-\left(x+\tfrac{1}{2}\right) \ln\frac{x}{n} - \left(n-x+\tfrac{1}{2}\right) \ln\frac{n-x}{n} \right. \\ &\qquad\qquad \left. + x \ln p + (n-x)\ln(1-p) \right]. \end{aligned}$$

By expanding the argument of the exponential in terms of $y = x - np$, where $1 \ll y \ll np$ and keeping only the dominant terms, it can be shown that

$$f(x) \approx \frac{1}{\sqrt{2\pi n}} \frac{1}{\sqrt{p(1-p)}} \exp\left[-\frac{1}{2} \frac{(x-np)^2}{np(1-p)} \right],$$

which is of Gaussian form with $\mu = np$ and $\sigma = \sqrt{np(1-p)}$.

Thus we see that the *value* of the Gaussian *probability density function* $f(x)$ is a good approximation to the *probability* of obtaining x successes in n trials. This approximation is actually very good even for relatively small n. For example, if $n = 10$ and $p = 0.6$ then the Gaussian approximation to the binomial distribution is (16.93) with $\mu = 10 \times 0.6 = 6$ and $\sigma = \sqrt{10 \times 0.6(1-0.6)} = 1.549$. The probability functions $f(x)$ for the binomial and associated Gaussian distributions for these parameters are given in Table 16.4, and it can be seen that the Gaussian approximation is a good one.[22]

Strictly speaking, however, since the Gaussian distribution is continuous and the binomial distribution is discrete, we should use the integral of $f(x)$ for the Gaussian distribution in the calculation of approximate binomial probabilities. More specifically, we should apply a *continuity correction* so that the discrete integer x in the binomial distribution

22 What are the chances of throwing 60 or more heads in 100 tosses of an unbiased coin?

Table 16.4 *Comparison of the binomial distribution for* $n = 10$ *and* $p = 0.6$ *with its Gaussian approximation*

x	$f(x)$ (binomial)	$f(x)$ (Gaussian)
0	0.0001	0.0001
1	0.0016	0.0014
2	0.0106	0.0092
3	0.0425	0.0395
4	0.1115	0.1119
5	0.2007	0.2091
6	0.2508	0.2575
7	0.2150	0.2091
8	0.1209	0.1119
9	0.0403	0.0395
10	0.0060	0.0092

becomes the interval $[x - 0.5,\ x + 0.5]$ in the Gaussian distribution. Explicitly,

$$\Pr(X = x) \approx \frac{1}{\sigma\sqrt{2\pi}} \int_{x-0.5}^{x+0.5} \exp\left[-\frac{1}{2}\left(\frac{u-\mu}{\sigma}\right)^2\right] du.$$

The Gaussian approximation is particularly useful for estimating the binomial probability that X lies between the (integer) values x_1 and x_2,

$$\Pr(x_1 < X \le x_2) \approx \frac{1}{\sigma\sqrt{2\pi}} \int_{x_1-0.5}^{x_2+0.5} \exp\left[-\frac{1}{2}\left(\frac{u-\mu}{\sigma}\right)^2\right] du,$$

as we now illustrate with a worked example.

Example A manufacturer makes computer chips of which 10% are defective. For a random sample of 200 chips, find the approximate probability that more than 15 are defective.

We first define the random variable

$$X = \text{number of defective chips in the sample},$$

which has a binomial distribution $X \sim \text{Bin}(200, 0.1)$. Therefore, the mean and variance of this distribution are

$$E[X] = 200 \times 0.1 = 20 \qquad \text{and} \qquad V[X] = 200 \times 0.1 \times (1 - 0.1) = 18,$$

and we may approximate the binomial distribution with a Gaussian distribution such that $X \sim N(20, 18)$. The standard variable is

$$Z = \frac{X - 20}{\sqrt{18}},$$

and so, using $X = 15.5$ to allow for the continuity correction,

$$\Pr(X > 15.5) = \Pr\left(Z > \frac{15.5 - 20}{\sqrt{18}}\right) = \Pr(Z > -1.06)$$
$$= \Pr(Z < 1.06) = 0.86.$$

The final probability in this calculation was obtained from Table 16.3. ◀

Gaussian approximation to the Poisson distribution

We first met the Poisson distribution as the limit of the binomial distribution for $n \to \infty$ and $p \to 0$, taken in such a way that $np = \lambda$ remains finite. Further, in the previous subsection, we considered the Gaussian distribution as the limit of the binomial distribution when $n \to \infty$ but p remains finite, so that $np \to \infty$ also. It should come as no surprise, therefore, that the Gaussian distribution can also be used to approximate the Poisson distribution when the mean λ becomes large. The probability function for the Poisson distribution is

$$f(x) = e^{-\lambda}\frac{\lambda^x}{x!},$$

which, on taking the logarithm of both sides, gives

$$\ln f(x) = -\lambda + x\ln\lambda - \ln x!. \qquad (16.103)$$

Stirling's approximation for large x gives

$$x! \approx \sqrt{2\pi x}\left(\frac{x}{e}\right)^x$$

implying that

$$\ln x! \approx \ln\sqrt{2\pi x} + x\ln x - x,$$

which, on substituting into (16.103), yields

$$\ln f(x) \approx -\lambda + x\ln\lambda - (x\ln x - x) - \ln\sqrt{2\pi x}.$$

Since we expect the Poisson distribution to peak around $x = \lambda$, we substitute $\epsilon = x - \lambda$ to obtain

$$\ln f(x) \approx -\lambda + (\lambda + \epsilon)\left\{\ln\lambda - \ln\left[\lambda\left(1 + \frac{\epsilon}{\lambda}\right)\right]\right\} + (\lambda + \epsilon) - \ln\sqrt{2\pi(\lambda + \epsilon)}.$$

Using the expansion $\ln(1 + z) = z - z^2/2 + \cdots$, we find

$$\ln f(x) \approx \epsilon - (\lambda + \epsilon)\left(\frac{\epsilon}{\lambda} - \frac{\epsilon^2}{2\lambda^2}\right) - \ln\sqrt{2\pi\lambda} - \frac{1}{2}\left(\frac{\epsilon}{\lambda} - \frac{\epsilon^2}{2\lambda^2}\right)$$
$$\approx -\frac{\epsilon^2}{2\lambda} - \ln\sqrt{2\pi\lambda},$$

when only the dominant terms are retained, after using the fact that ϵ is of the order of the standard deviation of x, i.e. of order $\lambda^{1/2}$. On exponentiating this result we obtain

$$f(x) \approx \frac{1}{\sqrt{2\pi\lambda}} \exp\left[-\frac{(x-\lambda)^2}{2\lambda}\right],$$

which is the Gaussian distribution with $\mu = \lambda$ and $\sigma^2 = \lambda$.

The larger the value of λ, the better is the Gaussian approximation to the Poisson distribution; the approximation is reasonable even for $\lambda = 5$, but $\lambda \geq 10$ is safer. As in the case of the Gaussian approximation to the binomial distribution, a continuity correction is necessary since the Poisson distribution is discrete. The next worked example incorporates this correction.

Example E-mail messages are received by an author at an average rate of one per hour. Find the probability that in a day the author receives 24 messages or more.

We first define the random variable

$$X = \text{number of messages received in a day.}$$

Thus $E[X] = 1 \times 24 = 24$, and so $X \sim \text{Po}(24)$. Since $\lambda > 10$ we may safely approximate the Poisson distribution by $X \sim \text{N}(24, 24)$. Now the standard variable is

$$Z = \frac{X - 24}{\sqrt{24}},$$

and, using the continuity correction, we find

$$\begin{aligned}\Pr(X > 23.5) &= \Pr\left(Z > \frac{23.5 - 24}{\sqrt{24}}\right)\\ &= \Pr(Z > -0.102) = \Pr(Z < 0.102) = 0.54.\end{aligned}$$

Again, the final probability was taken from Table 16.3. ◀

In fact, almost all probability distributions tend towards a Gaussian when the numbers involved become large – that this should happen is required by the central limit theorem, which we discuss in Section 16.10.

Multiple Gaussian distributions

Suppose X and Y are *independent* Gaussian-distributed random variables, so that $X \sim N(\mu_1, \sigma_1^2)$ and $Y \sim N(\mu_2, \sigma_2^2)$.

Let us now consider the random variable $Z = X + Y$. The PDF for this random variable may be found directly using (16.53), but it is easier to use the MGF. From (16.102), the MGFs of X and Y are

$$M_X(t) = \exp\left(\mu_1 t + \tfrac{1}{2}\sigma_1^2 t^2\right), \qquad M_Y(t) = \exp\left(\mu_2 t + \tfrac{1}{2}\sigma_2^2 t^2\right).$$

Using (16.78), since X and Y are independent RVs, the MGF of $Z = X + Y$ is simply the product of $M_X(t)$ and $M_Y(t)$. Thus, we have

$$M_Z(t) = M_X(t)M_Y(t) = \exp\left(\mu_1 t + \tfrac{1}{2}\sigma_1^2 t^2\right)\exp\left(\mu_2 t + \tfrac{1}{2}\sigma_2^2 t^2\right)$$
$$= \exp\left[(\mu_1 + \mu_2)t + \tfrac{1}{2}(\sigma_1^2 + \sigma_2^2)t^2\right],$$

which we recognize as the MGF for a Gaussian with mean $\mu_1 + \mu_2$ and variance $\sigma_1^2 + \sigma_2^2$. Thus, Z is also Gaussian distributed: $Z \sim N(\mu_1 + \mu_2,\ \sigma_1^2 + \sigma_2^2)$. One obvious extension of this is to the sum, and hence the arithmetic average, of n RVs all drawn from the same population; if the population is $\sim N(\mu,\ \sigma^2)$, then the average is $\sim N(\mu,\ \sigma^2/n)$.

A similar calculation may be performed to calculate the PDF of the random variable $W = X - Y$. If we introduce the variable $\tilde{Y} = -Y$ then $W = X + \tilde{Y}$, where $\tilde{Y} \sim N(-\mu_1,\ \sigma_1^2)$. Thus, using the result above, we find $W \sim N(\mu_1 - \mu_2,\ \sigma_1^2 + \sigma_2^2)$.

The following example makes no allowance for "(the wrong sort of) leaves on the line" or for flat tires!

Example An executive travels home from her office every evening. Her journey consists of a train ride, followed by a bicycle ride. The time spent on the train is Gaussian distributed with mean 52 minutes and standard deviation 1.8 minutes, while the time for the bicycle journey is Gaussian distributed with mean 8 minutes and standard deviation 2.6 minutes. Assuming these two factors are independent, estimate the percentage of occasions on which the *whole* journey takes more than 65 minutes.

We first define the random variables

$$X = \text{time spent on train}, \qquad Y = \text{time spent on bicycle},$$

so that $X \sim N(52, (1.8)^2)$ and $Y \sim N(8, (2.6)^2)$. Since X and Y are independent, the total journey time $T = X + Y$ is distributed as

$$T \sim N(52 + 8,\ (1.8)^2 + (2.6)^2) = N(60, (3.16)^2).$$

The standard variable is thus

$$Z = \frac{T - 60}{3.16},$$

and the required probability is given by

$$\Pr(T > 65) = \Pr\left(Z > \frac{65 - 60}{3.16}\right) = \Pr(Z > 1.58) = 1 - 0.943 = 0.057.$$

Thus the total journey time exceeds 65 minutes on 5.7% of occasions. ◀

The above results may be extended. For example, if the random variables X_i, $i = 1, 2, \ldots, n$, are distributed as $X_i \sim N(\mu_i, \sigma_i^2)$ then the random variable $Z = \sum_i c_i X_i$ (where the c_i are constants) is distributed as $Z \sim N(\sum_i c_i\mu_i, \sum_i c_i^2\sigma_i^2)$.

16.9.3 The exponential and gamma distributions

The exponential distribution with positive parameter λ is given by

$$f(x) = \begin{cases} \lambda e^{-\lambda x} & \text{for } x > 0, \\ 0 & \text{for } x \leq 0 \end{cases} \tag{16.104}$$

and satisfies $\int_{-\infty}^{\infty} f(x)\,dx = 1$ as required. The exponential distribution occurs naturally if we consider the distribution of the length of intervals between successive events in a Poisson process or, equivalently, the distribution of the interval (i.e. the waiting time) before the first event. If the average number of events per unit interval is λ then on average there are λx events in interval x, so that from the Poisson distribution the probability that there will be no events in this interval is given by

$$\Pr(\text{no events in interval } x) = e^{-\lambda x}.$$

The probability that an event occurs in the next infinitesimal interval $[x,\ x + dx]$ is given by $\lambda\,dx$, so that

$$\Pr(\text{the first event occurs in interval } [x, x + dx]) = e^{-\lambda x}\lambda\,dx.$$

Hence the required probability density function is given by

$$f(x) = \lambda e^{-\lambda x}.$$

The expectation and variance of the exponential distribution can be evaluated as $1/\lambda$ and $(1/\lambda)^2$ respectively. The MGF is given by

$$M(t) = \frac{\lambda}{\lambda - t}. \tag{16.105}$$

We may generalize the above discussion to obtain the PDF for the interval between every rth event in a Poisson process or, equivalently, the interval (waiting time) before the rth event. We begin by using the Poisson distribution to give

$$\Pr(r - 1 \text{ events occur in interval } x) = e^{-\lambda x}\frac{(\lambda x)^{r-1}}{(r-1)!},$$

from which we obtain

$$\Pr(r\text{th event occurs in the interval } [x, x + dx]) = e^{-\lambda x}\frac{(\lambda x)^{r-1}}{(r-1)!}\lambda\,dx.$$

Thus the required PDF is

$$f(x) = \frac{\lambda}{(r-1)!}(\lambda x)^{r-1}e^{-\lambda x}. \tag{16.106}$$

This is known as the *gamma distribution* of order r with parameter λ. Although our derivation applies only when r is a positive integer, the gamma distribution is defined for all positive r by replacing $(r-1)!$ by $\Gamma(r)$ in (16.106); see Section 9.10 for a discussion of the gamma function $\Gamma(x)$. If a random variable X is described by a gamma distribution of order r with parameter λ, we write $X \sim \gamma(\lambda, r)$; we note that the exponential distribution is the special case $\gamma(\lambda, 1)$. The gamma distribution $\gamma(\lambda, r)$ is plotted in Figure 16.14

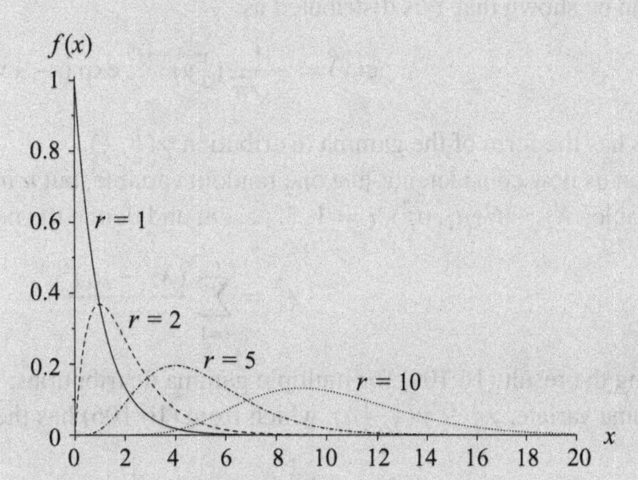

Figure 16.14 The PDF $f(x)$ for the gamma distributions $\gamma(\lambda, r)$ with $\lambda = 1$ and $r = 1, 2, 5, 10$.

for $\lambda = 1$ and $r = 1, 2, 5, 10$. For large r, the gamma distribution tends to the Gaussian distribution whose mean and variance are specified by (16.108) below.

The MGF for the gamma distribution is obtained from that for the exponential distribution, by noting that we may consider the interval between every rth event in a Poisson process as the sum of r intervals between successive events. Thus the rth-order gamma variate is the sum of r independent exponentially distributed random variables. From (16.105) and (16.79), the MGF of the gamma distribution is therefore given by

$$M(t) = \left(\frac{\lambda}{\lambda - t}\right)^r, \tag{16.107}$$

from which the mean and variance are found to be

$$E[X] = \frac{r}{\lambda}, \qquad V[X] = \frac{r}{\lambda^2}. \tag{16.108}$$

We may also use the above MGF to prove another useful theorem regarding multiple gamma distributions. If $X_i \sim \gamma(\lambda, r_i)$, $i = 1, 2, \ldots, n$, are independent gamma variates then the random variable $Y = X_1 + X_2 + \cdots + X_n$ has MGF

$$M(t) = \prod_{i=1}^{n} \left(\frac{\lambda}{\lambda - t}\right)^{r_i} = \left(\frac{\lambda}{\lambda - t}\right)^{r_1 + r_2 + \cdots + r_n}. \tag{16.109}$$

Thus Y is also a gamma variate, distributed as $Y \sim \gamma(\lambda, r_1 + r_2 + \cdots + r_n)$.

16.9.4 The chi-squared distribution

In Subsection 16.6.1 we considered only functions of random variables that had single-valued inverses. Functions that have multiple-valued inverses can be treated, but only by identifying the inverses individually. One such derived random variable is $Y = (X - \mu)^2/\sigma^2$ when X is Gaussian distributed with mean μ and variance σ^2, i.e. $X \sim N(\mu, \sigma^2)$.

It can be shown that Y is distributed as

$$g(y) = \frac{1}{2\sqrt{\pi}}\left(\tfrac{1}{2}y\right)^{-1/2}\exp\left(-\tfrac{1}{2}y\right).$$

This has the form of the gamma distribution $\gamma(\frac{1}{2}, \frac{1}{2})$.

Let us now consider, not just one random variable, but n independent Gaussian random variables $X_i \sim N(\mu_i, \sigma_i^2)$, $i = 1, 2, \ldots, n$, and define the new variable

$$\chi_n^2 = \sum_{i=1}^{n} \frac{(X_i - \mu_i)^2}{\sigma_i^2}. \tag{16.110}$$

Using the result (16.109) for multiple gamma distributions, χ_n^2 must be distributed as the gamma variate, $\chi_n^2 \sim \gamma(\frac{1}{2}, \frac{1}{2}n)$, which from (16.106) has the PDF

$$\begin{aligned} f(\chi_n^2) &= \frac{\frac{1}{2}}{\Gamma\left(\frac{1}{2}n\right)}\left(\tfrac{1}{2}\chi_n^2\right)^{(n/2)-1}\exp\left(-\tfrac{1}{2}\chi_n^2\right) \\ &= \frac{1}{2^{n/2}\Gamma\left(\frac{1}{2}n\right)}\left(\chi_n^2\right)^{(n/2)-1}\exp\left(-\tfrac{1}{2}\chi_n^2\right). \end{aligned} \tag{16.111}$$

This is known as the *chi-squared distribution* of order n and has numerous applications in statistics (see Chapter 17). Setting $\lambda = \frac{1}{2}$ and $r = \frac{1}{2}n$ in (16.108), we find that

$$E[\chi_n^2] = n, \qquad V[\chi_n^2] = 2n.$$

An important generalization occurs when the n Gaussian variables X_i are *not* linearly independent but are instead required to satisfy a linear constraint of the form

$$c_1 X_1 + c_2 X_2 + \cdots + c_n X_n = 0, \tag{16.112}$$

in which the constants c_i are not all zero. In this case, it may be shown that the variable χ_n^2 defined in (16.110) is still described by a chi-squared distribution, but one of order $n - 1$. Indeed, this result may be extended to show that if the n Gaussian variables X_i satisfy m linear constraints of the form (16.112) then the variable χ_n^2 defined in (16.110) is described by a chi-squared distribution of order $n - m$.

16.9.5 The Cauchy and Breit–Wigner distributions

A random variable X (in the range $-\infty$ to ∞) that obeys the *Cauchy distribution* is described by the PDF

$$f(x) = \frac{1}{\pi}\frac{1}{1+x^2}.$$

This is a special case of the *Breit–Wigner distribution*

$$f(x) = \frac{1}{\pi}\frac{\frac{1}{2}\Gamma}{\frac{1}{4}\Gamma^2 + (x - x_0)^2},$$

which is encountered in the study of nuclear and particle physics. In Figure 16.15, we plot some examples of the Breit–Wigner distribution for several values of the parameters x_0 and Γ.

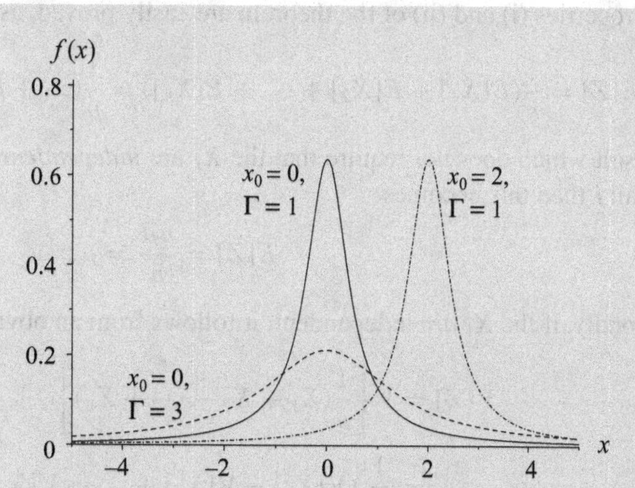

Figure 16.15 The PDF $f(x)$ for the Breit–Wigner distribution for different values of the parameters x_0 and Γ.

We see from the figure that the peak (or mode) of the distribution occurs at $x = x_0$. It is also straightforward to show that the parameter Γ is equal to the width of the peak at half the maximum height. Although the Breit–Wigner distribution is symmetric about its peak, it does not formally possess a mean since the integrals $\int_{-\infty}^{0} xf(x)\,dx$ and $\int_{0}^{\infty} xf(x)\,dx$ both diverge. Similar divergences occur for all higher moments of the distribution.

16.10 The central limit theorem

In Subsection 16.9.2 we discussed approximating the binomial and Poisson distributions by the Gaussian distribution when the number of trials is large. We now discuss why the Gaussian distribution is so common and therefore so important. The *central limit theorem* may be stated as follows.

Central limit theorem. *Suppose that X_i, $i = 1, 2, \ldots, n$, are independent random variables, each of which is described by a probability density function $f_i(x)$ (these may all be different) with a mean μ_i and a variance σ_i^2. The random variable $Z = \left(\sum_i X_i\right)/n$, i.e. the "mean" of the X_i, has the following properties:*

(i) *its expectation value is given by $E[Z] = \left(\sum_i \mu_i\right)/n$;*
(ii) *its variance is given by $V[Z] = \left(\sum_i \sigma_i^2\right)/n^2$;*
(iii) *as $n \to \infty$ the probability function of Z tends to a Gaussian with corresponding mean and variance.*

We note that for the theorem to hold, the probability density functions $f_i(x)$ must possess formal means and variances. Thus, for example, if any of the X_i were described by a Cauchy distribution then the theorem would not apply.

Properties (i) and (ii) of the theorem are easily proved, as follows. Firstly

$$E[Z] = \frac{1}{n}(E[X_1] + E[X_2] + \cdots + E[X_n]) = \frac{1}{n}(\mu_1 + \mu_2 + \cdots + \mu_n) = \frac{\sum_i \mu_i}{n},$$

a result which does *not* require that the X_i are *independent* random variables. If $\mu_i = \mu$ for all i then this becomes

$$E[Z] = \frac{n\mu}{n} = \mu.$$

Secondly, if the X_i *are* independent, it follows from an obvious extension of (16.57) that

$$V[Z] = V\left[\frac{1}{n}(X_1 + X_2 + \cdots + X_n)\right]$$
$$= \frac{1}{n^2}(V[X_1] + V[X_2] + \cdots + V[X_n]) = \frac{\sum_i \sigma_i^2}{n^2}.$$

Let us now consider property (iii), which is the reason for the ubiquity of the Gaussian distribution and is most easily proved by considering the moment generating function $M_Z(t)$ of Z. From (16.79), this MGF is given by

$$M_Z(t) = \prod_{i=1}^{n} M_{X_i}\left(\frac{t}{n}\right),$$

where $M_{X_i}(t)$ is the MGF of $f_i(x)$. Now

$$M_{X_i}\left(\frac{t}{n}\right) = 1 + \frac{t}{n}E[X_i] + \frac{1}{2}\frac{t^2}{n^2}E[X_i^2] + \cdots$$
$$= 1 + \mu_i\frac{t}{n} + \frac{1}{2}(\sigma_i^2 + \mu_i^2)\frac{t^2}{n^2} + \cdots,$$

and as n becomes large

$$M_{X_i}\left(\frac{t}{n}\right) \approx \exp\left(\frac{\mu_i t}{n} + \frac{1}{2}\sigma_i^2\frac{t^2}{n^2}\right),$$

as may be verified by expanding the exponential up to terms including $(t/n)^2$. Therefore

$$M_Z(t) \approx \prod_{i=1}^{n} \exp\left(\frac{\mu_i t}{n} + \frac{1}{2}\sigma_i^2\frac{t^2}{n^2}\right) = \exp\left(\frac{\sum_i \mu_i}{n}t + \frac{1}{2}\frac{\sum_i \sigma_i^2}{n^2}t^2\right).$$

Comparing this with the form of the MGF for a Gaussian distribution, (16.102), we can see that the probability density function $g(z)$ of Z tends to a Gaussian distribution with mean $\sum_i \mu_i/n$ and variance $\sum_i \sigma_i^2/n^2$. We emphasize that this result does not require the distributions of the individual X_i to be Gaussian, but merely that they have formal means and variances.

For the particular case in which we consider Z to be the mean of n *independent* measurements of the *same* random variable X (so that $X_i = X$ for $i = 1, 2, \ldots, n$) we have that, as $n \to \infty$, Z has a Gaussian distribution with mean μ and variance σ^2/n.

16.11 Joint distributions

As mentioned briefly in Subsection 16.4.3, it is common in the physical sciences to consider simultaneously two or more random variables that are not independent, in general, and are thus described by *joint probability density functions*. We will return to the subject of the interdependence of random variables after first presenting some of the general ways of characterizing joint distributions. We will concentrate mainly on *bivariate* distributions, i.e. distributions of only two random variables, though the results may be extended readily to multivariate distributions.

The subject of multivariate distributions is large and a detailed study is beyond the scope of this book; in particular, we do not have the space to discuss generating functions for joint distributions or how they behave under a transformation of their variables. Another important topic that will have to be omitted is the multivariate Gaussian distribution. The interested reader will therefore need to consult one of the many specialised texts on these subjects.

The first thing to note when dealing with bivariate distributions is that the distinction between discrete and continuous distributions may not be as clear as for the single variable case; the random variables can both be discrete, or both continuous, or one discrete and the other continuous. In general, for the random variables X and Y, the joint distribution will take an infinite number of values unless both X and Y have only a finite number of values. In this chapter we will consider only the cases where X and Y are either both discrete or both continuous random variables.

16.11.1 Discrete bivariate distributions

In direct analogy with the one-variable (univariate) case, if X is a discrete random variable that takes the values $\{x_i\}$ and Y one that takes the values $\{y_j\}$ then the probability function of the joint distribution is defined as

$$f(x, y) = \begin{cases} \Pr(X = x_i,\ Y = y_j) & \text{for } x = x_i,\ y = y_j, \\ 0 & \text{otherwise.} \end{cases}$$

We may therefore think of $f(x, y)$ as a set of spikes at valid points in the xy-plane, whose height at (x_i, y_i) represents the probability of obtaining $X = x_i$ and $Y = y_j$. The normalization of $f(x, y)$ implies

$$\sum_i \sum_j f(x_i, y_j) = 1, \tag{16.113}$$

where the sums over i and j take all valid pairs of values. We can also define the cumulative probability function

$$F(x, y) = \sum_{x_i \leq x} \sum_{y_j \leq y} f(x_i, y_j), \tag{16.114}$$

from which it follows that the probability that X lies in the range $[a_1, a_2]$ and Y lies in the range $[b_1, b_2]$ is given by

$$\Pr(a_1 < X \leq a_2,\ b_1 < Y \leq b_2) = F(a_2, b_2) - F(a_1, b_2) - F(a_2, b_1) + F(a_1, b_1).$$

Finally, we define X and Y to be *independent* if we can write their joint distribution in the form

$$f(x, y) = f_X(x) f_Y(y), \qquad (16.115)$$

i.e. as the product of two univariate distributions.

16.11.2 Continuous bivariate distributions

In the case where both X and Y are continuous random variables, the PDF of the joint distribution is defined by

$$f(x, y)\, dx\, dy = \Pr(x < X \le x + dx,\ y < Y \le y + dy), \qquad (16.116)$$

so $f(x, y)\, dx\, dy$ is the probability that x lies in the range $[x, x + dx]$ and y lies in the range $[y, y + dy]$. It is clear that the two-dimensional function $f(x, y)$ must be everywhere non-negative and that normalization requires

$$\int_{-\infty}^{\infty} \int_{-\infty}^{\infty} f(x, y)\, dx\, dy = 1.$$

It follows further that

$$\Pr(a_1 < X \le a_2,\ b_1 < Y \le b_2) = \int_{b_1}^{b_2} \int_{a_1}^{a_2} f(x, y)\, dx\, dy. \qquad (16.117)$$

We can also define the cumulative probability function by

$$F(x, y) = \Pr(X \le x,\ Y \le y) = \int_{-\infty}^{x} \int_{-\infty}^{y} f(u, v)\, du\, dv,$$

from which we see that (as for the discrete case),

$$\Pr(a_1 < X \le a_2,\ b_1 < Y \le b_2) = F(a_2, b_2) - F(a_1, b_2) - F(a_2, b_1) + F(a_1, b_1).$$

Finally we note that the definition of independence (16.115) for discrete bivariate distributions also applies to continuous bivariate distributions.

Two random variables are needed to describe the outcome of any one trial in the simple experiment analyzed in the following worked example.

Example A flat table is ruled with parallel straight lines a distance D apart, and a thin needle of length $l < D$ is tossed onto the table at random. What is the probability that the needle will cross a line?

Let θ be the angle that the needle makes with the lines, and let x be the distance from the center of the needle to the nearest line. Since the needle is tossed "at random" onto the table, the angle θ is uniformly distributed in the interval $[0, \pi]$, and the distance x is uniformly distributed in the interval $[0, D/2]$. Assuming that θ and x are independent, their joint distribution is just the product of their individual distributions, and is given by

$$f(\theta, x) = \frac{1}{\pi} \frac{1}{D/2} = \frac{2}{\pi D}.$$

The needle will cross a line if the distance x of its center from that line is less than $\frac{1}{2} l \sin\theta$. Thus the required probability is

$$\frac{2}{\pi D}\int_0^{\pi}\int_0^{\frac{1}{2}l\sin\theta} dx\, d\theta = \frac{2}{\pi D}\frac{l}{2}\int_0^{\pi}\sin\theta\, d\theta = \frac{2l}{\pi D}.$$

This gives an experimental (but cumbersome) method of determining π. ◀

16.11.3 Marginal and conditional distributions

Given a bivariate distribution $f(x,y)$, we may be interested only in the probability function for X *irrespective of the value of Y* (or vice versa). This *marginal* distribution of X is obtained by summing or integrating, as appropriate, the joint probability distribution over all allowed values of Y. Thus, the marginal distribution of X (for example) is given by

$$f_X(x) = \begin{cases} \sum_j f(x, y_j) & \text{for a discrete distribution,} \\ \int f(x,y)\, dy & \text{for a continuous distribution.} \end{cases} \tag{16.118}$$

It is clear that an analogous definition exists for the marginal distribution of Y.

Alternatively, one might be interested in the probability function of X *given that Y takes some specific value of $Y = y_0$*, i.e. $\Pr(X = x|Y = y_0)$. This *conditional* distribution of X is given by

$$g(x) = \frac{f(x, y_0)}{f_Y(y_0)},$$

where $f_Y(y)$ is the marginal distribution of Y. The division by $f_Y(y_0)$ is necessary in order that $g(x)$ is properly normalized.

16.12 Properties of joint distributions

The probability density function $f(x,y)$ contains all the information on the joint probability distribution of two random variables X and Y. In a similar manner to that presented for univariate distributions, however, it is conventional to characterize $f(x,y)$ by certain of its properties, which we now discuss. Once again, most of these properties are based on the concept of expectation values, which are defined for joint distributions in an analogous way to those for single-variable distributions (16.46). Thus, the expectation value of any function $g(X,Y)$ of the random variables X and Y is given by

$$E[g(X,Y)] = \begin{cases} \sum_i \sum_j g(x_i, y_j) f(x_i, y_j) & \text{for the discrete case,} \\ \int_{-\infty}^{\infty}\int_{-\infty}^{\infty} g(x,y) f(x,y)\, dx\, dy & \text{for the continuous case.} \end{cases}$$

16.12.1 Means

The means of X and Y are defined respectively as the expectation values of the variables X and Y. Thus, the mean of X is given by

$$E[X] = \mu_X = \begin{cases} \sum_i \sum_j x_i f(x_i, y_j) & \text{for the discrete case,} \\ \int_{-\infty}^{\infty} \int_{-\infty}^{\infty} x f(x, y)\, dx\, dy & \text{for the continuous case.} \end{cases} \tag{16.119}$$

$E[Y]$ is obtained in a similar manner and a simple, but not unexpected, result involving both is now proved.

Example Show that if X and Y are independent random variables then $E[XY] = E[X]E[Y]$.

Let us consider the case where X and Y are continuous random variables. Since X and Y are independent $f(x, y) = f_X(x) f_Y(y)$, so that

$$E[XY] = \int_{-\infty}^{\infty} \int_{-\infty}^{\infty} xy f_X(x) f_Y(y)\, dx\, dy = \int_{-\infty}^{\infty} x f_X(x)\, dx \int_{-\infty}^{\infty} y f_Y(y)\, dy = E[X]E[Y].$$

An analogous proof exists for the discrete case. ◀

16.12.2 Variances

The definitions of the variances of X and Y are analogous to those for the single-variable case (16.48), i.e. the variance of X is given by

$$V[X] = \sigma_X^2 = \begin{cases} \sum_i \sum_j (x_i - \mu_X)^2 f(x_i, y_j) & \text{for the discrete case,} \\ \int_{-\infty}^{\infty} \int_{-\infty}^{\infty} (x - \mu_X)^2 f(x, y)\, dx\, dy & \text{for the continuous case.} \end{cases} \tag{16.120}$$

Equivalent definitions exist for the variance of Y.

16.12.3 Covariance and correlation

Means and variances of joint distributions provide useful information about their marginal distributions, but we have not yet given any indication of how to measure the relationship between the two random variables. Of course, it may be that the two random variables are independent, but often this is not so. For example, if we measure the heights and weights of a sample of people we would not be surprised to find a tendency for tall people to be heavier than short people and vice versa. We will show in this section that two functions, the *covariance* and the *correlation*, can be defined for a bivariate distribution and that these are useful in characterizing the relationship between the two random variables.

The *covariance* of two random variables X and Y is defined by

$$\text{Cov}[X, Y] = E[(X - \mu_X)(Y - \mu_Y)], \tag{16.121}$$

where μ_X and μ_Y are the expectation values of X and Y respectively. Clearly related to the covariance is the *correlation* of the two random variables, defined by

$$\text{Corr}[X, Y] = \frac{\text{Cov}[X, Y]}{\sigma_X \sigma_Y}, \tag{16.122}$$

where σ_X and σ_Y are the standard deviations of X and Y respectively. It can be shown that the correlation function lies between -1 and $+1$. If the value assumed is negative, X and Y are said to be *negatively correlated*, if it is positive they are said to be *positively correlated* and if it is zero they are said to be *uncorrelated*. We will now justify the use of these terms.

One particularly useful consequence of its definition is that the covariance of two *independent* variables, X and Y, is zero. It immediately follows from (16.122) that their correlation is also zero, and this justifies the use of the term "uncorrelated" for two such variables. To show this extremely important property we first note that

$$\begin{aligned}\text{Cov}[X, Y] &= E[(X - \mu_X)(Y - \mu_Y)] \\ &= E[XY - \mu_{XY} - \mu_{YX} + \mu_X \mu_Y] \\ &= E[XY] - \mu_{XE}[Y] - \mu_{YE}[X] + \mu_X \mu_Y \\ &= E[XY] - \mu_X \mu_Y. \end{aligned} \tag{16.123}$$

Now, if X and Y are independent then $E[XY] = E[X]E[Y] = \mu_X \mu_Y$ and so $\text{Cov}[X,Y] = 0$.

It is important to note that the converse of this result is not necessarily true; two variables dependent on each other can still be uncorrelated. In other words, it is possible (and not uncommon) for two variables X and Y to be described by a joint distribution $f(x, y)$ that *cannot* be factorized into a product of the form $g(x)h(y)$, but for which $\text{Corr}[X, Y] = 0$. Indeed, from the definition (16.121), we see that for any joint distribution $f(x, y)$ that is symmetric in x about μ_X (or similarly in y) we have $\text{Corr}[X, Y] = 0$.

We have already asserted that if the correlation of two random variables is positive (negative) they are said to be positively (negatively) correlated. We have also stated that the correlation lies between -1 and $+1$. The terminology suggests that if the two RVs are identical (i.e. $X = Y$) then they are completely correlated and that their correlation should be $+1$. Likewise, if $X = -Y$ then the functions are completely anticorrelated and their correlation should be -1. Values of the correlation function between these extremes show the existence of some degree of correlation. In fact it is not necessary that $X = Y$ for $\text{Corr}[X, Y] = 1$; it is sufficient that Y is a linear function of X, i.e. $Y = aX + b$ (with a positive). If a is negative then $\text{Corr}[X, Y] = -1$. To show this we first note that $\mu_Y = a\mu_X + b$. Now

$$Y = aX + b = aX + \mu_Y - a\mu_X \quad \Rightarrow \quad Y - \mu_Y = a(X - \mu_X),$$

and so using the definition of the covariance (16.121)

$$\text{Cov}[X, Y] = aE[(X - \mu_X)^2] = a\sigma_X^2.$$

It follows from the properties of the variance (Subsection 16.5.3) that $\sigma_Y = |a|\sigma_X$ and so, using the definition (16.122) of the correlation,

$$\text{Corr}[X, Y] = \frac{a\sigma_X^2}{|a|\sigma_X^2} = \frac{a}{|a|},$$

which is the stated result.

It should be noted that, even if the possibilities of X and Y being non-zero are mutually exclusive, $\mathrm{Corr}[X, Y]$ need not have value ± 1.

The biased die of several previous worked examples is now used to demonstrate these definitions in action.

Example A biased die gives probabilities $\frac{1}{2}p, p, p, p, p, 2p$ of throwing 1, 2, 3, 4, 5, 6 respectively. If the random variable X is the number shown on the die and the random variable Y is defined as X^2, calculate the covariance and correlation of X and Y.

We have already calculated in Subsections 16.2.1 and 16.5.4 that

$$p = \frac{2}{13}, \quad E[X] = \frac{53}{13}, \quad E\left[X^2\right] = \frac{253}{13}, \quad V[X] = \frac{480}{169}.$$

Using (16.123), we obtain

$$\mathrm{Cov}[X, Y] = \mathrm{Cov}[X, X^2] = E[X^3] - E[X]E[X^2].$$

Now $E[X^3]$ is given by

$$\begin{aligned} E[X^3] &= 1^3 \times \tfrac{1}{2}p + (2^3 + 3^3 + 4^3 + 5^3)p + 6^3 \times 2p \\ &= \frac{1313}{2}p = 101, \end{aligned}$$

and so the covariance of X and Y is given by

$$\mathrm{Cov}[X, Y] = 101 - \frac{53}{13} \times \frac{253}{13} = \frac{3660}{169}.$$

The correlation is defined by $\mathrm{Corr}[X, Y] = \mathrm{Cov}[X, Y]/\sigma_X \sigma_Y$. The standard deviation of Y may be calculated from the definition of the variance. Letting $\mu_Y = E[X^2] = \frac{253}{13}$ gives

$$\begin{aligned} \sigma_Y^2 &= \frac{p}{2}\left(1^2 - \mu_Y\right)^2 + p\left(2^2 - \mu_Y\right)^2 + p\left(3^2 - \mu_Y\right)^2 + p\left(4^2 - \mu_Y\right)^2 \\ &\quad + p\left(5^2 - \mu_Y\right)^2 + 2p\left(6^2 - \mu_Y\right)^2 \\ &= \frac{187\,356}{169}p = \frac{28\,824}{169}. \end{aligned}$$

We deduce that

$$\mathrm{Corr}[X, Y] = \frac{3660}{169}\sqrt{\frac{169}{28\,824}}\sqrt{\frac{169}{480}} \approx 0.984.$$

Thus the random variables X and Y display a strong degree of positive correlation, as we would expect.[23] ◀

The covariance of two random variables plays a part in a variety of circumstances. For example, if X and Y are *not* independent then the variance of $X + Y$ is given by

$$\begin{aligned} V[X+Y] &= E[(X+Y)^2] - (E[X+Y])^2 \\ &= E[X^2] + 2E[XY] + E[Y^2] - \{(E[X])^2 + 2E[X]E[Y] + (E[Y])^2\} \\ &= V[X] + V[Y] + 2(E[XY] - E[X]E[Y]) \\ &= V[X] + V[Y] + 2\,\mathrm{Cov}[X, Y]. \end{aligned}$$

23 But note that they are not 100% correlated.

More generally, we find (for a, b and c constant)

$$V[aX + bY + c] = a^2 V[X] + b^2 V[Y] + 2ab\,\mathrm{Cov}[X, Y]. \qquad (16.124)$$

Note that if X and Y are in fact independent then $\mathrm{Cov}[X, Y] = 0$ and we recover the expression (16.57) in Subsection 16.6.3.

We may use (16.124) to obtain an approximate expression for $V[f(X, Y)]$ for any arbitrary function f, even when the random variables X and Y are correlated. Approximating $f(X, Y)$ by the linear terms of its Taylor expansion about the point (μ_X, μ_Y), we have

$$f(X, Y) \approx f(\mu_X, \mu_Y) + \left(\frac{\partial f}{\partial X}\right)(X - \mu_X) + \left(\frac{\partial f}{\partial Y}\right)(Y - \mu_Y), \qquad (16.125)$$

where the partial derivatives are evaluated at $X = \mu_X$ and $Y = \mu_Y$. Taking the variance of both sides, and using (16.124),[24] we find

$$V[f(X, Y)] \approx \left(\frac{\partial f}{\partial X}\right)^2 V[X] + \left(\frac{\partial f}{\partial Y}\right)^2 V[Y] + 2\left(\frac{\partial f}{\partial X}\right)\left(\frac{\partial f}{\partial Y}\right)\mathrm{Cov}[X, Y]. \qquad (16.126)$$

Clearly, if $\mathrm{Cov}[X, Y] = 0$, we recover the result (16.58) derived in Subsection 16.6.3. We note that (16.125), and hence also (16.126), are exact if $f(X, Y)$ is linear in X and Y.

For several variables X_i, $i = 1, 2, \ldots, n$, we can define the symmetric (positive definite)[25] *covariance matrix* whose elements are

$$V_{ij} = \mathrm{Cov}[X_i, X_j], \qquad (16.127)$$

and the symmetric (positive definite) *correlation matrix*

$$\rho_{ij} = \mathrm{Corr}[X_i, X_j].$$

The diagonal elements of the covariance matrix are the variances of the variables, whilst those of the correlation matrix are unity. For a function of several variables, (16.126) generalizes to

$$V[f(X_1, X_2, \ldots, X_n)] \approx \sum_i \left(\frac{\partial f}{\partial X_i}\right)^2 V[X_i] + \sum_i \sum_{j \neq i} \left(\frac{\partial f}{\partial X_i}\right)\left(\frac{\partial f}{\partial X_j}\right)\mathrm{Cov}[X_i, X_j],$$

where the partial derivatives are evaluated at $X_i = \mu_{X_i}$.

The following example is a little long, but is straightforward and included to illustrate the definitions introduced in this section.

---

24 With $a = \partial f/\partial X$, $b = \partial f/\partial Y$ and constant $c = f(\mu_X, \mu_Y) - \mu_X(\partial f/\partial X) - \mu_Y(\partial f/\partial Y)$.

25 Recall that this does *not* imply that all of its entries are positive; its diagonal elements are, but the off-diagonal ones can be zero or negative.

Example A card is drawn at random from a normal 52-card pack and its identity noted. The card is replaced, the pack shuffled and the process repeated. Random variables W, X, Y, Z are defined as follows:

$W = 2$ if the drawn card is a heart; $W = 0$ otherwise.
$X = 4$ if the drawn card is an ace, king, or queen; $X = 2$ if the card is a jack or ten; $X = 0$ otherwise.
$Y = 1$ if the drawn card is red; $Y = 0$ otherwise.
$Z = 2$ if the drawn card is black and an ace, king or queen; $Z = 0$ otherwise.

Establish the correlation matrix for W, X, Y, Z.

The means of the variables are given by

$$\mu_W = 2 \times \tfrac{1}{4} = \tfrac{1}{2}, \ \mu_X = \left(4 \times \tfrac{3}{13}\right) + \left(2 \times \tfrac{2}{13}\right) = \tfrac{16}{13},$$
$$\mu_Y = 1 \times \tfrac{1}{2} = \tfrac{1}{2}, \ \mu_Z = 2 \times \tfrac{6}{52} = \tfrac{3}{13}.$$

The variances, calculated from $\sigma_U^2 = V[U] = E\left[U^2\right] - (E[U])^2$, where $U = W, X, Y$ or Z, are

$$\sigma_W^2 = \left(4 \times \tfrac{1}{4}\right) - \left(\tfrac{1}{2}\right)^2 = \tfrac{3}{4}, \ \sigma_X^2 = \left(16 \times \tfrac{3}{13}\right) + \left(4 \times \tfrac{2}{13}\right) - \left(\tfrac{16}{13}\right)^2 = \tfrac{472}{169},$$
$$\sigma_Y^2 = \left(1 \times \tfrac{1}{2}\right) - \left(\tfrac{1}{2}\right)^2 = \tfrac{1}{4}, \ \sigma_Z^2 = \left(4 \times \tfrac{6}{52}\right) - \left(\tfrac{3}{13}\right)^2 = \tfrac{69}{169}.$$

The covariances are found by first calculating $E[WX]$, etc. and then forming $E[WX] - \mu_W\mu_X$, etc.

$$E[WX] = 2\,(4)\left(\tfrac{3}{52}\right) + 2\,(2)\left(\tfrac{2}{52}\right) = \tfrac{8}{13}, \quad \text{Cov}[W, X] = \tfrac{8}{13} - \tfrac{1}{2}\left(\tfrac{16}{13}\right) = 0,$$

$$E[WY] = 2(1)\left(\tfrac{1}{4}\right) = \tfrac{1}{2}, \quad \text{Cov}[W, Y] = \tfrac{1}{2} - \tfrac{1}{2}\left(\tfrac{1}{2}\right) = \tfrac{1}{4},$$

$$E[WZ] = 0, \quad \text{Cov}[W, Z] = 0 - \tfrac{1}{2}\left(\tfrac{3}{13}\right) = -\tfrac{3}{26},$$

$$E[XY] = 4(1)\left(\tfrac{6}{52}\right) + 2(1)\left(\tfrac{4}{52}\right) = \tfrac{8}{13}, \quad \text{Cov}[X, Y] = \tfrac{8}{13} - \tfrac{16}{13}\left(\tfrac{1}{2}\right) = 0,$$

$$E[XZ] = 4(2)\left(\tfrac{6}{52}\right) = \tfrac{12}{13}, \quad \text{Cov}[X, Z] = \tfrac{12}{13} - \tfrac{16}{13}\left(\tfrac{3}{13}\right) = \tfrac{108}{169},$$

$$E[YZ] = 0, \quad \text{Cov}[Y, Z] = 0 - \tfrac{1}{2}\left(\tfrac{3}{13}\right) = -\tfrac{3}{26}.$$

The correlations Corr$[W, X]$ and Corr$[X, Y]$ are clearly zero; the remainder are given by

$$\text{Corr}[W, Y] = \tfrac{1}{4}\left(\tfrac{3}{4} \times \tfrac{1}{4}\right)^{-1/2} = 0.577,$$

$$\text{Corr}[W, Z] = -\tfrac{3}{26}\left(\tfrac{3}{4} \times \tfrac{69}{169}\right)^{-1/2} = -0.209,$$

$$\text{Corr}[X, Z] = \tfrac{108}{169}\left(\tfrac{472}{169} \times \tfrac{69}{169}\right)^{-1/2} = 0.598,$$

$$\text{Corr}[Y, Z] = -\tfrac{3}{26}\left(\tfrac{1}{4} \times \tfrac{69}{169}\right)^{-1/2} = -0.361.$$

Finally, then, we can write down the correlation matrix:

$$\rho = \begin{pmatrix} 1 & 0 & 0.58 & -0.21 \\ 0 & 1 & 0 & 0.60 \\ 0.58 & 0 & 1 & -0.36 \\ -0.21 & 0.60 & -0.36 & 1 \end{pmatrix}.$$

As would be expected, X is uncorrelated with either W or Y, color and face-value being two independent characteristics. Positive correlations are to be expected between W and Y and between X and Z; both correlations are fairly strong. Moderate anticorrelations exist between Z and both W and Y, reflecting the fact that it is impossible for W and Y to be positive if Z is positive. ◄

Finally in this section, let us suppose that the random variables X_i, $i = 1, 2, \dots, n$, are related to a second set of random variables $Y_k = Y_k(X_1, X_2, \dots, X_n)$, $k = 1, 2, \dots, m$. By expanding each Y_k as a Taylor series as in (16.125) and inserting the resulting expressions into the definition of the covariance (16.121), we find that the elements of the covariance matrix for the Y_k variables are given by

$$\mathrm{Cov}[Y_k, Y_l] \approx \sum_i \sum_j \left(\frac{\partial Y_k}{\partial X_i}\right)\left(\frac{\partial Y_l}{\partial X_j}\right) \mathrm{Cov}[X_i, X_j]. \qquad (16.128)$$

It is straightforward to show that this relation is exact if the Y_k are linear combinations of the X_i. Equation (16.128) can then be written in matrix form as

$$\mathsf{V}_Y = \mathsf{S}\mathsf{V}_X\mathsf{S}^{\mathrm{T}}, \qquad (16.129)$$

where V_Y and V_X are the covariance matrices of the Y_k and X_i variables respectively and S is the rectangular $m \times n$ matrix with elements $S_{ki} = \partial Y_k / \partial X_i$.

SUMMARY

1. *Venn diagrams, unions* ($\cup$) *and intersections* ($\cap$)
   S is the sample space, $\emptyset$ is the empty event and the *complement* of event A is denoted by $\bar{A}$.
   - $A \cup \bar{A} = S$ and $A \cap \bar{A} = \emptyset$.
   - The operations $\cup$ and $\cap$ are commutative, associative, idempotent, and have distribution laws

$$A \cap (B \cup C) = (A \cap B) \cup (A \cap C),$$
$$A \cup (B \cap C) = (A \cup B) \cap (A \cup C).$$

   - De Morgan's laws: $\overline{A \cup B} = \bar{A} \cap \bar{B}$ and $\overline{A \cap B} = \bar{A} \cup \bar{B}$.
   - The Venn diagram for n events has 2^n regions, with ${}^nC_r$ r-fold intersection regions.

2. *Probability,* $\Pr(A)$
   (i) For a single event, $\Pr(\bar{A}) = 1 - \Pr(A)$.
   (ii) For two events, the *conditional* probability for event A, given that event B has already occurred, is $\Pr(A|B) = \Pr(A \cap B)/\Pr(B)$.

Relationship	$\Pr(A \cup B)$	$\Pr(A \cap B)$	$\Pr(A\|B)$
Statistically independent	$\Pr(A) + \Pr(B) - \Pr(A \cap B)$	$\Pr(A)\Pr(B)$	$\Pr(A)$
Mutually exclusive	$\Pr(A) + \Pr(B)$	0	0

(iii) For n events A_i

- The probability of their union, $\Pr(A_1 \cup A_2 \cup \cdots \cup A_n)$, is given by (sums run over all sets of unrepeated subscripts)

$$\sum_i \Pr(A_i) - \sum_{i,j} \Pr(A_i \cap A_j) + \sum_{i,j,k} \Pr(A_i \cap A_j \cap A_k) - \cdots + (-1)^{n+1} \Pr(A_1 \cap A_2 \cap \cdots \cap A_n).$$

- If a set of mutually exclusive events A_i exhaust S, then

$$\Pr(B) = \sum_i \Pr(A_i)\Pr(B|A_i).$$

- *Bayes' theorem*

$$\Pr(A|B) = \frac{\Pr(A)\Pr(B|A)}{\sum_i \Pr(A_i)\Pr(B|A_i)}.$$

Here (a) A may coincide with one of the A_i, but does not have to, and (b) in many applications $n = 2$ with $A_1 = A$ and $A_2 = \bar{A}$.

3. *Permutations, combinations and statistical counts*
   - The number of ways of selecting k objects (order immaterial) from n without replacement is ${}^nC_k = \dfrac{n!}{(n-k)!\,k!}$; with replacement it is ${}^{n+k-1}C_k = \dfrac{(n+k-1)!}{(n-1)!\,k!}$.
   - Multinomial distribution: with $\sum_{i=1}^{m} n_i = n$, the multinomial coefficient is

$$M(n; n_i) = \frac{n!}{n_1!\,n_2!\,\cdots n_m!}.$$

   (i) It gives the number of distinguishable permutations of the n objects when there are n_i objects of type i, for $i = 1, 2, \ldots, m$.
   (ii) It gives the number of ways n distinguishable objects can be divided into m piles, with n_i objects in the ith pile.
   (iii) It is the coefficient of $x_1^{n_1} x_2^{n_2} \cdots x_m^{n_m}$ in the (multinomial) expansion of $(x_1 + x_2 + \cdots + x_m)^n$.
   (iv) When x_i is the probability that a single Bernoulli trial will result in an outcome of type i, with $\sum_i x_i = 1$, $M(n; n_i) x_1^{n_1} x_2^{n_2} \cdots x_M^{n_m}$ is the probability that n trials will contain exactly n_i outcomes of type i for each i.

4. *Particle statistics*
A system of N particles occupies R energy levels, of which the ith has degeneracy g_i. The number of particles in any one quantum state (i.e. with both energy level and degenerate sub-state defined) may or may not be limited to not more than one. The number of ways the particles can be distributed amongst the quantum states such that the ith level contains n_i particles, where $\sum_i n_i = N$, are given below.

	No limit in each level	≤ 1 in each level
Distinguishable	$N!\prod_{i=1}^{R}\frac{g_i^{n_i}}{n_i!}$ (Maxwell–Boltzmann)	$N!\prod_{i=1}^{R}\frac{g_i!}{n_i!(g_i-n_i)!}$ (—)
Indistinguishable	$\prod_{i=1}^{R}\frac{(n_i+g_i-1)!}{n_i!(g_i-1)!}$ (Bose–Einstein)	$\prod_{i=1}^{R}\frac{g_i!}{n_i!(g_i-n_i)!}$ (Fermi–Dirac)

5. *Probability distributions*
In the table below, X is a random variable and $\Pr(X = x_i) = p_i$, $\Pr(x < X < x + dx) = f(x)\,dx$ and $E[U]$ is the expectation value of RV U. Only the continuous case is given; discrete PDFs can be replaced by $f(x) = \sum_{i=1}^{n} p_i\delta(x - x_i)$.

Property or definition	Formula	Conditions or notes
Normalization	$\int_{-\infty}^{\infty} f(x)\,dx = 1$	
Cumulative PF	$F(x) = \int_{-\infty}^{x} f(x)\,dx$	$F(\infty) = 1$
Independent RVs	$h(x, y) = f(x)g(y)$	
Mean, $E[X]$ or μ	$\mu = \int_{-\infty}^{\infty} xf(x)\,dx$	
Median	$M = F^{-1}(\frac{1}{2})$	M is such that $F(M) = \frac{1}{2}$
nth percentile	$P_n = F^{-1}(n/100)$	
Variance, $V[X]$	$\int_{-\infty}^{\infty} (x-\mu)^2 f(x)\,dx$	$V[aX + b] = a^2V[X]$
Standard deviation	$\sigma = +\sqrt{\text{variance}}$	$\sigma^2 = E[X^2] - \mu^2$
Moments, $E[x^k]$	$\mu_k = \int_{-\infty}^{\infty} x^k f(x)\,dx$	
Function, $Y(X)$	$g(y) = f(x(y))\left\lvert\frac{dx}{dy}\right\rvert$	Unique inverse $X = X(Y)$

(cont.)

(*Cont.*)

Property or definition	Formula	Conditions or notes
PDF for $Z = X + Y$	$p(Z) = f * g$	Convolution of $f(X)$ and $g(Y)$
$E[aX + bY]$	$aE[X] + bE[Y]$	
$V[aX + bY]$	$a^2V[X] + b^2V[Y]$	X and Y independent RVs
$E[W(X, Y)]$	$\approx W(\mu_X, \mu_Y)$	"
†$V[W(X, Y)]$	$\approx \left(\frac{\partial W}{\partial X}\right)^2 V[X] + \left(\frac{\partial W}{\partial Y}\right)^2 V[Y]$	"

†In the expression for $V[W(X, Y)]$ the two partial derivatives are both to be evaluated at $x = \mu_X$ and $y = \mu_Y$.

- For the parameters of important discrete distributions, see Table 16.1 on page 654
- For the parameters of important continuous distributions, see Table 16.2 on page 666
- For both tables, the PGF can be obtained from the MGF by replacing e^t by t.

6. *Generating functions*

	Probability GF	Moment GF
Applicability	discrete only	discrete and continuous
Definition	$\Phi_X(t) = E[t^X]$	$M_X(t) = E[e^{tX}]$
Relationship	$= \text{MGF}(\ln t)$	$= \text{PGF}(e^t)$
Formula	$\sum_{n=0}^{\infty} f_n t^n$	$\int_{-\infty}^{\infty} f(x)e^{tx}\,dx$
Particular probability	$f_n = \frac{1}{n!}\Phi_X^{(n)}(1)$	$E[X^n] = M_X^{(n)}(0)$
Mean μ	$\Phi'_X(1)$	$M'_X(0)$
Variance σ^2	$\Phi''_X(1) + \Phi'_X(1) - [\Phi'_X(1)]^2$	$M''_X(0) - [M'_X(0)]^2$
†Shifting and scaling	–	$M_{aX+b}(t) = e^{bt}M_X(at)$
Sum S_N of N IRVs; N fixed	$\Phi_{S_N}(t) = \prod_{i=1}^{N} \Phi_{X_i}(t)$	$M_{S_N}(t) = \prod_{i=1}^{N} M_{X_i}(t)$
Linear sum $L = \sum_i^N c_i X_i$ of N IRVs	–	$M_L(t) = \prod_{i=1}^{N} M_{X_i}(c_i t)$
††Sum S_N of N identical IRVs; N variable	$\Phi_{S_N}(t) = \chi_N(\Phi_X(t))$	$M_{S_N}(t) = \chi_N(M_X(t))$

IRV = independent random variable.
†Taking $a = 1$ and $b = -\mu$ gives the central moments.
††The PGF for N is $\chi_N(t)$.

7. *Central limit theorem*
   If X_i, $i = 1, 2, \dots, n$, are IRVs, each described by its own probability density function $f_i(x)$, with a mean μ_i and a variance σ_i^2, then the random variable $Z = \left(\sum_i X_i\right)/n$, has the following properties:
   (i) its expectation value is given by $E[Z] = \left(\sum_i \mu_i\right)/n$;
   (ii) its variance is given by $V[Z] = \left(\sum_i \sigma_i^2\right)/n^2$;
   (iii) as $n \to \infty$ the probability function of Z tends to a Gaussian with corresponding mean and variance.

8. *Bivariate distributions* $f(X, Y)$
   For a general function $g(X, Y)$, its expectation

$$E[g(X, Y)] = \int_{-\infty}^{\infty}\int_{-\infty}^{\infty} g(x, y) f(x, y)\, dx\, dy.$$

Parameter	Calculation
Normalization	$\int_{-\infty}^{\infty}\int_{-\infty}^{\infty} f(x, y)\, dx\, dy = 1$
Mean, μ_X	$E[X]$
Variance, $V[X]$ or σ_X^2	$E[(X - \mu_X)^2]$
$E[XY]$ for IRVs	$E[X]E[Y]$
Covariance, $\text{Cov}[X, Y]$	$E[(X - \mu_X)(Y - \mu_Y)] = E[XY] - \mu_X\mu_Y$
Covariance of IRVs	0
Correlation, $\text{Corr}[X, Y]$	$r = \dfrac{\text{Cov}[X, Y]}{\sigma_X\sigma_Y}, \quad -1 \le r \le +1$

9. *Multivariate distributions* $f(X_i)$ *for* $i = 1, 2, \dots, n$
   The *covariance matrix* V has entries $V_{ij} = \text{Cov}[X_i, X_j]$, and the *correlation matrix* ρ has entries $\rho_{ij} = \text{Corr}[X_i, X_j]$. The diagonal elements of V are all positive and V_{kk} gives the variance of X_k.
   The variance of a function $h(X_1, X_2, \dots, X_n)$ is given approximately by

$$\sum_i \left(\frac{\partial h}{\partial X_i}\right)^2 V[X_i] + \sum_i \sum_{j \ne i} \left(\frac{\partial h}{\partial X_i}\right)\left(\frac{\partial h}{\partial X_j}\right) \text{Cov}[X_i, X_j],$$

   where the partial derivatives are evaluated at $X_i = \mu_{X_i}$.

PROBLEMS

16.1. By shading or numbering Venn diagrams, determine which of the following are valid relationships between events. For those that are, prove the relationship using de Morgan's laws.
(a) $\overline{(\bar{X} \cup Y)} = X \cap \bar{Y}$.
(b) $\bar{X} \cup \bar{Y} = \overline{(X \cup Y)}$.

(c) $(X \cup Y) \cap Z = (X \cup Z) \cap Y$.
(d) $X \cup \overline{(Y \cap Z)} = (X \cup \bar{Y}) \cap \bar{Z}$.
(e) $X \cup \overline{(Y \cap Z)} = (X \cup \bar{Y}) \cup \bar{Z}$.

16.2. Given that events X, Y and Z satisfy

$$(X \cap Y) \cup (Z \cap X) \cup \overline{(\bar{X} \cup \bar{Y})} = \overline{(Z \cup \bar{Y})} \cup \{[\overline{(\bar{Z} \cup \bar{X})} \cup (\bar{X} \cap Z)] \cap Y\},$$

prove that $X \supset Y$, and that either $X \cap Z = \emptyset$ or $Y \supset Z$.

16.3. A and B each have two unbiased four-faced dice, the four faces being numbered 1, 2, 3, 4. Without looking, B tries to guess the sum x of the numbers on the bottom faces of A's two dice after they have been thrown onto a table. If the guess is correct B receives x^2 euros, but if not he loses x euros.

Determine B's expected gain per throw of A's dice when he adopts each of the following strategies:
(a) he selects x at random in the range $2 \leq x \leq 8$;
(b) he throws his own two dice and guesses x to be whatever they indicate;
(c) he takes your advice and always chooses the same value for x. Which number would you advise?

16.4. $X_1, X_2, \ldots, X_n$ are independent, identically distributed, random variables drawn from a uniform distribution on [0, 1]. The random variables A and B are defined by

$$A = \min(X_1, X_2, \ldots, X_n), \qquad B = \max(X_1, X_2, \ldots, X_n).$$

For any fixed k such that $0 \leq k \leq \frac{1}{2}$, find the probability, p_n, that both

$$A \leq k \qquad \text{and} \qquad B \geq 1 - k.$$

Check your general formula by considering directly the cases (a) $k = 0$, (b) $k = \frac{1}{2}$, (c) $n = 1$ and (d) $n = 2$.

16.5. Two duelists, A and B, take alternate shots at each other, and the duel is over when a shot (fatal or otherwise!) hits its target. Each shot fired by A has a probability α of hitting B, and each shot fired by B has a probability β of hitting A. Calculate the probabilities P_1 and P_2, defined as follows, that A will win such a duel: P_1, A fires the first shot; P_2, B fires the first shot.

If they agree to fire simultaneously, rather than alternately, what is the probability P_3 that A will win, i.e. hit B without being hit himself?

16.6. This problem shows that the odds are hardly ever "evens" when it comes to dice rolling.
(a) Gamblers A and B each roll a fair six-faced die, and B wins if his score is strictly greater than A's. Show that the odds are 7 to 5 in A's favor.
(b) Calculate the probabilities of scoring a total T from two rolls of a fair die for $T = 2, 3, \ldots, 12$. Gamblers C and D each roll a fair die twice and score

respective totals T_C and T_D, D winning if $T_D > T_C$. Realizing that the odds are not equal, D insists that C should increase her stake for each game. C agrees to stake £1.10 per game, as compared to D's £1.00 stake. Who will show a profit?

16.7. An electronics assembly firm buys its microchips from three different suppliers; half of them are bought from firm X, whilst firms Y and Z supply 30% and 20%, respectively. The suppliers use different quality-control procedures and the percentages of defective chips are 2%, 4% and 4% for X, Y and Z, respectively. The probabilities that a defective chip will fail two or more assembly-line tests are 40%, 60% and 80%, respectively, whilst all defective chips have a 10% chance of escaping detection. An assembler finds a chip that fails only one test. What is the probability that it came from supplier X?

16.8. As every student of probability theory will know, Bayesylvania is awash with natives, not all of whom can be trusted to tell the truth, and lost, and apparently somewhat deaf, travelers who ask the same question several times in an attempt to get directions to the nearest village.

One such traveler finds himself at a T-junction in an area populated by the Asciis and Bisciis in the ratio 11 to 5. As is well known, the Biscii always lie, but the Ascii tell the truth three quarters of the time, giving independent answers to all questions, even to immediately repeated ones.

(a) The traveler asks one particular native twice whether he should go to the left or to the right to reach the local village. Each time he is told "left". Should he take this advice, and, if he does, what are his chances of reaching the village?

(b) The traveler then asks the same native the same question a third time, and for a third time receives the answer "left". What should the traveler do now? Have his chances of finding the village been altered by asking the third question?

16.9. A boy is selected at random from amongst the children belonging to families with n children. It is known that he has at least two sisters. Show that the probability that he has $k-1$ brothers is

$$\frac{(n-1)!}{(2^{n-1}-n)(k-1)!(n-k)!},$$

for $1 \leq k \leq n-2$ and zero for other values of k. Assume that boys and girls are equally likely.

16.10. Villages A, B, C and D are connected by overhead telephone lines joining AB, AC, BC, BD and CD. As a result of severe gales, there is a probability p (the same for each link) that any particular link is broken.

(a) Show that the probability that a call can be made from A to B is

$$1 - p^2 - 2p^3 + 3p^4 - p^5.$$

(b) Show that the probability that a call can be made from D to A is

$$1 - 2p^2 - 2p^3 + 5p^4 - 2p^5.$$

16.11. A set of $2N + 1$ rods consists of one of each integer length $1, 2, \ldots, 2N, 2N + 1$. Three, of lengths a, b and c, are selected, of which a is the longest. By considering the possible values of b and c, determine the number of ways in which a non-degenerate triangle (i.e. one of non-zero area) can be formed (i) if a is even, and (ii) if a is odd. Combine these results appropriately to determine the total number of non-degenerate triangles that can be formed using three of the $2N + 1$ rods, and hence show that the probability that such a triangle can be formed from a random selection (without replacement) of three rods is

$$\frac{(N-1)(4N+1)}{2(4N^2-1)}.$$

16.12. A certain marksman never misses his target, which consists of a disc of unit radius with center O. The probability that any given shot will hit the target within a distance t of O is t^2, for $0 \le t \le 1$. The marksman fires n independent shots at the target, and the random variable Y is the radius of the smallest circle with center O that encloses all the shots. Determine the PDF for Y and hence find the expected area of the circle.

The shot that is furthest from O is now rejected and the corresponding circle determined for the remaining $n - 1$ shots. Show that its expected area is

$$\frac{n-1}{n+1}\pi.$$

16.13. The duration (in minutes) of a telephone call made from a public call-box is a random variable T. The probability density function of T is

$$f(t) = \begin{cases} 0 & t < 0, \\ \frac{1}{2} & 0 \le t < 1, \\ ke^{-2t} & t \ge 1, \end{cases}$$

where k is a constant. To pay for the call, 20 pence has to be inserted at the beginning, and a further 20 pence after each subsequent half-minute. Determine by how much the average cost of a call exceeds the cost of a call of average length charged at 40 pence per minute.

16.14. Kittens from different litters do not get on with each other, and fighting breaks out whenever two kittens from different litters are present together. A cage initially contains x kittens from one litter and y from another. To quell the fighting, kittens are removed at random, one at a time, until peace is restored. Show, by induction, that the expected number of kittens finally remaining is

$$N(x, y) = \frac{x}{y+1} + \frac{y}{x+1}.$$

16.15. A tennis tournament is arranged on a straight knockout basis for 2^n players, and for each round, except the final, opponents for those still in the competition are drawn at random. The quality of the field is so even that in any match it is equally likely that either player will win. Two of the players have surnames that begin with "Q". Find the probabilities that they play each other

(a) in the final,

(b) at some stage in the tournament.

16.16. A particle is confined to the one-dimensional space $0 \leq x \leq a$, and classically it can be in any small interval dx with equal probability. However, quantum mechanics gives the result that the probability distribution is proportional to $\sin^2(n\pi x/a)$, where n is an integer. Find the variance in the particle's position in both the classical and quantum mechanical pictures, and show that, although they differ, the latter tends to the former in the limit of large n, in agreement with the correspondence principle of physics.

16.17. This problem is about interrelated binomial trials.

(a) In two sets of binomial trials T and t, the probabilities that a trial has a successful outcome are P and p, respectively, with corresponding probabilities of failure of $Q = 1 - P$ and $q = 1 - p$. One "game" consists of a trial T, followed, if T is successful, by a trial t and then a further trial T. The two trials continue to alternate until one of the T-trials fails, at which point the game ends. The score S for the game is the total number of successes in the t-trials. Find the PGF for S and use it to show that

$$E[S] = \frac{Pp}{Q}, \qquad V[S] = \frac{Pp(1 - Pq)}{Q^2}.$$

(b) Two normal unbiased six-faced dice A and B are rolled alternately starting with A; if A shows a 6 the experiment ends. If B shows an odd number no points are scored, if it shows a 2 or a 4 then one point is scored, whilst if it records a 6 then two points are awarded. Find the average and standard deviation of the score for the experiment and show that the latter is the greater.

16.18. For a non-negative integer random variable X, in addition to the probability generating function $\Phi_X(t)$ defined in equation (16.60), it is possible to define the probability generating function

$$\Psi_X(t) = \sum_{n=0}^{\infty} g_n t^n,$$

where g_n is the probability that $X > n$.

(a) Prove that Φ_X and Ψ_X are related by

$$\Psi_X(t) = \frac{1 - \Phi_X(t)}{1 - t}.$$

(b) Show that $E[X]$ is given by $\Psi_X(1)$ and that the variance of X can be expressed as $2\Psi'_X(1) + \Psi_X(1) - [\Psi_X(1)]^2$.

(c) For a particular random variable X, the probability that $X > n$ is equal to α^{n+1}, with $0 < \alpha < 1$. Use the results in (b) to show that $V[X] = \alpha(1-\alpha)^{-2}$.

16.19. A point P is chosen at random on the circle $x^2 + y^2 = 1$. The random variable X denotes the distance of P from $(1, 0)$. Find the mean and variance of X and the probability that X is greater than its mean.

16.20. As assistant to a celebrated and imperious newspaper proprietor, you are given the job of running a lottery, in which each of his five million readers will have an equal independent chance, p, of winning a million pounds; you have the job of choosing p. However, if nobody wins it will be bad for publicity, whilst if more than two readers do so, the prize cost will more than offset the profit from extra circulation – in either case you will be sacked! Show that, however you choose p, there is more than a 40% chance you will soon be clearing your desk.

16.21. The number of errors needing correction on each page of a set of proofs follows a Poisson distribution of mean μ. The cost of the first correction on any page is α and that of each subsequent correction on the same page is β. Prove that the average cost of correcting a page is

$$\alpha + \beta(\mu - 1) - (\alpha - \beta)e^{-\mu}.$$

16.22. In the game of Blackball, at each turn Muggins draws a ball at random from a bag containing five white balls, three red balls and two black balls; after being recorded, the ball is replaced in the bag. A white ball earns him \$1, whilst a red ball gets him \$2; in either case, he also has the option of leaving with his current winnings or of taking a further turn on the same basis. If he draws a black ball the game ends and he loses all he may have gained previously. Find an expression for Muggins' expected return if he adopts the strategy of drawing up to n balls, provided he has not been eliminated by then.

Show that, as the entry fee to play is \$3, Muggins should be dissuaded from playing Blackball, but, if that cannot be done, what value of n would you advise him to adopt?

16.23. The probability distribution for the number of eggs in a clutch is $\text{Po}(\lambda)$, and the probability that each egg will hatch is p (independently of the size of the clutch). Show by direct calculation that the probability distribution for the number of chicks that hatch is $\text{Po}(\lambda p)$ and so justify the assumptions made in the worked example at the end of Subsection 16.7.1.

16.24. A shopper buys 36 items at random in a supermarket, where, because of the sales tax imposed, the final digit (the number of pence) in the price is uniformly and randomly distributed from 0 to 9. Instead of adding up the bill exactly, she rounds each item to the nearest 10 pence, rounding up or down with equal probability if the price ends in a "5". Should she suspect a mistake if the cashier asks her for 23 pence more than she estimated?

16.25. Under EU legislation on harmonization, all kippers are to weigh 0.2000 kg, and vendors who sell underweight kippers must be fined by their government. The weight of a kipper is normally distributed, with a mean of 0.2000 kg and a standard deviation of 0.0100 kg. They are packed in cartons of 100 and large quantities of them are sold.

Every day, a carton is to be selected at random from each vendor and tested according to one of the following schemes, which have been approved for the purpose.

(a) The entire carton is weighed, and the vendor is fined 2500 euros if the average weight of a kipper is less than 0.1975 kg.
(b) Twenty five kippers are selected at random from the carton; the vendor is fined 100 euros if the average weight of a kipper is less than 0.1980 kg.
(c) Kippers are removed one at a time, at random, until one has been found that weighs *more* than 0.2000 kg; the vendor is fined $4n(n-1)$ euros, where n is the number of kippers removed.

Which scheme should the Chancellor of the Exchequer be urging his government to adopt?

16.26. In a certain parliament, the government consists of 75 New Socialites and the opposition consists of 25 Preservatives. Preservatives never change their mind, always voting against government policy without a second thought; New Socialites vote randomly, but with probability p that they will vote for their party leader's policies.

Following a decision by the New Socialites' leader to drop certain manifesto commitments, N of his party decide to vote consistently with the opposition. The leader's advisors reluctantly admit that an election must be called if N is such that, at any vote on government policy, the chance of a simple majority in favor would be less than 80%. Given that $p = 0.8$, estimate the lowest value of N that would precipitate an election.

16.27. A practical-class demonstrator sends his 12 students to the storeroom to collect apparatus for an experiment, but forgets to tell each which type of component to bring. There are three types, A, B and C, held in the stores (in large numbers) in the proportions 20%, 30% and 50%, respectively, and each student picks a component at random. In order to set up one experiment, one unit each of A and B and two units of C are needed. Let $\Pr(N)$ be the probability that at least N experiments can be set up.

(a) Evaluate $\Pr(3)$.
(b) Find an expression for $\Pr(N)$ in terms of k_1 and k_2, the numbers of components of types A and B respectively selected by the students. Show

that Pr(2) can be written in the form

$$\Pr(2) = (0.5)^{12} \sum_{i=2}^{6} {}^{12}C_i \, (0.4)^i \sum_{j=2}^{8-i} {}^{12-i}C_j \, (0.6)^j .$$

(c) By considering the conditions under which no experiments can be set up, show that $\Pr(1) = 0.9145$.

16.28. A husband and wife decide that their family will be complete when it includes two boys and two girls – but that this would then be enough! The probability that a new baby will be a girl is p. Ignoring the possibility of identical twins, show that the expected size of their family is

$$2\left(\frac{1}{pq} - 1 - pq\right),$$

where $q = 1 - p$.

16.29. The continuous random variables X and Y have a joint PDF proportional to $xy(x-y)^2$ with $0 \le x \le 1$ and $0 \le y \le 1$. Find the marginal distributions for X and Y and show that they are negatively correlated with correlation coefficient $-\frac{2}{3}$.

16.30. A discrete random variable X takes integer values $n = 0, 1, \ldots, N$ with probabilities p_n. A second random variable Y is defined as $Y = (X - \mu)^2$, where μ is the expectation value of X. Prove that the covariance of X and Y is given by

$$\text{Cov}[X, Y] = \sum_{n=0}^{N} n^3 p_n - 3\mu \sum_{n=0}^{N} n^2 p_n + 2\mu^3.$$

Now suppose that X takes all of its possible values with equal probability, and hence demonstrate that two random variables can be uncorrelated, even though one is defined in terms of the other.

16.31. Two continuous random variables X and Y have a joint probability distribution

$$f(x, y) = A(x^2 + y^2),$$

where A is a constant and $0 \le x \le a$, $0 \le y \le a$. Show that X and Y are negatively correlated with correlation coefficient $-15/73$. By sketching a rough contour map of $f(x, y)$ and marking off the regions of positive and negative correlation, convince yourself that this (perhaps counter-intuitive) result is plausible.

HINTS AND ANSWERS

16.1. (a) Yes, (b) no, (c) no, (d) no, (e) yes.

16.3. Show that, if $p_x/16$ is the probability that the total will be x, then the corresponding gain is $[p_x(x^2+x) - 16x]/16$. (a) A loss of 0.36 euros; (b) a gain of 27/64 euros; (c) a gain of 46/16 euros, provided he takes your advice and guesses "6" each time.

16.5. $P_1 = \alpha(\alpha+\beta-\alpha\beta)^{-1}$; $P_2 = \alpha(1-\beta)(\alpha+\beta-\alpha\beta)^{-1}$; $P_3 = P_2$.

16.7. The relative probabilities are $X : Y : Z = 50 : 36 : 8$ (in units of 10^{-4}); 25/47.

16.9. Take A_j as the event that a family consists of j boys and $n-j$ girls, and B as the event that the boy has at least two sisters. Apply Bayes' theorem.

16.11. (i) For a even, the number of ways is $1+3+5+\cdots+(a-3)$, and (ii) for a odd it is $2+4+6+\cdots+(a-3)$. Combine the results for $a=2m$ and $a=2m+1$, with m running from 2 to N, to show that the total number of non-degenerate triangles is given by $N(4N+1)(N-1)/6$. The number of possible selections of a set of three rods is $(2N+1)(2N)(2N-1)/6$.

16.13. Show that $k=e^2$ and that the average duration of a call is 1 minute. Let p_n be the probability that the call ends during the interval $0.5(n-1) \le t < 0.5n$ and $c_n = 20n$ be the corresponding cost. Prove that $p_1 = p_2 = \frac{1}{4}$ and that $p_n = \frac{1}{2}e^2(e-1)e^{-n}$, for $n \ge 3$. It follows that the average cost is

$$E[C] = \frac{30}{2} + 20\frac{e^2(e-1)}{2}\sum_{n=3}^{\infty} ne^{-n}.$$

The arithmetico-geometric series has sum $(3e^{-1} - 2e^{-2})/(e-1)^2$ and the total charge is $5(e+1)/(e-1) = 10.82$ pence more than the 40 pence a uniform rate would cost.

16.15. If p_r is the probability that before the rth round both players are still in the tournament (and therefore have not met each other), show that

$$p_{r+1} = \frac{1}{4}\frac{2^{n+1-r}-2}{2^{n+1-r}-1}p_r \qquad \text{and hence that} \qquad p_r = \left(\frac{1}{2}\right)^{r-1}\frac{2^{n+1-r}-1}{2^n-1}.$$

(a) The probability that they meet in the final is $p_n = 2^{-(n-1)}(2^n-1)^{-1}$.
(b) The probability that they meet at some stage in the tournament is given by the sum $\sum_{r=1}^{n} p_r(2^{n+1-r}-1)^{-1} = 2^{-(n-1)}$.

16.17. (a) Use result (16.73) to show that the PGF for S is $Q/(1-Pq-Ppt)$. Then use equations (16.63) and (16.65).
(b) The PGF for the score is $6/(21-10t-5t^2)$ and the average score is 10/3. The variance is 145/9 and the standard deviation is 4.01.

16.19. Mean $= 4/\pi$. Variance $= 2 - (16/\pi^2)$. Probability that X exceeds its mean $= 1 - (2/\pi)\sin^{-1}(2/\pi) = 0.561$.

16.21. Consider separately, 0, 1 and ≥ 2 errors on a page.

16.23. $\Pr(k \text{ chicks hatching}) = \sum_{n=k}^{\infty} \text{Po}(n, \lambda)\, \text{Bin}(n, p)$.

16.25. There is not much to choose between the schemes. In (a) the critical value of the standard variable is -2.5 and the average fine would be 15.5 euros. For (b) the corresponding figures are -1.0 and 15.9 euros. Scheme (c) is governed by a geometric distribution with $p = q = \frac{1}{2}$, and leads to an expected fine of $\sum_{n=1}^{\infty} 4n(n-1)(\frac{1}{2})^n$. The sum can be evaluated by differentiating the result $\sum_{n=1}^{\infty} p^n = p/(1-p)$ with respect to p, and gives the expected fine as 16 euros.

16.27. (a) $[12!(0.5)^6(0.3)^3(0.2)^3]/(6!\,3!\,3!) = 0.0624$.

16.29. You will need to establish the normalization constant for the distribution (36), the common mean value (3/5) and the common standard deviation (3/10). The marginal distributions are $f(x) = 3x(6x^2 - 8x + 3)$, and the same function of y. The covariance has the value $-3/50$, yielding a correlation of $-2/3$.

16.31. $A = 3/(24a^4)$; $\mu_X = \mu_Y = 5a/8$; $\sigma_X^2 = \sigma_Y^2 = 73a^2/960$; $E[XY] = 3a^2/8$; $\text{Cov}[X, Y] = -a^2/64$.

17 Statistics

In this chapter, we turn to the study of statistics, which is concerned with the analysis of experimental data. In a book of this nature we cannot hope to do justice to such a large subject; indeed, many would argue that statistics belongs to the realm of experimental science rather than in a mathematics textbook. Nevertheless, physical scientists and engineers are regularly called upon to perform a statistical analysis of their data and to present their results in a statistical context. This justifies the inclusion of the subject in a book such as this, but we will concentrate on those aspects of direct relevance to the presentation of experimental data.[1]

17.1 Experiments, samples and populations

We may regard the product of any experiment as a set of N measurements of some quantity x or of some set of quantities $x, y, \ldots, z$. This set of measurements constitutes the *data*. Each measurement (or *data item*) consists accordingly of a single number x_i or a set of numbers $(x_i, y_i, \ldots, z_i)$, where $i = 1, \ldots, N$. For the moment, we will assume that each data item is a single number, although our discussion can be extended to the more general case.

As a result of inaccuracies in the measurement process, or because of intrinsic variability in the quantity x being measured, one would expect the N measured values $x_1, x_2, \ldots, x_N$ to be different each time the experiment is performed. We may therefore consider the x_i as a set of N random variables. In the most general case, these random variables will be described by some N-dimensional joint probability density function $P(x_1, x_2, \ldots, x_N)$.[2] In other words, an experiment consisting of N measurements is considered as a single

1 There are, in fact, two separate schools of thought concerning statistics: the frequentist approach and the Bayesian approach. Which of these approaches is the more fundamental is still a matter of heated debate. Here we will concentrate primarily on the more traditional frequentist approach (despite the preference of one of the authors for the Bayesian viewpoint!). For a fuller discussion of the frequentist approach one could refer to, for example, A. Stuart and K. Ord, *Kendall's Advanced Theory of Statistics, vol. 1* (London: Edward Arnold, 1994) or J. F. Kenney and E. S. Keeping, *Mathematics of Statistics* (New York: Van Nostrand, 1954). For a discussion of the Bayesian approach one might consult, for example, D. S. Sivia, *Data Analysis: A Bayesian Tutorial* (Oxford: Oxford University Press, 1996).

2 In this chapter, we will adopt the common convention that $P(x)$ denotes the particular probability density function that applies to its argument, x. This obviates the need to use a different letter for the PDF of each new variable. For example, if X and Y are random variables with different PDFs, then properly one should denote these distributions by $f(x)$ and $g(y)$, say. In our shorthand notation, these PDFs are denoted by $P(x)$ and $P(y)$, where it is understood that the functional form of the PDF may be different in each case.

Table 17.1 *Experimental data giving eight measurements of the round trip time in milliseconds for a computer "packet" to travel from Cambridge UK to Cambridge MA*

188.7	204.7	193.2	169.0
168.1	189.8	166.3	200.0

random *sample* from the joint distribution (or *population*) $P(\mathbf{x})$, where $\mathbf{x}$ denotes a point in the N-dimensional data space having coordinates $(x_1, x_2, \ldots, x_N)$.

The situation is simplified considerably if the sample values x_i are *independent*. In this case, the N-dimensional joint distribution $P(\mathbf{x})$ factorizes into the product of N one-dimensional distributions,

$$P(\mathbf{x}) = P(x_1)P(x_2)\cdots P(x_N). \tag{17.1}$$

In the general case, each of the one-dimensional distributions $P(x_i)$ may be different. A typical example of this occurs when N independent measurements are made of some quantity x but the accuracy of the measuring procedure varies between measurements.

It is often the case, however, that each sample value x_i is drawn independently from the *same* population. In this case, $P(\mathbf{x})$ is of the form (17.1), but, in addition, $P(x_i)$ has the same form for each value of i. The measurements $x_1, x_2, \ldots, x_N$ are then said to form a *random sample of size* N from the one-dimensional population $P(x)$. This is the most common situation met in practice and, unless stated otherwise, we will assume from now on that this is the case.

17.2 Sample statistics

Suppose we have a set of N measurements $x_1, x_2, \ldots, x_N$. Any function of these measurements (that contains no unknown parameters) is called a *sample statistic*, or often simply a *statistic*. Sample statistics provide a means of characterizing the data. Although the resulting characterization is inevitably incomplete, it is useful to be able to describe a set of data in terms of a few pertinent numbers. We now discuss the most commonly used sample statistics.

17.2.1 Averages

The simplest number used to characterize a sample is its *mean*, which for N values x_i, $i = 1, 2, \ldots, N$, is defined by

$$\bar{x} = \frac{1}{N}\sum_{i=1}^{N} x_i. \tag{17.2}$$

In words, the *sample mean* is the sum of the sample values divided by the number of values in the sample. As an almost trivial example based on data that we will use for illustration on a number of occasions, consider the following.

Example Table 17.1 gives eight values for the round trip time in milliseconds for a computer "packet" to travel from Cambridge UK to Cambridge MA. Find the sample mean.

Using (17.2) the sample mean in milliseconds is given by

$$\bar{x} = \tfrac{1}{8}(188.7 + 204.7 + 193.2 + 169.0 + 168.1 + 189.8 + 166.3 + 200.0)$$
$$= \frac{1479.8}{8} = 184.975.$$

Since the sample values in Table 17.1 are quoted to an accuracy of one decimal place, it is usual to quote the mean to the same accuracy, i.e. as $\bar{x} = 185.0$. ◀

Strictly speaking the mean given by (17.2) is the *arithmetic mean* and this is by far the most common definition used for a mean. Other definitions of the mean are possible, though less common, and include

(i) the *geometric mean*,

$$\bar{x}_{\rm g} = \left(\prod_{i=1}^{N} x_i\right)^{1/N}, \tag{17.3}$$

(ii) the *harmonic mean*,

$$\bar{x}_{\rm h} = \frac{N}{\sum_{i=1}^{N} 1/x_i}, \tag{17.4}$$

(iii) the *root mean square*,

$$\bar{x}_{\rm rms} = \left(\frac{\sum_{i=1}^{N} x_i^2}{N}\right)^{1/2}. \tag{17.5}$$

It should be noted that, $\bar{x}$, $\bar{x}_{\rm h}$ and $\bar{x}_{\rm rms}$ would remain well defined even if some sample values were negative, but the value of $\bar{x}_{\rm g}$ could then become complex. The geometric mean should not be used in such cases.

Example Calculate $\bar{x}_{\rm g}$, $\bar{x}_{\rm h}$ and $\bar{x}_{\rm rms}$ for the sample given in Table 17.1.

The geometric mean is given by (17.3) to be

$$\bar{x}_{\rm g} = (188.7 \times 204.7 \times \cdots \times 200.0)^{1/8} = 184.4.$$

The harmonic mean is given by (17.4) to be

$$\bar{x}_{\rm h} = \frac{8}{(1/188.7) + (1/204.7) + \cdots + (1/200.0)} = 183.9.$$

Finally, the root mean square is given by (17.5) to be

$$\bar{x}_{\rm rms} = \left[\tfrac{1}{8}(188.7^2 + 204.7^2 + \cdots + 200.0^2)\right]^{1/2} = 185.5.$$

Because the spread of measured values is not very large, the three means correspondingly have a small spread.[3] ◀

Two other measures of the "average" of a sample are its *mode* and *median*. The mode is simply the most commonly occurring value in the sample. A sample may possess several modes, however, and thus it can be misleading in such cases to use the mode as a measure of the average of the sample. The median of a sample is the halfway point when the sample values x_i $(i = 1, 2, \ldots, N)$ are arranged in ascending (or descending) order. Clearly, this depends on whether the size of the sample, N, is odd or even. If N is odd then the median is simply equal to $x_{(N+1)/2}$, whereas if N is even the median of the sample is usually taken to be $\frac{1}{2}(x_{N/2} + x_{(N/2)+1})$.

Example Find the mode and median of the sample given in Table 17.1.

From the table we see that each sample value occurs exactly once, and so any value may be called the mode of the sample.

To find the sample median, we first arrange the sample values in ascending order and obtain

$$166.3, 168.1, 169.0, 188.7, 189.8, 193.2, 200.0, 204.7.$$

Since the number of sample values $N = 8$, which is even, the median of the sample is

$$\tfrac{1}{2}(x_4 + x_5) = \tfrac{1}{2}(188.7 + 189.8) = 189.25.$$

Note that with this treatment of samples containing even numbers of sample values, the median is not necessarily equal to any of the actual values. ◀

17.2.2 Variance and standard deviation

The variance and standard deviation both give a measure of the spread of values in a sample about the sample mean $\bar{x}$. The *sample variance* is defined by

$$s^2 = \frac{1}{N}\sum_{i=1}^{N}(x_i - \bar{x})^2, \tag{17.6}$$

and the *sample standard deviation* is the positive square root of the sample variance, i.e.

$$s = \sqrt{\frac{1}{N}\sum_{i=1}^{N}(x_i - \bar{x})^2}. \tag{17.7}$$

3 By considering $(a \pm b)^2$, show that for two positive measurements their r.m.s. mean is greater than or equal to their geometric mean, and that $\bar{x}^2 \geq \bar{x}_{\rm rms}\,\bar{x}_{\rm g}$. Deduce that $\bar{x} \geq \bar{x}_{\rm g}$.

Example Find the sample variance and sample standard deviation of the data given in Table 17.1.

We have already found that the sample mean is 185.0 to one decimal place. However, when the mean is to be used in the subsequent calculation of the sample variance it is better to use the most accurate value available. In this case the exact value is 184.975, and so using (17.6),

$$s^2 = \frac{1}{8}\left[(188.7 - 184.975)^2 + \cdots + (200.0 - 184.975)^2\right]$$
$$= \frac{1608.36}{8} = 201.0,$$

where once again we have quoted the result to one decimal place. The sample standard deviation is then given by $s = \sqrt{201.0} = 14.2$. As it happens, in this case the difference between the true mean and the rounded value is very small compared with the variation of the individual readings about the mean and using the rounded value has a negligible effect; however, this would not be so if the difference were comparable to the sample standard deviation. ◀

Using the definition (17.7), it is clear that in order to calculate the standard deviation of a sample we must first calculate the sample mean. One nearly always wants to know the mean, but, by using an alternative form for s^2, a more efficient means of finding both the mean and the standard deviation is available. From (17.6), we see that

$$s^2 = \frac{1}{N}\sum_{i=1}^{N}(x_i - \bar{x})^2$$
$$= \frac{1}{N}\sum_{i=1}^{N}x_i^2 - \frac{1}{N}\sum_{i=1}^{N}2x_i\bar{x} + \frac{1}{N}\sum_{i=1}^{N}\bar{x}^2$$
$$= \overline{x^2} - 2\bar{x}^2 + \bar{x}^2 = \overline{x^2} - \bar{x}^2.$$

We may therefore write the sample variance s^2 as

$$s^2 = \overline{x^2} - \bar{x}^2 = \frac{1}{N}\sum_{i=1}^{N}x_i^2 - \left(\frac{1}{N}\sum_{i=1}^{N}x_i\right)^2, \qquad (17.8)$$

from which the sample standard deviation is found by taking the positive square root. Thus, by evaluating the quantities $\sum_{i=1}^{N} x_i$ and $\sum_{i=1}^{N} x_i^2$ for our sample, we can calculate the sample mean and sample standard deviation at the same time.

Example Calculate $\sum_{i=1}^{N} x_i$ and $\sum_{i=1}^{N} x_i^2$ for the data given in Table 17.1 and hence find the mean and standard deviation of the sample.

From Table 17.1, we obtain

$$\sum_{i=1}^{N} x_i = 188.7 + 204.7 + \cdots + 200.0 = 1479.8,$$

$$\sum_{i=1}^{N} x_i^2 = (188.7)^2 + (204.7)^2 + \cdots + (200.0)^2 = 275\,334.36.$$

Since $N = 8$, we find, as before, that

$$\bar{x} = \frac{1479.8}{8} = 185.0, \qquad s = \sqrt{\frac{275\,334.36}{8} - \left(\frac{1479.8}{8}\right)^2} = 14.2,$$

where the results have been quoted to one decimal place. ◀

17.2.3 Moments

By analogy with our discussion of probability distributions in Section 16.5, the sample mean and variance may also be described in terms of the first and second moments of the sample. In general, for a sample x_i, $i = 1, 2, \ldots, N$, we define the rth moment m_r by

$$m_r = \frac{1}{N} \sum_{i=1}^{N} x_i^r. \tag{17.9}$$

Thus the sample mean $\bar{x}$ and variance s^2 may also be written as m_1 and $m_2 - m_1^2$ respectively.[4]

17.2.4 Covariance and correlation

So far we have assumed that each data item of the sample consists of a single number. Now let us suppose that each item of data consists of a pair of numbers, so that the sample is given by (x_i, y_i), $i = 1, 2, \ldots, N$.

We may calculate the sample means, $\bar{x}$ and $\bar{y}$, and sample variances, s_x^2 and s_y^2, of the x_i and y_i values individually but these statistics do not provide any measure of the relationship between the x_i and y_i. By analogy with our discussion in Subsection 16.12.3 we measure any interdependence between the x_i and y_i in terms of the *sample covariance*,

4 As is common practice, we use a convention in which a sample statistic is denoted by the Roman letter corresponding to whichever Greek letter is used to describe the corresponding population statistic. We have not discussed central moments to any significant extent, but the convention applies to them as well. Thus, we use m_r to denote the rth moment of a sample, since in Section 16.5 we denoted the rth moment of a population by μ_r; similarly, n_r and ν_r are used to denote the rth central moments of the sample and population, respectively.

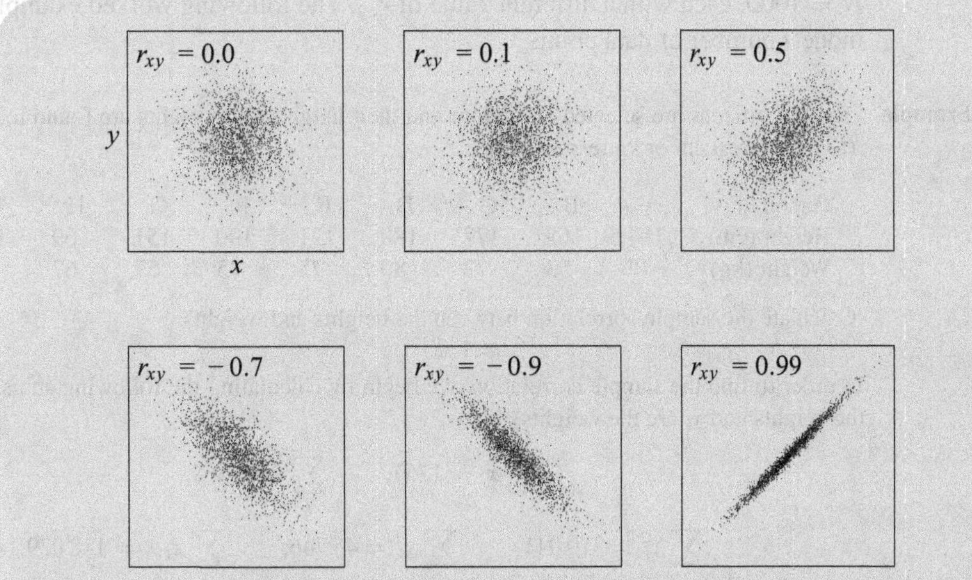

Figure 17.1 Scatter plots for two-dimensional data samples of size $N = 1000$, with various values of the correlation r. No scales are plotted, since the value of r is unaffected by shifts of origin or changes of scale in x and y.

which is given by

$$\begin{aligned} V_{xy} &= \frac{1}{N}\sum_{i=1}^{N}(x_i - \bar{x})(y_i - \bar{y}) \\ &= \overline{(x - \bar{x})(y - \bar{y})} \\ &= \overline{xy} - \bar{x}\bar{y} - \bar{x}\bar{y} + \bar{x}\bar{y} \\ &= \overline{xy} - \bar{x}\bar{y}. \end{aligned} \tag{17.10}$$

Writing out the last expression in full, we obtain the form most useful for calculations, which reads

$$V_{xy} = \frac{1}{N}\left(\sum_{i=1}^{N} x_i y_i\right) - \frac{1}{N^2}\left(\sum_{i=1}^{N} x_i\right)\left(\sum_{i=1}^{N} y_i\right).$$

We may also define the closely related *sample correlation* by

$$r_{xy} = \frac{V_{xy}}{s_x s_y},$$

which can take values between -1 and $+1$. If the x_i and y_i are independent then $V_{xy} = 0 = r_{xy}$, and from (17.10) we see that $\overline{xy} = \bar{x}\bar{y}$. It should also be noted that the value of r_{xy} is not altered by shifts in the origin or by changes in the scale of the x_i or y_i. In other words, if $x' = ax + b$ and $y' = cy + d$, where a, b, c, d are constants, then $r_{x'y'} = r_{xy}$. Figure 17.1 shows scatter plots for several two-dimensional random samples x_i, y_i of size

$N = 1000$, each with a different value of r_{xy}. The following worked example has a more modest number of data points.

Example Ten UK citizens are selected at random and their heights and weights are found to be as follows (to the nearest cm or kg respectively):

Person	A	B	C	D	E	F	G	H	I	J
Height (cm)	194	168	177	180	171	190	151	169	175	182
Weight (kg)	75	53	72	80	75	75	57	67	46	68

Calculate the sample correlation between the heights and weights.

In order to find the sample correlation, we begin by calculating the following sums (where x_i are the heights and y_i are the weights)

$$\sum_i x_i = 1757, \qquad \sum_i y_i = 668,$$

$$\sum_i x_i^2 = 310\,041, \qquad \sum_i y_i^2 = 45\,746, \qquad \sum_i x_i y_i = 118\,029.$$

The sample consists of $N = 10$ pairs of numbers, so the means of the x_i and of the y_i are given by $\bar{x} = 175.7$ and $\bar{y} = 66.8$. Also, $\overline{xy} = 11\,802.9$. Similarly, the standard deviations of the x_i and y_i are calculated, using (17.8), as

$$s_x = \sqrt{\frac{310\,041}{10} - \left(\frac{1757}{10}\right)^2} = 11.6,$$

$$s_y = \sqrt{\frac{45\,746}{10} - \left(\frac{668}{10}\right)^2} = 10.6.$$

Thus the sample correlation is given by

$$r_{xy} = \frac{\overline{xy} - \bar{x}\bar{y}}{s_x s_y} = \frac{11\,802.9 - (175.7)(66.8)}{(11.6)(10.6)} = 0.54.$$

Thus there is a moderate positive correlation between the heights and weights of the people measured. ◀

It is straightforward to generalize the above discussion to data samples of arbitrary dimension, the only complication being one of notation. We choose to denote the ith data item from an n-dimensional sample as $(x_i^{(1)}, x_i^{(2)}, \ldots, x_i^{(n)})$, where the bracketed superscript runs from 1 to n and labels the elements within a given data item whereas the subscript i runs from 1 to N and labels the data items within the sample. In this n-dimensional case, we can define the *sample covariance matrix* whose elements are

$$V_{kl} = \overline{x^{(k)}x^{(l)}} - \overline{x^{(k)}}\;\overline{x^{(l)}}$$

and the *sample correlation matrix* with elements

$$r_{kl} = \frac{V_{kl}}{s_k s_l}.$$

Both these matrices are clearly symmetric but are *not* necessarily positive definite.

17.3 Estimators and sampling distributions

In general, the population $P(\mathbf{x})$ from which a sample $x_1, x_2, \ldots, x_N$ is drawn is *unknown*. The *central aim* of statistics is to use the sample values x_i to infer certain properties of the unknown population $P(\mathbf{x})$, such as its mean, variance and higher moments. To keep our discussion in general terms, let us denote the various parameters of the population by $a_1, a_2, \ldots$, or collectively by $\mathbf{a}$. Moreover, we make the dependence of the population on the values of these quantities explicit by writing the population as $P(\mathbf{x}|\mathbf{a})$. For the moment, we are assuming that the sample values x_i are independent and drawn from the same (one-dimensional) population $P(x|\mathbf{a})$, in which case

$$P(\mathbf{x}|\mathbf{a}) = P(x_1|\mathbf{a})P(x_2|\mathbf{a})\cdots P(x_N|\mathbf{a}).$$

Suppose, we wish to *estimate* the value of one of the quantities $a_1, a_2, \ldots$, which we will denote simply by a. Since the sample values x_i provide our only source of information, any estimate of a must be some function of the x_i, i.e. some sample statistic. Such a statistic is called an *estimator* of a and is usually denoted by $\hat{a}(\mathbf{x})$, where $\mathbf{x}$ denotes the sample elements $x_1, x_2, \ldots, x_N$.

Since an estimator $\hat{a}$ is a function of the sample values of the random variables $x_1, x_2, \ldots, x_N$, it too must be a random variable. In other words, if a number of random samples, each of the same size N, are taken from the (one-dimensional) population $P(x|\mathbf{a})$ then the value of the estimator $\hat{a}$ will vary from one sample to the next and in general will not be equal to the true value a. The form of the sampling distribution generally depends upon the estimator under consideration and upon the form of the population from which the sample was drawn, including the true values of the quantities $\mathbf{a}$. It is also usually dependent on the sample size N, as we now illustrate.

Example The sample values $x_1, x_2, \ldots, x_N$ are drawn independently from a Gaussian distribution with mean μ and variance σ. Suppose that we choose the sample mean $\bar{x}$ as our estimator $\hat{\mu}$ of the population mean. Find the sampling distributions of this estimator.

The sample mean $\bar{x}$ is given by

$$\bar{x} = \frac{1}{N}(x_1 + x_2 + \cdots + x_N),$$

where the x_i are independent random variables distributed as $x_i \sim N(\mu, \sigma^2)$. From our discussion of multiple Gaussian distributions on p. 676, we see immediately that $\bar{x}$ will also be Gaussian distributed as $N(\mu, \sigma^2/N)$. In other words, the sampling distribution of $\bar{x}$ is given by

$$P(\bar{x}|\mu, \sigma) = \frac{1}{\sqrt{2\pi\sigma^2/N}} \exp\left[-\frac{(\bar{x}-\mu)^2}{2\sigma^2/N}\right]. \tag{17.11}$$

Note that the variance of this distribution is σ^2/N, i.e. a factor of $1/N$ smaller than that of the population from which the sample is drawn. ◀

17.3.1 Consistency, bias and efficiency of estimators

For any particular quantity a, we may in fact define any number of different estimators, each of which will have its own sampling distribution. The quality of a given estimator $\hat{a}$ may be assessed by investigating certain properties of its sampling distribution $P(\hat{a}\,|\,\mathbf{a})$. In particular, an estimator $\hat{a}$ is usually judged on the three criteria of *consistency*, *bias* and *efficiency*, each of which we now discuss.

Consistency

An estimator $\hat{a}$ is *consistent* if its value tends to the true value a in the large-sample limit, i.e.

$$\lim_{N\to\infty} \hat{a} = a.$$

Consistency is usually a minimum requirement for a useful estimator. An equivalent statement of consistency is that in the limit of large N the sampling distribution $P(\hat{a}\,|\,\mathbf{a})$ of the estimator must satisfy

$$\lim_{N\to\infty} P(\hat{a}\,|\,\mathbf{a}) \to \delta(\hat{a} - a).$$

Bias

The expectation value of an estimator $\hat{a}$ is given by

$$E[\hat{a}] = \int \hat{a} P(\hat{a}\,|\,\mathbf{a})\, d\hat{a} = \int \hat{a}(\mathbf{x}) P(\mathbf{x}\,|\,\mathbf{a})\, d^N\mathbf{x}, \qquad (17.12)$$

where the second integral extends over all possible values that can be taken by the sample elements $x_1, x_2, \ldots, x_N$. This expression gives the expected mean value of $\hat{a}$ from an infinite number of samples, each of size N. The *bias* of an estimator $\hat{a}$ is then defined as

$$b(\mathbf{a}) = E[\hat{a}] - a. \qquad (17.13)$$

We note that the bias b is a "theoretical" property dependent on the formulation of $\hat{a}$; its value depends on an integral over all of the *possible* values for the x_i, but does *not* depend on the actual measured sample values $x_1, x_2, \ldots, x_N$. In general, though, it will depend on the sample size N, the functional form of the estimator $\hat{a}$ and, as indicated, on the true properties $\mathbf{a}$ of the population, including the true value of a itself. If $b = 0$ then $\hat{a}$ is called an *unbiased* estimator of a. If the form of the bias is known, then a simple correction can be made. The following example covers the cases of a scaling plus shift, expressed by the constants b_1 and b_2; the former will, in general, be a function of N.

Example An estimator $\hat{a}$ is biased in such a way that $E[\hat{a}] = a + b(a)$, where the bias $b(a)$ is given by $(b_1 - 1)a + b_2$ and b_1 and b_2 are known constants. Construct an unbiased estimator of a.

Let us first write $E[\hat{a}]$ in the clearer form

$$E[\hat{a}] = a + (b_1 - 1)a + b_2 = b_1 a + b_2.$$

The task of constructing an unbiased estimator is now trivial, and an appropriate choice is $\hat{a}' = (\hat{a} - b_2)/b_1$, which has the expectation value

$$E[\hat{a}'] = \frac{E[\hat{a}] - b_2}{b_1} = a,$$

i.e. has the required value. ◀

Efficiency

The variance of an estimator is given by

$$V[\hat{a}] = \int (\hat{a} - E[\hat{a}])^2 P(\hat{a}|\mathbf{a})\, d\hat{a} = \int (\hat{a}(\mathbf{x}) - E[\hat{a}])^2 P(\mathbf{x}|\mathbf{a})\, d^N\mathbf{x} \qquad (17.14)$$

and describes the spread of values $\hat{a}$ about $E[\hat{a}]$ that would result from a large number of samples, each of size N. An estimator with a smaller variance is said to be more *efficient* than one with a larger variance. It can be shown[5] that for any given quantity a of the population there exists a theoretical *lower limit* on the variance of *any* estimator $\hat{a}$. This result is known as *Fisher's inequality* (or the *Cramér–Rao inequality*) and reads

$$V[\hat{a}] \geq \left(1 + \frac{\partial b}{\partial a}\right)^2 \Big/ E\left[-\frac{\partial^2 \ln P}{\partial a^2}\right], \qquad (17.15)$$

where P stands for the population $P(\mathbf{x}|\mathbf{a})$ and b is the bias of the estimator. Denoting the quantity on the RHS of (17.15) by V_{min}, the *efficiency* e of an estimator is defined as

$$e = V_{\text{min}}/V[\hat{a}].$$

An estimator for which $e = 1$ is called a *minimum-variance* or *efficient* estimator. Otherwise, if $e < 1$, $\hat{a}$ is called an *inefficient* estimator.

It should be noted that, in general, there is no unique "optimal" estimator $\hat{a}$ for a particular property a. To some extent, there is always a trade-off between bias and efficiency. One must often weigh the relative merits of an unbiased, inefficient estimator against another that is more efficient but slightly biased. Nevertheless, a common choice is the *best unbiased estimator* (BUE), which is simply the unbiased estimator $\hat{a}$ having the smallest variance $V[\hat{a}]$.

Finally, we note that some qualities of estimators are related. For example, suppose that $\hat{a}$ is an unbiased estimator, so that $E[\hat{a}] = a$ and $V[\hat{a}] \to 0$ as $N \to \infty$. Using inequalities similar to those discussed in Subsection 16.5.3, it follows immediately that $\hat{a}$

5 See, for example, K. F. Riley, M. P. Hobson and S. J. Bence, *Mathematical Methods for Physics and Engineering*, 3rd edn. (Cambridge: Cambridge University Press, 2006), subsection 31.3.2.

is also a consistent estimator. Nevertheless, it does *not* follow that a consistent estimator is unbiased.

The next worked example shows that the "natural" estimator of the mean of a Gaussian distribution is, in fact, the best possible.

Example The sample values $x_1, x_2, \ldots, x_N$ are drawn independently from a Gaussian distribution with mean μ and variance σ. Show that the sample mean $\bar{x}$ is a consistent, unbiased, minimum-variance estimator of μ.

We found earlier that the sampling distribution of $\bar{x}$ is given by

$$P(\bar{x}|\mu,\sigma) = \frac{1}{\sqrt{2\pi\sigma^2/N}} \exp\left[-\frac{(\bar{x}-\mu)^2}{2\sigma^2/N}\right],$$

from which we see immediately that $E[\bar{x}] = \mu$ and $V[\bar{x}] = \sigma^2/N$. Thus $\bar{x}$ is an unbiased estimator of μ, i.e. $b = 0$. Moreover, since it is also true that $V[\bar{x}] \to 0$ as $N \to \infty$, $\bar{x}$ is a consistent estimator of μ.

In order to determine whether $\bar{x}$ is a minimum-variance estimator of μ, we must use Fisher's inequality (17.15). Since the sample values x_i are independent and drawn from a Gaussian of mean μ and standard deviation σ, we have, from taking the logarithm of the product of N factors like (16.93), that

$$\ln P(\mathbf{x}|\mu,\sigma) = -\frac{1}{2}\sum_{i=1}^{N}\left[\ln(2\pi\sigma^2) + \frac{(x_i-\mu)^2}{\sigma^2}\right],$$

and, on differentiating twice with respect to μ, we find

$$\frac{\partial^2 \ln P}{\partial \mu^2} = -\frac{N}{\sigma^2}.$$

This is independent of the x_i and so its expectation value is also equal to $-N/\sigma^2$. With b set equal to zero in (17.15), Fisher's inequality thus states that, for *any* unbiased estimator $\hat{\mu}$ of the population mean,

$$V[\hat{\mu}] \geq \frac{\sigma^2}{N}.$$

Since $V[\bar{x}] = \sigma^2/N$, the sample mean $\bar{x}$ is a minimum-variance estimator of μ. ◀

17.3.2 Standard errors on estimators

For a given sample $x_1, x_2, \ldots, x_N$, we may calculate the value of an estimator $\hat{a}(\mathbf{x})$ for the quantity a. It is also necessary, however, to give some measure of the statistical uncertainty in this estimate. One way of characterizing this uncertainty is with the standard deviation of the sampling distribution $P(\hat{a}|\mathbf{a})$, which is given simply by

$$\sigma_{\hat{a}} = (V[\hat{a}])^{1/2}. \tag{17.16}$$

If the estimator $\hat{a}(\mathbf{x})$ were calculated for a large number of samples, each of size N, then the standard deviation of the resulting $\hat{a}$ values would be given by (17.16). Consequently, $\sigma_{\hat{a}}$ is called the *standard error* on our estimate.

In general, however, the standard error $\sigma_{\hat{a}}$ depends on the true values of some or all of the quantities $\mathbf{a}$ and they may be unknown. When this occurs, one must substitute estimated values of any unknown quantities into the expression for $\sigma_{\hat{a}}$ in order to obtain an estimated standard error $\hat{\sigma}_{\hat{a}}$. One then quotes the result as

$$a = \hat{a} \pm \hat{\sigma}_{\hat{a}}.$$

The next worked example assumes that the variance is known, and does not have to be estimated.

Example Ten independent sample values x_i, $i = 1, 2, \ldots, 10$, are drawn at random from a Gaussian distribution with standard deviation $\sigma = 1$. The sample values are as follows (to two decimal places):

2.22 2.56 1.07 0.24 0.18 0.95 0.73 −0.79 2.09 1.81

Estimate the population mean μ, quoting the standard error on your result.

We have shown in the final worked example of Subsection 17.3.1 that, in this case, $\bar{x}$ is a consistent, unbiased, minimum-variance estimator of μ and has variance $V[\bar{x}] = \sigma^2/N$. Thus, our estimate of the population mean with its associated standard error is

$$\hat{\mu} = \bar{x} \pm \frac{\sigma}{\sqrt{N}} = 1.11 \pm 0.32.$$

If the true value of σ had not been known, we would have needed to use an estimated value $\hat{\sigma}$ in the expression for the standard error. Useful basic estimators of σ are discussed in Subsection 17.4.2. ◀

It should be noted that the above approach is most meaningful for unbiased estimators. In this case, $E[\hat{a}] = a$ and so $\sigma_{\hat{a}}$ describes the spread of $\hat{a}$-values about the true value a. For a biased estimator, however, the spread about the true value a is given by the *root mean square error* $\epsilon_{\hat{a}}$, which is defined by

$$\begin{aligned}
\epsilon_{\hat{a}}^2 &= E[(\hat{a} - a)^2] \\
&= E[(\hat{a} - E[\hat{a}])^2] + (E[\hat{a}] - a)^2 \\
&= V[\hat{a}] + b(\mathbf{a})^2.
\end{aligned}$$

We see that $\epsilon_{\hat{a}}^2$ is the sum of the variance of $\hat{a}$ and the square of the bias and so can be interpreted as the sum of squares of statistical and systematic errors.[6] For a biased estimator, it is often more appropriate to quote the result as

$$a = \hat{a} \pm \epsilon_{\hat{a}}.$$

As above, it may be necessary to use estimated values $\hat{\mathbf{a}}$ in the expression for the root mean square error and thus to quote only an estimate $\hat{\epsilon}_{\hat{a}}$ of the error.

6 Carry through this calculation, giving the reasons for the various steps. The notation will probably be less confusing if $E[\hat{a}]$ is denoted by a different symbol, say $\hat{\mu}$. Write $\hat{a} - a$ as $(\hat{a} - \hat{\mu}) + (\hat{\mu} - a)$.

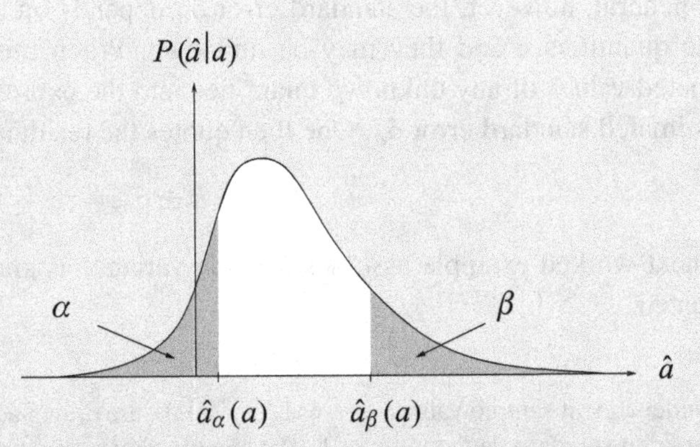

Figure 17.2 The sampling distribution $P(\hat{a}|a)$ of some estimator $\hat{a}$ for a given value of a. The shaded regions indicate the two probabilities $\Pr(\hat{a} < \hat{a}_\alpha(a)) = \alpha$ and $\Pr(\hat{a} > \hat{a}_\beta(a)) = \beta$.

17.3.3 Confidence limits on estimators

An alternative (and often equivalent) way of quoting a statistical error is with a *confidence interval*. Let us assume that, *other* than the quantity of interest a, the quantities $\mathbf{a}$ have known fixed values. Thus we denote the sampling distribution of $\hat{a}$ by $P(\hat{a}|a)$. For any particular value of a, one can determine the two values $\hat{a}_\alpha(a)$ and $\hat{a}_\beta(a)$ such that

$$\Pr(\hat{a} < \hat{a}_\alpha(a)) = \int_{-\infty}^{\hat{a}_\alpha(a)} P(\hat{a}|a)\, d\hat{a} = \alpha, \tag{17.17}$$

$$\Pr(\hat{a} > \hat{a}_\beta(a)) = \int_{\hat{a}_\beta(a)}^{\infty} P(\hat{a}|a)\, d\hat{a} = \beta. \tag{17.18}$$

This is illustrated in Figure 17.2. Thus, for any particular value of a, the probability that the estimator $\hat{a}$ lies within the limits $\hat{a}_\alpha(a)$ and $\hat{a}_\beta(a)$ is given by

$$\Pr(\hat{a}_\alpha(a) < \hat{a} < \hat{a}_\beta(a)) = \int_{\hat{a}_\alpha(a)}^{\hat{a}_\beta(a)} P(\hat{a}|a)\, d\hat{a} = 1 - \alpha - \beta.$$

Now, let us suppose that from our sample $x_1, x_2, \ldots, x_N$, we actually obtain the value $\hat{a}_{\text{obs}}$ for our estimator. If $\hat{a}$ is a good estimator of a then we would expect $\hat{a}_\alpha(a)$ and $\hat{a}_\beta(a)$ to be monotonically increasing functions of a (i.e. $\hat{a}_\alpha$ and $\hat{a}_\beta$ *both* change in the *same* sense as a when the latter is varied). Assuming this to be the case, we can uniquely define the two numbers a_- and a_+ by the relationships

$$\hat{a}_\alpha(a_+) = \hat{a}_{\text{obs}} \qquad \text{and} \qquad \hat{a}_\beta(a_-) = \hat{a}_{\text{obs}}.$$

From (17.17) and (17.18) it follows that

$$\Pr(a_+ < a) = \alpha \qquad \text{and} \qquad \Pr(a_- > a) = \beta,$$

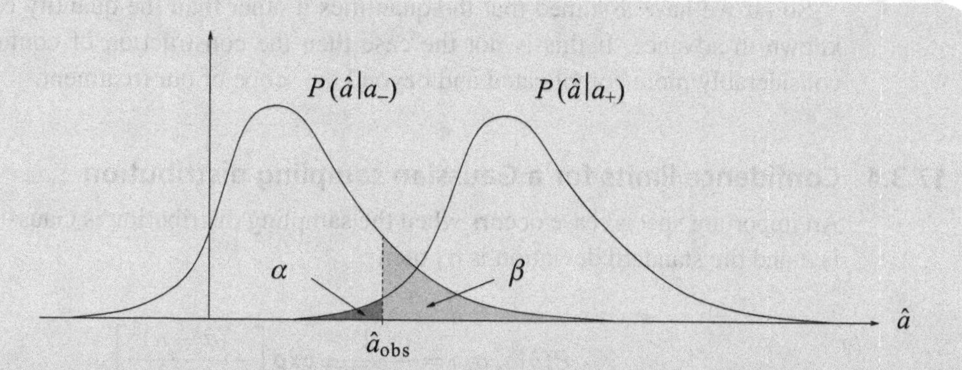

Figure 17.3 An illustration of how the observed value of the estimator, $\hat{a}_{\rm obs}$, and the given values α and β determine the two confidence limits a_- and a_+, which are such that $\hat{a}_\alpha(a_+) = \hat{a}_{\rm obs} = \hat{a}_\beta(a_-)$.

which when taken together imply

$$\Pr(a_- < a < a_+) = 1 - \alpha - \beta. \tag{17.19}$$

Thus, from our estimate $\hat{a}_{\rm obs}$, we have determined two values a_- and a_+ such that this interval contains the true value of a with probability $1 - \alpha - \beta$. It should be emphasized that a_- and a_+ are random variables. If a large number of samples, each of size N, were analyzed then the interval $[a_-, a_+]$ would contain the true value a on a fraction $1 - \alpha - \beta$ of the occasions.

The interval $[a_-, a_+]$ is called a *confidence interval* on a at the *confidence level* $1 - \alpha - \beta$. The values a_- and a_+ themselves are called respectively the *lower confidence limit* and the *upper confidence limit* at this confidence level. In practice, the confidence level is often quoted as a percentage. A convenient way of presenting our results is

$$\int_{-\infty}^{\hat{a}_{\rm obs}} P(\hat{a}|a_+)\, d\hat{a} = \alpha, \tag{17.20}$$

$$\int_{\hat{a}_{\rm obs}}^{\infty} P(\hat{a}|a_-)\, d\hat{a} = \beta. \tag{17.21}$$

The confidence limits may then be found by solving these equations for a_- and a_+ either analytically or numerically. The situation is illustrated graphically in Figure 17.3.

Occasionally one might not combine the results (17.20) and (17.21) but use either one or the other to provide a *one-sided* confidence interval on a. Whenever the results are combined to provide a *two-sided* confidence interval, however, the interval is *not* specified uniquely by the confidence level $1 - \alpha - \beta$. In other words, there are generally an infinite number of intervals $[a_-, a_+]$ for which (17.19) holds. To specify a unique interval, one often chooses $\alpha = \beta$, resulting in the *central confidence interval* on a. All cases can be covered by calculating the quantities $c = \hat{a} - a_-$ and $d = a_+ - \hat{a}$ and quoting the result of an estimate as

$$a = \hat{a}^{+d}_{-c}.$$

So far we have assumed that the quantities **a** other than the quantity of interest a are known in advance. If this is not the case then the construction of confidence limits is considerably more complicated and beyond the scope of our treatment.

17.3.4 Confidence limits for a Gaussian sampling distribution

An important special case occurs when the sampling distribution is Gaussian; if the mean is a and the standard deviation is $\sigma_{\hat{a}}$ then

$$P(\hat{a}|a, \sigma_{\hat{a}}) = \frac{1}{\sqrt{2\pi\sigma_{\hat{a}}^2}} \exp\left[-\frac{(\hat{a}-a)^2}{2\sigma_{\hat{a}}^2}\right]. \tag{17.22}$$

For almost any (consistent) estimator $\hat{a}$, the sampling distribution will tend to this form in the large-sample limit $N \to \infty$, as a consequence of the central limit theorem. For a sampling distribution of the form (17.22), the above procedure for determining confidence intervals becomes straightforward. Suppose, from our sample, we obtain the value $\hat{a}_{\text{obs}}$ for our estimator. In this case, equations (17.20) and (17.21) become

$$\Phi\left(\frac{\hat{a}_{\text{obs}} - a_+}{\sigma_{\hat{a}}}\right) = \alpha,$$
$$1 - \Phi\left(\frac{\hat{a}_{\text{obs}} - a_-}{\sigma_{\hat{a}}}\right) = \beta,$$

where $\Phi(z)$ is the cumulative probability function for the standard Gaussian distribution, discussed in Subsection 16.9.2. Solving these equations for a_- and a_+ gives

$$a_- = \hat{a}_{\text{obs}} - \sigma_{\hat{a}}\Phi^{-1}(1-\beta), \tag{17.23}$$
$$a_+ = \hat{a}_{\text{obs}} + \sigma_{\hat{a}}\Phi^{-1}(1-\alpha); \tag{17.24}$$

we have used the fact that $\Phi^{-1}(\alpha) = -\Phi^{-1}(1-\alpha)$ to make the equations symmetric. The value of the inverse function $\Phi^{-1}(z)$ can be read off directly from Table 16.3, given in Subsection 16.9.2. For the normally used central confidence interval one has $\alpha = \beta$. In this case, we see that quoting a result using the standard error, as

$$a = \hat{a} \pm \sigma_{\hat{a}}, \tag{17.25}$$

is equivalent to taking $\Phi^{-1}(1-\alpha) = 1$. From Table 16.3, we find $\alpha = 1 - 0.8413 = 0.1587$, and so this corresponds to a confidence level of $1 - 2(0.1587) \approx 0.683$. Thus, the standard error limits give the 68.3% central confidence interval. The following example reverses this procedure and finds the central confidence interval needed in order to have a given degree of confidence.

Example Ten independent sample values x_i, $i = 1, 2, \ldots, 10$, are drawn at random from a Gaussian distribution with standard deviation $\sigma = 1$. The sample values are as follows (to two decimal places):

$$2.22 \quad 2.56 \quad 1.07 \quad 0.24 \quad 0.18 \quad 0.95 \quad 0.73 \quad -0.79 \quad 2.09 \quad 1.81$$

Find the 90% central confidence interval on the population mean μ.

Our estimator $\hat{\mu}$ is the sample mean $\bar{x}$. As shown towards the end of Section 17.3, the sampling distribution of $\bar{x}$ is Gaussian with mean $E[\bar{x}]$ and variance $V[\bar{x}] = \sigma^2/N$. Since $\sigma = 1$ in this case, the standard error is given by $\sigma_{\bar{x}} = \sigma/\sqrt{N} = 0.32$. Moreover, in Subsection 17.3.2, we found the mean of the above sample to be $\bar{x} = 1.11$.

For the 90% central confidence interval, we require $\alpha = \beta = 0.05$. From Table 16.3, we find

$$\Phi^{-1}(1 - \alpha) = \Phi^{-1}(0.95) = 1.65,$$

and using (17.23) and (17.24) we obtain

$$a_- = \bar{x} - 1.65\sigma_{\bar{x}} = 1.11 - (1.65)(0.32) = 0.58,$$
$$a_+ = \bar{x} + 1.65\sigma_{\bar{x}} = 1.11 + (1.65)(0.32) = 1.64.$$

Thus, the 90% central confidence interval on μ is $[0.58, 1.64]$. For comparison, the true value used to create the sample was $\mu = 1$. ◀

In the case where the standard error $\sigma_{\hat{a}}$ in (17.25) is not known in advance, one must use a value $\hat{\sigma}_{\hat{a}}$ estimated from the sample. In principle, this complicates somewhat the construction of confidence intervals, since properly one should consider the two-dimensional joint sampling distribution $P(\hat{a}, \hat{\sigma}_{\hat{a}}|a)$. Nevertheless, in practice, provided $\hat{\sigma}_{\hat{a}}$ is a fairly good estimate of $\sigma_{\hat{a}}$ the above procedure may be applied with reasonable accuracy. In the special case where the sample values x_i are drawn from a Gaussian distribution with unknown μ and σ, it is in fact possible to obtain *exact* confidence intervals on the mean μ, for a sample of any size N, using Student's t-distribution. This is discussed in Subsection 17.6.4.

17.4 Some basic estimators

In many cases, one does not even know the functional *form* of the population from which a sample is drawn. Nevertheless, in a case where the sample values $x_1, x_2, \ldots, x_N$ are each drawn *independently* from a one-dimensional population $P(x)$, it is possible to construct some basic estimators for the moments and central moments of $P(x)$. In this section, we investigate the estimating properties of the common sample statistics presented in Section 17.2. In fact, expectation values and variances of these sample statistics can be calculated *without* prior knowledge of the functional form of the population; they depend only on the sample size N and certain moments and central moments of $P(x)$.

17.4.1 Population mean μ

Let us suppose that the parent population $P(x)$ has mean μ and variance σ^2. An obvious estimator $\hat{\mu}$ of the population mean is the sample mean $\bar{x}$. Provided μ and σ^2 are both finite, we may apply the central limit theorem directly to obtain exact expressions, valid for samples of any size N, for the expectation value and variance of $\bar{x}$. From parts (i) and (ii) of the central limit theorem, discussed in Section 16.10, we immediately obtain

$$E[\bar{x}] = \mu, \qquad V[\bar{x}] = \frac{\sigma^2}{N}. \tag{17.26}$$

Thus we see that $\bar{x}$ is an unbiased estimator of μ. Moreover, we note that the standard error in $\bar{x}$ is $\sigma/\sqrt{N}$, and so the sampling distribution of $\bar{x}$ becomes more tightly centered around μ as the sample size N increases. Indeed, since $V[\bar{x}] \to 0$ as $N \to \infty$, $\bar{x}$ is also a consistent estimator of μ.

In the limit of large N, we may in fact obtain an *approximate* form for the full sampling distribution of $\bar{x}$. Part (iii) of the central limit theorem (see Section 16.10) tells us immediately that, for large N, the sampling distribution of $\bar{x}$ is given approximately by the Gaussian form

$$P(\bar{x}|\mu,\sigma) \approx \frac{1}{\sqrt{2\pi\sigma^2/N}} \exp\left[-\frac{(\bar{x}-\mu)^2}{2\sigma^2/N}\right].$$

Note that this does *not* depend on the form of the original parent population. If, however, the parent population is in fact Gaussian then this result is *exact* for samples of *any* size N (as is immediately apparent from our discussion of multiple Gaussian distributions in Subsection 16.9.2).

17.4.2 Population variance σ^2

An estimator for the population variance σ^2 is not so straightforward to define as one for the mean. Complications arise because, in many cases, the true mean of the population μ is not known. Nevertheless, let us begin by considering the case where in fact μ is known. In this event, a useful estimator is

$$\widehat{\sigma^2} = \frac{1}{N}\sum_{i=1}^{N}(x_i-\mu)^2 = \left(\frac{1}{N}\sum_{i=1}^{N}x_i^2\right) - \mu^2, \tag{17.27}$$

and we now prove two of its qualities.

Example Show that $\widehat{\sigma^2}$ is an unbiased and consistent estimator of the population variance σ^2.

The expectation value of $\widehat{\sigma^2}$ is given by

$$E[\widehat{\sigma^2}] = \frac{1}{N}E\left[\sum_{i=1}^{N}x_i^2\right] - \mu^2 = E\left[x_i^2\right] - \mu^2 = \mu_2 - \mu^2 = \sigma^2,$$

from which we see that the estimator is unbiased. The variance of the estimator is

$$V[\widehat{\sigma^2}] = \frac{1}{N^2}V\left[\sum_{i=1}^{N}x_i^2\right] + V\left[\mu^2\right] = \frac{1}{N}V\left[x_i^2\right] = \frac{1}{N}(\mu_4 - \mu_2^2),$$

in which we have used that fact that $V[\mu^2] = 0$ and $V[x_i^2] = E[x_i^4] - (E[x_i^2])^2 = \mu_4 - \mu_2^2$, where μ_r is the rth population moment.[7] Since $\mu_4 - \mu_2^2$ is a fixed quantity, $V[\widehat{\sigma^2}] \to 0$ as $N \to \infty$, showing that $\widehat{\sigma^2}$ is a consistent estimator of σ^2. As it is also unbiased, the full result is established.

◀

If the true mean of the population is unknown, however, a natural alternative is to replace μ by $\bar{x}$ in (17.27), so that our estimator is simply the sample variance s^2 given by

$$s^2 = \frac{1}{N}\sum_{i=1}^{N} x_i^2 - \left(\frac{1}{N}\sum_{i=1}^{N} x_i\right)^2.$$

In order to determine the properties of this estimator, we must calculate $E[s^2]$ and $V[s^2]$. This task is straightforward but lengthy. However, for the investigation of the properties of a *central* moment (i.e. a moment about the mean of the population, rather than about zero) of the sample, there exists a useful trick that simplifies the calculation. It turns out that we can assume, with no loss of generality, that the mean μ_1 of the population from which the sample is drawn is equal to zero. With this assumption, the population central moments, ν_r, are identical to the corresponding moments μ_r, and we may perform our calculation in terms of the latter. At the end, however, we replace μ_r by ν_r in the final result and so obtain a general expression that is valid even in cases where $\mu_1 \neq 0$.

Example Calculate $E[s^2]$ for a sample of size N.

The expectation value of the sample variance s^2 for a sample of size N is given by

$$E[s^2] = \frac{1}{N}E\left[\sum_i x_i^2\right] - \frac{1}{N^2}E\left[\left(\sum_i x_i\right)^2\right]$$

$$= \frac{1}{N}NE\left[x_i^2\right] - \frac{1}{N^2}E\left[\sum_i x_i^2 + \sum_{\substack{i,j\\ j\neq i}} x_i x_j\right]. \tag{17.28}$$

The number of terms in the double summation in (17.28) is $N(N-1)$, so we find

$$E[s^2] = E\left[x_i^2\right] - \frac{1}{N^2}(NE\left[x_i^2\right] + N(N-1)E[x_i x_j]).$$

Now, since the sample elements x_i and x_j are independent, $E[x_i x_j] = E[x_i]E[x_j] = 0$, assuming the mean μ_1 of the parent population to be zero. Denoting the rth moment of the population by μ_r, we thus obtain

$$E[s^2] = \mu_2 - \frac{\mu_2}{N} = \frac{N-1}{N}\mu_2 = \frac{N-1}{N}\sigma^2, \tag{17.29}$$

where in the last line we have used the fact that the population mean is zero, and so $\mu_2 = \nu_2 = \sigma^2$. However, the final result is also valid in the case where $\mu_1 \neq 0$.

7 We have not explicitly proved the second equality, but it is established in the same way as $V[x_i] = \mu_2 - \mu_1^2$, by considering $E[(x_i^2 - \sigma^2)^2]$.

Using the above method, we can also find the variance of s^2, although the algebra is rather heavy going and we will not reproduce it here. The final expression, expressed in terms of central moments ν_r, is

$$V[s^2] = \frac{N-1}{N^3}\left[(N-1)\nu_4 - (N-3)\nu_2^2\right]; \qquad (17.30)$$

a result that is valid whether or not the mean of the population is zero. ◀

From (17.29), we see that s^2 is a *biased* estimator of σ^2, although the bias becomes negligible for large N. However, it immediately follows that an unbiased estimator of σ^2 is given simply by

$$\widehat{\sigma^2} = \frac{N}{N-1}s^2, \qquad (17.31)$$

where the multiplicative factor $N/(N-1)$ is often called *Bessel's correction*. Thus in terms of the sample values x_i, $i = 1, 2, \ldots, N$, an unbiased estimator of the population variance σ^2 is given by

$$\widehat{\sigma^2} = \frac{1}{N-1}\sum_{i=1}^{N}(x_i - \bar{x})^2. \qquad (17.32)$$

Using (17.30), we find that the variance of the estimator $\widehat{\sigma^2}$ is

$$V[\widehat{\sigma^2}] = \left(\frac{N}{N-1}\right)^2 V[s^2] = \frac{1}{N}\left(\nu_4 - \frac{N-3}{N-1}\nu_2^2\right),$$

where ν_r is the rth central moment of the parent population. We note that, since $E[\widehat{\sigma^2}] = \sigma^2$ and $V[\widehat{\sigma^2}] \to 0$ as $N \to \infty$, the statistic $\widehat{\sigma^2}$ is also a consistent estimator of the population variance.

17.4.3 Population standard deviation σ

The standard deviation σ of a population is defined as the positive square root of the population variance σ^2 (as, indeed, our notation suggests). Thus, it is common practice to take the positive square root of the variance estimator as our estimator for σ. Thus, we take

$$\hat{\sigma} = \left(\widehat{\sigma^2}\right)^{1/2}, \qquad (17.33)$$

where $\widehat{\sigma^2}$ is given by either (17.27) or (17.31), depending on whether the population mean μ is known or unknown. Because of the square root in the definition of $\hat{\sigma}$, it is not possible in either case to obtain an exact expression for $E[\hat{\sigma}]$ and $V[\hat{\sigma}]$. Indeed, although in each case the estimator is the positive square root of an unbiased estimator of σ^2, it is *not* itself an unbiased estimator of σ. However, the bias does becomes negligible for large N.

It is possible to obtain approximate expressions for $E[\hat{\sigma}]$ and $V[\hat{\sigma}]$ for a sample of size N. For the case where the population mean μ is unknown, i.e. with $\hat{\sigma} = [N/(N-1)]^{1/2}s$, we quote them without proofs:

$$E[\hat{\sigma}] \approx \sigma \quad \text{and} \quad V[\hat{\sigma}] \approx \frac{1}{4N\nu_2}\left(\nu_4 - \frac{N-3}{N-1}\nu_2^2\right).$$

17.4.4 Population moments μ_r

We may straightforwardly generalize our discussion of estimation of the population mean μ $(= \mu_1)$ in Subsection 17.4.1 to the estimation of the rth population moment μ_r. An obvious choice of estimator is the rth sample moment m_r. The expectation value of m_r is given by

$$E[m_r] = \frac{1}{N}\sum_{i=1}^{N} E[x_i^r] = \frac{N\mu_r}{N} = \mu_r,$$

and so it is an unbiased estimator of μ_r.

The variance of m_r may be found in a similar manner, although the calculation is a little more complicated. We find that

$$\begin{aligned} V[m_r] &= E[(m_r - \mu_r)^2] \\ &= \frac{1}{N^2}E\left[\left(\sum_i x_i^r - N\mu_r\right)^2\right] \\ &= \frac{1}{N^2}E\left[\sum_i x_i^{2r} + \sum_i\sum_{j\neq i} x_i^r x_j^r - 2N\mu_r\sum_i x_i^r + N^2\mu_r^2\right] \\ &= \frac{1}{N}\mu_{2r} + \frac{1}{N^2}\sum_i\sum_{j\neq i} E[x_i^r x_j^r] - 2\mu_r^2 + \mu_r^2. \end{aligned} \tag{17.34}$$

However, since the sample values x_i are assumed to be independent, we have

$$E\left[x_i^r x_j^r\right] = E\left[x_i^r\right]E\left[x_j^r\right] = \mu_r^2. \tag{17.35}$$

The number of terms in the double sum on the RHS of (17.34) is $N(N-1)$, and so we find

$$V[m_r] = \frac{1}{N}\mu_{2r} + \frac{N-1}{N}\mu_r^2 - \mu_r^2 = \frac{\mu_{2r} - \mu_r^2}{N}. \tag{17.36}$$

Since $E[m_r] = \mu_r$ and $V[m_r] \to 0$ as $N \to \infty$, the rth sample moment m_r is also a consistent estimator of μ_r.

Having considered the variance of the rth moment of the sample[8] we now turn to the covariance of two different moments.

8 Which might be thought of as the covariance of the rth moment with itself.

Example Find the covariance of the sample moments m_r and m_s for a sample of size N.

We obtain the covariance of the sample moments m_r and m_s in a similar manner to that used above to obtain the variance of m_r. From the definition of covariance, we have

$$\begin{aligned}
\text{Cov}[m_r, m_s] &= E[(m_r - \mu_r)(m_s - \mu_s)] \\
&= \frac{1}{N^2} E\left[\left(\sum_i x_i^r - N\mu_r\right)\left(\sum_j x_j^s - N\mu_s\right)\right] \\
&= \frac{1}{N^2} E\left[\sum_i x_i^{r+s} + \sum_i \sum_{j\neq i} x_i^r x_j^s - N\mu_r \sum_j x_j^s - N\mu_s \sum_i x_i^r + N^2 \mu_r \mu_s\right].
\end{aligned}$$

Assuming the x_i to be independent, we may again use result (17.35) to obtain

$$\begin{aligned}
\text{Cov}[m_r, m_s] &= \frac{1}{N^2}[N\mu_{r+s} + N(N-1)\mu_r\mu_s - N^2\mu_r\mu_s - N^2\mu_s\mu_r + N^2\mu_r\mu_s] \\
&= \frac{1}{N}\mu_{r+s} + \frac{N-1}{N}\mu_r\mu_s - \mu_r\mu_s \\
&= \frac{\mu_{r+s} - \mu_r\mu_s}{N}.
\end{aligned}$$

We note that by setting $r = s$, we recover the expression (17.36) for $V[m_r]$. ◀

The discussion of estimators for the second central moment ν_2 (or equivalently σ^2) given in Subsection 17.4.2 could be extended to the estimation of a general rth central moment ν_r. However, to do so adds little to the general principles involved; the results are complicated and not easily memorized. If they are needed, a more specialized text or a formula reference book should be consulted.

17.4.5 Population covariance Cov[x, y] and correlation Corr[x, y]

So far we have assumed that each of our N independent samples consists of a single number x_i. Let us now extend our discussion to a situation in which each sample consists of two numbers x_i, y_i, which we may consider as being drawn randomly from a two-dimensional population $P(x, y)$. In particular, we now consider estimators for the population covariance $\text{Cov}[x, y]$ and for the correlation $\text{Corr}[x, y]$.

When μ_x and μ_y are *known*, an appropriate estimator of the population covariance is

$$\widehat{\text{Cov}}[x, y] = \overline{xy} - \mu_x\mu_y = \left(\frac{1}{N}\sum_{i=1}^{N} x_i y_i\right) - \mu_x\mu_y. \tag{17.37}$$

This estimator is unbiased since

$$E\left[\widehat{\text{Cov}}[x, y]\right] = \frac{1}{N}E\left[\sum_{i=1}^{N} x_i y_i\right] - \mu_x\mu_y = E[x_i y_i] - \mu_x\mu_y = \text{Cov}[x, y].$$

Alternatively, if μ_x and μ_y are *unknown*, it is natural to replace μ_x and μ_y in (17.37) by the sample means $\bar{x}$ and $\bar{y}$ respectively, in which case we recover the sample covariance

$V_{xy} = \overline{xy} - \bar{x}\bar{y}$ discussed in Subsection 17.2.4. This estimator is biased, but an unbiased estimator of the population covariance is obtained by forming

$$\widehat{\text{Cov}}\,[x, y] = \frac{N}{N-1} V_{xy}, \tag{17.38}$$

as is proved in the following worked example.

Example Calculate the expectation value of the sample covariance V_{xy} for a sample of size N.

The sample covariance is given by

$$V_{xy} = \left(\frac{1}{N}\sum_i x_i y_i\right) - \left(\frac{1}{N}\sum_i x_i\right)\left(\frac{1}{N}\sum_j y_j\right).$$

Thus its expectation value is given by

$$E[V_{xy}] = \frac{1}{N} E\left[\sum_i x_i y_i\right] - \frac{1}{N^2} E\left[\left(\sum_i x_i\right)\left(\sum_j x_j\right)\right]$$

$$= E[x_i y_i] - \frac{1}{N^2} E\left[\sum_i x_i y_i + \sum_{\substack{i,j\\ j\neq i}} x_i y_j\right].$$

Since the number of terms in the double sum on the RHS is $N(N-1)$, we have

$$E[V_{xy}] = E[x_i y_i] - \frac{1}{N^2}(NE[x_i y_i] + N(N-1)E[x_i y_j])$$

$$= E[x_i y_i] - \frac{1}{N^2}(NE[x_i y_i] + N(N-1)E[x_i]E[y_j])$$

$$= E[x_i y_i] - \frac{1}{N}\left(E[x_i y_i] + (N-1)\mu_x \mu_y\right) = \frac{N-1}{N}\text{Cov}\,[x, y],$$

where we have used the fact that, since the samples are independent, $E[x_i y_j] = E[x_i]E[y_j]$. ◀

It is possible to obtain expressions for the variances of the estimators (17.37) and (17.38) but these quantities depend upon higher moments of the population $P(x, y)$ and are extremely lengthy to calculate.

Whether the means μ_x and μ_y are known or unknown, an estimator of the population correlation $\text{Corr}\,[x, y]$ is given by

$$\widehat{\text{Corr}}\,[x, y] = \frac{\widehat{\text{Cov}}\,[x, y]}{\hat{\sigma}_x \hat{\sigma}_y}, \tag{17.39}$$

where $\widehat{\text{Cov}}\,[x, y]$, $\hat{\sigma}_x$ and $\hat{\sigma}_y$ are the appropriate estimators of the population covariance and standard deviations. Although this estimator is unbiased only asymptotically, i.e. for large N, it is widely used because of its simplicity. Once again the variance of the estimator depends on the higher moments of $P(x, y)$ and is difficult to calculate.

In the case in which the means μ_x and μ_y are unknown, a suitable (but biased) estimator is

$$\widehat{\text{Corr}}\,[x, y] = \frac{N}{N-1}\frac{V_{xy}}{s_x s_y} = \frac{N}{N-1} r_{xy}, \tag{17.40}$$

where s_x and s_y are the sample standard deviations of the x_i and y_i respectively and r_{xy} is the sample correlation. In the special case when the parent population $P(x, y)$ is Gaussian, it may be shown that, if $\rho = \text{Corr}\,[x, y]$,

$$E[r_{xy}] = \rho - \frac{\rho(1-\rho^2)}{2N} + O(N^{-2}), \tag{17.41}$$

$$V[r_{xy}] = \frac{1}{N}(1-\rho^2)^2 + O(N^{-2}), \tag{17.42}$$

from which the expectation value and variance of the estimator $\widehat{\text{Corr}}\,[x, y]$ may be found immediately.

We note finally that our discussion may be extended, without significant alteration, to the general case in which each data item consists of n numbers $x_i, y_i, \ldots, z_i$.

17.4.6 A worked example

To conclude our discussion of basic estimators, we reconsider the set of experimental data given in Subsection 17.2.4. We carry the analysis as far as calculating the standard errors in the estimated population parameters, including the population correlation.

Example Ten UK citizens are selected at random and their heights and weights are found to be as follows (to the nearest cm or kg respectively):

Person	A	B	C	D	E	F	G	H	I	J
Height (cm)	194	168	177	180	171	190	151	169	175	182
Weight (kg)	75	53	72	80	75	75	57	67	46	68

Estimate the means, μ_x and μ_y, and standard deviations, σ_x and σ_y, of the two-dimensional joint population from which the sample was drawn, quoting the standard error on the estimate in each case. Estimate also the correlation $\text{Corr}[x, y]$ of the population, and quote the standard error on the estimate under the assumption that the population is a multivariate Gaussian.

In Subsection 17.2.4, we calculated various sample statistics for these data. In particular, we found that for our sample of size $N = 10$,

$$\bar{x} = 175.7, \qquad \bar{y} = 66.8,$$
$$s_x = 11.6, \qquad s_y = 10.6, \qquad r_{xy} = 0.54.$$

Let us begin by estimating the means μ_x and μ_y. As discussed in Subsection 17.4.1, the sample mean is an unbiased, consistent estimator of the population mean. Moreover, the standard error on $\bar{x}$ (say) is $\sigma_x/\sqrt{N}$. In this case, however, we do not know the true value of σ_x and we must estimate it using $\widehat{\sigma_x} = \sqrt{N/(N-1)}s_x$. Thus, our estimates of μ_x and μ_y, with associated standard errors, are

$$\hat{\mu}_x = \bar{x} \pm \frac{s_x}{\sqrt{N-1}} = 175.7 \pm 3.9,$$

$$\hat{\mu}_y = \bar{y} \pm \frac{s_y}{\sqrt{N-1}} = 66.8 \pm 3.5.$$

We now turn to estimating σ_x and σ_y. As just mentioned, our estimate of σ_x (say) is $\widehat{\sigma_x} = \sqrt{N/(N-1)}s_x$. Its variance (see the final line of Subsection 17.4.3) is given approximately by

$$V[\hat{\sigma}] \approx \frac{1}{4N\nu_2}\left(\nu_4 - \frac{N-3}{N-1}\nu_2^2\right).$$

Since we do not know the true values of the population central moments ν_2 and ν_4, we must use their estimated values in this expression. We may take $\hat{\nu}_2 = \widehat{\sigma_x^2} = (\hat{\sigma})^2$, which we have already calculated. It still remains, however, to estimate ν_4. It can be shown[9] that to $O(N^{-1})$, it is acceptable to take $\hat{\nu}_4 = n_4$. Thus for the x_i and y_i values, we have

$$(\hat{\nu}_4)_x = \frac{1}{N}\sum_{i=1}^{N}(x_i - \bar{x})^4 = 53\,411.6,$$

$$(\hat{\nu}_4)_y = \frac{1}{N}\sum_{i=1}^{N}(y_i - \bar{y})^4 = 27\,732.5.$$

Substituting these values into (17.33), we obtain

$$\hat{\sigma}_x = \left(\frac{N}{N-1}\right)^{1/2} s_x \pm (\hat{V}[\hat{\sigma}_x])^{1/2} = 12.2 \pm 6.7, \qquad (17.43)$$

$$\hat{\sigma}_y = \left(\frac{N}{N-1}\right)^{1/2} s_y \pm (\hat{V}[\hat{\sigma}_y])^{1/2} = 11.2 \pm 3.6. \qquad (17.44)$$

Finally, we estimate the population correlation $\text{Corr}[x, y]$, which we shall denote by ρ. From (17.40), we have

$$\hat{\rho} = \frac{N}{N-1}r_{xy} = 0.60.$$

Under the *assumption* that the sample was drawn from a two-dimensional Gaussian population $P(x, y)$, the variance of our estimator is given by (17.42). Since we do not know the true value of ρ, we must use our estimate $\hat{\rho}$. We find that the standard error $\Delta\rho$ in our estimate is given approximately by

$$\Delta\rho \approx \frac{10}{9}\left(\frac{1}{10}\right)[1 - (0.60)^2]^2 = 0.05.$$

Thus, in the standard notation, the estimate is that $\rho = 0.60 \pm 0.05$. ◀

9 The full relationship is $E[n_4] = N^{-3}(N-1)[(N^2 - 3N + 3)\nu_4 + 3(2N-3)\nu_2^2]$.

17.5 Data modeling

In the previous sections, we assumed that the populations $P(x)$ or $P(x, y, \ldots)$, from which our independent experimental samples were drawn, had unknown functional forms and used estimators of population statistics that made no assumptions about the actual functional forms. We now briefly discuss the process of *data modeling*, in which a specific form for the population is assumed, but with that form containing one or more parameters. The purpose of the analysis is to extract the best values, with errors, for those parameters.

The most general approach to this type of problem is the method known as *maximum likelihood* in which the probability of obtaining the observed set of data points is calculated as a function of the parameter values. For independent sample values, all drawn from the assumed population, the probability (the likelihood function) is simply the product of the individual single sample probabilities. The values of the parameters that maximize the likelihood function are then determined, either analytically or by numerical methods, and accepted as the best estimate of those parameters.

A full discussion of the maximum likelihood method, including the associated biases, efficiencies, errors and confidence limits, is beyond the level and scope of this book, and we will restrict our attention to just one particular aspect of it known as the *method of least squares*.

17.5.1 The method of least squares

This approach is so widely used as a method of parameter estimation that it has acquired this special name of its own. Its principal use in an experimental setting is the determination of the parameters a_i in a functional form f that connects two measured experimental quantities, x and y, through $y = f(x)$. For example, we might have

$$y = a_2x^2 + a_1x + a_0 \quad \text{or} \quad y = a_1 \exp\left(-\frac{a_2}{x}\right).$$

At the outset, let us suppose that a data sample consists of a set of pairs (x_i, y_i), $i = 1, 2, \ldots, N$. For example, these data might correspond to the temperature y_i measured at various points x_i along some metal rod.

We will suppose that the x_i are known with negligible error, whereas there exists a measurement error (or *noise*) n_i on each of the values y_i.[10] Moreover, let us assume that the true value of y at any position x is given by some function $y = f(x; \mathbf{a})$ that depends on the M unknown parameters $\mathbf{a}$. Then

$$y_i = f(x_i; \mathbf{a}) + n_i.$$

Our aim is to estimate the values of the parameters $\mathbf{a}$ from the data sample.

Bearing in mind the central limit theorem, let us suppose that the n_i are drawn from a *Gaussian* distribution with no systematic bias and hence zero mean. And, for the moment, let us further suppose that the measurement errors are independent, though not necessarily all with the same variance. Denoting the variance of n_i by σ_i^2, the likelihood function

10 The treatment of situations in which there are significant errors in both x and y is beyond the scope of this book.

takes the form

$$L(\mathbf{x}, \mathbf{y}; \mathbf{a}) = \frac{1}{(2\pi)^{N/2} Q} \exp\left[-\tfrac{1}{2}\chi^2(\mathbf{a})\right],$$

where Q is a normalizing constant equal to $\prod_{i=1}^{N} \sigma_i$. The vectors $\mathbf{x}$, $\mathbf{y}$ and $\mathbf{a}$ are shorthand for the sets of measurements x_i and y_i ($i = 1, \ldots, N$), and the parameters a_j ($j = 1, \ldots, M$). The quantity denoted by χ^2 takes the form

$$\chi^2(\mathbf{a}) = \sum_{i=1}^{N} \left[\frac{y_i - f(x_i; \mathbf{a})}{\sigma_i}\right]^2.$$

In words, χ^2 is the sum of the squares of the differences between the measured and predicted values of y_i, inversely weighted according to the variance of the corresponding measurement error; the smaller the error, the greater the emphasis laid upon the corresponding deviation from the predicted value of y.

The least-squares (LS) estimators $\hat{\mathbf{a}}_{\text{LS}}$ of the parameter values are defined as those that maximize the value of L or, equivalently in this case, minimize the value of $\chi^2(\mathbf{a})$; they are usually determined by solving the M equations

$$\left.\frac{\partial \chi^2}{\partial a_i}\right|_{\mathbf{a}=\hat{\mathbf{a}}_{\text{LS}}} = 0 \qquad \text{for } i = 1, 2, \ldots, M. \tag{17.45}$$

If the measurement errors n_i are Gaussian distributed, as assumed above, then the least squares and maximum likelihood estimators of the parameters $\mathbf{a}$ coincide. However, in practice, because of its relative simplicity, the method of least squares is often applied to cases in which the n_i are *not* Gaussian distributed. The resulting estimators $\hat{\mathbf{a}}_{\text{LS}}$ are then *not* the ML estimators, and the best that can be said in justification is that the method, whilst not optimal, is a reasonable procedure for parameter estimation.

In the most general case, the measurement errors n_i will *not* be independent but be described by an N-dimensional multivariate Gaussian with a non-trivial covariance matrix N, whose elements $N_{ij} = \text{Cov}\,[n_i, n_j]$ we assume to be known. Under these assumptions the likelihood function is

$$L(\mathbf{x}, \mathbf{y}; \mathbf{a}) = \frac{1}{(2\pi)^{N/2} |\mathsf{N}|^{1/2}} \exp\left[-\tfrac{1}{2}\chi^2(\mathbf{a})\right],$$

where the quantity denoted by χ^2 is now given by the quadratic form

$$\chi^2(\mathbf{a}) = \sum_{i,j=1}^{N} [y_i - f(x_i; \mathbf{a})](\mathsf{N}^{-1})_{ij}[y_j - f(x_j; \mathbf{a})] = (\mathsf{y} - \mathsf{f})^{\mathrm{T}} \mathsf{N}^{-1} (\mathsf{y} - \mathsf{f}). \tag{17.46}$$

In the last equality, we have rewritten the expression in matrix notation by defining the column vector f with elements $f_i = f(x_i; \mathbf{a})$.[11]

Finally, we note that the method of least squares is easily extended to the case in which each measurement y_i depends on several variables, which we denote by $\mathbf{x}_i$. For example,

11 In the common but special case in which the measurement errors n_i *are* independent, their covariance matrix takes the diagonal form $\mathsf{N} = \text{diag}(\sigma_1^2, \sigma_2^2, \ldots, \sigma_N^2)$. In this case, the quadratic expression (17.46) for χ^2 reduces to that given previously.

y_i might represent the temperature measured at the (three-dimensional) position $\mathbf{x}_i$ in a room. In this case, the data is modeled by a function $y = f(\mathbf{x}_i; \mathbf{a})$, and the remainder of the above discussion carries through unchanged.

17.5.2 Linear least squares

We have so far made no restriction on the form of the function $f(x; \mathbf{a})$. It so happens, however, that, for a model in which $f(x; \mathbf{a})$ is a *linear* function of the parameters $a_1, a_2, \ldots, a_M$, one can always obtain analytic expressions for the LS estimators $\hat{\mathbf{a}}_{\text{LS}}$ and their variances. The general form of this kind of model is

$$f(x_i; \mathbf{a}) = \sum_{j=1}^{M} a_j h_j(x_i) = \sum_{j=1}^{M} R_{ij} a_j, \tag{17.47}$$

where $\{h_1(x), h_2(x), \ldots, h_M(x)\}$ is some set of linearly independent fixed functions of x, often called the *basis functions*, and $R_{ij} = h_j(x_i)$; the matrix R is known as the *response matrix*. Note that the functions $h_i(x)$ themselves may be highly non-linear functions of x. The "linear" nature of the model (17.47) refers only to its dependence on the *parameters* a_i. Furthermore, in this case, it may be shown that the LS estimators $\hat{a}_i$ have zero bias and are minimum-variance, irrespective of the probability density function from which the measurement errors n_i are drawn.

Because the proofs of general results for linear least squares are rather complicated, they have been placed in an appendix (Appendix G) and here, in the main text, we will only quote those results and apply them to straight-line fits to experimental data. However it should be emphasized that the results apply to a wide range of sets of basis functions; the only requirement is that $f(x; \mathbf{a})$ can be written as a function linear in the a_i, as in (17.47).

In order to obtain analytic expressions for the LS estimators $\hat{\mathbf{a}}_{\text{LS}}$, we consider the expression for χ^2 given in (17.46). Written in matrix notation, it is

$$\chi^2(\mathbf{a}) = (\mathsf{y} - \mathsf{Ra})^{\mathrm{T}} \mathsf{N}^{-1} (\mathsf{y} - \mathsf{Ra}), \tag{17.48}$$

where, as previously, N is the covariance matrix of the measurement errors. The LS estimates of the parameters $\mathbf{a}$ are now found, as shown in (17.45), by differentiating (17.48) with respect to the a_i and setting the resulting expressions equal to zero.

Provided the matrix $\mathsf{R}^{\mathrm{T}}\mathsf{N}^{-1}\mathsf{R}$ is not singular, this gives, as the solution for $\hat{\mathbf{a}}$, that

$$\hat{\mathsf{a}} = (\mathsf{R}^{\mathrm{T}}\mathsf{N}^{-1}\mathsf{R})^{-1}\mathsf{R}^{\mathrm{T}}\mathsf{N}^{-1}\mathsf{y} \equiv \mathsf{Sy}, \tag{17.49}$$

thus defining the $M \times N$ matrix S. It follows that the LS estimates $\hat{a}_i$, $i = 1, 2, \ldots, M$, are linear functions of the original measurements y_j, $j = 1, 2, \ldots, N$. Moreover, the covariance matrix of the estimators $\hat{a}_i$ is given by

$$\mathsf{V} \equiv \mathrm{Cov}[\hat{a}_i, \hat{a}_j] = \mathsf{SNS}^{\mathrm{T}} = (\mathsf{R}^{\mathrm{T}}\mathsf{N}^{-1}\mathsf{R})^{-1}. \tag{17.50}$$

The two equations (17.49) and (17.50) contain the complete method of least squares. Further, if the LS estimates are calculated using (17.49), then their covariance matrix (17.50) has already been determined as a by-product.

We now give an example of the simplest use of linear least squares analysis, that of fitting the best straight line $y = mx + c$ to a set of N pairs of x, y values. The formulae we will derive for $\hat{m}$ and $\hat{c}$ are standard ones; they are to be found in many books of mathematical formulae as well as being "hard-wired" into most calculators under the title "linear regression".

Example An experiment produces the following data sample pairs (x_i, y_i):

x_i:	1.85	2.72	2.81	3.06	3.42	3.76	4.31	4.47	4.64	4.99
y_i:	2.26	3.10	3.80	4.11	4.74	4.31	5.24	4.03	5.69	6.57

where the x_i-values are known exactly but each y_i-value is measured only to an accuracy of $\sigma = 0.5$. Assuming the underlying model for the data to be a straight line $y = mx + c$, find the LS estimates of the slope m and intercept c and quote the standard error on each estimate.

The data are plotted in Figure 17.4, together with error bars indicating the uncertainty in the y_i-values. Our model of the data is a straight line, and so we have

$$f(x; c, m) = c + mx.$$

In the language of (17.47), our basis functions are $h_1(x) = 1$ and $h_2(x) = x$ and our model parameters are $a_1 = c$ and $a_2 = m$. From (17.47) the elements of the response matrix are $R_{ij} = h_j(x_i)$, and so

$$\mathsf{R} = \begin{pmatrix} 1 & x_1 \\ 1 & x_2 \\ \vdots & \vdots \\ 1 & x_N \end{pmatrix}, \qquad (17.51)$$

where x_i are the data values and $N = 10$ in our case. Further, since the standard deviation on each measurement error is σ, we have $\mathsf{N} = \sigma^2\mathsf{I}$, where I is the $N \times N$ identity matrix. Because of this simple form for N, the expression (17.49) for the LS estimates reduces to

$$\hat{\mathsf{a}} = \sigma^2(\mathsf{R}^{\mathrm{T}}\mathsf{R})^{-1}\frac{1}{\sigma^2}\mathsf{R}^{\mathrm{T}}\mathsf{y} = (\mathsf{R}^{\mathrm{T}}\mathsf{R})^{-1}\mathsf{R}^{\mathrm{T}}\mathsf{y}. \qquad (17.52)$$

Note that we cannot expand the inverse in this equation, since R itself is not square and hence does not possess an inverse. Inserting the form for R in (17.51) into the expression (17.52), we find

$$\begin{aligned} \begin{pmatrix} \hat{c} \\ \hat{m} \end{pmatrix} &= \begin{pmatrix} \sum_i 1 & \sum_i x_i \\ \sum_i x_i & \sum_i x_i^2 \end{pmatrix}^{-1} \begin{pmatrix} \sum_i y_i \\ \sum_i x_i y_i \end{pmatrix} \\ &= \frac{1}{N(\overline{x^2} - \bar{x}^2)} \begin{pmatrix} \overline{x^2} & -\bar{x} \\ -\bar{x} & 1 \end{pmatrix} \begin{pmatrix} N\bar{y} \\ N\overline{xy} \end{pmatrix}. \end{aligned}$$

We thus obtain the LS estimates

$$\hat{m} = \frac{\overline{xy} - \bar{x}\,\bar{y}}{\overline{x^2} - \bar{x}^2} \quad \text{and} \quad \hat{c} = \frac{\overline{x^2}\bar{y} - \bar{x}\,\overline{xy}}{\overline{x^2} - \bar{x}^2} = \bar{y} - \hat{m}\bar{x}, \qquad (17.53)$$

where the last expression for $\hat{c}$ shows that the best-fit line passes through the "center of mass" $(\bar{x}, \bar{y})$ of the data sample. To find the standard errors on our results, we must calculate the covariance

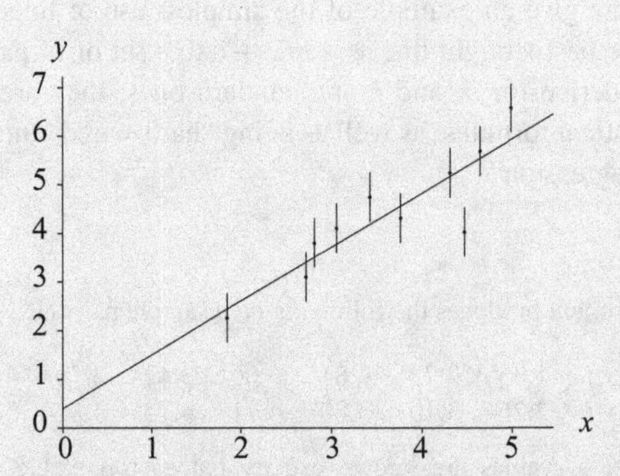

Figure 17.4 A set of data points with error bars indicating the uncertainty $\sigma = 0.5$ on the y-values. The straight line is $y = \hat{m}x + \hat{c}$, where $\hat{m}$ and $\hat{c}$ are the least-squares estimates of the slope and intercept.

matrix of the estimators. This is given by (17.50), and has effectively already been calculated above as

$$\mathsf{V} = \sigma^2(\mathsf{R}^\mathrm{T}\mathsf{R})^{-1} = \frac{\sigma^2}{N(\overline{x^2} - \bar{x}^2)}\begin{pmatrix} \overline{x^2} & -\bar{x} \\ -\bar{x} & 1 \end{pmatrix}. \tag{17.54}$$

The standard error on each estimator is simply the positive square root of the corresponding diagonal element, i.e. $\sigma_{\hat{c}} = \sqrt{V_{11}}$ and $\sigma_{\hat{m}} = \sqrt{V_{22}}$, and the covariance of the estimators $\hat{m}$ and $\hat{c}$ is given by $\mathrm{Cov}\,[\hat{c}, \hat{m}] = V_{12} = V_{21}$. Inserting the data sample averages and moments into (17.53) and (17.54), we find

$$c = \hat{c} \pm \sigma_{\hat{c}} = 0.40 \pm 0.62 \qquad \text{and} \qquad m = \hat{m} \pm \sigma_{\hat{m}} = 1.11 \pm 0.17.$$

The "best-fit" straight line $y = \hat{m}x + \hat{c}$ is plotted in Figure 17.4. For comparison, the true values used to create the data were $m = 1$ and $c = 1$. ◀

The extension of the method to fitting data to a higher-order polynomial, such as $f(x; \mathbf{a}) = a_1 + a_2x + a_3x^2$, is obvious. However, as the order of the polynomial increases the matrix inversions become rather complicated. Indeed, even when the matrices are inverted numerically, the inversion is prone to numerical instabilities. A better approach is to replace the basis functions $h_m(x) = x^m$, $m = 1, 2, \ldots, M$, with a set of polynomials that are "orthogonal over the data", i.e. such that

$$\sum_{i=1}^{N} h_l(x_i)h_m(x_i) = 0 \qquad \text{for } l \neq m.$$

Such a set of polynomial basis functions can always be found by using the Gram–Schmidt orthogonalization procedure presented in Section 8.1. The details of this approach are beyond the scope of our discussion but we note that, in this case, the matrix $\mathsf{R}^\mathrm{T}\mathsf{R}$ is diagonal and may be inverted easily.

If the function $f(x; \mathbf{a})$ is *not* linear in the parameters $\mathbf{a}$ then, in general, it is not possible to obtain an explicit expression for the LS estimates $\hat{\mathbf{a}}$ and one must employ an iterative (numerical) procedure. In practice, however, such problems are best solved using one of the many commercially available software packages.

17.6 Hypothesis testing

So far we have concentrated on using a data sample to obtain a number or a set of numbers. These numbers may be estimated values for the moments or central moments of the population from which the sample was drawn or, more generally, the values of some parameters $\mathbf{a}$ in an assumed model for the data. Sometimes, however, one wishes to use the data to give a "yes" or "no" answer to a particular question. For example, one might wish to know whether some assumed model does, in fact, provide a good fit to the data, or whether two parameters have the same value.

17.6.1 Simple and composite hypotheses

In order to use data to answer questions of this sort, the question must be posed precisely. This is done by first asserting that some *hypothesis* is true. The hypothesis under consideration is traditionally called the *null hypothesis* and is denoted by H_0. In particular, this usually specifies some form $P(\mathbf{x}|H_0)$ for the probability density function from which the data $\mathbf{x}$ are drawn. If the hypothesis determines the PDF uniquely, then it is said to be a *simple hypothesis*. If, however, the hypothesis determines the functional form of the PDF but not the values of certain parameters $\mathbf{a}$ on which it depends then it is called a *composite hypothesis*.

One decides whether to *accept* or *reject* the null hypothesis H_0 by performing some *statistical test*, as described below in Subsection 17.6.2. In fact, formally one uses a statistical test to decide between the null hypothesis H_0 and the *alternative hypothesis* H_1. We define the latter to be the complement $\overline{H_0}$ of the null hypothesis *within some restricted hypothesis space known (or assumed) in advance.* Hence, rejection of H_0 implies acceptance of H_1, and vice versa.

As an example, let us consider the case in which a sample $\mathbf{x}$ is drawn from a Gaussian distribution with a known variance σ^2 but with an unknown mean μ. If one adopts the null hypothesis H_0 that $\mu = 0$, which we write as $H_0 : \mu = 0$, then the corresponding alternative hypothesis must be $H_1 : \mu \neq 0$. Note that, in this case, H_0 is a simple hypothesis whereas H_1 is a composite hypothesis. If, however, one adopted the null hypothesis $H_0 : \mu < 0$ then the alternative hypothesis would be $H_1 : \mu \geq 0$, so that both H_0 and H_1 would be composite hypotheses. Very occasionally both H_0 and H_1 will be simple hypotheses. In our illustration, this would occur, for example, if one knew in advance that the mean μ of the Gaussian distribution were equal to either zero or unity. In this case, if one adopted the null hypothesis $H_0 : \mu = 0$ then the alternative hypothesis would be $H_1 : \mu = 1$.

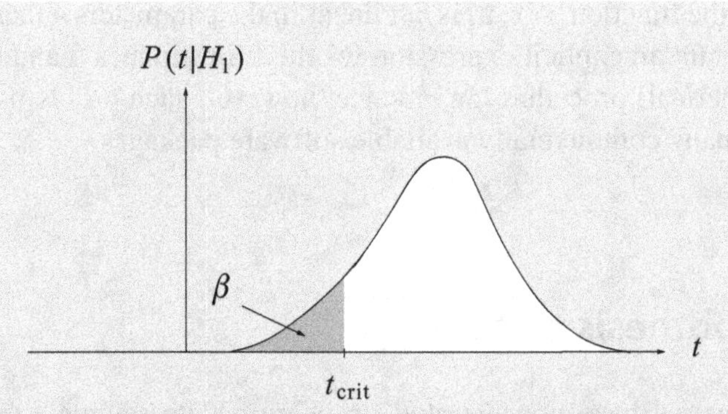

Figure 17.5 The sampling distributions $P(t|H_0)$ and $P(t|H_1)$ of a test statistic t. The shaded areas indicate the (one-tailed) regions for which $\Pr(t > t_{\rm crit}|H_0) = \alpha$ and $\Pr(t < t_{\rm crit}|H_1) = \beta$ respectively.

17.6.2 Statistical tests

In our discussion of hypothesis testing we will restrict our attention to cases in which the null hypothesis H_0 is *simple* (see above). We begin by constructing a *test statistic* $t(\mathbf{x})$ from the data sample. Although, in general, the test statistic need not be just a (scalar) number, and could be a multi-dimensional (vector) quantity, we will restrict our attention to the former case. Like any statistic, $t(\mathbf{x})$ will be a random variable. Moreover, given the simple null hypothesis H_0 concerning the PDF from which the sample was drawn, we may determine (in principle) the sampling distribution $P(t|H_0)$ of the test statistic. A typical example of such a sampling distribution is shown in Figure 17.5. One defines for t a *rejection region* containing some fraction α of the total probability. For example, the (one-tailed) rejection region could consist of values of t greater than some value $t_{\rm crit}$, for which

$$\Pr(t > t_{\rm crit}|H_0) = \int_{t_{\rm crit}}^{\infty} P(t|H_0)\,dt = \alpha; \qquad (17.55)$$

this is indicated by the shaded region in the upper half of Figure 17.5. Equally, a (one-tailed) rejection region could consist of values of t less than some value $t_{\rm crit}$. Alternatively, one could define a (two-tailed) rejection region by two values t_1 and t_2 such that $\Pr(t_1 < t < t_2|H_0) = \alpha$. In all cases, if the observed value of t lies in the rejection region then H_0 is *rejected* at *significance level* α; otherwise H_0 is *accepted* at this same level.

It is clear that there is a probability α of rejecting the null hypothesis H_0 even if it is true. This is called an *error of the first kind*. Conversely, an *error of the second kind* occurs when the hypothesis H_0 is accepted even though it is false (in which case H_1 is true). The probability β (say) that such an error will occur is, in general, difficult to calculate, since the alternative hypothesis H_1 is usually composite. However, in those cases where H_1 is a simple hypothesis, it is straightforward to calculate β (at least in principle).

Denoting the corresponding sampling distribution of t by $P(t|H_1)$, the probability β is the integral of $P(t|H_1)$ over the *complement* of the rejection region, called the *acceptance region*. For example, in the case corresponding to (17.55) this probability is given by

$$\beta = \Pr(t < t_{\text{crit}}|H_1) = \int_{-\infty}^{t_{\text{crit}}} P(t|H_1)\,dt.$$

This is illustrated in Figure 17.5. The quantity $1 - \beta$ is called the *power* of the statistical test to reject the wrong hypothesis.

In the case where H_0 and H_1 are both simple hypotheses, there exists a prescription (the *Neyman–Pearson lemma*) which allows one to determine the "best" rejection region and test statistic to use. However, as mentioned earlier, this situation rarely occurs, and so we will not pursue it here, but move on to the much more common case where, although H_0 is simple, H_1 is composite.

17.6.3 The generalized likelihood-ratio test

If the alternative hypothesis H_1 is composite then the corresponding distribution $P(\mathbf{x}|H_1)$ is not uniquely determined, in general, and a practical, rather than optimal approach has to be taken.

In line with our intention to concentrate on cases in which H_0 is simple, suppose that H_0 is the simple hypothesis $H_0 : \mathbf{a} = \mathbf{a}_0$. Then, a suitable test statistic $t(\mathbf{x})$ is provided by the ratio of the likelihood of the observed measurements taking $\mathbf{a}$ to be equal to $\mathbf{a}_0$, to the same likelihood with $\mathbf{a} = \hat{\mathbf{a}}$, where $\hat{\mathbf{a}}$ is the usual "best" maximum-likelihood estimate of $\mathbf{a}$; as a formula

$$t(\mathbf{x}) = \frac{L(\mathbf{x}; \mathbf{a}_0)}{L(\mathbf{x}; \hat{\mathbf{a}})}. \tag{17.56}$$

It is clear that t is a function of the sample values only and must lie between 0 and 1. Its sampling distribution $P(t|H_0)$ can now be determined (in principle).

As discussed in the previous subsection, the rejection region to be used for a given significance α is simply $t < t_{\text{crit}}$, where the value t_{crit} depends on α, but may not be optimal. Clearly, an equivalent procedure is to use as a test statistic $u = f(t)$, where $f(t)$ is any monotonically increasing function of t; the corresponding rejection region is then $u < f(t_{\text{crit}})$. Similarly, one may use a test statistic $v = g(t)$, where $g(t)$ is any monotonically decreasing function of t; the rejection region then becomes $v > g(t_{\text{crit}})$.

As the distributions of many sampling statistics turn out to be χ^2 distributions with various numbers of degrees of freedom and there are a number of significance levels that are commonly used, e.g. 10%, 5%, 1% and 0.1%, Table 17.2 is provided as a convenient reference. It is used in the following example.

Table 17.2 *The tabulated values are those which a variable distributed as χ^2 with n degrees of freedom exceeds with the given percentage probability. For example, a variable having a χ^2 distribution with 14 degrees of freedom takes values in excess of 21.06 on 10% of occasions*

%	99	95	10	5	2.5	1	0.5	0.1
$n = 1$	$1.57\ 10^{-4}$	$3.93\ 10^{-3}$	2.71	3.84	5.02	6.63	7.88	10.83
2	$2.01\ 10^{-2}$	0.103	4.61	5.99	7.38	9.21	10.60	13.81
3	0.115	0.352	6.25	7.81	9.35	11.34	12.84	16.27
4	0.297	0.711	7.78	9.49	11.14	13.28	14.86	18.47
5	0.554	1.15	9.24	11.07	12.83	15.09	16.75	20.52
6	0.872	1.64	10.64	12.59	14.45	16.81	18.55	22.46
7	1.24	2.17	12.02	14.07	16.01	18.48	20.28	24.32
8	1.65	2.73	13.36	15.51	17.53	20.09	21.95	26.12
9	2.09	3.33	14.68	16.92	19.02	21.67	23.59	27.88
10	2.56	3.94	15.99	18.31	20.48	23.21	25.19	29.59
11	3.05	4.57	17.28	19.68	21.92	24.73	26.76	31.26
12	3.57	5.23	18.55	21.03	23.34	26.22	28.30	32.91
13	4.11	5.89	19.81	22.36	24.74	27.69	29.82	34.53
14	4.66	6.57	21.06	23.68	26.12	29.14	31.32	36.12
15	5.23	7.26	22.31	25.00	27.49	30.58	32.80	37.70
16	5.81	7.96	23.54	26.30	28.85	32.00	34.27	39.25
17	6.41	8.67	24.77	27.59	30.19	33.41	35.72	40.79
18	7.01	9.39	25.99	28.87	31.53	34.81	37.16	42.31
19	7.63	10.12	27.20	30.14	32.85	36.19	38.58	43.82
20	8.26	10.85	28.41	31.41	34.17	37.57	40.00	45.31
21	8.90	11.59	29.62	32.67	35.48	38.93	41.40	46.80
22	9.54	12.34	30.81	33.92	36.78	40.29	42.80	48.27
23	10.20	13.09	32.01	35.17	38.08	41.64	44.18	49.73
24	10.86	13.85	33.20	36.42	39.36	42.98	45.56	51.18
25	11.52	14.61	34.38	37.65	40.65	44.31	46.93	52.62
30	14.95	18.49	40.26	43.77	46.98	50.89	53.67	59.70
40	22.16	26.51	51.81	55.76	59.34	63.69	66.77	73.40
50	29.71	34.76	63.17	67.50	71.42	76.15	79.49	86.66
60	37.48	43.19	74.40	79.08	83.30	88.38	91.95	99.61
70	45.44	51.74	85.53	90.53	95.02	100.4	104.2	112.3
80	53.54	60.39	96.58	101.9	106.6	112.3	116.3	124.8
90	61.75	69.13	107.6	113.1	118.1	124.1	128.3	137.2
100	70.06	77.93	118.5	124.3	129.6	135.8	140.2	149.4

Example Ten independent sample values x_i, $i = 1, 2, \ldots, 10$, are drawn at random from a Gaussian distribution with standard deviation $\sigma = 1$. The sample values are as follows:

2.22 2.56 1.07 0.24 0.18 0.95 0.73 −0.79 2.09 1.81

Test the null hypothesis $H_0 : \mu = 0$ at the 10% significance level.

We must test the (simple) null hypothesis $H_0 : \mu = 0$ against the (composite) alternative hypothesis $H_1 : \mu \neq 0$. Since the Gaussian distribution has unit standard deviation, the likelihood function takes the form

$$L(\mathbf{x}; \mu) = \frac{1}{(2\pi)^{N/2}} \exp\left[-\tfrac{1}{2}\textstyle\sum_i (x_i - \mu)^2\right],$$

which has its global maximum at $\mu = \bar{x}$. The test statistic t is therefore given by

$$\begin{aligned} t(\mathbf{x}) = \frac{L(\mathbf{x}; 0)}{L(\mathbf{x}; \bar{x})} &= \frac{\exp\left[-\frac{1}{2}\sum_i (x_i - 0)^2\right]}{\exp\left[-\frac{1}{2}\sum_i (x_i - \bar{x})^2\right]} \\ &= \frac{\exp\left[-\frac{1}{2}\sum_i x_i^2\right]}{\exp\left[-\frac{1}{2}\sum_i (x_i^2 - 2\bar{x}x_i + \bar{x}^2)\right]} \\ &= \frac{\exp\left[-\frac{1}{2}\sum_i x_i^2\right]}{\exp\left[-\frac{1}{2}\sum_i (x_i^2) + \frac{2}{2}N\bar{x}^2 - \frac{1}{2}N\bar{x}^2\right]} \\ &= \exp\left(-\tfrac{1}{2}N\bar{x}^2\right). \end{aligned}$$

It is in fact more convenient to consider the test statistic

$$v = -2 \ln t = N\bar{x}^2.$$

Since $-2 \ln t$ is a monotonically decreasing function of t, the rejection region now becomes $v > v_{\rm crit}$, where

$$\int_{v_{\rm crit}}^{\infty} P(v|H_0)\, dv = \alpha, \qquad (17.57)$$

α being the significance level of the test. Thus it only remains to determine the sampling distribution $P(v|H_0)$. Under the null hypothesis H_0, we expect $\bar{x}$ to be Gaussian distributed, with mean zero and variance $1/N$. Thus, from Subsection 16.9.4, v will follow a *chi-squared* distribution of order 1. Substituting the appropriate form for $P(v|H_0)$ in (17.57) and setting $\alpha = 0.1$, we find by numerical integration (or from Table 17.2) that $v_{\rm crit} = N\bar{x}_{\rm crit}^2 = 2.71$. Since $N = 10$, $\bar{x}_{\rm crit} = \sqrt{2.71/10} = 0.52$, and the rejection region on $\bar{x}$ at the 10% significance level is

$$\bar{x} < -0.52 \quad \text{and} \quad \bar{x} > 0.52.$$

As noted before, for this sample $\bar{x} = 1.11$, and so we may reject the null hypothesis $H_0 : \mu = 0$ at the 10% significance level. ◀

The above example illustrates the general situation that if the maximum-likelihood estimates $\hat{\mathbf{a}}$ of the parameters fall close to the values $\mathbf{a}_0$ assumed in the null hypothesis, then the sample will be considered consistent with H_0 and the value of t will be near unity. If $\hat{\mathbf{a}}$ is distant from $\mathbf{a}_0$ then the sample will not be in accord with H_0 and ordinarily t will have a small (positive) value.

It is clear that in order to prescribe the rejection region for t, or for a related statistic u or v, it is necessary to know the sampling distribution $P(t|H_0)$. If H_0 is simple then one can in principle determine $P(t|H_0)$, although this may prove difficult in practice. Moreover, if H_0 is composite, then it may not be possible to obtain $P(t|H_0)$, even in principle. Nevertheless, a useful approximate form for $P(t|H_0)$ exists in the large-sample limit. Consider a null hypothesis in which any R of the M parameters a_i are fixed. For discussion purposes we take the first R:

$$H_0 : \left(a_1 = a_1^0, a_2 = a_2^0, \ldots, a_R = a_R^0\right), \quad \text{with } R \le M$$

and where the a_i^0 are fixed numbers. If H_0 is true then it can be shown (although we will not prove it) that, when the sample size N is large, the quantity $-2\ln t$ follows an approximate *chi-squared* distribution of order R.

17.6.4 Student's *t*-test

Student's t-test is just a special case of the generalized likelihood ratio test applied to a sample $x_1, x_2, \ldots, x_N$ drawn independently from a Gaussian distribution for which *both* the mean μ and variance σ^2 are unknown, and for which one wishes to distinguish between the hypotheses

$$H_0 : \mu = \mu_0, \quad 0 < \sigma^2 < \infty \qquad \text{and} \qquad H_1 : \mu \neq \mu_0, \quad 0 < \sigma^2 < \infty,$$

where μ_0 is a given number. Here, the allowed parameter space $\mathcal{A}$ is the half-plane $-\infty < \mu < \infty, 0 < \sigma^2 < \infty$, whereas the subspace $\mathcal{S}$ characterized by the null hypothesis H_0 is only the line $\mu = \mu_0, 0 < \sigma^2 < \infty$.

The likelihood function for this situation is given by

$$L(\mathbf{x}; \mu, \sigma^2) = \frac{1}{(2\pi\sigma^2)^{N/2}} \exp\left[-\frac{\sum_i (x_i - \mu)^2}{2\sigma^2}\right]. \tag{17.58}$$

On the one hand, the values of μ and σ^2 that maximize L in $\mathcal{A}$ are $\mu = \bar{x}$ and $\sigma^2 = s^2$, where $\bar{x}$ is the sample mean and s^2 is the sample variance.[12] On the other hand, to maximize L in the subspace $\mathcal{S}$ we have to set $\mu = \mu_0$, and the only remaining parameter is σ^2; the value of σ^2 that maximizes L is then found to be[13]

$$\widehat{\sigma^2} = \frac{1}{N} \sum_{i=1}^{N} (x_i - \mu_0)^2.$$

To retain, in due course, the standard notation for Student's t-test, in this section we will denote the generalized likelihood ratio by λ (rather than t); it is thus given by

$$\begin{aligned}
\lambda(\mathbf{x}) &= \frac{L(\mathbf{x}; \mu_0, \widehat{\sigma^2})}{L(\mathbf{x}; \bar{x}, s^2)} \\
&= \frac{[(2\pi/N)\sum_i (x_i - \mu_0)^2]^{-N/2} \exp(-N/2)}{[(2\pi/N)\sum_i (x_i - \bar{x})^2]^{-N/2} \exp(-N/2)} \\
&= \left[\frac{\sum_i (x_i - \bar{x})^2}{\sum_i (x_i - \mu_0)^2}\right]^{N/2}.
\end{aligned} \tag{17.59}$$

12 This can be shown by taking the partial derivatives of the logarithm of (17.58) with respect to μ and σ, setting each equal to zero, and solving the resulting simultaneous equations.

13 Carry out this calculation.

Normally, our next step would be to find the sampling distribution of λ under the assumption that H_0 were true. It is more conventional, however, to work in terms of a related test statistic t, which was first devised by William Gossett, who wrote under the pen name of "Student".

The sum of squares in the denominator of (17.59) may be put into the form[14]

$$\sum_i (x_i - \mu_0)^2 = N(\bar{x} - \mu_0)^2 + \sum_i (x_i - \bar{x})^2. \tag{17.60}$$

Thus, on dividing the numerator and denominator in (17.59) by $\sum_i (x_i - \bar{x})^2$ (equal to Ns^2) and rearranging, the generalized likelihood ratio λ can be written

$$\lambda = \left(1 + \frac{t^2}{N-1}\right)^{-N/2},$$

where we have defined the new variable

$$t = \frac{\bar{x} - \mu_0}{s/\sqrt{N-1}}. \tag{17.61}$$

Since t^2 is a monotonically decreasing function of λ (and vice versa), the corresponding rejection region is $t^2 > c$, where c is a positive constant depending on the required significance level α. It is conventional, however, to use t, rather than t^2, as our test statistic, in which case our rejection region becomes two-tailed and is given by

$$t < -t_{\rm crit} \quad \text{and} \quad t > t_{\rm crit}, \tag{17.62}$$

where $t_{\rm crit}$ is the positive square root of the constant c.

The definition (17.61) and the rejection region (17.62) form the basis of Student's t-test. It only remains to determine the sampling distribution $P(t|H_0)$. At the outset, it is worth noting that if we write the expression (17.61) for t in terms of the standard estimator $\hat{\sigma} = \sqrt{Ns^2/(N-1)}$ of the standard deviation then we obtain

$$t = \frac{\bar{x} - \mu_0}{\hat{\sigma}/\sqrt{N}}. \tag{17.63}$$

If, in fact, we knew the true value of σ and used it in this expression for t then it is clear from our discussion in Section 17.3 that t would follow a Gaussian distribution with mean 0 and variance 1, i.e. $t \sim N(0, 1)$. When σ is not known, however, we have to use our estimate $\hat{\sigma}$ in (17.63), with the result that t is no longer distributed as the standard Gaussian. As one might expect from the central limit theorem, however, the distribution of t does tend towards the standard Gaussian for large values of N.

As noted earlier, the exact distribution of t, valid for any value of N, was first discovered by William Gossett. Starting from result (17.60), it can be shown[15] that, if the hypothesis H_0 is true, then $\bar{x}$ and s are independent variables and their joint sampling distribution is given by

$$P(\bar{x}, s|H_0) = Cs^{N-2} \exp\left(-\frac{Ns^2}{2\sigma^2}\right) \exp\left[-\frac{N(\bar{x} - \mu_0)^2}{2\sigma^2}\right], \tag{17.64}$$

14 Prove that this is so.

15 Though the proof is beyond the scope of this book.

where C is a normalization constant. We can use this result to obtain the joint sampling distribution of s and t by demanding that

$$P(\bar{x}, s|H_0)\, d\bar{x}\, ds = P(t, s|H_0)\, dt\, ds.$$

Using (17.61) to substitute for $\bar{x} - \mu_0$ in (17.64), and noting that $d\bar{x} = (s/\sqrt{N-1})\, dt$, we find

$$P(\bar{x}, s|H_0)\, d\bar{x}\, ds = As^{N-1} \exp\left[-\frac{Ns^2}{2\sigma^2}\left(1 + \frac{t^2}{N-1}\right)\right] dt\, ds,$$

where A is another normalization constant. In order to obtain the sampling distribution of t alone, we must integrate $P(t, s|H_0)$ with respect to s over its allowed range, from 0 to ∞. Thus, the required distribution of t alone is given by

$$P(t|H_0) = \int_0^\infty P(t, s|H_0)\, ds = A \int_0^\infty s^{N-1} \exp\left[-\frac{Ns^2}{2\sigma^2}\left(1 + \frac{t^2}{N-1}\right)\right] ds. \quad (17.65)$$

To carry out this integration, we set $y = s\{1 + [t^2/(N-1)]\}^{1/2}$, which on substitution into (17.65) yields

$$P(t|H_0) = A\left(1 + \frac{t^2}{N-1}\right)^{-N/2} \int_0^\infty y^{N-1} \exp\left(-\frac{Ny^2}{2\sigma^2}\right) dy.$$

Since the integral over y does not depend on t, it is simply a constant. We thus find that the sampling distribution of the variable t is

$$P(t|H_0) = \frac{1}{\sqrt{(N-1)\pi}} \frac{\Gamma\left(\frac{1}{2}N\right)}{\Gamma\left(\frac{1}{2}(N-1)\right)} \left(1 + \frac{t^2}{N-1}\right)^{-N/2}, \quad (17.66)$$

where we have used the normalization condition $\int_{-\infty}^{\infty} P(t|H_0)\, dt = 1$ to determine[16] the value of the constant.

The distribution (17.66) is called *Student's t-distribution with $N - 1$ degrees of freedom*. A plot of Student's t-distribution is shown in Figure 17.6 for various values of N. For comparison, we also plot the standard Gaussian distribution, to which the t-distribution tends for large N. As is clear from the figure, the t-distribution is symmetric about $t = 0$.

In Table 17.3 we list some critical points of the cumulative probability function $C_n(t)$ of the t-distribution, which is defined by

$$C_n(t) = \int_{-\infty}^{t} P(t'|H_0)\, dt',$$

where $n = N - 1$ is the number of degrees of freedom. Clearly, $C_n(t)$ is analogous to the cumulative probability function $\Phi(z)$ of the Gaussian distribution, discussed in Subsection 16.9.2. For comparison purposes, we also list the critical points of $\Phi(z)$, which corresponds to the t-distribution for $N = \infty$.

16 The calculation is quite lengthy.

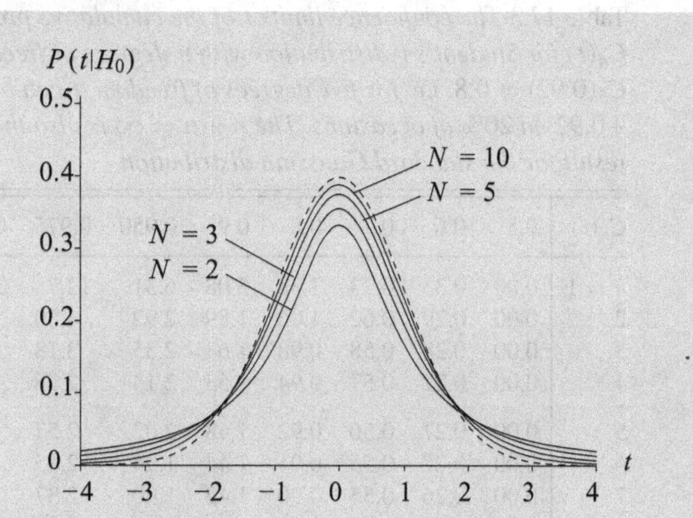

Figure 17.6 Student's t-distribution for various values of N. The broken curve shows the standard Gaussian distribution for comparison.

A straightforward application of Student's t-test now follows.

Example Ten independent sample values x_i, $i = 1, 2, \ldots, 10$, are drawn at random from a Gaussian distribution with unknown mean μ and unknown standard deviation σ. The sample values are as follows:

$$2.22 \quad 2.56 \quad 1.07 \quad 0.24 \quad 0.18 \quad 0.95 \quad 0.73 \quad -0.79 \quad 2.09 \quad 1.81$$

Test the null hypothesis $H_0 : \mu = 0$ at the 10% significance level.

For our null hypothesis, $\mu_0 = 0$. Since for this sample $\bar{x} = 1.11$, $s = 1.01$ and $N = 10$, it follows from (17.61) that

$$t = \frac{\bar{x}}{s/\sqrt{N-1}} = 3.33.$$

The rejection region for t is given by (17.62) where t_{crit} is such that

$$C_{N-1}(t_{\text{crit}}) = 1 - \alpha/2,$$

and α is the required significance of the test. In our case $\alpha = 0.1$ and $N = 10$, and from Table 17.3 we find $t_{\text{crit}} = 1.83$. Thus our rejection region for H_0 at the 10% significance level is

$$t < -1.83 \quad \text{and} \quad t > 1.83.$$

For our sample $t = 3.30$ and so we can clearly reject the null hypothesis $H_0 : \mu = 0$ at this level. ◀

It is worth noting the connection between the t-test and the classical confidence interval on the mean μ. The central confidence interval on μ at the confidence level $1 - \alpha$ is the

Table 17.3 *The confidence limits t of the cumulative probability function $C_n(t)$ for Student's t-distribution with n degrees of freedom. For example, $C_5(0.92) = 0.8$, i.e. for five degrees of freedom, t can be expected to exceed +0.92 on 20% of occasions. The row $n = \infty$ is also the corresponding result for the standard Gaussian distribution*

$C_n(t)$	0.5	0.6	0.7	0.8	0.9	0.950	0.975	0.990	0.995	0.999
$n = 1$	0.00	0.33	0.73	1.38	3.08	6.31	12.7	31.8	63.7	318.3
2	0.00	0.29	0.62	1.06	1.89	2.92	4.30	6.97	9.93	22.3
3	0.00	0.28	0.58	0.98	1.64	2.35	3.18	4.54	5.84	10.2
4	0.00	0.27	0.57	0.94	1.53	2.13	2.78	3.75	4.60	7.17
5	0.00	0.27	0.56	0.92	1.48	2.02	2.57	3.37	4.03	5.89
6	0.00	0.27	0.55	0.91	1.44	1.94	2.45	3.14	3.71	5.21
7	0.00	0.26	0.55	0.90	1.42	1.90	2.37	3.00	3.50	4.79
8	0.00	0.26	0.55	0.89	1.40	1.86	2.31	2.90	3.36	4.50
9	0.00	0.26	0.54	0.88	1.38	1.83	2.26	2.82	3.25	4.30
10	0.00	0.26	0.54	0.88	1.37	1.81	2.23	2.76	3.17	4.14
11	0.00	0.26	0.54	0.88	1.36	1.80	2.20	2.72	3.11	4.03
12	0.00	0.26	0.54	0.87	1.36	1.78	2.18	2.68	3.06	3.93
13	0.00	0.26	0.54	0.87	1.35	1.77	2.16	2.65	3.01	3.85
14	0.00	0.26	0.54	0.87	1.35	1.76	2.15	2.62	2.98	3.79
15	0.00	0.26	0.54	0.87	1.34	1.75	2.13	2.60	2.95	3.73
16	0.00	0.26	0.54	0.87	1.34	1.75	2.12	2.58	2.92	3.69
17	0.00	0.26	0.53	0.86	1.33	1.74	2.11	2.57	2.90	3.65
18	0.00	0.26	0.53	0.86	1.33	1.73	2.10	2.55	2.88	3.61
19	0.00	0.26	0.53	0.86	1.33	1.73	2.09	2.54	2.86	3.58
20	0.00	0.26	0.53	0.86	1.33	1.73	2.09	2.53	2.85	3.55
25	0.00	0.26	0.53	0.86	1.32	1.71	2.06	2.49	2.79	3.46
30	0.00	0.26	0.53	0.85	1.31	1.70	2.04	2.46	2.75	3.39
40	0.00	0.26	0.53	0.85	1.30	1.68	2.02	2.42	2.70	3.31
50	0.00	0.26	0.53	0.85	1.30	1.68	2.01	2.40	2.68	3.26
100	0.00	0.25	0.53	0.85	1.29	1.66	1.98	2.37	2.63	3.17
200	0.00	0.25	0.53	0.84	1.29	1.65	1.97	2.35	2.60	3.13
∞	0.00	0.25	0.52	0.84	1.28	1.65	1.96	2.33	2.58	3.09

set of values for which

$$-t_{\rm crit} < \frac{\bar{x} - \mu}{s/\sqrt{N-1}} < t_{\rm crit},$$

where $t_{\rm crit}$ satisfies $C_{N-1}(t_{\rm crit}) = \alpha/2$. Thus the required confidence interval is

$$\bar{x} - \frac{t_{\rm crit} s}{\sqrt{N-1}} < \mu < \bar{x} + \frac{t_{\rm crit} s}{\sqrt{N-1}}.$$

Hence, in the above example, the 90% classical central confidence interval on μ is

$$0.49 < \mu < 1.73,$$

which, as expected, is significantly removed from the hypothesis value of $\mu = 0$.

The t-distribution may also be used to compare different samples from Gaussian distributions. In particular, let us consider the case where we have two independent samples of sizes N_1 and N_2, drawn respectively from Gaussian distributions with a common variance σ^2 but with possibly different means μ_1 and μ_2. On the basis of the samples, one wishes to distinguish between the hypotheses

$$H_0 : \mu_1 = \mu_2, \quad 0 < \sigma^2 < \infty \qquad \text{and} \qquad H_1 : \mu_1 \neq \mu_2, \quad 0 < \sigma^2 < \infty.$$

In other words, we wish to test the null hypothesis that the samples are drawn from populations having the same mean. Suppose that the measured sample means and standard deviations are $\bar{x}_1$, $\bar{x}_2$ and s_1, s_2 respectively. In an analogous way to that presented above, one may show that the generalized likelihood ratio can be written as

$$\lambda = \left(1 + \frac{t^2}{N_1 + N_2 - 2}\right)^{-(N_1+N_2)/2}.$$

In this case, the variable t is given by

$$t = \frac{\bar{w} - \omega}{\hat{\sigma}}\left(\frac{N_1 N_2}{N_1 + N_2}\right)^{1/2}, \tag{17.67}$$

where $\bar{w} = \bar{x}_1 - \bar{x}_2$, $\omega = \mu_1 - \mu_2$ and

$$\hat{\sigma} = \left[\frac{N_1 s_1^2 + N_2 s_2^2}{N_1 + N_2 - 2}\right]^{1/2}.$$

It is straightforward (albeit with complicated algebra) to show that the variable t in (17.67) follows Student's t-distribution with $N_1 + N_2 - 2$ degrees of freedom, and so we may use an appropriate form of Student's t-test to investigate the null hypothesis $H_0 : \mu_1 = \mu_2$ (or equivalently $H_0 : \omega = 0$). As above, the t-test can be used to place a confidence interval on $\omega = \mu_1 - \mu_2$; this is done in the following worked example.

Example Suppose that two classes of students take the same mathematics examination and the following percentage marks are obtained:

Class 1:	66	62	34	55	77	80	55	60	69	47	50
Class 2:	64	90	76	56	81	72	70				

Assuming that the two sets of examination marks are drawn from Gaussian distributions with a common variance, test the hypothesis $H_0 : \mu_1 = \mu_2$ at the 5% significance level. Use your result to obtain the 95% classical central confidence interval on $\omega = \mu_1 - \mu_2$.

We begin by calculating the mean and standard deviation of each sample. The number of values in each sample is $N_1 = 11$ and $N_2 = 7$ respectively, and we find

$$\bar{x}_1 = 59.5, \; s_1 = 12.8 \quad \text{and} \quad \bar{x}_2 = 72.7, \; s_2 = 10.3,$$

leading to $\bar{w} = \bar{x}_1 - \bar{x}_2 = -13.2$ and $\hat{\sigma} = 12.6$. Setting $\omega = 0$ in (17.67), we thus find $t = -2.17$.

The rejection region for H_0 is given by (17.62), where $t_{\rm crit}$ satisfies

$$C_{N_1+N_2-2}(t_{\rm crit}) = 1 - \alpha/2, \tag{17.68}$$

where α is the required significance level of the test. In our case we set $\alpha = 0.05$, and from Table 17.3 with $n = 16$ we find that $t_{\rm crit} = 2.12$. The rejection region is therefore

$$t < -2.12 \quad \text{and} \quad t > 2.12.$$

Since $t = -2.17$ for our samples, we can reject the null hypothesis $H_0 : \mu_1 = \mu_2$, although only by a small margin.[17] The 95% central confidence interval on $\omega = \mu_1 - \mu_2$ is given by

$$\bar{w} - \hat{\sigma}\, t_{\rm crit}\left(\frac{N_1+N_2}{N_1N_2}\right)^{1/2} < \omega < \bar{w} + \hat{\sigma}\, t_{\rm crit}\left(\frac{N_1+N_2}{N_1N_2}\right)^{1/2},$$

where $t_{\rm crit}$ is given by (17.68). Thus, we find

$$-26.1 < \omega < -0.28,$$

which, as expected, does not (quite) contain $\omega = 0$. ◀

In order to apply Student's t-test in the above example, we had to make the assumption that the samples were drawn from Gaussian distributions possessing a common variance, which is clearly unjustified *a priori*. We can, however, perform another test on the data to investigate whether the additional hypothesis $\sigma_1^2 = \sigma_2^2$ is reasonable; this test is discussed in the next subsection. If this additional test shows that the hypothesis $\sigma_1^2 = \sigma_2^2$ may be accepted (at some suitable significance level), then we may indeed use the analysis in the above example to infer that the null hypothesis $H_0 : \mu_1 = \mu_2$ may be rejected at the 5% significance level. If, however, we find that the additional hypothesis $\sigma_1^2 = \sigma_2^2$ must be rejected, then we can only infer from the above example that the hypothesis that the two samples were drawn from the same Gaussian distribution may be rejected at the 5% significance level.

Throughout the above discussion, we have assumed that samples are drawn from a Gaussian distribution. Although this is true for many random variables, in practice it is usually impossible to know *a priori* whether this is the case. It can be shown, however, that Student's t-test remains reasonably accurate even if the sampled distribution(s) differ considerably from a Gaussian. Indeed, for sampled distributions that differ only slightly from a Gaussian form, the accuracy of the test is remarkably good. Nevertheless, when applying the t-test, it is always important to remember that the assumption of a Gaussian parent population is central to the method.

17.6.5 Fisher's F-test

Having concentrated on tests for the mean μ of a Gaussian distribution, we now consider tests for its standard deviation σ. Before discussing Fisher's F-test for comparing the standard deviations of two samples, we begin by considering the case when an independent sample $x_1, x_2, \ldots, x_N$ is drawn from a Gaussian distribution with unknown μ and σ, and

17 Indeed, it is easily shown that one cannot reject H_0 at the 2% significance level.

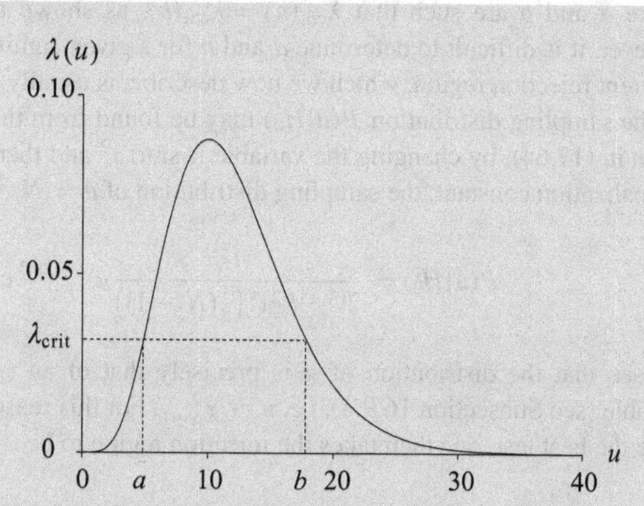

Figure 17.7 The sampling distribution $P(u|H_0)$ for $N = 10$; this is a chi-squared distribution for $N - 1$ degrees of freedom.

we wish to distinguish between the two hypotheses

$$H_0 : \sigma^2 = \sigma_0^2, \quad -\infty < \mu < \infty \qquad \text{and} \qquad H_1 : \sigma^2 \neq \sigma_0^2, \quad -\infty < \mu < \infty,$$

where σ_0^2 is a given number. Here, the whole possible parameter space $\mathcal{A}$ is the half-plane $-\infty < \mu < \infty$, $0 < \sigma^2 < \infty$, whereas the subspace $\mathcal{S}$ characterized by the null hypothesis H_0 is the line $\sigma^2 = \sigma_0^2$, $-\infty < \mu < \infty$.

The likelihood function for this situation is, as in the previous subsection, given by

$$L(\mathbf{x}; \mu, \sigma^2) = \frac{1}{(2\pi\sigma^2)^{N/2}} \exp\left[-\frac{\sum_i (x_i - \mu)^2}{2\sigma^2}\right].$$

The maximum of L in $\mathcal{A}$ occurs at $\mu = \bar{x}$ and $\sigma^2 = s^2$, whereas the maximum of L in $\mathcal{S}$ is at $\mu = \bar{x}$ and $\sigma^2 = \sigma_0^2$. Thus, the generalized likelihood ratio is given by[18]

$$\lambda(\mathbf{x}) = \frac{L(\mathbf{x}; \bar{x}, \sigma_0^2)}{L(\mathbf{x}; \bar{x}, s^2)} = \left(\frac{u}{N}\right)^{N/2} \exp\left[-\tfrac{1}{2}(u - N)\right],$$

where we have introduced the variable

$$u = \frac{Ns^2}{\sigma_0^2} = \frac{\sum_i (x_i - \bar{x})^2}{\sigma_0^2}. \tag{17.69}$$

An example of this distribution is plotted in Figure 17.7 for $N = 10$. From the figure, we see that the rejection region $\lambda < \lambda_{\text{crit}}$ corresponds to a two-tailed rejection region on u given by

$$0 < u < a \qquad \text{and} \qquad b < u < \infty,$$

18 Carry through this evaluation of λ in terms of u for yourself; it should take about five lines of working.

where a and b are such that $\lambda_{\rm crit}(a) = \lambda_{\rm crit}(b)$, as shown in Figure 17.7. In practice, however, it is difficult to determine a and b for a given significance level α, so a slightly different rejection region, which we now describe, is usually adopted.

The sampling distribution $P(u|H_0)$ may be found from the sampling distribution of s given in (17.64), by changing the variable first to s^2 and then to u. Including the correct normalization constant, the sampling distribution of $u = Ns^2/\sigma_0^2$ is given by

$$P(u|H_0) = \frac{1}{2^{(N-1)/2}\Gamma\left(\frac{1}{2}(N-1)\right)} u^{(N-3)/2} \exp\left(-\tfrac{1}{2}u\right).$$

We see that the distribution of u is precisely that of an $(N-1)$th-order chi-squared variable (see Subsection 16.9.4), i.e. $u \sim \chi^2_{N-1}$. For this reason, although it does not give quite the best test, one then takes the rejection region to be

$$0 < u < a \qquad \text{and} \qquad b < u < \infty,$$

with a and b chosen such that the two tails have *equal areas*; the advantage of this choice is that tabulations of the chi-squared distribution make the size of this region relatively easy to estimate. Thus, for a given significance level α, we have

$$\int_0^a P(u|H_0)\,du = \alpha/2 \qquad \text{and} \qquad \int_b^\infty P(u|H_0)\,du = \alpha/2.$$

In the following example $\alpha = 0.10$ and the number of degrees of freedom is 9.

Example Ten independent sample values x_i, $i = 1, 2, \ldots, 10$, are drawn at random from a Gaussian distribution with unknown mean μ and standard deviation σ. The sample values are as follows:

2.22 2.56 1.07 0.24 0.18 0.95 0.73 −0.79 2.09 1.81

Test the null hypothesis $H_0 : \sigma^2 = 2$ at the 10% significance level.

For our null hypothesis $\sigma_0^2 = 2$. Since for this sample $s = 1.01$ and $N = 10$, from (17.69) we have $u = 5.10$. For $\alpha = 0.1$ we find, either numerically or using Table 17.2, that $a = 3.33$ and $b = 16.92$. Thus, our rejection region is

$$0 < u < 3.33 \qquad \text{and} \qquad 16.92 < u < \infty.$$

The value $u = 5.10$ from our sample does not lie in the rejection region, and so we cannot reject the null hypothesis $H_0 : \sigma^2 = 2$. ◀

We now turn to Fisher's F-test. Let us suppose that two independent samples of sizes N_1 and N_2 are drawn from Gaussian distributions with means and variances μ_1, σ_1^2 and μ_2, σ_2^2 respectively, and we wish to distinguish between the two hypotheses

$$H_0 : \sigma_1^2 = \sigma_2^2 \qquad \text{and} \qquad H_1 : \sigma_1^2 \neq \sigma_2^2.$$

In this case, the generalized likelihood ratio is found to be

$$\lambda = \frac{(N_1+N_2)^{(N_1+N_2)/2}}{N_1^{N_1/2}N_2^{N_2/2}} \frac{[F(N_1-1)/(N_2-1)]^{N_1/2}}{[1+F(N_1-1)/(N_2-1)]^{(N_1+N_2)/2}},$$

where F is given by the variance ratio

$$F = \frac{N_1 s_1^2/(N_1-1)}{N_2 s_2^2/(N_2-1)} \equiv \frac{u^2}{v^2} \tag{17.70}$$

and s_1 and s_2 are the standard deviations of the two samples.

In fact, it does not matter whether the ratio F given in (17.70) is defined as u^2/v^2 or as v^2/u^2. It is conventional to put the larger sample variance on the top, so that F is always greater than or equal to unity. A large value of F indicates that the sample variances u^2 and v^2 are very different whereas a value of F close to unity means that they are very similar. One desirable consequence of this convention is that the rejection region corresponds to a one-tailed test, rather than the two-tailed test that would be necessary if F could be both greater and smaller than unity.

With this convention and a given significance level α, it is natural to define the rejection region on F as $F > F_{\text{crit}}$, where

$$C_{n_1,n_2}(F_{\text{crit}}) = \int_1^{F_{\text{crit}}} P(F|H_0)\,dF = 1-\alpha,$$

and $n_1 = N_1 - 1$ and $n_2 = N_2 - 1$ are the numbers of degrees of freedom. Table 17.4 lists values of F_{crit} corresponding to the 5% significance level (i.e. $\alpha = 0.05$) for various values of n_1 and n_2.

The distribution of F may be obtained in a reasonably straightforward manner from the distribution of the sample variance s^2 under the null hypothesis, H_0, that the two Gaussian distributions share a common variance. Firstly, the distributions of u^2 and v^2 are determined from that of s^2, using the definitions provided by (17.70), then u^2 is set equal to Fv^2, and finally the joint F-v^2 distribution is integrated over all v^2, from 0 to ∞, to yield $P(F|H_0)$. The final result (which we just quote) is that

$$P(F|H_0) = AF^{(N_1-3)/2}\left(1+\frac{N_1-1}{N_2-1}F\right)^{-(N_1+N_2-2)/2}, \tag{17.71}$$

where the normalization constant A is given by

$$A = \left(\frac{N_1-1}{N_2-1}\right)^{(N_1-1)/2} \frac{\Gamma[\frac{1}{2}(N_1+N_2-2)]}{\Gamma[\frac{1}{2}(N_1-1)]\,\Gamma[\frac{1}{2}(N_2-1)]}.$$

$P(F|H_0)$ is called the *F-distribution* (or occasionally the *Fisher distribution*) with $(N_1-1,\ N_2-1)$ degrees of freedom.

Although it may not look it at first sight, it can be verified that the F-distribution $P(F)$ is symmetric between the two data samples, i.e. that it retains the same form, but with

Table 17.4 *Values of F for which the cumulative probability function $C_{n_1,n_2}(F)$ of the F-distribution with (n_1, n_2) degrees of freedom has the value 0.95. For example, for $n_1 = 10$ and $n_2 = 6$, $C_{n_1,n_2}(4.06) = 0.95$*

$C_{n_1,n_2}(F)$	$n_1 = 1$	2	3	4	5	6	7	8
$n_2 = 1$	161	200	216	225	230	234	237	239
2	18.5	19.0	19.2	19.2	19.3	19.3	19.4	19.4
3	10.1	9.55	9.28	9.12	9.01	8.94	8.89	8.85
4	7.71	6.94	6.59	6.39	6.26	6.16	6.09	6.04
5	6.61	5.79	5.41	5.19	5.05	4.95	4.88	4.82
6	5.99	5.14	4.76	4.53	4.39	4.28	4.21	4.15
7	5.59	4.74	4.35	4.12	3.97	3.87	3.79	3.73
8	5.32	4.46	4.07	3.84	3.69	3.58	3.50	3.44
9	5.12	4.26	3.86	3.63	3.48	3.37	3.29	3.23
10	4.96	4.10	3.71	3.48	3.33	3.22	3.14	3.07
20	4.35	3.49	3.10	2.87	2.71	2.60	2.51	2.45
30	4.17	3.32	2.92	2.69	2.53	2.42	2.33	2.27
40	4.08	3.23	2.84	2.61	2.45	2.34	2.25	2.18
50	4.03	3.18	2.79	2.56	2.40	2.29	2.20	2.13
100	3.94	3.09	2.70	2.46	2.31	2.19	2.10	2.03
∞	3.84	3.00	2.60	2.37	2.21	2.10	2.01	1.94
	$n_1 = 9$	10	20	30	40	50	100	∞
$n_2 = 1$	241	242	248	250	251	252	253	254
2	19.4	19.4	19.4	19.5	19.5	19.5	19.5	19.5
3	8.81	8.79	8.66	8.62	8.59	8.58	8.55	8.53
4	6.00	5.96	5.80	5.75	5.72	5.70	5.66	5.63
5	4.77	4.74	4.56	4.50	4.46	4.44	4.41	4.37
6	4.10	4.06	3.87	3.81	3.77	3.75	3.71	3.67
7	3.68	3.64	3.44	3.38	3.34	3.32	3.27	3.23
8	3.39	3.35	3.15	3.08	3.04	3.02	2.97	2.93
9	3.18	3.14	2.94	2.86	2.83	2.80	2.76	2.71
10	3.02	2.98	2.77	2.70	2.66	2.64	2.59	2.54
20	2.39	2.35	2.12	2.04	1.99	1.97	1.91	1.84
30	2.21	2.16	1.93	2.69	1.79	1.76	1.70	1.62
40	2.12	2.08	1.84	1.74	1.69	1.66	1.59	1.51
50	2.07	2.03	1.78	1.69	1.63	1.60	1.52	1.44
100	1.97	1.93	1.68	1.57	1.52	1.48	1.39	1.28
∞	1.88	1.83	1.57	1.46	1.39	1.35	1.24	1.00

N_1 and N_2 interchanged, if F is replaced by $F' = F^{-1}$. Symbolically, if $P'(F')$ is the distribution of F' and $P(F) = \eta(F, N_1, N_2)$, then $P'(F') = \eta(F', N_2, N_1)$.

We now illustrate the use of the F-test to determine whether two samples could have come from distributions with the same variance (including the possibility of the same distribution).

Example Suppose that two classes of students take the same mathematics examination and the following percentage marks are obtained:

Class 1:	66	62	34	55	77	80	55	60	69	47	50
Class 2:	64	90	76	56	81	72	70				

Assuming that the two sets of examination marks are drawn from Gaussian distributions, test the hypothesis $H_0 : \sigma_1^2 = \sigma_2^2$ at the 5% significance level.

The variances of the two samples are $s_1^2 = (12.8)^2$ and $s_2^2 = (10.3)^2$ and the sample sizes are $N_1 = 11$ and $N_2 = 7$. Thus, we have

$$u^2 = \frac{N_1 s_1^2}{N_1 - 1} = 180.2 \quad \text{and} \quad v^2 = \frac{N_2 s_2^2}{N_2 - 1} = 123.8,$$

where we have taken u^2 to be the larger value. Thus, $F = u^2/v^2 = 1.46$ to two decimal places. Since the first sample contains 11 values and the second contains seven values, we take $n_1 = 10$ and $n_2 = 6$. Consulting Table 17.4, we see that, at the 5% significance level, $F_{\text{crit}} = 4.06$. Since our value lies comfortably below this, we conclude that there is no statistical evidence for rejecting the hypothesis that the two samples were drawn from Gaussian distributions with a common variance. ◀

It is also common to define the variable $z = \frac{1}{2} \ln F$, the distribution of which can be found straightforwardly from (17.71). This is a useful change of variable since it can be shown that, for large values of n_1 and n_2, the variable z is distributed approximately as a Gaussian with mean $\frac{1}{2}(n_2^{-1} - n_1^{-1})$ and variance $\frac{1}{2}(n_2^{-1} + n_1^{-1})$.[19]

17.6.6 Goodness of fit in least-squares problems

We conclude our discussion of hypothesis testing with an example of a goodness-of-fit test. In Subsection 17.5.1, we discussed the use of the method of least squares in estimating the best-fit values of a set of parameters $\mathbf{a}$ in a given model $y = f(x; \mathbf{a})$ for a data set (x_i, y_i), $i = 1, 2, \ldots, N$. We have not addressed, however, the question of whether the best-fit model $y = f(x; \hat{\mathbf{a}})$ does, in fact, provide a good fit to the data. In other words, we have not considered thus far how to verify that the functional form f of our assumed model is indeed correct. In the language of hypothesis testing, we wish to distinguish between the two hypotheses

$$H_0 : \text{model is correct} \quad \text{and} \quad H_1 : \text{model is incorrect}.$$

Given the vague nature of the alternative hypothesis H_1, we clearly cannot use the generalized likelihood-ratio test. Nevertheless, it is still possible to test the null hypothesis H_0 at a given significance level α.

19 Assuming that the sample sizes are large enough for this approximation to hold, determine how many standard deviations z is from its mean in the previous worked example.

The least-squares estimates of the parameters $\hat{a}_1, \hat{a}_2, \ldots, \hat{a}_M$, as discussed in Subsection 17.5.1, are those values that minimize the quantity

$$\chi^2(\mathbf{a}) = \sum_{i,j=1}^{N} [y_i - f(x_i;\mathbf{a})](\mathsf{N}^{-1})_{ij}[y_j - f(x_j;\mathbf{a})] = (\mathsf{y}-\mathsf{f})^{\mathrm{T}}\mathsf{N}^{-1}(\mathsf{y}-\mathsf{f}).$$

In the last equality, we rewrote the expression in matrix notation by defining the column vector f with elements $f_i = f(x_i;\mathbf{a})$. The value $\chi^2(\hat{\mathbf{a}})$ at this minimum can be used as a statistic to test the null hypothesis H_0, as follows. The N quantities $y_i - f(x_i;\mathbf{a})$ are Gaussian distributed. However, provided the function $f(x_j;\mathbf{a})$ is linear in the parameters $\mathbf{a}$, the equations (17.49) that determine the least-squares estimate $\hat{\mathbf{a}}$ constitute a set of M linear constraints on these N quantities. Thus, as discussed in Subsection 16.9.4, the sampling distribution of the quantity $\chi^2(\hat{\mathbf{a}})$ will be a *chi-squared distribution with* $N-M$ *degrees of freedom* (d.o.f.), which has the expectation value and variance

$$E[\chi^2(\hat{\mathbf{a}})] = N - M \qquad \text{and} \qquad V[\chi^2(\hat{\mathbf{a}})] = 2(N-M).$$

Thus we would expect the value of $\chi^2(\hat{\mathbf{a}})$ to lie typically in the range $(N-M) \pm \sqrt{2(N-M)}$. A value lying outside this range may suggest that the assumed model for the data is incorrect. A very small value of $\chi^2(\hat{\mathbf{a}})$ is usually an indication that the model has too many free parameters and has "over-fitted" the data. More commonly, the assumed model is simply incorrect, and this usually results in a value of $\chi^2(\hat{\mathbf{a}})$ that is larger than expected.

One can choose to perform either a one-tailed or a two-tailed test on the value of $\chi^2(\hat{\mathbf{a}})$. It is usual, for a given significance level α, to define the one-tailed rejection region to be $\chi^2(\hat{\mathbf{a}}) > k$, where the constant k satisfies

$$\int_k^{\infty} P(\chi_n^2)\, d\chi_n^2 = \alpha \tag{17.72}$$

and $P(\chi_n^2)$ is the PDF of the chi-squared distribution with $n = N - M$ degrees of freedom (see Subsection 16.9.4).

We now re-analyze the data of a previous worked example to examine whether it is a satisfactory fit to an assumed model.

Example An experiment produces the following data sample pairs (x_i, y_i):

x_i:	1.85	2.72	2.81	3.06	3.42	3.76	4.31	4.47	4.64	4.99
y_i:	2.26	3.10	3.80	4.11	4.74	4.31	5.24	4.03	5.69	6.57

where the x_i-values are known exactly but each y_i-value is measured only to an accuracy of $\sigma = 0.5$. At the one-tailed 5% significance level, test the null hypothesis H_0 that the underlying model for the data is a straight line $y = mx + c$.

These data are the same as those investigated in Subsection 17.5.2 and plotted in Figure 17.4. As shown previously, the least squares estimates of the slope m and intercept c are given by

$$\hat{m} = 1.11 \qquad \text{and} \qquad \hat{c} = 0.4. \tag{17.73}$$

Since the error on each y_i-value is drawn independently from a Gaussian distribution with standard deviation σ, we have

$$\chi^2(\mathbf{a}) = \sum_{i=1}^{N} \left[\frac{y_i - f(x_i; \mathbf{a})}{\sigma} \right]^2 = \sum_{i=1}^{N} \left[\frac{y_i - mx_i - c}{\sigma} \right]^2. \tag{17.74}$$

Inserting the values (17.73) into (17.74), we obtain $\chi^2(\hat{m}, \hat{c}) = 11.5$. In our case, the number of data points is $N = 10$ and the number of fitted parameters is $M = 2$. Thus, the number of degrees of freedom is $n = N - M = 8$. Setting $n = 8$ and $\alpha = 0.05$ in (17.72) we find from Table 17.2 that $k = 15.51$. Hence our rejection region is

$$\chi^2(\hat{m}, \hat{c}) > 15.51.$$

Since above we found $\chi^2(\hat{m}, \hat{c}) = 11.5$, we cannot reject the null hypothesis that the underlying model for the data is a straight line $y = mx + c$. ◀

As mentioned above, our analysis is only valid if the function $f(x; \mathbf{a})$ is linear in the parameters $\mathbf{a}$. Nevertheless, it is so convenient that it is sometimes applied in non-linear cases, provided the non-linearity is not too severe.

17.6.7 Elementary contingency analysis

A further way in which a χ^2-test can be employed is to test for correlations between two "qualities", i.e. classifications in which each class cannot meaningfully be given a numerical value, but can be indexed. For example, we may ask, whether, amongst a student population of N students studying natural sciences, there is any significant correlation between the general area, biological ($i = 1$) or physical ($i = 2$), that a student opts to study and the sex of the student ($j = 1$ or 2 for male or female, respectively).

The general procedure is to adopt as the null hypothesis H_0 that there is no such correlation and that the two factors, here subject choice and gender, are independent. The expected numbers $n_{ij}^{(0)}$ in each doubly indexed category are then determined by direct multiplication of the corresponding fractions of the whole that those classes represent. For example, $n_{12}^{(0)}$, the expected number of females studying biological sciences, is given by $N \times f_1 \times g_2$, where f_1 is the actual fraction of all science students who are reading biological sciences and, similarly, g_2 is the actual fraction of all students who are female.

A 2×2 table showing these expected values is called a *contingency table* and represents the theoretical model; a similar table of the actual data is also required. There are a number of tests that measure the likelihood of obtaining the observed data if the theoretical model is valid. The most straightforward, and the normal one to apply unless either the actual or predicted values are small,[20] is *Pearson's* χ^2*-test*. This test computes the quantity

$$\chi^2 = \sum_i \sum_j \frac{\left(n_{ij} - n_{ij}^{(0)}\right)^2}{n_{ij}^{(0)}} \tag{17.75}$$

which should be distributed as a χ^2-distribution with $(2-1) \times (2-1) = 1$ degree of freedom. Comparison with a χ^2 probability table then enables the null hypothesis of no correlation to be accepted or rejected at any particular level.

20 The method is normally considered safe if each table entry is greater than about 10. For significantly smaller entries, a more refined test, such as Fisher's exact test which is based on the hypergeometric function and described in more specialized books, should be used.

If the number of possible categories in each or both of the two classifications is more than 2, say n and m, then the double sum in (17.75) contains nm terms, but since one entry in each row and one in each column is determined by the corresponding class total, there are only $(n-1)(m-1)$ degrees of freedom.[21] In our final worked example $n = m = 3$.

Example In a certain Cambridge college, the class of degree obtained by each of its students tabulated against the type of school they had previously attended was as follows.

	1st	2nd	3rd
State	60	127	23
Independent	32	98	10
Overseas	8	35	7

Is there any significant correlation between these two classifications?

We first copy out the table, augmenting it with class totals and the corresponding class fractions

	1st	2nd	3rd	Total	Fraction
State	60	127	23	210	0.525
Independent	32	98	10	140	0.350
Overseas	8	35	7	50	0.125
Total	100	260	40	400	
Fraction	0.250	0.650	0.010		

The table of expected values, on the basis H_0 that there is no correlation between degree class and school type, can now be found from the observed fractions:

	1st	2nd	3rd	Total	Fraction
State	52.5	136.5	21	210	0.525
Independent	35	91	14	140	0.350
Overseas	12.5	32.5	5	50	0.125
Total	100	260	40	400	
Fraction	0.250	0.650	0.010		

Although two entries in the final row are marginally too small to be entirely satisfactory, we next evaluate χ^2 as given by (17.75):

$$\chi^2 = \frac{(7.5)^2}{52.5} + \frac{(9.5)^2}{136.5} + \frac{(2)^2}{21} + \frac{(3)^2}{35} + \frac{(7)^2}{91} + \frac{(4)^2}{14} + \frac{(4.5)^2}{12.5} + \frac{(2.5)^2}{32.5} + \frac{(2)^2}{5}$$
$$= 1.07 + 0.67 + 0.19 + 0.26 + 0.54 + 1.14 + 1.62 + 0.19 + 0.80$$
$$= 6.47.$$

21 This reduction can be thought of as "the price" for having chosen the most favorable weighted mean for each row/column when compiling the table of expected values.

The number of degrees of freedom for this comparison is $(3-1)\times(3-1)=4$ and reference to Table 17.2 shows that for 4 d.o.f. the 10% significance level for χ^2 is 7.78. Since our calculated value is less than this, we cannot reject our null hypothesis and are led to the conclusion that, at this level, there is no evidence for any correlation between degree result and school type. A more detailed cumulative χ^2-table would show that for 4 d.o.f. the observed value of 6.47 is exceeded by chance on more than 16% of occasions.

As two of the three largest contributions to the χ^2 total come from the overseas entries that are in single figures, it might appear that these are distorting the results. However, an analysis excluding overseas students produces essentially the same conclusion.[22] ◀

SUMMARY

1. *Sample statistics for sample elements* $\mathbf{x}=(x_1, x_2, \ldots, x_N)$

Statistic	Formula
Mean, $\bar{x}$	$\frac{1}{N}\sum_{i=1}^{N} x_i$
Variance, s^2	$\frac{1}{N}\sum_{i=1}^{N}(x_i-\bar{x})^2=\overline{x^2}-(\bar{x})^2$
Standard deviation, s	$+\sqrt{\text{variance}}$
Covariance, V_{xy}	$\frac{1}{N}\sum_{i=1}^{N}(x_i-\bar{x})(y_i-\bar{y})=\overline{xy}-\bar{x}\bar{y}$
Correlation, r_{xy}	$\frac{V_{xy}}{s_x\, s_y}$

2. *Estimators and errors*
Here $P(y|a)$ is the probability distribution for obtaining sample result y if the parent population has parameter a.
   - Desirable characteristics of estimator $\hat{a}$ of population parameter a.
     (i) Consistent: $\lim_{N\to\infty}\hat{a}=a$.
     (ii) Unbiased: the expectation value of $\hat{a}$, $E[\hat{a}]=\int \hat{a}(\mathbf{x})P(\mathbf{x}|a)\,d^N\mathbf{x}$, a "theoretical" quantity that does not depend on the actual samples used, should be equal to a.

22 Show that such an analysis gives $\chi^2=3.52$ for 2 d.o.f. This value has a probability of being exceeded on 17% of occasions.

(iii) Efficient: the variance of $\hat{a}$, $V[\hat{a}] = \int (\hat{a}(\mathbf{x}) - E[\hat{a}])^2 P(\mathbf{x}|a)\, d^N\mathbf{x}$ should be as close as possible to the (theoretical) Fisher lower limit.

- The root mean square error $\epsilon_{\hat{a}}$ is given by

$$\epsilon_{\hat{a}}^2 = V[\hat{a}] + (E[\hat{a}] - a)^2 = (\text{sample variance}) + (\text{bias})^2.$$

- The (usual) central confidence *interval* at the *confidence level* $1 - 2\alpha$ for an observed value $\hat{a}_o$ is reported as

$$a = (\hat{a}_o)^{+d}_{-c},$$

where $c = \hat{a} - a_-$, $d = a_+ - \hat{a}$, and a_+ and a_- are such that

$$\int_{-\infty}^{\hat{a}_o} P(\hat{a}|a_+)\, d\hat{a} = \alpha = \int_{\hat{a}_o}^{\infty} P(\hat{a}|a_-)\, d\hat{a}.$$

- For the Gaussian distribution, this normally becomes a symmetric interval of two standard deviations (known or estimated), quoted as $a = \hat{a}_o \pm \sigma_{\hat{a}}$, corresponding to a confidence level of 0.683.

3. *Practical (consistent) estimators*

Here, $\mu_k = \int_{-\infty}^{\infty} x^k P(x)\, dx$ is the kth moment of the population distribution; ν_k is the corresponding central moment. An estimator marked "U" is unbiased. The final four entries refer to two-dimensional population distributions $P(x, y)$.

Population parameter, a	Estimator, $\hat{a}$	$E[\hat{a}]$	$V[\hat{a}]$	U
Mean, μ	$\bar{x}$	μ	$\frac{\sigma^2}{N}$	✓
Variance, σ^2, μ known	$\frac{1}{N}\sum_{i=1}^{N} x_i^2 - \mu^2$	σ^2	$\frac{1}{N}\left(\mu_4 - \mu_2^2\right)$	✓
μ unknown	s^2	$\frac{N-1}{N}\sigma^2$	see (17.30) p. 724	×
	$\frac{N}{N-1}s^2$	σ^2	$g(N, \nu_2, \nu_4)$	✓
Standard deviation, σ; μ unknown	$\left(\widehat{\sigma^2}\right)^{1/2}$	$\approx \sigma$	$\frac{g(N, \nu_2, \nu_4)}{4\nu_2}$	×
rth population moment, μ_r	$m_r = \frac{1}{N}\sum_{i=1}^{N} x_i^r$	μ_r	$\frac{1}{N}\left(\mu_{2r} - \mu_r^2\right)$	✓

(cont.)

Population parameter, a	Estimator, $\hat{a}$	$E[\hat{a}]$	$V[\hat{a}]$	U
Covariance, means known	$\overline{xy} - \mu_x\mu_y$	Cov $[x, y]$	–	✓
means unknown	V_{xy} (see item 1)	$\frac{N-1}{N}$Cov $[x, y]$	–	×
	$\frac{N}{N-1}V_{xy}$	Cov $[x, y]$	–	✓
Correlation, ρ, means unknown	$\frac{N}{N-1}\frac{V_{xy}}{s_x s_y}$	$\approx \rho - \frac{\rho(1-\rho^2)}{2N}$ for Gaussian	$\approx \frac{1}{N}(1-\rho^2)^2$ for Gaussian	×

In the fourth and fifth entries, $g(N, \nu_2, \nu_4) = \frac{1}{N}\left(\nu_4 - \frac{N-3}{N-1}\nu_2^2\right)$.

4. *Data modeling*
To determine the M model parameters $\mathbf{a}$ from paired measurements (x_i, y_i) connected by $y = f(x; \mathbf{a})$, when the distribution of independent measurement errors n_i contributing to measurements y_i is Gaussian, with zero mean and variance σ_i^2, find the a_j that minimize

$$\chi^2(\mathbf{a}) = \sum_{i=1}^{N}\left[\frac{y_i - f(x_i; \mathbf{a})}{\sigma_i}\right]^2$$

by setting $\frac{\partial \chi^2}{\partial a_j} = 0$ for $j = 1, 2, \ldots, M$.

- If the measurement errors are not independent, the expression for $\chi^2(\mathbf{a})$ becomes a quadratic form $(\mathbf{y} - \mathbf{f})^{\mathrm{T}}\mathsf{N}^{-1}(\mathbf{y} - \mathbf{f})$, where N is the covariance matrix for the errors, $N_{ij} = \text{Cov}\,[n_i, n_j]$, and $f_i = f(x_i; \mathbf{a})$.
- *Linear least squares*: If $f(x; \mathbf{a})$ is *linear* in the a_j, then define the components of the $N \times M$ *response matrix* R by

$$f(x_i; \mathbf{a}) = \sum_{j=1}^{M} a_j h_j(x_i) \equiv \sum_{j=1}^{M} R_{ij} a_j.$$

The set of base functions, $h_j(x)$, need not be linear in x, but are best chosen, so far as is possible, to be orthogonal over the data, i.e.

$$\sum_{i=1}^{N} h_k(x_i) h_m(x_i) = 0 \qquad \text{for } k \neq m.$$

- The best linear least squares fit to the parameters $\mathbf{a}$ is given by

$$\hat{\mathbf{a}} = (\mathsf{R}^{\mathrm{T}}\mathsf{N}^{-1}\mathsf{R})^{-1}\mathsf{R}^{\mathrm{T}}\mathsf{N}^{-1}\mathbf{y} \quad \text{with} \quad \mathsf{V} \equiv \text{Cov}[\hat{a}_i, \hat{a}_j] = (\mathsf{R}^{\mathrm{T}}\mathsf{N}^{-1}\mathsf{R})^{-1}.$$

If the measurement errors are independent, N^{-1} is $\text{diag}(1/\sigma_i^2)$.

5. *Goodness of least-squares fit*
   To test the appropriateness of an assumed model $y = f(x; \mathbf{a})$, where f is linear in the M parameters a_j, form the χ^2 variable

$$\chi^2(\hat{\mathbf{a}}) = \sum_{i=1}^{N} \left[\frac{y_i - f(x_i; \hat{\mathbf{a}})}{\sigma_i} \right]^2,$$

   where the components of $\hat{\mathbf{a}}$ are the calculated best-fit parameters to the model. $\chi^2(\hat{\mathbf{a}})$ can then be tested against the χ^2-distribution for $N - M$ degrees of freedom which has expectation $N - M$ and variance $2(N - M)$; too large a value for $\chi^2(\mathbf{a})$ indicates an incorrect model, whilst too small a value suggests that the model has too many adjustable parameters a_j.

6. *Hypothesis testing for actual, approximate and assumed Gaussian populations*
   - A hypothesis H_0 is *simple* if it determines the PDF $P(x|H_0)$ of the measurements uniquely.
   - A test statistic $\lambda(\mathbf{x})$, computed from the measured sample and based on a likelihood ratio, can be used to accept or reject a simple hypothesis H_0 at some predetermined confidence level α, by comparing λ with a critical value calculated from $P(\lambda|H_0)$ and α.
   - Any function $f(\lambda)$ of λ will be an equally valid statistic, provided f is a monotonic function, either increasing or decreasing. In this case $P(f(\lambda)|H_0)$ must be used to calculate the critical value.

   In the following table of tests, only the final simplest form of the test statistic is given; where a quantity A appears in $P(f|H_0)$, it represents a normalization constant. In several cases, in order to produce a compact layout, additional symbols, as given in the notes below the table, are introduced.

Case	H_0	Test statistic $f(\lambda)$	$P(f\|H_0)$	Test
σ^2 known	$\mu = \mu_0$	$f = N(\bar{x} - \mu_0)^2$	$\chi_1^2(f)$	χ^2, 1 d.o.f.
[a]μ and σ^2 unknown	$\mu = \mu_0$	$t = \dfrac{\bar{x} - \mu_0}{s/\sqrt{N-1}}$	T(t, N, 1)	Student's t, $N - 1$ d.o.f.
[a,b]Two populations, common σ^2	$\mu_1 = \mu_2$	$t = \dfrac{\bar{w} - \omega}{\hat{\sigma}} N_H^{1/2}$	T(t, N, 2)	Student's t, $N_1 + N_2 - 2$ d.o.f.
Sample with unknown μ	$\sigma^2 = \sigma_0^2$	$u = \dfrac{Ns^2}{\sigma_0^2}$	$\chi_{N-1}^2(u)$	χ^2, $N - 1$ d.o.f.[c]

(*cont.*)

Case	H_0	Test statistic $f(\lambda)$	$P(f\|H_0)$	Test
[d]Two samples, different μ_i	$\sigma_1^2 = \sigma_2^2$	$F = \dfrac{N_1 s_1^2/n_1}{N_2 s_2^2/n_2}$	$g(F, n_1, n_2)$	Fisher's F, (n_1, n_2) d.o.f.
[e]As above, $N_1, N_2 \gg 1$	$\sigma_1^2 = \sigma_2^2$	$z = \frac{1}{2} \ln F$	$N[m_-, m_+]$	Gaussian

(a) $T(t, N, M) = A\left(1 + \dfrac{t^2}{N-M}\right)^{-N/2}$.

(b) $\omega = \mu_1 - \mu_2,\ \bar{w} = \bar{x}_1 - \bar{x}_2,\ N_H = \dfrac{N_1 N_2}{N},\ \hat{\sigma} = \left[\dfrac{N_1 s_1^2 + N_2 s_2^2}{N-2}\right]^{1/2}$ and $N = N_1 + N_2$.

(c) With symmetric rejection regions.

(d) Labels i chosen so that $F \geq 1$, $n_i = N_i - 1$ and

$$g(F, n_1, n_2) = AF^{(n_1-2)/2}\left(1 + \frac{n_1}{n_2}F\right)^{-(n_1+n_2)/2}.$$

(e) $n_i = N_i - 1,\ m_\pm = \frac{1}{2}(n_1^{-1} \pm n_2^{-1})$.

7. *Contingency analysis*

Contingency analysis tests for correlations between two "qualities", i.e. classifications in which each class cannot meaningfully be given a numerical value, but can be indexed, $i = 1, 2, \ldots, N$ and $j = 1, 2, \ldots, M$; the null hypothesis H_0 is that there is no such correlation. The expected numbers $n_{ij}^{(0)}$ in each doubly indexed category are then determined by direct multiplication of the corresponding fractions of the whole that those classes represent. A $N \times M$ *contingency table* showing these expected values, and a similar table of the actual data, n_{ij}, are drawn up. Then, unless some of the actual or predicted values are less than about 10, *Pearson's* χ^2*-test* is applied.

This test computes the quantity

$$\chi^2 = \sum_{i=1}^{N}\sum_{j=1}^{M} \frac{\left(n_{ij} - n_{ij}^{(0)}\right)^2}{n_{ij}^{(0)}},$$

which should be distributed as a χ^2-distribution with $(N-1) \times (M-1)$ degrees of freedom. Comparison with a χ^2 probability table then enables the null hypothesis of no correlation to be accepted or rejected at any particular level.

PROBLEMS

17.1. A group of students uses a pendulum experiment to measure g, the acceleration of free fall, and obtains the following values (in m s^{-2}): 9.80, 9.84, 9.72, 9.74,

9.87, 9.77, 9.28, 9.86, 9.81, 9.79, 9.82. What would you give as the best value and standard error for g as measured by the group?

17.2. Measurements of a certain quantity gave the following values: 296, 316, 307, 278, 312, 317, 314, 307, 313, 306, 320, 309. Within what limits would you say there is a 50% chance that the correct value lies?

17.3. The following are the values obtained by a class of 14 students when measuring a physical quantity x: 53.8, 53.1, 56.9, 54.7, 58.2, 54.1, 56.4, 54.8, 57.3, 51.0, 55.1, 55.0, 54.2, 56.6.

(a) Display these results as a histogram and state what you would give as the best value for x.

(b) Without calculation, estimate how much reliance could be placed upon your answer to (a).

(c) Data books give the value of x as 53.6 with negligible error. Are the data obtained by the students in conflict with this?

17.4. Prove that the sample mean is the best *linear unbiased estimator* of the population mean μ as follows.

(a) If the real numbers $a_1, a_2, \ldots, a_n$ satisfy the constraint $\sum_{i=1}^{n} a_i = C$, where C is a given constant, show that $\sum_{i=1}^{n} a_i^2$ is minimized by $a_i = C/n$ for all i.

(b) Consider the linear estimator $\hat{\mu} = \sum_{i=1}^{n} a_i x_i$. Impose the conditions (i) that it is *unbiased* and (ii) that it is as *efficient* as possible.

17.5. A population contains individuals of k types in equal proportions. A quantity X has mean μ_i amongst individuals of type i and variance σ^2, which has the same value for all types. In order to estimate the mean of X over the whole population, two schemes are considered; each involves a total sample size of nk. In the first the sample is drawn randomly from the whole population, whilst in the second (*stratified sampling*) n individuals are randomly selected from each of the k types.

Show that in both cases the estimate has expectation

$$\mu = \frac{1}{k}\sum_{i=1}^{k} \mu_i,$$

but that the variance of the first scheme exceeds that of the second by an amount

$$\frac{1}{k^2 n}\sum_{i=1}^{k} (\mu_i - \mu)^2.$$

17.6. This problem is intended to illustrate the dangers of applying formalized estimator techniques to distributions that are not well behaved in a statistical sense.

The following are five sets of 10 values, all drawn from the same Cauchy distribution with parameter a.

(i)	4.81	−1.24	1.30	−0.23	2.98
	−1.13	−8.32	2.62	−0.79	−2.85
(ii)	0.07	1.54	0.38	−2.76	−8.82
	1.86	−4.75	4.81	1.14	−0.66
(iii)	0.72	4.57	0.86	−3.86	0.30
	−2.00	2.65	−17.44	−2.26	−8.83
(iv)	−0.15	202.76	−0.21	−0.58	−0.14
	0.36	0.44	3.36	−2.96	5.51
(v)	0.24	−3.33	−1.30	3.05	3.99
	1.59	−7.76	0.91	2.80	−6.46

Ignoring the fact that the Cauchy distribution does not have a finite variance (or even a formal mean), show that $\hat{a}$, the estimator of a, has to satisfy

$$s(\hat{a}) = \sum_{i=1}^{10} \frac{1}{1 + x_i^2/\hat{a}^2} = 5. \quad (*)$$

Using a programmable calculator, spreadsheet or computer, find the value of $\hat{a}$ that satisfies $(*)$ for each of the data sets and compare it with the value $a = 1.6$ used to generate the data. Form an opinion regarding the variance of the estimator.

Show further that if it is assumed that $(E[\hat{a}])^2 = E[\hat{a}^2]$, then $E[\hat{a}] = \nu_2^{1/2}$, where ν_2 is the second (central) moment of the distribution, which for the Cauchy distribution is infinite!

17.7. According to a particular theory, two dimensionless quantities X and Y have equal values. Nine measurements of X gave values of 22, 11, 19, 19, 14, 27, 8, 24 and 18, whilst seven measured values of Y were 11, 14, 17, 14, 19, 16 and 14. Assuming that the measurements of both quantities are Gaussian distributed with a common variance, are they consistent with the theory? An alternative theory predicts that $Y^2 = \pi^2 X$; are the data consistent with this proposal?

17.8. On a certain (testing) steeplechase course there are 12 fences to be jumped, and any horse that falls is not allowed to continue in the race. In a season of racing a total of 500 horses started the course and the following numbers fell at each fence:

Fence:	1	2	3	4	5	6	7	8	9	10	11	12
Falls:	62	75	49	29	33	25	30	17	19	11	15	12

Use this data to determine the overall probability of a horse's falling at a fence, and test the hypothesis that it is the same for all horses and fences as follows.

(a) Draw up a table of the expected number of falls at each fence on the basis of the hypothesis.

(b) Consider for each fence i the standardized variable

$$z_i = \frac{\text{estimated falls} - \text{actual falls}}{\text{standard deviation of estimated falls}},$$

and use it in an appropriate χ^2 test.

(c) Show that the data indicates that the odds against all fences being equally testing are about 40 to 1. Identify the fences that are significantly easier or harder than the average.

17.9. During an investigation into possible links between mathematics and classical music, pupils at a school were asked whether they had preferences (a) between mathematics and English, and (b) between classical and pop music. The results are given below.

	Classical	None	Pop
Mathematics	23	13	14
None	17	17	36
English	30	10	40

Determine whether there is any evidence for
(a) a link between academic and musical tastes, and
(b) a claim that pupils either had preferences in both areas or had no preference.
You will need to consider the appropriate value for the number of degrees of freedom to use when applying the χ^2 test.

17.10. Three candidates X, Y and Z were standing for election to a vacant seat on their college's Student Committee. The members of the electorate (current first-year students, consisting of 150 men and 105 women) were each allowed to cross out the name of the candidate they least wished to be elected, the other two candidates then being credited with one vote each. The following data are known.
(a) X received 100 votes from men, whilst Y received 65 votes from women.
(b) Z received five more votes from men than X received from women.
(c) The total votes cast for X and Y were equal.
Analyze this data in such a way that a χ^2 test can be used to determine whether voting was other than random (i) amongst men and (ii) amongst women.

17.11. A particle detector consisting of a shielded scintillator is being tested by placing it near a particle source whose intensity can be controlled by the use of absorbers. It might register counts even in the absence of particles from the source because of the cosmic ray background.

The number of counts n registered in a fixed time interval as a function of the source strength s is given as:

source strength s:	0	1	2	3	4	5	6
counts n:	6	11	20	42	44	62	61

At any given source strength, the number of counts is expected to be Poisson distributed with mean

$$n = a + bs,$$

where a and b are constants. Analyze the data for a fit to this relationship and obtain the best values for a and b together with their standard errors.

(a) How well is the cosmic ray background determined?

(b) What is the value of the correlation coefficient between a and b? Is this consistent with what would happen if the cosmic ray background were imagined to be negligible?

(c) Do the data fit the expected relationship well? Is there any evidence that the reported data "are too good a fit"?

17.12. The function $y(x)$ is known to be a quadratic function of x. The following table gives the measured values and uncorrelated standard errors of y measured at various values of x (in which there is negligible error):

x	1	2	3	4	5
$y(x)$	3.5 ± 0.5	2.0 ± 0.5	3.0 ± 0.5	6.5 ± 1.0	10.5 ± 1.0

Construct the response matrix R using as basis functions $1, x, x^2$. Calculate the matrix $\mathsf{R}^{\mathrm{T}}\mathsf{N}^{-1}\mathsf{R}$ and show that its inverse, the covariance matrix V, has the form

$$\mathsf{V} = \frac{1}{9184}\begin{pmatrix} 12\,592 & -9708 & 1580 \\ -9708 & 8413 & -1461 \\ 1580 & -1461 & 269 \end{pmatrix}.$$

Use this matrix to find the best values, and their uncertainties, for the coefficients of the quadratic form for $y(x)$.

17.13. The following are the values and standard errors of a physical quantity $f(\theta)$ measured at various values of θ (in which there is negligible error):

θ	0	$\pi/6$	$\pi/4$	$\pi/3$
$f(\theta)$	3.72 ± 0.2	1.98 ± 0.1	-0.06 ± 0.1	-2.05 ± 0.1

θ	$\pi/2$	$2\pi/3$	$3\pi/4$	π
$f(\theta)$	-2.83 ± 0.2	1.15 ± 0.1	3.99 ± 0.2	9.71 ± 0.4

Theory suggests that f should be of the form $a_1 + a_2 \cos\theta + a_3 \cos 2\theta$. Show that the normal equations for the coefficients a_i are

$$481.3a_1 + 158.4a_2 - 43.8a_3 = 284.7,$$
$$158.4a_1 + 218.8a_2 + 62.1a_3 = -31.1,$$
$$-43.8a_1 + 62.1a_2 + 131.3a_3 = 368.4.$$

(a) If you have matrix inversion routines available on a computer, determine the best values and variances for the coefficients a_i and the correlation between the coefficients a_1 and a_2.

(b) If you have only a calculator available, solve for the values using a Gauss–Seidel iteration[23] and start from the approximate solution $a_1 = 2$, $a_2 = -2$, $a_3 = 4$.

17.14. It is claimed that the two following sets of values were obtained (a) by randomly drawing from a normal distribution that is $N(0, 1)$ and then (b) randomly assigning each reading to one of two sets A and B:

Set A:	−0.314	0.603	−0.551	−0.537	−0.160	−1.635	0.719
	0.610	0.482	−1.757	0.058			
Set B:	−0.691	1.515	−1.642	−1.736	1.224	1.423	1.165

Make tests, including t- and F-tests, to establish whether there is any evidence that either claims is, or both claims are, false.

HINTS AND ANSWERS

17.1. Note that the reading of 9.28 m s^{-2} is clearly in error, and should not be used in the calculation; 9.80 ± 0.02 m s^{-2}.

17.3. (a) 55.1. (b) Note that two-thirds of the readings lie within ± 2 of the mean and that 14 readings are being used. This gives a standard error in the mean ≈ 0.6. (c) Student's t has a value of about 2.5 for 13 d.o.f. (degrees of freedom), and therefore it is likely at the 3% significance level that the data are in conflict with the accepted value.

17.5. Recall that, because of the equal proportions of each type, the *expected* number of each type in the first scheme is n. Show that the variance of the estimator for the second scheme is $\sigma^2/(kn)$. When calculating that for the first scheme, recall that $\overline{x_i^2} = \mu_i^2 + \sigma^2$ and note that μ_i^2 can be written as $(\mu_i - \mu + \mu)^2$.

17.7. $\bar{X} = 18.0 \pm 2.2$, $\bar{Y} = 15.0 \pm 1.1$. $\hat{\sigma} = 4.92$ giving $t = 1.21$ for 14 d.o.f., and is significant only at the 75% level. Thus there is no significant disagreement between the data and the theory. For the second theory, only the mean values can be tested as Y^2 will not be Gaussian distributed. The difference in the means is $\bar{Y^2} - \pi^2\bar{X} = 47 \pm 36$ and is only significantly different from zero at the 82% level. Again the data is consistent with the proposed theory.

17.9. Consider how many entries may be chosen freely in the table if all row and column totals are to match the observed values. It should be clear that for an $m \times n$ table the number of degrees of freedom is $(m - 1)(n - 1)$.

23 If you are not familiar with this consult a text on numerical methods.

(a) In order to make the fractions expressing each preference or lack of preference correct, the expected distribution, if there were no correlation, must be

	Classical	None	Pop
Mathematics	17.5	10	22.5
None	24.5	14	31.5
English	28	16	36

This gives a χ^2 of 12.3 for four d.o.f., making it less than 2% likely, that no correlation exists.

(b) The expected distribution, if there were no correlation, is

	Music preference	No music preference
Academic preference	104	26
No academic preference	56	14

This gives a χ^2 of 1.2 for one d.o.f. and no evidence for the claim.

17.11. As the distribution at each value of s is Poisson, the best estimate of the measurement error is the square root of the number of counts, i.e. $\sqrt{n(s)}$. Linear regression gives $a = 4.3 \pm 2.1$ and $b = 10.06 \pm 0.94$.

(a) The cosmic ray background must be present, since $n(0) \neq 0$ but its value of about 4 is uncertain to within a factor 2.

(b) The correlation coefficient between a and b is -0.63. Yes; if a were reduced towards zero then b would have to be increased to compensate.

(c) Yes, $\chi^2 = 4.9$ for five d.o.f., which is almost exactly the "expected" value, neither too good nor too bad.

17.13. $a_1 = 2.02 \pm 0.06$, $a_2 = -2.99 \pm 0.09$, $a_3 = 4.90 \pm 0.10$; $r_{12} = -0.60$.

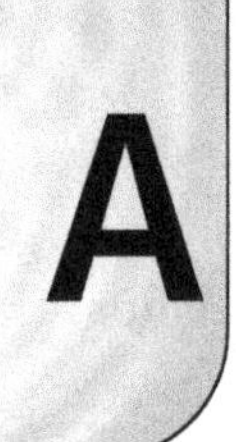

A Review of background topics

A.1 Arithmetic and geometry

1. *Logarithms and the exponential function*
   - For a logarithm to any base a (> 0), $x = a^{\log_a x}$ and

$$\log_a x^n = n \log_a x, \text{ where } n \text{ is any real number.}$$

   - For $x > 0$, its natural logarithm, $\log_e x \equiv \ln x$ is defined by $x = e^{\ln x}$, where $e = \exp(1)$ and

$$e^x = \exp(x) = \sum_{n=0}^{\infty} \frac{x^n}{n!}.$$

   - The exponential function and natural logarithm have the properties

$$\frac{d}{dx}(e^x) = e^x, \qquad \frac{d}{dx}(\ln x) = \frac{1}{x}, \qquad \ln x = \int_1^x \frac{1}{u}\, du,$$

$$\ln(xy) = \ln x + \ln y, \qquad \ln\left(\frac{x}{y}\right) = \ln x - \ln y.$$

2. *Rational and irrational numbers*
   - If $\sqrt{p}$ is irrational, $a + b\sqrt{p} = c + d\sqrt{p}$ implies that $a = c$ and $b = d$.
   - To rationalize $(a + b\sqrt{p})^{-1}$, write it as

$$\frac{(a - b\sqrt{p})}{(a + b\sqrt{p})(a - b\sqrt{p})} = \frac{a - b\sqrt{p}}{a^2 - b^2 p}.$$

3. *Physical dimensions*

   The base units are [length] $= L$, [mass] $= M$, [time] $= T$, [current] $= I$, [temperature] $= \Theta$.
   - The dimensions of any one physical quantity can contain only integer powers of the base units.
   - The dimension of a constant is zero for each base unit.
   - The dimensions of a product ab are the sums of the dimensions of a and b, separately for each base unit.
   - The dimensions of $1/a$ are the negatives of the dimensions of a for each base unit.
   - All terms in any physically acceptable equation (including the individual terms in any implied series) must have the same dimension for each base unit.

4. *Binomial expansion*
   - For any integer $n > 0$,

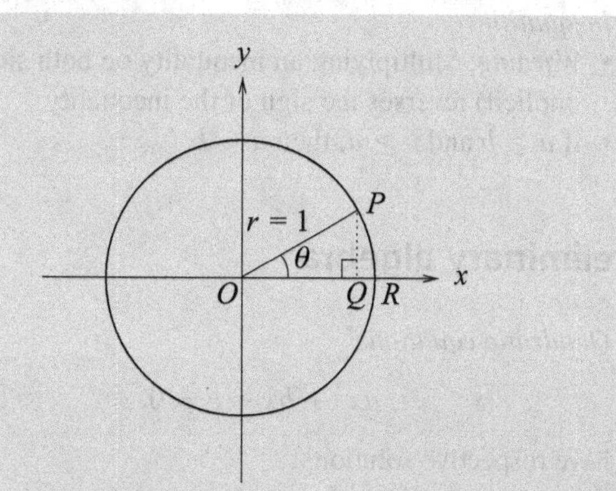

Figure A.1 The geometric definitions of the basic trigonometric functions.

(i) $(x+y)^n = \sum_{k=0}^{n} {}^nC_k x^{n-k} y^k$, with ${}^nC_k = \dfrac{n!}{k!\,(n-k)!}$.

(ii) $(x+y)^{-n} = x^{-n} \sum_{k=0}^{\infty} {}^{-n}C_k \left(\dfrac{y}{x}\right)^k$, with ${}^{-n}C_k = (-1)^k \times {}^{n+k-1}C_k$ and $|x| > |y|$.

- For any n, positive or negative, integer or non-integer, and $|x| > |y|$,

$$(x+y)^n = x^n \sum_{k=0}^{\infty} {}^nC_k \left(\frac{y}{x}\right)^k, \text{ with } {}^nC_0 = 1 \text{ and } {}^nC_{k+1} = \frac{n-k}{k+1}\,{}^nC_k.$$

5. *Trigonometry*
   - With θ measured in radians, in Figure A.1,

$$\sin\theta = \frac{QP}{OP} = \sum_{n=0}^{\infty} \frac{(-1)^n \theta^{2n+1}}{(2n+1)!}, \qquad \cos\theta = \frac{OQ}{OP} = \sum_{n=0}^{\infty} \frac{(-1)^n \theta^{2n}}{(2n)!}.$$

   - $\cos^2\theta + \sin^2\theta = 1, \qquad 1+\tan^2\theta = \sec^2\theta, \qquad 1+\cot^2\theta = \operatorname{cosec}^2\theta.$
   - $$\sin(A \pm B) = \sin A \cos B \pm \cos A \sin B,$$
     $$\cos(A \pm B) = \cos A \cos B \mp \sin A \sin B.$$
   - $$\sin 2\theta = 2\sin\theta\cos\theta,$$
     $$\cos 2\theta = \cos^2\theta - \sin^2\theta = 2\cos^2\theta - 1 = 1 - 2\sin^2\theta.$$
   - If $t = \tan(\theta/2)$, then

$$\sin\theta = \frac{2t}{1+t^2}, \qquad \cos\theta = \frac{1-t^2}{1+t^2}, \qquad \tan\theta = \frac{2t}{1-t^2}.$$

6. *Inequalities*
   - *Warning*: Multiplying an inequality on both sides by a negative quantity (explicit or implicit) reverses the sign of the inequality.
   - If $a \geq b$ and $b \geq a$, then $a = b$.

A.2 Preliminary algebra

1. *Quadratic equations*

$$ax^2 + bx + c = 0, \qquad ax^2 + 2bx + c = 0$$

have respective solutions

$$\alpha_{1,2} = \frac{-b \pm \sqrt{b^2 - 4ac}}{2a}, \qquad \alpha_{1,2} = \frac{-b \pm \sqrt{b^2 - ac}}{a}.$$

2. *Polynomial equations with real coefficients*
   - An nth-degree polynomial has exactly n zeros, but they are not necessarily real, nor necessarily distinct.
   - An nth-degree polynomial has an odd or even number of real zeros according to whether n is odd or even, respectively.
   - For the nth-degree polynomial equation

$$a_n x^n + a_{n-1} x^{n-1} + \cdots + a_1 x + a_0 = 0, \qquad a_n \neq 0$$

with roots $\alpha_1, \alpha_2, \ldots, \alpha_n$,

$$\prod_{k=1}^{n} \alpha_k = (-1)^n \frac{a_0}{a_n}, \qquad \sum_{k=1}^{n} \alpha_k = -\frac{a_{n-1}}{a_n}.$$

3. *Coordinate geometry*
   - Straight line: $y = mx + c$ or $ax + by + k = 0$.
   - The condition for two straight lines to be orthogonal is $m_1 m_2 = -1$.
   - Conic sections "centered" on (α, β) and their parameterizations:

Conic	Equation	x	y
circle	$(x-\alpha)^2 + (y-\beta)^2 = a^2$	$\alpha + a\cos\phi$	$\beta + a\sin\phi$
ellipse	$\frac{(x-\alpha)^2}{a^2} + \frac{(y-\beta)^2}{b^2} = 1$	$\alpha + a\cos\phi$	$\beta + b\sin\phi$
parabola	$(y-\beta)^2 = 4a(x-\alpha)$	$\alpha + at^2$	$\beta + 2at$
hyperbola	$\frac{(x-\alpha)^2}{a^2} - \frac{(y-\beta)^2}{b^2} = 1$	$\alpha + a\cosh\phi$	$\beta + b\sinh\phi$

4. *Plane polar coordinates*

$$x = \rho\cos\phi, \qquad y = \rho\sin\phi,$$

$$\rho = \sqrt{x^2 + y^2}, \qquad \cos\phi = \frac{x}{\sqrt{x^2 + y^2}}, \qquad \sin\phi = \frac{y}{\sqrt{x^2 + y^2}}.$$

5. *Partial fractions expansion*
For the representation of $f(x) = g(x)/h(x)$, with $g(x)$ a polynomial and $h(x) = (x - \alpha_1)(x - \alpha_2)\cdots(x - \alpha_n)$:
   - With the α_i all different,

$$f(x) = \sum_{k=1}^{n} \frac{A_k}{x - \alpha_k}, \quad \text{where } A_k = \frac{g(\alpha_k)}{\prod_{j \neq k}^{n}(\alpha_k - \alpha_j)}.$$

   - If the degree of $g(x)$ is $\geq m$, the degree of $h(x)$, then $f(x)$ must first be written as

$$f(x) = s(x) + \frac{r(x)}{h(x)}, \quad \text{where} \begin{cases} s(x) \text{ is a polynomial,} \\ \text{the degree of } r(x) \text{ is } < m. \end{cases}$$

   - If $h(x)$ contains a factor $a^2 + x^2$, then the corresponding term in the expansion takes the form $(Ax + b)/(a^2 + x^2)$.
   - If $h(x) = 0$ has a repeated root α, i.e. $h(x)$ contains a factor $(x - \alpha)^p$, then the expansion must contain

$$\text{either} \quad \frac{A_0 + A_1 x + \cdots + A_{p-1}x^{p-1}}{(x - \alpha)^p}$$

$$\text{or} \quad \frac{B_1}{x - \alpha} + \frac{B_2}{(x - \alpha)^2} + \cdots + \frac{B_p}{(x - \alpha)^p}.$$

6. *Proof by induction* (on n)
   (i) Assume the proposition is true for $n = N$ (or for all $n \leq N$).
   (ii) Use (i) to prove the proposition is then true for $n = N + 1$ (or for all $n \leq N + 1$).
   (iii) Show by observation, or by direct calculation without assumptions, that the proposition is true for the lowest n in its range.
   (iv) Conclude that the proposition is true for all n in its range.
7. *Proof by contradiction*
   (i) Assume the proposition is *not* true.
   (ii) Show, using *only* conclusions that *necessarily* follow from their predecessors and the assumption, that this leads to a contradiction.
   (iii) Conclude that the proposition is true.

   Warning: Failure to find a contradiction gives no information as to whether or not the proposition is true.
8. *Necessary and sufficient conditions*
   - A if B — B is a sufficient condition for A — $B \Rightarrow A$
     A only if B — B is a necessary consequence of A — $A \Rightarrow B$
     A IFF B — A and B necessarily imply each other — $A \Leftrightarrow B$
   - *Warning*: Necessary and sufficient condition proofs nearly always require two separate chains of argument. The second part of the proof is usually *not* the lines of the first part written in reverse order.

A.3 Differential calculus

1. *Definitions*

$$\frac{df(x)}{dx} \equiv f^{(1)} \equiv f'(x) \equiv \lim_{\Delta x \to 0} \frac{f(x+\Delta x) - f(x)}{\Delta x},$$

$$f^{(n+1)}(x) \equiv \lim_{\Delta x \to 0} \frac{f^{(n)}(x+\Delta x) - f^{(n)}(x)}{\Delta x}.$$

2. *Standard derivatives*

- $(e^{ax})' = ae^{ax}, \qquad (\ln ax)' = \frac{1}{x}, \qquad (a^x)' = a^x \ln a, \qquad (x^n)' = nx^{n-1}.$
- The derivatives of sinusoidal and inverse sinusoidal functions.

$$\frac{d}{dx}(\sin ax) = a\ \cos ax, \qquad \frac{d}{dx}(\cos ax) = -a\ \sin ax,$$

$$\frac{d}{dx}(\tan ax) = a\ \sec^2 ax, \qquad \frac{d}{dx}(\sec ax) = a\ \sec ax \tan ax,$$

$$\frac{d}{dx}(\cot ax) = -a\ \operatorname{cosec}^2 ax, \qquad \frac{d}{dx}(\operatorname{cosec} ax) = -a\ \operatorname{cosec} ax \cot ax,$$

$$\frac{d}{dx}\left(\sin^{-1}\frac{x}{a}\right) = \frac{1}{\sqrt{a^2-x^2}}, \qquad \frac{d}{dx}\left(\cos^{-1}\frac{x}{a}\right) = \frac{-1}{\sqrt{a^2-x^2}},$$

$$\frac{d}{dx}\left(\tan^{-1}\frac{x}{a}\right) = \frac{a}{a^2+x^2}.$$

- The derivatives of hyperbolic and inverse hyperbolic functions.

$$\frac{d}{dx}(\cosh x) = \sinh x, \qquad \frac{d}{dx}(\sinh x) = \cosh x,$$

$$\frac{d}{dx}(\tanh x) = \operatorname{sech}^2 x, \qquad \frac{d}{dx}(\operatorname{sech} x) = -\operatorname{sech} x \tanh x,$$

$$\frac{d}{dx}(\operatorname{cosech} x) = -\operatorname{cosech} x \coth x, \qquad \frac{d}{dx}(\coth x) = -\operatorname{cosech}^2 x.$$

$$\frac{d}{dx}\left(\cosh^{-1}\frac{x}{a}\right) = \frac{\pm 1}{\sqrt{x^2-a^2}}, \qquad \frac{d}{dx}\left(\sinh^{-1}\frac{x}{a}\right) = \frac{1}{\sqrt{x^2+a^2}},$$

$$\frac{d}{dx}\left(\tanh^{-1}\frac{x}{a}\right) = \frac{a}{a^2-x^2}, \qquad \text{for } x^2 < a^2,$$

$$\frac{d}{dx}\left(\coth^{-1}\frac{x}{a}\right) = \frac{-a}{x^2-a^2}, \qquad \text{for } x^2 > a^2.$$

3. *Derivatives of compound functions*
   If $u, v, \ldots, w$ are all functions of x, then
   - $(uv)' = u'v + uv'$.
   - $(uv \ldots w)' = u'v \ldots w + uv' \ldots w + \cdots + uv \ldots w'$.
   - $\left(\dfrac{u}{v}\right)' = \dfrac{vu' - uv'}{v^2}$.
   - If $f = uv$, then $f^{(n)} = \sum_{r=0}^{n} {}^nC_r u^{(r)} v^{(n-r)}$ (Leibnitz).
4. *Change of variable*
$$\frac{dx}{df} = \left(\frac{df}{dx}\right)^{-1}, \qquad \frac{df}{dx} = \frac{df}{du}\frac{du}{dx} \qquad \text{(chain rule).}$$
5. *Stationary points*
   - For a stationary point of $f(x)$, $f'(x) = 0$ and for
     maximum $f'' < 0$,
     minimum $f'' > 0$,
     point of inflection $f'' = 0$ and changes sign through the point.
   - If $a \leq x \leq c$, then for some b in $a < b < c$,
$$\frac{f(c) - f(a)}{c - a} = f'(b), \qquad \text{mean value theorem.}$$
   Rolle's theorem is a special case of this in which $f(c) = f(a)$ and $f'(b) = 0$.
6. *Radius of curvature* of $f(x)$
$$\rho = \frac{[1 + (f')^2]^{3/2}}{f''}.$$
7. *Graphs*
   Aspects that may help in sketching a graph: symmetry or antisymmetry about the x- or y-axis; zeros; particular simply calculated values; vertical and horizontal asymptotes; other asymptotes; stationary points.

A.4 Integral calculus

1. *Elementary properties of integrals*
$$\int_a^b 0\, dx = 0, \qquad \int_a^a f(x)\, dx = 0, \qquad \int_a^b f(x)\, dx = -\int_b^a f(x)\, dx,$$
$$\int_a^c f(x)\, dx = \int_a^b f(x)\, dx + \int_b^c f(x)\, dx,$$
$$\int_a^b [f(x) + g(x)]\, dx = \int_a^b f(x)\, dx + \int_a^b g(x)\, dx,$$
$$\frac{d}{dx} F(x) \equiv \frac{d}{dx}\left[\int_{x_0}^x f(u)\, du\right] = f(x).$$

2. *Standard integrals*
   - The integrals of elementary functions.

$$\int a\,dx = ax + c, \qquad \int ax^n\,dx = \frac{ax^{n+1}}{n+1} + c,$$
$$\int e^{ax}\,dx = \frac{e^{ax}}{a} + c, \qquad \int \frac{a}{x}\,dx = a\ln x + c,$$
$$\int a\cos bx\,dx = \frac{a\sin bx}{b} + c, \qquad \int a\sin bx\,dx = \frac{-a\cos bx}{b} + c,$$
$$\int a\tan bx\,dx = \frac{-a\ln(\cos bx)}{b} + c, \qquad \int a\cos bx\sin^n bx\,dx = \frac{a\sin^{n+1} bx}{b(n+1)} + c,$$
$$\int \frac{a}{a^2+x^2}\,dx = \tan^{-1}\left(\frac{x}{a}\right) + c, \qquad \int a\sin bx\cos^n bx\,dx = \frac{-a\cos^{n+1} bx}{b(n+1)} + c,$$
$$\int \frac{-1}{\sqrt{a^2-x^2}}\,dx = \cos^{-1}\left(\frac{x}{a}\right) + c, \qquad \int \frac{1}{\sqrt{a^2-x^2}}\,dx = \sin^{-1}\left(\frac{x}{a}\right) + c,$$

   where the integrals that depend on n are valid for all $n \neq -1$ and where a and b are constants. In the two final results $|x| \leq a$.
   - Some particularly important cases for physical science:

$$\int \sin x\,dx = -\cos x + c, \qquad \int \cos x\,dx = \sin x + c,$$
$$\int \frac{1}{a^2+x^2}\,dx = \frac{1}{a}\tan^{-1}\left(\frac{x}{a}\right) + c,$$
$$\int_0^{n\pi/2} \cos^2 x\,dx = \frac{n\pi}{4} = \int_0^{n\pi/2} \sin^2 x\,dx,$$
$$\int_{x_0}^{x_0+(n\pi/\alpha)} \cos^2(\alpha x)\,dx = \frac{n\pi}{2\alpha} = \int_{x_0}^{x_0+(n\pi/\alpha)} \sin^2(\alpha x)\,dx.$$

3. *Common substitutions*
   With $t = \tan\theta/2$,

$$\sin\theta = \frac{2t}{1+t^2}, \qquad \cos\theta = \frac{1-t^2}{1+t^2}, \qquad d\theta = \frac{2}{1+t^2}\,dt.$$

Integrand contains	Substitution	Differential
$\sqrt{a^2-x^2}$	$x = a\sin u$	$dx = a\cos u\,du$
$\sqrt{a^2+x^2}$	$x = a\sinh u$	$dx = a\cosh u\,du,$
$\sqrt{x^2-a^2}$	$x = a\cosh u$	$dx = a\sinh u\,du$

4. *Integration by parts*

$$\int u \frac{dv}{dx}\, dx = uv - \int \frac{du}{dx} v\, dx$$

$$\text{or} \qquad \int uw\, dx = u \int^x w\, dx' - \int \frac{du}{dx} \left(\int^x w\, dx' \right) dx.$$

It is sometimes helpful to use the second form with w as (a hidden) unity.

5. *Infinite and improper integrals*
   - $\displaystyle\int_a^\infty f(x)\, dx = \lim_{b\to\infty} \int_a^b f(x)\, dx = \lim_{b\to\infty} F(b) - F(a).$
   - If $\lim_{x\to c} f(x) = \infty$ with $a \le c \le b$, then

$$\int_a^b f(x)\, dx = \lim_{\delta\to 0} \int_a^{c-\delta} f(x)\, dx + \lim_{\delta\to 0} \int_{c+\delta}^b f(x)\, dx,$$

   provided both limits exist.

6. *Curve lengths, and areas and volumes of revolution*
   - Curve length

$$s = \int_a^b \sqrt{1 + \left(\frac{dy}{dx}\right)^2}\, dx \qquad \text{or} \qquad s = \int_c^d \sqrt{1 + \left(\frac{dx}{dy}\right)^2}\, dy.$$

   - Area of solid of revolution

$$S = 2\pi \int_a^b y \sqrt{1 + \left(\frac{dy}{dx}\right)^2}\, dx \qquad \text{or} \qquad S = 2\pi \int_c^{dx} \sqrt{1 + \left(\frac{dx}{dy}\right)^2}\, dy.$$

   - Volume of solid of revolution

$$V = \pi \int_a^b y^2\, dx \qquad \text{or} \qquad V = \pi \int_c^d x^2\, dy.$$

A.5 Complex numbers and hyperbolic functions

1. *Real and imaginary parts*
   - $a + ib = c + id \quad \Rightarrow \quad a = c$ and $b = d$.
   - With $z = x + iy$, and $z^* = x - iy$,

$$x = \text{Re}\, z = \frac{(z + z^*)}{2}, \qquad y = \text{Im}\, z = \frac{(z - z^*)}{2i}.$$

   - In the Argand diagram, $z = re^{i\theta}$ and $z^* = re^{-i\theta}$ with
     (a) $|z| = r = \sqrt{x^2 + y^2} = \sqrt{zz^*}$,
     (b) $\arg z = \theta = \tan^{-1}\left(\frac{y}{x}\right)$, taking account of the signs of x and y,
     (c) $x = r\cos\theta$, $y = r\sin\theta$.

2. *Complex algebra*
With $z_k = x_k + iy_k = r_k e^{i\theta_k}$,

$$z_1 \pm z_2 = (x_1 \pm x_2) + i(y_1 \pm y_2),$$
$$z_1 z_2 = (x_1 x_2 - y_1 y_2) + i(x_1 y_2 + y_1 x_2) = r_1 r_2 e^{i(\theta_1+\theta_2)},$$
$$|z_1 z_2| = |z_1|\,|z_2|, \quad \arg z_1 z_2 = \arg z_1 + \arg z_2,$$
$$\frac{z_1}{z_2} = \frac{r_1}{r_2} e^{i(\theta_1-\theta_2)},$$
$$\left|\frac{z_1}{z_2}\right| = \frac{|z_1|}{|z_2|}, \quad \arg\frac{z_1}{z_2} = \arg z_1 - \arg z_2.$$

3. *The unit circle*
   - $e^{i\theta} = \cos\theta + i\sin\theta$ (Euler's equation).
   - $(\cos\theta + i\sin\theta)^n = \cos n\theta + i\sin n\theta$ (de Moivre's theorem).
   - The nth roots of unity are $e^{2\pi ik/n}$ for $k = 0, 1, \ldots, n-1$.
4. *Hyperbolic functions*
   - $\cosh x = \frac{1}{2}(e^x + e^{-x})$, $\sinh x = \frac{1}{2}(e^x - e^{-x})$.
   - $\cos ix = \cosh x$, $\cosh ix = \cos x$, $\sin ix = i\sinh x$, $\sinh ix = i\sin x$.
   - $\cosh^2 x - \sinh^2 x = 1$.
   - $\sinh 2x = 2\sinh x\cosh x$, $\cosh 2x = \cosh^2 x + \sinh^2 x$.
   - $\cosh^{-1} x = \ln(x \pm \sqrt{x^2-1})$, $\sinh^{-1} x = \ln(x + \sqrt{x^2+1})$.
   - $\frac{d}{dx}(\cosh x) = \sinh x$, $\frac{d}{dx}(\sinh x) = \cosh x$.
   - $\frac{d}{dx}(\cosh^{-1}\frac{x}{a}) = \frac{\pm 1}{\sqrt{x^2-a^2}}$, $\frac{d}{dx}(\sinh^{-1}\frac{x}{a}) = \frac{1}{\sqrt{x^2+a^2}}$.

A.6 Series and limits

1. *Finite and infinite series*

Definitions: $S_N = \sum_{n=0}^{N-1} u_n$ and $S_\infty = \sum_{n=0}^{\infty} u_n$ with $|r| < 1$ where relevant.

Type	u_n	S_N	S_∞
Arithmetic	$a + nd$	$\frac{1}{2}N(u_0 + u_{N-1})$	∞
Geometric	ar^n	$\frac{a(1-r^N)}{1-r}$	$\frac{a}{1-r}$
Arithmetico-geometric	$(a+nd)r^n$	see below *	$\frac{a}{1-r} + \frac{rd}{(1-r)^2}$

$$^*S_N = \frac{a - [a + (N-1)d]\,r^N}{1-r} + \frac{rd(1-r^{N-1})}{(1-r)^2}.$$

- Difference method: If a function $f(n)$ can be found such that $u_n = f(n) - f(n-1)$, then $\sum_{n=1}^{N} u_n = f(N) - f(0)$.
- Powers of the natural numbers

$$\sum_{n=1}^{N} n = \frac{1}{2}N(N+1), \qquad \sum_{n=1}^{N} n^2 = \frac{1}{6}N(N+1)(2N+1),$$

$$\sum_{n=1}^{N} n^3 = \frac{1}{4}N^2(N+1)^2.$$

2. *Tests for the convergence of infinite series* $\sum u_n$
In all tests only the *ultimate* behavior matters; any finite number of terms can be disregarded. More symbolically, the criteria need only be satisfied for all $n > N$ where N can be as large as necessary, but must be finite.

In all cases, a necessary (but not sufficient) requirement for convergence is that $\lim_{n\to\infty} u_n = 0$.

- Alternating sign test: If successive terms alternate in sign and $|u_n| \to 0$ as $n \to \infty$, then $\sum u_n$ converges.
- Integral test: If $f(n) = u_n$ when n is an integer and $\lim_{N\to\infty} \int^N f(x)\,dx$ exists, then $\sum u_n$ is convergent.
- Other tests, based on quantitative comparisons, are given in the following table.

Test	Test quantity	Conclusion
Comparison	$u_n \le v_n$	$\sum v_n$ conv. $\Rightarrow$ $\sum u_n$ conv.
Ratio	$\rho = \lim_{n\to\infty}\left(\frac{u_{n+1}}{u_n}\right)$	< 1 the series converges, > 1 the series diverges, $= 1$ the test is inconclusive.
Ratio comparison	$\frac{u_{n+1}}{u_n} \le \frac{v_{n+1}}{v_n}$	$\sum v_n$ conv. $\Rightarrow$ $\sum u_n$ conv.
Quotient	$\rho = \lim_{n\to\infty}\left(\frac{u_n}{v_n}\right)$	$\ne 0$ and $\ne \infty$, then $\sum u_n$ and $\sum v_n$ converge or diverge together, $= 0$ and $\sum v_n$ converges, then so does $\sum u_n$, $= \infty$ and $\sum v_n$ diverges, then so does $\sum u_n$.
Cauchy's root	$\rho = \lim_{n\to\infty}(u_n)^{1/n}$	< 1 the series converges, > 1 the series diverges, $= 1$ the test is inconclusive.

3. *Power series* $P(z) = \sum_{n=0}^{\infty} a_n (z^m)^n$, *with m usually equal to unity*
   - Radius of the circle of convergence: $R = \left(\lim_{n\to\infty} \left| \frac{a_{n+1}}{a_n} \right| \right)^{-1/m}$.
   - Within its circle of convergence, a power series can be integrated or differentiated to produce another power series convergent in the same region.
   - Taylor series for $f(x)$ about the point $x = a$;

$$f(x) = f(a) + (x-a)f'(a) + \frac{(x-a)^2}{2!} f''(a) + \cdots$$
$$\cdots + \frac{(x-a)^{n-1}}{(n-1)!} f^{(n-1)}(a) + \frac{(x-a)^n}{n!} f^{(n)}(\xi),$$

   where $a \le \xi \le x$.
   - Maclaurin series for common functions,

$$\sin x = x - \frac{x^3}{3!} + \frac{x^5}{5!} - \frac{x^7}{7!} + \cdots \quad \text{for } -\infty < x < \infty,$$
$$\cos x = 1 - \frac{x^2}{2!} + \frac{x^4}{4!} - \frac{x^6}{6!} + \cdots \quad \text{for } -\infty < x < \infty,$$
$$\tan x = x + \frac{x^3}{3} + \frac{2x^5}{15} + \frac{17x^7}{315} + \cdots \quad \text{for } -\pi/2 < x < \pi/2,$$
$$\tan^{-1} x = x - \frac{x^3}{3} + \frac{x^5}{5} - \frac{x^7}{7} + \cdots \quad \text{for } -1 < x < 1,$$
$$e^x = 1 + x + \frac{x^2}{2!} + \frac{x^3}{3!} + \frac{x^4}{4!} + \cdots \quad \text{for } -\infty < x < \infty,$$
$$\sinh x = x + \frac{x^3}{3!} + \frac{x^5}{5!} + \frac{x^7}{7!} + \cdots \quad \text{for } -\infty < x < \infty,$$
$$\cosh x = 1 + \frac{x^2}{2!} + \frac{x^4}{4!} + \frac{x^6}{6!} + \cdots \quad \text{for } -\infty < x < \infty,$$
$$\ln(1+x) = x - \frac{x^2}{2} + \frac{x^3}{3} - \frac{x^4}{4} + \cdots \quad \text{for } -1 < x \le 1,$$
$$(1+x)^n = 1 + nx + n(n-1)\frac{x^2}{2!} + n(n-1)(n-2)\frac{x^3}{3!} + \cdots \quad \text{for } -\infty < x < \infty.$$

4. *Evaluation of limits of* $f(x)$ *as* $x \to a$
   - The limits obtained when $x \to a^+$ and $x \to a^-$ are not necessarily equal.
   - The fractions $0/0$ and ∞/∞ are indeterminate.
   - L'Hôpital's rule for determining the limit as $x \to a$ of $f(x)/g(x)$ when an indeterminate form is encountered:

$$\lim_{x\to a} \frac{f(x)}{g(x)} = \lim_{x\to a} \frac{f^{(n)}(x)}{g^{(n)}(x)},$$

   where n is the lowest value of m for which $f^{(m)}(a)/g^{(m)}(a)$ is not an indeterminate form.

- If the indeterminate form $0 \times \infty$ is encountered, write it as $0/0$ or ∞/∞ by using the inverse of one of the factors involved.

A.7 Partial differentiation

1. *Definitions and notation based on $f = f(x, y)$*
   - Partial derivative definition:

$$f_x \equiv \left(\frac{\partial f}{\partial x}\right)_y = \lim_{\Delta x \to 0} \frac{f(x + \Delta x, y) - f(x, y)}{\Delta x},$$

   i.e. y is held fixed.
   - Second derivatives:

$$f_{xx} = \frac{\partial^2 f}{\partial x^2} = \frac{\partial}{\partial x}\left(\frac{\partial f}{\partial x}\right), \qquad f_{yy} = \frac{\partial^2 f}{\partial y^2} = \frac{\partial}{\partial y}\left(\frac{\partial f}{\partial y}\right),$$

$$f_{xy} = \frac{\partial}{\partial x}\left(\frac{\partial f}{\partial y}\right) = \frac{\partial^2 f}{\partial x \partial y} = \frac{\partial^2 f}{\partial y \partial x} = \frac{\partial}{\partial y}\left(\frac{\partial f}{\partial x}\right) = f_{yx}.$$

   - Total differential: $df = \left(\frac{\partial f}{\partial x}\right)_y dx + \left(\frac{\partial f}{\partial y}\right)_x dy.$
   - $\left(\frac{\partial x}{\partial y}\right)_f = \left(\frac{\partial y}{\partial x}\right)_f^{-1}$ and $\left(\frac{\partial y}{\partial f}\right)_x \left(\frac{\partial f}{\partial x}\right)_y \left(\frac{\partial x}{\partial y}\right)_f = -1.$

2. *Differentials*
   - If $df = A(x, y)\,dx + B(x, y)\,dy$, then df is exact $\Leftrightarrow \frac{\partial A}{\partial y} = \frac{\partial B}{\partial x}$.
   - Chain rule: If $x = x(u)$ and $y = y(u)$, then

$$\frac{df}{du} = \left(\frac{\partial f}{\partial x}\right)_y \frac{dx}{du} + \left(\frac{\partial f}{\partial y}\right)_x \frac{dy}{du}.$$

   - Taylor's theorem for $f(x, y)$:

$$f(x, y) = f(x_0, y_0) + \frac{\partial f}{\partial x}\Delta x + \frac{\partial f}{\partial y}\Delta y + \frac{1}{2!}\left[\frac{\partial^2 f}{\partial x^2}(\Delta x)^2 + 2\frac{\partial^2 f}{\partial x \partial y}\Delta x \Delta y + \frac{\partial^2 f}{\partial y^2}(\Delta y)^2\right] + \cdots,$$

   where $\Delta x = x - x_0$ and $\Delta y = y - y_0$ and all derivatives are evaluated at (x_0, y_0).

3. *Stationary values for $f(x, y)$*
   - A necessary condition is $f_x = f_y = 0$, and then
     (i) minimum if both f_{xx} and f_{yy} are positive *and* $f_{xy}^2 < f_{xx} f_{yy}$,
     (ii) maximum if both f_{xx} and f_{yy} are negative *and* $f_{xy}^2 < f_{xx} f_{yy}$,
     (iii) saddle point if f_{xx} and f_{yy} have opposite signs *or* if $f_{xy}^2 \geq f_{xx} f_{yy}$.
   - Under a single constraint $g(x, y) = 0$, consider $h(x, y) = f(x, y) + \lambda g(x, y)$ and apply $h_x = 0$, $h_y = 0$, together with $g(x, y) = 0$, to solve for x, y and λ.

- General procedure for $f(x_i)$ with $i = 1, 2, \ldots, N$ subject to constraints $g_j(x_i) = 0$ with $j = 1, 2, \ldots, M$ and $M < N$: form $h(x_i) = f(x_i) + \sum_j \lambda_j g_j(x_i)$ and then solve $\partial h/\partial x_i = 0$, together with $g_j(x_i) = 0$, for the $N + M$ quantities x_i and λ_j.

4. *Envelopes*
The family of curves in the xy-plane given by $f(x, y, \alpha) = 0$, where α is a parameter, has an envelope given by eliminating α between the two equations

$$f(x, y, \alpha) = 0 \quad \text{and} \quad \frac{\partial}{\partial \alpha} f(x, y, \alpha) = 0.$$

5. *Differentiating integrals* (Leibnitz' rule)
   - For fixed limits $\dfrac{d}{dx}\displaystyle\int_u^v f(x, t)\, dt = \int_u^v \frac{\partial f(x, t)}{\partial x}\, dt.$
   - For x-dependent limits $u = u(x)$, $v = v(x)$,

$$\frac{d}{dx}\int_u^v f(x, t)\, dt = f(x, v(x))\frac{dv}{dx} - f(x, u(x))\frac{du}{dx} + \int_{u(x)}^{v(x)} \frac{\partial f(x, t)}{\partial x} dt.$$

A.8 Multiple integrals

The value of a multiple integral is independent of the order in which the integrations are carried out, though the difficulty of finding it may not be.

1. *Areas, volumes and masses*

$$A = \iint dx\, dy, \qquad V = \iiint dx\, dy\, dz, \qquad M = \int dM.$$

2. *Average values*
   - Center of gravity: $\bar{x} = \dfrac{\int x\, dM}{\int dM}$, and similarly for $\bar{y}$ and $\bar{z}$.
   - Mean value of $f(x_i) = \dfrac{\iint\cdots\int f(x_i)\, dx_1\, dx_2 \ldots dx_n}{\iint\cdots\int dx_1\, dx_2 \ldots dx_n}$.

3. *Pappus' theorems*
For volumes or areas of revolution formed by rotating a plane area or plane line segment, respectively, about an axis that *does not* intersect it.
   (i) The volume of revolution = plane area × the distance moved by its centroid.
   (ii) The area of revolution = segment length × the distance moved by its centroid.

4. *Change of variables*
   - The integral $I = \displaystyle\iint\cdots\int_R f(x_1, x_2, \ldots, x_n)\, dx_1\, dx_2 \ldots dx_n$ can be written as $\displaystyle\iint\cdots\int_{R'} g(y_1, y_2, \ldots, y_n)|J_{xy}|\, dy_1\, dy_2 \ldots dy_n$, where the volume elements are related by $dx_1\, dx_2 \ldots dx_n = J_{xy}\, dy_1\, dy_2 \ldots dy_n$, and the Jacobian J_{xy} is given by

the determinant

$$J_{xy} \equiv \frac{\partial(x_1, x_2, \ldots, x_n)}{\partial(y_1, y_2, \ldots, y_n)} \equiv \begin{vmatrix} \frac{\partial x_1}{\partial y_1} & \frac{\partial x_2}{\partial y_1} & \cdots & \frac{\partial x_n}{\partial y_1} \\ \frac{\partial x_1}{\partial y_2} & \frac{\partial x_2}{\partial y_2} & \cdots & \frac{\partial x_n}{\partial y_2} \\ \vdots & & \ddots & \vdots \\ \frac{\partial x_1}{\partial y_n} & \frac{\partial x_2}{\partial y_n} & \cdots & \frac{\partial x_n}{\partial y_n} \end{vmatrix}.$$

- The rows and columns of J_{xy} can be interchanged without changing its value.
- $J_{xy}J_{yx} = 1$ and $J_{xz} = J_{xy}J_{yz}$.

A.9 Vector algebra

1. *Vector algebra*
   - Addition, subtraction and scalar multiplication

$$\mathbf{a}+\mathbf{b} = \mathbf{b}+\mathbf{a}, \qquad \mathbf{a}+(\mathbf{b}+\mathbf{c}) = (\mathbf{a}+\mathbf{b})+\mathbf{c},$$
$$\mathbf{a}+(-\mathbf{a}) = \mathbf{0}, \qquad \mathbf{a}-\mathbf{b} = \mathbf{a}+(-\mathbf{b}),$$
$$\lambda(\mu\mathbf{a}+\nu\mathbf{b}) = \lambda\mu\mathbf{a}+\lambda\nu\mathbf{b}.$$

   - A unit vector in the direction of $\mathbf{a}$ is $\hat{\mathbf{a}} = \mathbf{a}/|\mathbf{a}|$ where $|\mathbf{a}|$ is the magnitude of $\mathbf{a}$.
   - The set of vectors $\{\mathbf{e}_i\}$ are linearly independent only if $\sum_i c_i\mathbf{e}_i = \mathbf{0}$ implies that $c_i = 0$ for all i.
2. *Scalar product*
   - Definition: Scalar $s = \mathbf{a}\cdot\mathbf{b} = |\mathbf{a}||\mathbf{b}|\cos\theta = \mathbf{b}\cdot\mathbf{a}$ with $0 \le \theta \le \pi$.
   - $\mathbf{a}\cdot(\mathbf{b}+\mathbf{c}) = \mathbf{a}\cdot\mathbf{b}+\mathbf{a}\cdot\mathbf{c}$.
   - $\mathbf{a}\cdot\mathbf{a} = |\mathbf{a}|^2$.
   - *Warning*: If the vectors may have complex components, then $\mathbf{a}\cdot\mathbf{b} = (\mathbf{b}\cdot\mathbf{a})^*$ and $(\lambda\mathbf{a})\cdot\mathbf{b} = \lambda^*(\mathbf{a}\cdot\mathbf{b})$.
3. *Vector product*
   - Definition: Vector $\mathbf{v} = \mathbf{a}\times\mathbf{b}$ with $\mathbf{a}$, $\mathbf{b}$ and $\mathbf{v}$ (in that order) forming a right-handed set.

$$|\mathbf{v}| = |\mathbf{a}||\mathbf{b}|\sin\theta \text{ with } 0 \le \theta \le \pi.$$

   - Properties

$$\mathbf{a}\times\mathbf{a} = \mathbf{0}, \qquad \mathbf{b}\times\mathbf{a} = -(\mathbf{a}\times\mathbf{b}),$$
$$(\mathbf{a}+\mathbf{b})\times\mathbf{c} = (\mathbf{a}\times\mathbf{c})+(\mathbf{b}\times\mathbf{c}),$$
$$(\mathbf{a}\times\mathbf{b})\times\mathbf{c} \ne \mathbf{a}\times(\mathbf{b}\times\mathbf{c}) \quad \text{(see below)}.$$

   - In Cartesian components

$$\mathbf{a}\times\mathbf{b} = (a_yb_z - a_zb_y)\mathbf{i} + (a_zb_x - a_xb_z)\mathbf{j} + (a_xb_y - a_yb_x)\mathbf{k}.$$

4. *Scalar triple product*
   - Definition: Scalar $[\mathbf{a}, \mathbf{b}, \mathbf{c}] \equiv \mathbf{a} \cdot (\mathbf{b} \times \mathbf{c})$. The product $[\mathbf{a}, \mathbf{b}, \mathbf{c}]$ is equal to the volume of the parallelepiped with edges **a**, **b** and **c**. In Cartesian coordinates

$$\mathbf{a} \cdot (\mathbf{b} \times \mathbf{c}) = a_x(b_y c_z - b_z c_y) + a_y(b_z c_x - b_x c_z) + a_z(b_x c_y - b_y c_x).$$

   - Properties

$$[\mathbf{a}, \mathbf{b}, \mathbf{c}] = [\mathbf{b}, \mathbf{c}, \mathbf{a}] = [\mathbf{c}, \mathbf{a}, \mathbf{b}] = -[\mathbf{a}, \mathbf{c}, \mathbf{b}] = -[\mathbf{b}, \mathbf{a}, \mathbf{c}] = -[\mathbf{c}, \mathbf{b}, \mathbf{a}],$$
$$(\mathbf{a} \times \mathbf{b}) \cdot (\mathbf{c} \times \mathbf{d}) = (\mathbf{a} \cdot \mathbf{c})(\mathbf{b} \cdot \mathbf{d}) - (\mathbf{a} \cdot \mathbf{d})(\mathbf{b} \cdot \mathbf{c}).$$

5. *Vector triple product*
   - Vector $\mathbf{a} \times (\mathbf{b} \times \mathbf{c})$ is perpendicular to **a** and lies in the plane defined by vectors **b** and **c**.
   - Non-associativity

$$\mathbf{a} \times (\mathbf{b} \times \mathbf{c}) = (\mathbf{a} \cdot \mathbf{c})\mathbf{b} - (\mathbf{a} \cdot \mathbf{b})\mathbf{c},$$
$$(\mathbf{a} \times \mathbf{b}) \times \mathbf{c} = (\mathbf{a} \cdot \mathbf{c})\mathbf{b} - (\mathbf{b} \cdot \mathbf{c})\mathbf{a}.$$

6. *Lines, planes and spheres*
   - The point P that divides AB in the ratio $\lambda : \mu$ is given by

$$\mathbf{p} = \frac{\mu}{\lambda + \mu}\mathbf{a} + \frac{\lambda}{\lambda + \mu}\mathbf{b}.$$

   - The centroid of the triangle ABC is given by $\mathbf{g} = \frac{1}{3}(\mathbf{a} + \mathbf{b} + \mathbf{c})$.
   - The line in the direction of **f** passing through the point A is

$$\mathbf{r} = \mathbf{a} + \lambda\mathbf{f} \qquad \text{or} \qquad (\mathbf{r} - \mathbf{a}) \times \mathbf{f} = \mathbf{0}.$$

   - The line passing through A and C is $\mathbf{r} = \mathbf{a} + \lambda(\mathbf{c} - \mathbf{a})$.
   - The plane with a normal in the direction of unit vector $\hat{\mathbf{n}}$ and containing the point A is

$$(\mathbf{r} - \mathbf{a}) \cdot \hat{\mathbf{n}} = 0 \qquad \text{or} \qquad \hat{\mathbf{n}} \cdot \mathbf{r} = p,$$

     where p is the perpendicular distance from the origin to the plane.
   - The plane containing points A, B and C is $\mathbf{r} = \alpha\mathbf{a} + \beta\mathbf{b} + \gamma\mathbf{c}$ with $\alpha + \beta + \gamma = 1$.
   - The sphere with center C and radius R is $(\mathbf{r} - \mathbf{c}) \cdot (\mathbf{r} - \mathbf{c}) = R^2$.
7. *Distances using vectors*
   - The distance of a point P from the line with direction **f** that passes through A is $d = |(\mathbf{a} - \mathbf{p}) \times \hat{\mathbf{f}}|$.
   - The distance of a point P from the plane with unit normal $\hat{\mathbf{n}}$ that contains A is $d = (\mathbf{a} - \mathbf{p}) \cdot \hat{\mathbf{n}}$, with the sign of d indicating which side of the plane P lies on.
   - The distance between the lines with directions **f** and **g**, passing through the points A and B respectively, is

$$d = |(\mathbf{a} - \mathbf{b}) \cdot \hat{\mathbf{n}}|, \text{ where } \hat{\mathbf{n}} = \frac{\mathbf{f} \times \mathbf{g}}{|\mathbf{f} \times \mathbf{g}|}.$$

   - The distance between a line through A and a plane (to which it is parallel) with unit normal $\hat{\mathbf{n}}$ is $d = |(\mathbf{r} - \mathbf{a}) \cdot \hat{\mathbf{n}}|$, where **r** is any point on the plane.

8. *Reciprocal vectors to the non-coplanar set* $\{\mathbf{a}, \mathbf{b}, \mathbf{c}\}$

$$\mathbf{a}' = \frac{\mathbf{b} \times \mathbf{c}}{[\mathbf{a}, \mathbf{b}, \mathbf{c}]}, \qquad \mathbf{b}' = \frac{\mathbf{c} \times \mathbf{a}}{[\mathbf{a}, \mathbf{b}, \mathbf{c}]}, \qquad \mathbf{c}' = \frac{\mathbf{a} \times \mathbf{b}}{[\mathbf{a}, \mathbf{b}, \mathbf{c}]},$$

have the properties

- $\mathbf{a} \cdot \mathbf{a}' = \mathbf{b} \cdot \mathbf{b}' = \mathbf{c} \cdot \mathbf{c}' = 1.$
- $\mathbf{a}' \cdot \mathbf{b} = \mathbf{a}' \cdot \mathbf{c} = \mathbf{b}' \cdot \mathbf{a} = \mathbf{b}' \cdot \mathbf{c} = \mathbf{c}' \cdot \mathbf{a} = \mathbf{c}' \cdot \mathbf{b} = 0.$

A.10 First-order ordinary differential equations

Equation types and their solution methods

$$\text{General form} \quad p = \frac{dy}{dx} = F(x, y) \quad \text{or} \quad A(x, y)\,dx + B(x, y)\,dy = 0.$$

Name	Typical form	Solution method
Separable	$F = f(x)g(y)$	$\int \frac{dy}{g(y)} = \int f(x)\,dx.$
Exact	$h(x, y) = \frac{\partial A}{\partial y} - \frac{\partial B}{\partial x} = 0$	$U(x, y) = \int A(x, y)\,dx + V(y)$; $V(y)$ is such that $\frac{\partial U}{\partial y} = B(x, y).$
Inexact	$h(x, y) = \frac{\partial A}{\partial y} - \frac{\partial B}{\partial x} \neq 0$	If $f = \frac{h}{B} \neq f(y)$, then $\mu(x) = \int f(x)\,dx$ is an IF. If $g = \frac{h}{A} \neq g(x)$, then $\mu(y) = -\int g(y)\,dy$ is an IF.
Linear	$\frac{dy}{dx} + P(x)y = Q(x)$	$\mu(x) = \exp\{\int P(x)\,dx\}$ is an IF.
Homogeneous	$\frac{dy}{dx} = H\left(\frac{y}{x}\right)$	Put $y = vx$ to obtain separated $\int \frac{dv}{H(v) - v} = \ln x + c.$
Bernoulli	$\frac{dy}{dx} + P(x)y = Q(x)y^n$ $(n \neq 0,\ n \neq 1)$	Put $v = y^{1-n}$ and obtain linear $\frac{dv}{dx} + (1 - n)P(x)v = (1 - n)Q(x).$
Higher degree, soluble for p	$\prod_{i=1}^{n}(p - F_i) = 0$	Solve $p - F_i = 0$ for $G_i(x, y) = 0$. Then $\prod_{i=1}^{n} G_i(x, y) = 0.$
*Higher degree, soluble for x	$x = H(y, p)$	Solve $\frac{1}{p} = \frac{\partial H}{\partial y} + \frac{\partial H}{\partial p}\frac{dp}{dy}$ for $G(y, p) = 0$. Eliminate p between this and $x = H(y, p)$.
*Higher degree, soluble for y	$y = H(x, p)$	Solve $p = \frac{\partial H}{\partial x} + \frac{\partial H}{\partial p}\frac{dp}{dx}$ for $G(x, p) = 0$. Eliminate p between this and $y = H(x, p)$.

- The higher degree equations marked * may give rise to singular solutions.
- One boundary condition is needed for each equation.

B Inner products

In the main text the idea of an inner product is introduced and its main properties are summarized. That summary is repeated here, but we also consider the case in which the basis vectors are not orthonormal.

The *inner product* of two vectors, denoted in general by $\langle \mathbf{a}|\mathbf{b}\rangle$, is a scalar function of $\mathbf{a}$ and $\mathbf{b}$ and has the following properties:

(i) $\langle \mathbf{a}|\mathbf{b}\rangle = \langle \mathbf{b}|\mathbf{a}\rangle^*$,
(ii) $\langle \mathbf{a}|\lambda\mathbf{b} + \mu\mathbf{c}\rangle = \lambda\,\langle \mathbf{a}|\mathbf{b}\rangle + \mu\,\langle \mathbf{a}|\mathbf{c}\rangle$.

For vectors in a complex vector space, (i) and (ii) imply that

(iii) $\langle \lambda\mathbf{a} + \mu\mathbf{b}|\mathbf{c}\rangle = \lambda^*\,\langle \mathbf{a}|\mathbf{c}\rangle + \mu^*\,\langle \mathbf{b}|\mathbf{c}\rangle$,
(iv) $\langle \lambda\mathbf{a}|\mu\mathbf{b}\rangle = \lambda^*\mu\,\langle \mathbf{a}|\mathbf{b}\rangle$.

Two vectors in a general vector space are defined to be *orthogonal* if and only if $\langle \mathbf{a}|\mathbf{b}\rangle = 0$.

The *norm* of a vector $\mathbf{a}$ is defined by $||\mathbf{a}|| = \langle \mathbf{a}|\mathbf{a}\rangle^{1/2}$. In a general vector space $\langle \mathbf{a}|\mathbf{a}\rangle$ can be positive or negative, but in this book only spaces in which $\langle \mathbf{a}|\mathbf{a}\rangle \geq 0$ will be considered further. Such spaces are said to have a *positive semi-definite norm*, and in them $\langle \mathbf{a}|\mathbf{a}\rangle = 0$ implies $\mathbf{a} = \mathbf{0}$.

The simplest basis for the N-dimensional vector space is a set of vectors, $\hat{\mathbf{e}}_1, \hat{\mathbf{e}}_2, \ldots, \hat{\mathbf{e}}_N$, that are *orthonormal*, i.e. the basis vectors are mutually orthogonal and each has unit norm. Expressed mathematically, the basis has the property

$$\left\langle \hat{\mathbf{e}}_i \,|\, \hat{\mathbf{e}}_j \right\rangle = \delta_{ij}. \tag{B.1}$$

Here δ_{ij} is the *Kronecker delta* symbol which is defined by

$$\delta_{ij} = \begin{cases} 1 & \text{for } i = j, \\ 0 & \text{for } i \neq j. \end{cases}$$

In this basis any two vectors, $\mathbf{a}$ and $\mathbf{b}$, may be expressed as

$$\mathbf{a} = \sum_{i=1}^{N} a_i\,\hat{\mathbf{e}}_i \qquad \text{and} \qquad \mathbf{b} = \sum_{i=1}^{N} b_i\,\hat{\mathbf{e}}_i.$$

Furthermore, for any $\mathbf{a}$,

$$\left\langle \hat{\mathbf{e}}_j|\mathbf{a}\right\rangle = \sum_{i=1}^{N} \left\langle \hat{\mathbf{e}}_j|a_i\,\hat{\mathbf{e}}_i\right\rangle = \sum_{i=1}^{N} a_i\left\langle \hat{\mathbf{e}}_j|\,\hat{\mathbf{e}}_i\right\rangle = a_j. \tag{B.2}$$

Thus the components of **a** can be calculated as $a_i = \langle \hat{\mathbf{e}}_i | \mathbf{a} \rangle$. It should be noted that this is *not* true unless the basis is orthonormal.

The inner product of **a** and **b** can be written, in terms of their components in an orthonormal basis, as

$$\begin{aligned}
\langle \mathbf{a}|\mathbf{b} \rangle &= \langle a_1\, \hat{\mathbf{e}}_1 + a_2\, \hat{\mathbf{e}}_2 + \cdots + a_N\, \hat{\mathbf{e}}_N | b_1\, \hat{\mathbf{e}}_1 + b_2\, \hat{\mathbf{e}}_2 + \cdots + b_N\, \hat{\mathbf{e}}_N \rangle \\
&= \sum_{i=1}^{N} a_i^* b_i \, \langle \hat{\mathbf{e}}_i | \hat{\mathbf{e}}_i \rangle + \sum_{i=1}^{N} \sum_{j \neq i}^{N} a_i^* b_j \left\langle \hat{\mathbf{e}}_i | \hat{\mathbf{e}}_j \right\rangle \\
&= \sum_{i=1}^{N} a_i^* b_i,
\end{aligned}$$

where the second equality follows from inner-product property (iv) and the third from (B.1).

The above may be generalized to the case where the base vectors $\mathbf{e}_1, \mathbf{e}_2, \ldots, \mathbf{e}_N$ are *not* orthonormal, as follows. Define N^2 numbers G_{ij} by

$$G_{ij} = \left\langle \mathbf{e}_i | \mathbf{e}_j \right\rangle . \tag{B.3}$$

Then, if $\mathbf{a} = \sum_{i=1}^{N} a_i \mathbf{e}_i$ and $\mathbf{b} = \sum_{i=1}^{N} b_i \mathbf{e}_i$, the inner product of **a** and **b** is given by

$$\begin{aligned}
\langle \mathbf{a}|\mathbf{b} \rangle &= \left\langle \sum_{i=1}^{N} a_i \mathbf{e}_i \middle| \sum_{j=1}^{N} b_j \mathbf{e}_j \right\rangle = \sum_{i=1}^{N} \sum_{j=1}^{N} a_i^* b_j \left\langle \mathbf{e}_i | \mathbf{e}_j \right\rangle \\
&= \sum_{i=1}^{N} \sum_{j=1}^{N} a_i^* G_{ij} b_j .
\end{aligned} \tag{B.4}$$

In matrix notation this result is

$$\langle \mathbf{a}|\mathbf{b} \rangle = \sum_{i=1}^{N} \sum_{j=1}^{N} a_i^* G_{ij} b_j = \mathsf{a}^\dagger \mathsf{G} \mathsf{b},$$

where $\mathsf{a}^\dagger$ is the Hermitian conjugate of the column matrix a.

From (B.3) and the properties of the inner product it follows that $G_{ij} = G_{ji}^*$. This ensures that $||\mathbf{a}|| = \langle \mathbf{a}|\mathbf{a} \rangle$ is real, since then

$$\langle \mathbf{a}|\mathbf{a} \rangle^* = \sum_{i=1}^{N} \sum_{j=1}^{N} a_i G_{ij}^* a_j^* = \sum_{j=1}^{N} \sum_{i=1}^{N} a_j^* G_{ji} a_i = \langle \mathbf{a}|\mathbf{a} \rangle .$$

C Inequalities in linear vector spaces

For a set of objects (vectors) forming a linear vector space in which $\langle \mathbf{a}|\mathbf{a}\rangle \geq 0$ for all $\mathbf{a}$, the following inequalities are often useful.

(i) *Schwarz's inequality* is the most basic result and states that

$$|\langle \mathbf{a}|\mathbf{b}\rangle| \leq ||\mathbf{a}||\,||\mathbf{b}||, \tag{C.1}$$

where the equality holds when $\mathbf{a}$ is a scalar multiple of $\mathbf{b}$, i.e. when $\mathbf{a} = \lambda\mathbf{b}$.
Schwarz's inequality may be proved by considering

$$\begin{aligned} ||\mathbf{a}+\lambda\mathbf{b}||^2 &= \langle \mathbf{a}+\lambda\mathbf{b}|\mathbf{a}+\lambda\mathbf{b}\rangle \\ &= \langle \mathbf{a}|\mathbf{a}\rangle + \lambda\langle \mathbf{a}|\mathbf{b}\rangle + \lambda^*\langle \mathbf{b}|\mathbf{a}\rangle + \lambda\lambda^*\langle \mathbf{b}|\mathbf{b}\rangle. \end{aligned}$$

If we write $\langle \mathbf{a}|\mathbf{b}\rangle$ as $|\langle \mathbf{a}|\mathbf{b}\rangle|e^{i\alpha}$ then

$$||\mathbf{a}+\lambda\mathbf{b}||^2 = ||\mathbf{a}||^2 + |\lambda|^2\,||\mathbf{b}||^2 + \lambda|\langle \mathbf{a}|\mathbf{b}\rangle|e^{i\alpha} + \lambda^*|\langle \mathbf{a}|\mathbf{b}\rangle|e^{-i\alpha}.$$

However, $||\mathbf{a}+\lambda\mathbf{b}||^2 \geq 0$ for all λ, so we may choose $\lambda = re^{-i\alpha}$ and require that, for all r,

$$0 \leq ||\mathbf{a}+\lambda\mathbf{b}||^2 = ||\mathbf{a}||^2 + r^2\,||\mathbf{b}||^2 + 2r|\langle \mathbf{a}|\mathbf{b}\rangle|.$$

This means that the quadratic equation in r formed by setting the RHS equal to zero must have no real roots. This, in turn, implies that

$$4|\langle \mathbf{a}|\mathbf{b}\rangle|^2 \leq 4\,||\mathbf{a}||^2\,||\mathbf{b}||^2,$$

which, on taking the square root (all factors are necessarily positive) of both sides, gives Schwarz's inequality.

(ii) The *triangle inequality* states that

$$||\mathbf{a}+\mathbf{b}|| \leq ||\mathbf{a}|| + ||\mathbf{b}|| \tag{C.2}$$

and may be derived from the properties of the inner product and Schwarz's inequality as follows. Let us first consider

$$||\mathbf{a}+\mathbf{b}||^2 = ||\mathbf{a}||^2 + ||\mathbf{b}||^2 + 2\,\mathrm{Re}\,\langle \mathbf{a}|\mathbf{b}\rangle \leq ||\mathbf{a}||^2 + ||\mathbf{b}||^2 + 2|\langle \mathbf{a}|\mathbf{b}\rangle|.$$

Using Schwarz's inequality we then have

$$||\mathbf{a}+\mathbf{b}||^2 \leq ||\mathbf{a}||^2 + ||\mathbf{b}||^2 + 2\,||\mathbf{a}||\,||\mathbf{b}|| = (||\mathbf{a}|| + ||\mathbf{b}||)^2,$$

which, on taking the square root, gives the triangle inequality (C.2).

(iii) *Bessel's inequality* requires the introduction of an orthonormal basis $\hat{\mathbf{e}}_i$, $i = 1, 2, \ldots, N$ into the N-dimensional vector space; it states that

$$||\mathbf{a}||^2 \geq \sum_i |\langle \hat{\mathbf{e}}_i|\mathbf{a}\rangle|^2, \tag{C.3}$$

where the equality holds if the sum includes all N basis vectors. If not all the basis vectors are included in the sum then the inequality results (though of course the equality remains if those basis vectors omitted all have $a_i = 0$). Bessel's inequality can also be written

$$\langle \mathbf{a}|\mathbf{a}\rangle \geq \sum_i |a_i|^2,$$

where the a_i are the components of $\mathbf{a}$ in the orthonormal basis. From (1.17) these are given by $a_i = \langle \hat{\mathbf{e}}_i|\mathbf{a}\rangle$. The above may be proved by considering

$$\left\|\mathbf{a} - \sum_i \langle \hat{\mathbf{e}}_i|\mathbf{a}\rangle\, \hat{\mathbf{e}}_i\right\|^2 = \left\langle \mathbf{a} - \sum_i \langle \hat{\mathbf{e}}_i|\mathbf{a}\rangle\, \hat{\mathbf{e}}_i \,\middle|\, \mathbf{a} - \sum_j \langle \hat{\mathbf{e}}_j|\mathbf{a}\rangle\, \hat{\mathbf{e}}_j \right\rangle.$$

Expanding out the inner product and using $\langle \hat{\mathbf{e}}_i|\mathbf{a}\rangle^* = \langle \mathbf{a}|\hat{\mathbf{e}}_i\rangle$, we obtain

$$\left\|\mathbf{a} - \sum_i \langle \hat{\mathbf{e}}_i|\mathbf{a}\rangle\, \hat{\mathbf{e}}_i\right\|^2 = \langle \mathbf{a}|\mathbf{a}\rangle - 2\sum_i \langle \mathbf{a}|\hat{\mathbf{e}}_i\rangle \langle \hat{\mathbf{e}}_i|\mathbf{a}\rangle + \sum_i \sum_j \langle \mathbf{a}|\hat{\mathbf{e}}_i\rangle \langle \hat{\mathbf{e}}_j|\mathbf{a}\rangle \langle \hat{\mathbf{e}}_i|\hat{\mathbf{e}}_j\rangle.$$

Now $\langle \hat{\mathbf{e}}_i|\hat{\mathbf{e}}_j\rangle = \delta_{ij}$, since the basis is orthonormal, and so we find

$$0 \leq \left\|\mathbf{a} - \sum_i \langle \hat{\mathbf{e}}_i|\mathbf{a}\rangle\, \hat{\mathbf{e}}_i\right\|^2 = ||\mathbf{a}||^2 - \sum_i |\langle \hat{\mathbf{e}}_i|\mathbf{a}\rangle|^2,$$

which is Bessel's inequality.

D Summation convention

In the main text, particularly when dealing with matrices and vector calculus in three dimensions, we often need to take a sum over a number of terms which are all of the same general form, and differ only in the value of an indexing subscript or the symbol associated with each of the three different Cartesian coordinates. Thus the prescription for the elements of the product P of two matrices, A and B, might be written as

$$P_{ij} = \sum_{k=1}^{N} A_{ik} B_{kj} \tag{D.1}$$

or the expression for the divergence of a vector **a** (see Chapter 2) given in the form

$$\nabla \cdot \mathbf{a} = \frac{\partial a_x}{\partial x} + \frac{\partial a_y}{\partial y} + \frac{\partial a_z}{\partial z}. \tag{D.2}$$

Sometimes, for example in a multiple matrix product, several sums appear in an expression, each with its own explicit summation sign.

Such calculations can be significantly compacted, and in some cases simplified, if the Cartesian coordinates x, y and z are replaced symbolically by the indexed coordinates x_i, where i takes the values 1, 2 and 3, and the so-called *summation convention* is adopted. We will also find that when the convention is used, together with two particular indexed quantities, the Kronecker delta δ_{ij} and the Levi–Civita symbol ϵ_{ijk},[1] many of the equalities appearing in vector algebra and calculus can be established very straightforwardly. Further discussion of δ_{ij} and ϵ_{ijk} is given in Appendix E.

The convention is that any *lower-case* alphabetic subscript that appears *exactly* twice in any term of an expression is understood to be summed over all the values that a subscript in that position can take (unless the contrary is specifically stated). The subscripted quantities may appear in the numerator and/or the denominator of a term in an expression. This naturally implies that any such pair of repeated subscripts must occur only in subscript positions that have the same range of values. Sometimes the ranges of values have to be specified but usually they are apparent from the context. As specific examples, for vector calculus in ordinary three-dimensional space the range is 1 to 3, and for the multiplication of $n \times n$ matrices it is 1 to n.

In this notation (D.1) becomes

$$P_{ij} = A_{ik} B_{kj} \qquad \text{i.e. without the explicit summation sign,}$$

1 These two quantities are technically isotropic Cartesian tensors, though this aspect will not be part of our considerations here.

and (D.2) is even more compacted to

$$\nabla \cdot \mathbf{a} = \frac{\partial a_i}{\partial x_i}.$$

The following simple examples illustrate further what is meant (in the three-dimensional case):

(i) $a_i x_i$ stands for $a_1 x_1 + a_2 x_2 + a_3 x_3$, the scalar product $\mathbf{a} \cdot \mathbf{x}$;
(ii) a_{ii} stands for $a_{11} + a_{22} + a_{33}$, the trace of A;
(iii) $a_{ij} b_{jk} c_k$ stands for $\sum_{j=1}^{3} \sum_{k=1}^{3} a_{ij} b_{jk} c_k$, the ith component of the vector $\mathsf{A}\mathsf{B}\mathbf{c}$;
(iv) $\dfrac{\partial^2 \phi}{\partial x_i \partial x_i}$ stands for $\dfrac{\partial^2 \phi}{\partial x_1^2} + \dfrac{\partial^2 \phi}{\partial x_2^2} + \dfrac{\partial^2 \phi}{\partial x_3^2}$, ∇^2 of scalar ϕ;
(v) $\dfrac{\partial^2 v_j}{\partial x_i \partial x_i}$ stands for $\dfrac{\partial^2 v_j}{\partial x_1^2} + \dfrac{\partial^2 v_j}{\partial x_2^2} + \dfrac{\partial^2 v_j}{\partial x_3^2}$, ∇^2 of the jth component of vector $\mathbf{v}$.

Subscripts that are summed over are called *dummy subscripts* and the others *free subscripts*. For example, in (v) above, i is a dummy subscript, but j is a free subscript. It is worth remarking that when introducing a dummy subscript into an expression, care should be taken not to use one that is already present, either as a free or as a dummy subscript. For example, $a_{ij} b_{jk} c_{kl}$ cannot, and must not, be replaced by $a_{ij} b_{jj} c_{jl}$ or by $a_{il} b_{lk} c_{kl}$, but could be replaced by $a_{im} b_{mk} c_{kl}$ or by $a_{im} b_{mn} c_{nl}$. Naturally, free subscripts must not be changed at all unless the working calls for it.

In Appendix E the Kronecker delta δ_{ij} is discussed in some detail. For present purposes we only need its definition, which is

$$\delta_{ij} = \begin{cases} 1 & \text{if } i = j, \\ 0 & \text{otherwise.} \end{cases}$$

When the summation convention has been adopted, the main effect of δ_{ij} is to replace one subscript by another in certain expressions. Examples might include

$$b_j \delta_{ij} = b_i,$$

and

$$a_{ij} \delta_{jk} = a_{ij} \delta_{kj} = a_{ik}. \tag{D.3}$$

In the second of these the dummy index shared by both terms on the left-hand side (namely j) has been replaced by the free index carried by the Kronecker delta (namely k), and the delta symbol has disappeared. In matrix language, (D.3) can be written as $\mathsf{A}\mathsf{I} = \mathsf{A}$, where A is the matrix with elements a_{ij} and I is the unit matrix having the same dimensions as A – its elements are simply δ_{ij}, i.e. ones on the leading diagonal where $i = j$, and zero elsewhere where $i \neq j$.

In some expressions we may use the Kronecker delta to replace indices in a number of different ways, e.g.

$$a_{ij} b_{jk} \delta_{ki} = a_{ij} b_{ji} \quad \text{or} \quad a_{kj} b_{jk},$$

where the two expressions on the RHS are totally equivalent to one another, both being equal to the trace of AB.

Use of the summation convention and the Levi–Civita symbol, defined by

$$\epsilon_{ijk} = \begin{cases} +1 & \text{if } i, j, k \text{ is an even permutation of } 1, 2, 3, \\ -1 & \text{if } i, j, k \text{ is an odd permutation of } 1, 2, 3, \\ 0 & \text{otherwise,} \end{cases}$$

allow very compact representations of some expressions that arise frequently in matrix and vector algebras. For example, the ith component of a vector product given in standard Cartesian coordinates as

$$\mathbf{a} \times \mathbf{b} = (a_y b_z - a_z b_y)\mathbf{i} + (a_z b_x - a_x b_z)\mathbf{j} + (a_x b_y - a_y b_x)\mathbf{k},$$

can be written using the summation convention as $\epsilon_{ijk} a_j b_k$, whilst the whole of the RHS can be written as $\epsilon_{ijk} a_j b_k \hat{\mathbf{e}}_i$ in a notation in which $\hat{\mathbf{e}}_i$ represents $\mathbf{i}$, $\mathbf{j}$ and $\mathbf{k}$ for i equal to 1, 2 and 3, respectively.

In a similar way, the scalar triple product $[\,\mathbf{a} \cdot (\mathbf{b} \times \mathbf{c})]$ can be expressed as $\epsilon_{ijk} a_i b_j c_k$ and the determinant given in (1.46) as $\epsilon_{ijk} a_{1i} a_{2j} a_{3k}$. It should be noted that whilst the expression for a vector component contains one free subscript, namely i, the other three examples given here contain only dummy subscripts; the 1, 2 and 3 appearing in the expression for a determinant are not lower-case alphabetic subscripts and are not classed as either.

E The Kronecker delta and Levi–Civita symbols

In a number of places in the main text and in Appendix D we have encountered and used the *Kronecker delta*, a two-subscript quantity δ_{ij} defined by

$$\delta_{ij} = \begin{cases} 1 & \text{if } i = j, \\ 0 & \text{otherwise.} \end{cases}$$

As indicated in Appendix D, when used with the summation convention δ_{ij} has the property $a_{ij}\delta_{jk} = a_{ik}$ for a general matrix A with elements a_{ij}. A particular case of this occurs when A is the unit matrix; then $a_{ij} = \delta_{ij}$ and the result

$$\delta_{ij}\delta_{jk} = \delta_{ik} \tag{E.1}$$

follows. One further important property is that $\delta_{ii} = 3$, the trace of the unit matrix in three dimensions.[1]

For our purposes there is not much to add to these basic properties, but it is important to note that the derivatives of the (three) Cartesian components with respect to each other can be written in terms of the Kronecker delta. Two obvious examples are $\partial x/\partial x = 1$ and $\partial x/\partial y = 0$, but all nine relationships can be summarized, ready for automatic implementation in summation convention formulae, by

$$\frac{\partial x_i}{\partial x_j} = \delta_{ij}. \tag{E.2}$$

At this point we introduce the three-subscript *Levi–Civita* symbol, ϵ_{ijk}, the value of which is given by

$$\epsilon_{ijk} = \begin{cases} +1 & \text{if } i, j, k \text{ is an even permutation of } 1, 2, 3, \\ -1 & \text{if } i, j, k \text{ is an odd permutation of } 1, 2, 3, \\ 0 & \text{otherwise.} \end{cases}$$

We see that, whilst δ_{ij} is symmetric, in that $\delta_{ij} = \delta_{ji}$, ϵ_{ijk} is totally antisymmetric, i.e. it changes sign under the interchange of any pair of subscripts. In fact ϵ_{ijk}, or any scalar multiple of it, is the *only* three-subscript quantity with this property.

It should be noted that both δ_{ij} and ϵ_{ijk} are simply numbers, and that, in fact, there are only three possible numbers, 0 and ± 1. These are such simple numbers that it may seem hardly worthwhile introducing a new notation to represent them, especially as including them greatly increases the formal number of terms in a multiple summation (even though the added terms all contain a factor zero and contribute nothing). However, as we will see, when using the summation convention we do not usually need to write out the terms

1 $\delta_{ii} = N$ for δ_{ij} defined in an N-dimensional space.

explicitly, and the form of the summation used for calculation is actually much more compact and workable than an explicit sum of the non-zero terms.

We begin our study of the Levi–Civita symbol ϵ_{ijk} by noting that we may use it to formulate an alternative prescription for the determinant of a 3×3 matrix A by means of the identity

$$|\mathsf{A}|\epsilon_{lmn} = A_{li}A_{mj}A_{nk}\epsilon_{ijk}. \tag{E.3}$$

By considering all possible cases for the values (1, 2 or 3) for each of l, m, n and i, j, k, (E.3) may be shown to be equivalent to the Laplace expansion (see Chapter 1) of the determinant of the matrix A.[2] The following is a simple example.

Example Evaluate the determinant of the matrix

$$\mathsf{A} = \begin{pmatrix} 2 & 1 & -3 \\ 3 & 4 & 0 \\ 1 & -2 & 1 \end{pmatrix}.$$

Setting $l = 1$, $m = 2$ and $n = 3$ in (E.3) we find

$$\begin{aligned} |\mathsf{A}| &= \epsilon_{ijk}A_{1i}A_{2j}A_{3k} \\ &= (2)(4)(1) - (2)(0)(-2) - (1)(3)(1) + (-3)(3)(-2) \\ &\quad + (1)(0)(1) - (-3)(4)(1) = 35, \end{aligned}$$

a result that may be verified using the Laplace expansion method. ◀

Many of the properties of determinants discussed in Chapter 1 can be proved very efficiently using (E.3) (see Problem 1.37).

In addition to providing a convenient notation for the determinant of a matrix, δ_{ij} and ϵ_{ijk} can be used to rewrite many of the familiar expressions of vector algebra and calculus. For example, provided we are using right-handed Cartesian coordinates, the vector product $\mathbf{a} = \mathbf{b} \times \mathbf{c}$ has as its ith component $a_i = \epsilon_{ijk}b_jc_k$. Below we give some examples of how expressions occurring in the algebra and calculus of vectors can be written compactly in terms of δ_{ij} and ϵ_{ijk}. Those involving the differential vector operator ∇ should be returned

2 This may be readily extended to an $N \times N$ matrix A, i.e.

$$|\mathsf{A}|\epsilon_{i_1i_2\cdots i_N} = A_{i_1j_1}A_{i_2j_2}\cdots A_{i_Nj_N}\epsilon_{j_1j_2\cdots j_N},$$

where $\epsilon_{i_1i_2\cdots i_N}$ equals 1 if $i_1i_2\cdots i_N$ is an even permutation of $1, 2, \ldots, N$ and equals -1 if it is an odd permutation; otherwise it equals zero.

to after Chapter 2 has been studied.

$$\mathbf{a}\cdot\mathbf{b} = a_i b_i = \delta_{ij} a_i b_j,$$

$$\nabla^2\phi = \frac{\partial^2\phi}{\partial x_i \partial x_i} = \delta_{ij}\frac{\partial^2\phi}{\partial x_i \partial x_j},$$

$$(\nabla\times\mathbf{v})_i = \epsilon_{ijk}\frac{\partial v_k}{\partial x_j},$$

$$[\nabla(\nabla\cdot\mathbf{v})]_i = \frac{\partial}{\partial x_i}\left(\frac{\partial v_j}{\partial x_j}\right) = \delta_{jk}\frac{\partial^2 v_j}{\partial x_i \partial x_k},$$

$$[\nabla\times(\nabla\times\mathbf{v})]_i = \epsilon_{ijk}\frac{\partial}{\partial x_j}\left(\epsilon_{klm}\frac{\partial v_m}{\partial x_l}\right) = \epsilon_{ijk}\epsilon_{klm}\frac{\partial^2 v_m}{\partial x_j \partial x_l},$$

$$(\mathbf{a}\times\mathbf{b})\cdot\mathbf{c} = \delta_{ij} c_i \epsilon_{jkl} a_k b_l = \epsilon_{ikl} c_i a_k b_l.$$

An important relationship between the Kronecker delta and Levi–Civita symbol is expressed by the identity

$$\epsilon_{ijk}\epsilon_{klm} = \delta_{il}\delta_{jm} - \delta_{im}\delta_{jl}. \qquad \text{(E.4)}$$

This result can be established as a special case of a more general one, that is, however, more straightforward to prove. Consider the determinant on the RHS of the following equation and recall that any determinant is equal to zero if any two of its columns or any two of its rows are equal:

$$\epsilon_{ijk}\epsilon_{pqr} = \begin{vmatrix} \delta_{ip} & \delta_{iq} & \delta_{ir} \\ \delta_{jp} & \delta_{jq} & \delta_{jr} \\ \delta_{kp} & \delta_{kq} & \delta_{kr} \end{vmatrix}. \qquad \text{(E.5)}$$

For a non-zero RHS, i, j and k must all take different values selected from 1, 2 and 3, and the same is true of p, q and r. Exactly the same conditions apply for a non-zero value of the LHS. Now, recalling that the value of a determinant changes sign if any two of its rows or any two of its columns are interchanged, and that the same is true of the Levi–Civita symbol if any two of its subscripts are interchanged, we see that both sides of (E.5) have the same subscript-interchange properties. Finally, if $i = p = 1$, $j = q = 2$ and $k = r = 3$ then both sides have unit value. Taken together these observations establish the validity of (E.5) for all combinations of all six subscripts.

To obtain relationship (E.4), we use (E.5) in the particular case of $p = k$. The LHS becomes the LHS of (E.4), apart from an irrelevant change in the naming of the free subscripts. The determinant on the RHS can be written as a Laplace expansion based on the first row, as in equation (1.45):

$$\begin{vmatrix} \delta_{ip} & \delta_{iq} & \delta_{ir} \\ \delta_{jp} & \delta_{jq} & \delta_{jr} \\ \delta_{kp} & \delta_{kq} & \delta_{kr} \end{vmatrix} = \delta_{ik}(\delta_{jq}\delta_{kr} - \delta_{kq}\delta_{jr}) - \delta_{iq}(\delta_{jk}\delta_{kr} - \delta_{kk}\delta_{jr}) + \delta_{ir}(\delta_{jk}\delta_{kq} - \delta_{kk}\delta_{jq}).$$

Now, we recall that, for any dummy subscript m, $\delta_{lm}\delta_{mn} = \delta_{ln}$ and $\delta_{mm} = 3$. These properties enable the RHS to be simplified to

$$(\delta_{jq}\delta_{ir} - \delta_{iq}\delta_{jr}) - (\delta_{iq}\delta_{jr} - 3\delta_{iq}\delta_{jr}) + (\delta_{ir}\delta_{jq} - 3\delta_{ir}\delta_{jq}) = \delta_{iq}\delta_{jr} - \delta_{ir}\delta_{jq},$$

thus establishing identity (E.4).

Clearly there are many ways to expand the determinant in (E.5) and several choices of pairs of subscripts to set equal. For example, one alternative form of (E.4) is

$$\epsilon_{ijk}\epsilon_{ilm} = \delta_{jl}\delta_{km} - \delta_{jm}\delta_{kl}. \qquad \text{(E.6)}$$

The pattern of subscripts in these identities is most easily remembered by noting that the subscripts on the first δ on the RHS are those that immediately follow (cyclically, if necessary) the common subscript, here i, in each ϵ-term on the LHS; the remaining combinations of j, k, l, m as subscripts in the other δ-terms on the RHS can then be filled in automatically.

One useful application of (E.4) is in obtaining alternative expressions for vector quantities that arise from the vector product of a vector product. As a particular example involving the differential operator ∇, we now obtain an alternative expression for the curl of the curl of a vector – in operator notation, $\nabla \times (\nabla \times \mathbf{v})$.

Amongst the examples given earlier was the summation convention form for this quantity:

$$[\nabla \times (\nabla \times \mathbf{v})]_i = \epsilon_{ijk}\epsilon_{klm}\frac{\partial^2 v_m}{\partial x_j \partial x_l}.$$

Using (E.4) this can be rewritten as

$$\begin{aligned}
[\nabla \times (\nabla \times \mathbf{v})]_i &= \epsilon_{ijk}\epsilon_{klm}\frac{\partial^2 v_m}{\partial x_j \partial x_l} \\
&= (\delta_{il}\delta_{jm} - \delta_{im}\delta_{jl})\frac{\partial^2 v_m}{\partial x_j \partial x_l} \\
&= \frac{\partial}{\partial x_i}\left(\frac{\partial v_j}{\partial x_j}\right) - \frac{\partial^2 v_i}{\partial x_j \partial x_j} \\
&= [\nabla(\nabla \cdot \mathbf{v})]_i - \nabla^2 v_i.
\end{aligned}$$

This result is used in Chapter 2 and the reader is referred there for a discussion of its applicability.

F Gram-Schmidt orthogonalization

An eigenvalue corresponding to two or more different eigenvectors (i.e. no two are simply multiples of one another) of a normal matrix A is said to be *degenerate*. Suppose that λ_1 is k-fold degenerate, i.e.

$$\mathsf{A}\mathsf{x}^i = \lambda_1 \mathsf{x}^i \quad \text{for } i = 1, 2, \ldots, k, \tag{F.1}$$

but that it is different from any of λ_{k+1}, λ_{k+2}, etc. Then any linear combination of these x^i is also an eigenvector with eigenvalue λ_1, since, for $\mathsf{z} = \sum_{i=1}^{k} c_i \mathsf{x}^i$,

$$\mathsf{A}\mathsf{z} \equiv \mathsf{A} \sum_{i=1}^{k} c_i \mathsf{x}^i = \sum_{i=1}^{k} c_i \mathsf{A}\mathsf{x}^i = \sum_{i=1}^{k} c_i \lambda_1 \mathsf{x}^i = \lambda_1 \mathsf{z}. \tag{F.2}$$

If the x^i defined in (F.1) are not already mutually orthogonal then we can construct new eigenvectors z^i that are orthogonal by the following procedure:

$$\begin{aligned}
\mathsf{z}^1 &= \mathsf{x}^1, \\
\mathsf{z}^2 &= \mathsf{x}^2 - \left[(\hat{\mathsf{z}}^1)^\dagger \mathsf{x}^2\right] \hat{\mathsf{z}}^1, \\
\mathsf{z}^3 &= \mathsf{x}^3 - \left[(\hat{\mathsf{z}}^2)^\dagger \mathsf{x}^3\right] \hat{\mathsf{z}}^2 - \left[(\hat{\mathsf{z}}^1)^\dagger \mathsf{x}^3\right] \hat{\mathsf{z}}^1, \\
&\vdots \\
\mathsf{z}^k &= \mathsf{x}^k - \left[(\hat{\mathsf{z}}^{k-1})^\dagger \mathsf{x}^k\right] \hat{\mathsf{z}}^{k-1} - \cdots - \left[(\hat{\mathsf{z}}^1)^\dagger \mathsf{x}^k\right] \hat{\mathsf{z}}^1.
\end{aligned}$$

In this procedure, known as *Gram–Schmidt orthogonalization*, each new eigenvector z^i is normalized to give the unit vector $\hat{\mathsf{z}}^i$ before proceeding to the construction of the next one; the normalization is carried out by dividing each element of the vector z^i by $[(\mathsf{z}^i)^\dagger \mathsf{z}^i]^{1/2}$.

It should be noted that each factor $(\hat{\mathsf{z}}^m)^\dagger \mathsf{x}^n$ in square brackets is a scalar product and thus only a number. It follows that, as shown in (F.2), each vector z^i so constructed is an eigenvector of A with eigenvalue λ_1 and will remain so on normalization. It is straightforward to check that, provided the previous new eigenvectors have been normalized as prescribed, each z^i is orthogonal to all its predecessors.

Therefore, even if A has some degenerate eigenvalues we can *by construction* obtain a set of N mutually orthogonal eigenvectors. Moreover, it may be shown (although the proof is beyond the scope of this book) that these eigenvectors are *complete* in that they form a basis for the N-dimensional vector space. As a result any arbitrary vector y can be expressed as a linear combination of the eigenvectors x^i:

$$\mathsf{y} = \sum_{i=1}^{N} a_i \mathsf{x}^i, \tag{F.3}$$

where $a_i = (\mathsf{x}^i)^\dagger \mathsf{y}$. Thus, the eigenvectors form an orthogonal basis for the vector space. By normalizing the eigenvectors so that $(\mathsf{x}^i)^\dagger \mathsf{x}^i = 1$ this basis is made orthonormal.

G Linear least squares

As stated in the main text, we write (17.47) in the form

$$f(x_i; \mathbf{a}) = \sum_{j=1}^{M} R_{ij} a_j, \tag{G.1}$$

where $R_{ij} = h_j(x_i)$ is an element of the *response matrix* R of the experiment. The expression for χ^2 given in (17.46) and written in matrix notation, is

$$\chi^2(\mathbf{a}) = (\mathsf{y} - \mathsf{Ra})^{\mathrm{T}} \mathsf{N}^{-1} (\mathsf{y} - \mathsf{Ra}). \tag{G.2}$$

The LS estimates of the parameters $\mathbf{a}$ are found, as stated in (17.45), by differentiating (G.2) with respect to the a_i and setting the resulting expressions equal to zero.

This is most conveniently done in component form, using the summation convention (see Appendix D). We start with

$$\chi^2(\mathbf{a}) = (y_i - R_{ik} a_k)(N^{-1})_{ij}(y_j - R_{jl} a_l).$$

Differentiating with respect to a_p gives

$$\begin{aligned}\frac{\partial \chi^2}{\partial a_p} &= -R_{ik}\delta_{kp}(N^{-1})_{ij}(y_j - R_{jl}a_l) + (y_i - R_{ik}a_k)(N^{-1})_{ij}(-R_{jl}\delta_{lp}) \\ &= -R_{ip}(N^{-1})_{ij}(y_j - R_{jl}a_l) - (y_i - R_{ik}a_k)(N^{-1})_{ij}R_{jp},\end{aligned} \tag{G.3}$$

where δ_{ij} is the Kronecker delta symbol discussed in Appendix E. By swapping the indices i and j in the second term on the RHS of (G.3) and using the fact that the matrix N^{-1} is symmetric, we obtain

$$\begin{aligned}\frac{\partial \chi^2}{\partial a_p} &= -2R_{ip}(N^{-1})_{ij}(y_j - R_{jk}a_k) \\ &= -2(R^{\mathrm{T}})_{pi}(N^{-1})_{ij}(y_j - R_{jk}a_k).\end{aligned} \tag{G.4}$$

If we denote the vector with components $\partial\chi^2/\partial a_p$, $p = 1, 2, \ldots, M$, by $\nabla\chi^2$ and write the RHS of (G.4) in matrix notation, we have

$$\nabla\chi^2 = -2\mathsf{R}^{\mathrm{T}}\mathsf{N}^{-1}(\mathsf{y} - \mathsf{Ra}). \tag{G.5}$$

Setting the expression (G.5) equal to zero at $\mathbf{a} = \hat{\mathbf{a}}$, we find

$$-2\mathsf{R}^{\mathrm{T}}\mathsf{N}^{-1}\mathsf{y} + 2\mathsf{R}^{\mathrm{T}}\mathsf{N}^{-1}\mathsf{R}\hat{\mathsf{a}} = 0.$$

Provided the matrix $\mathsf{R}^{\mathrm{T}}\mathsf{N}^{-1}\mathsf{R}$ is not singular, we may solve this equation for $\hat{\mathbf{a}}$ to obtain

$$\hat{\mathsf{a}} = (\mathsf{R}^{\mathrm{T}}\mathsf{N}^{-1}\mathsf{R})^{-1}\mathsf{R}^{\mathrm{T}}\mathsf{N}^{-1}\mathsf{y} \equiv \mathsf{Sy}, \tag{G.6}$$

thus defining the $M \times N$ matrix S. As noted in the main text, the LS estimates $\hat{a}_i$, $i = 1, 2, \ldots, M$, are linear functions of the original measurements y_j, $j = 1, 2, \ldots, N$.

The error propagation formula (16.129) derived in Subsection 16.12.3 shows that the covariance matrix of the estimators $\hat{a}_i$ is given by

$$\mathsf{V} \equiv \text{Cov}\,[\hat{a}_i, \hat{a}_j] = \mathsf{SNS}^{\mathrm{T}} = (\mathsf{R}^{\mathrm{T}}\mathsf{N}^{-1}\mathsf{R})^{-1}, \tag{G.7}$$

the final equality in this result being proved as follows.

Using the definition of S given in (G.6), the covariance matrix (G.7) becomes

$$\begin{aligned}\mathsf{V} &= \mathsf{SNS}^{\mathrm{T}} \\ &= [(\mathsf{R}^{\mathrm{T}}\mathsf{N}^{-1}\mathsf{R})^{-1}\mathsf{R}^{\mathrm{T}}\mathsf{N}^{-1}]\mathsf{N}[(\mathsf{R}^{\mathrm{T}}\mathsf{N}^{-1}\mathsf{R})^{-1}\mathsf{R}^{\mathrm{T}}\mathsf{N}^{-1}]^{\mathrm{T}}.\end{aligned}$$

Using the result $(\mathsf{AB}\cdots\mathsf{C})^{\mathrm{T}} = \mathsf{C}^{\mathrm{T}}\cdots\mathsf{B}^{\mathrm{T}}\mathsf{A}^{\mathrm{T}}$ for the transpose of a product of matrices and noting that, for any non-singular matrix, $(\mathsf{A}^{-1})^{\mathrm{T}} = (\mathsf{A}^{\mathrm{T}})^{-1}$ we find

$$\begin{aligned}\mathsf{V} &= (\mathsf{R}^{\mathrm{T}}\mathsf{N}^{-1}\mathsf{R})^{-1}\mathsf{R}^{\mathrm{T}}\mathsf{N}^{-1}\mathsf{N}(\mathsf{N}^{\mathrm{T}})^{-1}\mathsf{R}[(\mathsf{R}^{\mathrm{T}}\mathsf{N}^{-1}\mathsf{R})^{\mathrm{T}}]^{-1} \\ &= (\mathsf{R}^{\mathrm{T}}\mathsf{N}^{-1}\mathsf{R})^{-1}\mathsf{R}^{\mathrm{T}}\mathsf{N}^{-1}\mathsf{R}(\mathsf{R}^{\mathrm{T}}\mathsf{N}^{-1}\mathsf{R})^{-1} \\ &= (\mathsf{R}^{\mathrm{T}}\mathsf{N}^{-1}\mathsf{R})^{-1},\end{aligned}$$

where we have also used the fact that N is symmetric and so $\mathsf{N}^{\mathrm{T}} = \mathsf{N}$.

It is worth noting that one may also write the elements of the (inverse) covariance matrix as

$$(V^{-1})_{ij} = \frac{1}{2}\left(\frac{\partial^2\chi^2}{\partial a_i \partial a_j}\right)_{\mathbf{a}=\hat{\mathbf{a}}}.$$

Further, since $f(x;\mathbf{a})$ is linear in the parameters $\mathbf{a}$, one can write χ^2 *exactly* as

$$\chi^2(\mathbf{a}) = \chi^2(\hat{\mathbf{a}}) + \frac{1}{2}\sum_{i,j=1}^{M}\left(\frac{\partial^2\chi^2}{\partial a_i \partial a_j}\right)_{\mathbf{a}=\hat{\mathbf{a}}}(a_i - \hat{a}_i)(a_j - \hat{a}_j),$$

which is quadratic in the parameters a_i. Hence the form of the likelihood function $L \propto \exp(-\chi^2/2)$ is Gaussian.

H Footnote answers

This appendix contains short answers to those footnotes that are in the form of a question. They have been deliberately placed away from the questions so as to encourage the reader to formulate their own response, as they would be expected to do in a supervision or tutorial, before seeking confirmation. It should be remembered that the questions, typically requiring only brief answers, are normally designed to test whether a particular point in the main text, which may in itself be a relatively small one, has been correctly grasped. Thus some answers may seem trivial to the reader – if they do, so much the better!

1. Matrices and vector spaces

3 Denoting equation (1.12) by (i), equation (1.13) by (ii), and the general properties of complex conjugation by (iii):

$$\begin{aligned}\langle\lambda\mathbf{a}+\mu\mathbf{b}|\mathbf{c}\rangle &\overset{\text{(i)}}{=} \langle\mathbf{c}|\lambda\mathbf{a}+\mu\mathbf{b}\rangle^* \overset{\text{(ii)}}{=} (\langle\mathbf{c}|\lambda\mathbf{a}\rangle+\langle\mathbf{c}|\mu\mathbf{b}\rangle)^*\\ &\overset{\text{(iii)}}{=} \langle\mathbf{c}|\lambda\mathbf{a}\rangle^*+\langle\mathbf{c}|\mu\mathbf{b}\rangle^* \overset{\text{(ii)}}{=} (\lambda\langle\mathbf{c}|\mathbf{a}\rangle)^*+(\mu\langle\mathbf{c}|\mathbf{b}\rangle)^*\\ &\overset{\text{(iii)}}{=} \lambda^*\langle\mathbf{c}|\mathbf{a}\rangle^*+\mu^*\langle\mathbf{c}|\mathbf{b}\rangle^* \overset{\text{(i)}}{=} \lambda^*\langle\mathbf{a}|\mathbf{c}\rangle+\mu^*\langle\mathbf{b}|\mathbf{c}\rangle. \quad \text{(iv)}\end{aligned}$$

$$\langle\lambda\mathbf{a}|\mu\mathbf{b}\rangle \overset{\text{(iv)}}{=} \lambda^*\langle\mathbf{a}|\mu\mathbf{b}\rangle \overset{\text{(ii)}}{=} \lambda^*\mu\langle\mathbf{a}|\mathbf{b}\rangle.$$

4 With $\mathcal{A}\mathbf{x}=(2x_1+x_2,x_2)$, $\mathcal{B}\mathbf{x}=(x_1,x_1+2x_2)$ and $\mathcal{C}\mathbf{x}=(x_1-x_2,2x_2)$, we have, taking a general vector as $\mathbf{x}=(x,y)$,

$$\mathcal{A}\mathcal{C}\mathbf{x}=\mathcal{A}(x-y,2y)=[2(x-y)+2y,2y]=(2x,2y),$$
$$\mathcal{C}\mathcal{A}\mathbf{x}=\mathcal{C}(2x+y,y)=[(2x+y)-y,2y]=(2x,2y),$$

i.e. $\mathcal{A}$ and $\mathcal{C}$ commute. However,

$$\mathcal{A}\mathcal{B}\mathbf{x}=\mathcal{A}(x,x+2y)=[2x+(x+2y),x+2y]=(3x+2y,x+2y),$$
$$\mathcal{B}\mathcal{A}\mathbf{x}=\mathcal{B}(2x+y,y)=[2x+y,(2x+y)+2y]=(2x+y,2x+3y),$$

showing that $\mathcal{A}$ and $\mathcal{B}$ do not commute.

5 The relevant matrices are

$$\mathsf{A}=\begin{pmatrix}2&1\\0&1\end{pmatrix},\quad \mathsf{B}=\begin{pmatrix}1&0\\1&2\end{pmatrix},\quad \mathsf{C}=\begin{pmatrix}1&-1\\0&2\end{pmatrix}.$$

$$\mathsf{AC}=\begin{pmatrix}2&1\\0&1\end{pmatrix}\begin{pmatrix}1&-1\\0&2\end{pmatrix}=\begin{pmatrix}2&0\\0&2\end{pmatrix}=\begin{pmatrix}1&-1\\0&2\end{pmatrix}\begin{pmatrix}2&1\\0&1\end{pmatrix}=\mathsf{CA}.$$

$$\mathsf{AB} = \begin{pmatrix} 2 & 1 \\ 0 & 1 \end{pmatrix}\begin{pmatrix} 1 & 0 \\ 1 & 2 \end{pmatrix} = \begin{pmatrix} 3 & 2 \\ 1 & 2 \end{pmatrix}, \quad \text{whilst}$$

$$\mathsf{BA} = \begin{pmatrix} 1 & 0 \\ 1 & 2 \end{pmatrix}\begin{pmatrix} 2 & 1 \\ 0 & 1 \end{pmatrix} = \begin{pmatrix} 2 & 1 \\ 2 & 3 \end{pmatrix}, \quad \text{which is not the same.}$$

$$\mathsf{BC} = \begin{pmatrix} 1 & 0 \\ 1 & 2 \end{pmatrix}\begin{pmatrix} 1 & -1 \\ 0 & 2 \end{pmatrix} = \begin{pmatrix} 1 & -1 \\ 1 & 3 \end{pmatrix}, \quad \text{but}$$

$$\mathsf{CB} = \begin{pmatrix} 1 & -1 \\ 0 & 2 \end{pmatrix}\begin{pmatrix} 1 & 0 \\ 1 & 2 \end{pmatrix} = \begin{pmatrix} 0 & -2 \\ 2 & 4 \end{pmatrix}, \quad \text{which is not the same.}$$

So, $\mathcal{BC} \neq \mathcal{CB}$.

6 (i) If A and B commute, then

$$(\mathsf{A}+\mathsf{B})(\mathsf{A}-\mathsf{B}) = \mathsf{A}^2 + \mathsf{BA} - \mathsf{AB} - \mathsf{B}^2 = \mathsf{A}^2 + \mathsf{BA} - \mathsf{BA} - \mathsf{B}^2 = \mathsf{A}^2 - \mathsf{B}^2.$$

(ii) If $(\mathsf{A}+\mathsf{B})(\mathsf{A}-\mathsf{B}) = \mathsf{A}^2 - \mathsf{B}^2$, then

$$\mathsf{A}^2 - \mathsf{B}^2 = (\mathsf{A}+\mathsf{B})(\mathsf{A}-\mathsf{B}) = \mathsf{A}^2 + \mathsf{BA} - \mathsf{AB} - \mathsf{B}^2 \quad \Rightarrow \quad \mathsf{BA} - \mathsf{AB} = 0,$$

i.e. A and B commute.

7 If $\mathsf{A} = \begin{pmatrix} 1 & 0 & 0 \\ 0 & -1 & 0 \\ 0 & 0 & 1 \end{pmatrix}$, then $\mathsf{A}^{2m} = \begin{pmatrix} 1 & 0 & 0 \\ 0 & 1 & 0 \\ 0 & 0 & 1 \end{pmatrix}$ and $\mathsf{A}^{2m+1} = \begin{pmatrix} 1 & 0 & 0 \\ 0 & -1 & 0 \\ 0 & 0 & 1 \end{pmatrix}$.

$$\begin{aligned}\exp(i\mathsf{A}) &= \sum_{n=0}^{\infty} \frac{1}{n!}(i^n \mathsf{A}^n) = \sum_{m=0}^{\infty} \frac{(-1)^m}{(2m)!}\mathsf{A}^{2m} + i\sum_{m=0}^{\infty} \frac{(-1)^m}{(2m+1)!}\mathsf{A}^{2m+1} \\ &= \cos 1 \begin{pmatrix} 1 & 0 & 0 \\ 0 & 1 & 0 \\ 0 & 0 & 1 \end{pmatrix} + i \sin 1 \begin{pmatrix} 1 & 0 & 0 \\ 0 & -1 & 0 \\ 0 & 0 & 1 \end{pmatrix},\end{aligned}$$

$$\text{Tr } \exp(i\mathsf{A}) = \cos 1 \times 3 + i \sin 1 \times 1 = 3\cos 1 + i \sin 1.$$

12 A^{-1} is defined by $\mathsf{A}^{-1}\mathsf{A} = \mathsf{I}$. Define $\mathsf{A}_{\mathrm{R}}^{-1}$ by $\mathsf{A}\mathsf{A}_{\mathrm{R}}^{-1} = \mathsf{I}$ and consider

$$\mathsf{A}^{-1} = \mathsf{A}^{-1}\mathsf{I} = \mathsf{A}^{-1}(\mathsf{A}\mathsf{A}_{\mathrm{R}}^{-1}) = (\mathsf{A}^{-1}\mathsf{A})\mathsf{A}_{\mathrm{R}}^{-1} = \mathsf{I}\mathsf{A}_{\mathrm{R}}^{-1} = \mathsf{A}_{\mathrm{R}}^{-1},$$

thus establishing that the left inverse is also a right inverse.

13 For example, the cofactor of the "3" in the a_{23} position is given by $(-1)^{3+2}[(2)(-2) - (1)(3)] = +7$.

15 A matrix all of whose entries are equal has rank 1 (whatever the value of N).

(a) As all columns are the same, any two are linearly dependent and so the matrix has only *one* linearly independent column. Thus, the matrix has rank 1.

(b) Any 2×2 submatrix has a determinant of the form $(\lambda \times \lambda) - (\lambda \times \lambda) = 0$. The determinant of any larger submatrix can be expressed as a sum of the determinants of 2×2 submatrices; thus all such larger submatrices have zero determinant. The largest submatrix with a non-zero determinant is a 1×1 submatrix with determinant λ, and so the matrix has rank 1.

19 The determinant (= 0) is equal to the product of the diagonal elements; the trace (= 0) is equal to their sum. Thus, one element must be zero and the two remaining (non-zero) elements must have opposite signs, i.e. it has the form $\mathsf{A} = \text{diag}\,(0, \lambda, -\lambda)$ but, of course, the zero could appear as any one of the three entries.

20 A has determinant 1. As this $\neq 0$, A has an inverse:

$$\mathsf{A} = \begin{pmatrix} 1 & 0 & a \\ 0 & 1 & b \\ 0 & 0 & 1 \end{pmatrix} \qquad \mathsf{A}^{-1} = \begin{pmatrix} 1 & 0 & -a \\ 0 & 1 & -b \\ 0 & 0 & 1 \end{pmatrix},$$

and $\mathsf{A} + \mathsf{A}^{-1} = 2\mathsf{I}$. Multiplying on the right by A gives $\mathsf{A}^2 + \mathsf{A}^{-1}\mathsf{A} = 2\mathsf{IA}$; hence, $\mathsf{A}^2 - 2\mathsf{A} + \mathsf{I} = 0$. This can be written $(\mathsf{A} - \mathsf{I})^2 = 0$. However this does *not* imply that $\mathsf{A} - \mathsf{I} = 0$; two $N \times N$ matrices can multiply to give the zero matrix without either being the zero matrix. Explicitly, in this case,

$$(\mathsf{A} - \mathsf{I})^2 = \begin{pmatrix} 0 & 0 & a \\ 0 & 0 & b \\ 0 & 0 & 0 \end{pmatrix} \begin{pmatrix} 0 & 0 & a \\ 0 & 0 & b \\ 0 & 0 & 0 \end{pmatrix} = \begin{pmatrix} 0 & 0 & 0 \\ 0 & 0 & 0 \\ 0 & 0 & 0 \end{pmatrix},$$

yet neither matrix in the central expression is the zero matrix.

22 The matrices are

$$\mathsf{S}_x = \begin{pmatrix} 0 & 1 \\ 1 & 0 \end{pmatrix}, \quad \mathsf{S}_y = \begin{pmatrix} 0 & -i \\ i & 0 \end{pmatrix}, \quad \mathsf{S}_z = \begin{pmatrix} 1 & 0 \\ 0 & -1 \end{pmatrix}.$$

S_x and S_z: real, symmetric, Hermitian, orthogonal, unitary.
S_y: antisymmetric, Hermitian, unitary.

26 If one of the eigenvalues of A is 0, then the determinant of A (equal to the product of the eigenvalues) is also 0 and A is singular. Therefore, the inverse of A does not exist.

27 If A is real and orthogonal, i.e. $\mathsf{A}^{\mathrm{T}}\mathsf{A} = \mathsf{I}$, and has eigenvalue λ, then $\mathsf{A}\mathbf{x} = \lambda\mathbf{x}$ and $\mathbf{x}^{\mathrm{T}}\mathsf{A}^{\mathrm{T}} = \lambda\mathbf{x}^{\mathrm{T}}$ for some non-zero $\mathbf{x}$. Thus,

$$(\mathbf{x}^{\mathrm{T}}\mathsf{A}^{\mathrm{T}})(\mathsf{A}\mathbf{x}) = (\lambda\mathbf{x}^{\mathrm{T}})(\lambda\mathbf{x}) \quad \Rightarrow \quad \mathbf{x}^{\mathrm{T}}\mathsf{I}\mathbf{x} = \lambda^2\mathbf{x}^{\mathrm{T}}\mathbf{x} \quad \Rightarrow \quad \mathbf{x}^{\mathrm{T}}\mathbf{x}(1 - \lambda^2) = 0.$$

But $\mathbf{x}^{\mathrm{T}}\mathbf{x} \neq 0$ and so $\lambda^2 = 1$ and $\lambda = \pm 1$.

28 Following the suggested procedure:

$$|\mathsf{A} - \lambda\mathsf{I}| = \begin{vmatrix} 1-\lambda & 1 & 3 \\ 1 & 1-\lambda & -3 \\ 3 & -3 & -3-\lambda \end{vmatrix} = \begin{vmatrix} 2-\lambda & 1 & 3 \\ 2-\lambda & 1-\lambda & -3 \\ 0 & -3 & -3-\lambda \end{vmatrix}$$

$$= (2-\lambda)\begin{vmatrix} 1 & 1 & 3 \\ 1 & 1-\lambda & -3 \\ 0 & -3 & -3-\lambda \end{vmatrix} = (2-\lambda)\begin{vmatrix} 1 & 1 & 3 \\ 0 & -\lambda & -6 \\ 0 & -3 & -3-\lambda \end{vmatrix}$$

$$= (2-\lambda)(1)[(-\lambda)(-3-\lambda) - 18]$$

$$= (2-\lambda)(\lambda^2 + 3\lambda - 18) = (2-\lambda)(\lambda+6)(\lambda-3).$$

29 If all three eigenvectors are equal, then *any* vector is an eigenvector. Therefore choose the simplest orthonormal set, (1, 0, 0), (0, 1, 0) and (0, 0, 1).

32 If two of the eigenvalues are zero, the equation of the quadratic surface will contain only one of its coordinates, e.g. $\lambda_3 x_3^2 = 1$. This means $x_3 = \pm\lambda^{-1/2}$ and the surface consists of two parallel planes with their common normal in the direction of the eigenvector corresponding to the non-zero eigenvalue. The same situation can also be thought of formally as an ellipsoid with two infinitely large perpendicular semi-axes.
35 The various currents flowing in the circuit will be equal to the sums and differences of the rates at which charge flows onto or off the various capacitors in the circuit, and can therefore be expressed in terms of the $\dot{q}_i$ as, typically, $I = \sum_j s_j \dot{q}_j$ with the s_j equal to 1, 0 or -1 (according to Kirchhoff's first law). The current-dependent component of the energy in the circuit is that stored in the magnetic field of the inductors (the charge-dependent part is that stored in the electric fields of the capacitors). The energy stored in a typical inductor of self-inductance L is $\frac{1}{2}LI^2$. When the expression for a typical current is squared it will generate both terms such as $L\dot{q}_i\dot{q}_i$ and terms such as $L\dot{q}_i\dot{q}_j$ with $i \neq j$; the entry a_{ij} in T will be the sum (taking account of sign) of the coefficients in all terms containing $\dot{q}_i\dot{q}_j$ when all the inductors in the circuit have been accounted for.
36 $\mathsf{B} = (3,\ \frac{3}{2},\ 0,\ 2;\ \frac{3}{2},\ 5,\ -1,\ -3;\ 0,\ -1,\ -1,\ 4;\ 2,\ -3,\ 4,\ 0)$.
39 No. Any other configuration must be a linear combination of the two, and therefore contain some non-zero element of each. For repetition after time T, we would require $\omega_1 T = 2\pi n$ and $\omega_2 T = 2\pi m$ for some *integers* n and m. Irrespective of the value of T, this implies that $(5+\sqrt{19})n^2 = (5-\sqrt{19})m^2$; this, in turn, implies that $\sqrt{19}$ is rational – which it is not! So, repetition cannot occur.

2. Vector calculus

1 With $\mathbf{r}(t) = R\,\hat{\mathbf{e}}_\rho$, we have $\mathbf{v}(t) = R\dot{\phi}\,\hat{\mathbf{e}}_\phi$ and $\mathbf{a}(t) = R\ddot{\phi}\,\hat{\mathbf{e}}_\phi - R\dot{\phi}^2\,\hat{\mathbf{e}}_\rho$. With $\phi = \omega t + \phi_0$, $\ddot{\phi} = 0$ and the (inward) acceleration has magnitude $R\omega^2$.
2 The derivative of $[(\mathbf{a}\times\mathbf{b})\cdot\mathbf{c}]$ is

$$(\mathbf{a}\times\mathbf{b})\cdot\frac{d\mathbf{c}}{du} + \left(\mathbf{a}\times\frac{d\mathbf{b}}{du}\right)\cdot\mathbf{c} + \left(\frac{d\mathbf{a}}{du}\times\mathbf{b}\right)\cdot\mathbf{c} = (\mathbf{a}\times\mathbf{b})\cdot\frac{d\mathbf{c}}{du} + (\mathbf{c}\times\mathbf{a})\cdot\frac{d\mathbf{b}}{du} + (\mathbf{b}\times\mathbf{c})\cdot\frac{d\mathbf{a}}{du}.$$

4 $\mathbf{r}(\rho,\phi) = \rho\cos\phi\,\mathbf{i} + \rho\sin\phi\,\mathbf{j} + (\rho^2/4a)\,\mathbf{k}$, leading to $\mathbf{n} = (\partial\mathbf{r}/\partial\rho)\times(\partial\mathbf{r}/\partial\phi) = -(\rho/2a)[\cos\phi\,\mathbf{i} + \sin\phi\,\mathbf{j}] + \mathbf{k}$, which has magnitude $[(\rho/2a)^2 + 1]^{1/2}$. Hence $dS = (2a)^{-1}[\rho^2 + 4a^2]^{1/2}\,d\rho\,d\phi$.
5 (i) $2x^3y^2z^2$, (ii) $2xy^2z^2(x^2+y^2+z^2)$, (iii) $2xy^2z^2(2x^2+y^2+z^2)$.
6 $\nabla\phi = -\dfrac{nA}{r^{n+1}}\left(\dfrac{\partial r}{\partial x}, \dfrac{\partial r}{\partial y}, \dfrac{\partial r}{\partial z}\right) = -\dfrac{nA}{r^{n+1}}\left(\dfrac{x}{r}, \dfrac{y}{r}, \dfrac{z}{r}\right)$, i.e. directly towards the origin.

$$|\nabla\phi\cdot\hat{\mathbf{r}}| = \left|-\frac{nA}{r^{n+1}}\frac{x^2+y^2+z^2}{r^2}\right| = \frac{nA}{r^{n+1}}.$$

7 $2z(x^2+y^2)[2xz,\ 2yz,\ x^2+y^2]$.
8 If $z = 0$ or $x = y = 0$, i.e. on the xy-plane and on the z-axis.
9 $\displaystyle\sum_{x,y,z}\frac{\partial(\phi x)}{\partial x} = \sum_{x,y,z}\left(\phi + x\frac{\partial\phi}{\partial r}\frac{\partial r}{\partial x}\right) = \sum_{x,y,z}\left(\phi + x\frac{\partial\phi}{\partial r}\frac{x}{r}\right) = 3\phi + \frac{r^2}{r}\frac{\partial\phi}{\partial r}.$

14
$$\frac{1}{r^2}\frac{\partial}{\partial r}\left(r^2\frac{\partial \Phi}{\partial r}\right)=\frac{1}{r^2}\left[2r\frac{\partial \Phi}{\partial r}+r^2\frac{\partial^2 \Phi}{\partial r^2}\right]=\frac{\partial^2 \Phi}{\partial r^2}+\frac{2}{r}\frac{\partial \Phi}{\partial r},$$
$$\frac{1}{r}\frac{\partial^2}{\partial r^2}(r\phi)=\frac{1}{r}\frac{\partial}{\partial r}\left[\Phi+r\frac{\partial \Phi}{\partial r}\right]=\frac{1}{r}\frac{\partial \Phi}{\partial r}+\frac{1}{r}\frac{\partial \Phi}{\partial r}+\frac{\partial^2 \Phi}{\partial r^2}.$$

15 Circular cylinders centered on the z-axis; half-planes containing the z-axis; planes with normals along the z-axis.

3. Line, surface and volume integrals

1 $\mathbf{i}\left(\int_C a_y\,dz-\int_C a_z\,dy\right)+\mathbf{j}\left(\int_C a_z\,dx-\int_C a_x\,dz\right)+\mathbf{k}\left(\int_C a_x\,dy-\int_C a_y\,dx\right).$

2 The integral is $\int_1^4 (x+\frac{1}{2}\sqrt{x}+\frac{1}{2})\,dx.$

4 $ds=\phi\sqrt{4k^2+\phi^2(k^2+9\mu^2)}\,d\phi.$

5 (a) Simply, (b) multiply – consider a closed curve in the form of a "D" with the curved part lying within the handle and the straight part within the cylindrical part of the cup, (c) simply – though the physics of the cup is more interesting than its connectivity!

7
$$I=\int_R \sigma(x^2+y^2)\,dx\,dy=\int_{r=a}\sigma(xy^2\,dy-x^2y\,dx)-\int_{r=b}\sigma(xy^2\,dy-x^2y\,dx).$$

Now set $x=r\cos\phi$ and $y=r\sin\phi$ leading to $I=\sigma(a^4-b^4)\int_0^{2\pi}\frac{1}{2}\sin^2(2\phi)\,d\phi$. The mass $m=\sigma\pi(a^2-b^2)$.

8 The integrand is $r^4(\sin^2\phi\cos^2\phi-\cos^2\phi\sin^2\phi)=0$. The potential function $\psi(x,y)=\frac{1}{2}x^2y^2$.

10 Since the vector area of the closed hemisphere is $\mathbf{0}$ and that of its base is clearly $-\pi a^2\,\mathbf{k}$, the vector area of its curved surface must be $+\pi a^2\,\mathbf{k}$.

14
$$\Omega=\int_S\frac{r^2\sin\theta\,d\theta\,d\phi}{r^2}\hat{\mathbf{r}}\cdot\hat{\mathbf{r}}=\int_0^{2\pi}d\phi\int_0^{\alpha}\sin\theta\,d\theta=2\pi(1-\cos\alpha).$$

18 $\mathbf{c}\cdot\mathbf{P}=\mathbf{c}\cdot\mathbf{Q}\;\Rightarrow\;\mathbf{c}\cdot(\mathbf{P}-\mathbf{Q})=0$. Since $\mathbf{c}$ is arbitrary, take it as $\mathbf{P}-\mathbf{Q}$. Then $0=(\mathbf{P}-\mathbf{Q})\cdot(\mathbf{P}-\mathbf{Q})=|\mathbf{P}-\mathbf{Q}|^2$. The only vector with zero modulus is the zero vector $\mathbf{0}$. Thus $\mathbf{P}-\mathbf{Q}=\mathbf{0}$ and hence $\mathbf{P}=\mathbf{Q}$.

4. Fourier series

2 Either (i) using $\sin(2x+\alpha)=\sin 2x\cos\alpha+\cos 2x\sin\alpha$ and inspection, or (ii)
$$\text{even}=\frac{1}{2}[\sin(2x+\alpha)+\sin(-2x+\alpha)]=\frac{1}{2}2\sin\left(\frac{2\alpha}{2}\right)\cos\left(\frac{4x}{2}\right)=\sin\alpha\cos 2x,$$
$$\text{odd}=\frac{1}{2}[\sin(2x+\alpha)-\sin(-2x+\alpha)]=\frac{1}{2}2\cos\left(\frac{2\alpha}{2}\right)\sin\left(\frac{4x}{2}\right)=\cos\alpha\sin 2x.$$

3 All the cosines have unit value and so
$$0=\frac{4}{3}+\frac{16}{\pi^2}\sum_{r=1}^{\infty}\frac{(-1)^r}{r^2},$$
from which it follows that the given sum has the value $\pi^2/12$.

4

$$b_r = \frac{8}{\pi r}(-1)^{r-1} - 2\left(\frac{2}{\pi r}\right)^3 \left[1 - (-1)^{r-1}\right].$$

The stated conclusion follows from the first term.

5 (a) No, (b) Yes.

7 c_0 represents the average value of $f(x)$ over any number of complete cycles.

8 Taking both $f(x)$ and $g(x)$ as the function x, $c_r = 2i(-1)^r/\pi r$ and $\gamma_r^* = -2i(-1)^r/\pi r$ for $r \neq 0$; $c_0 = \gamma_0 = 0$. The equation reads

$$\frac{1}{4}\int_{-2}^{2} x^2\,dx = \sum_{\substack{r=-\infty \\ r\neq 0}}^{r=\infty} \frac{4}{\pi^2 r^2} = \frac{8}{\pi^2}\sum_{r=1}^{\infty}\frac{1}{r^2},$$

from which the stated result follows.

5. Integral transforms

4 A maximum, the limit as $q \to 0$ of $\sin^2 qb/q^2$ is b^2 and the forward intensity is $16b^2/r_0^2$, independent of a but quadratic in b.

6 $f^{(n)}(a) = (-1)^n \int_{-\infty}^{\infty} f(t)\delta^{(n)}(t-a)\,dt.$

7 $\sum_i \delta(t-a_i) + \sum_j H(t-b_j) - \sum_k H(t-c_k)$, where $a_i = 1, 2, 3, 20, 21, 22$, $b_j = 6, 10, 14$ and $c_k = 9, 13, 17$. Technically, there should be no spaces between the letters, as the signal is meant to be sent as a continuous sequence of nine sounds.

11 The transform of e^{at} is $(s-a)^{-1}$. But we also have

$$\mathcal{L}\left[\sum_{n=0}^{\infty}\frac{a^n t^n}{n!}\right] = \sum_{n=0}^{\infty}\frac{a^n\, n!}{n!\, s^{n+1}} = \frac{1}{s}\sum_{n=0}^{\infty}\frac{a^n}{s^n} = \frac{1}{s}\left(1-\frac{a}{s}\right)^{-1} = \frac{1}{s-a}.$$

The Laplace transformations require $s > a$ and $s > 0$; the binomial expansion requires $s > a$.

13 At $t = 0$ the integral has zero range and so has zero value. At $t = \infty$ the factor $e^{-st} \to 0$ for $s > 0$.

6. Higher-order ordinary differential equations

6 At $x = 1, a + b = 0$, whilst at $x = -1, -a + b = 0$; hence $a = b = 0$. $W = \pm(5x^9 - 5x^9) = 0$.

11 First determine the combination of functions that multiply each constant, e.g. for a it is $-4x^3 \sin 2x + 12x^2 \cos 2x + 6x \sin 2x + 4x^3 \sin 2x$. Then regroup all such terms, collecting together the combination of constants that multiply any particular function; the first two equations given are those corresponding to $x^2 \cos 2x$ and $x \cos 2x$.

12 The sum of -2 and -3 is -5 and their product is 6; the CE is therefore $\lambda^2 + 5\lambda + 6 = 0$ and the RR is $u_{n+1} + 5u_n + 6u_{n-1} = 0$ for $n \geq 1$, with $u_0 = 1$ and $u_1 = 0$. (a) $u_2 = -6$, $u_3 = 30$ and $u_4 = -114$. (b) $u_4 = 48 - 162 = -114$.

14 For the equation as given, $a_0 - a_1' + a_2'' = 1 - 4x + 1 + 4 \neq 0$. After dividing by λx^2, $a_0 - a_1' + a_2'' = \lambda^{-1}(x^{-2} - x^{-2} + 0) = 0$, and the equation is exact. [Its first integral is $dy/dx + (1 - \frac{1}{2}x^{-1})y = A$ and its solution $y = B\sqrt{x}e^{-x} + A\sqrt{x}e^{-x}F(x)$, where $F(x)$ is the indefinite integral of $e^x/\sqrt{x}$.]

15 Setting $y = xv(x)$ yields an equation for dv/dx and its derivative:

$$\frac{v''}{v'} = -\frac{2}{x} + \frac{1}{1-x} - \frac{1}{1+x},$$

which integrates to $v' = [x^2(1-x)(1+x)]^{-1}$.
[The second solution is

$$y(x) = \frac{x}{2}\ln\left(\frac{1+x}{1-x}\right) - 1,$$

which is known as a Legendre function of the second kind.]
17 The integrating factor is $\exp(\int^x 2\,du) = e^{2x}$ and so the equation reads

$$\frac{d}{dx}\left(pe^{2x}\right) = \int^x 4u\,e^{2u}\,du = 2xe^{2x} - e^{2x} + A \quad \Rightarrow \quad p = 2x - 1 + Ae^{-2x}.$$

18 (a) Circles with centers on the x-axis. (b) For a circle of radius a, the relationship is $y = a\cos\psi$, where ψ is as defined in the question. The identity $\sec^2\psi = 1 + \tan^2\psi$ is needed to give $d^2y/dx^2 = a^{-1}\sec^3\psi$.

7. Series solutions of ordinary differential equations

1 Writing $\tan x = t$ and $\sec x = s$, we have

$$\begin{vmatrix} t & s & 1 \\ s^2 & st & 0 \\ 2s^2t & s^3 + s^2t & 0 \end{vmatrix} = s^3(t^2 - 1 - t\sqrt{1+t^2}) \neq 0 \quad \Rightarrow \text{ functions linearly independent.}$$

But,

$$\begin{vmatrix} t^2 & s^2 & 1 \\ 2ts^2 & 2s^2t & 0 \\ 2s^4 + 4s^2t^2 & 2s^4 + 4s^2t^2 & 0 \end{vmatrix} = 0 \quad \Rightarrow \quad \text{functions linearly dependent.}$$

2 The differential equation must have $m^2 - [3 + (-2)]m + (3)(-2) = 0$ as its auxiliary equation, and so it is $y'' - y' - 6y = 0$. The Wronskian is $W = (e^{3x})(-2e^{-2x}) - (e^{-2x})(3e^{3x}) = -5e^x$. From the equation, it is $W = C\exp(-\int^x(-1)\,du) = Ce^x$, as expected.
4 Three, ±1 and ∞. The ones at $w = \pm1$ are the same as those at $z = \pm1$. All three are regular singular points.
5 Putting $z - 1 = Z$ and $y(z) = Y(Z)$ produces $Z(2+Z)Y'' + (Z+1)Y' - \nu^2Y = 0$, which has regular singularities at $Z = 0$ and $Z = -2$. Now setting $Z = 1/W$ and $Y(Z) = V(W)$ gives

$$\frac{d^2V}{dW^2} + \frac{3W+1}{W(2W+1)}\frac{dV}{dW} - \nu^2V = 0,$$

which has a regular singularity at $W = 0$, i.e. at $z = \infty$. The singularity at $W = -\frac{1}{2}$ is the same as that at $Z = -2$.
11 With $y = z^\sigma$, we have $\ln y = \sigma\ln z$ and hence $y^{-1}dy/d\sigma = \ln z$. Therefore, $d(z^\sigma)/d\sigma = dy/d\sigma = y\ln z = z^\sigma\ln z$.

8. Eigenfunction methods for differential equations

4 Any set of functions that are periodic with period 2π, for example $\{\cos nx, \sin nx\}$ for all integers n. Any sub-set of these would suffice to make a self-adjoint operator Hermitian.
7 e^{ik_jx} and e^{-ik_jx} are degenerate eigenfunctions of the Hermitian operator d^2/dx^2, with $\rho = 1$ and eigenvalue $-k^2$. The real linear combinations are $e^{-ik_jx} + e^{ik_jx} = 2\cos k_jx$ and $i(e^{-ik_jx} - e^{ik_jx}) = 2\sin k_jx$. Note that d/dx is not an acceptable example, as it is not Hermitian, and neither is $\pm i d/dx$ because, although it is Hermitian, its eigenvalue $\mp k$ is not degenerate.

9. Special functions

3 After the first integration, $y_2(x) = x\int^x u^{-2}\exp[-\ln(1-u^2)]\,du = x\int^x [u^{-2} + \frac{1}{2}(1-u)^{-1} + \frac{1}{2}(1+u)^{-1}]\,du$. This integrates to give $Q_1(x)$. There can be no u^{-1} term in the partial fraction expansion of $u^{-2}(1-u^2)^{-1}$ because the latter is even in u whilst the former is an odd function of u.
6 Either by the construction of multiples of the Legendre polynomials by the addition and subsequent subtraction of an appropriate multiple of the highest unaccounted-for power of x, starting with $P_3(x)$, or by the use of (9.14), we find $a_0 = \frac{4}{3}$, $a_1 = \frac{8}{5}$, $a_2 = \frac{2}{3}$ and $a_3 = \frac{2}{5}$. Knowing that $P_\ell(1) = 1$ and $P_\ell(-1) = (-1)^\ell$, the values of $f(1)$ and $f(-1)$ could have been used to find a_2 and a_3, once a_0 and a_1 were known, through the simultaneous equations $a_0 + a_1 + a_2 + a_3 = f(1)$ and $a_0 - a_1 + a_2 - a_3 = f(-1)$.
7 Setting $x = 0$ gives $(1+h^2)^{-1/2} = \sum P_n(0)h^n$. The binomial expansion will contain no odd powers of h and so $P_\ell(0) = 0$ if ℓ is odd; for any ℓ, $P_\ell(0)$ is equal to the constant term. For $\ell = 2r$, the coefficient of h^{2r} in the expansion is $(r!)^{-1}(-\frac{1}{2})(-\frac{3}{2})(-\frac{5}{2})\cdots(-\frac{2r-1}{2})$, which can be rearranged into the given form.
8 All the intermediate results are given on p. 324.
12 Here $m = 1$ and $n = 2$ in (9.40) and we need the coefficient of h^2 in

$$\frac{2!\,(1-x^2)^{1/2}}{2\,(1!)\,(1-2hx+h^2)^{3/2}} = (1-x^2)^{1/2}\left[1 - \frac{3}{2}(-2hx+h^2) + \frac{(-\frac{3}{2})(-\frac{5}{2})}{2!}(-2hx+h^2)^2 + \cdots\right].$$

This coefficient is

$$P_3^1(x) = (1-x^2)^{1/2}\left[-\frac{3}{2} + \frac{15}{8}4x^2\right] = \frac{3}{2}(1-x^2)^{1/2}(5x^2-1).$$

13 Note that $e^{\pm i\phi}\left[e^{\pm i(\phi+\pi)}\right]^* = e^{\mp i\pi} = -1$ and that $\cos(\pi-\theta) = -\cos\theta$. Equation (9.50) gives $\cos\gamma = -1$ and $P_1(-1) = -1$. The RHS of the equation is

$$\frac{4\pi}{3}\left[\frac{3}{8\pi}\sin^2\theta e^{-i\pi} + \frac{3}{4\pi}\cos\theta(-\cos\theta) + \frac{3}{8\pi}\sin^2\theta e^{i\pi}\right] = -1.$$

15 The formal method is to find the a_n using the given formula, evaluating it for each n by substituting $x = \cos\theta$ and integrating from π to 0. However, it is much easier to match coefficients between the given expression and a sum of multiples of $T_n(x)$, starting with $T_3(x)$ [because x^3 is the highest power of x present] and working downwards towards

$n = 0$:

$$1 + x + x^2 + x^3 = \tfrac{1}{4}(4x^3 - 3x) + \tfrac{1}{2}(2x^2 - 1) + (1 + \tfrac{3}{4})x + 1 + \tfrac{1}{2}$$
$$= \tfrac{1}{4}T_3(x) + \tfrac{1}{2}T_2(x) + \tfrac{7}{4}T_1(x) + \tfrac{3}{2}T_0(x).$$

17 For ν an integer ≥ 1,

$$J_{-\nu}(x) = \sum_{n=0}^{\infty} \frac{(-1)^n}{n!\,\Gamma(-\nu + n + 1)} \left(\frac{x}{2}\right)^{\nu+2n}.$$

However, for $-\nu + n + 1 \leq 0$, $\Gamma(-\nu + n + 1) = \infty$ and so all terms are zero for $n \leq \nu - 1$. Re-indexing using $n = \nu + s$ with $s \geq 0$ gives

$$J_{-\nu}(x) = \sum_{s=0}^{\infty} \frac{(-1)^{\nu+s}}{(\nu + s)!\,\Gamma(s + 1)} \left(\frac{x}{2}\right)^{-\nu+2\nu+2s}$$
$$= (-1)^{\nu} \sum_{s=0}^{\infty} \frac{(-1)^s}{\Gamma(\nu + s + 1)\,s!} \left(\frac{x}{2}\right)^{\nu+2s} = (-1)^{\nu} J_{\nu}(x).$$

21 Denoting $\int_0^{\infty} x^n e^{-x}\,dx$ by I_n and noting that $I_n = n!$,

$$\int_0^{\infty} L_1(x)L_2(x)e^{-x}\,dx = \frac{1}{2!}\int_0^{\infty} (1 - x)(x^2 - 4x + 2)e^{-x}\,dx$$
$$= \tfrac{1}{2}(-I_3 + 5I_2 - 6I_1 + 2I_0) = \tfrac{1}{2}(-6 + 10 - 6 + 2) = 0.$$

22 Setting $x = 0$ gives

$$G(0, h) = \frac{e^0}{1 - h} = 1 + h + h^2 + \cdots = \sum_{n=0}^{\infty} L_n(0)h^n \quad \Rightarrow \quad L_n(0) = 1 \text{ for all } n.$$

27 With $u = v^2$ and $du = 2v\,dv$,

$$\frac{1}{\sqrt{\pi}}\gamma\left(\frac{1}{2}, x^2\right) = \frac{1}{\sqrt{\pi}}\int_0^{x^2} u^{-1/2}e^{-u}\,du = \frac{1}{\sqrt{\pi}}\int_0^{x} \frac{1}{v}e^{-v^2}2v\,dv = \frac{2}{\sqrt{\pi}}\int_0^{x} e^{-v^2}\,dv.$$

10. Partial differential equations

2

$$x\left(\frac{2x}{5} - \frac{3f}{x^4}\right) + \frac{3x^2}{5} + \frac{3f}{x^3} = x^2. \quad \checkmark$$

5 Hyperbolic.

8 Since $A + B\lambda + C\lambda^2 = 0$, we have $\lambda_1 + \lambda_2 = -B/C$ and $\lambda_1\lambda_2 = A/C$. Thus, the condition is

$$2A + B\left(-\frac{B}{C}\right) + 2C\left(\frac{A}{C}\right) \neq 0 \quad \Rightarrow \quad B^2 \neq 4AC.$$

11. Separation of variables and other methods for PDEs

2 With separation constant μ^2, $u(x,t) = Ae^{\mu x+\mu^2\kappa t} + Be^{-\mu x+\mu^2\kappa t}$. With $A = 0$ and (speed) $c = \mu\kappa$, $u(x,t) = Be^{-\mu(x-ct)}$; for any x, the amplitude will be the common value $Be^{-\mu x_0}$ at time $c^{-1}(x - x_0)$.
3 For $u(0,y) = T_0\sin(3\pi y/b)$,

$$B_3 = \frac{2}{b}\int_0^b T_0\sin\left(\frac{3\pi y}{b}\right)\sin\left(\frac{3\pi y}{b}\right)dy = \frac{2T_0}{b}\frac{1}{2}b = T_0$$

and $B_n = 0$ for $n \neq 3$. Thus $u(x,y) = T_0e^{-3\pi x/b}\sin(3\pi y/b)$.
9 For large r, the ring appears as a point object at the origin; $u \sim -GM/r$.
For $r = 0$, all points of the ring are a distance a from the origin; $u \sim -GM/a$.
10 Using Br^n for the CF: $n(n-1) + 2n - 2 = 0 \quad\Rightarrow\quad n = 1, -2$.
Using Cr^3 for the PI: $Cr^3[(3)(2) + 2(3) - 2] = -Ar^3/\epsilon_0 \quad\Rightarrow\quad C = -A/10\epsilon_0$.
16

$$\begin{aligned}u(0,0,z_0) &= \frac{z_0}{2\pi}\iint\frac{f(x,y)}{[x^2+y^2+z_0^2]^{3/2}}\,dx\,dy\\ &= \frac{z_0V_0}{2\pi}\int_0^a\frac{2\pi\rho}{[\rho^2+z_0^2]^{3/2}}\,d\rho = V_0\left[1 - \frac{z_0}{(a^2+z_0^2)^{1/2}}\right].\end{aligned}$$

As $a \to \infty$, $u(0,0,z_0) \to V_0$ for all z_0.

12. Calculus of variations

3 The standard first integral of (12.8) reads

$$n(y)(1+y'^2)^{1/2} - \frac{y'\,n(y)\,y'}{(1+y'^2)^{1/2}} = k.$$

6

$$\lambda_0 = \frac{\int_0^1\left(\frac{\pi}{2}\cos\frac{\pi x}{2}\right)^2dx}{\int_0^1\sin^2\left(\frac{\pi x}{2}\right)dx} = \frac{\pi^2}{4}\frac{\int_0^1\cos^2\frac{\pi x}{2}\,dx}{\int_0^1\sin^2\frac{\pi x}{2}\,dx} = \frac{\pi^2}{4}. \quad \checkmark$$

13. Integral equations

2 $\mathcal{K}^\dagger y = \int_0^x (z^2 - x^2)e^{-ikx}y(z)\,dz$. **3** (a), (c) and (e). **7** (b) and (c).

14. Complex variables

2 In the standard notation $u = x$ and $v = \pm y$. The CR relations show that $x + iy$ is analytic; and that $x - iy$ is not.
4 (a) $|z|^2 = zz^*$; not analytic. (b) $(x-iy)/(x^2+y^2) = z^*/zz^* = 1/z$; analytic. (c) $\cos x\cosh y \pm i\sin x\sinh y = \cos(x \mp iy)$; $\cos x\cosh y + i\sin x\sinh y = \cos z^*$, not analytic; $\cos x\cosh y - i\sin x\sinh y = \cos z$, analytic.
5 $z^3 = (x^3 - 3xy^2) + i(3x^2y - y^3)$. Thus, $\nabla^2u = 6x - 6x = 0$ and $\nabla^2v = 6y - 6y = 0$.
7 The ratio of successive terms is

$$-\frac{z^{2n+2}}{2^{n+1}}\frac{2^n}{z^{2n}} = -\frac{z^2}{2} \quad\Rightarrow\quad \text{convergence for } |z| < \sqrt{2}.$$

11 (a) Imaginary. (b) $\exp(-\pi/2 - 2\pi k)$ $(\equiv a_k)$, real. (c) $\exp(-2\pi m)\exp(i \ln a_k)$, complex.

14
$$\frac{1}{z^3}\left[\left(z - \frac{z^3}{3!} + \cdots\right) + z\left(1 - \frac{z^2}{2!} + \cdots\right)\right] = \frac{2}{z^2} - \cdots \text{ pole of order 2,}$$
$$\frac{1}{z^3}\left[\left(z - \frac{z^3}{3!} + \cdots\right) - z\left(1 - \frac{z^2}{2!} + \cdots\right)\right] = \frac{1}{3} + \cdots \text{ removable singularity.}$$

15 $z = 1$. Re-arranging $w = (z - i)/(z + i)$ gives $z = i(1 + w)/(1 - w)$.

15. Applications of complex variables

1
$$E_x + iE_y = -\left[-\frac{q}{2\pi\epsilon_0}\frac{1}{z}\right]^* = \frac{q}{2\pi\epsilon_0}\frac{1}{z*} = \frac{q}{2\pi\epsilon_0}\frac{\cos\theta + i\sin\theta}{r}.$$

2 $V^2 = |f'(z)|^2$ is given by

$$V_0^2\left(1 - \frac{a^2}{z^2}\right)\left(1 - \frac{a^2}{z^{*2}}\right) = V_0^2\frac{|z|^4 - a^2(z^2 + z^{*2}) + a^4}{|z|^4} = V_0^2\frac{r^4 - 2a^2r^2\cos 2\theta + a^4}{r^4}.$$

Thus, $V = V_0(r^2 - a^2)/r^2$ when $\theta = 0$ or π, and $V = V_0(r^2 + a^2)/r^2$ when $\theta = \pm\pi/2$.
7 For $\theta \sim 0$, $\cos 2\theta \sim 1$ and $a^2 + b^2 - 2ab\cos\theta$ is close to $(a - b)^2$. If b is only a little greater than a the integrand is large but finite in this region – and so therefore is I.
9 The substitution $x = a\tan\theta$ reduces the integral to $a^{-7}\int_0^{\pi/2}\cos^6\theta\, d\theta$, which is equal to $\frac{1}{4}$ of the same integral from 0 to 2π.
13
$$\cot\pi z = \frac{\cos\pi z}{\sin\pi z} = \frac{\frac{1}{2}\left(e^{i\pi z} + e^{-i\pi z}\right)}{\frac{1}{2i}\left(e^{i\pi z} - e^{-i\pi z}\right)} = i\frac{e^{2\pi i(x+iy)} + 1}{e^{2\pi i(x+iy)} - 1}.$$

This $\to i(0 + 1)/(0 - 1) = -i$ for $y > 0$ and $\to i(\infty + 1)/(\infty - 1) = i$ for $y < 0$.

16. Probability

1 $A - B = A \cap \bar{B} = \emptyset$, because $B \supset A$ and so A and $\bar{B}$ do not intersect.
4 (a) $16N/52$. (b) $17N/52$.
5 Yes, the replaced card would increase the number that could be drawn, but would not count as a success. The probability would now be $(4/52) \times (3/52) = 3/676$.
7 $({}^4C_0\ {}^{48}C_{13})/{}^{52}C_{13} = 0.304$, but $({}^4C_1\ {}^{48}C_{12})/{}^{52}C_{13} = 0.439$ and so one king is more likely.
8 In the order of the sequences given, the factors are (3/10)(2/9)(7/8), (3/10)(7/9)(2/8) and (7/10)(3/9)(2/8).
10 With $0 \le Y \le 1$, $dY/dX = f(X)$ and hence, by (16.52), $g(Y) = f(X) \times |dX/dY| = f(X) \times [1/f(X)] = 1$.
13 $\Phi_n(t) = (q + pt)^n$. From which $\Phi_n'(t) = np(q + pt)^{n-1}$ and $\Phi_n''(t) = n(n - 1)p^2(q + pt)^{n-2}$. Then $\mu = \Phi_n'(1) = np$ and $V = \Phi_n''(1) + \Phi_n'(1) - [\Phi_n'(1)]^2 = -np^2 + np = npq$.
15 The second derivative of $e^{-\mu t}M_X(t)$ is $\mu^2 e^{-\mu t}M - 2\mu e^{-\mu t}M' + e^{-\mu t}M''$. At $t = 0$ this has the value $\mu^2 - 2\mu M'(0) + M''(0) = M''(0) - \mu^2$.

18 Writing $p_1p_2e^{2t} + (p_1q_2 + p_2q_1)e^t + q_1q_2$ as $f(t)$, $M \equiv M_{X+Y}(t) = [f(t)]^n$ and $M'(0) = n(2p_1p_2 + p_1q_2 + p_2q_1)[f(0)]^{n-1} = n(p_1 + p_2)$. Similarly, $M''(0) = n(n-1)(p_1 + p_2)^2 + n(2p_1p_2 + p_1 + p_2)$ leading to $V = n(p_1q_1 + p_2q_2)$.
22 The mean is 50 and, using the Gaussian approximation, $\sigma = \sqrt{100 \times 0.5 \times 0.5} = 5$. So 60 is 2 standard deviations from the mean and the chances of equalling or exceeding it is, $1 - \Phi(2) = 0.0228$, i.e. about 2.3%.

17. Statistics

3 $(a-b)^2 \geq 0 \quad \Rightarrow \quad \frac{1}{2}(a^2 + b^2) \geq ab$. Taking square roots shows that $\bar{x}_{\rm rms} \geq \bar{x}_{\rm g}$. From $[\frac{1}{2}(a+b)]^2 = \frac{1}{4}(a^2 + b^2 + 2ab)$ we have

$$\bar{x}^2 = \tfrac{1}{2}(\bar{x}_{\rm rms}^2 + \bar{x}_{\rm g}^2) \geq \tfrac{1}{2}\, 2\bar{x}_{\rm rms}\bar{x}_{\rm g} \geq \bar{x}_{\rm g}^2,$$

and hence the stated result. The final inequality uses the first result proved.
6 Writing $E[\hat{a}] = \hat{\mu}$,

$$\begin{aligned} \epsilon_0^2 &= E\left[(\hat{a} - a)^2\right] \\ &= E\left[(\hat{a} - \hat{\mu} + \hat{\mu} - a)^2\right] \\ &= E\left[(\hat{a} - \hat{\mu})^2 + 2(\hat{a} - \hat{\mu})(\hat{\mu} - a) + (\hat{\mu} - a)^2\right]. \end{aligned}$$

Now $\hat{\mu} - a$ is a constant, and so $E[\hat{\mu} - a] = \hat{\mu} - a$. Further, $E[\hat{a} - \hat{\mu}] = 0$, by definition, and so

$$\epsilon_0^2 = E\left[(\hat{a} - \hat{\mu})^2\right] + 0 + (\hat{\mu} - a)^2 = V[\hat{a}] + [b(\mathbf{a})]^2.$$

13 Setting the derivative of the logarithm of the likelihood function to zero, we have

$$\begin{aligned} L &= \frac{1}{(2\pi\sigma^2)^{N/2}} \exp\left[-\frac{\sum(x_i - \mu_0)^2}{2\sigma^2}\right], \\ \ln L &= k - N\ln\sigma - \frac{\sum(x_i - \mu_0)^2}{2\sigma^2}, \\ 0 = \frac{\partial \ln L}{\partial \sigma} &= -\frac{N}{\sigma} - \frac{(-2)\sum(x_i - \mu_0)^2}{2\sigma^3}, \\ \Rightarrow \quad \sum(x_i - \mu_0)^2 &= \frac{N}{\sigma}\sigma^3 = N\sigma^2. \end{aligned}$$

Hence the stated result.

14 $$N(\bar{x}-\mu_0)^2+\sum_i (x_i-\bar{x})^2 = N\bar{x}^2-2N\bar{x}\mu_0+N\mu_0^2+\sum_i x_i^2-2\bar{x}N\bar{x}+N\bar{x}^2$$
$$=\sum_i x_i^2-2N\bar{x}\mu_0+N\mu_0^2$$
$$=\sum_i x_i^2-2\mu_0\sum_i x_i+\sum_i \mu_0^2$$
$$=\sum_i (x_i-\mu_0)^2.$$

18
$$\lambda=\frac{(2\pi s^2)^{N/2}}{(2\pi\sigma_0^2)^{N/2}}\exp\left[-\frac{\sum(x_i-\bar{x})^2}{2\sigma_0^2}\right]\exp\left[\frac{\sum(x_i-\bar{x})^2}{2s^2}\right]$$
$$=\left(\frac{u}{N}\right)^{N/2}\exp\left[-\frac{\sum(x_i-\bar{x})^2}{2}\left(\frac{1}{\sigma_0^2}-\frac{1}{s^2}\right)\right]$$
$$=\left(\frac{u}{N}\right)^{N/2}\exp\left[-\frac{Ns^2}{2}\left(\frac{1}{\sigma_0^2}-\frac{1}{s^2}\right)\right]$$
$$=\left(\frac{u}{N}\right)^{N/2}\exp\left[-\frac{1}{2}(u-N)\right].$$

19 $z=\frac{1}{2}\ln F=\frac{1}{2}\ln 1.46=0.189$. The mean $=\frac{1}{2}(n_2^{-1}-n_1^{-1})=0.026$; the s.d. $=\sqrt{\frac{1}{2}(n_2^{-1}+n_1^{-1})}=0.341$. And so z is 0.48 s.d. from its mean.

22 Excluding the final row gives expected frequencies of [55.2, 135, 19.8; 36.8, 90, 13.2] and corresponding contributions to χ^2 of 0.417, 0.474, 0.517; 0.626, 0.711, 0.776. The d.o.f. are $(3-1)\times(2-1)=2$.

Index

ı the United States
ɔookmasters